Solar Energy Systems Design

Solar Energy Systems Design

NORMAN C. HARRIS

Emeritus Professor of Higher Education
The University of Michigan

CYDNEY E. MILLER

Solar Energy Consultant

IRVING E. THOMAS

Solar Engineer

John Wiley & Sons

New York   Chichester   Brisbane   Toronto   Singapore

Cover and text designer: Karia Gerdes Kincheloe
Cover photo: Gabe Palmer/The Image Bank

Library of Congress Cataloging in Publication Data:

Harris, Norman C.
Solar energy systems design.

Includes index.
1. Solar energy.   I. Miller, Cydney E. (Cydney Elizabeth), 1953-
II. Thomas, Irving E., 1948-
III. Title.

TJ810.H3618   1985   621.47   84-19679
ISBN 0-471-87104-4
Printed in the U.S. of A.
10 9 8 7 6 5 4 3 2 1

Contents

Preface   VII

PART ONE
SOLAR ENERGY: BASIC PRINCIPLES
1

Chapter 1   Introduction and Overview   3
Chapter 2   Basic Physics Review   27
Chapter 3   Heat Energy and Heat Transfer   53
Chapter 4   Solar Radiation   89
Chapter 5   Building Heating Loads   149

PART TWO
PASSIVE SOLAR ENERGY SYSTEMS
181

Chapter 6   Passive Solar Design Concepts   183
Chapter 7   Direct-Gain Passive Systems   227
Chapter 8   Indirect-Gain and Isolated-Gain Passive Systems   281
Chapter 9   Advanced Passive Methods—Selected Applications   329

PART THREE
ACTIVE SOLAR ENERGY SYSTEMS
381

Chapter 10   Introduction to Active Solar Energy Systems   383
Chapter 11   Flat-Plate Solar Collectors   409
Chapter 12   Performance of Flat-Plate Collectors   443
Chapter 13   Solar Heating of Domestic Hot Water   467
Chapter 14   Solar Heating of Swimming Pools and Spas   533
Chapter 15   Solar Energy Systems for Space Heating   569
Chapter 16   Summer Cooling of Solar Buildings   627
Chapter 17   Commercial and Industrial Applications of Solar Energy   667

PART FOUR
ELECTRICITY FROM THE SUN
689

Chapter 18   Solar-Energized Electric Power   691

APPENDIX I
SELECTED REFERENCES
737

APPENDIX II
SOLAR ECONOMICS
741

GLOSSARY
753

INDEX
765

Preface

Solar energy gives every promise of putting its stamp on the 1980s and 1990s just as electronics did on the 1960s and 1970s. Spurred by dwindling reserves, escalating costs, and an uncertain future for fossil fuels, research and development in the solar energy field is proceeding at a steady pace. New breakthroughs in theory and new industrial applications occur with increasing frequency, and they bring with them new demands for trained technical and professional personnel—demands that will have to be met by colleges, technical institutes, and community colleges through organized programs of solar technology.

Solar Energy Systems Design is a basic text for the education and training of solar systems designers, sales and application engineers, technicians, contractors, and installers in the solar energy industry. Design practices for both passive solar and active solar are developed in detail. All of the usual applications of solar energy are treated—space heating and cooling, domestic water heating, pool and spa heating, industrial process heating, and photovoltaics.

The book is intended for use in organized educational programs in four-year colleges and universities, community colleges, technical institutes, and technical–vocational schools, and it is well suited for use as a reference for architects, solar consultants, and contractors in the industry. It also meets the need for a basic text in related training for apprentices and for home-study programs.

The content, scope, and sequence of *Solar Energy Systems Design* have been planned for a basic one-year course. However, many other time sequences are feasible, and by judicious selection of topics, the instructor can easily accommodate the text to shorter or longer courses. In order to maximize learning and bring students' competencies up to industry requirements, the basic course should be accompanied by hands-on projects in a solar energy laboratory.

Instructors and students alike will find that the generous use of sketches, diagrams, and photographs in the text aids understanding of design theory and practice. The mathematics requirements have purposely been kept at about the level of high school algebra. Elementary physics and the basic principles of solar radiation and heat transfer are reviewed extensively in Chapters 2 and 3, since it is the authors' experience that most students will need such a review before beginning the study of solar energy theory and practice. Students already well grounded in basic physics can turn directly to Chapter 4, where the study of solar energy really begins.

Architects, designers, solar technicians, contractors, and mechanics will all appreciate the book's balanced approach to theory and practice, drawing on the authors' combined experience of many years in the design of solar energy systems, in air conditioning and refrigeration, and in college and university teaching.

Engineers, designers, and technicians are directly and continually involved with problem solving, no less in solar energy than in any other technological field. Recognizing this, the authors have included many sample problems, fully worked out, to illustrate proper approaches to system design. Many of these illustrative problems are adaptations of design problems from actual jobs in process. The step-by-step solutions to these examples, if carefully followed, should develop confidence and remove the fear that many students have when faced with the need to set up and solve design problems.

For further practice and review, there are end-of-chapter problems, ranging from simple and basic to advanced and system-specific.

Anticipating increased use of the SI–metric system by U.S. industry in the future, the text provides an introduction to metrics in the first four chapters. Space does not allow using duplicate systems of measurement in a basic book, however, and SI units are not carried on into later chapters where design practices are emphasized. The major emphasis throughout is on English engineering units of measurement, since that is the system still in daily use in the U.S. solar industry.

Authors alone do not create a book. We are most appreciative of the contributions to this volume made by many others. Many corporations, businesses, and public utilities have provided data, information on new products and installations, and the opportunity to observe solar energy plants and manufacturing operations. Naming them all would be impossible, but special mention must go to ARCO Solar, Inc. Phelps Dodge Solar Enterprises, Southern California Edison Company, and Solarwest Electric. Santa Barbara Solar Energy Systems supplied many photographs and provided data and examples from recently designed jobs. Solar equipment manufacturers were generous with technical data and photographs. They are recognized in the credit lines accompanying illustrations and tabular data.

Professional and trade associations were also extremely helpful, providing copies of research papers, engineering studies, and tabular data. Our gratitude is due especially to the American Society of Heating, Refrigerating, and Air Conditioning Engineers (ASHRAE); the International Solar Energy Society and its divisions; and the Solar Energy Research Institute (SERI).

Government agencies, institutes, and laboratories also provided a wealth of information and tabular data. Among these we mention especially the U.S. Department of Energy (DOE), the National Aeronautics and Space Administration (NASA); the National Oceanic and Atmospheric Administration (NOAA); the Los Alamos National Laboratory; and the U.S. Department of Housing and Urban Development (HUD).

For her highly competent manuscript typing and her genuine concern with our deadlines, we express sincere thanks to Elizabeth E. Auchincloss.

We acknowledge and are grateful for the interest of many associates—current and former colleagues—who took the time to read and criticize the developing manuscript and to suggest changes that would make it better. The following reviewers provided helpful comments on the manuscript: Jon Klima, Community College of Denver, Red Rocks Campus; George A. Cleland, Miami; Earl D. Stedman, Mississippi County Community College; Claude C. Hartman, Cerritos College; Bernard Jenkins, Lansing Community College; Stan Lowery, Pensacola Junior College; Clarence Tresler, Central Texas College; Marvin R. Peters, Butte College; Paul Hering, Universal Technical Institute; Richard Kelso, University of Tennessee; Biswa N. Dey, State University of New York, Utica; Frances J. Callahan, Wentworth Institute of Technology; William D. Lloyd, Lloyd–Vondada Architects, Portland, Oregon; H. D. Mol, Auburn University; Hal Barcus, Miami University; and Mark Mrohs, Training Manager, ARCO-Solar, Incorporated.

Both practitioners and teachers have contributed to *Solar Energy Systems Design*. We dedicate the book to them and to students of solar energy.

Santa Barbara, California

Norman C. Harris
Cydney E. Miller
Irving E. Thomas

PART ONE

Solar Energy: Basic Principles

Introduction and Overview

*I*n spite of what you read in the papers, the much-heralded energy shortage of the future need never reach crisis proportions. The sun—the ultimate source of nearly all energy on earth—shines every day, providing solar power to the earth's surface at the stupendous rate of 1.7×10^{14} kilowatts (kW). In more familiar terms, this flow of power is equivalent to that which could be obtained from burning 6 million tons of coal every second, around the clock, every day in the year—roughly 100 tons of coal per day for every person on earth.

Energy from the sun—*solar energy*—is a renewable resource. It will not be depleted for millions of years, it does not have to be imported from abroad, it does not have to be hauled or piped to the point of use, and it is readily available at any location where the sun shines. It is nonpolluting, has little if any adverse environmental impact, and leaves no residue or industrial waste. Taking into account night and day, the seasons of the year, cloudy and sunny days, and the latitude on the earth's surface, the average solar power at ground level is about 17 watts per square foot (W/ft^2)—enough, if it could be collected, stored, and used at 100 percent efficiency, to supply the total world demand for energy many times over.

The solar radiation energy that annually strikes the roof and walls of a typical one- or two-story residence is several times as great as the home's annual heating load energy demand. Why, then, does the specter of a worldwide energy shortage confront us as we look to the year 2000 and beyond? The truth is that we have not, until very recently, made a concerted, systematic, research-based attempt to harness solar power.

SOLAR ENERGY—PAST AND PRESENT

Although solar energy is just now at the threshold of massive industrial development, the use of the sun's rays for heat and energy is not new. A brief look at the history of solar energy follows.

1-1 Prehistoric and Ancient Uses of Energy from the Sun

Archaeology provides us with numerous examples of the use of the sun's rays for heating the dwellings of prehistoric and ancient peoples. The use of solar energy for home heating and for greenhouses was quite common in ancient Greece and Rome. These early solar energy systems consisted simply of facing the dwelling in a southerly direction, providing ample openings for the direct rays of the sun to enter, using interior materials that could store up large quantities of heat, and providing roof overhangs or shading to allow the low winter sun to shine in and to block out the rays of the high summer sun. Today, we refer to this general plan as a *passive solar* system. Native American Indians in the Southwest designed their dwellings with the sun's heat in mind. Colonial houses in the Northeast—the saltbox design—also took advantage of the sun and other climatic factors.

Whether it is legend or historical fact is not actually known, but ancient Greek literature records the use of solar power for military purposes in 212 B.C. A fleet of Roman warships appeared off the Greek coast of Syracuse and made preparations for the conquest of the island. Archimedes, the renowned mathematician and natural philosopher of the era, devised an array of solar energy-reflecting mirrors that focused the sun's rays to a narrow, high-temperature beam. As the Roman ships came within a range of about 200 ft offshore, the beam was directed at each in turn, setting them afire. According to the legend, the entire fleet was destroyed.

Wind power is, of course, a direct result of solar energy. The use of the wind for ship propulsion antedates even the most ancient records. By the time of the Greek and Roman wars referred to above, sailing ships powered by the wind were plying all of the seas of the ancient world. In ancient and medieval times wind power was also a basic source of energy for the windmills that ground grain and pumped water. These venerable engines are still turning in the wind today in many regions of the world, using solar energy to do some of the world's work.

1-2 Solar Energy and the Industrial Age

Through the period of the Roman Empire and on into the Middle Ages, history tells us little with respect to the use or further development of solar energy devices. Much of the solar knowledge developed earlier in Greece and Rome appears to have vanished from European culture. The only buildings in medieval Europe to have glass windows were the churches, and these were made of stained glass. The Church denounced the concept of growing plants out of their natural habitat or out of season, and as a result the use of greenhouses declined rapidly throughout Europe. Sometime during the thirteenth century, a Latin translation of an Arab treatise on the mathematics of mirrors was circulated in several European universities, and this text stimulated an interest in the construction of parabolic burning mirrors. In the late 1600s, a German noble named Baron Tchirnhausen constructed a copper mirror 5½ ft in diameter, which was the largest mirror constructed up to that time. Sometime in the 1500s, the use of greenhouses for growing plants was revived, and the Dutch and Flemish had soon developed greenhouse horticulture to a level equivalent to that of the ancient Greeks and Romans. The practice spread to France and England, and soon greenhouses became so popular that the 1700s have been called the Age of the Greenhouse by some historians of science.

In about 1615, as the Middle Ages phased into the early years of the Industrial Revolution, a French scientist–mechanic named Salomon de Caux devised and built a solar energy machine that heated air. The hot air was then allowed to expand in the machine, providing energy to pump water. A century-and-a-half later, just before the American Revolution, Antoine Lavoisier, another Frenchman, invented a solar furnace that produced temperatures high enough (1750°C) to melt several different metals. After another century had passed, Professor August Mouchot, also of France, developed a solar-energized steam boiler and engine that was put on display at the Paris World's Fair in 1878.

Also in the 1870s, John Ericsson, an American (designer of the steamship *Monitor*, the ironclad of U.S. Civil War fame), built several models of solar-powered steam and hot-air engines. Although they operated successfully, they could not compete economically with coal-fired boilers and did not undergo further development at that time.

At about this time (1870) the potential of solar energy for distilling seawater was recognized. Among several seawater–freshwater conversion plants built in the late 1800s, two are worthy of mention—one in Egypt and one in Chile. The Chile installation covered an area of more than 50,000 sq ft, and was reported to have produced 6000 gallons of fresh water per day.

As the twentieth century began, a solar-powered engine delivering 3.2 kW (4.5 hp) was built and installed at the Pasadena Ostrich Farm in southern California by a Boston engineer named Aubrey Eneas. More than 1700 mirrors were used to collect and focus the sun's rays on the steam boiler. This engine was capable of pumping water at a rate of 1400 gal/min. In about 1908, there was an early attempt to use flat-plate solar collectors to heat water; and in 1913, parabolic (concentrating) collectors were used to furnish solar energy for irrigation pumps along the Nile River south of Cairo. An American named Shuman and an Englishman named Boys were the organizers and engineers for this project. It is reported that their solar-driven engine developed nearly 100 horsepower (hp), an output (for solar) unheard of up to that time.

From about 1905 until 1930, there was an intensive development of small, roof-mounted solar water heaters for home use in southern California, Arizona, and Florida. The solar water-heating industry was just beginning to enter the mass-market phase in these states when low-priced natural gas made its appearance on the scene, with piped distribution to urban and suburban areas. After about 1940, solar water heaters for domestic use gradually disappeared until the energy "crisis" of 1973–74 spurred new interest in solar technology.

One of the first solar energy space-heating systems was designed and installed in Scotty's Castle at the Death Valley Ranch in California about 1930. It was built around flat-plate solar collectors, which provided heat for both space heating and domestic hot water.

Probably the first scientifically oriented, research-based program for solar energy development in the United States was conducted in the *Solar House* at the Massachusetts Institute of Technology, beginning in 1939. The *Solar House* was used as a laboratory, and a careful, instrumented, theory-based approach to the study of solar energy was begun. Solar collectors, heat exchangers, thermal storage tanks, and control systems were all involved in the research, as well as construction elements involved in the design of passive solar heating systems, such as direct-gain, south-facing windows, and thermal storage materials. With the approach of World War II, this and other solar research activities had to be set aside, and intensive solar research did not begin again until about 1950.

1-3   More Recent Developments—1950 to the Present

Assisted to some extent by research breakthroughs made during World War II, especially in the fields of solid-state physics, metallurgy, heat transfer, and radiation theory, solar energy research and development moved ahead at an accelerated pace in the 1950s. Improved solar collectors, both water-heating and air-heating, were soon developed. Better heat transfer materials for collector absorber plates and heat exchangers allowed for increased efficiency in converting solar radiation energy to usable heat energy, for both water-heating and air-heating systems. Research data on thermal storage materials and on time lag in thermal conduction provided the basis for improved design of passive solar structures.

In the 1950s, the MIT project resumed, this time featuring actual houses in which families lived. Also during this period, the Dover House (also in Massachusetts) was designed by Dr. Maria Telkes and associates to demonstrate the feasibility of complete solar heating in the Boston area. Dr. George Löf and his associates designed and built solar houses in Boulder and Denver, Colorado, in the 1950s. The solar heating systems (space heating and domestic hot water) of these houses are reported to be still in operation to this day. Among other researchers during this period was Harry Thomason of Washington, D.C., who built several solar houses equipped with liquid-heating solar collectors in which water trickled down in channels formed by corrugated (and blackened) aluminum sheeets as absorber plates.

By 1975, some 100 solar-heated houses had been built in the United States, and many others were reported from Australia, Canada, Great Britain, and France. Solar water heaters were in extensive use in Israel, Australia, and Japan by that date. These three countries, along with France, have been very active in solar energy development in recent years. Various combinations of solar water heaters and electric heat pumps as systems for space heating had been installed and tested by 1978, with excellent results in terms of system efficiency.

Spurred by the cutoff of oil from the Arabian Gulf in 1973–74, the U.S. government initiated an extensive program of energy research under the Energy Research and Development Administration (ERDA). By 1975, some $55 million had been allocated to solar energy research and development. At that time, ERDA planners suggested that attainable targets for the solar share of total U.S. energy needs would be 0.8 percent by 1985, 7 percent by 2000, and 25 percent by 2020.[1]

The 1973–74 oil embargo lasted only a short time, and as the flow of Middle East oil resumed, the urgency of the mid-1970s subsided somewhat. However, there has been continued federal support for solar energy research, and a number of centers for such research have been established throughout the United States. Some of these are associated with major universities, and some operate as units of national research laboratories (Fig. 1-1). We shall have occasion to mention the work of several of these laboratories in later chapters. Much of their work has been funded by various agencies of the U.S. government, including the Department of Energy (DOE), the National Aeronautics and Space Administration (NASA), and the National Oceanographic and Atmospheric Administration (NOAA). In some recent years, federal funding for research and development has been in the hundreds of millions of dollars. Many large corporations, too, have allocated large sums of money to solar energy development. Some of these are primarily "energy companies" (i.e., producers and marketers of oil, natural gas, or coal) that see ventures into the solar energy field as natural extensions of their present operations; others are aerospace corporations that see the solar energy industry as a profitable way to

---

[1] Richard C. Jordan and Benjamin Y. H. Liu, eds., *Applications of Solar Energy for Heating and Cooling of Buildings* (Atlanta: American Society of Heating, Refrigerating, and Air Conditioning Engineers, 1977), p. I-3.

Figure 1-1 One of the modular test buildings at the Los Alamos Scientific Laboratory in Los Alamos, New Mexico. The building is a solar energy laboratory equipped and instrumented to study every facet of active solar heating. (U.S. Department of Energy.)

use their high-tech research findings, manufacturing facilities, and specialized personnel.

1-4   Some Limitations on Sun Power

Plentiful and inexhaustible though it is, solar power has some limitations that must be recognized at the outset. Some of these are listed here, and others will be examined as we develop the principles of solar energy systems design in subsequent chapters.

1. The sun's energy is not concentrated on the earth's surface to provide large amounts of energy *at high temperature* in a small area. Large collection areas are required.
2. Solar energy is intermittent, because night and day, cloudy days, and the seasons of the year affect its reception on earth. Consequently, it must be collected when available, stored in large quantities, and used later as energy demands dictate.
3. As with all heat energy processes, the conversion of solar energy to usable heat or mechanical energy is accompanied by significant losses. Depending on the type and design of the solar energy system, efficiencies range from about 12 percent to a high of about 60 percent.
4. Capital investment for equipment is large. Although the sun's energy itself is free, the equipment and systems necessary to capture it and put it to work are expensive. This factor may well become less inhibiting, however, as mass production of solar energy components becomes commonplace.
5. Generally speaking, at locations and times of greatest energy need, solar energy is least available—for example, in winter in extreme northern (and southern) latitudes, in urban areas where high-rise and closely spaced buildings restrict solar access, and on cloudy or overcast days.

Despite these limitations, as the world's population climbs toward 5 billion, and as other energy sources become scarcer and higher priced, the era of solar energy has now really begun. Just as other industries (electronics, communications, transportation, and aerospace) have mushroomed in response to demand and as the direct result of systematic research and development programs, so will the solar energy industry grow in the years ahead. Some of the limitations listed above, and others to be dealt with later, will be overcome as research breakthroughs and mass production methods combine to create better and less expensive solar components. Moreover, the era of cheap fossil-fuel energy is over, and market forces alone will cause a major shift to renewable sources of energy. Of these, solar energy—energy from the primary source of all energy in our planetary system—holds the greatest promise.

1-5   Types of Solar Energy Systems

The use of solar energy takes many forms as it contributes to the energy needs of homes, commercial and institutional buildings, and industrial plants. The following paragraphs provide a brief introduction to the general classification of solar energy systems.

Solar Water Heating.   Of all fluids, water is the most suitable heat transfer and heat storage material, not only because it is readily available and cheap but also because of its excellent thermal properties. Consequently, even though the end output of a solar energy system may be mechanical energy, warm air, or heated industrial fluids, the process most often begins with the absorption of solar radiation by water. Moreover, heated water is often the desired end point, as in domestic and industrial hot-water systems, hot water for space heating, and heated water for swimming pools, spas, or hot tubs. Flat-plate solar collectors or parabolic (concentrating or focusing) collectors are used to heat the water—flat-plate collectors for water temperatures up to 200°F, and concentrating and/or tracking collectors for higher temperatures.

Thousands of solar hot-water heaters are operating in the United States and abroad at present (Fig. 1-2). Most of these are for residential hot-water systems, but many also serve industrial uses, in laundries and in plants for processed-food manufacturing. As the cost of conventional fuels continues to rise, more and more owners of swimming pools and spas are installing solar heating systems to provide comfortable water temperatures.

Space Heating and Cooling.   For the heating of interior spaces, hot water from solar collectors is stored in large thermal storage tanks, from which heat may be supplied to the space by conventional radiators or baseboard units, or by hot-water coils in a forced-air heating system (Fig. 1-3). Air-heating solar collectors also may be used, with the heated air being circulated to a rockbed thermal storage, from which it is then circulated to living spaces as needed to satisfy the heating load.

Solar-energized cooling systems use solar energy instead of mechanical or electrical energy to operate conventional refrigeration and air-conditioning equipment. Absorption refrigeration, rather than compression refrigeration, is most often used when solar radiation is the system energizer.

The space-heating systems referred to so far are called *active solar energy systems*, since they involve the active participation of mechanical equipment. However, given proper siting and architectural design, structures in many locations can be adequately heated by the sun's rays alone. The architectural design of the structure is such that it "lives with the sun." Such structures feature large areas of south-facing glass, interior materials that are excellent heat absorbers, ample insulation along with weathertight construction, and generous roof overhangs or shading allowing the low winter sun to shine directly into the space but blocking out the high summer sun. Designs of this type are called *passive solar energy systems* (Fig. 1-4).

(a)

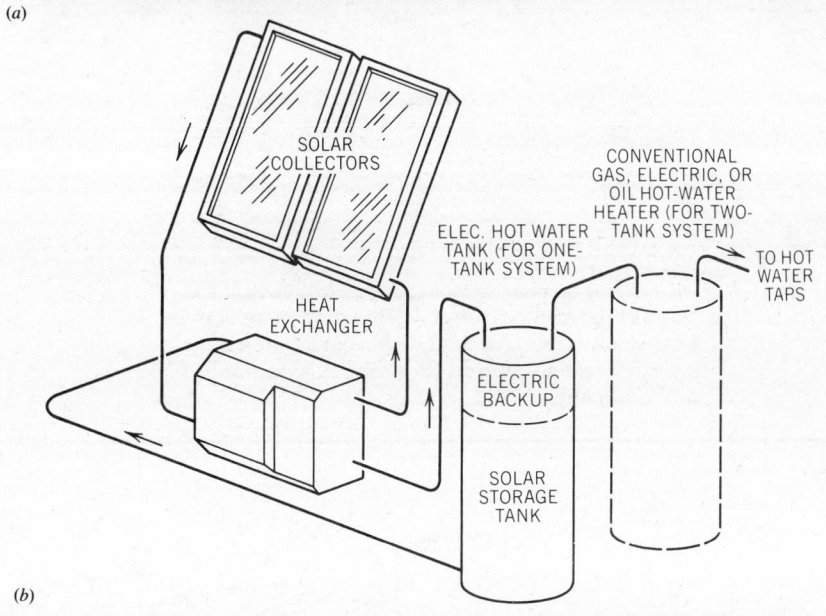

(b)

Figure 1-2 Domestic water heating involves the largest number of active solar energy installations. (a) Water-heating solar collectors on a house in New England. (b) Schematic flow diagram of a solar-energized domestic hot-water system. (Grumman Energy Systems Company.)

Electricity from Solar Energy—Photovoltaics.   The photoelectric cell was originally developed in the 1870s, using selenium as the photosensitive material. (A *photosensitive material*, as the term is used here, is one that gives off electrons when its atoms are struck by light.) As the science of solid-state physics advanced, silicon emerged as the favored semiconductor, and *silicon solar cells* or photovoltaic cells (PVCs) became useful generators for small amounts of electric power. By 1955, the silicon solar cell had reached an efficiency (solar energy to electric energy conversion) of more than 10 percent. Currently (1985), mass-produced, single-silicon solar cells routinely convert sunlight to electric energy at efficiencies of 12 to 15 percent. Materials other than silicon (cadmium, gallium, etc.) are also undergoing test and development, but silicon remains the dominant factor in the PVC industry at present.

Space Heating

There are two basic types of solar space heating systems—
those that use liquids to collect the solar energy and those
using air. The decision to use either a liquid or air system
depends upon design and application.

Also to heat a home adequately through hours of darkness
and during overcast or cloudy weather, heat energy must be
stored in a large reservoir during sunny weather and be
available for use when needed.

The first system circulates liquid, such as water or antifreeze
solution, through the collectors and into an insulated storage
tank as shown. A tank may hold anywhere from 500 to
2,000 gallons of water, depending on the area to be heated.

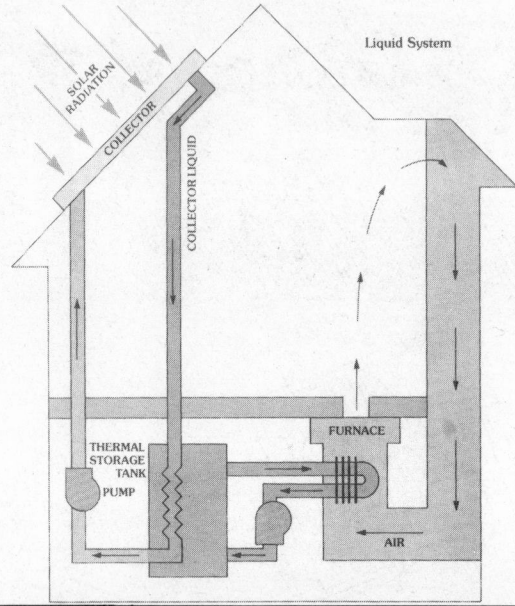

Figure 1-3 Schematic representation of an active
solar space-heating system with a conventional furnace as a backup heater. (Southern California Edison Company.)

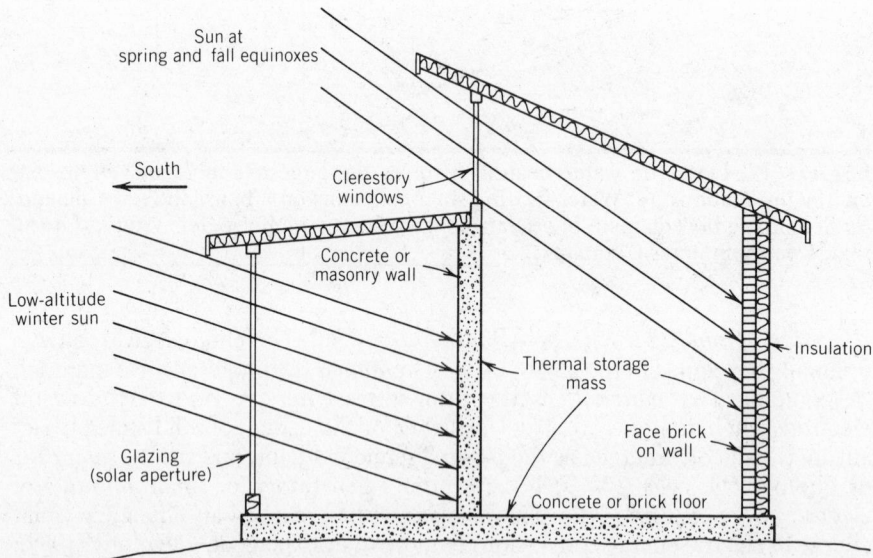

Figure 1-4 Cross-section diagram of a passive solar house. Note the south-facing
glass, the distribution of thermal storage mass in the heated space, and the use of
ample insulation to prevent conduction heat losses.

Figure 1-5 A solar-powered service station in San Diego, California. The roofs are covered with photovoltaic cells arranged in modules and arrays. A tie-in to utility power is necessary for night operation. (ARCO Solar, Inc.)

Panels and arrays of solar cells can be connected to deliver small or large amounts of electric power for use in remote locations or to energize signaling and communications equipment. Recently, a few very large photovoltaic generating plants, featuring PVC arrays many acres in extent, have been built to feed electric power into the regional network of a public utility. Photovoltaic power is finding ready acceptance in many homes and commercial establishments, often tied in with public utility power as a backup (Fig. 1-5).

Solar cell arrays power many satellites and provide onboard power for space vehicles as well, and some "futurists" envision an era when huge satellites in synchronous orbits 22,500 miles above the earth will beam power to ground receiving stations from solar cell arrays measuring thousands of acres in extent.

Solar-Thermal Power. Concentrating (sun-tracking) reflectors in large arrays hundreds of acres in extent can collect vast amounts of solar energy. Several such systems have been installed in the American Southwest, some for test and research purposes and others for supplying energy to specific installations. In one design format, the reflectors all focus on a receiver-steam generator, from which high-pressure steam flows to turbines to generate electric power. At least one public utility already (1984) has such a plant feeding solar-generated electricity into its distribution network (Fig. 1-6).

In this book you will find an in-depth treatment of solar water-heating systems, including residential and industrial water heating, swimming pools, and spas; space-heating and -cooling systems, both active and passive; and photovoltaic systems. All of these will be dealt with in great detail from the point of view of system design and operation. A descriptive but not design-oriented treatment will be accorded the topic of solar-thermal power.

The four topics that follow, although briefly described here to complete the orderly classification of solar energy systems, are not given any further treatment in the book.

Wind Energy Conversion. Wind is the result of the sun's uneven heating of the earth's surface. The wind systems of the earth are predictable within certain limits, and specific locations (mountain passes or across the broad reaches of the open plains) are known to have fairly steady winds at velocities consistent with significant amounts of power. Ordinary windmills have used wind energy for

Figure 1-6 Solar One—the world's largest solar-thermal electric power plant, located near Barstow, California. The heliostats in the huge array field all focus on the receiver–boiler at the top of the tower. High-pressure steam is generated in the boiler for the operation of conventional turbine generators. At full capacity, the plant generates 10 megawatts (MW) of alternating current (ac) power, enough for 5000 typical family homes. (Southern California Edison Company.)

centuries as they pumped water and ground grain. Today, huge wind-driven generators are being erected in suitable locations, and the resulting electric power output is fed into a public utility grid (Fig. 1-7). There is also a considerable expansion in the small wind-generator market for use on ranches, at resorts, or in other remote locations.

Figure 1-7 Wind turbine generators offer promise of significant amounts of electric energy. The large wind turbine unit shown produces 1300 kW at full power. (Southern California Edison Company.)

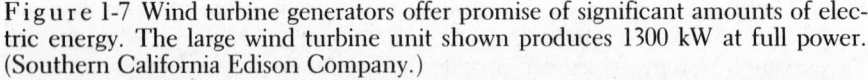

Ocean-Thermal Energy Conversion.   Heat from the sun is also the cause of the major currents in the world's oceans. Warm water from the tropics flows generally northward and southward to the poles. There, as the water chills, it sinks to great depths because of its greater density and slowly flows back toward the equator in deep ocean currents hundreds or even thousands of feet below the surface. At certain locations (in the Caribbean Sea and off Hawaii, for example), the temperature *difference* between surface water and water at the 2000-ft level is in excess of 40 Fahrenheit degrees (F°). With this temperature difference and hundreds of cubic miles of ocean water to work with, all that is required to extract incalculable amounts of energy is a suitable engine. No such energy conversion system has yet been perfected, but pilot models are currently being developed, and at least one model is currently being tested off Hawaii (1984). The engine itself is a low-pressure turbine that operates with fluorocarbons as the working medium instead of steam. Figure 1-8 shows a mockup of an ocean thermal generator.

Figure 1-8 Model of an ocean thermal energy conversion (OTEC) plant. Intended for installation on free-floating ocean platforms, these units operate on a Rankine cycle, using ammonia as the working fluid in the turbine. This cycle operates on the temperature difference (about 40°F) between surface water and water at 2000 to 4000 ft deep. The design concept for this particular configuration envisions a platform 340 ft across and 17 stories high. Expected output at full power is 100 MW, enough for a city of 60,000 population. Note the relative size of the "workers" shown at the several levels. (U.S. Department of Energy and TRW, Inc.)

Seawater-to-Freshwater Conversion.   Distillation of seawater is one of the very early uses to which solar energy was successfully applied. Throughout the world, in hot and arid coastal areas, solar-energized seawater conversion plants are in use or being planned for both domestic (potable) water and irrigation water. As conventional fuels become scarcer and more costly, solar-energy seawater conversion will become an absolute necessity in many parts of the world.

Biomass Energy Conversion.   The sun is the primary agent in the production of all life on earth. Plant life on earth and in the oceans grows and produces organic matter in quantities so large that their magnitudes can hardly be expressed. This *biomass* is the result of the sun's energy having arrived on the earth in the recent past. The production of heat, electricity, or fuel from the processing of biomass and/or organic waste materials, including garbage, is called biomass energy conversion. Many urban localities throughout the world now have biomass conversion facilities generating heat and electricity for local use. Although at first thought it seems that such installations produce energy from organic matter, in reality they are *recovering solar energy* that arrived on earth in the relatively recent past.

ENERGY TODAY AND FOR THE FUTURE

What is the actual world energy situation today, and, more specifically, what are the best estimates of the energy future of the United States? Since solar energy development will depend in part on the availability and price of other forms of energy, a brief look at the total energy picture is in order.

1-6 Primary Energy Sources of the Twentieth Century

In 1900, horses and mules supplied about 40 percent of the energy demand in the United States. Human muscle supplied about 25 percent, and machines (steam, internal combustion, and electric) contributed the remaining 35 percent. By 1950, machines had a 95 percent share, with the remaining 5 percent about equally divided between animals and humans. Today, with an electric or small engine-powered device for every purpose (chainsaws, eggbeaters, even toothbrushes), the machine share is close to 100 percent.

Conventional Energy Sources—The Fossil Fuels.   Coal, oil, and natural gas are called *fossil fuels* (from the Latin meaning "to dig up from the earth").
    Coal was the energizer of the Industrial Revolution, and it has remained an important fuel to this day. In fact, as a result of steps taken to reduce dependence on foreign oil, the use of coal in the United States has increased in recent years until, in 1984, it supplied about one fourth of the total energy used by the nation's economy. Estimates of the world coal supply vary widely, but there is some general agreement that, even with sharply rising demand, reserves are sufficient for 200 to 400 years. The United States is fortunate to have extensive coal reserves, making coal our most abundant energy source given present technologies.[2] However, coal has many disadvantages. It is hazardous and environmentally damaging to mine and therefore quite expensive; burning it pollutes the atmosphere unless high-cost control measures are introduced; it is difficult and expensive to handle and transport; and as demand increases and the more easily mined deposits are depleted, its price will continue to rise.
    The world oil supply–demand equation is well known, being in the news headlines on a week-to-week basis. Although there is no actual shortage at

[2] It was estimated in 1981 that the United States had more than 20 times as much energy in coal reserves as in oil and natural gas reserves combined. (*The National Energy Policy Plan*, U.S. Department of Energy, July 1981.)

present (1984), political instability in any one or several of the OPEC consortium nations could trigger a crisis at any time. Domestic oil supplied about 22 percent of the 1983 energy demand (about 10 million bbl per day), while imported oil supplied about 15 percent (7 million bbl per day). Petroleum—domestic and imported—provided nearly 40 percent of total U.S. energy in 1984, and imported oil constituted 40 percent of total oil consumption.

Estimates of world oil reserves vary widely. So-called "known reserves" are supposed to be about 300 billion bbl. Knowledgeable experts calculate that, at even the present rate of consumption, these reserves will be seriously depleted by about the year 2030. The year 2040 may signal the end of the petroleum era. With this relatively short time frame in mind, and with the havoc created in the national economy by exporting dollars in order to import oil, every possible effort must be made to develop indigenous energy sources for the immediate future. This is not to mention the ever-present possibility that a major share of foreign oil might be cut off at any time because of war or political instability in the Middle East or Latin America, leaving the United States with a catastrophic energy gap.

Natural gas supplied about one fourth of the U.S. energy demand in 1984. The peak of domestic natural gas production probably occurred about 1980, and it is doubtful that new discoveries in the future will match the rate of depletion of known reserves. Natural gas is an ideal fuel, rich in energy (1100 Btu/cu ft), easily transported in pipelines, clean-burning, and nonpolluting. Again, it is impossible to make accurate projections of future supply, but it is expected that by the year 2030 reserves will be depleted to the point where sustained production at anything like present rates will be impossible.

The Unique Values of Fossil Fuels.   Petroleum, coal, and natural gas are remarkable storehouses of energy. One gallon of gasoline, for example, will operate a 15-hp engine under full load for nearly an hour; or, in more familiar terms, it will propel a 4000-lb automobile along a level highway at 60 mph for 20 mi. The unique value of petroleum as a fuel is that it is so well suited to providing *mobile power*, power for propulsion where the *propellant must be carried along with the vehicle being propelled*. Coal and natural gas are not nearly as well suited to this purpose, unless they are first liquefied at considerable expense. Fossil fuels are so extremely valuable as primary sources of mobile power that it is folly to burn them in stationary plants merely to provide Btus or to generate electricity. The use of oil and natural gas especially, for the central station generation of electricity, is nothing short of tragic. Virtually all sea and air transportation is dependent on fossil fuels. So, too, is a major share of land transportation, although electrification of land transport, or a major share of it, is technically feasible. The mobility of the future is being traded for heat and electricity now. Those of us alive today may not live to regret it, but generations to come will pay the penalty for our unwise use of precious fossil fuels. It is not inconceivable, in this regard, to envision a twenty-first century without commercial air travel, for example.

Fossil fuels also are our most important source of chemical feedstocks, with end products ranging from fertilizers (absolutely essential to feed the world's population) to plastics, textiles, and pharmaceuticals. With proper planning and development, fossil fuels as heat producers can be largely replaced by other forms of energy, including solar energy, thus allowing oil, coal, and gas to serve far more important uses than the mere production of heat.

1-7   New, Promising, and Alternative Energy Sources

There are numerous alternatives to continued escalation in the use of oil and natural gas. Discussion of some of these alternatives follows.

Hydro and Geothermal Energy.   Presently, hydro power provides about 4 percent of the nation's energy, most of this output in the form of electricity.

Geothermal power accounts for only a fraction of 1 percent. Possibilities exist for a significant increase in hydro power, but environmental considerations inhibit plans for damming up many more rivers to create hydro reservoirs. "Small hydro"—the installation of small turbogenerators on creeks and rivers without creating large reservoirs—is being encouraged by federal policy, and power companies in many states are required by federal law to purchase the output of small hydro plants at rates set by state public utilities commissions.

Only a few localities are suited to geothermal power generation, three promising areas at present being the Geyserville (California) region, the Imperial Valley (California) region, and a large area in north-central Wyoming. Many other localities are known to exhibit geothermal activity, and the U.S. Geological Survey continues to encourage exploration for possible sites. This agency estimates that there are as many as 100 geothermal sites in the United States that have sufficient long-term energy potential to justify a geothermal electric plant. Some of these sites may have capacities of less than 50 megawatts (MW), but others are estimated at 3000 MW or more. A full development of the geothermal resource could, according to knowledgeable estimates, result in a 5 to 10 percent share of the total U.S. electrical energy demand, continuing for hundreds of years.

Nuclear Energy.   As of 1970, nuclear (fission) power was considered as the probable major energy resource of the future. By 1982, some 77 nuclear fission plants were in operation, producing about 13 percent of all electric power then being generated in the United States. This output is equivalent to that available from 1.4 million bbl of oil per day. Forecasts in the early 1970s by the Atomic Energy Commission (AEC) identified targets as follows:

By 1985	140 nuclear plants on line	20% of U.S. output
By 2000	1000 nuclear plants on line	60% of U.S. output

The so-called fast-breeder reactor (FBR) was scheduled for operation by 1990 to provide even greater generating capacity. It was to operate on plutonium, which is continually produced ("bred") in the reactor itself from ordinary uranium, thus bypassing the need for U-235, a very scarce and expensive isotope of uranium.

For a variety of reasons, however, these targets for nuclear power have proved to be unrealistic. Cost overruns, problems of disposing of radioactive waste, inadequate future supplies of U-235, concerns over operational safety, fear of thefts of nuclear material by terrorist organizations, and litigation resulting from sociopolitical opposition have all combined to relegate nuclear fission to a position that is currently not in the front rank of U.S. energy options. New plants are not being planned, contracts are being canceled, some plants under construction are not being completed, and there are even a few cases where fully completed plants have not been licensed for full-power operation. For the present, at least, it appears that nuclear fission reactors will not increase their share of the nation's electrical energy output much beyond the 12 percent share they now have (1984).

Nuclear fusion is a promise for the future, not a reality. The thermonuclear processes that provide the sun's energy and that operate in the hydrogen bomb are not yet subject to controlled operation for the production of usable energy. Research continues on the seemingly insurmountable problems of controlled nuclear fusion, both in the United States and abroad. Occasional breakthroughs are announced, but the consensus of scientists is that actual operating nuclear fusion plants will not be a reality before about 2030. There is no unanimity of opinion even on this date, and the possibility must be admitted that controlled nuclear fusion might elude us for a century or more.

If and when thermonuclear electric-generating plants are put on line, and if they prove to be safe and cost-effective, the world energy shortage will disappear, although some problems related to what we have referred to as *mobile energy sources* could still be troublesome. With thermonuclear fusion,

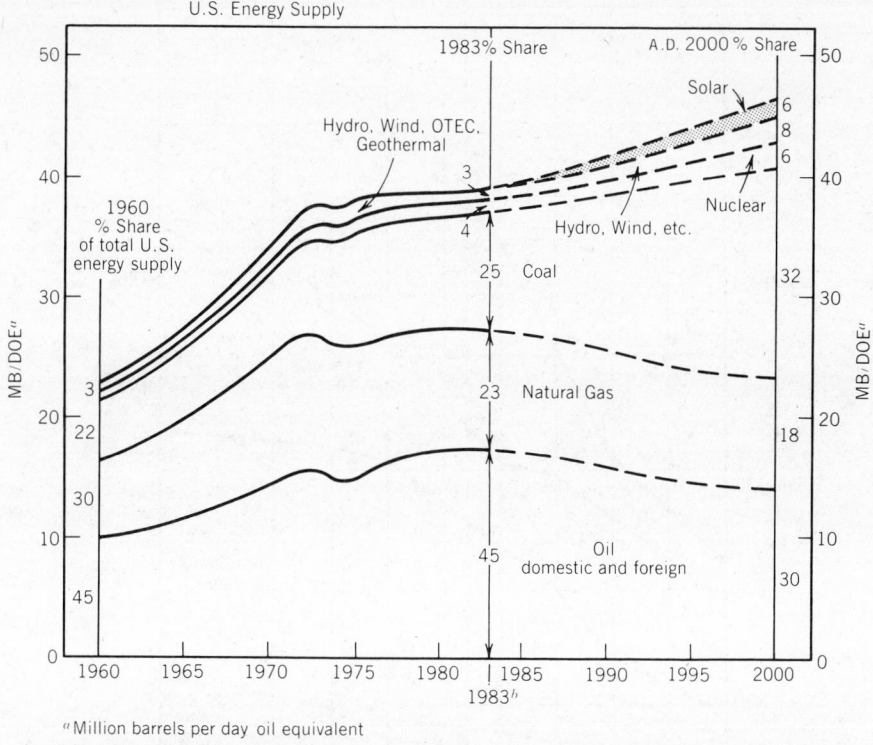

U.S. Energy Supply

1983 % Share

A.D. 2000 % Share

Solar

6

8

6

Hydro, Wind, OTEC,
Geothermal

3

Nuclear

1960
% Share
of total U.S.
energy supply

4

Hydro, Wind, etc.

25   Coal   32

3

23   Natural Gas   18

22

30

45   Oil
domestic and foreign

45   30

1960 1965 1970 1975 1980 | 1985 1990 1995 2000
1983[b]

MB/DOE[a]

a Million barrels per day oil equivalent
b Data after 1983 are estimates.

Figure 1-9 Chart showing the supply levels and percent share for various fuels and
other energy sources in the United States. Supplies are expressed in millions of
barrels per day of oil equivalent (MB/DOE). Coal usage is expected to increase
sharply even though its environmental problems are serious. Nuclear energy is
currently in a slump because of environmental and socio-political opposition, and its
share may not increase appreciably in this century. The growth shown for solar
energy is a very conservative estimate. Some forecasts by solar enthusiasts predict a
20 percent share for solar by A.D. 2000.
Sources: U.S. Department of Energy Information Administration and solar energy
industry forecasts.

we would be, as the saying goes, "back to square one," tapping the energy of the
atomic nucleus in the same primeval process that occurs in the interior of the
sun. We would be generating solar power on earth.

Figure 1-9 portrays the supply levels of the major sources of energy in the
United States from 1960 to 1983, with extrapolations to the year 2000. Figure
1-10 indicates actual and projected total energy demand, based on assumptions
that the population and the economy will grow at about the same rate in the
1985–2010 period as it did in the 1960–1985 period. Note that the energy
(vertical) scales are expressed in millions of barrels of oil per day equivalent
(Million B/DOE) and also in Quads (Q) (1Q = 10^{15} Btu).[3]

An alarming trend is readily noted from the chart—the gap that opens up
after about 1990 between projected demand and total possible supply *from all
currently known domestic sources.* Obviously, if these projections are even
approximately correct, only three options or courses of action are open:
(1) Scale down demand, with all of the difficulties associated therewith (economic stagnation, unemployment, falling standards of living); (2) return to the
nuclear option and get back on the schedule set by the AEC in the 1970s; or

[3] One *Quad* is 1 quadrillion (1000 trillion) Btu, an almost incomprehensible amount of energy.
In 1983, the total energy demand in the United States for the entire year was about 110 Quads,
equivalent to 52 million B/DOE.

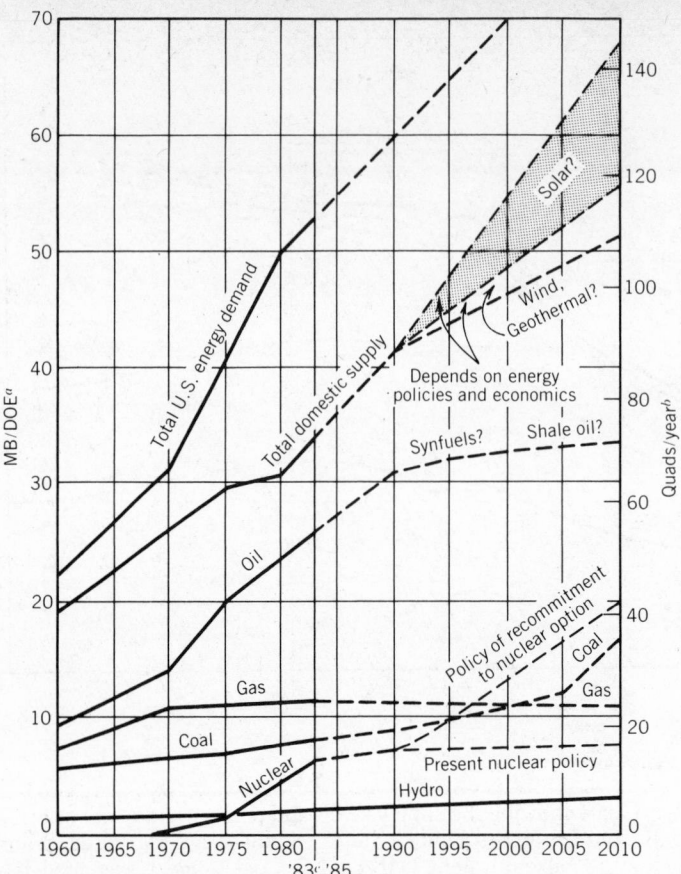

[a]Million barrels per day oil equivalent.

[b]1 Quad equals 10^{15} Btu.

[c]Curves beyond 1983 represent estimated values.

Figure 1-10 Chart showing the actual U.S. demand in MB/ DOE and in Quads per year of energy from various fuels and energy sources. The extrapolations beyond 1983 are subject to many uncertainties, including such factors as international trade (oil), sociopolitical attitudes (nuclear), environmental concerns (coal and geothermal), and research and cost breakthroughs (solar). Of all domestic energy sources, only solar and nuclear have more or less unlimited possibilities. As the chart indicates, by the year 2000, solar energy could well be a critical factor in narrowing the energy gap.

Sources: U.S. Department of Energy and solar industry analysts.

(3) develop alternative sources of energy, including energy from that greatest of all sources, the sun. If we continue on the present course, the energy gap will have widened by the turn of the century to a level of about 25 million B/DOE.

Coal Gasification and Liquefaction. As noted above, coal is our most plentiful energy resource, but the problems of mining it, transporting it, preparing it for use, reducing pollution, and repairing environmental damage all work against its massive use to close the energy gap. Some of these negative factors could be eliminated by the gasification and/or liquefaction of coal at the mine site. The resulting gaseous or liquid fuels could then be transported in pipelines and used as natural gas and oil are today. Pilot plants are already in operation developing and testing the technologies for these conversions. They may be successful, but they certainly will be expensive.

Oil Shale.   Billions of barrels of oil are locked up in oil shales, mostly in the Rocky Mountains. Thus far, although a few pilot plants have been in operation for several years, methods of extracting oil from these shales have not been cost-effective in the present energy market. Any significant interruption in the flow of imported oil or a substantial price increase by the OPEC consortium could change this situation overnight, however. The technology already exists for shale oil extraction, but there remain serious problems of cost and environmental impact yet to be addressed.

Tidal Power.   The enormous potential energy available in ocean tides is not the result of the sun's *heat*, but of forces of *gravitational attraction* among and between the earth, the sun, and the moon. This possible energy source, though significant in some parts of the world, is not one of great promise in the United States. There are few sites along our coasts that are well suited to the development of tidal power, and, further, the technology for converting tidal power to usable forms of energy has not been intensively developed, except for a few sites where the tidal changes result in hydraulic heads (i.e., difference in water levels) of 20 ft and more, as at some locations on the coast of France and in the Bay of Fundy between Nova Scotia and New Brunswick in Canada. It is doubtful that tidal power will make any appreciable contribution to U.S. energy needs in the foreseeable future.

In addition to the alternative sources of energy briefly identified above, other solutions to the energy problem are often subjects of scientific discussion or, more often, media speculation. Some of these are:

1. Attempting to harness the incalculable energy generated by storms—hurricanes, tornadoes, lightning, and so on. As yet, no feasible system has been proposed to absorb and collect such fantastic amounts of energy from the fury of the storm, let alone store the energy for orderly use.
2. Drilling extremely deep holes into the hot core of the earth to bring the earth's residual heat to the surface.
3. Drastic curtailment of energy demands by strict conservation measures and by mandated changes in societal life-styles and the economies of the developed nations. Proponents of this "limits-to-growth" alternative envision cottage industries powered by solar collectors or wind turbines, organic farming with a return to animal and human labor, increased use of sailing ships, decentralized energy production, a scaled-down economy, and a far less mobile population. Attractive as this alternative may sound at first—a return to a simpler, less hectic, closer-to-nature life-style—the concept fails to cope with the economic and demographic realities of the world as it is.

Nearly all of these alternatives have a contribution to make to world energy supplies; indeed, a number of them, taken singly or in combination, could narrow the energy gap appreciably. We have purposely explored these limited-potential alternatives first, leaving the really viable and ultimate energy source—the sun—for final emphasis as an introduction to the theme of the book itself.

1-8   Solar Energy—The Viable, Renewable, Inexhaustible Alternative

Energy from the sun represents the earth's largest energy resource. Furthermore, its limits are not finite; that is, it is not a reserve in storage where withdrawals deplete the reserve. Solar energy is *continuously* being produced, and it arrives on earth in a continuous, inexhaustible stream. Using untold megawatts of solar power today does not in any way diminish tomorrow's supply.

If large polar regions of the earth are omitted, the average solar power input to the earth's surface is about 20 W/ft² (215 W/m²). In very sunny, arid regions, such as the Sahara Desert or parts of Arizona and New Mexico, the solar power at ground level is 260 W/m² (24 W/ft²) or more. Direct conversion to electricity by arrays of *photovoltaic cells* (PVCs) can now be accomplished (1984) at 10 to 15 percent efficiency. From these data a simple calculation reveals that 10 acres of sun-bathed Arizona earth surface completely covered with PVCs could produce an electrical output of more than 1 MW (1 million watts). Now 10 acres is scarcely more than an average city block, so the energy potential of large areas hundreds of acres in extent can readily be appreciated. Moreover, if the energy output desired is hot water rather than electricity, the conversion can be effected at efficiencies of up to 55 or 60 percent by *flat-plate solar collectors*, instead of the 10 to 15 percent now possible from PVCs.

Covering every square foot of a given site with solar cells is impossible, of course. The solar arrays must be spaced widely enough to allow for unobstructed solar access to all of the PVCs, and to facilitate service and maintenance. A 1-MW plant actually requires a total site area of about 15 to 20 acres. And farther north or in regions with frequent cloud cover or hazy atmospheres, perhaps 20 to 30 acres would be required.

In even more familiar terms for the residential user, 500 to 800 ft² of south-facing roof fitted with flat-plate solar collectors will collect enough solar energy at most U.S. locations to supply 50 percent or more of the total energy needs of a typical residence, winter and summer.

Solar energy represents a bountiful energy supply from a continuous, renewable, and inexhaustible source; moreover, it does not require transportation, it is available at the site where needed, it is not subject to the control of foreign powers or cartels, it allows for decentralization of energy if that is desired, and, of all energy sources, it has minimum adverse environmental impact.

Solar Power in the year 2000.   Long-term estimates of the contribution of solar power to the national energy supply have recently been made by a number of federal agencies, departments, councils, and commissions. During the period from 1977 to 1982, the President's Council on Environmental Quality, the Department of Energy (Office of Solar Applications and the Energy Information Administration), and the Executive Office of the President (National Energy Plan) all submitted major study reports dealing with solar energy. Nongovernment research groups, business firms, and "think tanks" were also commissioned to submit forecasts, among these being the Stanford Research Institute, the Mitre Corporation, and the Bechtel Corporation.

There was, understandably, no close agreement on the future contributions of solar energy, and yet there was enough clustering to justify averaging some of the forecasts. Table 1-1 attempts to present a sort of median-case view of these forecasts, grouped in seven classifications of solar technologies.

The solar energy contributions indicated in Table 1-1 will, if met, cast solar energy in a key role in the attainment of energy independence for the United States by the turn of the century. With the possibility of 10 to 15 Quads from solar energy by the year 2000, solar would be supplying nearly 15 percent of the national energy demand. If, indeed, 35 Quads can be extracted from the sun's energy by the year 2030, the energy gap foreseen by that date can be narrowed appreciably. By that time, solar-related energy technologies should be supplying nearly one fourth of the total energy demand of the United States.

What will be the nature and magnitude of the industry that will develop over the next 20 years—an industry that may be called upon to support the production of one fourth of all U.S. energy? Predicting the future is best left to fortune-tellers and futurists, and we shall not attempt it. Instead, a brief assessment of the solar energy industry as it exists today is given here, with the reminder that hardly one Quad of energy now results from its efforts. If the targets listed in Table 1-1 are met, industry growth for the next several decades can readily by imagined.

Table 1-1 Contributions Expected from Various Solar Energy
Technologies to U.S. Energy Needs in Selected Years

Type of Solar Energy Technology	Selected Target Dates (Quads/Year)		
	1985	2000[a]	2030[a]
Direct solar-thermal conversion—active and passive (space heating and domestic hot-water and pool heating)	0.30	2.8	6
Direct solar-thermal conversion (industrial process heat)	0.15	2.5	10
Solar-thermal electric power generation	trace	1.0	5
Photovoltaic electric power generation	0.05	2.0	6
Wind energy conversion	trace	2.0	3
Geothermal energy conversion	0.10	2.0	4
Ocean thermal conversion	none	0.5	1
Totals	0.60 Q	12.8 Q	35 Q

Sources: U.S. Department of Energy and study reports referred to on page 20.

[a] Estimates in these columns are quite speculative, being dependent on a number of unforeseeable events such as cost–benefit factors, decisions on nuclear fission power, the outcome of thermonuclear research and development, and world political and economic conditions.

THE SOLAR ENERGY INDUSTRY TODAY

As Table 1-1 indicates, if wind energy, geothermal energy, and ocean thermal conversion (which do not use *direct* solar radiation) are omitted, there are four main segments of the solar energy industry:

1. Direct solar-thermal conversion for space heating of buildings and for heating water for domestic hot-water systems, swimming pools, and spas.
2. Direct solar-thermal conversion for industrial process heat and for solar-energized cooling (high-temperature applications).
3. Solar-thermal electric power generation.
4. Photovoltaic electric power generation.

The first two of these industry segments are based either on the use of passive solar buildings, or on solar collectors and associated heat storage and distribution equipment. The third segment (solar-thermal electric power) usually involves solar reflectors (*heliostats*), complex sun-tracking equipment, and boiler–steam-turbine–generator equipment. The fourth (photovoltaics) is dependent on the use of semiconductors, with heavy involvement in metallurgy, chemistry, and electric circuitry. The manufacture, marketing, and installation of these sophisticated (high-tech) components and systems involves hundreds of millions of dollars of investment capital and thousands of professional, technical, and trade and craft workers.

Many system designs within the solar industry involve the interaction of the sun's radiant energy with heat-transfer equipment and heat-transfer fluids acting on or by means of mechanical or electrical equipment. Another equally important aspect of solar energy is known as *passive solar*—"living with the sun," without the use of mechanical or electrical devices. Homes and commercial buildings can be designed to use the sun's heat directly for space heating, storing heat in the mass of the structure itself. Space cooling can also be accomplished, to some extent, by structural elements that feature night ventilation and night-sky radiative cooling, without mechanical refrigeration. The passive-solar industry is also a dynamic and growing element of the national economy. It is closely allied with the construction industry and with architects and building designers. This book deals extensively with both passive solar and active solar systems design.

1-9 Present Status and Projected Growth of the Industry

The solar energy industry has been growing at a steady pace since the early 1970s. In the 1960s, most of the activity in the solar field was government-sponsored, centering around photovoltaic devices to power the onboard systems for satellites and spacecraft. After the 1973 OPEC oil embargo, however, the U.S. government initiated a great variety of activities to encourage the development of alternative sources of energy. Funds were made available, and commissions and research institutes were established for research and development in such fields as solar energy, wind energy, small hydro, geothermal, biomass, and ocean thermal energy conversion. In the late 1970s, the National Energy Plan was initiated. This plan included tax incentives and tax credits to encourage the installation of solar energy systems.

The use of solar energy, partly as a result of these incentives, has increased dramatically in the last several years, although the "oil glut" of the early 1980s has temporarily slowed growth in all alternative energy industries, including solar.

In 1980–81, the U.S. Department of Energy sponsored a nationwide survey of the solar industry to determine the approximate level of business activity. Among the findings were that in the 1980 calendar year, there were an estimated 113,600 active solar energy systems installed in the United States, with an estimated price tag of $437 million. By 1983, the figures had increased to 163,380 installed systems billed at $784 million.

Domestic hot-water heating accounts for the largest use of active solar energy systems, amounting to about 56 percent of the total systems installed in 1980 and nearly 70 percent of the systems installed in 1981. It is interesting to note that, in 1981, 86 percent of the solar hot-water systems installed for single-family residences were placed on *existing* structures. This market, called the retrofit market by the industry, offers almost limitless opportunity for solar energy consultants and contractors.

The use of solar heaters for swimming pool heating showed a decrease from about 30,000 installations in 1980 to about 21,000 in 1981. This drop in swimming pool solar installations is explained in part by the sluggish economy of 1981–82, and also by the fact that many of the federal and state tax incentives had by that time become less favorable for pool systems or had been eliminated altogether.

The number of active solar space-heating systems increased in 1981, and the dollar value of these systems showed a substantial increase, because many of them were located in colder climates where large systems with more solar collectors were involved. In 1981, the installed cost of an active solar space-heating system for typical residences ranged from a low of about $7000 up to $15,000.

The solar energy industry is expected to show substantial growth in the last half of the 1980s. Continually rising costs of natural gas, fuel oil, and electric energy will spur solar energy growth as homeowners and building managers see these costs for conventional energy sources doubling or perhaps tripling over the next decade. Also, there has been increased interest in passive solar buildings in recent years, as architects and designers have become knowledgeable about solar theory and practice.

The segment of the solar industry related to the generation of electric power—the photovoltaics industry—is in a steady growth pattern in the mid-1980s. The recent completion of several photovoltaic central power plants feeding electric power directly into public utility grids (see Chapter 18) and current plans for many more such plants indicate that the photovoltaics industry is just now entering a period of sustained growth. New plants with a total of more than 250 MW of electrical output from photovoltaic and solar thermal systems are in the planning process now, to go on line by 1990.

The President's Council on Environmental Quality, in a White House report of the future impact of solar energy, projected an annual photovoltaic output of 2 to 8 Quads (1 Quad = 10^{15} Btu) of energy (in the United States) by the year 2000. The total U.S. energy use in that year is projected to be 95 to 100 Quads.

Another assessment conducted under the aegis of the federal government was the *Domestic Policy Review of Solar Energy*, which was published in 1979. This study set a goal of 20 percent of national energy consumption to have its source in renewable energy by the year 2000. A major portion of this—perhaps two thirds—would come from solar in three technologies: active solar, passive solar, and photovoltaics.

In terms of market size, these estimates, assessments, and forecasts predict a U.S. solar energy market for the mid-1990s that could result in sales totaling anywhere from $10 billion to $50 billion. It is impossible to pinpoint an accurate figure because of the uncertainties about world oil and about the environmental results of an all-out move to coal. At one level of these uncertainties—an uninterrupted flow of oil (but at increasing prices) and progress on reducing pollution from burning coal—the growth of the solar industry would be steady but slow. On the other hand, if another "oil crisis" like that of 1973, or worse, occurs for any reason—political instability, terrorist acts, or war—then an explosive growth for all renewable energy industries, including solar energy, can be predicted with certainty.

Employment in the Industry. The U.S. Department of Commerce sponsored a study in 1977 by the Bechtel Corporation for the purpose of making energy forecasts to the year 2000.[4] A major section of the Bechtel report deals with labor force requirements in the energy industry. The expectation is that the solar energy industry will employ directly about 279,000 workers by the turn of the century. The annual growth rate of the solar industry work force for the period from 1985 to 2000 was estimated at 16 percent.

Another approach to work force estimation is through the dollar volume of industry sales. Let us assume a midlevel estimate of $30 billion in the above-cited range of $10 billion to $50 billion for the 1990s solar energy industry market. This kind of high-tech industry has higher than average labor costs since a larger share of its workers are well educated and highly trained personnel, many of them with college and graduate training in the sciences and technologies. An assumption that labor costs will be about 30 percent of the market price of the equipment and services provided by the industry is justifiable for this level of labor force. Thirty percent of $30 billion is $9 billion. At an average of $25,000 (1990) per year per industry worker, if we divide $9 billion by $25,000, an estimated 360,000 full-time employees is obtained. Many, if not most, of these persons will have to have either preemployment education in solar energy or extensive on-the-job training after employment.

Thousands of solar industry workers will be in professional occupations—scientists, engineers, architects, and business managers—but even more will be technologists, technicians, sales representatives, and highly skilled trade and craft workers. We turn now to a brief analysis of the careers that solar energy and its industry will provide.

CAREERS IN SOLAR ENERGY

Many readers of this book are students at colleges and technical schools enrolled in courses whose purpose is to provide education and training for careers in the solar industry. The analysis of sections 1-8 and 1-9 portray an industry

[4] Bechtel National, Inc., *Projected Annual Resource Requirements at the National and Regional Level* (San Francisco, 1977).

already organized on a sound technological and marketing base, ready for significant expansion and development in the 1980s and 1990s. Career opportunities will increase in numbers as the industry grows, and, just as in other industries, the best jobs in solar energy will be held by those men and women who have prepared themselves with a solid background in theory and practice.

1-10 The Work of Scientists, Engineers, and Architects

Any industry, in its developmental stages, requires intensive work in research and development, and the solar industry is no exception. Areas of solar energy needing further research include, but are not limited to, heat transfer, solar radiation absorption, selective coatings, semiconductor performance and the efficiency of solar cells, performance of solar components, controls and instrumentation, system design and long-term testing, design and performance of passive solar structures, climatology, computer simulation, and cost–benefit analysis.

Preparation for this level of work in the industry ordinarily requires, at the minimum, a college or university degree in one of the sciences or in architecture or engineering. The course of study should include or at least provide options for basic courses in environmental science and solar energy. Many workers in these professional-level jobs will have completed graduate studies to the level of the master's degree or the doctorate; others will have taken a professional degree, as in the case of architects and designers.

Engineering technologists and applications engineers may prepare for solar energy careers through bachelor of engineering technology (BET) and bachelor of industrial technology (BIT) programs in four-year technological universities and state colleges. Many of these institutions provide basic and advanced studies in solar energy.

1-11 Technicians in the Solar Industry

Engineering and industrial technicians will be in demand as the solar industry grows. Technicians will serve as small-system designers, specification writers, estimators, computer operators and programmers, job layout specialists, installation supervisors, startup and check-test supervisors, and operating engineers of large and complex systems.

In the manufacturing segment of the industry, technicians will find jobs as manufacturing process supervisors, research and test laboratory technicians, quality control technicians, foremen and supervisors, and sales and service representatives. There may emerge in the next decade the special job classification of *solar engineering technician*.

Engineering and industrial technicians are ordinarily graduates of two-year associate degree programs in community junior colleges, technical institutes, and technical colleges. Many of these institutions already offer high-quality programs and courses in solar energy technology, and the number of such programs is increasing rapidly. Persons already possessing basic technician training or experience can avail themselves of short-term training courses in solar technology at these colleges and technical schools.

1-12 Trade and Craft Workers in the Solar Field

After a solar energy system is designed and contracted for, the installation begins. Nearly all of this phase of the job—rigging, plumbing, electrical wiring, carpentry, masonry, insulating, glazing, sheet metal, auxiliary heating—fits into established job patterns in the building and mechanical trades. Training courses and apprenticeships for these careers are determined by the various trades themselves and by boards of apprenticeship standards in the several states.

One or more courses in the theory and practice of solar energy systems design will enable the journeyman mechanic or apprentice in any trade to be more effective on the job and to be more valuable to the builder or contractor who is his or her employer. The ability to work effectively on solar energy jobs will be a requirement of the future for many trade and craft occupations.

Short-term courses and programs for training in solar energy are available at more than 100 community junior colleges, public and private vocational–technical schools, and state-operated area vocational schools. Correspondence and home-study programs in solar energy are also offered by a number of recognized and accredited correspondence schools.

CONCLUSION

Solar energy is here—ready and waiting for us to use it. It can, in the short run, help close the energy gap; in the long run, it could be our major source of energy. The real tragedy of any future energy crisis is that there need not be a crisis at all. Solar energy can prevent that crisis if we proceed without delay to develop methods of using it efficiently and economically. This book is addressed to that challenge.

Basic Physics Review

S olar energy systems design is rapidly becoming a recognized engineering and technical field with its own body of knowledge and its own approved practices. However, the science of solar energy itself and the design and operation of solar systems equipment are based, for the most part, on well-known principles of physics. It is the purpose of this chapter to review such elementary physical principles and concepts as *temperature*; the *nature of heat*; *force, mass,* and *velocity*; *work, energy,* and *power*; *density*; and *pressure* and *pressure measurement.* Both the English (engineering) system and the SI–metric system of measurement will also be reviewed. Subsequent chapters will make use of somewhat more advanced physics concepts in developing theory and practice in solar systems design.

TEMPERATURE AND THE NATURE OF HEAT

The senses of touch and feeling tell us that a metal plate in direct sunlight is "hot" and that snow and ice are "cold." Temperature is a measure of hotness or coldness, but bodily sensations of hot and cold are not very accurate indicators. Instruments for the accurate measurement of temperature are called *thermometers,* and temperature is measured in units called *degrees.* Some common thermometers will be described and illustrated below.

2-1 The Fahrenheit Scale of Temperature

The first successful thermometer was a liquid-in-glass instrument devised by Gabriel Daniel Fahrenheit about 1715. He calibrated it in degrees to establish the Fahrenheit scale, and this scale is still used in many English-speaking countries, including the United States.

The Fahrenheit scale is based on two fixed points—the boiling point (BP) of pure water at atmospheric pressure, marked as 212 degrees Fahrenheit (212°F); and the freezing point (FP) of water, marked as 32°F. This allows for 180 equal divisions (Fahrenheit degrees) on the scale between the two fixed points. The scale is continued with the same intervals above the boiling point and below the freezing point for indications of more extreme temperatures (Fig. 2-1).

2-2 Types of Thermometers

Liquid-in-glass thermometers (often handheld) consist of a glass tube called the stem, with a small bulb at the bottom (Fig. 2-1). The slender bore in the stem must be of uniform diameter. The liquid that fills the bulb is usually either mercury or alcohol. The space in the stem above the liquid is evacuated. Changes in temperature of the substance in which the bulb is imbedded cause the liquid in the bulb to expand or contract, and since the bore of the stem is of very small diameter, only a small change in volume of the liquid in the bulb will result in a very significant movement of the liquid column in the stem. The expansion (or contraction) of the glass bulb and stem is very small compared to that of the mercury or alcohol in the instrument, and in ordinary industrial practice no correction need be made for this effect. Precision-grade thermometers for scientific research are compensated for this differential expansion, however.

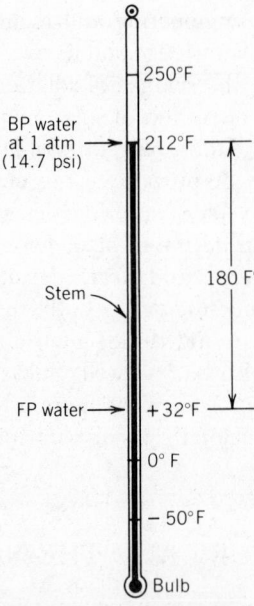

Figure 2-1 The Fahrenheit temperature scale. The fixed points are the boiling point (BP) and the freezing point (FP) of water at atmospheric pressure. The space between these two fixed points is divided into 180 equal degrees.

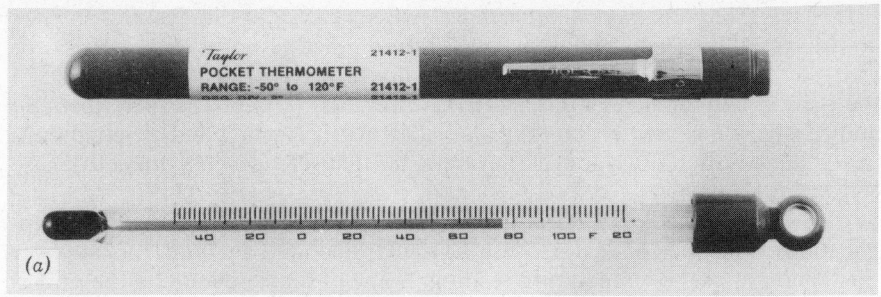

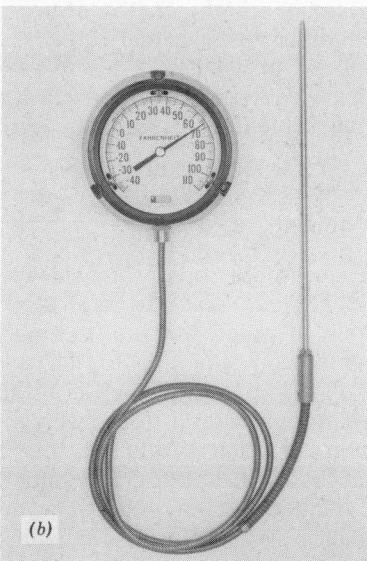

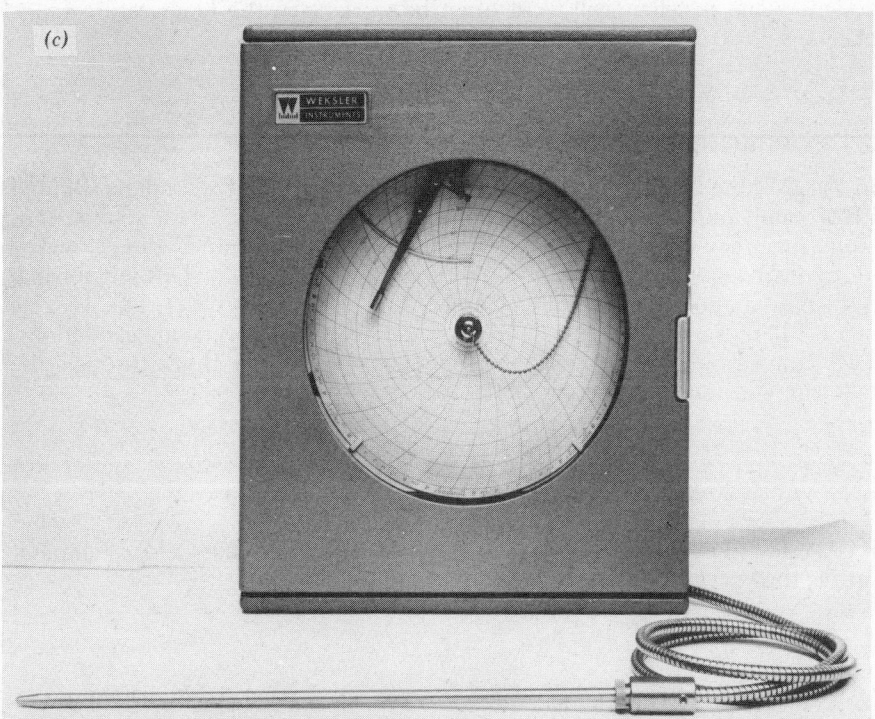

F i g u r e 2-2 Representative types of thermometers used in solar energy practice. (a) Engineer's pocket thermometer (Taylor Instruments). (b) Remote-reading dial thermometer. (c) Recording thermometer with remote sensing bulb (Weksler Instruments Corp.)

TEMPERATURE AND THE NATURE OF HEAT

Liquid-in-glass thermometers have two very evident disadvantages: They are subject to breakage, and their temperature range is somewhat limited. (For example, mercury freezes at about −38°F, and alcohol freezes at −200°F. Also, the boiling point of alcohol is +171°F.) However, they are reasonably accurate and relatively inexpensive, so they are widely used. Heating and air-conditioning engineers often use a pocket thermometer that screws into a protective case (Fig. 2-2a).

Dial thermometers are frequently used in industrial and engineering applications. The scale of degrees is on a circular dial, and a pointer moving over the dial registers the temperature reading. Some of these operate on the principle of bimetallic (differential) expansion, others are actuated by thermoelectricity produced by a difference in temperature at two metallic junctions (the *Seebeck effect*), and still another type is actuated by the changing pressure of a gas in a thermal bulb placed at a location where the temperature is to be monitored. Figure 2-2b shows a typical remote-reading dial thermometer. Figure 2-2c is a recording thermometer with a remote sensing bulb.

2-3 The Celsius Temperature Scale

Another temperature scale, currently used worldwide for scientific work, is the *Celsius* (formerly called centigrade) scale. Its fixed points are also the freezing point and the boiling point of water, but the freezing point is assigned zero degrees Celsius (0°C), and the boiling point is 100 degrees Celsius (100°C). With 100 equal degrees between the fixed points, the Celsius temperature scale lends itself conveniently to decimalization and to use with the SI–metric system of measurement. (See section 2-9 below.) Although the Fahrenheit scale is still widely used in industry in the United States (including the solar energy industry), the Celsius scale is gradually coming into use as the nation edges its way toward adoption of the metric system of measurement. Engineers and technicians must be able to use both temperature scales with ease and accuracy. Solar designers and installers will most often be working in the temperature range from −40°F to 240°F (−40°C to 116°C).

2-4 Comparison of Fahrenheit and Celsius Scales

In Fig. 2-3, the two temperature scales are aligned on the fixed points (BP and FP of water) with several other key points of each scale labeled for comparison. Note that 100 Celsius degrees span the same temperature range as 180 Fahrenheit degrees. Thus, 1 C° = 1.8 F°, and 1 F° = 1/1.8 C°. This relationship is the basis for converting from one temperature scale to the other. The freezing point (FP) of water is used as the base for temperature conversion calculations. Two conversion formulas will be given below, but first think through the following illustrative problems.

Illustrative Problem 2-1

It is desired to maintain an indoor temperature of 70°F. What Celsius temperature is this?

Solution

Note that 70°F is 38 Fahrenheit degrees (70 − 32 = 38) above the FP of water. Each Fahrenheit degree (F°) equals 1/1.8 Celsius degrees.
38/1.8 = 21.1C° above FP. Since FP water is zero degrees on the Celsius scale (0°C), the desired room temperature is

$$t = 21.1°C \qquad \textit{Ans.}$$

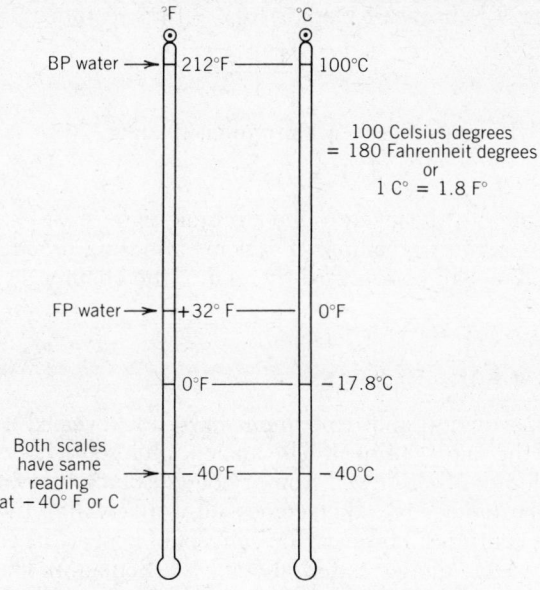

Figure 2-3 Fahrenheit and Celsius temperature scales compared. The reference points for both are the boiling point and the freezing point of water.

Illustrative Problem 2-2

The water flowing out of a flat-plate solar collector has a recorded temperature of 65°C. Convert this to a Fahrenheit temperature.

Solution

65°C is 65 C° above FP water, since FP = 0°C.
65 C° = 65 × 1.8 = 117 F°
Therefore, the Fahrenheit reading will be 117 F° above FP.
    But FP water (Fahrenheit) is 32°F.
    Therefore, the water temperature is

$$t = 117°F + 32°F = 149°F \qquad \qquad Ans.$$

Illustrative Problem 2-3

In a northern climate, the expected low temperature for the day is +8°F. What Celsius temperature is this?

Solution

8°F is 24 F° (32 − 8 = 24) below FP water.
24 F° = 24/1.8 C° = 13.3 C° below FP water.
    But FP water on the Celsius scale is 0°C. Therefore, the Celsius temperature is

$$t = 0°C - 13.3°C = -13.3°C \qquad \qquad Ans.$$

Two simple formulas are provided below for making these scale conversions, but it is stressed that the formulas should not be used without the understanding that the above analysis emphasizes. Be sure that you understand the actual relationships before "cranking out" a temperature conversion problem by formula.

TEMPERATURE AND THE NATURE OF HEAT

To change a Fahrenheit reading to a Celsius reading:

$$°C = \frac{°F - 32}{1.8} \qquad (2\text{-}1)$$

To change a Celsius reading to a Fahrenheit reading:

$$°F = 1.8°C + 32 \qquad (2\text{-}2)$$

Careful attention must be given to the algebraic sign (+ or −, above or below zero) of the temperature readings. Problems affording practice in the use of Eqs. (2-1) and (2-2) will be found at the end of the chapter.

2-5   Heat Is a Form of Energy

Centuries of speculation and experiment have not revealed a simple, precise explanation of the nature of heat. The ancient philosophers believed that heat was an invisible fluid that flowed from a hot substance to a colder one. They called this fluid *caloric* (from which comes our word *calorie*). In the seventeenth and eighteenth centuries, however, the concept of heat as the *energy of molecular motion* began to emerge. Later, about 1795, Benjamin Thompson, on the basis of careful measurements made during the boring of brass cannon, concluded that heat must be a form of energy that somehow flowed into the metal as a result of the mechanical work done by the drilling bits. Thus was introduced

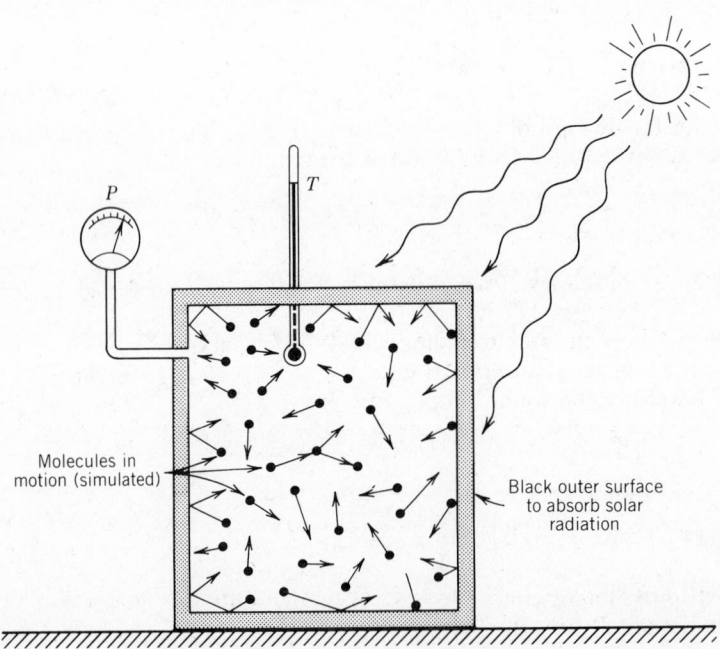

Figure 2-4 Schematic representation of the kinetic–molecular theory of gases. Imagine millions of molecules of a gas in a closed container fitted with a thermometer and a pressure gauge. As the container is heated (solar heating is shown, but any heat source would serve as well), heat energy increases molecular motion. Increased molecular velocities show up as a temperature increase on the thermometer, T. Also, higher velocities cause more frequent molecular collisions, both with other molecules and with the walls of the container. These violent impacts register as increased pressure on the gauge P. Motion, not rest, is the normal state of molecules. *Heat energy* is considered to be the translational kinetic energy of all of the molecules. *Temperature* is a manifestation of the average speed of the molecules.

the idea of some sort of equivalence between work and heat, an idea that later led to the entire science of thermodynamics. The actual determination of the *mechanical equivalent of heat* was undertaken some 50 years later by James Prescott Joule.

The caloric theory of the ancient philosophers was thus discredited, and the concept of heat as a form of energy began to be accepted. Other scientists—among them Faraday, Maxwell, and Kelvin—working in the nineteenth century produced experimental evidence and mathematical models that led to the theory that heat is the *energy of molecular motion* (Fig 2-4).

Since the late 1800s, the theory that heat is a manifestation of molecular motion (also called the *kinetic–molecular hypothesis*) has been modified somewhat to fit it in with other forms of energy that matter possesses. The concept of *total internal energy* is now used in explaining the nature of heat.

The *internal energy* of a given mass of matter is defined as the sum total of (1) the kinetic energy of its molecules in their random motions, (2) the rotational and/or vibrational energy of its atoms and subatomic particles, and (3) the potential energy of the forces of attraction and repulsion between molecules and of the binding forces between and among atoms and subatomic (elemental) particles. Of these three classes of internal energy, it is the first—molecular kinetic energy—that concerns us here.

2-6   Heat and Temperature Related

The total energy of a substance or a system depends on its mass, its temperature, its *state* (solid, liquid, or gas) and a variety of other characteristics such as pressure, magnetic and electric properties, and chemical properties. Temperature and heat are related in that they both seem to be manifestations of molecular motion.

It should be emphasized that the kinetic–molecular theory of heat is just that—a theory, still lacking in demonstrable proof. The theory, however, fits the observable facts, provides a basis for advanced scientific and mathematical analysis of energy phenomena, and affords the foundation for one of the most important branches of physics and engineering—the field of *thermodynamics*.

Two important definitions follow from the kinetic–molecular theory of heat:

1. Temperature is the property of a substance or system that determines whether or not heat energy will flow from it or into it, with respect to other systems. The temperature of a substance is a measure of the potential for heat flow from or to that substance and is proportional to the average kinetic energy of the molecules of the substance.
2. Heat energy is energy that is transferred from a hot substance or system to a colder one as a result of the temperature difference between the two substances or systems.

It should be noted, from the second of the above definitions, that *heat flow* occurs only when a *temperature difference* is present. Heat energy will flow (naturally) only from a hotter body to a cooler one—down a "temperature hill" (Fig. 2-5). Heat is somewhat analogous to water with respect to flow properties. Water flows downhill naturally. It will flow uphill only if forced or pumped—that is, if mechanical work is done on it. And so it is with heat. When heat flows naturally out of a hot body to a colder body, the hot body undergoes a temperature drop and the cold body's temperature rises. Heat will not flow from a colder body or substance to a hotter one unless mechanical work is done in the process. The *heat pump* is one device for mechanically forcing heat up a temperature hill. We shall deal at some length with the use of heat pumps in certain solar energy systems in Chapters 16 and 17.

TEMPERATURE AND THE NATURE OF HEAT

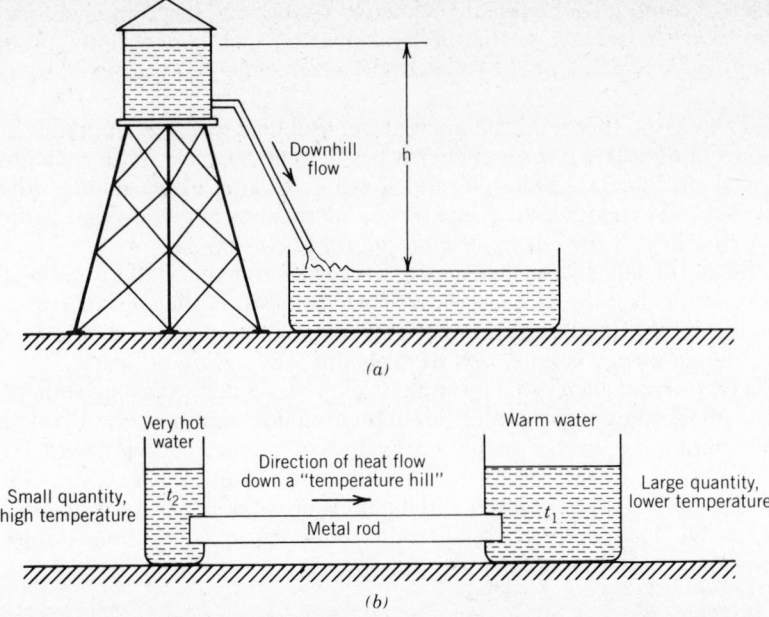

(a)

(b)

Figure 2-5 Heat flow compared to water flow. (*a*) Water flows naturally downhill, from a higher elevation to a lower elevation. It will not flow uphill unless it is pumped—that is, unless mechanical work is done on it. (*b*) Heat flows as a result of temperature difference. In the situation shown, the flow is "down a temperature hill" from a higher temperature, t_2, to a lower temperature, t_1. Note that it is *temperature*, not total heat quantity, that determines the direction of heat flow. The large quantity of warm water at temperature t_1 has more stored heat energy than does the smaller amount of hot water, but heat flow is from the higher temperature "source" to the lower temperature "sink."

SYSTEMS AND UNITS OF MEASUREMENT

At the present time, engineers and technicians in the United States must be familiar with and use two systems of measurement. The English (engineering) system is still the standard for most industrial and commercial activity, including the field of solar energy. However, the Metric Conversion Act passed by the Congress and signed by the president in 1975, calls for an orderly and voluntary conversion to the metric system, but no date was set for completion of the changeover. Although the "move to metrics" seems very slow, particularly in the construction industry, most professional engineering societies are revising their publications, data books, and technical manuals for use with SI–metric units. It may be a decade or more before metrication becomes complete, but in the interim all engineers and technicians must be at home with metric units. The following sections provide a review of both the English (engineering) system and the SI–metric system.

2-7   Fundamental Units and Derived Units

The *fundamental units* of measurement are those of mass (M), length (L), and time (T). Other frequently used units are known as *derived units*. *Volume*, *density*, and *velocity*, for example, are common derived units. The unit of volumetric flow rate is a slightly more complex derived unit. It is obtained by dividing a volume (L^3) by a time unit, obtaining the flow-rate unit in cubic feet per second (cu ft/sec). Many derived units are used in solar design work, and their definitions and correct usage will be given as they are introduced in the chapters to follow.

2-8 Review of English (Engineering) System Units

The defined standard of *length* in the English system is the *yard* (yd), but the *foot* (ft) and the *inch* (in.) are more commonly used in the construction industry and in solar energy system design.

The basic unit of *mass* in the English system is the *slug* (sl). Scientists and engineers use the slug in most of their research and design work. However, in the United States, industrial and commercial practices tend to deal more in terms of weight,[1] the standard unit of which is the pound (lb). A smaller unit, the ounce (oz), is used for smaller weights, and an even smaller unit, the grain (gr), is used to stipulate the weight of water vapor in air–water vapor mixtures. One gr equals 1/7000 lb.

The standard unit of time for engineers and scientists is the second (sec), but solar energy calculations more often use the hour (hr) and the minute (min). The day (24 hr) and the month are also used in connection with solar radiation measurement and in summing up heating and cooling loads.

Other English system units will be defined as they are introduced on later pages.

2-9 SI-Metric System Units

The standard unit of length in the SI–metric system is the *meter* (spelled *metre* in some countries[2]). The metric system has the advantage of being decimalized—all units of any measure of length or mass (but not of time) are related to one another by a power-of-10 multiplier. The SI–metric system is the official measuring system of most of the world's nations, including Canada and Great Britain.

Some multiples and submultiples of the meter are the *kilometer* (km, 1000 m), the *centimeter* (cm, 0.01 m), and the *millimeter* (mm, 0.001m). To the extent that the construction industry is "metricating" (and this is very slow at present), the meter and the millimeter are the units most often used.

The unit of mass in the metric system is the kilogram (kg). It has a base in a common substance—pure water. One kilogram is defined as the mass of a 10-cm cube (1000 cm^3, or 1 *liter*) of pure water at a temperature of 4°C.

The *gram* (g) is 0.001 kg, and the *milligram* (mg) is 0.000001 kg or 0.001 g. A *metric ton* is equal to 1000 kg.

Time has the same units in the metric system as it does in the English system—seconds (sec), minutes (min), hours (hr), and days.

2-10 Comparison of English and Metric Systems—Conversion Factors

One meter is approximately equal to 39.37 in.—about 3.4 in. longer than a yard (Fig. 2-6*a*). One in. is then about 2.54 cm. Fig. 2-6*b* shows the inch and the centimeter compared.

One kilogram is the mass equivalent of approximately 2.2 pounds (lb). Table 2-1 collects and defines some common units of measurement in both the English and metric systems, and Table 2-2 provides conversion factors or approximate equivalencies for some often-used units.

---

[1] Weight (*w*) is a measure of the pull of the earth's gravity on mass (*m*). Weight is really a *force*, not a mass. Mathematically, $w = mg$, where g is the acceleration of gravity at the earth's surface.

[2] The International Committee on Weights and Measures, meeting in a General Conference in Paris in 1960, restandardized the metric system and gave it a new name—*Système International d'Unités*. The term *SI–metric* comes from the initials of the first two words of the French name. In the interest of brevity, we shall often use the common term, *metric system*.

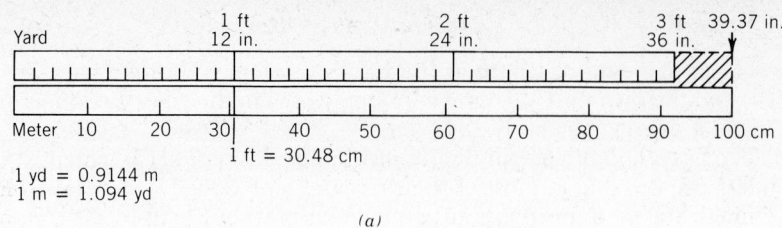

1 yd = 0.9144 m
1 m = 1.094 yd

(a)

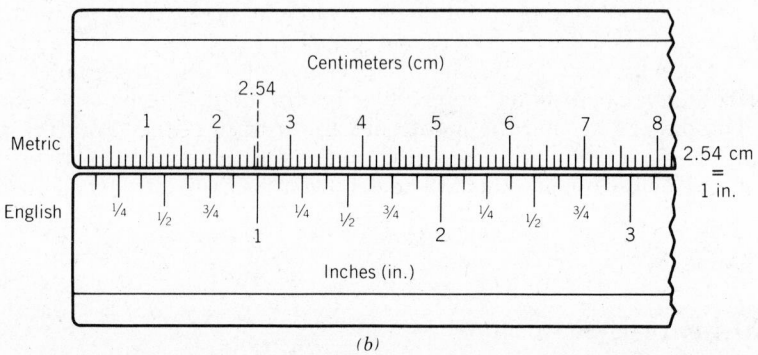

(b)

Figure 2-6 Comparisons between various units of the English and SI–metric systems of measurement. (*a*) The yard (yd) compared to the meter (m). (*b*) The inch (in.) compared to the centimeter (cm).

Table 2-1   Some Commonly Used Units of Measurement

English (Engineering) Units	*SI– Metric Units*
Length	
1 yard (yd) (the standard)	1 meter (m) (the standard)
1 foot (ft) = 1/3 yd	1 centimeter (cm) = 0.01 m
1 inch (in.) = 1/12 ft	1 millimeter (mm) = 0.001 m
1 microinch (μ in.) = 0.000001 in.	1 micron[a] (μ) = 0.000001 m
	1 micrometer[a] (μm) = 0.000001 m
Weight[b]	*Mass*
1 pound (lb) (the standard)	1 kilogram (kg) (the standard)
1 ounce (oz) = 1/16 lb	1 gram (g) = 0.001 kg
1 grain (gr) = 1/7000 lb	1 milligram (mg) = 0.001 g
1 ton (T) = 2000 lb	
Mass	
1 slug (sl) (The slug is the engineering unit of mass. Its weight equivalent is approximately 32.2 lb.)	1 metric ton = 1000 kg
Time	
60 seconds (sec) = 1 minute (min)	Same as English units
60 min = 1 hour (hr)	
24 hr = 1 day	

[a] Note that these are equal measures. *Micrometer* is the recently approved term, but *micron* is still preferred in much of the solar energy industry.

[b] Weight units are used as a measure of mass in U.S. commerce and industry. Weights are actually *forces*.

Table 2-2   English and Metric Equivalents of Length, Mass, Weight, and Volume.

English to Metric	Metric to English
Length	
1 yard (yd) = 0.9144 m	1 meter (m) = 39.37 in.
1 foot (ft) = 0.3048 m	= 3.28 ft
= 30.48 cm	= 1.094 yd
1 inch (in.) = 2.54 cm	1 cm = 0.394 in.
= 25.4 mm	1 mm = 0.0394 in.
Weight	*Mass*
1 pound (lb) = 454 g	1 kilogram (kg) = 2.205 lb
= 0.454 kg	1 gram (g) = 0.0353 oz
1 ounce (oz) = 28.35 g	1 milligram (mg) = 0.0154 gr
1 grain (gr) = 65 mg	1 kilogram (kg) = 0.0685 sl
= 0.065 g	
1 ton (U.S.) = 907 kg	
Mass	
1 slug (sl) = 14.6 kg	
≅ 32.2 lb wt	
Volume	
1 cu yd (yd³) = 0.765 m³	1 cu meter (m³) = 1.31 yd³
= 27 ft³	= 35.4 ft³
1 cu ft (ft³) = 0.037 yd³	= 1000 liters (L)
= 28.3 liters (L)	1000 cm³ = 1 liter (L)
= 7.48 U.S. gallons (gal)	1 liter (L) = 1.057 qt
1 quart (qt) = 0.946 liter (L)	= 0.264 gal
1 gallon (gal) = 3.785 liters (L)	
= 0.833 Imperial gal	
1 Imperial gal = 1.2 U.S. gal	

Notes: The English units are weights, and the metric units are masses. In U.S. industrial and commercial practice, the *pound*, the *ounce*, the *ton*, and the *grain*, although they are actually *weight* (force) units, are used interchangeably as *mass* units. The *slug* is the scientific and engineering unit of *mass*. Equivalencies in the table are approximate.

Illustrative Problem 2-4

A radiant heating installation requires 850 ft of copper pipe to be embedded in a slab floor. What is this length expressed in meters?

Solution

From Table 2-2, note that 1 ft = 0.3048 m.
    The length required is

$$850 \text{ ft} \times 0.3048 \text{ m/ft} = 259 \text{ m} \qquad \textit{Ans.}$$

Illustrative Problem 2-5

A rock bin for a solar-heated home in Canada is planned to contain 55,000 kg of stones. How many U.S. tons is this?

Solution

From Table 2-2, 1 kg = 2.205 lb (in weight equivalent).
    And from Table 2-1, 1 U.S. ton = 2000 lb.

The weight of stones is therefore given by

$$w = \frac{55{,}000 \text{ kg} \times 2.205 \text{ lb/kg}}{2000 \text{ lb/ton}} = 60.64 \text{ tons*}$$ *Ans.*

Illustrative Problem 2-6

A system of flat-plate solar collectors for a small office building has a total maximum water circulation rate of 1000 gal/hr. How many liters per second is this?

Solution

From Table 2-2,

$$1 \text{ gal} = 3.785 \text{ L}$$
$$1000 \text{ gal/hr} \times 3.785 \text{ L/gal} = 3785 \text{ L/hr}$$

Since 1 hr = 3600 sec, the flow rate is

$$\frac{3785 \text{ L/hr}}{3600 \text{ sec/hr}} = 1.05 \text{ L/sec}$$ *Ans.*

2-11 Some Important Derived Units

Area.   The standard unit of area for solar energy system design is the *square foot* (sq ft or ft²). Sometimes the *square inch* (sq in. or in.²) is used, especially when pressure measurements are being expressed. Metric units of area are the *square meter* (m²), the *square centimeter* (cm²), and the *square millimeter* (mm²). Area is a very important factor in solar system design, since the amount of solar radiation received on a surface is directly proportional to surface area exposed (perpendicularly) to the sun's rays. Also, the amount of heat a surface can radiate to its (colder) surroundings is a direct function of the area of the warm surface; and, finally, the amount of heat that will be conducted through solid substances as a result of temperature difference is directly proportional to the cross-section area of the material.

Volume.   Several volume units are commonly used in solar energy work. The *cubic foot* (cu ft or ft³) is often used for solids and gases (including air). The *gallon* (gal) is the most common unit for liquids. The *cubic yard* (yd³), often referred to in the industry as just "yards," is a common measure of sand, gravel, and rock. In the metric system, the *cubic meter* (m³) is used for solids and gases, including air, and the *liter* (L) is the standard volume unit for liquids.

Force.   *Force* is defined as a push or pull that tends to cause a change in motion. Force has the same units as weight in the English system—*pounds*, *ounces*, and *grains*. In U.S. industrial practice, it is the pound that is almost always used to express forces.

In the metric system, the basic force unit is the *newton* (**N**), named for the famous English physicist, Sir Isaac Newton. One newton (**N**) of force, when applied to 1 kg of mass (*in the absence of other forces*) will impart to the mass an acceleration of 1 m/sec/sec (1 m/sec²). The dimensions of the newton are therefore

$$1 \text{ } \mathbf{N} = 1 \frac{\text{kg-m}}{\text{sec}^2} \tag{2-3}$$

---

* Answers to illustrative problems have been computed with a pocket electronic calculator. Accuracy to three significant figures can be expected, with some degree of confidence in the fourth digit if it is supplied. It should be recalled that no calculated answer is ever any more accurate than the least accurate of the raw data that form the basis of the computation.

The earth's force of gravity gives a mass of 1 kg an acceleration of 9.81 m/sec² at sea level at the equator. This value (9.81 m/sec²) is the metric value of *g*, the acceleration of gravity. The English system value of *g* is 32.17 ft/sec²—that is, the force of earth's gravity on the mass equivalent of 1 lb will accelerate it at 32.17 ft/sec² (in the absence of all other forces).

These definitions also serve to define the *slug* (sl), referred to earlier as the English system science and engineering unit of *mass*. One *pound of force* (in the absence of any other forces) will accelerate one *slug of mass* at a rate of 1 ft/sec². Consequently, the dimensions of the slug are

$$1 \text{ sl} = 1\frac{\text{lb-sec}^2}{\text{ft}} \tag{2-4}$$

One slug of mass is equal to the mass equivalent of 32.17 lb.

Current practice in the U.S. solar industry is to consider the *pound* as the basic unit for *force*, *weight*, and *mass*. In *scientific* and *engineering* calculations, however, the slug is the officially approved unit of mass, and the pound is a force unit.

Some metric countries also still use a dual system of masses and weights (forces). Although the newton (**N**) is now the official unit of both force and weight in metric countries, there is still frequent use of the older unit, the *kilogram-force* (kg-f, or *kilo*) for everyday commercial transactions. One (*kilo*) equals 9.81 **N**.

Velocity.   Fluid flow (both liquids and gases) is an important factor in many solar heating systems. Often, the design must provide for a stipulated velocity and flow rate of fluids through pipes or ducts. The English system standard of velocity is the *foot per second* (ft/sec), but the *foot per minute* (ft/min) is often used in heating and air-conditioning design to express the velocity of air flow in ducts. Metric units of velocity are the *meter per second* (m/sec) and the *centimeter per second* (cm/sec).

Pressure.   Pressure is *force* acting on *unit area* (Fig. 2-7). The standard English system unit of pressure is the *pound per square inch* (lb/in.² or psi). Occasionally, the *pound per square foot* (lb/ft² or psf) will be encountered in the construction industry. Pressures are also expressed in *inches of mercury* (in. Hg) and, for air and other gases, in *inches of water* (in w.). *Atmospheric pressure* is the average pressure at sea level exerted by the weight of the earth's atmosphere. One atmosphere pressure (1 atm) equals 14.7 psi, or 29.92 in. Hg, or 407 in. w.

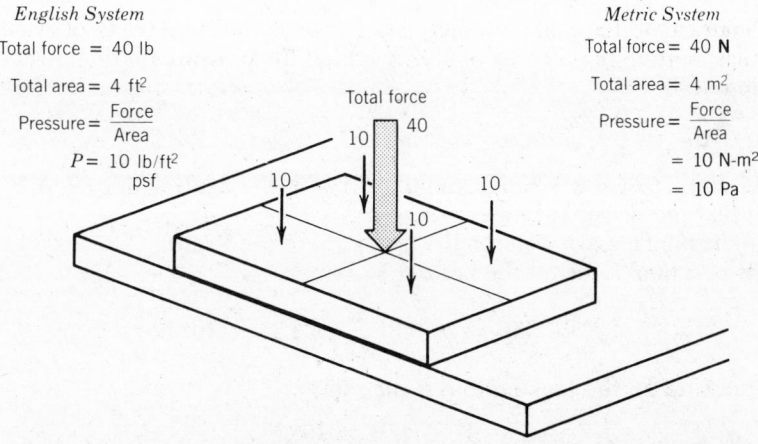

Figure 2-7 Pressure is force per unit area (*P = F/A*). The total force of 40 units, portrayed by the broad arrow, is assumed to be equally distributed over the entire area. Forces are in pounds or newtons, and area is in square feet or square meters.

Mathematically, pressure

$$p = \frac{\text{Force}}{\text{Area}} = \frac{F}{A} \tag{2-5a}$$

It follows, then, that the total force on a surface against which pressure acts perpendicularly is

$$F = \text{Pressure} \times \text{Area} = pA \tag{2-5b}$$

In the metric system, the standard unit of pressure is the pascal (Pa), but a more useful measure for solar energy system design is the kilopascal (1 kPa = 1000 Pa), since it is large enough for practical industrial problems. The pascal is a pressure of one newton per square meter (1 Pa = 1 N/m^2). Other metric units of pressure are the centimeter of mercury (cm Hg) and the millimeter of mercury (mm Hg).

Illustrative Problem 2-7

The hydrostatic pressure at the bottom of a cylindrical water storage tank is observed to be 5.2 psi. If the diameter of the tank is 6 ft, find the total force on the (flat) bottom.

Solution

Since pressure is given in pounds per square inch, the area of the tank bottom must be expressed in square inches in order to have consistent units. From simple geometry,

$$\text{Area of a circle, } A = \frac{\pi \times \text{diameter}^2}{4}$$

Substituting,

$$A = \frac{3.14 \times (6 \text{ ft})^2}{4} \times \frac{144 \text{ in.}^2}{1 \text{ ft}^2} = 4070 \text{ in.}^2$$

Now, from Eq. (2-5b),

$$F = pA = 5.2 \frac{\text{lb}}{\text{in.}^2} \times 4070 \text{ in.}^2 = 21{,}160 \text{ lb} \qquad \textit{Ans.}$$

Illustrative Problem 2-8 (SI–Metric)

A cylindrical water tank for thermal storage contains 5000 L of water. Its (circular) bottom has an area of 0.92 m^2. Find the pressure on the tank bottom in kilopascals.

Solution

Recall that a pascal is 1 N/m^2, and that 1 kg-f is the equivalent of 9.81 **N**. Also recall that the density of water is 1 kg/L.

   The total mass of water in the tank is therefore 5000 kg. This 5000-kg mass results in a total force on the bottom,

$$F = 5000 \text{ kg} \times 9.81 \frac{\textbf{N}}{\text{kg}} = 4.91 \times 10^4 \textbf{ N}$$

The pressure on the tank bottom is therefore

$$p = \frac{F}{A} = \frac{4.91 \times 10^4 \textbf{ N}}{0.92 \text{ m}^2}$$

$$= 5.33 \times 10^4 \frac{\textbf{N}}{\text{m}^2} \text{ (Pa)}$$

$$= 53.3 \text{ kPa} \qquad \textit{Ans.}$$

Density.   Engineers and scientists deal with two kinds of density. *Mass density* is defined as mass per unit volume. The symbol for mass density is the Greek letter rho (ρ). Mathematically, mass density

$$\rho = \frac{m}{V} \tag{2-6}$$

Common units of mass density are the slug per cubic foot (sl/ft³) in the English system and the kilogram per cubic meter (kg/m³) in the metric system. The gram per cubic centimeter (g/cm³) and the gram per liter (g/L) are also used occasionally.

   Weight density is the more commonly used kind of density in U.S. industrial practice. Weight density is defined as weight per unit volume. Its symbol is *D*, and by definition

$$D = \frac{w}{V} \tag{2-7}$$

Weight density is almost always expressed in pounds per cubic foot (lb/ft³). Most of the discussions and problems in this book will be involved with weight density, and for all English system discussions the word *density* will refer to weight density, *D*. The symbol ρ and the term *mass density* will be used only with SI–metric discussions.

   Table 2-3 gives a number of equivalencies and conversion factors (English to metric) that are often needed in solar system design work.

T a b l e   2 - 3   Definitions, Equivalencies, and Conversion Factors for Selected Derived Units

Area
1 ft² = 144 in.² = 0.111 yd² = 929 cm² = 0.093 m²
1 yd² = 9 ft² = 0.836 m²
1 m² = 10.76 ft² = 1.196 yd² = 1550 in.² = 10,000 cm²

Force
1 lb (force) = 4.45 **N** = 454 g-f = 0.454 kg-f (kilo)
1 **N** = 0.225 lb-f = 0.102 kg-f (kilo) = 102 g-f
1 kilonewton (k**N**) = 1000 **N** = 225 lb-f
1 ton (U.S.) = 2000 lb = 907 kilos (kg-f)

Pressure
1 lb/in.² (psi) = 144 lb/ft² = 6.9×10^3 Pa (**N**/m²)
1 atm = 14.7 psi = 29.92 in. Hg = 33.9 ft w = 407 in. w
       = 760 mm Hg = 1.013×10^5 Pa = 101.3 kPa
1 Pa = 1 **N**/m² = 1.45×10^{-4} psi
1000 Pa = 1 kPa = 0.145 psi
1 in. Hg = 3385 Pa = 3.385 kPa $\cong$ 1/30 atm
1 in. w = 1.86 mm Hg = 249 Pa $\cong$ 1/407 atm
1 ft w = 0.434 psi = 0.0294 atm
1 mm Hg = 133.3 Pa = 1/760 atm

Density of Selected Fluids

Substance	Weight Density, D	Mass Density, ρ
Water (14°C, 39°F)	62.4 lb/ft³	1000 kg/m³
	8.34 lb/gal	1 kg/L
		1 g/cm³
Dry air	0.0807 lb/ft³	1.293 kg/m³
	(32°F, 1 atm)	(0°C, 1 atm)
Mercury	850 lb/ft³	13.6 g/cm³

Illustrative Problem 2-9

A flat-bottomed thermal storage tank is circular in cross-section, with a diameter of 4 ft. (a) What total weight of water (39° F) will it contain when filled to a depth of 7 ft? (b) How many gallons does it contain?

Solution

(a) The volume must first be found. For a right circular cylinder,

$$V = \frac{\pi \, (\text{diameter})^2}{4} \times \text{height}$$

Substituting,

$$V = \frac{3.14 \times (4 \text{ ft})^2}{4} \times 7 \text{ ft}$$
$$= 87.9 \text{ ft}^3$$

Now use Eq. (2-7) and solve for w,

$$w = D \, V$$

From Table 2-3, the density of water at 39°F is 62.4 lb/ft³. Substituting,

$$w = 62.4 \, \frac{\text{lb}}{\text{ft}^3} \times 87.9 \text{ ft}^3$$
$$= 5485 \text{ lb} \qquad \qquad Ans.$$

(b) From Table 2-3, 1 ft³ = 7.48 gal. Therefore,

$$V_{\text{gal}} = 87.9 \text{ ft}^3 \times 7.48 \, \frac{\text{gal}}{\text{ft}^3}$$
$$= 657 \text{ gal} \qquad \qquad Ans.$$

Illustrative Problem 2-10 (SI–Metric)

A closed fluid-circulating system is held under a gauge pressure of 15 psi. Express this pressure in kilopascals.

Solution

From Table 2-3, note that 1 psi is equivalent to 6.9×10^3 Pa. Therefore,

$$p_{\text{kPa}} = \frac{15 \text{ psi} \times 6.9 \times 10^3 \, \frac{\text{Pa}}{\text{psi}}}{1000 \, \frac{\text{Pa}}{\text{kPa}}}$$
$$= 103.5 \text{ kPa} \qquad \qquad Ans.$$

2-12 Fluid Pressure and Its Measurement

Fluid pressure is measured by instruments called gauges (sometimes spelled *gages*). Two types of gauges are in common use in industrial practice—the U-tube or manometer gauge, and the Bourdon-type gauge.

U-tube (Manometer) Gauges.   These gauges make use of a glass tube bent in the shape of a U and filled with a heavy liquid such as water or mercury (Fig. 2-8a). Fluid pressure in pipe AB at point C exerts a force down on the liquid in the left side of the U tube, lowering it to point E on the left side and raising it to point F on the right side. In this version of the instrument, the right side of the tube is open to the atmosphere. The difference in elevation (deflection) FE (or h) is a measure of the gauge pressure at point C in the pipe. The gauge pressure

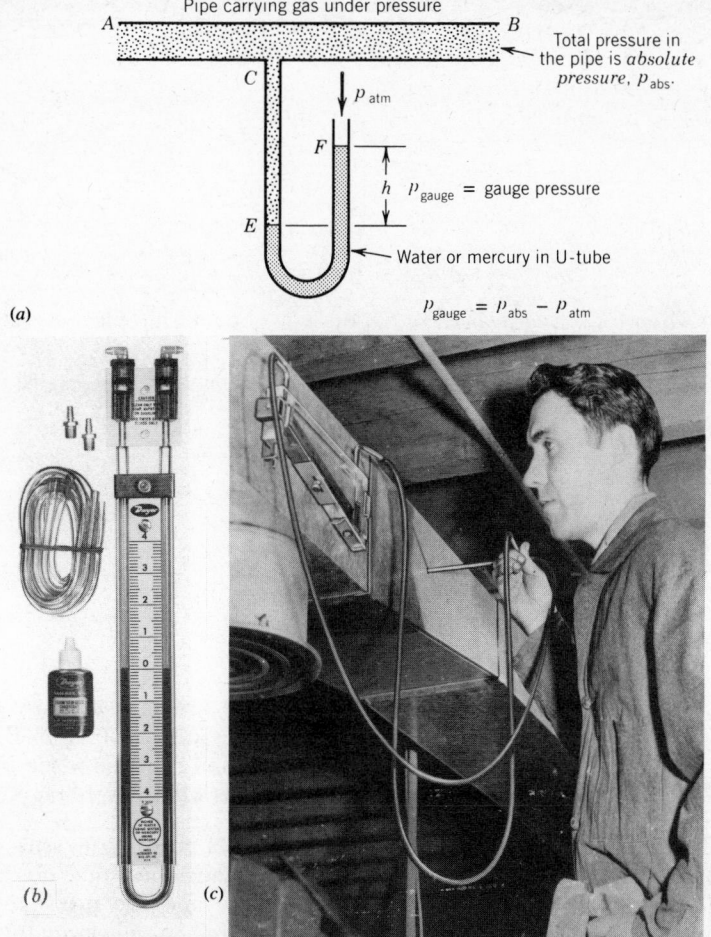

Pipe carrying gas under pressure

A ———————————————————————————————— B

Total pressure in the pipe is *absolute pressure*, p_{abs}.

C

$\downarrow p_{atm}$

F

h p_{gauge} = gauge pressure

E

← Water or mercury in U-tube

(a)

$p_{gauge} = p_{abs} - p_{atm}$

(b) (c)

Figure 2-8 U-tube manometers for the measurement of relatively low pressures. (*a*) Diagram of an open U-tube manometer. Water is used in the tube for very low pressures (such as air flowing in ducts or natural gas flowing in pipes to home appliances), and mercury is used for somewhat higher pressures. One end of this type of U tube is open to the atmosphere. (*b*) A commercial U-tube gauge. It reads in inches of water when filled with water, and in inches of mercury when filled with mercury. (*c*) An inclined manometer used to measure air flow in an air-conditioning duct. (Photos from Dwyer Instruments, Inc.)

can be read directly from a scale attached to the U tube. Readings on the scale are usually in inches or millimeters of mercury or inches of water, depending on the liquid used in the U tube. Water-filled U tubes are suited to the measurement of the low pressures associated with fuel gas flowing in pipes or air flowing in ducts. Mercury-filled U tubes are used for somewhat greater pressures. Scales can be graduated to read directly in pounds per square inch (psi) or in kilopascals (kPa); or the technician can resort to tables or charts for this conversion. The following problem illustrates one method of calculating such conversions.

Illustrative Problem 2-11

A water-filled U-tube gauge shows a deflection of 8.8 in. of water when attached to an industrial service gas line. Express this gauge pressure in (a) pounds per square inch, (b) millimeters of mercury, and (c) kilopascals.

Solution

From Table 2-3, (*Pressure* section), note that a pressure of 1 atm = 14.7 psi = 407 in. w = 760 mm Hg = 101.3 kPa.

(a) The psi equivalent of 8.8 in. w is obtained from the proportion

$$\frac{p_{psi}}{14.7 \text{ psi}} = \frac{8.8 \text{ in. w}}{407 \text{ in. w}}$$

From which,

$$p_{psi} = 0.32 \text{ psi} \qquad \text{Ans.}$$

(b) The mm Hg equivalent of 8.8 in. w is obtained, in like manner, from

$$\frac{p_{mm\,Hg}}{760 \text{ mm Hg}} = \frac{8.8 \text{ in. w}}{407 \text{ in. w}}$$

From which

$$p_{mm\,Hg} = 16.4 \text{ mm Hg} \qquad \text{Ans.}$$

(c) Finally,

$$p_{kPa} = 101.3 \text{ kPa} \times \frac{8.8 \text{ in. w}}{407 \text{ in. w}}$$
$$= 2.19 \text{ kPa} \qquad \text{Ans.}$$

As remarked above, only relatively low pressures are measured by U-tube manometers. Pressures in medium and higher ranges are measured by Bourdon gauges, described below. First, however, a distinction must be made between two kinds of pressure in order to allow for the effect of atmospheric pressure.

Gauge Pressure and Absolute Pressure. When a pressure measurement is read from an ordinary gauge, the reading is called gauge pressure. It is actually the difference between total or absolute pressure and the pressure of the atmosphere at that place and time. Reference to Fig. 2-8a will clarify this point. The total pressure in the pipe (p_{abs}) is the indicated (gauge) pressure ($h = p_{gauge}$) plus the pressure of the atmosphere exerting a downward force on the liquid in the right arm of the U tube. As an equation, gauge pressure,

$$p_{gauge} = p_{abs} - p_{atm} \qquad (2\text{-}8)$$
(gauge pressure = absolute pressure − atm pressure)

Absolute pressure is expressed in the same units as gauge pressure, with the word *absolute* added—as pounds per square inch, absolute (psia), or kilopascals, absolute (kPaa). Occasionally, for clarity, gauge pressures are expressed as pounds per square inch, gauge (psig), or inches of water, gauge (in. wg), or kilopascals, gauge (kPag). In solar energy field work, the word *gauge* is usually omitted, and pressure readings are to be understood as gauge values unless *absolute* is clearly indicated. In research and design, however—particularly with respect to refrigeration compressors, heat pumps, gaseous fuels, and so on—pressures (and, for that matter, temperatures) are ordinarily reported in absolute values.

Bourdon Gauges. For the medium to high pressures often encountered in solar energy systems, Bourdon-type gauges are used instead of U-tube manometer gauges. Bourdon gauges are constructed so that a needle or pointer moves across a scale (usually on a circular arc) that is calibrated to read directly in pressure units (pounds per square inch or kilopascals). Some Bourdon gauges read in inches or millimeters of mercury. Figure 2-9 shows a common type of Bourdon gauge and a cutaway view of the essential working parts.

Compound Gauges. Some Bourdon gauges are designed to allow pressure readings both above and below atmospheric pressure. These are called compound gauges. The zero point on the scale is zero pounds per square inch,

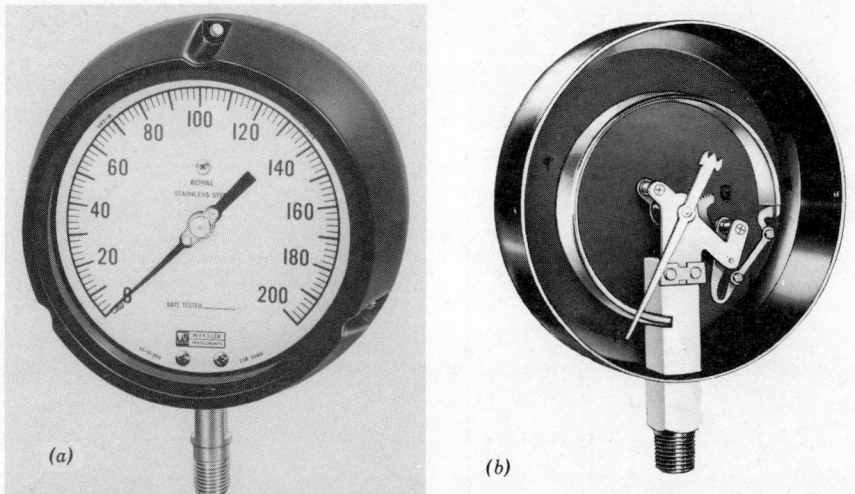

Figure 2-9 A typical Bourdon pressure gauge. (*a*) Front view, showing dial and range of pressures for the instrument. (Weksler Instruments, Inc.) (*b*) Cutaway, showing the working parts and operating principles of a Bourdon gauge.

gauge (psig), beginning at a pressure of 1 atm. Above zero, the typical compound gauge reads in pounds per square inch, but below zero the scale is calibrated in inches of mercury (English system gauges). Readings above zero are pressure readings, and readings below zero are vacuum readings. The vacuum scale also starts at zero and reads to 30 in. Hg vacuum. A 30-inch reading (actually 29.92 in.) would mean a perfect vacuum or zero psia.

Vacuum gauges are necessary in the design and testing of compressors and for evacuating piping systems before filling the system with refrigerant or other special gases.

2-13 Pressure Calculations in Nonflowing Liquids—Hydrostatics

The pressure at any point in a nonflowing liquid system is the result of the actual weight of liquid above that point. Since pressure in a nonflowing liquid is transmitted equally and without loss in all directions throughout the liquid (Pascal's law), the depth at the point being considered is the vertical distance from that point to the level of the liquid surface (Fig. 2-10). The study of the properties of nonflowing liquids is known as hydrostatics, and pressure in nonflowing liquids is called hydrostatic pressure.

The system shown in diagram form in Fig. 2-10 contains liquid of weight density D (or mass density ρ). The liquid surface is at an elevation h, measured vertically from the valve V, which is at street level. Note that valve V is closed—liquid not flowing.

The pressure in hydrostatic systems can be calculated (English system) from the equation

$$p = hD$$
$$\text{(pressure = depth} \times \text{density)}$$
(2-9a)

Equation (2-9a) gives the pressure in pounds per square foot (lb/ft^2 or psf) when the depth h is in feet and the density D is in pounds per cubic foot. To obtain pressure in pounds per square inch (lb/in.^{2} or psi), the basic equation can be modified as follows:

$$p = \frac{hD}{144}$$
(2-9b)

since there are 144 in.^{2} in 1 ft^2.

SYSTEMS AND UNITS OF MEASUREMENT

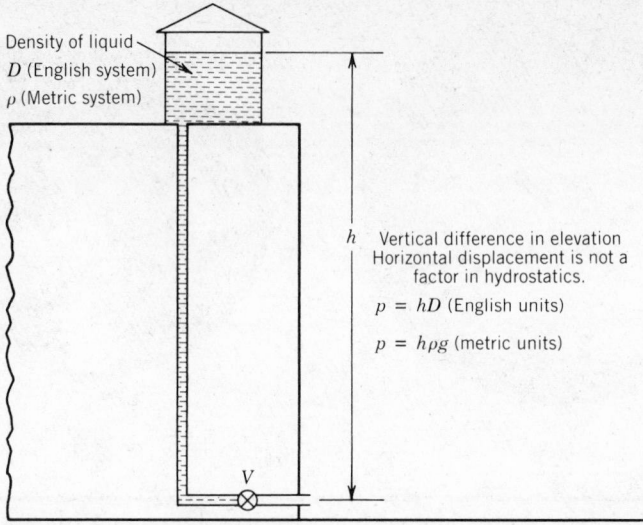

Density of liquid
D (English system)
ρ (Metric system)

h Vertical difference in elevation
Horizontal displacement is not a factor in hydrostatics.

$p = hD$ (English units)

$p = h\rho g$ (metric units)

V

Figure 2-10 Hydrostatic pressure—pressure in nonflowing liquid systems. This schematic diagram shows that, for nonflowing liquids, pressure at any point in the system is dependent only on the vertical height h of the liquid above that point (in this case the valve V), and on the density D (or ρ in the metric system) of the liquid.

For SI–metric system calculations, it must be recalled that mass density ρ is expressed in kilograms per cubic meter. But pressure is force per unit area, and kilograms are *masses*, not forces. The kilogram (mass) units must be converted to force units (newtons) by multiplying by g, the acceleration of gravity (see section 2-11). In SI units, therefore, the hydrostatic equation becomes

$$p = h\rho g \qquad (2\text{-}10)$$

where p is in pascals (N/m^2) when h is in meters, and ρ is in kilograms per cubic meter. The acceleration of gravity g has the metric value of 9.81 m/sec².

Illustrative Problem 2-12

The city water reservoir surface is 185 ft above the street level at a residential customer's water meter. Find the pressure in psi at the meter under conditions of nonflow.

Solution

Use Eq. (2-9b) and write

$$p = \frac{185 \text{ ft} \times 62.4 \, \frac{\text{lb}}{\text{ft}^3}}{144 \, \frac{\text{in.}^2}{\text{ft}^2}}$$

$$= 80.2 \, \frac{\text{lb}}{\text{in.}^2} \text{ (psi)} \qquad \textit{Ans.}$$

Illustrative Problem 2-13 (SI–Metric)

The water surface of a penthouse swimming pool and spa is 140 m above the pumps in the basement of the building. What hydrostatic pressure does this elevation cause in the basement? Answer in kilopascals.

Solution

Use Eq. (2-10), $p = h\rho g$. Substituting values,

$$p = 140 \text{ m} \times 1000 \frac{\text{kg}}{\text{m}^3} \times 9.81 \frac{\text{m}}{\text{sec}^2}$$

$$= 1.373 \times 10^6 \frac{\text{kg}}{\text{m-sec}^2}$$

$$= 1.373 \times 10^6 \frac{\text{N}}{\text{m}^2} \text{ (Pa)}*$$

$$= 1373 \text{ kPa} \qquad\qquad\qquad \textit{Ans.}$$

It is again emphasized that the above discussions and Eqs. (2-9a), (2-9b), and (2-10) apply only to nonflowing liquid systems. The analysis of pressure conditions and variations in systems where liquids are flowing (*hydrodynamics* or *hydraulics*) is beyond the scope of an introductory text.

2-14   Energy and Power

One definition of energy is that it is the capacity to do work. There are many forms of energy, such as heat energy, mechanical energy, electrical energy, nuclear energy, and of course solar energy, which arrives on earth in the form of electromagnetic radiations.

Since solar energy systems often involve mechanical and electrical devices, we shall review here some basic definitions and principles relating work, energy, and power.

Work is accomplished when a force is applied to an object, moving it through a measurable distance. The amount of work done is the product of the force applied and the distance moved in the direct line of the force (Fig. 2-11). As an equation,

$$W = Fs$$
$$\text{Work} = \text{Force} \times \text{distance} \qquad (2\text{-}11)$$

The unit of work in the English (engineering) system is the foot-pound (ft-lb). In the SI–metric system, the unit of work is the newton-meter (N-m). The name joule (**J**) has been given to the newton-meter to commemorate the research of James Prescott Joule, noted for his nineteenth-century discoveries in the field of mechanics and heat. Figure 2-11 illustrates the definition of mechanical work.

A body possesses energy if its position, its condition of motion, or its molecular, atomic, or nuclear arrangement is such that it could accomplish work under stipulated conditions. The units of energy are the same units (defined above) used for work—namely, the foot-pound and the joule.

Power is the time rate at which work is done (output power) or the rate at which energy is expended (input power).

$$P = \frac{W}{t} = \frac{Fs}{t} \qquad (2\text{-}12)$$

In engineering and technical fields, the basic unit of power is the foot-pound per second (ft-lb/sec). This is a rather small unit of power, however, and a more commonly used unit is the horsepower (hp), defined as follows:

$$1 \text{ hp} = 550 \frac{\text{ft-lb}}{\text{sec}} = 33,000 \frac{\text{ft-lb}}{\text{min}} \qquad (2\text{-}13)$$

The metric unit of power is the joule per second (**J**/sec). The joule per second is named the watt (**W**) in honor of James Watt, the Scotsman who developed the reciprocating steam engine.

---

*Recall that the dimensions of the newton are $\frac{\text{kg-m}}{\text{sec}^2}$ and that $1 \frac{\text{N}}{\text{m}^2} = 1$ Pa. (Review section 2-11.)

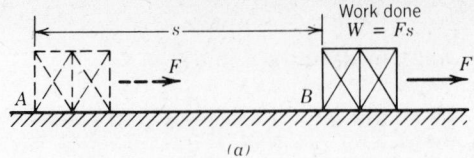

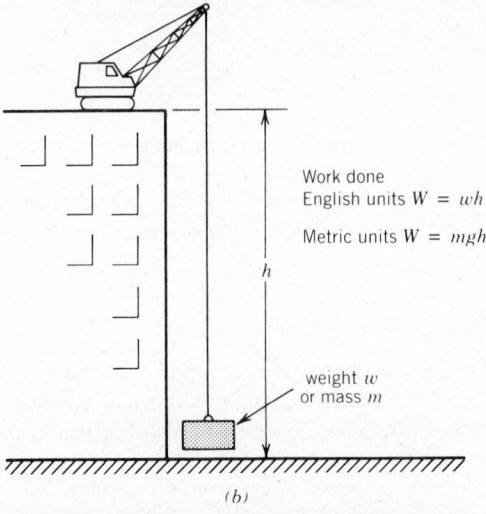

Figure 2-11 The definition of mechanical work. (*a*) A heavy crate of machinery is skidded over a floor from A to B for a distance *s*, by the application of a force *F*, in the direction or line of motion. The work done is $W = Fs$. (*b*) A crate of solar collectors, total weight *w* (or mass *m* in the metric system), is hoisted at a uniform speed to the roof of a building through an elevation difference *h*. *Weight* is the force of gravity that the hoist must overcome. The minimal work done is equal to the force acting times the vertical distance hoisted. $W = wh$ (English units). With metric units, it must be remembered that force $F = mg$, and consequently the work done in metric units is $W = mgh$.

Electric power is also expressed in watts and in kilowatts (kW); and electric energy in kilowatt-hours (kW-hr). The English and the metric systems use identical units for all electrical measurements. Electric power being used in a circuit can be measured by a wattmeter, and electric energy consumed is measured by kilowatt-hour meters.

It is often necessary to convert energy and power relationships from English units to metric units and vice versa. The following conversion factors are useful for this purpose:

$$1 \text{ ft-lb} = 1.356 \text{ J} \qquad (2\text{-}14)$$
$$1 \text{ J} = 0.738 \text{ ft-lb}$$
$$1 \text{ hp} = 746 \text{ W} = 0.746 \text{ kW}$$
$$1 \text{ kW} = 1.34 \text{ hp}$$

Illustrative Problem 2-14

Several solar collector panels whose total crated weight is 560 lb are to be hoisted from the ground to the roof of a building, 85 ft vertically upward. (a) Find the work done. (b) If the lift is accomplished in 42 sec, what is the horsepower output of the hoist?

Solution

Recall that weight is the force of gravity acting on mass. A force equal to the weight of the crated panel will be required to lift it vertically (no acceleration).
(a) From Eq. (2-11),

$$W = Fs = 560 \text{ lb} \times 85 \text{ ft}$$
$$= 47,600 \text{ ft-lb} \qquad Ans.$$

(b) From Eq. (2-12),

$$P = \frac{W}{t} = \frac{47,600 \text{ ft-lb}}{42 \text{ sec}}$$
$$= 1130 \frac{\text{ft-lb}}{\text{sec}}$$

$$P_{hp} = \frac{1130 \dfrac{\text{ft-lb}}{\text{sec}}}{550 \dfrac{\text{ft-lb}}{\text{sec-hp}}} = 2.06 \text{ hp} \qquad Ans.$$

Illustrative Problem 2-15

A pump lifts water from a swimming pool to solar collectors located on the roof of a hotel. The vertical lift is 210 ft, and the flow rate of water is 115 gal/min. (a) Find the net output horsepower required. (b) Assuming 100 percent efficiency of the pump and motor system, what is the electrical power used by the pump motor? (Neglect pipe friction and velocity head power requirements.)

Solution

(a) The net power output is the actual work done per minute in lifting the water to the roof. Substituting in Eq. (2-12),

$$P = 115 \frac{\text{gal}}{\text{min}} \times 8.33 \frac{\text{lb}}{\text{gal}} \times 250 \text{ ft} = 239,500 \frac{\text{ft-lb}}{\text{min}}$$

$$P_{hp} = \frac{239,500 \dfrac{\text{ft-lb}}{\text{min}}}{33,000 \dfrac{\text{ft-lb}}{\text{min-hp}}} = 7.26 \text{ hp} \qquad Ans.$$

(b) From (a), the horsepower is 7.26. Converting, using Eq. (2-14),

$$P_{kW} = 0.746 \frac{\text{kW}}{\text{hp}} \times 7.26 \text{ hp}$$
$$= 5.42 \text{ kW} \qquad Ans.$$

2-15 Units and Dimensions in Problem Solving

The design of even the simplest solar energy system involves calculations and solutions to problems. The ultimate performance of the system and your reputation as a solar designer, contractor, or technician will depend on the accuracy with which these problems are solved. As a solar designer, you will necessarily be a problem solver.

Two parts of the answer to any problem are of equal importance—the numerical part, such as 278.3, and the dimensions or units, such as Btu/ft³-F°. It hardly needs to be said that the answer cannot be correct unless both the numerical value and the dimensions are correct. There is a vast and potentially catastrophic difference between 278.3 Btu/ft³-F° and 278.3 Btu/lb-F°.

Be sure to form the habit of setting up problems with the proper units at the outset, and then carry them all the way through—step by step—to the final answer. Units are subject to multiplication and division (cancellation), they can acquire exponents, and they can tell you during the solution and at the end result whether or not you are on the right track. A quick check on the dimensions of the answer will tell you immediately if the answer is wrong. If the dimensions are correct, the answer is not necessarily correct; but if the dimensions are wrong, the answer is assuredly wrong. The "dimensions check" is, then, a necessary but not sufficient test of an answer's correctness.

You will have noted that the Illustrative Problems solved for you in the text carry the dimensions through all steps of the problem. Here is another example.

Illustrative Problem 2-16

A water-cooling tower is to be installed on a flat roof. Its total weight is 6.5 tons, distributed evenly over an area of 40 sq ft. Find the pressure ("roof loading") caused by this installation, in pounds per square inch.

Solution

From Tables 2-1 and 2-3,

$$1 \text{ ton} = 2000 \text{ lb}$$
$$1 \text{ ft}^2 = 144 \text{ in.}^2$$

From Eq. (2-5),

$$p = \frac{6.5 \text{ tons}}{40 \text{ ft}^2} \text{ (in tons/ft}^2)$$

To change tons/ft² into lb/in.²,

$$p = \frac{6.5 \text{ tons}}{40 \text{ ft}^2} \times \frac{2000 \text{ lb}}{\text{ton}} \times \frac{1 \text{ ft}^2}{144 \text{ in.}^2}$$

Note that the ton units and the ft² units cancel out, leaving lb/in.² units, as desired.

Now, performing the numerical calculation, the result is

$$p = 2.26 \frac{\text{lb}}{\text{in.}^2} \text{ (psi)} \hspace{3cm} Ans.$$

This chapter has dealt for the most part with simple, basic units common to everyday industrial practice. The next chapter will introduce many new units involved in heat energy and heat transfer, and in Chapter 4 specialized units and dimensions for solar energy will be introduced.

PROBLEMS

1. Hot water coming off a solar collector is at a temperature of 140°F. Express this temperature in degrees Celsius.

2. A solar space-heating system is designed to maintain a space temperature no lower than 20°C. What Fahrenheit temperature is this?

3. Find the Celsius temperature when the local weather bureau reports the low for the day as +12°F.

4. A hot tub owner desires a water temperature of 102°F. Express this in degrees Celsius.

5. A flat-plate solar collector measure 96 in. by 48 in. How many square feet is this?

6. A water tank for thermal storage is in the form of a right circular cylinder, diameter 4 ft, height 7 ft. How many gallons will it hold when full?

7. The flow rate into a bank of solar collectors is measured as 54 gal/min. If this flow rate remains steady, how many pounds of water per hour will flow through the collector bank? (The water is at a temperature where its density $D = 62.1$ lb/ft³).

8. A 10-ton crate containing machinery rests on the floor on its side—dimensions 6 ft × 12 ft. If the weight is evenly distributed, find the pressure exerted on the floor (called floor loading), in (a) pounds per square foot, and (b) pounds per square inch.

9. The gauge pressure in a gas tank is reported to be 6 atm. Express this pressure in (a) pounds per square inch, and (b) inches of mercury.

10. An industrial gas line is under a gauge pressure of 1.45 lb/in.^{2} Express this pressure in inches of water, gauge.

11. A hoist lifts a crate of solar panels and fittings whose total weight is 1600 lb from the ground to the roof level of a building—a vertical distance of 120 ft—in 0.75 min. Find the horsepower output of the hoist.

12. An electric motor is rated at 5 hp. If it runs for 8 hr at full load, how many kilowatt hours of electric energy are used? (Assume no losses.)

13. An open U-tube manometer gauge indicates a deflection of 14.2 in. of water when applied to an industrial service gas line. (a) Express this pressure in pounds per square inch, gauge. (b) What is the absolute pressure in the gas line?

14. A Bourdon gauge reads 105 psi. What is the absolute pressure where it is connected, expressed in atmospheres?

15. A ground-floor water line is supplied from a tank on the roof. The vertical distance to the water level in the tank is 210 ft. Find the gauge pressure (psi) in the ground-floor pipe (water not flowing).

16. The basement water pressure in a tall building is 115 psig. What pressure could be expected (water not flowing) on the tenth floor, 120 ft above the basement water meter?

SI–METRIC PROBLEMS

1. The roof of a house is known to have an area of 2800 sq ft. How many square meters is this?

2. A standard 2 × 4 framing stud has actual dimensions of 1.5 in. by 3.5 in. Express these dimensions in millimeters.

3. A shipment of copper tubing is certified to have a mass of 1420 kg. Convert this figure to pounds weight.

4. A solar heating installation has a steady-state water flow of 6.2 L/sec. How many U.S. gallons per hour is this?

5. A shipment from Mexico is labeled 750 kilos. How many newtons is this? How many pounds?

6. A heavy crate has a mass of 950 kg and rests flat on the floor on a side whose dimensions are 1.5 m by 3.5 m. Assuming even distribution of the force equivalent of this mass over the contact area, find the pressure on the floor (floor loading) in kilopascals.

7. If it is desired to maintain a gauge pressure of 30 psi in a hot-water circulating system, what would be the gauge pressure reading in kilopascals?

8. A hot-water heating system pump lifts 8.5 L/sec from a ground-floor location to the eighth floor, a 50-m gain in elevation. Take the mass density of the hot water as $\rho = 0.993$ kg/L. Neglect friction and other losses, and calculate the power being delivered by the pump in kilowatts. (Assume an open-circuit system.)

9. A pump and motor system is rated at 5 kW. What is the maximum rate (liters per second) at which it can lift water from a swimming pool at ground level to an open storage tank on the roof 50 m above the ground? (Take the mass density of the water as 0.997 kg/L.)

10. The water surface in a city reservoir is 200 m higher in elevation than the water main in a downtown street. What is the hydrostatic pressure (no flow) in the main in kilopascals?

CHAPTER 3

Heat Energy and Heat Transfer

The concept of heat as the energy of molecular motion was developed in the preceding chapter, and the relationship of heat and temperature was explained. Heat, it will be recalled, flows as a result of *temperature difference*, and it flows only in the direction of a descending temperature gradient.

The design of successful solar energy systems depends on a knowledge of many of the properties, attributes, and effects of heat. The purpose of the present chapter is to introduce and explain a number of these essential factors.

THE MEASUREMENT OF HEAT

3-1 Heat Quantity and Heat Measurement

At the outset, it is emphasized that a thermometer measures temperature, not heat. The amount of heat energy that a material substance can store, or that can flow into or out of a substance, cannot be directly measured by thermometers. And the amount of heat required to warm up, melt (liquefy), or vaporize a substance, or the amount of heat surrendered by a body as it cools (or solidifies or liquefies) cannot be measured directly by a thermometer either. However, temperature change is involved in determining heat quantity and in defining the basic unit of heat energy.

The engineering unit of heat quantity is the *British thermal unit* (Btu).

> *One Btu is the amount of heat energy required to raise the temperature of 1 lb of water 1 F°.*

Note that water is the substance used to standardize the unit of heat quantity. The reason for using water will become apparent in the definition of thermal capacity given below. Laboratory tests of heat-flow quantities are made by measuring with thermometers the temperature change that takes place in a given weight (mass) of water.

In like manner, in the metric system, the basic unit of heat is the calorie (cal).

> *One calorie is the amount of heat energy required to raise the temperature of 1 g of water 1 C°.*

In actual practice, in the metric system, the kilocalorie (kcal) is used, since the metric standard of mass is the kilogram, not the gram.

> One kilocalorie (1000 cal) *raises the temperature of 1 kg of water 1 C°.*

The relationship between the Btu and the kilocalorie can be obtained by recalling that the mass equivalent of a standard pound is about 454 g, and that $1\ C° = 1.8\ F°$.

Therefore,

$$1\ \text{Btu} = \frac{454}{1.8} = 252\ \text{cal}$$

$$= 0.252\ \text{kcal}$$

And

$$1\ \text{kcal} = 3.97\ \text{Btu}$$

3-2 Thermal Capacity and Specific Heat (The Basic Heat Equation)

Based on the above definition of the unit of heat energy, the heat required to warm up a given amount of water can be evaluated as follows. Let

> H = the heat quantity required, Btu or kcal
> w = weight (or m = mass) of the water, lb or kg
> t_1 = the initial temperature of the water
> t_2 = the final temperature

Then the water will absorb heat in accordance with this equation:

$$H_{\text{Btu}} = w(t_2 - t_1)\,F° \qquad (3\text{-}1a)$$

and, in SI units,

$$H_{\text{kcal}} = m(t_2 - t_1)\,C° \qquad (3\text{-}1b)$$

From these equations, it follows that the units of the Btu are lb-F°, and the units of the kcal are kg-C°.

Just as Eqs. (3-1a) and (3-1b) give the amount of heat required to warm up a given amount of water, they also predict the amount of heat that a known amount of water will give off as it cools through the temperature difference $(t_2 - t_1)$. Water is often used as a heat-storing substance in solar energy systems. When so used, it is referred to as *thermal storage*.

Thermal Capacity. The amount of heat that 1 lb of a substance requires to warm up 1 F° (and that it will give off as it cools 1 F°) is called its thermal capacity. From the above definitions, the thermal capacity of water is 1 Btu/lb-F°, or 1 kcal/kg-C°.

A few sample problems will serve to clarify the concepts discussed thus far.

Illustrative Problem 3-1

A thermal storage tank contains 1200 gal of water at 140°F. How much heat (Btu) can it surrender as its temperature drops to 95°F?

Solution

Use Eq. (3-1a) and recall from Table 2-3 that water weighs approximately 8.34 lb/gal.

Substituting,

$$H = 1200 \text{ gal} \times 8.34 \text{ lb/gal} \times (140\text{-}95) \text{ F}°$$
$$= 450{,}400 \text{ Btu} \qquad\qquad \textit{Ans.}$$

Illustrative Problem 3-2

The water flow rate through a bank of solar collectors is 18 gal/min. The water-in temperature is 85°F, and the water-off temperature is 150°F. How many Btu per hour of solar energy are being absorbed and carried away by the water? (The symbol for heat flow rate is Q).

Solution

First, find the hourly water flow, since it is Btu per hour that is desired.

$$w_{\text{lb/hr}} = 18 \text{ gal/min} \times 8.34 \text{ lb/gal} \times 60 \text{ min/hr}$$
$$= 9010 \text{ lb/hr}$$

Substituting in Eq. (3-1a), modified to give heat flow rate,

$$Q_{\text{Btu/hr}} = 9010 \text{ lb/hr} \times (150\text{-}85) \text{ F}°$$
$$= 585{,}600 \text{ lb-F}°/\text{hr or Btu/hr} \qquad \textit{Ans.}$$

Of all common substances, water has the highest thermal capacity. For this reason and many others—low cost, ready availability, excellent flow properties, nontoxicity, and relatively low corrosion effects—water is a universally used heat transfer and heat storage medium.

Other substances also absorb and store heat. Their thermal capacities are, for the most part, less than one. In other words, it takes less heat to warm up most substances than it takes to warm up an equal weight (mass) of water through the same temperature range, and these other substances give off less heat as they cool than water does. It follows that they cannot store as much heat per unit weight (or mass) as water can.

Specific Heat Capacity. The quantity of heat required to change the temperature of unit weight (mass) of any substance by one degree is called the specific heat capacity c (or simply specific heat) of the substance. The specific heat of water is 1.00 by definition. Specific heat capacity c has the units Btu/lb-F°, kcal/kg-C°, or cal/gm-C°. Specific heat may be regarded as a measure of the thermal inertia of a substance; and a substance's specific heat tells at a glance whether or not it is suited for use as a thermal storage medium.

Table 3-1 provides values of specific heat capacity for some common substances.

Table 3-1 Specific Heat Capacity *c* of Selected Substances

Substance (near 70°F except as noted)	Specific Heat, c (Btu/lb-F°, kcal/kg- C°, cal/gm-C°)	kJ/kg-C°
Water (the standard)	1.00	4.19
Adobe	0.23	0.96
Air	0.24	1.01
Aluminum	0.22	0.92
Brick (common)	0.20	0.84
Concrete	0.16 to 0.20	0.67 to 0.84
Copper	0.093	0.39
Earth (dry soil)	0.20	0.84
Glass	0.15 to 0.21	0.63 to 0.88
Ice	0.50	2.09
Iron (steel)	0.115	0.48
Marble	0.21	0.88
Steam (212°F, 100°C)	0.48	2.01
Stone (average)	0.20	0.84
Water vapor (70°F)	0.45	1.88
Wood		
Hardwoods (average)	0.30	1.25
Pine, fir, cedar, etc. (average)	0.33	1.38

In solar energy system design, it is essential to be able to predict in advance how much heat a fluid medium or a solid structure will absorb as its temperature rises (and how much heat it will give off as its temperature falls). For example, if water is heated in solar collectors and stored in a large tank in order to surrender its heat later to a living space, how much thermal storage water will be needed? Or, if air is heated in a solar collector, how much heat can it store and transfer to the space to be heated? Or, if the structure itself is designed to absorb, store, and ultimately surrender heat to the space (a passive solar design), what mass of structural elements (concrete, stone, brick, wood) will be necessary? Some of these factors will now be analyzed, and two elementary problems will be set up and solved as illustrations of thermal storage practice.

Let H be the heat to be absorbed, w the weight of the material that will absorb and store heat, and $(t_2 - t_1)$ the temperature change that occurs in the process. The fundamental heat equation can then be written:

$$H = wc \, (t_2 - t_1) \quad \text{(English–engineering)} \qquad \text{(3-2a)}$$

or

$$H = mc \, (t_2 - t_1) \quad \text{(SI–metric)} \qquad \text{(3-2b)}$$

The following problems illustrate the nature of predictive calculations that can be made with the fundamental heat equation.

Illustrative Problem 3-3

A brick wall is to be used as thermal storage mass. It is erected across the entire south side of a house, behind glass where it can absorb direct solar radiation. It is desired to store 600,000 Btu of heat in the wall as it warms up from 75°F at 8 A.M. to 100°F at 4 P.M. Find the weight of brick required for the wall.

Solution

Solve Eq. (3-2a) for w:

$$w = \frac{H}{c(t_2 - t_1)}$$

From Table 3-1, the value of c for brick is 0.20 Btu/lb-F°.

Substituting values,

$$w = \frac{600,000 \text{ Btu}}{0.20 \dfrac{\text{Btu}}{\text{lb-F°}} \times (100 - 75) \text{ F°}}$$

$$= 120,000 \text{ lb (or 60 tons)} \qquad\qquad Ans.$$

Illustrative Problem 3-4 (SI–metric)

Stones, contained in a rock bin excavated under a building, are often used as thermal storage. How much heat, in kilocalories, could be stored in 80 metric tons of stones as they are warmed up by hot air from a solar collector, from 20° C to 48° C?

Solution

From Table 3-1, c for stone is 0.20 kcal/kg-C°. Substituting in Eq. (3-2b),

$$H = 80,000 \text{ kg} \times 0.20 \frac{\text{kcal}}{\text{kg-C°}} \times (48 - 20) \text{ C°}$$

$$= 448,000 \text{ kcal} \qquad\qquad Ans.$$

The basic principles just discussed and the methods of problem solution illustrated will be intensively used in later chapters as the actual practice of solar space heating is dealt with in detail.

CONSERVATION OF ENERGY— HEAT ENERGY AND MECHANICAL ENERGY

One of the best known principles in all of science is the *Law of Conservation of Energy*. Briefly stated, this law says that energy cannot be created or destroyed, but it can be changed in form. Energy transformations take place around us all the time. Electric energy is converted to heat by resistance (strip) heaters, or into mechanical energy of rotation by an electric motor. Heat energy is converted to mechanical energy by a turbine or an internal combustion engine. Mechanical energy is converted to heat energy by friction, and to electric energy by a generator or alternator. In all such energy transformations, no energy is gained or lost; it is merely changed from one form to another.[1] [There always *seems* to be lost energy, but it is not lost—it is merely wasted, not useful for the purpose at hand. See Eqs. (3-4) and (3-5).]

Although heat energy is always the primary consideration in solar energy systems design, mechanical energy and electric energy are also involved in all but the simplest passive solar heating systems. It is essential, therefore, to be able to convert heat energy measurements and units into mechanical and electric energy units, and vice versa.

3-3   The Mechanical Equivalent of Heat

The heat equivalent of mechanical work was first investigated by James Prescott Joule, a nineteenth-century British scientist. He arranged an experiment using a device whose essential elements are diagrammed in Fig. 3-1. The churnlike

[1] Since Einstein's formulation of his famous mass–energy equation, $E = mc^2$, in 1905, and reflecting subsequent developments in atomic and nuclear energy in which it is apparent that energy is liberated as mass disappears, the basic law has been modified to become the law of conservation of mass–energy.

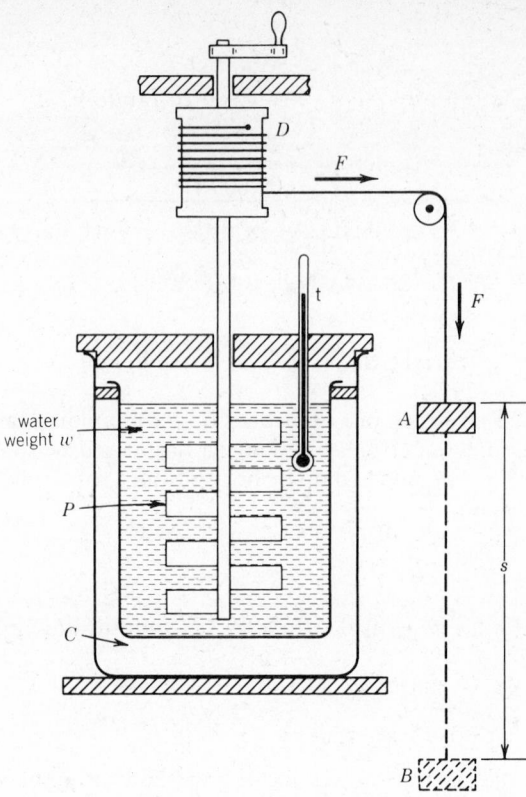

Figure 3-1 Schematic diagram of an early form of Joule's apparatus. A known weight of water w is placed in the calorimeter cup C, completely covering the paddle arrangement P. A thermometer t is inserted through the calorimeter lid and into the water. A heavy weight is allowed to fall a distance s under the pull of gravity. It moves from A to B, exerting a steady force F (equal to its weight) on the cord that is wound around drum D. The amount of mechanical work done is therefore $W = F s$. This amount of mechanical work is done on the water by the rotating paddles, and the water temperature rises. Assuming no losses (calorimeters are well insulated), we can equate the work done to the heat produced:

$$W = J H$$

But from Eq. (3–1a), H can be calculated from the known weight of water and its temperature rise $(t_2 - t_1)$. Joule's constant J can thus be determined.

apparatus featured a system of rotating paddles that could expend a measured amount of mechanical work on a known weight (mass) of water contained in an insulated cup called a calorimeter. As mechanical work was done on the water by the paddles, the temperature of the water increased. The total amount of heat produced was calculated from the fundamental heat equation, Eq. (3-2), where $c = 1.00$ for water. Joule defined the mechanical equivalent of heat as the ratio of mechanical work input to the heat output produced. This ratio has since come to be called *Joule's constant*, J. As defined,

Joule's constant,

$$J = \frac{W}{H} \qquad\qquad (3\text{-}3a)$$

and

$$W = JH \qquad (3\text{-}3b)$$

Joule's original experiment has been repeated numberless times with the most careful research methods and modern laboratory equipment. The presently accepted values of the constant J are

$$J = 778 \text{ ft-lb/Btu}$$
$$= 4.186 \times 10^3 \text{ joules/kcal}$$
$$= 4.186 \text{ joules/cal}$$

NOTE *Joule's constant* is designated by J, while the symbol for the joule (SI–metric unit of energy) is **J** in boldface type.

Energy transformations are rarely, if ever, 100 percent efficient—that is, not all of the energy input results in output of the kind of energy desired. In heat engines, for example, only about 20 to 35 percent of the heat energy input is converted to mechanical energy; the rest is given off as wasted heat. Even electric motors and generators do not convert all of the input energy into usable output. The ratio of the useful output of a machine (energy in the form desired) to the input energy is called efficiency.

$$\% \text{ efficiency}_{(\text{machine})} = \frac{\text{Useful energy output}}{\text{Energy input}} \times 100 \qquad (3\text{-}4)$$

Electrical equipment (motors, generators, transformers) usually operates in the 90 to 96 percent range of efficiency. Solar collectors (see Chapters 11 and 12) can attain efficiencies in the range from 30 to 70 percent. Photovoltaic (solar) cells of current manufacture (1984) can convert about 15 to 18 percent of the solar radiation energy that falls on them into electric energy.

In order to incorporate the idea of wasted energy into the law of conservation of energy, the following equations can be written:

$$\begin{matrix} W & - & W_w & = & JH \\ \text{mechanical} & & \text{wasted} & & \text{energy converted} \\ \text{energy input} & & \text{work} & & \text{to heat} \end{matrix} \qquad (3\text{-}5)$$

$$\begin{matrix} H & - & H_w & = & \dfrac{W}{J} \\ \text{heat energy} & & \text{wasted} & & \text{heat converted} \\ \text{input} & & \text{heat} & & \text{to work} \end{matrix} \qquad (3\text{-}6)$$

The branch of science that deals specifically with the relationships between heat and mechanical work is called thermodynamics. Eqs. (3-5) and (3-6) are mathematical statements of the *First Law of Thermodynamics*.

Illustrative Problem 3-5

A bank of electric strip (resistance) heaters is used as a backup to provide supplemental heat to a solar heating system. For each kilowatt-hour of electric energy used, how many Btu of heat will be supplied to the system? (Assume 100 percent efficiency.)

Solution

Recall that

$$1 \text{ W} = 1 \text{ J/sec}$$

Therefore

$$1 \text{ kW} = 1000 \text{ J/sec} = 1 \text{ kJ/sec}$$
$$= 3.6 \times 10^3 \text{ kJ/hr}$$

But

$$J = 4.186 \text{ kJ/kcal}$$

Therefore, 1 kW produces

$$\frac{3.6 \times 10^3 \text{ kJ/hr}}{4.186 \text{ kJ/kcal}} \text{ or } 860 \text{ kcal/hr.}$$

But (from Section 3-1),

$$1 \text{ Btu} = 0.252 \text{ kcal}$$

Therefore

$$1 \text{ kW} = \frac{860 \text{ kcal/hr}}{0.252 \text{ kcal/Btu}}$$

$$= 3410 \text{ Btu/hr} \qquad\qquad Ans.$$

This answer is an important and convenient relationship to remember since there is no waste in the electric energy-to-heat conversion. The heat produced by electric resistance heaters is determined from the relation:

$$\begin{matrix} 1 \text{ kW} & = & 3410 \text{ Btu/hr} \\ \text{(electric power)} & = & \text{(heat flow rate)} \end{matrix} \qquad (3\text{-}7)$$

Electric strip (resistance) heater elements are often supplied in 2-kW units. These can be assembled in banks and controlled by a sequencing programmer to "kick in" one at a time as the need for supplemental heat increases.

In air-conditioning practice it is more common to express heat quantities in kilojoules rather than in kilocalories, and heat flow rates in the same units as electric power (i.e., in kilowatts) rather than in kilocalories per second.

HEAT AND CHANGE OF STATE

At the ordinary temperatures and pressures encountered on earth, different kinds of matter exist in three different conditions, called *states*. There are *solids* like wood, metal, stone, and metallic salt-hydrates; *liquids* like water, mercury, and gasoline; and *gases* like air, oxygen, and carbon dioxide. However, even with relatively small changes in temperature and/or pressure, the state of a substance may change. Water, for example, is a liquid as we ordinarily think of it, but it becomes a solid at 32°F, and a gas (vapor) at 212°F (at atmospheric pressure). Metals melt to a liquid at temperatures ranging upward from 500°F. Several metallic salt-hydrates are solids at normal room temperature, but they melt to liquids at temperatures in the range from 80° to 120°F. And gases condense to liquids when the pressure–temperature conditions are right. Perhaps the most common example of the latter is the condensation of water vapor in the atmosphere to form rain or dew, often when a temperature drop of only a few degrees occurs.

In summary, all matter exists in one of three states—solid, liquid, or gas.[2] The particular state at the time depends first on the nature of the substance itself and also on its temperature, its heat content, and the pressure to which it is subjected.

Conditions encountered in solar energy work sometimes result in changes of state for water, and they routinely cause changes of state for the refrigerants used in solar-energized cooling systems and in solar-assisted heat-pump systems. Moreover, the metallic salt-hydrates that are sometimes used for thermal stor-

---

[2] A fourth state of matter (called *plasmas*) has been proposed by some scientists to characterize matter that has become intensely ionized at temperatures of millions of degrees—conditions that exist on the sun, for example.

age in solar energy systems undergo changes of state as they store heat and give off heat. These substances are sometimes referred to as *eutectic mixtures*.

At this point, some of the change-of-state properties of water will be described. The behavior of refrigerants will be dealt with briefly in Chapter 16.

3-4   The Change-of-State Behavior of Water

In heating system design, water is a commonly used medium of heat transfer and heat storage. Water is often the fluid used in solar collectors, and hot water is circulated to radiators and convectors where it gives off its heat to living spaces. Depending on the type of heating system and the location of its component equipments, the water may be in liquid or gaseous (vapor or steam) form; or, if proper precautions are not taken, the water may freeze to a solid. It is necessary, therefore, to be able to predict the change-of-state behavior of water.

Figure 3-2 is a temperature–heat (T–H) diagram for 1 lb of water at a constant pressure of 1 atm (14.7 psi). Note that the starting point is 1 lb of ice at $-20°F$ under 1 atm pressure (see lower left corner of the diagram). Suppose that heat is now added slowly to the ice. Its temperature rises 1 F° for each 0.5 Btu of heat added. (The specific heat capacity c of ice is 0.50 Btu/lb F°.) Successive temperature–heat content readings plot along the line AB as ice is warmed up while it is still a solid. The T–H curve at B shows that 26 Btu of heat have been added by the time the 1 lb of ice reaches a temperature of 32°F, ready to melt but not yet melting. Since this heat added in the AB region results in a rising temperature that can be sensed either by the human senses or by a thermometer, it is called *sensible heat*.

Adding more heat at point B does not immediately increase the temperature of the ice, but a change of state (melting) begins. Laboratory tests show that 144 Btu of heat are required to melt 1 lb of ice already at 32°F to liquid water, still at 32°F. No temperature change occurs during the change-of-state process, and since the heat absorbed by the ice does not show up in any way evident to the senses (or to a thermometer), it is called *latent heat*. The melting process is technically called *fusion*, and it is shown on the diagram as line BC.

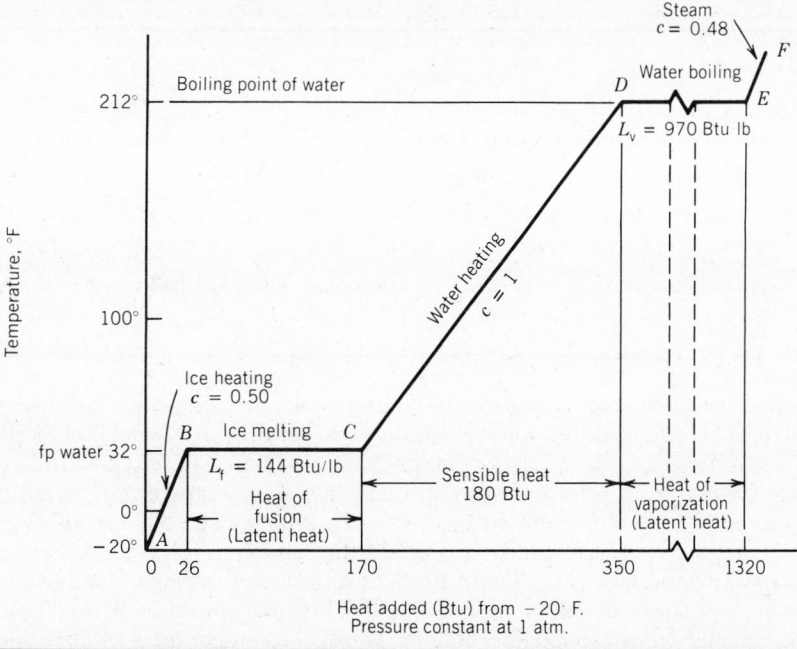

Figure 3-2 Temperature–heat (T–H) diagram for 1 lb of pure water at atmospheric pressure, in English units. Note the break at the right-hand side of the heat-content (horizontal axis) scale.

The *latent heat of fusion of water* is

$$L_f = 144 \text{ Btu/lb}$$

At point C, the melting (fusion) process is complete, the original pound of ice is now 1 lb of liquid water at 32°F, and a total of 170 Btu (26 + 144) of heat has been added to the original pound of ice.

From point C, the addition of more heat results once more in rising temperature, as the liquid water warms at the rate of 1 F° for every Btu added (c for liquid water = 1.00 Btu/lb-F°). After absorbing an additional 180 Btu of heat, the water will have attained a temperature of 212°F at point D. This temperature (212°F) is the boiling point of water at 1 atm pressure. This amount of heat (process CD) is again sensible heat, since it causes a temperature change. Total heat added to this point is 350 Btu.

From point D, the addition of more heat does not cause a temperature increase but initiates another change of state—*vaporization*, or boiling. To vaporize the entire pound of water at 1 atm pressure, 970 Btu of heat are required. This process occurs along line DE, and the heat absorbed by the water is latent heat. (Note that it is necessary to have a scale break in the diagram between D and E, since the textbook page size will not permit retention of the original scale).

The *latent heat of vaporization of water* (at 1 atm pressure) is

$$L_v = 970 \text{ Btu/lb}$$

At state point E of the diagram, there is now 1 lb of (saturated or wet) steam. Up to this point, the total heat added is 1320 Btu. If more heat is supplied to the pound of steam, its temperature rises once more. The process now follows the line EF at a slope dictated by the fact that for steam at 1 atm pressure, $c = 0.48$ Btu/lb-F°.

Illustrative Problem 3-6

A low-pressure steam boiler is supplied with water at 60°F. It is desired to produce 300 lb of (saturated) steam per hour from this feed water. Calculate the heat output of the boiler in Btu per hour.

Solution

Set up the problem in two parts—the first phase involving heating the water to the boiling point and the second phase involving the water-to-steam conversion.

$$H_1 = 300 \text{ lb} \times 1.00 \text{ Btu/lb-F}° \times (212 - 60)\text{F}°$$
$$= 45,600 \text{ Btu (to heat the water to the boiling point)}$$
$$H_2 = 300 \text{ lb} \times 970 \text{ Btu/lb} = 291,000 \text{ Btu (for the change of state)}$$

Boiler heat output = 291,000 + 45,600 = 336,600 Btu/hr   *Ans.*

3-5   Phase Change in Metallic Salt-Hydrates

Certain metallic salt-hydrates—including calcium chloride hexahydrate ($CaCl_2 \cdot 6\ H_2O$), sodium sulfate decahydrate ($Na_2SO_4 \cdot 10\ H_2O$) (Glauber's salt), and disodium phosphate dodecahydrate $Na_2HPO_4 \cdot 12\ H_2O$)—undergo a phase change from solid to liquid and back again at temperatures commonly encountered on a daily basis in most solar heating systems. These substances (when prepared as phase-change materials, they are often referred to as *eutectic mixtures*) exhibit quite high latent heats of fusion. For example, Glauber's salt absorbs and stores as latent heat about 100 Btu per pound as it undergoes a phase change from a crystalline solid to liquid at a temperature of about 90°F. The density of Glauber's salt (liquid phase) is about 85 lb/ft³, and 1 cu ft will store more than 8000 Btu as it melts and give this heat back to the system when it solidifies again as the surrounding temperature drops below 90°F.

The phase-change behavior of such salts makes them useful as thermal storage materials in solar heating systems. Further discussion of this topic is found in Chapter 7.

THERMAL EXPANSION

Everyday observation provides evidence that as materials are heated they expand, and as they are cooled they contract. Steel bridges and steel rails are fitted with expansion joints to allow for expansion and contraction as daily temperature changes occur. Pistons are fitted to engine cylinders and compressors with tolerances that allow for thermal expansion. A piping system full of liquid at a given temperature will increase in length significantly if the temperature of the fluid it is handling rises. Expansion joints are usually provided in piping circuits if operating temperature variations of some magnitude are expected.

Thermal expansion actually occurs on a three-dimensional basis, but, to simplify the treatment here, we list three types of expansion as though they took place independently.

1. Variation in length, or *linear expansion*. Linear expansion occurs in metal rods, pipes, or wires (solids only).
2. Expansion in a two-dimensional plane, or *area expansion*. This kind of expansion occurs with flat plates or sheets whose variation in thickness is not a factor of importance (solids only).
3. *Volume expansion* (sometimes called *cubical expansion*). Solids, liquids, and gases are all involved in volume expansion, although the effects are of practical significance mostly with liquids and gases.

Only linear expansion will be reviewed here. Any standard physics text may be consulted for a treatment of area expansion and volume expansion. Be sure to keep in mind that the word *expansion* also implies the opposite effect—contraction—which occurs as a substance undergoes a drop in temperature.

3-6 Linear Expansion

Consider, for example, a straight copper pipe carrying water from a solar collector to a storage tank. At dawn, the water temperature might be 40°F and at 4 P.M. 140°F. As it heats up, each unit length of the pipe expands a small amount as its temperature increases degree by degree. The small increase of each unit length for each degree rise in temperature is called the *coefficient of linear expansion*. Stated another way, the coefficient of linear expansion is the change in length per unit of original length per degree change in temperature (Fig. 3-3). As an equation,

$$\alpha = \frac{\Delta L}{L_0 \Delta t} \tag{3-8}$$

in which

α = the coefficient of linear expansion
ΔL = the change in length caused by a temperature change Δt
L_0 = the length at the initial temperature

NOTE   The Greek letter delta (Δ) is used to indicate a small increment of change in a measurement or variable. ΔL is read "delta L," not "delta times L."

The coefficient of linear expansion α has widely differing values for different materials. It varies to some extent at extreme temperature ranges even for the same material. The values are quite small for most materials but large enough to be a design factor in many manufacturing, assembly, and piping operations. For

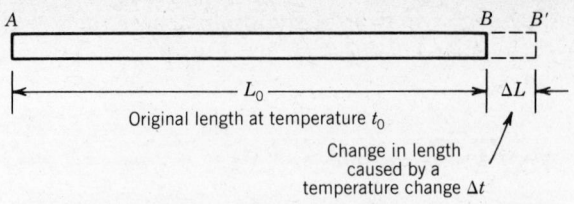

Original length at temperature t_0

Change in length
caused by a
temperature change Δt

Figure 3-3 Defining the coefficient of linear expansion. AB is a rod or pipe with a length L_0 at temperature t_0. If the temperature increases by an amount Δt, the rod expands to B', a net expansion in length of ΔL.

copper, as in the above example, $\alpha = 9.5 \times 10^{-6}$ ft/ft-F°. In metric units, the coefficient is 1.8 times larger, or 17.1×10^{-6} mm/mm-C°, since the Celsius degree is 1.8 times as large as the Fahrenheit degree.

Note from Eq. (3-8) that length units cancel out, and the coefficients can be used as pure numbers per Fahrenheit degree or Celsius degree.

Table 3-2 lists values of α for some materials commonly encountered in the solar energy industry.

Table 3-2 Coefficients of Linear Expansion α for Selected Industrial Materials

Material	α per F°	α per C°
Aluminum	13×10^{-6}	24×10^{-6}
Brass	10×10^{-6}	18×10^{-6}
Concrete	6.0×10^{-6}	11×10^{-6}
Copper	9.5×10^{-6}	17.1×10^{-6}
Glass		
Common	4 to 5×10^{-6}	7 to 9.5×10^{-6}
Pyrex	1.7×10^{-6}	3.0×10^{-6}
Steel	6.7×10^{-6}	12×10^{-6}

Illustrative Problem 3-7

A 50-ft length of copper pipe runs from the header of a solar collector panel to a thermal storage (water) tank. At 7 A.M. the copper temperature is 35°F, and by 4 P.M. it increases to 150°F. How much has it expanded in length (inches)?

Solution

Use Eq. (3-8) and solve for ΔL, the increase in length:

$$\Delta L = \alpha L_0 \Delta t$$

From Table 3-2,

$$\alpha_{copper} = 9.5 \times 10^{-6}/F°$$

Substituting,

$$L = 9.5 \times 10^{-6}/F° \times 50 \text{ ft} \times 115 \text{ F°}$$
$$= 5.46 \times 10^{-2} \text{ ft}$$
$$= 0.655 \text{ in.} \qquad\qquad Ans.$$

Unless a suitable allowance for this daily expansion and contraction is made in the piping installation, structural failure is likely to occur. As the saying goes, "something has to give."

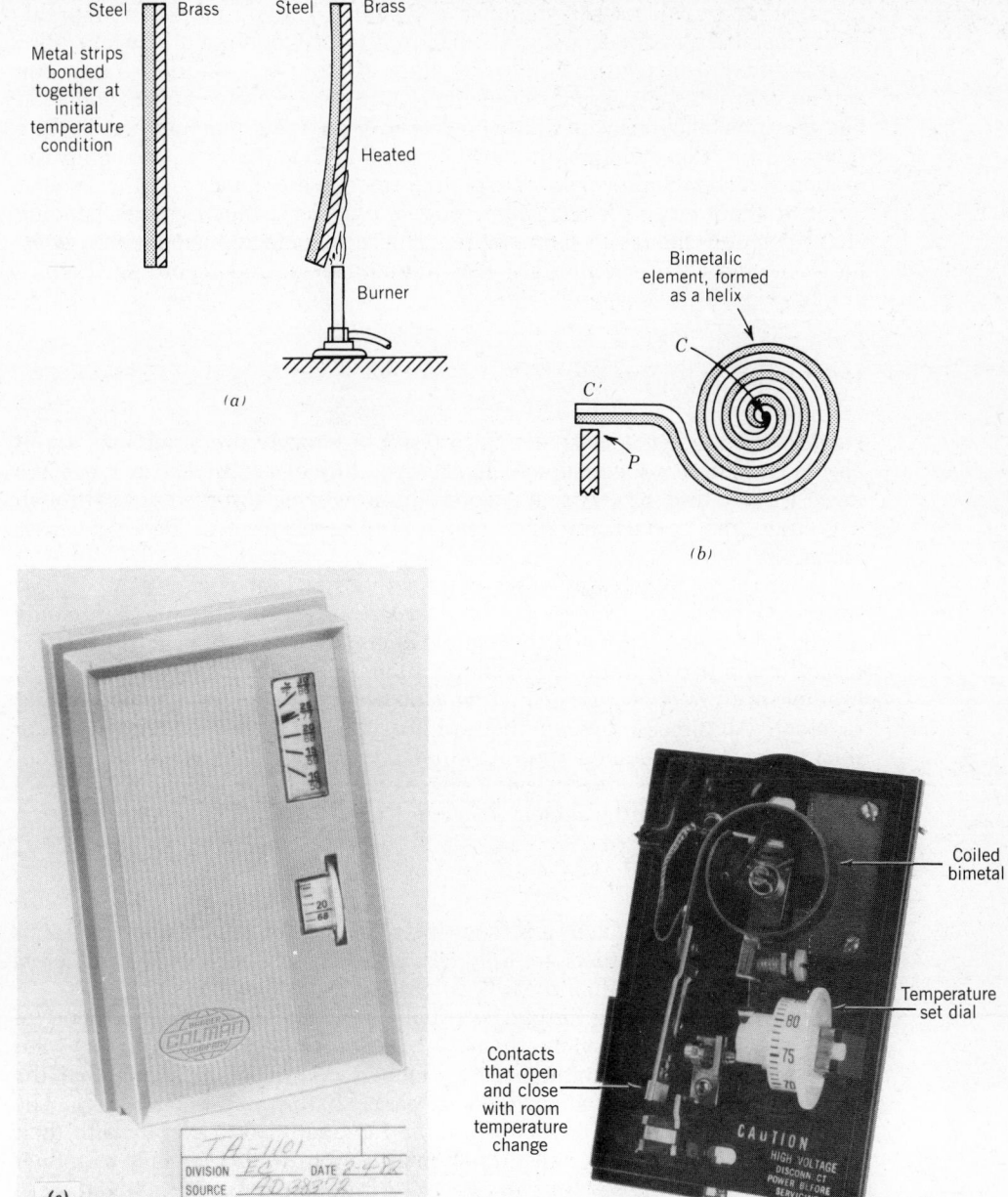

Figure 3-4 Differential expansion of bimetallic elements. (*a*) A strip of steel and a strip of brass are shown bonded together in the left diagram. The two strips are exactly the same length at a given initial temperature. When heated (right diagram), the brass expands at a much greater rate than the steel (see Table 3–2). Since the two strips cannot slide along each other, the result of the differential expansion is a bending of the bimetallic element. (*b*) A bimetallic element formed in the shape of a helix to magnify the movement caused by small temperature changes. The diagram illustrates the (much simplified) action in a common thermostat. Points *P* and *C* are connected to a low-voltage circuit. The contacts at *C'–P* are normally open for heating. As room temperature falls, the bimetal helix brings *C'* in contact with *P*. The low-voltage circuit is thus energized, and it in turn actuates controls, switches, and valves that operate the heating system. Ordinarily, there is a magnet (not shown in the diagam) to produce a snap action as *C'* and *P* come close together. (*c*) A typical air-conditioning thermostat, showing the thermometer that indicates the room temperature and the dial for setting the desired temperature. (*d*) Same thermostat with cover removed, showing the bimetallic expansion element and the contact points that open and close the control circuit. (Both photos from Barber-Colman Company.)

Differential Expansion—Bimetallic Elements. As Table 3-2 shows, different metals expand and contract at different rates. Consequently, when strips of two different metals are bonded together at room temperature so that one cannot slide along the surface of the other, if the temperature changes appreciably, the bimetallic element will bend one way on rising temperature and the other way on falling temperature, as shown in Fig. 3-4a. In order to magnify the motion of bimetallic strips, they are often formed in the shape of a helical coil in order to attain greater length. This scheme provides a much greater bending effect per unit change in temperature. Helical bimetallic elements are commonly employed in thermostats, dial thermometers, and recording thermographs (Figs. 3-4b, 3-4c, and 3-4d).

HEAT TRANSFER

Heat flows in material substances as a result of temperature difference, and it always flows from a source at a higher temperature to a substance or space at a lower temperature. Energy that becomes heat on being absorbed flows through a vacuum and through interplanetary space, in the form of *electromagnetic radiations.*

There are three easily observable methods of heat energy flow, or *heat transfer: conduction, convection,* and *radiation.* Heat entering a building through the walls and roof is an example of conduction, heat carried from one place to another in a moving medium such as hot water or warm air is an example of convection, and heat arriving on earth from the sun is an example of radiation. All three of these methods of heat transfer are basic to the design of solar energy systems, and a thorough understanding of them is essential. Problems in heat transfer arise in every phase of solar energy system design, so a detailed treatment of the subject is necessary.

3-7 Conduction

In solid materials, conduction is the primary method of heat flow. Some solids are good *conductors*, others are only fair, and some are such poor conductors that they are used to inhibit heat flow and are called *insulators*.

Conduction of heat in solids is usually explained by the kinetic theory of heat. According to the kinetic theory, as substances absorb heat energy from any source, their temperature rises, which is equivalent to saying that the average velocity of their molecules increases sharply (see Section 2-5). The higher-energy molecules now collide more frequently and violently with their neighbors, setting them in violent motion also. In turn, these collide with their neighbors, and energy is thus passed along through the solid material, as is depicted in Fig. 3-5.

Examples of conduction processes encountered in solar energy systems include, but are by no means limited to, the following:

1. Heat flow in and through metals as in the base (absorber) plate and tubes of solar collectors.
2. Heat flow from hot water through the metal coils and fins of a *heat exchanger* in order to heat a stream of air or a liquid in a different circuit.
3. Heat flow into and through the walls, roof, and windows of buildings. This flow is termed *heat gain* if it is inward to the building and *heat loss* if it is outward from the building.

3-8 Heat Conductivity

Solids conduct heat at widely differing rates. Liquids conduct heat very poorly, but they transfer heat very effectively by means of convection (discussed below). Gases are also very poor conductors of heat.

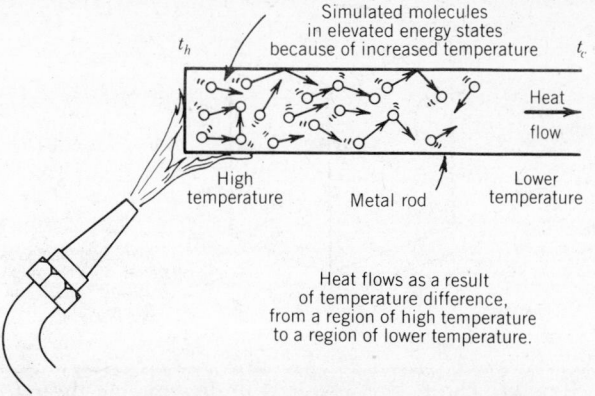

Simulated molecules
in elevated energy states
because of increased temperature

t_h t_c

Heat
flow

High
temperature   Metal rod   Lower
temperature

Heat flows as a result
of temperature difference,
from a region of high temperature
to a region of lower temperature.

Figure 3-5 A schematic representation of the kinetic–molecular hypothesis as the explanation of heat flow by conduction along a metal rod.

Among good conductors of heat are most of the metals. Falling into the "fair conductor" category are such materials as brick, wood, concrete, stucco, and plaster. Poor conductors (insulating materials) include cork, fiberglass, mineral wool, polyurethane and polystyrene boards, cellulose board, sawdust, wood chips, and ground bark. Air is a very poor conductor of heat, so any process of fabrication or the use of any material or structural element that will entrap air cells by the millions in an otherwise solid material will result in a good insulator. Fiberglass, mineral wool, and the various foam materials are examples of such insulators.

The rate at which a particular solid will conduct heat is determined by experiment, in a manner suggested by the sketch in Fig. 3-6. A slab of the material to be tested is indicated by the block with its parallel shaded surfaces W and C. The thickness of the slab L and the cross-sectional area A must be known. The two surfaces W and C are provided with thermometers so that their temperatures can be accurately determined. The test slab must be enclosed in a well-insulated container (not shown) so that heat losses to (or gains from) the surroundings are negligible. Heat (perhaps from steam) is supplied to the warm side W and removed (usually by flowing water) from the cold side C. Heat flow is maintained until a steady-state condition is reached—that is, until the thermometer readings and the water flow rate are holding constant. Then readings are taken over a period of about an hour. The heat flow can be calculated from the mass flow of water and its temperature rise, using the fundamental heat equation, Eq. (3-1a) or (3-1b). Such experiments have demonstrated that the quantity of heat that will be conducted through solids depends on the following factors:

1. The temperature difference $(t_2 - t_1)$ between the two parallel surfaces.
2. The thickness of the material (L).
3. The time the heat flows (T).
4. The cross-sectional area of the slab of material (A).
5. The nature of the material itself. This factor is called *conductivity*, and is designated by K.

These factors are combined in the following manner to result in the basic equation for heat conduction in solids:

$$H = \frac{K\,A\,T\,(t_2 - t_1)}{L} \qquad (3\text{-}9)$$

The quantity $(t_2 - t_1)/L$ is often written $\Delta t/L$ and is called the *temperature gradient*.

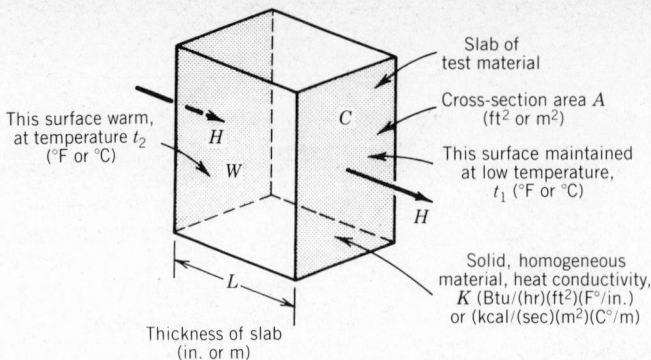

This surface warm, at temperature t_2 (°F or °C)

Slab of test material

Cross-section area A (ft² or m²)

This surface maintained at low temperature, t_1 (°F or °C)

Solid, homogeneous material, heat conductivity, K (Btu/(hr)(ft²)(F°/in.) or (kcal/(sec)(m²)(C°/m)

Thickness of slab (in. or m)

Figure 3-6 The factors involved in determining the rate of heat flow by conduction in a solid slab of material. All of these factors are found in the basic equation for heat conduction, Eq. (3–9). Note: Entire slab of test material must be encased in heavy insulation to prevent losses to surroundings. (Source of heat for side W, and means of removing heat from side C, not shown.)

In Eq. (3-9), H is the total heat (Btu) conducted in T hours through a slab of cross-sectional area A sq ft and thickness L in., when the temperature difference between its two surfaces is $(t_2 - t_1)$ F°. K is the conductivity of the material, and its meaning and dimensions can be determined by solving Eq. (3-9) for K and noting the units of each (measured) variable in the equation.

$$K = \frac{H \text{ (Btu)}}{T\text{(hr) } A\text{(ft}^2) \dfrac{(t_2 - t_1) \text{ F}°}{L\text{(in.)}}}$$

From this expression, it is readily evident that K has dimensions as follows:

$$K_{\text{(dimensions)}} = \frac{\text{Btu}}{\text{(hr)(ft}^2)\text{(F}°/\text{in.)}} \text{ (English system)}$$

These units are often written as Btu/hr-ft²-F°/in.

Rates of heat flow by conduction are ordinarily expressed in Btu per hour. For conduction,

$$Q_{\text{Btu/hr}} = \frac{K A (t_2 - t_1)}{L} = \frac{K A \Delta T}{L} \qquad (3\text{-}10)$$

SI–Metric Units for Conduction Heat Flow. Based on the same principles discussed above and illustrated in Fig. 3-6, the situation for conduction heat flow measured in metric units may be analyzed as follows.

H is the heat in joules J conducted in T sec through a slab of solid material whose cross-sectional area is A m² and whose thickness is L m, when the temperature difference between the two surfaces is $(t_2 - t_1)$C°. K is then expressed as follows, in metric units:

$$K = \frac{H \text{ (joules)}}{T \text{ (sec) } A \text{ (m}^2) \dfrac{(t_2 - t_1)\text{C}°}{L \text{ (m)}}}$$

The metric dimensions of K are therefore joules per (sec)(m²)(C°/m). Noting that 1 joule/sec = 1 watt, and rearranging,

$$K_{\text{(dimensions)}} = \frac{\text{W}}{\text{(m}^2)\text{(C}°/\text{m)}} \text{ (SI–metric system)}$$

Or, written out and simplified, W/m-C°.

Heat flow rates are most often expressed in watts (or kilowatts) in the metric system rather than in kilocalories per second or kilojoules per second. The conduction rate equation for the metric system is

$$Q_{kW} = \frac{KA\,(t_2 - t_1)}{10^3 L}$$

(3-11)

Table 3-4 provides values for the heat conductivity K of selected industrial materials.

Table 3-4 Heat Conductivity K of Selected Industrial Materials in Normal Temperature Ranges

Material	K Btu/(hr)(ft²)(F°/in.)	K W/(m²)(C°/m)
Air (at rest)	0.168	0.024
Aluminum	1480	212
Brick (common)	5.0	0.72
Celotex board	0.30	0.045
Concrete (average)	9–12	1.7
Copper	2640	380
Fiberglass insulation	0.26	0.037
Glass (average)	5.5	0.80
Insulating boards (range)	0.30–0.35	0.04–0.05
Plaster and gypsum lath (drywall)	4.0	0.57
Steel	312	45
Water Liquid	4.28	0.61
Ice	15.6	2.25
Wood Hardwoods (average)	1.15	0.165
Softwoods (average)	0.80	0.115

Illustrative Problem 3-8

Solar radiation strikes one side of a common brick wall, maintaining its surface temperature at 125°F. The wall is 40 ft by 10 ft by 14 in. thick. Calculate the steady-state rate of heat gain through the wall into the space beyond, in Btu per hour, when the inner wall surface temperature is 80°F.

Solution

Substitute given values in Eq. (3-12), noting from Table 3-3 that K for common brick is 5.0 Btu/(hr)(ft²)(F°/in.).

$$Q = \frac{5.0\ \text{Btu/(hr)(ft}^2\text{)(F°/in.)} \times 400\ \text{ft}^2 \times (125 - 80)\text{F°}}{14\ \text{in.}}$$

$$= 6430\ \text{Btu/hr}$$

Ans.

Illustrative Problem 3-9

A flat steel plate 3 ft by 8 ft is exposed to direct solar radiation, and its top surface temperature is noted as 155°F. The plate is 0.18 in. thick. The temperature of its lower surface is maintained at 154.85°F by water flowing across it. At what rate is heat being conducted through the plate?

Solution

Note in Table 3-3 that K for steel is 312 Btu/(hr)(ft²)(F°/in.). Substituting in Eq. (3-10),

$$Q = \frac{312 \text{ Btu/(hr)(ft}^2)(\text{F}°/\text{in.}) \times 24 \text{ ft}^2 \times 0.15 \text{ F}°}{0.18 \text{ in.}}$$

$$= 6240 \text{ Btu/hr} \qquad \qquad Ans.$$

Illustrative Problem 3-10 (SI–Metric)

A large view window of plate glass measures 8 m by 3 m by 0.085 m thick. Its inner surface temperature is 15°C. What is the rate of heat conduction loss (in kilowatts) through this window when its outer surface is at −8°C?

Solution

From Table 3-3, K for glass is 0.80 W/(m²)(C°/m). Substituting in Eq. (3-11),

$$Q_{kW} = \frac{0.80 \text{ W/(m}^2)(\text{C}°/\text{m}) \times 24 \text{ m}^2 \times 23 \text{ C}°}{0.085 \text{ m} \times 10^3}$$

$$= 5.2 \text{ W} \qquad \qquad Ans.$$

3-9 Other Measures Related to Heat Conduction— Conductance, Thermal Resistance, and Overall Heat Transmission Coefficients

Although conductivity K is the basic scientific measure of the heat-conducting properties of materials, three other measures are more common in industrial practice. They will be briefly discussed here and will be used extensively in later chapters.

Conductance (C) is similar to K, differing only in that it refers to the thickness of the materials *as stated, not per inch thick*. Both K and C are used only with homogeneous materials. For example, if a certain mix of concrete (when poured and set) has a conductivity

$$K = 12.0 \text{ Btu/(hr)(ft}^2)(\text{F}°/\text{in.}),$$

then a concrete wall 10 in. thick would have a conductance

$$C = 1.2 \text{ Btu/(hr)(ft}^2)(\text{F}°),$$

for the thickness stated, 10 in. Manufacturers of sheathing, building boards, and insulating boards provide values of the conductance C for their boards in the thicknesses as manufactured. If a material of conductivity K is x in. thick, its conductance,

$$C = \frac{K}{x}$$

Thermal resistance (R) is a measure of the resistance that materials have to heat flow. R values can be expressed either per inch of thickness or for the thickness stated.

$$R \text{ value per inch} = \frac{1}{K}$$

$$R \text{ value for thickness } x = \frac{x}{K}$$

$$R \text{ value per thickness stated} = \frac{1}{C}$$

R values are currently the federally mandated standard designation (in the United States) for the rating of insulating materials (mineral wool, fiberglass,

Figure 3-7 Insulating materials must be clearly labeled with applicable R values. R-30 blanket insulation of fiberglass is being installed over a ceiling. (Owens/Corning *Fiberglas.*)

polyurethane and polystyrene boards, pressed-fiber boards, etc.). Manufacturers must label the product with the R value for the thickness stated (Fig. 3-7). For example, a pressed-cellulose insulating board 3/4-in. thick would have an R value of about 2.0, labeled R-2. A 6-in. thick fiberglass or mineral wool blanket would have an R value of about 19 or 20, labeled R-19. The higher the R value, the greater the resistance to heat transmission by conduction. R-19 in walls and R-30 in ceilings are recommended values for solar buildings.

The problem of heat transmission gains and losses to and from structures is complicated by the fact that walls, roofs, and building components generally are not made of one homogeneous material. A frame wall for a typical house, for example, may have 1-in.- thick board-and-batt siding on fiberboard or wood sheathing, supported by a 2 × 4 Douglas-fir stud wall, which is lined inside with lath and plaster or drywall (gypsum board) (Fig. 3-8). Spaces between the studs are usually filled with blanket insulation 3 to 4 in. thick. The direct use of K-values, C-values, or R-values to determine heat gains and losses through such composite walls is extremely cumbersome, and as a result the concept of *overall heat transmission coefficients* (U-values) has been adopted.

A composite wall, such as that illustrated in Fig. 3-8, can be looked upon as a *series resistance.* The heat flow through the wall is analogous to the flow of electric current through a series circuit. Heat flow (current) passes through each of the thermal resistances, one after another in series. Since heat flow is the result of temperature difference, $t_2 - t_1 = \Delta t$ is analogous to potential difference (voltage) in an electrical circuit.

The total resistance to heat flow (i.e., the thermal resistance) of a composite wall is equal to the sum of the separate resistances in series. That is,

$$R_T = R_1 + R_2 + \text{--------} R_n \qquad (3\text{-}12)$$

U, on the other hand, is the *thermal transmittance* of the composite wall, the reciprocal of the thermal resistance:

$$U = \frac{1}{R_T} = \frac{1}{R_1 + R_2 + \text{----} R_n} \qquad (3\text{-}13)$$

U values are either calculated by computer analysis or actually measured in laboratories. Tables of U values for stipulated built-up structural elements (walls, floors, ceilings, partitions, etc.) are available in handbooks for the use of

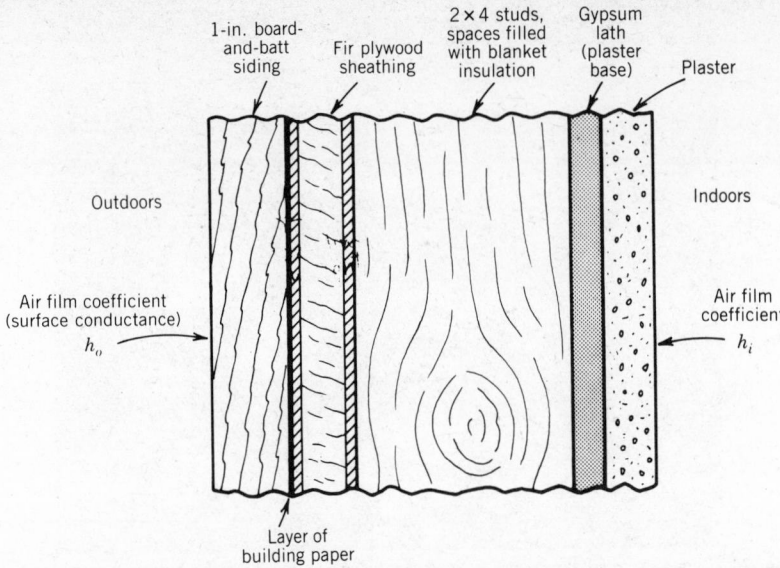

Figure 3-8 A vertical section through a composite wall—typical frame construction for a residence.

engineers and designers. As an example, the overall heat transmission coefficient for the typical frame wall described above (Fig. 3-8) would be about

$$U = 0.24 \text{ Btu/(hr)(ft}^2)(\text{F}°)$$

without insulation between the studs. With (nominal) 4 in. (actually about 3½ in.) of mineral wool or fiberglass blanket insulation in the stud wall, U would be reduced to about

$$0.07 \text{ Btu/(hr)(ft}^2)(\text{F}°)$$

R values and U values will be used extensively in Chapters 5 and 16, where the calculation of heating and cooling loads of buildings is discussed. A brief table of U values is provided in Chapter 5.

3-10 Convection

Gases and liquids generally are rather poor conductors of heat, but heat transfer is readily accomplished in these fluids by moving currents called *convection currents*. The fluid (liquid or gas) of a convection current absorbs heat at one location, and, after flowing to another location, it gives up heat as it mixes with cooler portions of the fluid. In the process of convection, high-energy molecules actually move from one location to another in the fluid, carrying their energy (heat) with them. Examples of convection are numerous and easily observed. Winds move across the earth, bringing heat waves or cold waves with them. The oceans of the earth incorporate massive convection currents of warm water in some regions and cold water in others, which warm up or cool off vast areas of certain continents. Hot water rises through pipes from boilers to heating units (convectors) in the living spaces of buildings. Warm air or cold air flows in ducts and emerges from registers to heat or cool homes, offices, and institutions. Warm air is less dense than cold air, so warm air rises in a room and cold air "falls" as a result of gravitational action (Fig. 3-9).

Two kinds of convection are identified: *natural convection*, which occurs as a result of differences in density caused by temperature variations (winds, ocean currents, fluid currents rising or falling as a result of being heated or cooled, etc.), and *forced convection*, which is mechanically induced by pumps or fans. Solar heating and cooling systems, depending on their design, may involve either natural or forced convection.

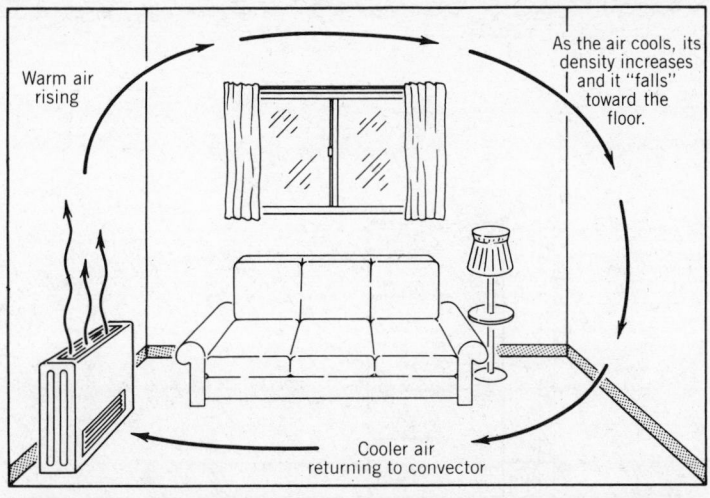

Figure 3-9 Convection currents in a room heated by a convector supplied with hot water.

Air is an ideal medium for convection, though it is a very poor conductor. Consequently, to minimize heat gains and losses in buildings, the walls, ceilings and other structural units should be fabricated in such a way that there are no large air spaces in which convection currents might be established. If air can move, it will carry heat with it (a convection current). If air is trapped and cannot move, any heat transfer would have to be by conduction, and air, as we have seen, is one of the poorest conductors known.

The factors involved in heat transfer by convection do not fit into an easily evaluated formula like that of Eq. (3-10) for conduction. For the purposes of solar energy, two cases are commonly encountered: flat plates or walls, along the surfaces of which a fluid (liquid or air) is free to move; and pipes or ducts, inside whose formed or curved surfaces fluids move, usually forced by pumps or fans. In both cases there is a temperature difference Δt between the fluid and the surface of the plate, wall, or pipe. Even though the fluid is in motion along the solid surface, there is a thin boundary film of the fluid in direct contact with the wall that either does not move at all or moves very slowly; it is said to be stagnant, and the boundary film is referred to as the *stagnant layer*. The thickness of this film depends on the turbulence of the fluid motion; the more turbulence, the thinner the film.

Consider a warm wall, for example, with a current of air moving slowly across it at a velocity of less than 15 ft/min. The wall could be a thermal storage wall in a solar-heated building, or it could be a wall in which hot-water pipes are imbedded for radiant heating. Heat flows from the wall into the room air by a combined conduction–convection process—conduction through the stagnant film of air in direct contact with the wall and convection in the air that moves along the wall (Fig. 3-10). There is also heat transfer by radiation, but that effect is neglected in the present discussion. (Radiation will be treated in Section 3-11 below.)

The combined effect of conduction through the stagnant film and convection in the moving air can be expressed by a convection coefficient or *film coefficient h* (sometimes called *surface conductance*.) An equation for the rate of heat transfer along a wall or flat plate can be written as

$$Q = h\,A\,\Delta t \qquad\qquad (3\text{-}14)$$

Where

Q = the rate of convection heat transfer, Btu/hr (or W)
A = the area of the wall or plate, ft^2 (or m^2)
Δt = the difference in temperature between the wall (plate) surface

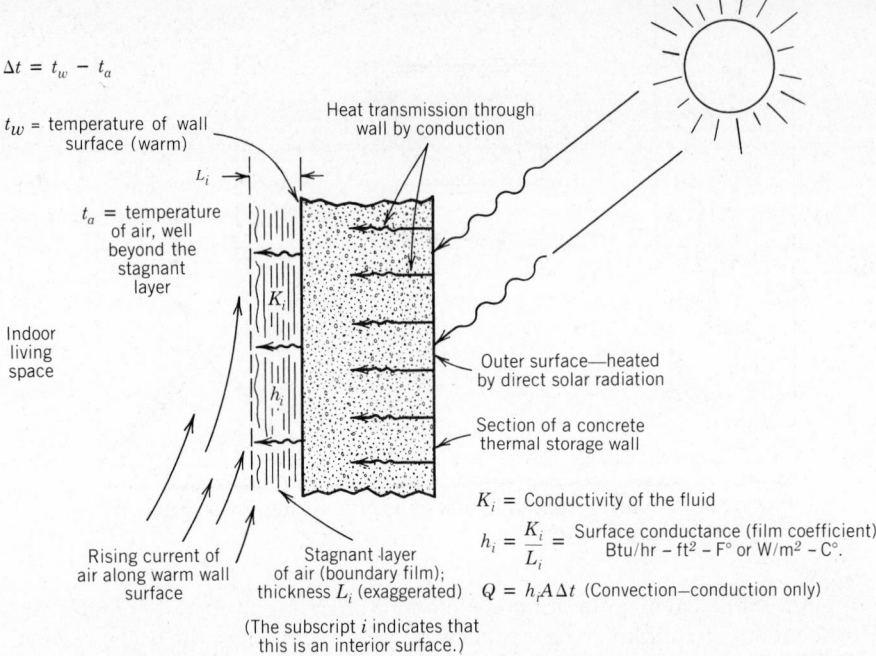

$\Delta t = t_w - t_a$

t_w = temperature of wall surface (warm)

L_i

t_a = temperature of air, well beyond the stagnant layer

Indoor living space

K_i

h_i

Heat transmission through wall by conduction

Outer surface—heated by direct solar radiation

Section of a concrete thermal storage wall

Rising current of air along warm wall surface

Stagnant layer of air (boundary film); thickness L_i (exaggerated)

K_i = Conductivity of the fluid

$h_i = \dfrac{K_i}{L_i} = $ Surface conductance (film coefficient) Btu/hr – ft² – F° or W/m² – C°.

$Q = h_i A \Delta t$ (Convection–conduction only)

(The subscript i indicates that this is an interior surface.)

Figure 3-10 Heat transfer by convection–conduction from the indoor surface of a warm wall. The film coefficient h is portrayed.

and the main fluid (air) flow, well beyond the stagnant region, F° (or C°)

h = the convection (film) coefficient, whose dimensions are K/L from Eq. (3-10)—the conductivity of the fluid divided by the effective thickness of the stagnant film; and h is the heat transmission per unit time, to or from a unit area of solid surface contact, for unit temperature difference between the surface and the main body of the fluid flowing along the surface, in Btu/hr-ft²-F° or W/m²-C°

Convection (film) coefficients are not susceptible to easy calculation. Their values depend on (1) the nature of the wall–liquid interface and whether the surface is flat or curved, horizontal or vertical; (2) the properties and flow velocity of the fluid; and (3) whether or not evaporation or condensation occurs at the boundary.

One frequently encountered problem involving film coefficients arises when the heat transmission loss of a building is being estimated and overall U values for the outer walls are needed. In addition to the separate conductances of the various elements in the wall structure itself, there are the inside and outside surface conductances (film coefficients) to consider. Since values of h are *conductances*, the *resistances* at the inner and outer wall surfaces are:

$$R_i = \frac{1}{h_i} \text{ (inside)}$$

and

$$R_o = \frac{1}{h_o} \text{ (outside)}$$

Eq. (3-13) can be modified to include surface convection effects as follows:

$$R_T = \frac{1}{h_i} + R_1 + R_2 + \text{------} R_n + \frac{1}{h_0} \tag{3-15}$$

When R_T is obtained from Eq. (3-15), U values are computed as before from Eq. (3-13), $U = \dfrac{1}{R_T}$.

The U values for composite walls listed in engineering tables take into account the surface conductances (film coefficients) of indoor and outdoor surfaces. Still air is usually assumed for indoors, and wind velocities of 15 to 25 mph are usually stipulated for outdoors for these tabulated values.

Figure 3-11 is a chart giving surface conductances (film coefficients) h for several different wall surfaces as the air velocity along the wall surface varies. Note that for white paint on pine, h_i (still air) is about 1.5 Btu/hr-ft²-F°, and h_0 (20 mph wind velocity, outdoors) is about 6 Btu/hr-ft²-F°.

It should be noted that the discussion given here is for natural convection only. In forced convection, turbulent flow is involved, the mathematical treatment of which is beyond the scope of an elementary text.

Illustrative Problem 3-11

A solar energy storage wall of common brick is 20 ft long by 10 ft high. It is warmed by direct solar radiation striking its outer surface. The inner surface forms one wall of a living space. If the inner surface temperature of the wall is 85°F and air movement along the wall is negligible, calculate the instantaneous rate of heat transfer by convection from the wall to the inside living space when the air temperature 1 ft away from the wall is 68°F.

Solution

From the chart of Fig. 3-11, determine the film coefficient for still air and common brick:

$$h_i = 2.0 \text{ Btu/hr-ft}^2\text{-F}°$$

Substituting values in Eq. (3-14):

$$Q = 2.0 \text{ Btu/hr-ft}^2\text{-F}° \times 20 \text{ ft} \times 10 \text{ ft} \times (85 - 68)\text{F}°$$
$$= 6800 \text{ Btu/hr} \qquad\qquad Ans.$$

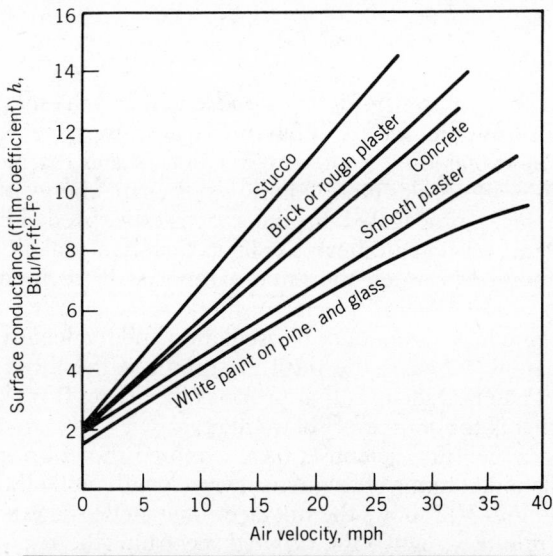

Figure 3-11 Chart showing how values of the film coefficient h vary with the velocity of air across wall surfaces. (Adapted with permission from *ASHRAE Handbook, 1977 Fundamentals*.)

Illustrative Problem 3-12

Calculate the U value for a frame-construction wall if the inside surface (still air) is smooth plaster and the outside surface (20 mph wind) is white paint on pine siding. The wall construction is as follows, from the inside: 1-in. plaster and lath on 2 × 4 studs with 3½-in. fiberglass blanket insulation between studs, ¾-in. fir plywood sheathing outside the studs, on which is 1-in. pine board-and-batt vertical siding, painted white. (Neglect stud area and assume fiberglass insulation over entire area.)

Solution

From the chart of Fig. 3-11, determine:

$$h_i = 1.8 \text{ Btu/hr-ft}^2\text{-F}°$$
$$h_o = 6.2 \text{ Btu/hr-ft}^2\text{-F}°$$

From Table 3-3, determine:

$$K_{\text{plaster-and-lath}} = 4.0 \text{ Btu/hr-ft}^2\text{-F}°/\text{in.}$$
$$K_{\text{fiberglass}} = 0.26$$
$$K_{\text{fir plywood}} = 0.80$$
$$K_{\text{pine siding}} = 0.80$$

Using Eq. (3-15), substituting values:

$$R_T = \frac{1}{1.8} + \frac{1}{4.0} + \frac{3.5}{0.26} + \frac{0.75}{0.80} + \frac{1}{0.80} + \frac{1}{6.2}$$
$$= 16.6$$

And, from Eq. (3-13):

$$U = \frac{1}{R_T} = 0.06 \text{ Btu/hr-ft}^2\text{-F}° \qquad \textit{Ans.}$$

In summary, convection is a method of heat transfer in which heat energy moves from one location to another carried by the actual movement of a fluid medium. Solar energy system design must always take convection into account—to use it when heat transfer is desired and to minimize it when heat flow is not wanted.

3-11 Radiation

The third method of heat transfer is called *radiation*. It is the method by which energy reaches us from the sun, 93 million miles away in space. Radiation heat transfer cannot be explaind by such simple hypotheses as the molecular collision idea in solids or molecular motion in fluids. Radiant energy is the energy of electromagnetic waves, propagated through space at the speed of light—186,000 mi/sec or 3×10^8 m/sec. Radiant heat, like light, travels in straight lines, and it is reflected and absorbed in much the same manner as light. Radiant heat and light are both manifestations of electromagnetic wave energy, the difference between them being in the frequency of oscillation and the length of the waves. Radiation is, in some respects, the most important of the three heat transfer methods, since it is solar radiation that provides for life on this planet, and it is solar radiation that is the source of solar energy.

Over the past century, scientists have explored the entire range of electromagnetic radiation in terms of frequency, wavelength, and other characteristics. The chart in Fig. 3-12 shows the full electromagnetic spectrum and locates the *visible spectrum* (i.e., light), the *infrared* spectrum (the region of so-called heat waves), and the *ultraviolet* region. It is interesting to note how small these spectrum regions—infrared and visible—are, compared to the entire electromagnetic spectrum. Most of the solar heating effect in solar radiation comes from these two regions.

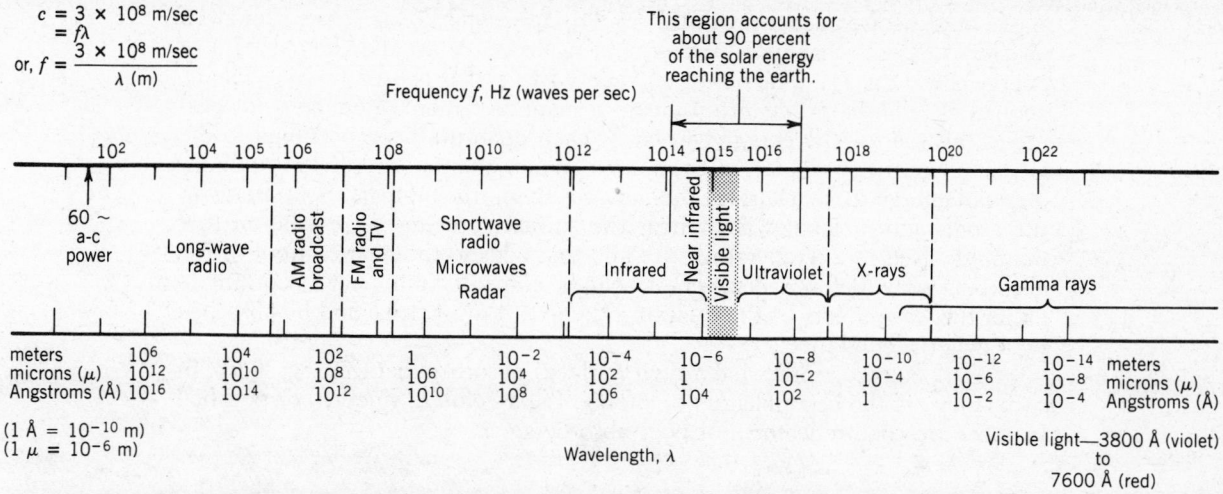

Figure 3-12 A chart of the full electromagnetic spectrum. The approximate relationship of wavelength to frequency is shown. Some of the key regions or spectral bands are identified. Boundaries between regions are not sharp—one region merges into the next.

Visible light ranges in wavelength from about 40×10^{-8} m (violet) to about 70×10^{-8} m (red). Infrared or heat waves occupy a wavelength range from about 10^{-4} m to the beginning of the red end of the visible spectrum at 70×10^{-8} m.

Besides wavelength, the other essential characteristic of wave motion is its frequency. Any wave motion travels at a velocity equal to the product of its wave length λ and its frequency f. As an equation,

$$v = f\lambda \qquad (3\text{-}16)$$

Since all electromagnetic radiations travel at the same velocity, namely the velocity of light, the basic relation for solar radiation is

$$c = f\lambda \qquad (3\text{-}17)$$

where c is the velocity of light (3×10^8 m/sec), f is the frequency of the radiation in cycles per second (*hertz*), and λ is the length of a complete wave in meters. Since c (the velocity of light) is a constant, f and λ are inversely proportional to each other —their product must always equal the velocity of light. Short waves are of high frequency, and long waves are of low frequency.

Units for Wavelength Measurement. The wavelength of electromagnetic radiation is commonly expressed in meters, but when so expressed the use of negative exponents is unavoidable—for example, the red waves cited above ($\lambda = 70 \times 10^{-8}$ m). The micron, recently renamed the micrometer (10^{-6} m), is also used. The same red waves would have a wavelength of 70×10^{-2} microns (μ). Another unit still used by solar designers is the *angstrom unit* (Å); $1\text{Å} = 10^{-10}$ m. The same red waves would have a wavelength of 7000 Å. Still another measuring unit is the nanometer (nm); 1 nm = 10^{-9} m. This unit was only recently adopted and has not yet become universally used in solar energy practice.

3-12 The Nature of Radiant Energy

There is as yet no neat unifying theory explaining radiant energy that approaches the simplicity of the hypotheses for conduction and convection. There are, in fact, two theories, more than a little bit at odds with each other. One theory (that of classical physics) supposes light and heat to be forms of wave energy—a portion of the total electromagnetic spectrum—and the discussions above have been in accord with this theory. A second theory (that of modern physics) supposes that radiant energy is made up of bundles or particles of energy called *quanta*. The quantum theory was first proposed by Max Planck about 1901. (The word *quantum* means a small increment of energy.) According to this theory, the energy carried by each quantum is proportional to the frequency of the oscillation of the source. When quanta or particles (or "bundles") of energy strike a surface that absorbs them, they impart their energy to the atoms and molecules at or near the surface, creating an excitation that manifests itself in a temperature rise and that adds to the heat content of the absorbing body. Before radiant energy can be converted into heat (according to either the wave theory or the quantum theory), it must strike and be absorbed by a material substance.

Dark-colored, somewhat rough surfaces absorb radiant energy best. In turn, they are the best radiators of energy. Light-colored, smooth, or polished surfaces are good reflectors but poor absorbers.

SOME LAWS OF RADIATION

There are a number of laws of radiation that are critically important to the study of solar energy. Three such laws or principles—the concept of blackbody radiation, the Stefan–Boltzmann law, and Wien's displacement law—will be presented in the following sections. First, however, a digression is in order to review the concept of *absolute temperature* and to explain the Kelvin and Rankine (absolute) temperature scales.

3-13   Absolute Temperature

Both the Fahrenheit and Celsius temperature scales use the freezing point and boiling point of water as fixed points. The Celsius scale, in fact, designates the freezing point of water as zero. But temperatures colder than this are commonplace, and the use of below-zero temperatures is often necessary (see Section 2-4 and Fig. 2-4).

As a result of extensive experimental research on the laws of gases during the late eighteenth century, French scientist Jacques Charles and others found that as a given mass of gas at 0°C was cooled at constant pressure, its volume decreased by 1/273 of the zero-degree volume for each decrease of 1C° in temperature. This finding led to speculation that at −273°C any gas would occupy zero volume, that all molecular motion would cease, and that a zero level of energy would be reached. These hypotheses could not be tested, however, because all gases liquefy (and subsequently solidify) before reaching −273°C. In any event, the concept of absolute zero was adopted, and its value was pegged at −273°C.[3]

The Kelvin Scale.   Later, in the nineteenth century, Kelvin proposed an absolute temperature scale, with its zero at absolute zero (−273°C) and having degrees equal in temperature difference to those of the Celsius scale. Since the late nineteenth century, this scale has been the standard for temperature measurement in all scientific and much engineering research work. It is called the *Kelvin scale* and its degrees are called *kelvins* (not capitalized). Any Kelvin temperature may be obtained by adding 273 (algebraically) to the Celsius temperature.

[3] Twentieth-century research in low-temperature physics has resulted in reaching and maintaining a temperature in the laboratory of about 0.001C° above absolute zero. On a transient basis, a temperature within 0.000001°C of absolute zero has been attained for very brief time periods.

$$T_K = t_C + 273$$

Kelvin temperatures use no degree symbol. A temperature of 15°C is 288 kelvins, or 288 K.

The Rankine Scale.   In like manner, a similar absolute scale is used for Fahrenheit degree intervals. It is called the *Rankine scale,* named for the Scotsman who suggested it. Absolute zero occurs at −460°F, and any Rankine temperature can be obtained by adding 460 (algebraically) to the Fahrenheit temperature.

$$T_R = t_F + 460$$

The degree symbol is used with Rankine temperatures, as for example 492° R. Absolute temperatures, either Kelvin or Rankine, are always indicated by the capital T. All four temperature scales are shown side by side in Fig. 3-13 for ready comparison.

Illustrative Problem 3-13

The surface temperature of a black metal plate in direct sunlight can reach 230°F. Express this temperature in (a) degrees Rankine, and (b) kelvins.

Solution

(a) $T_R = 230 + 460 = 690°$ R   *Ans.*

(b) $t_C = \dfrac{F - 32}{1.8} = \dfrac{230 - 32}{1.8} = 110°$ C

$T_K = 110 + 273 = 383$ K   *Ans.*

3-14   Ideal or Blackbody Radiation

It was pointed out above that dark-colored, rough surfaces make the best absorbers and the best radiators. A theoretical substance that could absorb all of the radiation that falls on it (i.e., a perfect absorber) and reflect none is called an

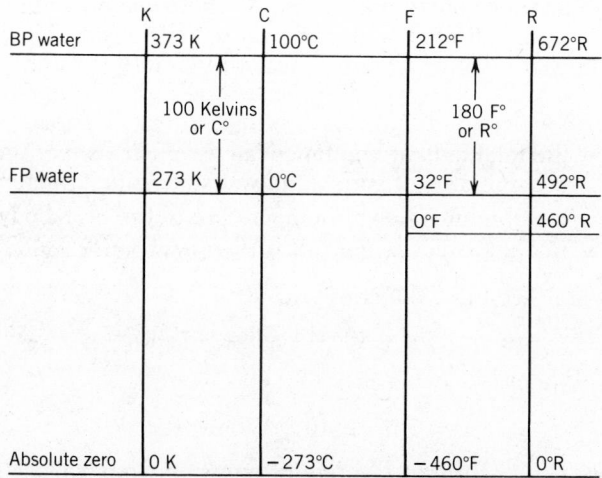

Figure 3-13 Schematic representation of the four temperature scales encountered in solar energy practice. The Kelvin, Celsius, Fahrenheit, and Rankine scales are placed side by side for ready comparison. Note the fixed points—BP of water, FP of water, and absolute zero.

SOME LAWS OF RADIATION

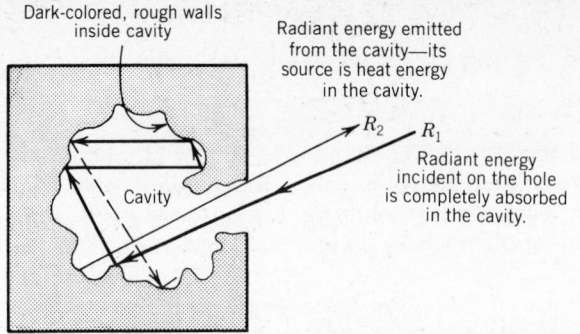

Figure 3-14 Cavity with total absorption of radiant energy. All entering energy is absorbed by the dark, rough walls of the cavity. Any leaving energy is from reradiation within the cavity itself. An ideal blackbody.

ideal blackbody. In reality, there is no perfect absorber, and therefore no ideal blackbody, but some substances come very close to the ideal. Black velvet and lampblack, for example, are near-perfect absorbers. Other substances can have their surfaces treated so that an approximation to a blackbody is attained. Metal plates are coated with flat-black paint, for example, to bring their absorption characteristics to within a few percentage points of that of a blackbody.

The best (theoretical) approach to an ideal blackbody is a small hole leading into a cavity having rough and very dark interior walls (Fig. 3-14). All radiation from any outside source that strikes the hole enters the cavity and is completely absorbed therein, perhaps after several internal reflections. None of the incident radiant energy is reflected back out again, so the definition of an ideal blackbody is satisfied. As the cavity absorbs radiation, its temperature rises, and the cavity itself becomes a source of radiation. Any of this internal *cavity radiation* that is emitted from the hole has a frequency and an energy level corresponding to the Kelvin temperature of the interior.

3-15   The Stefan–Boltzmann Radiation Law

About 1880, Josef Stefan determined experimentally that the rate of emission of radiant energy from an ideal blackbody was proportional to the fourth power of the Kelvin (absolute) temperature. Later, L. Boltzmann reached the same conclusion by mathematical analysis. The actual relationship is now known as the *Stefan–Boltzmann law of radiation*. It is written in this form:

$$R = \sigma T^4 \tag{3-18}$$

where R = the total radiant emittance (all wavelengths) per unit time and per unit area of surface, in watts per square meter

T = the absolute (Kelvin) temperature of the blackbody, in kelvins

σ = the radiation constant (σ is the Greek letter *sigma*)

When R is in kcal/(sec)(m²), the constant

$$\sigma = 1.36 \times 10^{-11} \text{ kcal/(sec)(m}^2\text{)(K}^4\text{)}$$

When R is in W/m²,

$$\sigma = 5.67 \times 10^{-8} \text{ W/(m}^2\text{)(K}^4\text{)}$$

It follows from Eq. (3-18) that the rate of emission from a blackbody source increases very rapidly with increasing temperature. For example, doubling the Kelvin temperature increases the radiant emittance rate by a factor of 16.

The Stefan–Boltzmann relationship holds generally for any radiating body, but R will be some fraction of the ideal value given by Eq. (3-18), the actual value varying with the nature of the radiating surface. Since, at very high temperatures, a number of actual surfaces (for example, the surface of the sun) do come close to being ideal blackbodies, Eq. (3-18) can be used to estimate the surface temperature of a body after determining the total radiant emittance R by measurement. This is one method by which the surface temperatures of the sun (about 6200 K) and the stars have been estimated.

3-16 Wien's Displacement Law

The radiant energy emitted by a surface depends on the kind of surface and on its absolute (or Kelvin) temperature. At low surface temperatures, the rate of emission is small and the radiation is mainly of long wavelength (low frequency). On rising temperature, the rate of emission (power) goes up sharply, not in a linear fashion but proportional to the fourth power of the Kelvin temperature. Also, at higher temperatures, the energy emitted contains more and more short-wavelength (high-frequency) radiations. When a hot body radiates energy, there is a distribution of wavelengths and their associated frequencies, and most of the energy is radiated at wavelengths near the peak of the radiation curve for a given temperature. Figure 3-15 shows radiation curves for four different Kelvin temperatures, with wavelengths plotted horizontally against energy radiated plotted vertically, for idealized blackbodies. Note from the curves that as the temperature of the radiating source rises, two events occur: (1) A great deal more total radiation energy is emitted, in accord with the fourth-power relationship of the Stefan–Boltzmann law; and (2) there is a shift or displacement of the curves to the left toward shorter wavelengths (higher frequencies). For each temperature, the radiation power emitted varies with wavelength, and each curve shows maximum emittance at a specific wavelength. As temperature increases, the wavelength at which maximum radiated power occurs is displaced in the direction of smaller wavelengths.

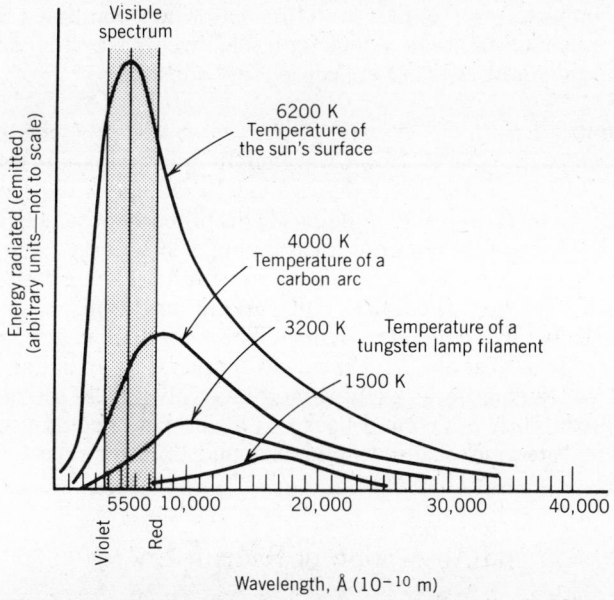

Figure 3-15 Radiation curves for an ideal blackbody for different temperatures. Total radiation emitted is plotted against wavelength for the separate temperatures. The curves are approximate only, intended for illustrative purposes and not for use in calculations. Note that as the temperature of the radiating source rises, total radiant energy emitted increases markedly; and also that the emitted-energy maximums of the curves are shifted toward shorter wavelengths and associated higher frequencies (Wien's displacement law).

The German physicist Wilhelm Wien, after intensive research on these phenomena, proposed a simple equation that expresses this shift or displacement of the radiation curves. His equation, now known as *Wien's displacement law*, is:

$$\lambda_{max} T = (\text{constant}) = 2.898 \times 10^{-3} \qquad (3\text{-}19)$$

Where

λ_{max} is the wavelength (meters) at the point of maximum radiated power (peak of the curve)

T is the absolute (Kelvin) temperature of the radiating source

The constant 2.898×10^{-3} is in meters-degree Kelvin

Illustrative Problem 3-14

Extraterrestrial photographs of the solar spectrum indicate that the peak solar radiation occurs at a wavelength (λ_{max}) of about 4.64×10^{-7} m. Using Wien's law, find the temperature of the surface of the sun.

Solution

Substituting values in Eq. (3-19),

$$4.64 \times 10^{-7} \text{ m} \times T = 2.898 \times 10^{-3} \text{ m-K}$$

From which,

$$T = 6250 \text{ K} \qquad\qquad Ans.$$

Wien's law predicts that real (i.e., nonblackbody) objects at normal temperatures experienced on earth will emit radiations of much longer wavelength than those that comprise solar radiation. This fact, when combined with the capability of some materials to be wavelength selective, makes it possible to "trap" solar heat in greenhouses, solar collectors, and sunspaces.

The Greenhouse Effect. A pane of glass is wavelength selective in that it transmits a great deal of the short-wavelength solar radiation that strikes its surface but absorbs most of the long-wavelength radiations that emanate from low-temperature surfaces inside buildings. This property of selectivity possessed by glass makes it an ideal material for trapping solar energy. The short-wavelength, high-frequency solar radiation is transmitted through glass with little loss, and strikes the floor, benches, walls, and other thermal storage materials where its wave energy is converted to heat. These thermal masses, as they warm up, begin to radiate heat also, but their low-temperature radiations are of long wavelength, which glass does not transmit. Since very little of the radiant energy entering a greenhouse is radiated back out again, the greenhouse serves as a heat trap. The "greenhouse effect" and its applications for solar heating will be treated in detail in later chapters.

3-17 Emission and Absorption of Radiant Energy

Radiant energy is emitted from the surfaces of hot bodies, and when it strikes a surface the energy of the wave is absorbed and/or reflected (or transmitted), depending on the nature of the surface. Some opaque, dark-colored materials (almost blackbodies) will absorb nearly all of the radiant energy that strikes them, while other substances, like glass, absorb some energy, reflect some, and transmit some (Fig. 3-16). All three of these characteristics of surfaces—absorption, emission, and reflection—are of great importance in solar energy work.

Emissivity. The behavior of nonblackbody surfaces is expressed by introducing a factor called *emissivity, e*. The emissivity *e* of a surface is defined as the

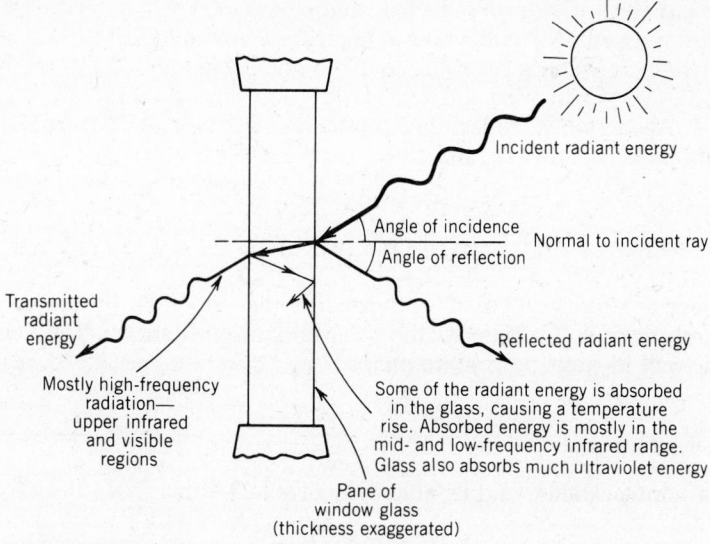

Figure 3-16 Solar radiation striking a pane of glass. Some of the incident radiation energy is reflected, some transmitted, and some absorbed. Glass is selective with respect to the wavelengths it transmits and absorbs.

ratio of the total radiant emittance R, for that surface, to the total radiant emittance from an ideal blackbody of equal area and at the same temperature. It follows that $e_{\text{blackbody}} = 1.00$. Values for other surfaces range downward to near zero. (See Table 3-4.) For nonblackbody surfaces, Eq. (3-18) becomes

$$R = e\sigma T^4 \qquad (3-20)$$

Where e is the emissivity of the surface.

Absorptance. In like manner, the *absorptance* α of a surface is the ratio of the radiant energy absorbed by it to the total radiant energy incident on the same area. The experiments of G. Kirchoff (about 1875) established the fact that the absorptance α of a surface is positively related to its emissivity e. In other words, good emitters of radiation are good absorbers, and poor emitters are poor absorbers.

Reflectance. The *reflectance* r of a surface is also a ratio—the ratio of the radiant energy reflected from the surface to the total radiation falling on the same surface area. For perfectly opaque bodies, all incident radiation energy

Table 3-4 Emissivities and Reflectances of Selected Opaque Surfaces

Surfaces	Temperature °C	K	e	r
Ideal blackbody	At any temperature		1.00	0.00
Brick, common	40	313	0.93	0.07
Lampblack	100	373	0.97	0.03
Flat-black paint	100	373	0.96	0.04
Human body surface (any pigmentation)	30	303	0.96	0.04
Polished metals (bright)	100	373	0.05	0.95
White canvas	100	373	0.84	0.16

Note: Both e and r are ratios, and consequently their values are dimensionless numbers.

must be either absorbed or reflected, and $e + r = 1.00$. A good reflector is therefore a poor emitter, and a poor reflector is a good emitter. A good absorber is a poor reflector, and a poor absorber is a good reflector.

Values of emissivities and reflectances for various surfaces are given in Table 3-4. Absorptance, reflectance, and emissivity will all be treated in more detail in Chapters 7, 10, 11, and 12.

Illustrative Problem 3-15

The surface of a brick wall used for thermal storage in a solar-heated house is at a temperature of 40°C. What is the total radiant emittance (all wavelengths) from this wall in watts per square meter?

Solution

This is a nonblackbody, and its emissivity (Table 3-4) is

$$e = 0.93$$

Substituting in Eq. (3-20),

$$R = 0.93 \times 5.67 \times 10^{-8} \frac{W}{m^2\text{-}K^4} \times (313K)^4$$

$$= 506 \ W/m^2 \qquad\qquad Ans.$$

All bodies are continually radiating and absorbing energy. A body that is absorbing radiant energy will (in the absence of conduction and convection effects) undergo a temperature rise until its surface is hot enough to radiate energy at a rate equal to the rate of absorption. The body is then said to be in equilibrium with its surroundings, and its temperture will hold constant as long as this condition is maintained. If a body radiates energy in excess of that absorbed (conduction and convection absent), its temperature will drop. A continual exchange of energy is thus going on among objects at different temperatures, with the hotter surfaces losing energy to the colder ones.

One must realize that radiation gains and losses occur regardless of the surrounding (ambient) air temperature. The human body, for example, will radiate heat to cold walls or floors, and a feeling of chill will result even though the air in the room may be a comfortable 70°F. Radiation to the cold night sky causes frost on an auto top even when the nighttime air temperature stays above 32°F. Solar collectors can freeze up even in the mild winter climates of California and Florida, as a result of radiation losses to the cold night sky. Freeze protection (automatic draining of water from the collectors at night or the use of an antifreeze) should be provided in all locations where freezing caused by radiation losses could conceivably occur.

At the other extreme, on summer nights radiation from warm ceilings and walls often causes extreme discomfort to people even though the air temperature itself is judged to be comfortable. Large areas of undraped glass may cause much discomfort, because of radiation effects under both summer and winter conditions.

3-18 Body Heat Gain and Loss Caused by Radiation

The human body is constantly radiating heat energy, based on the average body-surface (skin) temperature of about 88 to 90°F. There are two reasons why this heat loss does not result in a continually falling body temperature: (1) Metabolic processes within the body produce heat; and (2) the body may be gaining heat by radiation from the surroundings. Considering only radiation gains and losses, let us assume that the body surface temperature is T_1 kelvins, and the temperature of the surrounding walls, floor, and ceiling is T_2 kelvins.

The net rate of body heat gain or loss, per unit area, is then

$$R_{net} = e \sigma T_1^4 - e \sigma T_2^4 = e \sigma (T_1^4 - T_2^4) \qquad (3\text{-}21)$$

Equation (3-21) applies, of course, not only to persons but also to any objects that are simultaneously radiating and absorbing radiation.

Illustrative Problem 3-16

A person with a total body surface area of 1.35 m² and an average skin temperature of 32°C is standing unclothed in a room in which the walls, ceiling, and floor are at a temperature of 15°C. Find the net rate of heat loss from the body (radiation effects only) in watts.

Solution

Note from Table 3-4 that e for the human body surface is 0.96 in the temperature range of the problem.

Converting to Kelvin temperatures,

$$T_1 = 32 + 273 = 305 \text{ K}$$
$$T_2 = 15 + 273 = 288 \text{ K}$$

Substituting in Eq. (3-21),

$$\text{Radiant emittance, } R_{net} = 0.96 \times 5.67 \times 10^{-8} \frac{W}{m^2\text{-}K^4} \times [(305 \text{ K})^4 - (288 \text{ K})^4]$$

$$= 96.6 \text{ W/m}^2$$

$$\text{Total net radiation loss} = 96.6 \text{ W/m}^2 \times 1.35 \text{ m}^2 = 130 \text{ W} \qquad \textit{Ans.}$$

Illustrative Problem 3-17

Calculate the net total radiant energy emitted from a spherical blackbody 50 cm in diameter when its surface temperature is 700°C. Assume that it is surrounded by surfaces at a temperature of 20°C.

Solution

By definition, for a blackbody, $e = 1.00$. Converting to Kelvin temperatures,

$$700°C = 973 \text{ K}$$
$$20°C = 293 \text{ K}$$

Substituting in Eq. (3-21),

$$\text{Radiant emittance, } R_{net} = 1.00 \times 5.67 \times 10^{-8} \frac{W}{m^2\text{-}K^4}$$

$$\times [(973 \text{K})^4 - (293 \text{K})^4]$$

$$= 50,400 \text{ W/m}^2$$

$$= 50.4 \text{ kW/m}^2$$

$$\text{Total radiation emitted} = 50.4 \text{ kW/m}^2 \times 4\pi \times (0.25 \text{ m})^2$$

$$= 39.6 \text{ kW} \qquad \textit{Ans.}$$

CONCLUSION

In this chapter, we have dealt with some of the basic principles of heat energy and heat transfer. Understanding these principles is essential to successful solar energy systems design. Further, in addition to an understanding of the principles, the designer must be able to make meaningful and accurate calculations involving heat energy and heat transfer. In the next chapter, the specific topic of solar radiation will be introduced along with basic solar energy principles.

PROBLEMS

1. A thermal storage tank holds 1000 gal of water. How many Btu of heat must be supplied to warm this water from 55°F to 150°F?

2. The design of a solar heating system calls for a water-in temperature to the solar collectors of 80°F and a water-out temperature of 145°F. If 100,000 Btu/hr of heat is to be removed from the collectors by circulating water, what must be the water flow rate, in gallons per minute (gpm)?

3. A rock bin contains 300 tons of stones. At 5 P.M., the entire mass of stones is at a temperature of 110°F. What steady flow of heat (in Btu per hour) could this rock-bin thermal storage provide, averaged over the next 15 hr, until 8 A.M. the next day?

4. A brick wall is designed for thermal storage. It is behind glass on a south-facing wall and absorbs solar radiation all day. Its dimensions are 8 ft by 80 ft by 14 in. thick. Find the total heat storage capacity of the wall in Btu if it operates between 110°F and 80°F.

5. A thermal storage wall of adobe is 50 ft by 10 ft by 14 in. thick. It is behind glass on the south side of the house. Its average temperature at 8 A.M. is 72°F. If it warms up to a uniform 108°F by 4 P.M., how much heat will it have stored up for release during the night?

6. An electric water heater has an input power of 15 kW. If the water-in temperature is 60°F and the water-out temperature is to be 140°F, what maximum continuous flow (in gallons per minute) could this water heater handle?

7. A 6-kW bank of electric strip heaters is installed in a supply air duct to serve as back up heat for a solar system. A blower moves return air from the living spaces (at 65°F) into the heaters and then circulates the warmed air to the spaces. If the air off the heater is to be maintained at 120°F, what flow volume of return air (cubic feet per minute) should the blower supply? (Note: The density of air at 65°F is 0.0755 lb/ft³.)

8. Low-pressure (saturated) steam at 212°F is supplied to a home heating system at a rate of 300 lb/hr. It returns to the steam boiler as hot water at 130°F. How much heat (Btu per hour) has been transferred to the conditioned spaces?

9. A metallic salt-hydrate (Glauber's salt) is used as a heat storage material. How many cubic feet of Glauber's salt (liquid phase) would be required to store up and give off 65,000 Btu as it solidifies at a temperature of about 88°F?

10. A copper pipe is nominally 80 ft long. How much does it expand (in inches) as it warms from 50°F to 160°F?

11. A straight run of steel pipe 85 ft long carries hot water from a bank of solar collectors on the roof of an office building to a heat exchanger in the ground-floor equipment room. By what total length will this pipe expand as its temperature rises from 40°F at 7 A.M. to 180°F at 2 P.M.?

12. A steel tank has a rated volume of exactly 1200 gal at 60°F. If the tank is completely full of water at that temperature, how much water will overflow into the expansion tank when the temperature of both the water and the tank reaches 150°F?

13. A concrete wall is 8 in. thick, 50 ft long, and 10 ft high. Its outer surface temperature is 115°F. If the inner wall surface temperature is steady at 78°F, find the steady-state heat transmission gain through the wall by conduction, in Btu per hour.

14. The coefficient of conductivity K for mineral wool insulation is 0.27 Btu/hr(ft^2)(F°/in.). What is the conductance C of a 6-in.-thick blanket? What is the R value of a 10-in. thick blanket?

15. A composite wall is made up as follows, from the outside in: One-inch-thick cedar siding, ½-in.-thick insulating board; 3½-in. of fiberglass insulation between studs, ¾-in-thick gypsum drywall inside finish. Calculate the overall coefficient of heat transmission U, including surface film effects. (Neglect effect of the studs.)

16. A warm, solar-heated brick wall has an indoor surface temperature of 92°F. The air temperature, measured 6 to 8 in. away from the wall, is 70°F. Air movement along the wall is estimated to be about 7.5 mhp. Calculate the approximate rate of heat transfer per square foot of wall area from this wall to the room air as a result of the combined convection–conduction process.

17. Radiation in the far infrared region is determined to have a frequency of 3×10^{12} hertz (Hz). Find the wavelength of these rays in meters.

18. A hot-water radiator has a black, rough surface, and its operating surface temperature is 180°F. Its effective surface area is 20 ft^2. At what rate (Btu per hour) will it radiate heat to the room? Take room temperature as 70°F. All units must be converted to SI-metric foor use in Eq. (3-21).

19. A hopper of molten metal is observed through a radiation pyrometer, and the wavelength at the point of maximum radiation intensity is found to be 1.8×10^{-6} m. What is the approximate Kelvin temperature of the molten surface?

20. A copper plate whose surface is coated with flat black paint is in direct sunlight and reaches a temperature of 85°C. Calculate its total radiant emittance R (all wavelengths) in watts per square meter.

21. A person whose total body surface area is 14 ft^2 and whose average skin temperature is 89°F is standing unclothed in a room where all surrounding objects and surfaces are at a temperature of 50°F. What is the net radiation heat loss from the body, in Btu per hour? (Remember to convert units.)

SI–METRIC PROBLEMS

1. A hot-water tank holds 400 L. How much heat (kilocalories) will be required to warm this water from 10°C to 60°C?

2. An auxiliary (booster) electric heater for a domestic hot-water system is rated at 5 kW. How many kilocalories per second can it supply to the water in the tank?

3. A glass view window is 8 mm thick and measures 5 m by 2 m. If its outside surface temperature is 45°C and its inside surface temperature is 27°C, find the conduction heat gain to the space through the window, in kilowatts.

4. A furnace firebox has a hole in it whose area is 10 cm^2. The furnace interior is at a temperature of 550°C. Assume blackbody radiation, and calculate the radiant heat loss from the hole in kilocalories per second.

5. A black metal cube whose edges are 30 cm long has a surface temperature of 600°C. It is surrounded by surfaces whose temperature is 30°C. Calculate the net radiant energy loss from the cube in kilowatts.

6. The interior surface of a dark, rough wall is maintained at a temperature of 38°C by conduction from its outer surface, which is exposed to direct solar radiation. Its dimensions are 2.5 m by 12 m. The emissivity e of the inner wall surface is 0.84. Calculate the total radiant emittance from this wall, in kilowatts.

Solar Radiation

The basic principles of heat energy and heat transfer were presented in the preceding chapter. In designing solar energy systems, however, it is usually necessary to go beyond basic ideas to more advanced concepts in the science of heat energy. Consequently, in this and later chapters further and more detailed discussions of such topics as thermal capacity, heat power, change of state, conduction, convection, and radiation will be presented as they are needed in the development of solar industry practice.

This chapter is devoted to an in-depth discussion of solar radiation, the energy that arrives at the earth's surface from the sun.

THE SUN AND ITS ENERGY

The sun is just one of the more than 100 billion stars in the Milky Way galaxy. It is no more than average in size, not especially hot compared to other stars, and is therefore not a very bright star. From a vantage point in some far-off constellation, astronomers would probably catalog our sun as an "insignificant" star.

For life on earth, however, the sun is all important. Its mass contributes directly to the force of mutual attraction that holds the earth and the other eight planets in the orbital geometry that we call the solar system. Nearly all of the energy on earth comes from the sun, either directly in the form of present-day solar radiation or in the form of fossil fuels deposited in past ages as a product of the sun's energy in that time. Only atomic fission, thermonuclear energy, geothermal energy, and that portion of tidal energy caused by the moon's gravitational pull are energy sources that do not originate in the sun.

4-1   The Sun as a Giant Furnace

The sun is a gigantic sphere of high-temperature gases, about 865,000 mi (1,392,000 km) in diameter. Its mass is on the order of 2.2×10^{27} tons (2.0×10^{30} kg), or some 330,000 times that of the earth. Taken alone, the sun's mass is more than 99 percent of the mass of the entire solar system—that is, of the sun and the nine planets. Nearly 75 percent of the sun's mass is hydrogen, about 24 percent is helium, and the remainder is comprised of small traces of all of the known elements. As a gaseous sphere, its average density is not great—about 1.4 times that of water, or 87 lb/ft^3, or 1.4 kg/L. However, the sun's size and mass are so great that the resultant gravitational force (the gravity force on the surface of the sun is about 28 times that at the earth's surface) causes the density to increase steadily from the surface to the center, where it may be on the order of 160 kg/L (that is, 160 times as dense as water) at a pressure on about 70×10^9 (earth) atm.

Layers in the Outer Regions.   The surface of the sun that we see is called the *photosphere* (*photo* from the Greek, meaning "light"), and it is from this region that most of the radiant energy that reaches the earth emanates (Fig.

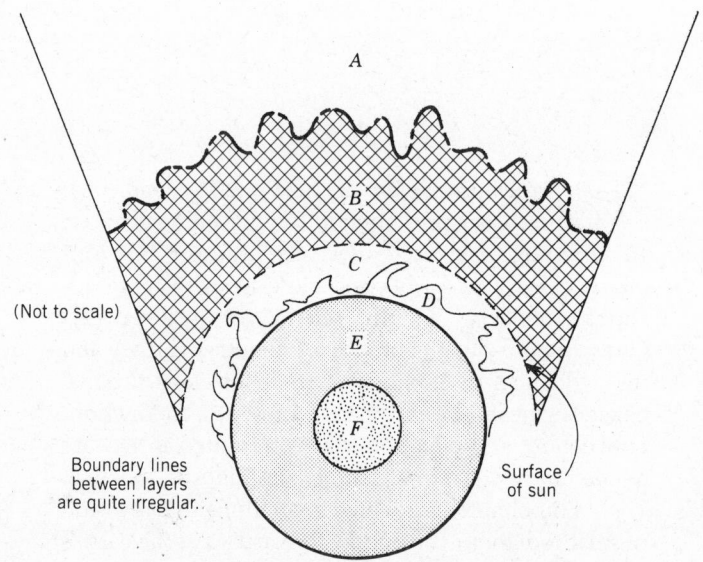

A—*Corona*, several hundred thousand to perhaps a million miles
thick. Very low-density gas; shines with pearly-white light.
B—*Chromosphere*, some 2000 to 6000 miles thick. Low-density
hydrogen gas, temperature 5000 to 30,000 °C. Visible
only during solar eclipse. Shines with reddish light.
C—*Photosphere*, the visible "surface" of the sun. Most of
the radiant energy reaching the earth emanates from this
layer. Several hundred miles thick—temperature 6000°C.
D—*Solar prominences* and *solar flares*.
E—The sun's *interior*. Radiation energy from the core makes its
way to the surface through 350,000 mi of the sun's mass.
F—The *core*. Here thermonuclear fusion occurs. Enormous
pressure and density. Temperature, 12 to 18 million °C.

Figure 4-1 The various layers of the sun from the inner core outward. The photosphere, the chromosphere, and the corona are shown (not to scale). The sun's "surface"—the photosphere—is at a temperature of about 5800 to 6000°C.

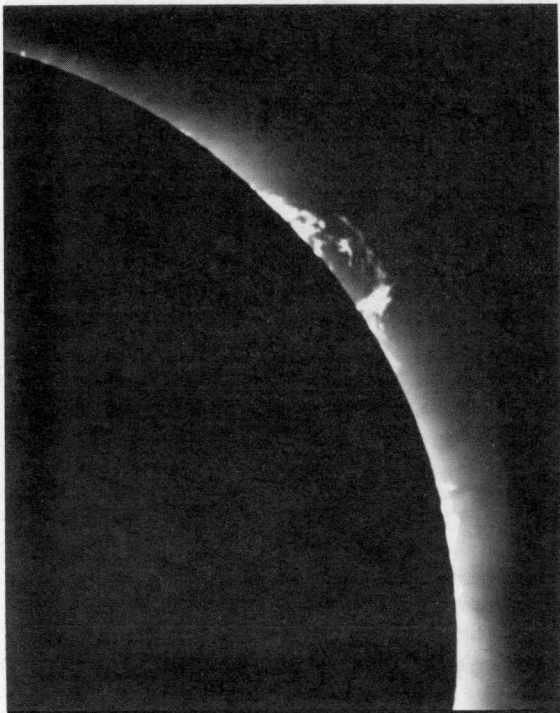

Figure 4-2 Photo of the edge of the sun, showing solar flares and prominences. The photosphere has been almost entirely blanked out by an eclipse, allowing the edge of the sun to be recorded on film. The flares shown are thousands of miles in length and width. (Lick Observatory.)

4-1). The photosphere has an average temperature of about 5800 to 6000°C (10,000°F) at a pressure of about 0.01 earth atm (0.147 psi). It is several hundred miles thick.

Outside the photosphere is a region called the *chromosphere*, a layer some 2000 to 6000 mi thick, made up of glowing red gaseous hydrogen. Beyond the chromosphere is the *corona*, an extensive envelope of low-density gases, varying in thickness from a few hundred thousand miles to more than a million miles. When observable at all, it gives off a white, pearly light. Both the chromosphere and the corona are visible only during a total eclipse, since they are otherwise blanked out by the strong glare of the photosphere.

The sun's surface should not be thought of in the same context as the surface of the earth. First, it is gaseous and therefore has no rigid form or boundary. Second, it is in a state of constant and violent agitation, with eruptions of inconceivable fury boiling up from the interior and out into immense solar flares.[1] Magnetic storms covering thousands of miles add to the indescribable turbulence caused by the solar flares. These storms are thought to be the cause of sunspots, which are regions that are somewhat cooler than the surrounding photosphere. During an eclipse, and with the aid of camera-equipped telescopes, clear color pictures of solar flares and sunspots can be obtained, showing the long tongues of flame surging out into space for thousands of miles (Fig. 4-2).[2] Scientists have calculated that in a typical solar flare there is enough energy, if it could be captured and stored, to meet all of the earth's energy needs for thousands of years.

[1] Gases in such violent, high-temperature conditions are almost totally ionized—that is, their orbital electrons are no longer bound in the atoms. Such gases are known as *plasmas*, and scientists regard them as a fourth state of matter, the other three being solids, liquids, and gases.

[2] *Warning!* Never look directly at the sun with unprotected eyes. Serious retinal damage can occur unless proper filters are used.

Origin of the Sun's Energy. Deep in the interior of the sun, solar energy is produced by thermonuclear reactions.[3] Temperatures of 15 million K (27 million °F) and pressures of 70×10^9 earth atmospheres provide ideal conditions for the fusion of lighter particles (hydrogen nuclei) to form heavier particles (helium nuclei). Hydrogen is the fuel in the sun's "nuclear furnace." The fusion process is probably one in which four atoms of hydrogen come together to form one atom of helium (under the extremely high temperature conditions of the sun's interior), and in the process a small amount of mass is lost. This lost mass reappears as energy, in accordance with Einstein's classic mass–energy equation,

$$E = m\,c^2 \qquad\qquad (4\text{-}1)$$

where

E = the energy produced, joules
m = the lost mass, kilograms
c = the velocity of light, 3×10^8 m/sec

The sun's internal energy is so great that it is almost incomprehensible by earth standards. Most of it is in the form of high-frequency electromagnetic radiation, primarily *gamma rays*, and it slowly makes its way some 350,000 to 400,000 mi to the photosphere, impeded by innumerable interactions with the elemental particles making up the dense interior matter of the sun. This energy degradation en route to the surface results in both a decreasing temperature and a decreasing radiation frequency, with the consequence that solar radiation at the photosphere is that associated with about 5800°C (10,000°F). The greater part of the energy radiated from the photosphere has frequencies in the visible light and infrared regions of the electromagnetic spectrum—that is, sunlight and heat (see Figs. 3-12, 4-5, and 4-6). Some of the energy from the surface plasma is dissipated in the form of cosmic rays, the solar wind, and other radiations, including *neutrinos* and other high-energy particles.

The total energy streaming out into space from the sun's surface is estimated to be about 3.8×10^{20} MW, radiating outward in all directions. This represents an energy release of about 63 MW/m² (8000 hp/ft²) at the sun's (photosphere) surface (Fig. 4-3). Only a small amount of this immense total is captured by the earth, since the solid angle in space subtended by the earth at its 93-million-mile distance from the sun is small indeed (see section 4-3). At a point just outside the earth's atmosphere, the solar power is estimated to be about 1.35 kW/m² (429 Btu/hr-ft², or 0.17 hp/ft²).

As mentioned above, the sun's nuclear furnace produces energy at the expense of lost mass. Eventually, the sun will "waste away," but its mass is so great that we hardly need to worry. Rough calculations indicate that the sun's present rate of loss of mass (conversion of mass to energy) is of the order of magnitude of 4.5 million tons/sec. At this rate, several billion years of satisfactory solar furnace operation can be expected before the sun's energy output would be appreciably depleted.

In summary, the sun is a fantastic nuclear furnace producing energy at temperatures in the millions of degrees. The high-frequency gamma radiation produced by nuclear fusion in the interior of the sun is tempered by its passage to the surface, and the energy that we call solar radiation emanates from the photosphere mostly as visible light and infrared waves characteristic of a radiating surface at a temperature of about 5800°C (6000 K or 10,000°F).

4-2 The Solar Spectrum

Solar energy arrives on earth in the form of electromagnetic radiations differing in wavelength and frequency of vibration. Wavelength and frequency are inversely proportional to each other, being related by the basic wave equation (see section 3-12),

[3] Thermonuclear fusion is currently the subject of intensive research in laboratories all over the world, with the hope of attaining controllable fusion energy for industrial uses.

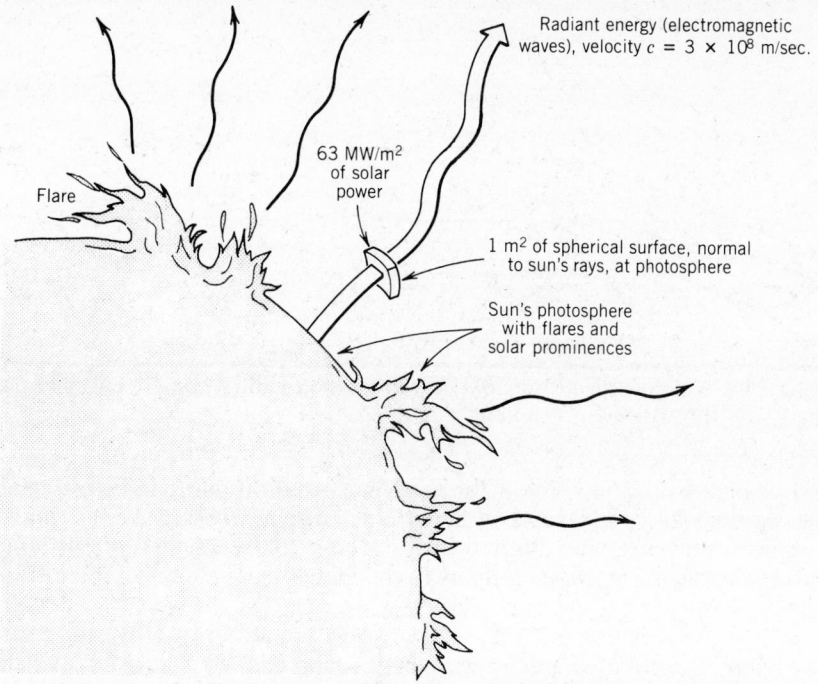

Figure 4-3 Radiant energy streams out in all directions from the surface of the sun. These electromagnetic waves have a radiation intensity of about 63 MW per square meter of surface normal to the rays, at a point near the sun's surface.

In the figure, the following labels appear:
- Radiant energy (electromagnetic waves), velocity $c = 3 \times 10^8$ m/sec.
- Flare
- 63 MW/m^2 of solar power
- 1 m^2 of spherical surface, normal to sun's rays, at photosphere
- Sun's photosphere with flares and solar prominences

$$c = f\lambda \qquad\qquad (3\text{-}19)$$

where

f = frequency, waves per second (hertz)
λ = wavelength, meters
c = the velocity of light, 3×10^8 m/sec

The unit of frequency, the hertz (Hz), is named for the German physicist who is credited with first producing electromagnetic waves in the laboratory. One hertz equals one vibration (complete wave) per second.

The term *extraterrestrial solar radiation* is used to describe the nature of solar radiation at the extreme outer limits of the earth's atmosphere. *Sea-level solar radiation* is radiation actually reaching the earth's surface, and its value is considerably less than that of extraterrestrial solar radiation because of the absorption and scattering effects of the earth's atmosphere.

Much information about the sun's energy can be obtained from a study of the *solar spectrum*. A spectrum is defined as a band of images formed when a beam of radiant energy is dispersed and then brought to a focus with the waves arranged in order of their wavelengths. An instrument called a *spectroscope* is used to obtain a visual representation of a solar spectrum (Fig. 4-4). In its basic form, a spectroscope consists of a prism (for dispersing the light into its constituent wavelengths or colors), a narrow slit through which a beam of sunlight is allowed to enter the prism, a collimator lens to render the rays parallel as they approach the prism, and an objective lens to focus the dispersed rays (coming from the prism) onto a suitable viewing screen or photographic plate. The result—either on the screen or in a photograph—is a band of color, the solar (visible) spectrum, ranging from deep violet through blue, green, yellow, and orange, into red. Figure 4-5 is a chart of the solar spectrum with the colors indicated along an approximate scale of wavelengths.

It should be noted that the solar spectrum is an expanded version of the narrow band labeled "visible light" in the complete electromagnetic spectrum

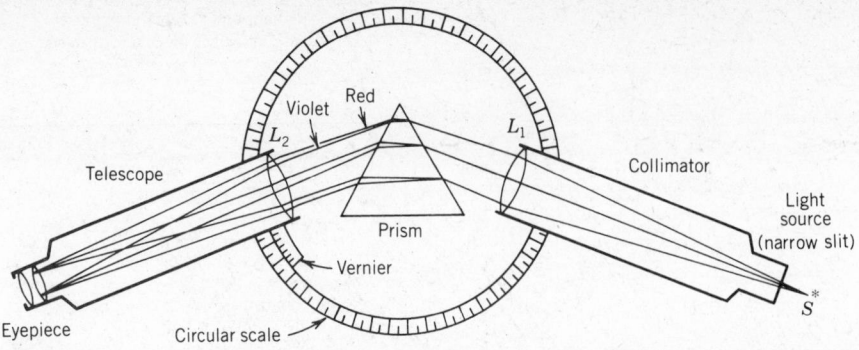

Figure 4-4 A simple spectroscope. Diagram of the essential components and path of light rays through the instrument.

chart of Fig. 3-12. The visible light band is flanked on one side by ultraviolet light (shorter wavelengths) and on the other by infrared light (waves too long to be visible to the eye). The infrared region closest to the visible is often termed *near-infrared*, while *far-infrared* refers to the longer wavelengths of the infrared region.

The sun's radiation intensity (solar power) at a level just above the earth's atmosphere is distributed among wavelengths approximately as listed in Table 4-1. It will be noted that about 85 percent of the sun's usable energy falls in the spectrum range that includes the visible and the near-infrared regions, to a wavelength of about 2.0 μm. Although some solar energy is available for collection in the far-infrared region, it has been found advantageous to screen out incoming energy at these wavelengths in order that reradiation losses at these same wavelengths (low-frequency radiation from low-temperature sources) can be prevented or minimized. Glass, for example, readily transmits high-frequency, short-wavelength radiation and rather effectively blocks low-frequency; long-wavelength radiation (see section 3-17). Selective coatings and special glazings are used with flat-plate solar collectors to maximize the transfer of solar energy to the space or fluid being heated (see Chapters 11 and 12).

Since only about 7 percent of total incoming solar energy is in the far-infrared spectrum range and beyond, no great loss is incurred when solar collectors and/or structures are designed to block out these wavelengths. On the other hand, reradiation from warm walls and floors of structures and from the hot absorber plates of solar collectors is nearly all in the infrared region, and unacceptably high losses of heat energy already collected would occur if materials that selectively block the passage of these longer wavelengths were not used. (Further discussion of this topic is contained in Chapters 8 and 12).

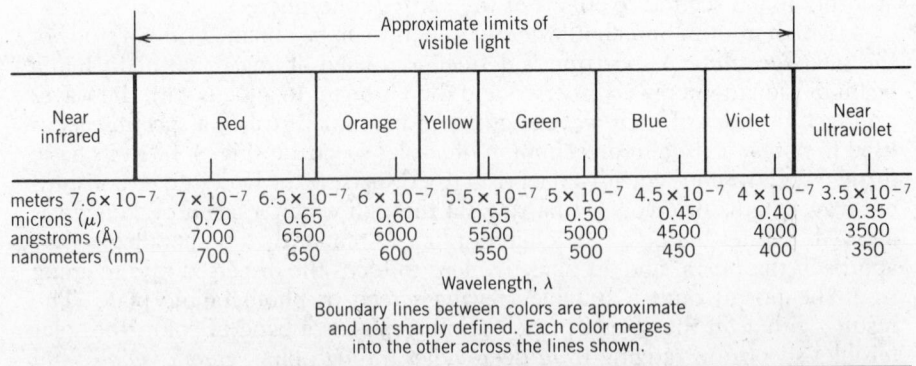

	Near infrared	Red	Orange	Yellow	Green	Blue	Violet	Near ultraviolet
meters	7.6×10^{-7}	7×10^{-7}	6.5×10^{-7}	6×10^{-7}	5.5×10^{-7}	5×10^{-7}	4.5×10^{-7}	4×10^{-7}	3.5×10^{-7}
microns (μ)		0.70	0.65	0.60	0.55	0.50	0.45	0.40	0.35
angstroms (Å)		7000	6500	6000	5500	5000	4500	4000	3500
nanometers (nm)		700	650	600	550	500	450	400	350

Wavelength, λ
Boundary lines between colors are approximate and not sharply defined. Each color merges into the other across the lines shown.

Figure 4-5 A chart of the visible region of the solar spectrum showing the separation (dispersion) of white light (sunlight) into bands of color. Approximate wavelengths are indicated along the horizontal scale in various units common to physics, astronomy, and solar energy.

Table 4-1 Approximate Distribution of Extraterrestrial Solar Energy among Wavelength Ranges

Wavelength Range μm	Radiation Intensity		Approximate Percent of Total Radiation Intensity
	Btu/hr-ft²	W/m²	
0.00 to 0.40 (gamma through ultraviolet)	34	108	8.0
0.40 to 0.70 (visible)	168	528	39
0.70 to 2.0 (near-infrared)	197	622	46
2.0 and above (far-infrared and beyond)	30	95	7.0
Totals	429	1353	100

Source: Richard C. Jordan and Benjamin Y. U. Liu, eds., *Applications of Solar Energy for Heating and Cooling.* (Atlanta: American Society of Heating, Refrigerating and Air Conditioning Engineers, 1977). Adapted with permission.

Figure 4-6 is a graphical presentation of data similar to the findings presented in Table 4-1. The curves shown are those for extraterrestrial solar radiation and for sea-level solar radiation. For comparison, a smoothed curve for blackbody radiation at 5800 K is also included. In all three cases, the curves are rough approximations only, intended to illustrate relationships, not to provide values for actual calculations.

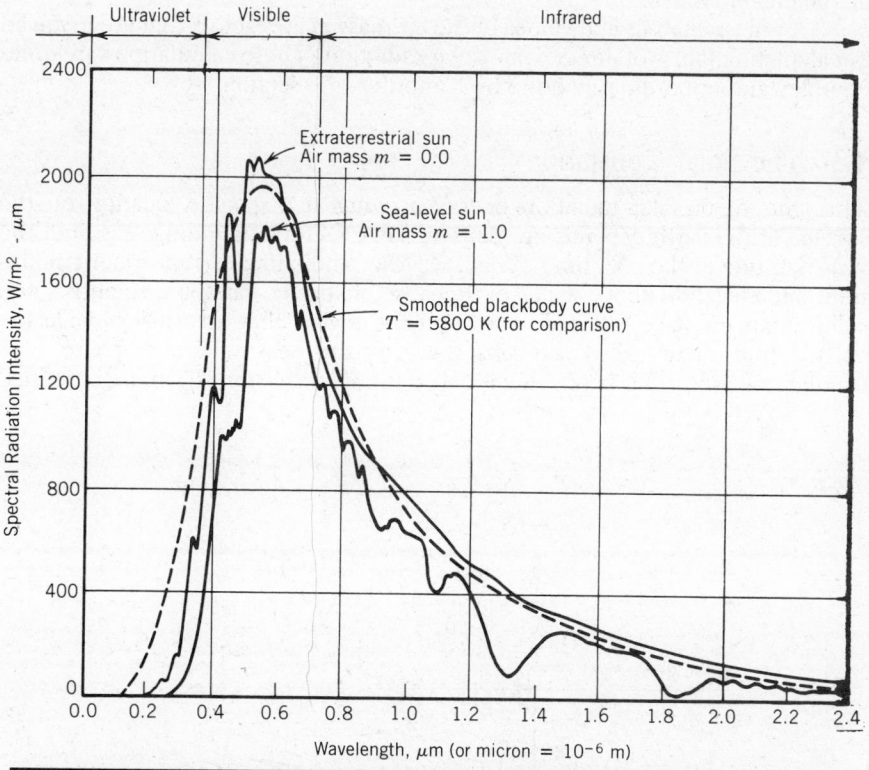

Figure 4-6 Spectral radiation intensity for the sun and for an ideal blackbody, plotted against wavelength. Note that most of the sun's energy occurs in radiations whose wavelength falls between 0.4 and 2.0 microns. Adapted from data compiled by NASA and published in *Applications of Solar Energy for Heating and Cooling of Buildings*, Richard C. Jordan and Benjamin Y. H. Liu, eds., Atlanta: ASHRAE, 1977, p. III-6. Used with permission.

Air Mass Attenuation.   The curves of Fig. 4-6 indicate significant differences between extraterrestrial solar energy and sea-level solar energy. These differences are caused by the absorption and scattering effects of the atmosphere. The amount of the decrease, or attenuation, depends on the mass of air directly between an earth location and the sun. Thus, the term *air mass attenuation* is used to describe the effect. Also, in addition to the air itself, the presence in the atmosphere of water vapor, dust, industrial contaminants, and ozone all contribute to the attenuation of radiant energy.

Air mass (m) at the top of the atmosphere before attenuation begins is assigned a value of zero. At sea level, with the sun directly overhead and the sky perfectly clear, air mass is assigned a value of 1.00. When the sun is not directly overhead but has a *solar altitude angle* β (Fig. 4-7), the sea-level air mass can be calculated from the relation $m = 1/\sin β$. In summary, and referring to Fig. 4-7:

Case 1 (extraterrestrial) $m = 0.0$
Case 2 (sea level, clear, β = 90°) $m = 1.00$
Case 3 (sea level, clear, β < 90°) $m = 1/\sin β$ (4-2)

The effect of air mass attenuation on solar radiation is considerable. In fact, sea-level solar radiation intensity, on the average, is roughly only about half of the extraterrestrial radiation intensity, even in the case when the sun is directly overhead (β = 90°). Selectively, air mass attenuation effectively eliminates ultraviolet radiation of wavelength shorter than about 0.3 μm and infrared radiation beyond about 2.5 μm (see Fig. 4-6 and Table 4-1).

The altitude above sea level of the site affects the value of the air mass appreciably, since there is less atmosphere for the sun's rays to traverse. At an altitude of 3000 ft with β = 90°, $m \cong 0.88$; at an altitude of 1 mi, $m \cong 0.81$; and at 7000 ft, $m \cong 0.76$.

Air mass analysis and values of the air mass m are factors that are involved in calculating tables of direct beam solar radiation. These calculations are quite complex and are ordinarily done by computer (see section 4-6).

4-3   The Solar Constant

The amount of solar radiation energy arriving at a specific location on the surface of the earth depends on many variable factors, including season of the year, latitude of the site, time of day, weather and climate, atmospheric pollution, and elevation above sea level. Because of surface variables, scientists and solar engineers have selected as a reference region that area just outside the earth's atmosphere, called *near-earth space*, for the measurement of a base value of solar radiation. This base value is called the *solar constant I_0*, and it is defined

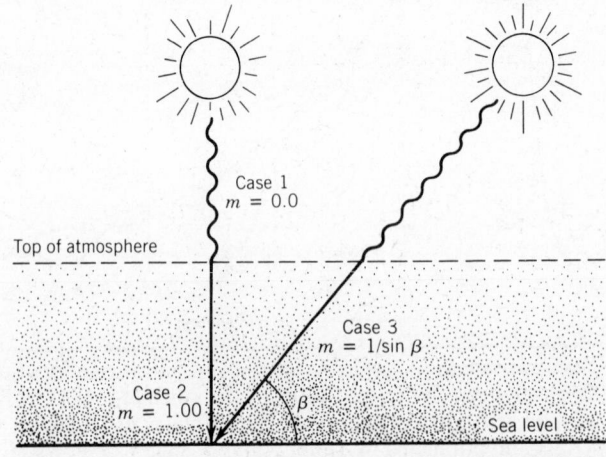

Figure 4-7 Relationships involved in air mass attenuation. The general case is represented by m = 1/sin β.

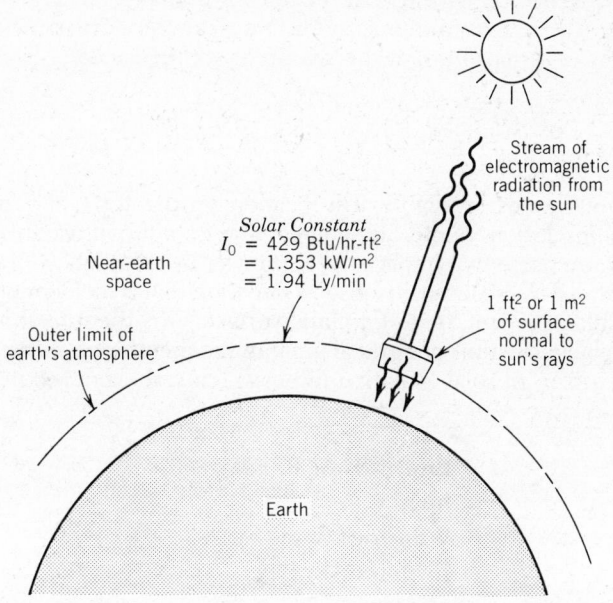

Figure 4-8 The solar constant, I_0. At a point in near-earth space, just before the earth's atmosphere becomes appreciable, the sun's radiant energy has an intensity (solar power) of 429 Btu/hr-ft², or 1.353 kW/m².

as the solar radiation energy per unit of area perpendicular to the sun's rays, per unit time, received at the top of the earth's atmosphere. Stated in another way, the solar constant is the solar radiation power per square foot (or per square meter) in near-earth space (Fig. 4-8). This near-earth space radiation is often referred to as extraterrestrial radiation.

The solar constant is a reference point used as a basis for comparison with other solar radiation intensity values. Nowhere on the earth's surface can solar power be as great as the value of the solar constant, because of the air mass attenuation discussed above.

After many years of careful observations and measurements, lately with input from instruments carried by satellites and spacecraft, scientists have agreed on an average value for the solar constant as:

$$I_0 = 429 \text{ Btu/hr-ft}^2$$
$$= 1.353 \text{ kW/m}^2$$

Table 4-2 gives values of I_0 in units commonly used by both science and industry.

We speak of an average value for the solar constant because it does vary somewhat, partly because of sunspot activity, which reduces solar radiation intensity. The major variation, however, is an annual one caused by the elliptical orbit of the earth around the sun. The sun and the earth *are closest to each other in January and farthest apart in July*. This distance variation causes the

Table 4-2 Values of the Solar Constant, I_0, in Selected Units

$I_0 =$	429 Btu/hr-ft²
	1.52 hp/yd²
	1353 W/m²
	1.353 kW/m²
	4870 kJ/hr-m²
	1.94 langleys/minute (Ly/min)

Note: The langley (Ly) is a unit used primarily by meteorologists. One Ly equals 1 cal/cm². Solar radiation data are often reported in langley terminology.

THE SUN AND ITS ENERGY

solar constant to be about 6 percent greater in January than it is in July. For solar heating, this is a fortuitous circumstance for the northern hemisphere, since solar power is maximum at the season of greatest need.

4-4 Intensity of Solar Radiation at the Earth's Surface—Insolation

The usable solar power available at a location on the surface of the earth is referred to as *insolation*. (Note the spelling, and do not confuse this word with *insulation*, a substance that impedes heat transfer by conduction.) Insolation is accurately defined as the intensity of solar radiation that is incident normal (that is, perpendicular) on unit area of a plane surface. It is the *rate* at which solar energy is available per unit of surface area and is therefore a measure of *solar power*. Solar power, or solar radiation intensity, can be expressed in any of the following units:

$$Btu/hr\text{-}ft^2 \text{ or } Btu/day\text{-}ft^2$$
$$hp/ft^2$$
$$W/m^2 \text{ or } kW/m^2$$
$$Ly/min \text{ or } Ly/day$$
$$kJ/hr\text{-}m^2$$
$$kcal/sec\text{-}m^2$$

Insolation at the earth's surface consists of wavelengths in the general range from about 0.32 μm to about 2.5 μm, with nearly all of the usable energy in the visible and near-infrared regions of the spectrum.

Total solar radiation intensity (insolation) received on a surface is denoted by the symbol I. It is made up of three components, defined as follows, and its value depends on the angle between the incoming solar rays and the normal to the surface (Fig. 4-9).

I_{DN} is the intensity of direct-beam solar radiation on a surface perpendicular (normal) to the sun's rays. Direct-beam solar radiation follows a direct path from the sun to the receiving surface and forms an angle with the normal to that surface. The greatest amount of usable solar power is provided by this component.

I_d is the component of total insolation contributed by diffuse radiation from the sky, caused by the scattering effects of air molecules, water vapor, or contaminants in the atmosphere. On clear days, diffuse radiation amounts to only about 15 percent of total insolation, while on days with a complete cloud

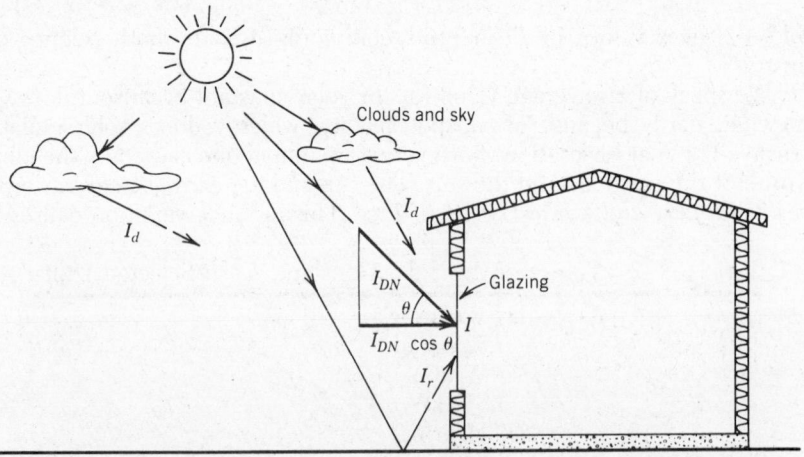

Figure 4-9 Sketch illustrating the components of total insolation.
$I = I_{DN}\cos \theta + I_d + I_r.$

overcast, although total insolation will be small, I_d could account for a large part of the insolation received.

I_r is the component of total insolation contributed by reflection from bodies on the earth's surface—buildings, paved surfaces, the surrounding ground, and bodies of water.

θ is the angle between the incoming solar rays and the normal to the receiving surface. It is called the *angle of incidence*.

In summary, then, the total insolation incident on a flat surface is

$$I = I_{DN} \cos \theta + I_d + I_r \qquad (4\text{-}3)$$

The term $I_{DN} \cos \theta$ is the *direct-beam radiation* incident on a surface that is tilted in such a way that its normal forms an angle θ with the incoming solar radiation. Figure 4-9 illustrates the factors involved in Eq. (4-3).

Since solar radiation intensity varies markedly hour by hour during the day and falls to zero for the night hours, it is a rather common practice to report insolation values on a per-day basis. *Average daily insolation* for January may vary from 450 Btu/ft² in Canada and northern Europe, to 625 Btu/ft² across much of the central belt of the United States, to as much as 1200 to 1600 Btu/ft² in the U.S. Southwest and other arid regions. At solar noon on a clear day in July at 35°N Lat, insolation on a horizontal surface may reach a rate of 325 Btu/hr-ft². For locations at 35° to 40°N Lat, typical insolation values for a solar altitude β of 50° and a cloudless atmosphere range from 230 to 300 Btu/hr-ft².

This amount of energy, streaming from the sky on clear, sunny days, is a resource of tremendous value in an energy-short world. Some familiar examples will indicate both the potential and the limitations of solar power.

First, consider a flat roof on a typical residence of about 1600 ft² of floor area, located at 35°N Lat in the United States. On a clear day in July, the total insolation received (direct and diffuse) would be on the order of 1500 Btu/day-ft². For the entire roof area of 1600 ft², the daily total energy available would be about 2.4 million Btu. For the 14 hours of daylight in July, the solar power incident on the roof would average about 170,000 Btu/hr. If it could be converted to mechanical or electrical power at 100 percent efficiency, the output would be nearly 67 hp or 50 kW—a truly significant amount of power, enough to satisfy the energy requirements of almost any family—all of the appliances, the heating system for winter, air conditioning for the summer, lights, and enough left over for a well-equipped home shop.

However, *available* energy and *utilizable* energy are not the same. Converting solar power to usable heat can be accomplished at efficiencies ranging from 35 to 60 percent, and the current state of the technology for conversion of solar radiation directly into electric power provides efficiencies of about 12 to 15 percent.

In order to receive and capture significant amounts of solar energy, large collection areas are required. Practical machines for individual and commercial use are not generally operable on the amount of solar energy that their unextended surfaces can intercept. An automobile, for example, cannot be powered by the solar energy incident on its surface, since less than 2 hp would be available and utilizable (at 15 percent efficiency) from the solar radiation incident on its approximately 60 ft² of horizontal surface (averaged for season of the year, day and night, clear and cloudy days, etc.).

Solar energy plants using collector areas of many acres in extent are currently being constructed. Some are already in operation, generating electric energy from solar power. Several such plants will be discussed in Chapter 18.

Units for Expressing Insolation Values. Insolation, or solar power, is expressed in several different kinds of units—heat units, electrical (power) units, and the special unit referred to above, the langley. Table 4-3 provides some conversion factors for dealing with insolation values and their electrical and mechanical energy equivalents.

Table 4-3 Useful Conversion Factors for Insolation Calculations

$$1 \text{ Btu} = 252 \text{ cal} = 0.252 \text{ kcal}$$
$$= 0.293 \text{ W-hr}$$
$$= 2.93 \times 10^{-4} \text{ kW-hr}$$
$$1 \text{ W} = 3.41 \text{ Btu/hr}$$
$$1 \text{ kW} = 3410 \text{ Btu/hr}$$
$$= 0.239 \text{ kcal/sec}$$
$$1 \text{ Btu/hr-ft}^2 = 3.155 \text{ W/m}^2 = 3.155 \times 10^{-3} \text{ kW/m}^2$$
$$= 0.293 \text{ W/ft}^2$$
$$= 6.5 \text{ Ly/day}$$
$$1 \text{ W/m}^2 = 0.317 \text{ Btu/hr-ft}^2$$
$$1 \text{ kW/m}^2 = 317 \text{ Btu/hr-ft}^2$$
$$= 2060 \text{ Ly/day}$$
$$1 \text{ Ly/day} = 0.154 \text{ Btu/hr-ft}^2$$
$$= 0.485 \text{ W/m}^2 = 4.85 \times 10^{-4} \text{ kW/m}^2$$
$$= 3.69 \text{ Btu/day-ft}^2$$

4-5   The Measurement of Insolation

Solar radiation is continually being measured on a daily basis at research laboratories of the National Weather Service and by special project teams of such government agencies as the National Aeronautics and Space Administration (NASA) and the National Oceanic and Atmospheric Administration (NOAA). Many other nations around the world also conduct continuing studies of solar radiation intensity. Many of the U.S. stations measure total insolation, I, and some are concerned primarily with extremely accurate measurements of direct normal radiation, I_{DN}. Most also maintain records of sunshine duration (total hours and minutes per day) and of "percent possible sunshine". Cloud-cover data are also collected and recorded. From these and similar data, insolation tables for selected latitudes and for specific cities and localities are prepared. Selections from such tables are included at the end of this chapter.

Insolation Measuring Instruments.   Solar radiation measurement techniques are based on the use of thermoelectric instruments that incorporate integrating capabilities so that, not only can instantaneous values be read, but so can hourly and daily totals. Daily totals can be aggregated into monthly totals, from which average daily totals for each month of the year can be compiled. The two most commonly used instruments for these measurements are the *pyrheliometer* and the *pyranometer*.[4]

Pyrheliometers measure direct-beam solar radiation at normal incidence. Since it is only the direct rays of the sun that are desired, a long collimating tube is fitted with a small aperture so that the angular acceptance of the instrument is limited to a solid angle of less than 5°. Pyrheliometers are mounted on a mechanism that moves them exactly 15°/hr so that they track the sun across the sky. Figure 4-10 illustrates a typical pyrheliometer, less mounting.

Pyranometers measure the total solar radiation from the entire sky hemisphere (I_{DN} and I_d) and, unless the researcher blocks it out by some means, the reflected component, I_r, as well. Precise positioning of the instrument in a horizontal plane is an absolute requirement. A precision spectral pyranometer is shown in Fig. 4-11.

Local conditions have a significant effect on the amount of solar radiation received on a surface and on the amount that can be captured for solar energy purposes. Elevation above sea level and percent cloud cover have already been

[4] For an in-depth treatment of solar radiation measurement, see Chapter Three by John I. Yellot in *Applications of Solar Energy for Heating and Cooling of Buildings*, Richard C. Jordan and Benjamin Y. H. Liu, eds., ASHRAE GRP170. (Atlanta: ASHRAE, 1977).

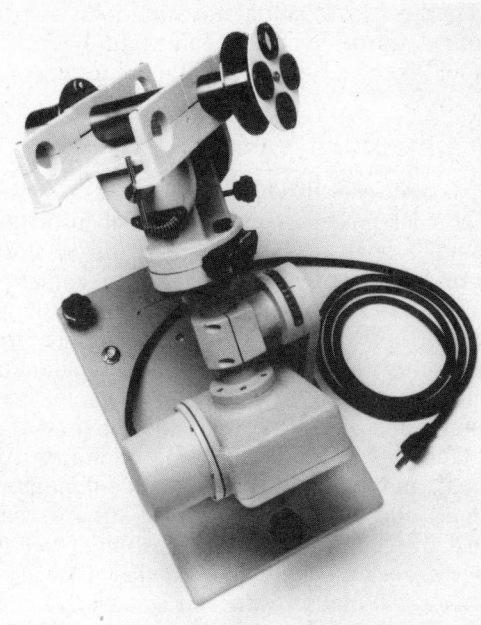

Figure 4-10 A normal-incidence pyrheliometer for measuring direct-beam solar radiation. Note the long collimating tube that limits the angle of acceptance to about 5°. A rotatable disk allows for three filters and one open aperture. (The Eppley Laboratory, Inc.)

mentioned, as have atmospheric conditions such as dust, humidity, pollen, pollution, haze, and fog. Site conditions such as paving, concrete, grass and green spaces, roof surfaces, water surfaces, nearby structures, and snow or ice all affect the radiation intensity on a given surface. And, with respect to transferring the available solar power into a structure or to a fluid medium, the ambient air temperature and the direction and velocity of the wind are factors that must be considered.

Although insolation tables (see Table 4-E in the Chapter Appendix) are a valuable source of data for estimating the total solar power available at the site of

Figure 4-11 A precision spectral pyranometer. This instrument measures total sun and sky radiation. Its active element is a circular multijunction thermopile. A spirit level and adjustable leveling screws are incorporated in the base of the instrument. (The Eppley Laboratory, Inc.)

a proposed solar energy system, actual measurements of available insolation by a precision pyranometer provide more reliable information. Whenever possible, tabular values should be checked by actual measurements at the site.

4-6   Calculating Insolation Values

Given the solar constant (429 Btu/hr-ft^2), which, as previously pointed out, is the extraterrestrial solar power or extraterrestrial insolation, it is possible to calculate by computer analysis values of I_{DN} (direct normal insolation) for specific locations and for specified days and times. Values for I_d (diffuse radiation, clear-day values) can also be incorporated in the calculation. I_r values (the reflection component) are not precalculated, however, since they vary with every locality and even with the specific site and building orientation.

Many variables enter into the computer program, including the location of the site, the month and day of the year, and the time of day. Climatological factors are important, so weather data are always inputs. Air mass attenuation has to be considered, and so must the effects of contaminants in the air—water vapor, dust, pollens, and industrial pollutants such as smog. Many weather stations now keep daily records of hours of sunshine, percent possible sunshine, and mean (i.e., average) cloud cover. These data can be applied (where known) to arrive at site-specific insolation values of considerable precision.

The basic equations, constants, and parameters that research laboratories use for these computer programs are too complex to be dealt with in an elementary text.[5] Furthermore, the practicing solar designer, engineer, or technician need not make such calculations, since insolation tables (measured values or computer-generated values) are already available and suitable for use in solar energy system design for most localities throughout the United States and Canada and for many other countries of the world.

SOLAR GEOMETRY

The sun is the center of our solar system, with the earth and the other eight planets revolving around it in slightly elliptical orbits. The earth's orbit has a mean radius of about 93 million mi, and the period of the earth's traverse around the sun determines our year (365.25 days). The earth rotates once each day about an axis that is tilted at an angle of about 23.5° with respect to the flat plane in which it revolves around the sun (Fig. 4-12). Because of gyroscopic effects, the earth's axis remains fixed on a point in space as the earth moves around the sun in its yearly journey. As a result, the north and south poles of the earth go through an annual cycle of pointing alternately toward and then away from the sun.

4-7   The Earth's Motion and the Seasons of the Year

The tilt of the earth's axis is what causes the seasons of the year—summer in the hemisphere that is tipped toward the sun, winter in the hemisphere that is tipped away from the sun, and spring and autumn at the halfway positions in between. In the northern hemisphere, the *winter solstice* occurs on December 21, the *summer solstice* on June 21, the *vernal equinox* on March 21, and the *autumnal equinox* on September 21.

The hemisphere that is temporarily tipped toward the sun has longer days and shorter nights (summer) and therefore receives more solar radiation. Both

[5] Interested readers are referred to the following sources for a discussion of these equations: Power Systems Group/Ametek, Inc. *Solar Energy Handbook, Theory and Applications* (Radnor, Pa.: Chilton Book Company, 1979), pp. 24–27. Richard C. Jordan and Benjamin Y. H. Liu, eds., *Applications of Solar Energy for Heating and Cooling of Buildings* ASHRAE Grp 170. (Atlanta: American Society of Heating, Refrigerating, and Air Conditioning Engineers, 1977), pp. IV–1 to IV–7.

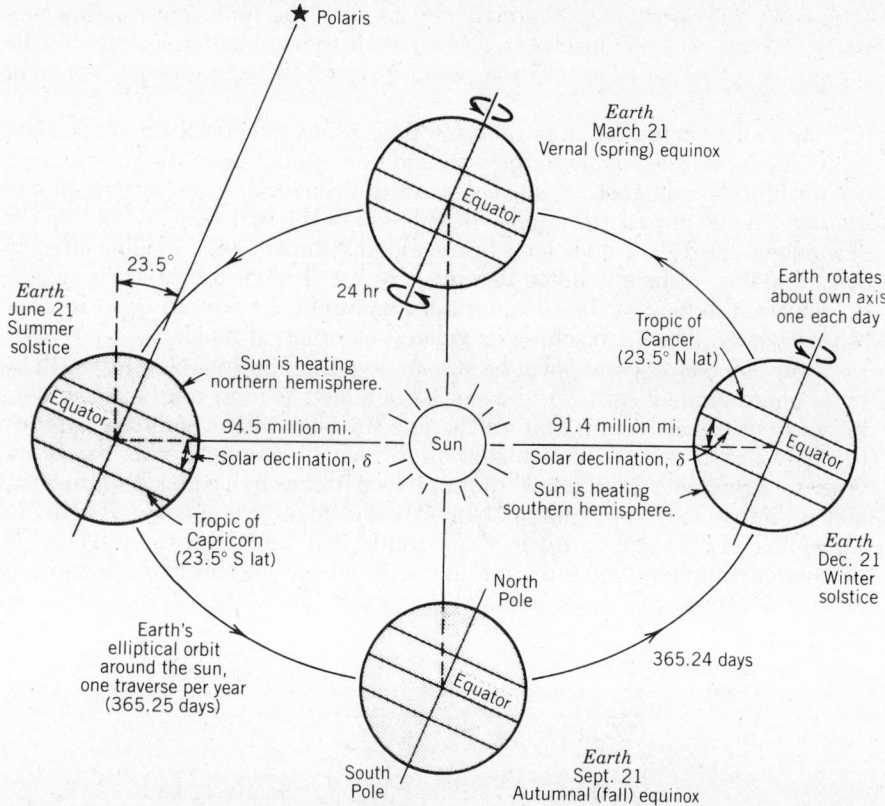

Figure 4-12 The earth in its yearly traverse around the sun. Note how the seasons of the year are determined by the tilt of the earth's axis. Note also that the earth is closer to the sun in winter than it is in summer (northern hemisphere).

the land mass and the water mass of that hemisphere warm up and they warm the air in turn. The opposite is true of the hemisphere that is tipped away from the sun.

On March 21 and again on September 21, the sun appears to be directly over the equator, and both hemispheres are receiving the same amount of solar radiation. On these dates, the duration of daylight is equal to the duration of darkness (*equinox* means "equal night").

Although land and water are warmer in summer than in winter, this is not to say that solar radiation is more intense in summer than in winter. In fact, as noted in section 4-3, the earth is closest to the sun in January, and for the *northern hemisphere* the solar constant I_0 is about 6 percent greater in January than it is in July. January, of course, is summer in the southern hemisphere, so "down under" hot weather coincides with increased solar radiation intensity and cold weather (July) with decreased solar radiation. It is important to note in this connection that the amount of heat energy *potentially available* in solar radiation is not affected by air temperature or the temperature of the land–water mass. Given perpendicular incidence and a clear sky and assuming equal elevation of the sun above the horizon, there is just as much instantaneous solar radiation power at a given location on a −20°F January day as there is on an 80°F day in July (in fact, 6 percent more in the northern hemisphere). To be sure, temperature, wind, and other climatic conditions markedly reduce the amount of the available power that can be captured—*usable solar radiation*.

4-8 The Sun's Apparent Motion

We have been discussing the actual motion of the earth with respect to the sun, with the sun as the centerpiece of the geometry. However, in order to design solar energy systems on earth, the earth must become the effective center of the action. We will regard it as stationary and speak of the sun's "daily path across the sky". Thus, we find ourselves reverting to the thought patterns of the Middle Ages, except that we keep reality in mind by using the terminology "apparent motion of the sun."

For solar energy design purposes, it is necessary to know the sun's position in the sky for any given day of the year and time of the day. This knowledge is essential for designing either active systems (for the location and orientation of solar collectors) or passive systems (orientation of the structure and placement of windows, fixed glass, and clerestories, and for estimating the shading effect of other buildings or nearby hills). Reference to Fig. 4-13a will help in visualizing the apparent motion of the sun (northern hemisphere) as it moves in its daily traverse across the sky, reaching its greatest elevation at midday.

An observer at point 0 will be able to note that, on June 21 (the northern hemisphere summer solstice), the sun has attained its most northerly track for its apparent passage around the earth. The sun rises earliest and sets latest on this day, and the sun's path is higher on this day than on any other day of the year. By September 21 (the fall equinox), its path has moved back south again and is directly over the equator. Now day and night are equal in length. By December 21, the sun's path is as far south as it will go, it is winter in the northern hemisphere, the sun is low in the sky all day, and the sun rises late and sets early, resulting in short days and long nights.

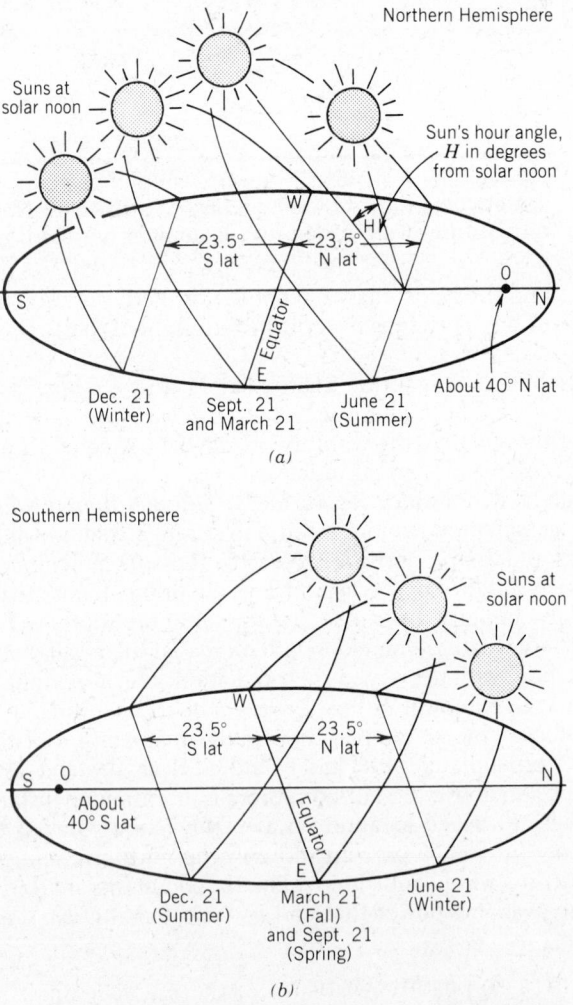

Figure 4-13 The *apparent motion* of the sun, as seen by an observer on earth (*a*) at about 40° N Lat and (*b*) at about 40° S Lat.

All of this is reversed south of the equator, as shown in Fig. 4-13b. Note that, in either hemisphere, if maximum solar radiation on a fixed structure is desired, its major axis should be east–west, so that a long wall faces due south in the northern hemisphere or due north in the southern hemisphere.

4-9 Solar Angles and Their Determination

The position of the sun in the sky with respect to a given site and with respect to a specified time of year and time of day, is expressed in terms of a number of *solar angles*. These angles are illustrated in Figs. 4-12, 4-13, and 4-14, and are defined as follows:

1. The *solar elevation angle*, or *solar altitude*, is the angular distance of the sun above the horizon, at a specified time of the year and time of day, from a particular location. This is the angle β of Fig. 4-14 (or angle *POH*).
2. The *solar azimuth* is the angular distance from due south to the sun's projection on the horizon, measured from due south (northern hemisphere). This is the angle φ of Fig. 4-14 (or angle *SOH*).
3. The *local latitude L*, is measured in degrees north or south of the equator.
4. The *solar declination* is the angular distance of the sun north (+) or south (−) of the equator, at a specified time of the year. This is angle δ of Fig. 4-12.
5. The *apparent solar time*, or the *sun's hour angle*. This is the angular distance (degrees) of the sun from its highest position at solar noon. It is shown as angle *H* in Fig. 4-13a.

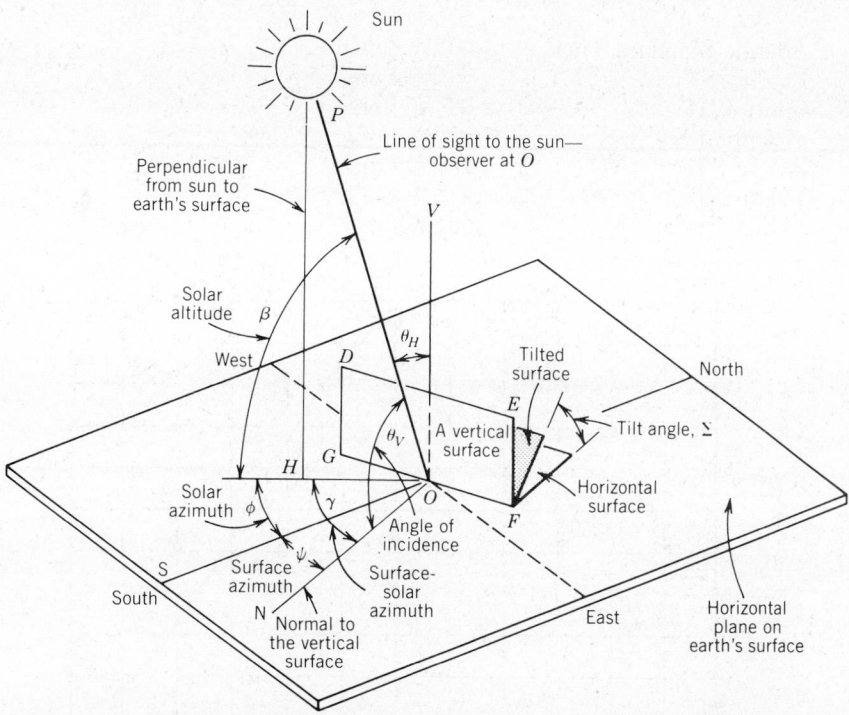

Figure 4-14 Diagram showing (and defining) many of the solar angles essential in solar energy design. Shown are the *solar azimuth* φ, the *solar altitude* β, the *angle of incidence* of solar radiation θ, the *surface azimuth* ψ, the *surface-solar azimuth* γ, and the *tilt angle* of a surface receiving radiation Σ. Adapted, with permission, from *ASHRAE 1977 Fundamentals*, Atlanta: ASHRAE, 1977, p. 26.3.

The solar altitude β and the solar azimuth φ are termed the *sun angles* in solar industry practice. They can be measured if suitable instruments are available, but they are ordinarily determined from prepared charts and tables, and methods of doing so will be explained in the next section. However, β and φ can also be calculated directly from the solar geometry, using the following equations:

$$\sin \beta = \cos L \cos \delta \cos H + \sin L \sin \delta \qquad (4\text{-}4)$$

$$\sin \phi = \frac{\cos \delta \sin H}{\cos \beta} \qquad (4\text{-}5)$$

To solve these equations, values must first be obtained for the various angles, as follows:

1. *L* is the latitude of the site, a known factor.
2. Angle δ is the solar declination for the particular time of year. Its value for any time of year may be obtained from the chart in Fig. 4-15.
3. *H* is the hour angle, and to evaluate it one must first know the *local solar time* (LSoT). Then obtain the difference between LSoT and solar noon in minutes of time. Finally, divide this time difference by 4 min (time) per degree of arc to obtain the hour angle as an angular distance in degrees.

Local Solar Time.   An explanation of local solar time (LSoT) is necessary at this point. Solar time is not the same as *clock time*, for a variety of reasons. Clock time is the same for all localities in the same standard time zone. Each standard time zone occupies 15° of longitude, although boundaries are often shifted one way or another to include specific population or commercial centers. The meridian of longitude that runs through the middle of each standard time zone is called the *standard time meridian*. These meridians for the time zones of the United States and Canada are listed below:

Atlantic Standard Time	60° W Long
Eastern ST	75° W Long
Central ST	90° W Long
Mountain ST	105° W Long
Pacific ST	120° W Long
Yukon ST	135° W Long
Alaska–Hawaii ST	150° W Long

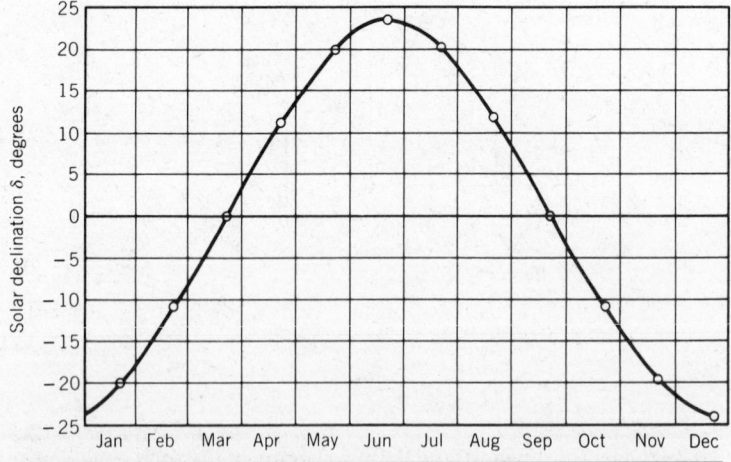

Figure 4-15 Solar declination for the 12 months of the year. Approximations for exact dates can be made from the curve with sufficient accuracy for ordinary solar design calculations. *Source: ASHRAE 1977 Fundamentals*, Atlanta: ASHRAE, 1977, p. 26.2.

When the sun is directly overhead at a standard time meridian, local solar time and standard time are very nearly the same (but see the discussion of the *equation of time* below). At all other times and places across the zone, local time and standard time differ. Local solar time (LSoT) differs from local standard time (LST) by four minutes of time per degree of longitude east or west of the standard time meridian for that zone. Stated another way, solar noon occurs four minutes earlier than standard time noon for each degree of longitude east of the meridian for that zone and four minutes later for each degree west of the zone's standard meridian. The time problem is further complicated by daylight saving time (DST) when it is in effect.

Converting Clock Time to Local Solar Time. The first step in determining sun angles is to obtain the local solar time. Converting clock time to local solar time is accomplished by the following procedure:

1. First, if daylight time (DT) is in effect at the locality, subtract one hour from clock time to obtain local standard time (LST).
2. Check the longitude of the locality (LL) and also the longitude of the standard time meridian (LSTM) for the time zone. Express the difference between these longitudes in degrees east or west of the standard meridian to the nearest tenth of a degree.
3. Multiply the difference by four minutes (of time) per degree. If the site is east of the standard meridian, add the correction to standard time; if west, subtract it from standard time.
4. A further correction is needed because of the earth's elliptical orbit and axis tilt. Both of these factors are incorporated into an equation called the *equation of time*. A plot of the equation of time is given in Fig. 4-16, from which values of the equation-of-time correction (ET), in minutes, can be read for each month of the year. Add this correction to the above result from step 3 to obtain the local solar time (LSoT).

The four steps above can be incorporated into the following equation:

$$LSoT = LST + 4(LSTM - LL) + ET \qquad (4\text{-}6)$$

where

LSoT = Local solar time
LSTM = Longitude of the local standard time meridian, degrees of arc
LL = Longitude of the locality, degrees
ET = The equation-of-time correction, minutes of time

The factor 4 is the minutes of elapsed time for 1° of earth rotation.

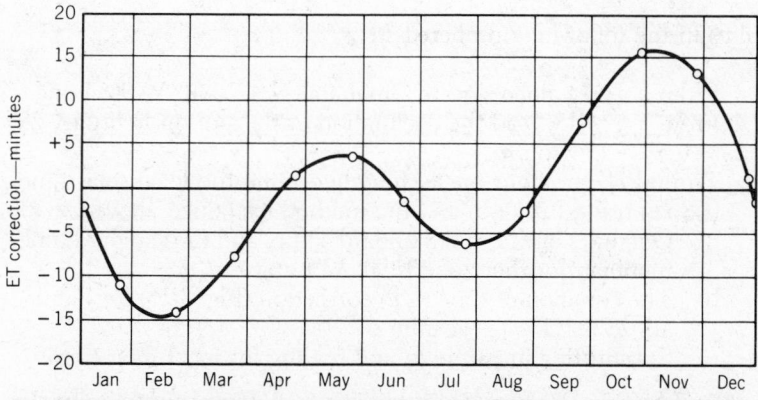

Figure 4-16 Curve of the equation of time (ET). The corrections are positive (+) or negative (−), as indicated on the vertical scale at left. *Source: ASHRAE 1977 Fundamentals*, Atlanta: ASHRAE, 1977, p. 26.2. Adapted with permission.

Equation (4-6) must be used with careful attention to the algebraic sign of the quantities involved. For example, the term (LSTM − LL) must be interpreted as the difference between the longitude of the local standard time meridian and the local longitude, *whichever is greater*. Also, the term 4(LSTM − LL) will be positive or negative depending on whether the site is east (positive) or west (negative) of the local standard meridian. The value of ET may be positive or negative, as read from the chart in Fig. 4-16.

An example will serve to clarify how local solar time is determined from clock time.

Illustrative Problem 4-1

Oklahoma City is located at 97.5° W Long in the Central Standard Time Zone (meridian at 90° W Long). Find the local solar time (LSoT) at 3:00 P.M. Central Daylight Time (CDT) on August 15.

Solution

First, convert DT (clock time) to LST by subtracting one hour.

$$\text{LST} = \text{3:00 P.M.} - 1 \text{ hr} = \text{2:00 P.M.}$$

From Fig. 4-16, read the equation-of-time (ET) correction for August 15 as

$$\text{ET} = -5 \text{ min}$$

Substituting in Eq. (4-6),

$$\begin{aligned} \text{LSoT} &= \text{2:00 P.M.} + 4 \text{ min/deg.} (90 - 97.5) \text{ deg} + (-5 \text{ min}) \\ &= \text{2:00} + (-30 \text{ min}) - 5 \text{ min} \\ &= \text{1:25 P.M.} \qquad\qquad\qquad\qquad\qquad\qquad \textit{Ans.} \end{aligned}$$

Now that the method of determining local solar time has been explained, we can proceed with an example of the calculation of solar altitude and solar azimuth, using Eqs. (4-4) and (4-5).

Illustrative Problem 4-2

Calculate the solar altitude and the solar azimuth at Omaha, Nebraska, at 2:00 P.M. local standard time on February 25.

Solution

Proceed as in the following numbered steps:

1. From a world globe, locate Omaha at N 42° Lat, W 96° Long.
2. From Fig. 4-15, read the declination of the sun on February 25 as $\delta = -10°$ (south).
3. To find H, the hour angle, first determine the local solar time (LSoT) using Eq. (4-6) and the method explained above.
   a) Omaha is on Central Standard Time, and the standard time meridian for that zone is 90° W Long.
   b) The equation-of-time (ET) correction (Fig. 4-16) for February 25 is ET = −13 min.
   c) Substituting in Eq. (4-6) and solving for local solar time,

$$\begin{aligned} \text{LSoT} &= \text{2:00 P.M.} + 4 \text{ min/deg} (-6 \text{ deg}) + (-13 \text{ min}) \\ &= \text{1:23 P.M. or 83 min past solar noon} \end{aligned}$$

   d) The hour angle, $H = \dfrac{+83 \text{ min}}{4 \text{ min/degree}} = +20.8°$

4. Substituting these values in Eq. (4-4) for solar altitude,

$$\sin \beta = \cos 42 \cos (-10) \cos 20.8 + \sin 42 \sin (-10)$$
$$= 0.68 + -0.116 = 0.564$$

and

$$\beta = 34.3° \qquad\qquad Ans.$$

5. For the azimuth angle, substitute in Eq. (4-5),

$$\sin \phi = \frac{\cos (-10) \sin 20.8}{\cos 34.3}$$
$$\phi = 25° \qquad\qquad Ans.$$

4-10 Sun Angles from Sun Charts

Trigonometric calculation of sun angles is not common practice in solar design work. These angles are ordinarily read directly from prepared graphs or charts that give values of β and ϕ for different latitudes and for each hour of the day and each month of the year. A limited set of sun charts is presented here as Fig. 4-17. To use the charts, consider the observer facing due south in the northern hemisphere. The solid, bell-shaped curves describe the sun's apparent path across the sky, in terms of its β and ϕ angular coordinates, at the designated times of the year. Interpolation between the month curves will give reasonable accuracy for intervening dates. The dashed lines give the time of day in local solar time. The solar altitude β is read off the vertical axis, and the solar azimuth ϕ off the horizontal axis.

The solar altitude β and the solar azimuth ϕ are very important angles for solar designers, since, for any specific day of the year and time of day, they determine whether or not the sun will shine on a particular site or location. In other words, they are essential angles in determining *solar access*.

The use of sun charts is illustrated by the following problems.

Illustrative Problem 4-3

Determine the solar altitude β and the solar azimuth ϕ at Savannah, Georgia, at 2:00 P.M. local solar time on August 8.

Solution

First, check the latitude of Savannah as being about 32° N Lat. Refer to the 32° N Lat sun chart of Fig. 4-17 and locate the (dashed) 2:00 P.M. line. August 8 is exactly halfway between July 21 and August 21, so locate a point on the 2:00 P.M. line halfway between the July and August (heavy) curves. Read the solar altitude β off the vertical scale at the right of the chart, and the solar azimuth ϕ off the scale at the bottom of the chart.

$$\beta = 59°$$
$$\phi = 67.5° \text{ (west of south)} \qquad\qquad Ans.$$

Illustrative Problem 4-4

Indianapolis is located at approximately 40°N Lat, 86°W Long. Find both sun angles at 4:00 P.M. Central Daylight Time (CDT) on January 21.

Solution

First, Central Daylight Time (CDT) (clock time) must be converted to local solar time (LSoT), as follows:

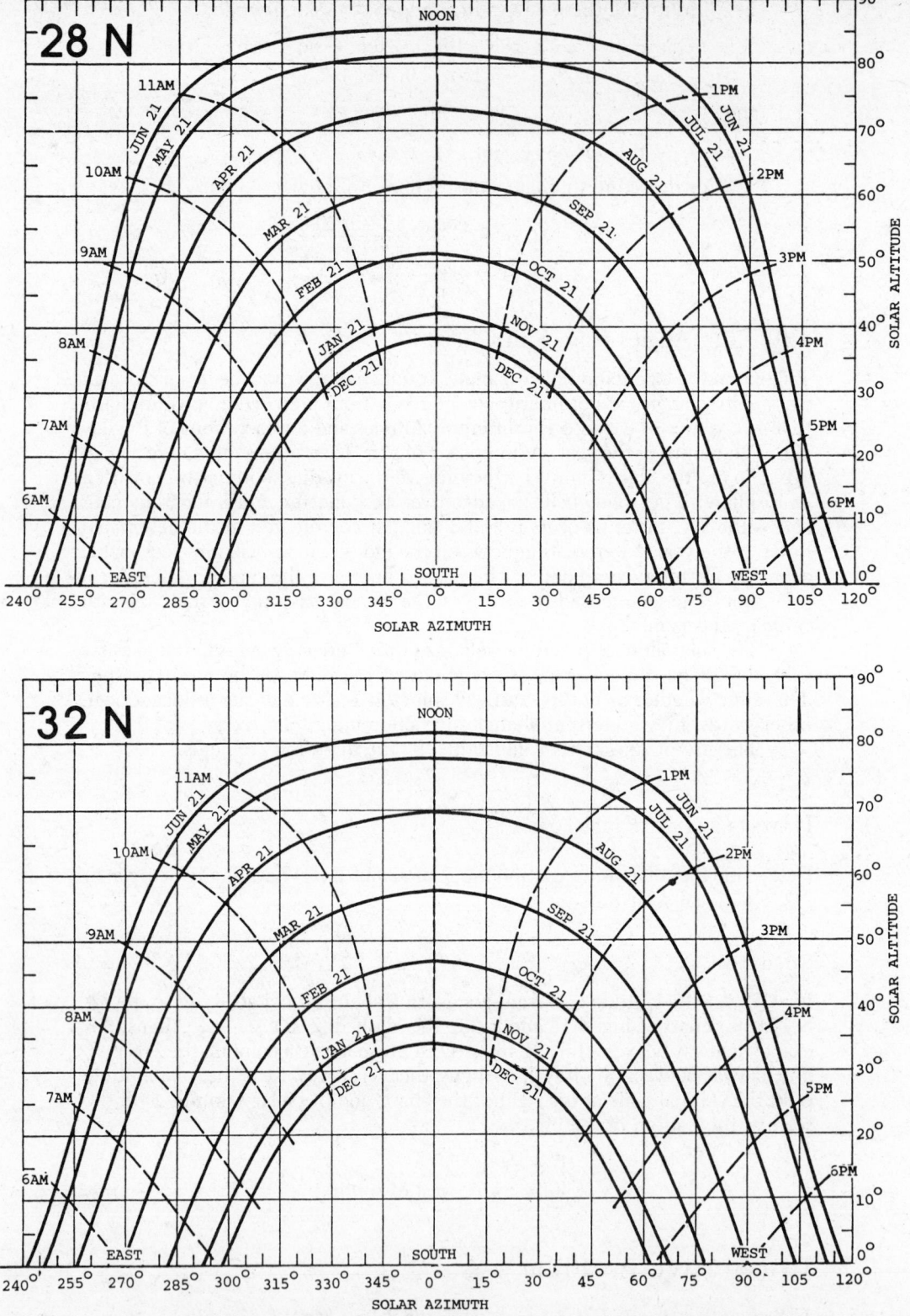

Figure 4-17 Sun charts for determining the sun angles or the sun's position in the sky for any day of the year and time of day. The solid bell-shaped lines show the sun's path across the sky. Read the solar azimuth ϕ from the horizontal scale and the solar altitude β from the vertical scale. The dashed lines give the time of day (in local solar time). *Source: Engineer's Guide To Solar Energy*, Bereny, J. A. and Howell, Yvonne. Sacramento: Solar Energy Information Services, 1979. Reprinted with permission.

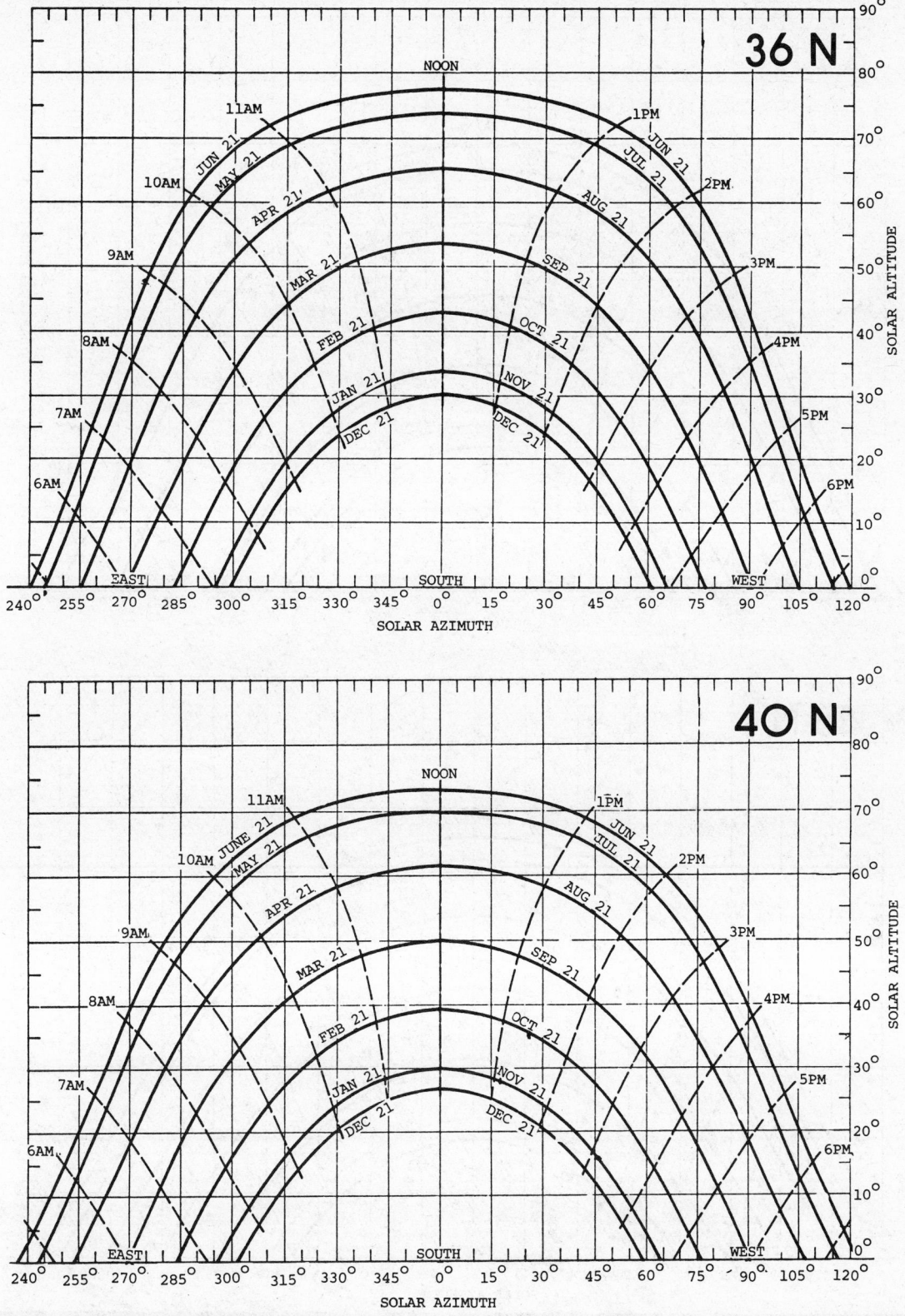

Figure 4-17 (*continued*)

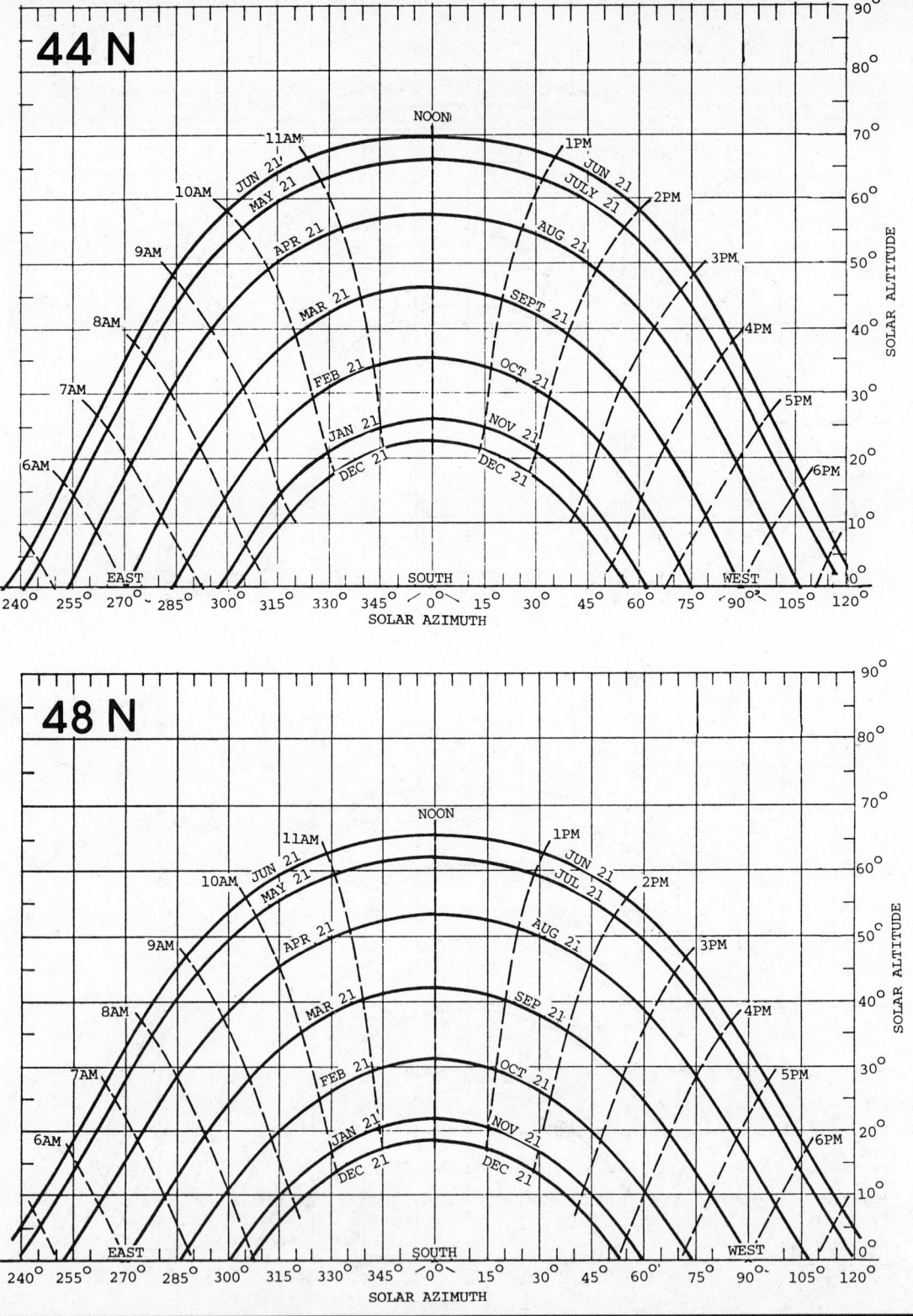

Figure 4-17 (*continued*)

Clock time − 1 hr = 4:00 P.M. − 1 hr = 3:00 P.M. Central Standard Time (CST)

The standard meridian for Central Standard Time at Indianapolis is 90°W Long. Therefore, from Eq. (4-6),

$$\text{Long. diff., LSTM} - \text{LL} = 90° - 86°$$
$$= 4° \text{ east of the meridian}$$

The correction will therefore be positive.
The equation-of-time correction (from Fig. 4-16) for Jan. 21,

$$ET = -12 \text{ min}$$

Solving Eq. (4-6),

$$\text{LSoT} = 3:00 \text{ P.M.} + 4 \text{ min/deg} (+4 \text{ deg}) + (-12 \text{ min})$$
$$\text{LSoT} = 3:04 \text{ P.M.}$$

Now refer to the sun chart for 40°N Lat (Fig. 4-17). On the January 21–November 21 (heavy line) curve, locate a point at your best estimate of 3:04 P.M. (interpolating between the 3:00 P.M. and the 4:00 P.M. line). Read solar altitude (vertical scale) and solar azimuth (horizontal scale) as,

$$\beta = 17°$$
$$\phi = 45° \hspace{3cm} Ans.$$

4-11 Some Other Important Angles of Solar Geometry

Surface Azimuth.   Surfaces on which the sun shines and from which solar radiation energy can be collected usually face due south, or they may be turned toward the east or west from due south, for any one of several reasons. In Fig. 4-14, vertical surface DEFG, for example, faces to the east of due south by an angle ψ. Angle SON (ψ) is called the surface azimuth. Note that it is the angle formed by the normal to the vertical surface and the due south line.

Surface-Solar Azimuth.   When the solar azimuth φ (angle HOS) and the surface azimuth ψ are added algebraically (they are both in a horizontal plane in Fig. 4-14), an angle γ, called the surface-solar azimuth, is obtained. The surface-solar azimuth is an indicator of the obliquity with which the sun's rays will strike a given surface at a stipulated time. In Fig. 4-14, γ is the angle NOH.

Angle of Incidence.   It was noted in section 4-4 that the total solar radiation intensity received on a surface depends on the angle of incidence—that is, the angle between the incoming solar rays and the normal to the surface. For the vertical surface DEFG in Fig. 4-14, the angle of incidence θ_V is angle PON. For the horizontal surface, θ_H is angle POV.

The angle of incidence for *any surface* can be calculated if the solar altitude β, the surface-solar azimuth γ, and the tilt angle Σ (see Fig. 4-14) are known. The equation for this calculation is

$$\cos \theta = \cos \beta \cos \gamma \sin \Sigma + \sin \beta \cos \Sigma \hspace{2cm} (4-7)$$

Angles of incidence are not ordinarily calculated by solar designers in actual field practice since they are readily available in tables. (See Table 4-A in the Chapter Appendix.) As is the case with many factors and parameters needed in solar design, tables of solar position and insolation typically provide values that already take into account the effects of tilt and orientation of the surface on the angle of incidence.

Tilt Angle.   Windows or glazings that admit solar radiation are often vertical, but this is not always the case. Likewise, skylights and solar collector surfaces may be horizontal, but more often they are installed on sloping roofs.

The angle between solar-receptive surfaces and a horizontal plane is called the *tilt angle* Σ, and it is illustrated in Fig. 4-14. Tables of insolation values will typically provide columns for several different tilt angles.

Profile (Shadow Line) Angles.   For centuries, it has been common practice to construct buildings in a manner that allows the winter sun to enter but effectively blocks out much of the summer sun. Various methods and construction elements, including roof overhangs, marquees, trellises, and awnings, are used for this purpose. The effect of a simple roof overhang, for example, is diagrammed in Fig. 4-18. Note that the high-solar-altitude summer sun does not enter through the window, since the shadow line falls outside the glazed surface, while the low-solar-altitude winter sun streams in under the overhang and shines into the living space.

The design of overhangs and the general problem of shadow lines affecting windows and glazed surfaces will be treated in detail in Chapter 7. At this point, however, we introduce one more important angle of solar geometry called the *profile angle* (Fig. 4-19). This angle is of critical importance when dealing with the problem of shading lines for windows and walls at particular times of the year and times of the day.

The profile angle is defined as the angle through which a horizontal plane must be rotated about a horizontal line located in the plane of a window or wall, in order for the plane being rotated to just include the position of the sun. In Fig. 4-19, *LMNO* is the (vertical) plane of the window, and line *DBE* is a horizontal line in the plane of the window. Plane *DEXY* is the horizontal plane that is to be rotated. Note that it will rotate about line *DE*. When it has been rotated sufficiently for line *XY* to just include the sun's position, *XY* will fall along *AF*, and angle *FDG* is the profile angle (sometimes called *shadow-line angle*).

Be sure to note the difference between the profile angle *FDG* and the sun's altitude angle *ABC*. These are not the same, either in meaning or in absolute value in the general case. Only when the window or wall faces due south (solar azimuth angle φ equals zero) and only at noon (12:00 LSoT) does the profile angle become equal to the sun's altitude.

Table 4-B (in the Chapter Appendix) provides values of solar position (solar altitude and solar azimuth), along with angles of incidence on *vertical surfaces* facing selected compass directions, and values of profile angles for walls

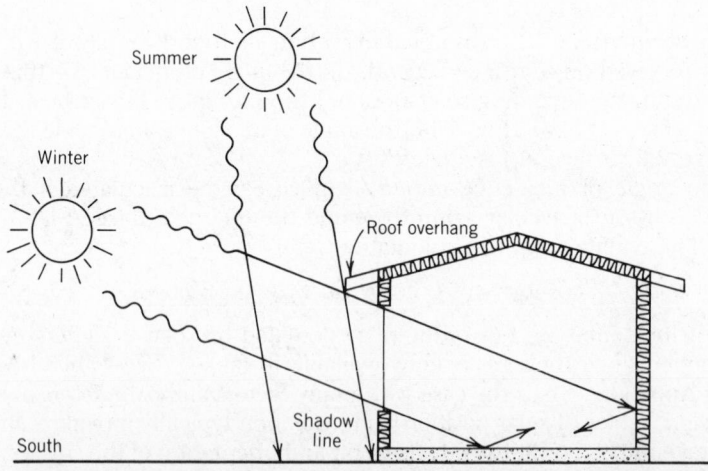

Figure 4-18 Effect of roof overhang on direct solar radiation entering a building. The low-altitude winter sun can shine under the overhang, and solar radiation enters the space nearly all day. In summer, however, the shadow line of the roof overhang is such that little, if any, solar radiation enters the space.

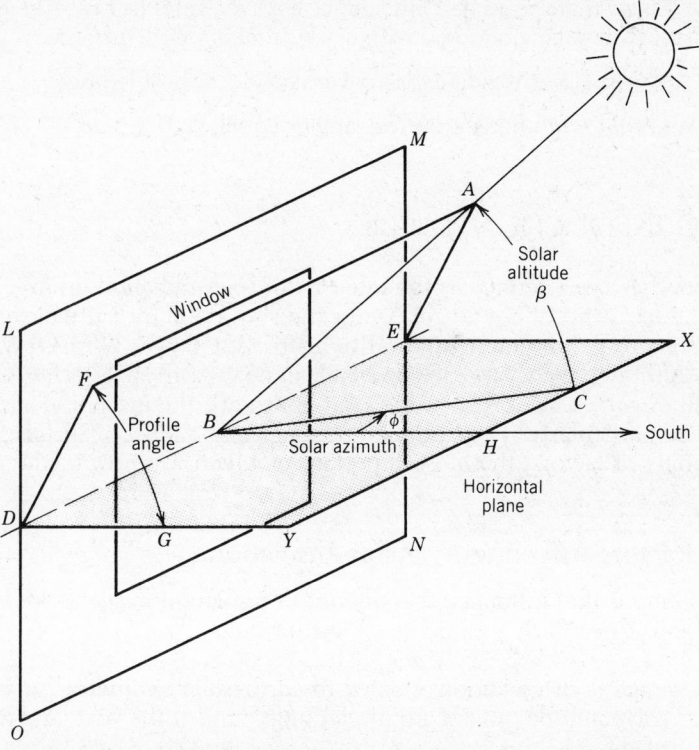

Figure 4-19 Sketch defining the *profile angle* and its relationship to solar altitude, solar azimuth, and the horizontal plane through a window.

or windows facing the same directions. Values are given for selected northern latitudes from 16°N Lat to 56°N Lat, and for each hour of daylight of the twenty-first day of each month of the year.

For example, at 32°N Lat on February 21 at 2:00 P.M. LSoT (see column at far right of table), the sun's altitude angle β is 38°, and its azimuth angle φ is 39°. The angle of incidence of direct-beam solar radiation on a vertical surface facing due south θ is 52°. For the same latitude, date, and time, the profile (shadow-line) angle for a wall or window facing due south is 46°.

Illustrative Problem 4-5

A site in Spokane, Washington, is being assessed for solar design purposes. Determine the solar position angles, the angle of incidence on a vertical surface facing due south, and the profile angle for a window facing due south, all at 12:00 noon LSoT on November 21.

Solution

From a world globe or other source, find the latitude of Spokane as approximately 48°N Lat. From the 48°N Lat section of Table 4-B, note that at 12:00 noon LSoT on November 21,

β = 22°

φ = 0° (at 12:00 noon LSoT, the sun is directly over the meridian)

Note that the profile angle is also 22°. (Recall that the sun's altitude angle and the profile angle on a wall facing due south are equal only at solar noon.)

Note also that the angle of incidence θ is 22°. Refer to Fig. 4-14 and satisfy yourself that, at solar noon and with a south-facing wall surface:

Angle of incidence = solar altitude = profile angle

Under no other conditions are these angles equal.

USING INSOLATION TABLES

Insolation has been defined as the intensity of solar radiation on unit area of a plane surface. Since it is a *rate* of energy transfer, it is a measure of solar power and it is expressed in power units—Btu/hr-ft^2, Btu/day-ft^2, W/m^2, kW/m^2, and Ly/day. Although the solar power available in near-earth space is fairly constant at 429 Btu/hr-ft^2 (1.353 kW/m^2) (the solar constant), the insolation available on earth, and particularly at a given site or on a given building or surface, depends on a variety of factors, the most important of which will now be discussed.

4-12 Factors Affecting Available Insolation

The conditions that influence the amount of insolation available at a specified site fall into three fairly well defined classifications:

1. Factors or conditions related to earth–solar geometry, such as solar altitude, solar azimuth, latitude, time of the year, and time of day.
2. Factors related to climatological and air mass conditions, such as the air mass attenuation m (section 4-2), sunny versus cloudy days, atmospheric clearness, industrial pollution, humidity, and the like.
3. Factors associated with the surface that is to receive the radiation. Is the surface facing directly south, or is it oriented in some other direction? Is it horizontal, vertical, or tilted?

Most of these factors have already been discussed, some briefly and some at considerable length. Additional comment about certain climatological factors is in order, however.

The term *clear-day values* means that the insolation values listed have been calculated with the assumption that the sun will shine brightly all day, with no cloud, fog, dust, smoke, or other impediment that would block out the sun, even for brief periods.

The term *percent possible sunshine* means the percentage of sun versus cloud cover for a given locality, determined from hourly and daily observations made over a long period of time at that location. The *average daily solar energy* delivered is obtained by multiplying the total clear-day insolation summed up for the entire day by the percent sunshine for that location. Table 4-C (in the Chapter Appendix) lists values of percent possible sunshine for selected localities in the United States.

The term *clearness factor* is related to atmospheric conditions like humidity, haze, dust, and industrial or biological contaminants—conditions that do not completely obscure the sun but appreciably reduce its radiation intensity. Clearness numbers for U.S. locations range from 1.00 (representing a good, average bright day) up to 1.15 (extremely clear atmosphere as on a mountain plateau) and down to about 0.85 (for locations with consistently high humidity and/or persistent hazy, smoky atmospheres from industrial pollution). Local conditions often govern this factor.

After applying the percent sunshine multiplier to clear-day values, the result should then be multiplied by the clearness number (if known) for the locality to obtain the net available insolation. Figure 4-20 indicates the range of clearness numbers for locations in the United States, for summer and winter,

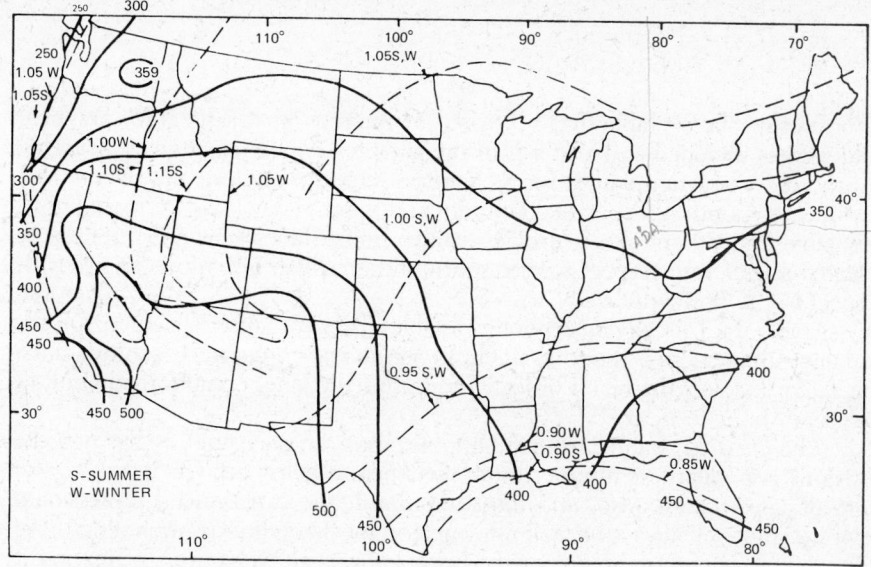

Figure 4-20 A "clearness number" map of the United States. The dashed lines are loci of equal clearness numbers for winter (W) and summer (S). Local conditions are highly important. This chart gives no indication of local haze, smog, or dust. Clear-day insolation values should be multiplied by the clearness number to get corrected values. The solid lines show the approximate average clear-day insolation in langleys per day. (Multiply Ly/day values by 3.69 to get Btu/ft²-day values). (Richard C. Jordan and Benjamin Y. H. Liu, eds. *op. cit.*, p. IV-2). Reprinted with permission.

and also shows the approximate average annual insolation for U.S. locations, in langleys per day.

Insolation on a surface is also affected markedly by the solar altitude, since at low sun angles the air mass attenuation increases geometrically (see section 4-2 and Fig. 4-7). However, since solar altitude is dictated by latitude, time of year, and time of day, all of which are the usual entry points into solar insolation tables, the air mass attenuation has already been taken into account in the calculation of tabular values of insolation.

4-13   Introduction to Insolation Tables

At this point, we shall introduce and illustrate the use of common types of insolation tables.

Table 4-D (in the Chapter Appendix) is included in order to show the variations in insolation values resulting from changes in the value of the solar altitude angle and from different conditions of the atmosphere, including sun versus cloud cover variations.

Illustrative Problem 4-7

At a day and time and latitude when the solar altitude β is 35°, find the following: (a) air mass m, (b) total insolation on a horizontal surface (clear-day value), (c) direct normal radiation for an industrial but cloudless atmosphere, and (d) total insolation received through an overcast of stratocumulus clouds.

Solution

Read values from Table 4-D, as follows:

(a) $m = 1.74$
(b) $I_h = 172$ Btu/hr-ft²

(c) $I_{DNi} = 148$ Btu/hr-ft^2
(d) $I_{oc} = 60$ Btu/hr-ft^2

A form of insolation table that is favored by many solar designers for practical field work is included as Table 4-E in the Chapter Appendix. It gives clear-day hourly total insolation values for a clearness factor of 1.00 for the twenty-first day of each month, for specified latitudes and times of day (LSoT). In addition, this table and others like it provide values for surfaces that are oriented at selected angles with respect to due south and are tilted at various angles with respect to the horizontal.

Such tables already incorporate variables like solar altitude, solar azimuth, and angle of incidence. They do not take *ground reflectance* into account, since this factor is solely the result of local (and unpredictable) conditions that often vary from hour to hour.

The clear-day values from such tables must be corrected by the percent sunshine and clearness number multipliers before being used as inputs to the solar design process. Also, and most important, the experienced professional solar designer will always be well informed on local conditions, including those that may be unique to the site itself. "Local knowledge" is a critically important input to solar energy systems design.

Illustrative Problem 4-8

A system featuring solar collectors mounted on a roof sloped at an angle of 30° with the horizontal is being designed for a Baltimore location. Because of site conditions, the roof surface is to face southwest (45°W of S). Find the hourly total insolation to be expected on a clear day at 2:00 P.M. LSoT on March 21.

Solution

Locate Baltimore at about 40°N Lat. Enter the 40°N Lat insolation table (Table 4-E) at March 21 in the PM column at 2:00 LSoT. In the 30°tilt section and in the PM/SW column, find

$$I_{\text{clear-day}} = 301 \text{ Btu/hr-ft}^2 \qquad \textit{Ans.}$$

Even on a clear day in Baltimore, however, there is some industrial pollution. The chart in Fig. 4-20 indicates an actual clearness number for Baltimore of about 0.98 (interpolating). Applying this multiplier,

$$I_{\text{net}} = 301 \times 0.98 = 295 \text{ Btu/hr-ft}^2 \qquad \textit{Ans.}$$

If it were desired now to sum up the hourly values to get a total for the day and then sum up the days for a monthly total, one would then apply the percent sunshine multiplier for Baltimore (Table 4-C gives 55 percent for Baltimore in March) to arrive at a realistic estimate of the total monthly insolation available to the solar collectors (expressed in Btu/month-ft^2).

Illustrative Problem 4-9

An active solar system for space heating is being installed at Midland, Texas. The solar collectors are to be mounted on the roof facing southwest, with a tilt angle of 52° from the horizontal. Determine the angle of incidence on the collector surfaces at solar noon on December 21. Also, determine the total insolation for clear-day conditions in Btu/hr-ft^2.

Solution

Verify the location of Midland at about 32°N Lat. To find the angle of incidence on the collectors, enter Table 4-A in the 32°N Lat section for December 21. At 12:00 noon LSoT, move across to the 52° tilt section and SW-facing column and read angle of incidence

$$\theta = 36.1° \qquad\qquad Ans.$$

To determine the insolation, enter the 32°N Lat section of Table 4-E at 12:00 noon LSoT for December 21. Move right across the table to the 52° tilt section and SW-facing column, and read

$$I_{\text{clear-day}} = 260 \text{ Btu/hr-ft}^2 \qquad\qquad Ans.$$

From Fig. 4-20, the clearness number for Midland is about 0.98 (interpolating). Applying this multiplier gives

$$I_{\text{net}} = 260 \times 0.98 = 255 \text{ Btu/hr-ft}^2 \qquad\qquad Ans.$$

Illustrative Problem 4-10

A passive solar house is being planned for a site near Traverse City, Michigan, at about 44°N Lat. The main solar collection area is to be a large vertical glazing facing due south. Find (a) the angle of incidence on the glazing surface at solar noon on November 21. Also determine, for solar noon, (b) the total hourly insolation on the glazing for clear-day conditions and (c) the hourly insolation, taking clearness factor into account.

Solution

Since a 44°N Lat table is not included, interpolation between 40°and 48°values will be necessary.
(a) Enter Table 4-A for 40°N Lat and November 21 at 12:00 solar noon. Note that, for a vertical surface (tilt angle 90°) facing south,

$$\theta_{40} = 30.2°$$

From the same table, for 48°N Lat, all conditions otherwise the same,

$$\theta_{48} = 22.2°$$

The 44° value is halfway between these two, or

$$\theta_{44} = 26.2° \qquad\qquad Ans.$$

(b) Enter Table 4-E for 40°N Lat and November 21 at 12:00 solar noon. For a vertical surface (tilt angle 90°) facing S, the clear-day hourly insolation is:

$$I_{\text{CD40}} = 258 \text{ Btu/hr-ft}^2 \qquad\qquad Ans.$$

From the same table, for 48°N Lat, all conditions otherwise the same,

$$I_{\text{CD48}} = 250 \text{ Btu/hr-ft}^2$$

The 44° value is then

$$I_{\text{CD44}} = 254 \text{ Btu/hr-ft}^2 \qquad\qquad Ans.$$

(c) Refer to Fig. 4-20 and note that for northern Michigan the clearness number is actually greater than 1.00—about 1.02.

$$I_{\text{net44}} = 1.02 \times 254 = 259 \text{ Btu/hr-ft}^2 \qquad\qquad Ans.$$

A third and very useful type of insolation table is one that lists *daily* or monthly total insolation values for south-facing surfaces, for various latitudes and selected tilt angles. The values in such tables are average values and are given for each day of the month or each month of the year. Table 4-F (in the Chapter Appendix) lists *daily total* insolation values for 32°N Lat and 40°N Lat, for selected angles of tilt. Space precludes the inclusion of tables for other latitudes.

A distinct advantage of this type of table is that it provides for a quick summation of the *total heating season* insolation Q_{IHS} and the *total cooling season* insolation Q_{ICS}. The heating season for latitudes 32°N and 40°N is usually taken as October through April, and the cooling season as May through

September. The percent possible sunshine for the locality (Table 4-C) and the clearness number should both be applied to the values from the table, since they are average clear-day values.

Q_{IHS} (total heating season insolation) is a primary design factor in balancing the *total season heating load*, Q_{HS}. The Q_{ICS} (total cooling season insolation) is of equal importance in balancing the *total season cooling load*, Q_{CS}. Both of these loads (Q_{HS} and Q_{CS}) will be treated at length in subsequent chapters.

Illustrative Problem 4-11

An office building near Little Rock, Arkansas (use the 32°N Lat insolation values), is to have a solar-energized cooling system installed. Calculate the expected total cooling season insolation Q_{ICS} from the data of Table 4-F. Assume that the solar collectors will be inclined at a tilt angle of 22° on a south-facing roof.

Solution

From the 22° tilt column of the 32°N Lat section of the table, sum up the daily values as follows, multiplying by the appropriate value of percent possible sunshine from Table 4-C.

May	2455 Btu/day-ft² × 31 days × 0.68 =	51,750 Btu/ft²
June	2435 Btu/day-ft² × 30 days × 0.73 =	53,325
July	2420 Btu/day-ft² × 31 days × 0.71 =	53,265
August	2390 Btu/day-ft² × 31 days × 0.73 =	54,085
September	2290 Btu/day-ft² × 30 days × 0.68 =	46,715
Total		259,140 Btu/ft²

Now, from Fig. 4-20, note that the summer clearness factor for Little Rock is about 0.96. Applying this factor, the expected total cooling season insolation per square foot of collector at Little Rock is

$$Q_{ICS} = 259{,}140 \text{ Btu/ft}^2 \times 0.96 = 248{,}800 \text{ Btu/ft}^2 \qquad \textit{Ans.}$$

CONCLUSION

The purpose of this chapter has been to develop a basic understanding of solar radiation and the geometry of the various sun angles. The concept of insolation on horizontal and tilted surfaces was also presented. Tables of solar position angles and of solar radiation (insolation) have been introduced, and elementary exercises in the use of these tables have been provided. Subsequent chapters, beginning with Chapter 6, will develop more advanced concepts, both in theory and practice, for collecting, storing, and distributing heat energy from solar radiation.

The next chapter makes a necessary digression to explain methods of determining the heating load for passive and active solar buildings.

PROBLEMS

1. At a location in near-earth space, what amount of solar power (kilowatts) would be intercepted by a satellite with a surface collection area of 1 acre, assuming normal incidence of solar radiation?
2. Calculate the frequency of solar radiations that are in the near-infrared region at a wavelength of 0.9 microns.
3. A table lists the daily insolation at a 40°N Lat location on a 44° tilted surface

for December as averaging 380 Ly/day. Express this value in Btu/day-ft².

4.   At a 32°N Lat location, the insolation is measured with a pyranometer as 295 Btu/hr-ft². Express this value in kW/ft². In W/m².

5.   The direct-beam radiation intensity at normal incidence on a surface is measured as 265 Btu/hr-ft². If the surface is tilted so that the angle of incidence is 40°, what is the radiation intensity?

6.   Calculate the theoretical solar power available on a flat roof of 2000 ft² area, averaged over a 12-hr daytime period, if the total daily insolation received (direct and diffuse) is 1600 Btu/day-ft². Answer in Btu/per hour and in kilowatts.

7.   It is 3:00 P.M. local daylight time on September 15th, at a location near Oklahoma City, Oklahoma, at approximately 97.2°W Long in the Central Standard Time zone. Calculate the local solar time.

8.   Near Minneapolis, Minnesota, on February 15 it is 11:00 A.M. local standard time. The location is at 92.5°W Long in the Central Standard Time zone. What is the local solar time?

9.   Near Billings, Montana, the longitude is 108.2°W, and the latitude is 46°N. Tables of sun angles are not available. Calculate the altitude angle β and the azimuth angle ϕ of the sun at 11:00 A.M. local standard time on February 28.

10.   At a location near Atlanta, Georgia, the longitude is 83.4°W, and the latitude is 33°N. It is 3:00 P.M. daylight time on July 15. Calculate the solar altitude and the solar azimuth.

11.   From the sun charts (Fig. 4-17), determine the solar altitude and solar azimuth for each of the following situations. Use a world globe to estimate latitudes, interpolating as necessary in the tables.

   (a)   Tampa, Florida, at 3:00 P.M. LSoT on September 1.
   (b)   Shreveport, Louisiana, at 10:00 A.M. LSoT on March 21.
   (c)   Las Vegas, Nevada, at 3:30 P.M. LSoT on October 21.
   (d)   Pierre, South Dakota, at 10:30 A.M. LSoT on May 21.

12.   A site near Columbus, Ohio, is being studied for a solar energy installation. From tables, determine the sun angles, the angle of incidence on a vertical surface facing due south, and the profile angle for a window facing due south, all at 11:00 A.M. LSoT on December 21.

13.   Determine the same angles indicated in problem 12 for a location near El Paso, Texas, at 3:00 P.M. LSoT on April 21.

14.   A roof for a building in Provo, Utah, has a slope of 40° with the horizontal, and it faces due south. Find the hourly total clear-day insolation to be expected at 3:00 P.M. LSoT on January 21.

15.   An office building roof near San Diego, California, has a 32° slope and faces southwest. Find the hourly total clear-day insolation at 11:00 A.M. LSoT on March 21.

16.   A residence near Peoria, Illinois, has a roof that faces due south with a 50° slope. Find (a) the hourly total clear-day insolation at 12:00 noon LSoT on February 21, and (b) the estimated monthly insolation Q_{Feb}, taking into account the clearness number and the percent possible sunshine for the locality.

17.   A university building in Reno, Nevada, is to have a solar space-heating installation. Calculate the expected total heating season insolation Q_{IHS} from the data of Table 4-F. The solar collectors will be tilted at an angle of 50° on a south-facing roof.

18.   What total cooling season insolation Q_{ICS} can be expected in Tucson, Arizona, if the solar collectors are mounted on a south-facing roof at a tilt angle of 22°? The percent possible sunshine averages 91 during the entire summer cooling season.

Table 4-A Solar Angle of Incidence on Inclined Surfaces for Selected Northern Latitudes

24°N

Solar Angle Of Incidence On Surfaces Tilted With Respect To The Horizontal

Date	Solar Time AM/PM	0°	14° E/W	14° SE/SW	14° S	14° SW/SE	14° W/E	24° E/W	24° SE/SW	24° S	24° SW/SE	24° W/E	34° E/W	34° SE/SW	34° S	34° SW/SE	34° W/E	44° E/W	44° SE/SW	44° S	44° SW/SE	44° W/E	90° E/W	90° SE/SW	90° S	90° SW/SE	90° W/E
Jan 21	7/5	85.2	72.5	72.2	79.6	90.0	90.0	63.6	62.9	75.9	90.0	90.0	54.8	53.8	72.6	90.0	90.0	46.3	45.0	69.8	90.0	90.0	24.8	21.1	65.7	90.0	90.0
	8/4	73.1	61.4	59.5	66.2	76.9	85.1	53.4	49.9	62.0	79.9	90.0	45.9	40.4	58.5	83.2	90.0	39.3	31.2	56.0	86.8	90.0	35.5	21.3	59.8	90.0	90.0
	9/3	62.1	52.1	48.1	53.5	63.9	72.9	45.8	38.2	48.4	66.2	81.0	40.6	28.2	44.5	69.2	89.1	37.0	18.3	42.2	72.8	90.0	48.4	28.1	54.4	89.5	90.0
	10/2	52.8	45.6	39.0	42.1	52.0	61.8	42.0	29.3	35.5	53.0	68.8	40.2	19.7	30.6	55.3	76.2	40.4	11.0	28.2	58.7	83.8	62.0	38.1	50.0	78.1	90.0
	11/1	46.4	43.2	34.1	33.4	41.9	52.4	43.2	26.3	24.8	40.8	57.9	44.9	20.3	17.6	41.8	64.1	48.3	18.1	14.1	44.6	70.9	75.9	49.2	47.0	66.7	90.0
	12	44.0	45.7	35.2	30.0	35.2	45.7	48.9	31.0	20.0	31.0	48.9	53.4	29.4	10.0	29.4	53.4	58.8	30.8	0.0	30.8	58.8	90.0	60.6	46.0	60.6	90.0
Feb 21	7/5	80.7	67.2	68.6	77.2	87.7	90.0	57.6	60.2	75.2	90.0	90.0	48.2	52.1	73.7	90.0	90.0	39.0	44.5	72.6	90.0	90.0	18.0	30.9	74.8	90.0	90.0
	8/4	67.7	55.0	54.9	62.9	73.5	80.7	46.1	46.0	60.5	78.2	80.7	37.7	37.5	59.0	83.2	85.4	30.1	29.8	58.5	83.6	90.0	31.5	31.0	69.0	90.0	90.0
	9/3	55.6	44.2	42.0	48.9	59.7	67.6	36.9	32.4	45.9	64.3	76.5	30.0	23.2	44.3	68.4	71.8	27.1	15.0	44.5	73.6	90.4	45.5	36.4	63.8	89.5	90.0
	10/2	44.9	36.2	30.9	35.9	46.4	55.4	32.1	20.9	31.5	49.4	63.5	30.4	10.9	29.5	53.7	58.7	31.6	1.0	30.6	58.9	80.4	60.5	45.1	59.6	88.9	90.0
	11/1	37.7	32.2	24.3	25.0	34.5	46.4	33.6	16.3	18.0	35.6	51.3	36.4	11.9	14.8	39.0	46.6	41.3	14.7	17.6	44.2	66.6	75.2	55.5	56.9	78.1	90.0
	12	34.0	36.4	25.8	20.0	25.8	36.4	40.8	23.3	10.0	23.3	40.8	46.6	24.7	0.0	24.7	46.6	53.4	29.4	10.0	29.4	53.4	90.0	66.7	56.0	66.7	90.0
Mar 21	7/5	76.3	62.4	65.6	75.2	85.3	90.0	52.5	58.4	75.0	90.0	90.0	42.6	51.7	75.2	90.0	90.0	32.7	45.9	75.9	90.0	90.0	15.0	40.8	84.0	90.0	90.0
	8/4	62.8	49.3	51.2	60.5	70.7	76.5	39.7	43.5	60.0	76.9	85.7	30.3	36.7	60.5	83.3	86.3	21.5	31.4	62.0	89.8	90.0	30.0	40.9	78.3	90.0	90.0
	9/3	49.8	37.1	37.2	45.9	56.3	62.9	28.6	28.8	45.0	62.0	72.0	21.4	21.7	45.9	68.3	72.4	17.1	17.6	48.4	75.6	81.9	45.0	45.3	73.3	89.1	90.0
	10/2	37.3	27.3	24.0	31.5	42.1	49.7	22.2	14.6	30.0	47.1	58.3	20.7	6.8	31.5	53.3	58.7	23.6	9.0	35.5	60.2	67.9	60.0	52.9	69.4	84.4	90.0
	11/1	28.1	23.2	14.6	18.0	28.5	37.5	24.3	6.5	15.0	32.4	45.5	28.8	8.4	18.0	38.3	45.5	35.5	17.3	24.8	45.5	54.1	75.0	62.6	66.9	80.7	90.0
	12	24.0	27.6	17.1	10.0	17.1	27.6	33.4	17.9	0.0	17.9	33.4	40.8	23.3	10.0	23.3	40.8	48.9	31.0	20.0	31.0	48.9	90.0	73.3	66.0	73.3	90.0
Apr 21	6/6	85.3	71.6	77.6	88.0	90.0	90.0	61.8	72.3	90.0	90.0	90.0	52.0	67.5	90.0	90.0	90.0	42.3	63.3	90.0	90.0	90.0	11.6	55.8	90.0	90.0	90.0
	7/5	71.7	57.8	63.1	73.5	82.6	85.7	47.8	57.6	75.3	83.7	85.7	37.9	53.0	76.0	83.7	90.0	28.0	49.4	80.2	90.0	90.0	18.9	52.3	90.0	90.0	90.0
	8/4	58.0	44.0	48.6	58.9	68.2	72.0	34.1	43.0	60.7	72.9	82.0	24.1	38.7	63.4	69.1	78.2	14.1	36.4	67.0	83.7	90.0	32.0	52.3	89.1	90.0	90.0
	9/3	44.4	30.6	34.0	44.2	53.7	58.3	20.8	28.3	46.2	61.2	68.2	11.6	25.2	49.7	54.5	68.2	5.6	25.5	54.4	69.1	78.2	46.2	56.0	84.4	88.1	90.0
	10/2	31.0	18.2	19.5	29.5	39.2	44.5	10.9	14.1	32.0	46.5	54.0	10.2	14.6	36.8	40.2	54.0	17.0	20.6	43.2	54.5	64.1	60.7	62.6	80.7	74.0	90.0
	11/1	18.9	11.6	5.2	14.8	24.6	31.0	14.6	5.6	18.9	31.9	40.4	22.2	15.4	26.2	26.5	40.4	31.0	25.3	34.9	40.2	50.0	75.3	71.2	78.4	59.7	90.0
	12	12.4	18.6	10.1	1.6	10.1	18.6	26.8	17.4	11.6	17.4	26.8	35.9	26.5	21.6	26.5	35.9	45.4	36.1	31.6	36.1	45.4	90.0	81.3	77.6	81.3	90.0
May 21	6/6	82.0	68.8	76.0	86.6	90.0	90.0	59.4	72.1	90.0	90.0	90.0	50.1	68.7	90.0	90.0	90.0	41.2	65.9	90.0	90.0	90.0	20.0	63.7	90.0	90.0	90.0
	7/5	68.8	52.2	62.0	72.6	80.9	82.5	45.6	58.0	75.9	89.6	82.5	36.2	55.0	79.6	84.2	90.0	27.1	53.1	83.6	90.0	90.0	24.8	60.6	90.0	90.0	90.0
	8/4	55.4	41.6	47.9	58.6	67.0	69.2	31.8	44.1	62.0	75.5	75.5	22.1	41.8	66.2	70.2	79.2	13.1	41.5	71.1	84.2	90.0	35.5	60.7	90.0	90.0	90.0
	9/3	41.7	27.8	33.9	44.5	52.9	55.7	17.8	30.6	48.4	61.4	61.4	8.1	30.2	53.1	56.4	65.7	3.3	32.6	59.5	70.2	75.7	48.4	63.9	88.9	89.1	90.0
	10/2	28.0	14.1	20.0	30.6	38.4	42.0	6.1	18.0	35.5	47.4	47.4	6.1	22.4	41.4	43.2	52.0	16.0	29.0	49.6	56.4	62.6	62.0	69.9	88.9	79.1	90.0
	11/1	14.5	3.3	7.8	16.8	24.8	28.3	10.4	13.9	24.8	33.8	33.8	20.2	22.9	33.4	31.3	38.2	30.1	32.4	42.6	43.2	48.2	75.9	77.8	86.7	72.0	90.0
	12	4.0	14.5	11.5	10.0	11.5	14.5	24.3	21.3	20.0	21.3	24.3	34.2	31.3	30.0	31.3	34.2	44.1	41.2	40.0	41.2	44.1	90.0	87.2	86.0	87.2	90.0
June 21	6/6	80.7	67.7	75.4	86.0	90.0	90.0	58.6	72.1	90.0	90.0	90.0	49.7	69.3	90.0	90.0	90.0	41.1	67.1	90.0	90.0	90.0	23.5	66.9	90.0	90.0	90.0
	7/5	67.7	54.4	61.7	72.4	86.6	81.2	45.1	58.4	76.3	89.4	81.2	35.9	56.0	80.5	84.4	87.9	27.4	54.8	85.0	90.0	90.0	27.6	64.1	90.0	90.0	90.0
	8/4	54.5	40.9	48.0	58.6	70.8	68.2	31.4	44.9	62.7	75.4	75.4	22.2	43.5	67.5	70.8	74.8	14.1	43.8	72.8	84.4	90.0	37.4	64.1	90.0	90.0	90.0
	9/3	41.0	27.3	34.3	44.9	52.8	54.9	17.6	32.1	49.6	61.7	61.7	8.8	32.6	55.3	55.3	64.8	6.6	35.7	61.7	70.8	74.8	49.6	67.2	88.9	89.6	90.0
	10/2	27.4	13.5	21.1	31.5	39.1	41.4	6.6	20.3	37.4	48.1	48.1	7.1	25.7	44.5	44.8	48.1	16.8	32.4	52.4	54.5	61.4	62.7	72.8	86.6	79.9	90.0
	11/1	13.7	0.3	10.7	19.7	25.7	27.7	10.3	17.3	27.6	35.1	35.1	20.3	26.1	36.5	33.6	47.7	30.3	35.5	45.9	45.9	47.7	76.3	80.5	86.0	74.7	90.0
	12	0.6	14.0	13.6	13.4	13.6	14.0	24.0	23.6	23.4	23.6	24.0	34.0	33.6	33.4	33.6	34.0	44.0	43.6	43.4	43.6	44.0	90.0	89.6	89.4	89.6	90.0

(Based on data in Table 1, pp. 387, 1972 ASHRAE HANDBOOK OF FUNDAMENTALS; 0% ground reflectance, 1.0 clearness factor)

Table 4-A Solar Angle of Incidence on Inclined Surfaces for Selected Northern Latitudes

24°N

Solar Angle of Incidence On Surfaces Tilted With Respect To The Horizontal

Date	AM	PM	0°	14° E/W	14° SE/SW	14° S	14° SW/SE	14° W/E	24° E/W	24° SE/SW	24° S	24° SW/SE	24° W/E	34° E/W	34° SE/SW	34° S	34° SW/SE	34° W/E	44° E/W	44° SE/SW	44° S	44° SW/SE	44° W/E	90° E/W	90° SE/SW	90° S	90° SW/SE	90° W/E
July 21	6	6	81.8	68.6	75.9	86.5	90.0	90.0	59.2	72.1	90.0	90.0	90.0	50.1	68.8	90.0	90.0	90.0	41.1	66.1	90.0	90.0	90.0	20.6	64.2	90.0	90.0	90.0
	7	5	68.6	55.1	62.0	72.6	80.7	82.0	45.5	58.1	76.0	89.5	90.0	36.1	55.1	79.8	89.5	90.0	27.1	53.4	83.8	90.0	90.0	25.3	61.2	90.0	90.0	90.0
	8	4	55.2	41.4	47.9	58.5	66.9	69.0	31.7	44.2	62.1	75.5	88.9	22.1	42.1	66.4	84.2	79.0	13.2	41.9	71.4	90.0	90.0	35.8	61.2	90.0	90.0	90.0
	9	3	41.6	27.7	34.0	44.5	52.9	55.6	17.8	30.9	48.6	61.5	75.5	8.1	30.6	53.8	70.3	65.5	3.9	33.1	59.9	79.2	85.5	48.6	64.3	90.0	90.0	90.0
	10	2	27.9	13.9	20.1	30.7	38.8	41.9	3.9	19.2	35.8	47.5	65.5	6.1	22.9	42.5	56.6	51.9	16.1	29.6	50.1	65.8	71.9	62.1	70.2	89.5	90.0	90.0
	11	1	14.3	2.7	8.2	17.9	24.9	28.2	10.3	14.5	25.3	34.0	51.9	20.1	23.4	34.0	43.5	38.1	30.1	33.0	43.2	53.1	58.1	76.0	78.2	87.3	89.5	90.0
	12		3.4	14.4	11.8	10.6	11.8	14.4	24.2	21.7	20.6	21.7	24.2	34.1	31.7	30.6	31.7	34.1	44.1	41.7	40.6	41.7	44.1	90.0	87.6	86.6	87.6	90.0
Aug 21	6	6	85.0	71.3	77.4	87.9	90.0	90.0	61.5	72.3	90.0	90.0	90.0	51.8	67.6	90.0	90.0	90.0	42.2	63.5	90.0	90.0	90.0	12.3	56.4	90.0	90.0	90.0
	7	5	71.5	57.6	63.0	73.4	82.5	85.4	47.6	57.6	75.4	85.4	90.0	37.7	53.1	77.8	90.0	90.0	27.9	49.7	80.5	89.8	90.0	19.3	53.0	90.0	90.0	90.0
	8	4	57.8	43.8	48.5	58.8	68.1	71.8	33.8	43.0	60.8	75.8	81.8	23.8	38.9	63.6	83.7	81.8	13.8	36.8	67.3	90.0	90.0	32.2	56.7	89.8	90.0	90.0
	9	3	44.1	30.3	34.0	44.2	53.6	58.0	20.5	28.4	46.3	61.2	68.0	11.1	25.5	50.0	69.1	68.0	5.0	26.1	54.8	77.3	87.9	46.3	56.2	85.0	90.0	89.1
	10	2	30.7	17.8	19.4	29.5	39.1	44.3	10.3	14.3	32.2	46.5	54.1	9.6	15.2	37.2	54.6	54.1	16.7	21.3	43.8	63.1	73.8	60.8	71.8	81.4	89.1	77.8
	11	1	18.4	10.9	5.0	14.9	24.5	30.7	14.2	6.3	19.3	32.0	40.2	21.9	16.0	26.8	40.4	40.2	30.9	25.9	35.5	49.4	59.5	75.4	71.8	79.1	81.8	73.8
	12		11.7	18.2	10.0	2.3	10.0	18.2	26.5	17.7	12.3	17.7	26.5	35.7	26.9	22.3	26.9	35.7	45.2	36.5	32.3	36.5	45.2	90.0	81.8	78.3	81.8	90.0
Sept 21	7	5	76.3	62.4	65.6	75.2	85.3	90.0	52.5	58.4	75.0	90.0	90.0	42.6	51.7	75.2	90.0	90.0	32.7	45.9	75.9	90.0	90.0	15.0	40.8	84.0	90.0	90.0
	8	4	62.8	49.3	51.2	60.5	70.7	76.5	39.7	43.5	60.0	76.9	86.3	30.3	36.7	60.5	83.3	86.3	21.5	31.4	62.0	89.8	90.0	30.0	40.9	78.3	90.0	90.0
	9	3	49.8	37.1	37.2	45.9	56.3	62.9	28.6	28.8	45.0	62.0	72.4	21.4	17.6	45.5	68.5	72.4	17.1	9.0	48.4	75.0	81.9	45.0	45.3	73.3	90.0	81.9
	10	2	37.7	27.3	24.0	31.5	42.1	49.7	22.2	14.6	30.0	47.1	58.7	23.6	8.4	31.5	53.3	58.7	23.6	17.3	35.5	60.2	67.2	60.0	52.9	69.4	84.6	77.2
	11	1	28.1	23.2	14.6	18.0	28.5	37.5	23.3	6.5	15.0	32.4	45.2	28.8	8.4	18.0	38.3	45.2	35.5	24.8	24.8	45.5	62.9	75.0	62.6	66.9	73.3	62.9
	12		24.0	27.6	17.1	10.0	17.1	27.6	33.4	17.9	0.0	17.9	33.4	48.9	23.3	10.0	23.3	48.9	48.9	31.0	20.0	31.0	48.9	90.0	73.3	66.0	73.3	90.0
Oct 21	7	5	80.9	67.3	68.8	77.4	87.9	90.0	57.9	60.3	75.3	90.0	90.0	48.5	52.2	73.6	90.0	90.0	39.3	44.5	72.5	90.0	90.0	18.2	30.4	74.3	90.0	90.0
	8	4	68.0	55.3	55.1	63.1	73.7	80.9	46.5	46.1	60.6	78.3	88.2	38.1	37.6	58.9	83.2	88.2	30.5	29.8	58.4	88.2	90.0	31.6	36.0	68.5	86.3	90.0
	9	3	55.9	44.6	41.3	49.2	59.8	67.0	37.3	32.7	46.0	63.8	73.6	31.4	23.4	44.2	68.4	76.7	27.6	15.0	44.3	73.6	80.5	46.0	44.7	63.3	72.6	80.5
	10	2	45.3	36.7	31.3	36.2	46.6	55.7	32.6	21.3	31.6	49.5	59.0	30.9	11.3	29.5	53.7	63.7	32.0	1.5	30.4	58.9	66.8	60.5	55.2	59.1	66.8	66.8
	11	1	37.5	33.7	24.8	25.4	34.8	44.9	34.1	16.8	18.2	35.8	49.5	36.9	12.3	14.8	39.1	51.6	41.6	14.7	17.3	44.2	53.6	75.3	55.2	56.4	59.1	53.6
	12		34.5	36.9	26.3	20.5	26.3	36.9	41.2	23.7	10.5	23.7	46.9	46.9	24.9	0.5	24.9	41.2	53.6	29.4	9.5	29.4	53.6	90.0	66.4	55.5	66.4	90.0
Nov 21	7	5	85.1	72.4	72.1	79.6	90.0	90.0	63.4	62.9	75.9	90.0	90.0	54.7	53.8	72.6	90.0	90.0	46.2	44.9	69.8	90.0	90.0	24.7	21.3	65.9	90.0	90.0
	8	4	73.0	61.3	59.4	66.1	76.7	85.1	52.0	49.8	61.9	79.8	90.0	45.8	40.3	58.5	86.8	90.0	39.1	31.1	56.1	86.8	90.0	35.4	21.5	59.9	86.3	90.0
	9	3	62.0	51.9	48.0	53.4	63.8	72.8	45.6	38.0	48.3	66.1	80.9	40.4	27.5	44.4	72.8	80.9	36.3	18.2	42.2	72.8	89.0	48.3	28.3	54.5	72.6	89.1
	10	2	52.7	45.4	38.9	42.0	51.7	61.7	41.8	29.1	35.4	52.9	68.7	40.1	19.6	30.5	58.7	68.4	40.3	10.8	28.2	58.7	76.1	61.9	38.2	50.1	58.7	83.1
	11	1	46.2	43.0	33.9	33.1	41.7	52.2	43.0	26.1	24.7	40.7	57.8	44.7	20.1	17.5	44.6	57.7	48.7	18.0	14.1	44.6	64.0	75.9	49.3	47.2	49.3	72.1
	12		43.8	45.5	35.0	29.8	35.0	45.5	48.7	30.9	19.8	30.9	48.7	53.2	29.3	9.8	30.8	53.2	58.7	30.8	0.2	30.8	58.7	90.0	60.7	46.2	60.7	90.0
Dec 21	7	5	86.8	74.5	73.5	80.5	90.0	90.0	65.7	64.1	76.3	90.0	90.0	57.2	54.7	72.4	90.0	90.0	49.1	45.3	68.9	90.0	90.0	27.6	17.8	62.6	90.0	90.0
	8	4	75.1	63.8	61.3	67.5	78.0	86.7	56.1	51.6	62.7	80.5	90.0	48.9	41.9	58.6	83.3	90.0	42.5	32.3	55.4	86.3	90.0	37.4	18.1	56.6	86.3	90.0
	9	3	64.5	54.9	50.5	55.5	65.5	74.9	48.9	40.5	49.6	67.5	82.5	44.0	30.5	44.9	69.6	82.5	40.6	20.5	41.8	72.6	85.0	49.6	25.5	51.1	72.6	85.0
	10	2	55.7	48.8	42.0	44.5	54.1	64.1	45.4	32.4	37.4	54.2	70.8	43.6	23.5	31.6	56.1	70.8	43.6	14.5	28.0	58.8	72.5	62.7	35.9	46.6	58.8	80.7
	11	1	49.6	46.7	37.3	36.4	44.6	55.2	46.5	29.8	27.6	43.0	60.3	48.0	23.5	19.6	43.2	60.3	50.9	20.3	14.3	45.2	66.2	76.3	47.2	43.6	47.2	69.9
	12		47.4	49.0	38.5	33.4	38.5	49.0	51.8	33.9	23.5	33.9	55.9	55.9	31.6	13.5	31.6	55.9	60.9	32.0	3.4	32.0	60.9	90.0	58.6	42.5	58.6	90.0

(Based on data in Table 1, pp. 387, 1972 ASHRAE HANDBOOK OF FUNDAMENTALS; 0% ground reflectance, 1.0 clearness factor)

Table 4-A   Solar Angle of Incidence on Inclined Surfaces for Selected Northern Latitudes

32°N

Solar Angle Of Incidence On Surfaces Tilted With Respect To The Horizontal

Angle of incidence (degrees). For each tilt, orientation sub‑columns are: E/W, SE/SW, S/S, SW/SE, W/E (AM value = E, SE, S, SW, W; PM value = W, SW, S, SE, E).

Date	AM	PM	0°	22° E/W	22° SE/SW	22° S	22° SW/SE	22° W/E	32° E/W	32° SE/SW	32° S	32° SW/SE	32° W/E	42° E/W	42° SE/SW	42° S	42° SW/SE	42° W/E	52° E/W	52° SE/SW	52° S	52° SW/SE	52° W/E	90° E/W	90° SE/SW	90° S	90° SW/SE	90° W/E
Jan. 21	7	5	88.6	68.7	68.0	79.6	90.0	90.0	59.9	58.8	75.9	90.0	90.0	51.2	49.7	72.6	90.0	90.0	43.1	41.0	69.8	90.0	90.0	24.8	20.3	65.2	90.0	90.0
	8	4	77.5	59.6	56.0	66.2	82.6	90.0	52.0	46.3	62.6	85.3	90.0	45.1	36.7	58.5	88.2	90.0	39.2	27.4	56.0	90.0	90.0	35.5	16.9	57.4	90.0	90.0
	9	3	67.5	52.9	45.5	53.5	69.6	83.9	47.4	35.5	48.4	71.6	90.0	43.2	25.5	44.5	74.1	90.0	42.9	15.5	42.2	77.1	90.0	48.4	22.5	50.0	79.8	90.0
	10	2	59.4	49.6	38.1	42.1	57.4	72.8	47.1	28.5	35.5	58.3	79.5	46.2	19.6	30.6	60.2	86.3	46.9	12.3	28.2	63.0	86.3	62.0	32.6	43.8	68.1	90.0
	11	1	53.9	50.4	35.5	33.4	46.7	63.0	51.1	28.5	24.8	45.8	68.3	53.1	23.5	17.6	46.6	74.1	56.4	22.0	14.1	49.0	80.2	75.9	44.2	39.6	56.1	90.0
	12		52.0	55.2	38.8	30.0	38.8	55.2	58.5	35.2	20.0	35.2	58.5	62.8	33.9	10.0	33.9	62.8	67.7	35.1	0.0	35.1	67.7	90.0	56.1	38.0	56.1	90.0
Feb. 21	7	5	82.9	61.9	63.8	77.2	90.0	90.0	52.5	55.4	75.2	90.0	90.0	43.2	47.5	73.7	90.0	90.0	34.3	40.2	72.6	90.0	90.0	18.0	29.3	73.6	90.0	90.0
	8	4	71.0	51.6	50.5	62.9	79.4	90.0	43.3	41.5	60.5	83.7	90.0	35.6	33.0	59.0	88.2	90.0	29.2	25.4	58.5	90.0	90.0	31.5	26.9	65.9	90.0	90.0
	9	3	60.1	43.7	38.4	49.0	65.5	78.4	37.2	28.7	45.9	69.2	86.9	33.2	19.2	44.3	73.4	90.0	31.2	10.7	44.5	78.1	90.0	45.9	31.0	58.9	85.7	90.0
	10	2	50.9	39.7	29.0	35.9	52.2	66.4	37.2	19.2	31.5	54.9	74.1	37.0	9.7	29.5	58.7	82.0	39.0	4.5	30.6	63.4	90.0	60.5	39.5	53.2	73.7	90.0
	11	1	44.4	40.7	25.5	25.0	39.9	55.4	42.2	19.0	18.0	41.0	61.9	45.0	16.3	14.8	44.1	68.9	50.1	19.1	17.6	48.6	76.2	75.2	50.1	49.4	61.8	90.0
	12		42.0	46.4	30.0	20.0	30.0	46.4	50.9	28.2	10.0	28.2	50.9	56.5	29.7	0.0	29.7	56.5	62.8	33.9	10.0	33.9	62.8	90.0	61.8	48.0	61.8	90.0
Mar. 21	7	5	77.3	55.6	60.3	75.2	90.0	90.0	45.7	53.2	75.0	90.0	90.0	36.0	46.8	75.2	90.0	90.0	26.3	41.4	75.9	90.0	90.0	15.0	38.7	82.1	90.0	90.0
	8	4	64.9	44.1	46.2	60.5	76.5	86.1	35.1	38.4	60.0	82.3	90.0	26.5	31.8	60.5	83.2	90.0	19.4	27.0	62.0	79.4	90.0	30.0	36.9	74.6	90.0	90.0
	9	3	53.2	34.8	32.6	45.9	62.1	73.1	28.0	23.9	45.0	67.4	82.3	23.3	16.8	45.5	69.0	87.0	22.1	13.6	48.4	64.6	86.7	45.0	40.1	68.0	90.0	90.0
	10	2	42.7	29.7	20.8	31.5	47.9	60.4	27.4	10.8	30.0	52.6	69.0	28.3	1.8	31.0	54.4	73.4	32.3	9.4	35.5	49.8	72.5	60.0	47.3	62.7	90.0	90.0
	11	1	35.0	31.1	15.5	18.0	34.3	48.5	33.7	10.4	15.0	37.9	56.1	38.6	13.2	17.0	39.7	59.6	44.9	21.0	24.8	35.2	58.5	75.0	57.0	59.2	79.7	90.0
	12		32.0	38.2	22.1	10.0	22.1	38.2	44.0	23.4	0.0	23.4	44.0	50.9	28.2	10.0	28.2	50.9	58.5	35.2	20.0	35.2	58.5	90.0	68.0	58.0	68.0	90.0
Apr. 21	6	6	83.9	62.2	71.8	88.0	90.0	90.0	52.4	66.8	90.0	90.0	90.0	42.7	62.5	90.0	90.0	90.0	33.1	58.9	90.0	90.0	90.0	11.6	55.1	90.0	90.0	90.0
	7	5	71.2	49.3	57.4	73.5	87.8	90.0	39.3	52.1	75.3	90.0	90.0	29.3	48.0	77.6	90.0	90.0	19.3	45.3	80.2	90.0	90.0	18.9	49.9	84.9	90.0	90.0
	8	4	58.5	36.7	42.9	58.9	71.9	80.4	26.9	37.5	60.7	80.9	87.0	17.8	33.8	63.4	88.4	90.0	8.2	32.5	67.0	81.3	90.0	32.0	48.5	78.7	90.0	90.0
	9	3	46.1	24.5	28.5	44.2	59.2	67.4	17.1	22.8	46.2	66.3	73.7	8.2	20.5	49.7	73.7	77.2	13.3	22.7	54.4	67.0	83.0	46.2	51.0	73.8	90.0	90.0
	10	2	34.3	18.7	14.2	29.5	44.8	54.4	16.3	8.7	32.0	51.6	63.8	19.7	12.2	36.8	59.1	63.8	26.6	20.5	43.2	52.9	68.9	60.7	57.1	70.7	86.9	90.0
	11	1	24.6	20.1	4.0	14.8	30.3	41.6	25.1	8.2	18.9	36.9	50.5	32.3	17.8	26.2	44.6	50.5	40.6	27.7	34.9	39.5	54.8	75.3	65.6	69.6	75.7	90.0
	12		20.4	29.7	16.0	1.6	16.0	29.7	37.4	22.3	11.6	22.3	37.4	45.9	30.5	21.6	30.5	45.9	54.8	39.5	31.6	39.5	54.8	90.0	75.7	69.6	75.7	90.0
May 21	6	6	79.6	58.7	70.1	86.6	90.0	90.0	49.3	66.6	90.0	90.0	90.0	40.2	63.8	90.0	90.0	90.0	31.5	61.7	90.0	90.0	90.0	20.0	62.7	90.0	90.0	90.0
	7	5	67.2	45.6	56.1	72.6	85.6	90.0	35.9	52.5	75.9	88.5	90.0	26.5	50.1	79.6	90.0	90.0	17.8	49.1	83.6	90.0	90.0	24.8	58.1	90.0	90.0	90.0
	8	4	54.6	32.6	42.0	58.5	71.9	76.5	22.7	38.6	62.0	74.4	86.5	12.8	37.2	66.2	80.1	90.0	3.5	38.1	71.1	82.9	90.0	38.5	56.9	86.4	90.0	90.0
	9	3	41.9	20.3	28.0	44.5	57.9	63.8	10.4	25.3	48.4	60.5	73.8	3.6	26.2	53.5	66.1	83.8	10.9	30.3	59.6	69.3	80.4	52.4	59.1	81.9	90.0	90.0
	10	2	29.4	10.3	14.1	30.6	43.8	50.8	8.9	14.3	35.0	47.0	60.7	15.9	20.5	42.1	52.0	70.5	24.9	28.7	49.6	56.2	66.8	64.4	64.4	79.0	86.4	90.0
	11	1	18.0	13.4	4.6	17.6	29.8	37.8	20.7	14.3	24.8	34.4	47.3	29.6	24.2	33.4	38.1	57.0	39.0	34.2	42.6	44.1	53.0	79.0	72.1	78.0	81.5	90.0
	12		12.0	24.9	15.8	10.0	15.8	24.9	34.0	24.8	20.0	24.8	34.0	43.4	34.4	30.0	34.4	43.4	53.0	44.1	40.0	44.1	53.0	90.0	81.5	78.0	81.5	90.0
June 21	6	6	77.8	57.4	69.6	86.0	90.0	90.0	48.3	66.7	90.0	90.0	90.0	39.6	64.5	90.0	90.0	90.0	31.5	63.1	90.0	90.0	90.0	23.4	65.8	90.0	90.0	90.0
	7	5	65.7	44.4	55.9	72.4	84.8	90.0	35.0	53.0	76.3	88.6	90.0	25.9	51.3	80.5	90.0	90.0	17.8	51.0	85.0	90.0	90.0	27.6	61.5	90.0	90.0	90.0
	8	4	53.1	31.4	42.1	58.6	71.3	75.0	21.6	39.0	62.3	74.8	85.0	12.2	39.0	67.5	79.9	90.0	5.5	40.6	72.8	83.7	90.0	37.4	60.3	89.6	90.0	90.0
	9	3	40.4	18.4	28.5	44.9	57.6	62.4	8.5	27.0	49.6	61.3	72.4	1.6	28.9	55.3	66.1	82.4	11.6	33.5	61.7	70.4	79.4	49.6	62.4	85.2	90.0	90.0
	10	2	27.8	7.2	15.0	31.6	43.9	49.6	6.5	17.6	37.4	48.3	59.5	15.3	23.9	44.5	52.4	69.5	25.0	32.1	52.4	57.7	66.1	62.7	67.5	82.4	89.6	90.0
	11	1	15.8	11.1	8.0	19.6	30.2	36.5	19.6	17.3	27.6	35.0	46.3	29.1	27.1	36.5	39.0	56.2	38.8	37.0	45.8	46.2	52.5	76.3	74.9	81.4	84.0	90.0
	12		8.6	23.5	17.0	13.4	17.0	23.5	33.0	26.6	23.4	26.6	33.0	42.7	36.4	33.4	36.4	42.7	52.5	46.2	43.4	46.2	52.5	90.0	84.0	81.4	84.0	90.0

(Based on data in Table 1, pp. 387, 1972 ASHRAE HANDBOOK OF FUNDAMENTALS; 0% ground reflectance, 1.0 clearness factor)

Table 4-A Solar Angle of Incidence on Inclined Surfaces for Selected Northern Latitudes

32°N

Solar Angle Of Incidence On Surfaces Tilted With Respect To The Horizontal

Date	Solar Time AM / PM	0°	22° E/W	22° SE/SW	22° S/S	22° SW/SE	22° W/E	32° E/W	32° SE/SW	32° S/S	32° SW/SE	32° W/E	42° E/W	42° SE/SW	42° S/S	42° SW/SE	42° W/E	52° E/W	52° SE/SW	52° S/S	52° SW/SE	52° W/E	90° E/W	90° SE/SW	90° S/S	90° SW/SE	90° W/E
Jul. 21	6 / 6	79.3	58.4	70.0	86.5	90.0	90.0	49.1	66.6	90.0	90.0	90.0	40.1	63.9	90.0	90.0	90.0	31.5	62.0	90.0	90.0	90.0	20.6	63.2	90.0	90.0	90.0
	7 / 5	66.9	45.4	56.1	72.6	85.5	88.6	35.8	52.6	76.0	90.0	88.6	26.3	50.3	79.8	90.0	90.0	17.5	49.4	83.8	90.0	90.0	25.3	58.7	90.0	90.0	90.0
	8 / 4	54.3	32.4	42.1	58.5	71.8	76.3	22.4	38.8	62.1	80.1	76.3	12.6	37.5	66.4	88.5	86.3	3.7	38.5	71.4	90.0	90.0	35.8	57.5	87.0	90.0	90.0
	9 / 3	41.6	19.7	28.1	44.5	57.9	59.6	10.0	25.6	48.6	66.1	60.5	13.0	26.7	53.8	60.7	83.5	10.9	29.3	59.9	83.1	83.5	48.6	59.7	82.5	87.3	90.0
	10 / 2	29.1	9.7	14.3	30.7	43.9	50.6	8.5	15.1	35.8	52.1	50.6	15.7	21.1	42.5	47.2	70.3	24.9	34.7	50.1	69.5	80.3	62.1	64.9	82.5	79.3	90.0
	11 / 1	17.6	12.9	5.2	17.9	29.8	37.5	20.5	14.8	25.3	38.2	37.5	29.2	24.7	34.0	34.7	56.9	38.9	34.7	43.2	56.5	66.7	76.0	72.6	79.6	71.4	90.0
	12	11.4	24.6	16.0	10.6	16.0	24.6	33.8	25.1	20.6	25.1	24.6	43.2	34.7	30.6	34.7	43.2	52.9	44.5	40.6	44.5	52.9	90.0	82.0	78.6	82.0	90.0
Aug. 21	6 / 6	83.5	61.9	71.6	87.9	90.0	90.0	52.2	66.8	90.0	90.0	90.0	42.5	62.6	90.0	90.0	90.0	32.9	59.1	90.0	90.0	90.0	12.3	55.7	90.0	90.0	90.0
	7 / 5	70.9	48.9	57.2	73.4	87.6	90.0	38.9	52.1	75.4	90.0	90.0	29.0	48.1	77.8	90.0	90.0	19.1	45.5	80.5	90.0	90.0	19.3	50.6	90.0	90.0	90.0
	8 / 4	58.2	36.3	42.8	58.8	73.4	80.1	26.4	37.5	60.8	80.8	80.1	16.7	34.0	63.6	88.4	86.7	7.5	32.9	67.3	90.0	90.0	32.2	51.7	85.5	90.0	90.0
	9 / 3	45.7	24.9	28.4	44.2	59.1	67.1	16.5	29.0	46.3	66.2	67.1	11.3	20.9	50.0	59.2	76.9	12.9	23.3	54.8	81.5	82.7	46.3	51.7	79.3	81.5	90.0
	10 / 2	33.9	17.6	14.0	29.5	44.6	54.1	15.6	9.0	32.2	51.6	54.1	19.3	12.8	37.2	59.2	59.2	26.3	21.1	43.8	67.2	82.7	60.8	57.7	74.5	67.2	90.0
	11 / 1	24.0	19.7	3.3	14.9	30.2	41.2	24.7	8.5	19.3	36.9	41.2	32.0	18.3	26.8	44.8	59.2	40.4	28.2	35.5	53.2	68.7	75.4	66.1	71.4	53.2	90.0
	12	19.7	29.2	15.8	2.3	15.8	29.2	37.0	22.4	12.3	22.4	29.2	45.6	30.8	22.3	30.8	45.6	54.6	39.9	32.3	39.9	54.6	90.0	76.2	70.3	76.2	90.0
Sept. 21	7 / 5	77.3	55.6	60.3	75.2	90.0	90.0	45.7	53.2	75.0	90.0	90.0	36.0	46.8	75.2	90.0	90.0	26.3	41.4	75.9	90.0	90.0	15.0	38.7	82.1	90.0	90.0
	8 / 4	64.9	44.1	46.2	60.5	76.5	86.1	35.1	38.4	60.0	82.3	86.1	26.5	31.8	59.0	88.2	90.0	19.4	27.0	62.0	88.2	90.0	30.0	36.9	74.6	90.0	90.0
	9 / 3	53.2	34.8	32.6	45.9	62.1	73.1	28.0	23.9	45.0	67.4	73.1	23.3	1.8	45.9	73.2	86.7	22.1	13.6	48.4	79.4	86.7	45.0	40.1	68.0	79.4	90.0
	10 / 2	42.7	29.7	20.8	31.5	47.9	60.4	27.4	10.4	30.0	52.6	60.4	28.6	13.2	31.5	58.2	77.8	32.2	9.4	35.5	64.6	86.7	60.0	47.3	62.7	64.6	90.0
	11 / 1	35.0	31.1	15.5	18.0	34.3	48.5	33.7	10.4	15.0	37.9	48.5	38.6	13.2	18.0	43.2	64.2	44.9	21.0	24.8	49.8	72.5	75.0	57.0	59.2	49.8	90.0
	12	32.0	33.2	22.1	10.0	22.1	38.2	44.0	23.4	90.0	23.4	38.2	50.9	28.2	10.0	28.2	50.9	58.5	35.2	20.0	35.2	58.5	90.0	68.0	58.0	68.0	90.0
Oct. 21	7 / 5	83.2	62.2	64.0	77.4	90.0	90.0	52.8	55.6	75.3	90.0	90.0	43.6	47.6	73.6	90.0	90.0	34.7	40.2	72.5	90.0	90.0	18.2	28.8	73.2	90.0	90.0
	8 / 4	71.3	52.0	50.8	63.1	79.5	90.0	43.7	41.7	60.6	83.8	90.0	36.1	33.1	58.9	88.2	90.0	29.7	25.4	58.4	78.0	90.0	31.6	26.4	65.5	85.4	90.0
	9 / 3	60.5	44.1	38.8	49.2	65.7	78.7	38.1	29.0	46.0	69.3	78.7	33.7	19.5	44.2	73.4	87.2	31.6	4.7	44.3	63.3	87.2	46.0	30.5	58.5	63.3	90.0
	10 / 2	51.3	41.2	29.5	36.2	52.4	66.7	37.7	19.7	31.6	55.0	66.7	37.5	10.2	29.5	58.7	74.3	39.4	19.1	30.4	48.6	76.4	60.6	39.1	52.7	48.6	90.0
	11 / 1	44.9	41.2	26.0	25.4	40.2	55.8	42.6	19.4	18.2	41.2	55.8	45.8	16.6	14.8	44.1	62.2	51.1	19.1	17.3	48.6	76.4	75.3	49.8	48.9	73.4	90.0
	12	42.5	46.9	30.4	20.5	30.4	46.9	51.3	28.6	10.5	28.6	46.9	56.8	29.8	0.5	29.8	56.8	63.0	33.9	9.5	33.9	63.0	90.0	61.5	47.5	61.5	90.0
Nov. 21	7 / 5	88.5	68.6	67.9	79.6	90.0	90.0	59.7	58.7	75.9	90.0	90.0	51.1	49.7	72.6	90.0	90.0	42.9	41.0	69.8	90.0	90.0	24.7	20.4	65.4	90.0	90.0
	8 / 4	77.3	59.4	55.8	66.1	82.5	90.0	51.8	46.2	61.9	85.3	90.0	44.9	36.6	58.5	88.2	90.0	39.0	27.4	56.1	77.1	90.0	35.4	17.1	57.6	88.2	90.0
	9 / 3	67.4	52.7	45.4	53.4	69.5	83.8	47.2	35.4	48.3	71.5	83.8	43.0	25.4	44.4	74.1	86.3	40.4	12.1	42.2	63.6	86.3	48.3	22.7	50.2	63.6	90.0
	10 / 2	59.2	49.4	37.9	42.0	57.3	72.7	46.9	28.3	35.4	58.2	72.7	46.7	19.4	30.5	60.1	74.4	46.7	15.6	28.2	48.9	80.1	61.9	32.7	44.0	48.9	90.0
	11 / 1	53.8	50.3	35.3	33.3	46.6	62.8	50.9	23.3	24.7	45.7	62.8	53.0	23.1	17.5	46.5	68.1	56.2	21.9	14.1	48.9	80.1	75.9	44.3	39.7	68.2	90.0
	12	51.8	55.0	38.6	29.8	38.6	55.0	58.4	35.0	19.8	35.0	55.0	62.6	33.8	9.8	33.8	62.6	67.6	35.1	0.2	35.1	67.6	90.0	56.2	38.2	56.2	90.0
Dec. 21	8 / 4	79.7	62.4	58.0	67.5	83.8	90.0	55.1	48.2	62.7	85.9	90.0	48.4	38.5	58.6	88.2	90.0	42.6	28.9	55.4	76.9	90.0	37.4	13.5	54.5	88.7	90.0
	9 / 3	70.2	56.1	48.2	55.3	71.1	85.9	50.8	38.2	49.6	72.5	85.9	46.7	28.2	44.9	74.5	87.9	43.9	18.2	41.8	76.9	90.0	49.6	19.9	47.1	77.8	90.0
	10 / 2	62.4	53.1	41.3	44.5	59.4	75.1	50.5	31.8	37.4	59.7	75.1	49.4	23.1	31.6	60.9	81.4	49.7	23.1	28.0	63.1	81.6	62.7	30.6	40.7	44.0	90.0
	11 / 1	57.3	53.8	38.9	36.5	49.3	65.7	54.2	31.8	27.6	47.8	65.7	55.9	26.4	19.6	47.8	75.9	58.7	23.8	14.3	49.4	81.6	76.3	42.4	36.2	66.2	90.0
	12	55.4	58.3	41.9	33.4	41.9	58.3	61.3	37.9	23.4	37.9	58.3	65.1	35.8	13.5	35.8	65.1	69.6	36.1	3.5	36.1	69.6	90.0	54.4	34.5	54.4	90.0

(Based on data in Table 1, pp. 387, 1972 ASHRAE HANDBOOK OF FUNDAMENTALS; 0% ground reflectance, 1.0 clearness factor)

Table 4-A   Solar Angle of Incidence on Inclined Surfaces for Selected Northern Latitudes

40°N

Solar Angle Of Incidence On Surfaces Tilted With Respect To The Horizontal

Within each tilt-angle group the five orientation columns are: **E/W, SE/SW, S, SW/SE, W/E**.

Date	Solar Time AM	Solar Time PM	0°	30° E/W	30° SE/SW	30° S	30° SW/SE	30° W/E	40° E/W	40° SE/SW	40° S	40° SW/SE	40° W/E	50° E/W	50° SE/SW	50° S	50° SW/SE	50° W/E	60° E/W	60° SE/SW	60° S	60° SW/SE	60° W/E	90° E/W	90° SE/SW	90° S	90° SW/SE	90° W/E
Jan. 21	8	4	81.9	58.1	52.5	66.2	88.1	90.0	50.9	42.8	62.0	90.0	90.0	44.5	33.2	58.5	90.0	90.0	39.2	24.0	56.0	90.0	90.0	35.5	13.0	55.7	90.0	90.0
	9	3	73.2	54.4	43.2	53.5	75.0	90.0	49.6	33.2	48.4	76.5	90.0	46.0	23.2	44.5	78.5	90.0	43.9	13.2	42.2	80.8	90.0	48.4	16.8	46.4	89.0	90.0
	10	2	66.2	54.2	37.5	42.1	62.6	83.4	52.3	28.4	35.5	63.1	89.6	51.7	20.1	30.6	64.5	90.0	52.5	14.0	28.4	66.7	90.0	62.0	27.5	38.3	77.1	90.0
	11	1	61.6	57.8	37.2	33.4	51.3	73.1	58.6	30.8	24.8	50.4	78.0	60.5	26.4	17.6	50.8	83.1	63.4	25.3	14.1	52.7	88.4	75.9	39.7	32.3	64.8	90.0
	12		60.0	64.3	42.3	30.0	42.3	64.3	67.5	39.0	20.0	39.0	67.5	71.3	37.8	10.0	37.8	71.3	75.5	38.7	0.0	38.7	75.5	90.0	52.2	30.0	52.2	90.0
Feb. 21	7	5	85.2	56.8	59.1	77.2	90.0	90.0	47.5	50.8	75.2	90.0	90.0	38.5	43.1	73.7	90.0	90.0	30.0	36.3	72.6	90.0	90.0	28.0	28.0	72.7	90.0	90.0
	8	4	74.6	49.0	46.3	62.9	85.0	90.0	41.3	37.3	60.5	88.0	90.0	34.5	28.8	59.0	88.8	90.0	29.4	21.5	58.5	90.0	90.0	31.5	22.9	63.3	90.0	90.0
	9	3	65.0	44.5	35.2	49.0	71.1	89.0	39.6	25.4	45.9	74.3	90.0	36.4	15.7	44.3	74.8	90.0	35.6	6.8	44.5	81.9	90.0	45.9	25.4	54.5	87.7	90.0
	10	2	57.2	44.3	27.9	35.9	57.6	77.1	43.0	18.5	31.5	59.9	84.3	43.5	10.3	29.5	63.2	90.0	45.8	8.3	30.6	67.3	90.0	60.5	33.9	47.1	82.3	90.0
	11	1	51.9	48.6	27.4	25.0	45.0	66.0	50.5	22.0	18.0	46.0	72.0	53.7	20.3	14.8	48.6	78.4	58.0	23.0	10.0	52.5	85.0	75.2	45.0	41.9	69.7	90.0
	12		50.0	56.2	34.2	20.0	34.2	56.2	60.5	32.8	10.0	32.8	60.5	65.6	34.1	0.0	34.1	65.6	71.3	37.8	10.0	37.8	71.3	90.0	57.2	40.0	57.2	90.0
Mar. 21	7	5	78.6	49.1	55.1	75.2	90.0	90.0	39.4	48.2	75.0	90.0	90.0	29.8	42.2	75.2	90.0	90.0	20.7	37.6	75.9	90.0	90.0	15.0	36.8	80.4	90.0	90.0
	8	4	67.5	40.1	41.3	60.5	82.0	90.0	31.8	33.6	60.0	87.4	90.0	24.5	27.2	60.5	90.0	90.0	19.7	23.3	62.0	90.0	90.0	30.0	32.9	71.3	90.0	90.0
	9	3	57.2	34.6	28.4	45.9	67.7	83.4	29.6	19.5	45.0	72.5	90.0	27.1	12.2	45.9	77.8	90.0	28.0	10.8	48.4	83.3	90.0	45.0	34.8	63.0	87.7	90.0
	10	2	48.4	34.5	18.5	31.5	53.5	71.1	33.9	8.7	30.0	57.5	79.2	36.0	2.8	31.5	62.8	87.5	40.1	11.8	30.0	68.5	90.0	60.0	41.7	56.2	83.0	90.0
	11	1	42.3	39.6	17.9	18.0	39.8	59.2	42.8	14.8	15.0	43.0	66.4	47.6	17.8	10.0	47.8	73.9	53.6	24.7	20.0	53.7	81.6	75.0	51.5	51.6	75.2	90.0
	12		40.0	48.4	27.0	10.0	27.0	48.4	54.1	28.5	0.0	28.5	54.1	60.5	32.8	18.0	32.8	60.5	67.5	39.0	20.0	39.0	67.5	90.0	63.0	50.0	63.0	90.0
Apr. 21	6	6	82.6	53.0	66.2	88.0	90.0	90.0	43.2	61.7	90.0	90.0	90.0	33.5	58.0	90.0	90.0	90.0	24.1	55.2	90.0	90.0	90.0	11.6	54.3	90.0	90.0	90.0
	7	5	71.1	41.1	51.8	73.5	90.0	90.0	31.1	47.0	75.3	90.0	90.0	21.1	43.5	77.6	90.0	90.0	11.1	41.7	80.2	90.0	90.0	18.9	47.5	89.5	90.0	90.0
	8	4	59.7	27.4	37.5	58.9	88.8	90.0	21.3	32.3	60.7	85.8	90.0	13.1	16.8	63.4	88.1	90.0	9.3	29.5	67.0	90.0	90.0	32.0	44.5	80.7	90.0	90.0
	9	3	48.7	23.1	23.2	44.2	64.6	77.0	18.1	17.7	46.2	71.2	86.5	17.3	4.2	49.7	76.5	90.0	21.6	21.2	54.4	85.2	90.0	46.2	45.9	73.1	85.3	90.0
	10	2	38.8	25.1	9.5	29.5	50.2	64.5	24.3	4.2	32.0	56.5	73.6	28.8	12.0	36.8	63.4	82.8	35.5	21.6	43.2	70.8	90.0	60.7	51.5	67.0	81.5	90.0
	11	1	31.3	30.0	8.2	14.8	35.8	52.2	35.2	12.6	18.9	41.8	60.6	42.0	21.2	26.2	48.9	69.2	49.7	30.6	34.9	56.6	78.0	75.3	60.0	63.0	74.8	90.0
	12		28.4	40.4	21.6	1.6	21.6	40.4	47.6	27.1	11.6	27.1	47.6	55.6	34.6	21.6	34.6	55.6	63.9	43.0	31.6	43.0	63.9	90.0	70.3	61.6	70.3	90.0
May 21	5	7	88.1	61.1	78.3	90.0	90.0	90.0	52.5	75.6	90.0	90.0	90.0	44.2	73.3	90.0	90.0	90.0	36.6	71.6	90.0	90.0	90.0	24.8	69.8	90.0	90.0	90.0
	6	6	77.3	48.7	64.5	86.6	90.0	90.0	39.4	61.5	75.9	90.0	90.0	30.6	59.4	79.6	90.0	90.0	22.5	58.3	83.6	90.0	90.0	20.0	61.4	89.5	90.0	90.0
	7	5	66.0	36.3	50.6	72.6	90.0	90.0	26.5	47.5	62.0	84.7	90.0	17.0	33.4	66.2	84.7	90.0	8.4	46.0	71.1	90.0	90.0	24.8	55.5	87.7	90.0	90.0
	8	4	54.6	24.6	36.5	58.5	76.8	90.0	14.7	33.7	48.4	70.6	90.0	5.1	23.4	53.5	70.6	90.0	5.9	35.6	59.5	86.6	90.0	35.5	54.1	80.5	90.0	90.0
	9	3	43.2	15.2	22.4	44.5	63.0	84.5	9.8	20.7	35.5	56.9	82.5	12.2	23.4	42.1	62.5	90.0	20.0	29.3	59.5	72.9	90.0	48.4	58.9	74.9	86.8	90.0
	10	2	32.5	15.2	8.6	30.6	49.0	72.6	18.6	12.0	24.8	42.6	69.9	25.6	21.3	33.4	50.9	79.5	34.1	29.6	42.6	59.6	89.2	62.0	58.9	71.2	78.9	90.0
	11	1	23.8	23.9	7.2	17.6	34.9	63.0	31.0	16.7	20.8	28.9	57.0	39.2	26.6	30.0	37.9	66.3	48.1	36.5	40.0	47.2	75.7	75.9	66.5	70.0	76.0	90.0
	12		20.0	33.5	20.8	10.0	20.8	44.0	44.0	28.9	20.8	28.9	44.0	52.8	37.9	30.0	37.9	52.8	62.0	47.2	40.0	47.2	62.0	90.0	76.0	70.0	76.0	90.0
June 21	5	7	85.8	59.5	77.6	86.0	90.0	90.0	51.2	75.4	90.0	90.0	90.0	43.4	73.8	90.0	90.0	90.0	36.4	72.6	90.0	90.0	90.0	27.6	72.4	90.0	90.0	90.0
	6	6	75.2	47.1	64.0	86.4	90.0	90.0	38.2	61.7	76.3	90.0	90.0	29.9	60.2	80.5	90.0	90.0	22.7	59.1	85.0	90.0	90.0	23.4	64.3	90.0	90.0	90.0
	7	5	64.0	34.7	50.3	72.4	89.3	90.0	25.2	48.0	62.7	84.3	90.0	16.2	47.2	67.5	84.3	90.0	9.5	48.1	72.8	90.0	90.0	27.6	58.7	90.0	90.0	90.0
	8	4	52.6	22.6	36.6	58.6	76.0	90.0	2.7	34.7	49.6	70.6	90.0	2.7	35.4	55.3	70.6	90.0	7.4	38.4	61.7	86.9	90.0	37.4	56.3	83.6	90.0	90.0
	9	3	41.2	12.5	22.9	44.9	62.5	82.9	6.5	22.7	37.4	56.8	80.8	24.9	23.7	44.5	65.2	90.0	20.3	32.9	61.7	73.9	90.0	49.6	57.5	76.0	87.3	90.0
	10	2	30.2	12.1	10.4	31.6	48.8	70.6	16.9	15.1	27.6	43.2	68.4	34.0	29.2	36.5	52.6	78.2	34.1	39.2	43.3	61.1	88.0	62.7	62.0	74.6	88.9	90.0
	11	1	20.8	23.9	9.3	19.6	35.0	58.7	29.7	19.2	23.4	30.3	55.7	38.5	29.2	36.5	39.6	65.2	47.7	39.2	43.3	49.2	74.8	76.3	69.2	74.6	78.4	90.0
	12		16.6	33.9	21.4	13.4	21.4	33.9	42.8	30.3	23.1	30.3	42.8	52.0	39.6	33.4	39.6	52.0	61.4	49.2	43.3	49.2	61.4	90.0	78.4	73.4	78.4	90.0

(Based on data in Table 1, pp. 387, 1972 ASHRAE HANDBOOK OF FUNDAMENTALS; 0% ground reflectance, 1.0 clearness factor)

Table 4-A Solar Angle of Incidence on Inclined Surfaces for Selected Northern Latitudes

40°N

Solar Angle of Incidence On Surfaces Tilted With Respect To The Horizontal

Date	AM	PM	0°	30° E/W	30° SE/SW	30° S/S	30° SW/SE	30° W/E	40° E/W	40° SE/SW	40° S/S	40° SW/SE	40° W/E	50° E/W	50° SE/SW	50° S/S	50° SW/SE	50° W/E	60° E/W	60° SE/SW	60° S/S	60° SW/SE	60° W/E	90° E/W	90° SE/SW	90° S/S	90° SW/SE	90° W/E
July 21	5	7	87.7	60.8	78.2	90.0	90.0	90.0	52.2	75.6	90.0	90.0	90.0	44.1	73.4	90.0	90.0	90.0	36.6	71.7	90.0	90.0	90.0	25.3	70.2	90.0	90.0	90.0
	6	6	76.9	48.4	64.4	86.5	90.0	90.0	39.2	61.6	90.0	90.0	90.0	30.4	59.6	90.0	90.0	90.0	22.5	58.6	90.0	90.0	90.0	20.6	61.9	90.0	90.0	90.0
	7	5	65.7	36.0	50.5	72.6	90.0	90.0	26.3	47.6	76.0	90.0	90.0	16.8	46.1	79.8	90.0	90.0	8.5	46.4	83.8	90.0	90.0	25.3	56.6	90.0	90.0	90.0
	8	4	54.2	24.3	36.5	58.5	76.7	84.2	14.3	33.8	62.1	84.6	90.0	4.6	33.7	66.4	90.0	90.0	6.1	36.1	71.4	90.0	90.0	35.8	53.5	88.2	90.0	90.0
	9	3	42.8	15.0	22.5	44.5	62.9	72.3	9.2	21.0	48.6	70.6	82.2	11.9	23.9	53.8	78.6	82.5	20.0	29.8	59.1	86.7	89.0	48.6	54.7	81.0	90.0	90.0
	10	2	32.1	14.6	8.9	30.7	48.9	60.0	18.7	12.6	35.8	56.6	69.8	25.4	20.9	42.5	64.7	69.0	34.0	30.2	50.1	73.1	75.5	62.1	59.4	75.4	87.2	90.0
	11	1	23.3	23.5	7.5	17.9	34.9	47.6	30.7	17.1	25.3	41.8	56.8	39.1	27.0	34.0	51.1	55.3	48.0	37.0	43.2	59.9	61.9	76.0	66.9	71.8	82.3	90.0
	12	12	19.4	35.2	20.9	10.6	20.9	35.2	43.7	29.1	20.6	29.1	43.7	52.7	38.2	30.6	38.2	52.7	61.9	47.5	40.6	47.5	61.9	90.0	76.4	70.6	76.4	90.0
Aug. 21	6	6	82.1	52.6	66.0	87.9	90.0	90.0	42.9	61.7	90.0	90.0	90.0	33.2	58.1	90.0	90.0	90.0	23.8	55.5	90.0	90.0	90.0	12.3	54.9	90.0	90.0	90.0
	7	5	70.7	40.7	51.7	73.4	90.0	90.0	30.7	47.0	75.4	90.0	90.0	20.7	43.7	77.8	90.0	90.0	10.7	42.0	80.5	90.0	90.0	19.3	48.2	90.0	90.0	90.0
	8	4	59.3	30.0	37.3	58.8	78.6	88.6	20.7	32.4	60.8	85.7	90.0	12.4	29.7	63.6	90.0	90.0	8.7	30.0	67.3	83.3	90.0	32.2	45.2	81.3	85.3	90.0
	9	3	48.2	22.7	23.0	44.2	64.4	76.5	17.4	17.8	46.3	71.1	84.6	16.8	17.3	50.0	78.1	85.2	21.3	21.8	54.8	68.5	88.4	46.3	46.6	73.7	73.7	90.0
	10	2	38.3	22.5	9.2	29.5	50.1	64.2	23.7	4.8	32.2	56.5	72.3	28.5	12.7	37.2	63.5	72.6	35.3	22.4	43.8	53.7	88.4	60.8	52.1	67.6	82.1	90.0
	11	1	30.7	29.4	7.7	14.9	35.7	51.8	34.8	12.8	19.3	41.8	60.3	41.7	21.6	26.8	49.0	66.1	49.5	31.1	35.5	39.0	77.9	75.4	60.5	62.3	69.5	89.1
	12	12	27.7	39.9	21.4	2.3	21.4	39.9	47.3	27.2	12.3	27.2	47.3	55.3	34.8	22.3	34.8	55.3	63.7	43.3	32.3	43.3	63.7	90.0	70.8	62.3	70.8	90.0
Sept. 21	7	5	78.6	49.1	55.1	75.2	90.0	90.0	39.4	48.2	75.0	90.0	90.0	29.5	42.2	75.2	90.0	90.0	20.7	37.6	75.9	90.0	90.0	15.0	36.8	80.4	90.0	90.0
	8	4	67.5	40.1	41.3	60.5	82.0	90.0	31.8	33.6	60.0	87.4	90.0	24.5	27.2	60.5	90.0	90.0	19.7	23.6	62.0	90.0	90.0	30.0	32.9	71.3	90.0	90.0
	9	3	57.2	34.6	28.4	45.9	67.7	83.4	29.6	19.5	45.0	72.5	89.1	27.1	12.2	45.9	77.8	90.0	28.0	11.8	48.4	83.3	90.0	45.0	34.8	63.0	87.7	90.0
	10	2	48.4	34.5	18.5	31.5	53.5	71.1	33.9	8.7	30.0	60.1	77.4	36.0	2.8	31.5	62.8	87.5	40.1	24.7	35.5	68.5	81.6	60.0	41.7	51.6	75.2	89.7
	11	1	42.3	33.6	17.9	18.0	39.8	59.2	42.8	14.8	15.0	46.1	66.3	47.6	17.8	18.0	47.8	73.9	53.6	24.7	24.8	53.7	73.9	75.0	51.5	50.0	63.0	90.0
	12	12	40.0	46.4	27.0	10.0	27.0	46.4	54.1	28.5	2.3	28.5	54.1	60.5	34.8	10.0	34.8	60.5	67.5	39.0	20.0	39.0	67.5	90.0	63.0	50.0	63.0	90.0
Oct. 21	7	5	85.5	57.2	59.3	77.4	90.0	90.0	47.9	51.0	75.3	90.0	90.0	39.0	43.2	73.6	90.0	90.0	30.5	36.3	72.5	90.0	90.0	18.2	27.6	72.4	90.0	90.0
	8	4	75.0	49.4	46.6	63.1	85.1	90.0	41.7	37.5	60.6	88.9	90.0	35.0	29.0	58.9	90.0	90.0	29.9	21.5	58.4	81.9	90.0	31.6	22.4	62.9	90.0	90.0
	9	3	65.0	45.0	35.6	46.2	71.3	89.3	40.1	25.7	46.0	74.4	90.0	36.9	16.0	46.0	78.0	90.0	35.9	6.9	44.3	67.2	90.0	46.0	25.0	46.6	69.5	90.0
	10	2	57.6	44.8	28.3	36.2	57.8	77.4	43.9	19.0	31.6	60.1	84.6	43.0	10.8	29.5	63.3	78.6	46.1	8.4	30.4	67.2	85.1	60.6	33.6	46.6	82.1	90.0
	11	1	52.4	49.0	27.9	25.4	45.3	66.3	50.9	22.5	18.2	46.1	72.3	54.0	20.5	14.8	48.6	78.6	58.3	23.0	17.3	52.5	85.1	75.3	44.7	39.5	69.5	90.0
	12	12	50.5	56.6	34.5	20.5	34.5	56.6	60.8	33.1	10.5	33.1	60.8	65.9	34.2	0.5	34.2	65.9	71.5	37.8	9.5	37.8	71.5	90.0	56.9	39.5	56.9	90.0
Nov. 21	8	4	81.8	57.9	52.4	66.1	88.0	90.0	50.7	42.7	61.9	90.0	90.0	44.3	33.1	58.5	90.0	90.0	39.0	23.9	56.1	90.0	90.0	35.4	13.2	55.8	90.0	90.0
	9	3	73.0	54.2	43.0	53.4	74.9	90.0	49.4	33.0	48.3	76.5	90.0	45.8	23.0	44.4	78.5	90.0	43.8	13.1	42.2	80.8	90.0	48.3	17.0	46.6	89.1	90.0
	10	2	66.0	54.0	37.3	42.0	62.4	83.3	52.1	28.2	35.4	63.0	89.5	51.6	19.9	30.5	64.5	88.4	52.4	13.8	28.2	66.7	88.4	61.9	27.6	38.4	77.2	90.0
	11	1	61.4	57.6	37.0	33.3	51.2	73.0	58.5	30.6	24.7	50.3	77.9	60.4	26.3	17.5	50.8	83.0	63.2	25.2	14.1	52.6	88.4	75.9	39.8	32.5	64.9	90.0
	12	12	59.8	64.2	42.2	29.8	42.2	64.2	67.3	38.9	19.8	38.9	67.3	71.1	37.7	9.8	37.7	71.1	75.4	38.7	0.2	38.7	75.4	90.0	52.3	30.2	52.3	90.0
Dec. 21	8	4	84.5	61.3	54.9	67.5	89.2	90.0	54.3	45.0	62.7	90.0	90.0	47.9	35.2	58.6	90.0	90.0	42.6	25.6	55.4	90.0	90.0	37.4	9.7	53.2	90.0	90.0
	9	3	76.0	57.8	46.1	55.3	76.4	90.0	53.0	36.1	49.6	77.4	90.0	49.3	26.2	44.9	78.8	90.0	47.0	16.3	41.8	80.5	90.0	49.6	14.3	43.8	87.0	90.0
	10	2	69.3	57.0	40.9	44.5	64.4	84.8	55.6	31.8	37.4	64.4	84.8	54.7	23.5	31.6	65.2	89.7	55.0	16.9	28.0	66.7	89.7	62.7	25.7	35.4	75.4	90.0
	11	1	65.0	61.0	40.6	36.5	53.7	75.7	61.5	34.0	27.6	52.2	80.1	63.0	29.0	19.6	51.9	84.8	65.3	26.8	14.3	53.0	89.7	76.3	38.2	35.4	64.9	90.0
	12	12	63.4	67.2	45.3	33.4	45.3	67.2	70.0	41.5	23.4	41.5	70.0	73.3	39.5	13.5	39.5	73.3	77.1	39.5	3.5	39.5	77.1	90.0	50.8	26.6	50.8	90.0

(Based on data in Table 1, pp. 387, 1972 ASHRAE HANDBOOK OF FUNDAMENTALS; 0% ground reflectance, 1.0 clearness factor)

Table 4-A Solar Angle of Incidence on Inclined Surfaces for Selected Northern Latitudes

48°N

Date	Solar Time		Solar Angle Of Incidence On Surfaces Tilted With Respect To The Horizontal																									
			0°	38°					48°					58°					68°					90°				
				AM: E	SE	S	SW	W	E	SE	S	SW	W	E	SE	S	SW	W	E	SE	S	SW	W	E	SE	S	SW	W
	AM	PM		PM: W	SW	S	SE	E	W	SW	S	SE	E	W	SW	S	SE	E	W	SW	S	SE	E	W	SW	S	SE	E
Jan 21	8	4	86.5	56.7	49.2	66.2	90.0	90.0	49.8	39.5	62.0	90.0	90.0	43.8	29.9	58.5	90.0	90.0	39.0	20.8	56.0	90.0	90.0	35.5	10.2	54.7	90.0	90.0
	9	3	79.0	56.0	41.1	53.5	79.9	90.0	51.6	31.1	48.4	80.9	90.0	48.4	21.1	44.5	82.2	90.0	46.6	11.3	42.2	83.7	90.0	48.4	11.2	43.7	87.6	90.0
	10	2	73.1	58.8	37.2	42.1	67.2	90.0	57.1	28.4	35.5	67.3	90.0	56.5	20.7	30.6	68.2	90.0	57.0	15.6	28.2	69.7	90.0	62.0	22.8	33.5	75.1	90.0
	11	1	69.3	64.7	39.0	33.4	55.5	82.6	65.4	33.0	24.8	54.3	86.8	66.8	29.0	17.6	54.3	90.0	69.0	27.9	14.1	55.6	90.0	75.9	35.8	25.4	62.2	90.0
	12		68.0	72.8	45.7	30.0	45.7	72.8	75.5	42.4	20.0	42.4	75.5	78.5	41.0	10.0	41.0	78.5	81.9	41.6	90.0	41.6	81.9	90.0	49.0	22.0	49.0	90.0
Feb 21	7	5	87.6	51.8	54.6	77.2	90.0	90.0	42.7	46.5	75.2	90.0	90.0	34.0	39.1	73.7	90.0	90.0	26.2	32.9	72.6	90.0	90.0	18.0	27.3	72.2	90.0	90.0
	8	4	78.4	46.9	42.3	62.9	90.0	90.0	39.8	33.3	60.5	90.0	90.0	34.0	24.9	59.0	90.0	90.0	30.0	18.1	58.5	90.0	90.0	31.5	19.3	61.2	90.0	90.0
	9	3	70.3	46.0	32.4	49.0	76.2	90.0	42.0	22.4	45.9	78.9	90.0	39.7	12.5	44.3	81.9	90.0	39.5	3.4	44.5	85.1	90.0	45.9	19.9	50.7	90.0	90.0
	10	2	63.8	49.4	27.3	35.9	62.6	87.4	48.6	18.5	31.5	64.5	90.0	49.3	11.7	29.5	67.1	90.0	51.5	11.5	30.6	70.5	90.0	60.5	28.5	41.4	79.5	90.0
	11	1	59.5	56.2	29.6	25.0	49.7	75.9	58.1	25.0	18.0	50.3	81.4	51.0	23.7	14.8	52.4	87.0	64.8	26.3	17.6	55.8	90.0	75.2	40.3	34.6	66.3	90.0
	12		58.0	65.3	38.1	20.0	38.1	65.3	69.2	36.8	10.0	36.8	69.2	73.7	37.9	90.0	37.9	73.7	78.5	41.0	10.0	41.0	78.5	90.0	53.2	32.0	53.2	90.0
Mar 21	7	5	80.0	43.0	50.2	75.2	90.0	90.0	33.5	43.6	75.0	90.0	90.0	24.4	38.2	75.2	90.0	90.0	16.2	34.5	75.9	90.0	90.0	15.0	35.0	78.9	90.0	90.0
	8	4	70.5	37.2	36.6	60.5	87.2	90.0	29.8	29.0	60.0	90.0	90.0	24.3	23.2	60.5	90.0	90.0	21.8	20.5	62.0	90.0	90.0	30.0	28.9	68.2	90.0	90.0
	9	3	61.8	36.1	24.6	45.9	72.9	90.0	32.6	15.4	45.0	77.2	90.0	31.7	8.2	45.9	81.8	90.0	33.6	9.8	48.4	86.7	90.0	45.0	29.4	58.3	90.0	90.0
	10	2	54.6	40.1	17.4	31.5	58.7	81.4	40.6	8.6	30.0	62.4	89.1	43.0	6.9	31.5	66.8	90.0	47.1	14.8	35.5	71.9	90.0	60.0	36.0	49.9	84.2	90.0
	11	1	49.7	48.0	20.9	18.0	44.8	69.5	51.3	19.0	15.0	47.6	76.1	55.8	21.9	18.0	51.8	82.9	61.2	28.1	24.8	57.1	89.9	75.0	46.3	44.1	71.1	90.0
	12		48.0	58.2	31.7	10.0	31.7	58.2	63.4	33.0	90.0	33.0	63.4	69.2	36.8	10.0	36.8	69.2	75.5	42.4	20.0	42.4	75.5	90.0	58.3	42.0	58.3	90.0
Apr 21	6	6	81.4	43.9	60.9	88.0	90.0	90.0	34.1	57.0	90.0	90.0	90.0	24.5	54.1	90.0	90.0	90.0	15.4	52.4	90.0	90.0	90.0	11.6	53.3	90.0	90.0	90.0
	7	5	71.4	33.5	46.6	73.5	90.0	90.0	23.6	42.4	75.3	90.0	90.0	13.7	39.8	77.6	90.0	90.0	4.6	39.2	80.2	90.0	90.0	18.9	45.0	86.9	90.0	90.0
	8	4	61.5	26.1	32.3	58.9	83.9	90.0	18.2	27.7	60.7	90.0	90.0	13.5	26.0	63.4	90.0	90.0	15.1	27.7	67.0	90.0	90.0	32.0	40.4	76.7	90.0	90.0
	9	3	52.2	24.6	18.2	44.2	69.7	86.8	22.4	13.1	46.2	75.8	90.0	24.2	14.5	49.7	82.1	90.0	29.3	21.1	54.4	88.6	90.0	46.2	40.7	67.7	90.0	90.0
	10	2	44.2	23.9	6.2	29.5	55.4	74.7	32.5	3.8	32.0	61.1	83.3	37.3	13.8	36.8	67.4	90.0	43.7	23.8	43.2	74.2	90.0	60.7	45.8	60.3	89.8	90.0
	11	1	38.5	39.4	12.9	14.8	41.1	62.6	44.6	17.1	18.9	46.4	70.4	51.0	24.8	26.2	52.9	78.5	58.1	33.7	34.9	60.0	86.7	75.3	54.5	55.3	77.1	90.0
	12		36.4	50.6	26.8	1.6	26.8	50.6	57.4	31.7	11.6	31.7	57.4	64.8	38.5	21.6	38.5	64.8	72.5	46.3	31.6	46.3	72.5	90.0	65.2	53.6	65.2	90.0
May 21	5	7	84.8	50.9	73.2	90.0	90.0	90.0	42.6	71.2	90.0	90.0	90.0	35.1	69.7	90.0	90.0	90.0	28.9	68.9	90.0	90.0	90.0	24.8	69.4	90.0	90.0	90.0
	6	6	75.3	38.8	59.4	86.6	90.0	90.0	29.7	57.1	90.0	90.0	90.0	21.3	55.9	90.0	90.0	90.0	14.9	55.9	90.0	90.0	90.0	20.0	59.8	90.0	90.0	90.0
	7	5	65.4	27.5	45.4	72.6	90.0	90.0	17.5	43.1	75.9	90.0	90.0	7.8	42.6	79.6	90.0	90.0	3.8	44.0	83.6	90.0	90.0	24.8	52.6	90.0	90.0	90.0
	8	4	55.4	18.4	31.3	58.5	81.6	90.0	9.9	29.4	62.0	89.1	90.0	7.5	30.5	66.2	90.0	90.0	14.6	34.4	71.1	90.0	90.0	35.5	48.7	83.1	90.0	90.0
	9	3	45.7	16.3	17.2	44.5	67.9	81.9	16.0	17.1	48.4	75.1	90.0	21.0	21.9	53.5	82.5	90.0	28.6	29.4	59.5	90.0	90.0	48.4	48.9	74.6	90.0	90.0
	10	2	37.0	23.3	4.0	30.6	53.9	70.1	27.9	11.8	35.5	61.0	79.3	34.7	21.5	42.1	68.5	88.6	42.7	31.4	49.6	76.3	90.0	62.0	53.3	67.9	90.0	90.0
	11	1	30.5	34.0	11.8	17.6	39.9	58.1	40.8	20.2	24.8	46.9	66.7	48.5	29.6	33.4	54.7	75.5	56.8	39.3	42.6	62.9	84.4	75.9	60.8	63.5	81.8	90.0
	12		28.0	45.9	25.8	10.0	25.8	45.9	53.8	33.1	20.0	33.1	53.8	62.1	41.5	30.0	41.5	62.1	70.7	50.3	40.0	50.3	70.7	90.0	70.6	62.0	70.6	90.0
June 21	5	7	82.1	49.2	72.5	90.0	90.0	90.0	41.4	71.1	90.0	90.0	90.0	34.5	70.2	90.0	90.0	90.0	29.2	70.0	90.0	90.0	90.0	27.6	71.7	90.0	90.0	90.0
	6	6	72.8	37.1	58.9	86.0	90.0	90.0	28.4	57.3	90.0	90.0	90.0	20.8	56.8	90.0	90.0	90.0	16.0	57.5	90.0	90.0	90.0	23.4	62.6	90.0	90.0	90.0
	7	5	63.0	25.3	45.2	72.4	90.0	90.0	15.7	43.7	76.3	90.0	90.0	7.1	44.1	80.5	90.0	90.0	7.3	46.2	85.0	90.0	90.0	27.6	55.7	90.0	90.0	90.0
	8	4	52.9	15.4	31.4	58.6	80.7	90.0	6.4	30.7	62.7	88.6	90.0	6.7	32.8	67.5	90.0	90.0	15.8	37.3	72.8	90.0	90.0	37.4	52.1	85.7	90.0	90.0
	9	3	43.1	13.0	17.9	44.9	67.3	79.9	14.0	19.5	49.6	74.9	89.6	20.5	25.2	55.3	82.7	90.0	29.0	32.8	61.7	90.0	90.0	49.6	52.3	77.5	90.0	90.0
	10	2	34.2	20.9	6.9	31.6	53.6	68.3	26.6	15.2	37.4	61.1	77.7	34.2	24.8	44.5	69.0	87.2	42.7	34.6	52.4	77.2	90.0	62.7	56.4	71.1	90.0	90.0
	11	1	27.3	32.2	13.0	19.6	39.9	56.4	39.5	22.2	27.6	47.5	65.3	47.8	31.9	36.5	55.7	74.4	56.4	41.8	45.8	64.3	83.5	76.3	63.6	66.9	83.7	90.0
	12		24.6	44.2	26.1	13.4	26.1	44.2	52.5	34.2	23.4	34.2	52.5	61.2	43.0	33.4	43.0	61.2	70.1	52.2	43.4	52.2	70.1	90.0	72.9	65.4	72.9	90.0

(Based on data in Table 1, pp. 387, 1972 ASHRAE HANDBOOK OF FUNDAMENTALS; 0% ground reflectance, 1.0 clearness factor)

Table 4-A Solar Angle of Incidence on Inclined Surfaces for Selected Northern Latitudes

48°N

Solar Angle Of Incidence On Surfaces Tilted With Respect To The Horizontal

Note on table structure: For each tilted plane (0°, 38°, 48°, 58°, 68°, 90°) five surface orientations are given. Each value serves the AM hour (first orientation letter) and the symmetric PM hour (second orientation letter): E/W, SE/SW, S/S, SW/SE, W/E. Solar Time is given as AM (E) / PM (W).

Date	AM	PM	0°	38° E·W	38° SE·SW	38° S	38° SW·SE	38° W·E	48° E·W	48° SE·SW	48° S	48° SW·SE	48° W·E	58° E·W	58° SE·SW	58° S	58° SW·SE	58° W·E	68° E·W	68° SE·SW	68° S	68° SW·SE	68° W·E	90° E·W	90° SE·SW	90° S	90° SW·SE	90° W·E
July 21	5	7	84.3	50.6	73.1	90.0	90.0	90.0	42.4	71.1	90.0	90.0	90.0	35.0	69.8	90.0	90.0	90.0	28.9	69.0	90.0	90.0	90.0	25.3	69.8	90.0	90.0	90.0
	6	6	74.8	38.5	59.3	86.5	90.0	90.0	29.5	57.1	90.0	90.0	90.0	21.2	56.0	90.0	90.0	90.0	15.0	56.1	90.0	90.0	90.0	20.6	60.3	90.0	90.0	90.0
	7	5	64.9	27.1	45.3	72.6	90.0	90.0	17.2	43.2	76.0	90.0	90.0	7.6	42.8	79.8	90.0	90.0	4.4	44.3	83.8	90.0	90.0	25.3	53.1	83.5	90.0	90.0
	8	4	54.9	17.8	31.3	58.5	81.4	90.0	9.3	29.6	62.1	89.0	90.0	7.3	30.9	66.4	90.0	90.0	14.8	34.9	71.4	90.0	90.0	35.8	49.3	75.1	90.0	90.0
	9	3	45.2	15.7	17.3	44.5	67.8	81.5	15.6	17.5	48.6	75.0	90.0	20.8	22.1	53.8	82.5	90.0	28.7	30.0	59.9	90.0	90.0	48.6	53.8	68.5	90.0	90.0
	10	2	36.5	22.8	4.4	30.7	53.9	69.3	27.7	12.4	35.8	61.0	79.0	34.6	22.2	42.5	68.6	88.3	42.7	31.9	50.1	76.4	90.0	62.1	63.8	64.1	90.0	90.0
	11	1	29.5	33.7	12.0	17.9	39.9	57.7	40.6	20.5	25.6	47.0	66.4	48.3	30.0	34.0	54.9	75.3	56.7	39.7	43.2	63.1	84.3	76.0	71.0	62.6	75.0	90.0
	12		27.4	45.6	25.8	10.6	25.8	45.6	53.6	33.3	20.6	33.3	53.6	61.9	41.7	30.6	41.7	61.9	70.6	50.6	40.6	50.6	70.6	90.0	71.0	62.6	71.0	90.0
Aug 21	6	6	80.9	43.4	60.8	87.9	90.0	90.0	33.7	57.0	90.0	90.0	90.0	24.2	54.2	90.0	90.0	90.0	15.2	52.7	90.0	90.0	90.0	12.3	53.8	87.4	90.0	90.0
	7	5	70.9	33.0	46.5	73.4	90.0	90.0	23.0	42.4	75.4	90.0	90.0	13.1	39.9	77.8	90.0	90.0	3.9	39.5	80.5	90.0	90.0	19.3	45.6	77.3	90.0	90.0
	8	4	61.0	25.4	32.1	58.8	83.7	90.0	17.5	27.7	60.8	90.0	90.0	12.9	26.3	63.6	90.0	90.0	14.9	28.2	67.3	90.0	90.0	32.2	41.3	68.2	90.0	90.0
	9	3	51.6	23.9	18.0	44.2	69.5	86.3	21.8	13.3	46.3	82.1	90.0	23.8	15.1	50.0	82.1	90.0	29.2	24.4	54.8	88.7	90.0	46.3	46.4	60.9	90.0	90.0
	10	2	43.6	29.3	5.6	29.5	55.3	74.3	32.0	4.4	32.2	61.1	83.0	37.0	14.4	37.2	67.5	90.0	43.6	34.1	43.8	74.3	90.0	60.8	55.0	56.0	90.0	90.0
	11	1	37.8	38.9	12.6	14.9	40.9	62.2	44.2	17.3	19.3	46.4	70.1	50.7	25.1	26.8	53.0	78.2	58.0	39.7	33.5	66.2	86.5	75.4	65.6	54.3	77.5	90.0
	12		35.7	50.2	26.6	2.3	26.6	50.2	57.1	31.8	12.3	31.8	57.1	64.5	38.7	23.3	38.7	64.5	72.3	46.6	32.3	46.6	72.3	90.0	65.6	54.3	65.6	90.0
Sept 21	7	5	80.0	43.0	50.2	75.2	90.0	90.0	33.5	43.6	75.0	90.0	90.0	24.4	38.2	75.2	90.0	90.0	16.2	34.5	75.9	90.0	90.0	15.0	35.0	78.9	90.0	90.0
	8	4	70.5	37.2	36.6	60.5	87.2	90.0	29.8	29.8	60.0	90.0	90.0	24.3	23.2	60.5	90.0	90.0	21.8	20.5	62.0	90.0	90.0	30.0	28.9	68.2	90.0	90.0
	9	3	61.8	36.1	24.6	45.9	72.9	89.1	32.6	15.4	45.0	77.2	90.0	31.7	8.2	45.9	81.8	90.0	33.6	9.8	48.4	86.7	90.0	45.0	29.4	58.3	90.0	90.0
	10	2	54.6	40.1	17.4	31.5	58.7	69.5	40.6	8.6	30.0	62.4	87.7	43.0	6.9	31.5	66.8	89.1	47.1	14.8	35.5	71.9	90.0	60.0	36.0	49.0	90.0	90.0
	11	1	49.7	49.0	20.9	18.0	44.8	58.2	51.3	19.0	15.0	47.6	76.3	55.8	21.9	18.0	51.8	82.9	61.2	28.1	24.8	57.1	89.9	75.0	46.3	44.1	77.1	90.0
	12		48.0	58.2	31.7	10.0	31.7	58.2	63.4	33.0	90.0	33.0	63.4	69.2	36.8	10.0	36.8	69.2	75.5	42.4	20.0	42.4	75.5	90.0	58.3	42.0	58.3	90.0
Oct 21	7	5	88.0	52.3	54.8	77.4	90.0	90.0	43.2	46.7	75.3	90.0	90.0	34.5	39.2	73.6	90.0	90.0	26.7	32.9	72.5	90.0	90.0	18.2	26.9	71.9	90.0	90.0
	8	4	78.8	47.4	42.7	63.1	90.0	90.0	40.3	33.6	60.6	90.0	90.0	34.4	25.1	58.9	90.0	90.0	30.5	18.1	58.4	90.0	90.0	31.6	18.8	60.3	90.0	90.0
	9	3	70.7	46.5	32.8	49.2	76.4	87.7	42.5	22.8	46.0	79.0	90.0	40.7	12.9	44.2	81.9	90.0	39.8	3.6	44.3	85.0	90.0	46.0	19.4	50.3	90.0	90.0
	10	2	64.3	49.8	27.8	36.2	62.4	76.3	49.0	19.0	31.6	64.6	87.7	51.8	12.2	29.5	67.2	90.0	51.8	11.5	30.4	70.4	90.0	60.6	28.2	41.0	90.0	90.0
	11	1	60.0	56.6	30.1	25.4	49.9	65.7	58.4	25.4	18.2	50.5	76.3	61.3	23.9	14.8	52.5	87.2	65.0	26.3	17.3	55.7	90.0	75.3	40.1	34.1	79.3	90.0
	12		58.5	65.7	38.5	20.5	38.5	65.7	69.5	37.1	10.5	37.1	69.5	73.9	38.0	0.5	38.0	73.9	78.7	41.0	9.5	41.0	78.7	90.0	52.9	31.5	52.9	90.0
Nov 21	8	4	86.4	56.5	49.1	66.1	90.0	90.0	49.6	39.4	61.9	90.0	90.0	43.6	29.8	58.5	90.0	90.0	38.8	20.7	56.1	90.0	90.0	35.4	10.4	54.8	90.0	90.0
	9	3	78.8	55.8	40.9	53.4	79.8	90.0	51.4	30.9	48.3	80.9	90.0	48.2	21.0	44.4	82.2	90.0	46.4	11.1	42.2	83.7	90.0	48.3	11.4	43.9	87.7	90.0
	10	2	72.9	58.6	37.0	42.0	67.1	90.0	56.9	28.2	35.4	67.3	90.0	56.3	20.5	30.5	68.1	90.0	56.9	15.4	28.2	69.7	90.0	61.9	22.9	33.7	75.2	90.0
	11	1	69.1	64.5	38.8	33.3	55.4	82.5	65.2	32.8	24.7	54.2	86.7	66.7	28.9	17.5	54.3	90.0	68.9	27.8	14.1	55.6	90.0	75.9	35.9	25.6	62.2	90.0
	12		67.8	72.7	45.5	29.8	45.5	72.7	75.4	42.3	19.8	42.3	75.4	78.4	40.9	9.8	40.9	78.4	81.9	41.5	0.2	41.5	81.9	90.0	49.1	22.2	49.1	90.0
Dec 21	9	3	82.0	59.4	44.2	55.3	81.2	90.0	54.9	34.2	49.6	81.7	90.0	51.4	24.3	44.9	82.4	90.0	49.2	14.6	41.8	83.3	90.0	49.6	8.9	41.6	86.0	90.0
	10	2	76.4	62.1	40.7	44.5	69.0	90.0	60.1	31.9	37.4	68.5	90.0	59.1	24.0	31.6	68.7	90.0	59.1	18.1	28.1	69.6	90.0	62.7	21.5	31.1	66.0	90.0
	11	1	72.7	67.6	42.3	36.5	57.8	84.9	68.0	35.9	27.6	55.9	88.7	69.0	31.3	19.6	55.3	90.0	70.6	29.2	14.3	55.3	90.0	76.3	34.7	22.4	60.9	90.0
	12		71.4	75.5	48.4	33.4	48.4	75.5	77.7	44.7	23.4	44.7	77.7	80.3	42.5	13.5	42.5	80.3	83.2	42.2	3.5	42.2	83.2	90.0	47.9	18.5	47.9	90.0

(Based on data in Table 1, pp. 387, 1972 ASHRAE HANDBOOK OF FUNDAMENTALS; 0% ground reflectance, 1.0 clearness factor)

Table 4-A Solar Angle of Incidence on Inclined Surfaces for Selected Northern Latitudes

56°N

Solar Angle of Incidence On Surfaces Tilted With Respect To The Horizontal

Sub-columns for each tilt angle: E/W (AM E, PM W), SE/SW, S, SW/SE, W/E (AM W, PM E)

Date	AM	PM	0° S	46° E/W	46° SE/SW	46° S	46° SW/SE	46° W/E	56° E/W	56° SE/SW	56° S	56° SW/SE	56° W/E	66° E/W	66° SE/SW	66° S	66° SW/SE	66° W/E	76° E/W	76° SE/SW	76° S	76° SW/SE	76° W/E	90° E/W	90° SE/SW	90° S	90° SW/SE	90° W/E
Jan. 21	9	3	85.0	57.4	39.0	53.5	84.2	90.0	53.1	29.1	48.4	84.6	90.0	50.1	19.2	44.5	85.1	90.0	48.2	9.5	42.2	85.7	90.0	48.4	6.0	42.1	86.9	90.0
	10	2	80.1	62.8	37.0	42.1	71.3	90.0	61.0	28.5	35.5	70.8	90.0	60.1	21.2	30.6	71.0	90.0	60.2	16.7	28.5	71.7	90.0	62.0	19.2	30.0	73.7	90.0
	11	1	77.1	70.7	40.6	33.4	59.2	90.0	70.9	34.8	24.8	57.6	90.0	71.7	30.9	17.6	57.1	90.0	73.1	29.7	14.1	57.7	90.0	75.9	32.9	19.3	60.3	90.0
	12		76.0	80.3	48.6	30.0	48.6	80.3	82.2	45.2	20.0	45.2	82.2	84.4	43.5	10.0	43.5	84.4	86.6	43.6	90.0	43.6	86.6	90.0	46.7	14.0	46.7	90.0
Feb. 21	8	4	82.5	45.2	38.5	62.9	90.0	90.0	38.7	29.6	60.5	90.0	90.0	33.6	21.4	59.0	90.0	90.0	30.8	15.5	58.5	90.0	90.0	31.1	16.2	59.6	90.0	90.0
	9	3	75.8	47.8	29.8	49.0	80.8	90.0	44.4	19.8	45.9	82.9	90.0	42.6	9.8	44.3	85.1	90.0	42.7	0.9	44.5	87.5	90.0	45.9	14.2	47.6	90.0	90.0
	10	2	70.6	54.2	27.0	35.9	67.1	90.0	53.5	18.9	31.5	68.3	90.0	54.2	13.4	29.5	70.3	90.0	56.1	14.0	30.6	72.9	90.0	60.5	23.5	36.5	77.3	90.0
	11	1	67.2	63.1	31.8	25.0	53.8	85.1	64.6	27.7	18.0	54.1	89.7	67.0	26.5	14.8	55.6	90.0	70.1	28.8	17.6	58.2	90.0	75.2	36.2	27.7	63.5	90.0
	12		66.0	73.6	41.6	20.0	41.6	73.6	76.9	40.3	10.0	40.3	76.9	80.5	40.9	90.0	40.9	80.5	84.4	43.5	10.0	43.5	84.4	90.0	49.8	24.0	49.8	90.0
Mar. 21	7	5	81.7	37.3	45.5	75.2	90.0	90.0	28.1	39.4	75.0	90.0	90.0	19.7	34.8	75.2	90.0	90.0	13.5	32.3	75.9	90.0	90.0	15.0	33.4	77.6	90.0	90.0
	8	4	73.8	35.2	32.3	60.5	90.0	90.0	29.0	24.9	60.0	90.0	90.0	25.2	19.8	60.5	90.0	90.0	24.8	18.9	62.0	90.0	90.0	30.0	25.1	65.5	90.0	90.0
	9	3	66.7	38.4	21.2	45.9	77.7	90.0	36.2	11.7	45.0	81.4	90.0	36.2	4.9	45.9	85.3	90.0	38.6	10.6	48.4	89.3	90.0	45.0	23.9	54.1	90.0	90.0
	10	2	61.0	45.8	17.1	31.5	63.4	90.0	46.7	10.3	30.0	66.5	90.0	49.2	10.3	31.5	70.3	90.0	53.0	10.7	35.5	74.5	90.0	60.0	30.5	44.1	81.1	90.0
	11	1	57.3	55.8	23.9	18.0	49.4	79.1	58.9	22.6	15.0	51.7	85.0	62.9	25.3	18.0	55.3	90.0	67.6	30.7	24.8	59.8	90.0	75.0	41.5	36.8	67.5	90.0
	12		56.0	67.1	35.9	10.0	35.9	67.1	71.8	37.0	90.0	37.0	71.8	76.9	40.3	10.0	40.3	76.9	82.2	45.2	20.0	45.2	82.2	90.0	54.1	34.0	54.1	90.0
Apr. 21	5	7	88.6	45.7	70.5	90.0	90.0	90.0	37.0	67.7	90.0	90.0	90.0	29.0	65.6	90.0	90.0	90.0	22.5	64.3	90.0	90.0	90.0	18.9	63.8	90.0	90.0	90.0
	6	6	80.4	34.9	56.2	88.0	90.0	90.0	25.1	53.0	88.0	90.0	90.0	15.7	51.1	90.0	90.0	90.0	7.8	50.6	90.0	90.0	90.0	11.6	52.2	90.6	90.0	90.0
	7	5	72.0	26.5	41.8	73.5	88.6	90.0	16.9	38.3	75.3	90.0	90.0	8.2	36.9	77.6	90.0	90.0	6.9	37.7	80.2	90.0	90.0	18.9	42.4	84.4	90.0	90.0
	8	4	63.9	23.6	27.5	58.9	74.5	90.0	18.3	23.7	60.7	90.0	90.0	17.4	23.5	63.4	90.0	90.0	21.6	27.1	67.0	90.0	90.0	32.0	36.1	72.9	90.0	90.0
	9	3	56.4	28.0	13.6	44.2	60.2	84.7	27.9	9.4	46.2	80.0	90.0	30.9	13.8	49.7	85.6	90.0	36.3	22.2	54.4	90.0	90.0	46.2	35.3	62.5	90.0	90.0
	10	2	50.1	37.1	5.7	29.5	45.9	72.5	40.1	7.3	32.0	65.3	84.7	44.9	16.5	36.2	71.0	90.0	50.9	26.3	43.2	77.0	90.0	60.7	40.2	53.8	85.9	90.0
	11	1	45.9	48.3	17.4	14.8	31.5	60.2	53.2	21.3	18.9	50.6	79.7	59.0	28.3	26.2	56.4	79.7	65.5	36.6	34.9	62.9	90.0	75.3	49.1	47.8	72.8	90.0
	12		44.4	60.2	31.5	1.6	31.5	60.2	66.5	35.9	11.6	35.9	66.5	73.1	42.0	21.6	42.0	73.1	80.0	49.2	31.6	49.2	80.0	90.0	60.3	45.6	60.3	90.0
May 21	4	8	88.8	53.3	82.4	90.0	90.0	90.0	46.7	81.5	90.0	90.0	90.0	41.2	80.9	90.0	90.0	90.0	37.4	80.5	90.0	90.0	90.0	35.5	80.5	90.0	90.0	90.0
	5	7	81.5	40.9	68.6	90.0	90.0	90.0	33.4	67.4	90.0	90.0	90.0	27.2	66.9	90.0	90.0	90.0	23.6	67.1	90.0	90.0	90.0	24.8	68.7	90.0	90.0	90.0
	6	6	73.5	29.2	54.7	86.6	90.0	90.0	20.3	53.3	90.0	90.0	90.0	13.1	53.2	90.0	90.0	90.0	11.4	54.4	90.0	90.0	90.0	20.0	58.0	89.4	90.0	90.0
	7	5	65.2	19.2	40.7	72.6	90.0	90.0	9.2	39.4	75.9	90.0	90.0	1.0	40.2	79.6	90.0	90.0	10.8	43.0	83.6	90.0	90.0	24.8	49.5	89.4	90.0	90.0
	8	4	56.9	15.2	26.6	58.5	86.2	90.0	11.4	26.0	62.0	86.0	90.0	14.3	28.8	66.2	90.0	90.0	22.8	34.3	71.1	90.0	90.0	35.5	43.6	78.6	90.0	90.0
	9	3	49.1	21.1	12.6	44.9	72.9	90.0	23.5	14.8	48.4	79.2	90.0	29.1	21.9	53.5	86.0	90.0	36.6	33.0	59.5	86.0	90.0	48.4	47.6	68.9	89.5	90.0
	10	2	42.4	31.7	3.7	30.6	58.7	79.9	36.6	13.6	35.5	65.1	88.7	43.1	23.6	42.1	72.0	90.0	50.6	33.6	49.6	79.2	90.0	62.0	55.3	61.1	77.0	90.0
	11	1	37.7	43.6	16.5	17.6	44.5	68.0	49.9	23.7	24.8	51.0	76.1	57.0	32.7	33.3	58.2	84.3	64.7	42.0	42.6	65.3	90.0	75.9	65.4	54.0	65.4	90.0
	12		36.0	55.8	30.6	10.0	30.6	55.8	63.1	37.2	20.0	37.2	63.1	70.8	44.9	30.0	44.9	70.8	78.7	53.2	40.0	53.2	78.7	90.0	65.4	54.0	65.4	90.0
June 21	4	8	85.8	51.5	81.5	90.0	90.0	90.0	45.6	81.2	90.0	90.0	90.0	40.9	81.2	90.0	90.0	90.0	37.9	81.4	90.0	90.0	90.0	37.4	82.2	90.0	90.0	90.0
	5	7	78.6	39.3	68.0	90.0	90.0	90.0	32.3	67.4	90.0	90.0	90.0	27.6	67.0	90.0	90.0	90.0	24.8	68.4	90.0	90.0	90.0	27.6	70.7	90.0	90.0	90.0
	6	6	70.7	27.2	54.4	86.0	90.0	90.0	19.1	53.7	90.0	90.0	90.0	13.5	54.3	90.0	90.0	90.0	14.1	56.2	88.0	90.0	90.0	23.4	60.6	90.0	90.0	90.0
	7	5	62.4	16.5	40.6	72.4	85.2	90.0	9.4	40.3	76.3	90.0	90.0	3.9	41.9	80.5	90.0	90.0	13.6	45.4	72.8	90.0	90.0	27.6	52.5	80.9	90.0	90.0
	8	4	54.1	11.8	26.9	58.6	85.2	89.2	9.4	27.6	62.7	90.0	90.0	15.3	31.4	67.5	90.0	90.0	24.1	37.0	72.8	90.0	90.0	37.4	47.7	80.9	90.0	90.0
	9	3	46.2	18.5	13.7	44.9	71.8	78.0	22.3	17.7	49.6	78.9	90.0	29.0	25.3	55.3	86.2	90.0	37.2	34.0	61.7	86.2	90.0	49.6	47.0	68.8	89.5	90.0
	10	2	39.3	29.8	6.8	31.6	58.2	66.3	35.6	16.8	37.4	65.2	87.0	42.8	26.8	44.5	72.5	90.0	50.8	36.8	52.4	80.1	90.0	62.7	50.7	64.1	77.0	90.0
	11	1	34.4	41.9	17.3	19.6	44.5	54.2	48.8	25.7	27.6	51.5	74.6	56.5	34.9	36.5	59.1	83.2	64.5	44.4	45.8	67.2	90.0	76.3	58.0	59.2	78.8	90.0
	12		32.6	54.2	30.8	13.4	30.8	54.2	61.9	38.1	23.4	38.1	61.9	69.9	46.3	33.4	46.3	69.9	78.2	55.0	43.4	55.0	78.2	90.0	67.6	57.4	67.6	90.0

(Based on data in Table 1, pp. 387, 1972 ASHRAE HANDBOOK OF FUNDAMENTALS; 0% ground reflectance, 1.0 clearness factor)

Table 4-A   Solar Angle of Incidence on Inclined Surfaces for Selected Northern Latitudes

56°N

Solar Angle Of Incidence On Surfaces Tilted With Respect To The Horizontal

Date	AM	PM	0°	46° E/W	46° SE/SW	46° S	46° SW/SE	46° W/E	56° E/W	56° SE/SW	56° S	56° SW/SE	56° W/E	66° E/W	66° SE/SW	66° S	66° SW/SE	66° W/E	76° E/W	76° SE/SW	76° S	76° SW/SE	76° W/E	90° E/W	90° SE/SW	90° S	90° SW/SE	90° W/E
July 21	4	8	88.0	52.8	82.2	90.0	90.0	90.0	46.5	81.4	90.0	90.0	90.0	41.2	80.9	90.0	90.0	90.0	37.5	80.7	90.0	90.0	90.0	35.8	80.8	90.0	90.0	90.0
	5	7	81.0	40.6	68.5	90.0	90.0	90.0	33.2	67.4	90.0	90.0	90.0	27.2	67.0	90.0	90.0	90.0	23.8	67.3	90.0	90.0	90.0	25.3	69.0	90.0	90.0	90.0
	6	6	73.0	28.8	54.6	86.5	90.0	90.0	20.1	53.4	90.0	90.0	90.0	13.2	53.4	90.0	90.0	90.0	11.8	54.7	90.0	90.0	90.0	20.6	58.5	90.0	90.0	90.0
	7	5	64.7	18.7	40.6	72.6	90.0	90.0	11.0	39.5	76.0	90.0	90.0	1.3	40.5	79.8	90.0	90.0	11.3	43.4	83.8	90.0	90.0	25.3	50.0	89.8	90.0	90.0
	8	4	56.4	14.6	26.6	58.5	86.0	90.0	11.0	26.3	62.1	90.0	90.0	15.1	29.3	64.4	90.0	90.0	23.0	34.8	71.4	90.0	90.0	35.8	44.9	79.0	90.0	90.0
	9	3	48.6	20.6	12.7	44.5	72.4	90.0	23.4	15.3	48.6	79.1	90.0	29.1	22.5	53.8	86.0	90.0	36.6	31.2	59.9	90.0	90.0	48.6	44.2	69.4	90.0	90.0
	10	2	41.8	31.3	4.2	30.7	58.6	79.6	36.4	14.2	35.8	59.1	88.4	43.1	24.5	42.5	72.1	88.4	50.6	34.2	50.1	79.3	90.0	62.1	48.2	61.6	89.7	90.0
	11	1	37.1	43.3	16.6	17.9	44.6	67.7	49.7	24.2	25.3	51.1	75.8	56.9	33.1	34.0	58.3	75.8	64.7	42.4	43.2	66.6	84.1	76.0	55.8	56.4	77.3	90.0
		12	35.4	55.5	30.6	10.6	30.6	55.5	62.9	37.3	20.6	37.3	62.9	70.6	45.1	30.6	45.1	70.6	78.6	53.5	40.6	53.5	78.6	90.0	65.8	54.6	65.8	90.0
Aug. 21	5	7	88.0	45.3	70.3	90.0	90.0	90.0	36.7	67.7	90.0	90.0	90.0	28.8	65.7	90.0	90.0	90.0	22.4	64.5	90.0	90.0	90.0	19.3	64.2	90.0	90.0	90.0
	6	6	79.8	34.4	56.0	87.9	90.0	90.0	24.7	53.0	90.0	90.0	90.0	15.3	51.2	90.0	90.0	90.0	7.8	50.9	90.0	90.0	90.0	12.3	52.7	90.0	90.0	90.0
	7	5	71.5	25.9	41.7	73.4	90.0	90.0	16.2	38.4	75.4	90.0	90.0	7.8	37.1	77.8	90.0	90.0	6.9	38.1	80.5	90.0	90.0	19.3	43.0	84.4	90.0	90.0
	8	4	63.3	22.9	27.3	58.8	90.0	90.0	17.6	23.8	60.8	90.0	90.0	17.1	23.9	63.6	90.0	90.0	21.6	27.7	67.3	90.0	90.0	32.2	36.0	73.4	90.0	90.0
	9	3	55.7	27.4	13.3	44.2	74.3	90.0	27.4	9.7	46.3	79.9	90.0	30.7	14.5	50.0	85.7	90.0	36.2	22.9	54.8	90.0	90.0	46.3	36.8	63.0	90.0	90.0
	10	2	49.5	36.6	5.1	29.5	60.1	84.3	39.8	7.6	32.2	65.3	86.0	44.7	17.1	37.2	71.0	90.0	50.9	26.9	43.8	77.2	90.0	60.8	40.8	54.4	86.2	90.0
	11	1	45.2	47.8	17.2	14.9	45.7	72.1	52.9	21.4	19.3	50.6	73.1	58.8	28.6	28.8	56.5	79.4	65.4	37.0	35.5	63.1	86.8	75.4	49.6	48.4	73.1	90.0
		12	43.7	59.9	31.4	2.3	31.4	59.9	66.2	36.0	12.3	36.0	66.2	72.9	42.2	22.3	42.2	72.9	79.9	49.5	32.3	49.5	79.9	90.0	60.8	46.3	60.8	90.0
Sept. 21	7	5	81.7	37.3	45.5	75.2	90.0	90.0	28.1	39.4	75.0	90.0	90.0	19.7	34.8	75.2	90.0	90.0	13.5	32.3	75.9	90.0	90.0	15.0	33.4	77.6	90.0	90.0
	8	4	73.8	35.2	32.3	60.5	90.0	90.0	29.0	24.9	60.0	90.0	90.0	25.2	19.8	60.5	90.0	90.0	24.8	18.9	62.0	90.0	90.0	30.0	25.1	65.5	90.0	90.0
	9	3	66.7	38.4	21.2	45.9	77.7	90.0	36.2	11.7	45.0	81.4	90.0	36.2	4.9	45.9	85.3	90.0	38.6	10.7	48.4	89.3	90.0	45.0	23.9	54.1	90.0	90.0
	10	2	61.0	45.8	17.1	31.5	63.4	90.0	46.7	11.0	30.0	66.5	90.0	49.2	10.3	31.5	70.3	90.0	53.0	17.7	35.5	74.5	90.0	60.0	30.5	44.1	89.3	90.0
	11	1	57.3	55.8	23.9	18.0	49.4	79.1	58.9	22.6	15.0	51.7	85.0	62.9	25.3	18.0	55.3	86.8	67.6	30.9	24.8	59.8	90.0	75.0	41.5	36.8	74.5	90.0
		12	56.0	67.1	35.9	10.0	35.9	67.1	71.8	37.0	0.0	37.0	71.8	76.9	40.3	10.0	40.3	76.9	82.2	45.2	20.0	45.2	82.2	90.0	54.1	34.0	54.1	90.0
Oct. 21	8	4	82.9	45.7	38.9	63.1	90.0	90.0	39.2	29.9	60.6	90.0	90.0	34.1	21.6	58.9	90.0	90.0	31.1	15.5	58.4	90.0	90.0	31.6	15.8	59.4	90.0	90.0
	9	3	76.2	48.3	30.3	49.2	81.0	90.0	44.8	20.3	46.0	82.9	90.0	43.0	10.3	44.2	85.1	90.0	42.9	0.7	44.3	87.4	90.0	46.0	13.8	47.3	85.1	90.0
	10	2	71.0	54.6	27.5	36.2	67.3	90.0	53.9	19.3	31.6	68.5	90.0	54.5	13.7	29.5	70.3	90.0	56.2	14.0	30.4	72.8	90.0	60.6	23.2	36.1	70.8	90.0
	11	1	67.7	63.4	32.2	25.4	54.1	85.4	64.9	28.0	18.2	54.2	89.9	67.2	26.7	14.8	55.6	89.9	70.2	28.8	17.3	58.2	90.0	75.3	36.0	27.2	57.5	90.0
		12	66.5	73.9	42.0	20.5	42.0	73.9	77.1	40.5	10.5	40.5	77.1	80.7	41.0	0.5	41.0	80.7	84.5	43.5	9.5	43.5	84.5	90.0	49.6	23.5	49.6	90.0
Nov. 21	9	3	84.8	57.2	38.9	53.4	84.2	90.0	52.9	28.9	48.3	84.5	90.0	49.8	19.0	44.4	85.0	90.0	48.1	9.3	42.2	85.8	90.0	48.3	6.1	42.2	86.9	90.0
	10	2	79.9	62.6	36.8	42.4	71.2	90.0	60.8	28.3	35.4	70.8	90.0	59.9	21.0	30.5	71.0	90.0	60.1	16.6	28.2	71.8	90.0	61.9	19.2	30.1	73.8	90.0
	11	1	76.9	70.5	40.4	33.3	59.1	90.0	70.8	34.7	24.7	57.5	90.0	71.6	30.8	17.5	57.0	90.0	73.1	29.7	14.1	57.7	90.0	75.9	33.0	19.5	60.4	90.0
		12	75.8	80.2	48.4	29.8	48.4	80.2	82.1	45.1	19.8	45.1	82.1	84.3	43.4	9.8	43.4	84.3	86.5	43.6	0.2	43.6	86.6	90.0	46.7	14.2	46.7	90.0
Dec. 21	9	3	88.1	60.7	43.3	55.3	85.4	90.0	56.2	32.4	49.6	85.2	90.0	52.7	22.5	44.9	85.1	90.0	50.4	12.9	41.8	85.2	90.0	49.6	4.9	40.5	85.5	90.0
	10	2	83.4	65.8	40.7	44.5	72.9	90.0	63.6	31.9	37.4	71.8	90.0	62.3	24.2	31.6	71.4	90.0	61.8	18.7	28.0	71.5	90.0	62.7	18.7	28.2	72.6	90.0
	11	1	80.5	73.4	43.7	36.5	61.2	90.0	73.2	37.5	27.6	59.0	90.0	73.5	33.0	19.6	57.8	90.0	74.3	30.7	14.3	57.7	90.0	76.3	32.4	16.8	59.4	90.0
		12	79.4	82.7	51.2	33.4	51.2	82.7	84.1	47.3	23.4	47.3	84.1	85.7	44.8	13.5	44.8	85.7	85.7	44.0	3.4	44.0	87.5	90.0	46.0	10.5	46.0	87.5

Source: Richard C. Jordan and Benjamin Y. H. Liu, eds. Applications of Solar Energy for Heating and Cooling of Buildings. Atlanta: ASHRAE, 1977, Table 11, pp. IV-24–IV-33. Reprinted with permission.

Table 4-B Solar Position and Other Related Angles for Selected Northern Latitudes

16°N

	Solar Time	Solar Position		Profile (Shadow Line) Angle												Angles of Incidence Vertical Surfaces													Solar Time
Date	AM	ALT	AZ	N	NNE	NE	ENE	E	ESE	SE	SSE	S	SSW	SW	WSW	N	NNE	NE	ENE	E	ESE	SE	SSE	S	SSW	SW	WSW	HOR	PM
DEC	7	7	63			21	10	8	7	7	9	15	58					72	50	28	8	19	41	63	86			83	5
	8	19	57			59	32	23	20	20	23	33	63					78	57	37	22	23	39	59	80			71	4
	9	31	49			83	53	38	32	31	34	42	62					86	67	50	35	31	40	56	74			59	3
	10	41	37				74	55	45	41	42	47	60	81					79	63	49	41	43	53	68	84		49	2
	11	48	21					72	58	51	48	50	57	70	88					76	63	52	48	51	61	74	89	42	1
	12	51	0					90	73	60	53	51	53	60	73					90	76	63	54	51	54	63	76	39	12
JAN	7	8	66			21	11	9	8	9	11	19	83					69	47	25	8	23	45	67	89			82	5
+	8	21	61			55	32	24	21	22	26	38	73					75	55	36	22	26	43	63	84			69	4
NOV	9	33	52			79	52	39	34	33	37	47	68					84	65	48	36	34	43	59	77			57	3
	10	43	40				72	56	47	44	45	51	64	85					77	62	50	44	46	56	71	87		47	2
	11	51	23				90	73	60	53	51	53	60	73					90	76	63	54	51	55	64	76		39	1
	12	54	0					90	74	63	56	54	56	63	74					90	77	65	57	54	57	65	77	36	12
FEB	7	11	75		56	21	14	11	11	13	18	38					82	60	39	18	14	32	54	76				79	5
+	8	25	70		85	48	32	26	25	27	34	53					88	68	48	32	25	34	52	72				65	4
OCT	9	38	62			70	51	42	38	39	45	59	83					77	60	46	38	41	52	68	86			52	3
	10	50	50			86	69	57	51	50	53	62	76					87	73	61	52	50	55	66	79			40	2
	11	59	30				86	74	65	60	60	63	70	81					86	75	66	61	60	64	72	82		31	1
	12	63	0					90	79	70	65	63	65	70	79					90	80	71	65	63	65	71	80	27	12
MAR	7	14	86		39	21	16	14	15	19	30	74					72	51	30	15	23	43	64	86				76	5
+	8	29	81		67	43	33	29	29	34	46	74					78	59	42	30	31	45	63	82				61	4
SEP	9	43	75		82	62	50	44	43	47	56	74					85	69	55	45	43	50	63	79				47	3
	10	56	64			77	66	59	56	58	64	74	88					79	68	60	56	59	66	76	88			34	2
	11	68	44				82	74	70	68	70	74	81	90					82	75	70	68	70	75	82	90		22	1
	12	74	0					90	84	79	75	74	75	79	84					90	84	79	75	74	75	79	84	16	12
APR	6	3	101	16	6	4	3	3	4	6	16					79	56	34	12	12	34	56	79					87	6
+	7	17	97	68	32	22	18	18	20	27	50					83	62	41	23	19	34	54	76					73	5
AUG	8	32	94	84	54	39	33	32	35	43	63					87	68	50	36	32	40	56	74					58	4
	9	46	90	90	70	56	48	46	48	56	70					90	74	61	50	46	50	61	75					44	3
	10	61	85		80	70	63	61	62	67	76	87					81	71	64	61	62	68	77	88				29	2
	11	75	75		88	82	78	75	75	77	81	86					88	82	78	75	75	77	81	86				15	1
	12	86	0					90	88	87	86	86	86	87	88					90	88	87	86	86	86	87	88	4	12
MAY	6	5	109	16	8	6	5	6	7	12	59					71	48	26	6	20	42	64	87					85	6
+	7	19	106	51	29	22	19	20	24	36	72					75	54	34	20	25	42	63	84					71	5
JUL	8	33	104	70	48	37	33	34	39	51	77					79	60	44	34	36	47	64	83					57	4
	9	47	102	79	62	52	48	48	53	63	81					82	67	55	48	48	56	68	83					43	3
	10	61	103	83	73	65	62	62	66	74	85					84	74	66	62	62	67	75	85					29	2
	11	75	108	85	80	77	75	76	79	83	89					85	80	77	75	76	79	83	89					15	1
	12	86	180	86	86	87	88	90								86	86	87	88	90								4	12
JUN	6	6	113	16	9	7	6	7	9	16						68	45	23	6	23	45	68						84	6
	7	20	110	47	28	22	20	21	26	40	82					72	51	32	20	28	46	66	87					70	5
	8	33	108	65	46	37	33	35	41	55	83					75	57	42	34	37	51	68	86					57	4
	9	47	107	74	59	51	47	49	55	67	85					78	64	53	47	50	59	72	87					43	3
	10	61	110	79	69	63	61	62	68	77	88					81	71	64	61	63	69	78	89					29	2
	11	74	120	82	77	75	74	76	80	86						82	77	75	74	76	80	86						16	1
	12	83	180	83	83	85	87	90								83	83	85	87	90								7	12
				N	NNW	NW	WNW	W	WSW	SW	SSW	S	SSE	SE	ESE	N	NNW	NW	WNW	W	WSZ	SW	SSW	S	SSE	SE	ESE	HOR	PM

Dates vary year to year within plus or minus three days of the twenty-first day of the month

Table 4-B Solar Position and Other Related Angles for Selected Northern Latitudes

24°N

	Solar Time	Solar Position		Profile (Shadow Line) Angles												Angles of Incidence Vertical Surfaces													Solar Time
DATE	AM	ALT	AZ	N	NNE	NE	ENE	E	ESE	SE	SSE	S	SSW	SW	WSW	N	NNE	NE	ENE	E	ESE	SE	SSE	S	SSW	SW	WSW	HOR	PM
DEC	7	3	63			10	5	4	3	3	4	7	33					72	50	28	6	18	40	63	85			87	5
	8	15	55			56	26	18	15	15	18	25	52					80	58	37	19	18	36	57	78			75	4
	9	26	46			88	50	34	52									89	69	50	33	26	34	51	71			64	3
		34	34				74	51	39	35	35	39	51	74					81	63	47	36	36	47	63	81		56	2
	11	40	18					70	53	44	40	42	48	62	85					76	60	47	41	44	55	70	87	50	1
	12	43	0					90	67	52	45	43	45	52	67					90	74	59	47	43	47	59	74	47	12
JAN	7	5	66			13	7	5	5	5	7	11	69					69	47	25	5	21	43	66	88			85	5
+	8	17	58			53	27	20	17	17	20	30	62					77	56	36	19	21	39	60	81			73	4
NOV	9	28	49			83	50	35	29	28	31	39	59					87	67	48	33	28	38	54	73			62	3
	10	37	36				73	52	42	38	38	43	56	79					79	62	47	38	39	50	66	83		53	2
	11	44	20					71	55	47	44	45	52	66	87					76	61	49	44	47	58	72	88	46	1
	12	46	0					90	70	56	48	46	48	56	70					90	75	61	50	46	50	61	75	44	12
FEB	7	9	74		55	18	11	9	9	10	14	30					84	62	40	18	11	30	52	74				81	5
+	8	22	66			48	30	24	22	23	29	45	87					70	50	32	22	30	48	68	89			68	4
OCT	9	34	57			73	50	39	34	35	39	51	75					80	62	46	35	36	47	63	81			56	3
	10	45	44				70	55	47	45	46	54	68	88					75	61	49	45	48	59	73	89		45	2
	11	52	25				88	72	60	54	52	55	62	75					89	75	63	55	52	56	65	78		38	1
	12	55	0					90	75	64	57	55	57	64	75					90	77	66	58	55	58	66	77	35	12
MAR	7	14	84		41	21	16	14	14	17	27	66					74	53	32	15	21	41	62	84				76	5
+	8	27	77		73	44	32	28	27	31	41	66					82	62	44	30	29	41	59	78				63	4
SEP	9	40	68		90	65	50	42	40	43	50	66					90	73	57	45	40	45	58	73				50	3
	10	52	55			82	68	58	53	53	57	66	80					84	71	60	53	53	59	69	82			38	2
	11	62	33				84	74	66	62	62	66	73	84					85	75	67	63	62	67	75	85		28	1
	12	66	0					90	80	73	68	66	68	73	80					90	81	73	68	66	68	73	81	24	12
APR	6	5	101	24	9	6	5	5	6	8	22					79	57	35	13	12	33	56	78					85	6
+	7	18	95	76	36	23	19	18	20	27	47					85	64	43	25	19	33	52	73					72	5
AUG	8	32	89		60	42	34	32	34	41	57	88					72	54	39	32	38	52	70	89				58	4
	9	46	82		76	60	50	46	47	52	64	82					80	65	53	46	47	56	69	84				44	3
	10	59	72		87	75	65	60	59	62	69	79					88	77	67	61	59	63	70	81				31	2
	11	71	52			88	81	75	72	71	73	78	85					88	81	75	72	71	74	78	85			19	1
	12	78	0					90	85	81	79	78	79	81	85					90	85	81	79	78	79	81	85	12	12
MAY	6	8	108	24	12	9	8	8	11	17	63					72	50	28	9	20	42	64	86					82	6
+	7	21	103	59	34	24	21	22	26	36	67					78	57	38	23	25	41	61	81					69	5
JUL	8	35	98	78	53	41	35	35	39	49	71					83	65	49	37	36	45	61	79					55	4
	9	48	94	87	69	56	50	48	51	59	74					88	73	60	51	48	53	64	78					42	3
	10	62	88		80	70	64	62	63	69	77	89					81	71	65	62	64	70	79	89				28	2
	11	76	77		88	82	78	76	76	78	81	87					88	82	78	76	76	78	82	87				14	1
	12	86	0					90	88	87	86	86	86	87	88					90	88	87	86	86	86	87	88	4	12
JUN	6	9	112	24	13	10	9	10	13	22	85					69	47	25	9	23	45	67	89					81	6
	7	22	107	55	33	25	22	23	28	41	76					75	54	35	23	28	44	64	85					68	5
	8	36	103	73	51	40	36	36	41	53	76					80	62	47	37	37	48	64	82					54	4
	9	49	99	82	66	55	50	49	53	63	78					84	70	58	50	50	56	67	81					41	3
	10	63	95	87	77	68	64	63	65	72	81					88	78	69	64	63	66	73	82					27	2
	11	76	91	90	84	80	77	76	77	80	85					90	85	80	77	76	77	80	85					14	1
	12	89	0					90	90	90	89	89	89	90	90					90	90	90	89	89	89	90	90	1	12
				N	NNW	NW	WNW	W	WSW	SW	SSW	S	SSE	SE	ESE	N	NNW	NW	WNW	W	WSW	SW	SSW	S	SSE	SE	ESE	HOR	PM

Dates vary year to year within plus or minus three days of the twenty-first day of the month

Table 4-B Solar Position and Other Related Angles for Selected Northern Latitudes

32°N

	Solar Time	Solar Position		Profile (Shadow Line) Angles												Angles of Incidence Vertical Surfaces													Solar Time
DATE	AM	ALT	AZ	N	NNE	NE	ENE	E	ESE	SE	SSE	S	SSW	SW	WSW	N	NNE	NE	ENE	E	ESE	SE	SSE	S	SSW	SW	WSW	HOR	PM
DEC	8	10	54			50	19	13	11	10	12	17	37					81	59	37	17	14	33	55	77			80	4
	9	20	44				45	28	22	20	21	26	42	86					70	50	31	20	29	47	68	89		70	3
	10	28	31				74	45	33	28	28	31	41	65					82	63	44	31	29	41	58	78		62	2
	11	33	16					66	46	36	33	34	40	53	81					76	58	42	33	36	49	66	85	57	1
	12	35	0					90	61	44	37	35	37	44	61					90	72	54	40	35	40	54	72	55	12
JAN	7	1	65			4	2	2	1	2	2	3	32					70	47	25	3	20	43	65	88			89	5
+	8	13	56			48	22	15	13	13	15	22	49					79	57	36	17	17	36	57	79			77	4
NOV	9	22	46			88	46	30	24	22	24	31	48					89	68	48	31	22	32	50	70			68	3
	10	31	33				73	47	36	31	31	35	46	71					81	62	45	33	32	44	61	80		59	2
	11	36	18					68	49	39	36	37	44	58	83					76	59	44	36	40	52	68	86	54	1
	12	38	0					90	64	48	40	38	40	48	64					90	72	56	43	38	43	56	72	52	12
FEB	7	7	73		52	14	9	7	7	8	10	22					85	62	40	18	9	29	51	73				83	5
+	8	18	64			46	27	20	19	19	24	37	79					72	51	32	19	26	45	65	86			72	4
OCT	9	29	53			76	48	35	30	30	33	43	66					83	64	46	33	30	41	58	77			61	3
	10	38	39				70	52	42	39	40	46	59	82					77	61	47	39	41	52	68	85		52	2
	11	45	21					70	55	47	45	47	54	68	89					75	61	50	45	49	59	73	89	45	1
	12	47	0					90	70	57	49	47	49	57	70					90	75	61	51	47	61	61	75	43	12
MAR	7	13	82		42	21	15	13	13	16	24	58					76	54	33	15	19	39	60	82				77	5
+	8	25	73		78	45	31	26	25	28	36	58					85	65	46	30	26	37	55	75				65	4
SEP	9	37	62			69	50	40	37	38	44	58	83					76	59	45	37	40	52	68	86			53	3
	10	47	47			88	69	56	49	47	50	58	72					88	73	60	50	47	52	63	77			43	2
	11	55	27				87	72	62	56	55	58	65	78					88	75	64	57	55	59	68	80		35	1
	12	58	0					90	77	66	60	58	60	66	77					90	78	68	61	58	61	68	78	32	12
APR	6	6	100	32	11	7	6	6	7	11	26					80	58	36	14	12	33	55	77					84	6
+	7	19	92	84	39	25	20	19	20	27	44					88	67	46	27	19	31	50	71					71	5
AUG	8	31	84		65	44	35	32	33	38	52	80					76	58	41	32	35	48	66	85				59	4
	9	44	74		83	63	51	45	44	48	57	74					85	69	56	46	44	51	63	79				46	3
	10	56	60			80	67	59	56	57	62	71	85					81	70	61	56	57	64	74	86			34	2
	11	65	37				83	74	68	66	66	70	77	87					84	75	69	66	66	71	78	87		25	1
	12	70	0					90	82	75	71	70	71	75	82					90	82	76	71	70	71	76	82	20	12
MAY	6	10	107	32	16	12	10	11	13	22	63					73	51	30	12	20	41	63	85					80	6
+	7	23	100	67	38	27	23	23	27	36	63					81	60	41	26	25	39	58	79					67	5
JUL	8	35	93	86	59	44	37	35	38	47	65					88	70	53	40	36	43	57	74					55	4
	9	48	85		75	60	52	48	49	55	67	85					79	65	54	48	50	59	72	86				42	3
	10	61	73		87	75	66	62	61	64	70	81					87	77	68	62	61	64	72	82				29	2
	11	72	52			88	81	76	73	72	74	79	85					88	81	76	73	72	74	79	85			18	1
	12	78	0					90	85	81	79	78	79	81	85					90	85	82	79	78	79	82	85	12	12
JUN	5	1	118	1	1	1	1	1	1	2						62	40	17	5	28	50	73						89	7
	6	12	110	32	18	13	12	13	16	27	79					70	48	27	12	23	44	66	88					78	6
	7	24	103	63	38	28	25	25	29	41	71					78	58	39	26	28	42	62	82					66	5
	8	37	97	81	57	44	38	37	41	50	70					85	67	51	40	37	46	60	77					53	4
	9	50	89		72	59	52	50	52	59	72	90					76	63	53	50	53	62	75	90				40	3
	10	62	80		84	73	66	63	63	67	74	85					84	75	67	63	63	67	75	85				28	2
	11	74	61			86	80	76	74	75	78	82	88					86	80	76	74	75	78	82	88			16	1
	12	81	0					90	87	84	82	81	82	84	87					90	87	84	82	81	82	84	87	9	12
				N	NNW	NW	WNW	W	WSW	SW	SSW	S	SSE	SE	ESE	N	NNW	NW	WNW	W	WSW	SW	SSW	S	SSE	SE	ESE	HOR	PM

Dates vary year to year within plus or minus three days of the twenty-first day of the month.

Table 4-B Solar Position and Other Related Angles for Selected Northern Latitudes

40°N

	Solar Time	Solar Position		Profile (Shadow Line) Angles												Angles of Incidence Vertical Surfaces													Solar Time
DATE	AM	ALT	AZ	N	NNE	NE	ENE	E	ESE	SE	SSE	S	SSW	SW	WSW	N	NNE	NE	ENE	E	ESE	SE	SSE	S	SSW	SW	WSW	HOR	PM
DEC	8	5	53			35	11	7	6	6	6	9	21					82	60	37	16	10	31	53	76			85	4
	9	14	42				37	20	15	14	15	18	30	78					71	50	29	14	24	44	65	87		76	3
	10	21	29				72	38	26	21	21	23	31	54					84	63	43	26	22	35	55	75		69	2
	11	25	15					61	37	28	25	26	31	43	75					76	56	38	26	29	44	63	83	65	1
	12	27	0					90	53	35	28	27	28	35	53					90	70	51	34	27	34	51	70	63	12
JAN	8	8	55			38	15	10	8	8	10	14	34					80	58	36	15	13	34	56	78			82	4
+	9	17	44				40	24	18	17	18	23	37	87					70	48	29	17	27	46	68	89		73	3
NOV	10	24	31				72	41	29	24	24	27	36	61					82	62	43	27	25	38	57	77		66	2
	11	28	16					63	41	32	29	29	35	48	78					76	57	40	29	32	47	65	84	62	1
	12	30	0					90	56	39	32	30	32	39	56					90	71	52	37	30	37	52	71	60	12
FEB	7	4	72		43	9	6	4	4	5	7	14					85	63	41	18	6	27	50	72				86	5
+	8	15	62			43	23	17	15	15	19	29	69					74	52	32	16	22	41	63	84			75	4
OCT	9	24	50			80	45	31	25	24	27	35	56					86	65	46	30	25	36	54	74			66	3
	10	32	35				70	47	36	32	33	38	50	75					79	61	44	33	34	46	63	82		58	2
	11	37	19					67	49	40	37	39	45	60	85					75	58	45	37	41	53	69	87	53	1
	12	39	0					90	65	49	41	39	41	49	65					90	73	57	44	39	44	57	73	51	12
MAR	7	11	80		43	13	13	12	12	14	21	50					78	56	34	15	17	37	58	80				79	5
+	8	23	70		85	45	29	24	23	25	31	50					88	67	47	30	23	33	51	71				67	4
SEP	9	33	57			72	48	37	33	33	38	50	75					80	61	45	34	35	46	63	81			57	3
	10	42	42				69	53	45	42	43	50	64	87					76	60	48	42	45	56	71	88		48	2
	11	48	23				90	71	57	50	48	50	57	71					90	75	62	52	48	52	62	75		42	1
	12	50	0					90	72	59	52	50	52	59	72					90	76	63	54	50	54	63	76	40	12
APR	6	7	99	40	14	9	8	8	9	12	29					81	59	37	15	12	32	54	77					83	6
+	7	19	89		42	26	20	19	20	26	41	88					69	48	29	19	29	48	68	89				71	5
AUG	8	30	79		71	46	35	31	31	35	47	72					80	61	44	32	32	44	62	81				60	4
	9	41	67			67	51	44	41	43	51	66	90					74	58	46	41	46	58	73	90			49	3
	10	51	51			85	69	58	52	51	55	63	77					86	72	61	53	51	57	67	80			39	2
	11	59	29				86	73	64	60	59	62	69	81					87	75	66	60	59	63	71	82		31	1
	12	62	0					90	78	69	63	62	63	69	78					90	80	70	64	62	64	70	80	28	12
MAY	5	2	115	5	3	2	2	2	3	6						65	43	20	3	25	47	70						88	7
+	6	13	106	40	20	15	13	13	16	25	62					75	53	32	14	20	40	61	83					77	6
JUL	7	24	97	75	42	30	25	24	27	36	58					84	64	44	28	25	37	55	76					66	5
	8	35	87		65	47	38	35	37	44	59	86					74	57	43	36	40	53	70	88				55	4
	9	47	76		82	64	53	48	47	51	61	77					84	69	57	48	47	54	66	80				43	3
	10	57	61			80	68	61	58	58	63	73	86					82	70	62	58	59	65	75	86			33	2
	11	66	37				84	75	69	66	67	71	77	87					84	76	70	66	67	71	78	87		24	1
	12	70	0					90	82	76	71	70	71	76	82					90	82	76	72	70	72	76	82	20	12
JUN	5	4	117	9	6	4	4	5	7	14						63	40	18	6	28	50	72						86	7
	6	15	108	40	22	16	15	16	19	31	75					72	51	30	15	23	43	64	86					75	6
	7	26	100	71	42	31	27	26	30	40	66					81	61	43	29	28	40	59	79					64	5
	8	37	91	89	63	47	39	37	40	48	64					89	72	55	42	37	43	56	73					53	4
	9	49	80		79	63	54	49	50	54	65	82					82	68	56	50	50	57	69	84				41	3
	10	60	66			78	68	62	60	61	67	77	89					80	70	63	60	62	69	78	89			30	2
	11	69	42				83	76	71	69	70	74	81	89					83	76	71	69	70	75	81	89		21	1
	12	73	0					90	84	78	75	73	75	78	84					90	84	78	75	73	75	78	84	17	12
				N	NNW	NW	WNW	W	WSW	SW	SSW	S	SSE	SE	ESE	N	NNW	NW	WNW	W	WSW	SW	SSW	S	SSE	SE	ESE	HOR	PM

Dates vary year to year within plus or minus three days of the twenty-first day of the month.

Table 4-B Solar Position and Other Related Angles for Selected Northern Latitudes

48°N

Date	Solar Time AM	Solar Position ALT	AZ	Profile (Shadow Line) Angles N	NNE	NE	ENE	E	ESE	SE	SSE	S	SSW	SW	WSW	Angles of Incidence Vertical Surfaces N	NNE	NE	ENE	E	ESE	SE	SSE	S	SSW	SW	WSW	HOR	Solar Time PM
DEC	8	1	53			5	1	1	1	1	1	1	2					82	60	37	15	8	30	53	75			89	4
	9	8	41				24	12	9	8	8	10	17	63				72	50	28	9	20	42	64	86			82	3
	10	14	28				68	27	17	14	14	15	21	40				84	63	41	22	15	31	52	74	82		76	2
	11	17	14					51	27	20	17	18	21	31	66				76	55	35	19	22	40	61	82		73	1
	12	19	20					90	41	25	20	19	20	25	41				90	69	48	29	19	29	48	69	71	12
JAN	8	3	55			20	6	4	4	4	4	6	15					80	58	36	13	10	32	55	77			87	4
+	9	11	43				29	16	12	11	12	15	25	78				70	48	27	11	23	44	66	88			79	3
NOV	10	17	29				68	32	21	17	17	19	26	48				83	62	41	23	18	34	54	75			73	2
	11	21	15					55	32	24	21	21	25	37	71				76	55	36	22	25	42	62	83		69	1
	12	22	0					90	47	30	24	22	24	30	47				90	69	49	31	22	31	49	69	68	12
FEB	7	2	72		23	4	2	2	2	2	3	6					86	63	41	18	5	27	49	72			88	5
+	8	11	60			37	18	13	11	11	14	21	56				75	53	32	13	19	39	61	83			79	4
OCT	9	19	47			83	39	25	20	19	21	27	45				88	67	46	27	19	31	50	71			71	3
	10	25	33				69	41	30	26	26	30	40	66				81	61	42	28	27	41	59	79		65	2
	11	30	17					63	42	33	30	31	36	51	81				75	56	40	30	34	48	66	85	60	1
	12	31	0					90	58	41	33	31	33	41	58				90	71	53	38	31	38	53	71	59	12
MAR	7	10	79		42	18	12	10	10	12	18	42					79	57	35	15	15	35	57	79			80	5
+	8	20	67			44	27	21	20	21	26	42	88				70	49	30	20	29	48	68	89			70	4
SEP	9	28	53			75	46	34	29	28	32	42	66				83	63	45	31	29	41	58	78			62	3
	10	35	38				70	49	39	36	36	42	55	80				78	60	45	36	38	50	66	84		55	2
	11	40	20					68	52	43	40	42	49	63	87				75	59	46	40	44	56	71	88	50	1
	12	42	0					90	67	52	44	42	44	52	67				90	73	58	47	42	47	58	73	48	12
APR	6	9	98	48	17	11	9	9	10	14	31				82	60	38	17	12	31	53	75				81	6
+	7	19	87		46	27	20	19	20	24	38	80				72	51	31	19	27	45	66	87			71	5
AUG	8	29	75		77	47	34	29	29	32	42	64				84	64	46	32	29	40	58	77			61	4
	9	38	61			70	51	42	38	39	45	58	82				77	60	46	38	41	52	68	85		52	3
	10	46	45				70	56	48	46	48	55	69	90				75	61	50	46	50	60	74	90	44	2
	11	51	24					89	72	60	53	52	54	61	74				75	63	54	52	55	65	77	39	1
	12	54	0					90	74	62	56	54	56	62	74				90	77	65	57	54	57	65	77	36	12
MAY	5	5	114	13	7	6	5	6	8	15				66	43	21	6	25	47	69					85	7
+	6	15	104	48	24	17	15	15	18	27	60			77	55	34	17	20	39	60	81				75	6
JUL	7	25	93	83	47	32	26	25	27	34	54			87	67	47	31	25	35	53	72				65	5
	8	35	82		71	49	39	35	35	41	53	78			78	61	45	36	37	49	65	83			55	4
	9	44	68		89	68	54	46	44	47	54	69			89	74	59	48	44	49	60	75			46	3
	10	53	51			85	70	60	54	53	57	65	78			86	73	62	55	53	58	68	80		37	2
	11	59	29				86	74	65	61	60	63	70	81			87	76	67	61	60	64	71	82	31	1
	12	62	0					90	78	69	64	62	64	69	78			90	80	71	64	62	64	71	80	28	12
JUN	5	8	117	17	10	8	8	9	12	24				64	42	20	9	28	50	72					82	7
	6	17	106	48	26	19	17	18	22	33	70			75	53	33	18	23	42	63	84				73	6
	7	27	96	79	47	33	28	27	30	39	61			85	65	46	31	28	38	56	75				63	5
	8	37	85		69	50	41	37	38	44	58	83			76	59	45	37	40	52	68	86			53	4
	9	47	72		86	67	55	48	47	50	58	74			87	72	59	50	47	52	63	78			43	3
	10	56	55			83	70	61	56	56	60	69	81			85	73	63	57	56	62	71	83		34	2
	11	63	31				86	75	67	63	63	66	73	83			86	76	68	64	63	67	74	84	27	1
	12	65	0					90	80	72	67	65	67	72	80			90	81	73	67	65	67	73	81	25	12

Bottom axis labels: N NNW NW WNW W WSW SW SSW S SSE SE ESE | N NNW NW WNW W WSW SW SSW S SSE SE ESE | HOR | PM

Table 4-B Solar Position and Other Related Angles for Selected Northern Latitudes

56°N

Column groups: **Solar Time (AM)** = AM; **Solar Position** = ALT, AZ; **Profile (Shadow Line) Angles** = first block of N…WSW; **Angles of Incidence Vertical Surfaces** = second block of N…WSW plus HOR; **Solar Time (PM)** = PM.

Date	AM	ALT	AZ	N	NNE	NE	ENE	E	ESE	SE	SSE	S	SSW	SW	WSW	N	NNE	NE	ENE	E	ESE	SE	SSE	S	SSW	SW	WSW	HOR	PM
DEC	9	2	40				6	3	2	2	2	2	4	23					72	50	27	5	18	41	63	85			88	3
	10	7	27				53	14	9	7	7	7	10	21					85	63	40	19	8	28	50	73			88	2
	11	10	14					35	16	11	10	10	12	18	48					76	54	32	13	17	37	59	82	80	1
	12	11	0					90	26	15	11	11	11	15	26					90	68	46	25	11	25	46	68	79	12
JAN + NOV	9	5	42				15	8	6	5	5	5	7	12	58				71	48	26	6	20	42	64	87		85	3
	10	10	28				59	20	13	10	10	10	11	15	31				84	62	40	19	12	30	52	74		80	2
	11	13	14				43	21	15	13	13	13	16	24	59					76	54	33	15	19	39	60	82	77	1
	12	14	0					90	33	19	15	14	15	19	33					90	68	47	26	14	26	47	68	76	12
FEB + OCT	8	7	59			26	11	8	7	7	8	13	39					76	54	32	11	16	37	59	82			83	4
	9	13	46			88	31	19	14	13	15	19	33					89	68	46	26	13	27	47	69			77	3
	10	19	31				66	33	23	19	19	22	30	55					82	61	40	23	21	36	56	77		71	2
	11	22	16					56	33	25	22	23	27	40	74					75	55	36	23	27	43	63	84	68	1
	12	23	0					90	48	31	25	23	25	31	48					90	69	49	32	23	32	49	69	67	12
MAR + SEP	7	8	77		40	15	10	9	8	10	14	34					80	58	36	15	13	33	55	78				82	5
	8	16	64			41	24	18	16	16	17	21	34	80					71	50	30	17	25	44	66	87		74	4
	9	23	50				78	43	29	24	23	26	34	56					85	65	45	29	24	36	54	74		67	3
	10	29	35					69	44	33	29	30	34	46	72					79	60	43	31	41	62	81		61	2
	11	33	18				64	45	36	33	33	34	40	55	83					75	57	41	33	37	50	67	86	57	1
	12	34	0					90	60	44	36	34	36	44	60					90	72	54	40	34	40	54	72	56	12
APR + AUG	5	1	109	4	2	2	2	2	3	21						71	49	26	4	19	41	64	86					89	7
	6	10	97	56	19	12	10	10	11	15	32					84	61	39	19	12	30	52	74					80	6
	7	18	84		49	27	20	18	19	23	34	72					74	53	33	19	24	42	63	84				72	5
	8	26	71			48	33	27	26	29	36	56					87	67	48	32	26	36	53	73			64	4
	9	34	56				74	50	39	34	34	39	50	74					81	62	46	35	35	46	62	81	56	3
	10	40	40				71	53	43	41	41	47	61	84					61	47	40	43	54	69	86		50	2
	11	44	21					70	55	47	44	46	53	67	88					75	61	49	44	48	58	73	89	46	1
	12	46	0					90	69	55	48	46	48	55	69					90	74	60	50	46	50	60	74	44	12
MAY + JUL	4	1	126	2	1	1	1	1	2	7						54	32	10	13	36	58	81						89	8
	5	8	113		21	12	9	8	9	12	22				67	45	23	9	25	47	69						82	7
	6	16	102		56	28	20	17	17	20	28	57					79	58	37	20	20	37	58	79				74	6
	7	25	89			51	33	27	25	26	33	50	89					70	51	33	25	33	50	69	89			65	5
	8	33	76				77	51	39	34	33	37	48	70					83	64	47	36	34	44	60	79		57	4
	9	41	62					54	45	41	42	48	61	83						78	62	48	41	44	54	69	86	49	3
	10	48	44					71	58	50	48	50	57	70	89					76	62	52	48	51	61	75	89	42	2
	11	52	23					89	73	61	54	52	55	62	74					89	76	64	55	52	56	65	77	38	1
	12	54	0					90	74	63	56	54	56	63	74					90	77	65	57	54	57	65	77	36	12
JUN	4	4	127	7	5	4	4	5	8	28						53	31	9	15	37	60	82						86	8
	5	11	115		25	15	12	11	13	17	31				65	43	23	12	28	49	71						79	7
	6	28	92		87	52	36	29	28	30	37	56					89	69	50	34	28	36	53	72				62	6
	7	28	92		87	52	36	29	28	30	37	56					89	69	50	34	28	36	53	72				62	5
	8	36	79			75	52	41	36	36	41	53	75					81	63	48	37	37	48	63	81			54	4
	9	44	64				71	55	47	44	45	52	66	86					76	61	50	44	47	57	72	88	46	3
	10	51	46					89	72	59	53	51	53	61	74					89	75	63	54	51	55	64	77	39	2
	11	56	25					88	74	63	57	56	58	65	77					89	76	65	58	56	56	59	68	34	1
	12	57	0					90	76	68	60	57	60	68	76					90	78	68	60	57	60	68	78	33	12

Bottom (PM-reading) direction headers: Profile block — N NNW NW WNW W WSW SW SSW S SSE SE ESE; Incidence block — N NNW NW WNW W WSW SW SSW S SSE SE ESE HOR; PM.

Dates vary year to year within plus or minus three days of the twenty-first day of the month

Source: ASHRAE Handbook, 1977 Fundamentals. Atlanta: ASHRAE, 1977. pp. 26.5–26.7. Reprinted with permission.

Table 4-C Average Percent Possible Sunshine—Selected Cities, United States

State	Location	Jan	Feb	Mar	Apr	May	Jun	Jul	Aug	Sep	Oct	Nov	Dec	Year
AL	Montgomery	47	53	58	64	66	65	63	65	63	66	57	50	59
AK	Juneau	33	32	37	38	38	34	30	30	25	19	23	20	31
AZ	Phoenix	78	80	83	89	93	94	85	85	89	88	83	77	86
AR	Little Rock	46	54	57	61	68	73	71	73	68	69	56	48	63
CA	Los Angeles	69	72	73	70	66	65	82	83	79	73	74	71	73
	Sacramento	45	61	70	80	86	92	97	96	94	84	64	46	79
CO	Denver	72	71	69	66	64	70	70	72	75	73	65	68	70
CN	Hartford	58	57	56	57	58	58	61	63	59	58	46	48	57
DE	Wilmington	50	54	57	57	59	64	63	61	60	60	54	51	53
DC	Washington	48	51	55	56	58	64	62	62	62	60	53	47	57
FL	Jacksonville	57	61	66	71	69	61	59	58	53	56	61	56	61
GA	Atlanta	47	52	57	65	69	67	61	65	63	67	60	50	61
HI	Honolulu	63	65	69	67	70	71	74	75	75	67	60	59	68
ID	Boise	41	52	63	68	71	75	89	85	82	67	45	39	67
IL	Peoria	45	50	52	55	59	66	68	67	64	63	44	39	57
IN	Indianapolis	41	51	51	55	61	68	70	71	66	64	42	39	58
IO	Des Moines	51	54	54	55	60	67	71	70	64	64	49	45	59
KS	Wichita	59	59	60	62	64	69	74	73	65	66	59	56	65
KY	Louisville	41	47	50	55	62	67	66	68	65	63	47	39	57
LA	Shreveport	49	54	56	55	64	71	74	72	68	71	62	53	64
ME	Portland	55	59	56	56	56	60	64	65	61	58	47	53	58
MD	Baltimore	51	55	55	55	57	62	65	62	60	59	51	48	57
MA	Boston	54	56	57	56	58	63	66	67	63	61	51	52	59
MI	Detroit	32	43	49	52	59	65	70	65	61	56	35	32	54
MN	Duluth	49	54	56	54	55	58	67	61	52	48	34	39	54
	Minn.–St. Paul	51	57	54	55	58	63	70	67	61	57	39	40	58
MO	Kansas City	64	54	61	65	67	72	84	69	51	62	46	54	64
	St. Louis	52	51	54	56	62	68	71	66	63	62	49	41	58
MT	Great Falls	49	57	67	62	64	65	81	78	68	61	46	46	64
NE	Omaha	55	55	55	59	62	68	76	72	67	67	52	48	62
NV	Reno	66	68	74	80	81	85	92	93	92	83	70	63	80
NH	Concord	52	54	52	53	54	57	62	60	54	54	42	47	54
NJ	Atlantic City	49	48	51	53	54	58	60	62	59	57	50	42	54
NM	Albuquerque	73	73	74	77	80	83	76	76	80	79	78	72	77
NY	Albany	46	51	52	53	55	59	64	61	56	53	36	38	53
	New York	50	55	56	59	61	64	65	64	63	51	52	49	59
NC	Charlotte	55	59	63	70	69	71	68	70	68	69	63	58	66
ND	Bismarck	54	56	60	58	63	64	76	73	65	59	44	47	62
OH	Cincinnati	41	45	51	55	61	67	68	67	66	59	44	38	57
	Cleveland	32	37	44	53	59	65	68	64	60	55	31	26	52
OK	Okla. City	59	61	63	63	65	73	75	77	69	68	60	59	67
OR	Portland	24	35	42	48	54	51	69	64	60	40	27	20	47
PA	Philadelphia	50	53	56	56	57	63	63	63	60	60	53	49	58
	Pittsburgh	36	38	45	48	53	60	62	60	60	56	40	30	50
RI	Providence	57	56	55	55	57	57	59	59	58	60	49	51	56
SC	Columbia	56	59	64	67	66	65	64	65	65	66	64	60	63
SD	Rapid City	54	59	61	59	57	60	71	73	67	65	56	54	62
TN	Nashville	40	47	52	59	62	67	64	66	63	64	50	40	57
TX	El Paso	78	82	85	87	89	89	79	80	82	84	83	78	83
	Houston	41	54	48	51	57	63	68	61	57	61	58	69	56
UT	Salt Lake City	47	55	64	66	73	78	84	83	84	73	54	44	70
VT	Burlington	42	48	52	50	56	60	65	62	55	50	30	33	51
VA	Richmond	51	54	59	62	64	67	65	64	63	59	56	51	60
WA	Seattle	21	42	49	51	58	54	67	65	61	42	27	17	49
	Spokane	26	41	53	60	63	65	81	78	71	51	28	20	57
WV	Parkersburg	32	36	43	49	56	59	62	60	59	54	37	29	48
WI	Milwaukee	44	47	51	54	59	63	70	67	60	57	41	38	56
WY	Cheyenne	61	65	64	61	58	65	68	68	69	68	60	59	64

Source: Adapted from Comparative Climatic Data, U.S. National Oceanic and Atmospheric Administration. Used with permission.

Table 4-D Values of Direct Normal and Diffuse Solar Radiation for Clear and for Industrial Atmospheres (ASHRAE, 1956) and Average Total Insolation Through Complete Overcasts of Various Cloud Types (Haurnitz, 1948)

Solar altitude β degrees	Optical air mass[a] m csc β	Standard, Cloudless Atmosphere			Industrial, Cloudless Atmosphere			Through Complete Overcasts, Blue Hill, Average Total Insolation on Horizontal			
		Direct normal radiation I_{DN}	Diffuse on horizontal I_d	Total on horizontal I_h	Direct normal radiation I_{DNi}	Diffuse on horizontal I_{di}	Total on horizontal I_{hi}	Cirro-stratus I_{oc}	Alto-cumulus I_{oc}	Strato-cumulus I_{oc}	Fog I_{oc}
5	10.39	67	7	13	34	9	12	—	—	—	—
10	5.60	123	14	35	58	18	28	—	—	15	10
15	3.82	166	19	62	80	24	45	50	35	25	15
20	2.90	197	23	90	103	31	64	70	50	35	20
25	2.36	218	26	118	121	38	89	95	65	40	20
30	2.00	235	28	146	136	44	112	120	75	50	25
35	1.74	248	30	172	148	48	133	145	90	60	30
40	1.55	258	31	197	158	52	154	165	105	70	35
45	1.41	266	32	220	165	55	172	185	115	80	40
50	1.30	272	33	242	172	58	190	205	130	85	40
60	1.15	283	34	279	181	63	220	235	150	100	45
70	1.06	289	35	307	188	69	246	260	160	110	50
80	1.02	292	35	322	195	—	—	—	—	—	—
90	1.00	294	36	328	200	—	—	—	—	—	—

Source: ASHRAE Handbook of Fundamentals, 1956. Reproduced with permission

Note: All tabular values are in Btu/hr-ft².

[a] Smithsonian Meteorological Tables, 6th rev. ed., 1951, p. 422.

Table 4-E Insolation on Inclined Surfaces for Selected Northern Latitudes

24°N

BTU/FT² Hourly Total Insolation, $I_{D\theta} + I_{d}$, On Surfaces Tilted With Respect To The Horizontal

Date	AM	PM	0° S	14° E/W	14° SE/SW	14° S/S	14° SW/SE	14° W/E	24° E/W	24° SE/SW	24° S/S	24° SW/SE	24° W/E	34° E/W	34° SE/SW	34° S/S	34° SW/SE	34° W/E	44° E/W	44° SE/SW	44° S/S	44° SW/SE	44° W/E	90° E/W	90° SE/SW	90° S/S	90° SW/SE	90° W/E
Jan. 21	7	5	10	25	26	17	4	4	35	36	21	4	4	44	45	25	4	4	52	53	28	4	4	66	68	31	2	2
	8	4	83	128	135	110	68	34	156	167	126	55	13	179	195	137	41	13	197	216	145	25	12	201	230	127	7	7
	9	3	151	193	209	188	143	101	217	242	207	132	61	234	269	221	118	20	244	288	228	100	14	200	262	176	8	8
	10	2	204	233	257	246	208	163	246	302	268	203	128	252	307	282	192	90	250	318	287	176	49	154	252	207	47	9
	11	1	237	249	281	283	255	212	249	302	306	258	186	242	315	319	254	155	227	318	324	242	120	86	217	226	108	9
	12		249	242	280	296	280	242	228	292	319	292	288	208	296	332	296	208	182	291	336	291	182	9	167	232	167	9
Feb. 21	7	5	35	71	67	44	16	9	94	88	49	9	9	114	106	53	9	9	131	121	56	8	8	156	141	46	5	5
	8	4	116	167	167	135	90	58	198	198	145	69	15	223	223	150	46	14	241	242	151	22	14	233	234	102	8	8
	9	3	187	232	240	213	168	131	256	269	225	149	87	273	291	230	126	40	281	304	228	100	15	217	249	141	9	9
	10	2	241	272	288	273	235	197	284	312	286	223	158	288	326	291	203	115	284	330	287	178	69	164	231	168	12	9
	11	1	276	288	312	310	284	248	286	327	324	280	220	276	332	328	267	185	258	328	323	247	144	92	192	185	76	10
	12		288	279	310	323	310	279	264	316	337	316	264	240	312	341	312	240	210	299	335	299	210	10	138	191	138	10
Mar. 21	7	5	60	104	94	63	30	14	132	115	64	13	13	156	133	62	13	13	175	147	59	12	12	195	154	27	7	7
	8	4	141	193	186	150	107	81	224	212	152	79	36	248	232	149	49	17	265	244	142	17	16	241	212	64	9	9
	9	3	212	256	236	226	185	155	279	279	229	159	109	294	293	225	128	61	300	300	214	94	18	219	218	95	10	10
	10	2	266	296	304	285	251	221	307	320	288	231	181	309	327	283	204	136	302	324	270	172	87	165	197	120	11	11
	11	1	300	312	327	322	299	272	309	334	326	288	242	297	332	320	268	205	276	320	305	240	163	93	156	135	41	11
	12		312	303	325	334	325	303	286	323	339	323	286	261	312	333	312	261	228	291	317	291	228	11	102	140	102	11
Apr. 21	6	6	7	16	12	5	4	4	23	16	4	4	4	28	19	4	4	4	33	21	3	3	3	41	24	2	2	2
	7	5	83	127	111	77	45	35	155	127	70	19	19	178	140	62	18	18	196	149	51	17	17	202	134	10	10	10
	8	4	160	209	194	157	120	103	236	211	149	86	59	257	223	137	51	23	270	228	122	21	21	230	169	16	12	12
	9	3	227	280	259	227	192	174	287	272	220	161	130	299	278	206	125	82	302	276	186	85	33	207	170	41	14	14
	10	2	278	302	303	282	254	236	314	310	275	228	197	313	308	259	195	153	303	297	237	157	105	157	148	61	14	14
	11	1	310	320	325	316	299	283	316	324	309	280	254	302	313	293	254	218	280	294	269	220	175	90	110	74	14	14
	12		321	312	323	328	323	311	295	313	321	313	295	269	294	305	294	269	235	267	280	267	235	15	60	79	60	15
May 21	6	6	22	41	31	15	10	10	53	36	10	10	10	64	41	9	9	9	73	44	9	9	9	86	43	5	5	5
	7	5	98	140	120	85	57	51	166	131	73	25	24	187	139	59	23	23	202	143	44	21	21	197	112	12	12	12
	8	4	171	215	196	159	127	117	239	207	145	91	75	257	212	127	53	31	267	211	106	26	26	217	136	15	15	15
	9	3	233	270	256	224	194	184	287	263	210	160	142	296	262	190	121	96	297	255	165	79	48	195	135	16	16	16
	10	2	281	305	297	275	252	242	312	298	261	222	208	310	290	239	186	162	299	274	211	145	116	149	114	22	17	17
	11	1	311	319	317	307	294	286	314	311	293	271	258	300	295	270	240	222	277	271	240	203	181	87	78	34	17	17
	12		322	313	316	317	316	313	295	301	304	301	295	270	278	281	278	270	236	246	250	246	236	17	32	37	32	17
June 21	6	6	29	50	37	20	13	13	63	42	12	12	12	75	46	12	12	12	84	49	11	11	11	96	45	7	7	7
	7	5	103	143	122	87	61	57	168	131	73	29	26	187	137	58	25	25	201	139	41	23	23	191	101	13	13	13
	8	4	173	215	194	158	128	122	238	203	142	92	81	254	206	122	53	39	263	203	99	28	28	209	122	16	16	16
	9	3	234	268	252	221	194	186	284	256	204	158	146	292	254	182	119	101	291	244	155	76	55	188	120	18	18	18
	10	2	280	302	292	269	249	242	308	290	253	218	206	305	280	229	181	165	294	263	199	139	119	144	99	18	18	18
	11	1	309	316	311	300	289	284	311	303	283	264	257	296	285	259	233	222	273	260	227	194	181	85	65	19	19	19
	12		319	310	310	310	310	310	293	294	294	294	293	267	269	269	269	267	235	236	236	236	235	19	21	22	21	19

(Based on data in Table 1, pp. 387, 1972 ASHRAE HANDBOOK OF FUNDAMENTALS; 0% ground reflectance, 1.0 clearness factor)

Table 4-E Insolation on Inclined Surfaces for Selected Northern Latitudes

24°N

BTU/FT² Hourly Total Insolation, $I_{D\theta} + I_d$, On Surfaces Tilted With Respect To The Horizontal

Orientation key for each tilt: columns are (AM orientation / PM orientation): E/W, SE/SW, S/S, SW/SE, W/E

Date	AM	PM	0°	14° E/W	14° SE/SW	14° S	14° SW/SE	14° W/E	24° E/W	24° SE/SW	24° S	24° SW/SE	24° W/E	34° E/W	34° SE/SW	34° S	34° SW/SE	34° W/E	44° E/W	44° SE/SW	44° S	44° SW/SE	44° W/E	90° E/W	90° SE/SW	90° S	90° SW/SE	90° W/E
July 21	6	6	23	40	31	16	11	11	52	35	11	11	11	62	39	10	10	10	70	42	9	9	9	81	41	6	6	6
	7	5	98	138	118	85	57	52	162	128	73	27	25	182	136	59	24	24	196	139	44	23	23	190	107	13	13	13
	8	4	169	212	192	157	126	118	235	203	143	91	77	252	207	125	54	35	261	206	104	28	28	210	131	18	16	16
	9	3	231	266	251	221	192	182	282	258	207	159	142	291	257	187	120	98	291	249	161	79	51	190	130	18	18	18
	10	2	278	301	292	270	248	239	307	292	256	219	203	304	285	235	184	162	293	269	206	143	116	146	110	21	19	19
	11	1	307	315	312	302	289	282	310	305	287	266	255	295	289	265	236	220	273	266	235	199	179	86	76	32	19	19
	12		317	308	311	312	311	308	291	296	298	296	291	266	273	275	273	266	233	242	245	242	233	19	31	36	31	19
Aug. 21	6	6	7	15	12	5	4	4	20	15	4	4	4	25	17	4	4	4	29	19	4	4	4	36	21	2	2	2
	7	5	82	122	107	76	47	37	147	122	69	22	22	168	133	60	21	21	184	140	50	20	20	187	124	11	11	11
	8	4	158	203	188	154	119	104	228	204	146	87	62	247	214	134	53	27	259	218	118	25	25	218	159	16	15	15
	9	3	223	261	252	222	189	172	279	264	214	159	130	290	269	200	124	85	292	266	181	86	37	199	162	39	16	16
	10	2	273	298	295	275	249	232	306	302	268	224	195	305	299	252	192	153	295	288	230	155	107	153	142	58	17	17
	11	1	304	313	317	309	292	278	309	315	301	274	250	295	305	285	248	215	274	285	261	215	174	89	106	71	17	17
	12		315	306	316	320	316	306	289	306	313	306	289	264	287	296	287	264	231	260	272	260	231	17	58	75	58	17
Sept. 21	7	5	57	96	87	60	30	16	120	106	60	15	15	142	122	59	15	15	159	134	56	14	14	175	139	26	8	8
	8	4	136	184	178	144	104	80	212	201	146	78	38	235	219	143	50	21	250	231	136	20	20	226	199	62	11	11
	9	3	205	247	246	218	179	152	268	268	221	155	108	282	281	217	126	62	287	287	206	94	22	209	208	93	13	13
	10	2	258	286	293	275	243	215	298	308	278	224	177	298	314	273	199	134	291	311	261	168	88	159	189	116	13	13
	11	1	291	301	316	311	290	264	298	323	315	279	236	287	321	309	259	200	267	309	295	233	160	91	151	131	42	14
	12		302	294	315	323	315	294	277	313	327	313	277	253	301	321	301	253	221	281	306	281	221	14	100	136	100	14
Oct. 21	7	5	32	60	60	40	15	10	83	78	45	10	10	100	94	48	9	9	115	107	50	9	9	136	124	42	5	5
	8	4	111	158	159	129	87	51	187	188	139	67	17	211	212	144	46	16	228	230	145	23	15	219	221	99	9	9
	9	3	180	223	231	206	163	127	246	259	217	145	85	262	280	223	124	41	270	292	221	98	18	208	240	138	10	10
	10	2	234	263	279	265	228	191	275	302	277	216	154	279	315	282	198	113	274	320	279	174	69	159	225	165	16	11
	11	1	268	279	303	301	276	241	278	317	315	272	214	268	323	319	260	180	250	318	314	241	141	90	188	182	76	11
	12		279	271	301	314	301	271	256	307	328	307	256	233	303	332	303	233	204	291	327	291	204	11	136	188	136	11
Nov. 21	7	5	10	24	25	16	4	4	34	34	20	4	4	42	43	24	4	4	50	51	27	4	4	63	64	29	2	2
	8	4	82	126	133	108	68	34	153	164	123	55	14	175	190	135	41	13	193	212	142	26	13	197	223	124	7	7
	9	3	150	191	206	186	142	101	214	239	205	131	62	231	265	217	117	21	241	283	224	99	15	196	257	172	9	9
	10	2	203	231	255	244	206	163	244	283	265	201	128	249	303	278	190	90	247	314	283	174	50	152	247	204	46	10
	11	1	236	247	278	280	252	211	247	299	302	255	185	240	311	316	251	155	225	314	320	239	120	86	213	222	106	10
	12		247	240	277	293	277	240	227	289	315	289	227	207	293	328	293	207	181	288	332	288	181	10	164	228	164	10
Dec. 21	7	5	3	10	10	7	2	2	14	15	9	2	2	18	19	11	2	2	21	22	12	1	1	27	29	14	1	1
	8	4	71	112	121	99	59	26	138	152	116	49	12	160	179	129	38	12	177	201	139	26	11	185	220	130	6	6
	9	3	137	177	195	176	132	89	200	229	198	124	52	217	257	214	113	15	228	277	223	98	14	190	262	184	8	8
	10	2	189	217	243	234	195	150	230	273	258	193	116	236	296	275	185	80	235	309	283	172	41	148	255	217	58	9
	11	1	221	233	267	270	241	197	233	290	295	247	173	227	304	312	245	141	213	310	320	237	110	84	222	236	117	9
	12		232	226	266	282	266	226	213	280	308	280	213	194	287	325	287	194	170	284	332	284	170	9	174	243	174	9

(Based on data in Table 1, pp. 387, 1972 ASHRAE HANDBOOK OF FUNDAMENTALS; 0% ground reflectance, 1.0 clearness factor)

Table 4-E Insolation on Inclined Surfaces for Selected Northern Latitudes

32°N

BTU/FT² Hourly Total Insolation, $I_{D\theta} + I_d$, On Surfaces Tilted With Respect To The Horizontal

Orientation columns within each tilt: AM E / PM W (E/W), SE/SW, S/S, SW/SE, AM W / PM E (W/E)

Date	Solar Time AM/PM	0°	22° E/W	22° SE/SW	22° S	22° SW/SE	22° W/E	32° E/W	32° SE/SW	32° S	32° SW/SE	32° W/E	42° E/W	42° SE/SW	42° S	42° SW/SE	42° W/E	52° E/W	52° SE/SW	52° S	52° SW/SE	52° W/E	90° E/W	90° SE/SW	90° S	90° SW/SE	90° W/E
Jan. 21	7/5	0	1	1	0	0	0	1	1	0	0	0	1	1	0	0	0	167	190	0	0	0	171	200	1	1	0
	8/4	56	114	125	93	37	5	136	151	106	27	11	153	173	116	17	11	217	272	123	10	10	187	256	115	6	6
	9/3	118	177	203	175	104	43	196	233	193	99	14	210	256	206	87	14	216	302	212	73	13	147	257	181	8	8
	10/2	167	208	249	235	175	104	217	275	256	171	70	219	293	269	162	34	184	299	274	148	14	83	229	221	61	9
	11/1	198	212	267	273	227	156	209	286	295	230	130	199	297	308	226	100	132	268	312	215	67	9	182	245	123	9
	12	209	194	259	285	259	194	178	270	308	270	178	157	273	321	273	157	132	268	324	268	132	9	182	253	182	9
Feb. 21	7/5	22	64	60	34	7	7	80	75	37	7	7	94	88	40	6	6	106	98	42	6	6	118	109	38	4	4
	8/4	95	168	172	127	60	14	194	199	136	60	14	214	220	140	21	13	228	235	141	12	12	218	228	108	7	7
	9/3	161	225	243	206	136	75	244	269	217	118	31	256	287	222	97	15	261	297	220	74	14	209	256	158	9	9
	10/2	212	254	286	266	206	141	261	306	278	193	101	261	318	283	175	59	253	320	279	152	15	160	246	193	32	9
	11/1	244	257	302	304	260	197	251	315	317	255	166	237	319	321	243	130	217	313	315	223	91	90	211	214	98	10
	12	255	237	293	316	293	237	217	297	330	297	217	192	292	334	292	192	160	279	328	279	160	10	160	222	160	10
Mar. 21	7/5	54	117	104	60	13	13	141	123	60	12	12	161	138	59	11	11	175	149	56	11	11	185	151	32	7	7
	8/4	129	205	198	146	79	36	230	221	147	52	17	249	237	144	24	16	260	247	137	15	15	235	217	78	9	9
	9/3	194	258	264	222	155	104	275	284	224	130	58	284	295	220	102	18	285	298	209	70	17	215	232	119	10	10
	10/2	245	285	305	280	225	171	290	319	283	205	129	286	323	278	179	83	275	317	265	148	35	163	217	150	11	11
	11/1	277	287	321	317	278	227	279	326	321	266	194	262	322	315	246	155	238	308	300	218	111	91	180	170	67	11
	12	287	267	311	329	311	267	246	308	333	308	246	217	295	327	295	217	181	274	312	274	181	11	128	177	128	11
Apr. 21	6/6	14	37	27	9	6	6	46	32	6	6	6	54	36	6	6	6	61	40	5	5	5	68	41	3	3	3
	7/5	86	153	130	78	27	19	178	145	71	18	18	197	155	62	17	17	210	161	51	16	16	205	142	10	10	10
	8/4	158	228	211	156	96	66	250	225	148	63	23	265	234	136	29	22	272	235	120	20	22	229	181	35	12	12
	9/3	220	277	270	225	168	133	290	281	217	137	86	295	284	203	102	38	292	278	183	64	58	206	188	68	13	13
	10/2	267	305	308	279	233	196	304	312	272	206	154	297	307	256	173	107	282	294	234	136	129	156	171	95	14	14
	11/1	297	305	322	313	283	248	294	319	306	263	214	275	306	290	235	174	247	285	265	201	195	89	136	112	30	17
	12	307	286	313	325	313	286	263	302	318	302	263	232	281	301	281	232	195	252	276	252	195	14	88	118	88	14
May 21	6/6	36	76	54	21	14	14	91	60	13	13	13	103	65	13	13	13	113	68	12	12	12	119	62	7	7	7
	7/5	107	172	142	88	41	29	194	152	75	24	24	211	158	60	22	22	222	159	44	21	21	204	124	13	13	13
	8/4	175	239	214	159	107	87	258	223	145	71	43	270	225	127	33	26	274	221	105	24	26	218	151	33	15	15
	9/3	233	284	269	223	174	150	295	273	209	139	105	298	270	188	101	74	290	259	163	59	74	195	154	56	16	16
	10/2	277	308	304	273	234	209	307	302	259	203	168	298	291	237	167	140	281	272	208	126	126	148	138	72	17	17
	11/1	305	310	317	305	280	258	290	308	290	256	225	278	290	268	224	201	249	263	237	186	186	87	105	77	35	17
	12	315	293	309	315	309	293	270	292	301	292	270	238	267	278	267	238	201	234	247	234	201	17	59	77	59	17
June 21	6/6	45	87	62	26	17	17	103	68	16	16	16	116	72	15	15	15	125	73	14	14	14	128	62	9	9	9
	7/5	115	177	145	91	46	38	198	152	76	26	26	213	156	59	25	25	222	155	41	23	23	200	114	14	14	14
	8/4	180	241	213	159	116	95	258	219	143	73	52	268	219	122	35	29	271	213	99	27	29	211	138	16	16	16
	9/3	236	284	266	221	175	156	293	267	204	139	112	294	261	181	100	65	287	248	153	58	65	189	140	19	18	18
	10/2	279	307	299	268	233	213	306	295	251	201	173	296	282	227	163	128	278	261	197	121	121	144	123	41	18	18
	11/1	306	308	312	299	277	260	297	301	282	251	227	276	281	257	218	188	248	253	224	179	179	85	92	56	19	19
	12	315	293	304	309	304	293	270	285	292	285	270	239	259	267	259	239	201	224	234	224	201	19	48	60	48	19

(Based on data in Table 1, pp. 387, 1972 ASHRAE HANDBOOK OF FUNDAMENTALS; 0% ground reflectance, 1.0 clearness factor)

Table 4-E Insolation on Inclined Surfaces for Selected Northern Latitudes

32°N

BTU/FT² Hourly Total Insolation, $I_{D\theta} + I_d$, On Surfaces Tilted With Respect To The Horizontal

Date	AM	PM	0°	22° E/W	22° SE/SW	22° S/S	22° SW/SE	22° W/E	32° E/W	32° SE/SW	32° S/S	32° SW/SE	32° W/E	42° E/W	42° SE/SW	42° S/S	42° SW/SE	42° W/E	52° E/W	52° SE/SW	52° S/S	52° SW/SE	52° W/E	90° E/W	90° SE/SW	90° S/S	90° SW/SE	90° W/E
July 21	6	6	37	74	54	22	15	15	88	59	14	14	14	100	63	13	13	13	109	66	12	12	12	114	59	8	8	8
	7	5	107	169	140	87	43	32	190	149	75	25	25	206	154	60	24	24	216	154	44	22	22	197	119	14	14	14
	8	4	174	235	211	158	107	89	253	218	143	72	46	264	220	125	35	29	267	215	104	27	27	212	146	16	16	16
	9	3	231	280	264	220	173	150	290	268	205	139	107	291	264	185	101	60	285	253	159	60	29	190	149	31	18	18
	10	2	274	303	299	269	231	208	303	296	254	201	168	293	285	232	165	123	276	267	204	125	76	146	133	54	30	18
	11	1	302	306	312	300	276	256	294	302	285	252	223	274	284	262	221	184	246	258	232	183	140	86	102	69	58	19
	12		311	290	304	310	304	290	267	287	296	287	267	236	262	273	262	236	199	229	242	229	199	19	58	74	58	19
Aug. 21	6	6	14	35	26	9	7	7	43	30	7	7	7	50	34	6	6	6	55	36	5	5	5	61	37	4	4	4
	7	5	85	147	125	77	30	22	169	138	69	21	21	186	147	60	20	20	198	152	50	19	19	191	132	12	12	12
	8	4	156	221	204	152	97	69	242	217	144	65	27	255	224	132	32	25	261	225	116	24	24	217	171	33	15	15
	9	3	216	270	263	220	166	133	282	272	212	136	89	286	274	197	102	43	273	268	178	65	26	198	179	65	16	16
	10	2	262	295	300	272	228	194	296	303	264	202	154	289	298	249	171	109	274	276	226	134	62	151	164	91	17	17
	11	1	292	298	314	305	277	245	288	310	298	257	212	269	297	281	230	173	242	276	257	197	130	88	131	107	30	17
	12		302	281	306	317	306	281	258	294	309	294	258	229	274	292	274	229	192	246	268	246	192	17	85	113	85	17
Sept. 21	7	5	51	107	95	56	14	14	128	112	56	14	14	145	125	55	13	13	158	134	52	12	12	165	135	30	7	7
	8	4	124	194	188	140	78	38	217	209	141	53	20	234	224	138	27	19	245	232	131	18	18	219	203	75	11	11
	9	3	188	247	253	213	151	103	263	271	215	127	59	271	282	211	100	22	272	284	201	70	22	205	220	114	12	12
	10	2	237	274	294	270	218	167	279	306	273	199	127	276	310	268	174	84	264	304	255	144	84	157	208	145	13	13
	11	1	268	278	309	306	269	221	270	314	309	257	189	254	310	303	238	152	230	296	289	212	110	90	174	164	66	14
	12		278	259	301	318	301	259	238	297	321	297	238	210	285	315	285	210	177	264	300	264	177	14	125	171	125	14
Oct. 21	7	5	19	53	50	29	7	7	66	63	32	7	7	78	73	34	6	6	87	81	36	6	6	97	90	32	4	4
	8	4	90	157	161	120	58	16	181	187	128	40	16	200	207	133	22	15	213	221	134	14	14	204	214	104	8	8
	9	3	155	215	232	198	132	73	233	257	208	115	73	245	275	213	95	17	249	285	212	73	16	200	245	153	10	10
	10	2	204	244	275	257	199	136	251	295	269	188	99	251	307	273	171	56	243	309	270	149	17	155	238	188	34	11
	11	1	236	248	292	294	252	191	242	305	307	247	191	229	308	311	236	127	210	303	306	217	89	88	206	209	97	11
	12		247	229	284	306	284	229	211	288	320	288	211	186	283	324	283	186	156	271	318	271	156	11	156	217	156	11
Nov. 21	7	5	0	1	1	0	0	0	1	1	0	1	1	1	1	1	1	0	1	1	1	0	0	1	2	1	0	0
	8	4	55	112	122	91	37	12	133	147	104	27	11	150	168	113	17	11	163	184	119	10	10	166	194	111	6	6
	9	3	118	175	200	173	108	44	194	229	190	99	15	206	252	202	86	14	213	267	208	72	13	183	251	176	8	8
	10	2	166	206	246	233	174	104	214	271	252	169	70	217	289	265	160	35	213	297	270	146	15	145	252	217	60	9
	11	1	197	211	264	270	225	156	207	282	291	228	130	198	293	303	223	100	183	294	307	213	67	83	225	241	121	9
	12		207	193	256	282	256	193	177	267	304	267	177	156	270	316	270	156	131	264	320	264	131	10	179	249	179	10
Dec. 21	8	4	41	91	103	77	29	10	110	127	90	22	9	126	147	101	14	9	138	162	108	8	8	145	176	107	5	5
	9	3	102	158	186	161	97	33	176	216	180	91	14	189	240	195	82	13	197	256	204	70	12	174	249	183	13	7
	10	2	150	189	232	221	162	90	198	259	244	160	58	202	279	259	154	25	199	290	267	143	13	180	256	226	69	8
	11	1	180	194	250	258	213	140	192	271	282	218	116	184	284	298	217	88	170	289	305	209	58	80	231	251	130	9
	12		190	177	243	271	243	177	162	256	295	256	162	143	262	311	262	143	120	260	318	260	120	9	186	259	186	9

(Based on data in Table 1, pp. 387, 1972 ASHRAE HANDBOOK OF FUNDAMENTALS; 0% ground reflectance, 1.0 clearness factor)

Table 4-E   Insolation on Inclined Surfaces for Selected Northern Latitudes

40°N

BTU/FT² Hourly Total Insolation, $I_{D\theta} + I_d$, On Surfaces Tilted With Respect To The Horizontal

Date	AM	PM	Horiz.	30° E/W	30° SE/SW	30° S	30° SW/SE	30° W/E	40° E/W	40° SE/SW	40° S	40° SW/SE	40° W/E	50° E/W	50° SE/SW	50° S	50° SW/SE	50° W/E	60° E/W	60° SE/SW	60° S	60° SW/SE	60° W/E	90° E/W	90° SE/SW	90° S	90° SW/SE	90° W/E
Jan 21	8	4	28	82	94	65	12	8	96	111	74	7	7	108	125	81	7	7	116	135	85	6	6	119	142	84	4	4
	9	3	83	152	187	155	75	13	167	212	171	68	12	177	231	182	59	11	182	243	187	48	10	166	235	171	11	7
	10	2	127	175	232	218	141	46	182	255	237	138	16	183	271	249	131	13	179	278	254	120	12	137	251	223	69	8
	11	1	154	170	246	257	196	100	165	263	277	199	75	156	273	290	197	48	142	274	293	188	20	79	231	253	132	8
	12		164	143	233	270	233	143	127	243	291	243	127	108	246	303	246	108	86	242	306	242	86	9	188	263	188	9
Feb 21	7	5	10	42	39	19	4	4	50	47	21	4	4	57	54	23	3	3	63	59	24	3	3	68	63	22	2	2
	8	4	73	159	167	114	32	13	180	190	122	16	12	196	207	126	11	11	205	218	127	10	10	198	213	107	7	7
	9	3	132	211	239	195	104	20	225	262	205	89	14	234	277	209	71	13	235	284	208	51	12	199	255	167	8	8
	10	2	178	228	277	256	175	82	232	296	267	163	45	229	305	271	148	15	219	305	267	127	13	154	254	210	48	9
	11	1	206	219	288	293	233	141	210	299	306	228	110	195	301	310	217	76	175	294	304	199	40	87	225	236	115	9
	12		216	189	272	306	272	189	168	275	319	275	168	142	270	323	270	142	113	257	317	257	113	9	176	245	176	9
Mar 21	7	5	46	123	109	55	11	11	143	125	55	11	11	158	137	54	10	10	169	145	51	9	9	171	143	35	6	6
	8	4	114	208	205	140	51	17	228	224	141	27	16	242	237	138	15	15	249	243	131	13	13	226	219	89	10	9
	9	3	173	251	267	215	126	51	263	283	217	102	18	267	292	213	76	16	264	292	202	48	15	209	242	138	22	10
	10	2	218	265	301	273	196	116	265	312	276	177	74	258	314	271	153	30	243	307	258	125	16	159	233	176	89	11
	11	1	247	255	310	310	254	176	242	313	313	242	141	223	308	307	222	102	197	293	293	196	61	90	200	200	150	11
	12		257	224	294	322	294	224	199	289	326	289	199	169	276	320	276	169	134	255	305	255	134	11	150	208	150	11
Apr 21	6	6	20	62	44	11	8	8	73	50	8	8	8	82	55	7	7	7	88	57	7	7	7	92	57	4	4	4
	7	5	87	174	146	77	19	19	194	158	70	18	18	209	166	61	16	16	217	169	50	15	15	205	149	12	10	10
	8	4	152	240	223	153	72	26	256	235	145	40	22	266	240	133	20	20	267	238	117	18	18	226	192	53	12	12
	9	3	207	276	277	221	143	86	285	285	213	112	40	283	284	199	78	22	275	275	179	43	20	203	204	93	13	13
	10	2	250	289	308	275	209	149	285	309	267	182	105	273	302	252	151	59	253	286	229	115	21	154	192	126	56	14
	11	1	278	279	315	308	263	205	263	310	301	242	168	240	295	285	215	127	210	272	260	182	82	88	160	147	113	14
	12		287	250	299	320	299	250	223	286	313	286	223	189	265	296	265	189	150	236	271	236	150	14	113	154	113	14
May 21	5	7	0	1	0	0	0	0	1	0	0	0	0	1	0	0	0	0	1	0	0	0	0	1	0	0	0	0
	6	6	49	111	78	25	16	16	126	84	15	15	15	138	87	14	14	14	146	88	13	13	13	144	77	9	9	9
	7	5	114	198	162	89	24	24	216	169	76	24	23	228	172	60	21	21	233	170	44	20	20	209	136	13	13	13
	8	4	175	255	229	158	85	52	268	234	144	50	27	273	233	125	25	25	271	226	104	23	23	218	166	25	16	15
	9	3	227	288	277	221	152	110	292	279	206	117	64	288	272	186	80	27	276	258	160	40	24	194	173	60	17	16
	10	2	267	299	306	270	213	169	293	301	255	183	125	278	288	233	147	78	255	265	205	107	29	147	160	89	33	17
	11	1	293	290	312	301	264	221	272	301	287	238	184	247	281	264	206	142	214	253	234	168	95	86	130	108	86	17
	12		301	263	298	312	298	263	235	279	297	279	235	200	252	274	252	200	159	219	243	219	159	17	86	114	86	17
June 21	5	7	4	14	7	3	3	3	16	8	3	3	3	18	8	2	2	2	19	9	2	2	2	21	8	1	1	1
	6	6	60	125	87	30	19	19	140	92	18	18	18	151	94	17	17	17	158	93	16	16	16	152	77	10	10	10
	7	5	123	205	165	92	27	27	221	170	77	26	26	231	170	59	24	24	235	166	41	22	22	206	127	14	14	14
	8	4	182	258	228	159	90	62	269	231	142	54	29	273	228	121	27	27	269	218	97	25	25	212	153	16	16	16
	9	3	233	289	275	219	154	119	292	273	202	119	73	287	264	179	80	29	273	248	151	39	27	188	159	47	18	18
	10	2	272	300	302	266	213	175	293	295	248	181	132	277	279	224	144	86	253	256	194	103	37	143	146	74	24	18
	11	1	296	292	308	296	262	226	273	301	278	235	189	247	272	253	201	147	214	243	221	162	100	84	117	92	75	19
	12		304	266	294	306	294	263	238	274	289	274	238	202	245	263	245	202	162	210	230	210	162	19	75	98	75	19

(Based on data in Table 1, pp. 387, 1972 ASHRAE HANDBOOK OF FUNDAMENTALS; 0% ground reflectance, 1.0 clearness factor)

Table 4-E Insolation on Inclined Surfaces for Selected Northern Latitudes

40°N

BTU/FT² Hourly Total Insolation, $I_{D\theta} + I_d$, On Surfaces Tilted With Respect To The Horizontal

Orientation columns within each tilt group: AM = E / PM = W (E/W); SE/SW; S/S; SW/SE; W/E.

Date	AM	PM	0°	30° E/W	30° SE/SW	30° S/S	30° SW/SE	30° W/E	40° E/W	40° SE/SW	40° S/S	40° SW/SE	40° W/E	50° E/W	50° SE/SW	50° S/S	50° SW/SE	50° W/E	60° E/W	60° SE/SW	60° S/S	60° SW/SE	60° W/E	90° E/W	90° SE/SW	90° S/S	90° SW/SE	90° W/E
July 21	5	7	0	1	1	0	0	0	2	1	0	0	0	2	1	0	0	0	2	1	0	0	0	2	1	0	0	0
	6	6	50	109	77	26	17	17	123	82	17	17	17	134	85	15	15	15	141	86	14	14	14	138	74	9	9	9
	7	5	114	195	159	89	26	26	212	165	75	25	25	222	167	60	23	23	227	165	44	25	21	202	130	14	14	14
	8	4	174	251	225	157	86	55	263	230	142	52	29	268	228	124	27	27	265	220	102	41	25	212	160	24	16	16
	9	3	225	283	273	218	151	112	287	280	203	117	66	283	266	182	80	29	270	251	157	106	26	189	168	58	18	18
	10	2	265	295	300	266	211	169	288	295	251	181	126	273	282	229	145	80	251	260	200	166	32	144	155	86	18	18
	11	1	290	287	307	296	260	220	269	295	281	235	183	244	275	258	203	142	212	247	228	200	97	85	126	104	32	18
	12		298	261	293	307	293	261	233	274	292	274	233	198	248	269	248	198	158	215	238	215	158	19	84	111	84	19
Aug 21	6	6	21	58	42	12	9	9	68	47	9	9	9	76	51	8	8	8	81	53	7	8	8	84	51	5	5	5
	7	5	87	167	140	76	22	22	185	151	69	21	21	198	157	60	19	19	205	159	49	17	16	192	139	12	12	12
	8	4	150	232	215	150	74	32	247	226	141	43	26	255	229	129	24	24	256	227	113	22	18	215	181	50	14	14
	9	3	205	274	268	216	142	90	276	275	207	112	45	275	274	193	80	26	265	265	173	45	21	195	194	89	16	16
	10	2	246	282	299	267	205	149	278	300	259	179	107	266	292	244	148	63	246	276	221	114	45	149	184	120	17	17
	11	1	273	274	307	300	257	203	258	301	292	237	168	235	286	276	210	128	206	263	252	177	101	87	154	140	54	17
	12		282	246	292	311	292	246	220	279	303	279	220	187	258	287	258	187	149	229	262	229	149	17	109	147	109	17
Sept 21	7	5	43	111	98	51	13	13	128	112	51	12	12	141	122	49	11	11	150	129	47	10	10	151	127	32	7	7
	8	4	109	196	193	133	52	20	214	210	134	29	19	227	222	131	17	17	232	227	124	16	16	210	204	84	11	11
	9	3	167	239	254	206	123	52	250	270	208	100	21	254	277	203	76	20	251	277	193	49	18	198	228	132	12	12
	10	2	211	254	289	262	190	115	255	299	265	172	75	247	300	260	149	33	233	293	247	122	19	153	222	168	24	13
	11	1	239	246	298	298	246	172	234	301	301	234	138	215	295	295	215	101	191	281	281	190	62	88	192	192	87	13
	12		249	217	283	310	283	217	194	278	313	278	194	165	265	307	265	165	131	245	292	245	131	13	145	200	145	13
Oct 21	7	5	7	30	28	14	3	3	36	34	15	3	3	41	38	17	3	3	44	42	17	3	2	48	45	16	2	2
	8	4	68	147	154	106	31	14	165	175	113	17	13	179	191	117	12	12	188	201	118	11	11	181	196	100	7	7
	9	3	126	199	227	185	100	21	213	248	195	86	17	221	263	200	69	15	222	269	198	50	14	188	242	160	49	9
	10	2	170	218	266	245	168	80	222	283	257	158	44	219	292	261	143	17	210	293	257	124	15	148	244	203	113	10
	11	1	199	211	277	283	225	137	202	288	295	220	107	188	290	299	210	75	169	284	294	193	41	85	217	229	171	11
	12		208	182	262	295	262	182	162	265	308	265	162	138	261	312	261	138	110	248	306	248	110	11	171	238	171	11
Nov 21	8	4	28	80	91	63	13	8	94	108	72	8	8	105	121	78	7	7	112	131	82	6	6	115	137	81	4	4
	9	3	82	150	183	152	74	14	164	208	167	67	13	174	226	178	58	12	179	237	183	48	11	162	229	167	11	7
	10	2	126	173	229	215	140	47	180	251	233	137	17	181	266	245	129	14	176	273	249	119	13	135	246	219	68	8
	11	1	153	169	243	254	194	100	164	260	273	197	75	155	269	285	194	49	141	270	288	185	21	78	227	248	129	9
	12		163	142	230	267	230	142	127	240	287	240	127	108	243	298	243	108	86	238	301	238	86	9	185	258	185	9
Dec 21	8	4	14	47	56	39	6	5	56	67	45	4	4	63	76	50	4	4	69	84	54	4	4	73	90	56	3	3
	9	3	65	127	162	135	62	12	141	186	152	58	11	152	205	164	52	10	157	218	171	45	9	147	216	163	17	6
	10	2	107	154	212	200	127	34	161	235	221	126	17	163	252	235	122	12	161	261	242	114	11	127	243	221	73	7
	11	1	134	150	227	239	194	84	147	246	262	186	62	140	257	276	186	38	129	261	283	180	14	74	228	252	134	8
	12		143	125	215	253	215	125	112	227	275	227	112	95	233	290	233	95	76	232	296	232	76	8	188	263	188	8

(Based on data in Table 1, pp. 387, 1972 ASHRAE HANDBOOK OF FUNDAMENTALS; 0% ground reflectance, 1.0 clearness factor)

Table 4-E Insolation on Inclined Surfaces for Selected Northern Latitudes

48°N

BTU/FT² Hourly Total Insolation, $I_{D\theta} + I_d$, On Surfaces Tilted With Respect To The Horizontal

Date	AM	PM	0°	38° E/W	38° SE/SW	38° S	38° SW/SE	38° W/E	48° E/W	48° SE/SW	48° S	48° SW/SE	48° W/E	58° E/W	58° SE/SW	58° S	58° SW/SE	58° W/E	68° E/W	68° SE/SW	68° S	68° SW/SE	68° W/E	90° E/W	90° SE/SW	90° S	90° SW/SE	90° W/E
Jan. 21	8	4	4	22	26	17	2	2	26	30	19	2	2	28	34	21	2	2	30	36	22	1	1	31	37	22	1	1
	9	3	46	113	149	120	42	10	124	167	132	38	9	131	181	140	33	8	135	189	145	28	7	128	187	139	13	5
	10	2	83	136	203	190	105	12	142	222	206	104	12	143	234	216	100	11	140	240	220	93	10	119	227	206	68	7
	11	1	107	125	216	231	161	47	121	231	249	165	27	114	240	260	164	12	104	241	263	158	10	71	219	243	129	8
		12	115	93	200	245	200	93	80	210	264	213	80	65	213	275	213	65	48	210	278	210	48	8	183	255	183	8
Feb. 21	7	5	1	8	8	3	1	1	9	9	4	1	1	10	10	4	1	1	11	11	4	0	0	12	11	4	0	0
	8	4	49	138	149	95	10	10	154	166	102	9	9	164	179	105	9	9	170	186	106	8	8	166	183	96	6	6
	9	3	100	188	226	178	73	13	199	245	187	61	12	205	257	191	47	12	204	261	190	32	10	182	244	167	8	8
	10	2	139	196	262	240	143	27	198	277	251	134	14	194	285	255	121	13	184	284	251	104	11	145	252	217	59	8
	11	1	165	177	268	278	203	86	168	277	290	200	58	154	279	294	190	29	136	272	288	175	12	83	230	247	125	9
		12	173	138	247	291	247	138	119	249	304	249	119	96	245	307	245	96	70	233	301	233	70	8	185	258	185	9
Mar. 21	7	5	37	121	108	49	10	10	136	120	49	9	9	147	128	47	8	8	154	133	45	7	7	153	131	35	5	5
	8	4	96	203	204	131	26	15	219	220	132	14	14	228	230	129	13	13	230	232	122	12	12	213	215	96	8	8
	9	3	147	236	263	205	97	17	244	277	207	76	16	245	282	203	53	15	238	280	193	29	13	201	245	152	10	10
	10	2	187	238	292	263	167	61	235	301	266	150	22	226	301	261	129	16	210	292	248	103	14	154	242	195	39	10
	11	1	212	216	295	300	228	122	202	297	303	216	88	182	290	297	199	52	157	275	283	175	15	87	214	223	106	10
		12	220	176	272	312	272	176	151	267	315	267	151	122	254	309	254	122	89	234	294	234	89	11	167	232	167	11
Apr. 21	6	6	27	87	62	13	9	9	98	67	9	9	9	106	71	8	8	8	111	73	7	7	7	111	70	5	5	5
	7	5	85	189	158	76	18	18	204	168	69	17	15	214	173	59	15	15	218	172	48	14	14	204	155	21	10	10
	8	4	142	243	230	149	48	21	254	239	141	80	18	258	240	129	57	18	255	235	113	25	16	221	200	69	12	12
	9	3	191	267	278	216	114	38	270	283	208	88	20	265	293	194	128	20	252	275	174	95	18	199	217	115	13	13
	10	2	228	267	303	268	183	98	259	302	260	158	22	244	293	245	128	21	221	275	223	95	19	151	209	152	15	14
	11	1	252	246	304	301	240	157	227	297	294	220	55	201	281	278	194	78	170	257	254	162	36	86	180	177	78	14
		12	260	258	282	313	282	208	178	268	305	268	178	144	247	289	247	144	106	218	264	218	106	14	135	185	135	14
May 21	5	7	9	30	16	4	4	4	34	17	4	4	4	37	18	4	4	4	39	18	3	3	3	40	17	2	2	2
	6	6	61	144	100	27	18	18	157	104	16	16	16	166	106	15	15	15	170	104	13	13	13	162	91	10	10	10
	7	5	118	218	177	89	24	24	231	182	75	22	20	237	181	60	20	21	236	176	43	18	18	212	146	13	13	13
	8	4	171	262	239	156	63	27	269	241	142	29	25	270	237	123	23	24	261	225	101	21	22	212	179	45	15	15
	9	3	217	282	281	217	128	66	281	279	202	95	27	271	270	182	59	25	254	252	156	22	23	192	190	86	16	16
	10	2	252	281	303	265	191	123	270	296	251	160	78	250	280	229	128	32	224	257	200	88	50	145	180	120	17	17
	11	1	274	261	303	296	244	178	239	290	281	219	138	211	268	258	187	96	176	239	228	150	116	85	153	141	57	17
		12	281	225	283	306	283	225	194	263	292	263	194	157	236	269	236	157	116	202	238	202	116	17	110	149	110	17
June 21	5	7	21	60	32	9	9	9	66	34	9	9	9	71	34	8	8	8	74	33	7	7	7	74	29	5	5	5
	6	6	74	158	110	33	21	21	171	112	19	19	19	179	112	18	18	18	182	109	16	16	16	170	91	12	12	12
	7	5	129	225	181	93	26	26	236	183	77	25	25	241	180	59	23	23	238	172	39	20	20	209	138	15	15	15
	8	4	181	266	239	157	69	29	271	239	140	34	27	271	232	119	25	25	259	218	95	23	23	212	167	45	16	16
	9	3	225	285	279	216	132	77	282	275	198	97	31	271	263	175	60	27	252	243	147	24	24	186	177	86	17	17
	10	2	259	284	300	262	192	132	271	290	244	160	87	250	272	220	124	41	223	246	189	85	56	142	140	120	48	18
	11	1	280	265	300	291	243	185	242	284	273	216	145	212	261	248	183	102	177	230	216	144	119	83	99	1.5	99	18
		12	287	230	280	301	280	230	198	259	283	259	198	161	230	258	230	161	119	194	225	194	119	18	99	1.3	99	18

(Based on data in Table 1, pp. 387, 1972 ASHRAE HANDBOOK OF FUNDAMENTALS; 0% ground reflectance, 1.0 clearness factor)

Table 4-E Insolation on Inclined Surfaces for Selected Northern Latitudes

48°N

BTU/FT² Hourly Total Insolation, $I_{D\theta} + I_d$, On Surfaces Tilted With Respect To The Horizontal

Date	AM	PM	0°	38° E/W	38° SE/SW	38° S/S	38° SW/SE	38° W/E	48° E/W	48° SE/SW	48° S/S	48° SW/SE	48° W/E	58° E/W	58° SE/SW	58° S/S	58° SW/SE	58° W/E	68° E/W	68° SE/SW	68° S/S	68° SW/SE	68° W/E	90° E/W	90° SE/SW	90° S/S	90° SW/SE	90° W/E
July 21	5	7	10	32	18	5	5	5	36	19	5	5	5	40	19	4	4	4	42	19	4	4	4	42	18	3	3	3
	6	6	62	141	99	28	19	19	153	102	18	18	19	162	103	16	16	16	165	101	15	15	15	156	88	11	11	11
	7	5	118	214	174	89	26	26	226	178	75	24	24	231	177	59	22	22	230	171	42	20	20	205	141	14	14	14
	8	4	171	258	234	154	65	29	264	236	140	31	27	263	231	121	25	25	254	219	99	22	22	211	173	43	16	16
	9	3	215	278	276	214	128	69	276	274	199	95	45	266	264	178	60	35	249	251	153	24	24	187	184	83	17	17
	10	2	250	277	297	261	189	124	266	290	246	159	81	246	263	224	125	54	220	251	195	87	25	143	175	116	18	18
	11	1	272	258	298	291	241	177	237	284	276	215	139	208	232	253	184	97	174	234	223	148	52	84	148	137	56	18
		12	279	224	278	301	278	224	193	259	286	259	193	157	232	263	232	157	116	198	232	198	116	19	107	144	107	19
Aug. 21	6	6	28	82	59	14	11	11	92	64	10	10	10	99	67	9	9	9	102	68	8	8	8	102	64	6	6	6
	7	5	85	180	152	75	21	21	194	160	67	19	19	203	163	58	18	18	206	162	47	16	16	191	144	20	12	12
	8	4	141	235	222	145	51	25	245	229	137	24	24	248	230	125	22	22	244	224	109	19	19	210	189	65	14	14
	9	3	189	260	269	210	116	44	262	273	201	89	26	256	269	187	58	24	243	257	168	27	22	191	206	110	15	15
	10	2	225	261	294	260	181	101	261	292	252	156	59	237	282	237	127	25	215	264	214	94	24	146	200	146	16	16
	11	1	248	241	295	293	235	157	223	288	285	215	120	198	272	268	189	81	167	248	244	158	40	85	173	169	75	17
		12	256	205	275	304	275	205	177	261	296	261	177	144	239	279	239	144	106	211	255	211	106	17	130	177	130	17
Sept. 21	7	5	35	107	95	44	11	11	120	105	44	11	11	129	113	43	9	9	134	117	40	8	8	133	114	31	6	6
	8	4	92	189	190	124	28	18	203	205	124	17	18	211	213	121	15	15	213	215	115	14	14	196	198	90	10	10
	9	3	142	224	249	196	94	21	231	261	197	75	21	225	266	193	53	18	225	263	183	30	16	189	230	143	12	12
	10	2	181	228	279	251	162	62	225	287	254	145	25	216	286	248	125	19	200	277	236	101	17	147	230	185	40	12
	11	1	205	208	282	287	220	120	195	284	289	208	88	176	277	284	191	54	151	262	269	168	18	85	204	212	103	13
		12	213	171	261	299	261	171	147	256	302	256	147	119	244	296	244	119	88	224	281	224	88	13	160	221	160	13
Oct. 21	7	5	0	3	2	1	0	0	3	3	1	0	0	3	3	1	0	0	4	3	1	0	0	4	4	1	0	0
	8	4	44	123	132	86	11	11	136	148	91	10	11	146	159	95	9	9	151	165	95	8	8	147	163	87	8	6
	9	3	94	175	211	167	70	15	186	229	176	59	14	191	240	180	46	13	190	244	178	32	12	170	228	157	8	8
	10	2	133	186	248	228	137	27	187	263	239	128	16	184	270	242	116	15	175	269	239	101	13	138	240	207	58	10
	11	1	157	169	255	266	195	83	160	265	277	191	57	147	266	281	182	29	130	260	276	168	14	80	220	237	121	10
		12	166	133	236	279	236	133	114	239	291	239	114	93	235	294	235	93	68	224	288	224	68	10	178	247	178	10
Nov. 21	8	4	5	22	26	17	2	2	22	30	19	2	2	28	33	21	2	2	30	36	22	2	2	31	37	22	1	1
	9	3	46	111	146	117	42	10	121	163	129	38	10	128	176	137	33	9	131	184	141	27	8	125	181	135	13	6
	10	2	83	134	199	186	104	13	139	217	202	102	13	140	229	212	98	12	137	235	215	91	10	117	222	201	67	7
	11	1	107	124	213	227	159	48	120	228	245	162	28	113	235	255	161	28	103	236	258	155	11	70	214	238	127	8
		12	115	92	198	241	198	92	80	207	259	207	80	65	210	270	210	65	48	207	272	207	48	8	179	250	179	8
Dec. 21	9	3	27	78	108	87	28	7	87	122	98	27	7	93	134	105	25	6	97	141	110	22	5	95	142	109	14	4
	10	2	63	111	173	164	88	11	117	192	180	89	10	119	205	192	87	9	118	212	197	83	9	104	205	190	66	6
	11	1	86	105	192	207	142	34	103	208	226	148	34	98	218	239	149	11	90	221	244	146	9	64	206	231	125	7
		12	94	75	179	222	179	75	65	190	241	190	65	53	195	254	195	53	40	195	260	195	40	7	175	244	175	7

(Based on data in Table 1, pp. 387, 1972 ASHRAE HANDBOOK OF FUNDAMENTALS; 0% ground reflectance, 1.0 clearness factor)

Table 4-E Insolation on Inclined Surfaces for Selected Northern Latitudes

56°N

BTU/FT² Hourly Total Insolation, $I_{D\theta} + I_d$, On Surfaces Tilted With Respect To The Horizontal

Orientation key for each tilt angle block: **E/W** (AM = E, PM = W), **SE/SW**, **S/S**, **SW/SE**, **W/E** (AM = W, PM = E).

Date	AM	PM	0°	46° E/W	46° SE/SW	46° S/S	46° SW/SE	46° W/E	56° E/W	56° SE/SW	56° S/S	56° SW/SE	56° W/E	66° E/W	66° SE/SW	66° S/S	66° SW/SE	66° W/E	76° E/W	76° SE/SW	76° S/S	76° SW/SE	76° W/E	90° E/W	90° SE/SW	90° S/S	90° SW/SE	90° W/E
Jan 21	9	3	11	46	64	50	12	4	50	71	55	10	4	53	77	59	10	3	55	79	60	9	3	54	80	60	7	2
	10	2	39	86	144	135	63	8	90	158	146	64	8	92	166	154	62	7	91	169	156	60	6	85	166	153	53	5
	11	1	58	79	167	183	116	10	77	179	197	120	9	73	186	206	121	8	67	187	208	118	7	56	180	201	108	6
	12		65	47	154	198	154	47	39	162	214	162	39	30	166	222	166	30	21	165	225	165	21	6	155	217	155	6
Feb 21	8	4	25	97	107	65	7	7	106	118	69	6	6	113	125	72	5	5	115	129	72	5	5	114	127	69	4	4
	9	3	65	155	197	151	45	11	163	211	159	37	10	167	220	162	27	9	165	222	161	17	8	156	214	151	6	6
	10	2	98	159	235	215	110	13	160	248	225	104	12	157	254	228	95	11	149	252	224	83	10	131	237	208	63	7
	11	1	119	134	239	254	170	36	126	248	265	168	14	115	249	268	161	11	101	243	263	150	10	76	222	243	126	8
	12		126	90	216	268	216	90	74	219	279	219	74	56	216	282	216	56	37	206	276	206	37	8	183	255	183	8
Mar 21	7	5	28	109	97	40	8	8	120	106	40	7	7	127	112	39	6	6	130	114	37	6	6	128	111	32	5	5
	8	4	75	189	195	119	13	13	200	207	120	12	12	205	213	117	11	11	205	213	111	9	8	194	203	97	8	8
	9	3	118	214	252	192	69	16	219	262	193	52	14	217	265	189	34	13	209	260	180	14	11	188	241	158	9	9
	10	2	151	206	277	249	138	70	202	283	251	124	15	192	282	246	106	14	176	272	234	85	13	146	244	205	52	10
	11	1	172	175	274	285	200	128	161	276	288	190	40	143	269	282	175	15	120	254	268	154	15	83	221	236	118	10
	12		179	128	248	297	248	128	105	243	300	243	105	79	231	294	231	79	51	213	280	213	51	10	177	246	177	10
Apr 21	5	7	0	0	0	0	0	0	0	0	0	0	0	0	0	0	0	0	0	0	0	0	0	0	0	0	0	0
	6	6	32	110	78	14	10	10	120	83	9	9	8	129	85	8	8	7	129	85	7	7	7	126	81	6	6	6
	7	5	81	196	166	74	17	17	207	173	66	16	14	211	174	57	14	12	211	171	46	12	12	200	158	29	10	10
	8	4	129	239	232	143	25	20	245	237	135	21	16	245	236	123	16	14	237	227	108	14	14	215	205	82	12	12
	9	3	169	251	274	208	91	48	250	276	200	73	18	241	270	186	38	16	225	257	167	16	16	193	225	133	13	13
	10	2	201	239	293	259	157	106	228	290	251	138	21	211	279	236	107	18	188	260	214	77	18	146	221	174	33	13
	11	1	220	208	288	292	216	162	188	280	284	198	48	162	264	268	173	33	132	240	245	143	17	84	195	200	96	13
	12		227	162	261	303	261	162	133	248	295	248	133	100	227	279	227	100	65	200	255	200	65	14	152	209	152	14
May 21	4	8	0	0	0	0	0	0	0	0	0	0	0	0	0	0	0	0	0	0	0	0	0	0	0	0	0	0
	5	7	25	80	43	10	10	10	86	44	9	9	8	90	44	8	8	7	92	43	7	7	7	90	39	6	6	6
	6	6	71	171	119	28	18	18	181	121	19	16	15	186	120	15	15	13	185	119	13	13	13	175	103	11	11	11
	7	5	119	230	189	88	22	22	237	190	76	18	18	238	186	74	18	16	232	177	41	16	16	212	156	16	13	13
	8	4	163	261	244	153	41	25	263	243	138	21	21	257	235	119	20	18	244	220	98	18	18	214	190	63	15	15
	9	3	201	269	280	212	104	27	262	275	197	73	22	249	263	176	41	20	228	244	151	19	19	188	204	109	16	16
	10	2	231	256	295	259	167	75	241	286	244	138	23	219	269	222	107	21	191	244	194	71	20	142	197	146	19	16
	11	1	249	226	290	288	222	130	202	275	274	198	92	172	253	251	167	50	137	224	222	132	21	83	172	170	78	17
	12		255	183	265	299	265	183	150	245	284	245	150	114	218	261	218	114	74	185	231	185	74	17	131	178	131	17
June 21	4	8	4	16	6	2	2	2	17	5	2	2	2	18	5	2	2	2	18	5	2	2	2	18	4	1	1	1
	5	7	40	108	60	14	14	14	116	60	13	13	13	120	58	11	11	13	121	55	10	10	10	116	48	8	8	8
	6	6	86	186	129	34	21	21	195	129	25	19	19	198	126	17	17	19	195	119	15	15	15	182	103	12	12	12
	7	5	132	238	193	92	25	25	243	192	76	23	23	242	186	57	21	21	234	174	38	18	18	211	150	15	15	15
	8	4	175	266	245	154	48	28	265	241	137	25	25	257	231	116	23	23	242	214	92	20	20	210	180	55	16	16
	9	3	212	272	278	211	109	33	264	271	193	76	27	249	256	170	41	24	226	234	143	21	21	184	192	98	17	17
	10	2	240	260	293	255	169	85	243	281	238	139	42	219	261	214	104	25	189	234	184	68	22	139	185	133	18	17
	11	1	258	231	287	284	222	139	205	271	267	196	99	174	246	242	163	57	138	215	210	127	22	82	161	156	70	18
	12		264	189	263	294	263	189	156	241	276	241	156	118	212	251	212	118	78	178	219	178	78	18	121	164	121	18

(Based on data in Table 1, pp. 387, 1972 ASHRAE HANDBOOK OF FUNDAMENTALS; 0% ground reflectance, 1.0 clearness factor)

Table 4-E Insolation on Inclined Surfaces for Selected Northern Latitudes

56°N

BTU/FT² Hourly Total Insolation, $I_{D\theta} + I_d$, On Surfaces Tilted With Respect To The Horizontal

Date	AM	PM	0°	46° E/W	46° SE/SW	46° S	46° SW/SE	46° W/E	56° E/W	56° SE/SW	56° S	56° SW/SE	56° W/E	66° E/W	66° SE/SW	66° S	66° SW/SE	66° W/E	76° E/W	76° SE/SW	76° S	76° SW/SE	76° W/E	90° E/W	90° SE/SW	90° S	90° SW/SE	90° W/E
July 21	4	8	0	0	0	0	0	0	0	0	0	0	0	0	0	0	0	0	0	0	0	0	0	0	0	0	0	0
	5	7	27	80	44	11	11	11	86	45	10	10	10	90	44	9	9	9	91	43	8	8	8	89	39	6	6	6
	6	6	72	168	117	30	19	19	177	119	18	18	18	181	117	16	16	16	180	112	14	14	14	170	100	12	12	12
	7	5	119	225	185	88	24	24	232	186	74	22	22	232	181	58	20	20	226	172	41	18	18	206	150	15	14	14
	8	4	163	256	239	151	44	27	257	237	136	25	25	251	229	117	23	23	238	214	96	20	20	208	184	61	16	16
	9	3	201	264	274	208	105	29	258	269	193	74	27	244	257	173	41	24	223	237	147	21	21	184	197	106	17	17
	10	2	230	253	290	254	166	77	237	280	239	137	35	215	263	217	105	25	187	238	189	70	22	140	192	142	19	18
	11	1	248	224	285	283	219	131	200	270	268	195	93	170	248	245	165	53	136	218	216	130	22	82	167	165	76	18
		12	254	182	260	293	260	182	150	241	278	241	150	114	214	255	214	114	75	181	225	181	75	18	127	173	127	18
Aug 21	5	7	0	1	1	0	0	0	1	1	0	0	0	1	1	0	0	0	1	1	0	0	0	1	1	0	0	0
	6	6	34	104	75	16	12	12	113	78	11	11	11	118	80	10	10	10	120	80	9	9	9	117	75	7	7	7
	7	5	82	187	159	73	19	19	197	164	65	18	18	201	165	56	16	16	199	161	45	14	14	187	148	28	11	11
	8	4	128	230	223	140	29	23	235	227	131	21	21	234	225	119	19	19	226	216	104	17	17	204	193	78	14	14
	9	3	168	243	264	202	92	25	241	265	193	67	23	232	259	179	40	21	217	245	160	19	19	185	214	126	15	15
	10	2	199	233	283	251	155	52	222	280	242	137	25	205	268	227	106	22	182	249	206	77	20	142	211	166	33	16
	11	1	218	204	279	282	211	108	184	271	274	193	74	159	254	258	168	37	130	231	235	139	20	83	187	191	93	16
		12	225	161	254	293	254	161	133	240	285	240	133	101	220	269	220	101	75	193	245	193	67	16	146	200	146	16
Sept 21	7	5	25	94	84	36	8	8	102	91	36	8	8	108	95	34	7	7	111	97	32	6	6	109	95	28	5	5
	8	4	72	173	179	111	15	15	183	190	111	14	14	188	195	108	13	13	187	194	102	11	11	177	184	89	9	9
	9	3	114	207	236	181	68	18	205	245	182	52	17	203	248	178	34	16	196	243	168	16	13	176	224	147	11	11
	10	2	146	196	262	236	133	20	192	268	237	119	18	182	255	232	102	17	167	256	221	82	14	138	230	193	51	12
	11	1	166	168	261	271	192	70	155	262	273	182	42	137	241	267	167	78	115	241	254	147	15	80	209	223	113	12
		12	173	124	236	283	236	124	102	232	285	232	102	78	202	279	220	78	51	202	265	202	51	12	168	233	168	12
Oct 21	8	4	20	79	87	53	6	6	86	96	57	6	6	91	102	59	5	5	94	105	59	5	5	92	104	57	4	4
	9	3	60	140	178	138	42	12	148	192	145	35	11	151	200	148	26	12	150	202	147	18	9	141	194	138	7	7
	10	2	92	148	219	201	104	14	149	231	210	98	13	146	236	213	90	13	139	235	210	79	10	122	221	195	60	8
	11	1	112	126	225	240	161	35	119	233	250	159	15	109	234	253	153	15	95	229	248	142	11	72	210	230	120	9
		12	119	86	204	253	204	86	71	207	263	207	71	54	204	266	204	54	36	195	261	195	36	9	173	241	173	9
Nov 21	9	3	12	45	63	49	12	4	49	70	54	11	4	52	75	57	10	3	53	78	59	9	3	53	78	58	6	2
	10	2	39	85	141	132	62	9	89	153	143	62	8	90	161	149	61	7	89	165	152	58	6	83	161	148	51	5
	11	1	58	78	164	179	114	11	76	175	193	118	10	72	182	201	118	9	66	183	203	115	8	55	175	196	106	6
		12	65	47	151	194	151	47	39	159	209	159	39	30	162	217	162	30	21	161	219	161	21	7	151	211	151	7
Dec 21	9	3	0	3	4	3	1	0	3	5	4	1	0	3	5	4	1	0	3	5	4	1	0	4	5	4	1	0
	10	2	19	52	91	86	39	5	55	101	95	40	5	57	108	101	41	5	57	111	104	40	4	55	110	103	37	3
	11	1	37	55	128	141	88	8	55	139	154	93	8	54	146	163	95	7	51	148	167	94	6	44	145	164	89	5
		12	43	32	122	159	122	32	26	130	173	130	26	21	135	182	135	21	14	136	186	136	14	5	130	182	130	5

Source: Richard C. Jordan and Benjamin Y. H. Liu, *op. cit.*, pp. IV-8–IV-17. Reprinted with permission.

Table 4-F Average Daily Total Insolation on a South-Facing Surface for Each Month of the Year and for Selected Tilt Angles, Btu/day-ft² (Clear-Day Values)

32° N

Month	Tilt Angle from Horizontal			
	22°	32°	42°	52°
January	1840	2010	2120	2165
February	2190	2300	2345	2320
March	2380	2405	2360	2245
April	2445	2355	2205	1995
May	2455	2285	2065	1785
June	2435	2235	1990	1690
July	2420	2250	2030	1755
August	2390	2295	2145	1935
September	2290	2310	2265	2155
October	2100	2210	2250	2230
November	1815	1980	2085	2130
December	1705	1890	2015	2085

40° N

Month	Tilt Angle from Horizontal			
	30°	40°	50°	60°
January	1930	2100	2210	2250
February	2365	2480	2525	2495
March	2630	2655	2605	2480
April	2730	2635	2465	2225
May	2750	2560	2310	2005
June	2740	2515	2230	1905
July	2720	2520	2275	1965
August	2665	2560	2395	2155
September	2490	2545	2490	2370
October	2255	2365	2410	2380
November	1900	2065	2170	2205
December	1730	1910	2030	2090

Source: Calculated from the hourly values of Table 4-E, for the twenty-first day of each month. Table 4-E values are from 1972 ASHRAE *Handbook of Fundamentals*.

CHAPTER 5

Building Heating Loads

Structures in which people live and work are continually gaining and losing heat. Gains come from such sources as direct solar radiation; thermal conduction inward through the building envelope (walls, windows, ceiling or roof, and floor) because of indoor–outdoor temperature difference; infiltration of warm outside air and moisture; and internally generated heat from occupants, lights, appliances, and heat-producing machinery operating in the space. Losses result from thermal conduction outward through the building envelope, and heat is required to offset the infiltration of cold outside air. In commercial buildings, ventilation air may be purposely introduced, and the heat required to warm this air is also a part of the building heating load.

The term *building heating load* describes the heating requirement necessary to maintain a specified or design room temperature. For typical residences, the building heating load is usually considered equal to the building heat loss, with internally generated heat (a gain) assumed to be negligible. In commercial and industrial structures, however, internal heat gains are often of major proportions, and they are an important consideration in determining the building heating load.

The building heating load is an essential factor in designing either a passive solar or an active solar building. In passive systems, the solar collection area (*aperture*) and the amount and distribution of thermal storage mass will be directly influenced by the building heating load. In active systems, the magnitude of the heating load determines the number of solar collectors to be used, the capacity of the thermal storage system, and the nature of the heat distribution system within the structure.

5-1 Energy Conservation Measures

The capitalized and operational costs of any heating system—solar or conventional—bear a direct relationship to the required heat output. Consequently, the essential first step in a solar energy system design is *not the calculation of heat losses* but a consideration of *how to minimize them*. In new construction for solar buildings, all of the conservation measures listed below should be incorporated. For retrofit applications in existing buildings, as many as possible should be implemented.

Minimizing Heat Transmission Losses.

1. Insulate ceilings to R-19 or even to R-30 in colder climates (see Chapter 3, section 3-10) and walls to R-11. These values are now standard building code requirements for new construction in the United States.
2. Within the constraints imposed by owners' life style, architectural considerations, and so on, design the structure with the smallest possible exterior wall area.
3. Use double-glazed windows throughout.
4. Install fiberglass blanket or rigid-board insulation under floors with crawl space; use perimeter or edge insulation for slab floors.
5. Provide drapes or shades at all windows to minimize air circulation (and the resultant convection–conduction heat losses) near the glass surface.

Minimizing Infiltration Losses.   (See also section 5-4.)

1. Insist on tight construction—well-fitted windows and doors. All cracks are to be carefully caulked with approved sealant.
2. Window area on the north wall should be as small as possible, consistent with other priorities such as desirable views and/or openings for summer cooling.
3. Weatherstrip windows and outside doors.
4. Specify tight-fitting dampers in fireplaces, and keep them closed when the fireplace is not in use.
5. Use exhaust fans only when necessary.

In all of the above, *specifying* the conservation measure is only the first step. Equally important are the installation of insulation and weatherstripping, the proper fitting of doors and windows, and careful caulking of all cracks. Substandard workmanship on any of these jobs can result in unacceptable heat losses and failure of the system to meet design specifications.

The importance of energy conservation is reinforced by the following suggested standard for maximum allowable heat loss in solar-heated residential buildings.

The Space-Heating Rule of Thumb for Residences.   Unless the building heat loss of a residence can be reduced to not more than 8 to 10 Btu/day-ft² of building floor area per Fahrenheit degree of indoor–outdoor temperature difference, a solar energy heating system, either passive or active, is not indicated.

This rule of thumb has been proposed by the Los Alamos Research Laboratories and is intended to be applied to new construction only, not to existing buildings.

Recapitulating the above, the two main sources of building heat loss are (1) transmission through the building envelope or skin, and (2) the heat required to warm up cold air infiltrating into the building and/or outside air brought in for ventilation.

The three main considerations involved in determining the heating load are (1) the design indoor and outdoor temperatures; (2) the materials, thermal properties, and construction details of the building; and (3) the amount of internally generated heat (commercial buildings only).

THE DESIGN HEATING LOAD

This section will analyze the factors that enter into the design heating load for solar buildings and will develop methods for calculating building heating loads. First, however, it is necessary to define somewhat more exactly some of the terms that have been introduced and used so far only in a descriptive way.

Building heat loss is the rate at which heat energy is leaving a building, at the specified indoor and outdoor temperatures. It can be expressed in Btu per hour, Btu/day, or Btu/month.

Building heating load is the rate at which heat energy must be supplied to achieve thermal balance in the structure and maintain a specified (design) indoor temperature against a specified (design) outdoor temperature. For residences, building heating load is assumed to be equivalent to building heat loss.

Outdoor design temperature is a temperature somewhat above the extreme low for the season at the locality being considered. The extreme low may have been reached only once or twice over a period of 20 years, and then for only a period of a few hours. Most of the time, even during a severe winter, the outdoor ambient temperature is above the outdoor design temperature by several degrees. Table 5-1 lists outdoor design temperatures for a number of locations in the United States and Canada, and for selected cities throughout the world.[1] Since, for most of the heating season, the actual outdoor temperature is well above the design outdoor temperature, the *actual* building heating load is usually considerably less than the *design* building heat loss.

Dry-bulb and Wet-bulb Temperatures. It will be noted that Table 5-1 gives outside design temperatures in terms of dry-bulb temperature and wet-bulb temperature.

Dry-bulb temperature (DB) is the temperature registered by an ordinary thermometer held in the air with its bulb not in contact with water or moisture. Temperatures reported in routine daily weather reports are DB temperatures.

Wet-bulb temperature (WB) is the temperature registered by a thermometer held in air, whose bulb is surrounded by and in intimate contact with a cloth wick that is soaked with water. The water evaporates into the air, cooling the thermometer bulb, and the thermometer registers a lower temperature as a result of the cooling effect of the evaporating water.

In dry air, the rate of evaporation from a wet wick is rapid, and the wet-bulb temperature will be correspondingly low. On the other hand, if the surrounding air is already *saturated* with moisture (100 percent relative humidity), there will be no evaporation, no cooling, and the WB reading will be the same as the DB reading. *Wet-bulb equals dry-bulb at saturation.*

Wet-bulb temperature, then, is a measure of the moisture content of the air. If the DB–WB temperature difference is large, the surrounding air is very dry; and if the DB–WB temperature difference (called the WB *depression*) is small, the surrounding air contains a good deal of moisture. If the WB depression is zero, the surrounding air is saturated with moisture. In very dry air at 90°F DB, the WB depression may be as much as 30F°.

---

[1] Complete tables of climatic conditions for locations throughout the world are provided in *ASHRAE 1977 Fundamentals* (Atlanta: ASHRAE, 1977), pp. 23.3–23.22.

UNITED STATES		Winter (Heating), Dry Bulb, °F	Summer (Cooling)		UNITED STATES		Winter (Heating), Dry Bulb, °F	Summer (Cooling)	
State	City		Dry Bulb, °F	Wet Bulb, °F	State	City		Dry Bulb, °F	Wet Bulb, °F
Alabama	Birmingham	17	95	78		Sioux City	−11	92	78
	Mobile	25	95	80	Kansas	Topeka	0	97	78
Alaska	Anchorage	−23	68	60	Kentucky	Louisville	5	93	78
	Fairbanks	−50	78	64	Louisiana	New Orleans	29	93	80
Arizona	Flagstaff	−2	84	60	Maine	Augusta	−7	85	73
	Phoenix	30	107	76	Maryland	Baltimore	10	92	78
	Tucson	28	102	72	Massachusetts	Boston	6	88	74
	Yuma	36	109	78		Worcester	0	85	72
Arkansas	Little Rock	15	96	80	Michigan	Detroit	3	90	75
California	Bakersfield	30	104	73		Lansing	−3	88	75
	El Centro	35	110	80		Marquette	−12	81	70
	Fresno	28	102	72	Minnesota	Duluth	−22	83	70
	Los Angeles	37	90	70		Minneapolis	−16	90	75
	San Diego	40	85	70	Mississippi	Vicksburg	20	95	80
	San Francisco	36	80	65	Missouri	Kansas City	2	96	78
	Santa Barbara	36	80	67		St. Louis	2	95	78
Colorado	Denver	−5	92	64	Montana	Butte	−24	84	60
	Pueblo	−10	95	66		Miles City	−20	95	68
Connecticut	Bridgeport	6	85	74	Nebraska	Omaha	−8	92	78
	Hartford	3	88	75	Nevada	Reno	5	94	63
Delaware	Wilmington	10	90	76	New Hampshire	Concord	−8	90	73
District of					New Jersey	Newark	10	92	76
Columbia	Washington	14	92	78	New Mexico	Albuquerque	10	95	66
Florida	Jacksonville	29	95	79	New York	Buffalo	2	85	73
	Miami	44	91	79		New York	10	90	75
Georgia	Atlanta	17	92	76		Syracuse	−3	90	73
	Savannah	24	93	78	North Carolina	Asheville	10	88	74
Hawaii	Honolulu	62	85	75		Charlotte	18	93	76
Idaho	Boise	3	94	66		Raleigh	16	93	78
Illinois	Chicago	−5	93	75	North Dakota	Bismarck	−23	92	72
	Springfield	−3	92	77	Ohio	Akron	1	86	73
Indiana	Indianapolis	−2	90	76		Dayton	−1	89	76
	Terre Haute	−2	92	77		Toledo	−3	88	75
Iowa	Des Moines	−10	92	78	Oklahoma	Tulsa	8	100	78

(Table 5-1 continued on facing page)

Wet-bulb temperature is of critical importance in determining summer cooling loads, since outdoor air in summer in most U.S. locations contains a great deal of moisture. In winter, however, outdoor air has very little moisture, and the WB temperature is of much less importance in the calculation of winter heating loads.

Further discussion of moisture in air and its effects on human comfort and on cooling and heating loads will be found in Chapters 6 and 16.

The *design temperature difference* Δt is the difference between the indoor design temperature t_i and the outdoor design temperature t_o. $\Delta t = t_i - t_o$. If t_o is below zero, it becomes a negative number, and the effect is that below-zero outdoor temperatures are added to the indoor temperature to obtain the design temperature difference Δt.

The *indoor design temperature* may vary somewhat, depending on the application. For residences, however, it is now usually taken as 70°F (21°C), although recent energy conservation guidelines in the United States have recommended dialing down to 68°F (20°C). Most individuals feel comfortable at indoor dry-bulb temperatures (winter) ranging from 68°F to 76°F. Besides dry-

Table 5-1 (continued)

UNITED STATES State	City	Winter (Heating), Dry Bulb, °F	Summer (Cooling) Dry Bulb, °F	Summer (Cooling) Wet Bulb, °F	UNITED STATES State	City	Winter (Heating), Dry Bulb, °F	Summer (Cooling) Dry Bulb, °F	Summer (Cooling) Wet Bulb, °F
Oregon	Bend	−3	88	62	Utah	Salt Lake City	0	95	65
	Portland	17	85	68	Vermont	Burlington	−10	85	72
Pennsylvania	Philadelphia	10	90	76	Virginia	Richmond	14	94	78
	Pittsburgh	0	89	73	Washington	Seattle	21	82	66
Puerto Rico	San Juan	68	87	79		Spokane	−7	92	64
Rhode Island	Providence	5	87	74	West Virginia	Charleston	7	90	75
South Carolina	Charleston	24	92	80		Wheeling	1	86	72
South Dakota	Sioux Falls	−15	92	75	Wisconsin	LaCrosse	−13	88	75
Tennessee	Nashville	10	92	76		Milwaukee	−8	90	75
Texas	Austin	24	100	78	Wyoming	Casper	−11	90	61
	Dallas	18	100	78		Cheyenne	−10	86	62
	El Paso	20	100	69		Laramie	−14	81	60
	Houston	28	95	79					

CANADA Province	City	Winter (Heating), Dry Bulb, °C	Summer (Cooling) Dry Bulb °C	Summer (Cooling) Wet Bulb °C	CANADA Province	City	Winter (Heating), Dry Bulb, °C	Summer Cooling Dry Bulb, °C	Summer Cooling Wet Bulb, °C
Alberta	Calgary	−33	27	17	Nova Scotia	Halifax	−17	24	19
British Columbia	Penticton	−18	32	20	Ontario	Ottawa	−27	31	23
	Vancouver	−9	25	19		Toronto	−21	31	23
	Victoria	−7	23	17	Prince Edward	Charlottes-			
Manitoba	Winnipeg	−34	30	23	Island	town	−22	26	21
New Brunswick	Fredericton	−27	29	22	Quebec	Quebec	−28	29	22
Newfoundland	St. Johns	−16	24	19	Saskatchewan	Regina	−36	31	21
Northwest Territory	Yellowknife	−45	25	17	Yukon Territory	Whitehorse	−43	25	15

OTHER LOCATIONS	Winter, Dry Bulb, °C	Summer Dry Bulb, °C	Summer Wet Bulb, °C	OTHER LOCATIONS	Winter, Dry Bulb, °C	Summer Dry Bulb, °C	Summer Wet Bulb, °C
Athens, Greece	31	34	22	Panama City, Panama	22	33	27
Bogota, Columbia	7	21	16	Paris, France	−6	30	20
Bombay, India	18	34	28	Rio de Janeiro, Brazil	14	33	26
Buenos Aires, Argentina	0	32	24	Riyadh, Saudia Arabia	3	42	25
Cairo, Egypt	7	38	24	Rome, Italy	−1	33	23
Capetown, South Africa	4	32	22	Shanghai, China	5	32	28
Caracas, Venezuela	11	28	21	Singapore	22	33	27
Kuala Lumpur, Malaysia	21	34	28	Stockholm, Sweden	−15	24	17
London, England	−4	27	19	Sydney, Australia	4	29	23
Madrid, Spain	−4	33	21	Tel Aviv, Israel	4	34	23
Manila, Philippines	20	33	27	Tokyo, Japan	−3	32	27
Melbourne, Australia	2	33	21	Vienna, Austria	−14	30	21
Mexico City, Mexico	2	27	16				

Source: Norman C. Harris, Modern Air Conditioning Practice, 3rd ed. New York: McGraw-Hill, 1983, pp. 110–11.
Reprinted with permission.

bulb temperature, such factors as relative humidity, air motion, radiation to or from the human body, clothing worn, level of physical activity, and the season of the year all have an effect on whether or not people feel comfortable indoors. A full discussion of this topic, including the use of *Comfort Charts* in the design process, will be found in Chapter 6, for winter comfort. See Chapter 16 for a discussion of summer comfort.

5-2   Heat Loss Through the Building Envelope

Heat leaves a building principally as a result of conduction and convection. Radiant energy from the low-temperature sources inside buildings is quite effectively blocked by window glass (see section 3-16), and radiation losses from building interiors are usually considered to be negligible in determining building heat loads.

Conduction Heat Loss.   The building envelope contains walls, windows, ceilings, roofs, floors, and doors. Most of these are nonhomogeneous structural components, made up of two or more materials of differing conductivities K (see sections 3-8 and 3-9). Each discrete material exhibits its own thermal resistance R to heat transmission by conduction. Walls, ceilings, roofs, and floors, as composite structural elements, can be looked upon as *series resistances* with respect to a current of thermal energy flowing through them (see section 3-9 and Fig. 5-2). The total resistance to heat flow for a composite structural element is equal to the sum of the separate resistances in series (comparable to a series network in electricity). The relationship is repeated here for convenience, referred to Fig. 5-1:

$$R_T = R_1 + R_2 + \text{------- } R_n \qquad (3\text{-}12)$$

It will be recalled that, from this relation for total thermal resistance of a composite wall, the inverse—total thermal transmittance—was derived as

$$U = \frac{1}{R_T} = \frac{1}{R_1 + R_2 + \text{----- } R_n} \qquad (3\text{-}13)$$

U-values, it will be remembered, are also called *overall heat transmission coefficients*, and they are expressed in Btu/hr-ft^2-F$°$, or in W/m^2-C$°$.

The Effects of Convection.   An analysis of Fig. 3-6 reveals that heat flows through a material such as a wall by conduction as a result of the temperature difference between *wall surfaces*, not as a result of the temperature difference between indoor and outdoor air. As heat flows from a warm interior surface of a

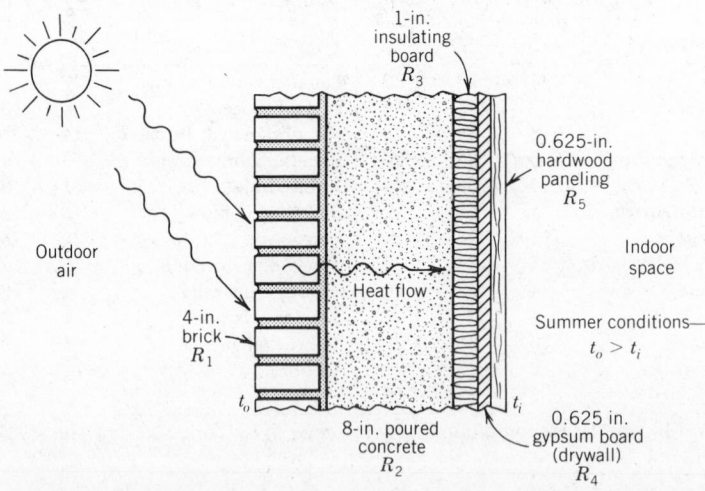

Figure 5-1 Heat flow by conduction through a series of structural elements with differing thermal resistances R. A composite wall is shown in vertical cross-section. The brick, concrete, insulating board, drywall, and wood paneling have different resistances to conduction heat flow. These resistances are R_1, R_2, R_3, R_4, and R_5 respectively. Heat flows by conduction through these resistances in series, because of temperature difference between the surfaces, $\triangle t = t_o - t_i$. $R_T = R_1 + R_2 + R_3 + R_4 + R_5$.

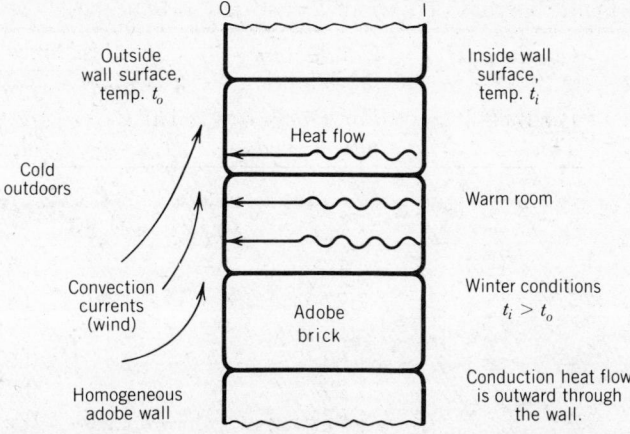

Figure 5-2 The effects of convection at a wall surface on the flow of heat through a wall by conduction. Heat flows to surface O by conduction because of the temperature difference (winter) $\triangle t = t_i - t_o$. If this arriving heat were not removed from the surface by convection currents (wind), t_o would increase until $t_o = t_i$ and conduction heat flow would cease.

wall through the wall material to its outer surface (winter), if the energy arriving there were not rapidly removed, the outer-surface temperature would rise, the temperature difference between the surfaces would decrease and eventually approach zero, and heat flow would cease, regardless of the indoor and outdoor air temperatures (Fig. 5-2). Since this result does not, in fact, occur, some process is obviously removing the heat from the outer surface. That process is convection, and its effects were described in section 3-10, in terms of the air-film coefficient h (also called *surface conductance*).

Tables of U-values prepared for designers and application engineers already incorporate the effect of the air-film coefficients at the inner and outer wall surfaces. For average conditions, values of the film coefficients used in preparing tables of U-values are, for indoor wall surfaces, $h_i = 1.50$ to 1.65; and, for outdoor wall surfaces, $h_o = 6.00$ Btu/hr-ft²-F° for 15 mph wind velocity. Values for h for a variety of wall surfaces and different velocities of air movement were given in Fig. 3-11.

Air Spaces in Walls. Some walls incorporate air spaces inside the wall structure. Examples are (1) frame walls in which the spaces between the studs are not filled with insulation, and (2) a masonry or built-up wall with the interior finish (plaster or drywall board) put on furring strips (usually 1 × 3 wood strips) nailed to the masonry. The thermal conductance a of such air spaces is also measured in Btu/hr-ft²-F° and is ordinarily given for the *thickness of the space*, not per inch thick. In fact, within a range of from 0.75 in. to as much as 4 in., the value of a changes very little with thickness of the air space (see Fig. 5-3).

For air spaces 0.75 to 3.5 in. thick, within wall structures with nonreflective surfaces, an average value of about 1.1 Btu/hr-ft²-F° may be used for a. There results an air-space R value of 1/1.1 or 0.91. For a few other situations, see the *air spaces* section of Table 5-2; and for complete data, refer to Chapter 22 of *ASHRAE Handbook, 1977 Fundamentals*.

Thermal Properties of Typical Construction Materials. In order to provide a few examples of the method of determining overall coefficients of transmission U, selected thermal properties of typical construction materials are included herein as Table 5-2. Some constructions are simplified and some values

Table 5-2 Selected Thermal Properties of Various Construction Materials

Material Described	Conductivity K Per Inch Thick	Conductance C For Thickness Stated	Resistance R Per Inch Thick (1/K)	Resistance R For Thickness Stated (1/C)	Specific Heat c Btu/lb-F°
Building Boards					
Gypsum or plaster board (drywall)					
0.5 in.	—	3.10	—	0.32	0.26
0.625 in.	—	2.22	—	0.45	—
Plywood (Douglas fir)					
0.5 in.	—	1.60	—	0.62	—
0.625 in.	—	1.29	—	0.77	—
Plywood subfloor, 0.75 in.	—	1.07	—	0.93	0.29
Vegetable fiberboard sheathing, 0.5 in.	—	0.76	—	1.32	0.31
Building Membrane					
Vapor barrier, 15 lb mopped felt, two layers	—	8.35	—	0.12	—
Finish Flooring					
Carpet and fibrous pad	—	0.48	—	2.08	0.34
Terrazzo, 1 in.	—	12.50	—	0.08	0.19
Tile, asphalt, vinyl, rubber, etc.	—	20.00	—	0.05	0.30
Wood, hardwood finish, 0.75 in.	—	1.47	—	0.68	0.30
Insulating Materials					
Blanket or batt					
Mineral or glass fiber					
3–3.5 in. thick	—	0.091	—	11.0	—
5.5–6.5 in. thick	—	0.053	—	19.0	—
8.5 in. thick	—	0.033	—	30.0	—
Boards or slabs					
Expanded polystyrene	0.20	—	5.00	—	0.29
Expanded polyurethane	0.16	—	6.25	—	0.38
Acoustical tile (cane fiberboard)					
0.5 in.	0.35	0.80	—	1.25	0.31
0.75 in.	—	0.53	—	1.89	—
Masonry Materials					
Cement concrete	5.0	—	0.20	—	—
Stucco	5.0	—	0.20	—	—
Masonry Units					
Brick					
Common	5.0	—	0.20	—	0.19
Face	9.0	—	0.11	—	—
Concrete block, three-cavity core, unfilled					
8 in.	—	0.9	—	1.11	—
12 in.	—	0.78	—	1.28	—
Stone, average	12.50	—	0.08	—	0.19

Table 5-2 (continued)

Material Described	Conductivity K Per Inch Thick	Conductance C For Thick- ness Stated	Resistance R		Specific Heat c Btu/lb-F°
			Per Inch Thick (1/K)	For Thick- ness Stated (1/C)	
Plastering Materials					
Cement plaster, 0.75 in.	—	6.66	—	0.15	0.20
Gypsum plaster, lightweight aggregate on metal lath					
0.625 in.	—	2.67	—	0.39	—
0.75 in.	—	2.13	—	0.47	—
Roofing					
Asphalt shingles, 70 lb	—	2.27	—	0.44	0.30
Built-up roofing, 0.37 in.	—	3.00	—	0.33	0.35
Wood shingles	—	1.06	—	0.94	0.31
Siding Materials					
Shingles					
Cedar, 16 in.					
7.5 in. exposure	—	1.15	—	0.87	0.31
Cedar, on 0.3125 in. insulating board	—	0.71	—	1.40	0.31
Siding					
Wood, board-and- batt, 1 × 8	—	1.27	—	0.79	0.28
Wood, plywood, 0.375 in., lapped	—	1.59	—	0.59	0.29
Aluminum					
Over sheathing	—	1.61	—	0.61	0.29
Backed by 0.375 in. insulating board and foil	—	0.34	—	2.96	—
Woods					
Hardwoods	1.10	—	0.91	—	0.30
Fir, pine, etc.	0.80	—	1.25	—	0.33
0.75 in.	—	1.06	—	0.94	0.33
2.5 in.	—	0.32	—	3.12	—
3.5 in.	—	0.23	—	4.35	—
Air Spaces in Vertical Walls (0.75 in. to 3.5 in. air space) Mean temp. 50°F; Temperature differential 20°F				0.91 (average)[a]	

[a] R values for air spaces vary rather widely with the mean temperature, the temperature difference, and the reflectance of the air-space surfaces. See Table 2 of source cited below for specific situations.

Source: Abstracted with permission from Tables 2 and 3A, Chapter 22, *ASHRAE Handbook 1977 Fundamentals* (Atlanta, ASHRAE 1977), pp. 22.13–22.17.

are averaged to avoid undue complexity of the data, since the purpose here is to illustrate the concept and method of calculating U values.[2]

Note that the table lists the materials described under several major headings, and in general gives values of conductance C and thermal resistance

[2] Complete tables of the thermal properties of materials may be found in *ASHRAE Handbook 1977 Fundamentals* (Atlanta: ASHRAE, 1977), pp. 22.13–22.16, and also in later editions.

R. Values of conductivity K and specific heat c are provided for some of the materials.

Sample Calculations of U-values—Walls.   In the next section, tables of overall transmission coefficients (U values) for typical construction elements will be provided as a data base for calculating building heat loss resulting from transmission through the building envelope. At this point, however, we digress briefly to show by two examples how the U values in such tables are calculated.

Example 1.
Frame wall, as for the exterior of a residence, cedar shingle siding on plywood sheathing, space between studs insulated with 3.5-in. fiberglass blanket insulation. Other details are shown in Fig. 5-4.

The thermal resistances of the several layers or elements of construction are obtained from Table 5-2 and are summed up from the outside in, as resistances in series:

Wall Component		R
Outside air film	$R = \dfrac{1}{6.0} =$	0.17
Cedar shingle siding		0.87
Plywood sheathing, 0.625 in.		0.77
Vapor barrier (membrane)		0.12
Fiberglass blanket, 3.5 in.		11.00
(Neglect effect of studs, but see below)		
Plaster board, 0.5 in.		0.32
Gypsum plaster, 0.625 in.		0.39
Inside air film	$R = \dfrac{1}{1.65} =$	0.61
	$R_T =$	14.25

$$U \text{ value} = \frac{1}{R_T} = 0.07 \text{ Btu/hr-ft}^2\text{-F}°$$

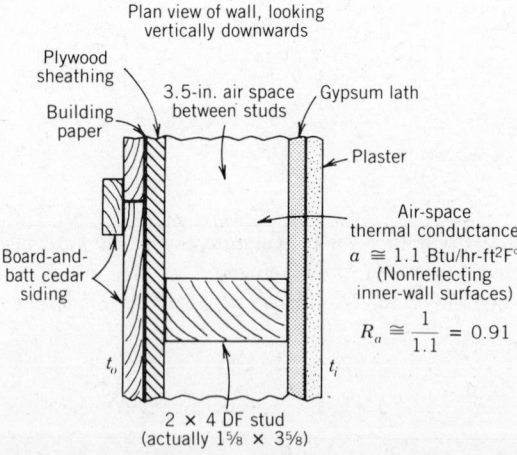

F i g u r e 5-3 Thermal conductance of air spaces in walls. A plan view of a residential frame wall, looking downward on a horizontal cross-section. The thermal conductance of the air space is a. For nonreflective wall surfaces and for air-space thicknesses of from 0.75 in. to 4.00 in., $a \cong 1.1$ Btu/hr-ft²-F°.

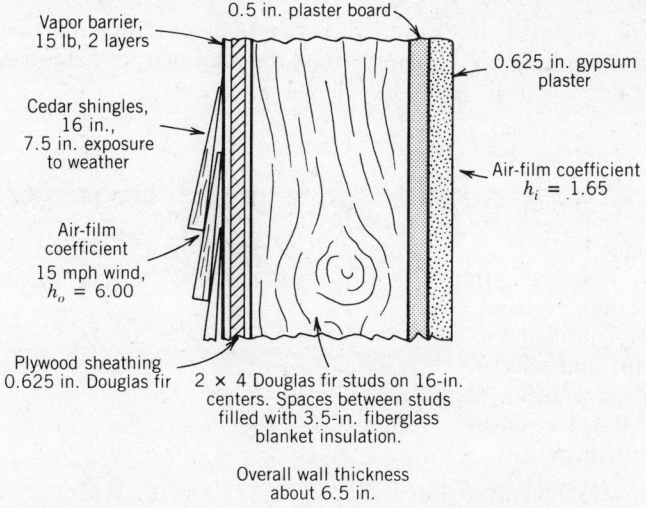

Vapor barrier,
15 lb, 2 layers

0.5 in. plaster board

0.625 in. gypsum
plaster

Cedar shingles,
16 in.,
7.5 in. exposure
to weather

Air-film coefficient
$h_i = 1.65$

Air-film
coefficient
15 mph wind,
$h_o = 6.00$

Plywood sheathing
0.625 in. Douglas fir

2 × 4 Douglas fir studs on 16-in.
centers. Spaces between studs
filled with 3.5-in. fiberglass
blanket insulation.

Overall wall thickness
about 6.5 in.

Figure 5-4 Typical frame wall for a residence in vertical section, with spaces between studs filled with 3.5-in. fiberglass blanket insulation. Cedar shingles on fir plywood sheathing with rolled-paper vapor barrier make up the outer wall structure. The inside is finished with gypsum plaster on plaster board.

Note that this wall is insulated in accordance with current standards, and its rate of heat loss by transmission would be relatively small. For a comparison with a noninsulated wall, see Example 2 below.

Effect of Studs in a Frame Wall. When heat transmission takes place through a wall by *parallel paths*—for example, some heat flowing through wood studs and some through the blanket insulation between the studs—a precise treatment of the heat flow problem requires the computation of a composite R value. With 2 × 4 studs (actually 1.625 in. by 3.5 in.) on 16-in. centers and blanket insulation between the studs, the total cross-sectional area through which conduction occurs is about 10 percent wood studs and 90 percent insulation. As a result,

$$R_{\text{composite}} = \frac{1}{\dfrac{0.10}{R_{\text{fir}}} + \dfrac{0.90}{R_{\text{insul.}}}}$$

From tables,

$$3.5 \text{ in. fir}—R = 4.35$$
$$3.5 \text{ in. insul.}—R = 11.0$$

And, substituting,

$$R_{\text{composite}} = \frac{1}{\dfrac{0.10}{4.35} + \dfrac{0.90}{11.0}}$$
$$= 9.54$$

Substituting this (composite) R value in place of the 11.0 R value (studs neglected) would result in a U value of 0.078 instead of the 0.07 obtained from assuming that the entire space is filled with insulation. Where a precise building heat loss estimate is indicated, the composite U value should be used, since wall heat loss usually accounts for a large share of the total structural heat loss.

Example 2.
Masonry wall, 4-in. face brick on 8-in. concrete block with cores hollow; other details as shown in Fig. 5-5. Obtain thermal resistances from Table 5-2, and sum them up as resistances in series, as follows:

Wall Component	*R*
Outside air film, $h = 6.0$	$R = \dfrac{1}{6.0} = 0.17$
4-in face brick	$R = 4 \times 0.11 = 0.44$
0.5-in. cement mortar	$R = 0.5 \times 0.20 = 0.10$
8-in. concrete block, cores unfilled	1.11
0.75-in. nonreflective air space; mean temp. 50° (Neglect effect of furring strips)	$R = \dfrac{1}{a} = \dfrac{1}{1.1} = 0.91$
Gypsum wallboard, 0.5-in.	0.32
Inside air film, $h_i = 1.65$	$R = \dfrac{1}{1.65} = 0.61$
	$R_T = \overline{3.66}$

$$U = \frac{1}{R_T} = 0.27 \ \text{Btu/hr-ft}^2\text{-F}°$$

Despite the sturdier construction of this wall and its greater overall thickness, the fact that it lacks insulating material makes its heat loss rate by transmission nearly four times as great as the wall of Example 1.

Windows and Fixed Glass.   For the purpose of calculating winter building heat loss, solar gains through glass are not considered since, for the heating season, days are short and nights long, and for that matter many days have no sunshine. Radiation losses outward through glass are negligible (section 3-16), so

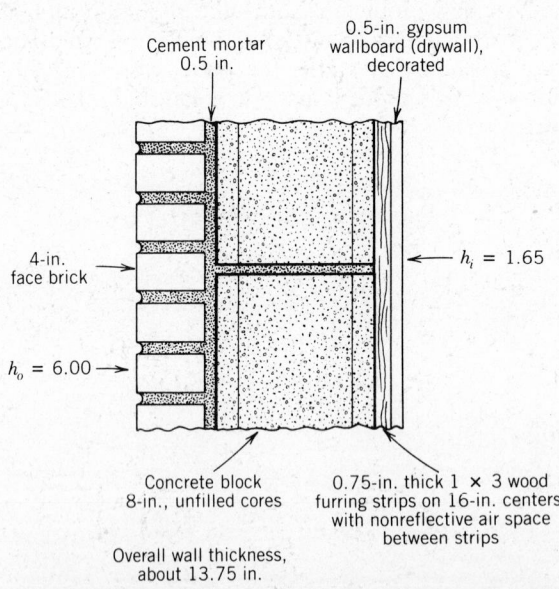

Cement mortar 0.5 in.

0.5-in. gypsum wallboard (drywall), decorated

4-in. face brick

$h_i = 1.65$

$h_o = 6.00$

Concrete block 8-in., unfilled cores

0.75-in. thick 1 × 3 wood furring strips on 16-in. centers, with nonreflective air space between strips

Overall wall thickness, about 13.75 in.

Figure 5-5 Masonry wall for light commercial construction. Four-in. face brick on hollow-core concrete block, with gypsum wallboard (drywall) on furring strips for the interior finish. Spaces between furring strips are air spaces.

the heat loss through glazing is assumed to be entirely from conduction. U values for various glass and clear plastic components are given in Table 5-3, which begins on page 162. Infiltration (cold air) losses will be discussed later.

Doors. Conduction losses through doors are calculated in the regular manner, with U values such as those listed in section 9 of Table 5-3. Infiltration losses around doors are discussed later in this chapter.

Ceilings and Roofs. Ceilings under flat roofs with space only for joists offer the simplest ceiling-heat-loss calculation. The method is similar to that for exterior walls.

Pitched roofs with sloping ceilings (beam ceilings or cathedral ceilings) also lend themselves to direct calculation of conduction heat losses based on the indoor–outdoor design temperature difference. See section 5 of Table 5-3 for U values and typical constructions of the above two types of roof–ceiling combinations.

Ceilings under unheated attic spaces involve one additional step in heat-loss calculations, because a fairly accurate estimate of the attic-space temperature must be made. The temperature difference in this case is $\Delta t = (t_i - t_{attic})$. See section 4 of Table 5-3 for U values.

Floors. For floors above a crawl space, assume that the temperature in the crawl space is the design outdoor temperature. See U values in section 7 of Table 5-3 for typical constructions.

For floors over basements, see section 4 of Table 5-3. If the basement temperature is not otherwise specified, use 60°F.

Concrete slab floors constitute a different kind of heat-loss problem. Since ground temperatures under slabs, even in cold climates, do not drop to very low levels, the outward heat loss at the perimeter of the slab is the overriding factor rather than losses down through the slab. Instead of being expressed as U values, heat-loss coefficients from concrete slab floors are expressed as F factors, in units of Btu/hr-F°-lineal foot of slab perimeter.

The heat loss Q from the small to medium sized slabs encountered in residential and small-commercial construction can be closely approximated by using a formula recommended by ASHRAE, as follows:

$$Q = P \times F \times \Delta t \qquad (5\text{-}1)$$

where

Q = the heat loss rate, Btu per hour
P = the outside perimeter of the slab, feet
Δt = the indoor–outdoor design temperature difference, degrees Fahrenheit
F = the perimeter heat-loss coefficient, an experimentally determined multiplier, Btu/hr-F°-lin. ft of edge

Recommended values of F are:

Slab with no edge insulation (not a recommended construction) $F = 0.81$ Btu/hr-F°-lin. ft

Slab with edge insulation 1 in. thick and 24 in. wide, extending 2 ft under slab edge. $F = 0.55$ Btu/hr-F°-lin. ft

As an example of the calculation, consider a 30 ft by 50 ft concrete slab for a house in a location where $t_o = 10°F$. The insulated slab heat loss would be

$$Q = 160 \text{ lin. ft} \times 0.55 \text{ Btu/hr-F°-lin. ft} \times (70 - 10)\text{F°}$$
$$= 5280 \text{ Btu/hr}$$

Table 5-3 Coefficients of Heat Transmission U for Selected Structural Elements—Btu/hr-ft²-F° Difference in Temperature Between the Air on the Two Sides

1. Frame Walls

A. Wood siding

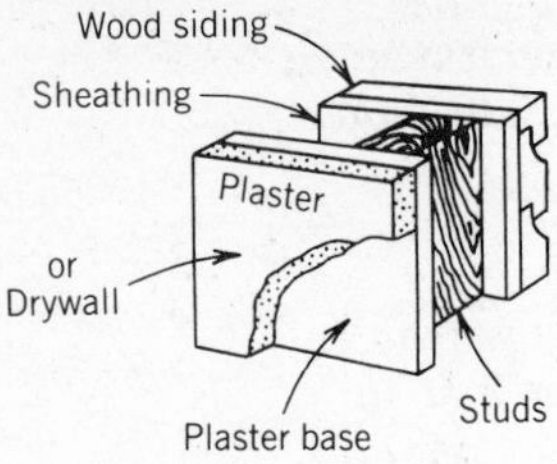

Wood siding
Sheathing
Plaster
or Drywall
Plaster base
Studs

Also applicable to:
Wood shingles
7 in. exposure
Board and batt siding
¾ in. thick

	Sheathing			
	0.625 in. Plywood		0.625 Insulation Board	
Interior finish	No insulation in wall	3.5 in. insulation	No insulation in wall	3.5 in. insulation
Gypsum board (drywall)	0.24	0.08	0.21	0.065
Plaster on gypsum lath	0.21	0.075	0.19	0.062

B. Stucco

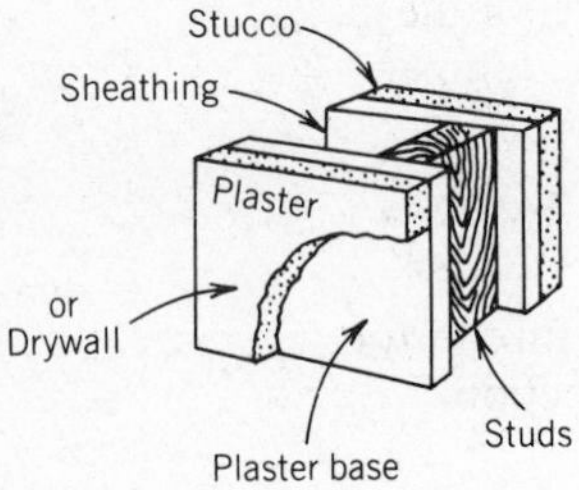

Stucco
Sheathing
Plaster
or Drywall
Plaster base
Studs

Gypsum board (drywall)	0.36	0.09	0.21	0.075
Plaster on gypsum lath	0.32	0.08	0.19	0.072

2. Masonry Walls

A. Solid brick

	Interior Finish		
Thickness inches	None	Plaster on gypsum lath on furring, 0.75 in. air space	Plaster on gypsum lath, furred, with insulation between the furring strips
8	0.625	0.26	0.13
12	0.42	0.22	0.12

B. Stone or poured concrete

8	0.70	0.36	0.16
12	0.58	0.33	0.15

C. Concrete block (no exterior facing)

8	0.56	0.32	0.15
12	0.50	0.30	0.14

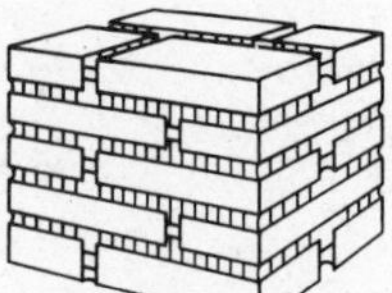

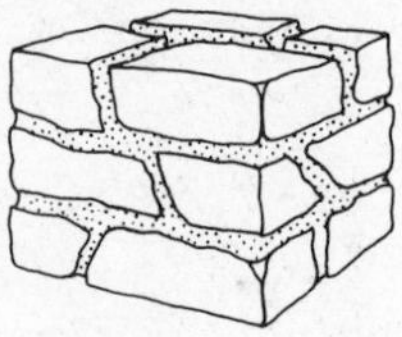

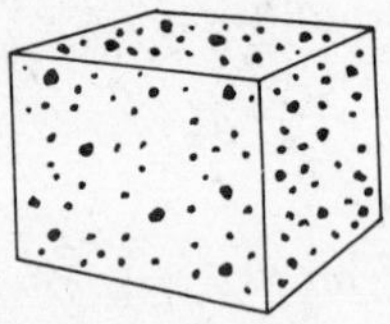

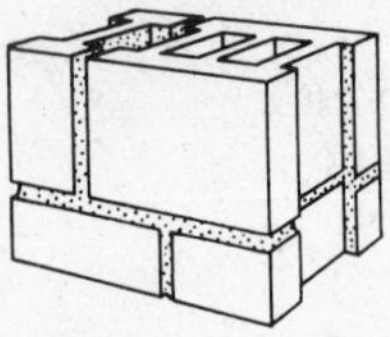

Masonry Walls (*continued*)

D. Concrete block (with 4-in. face brick)

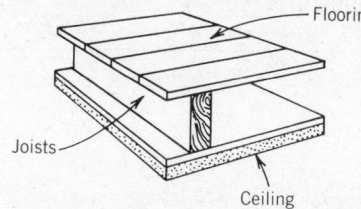

		Interior Finish	
Thickness inches	None	Plaster on gypsum lath on furring, 0.75 in. air space	Plaster on gypsum lath, furred, with insulation between the furring strips
12	0.33	0.18	0.13
16	0.31	0.16	0.12

3. Interior Walls (Partitions)

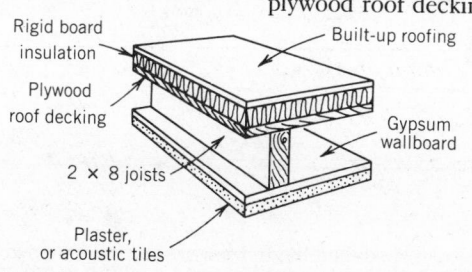

2 × 4 stud frame, with finish as indicated

Finish (Both Sides)	Insulation	
	None	3.5 in.
Gypsum lath and plaster	0.29	0.080
Drywall, decorated	0.30	0.084
Wood paneling, ⅜ in.	0.34	0.088

4. Ceiling/Floor Combinations

A. Unheated attic space above heated rooms. Gypsum lath and plaster on ceiling, 2 × 8 joists with R-19 blanket insulation, plywood subfloor, and wood floor. $\Delta t = t_i - t_{attic}$.

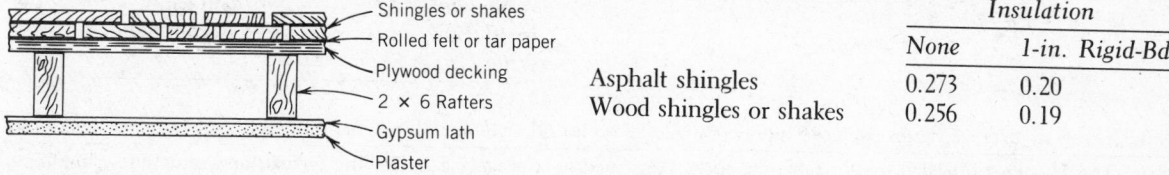

Average of such constructions	$U = 0.055$
With R-30 insulation	$U = 0.035$
Same, but no insulation	$U = 0.22$

B. Basement below heated rooms. If same construction, use same values, but $\Delta t = t_i - t_{basement}$.

5. Flat Roofs and Ceilings

Plaster or acoustical tile on gypsum lath or wallboard for ceiling. Wood 2 × 8 joists, plywood roof decking covered with rigid-board insulation, with built-up roof on top.

	Insulation Between Joists		
	None	R-19	R-30
$U =$	0.16	0.046	0.035

6. Pitched Roofs and Ceilings

A. Ceiling directly under pitched roof, plaster on gypsum lath, or decorated drywall, no attic space. Wood 2 × 4 or 2 × 6 rafters (as span requires), 45° slope. Plywood sheathing 0.625 in., covered with rolled felt and shingles or shakes.

Sectional view looking up the slope

	Insulation	
	None	1-in. Rigid-Bd.
Asphalt shingles	0.273	0.20
Wood shingles or shakes	0.256	0.19

Pitched Roofs and Ceilings (continued)

Sectional view looking up the slope

- Shingles or shakes
- Rolled felt or tar paper
- Plywood sheathing
- Rigid board insulation
- T & G fir roof decking, 2 × 4, nailed to beams

Douglas fir beam (exposed)

B. Beam ceiling directly under sloping roof, 2 × 4 tongue-and-groove fir roof decking nailed on top of wooden beams, no attic space. Rigid-board insulation on decking, plywood sheathing 0.625 in., covered with rolled felt paper and asphalt shingles or shakes.

	Insulation		
	None	*1-in.*	*2-in.*
Asphalt shingles	0.35	0.17	0.085
Wood shingles	0.32	0.16	0.083
Heavy wood shakes	0.29	0.15	0.080

7. *Floors*

Over crawl space only. Floors over basement and slab floors covered in other sections of this table.

Sectional view, end-on to joists

Hardwood flooring, or ¾-in. plywood with carpet and pad

- Fir plywood subfloor
- Floor joists

Type of Floor	Insulation	
	None	*R-5 Board*
Hardwood on 0.75 in. plywood subfloor on wood joists	0.37	0.15
Carpet and pad on 1-in. plywood on subfloor on wood joists	0.30	0.13
Linoleum or asphalt or vinyl tile on 1-in. plywood on subfloor on wood joists	0.55	0.17

8. *Windows and Skylights—Exterior Exposure*
   A. Vertical panes, sliding glass doors, etc.

Flat, single-pane	1.10
Double-pane insulating glass, 0.25 in. air space	0.58
Triple-pane insulating glass, 0.25 in. air spaces	0.39
Storm windows, 1 in. to 4 in. air space	0.50

   B. Horizontal panes, skylights, and clear plastic domes

Flat, single glass	1.23
Double-pane insulating glass, 0.25 in. air space	0.65
Clear plastic dome, single-walled	0.70

Note: For panes mounted at various tilt angles, see *ASHRAE Handbook.*

9. *Doors*	*No storm door*	*With storm door*
Thin wood hollow-core doors	1.13	0.65
Solid wood doors		
1 in. thick	0.64	0.30
1.5 in.	0.49	0.27
2.0 in.	0.43	0.24

Note: For doors with glass, the areas of wood and glass must be separately calculated, and the transmission loss determined accordingly.

10. *Grade-Level Concrete Slab Floors*

$$Q = PF(t_i - t_o)$$

Outdoor Design Temperature	*Value of F (Unheated Slab)*	
	With 1-in. Edge Insulation, 24-in. Wide Under Slab	*No Insulation (Not Recommended)*
+10 to −20°F	0.55	0.81

Note: These are average values. Refer to more complete tables for other temperatures and different insulation specifications.

Note: The *U* values provided in this table are to be construed as average values for the typical (and sometimes simplified) constructions described. They are not precise values for specific constructions. If such precise values are required, please consult engineering tables or manufacturers' specifications. Consult other tables also for constructions differing from those described.

Source: Adapted, with permission, from *ASHRAE Handbook, 1977 Fundamentals,* (Atlanta: ASHRAE, 1977), p. 22.18 ff; and from manufacturers' technical data.

5-3   Using Overall Coefficients of Heat Transmission

We are now ready to begin the actual calculation of building heat losses from transmission through the building envelope. Rarely does the solar designer have to carry out detailed calculations of heat-loss coefficients from tables of K values and R values, because overall coefficients of heat transmission (U values) are readily available in engineering tables for most of the structural components that will be encountered in solar design work.

Table 5-3 provides U values for structural elements *as described*. The coefficients are expressed in Btu/hr-ft²-F° difference in temperature between the air on the two sides (except as otherwise noted). Air-film and air-space coefficients have already been incorporated into the values given. The outdoor wind velocity is taken as 15 mph. Air velocity inside is assumed to be zero.

Space limitations in an introductory text do not permit the inclusion of complete tables covering all possible industry applications. Students and practicing designers needing a greater range of tabular data than that provided in Table 5-3 are referred to the handbooks of the American Society of Heating, Refrigerating, and Air Conditioning Engineers (ASHRAE) or to other engineering handbooks.

The basic equation for transmission heat loss through structural elements is

$$Q = UA \, \Delta t \qquad\qquad (5\text{-}2)$$

where

Q = the heat loss rate, Btu per hour
U = the coefficient of heat transmission of the composite structural element
A = the surface area of the structural element, square feet
$\Delta t = (t_i - t_o)$ is the design indoor–outdoor temperature difference

The following solved problems will illustrate the method of calculating transmission heat loss and afford practice in the use of tables of U values.

Illustrative Problem 5-1

A small commercial building has been constructed on a concrete slab with dimensions of 40 ft by 60 ft. The slab has edge insulation (1 in. thick and 24 in. wide under the slab) around the entire perimeter. The walls are 12 ft high, 8-in. hollow-core concrete block, with 4-in. face-brick exterior. Interior finish on the walls is plaster on gypsum lath over 1 × 3 furring strips, with ¾-in. rigid-board insulation between the strips.

Window areas total 700 ft² double-pane insulating glass. There are two entrances—the front one fitted with two 3 × 8 ft solid wood doors 2 in. thick, and the rear entrance with a single 3 × 8 ft 1.5-in.-thick solid wood door with a storm door.

The ceiling–roof structure is acoustic tile on gypsum wallboard on 2 × 10 joists, with R-19 insulation between the joists. Plywood roof decking with rigid-board insulation and built-up roof are on top. The roof is flat.

Calculate the building heat loss rate in Btu per hour, from transmission through the building envelope only, for a location where the outside design temperature is 15°F and inside design temperature is 70°F.

Solution

Use Eq. (5-2), $Q = UA \, (t_i - t_o)$. Refer to Table 5-3 for U values. Lay out the work as follows:

1. **Transmission through walls**

   Wall area = 2 (60 ft × 12 ft + 40 ft × 12 ft) − 700 ft^2 (windows)
   = 2400 ft^2 − 700 ft^2 = 1700 ft^2

   Wall U value (Table 5-3, section 2-D) = 0.13 Btu/hr-ft^2-F°

   Q_{walls} = 0.13 Btu/hr-ft^2-F° × 1700 ft^2 × (70 − 15)F°
   = 12,150 Btu/hr   12,150

2. **Transmission through windows**

   Window U value (Table 5-3, section 8-A) = 0.58 Btu/hr-ft^2-F°

   $Q_{windows}$ = 0.58 Btu/hr-ft^2-F° × 700 ft^2 × 55 F°
   = 22,300 Btu/hr   22,300

3. **Transmission through doors**

   From Table 5-3, section 9, $U_{front\,doors}$ = 0.43

   $U_{rear\,door}$ = 0.27

   Q_{doors} = 0.43 Btu/hr-ft^2-F° × 48 ft^2 × 55 F°
   + 0.27 Btu/hr-ft^2-F° × 24 ft^2 × 55 F°
   = 1490 Btu/hr   1,490

4. **Transmission through ceiling–roof**

   From Table 5-3, section 5, $U_{ceiling-roof}$ = 0.046 Btu/hr-ft^2-F°

   $Q_{ceiling-roof}$ = 0.046 Btu/hr-ft^2-F° × 2400 ft^2 × 55 F°
   = 6070 Btu/hr   6,070

5. **Heat loss at edge of slab**

   Use Eq. (5-1), $Q = PF(t_i - t_o)$

   From Table 5-3, section 10, F = 0.55 Btu/hr-F°-lin. ft of slab

   Q_{slab} = 200 lin. ft × 0.55 Btu/hr-F°-lin. ft. × 55 F°
   = 6050 Btu/hr   <u>6,050</u>

   Total building–envelope transmission loss   48,060 Btu/hr
   Ans.

We point out that this is the transmission loss through the building envelope only. It is not the *total heating load*. Infiltration of cold air and/or ventilation air purposely brought into the building will add to the total load, as will be discussed later. Since the structure is a commercial building, and, depending on its use, there may be internally generated heat from lights, machinery, occupants, and so on, that would help balance the heat transmission loss.

Illustrative Problem 5-2

A 2400 ft^2 residence (one floor with half basement) is being planned for Rapid City, South Dakota. It has large window areas for solar gain, and it is rectangular in shape (60 × 40 ft), with its major axis east–west. The solar collection area consists of 440 ft^2 of double-pane glass in the south wall. In addition, there are other double-pane windows totaling 210 ft^2, and a clear plastic dome skylight with an area of 36 ft^2.

The walls are frame construction, gypsum drywall interior finish, 3.5 in. insulation between studs, 0.625 in. insulating board sheathing with building paper, and 1-in. board-and-batt cedar siding on the exterior. The total wall area (less glass and doors) is 1270 ft².

The front (south) half of the ceiling is sloped upward to the rear and is a cathedral ceiling directly under the roof. It has an area of 1250 ft². Its construction is gypsum drywall on 2 × 6 wood rafters, with 1-in. rigid board insulation, rolled felt, and cedar shakes to the weather. The rear (north) half of the ceiling is horizontal at regular ceiling height (8 ft), with unheated attic space above; gypsum drywall ceiling, 2 × 6 joists with R-30 insulation. Local experience dictates that the attic design temperature should be 0°F.

The floor in the front half of the house is brick and ceramic tile on a concrete slab, with edge insulation on the three outdoor-facing sides. There is a basement under the rear half of the house, with the floor over the basement being carpet and pad on 1-in. plywood on subfloor on 2 × 10 in. joists. Basement design temperature is 60°F.

There are four 1.5-in., solid-wood exterior doors with storm doors, with a total door area of 96 ft².

Calculate the building heat loss rate in Btu per hour from transmission through the envelope only. Outside design temperature for Rapid City is −10°F. Inside design is 70°F. (Disregard solar gains. They offset building transmission losses, but they do not enter into the calculation of such losses.)

Solution

Use Eq. (5-2) as needed, and refer to Table 5-3 for U values.

1. **Transmission through glass and skylight**

$Q_{glass} = 0.58$ Btu/hr-ft²-F° $\times$ 650 ft² $\times$ [70 − (−10)] F°
$= 30,200$ Btu/hr   30,200

$Q_{skylight} = 0.70$ Btu/hr-ft²-F° $\times$ 36 ft² $\times$ [70 − (−10)] F°
$= 2,000$ Btu/hr   2,000

2. **Transmission through walls**

$Q_{walls} = 0.065$ Btu/hr-ft²-F° $\times$ 1270 ft² $\times$ [70 − (−10)] F°
$= 6600$ Btu/hr   6,600

3. **Transmission through ceilings and roof**

Front half
$Q_{ceiling} = 0.19$ Btu/hr-ft²-F° $\times$ 1250 ft² $\times$ [70 − (−10)] F°
$= 19,000$ Btu/hr   19,000

Rear half
$Q_{ceiling} = 0.035$ Btu/hr-ft²-F° $\times$ 1200 ft² $\times$ [70 − 0] F°
$= 2,950$ Btu/hr   2,950

4. **Transmission loss to basement, rear half of house**

$Q_{floor} = 0.22$ Btu/hr-ft²-F° $\times$ 1200 ft² $\times$ (70 − 60) F°
$= 2,650$ Btu/hr   2,650

5. **Transmission loss to ground through three edges of slab (front of house)**

Use Eq. (5-1), $Q_{slab} = PF(t_i - t_o)$

Substituting, $Q_{slab} = 0.55$ Btu/hr-F°-lin. ft $\times$ 100 ft $\times$ [70 − (−10)] F°
$= 4,400$ Btu/hr   4,400

6. Transmission loss through doors

$$Q_{\text{doors}} = 0.27 \text{ Btu/hr-ft}^2\text{-F}° \times 96 \text{ ft}^2 \times [70 - (-10)] \text{ F}°$$
$$= 2{,}070 \text{ Btu/hr}$$

$$\underline{2{,}070}$$

Total building envelope transmission loss   69,900 Btu/hr

Ans.

The discussions and sample problems above cover many of the typical situations involving building heat transmission ordinarily encountered by solar designers. Every application has its own unique attributes, however, and every location and building site has its own variations from average conditions. The general methods covered here constitute acceptable guidelines, but engineers and technicians should always be alert for specific local data and for indicated departures from the norm.

5-4 Heat Losses from Infiltration

Outside air enters buildings in two different ways—by *infiltration* and by *ventilation*. First we examine the effect of infiltrating air on the building heating load.

Assisted by winter winds, air infiltrates into buildings through cracks around windows and doors and through all of the other cracks and fissures that exist even in well-built structures. The cold air that infiltrates must be warmed up to the inside design temperature, and the heat required to warm it becomes a part of the building heating load.

Heat Required to Warm Air. If air is assumed dry (and cold winter air has very little moisture content), the amount of heat required to warm it from the outside design temperature to the inside design temperature is given by

$$Q_{\text{OA}} = wc(t_i - t_o) \qquad (5\text{-}3)$$

where

    Q_{OA} = the heat required to warm the outside air, Btu/per hour
    w = the weight of air heated, pounds per hour
    c = 0.24 Btu/lb-F°, the specific heat of dry air

But the weight of air heated per hour equals the volume per hour times the density of the air, or $w = V \times D$. For standard air (60°F and dry), density $D = 0.075 \text{ lb/ft}^3$. Substituting these values for c and w and simplifying, Eq. (5-3) becomes

$$Q_{\text{OA}} = 0.018 \, V(t_i - t_o) \qquad (5\text{-}4)$$

where

    V = volume rate of outside air entering the building, cubic feet per hour

You should check the dimensional analysis to verify the value of the units of the derived constant, 0.018. The units are Btu/ft³-F°.

Estimating the Volume of Infiltration Air. Infiltration air volume is usually estimated by the *air-change method*. Table 5-4 lists the number of complete air changes per hour caused by infiltration that would normally be expected in residences and small commercial buildings, for the conditions listed, assuming door usage is minimal to moderate.

The estimated hourly volume of infiltration air—V in Eq. (5-4)—is obtained by multiplying the proper air-change value from Table 5-4 by the volume of the room or space. Then use Eq. (5-4) to estimate the heating load contributed by the infiltration air. The method is illustrated as follows.

Table 5-4 Normally Expected Air Changes Caused by Infiltration—
Residences and Small Commercial Buildings with Limited Door Traffic
(Ventilation and Exhaust Air Not Considered)

Type of Space	Number of Complete Air Changes per Hour
Room or space with no windows or doors	0.5[a]
Room or space with windows or exterior door on one side only	1.0
Room with window(s) or door on two sides	1.5
Same, on three sides	2.0
Entrance halls and lobbies (moderate traffic)	2.3

Source: Adapted from *ASHRAE Handbook 1977 Fundamentals* (Atlanta: ASHRAE, 1977), Chapter 21.

[a] These are average values for good construction and moderate winter weather. For loose construction and winter wind velocities above 25 mph, these values should be increased by 30 to 50 percent.

Illustrative Problem 5-3

The living room of a residence measures 30 ft by 20 ft by 10 ft high. It has windows and doors on two sides. Construction is judged to be average to tight. Design conditions are $t_i = 70°F$, $t_o = 10°F$, wind velocity moderate. Calculate the heating load contributed by infiltration air.

Solution

The room volume is 30 ft × 20 ft × 10 ft = 6000 ft³.
From Table 5-4, the expected air-change rate is 1.5 changes per hour.

The volume of infiltration air per hour is therefore

$$V = 6000 \text{ ft}^3 \times 1.5 \text{ changes/hr}$$
$$= 9000 \text{ ft}^3/\text{hr}$$

Substitution in Eq. (5-4) gives

$$Q_{OA} = 0.018 \text{ Btu/ft}^3\text{-F}° \times 9000 \text{ ft}^3/\text{hr} \times (70 - 10)\text{F}°$$
$$= 9700 \text{ Btu/hr} \qquad\qquad Ans.$$

Effect of Exhaust Fans. In residences, exhaust fans are usually limited to kitchens and bathrooms, and their operation is relatively infrequent. For these applications, the air-change values in Table 5-4 need not be adjusted. However, air that is mechanically exhausted will be replaced by infiltration, and in buildings where exhaust fan usage is significant, the volume of exhaust air must be added to the normally expected infiltration air.

5-5 Ventilation Air

Air that is purposely brought into a building, usually by mechanical means, to satisfy fresh air requirements is called *ventilation air*. For residences with conventional heating systems, it is (regrettably) common practice not to bring in outside air for ventilation, but good engineering design requires it nevertheless. In commercial structures, current standards and building codes stipulate required amounts of ventilation air (see engineering handbooks or local codes), and this air also represents a heating load. Ventilation air and infiltration air are not added to determine the total outside air quantity, however, since, to the extent that ventilation air is mechanically introduced, it will develop a slight positive pressure inside the structure, thus reducing the amount of infiltration air that will enter by exactly the amount of ventilation air brought in.

5-6   Total Outside Air

The problem of estimating *total outside air* is approached as follows:

1. If exhaust air is not a factor, the total outside air is taken as the greater of infiltration air or ventilation air, *not both*.
2. If normally expected infiltration is greater than ventilation air, and there is appreciable exhaust air, add the exhaust air quantity to the infiltration air quantity to obtain total outside air.
3. If ventilation air is greater than infiltration plus exhaust air, ventilation air alone is taken as total outside air.

In most residences and small commercial buildings (except those with high occupancy rates or with air-polluting activities), infiltration will exceed mechanically induced ventilation by a large margin.

Illustrative Problem 5-4

A small dental clinic in Des Moines, Iowa, has windows and doors on two sides. Door traffic is estimated at 30 passages per hour (moderate). The building is relatively new and can be assumed to be of tight construction. Its total volume (50 ft by 60 ft by 10 ft) is 30,000 ft³. Code requirements necessitate the introduction of ventilation air at a rate of 12,000 ft³/hr (about 7 ft³/min per person for an average occupancy of 30 people). Exhaust fans operate continuously during office hours, exhausting 9000 ft³/hr.

The heat transmission loss through the building envelope has been calculated as 82,000 Btu/hr for the following design conditions: $t_i = 70°F$, $t_o = -7°F$, wind velocity 20 mph.

Make a careful estimate of the total building heating load at the design conditions.

Solution

From Table 5-4, since there are doors and windows on two sides and moderate traffic through the front door, we will use a factor of 1.7 air changes per hour.

$$\text{Expected normal infiltration} = 1.7 \times 30,000 = 51,000 \text{ ft}^3/\text{hr}$$

$$\text{Total infiltration} = \text{normal infiltration} + \text{exhaust air}$$
$$= 51,000 \text{ ft}^3/\text{hr} + 9,000 \text{ ft}^3/\text{hr} = 60,000 \text{ ft}^3/\text{hr}$$

Since total infiltration (including exhaust air) is greater than the ventilation air quantity, the infiltration air quantity is used.

$$\text{Total outside air} = 60,000 \text{ ft}^3/\text{hr}$$

From Eq. (5-4),

$$Q_{\text{outside air}} = 0.018 \text{ Btu/ft}^3\text{-F}° \times 60,000 \text{ ft}^3/\text{hr} \times [70 - (-7)] \text{ F}°$$
$$= 83,200 \text{ Btu/hr}$$

$$\text{Building heating load} = \text{transmission load} + \text{outside air load}$$
$$= 82,000 \text{ Btu/hr} + 83,200 \text{ Btu/hr}$$
$$= 165,200 \text{ Btu/hr} \qquad \textit{Ans.}$$

In some commercial buildings there might be significant amounts of internally generated heat to offset a part of this load. But in a dental clinic, little if any allowance would be made for internal heat.

Illustrative Problem 5-5

You have been asked for a preliminary estimate of the feasibility of heating the dental clinic described in Illustrative Problem 5-4 by means of a solar energy system. Using the Los Alamos Laboratories' rule of thumb stated in section 5-1, would you recommend a solar heating system?

Solution

From the given data, the *daily* heating load (assuming steady-state conditions),

$$Q_{daily} = 165,200 \text{ Btu/hr} \times 24 \text{ hr} = 3,965,000 \text{ Btu/day}$$

From this result, the load coefficient in Btu/day-ft²-F° is calculated as follows:

$$\text{Load coefficient} = \frac{3,965,000 \text{ Btu/day}}{(50 \text{ ft} \times 60 \text{ ft}) \times [70 - (-7)] \text{ F}°}$$

$$= 17.2 \text{ Btu/day-ft}^2 \text{ floor area-F}°$$

Since the Los Alamos rule of thumb recommends not exceeding 10 Btu/day-ft²-F°, a solar heating application for this clinic could not be strongly recommended. Given the design conditions at Des Moines, a solar heating system would supply only a part (probably not even 50 percent) of the clinic's space-heating needs. Steps could be taken to reduce transmission load or infiltration load or both. Also, it should be remembered that winter outdoor design temperatures are set only a few degrees above 20-year lows for the locality. For most winters, and for most days of any winter, the actual load would be much less than the design heating load.

In summary, under the given conditions a solar application is marginal at best. However, if improved construction could appreciably reduce the heating load, and if the client accepts the fact that only a part of the load will be met by solar (requiring a sizable auxiliary heating installation), further feasibility studies, including cost–benefit analysis could be recommended.

5-7 Design Heating Load

The sum of transmission loss through the building envelope and infiltration loss is termed the *design building heating load*. It can be represented by the following equation:

$$Q_D = U A \, \Delta t + 0.018 \, V \, \Delta t = (U A + 0.18 \, V) \, \Delta t \qquad (5\text{-}5)$$

where

$$Q_D = \text{the design building heating load, Btu per hour}$$
$$UA \, \Delta t = \text{the transmission heat loss, Btu per hour}$$
$$0.018 \, V \, \Delta t = \text{the infiltration–ventilation heat loss, Btu per hour}$$

HEATING LOADS AND SOLAR DESIGN

For solar energy applications, the design building heating load—Eq. (5-5)—based as it is on an instantaneous rate of heat flow (Btu per hour), is not a very suitable measure. A load measure averaged over a longer time period is necessary when the sun, rather than fossil fuels, is the source of energy for the system. Some reasons for this are elaborated here.

Conventional systems are sized for design (i.e., "worst case") conditions, but they can be controlled to deliver any desired instantaneous heating rate, from zero to the maximum design output. Their heat energy is "stored" as natural or bottled gas, oil, coal, or electrical energy.

Solar energy systems, on the other hand, collect heat energy only when the sun shines, and they store it in the same form—as heat energy. Storage capacity (thermal mass) is limited both by cost considerations and by the architectural design. On any night and during some days, no solar energy at all can be collected; and on some clear days under ideal conditions, far more energy than the instantaneous load requires is easily collected and put into storage. For solar energy, then, the concept of *average heating load* is necessary to take the place of the instantaneous or *design heating load*. The time period that has seemed most useful for averaging is a month, and the load parameter favored by solar designers is the *average monthly heating load*.

5-8 Degree-Days as a Measure of Heating Load

Structural heat loss occurs whenever the outdoor temperature falls below the indoor temperature, and, as we have seen, the building heat loss is proportional to the temperature difference, Δt. It has been assumed that space heating is desirable and perhaps required for many persons when the outside temperature falls below 65°F. Over time, say an entire day, it is essential to know how much below the base temperature of 65°F the average temperature has been. (Average temperature is the mean of the minimum and maximum temperatures for the day.) This difference, in degrees, is termed the "degree-days" for that day. For example, a day with an average temperature of 60°F will have five degree-days (DD). Heating degree-days have been recorded for years at many locations by the U.S. National Weather Service and have been aggregated for monthly totals. A more precise definition of degree-days follows:

Degree-Days (DD) Defined.   A degree-day is the difference between a fixed base temperature (usually 65°F) and the daily mean outdoor temperature, summed up for a specified period of time, such as a month or a year. Table 5-5 lists average monthly and yearly degree-days for selected locations in the United States. The higher the degree-day total for a given locality and month, the greater will be the heating load.

5-9 Average Monthly Structure Heating Load

Sections 5-3 to 5-5 developed in detail methods of calculating the transmission heat loss and the infiltration–ventilation heat loss. Section 5-7 noted that these two loads are combined—Eq. (5-5)—to express the design heating load.

We will now convert *design heating load* (an instantaneous rate in Btu per hour) to the *average monthly heating load*, expressed in Btu per month. This conversion is accomplished by making use of the degree-day concept.

From Eq. (5-5),

$$\frac{Q_D}{\Delta t} = U\,A + 0.018\,V$$

= the heat loss rate per degree of design temperature difference, Btu/hr-F°

The monthly Btu requirement for solar heating is obtained by substituting the proper number of degree-days per month at the locality (from Table 5-5) for the design temperature difference (Δt), and then multiplying by 24 hr/day. The result is

$$Q_{\text{month}} = (UA + 0.018\,V)\,\frac{\text{Btu}}{\text{hr-F°}} \times DD\,\frac{\text{F°-day}}{\text{month}} \times 24\,\frac{\text{hr}}{\text{day}} \qquad (5\text{-}6)$$

The term *UA* is the hourly transmission heat loss per degree of *design temperature difference*—Eq. (5-5)—and *the term* 0.018 V is the infiltration–ventilation hourly heat loss per degree of design temperature difference. (Check the dimensions on the right side of the equation to verify that the units of Q will be Btu per month.)

It should be recognized that, although the design (hourly rate) heating load is based on the design temperature difference (70°F − t_o), the average

Table 5-5 Monthly and Annual Heating Degree-Days, 65° Base—Selected States and Cities

State and City		Jan.	Feb.	Mar.	Apr.	May	June	July	Aug.	Sept.	Oct.	Nov.	Dec.	Annual
AL	Mobile	451	337	221	40	—	—	—	—	—	39	211	385	1684
AZ	Phoenix	428	292	185	60	—	—	—	—	—	17	182	388	1552
AR	Little Rock	791	619	470	139	21	—	—	—	5	143	441	725	3354
CA	Los Angeles	331	270	267	195	114	71	19	15	23	77	158	267	1819
	Sacramento	617	426	372	227	120	20	—	—	5	101	360	595	2843
	San Francisco	518	386	372	291	210	120	93	84	66	137	291	474	3042
CO	Denver	1088	902	868	525	253	80	—	—	120	408	768	1004	6016
DC	Washington	911	776	617	265	72	—	—	—	14	190	510	856	4211
FL	Jacksonville	348	282	176	24	—	—	—	—	—	19	161	317	1327
	Miami	53	67	17	—	—	—	—	—	—	—	13	56	206
GA	Atlanta	701	560	443	144	27	—	—	—	8	137	408	667	3095
ID	Boise	1116	826	741	480	252	97	—	12	127	406	756	1020	5833
IL	Chicago	1262	1053	874	453	208	26	—	8	57	316	738	1132	6127
IN	Indianapolis	1150	960	784	387	159	11	—	5	63	302	699	1057	5577
IO	Des Moines	1414	1142	965	465	186	26	—	13	94	350	816	1240	6710
KS	Wichita	1045	804	671	275	90	7	—	—	32	211	606	946	4687
KY	Louisville	983	818	661	286	105	5	—	—	35	241	600	911	4645
LA	New Orleans	403	299	188	29	—	—	—	—	—	40	179	327	1465
MA	Boston	1110	969	834	492	218	27	—	8	76	301	594	992	5621
MI	Detroit	1225	1067	918	507	238	26	—	11	80	342	717	1097	6228
MN	Duluth	1751	1481	1287	792	484	194	67	104	318	611	1098	1569	9756
	Minneapolis–St. Paul	1637	1358	1138	587	271	65	11	21	173	472	978	1438	8159
MO	Kansas City	1153	893	745	314	111	12	—	—	42	235	642	1014	5161
	St. Louis	1045	837	682	272	103	10	—	—	35	224	600	942	4750
MT	Great Falls	1380	1075	1070	648	367	162	18	42	260	524	912	1194	7652
NE	Omaha	1314	1036	865	391	148	20	—	6	71	301	750	1147	6049
NV	Reno	1026	781	766	546	328	145	17	50	168	456	747	992	6022
NM	Albuquerque	924	700	595	282	58	—	—	—	7	218	615	893	4292
NY	Albany	1349	1162	980	543	253	39	9	22	135	422	762	1212	6888
	New York	1017	885	741	387	137	—	—	—	29	209	528	915	4848
NC	Charlotte	710	588	461	145	34	—	—	—	10	152	420	698	3218
ND	Bismarck	1761	1442	1237	660	339	122	18	35	252	564	1083	1531	9044
OH	Columbus	1135	972	800	418	176	13	—	8	76	342	699	1063	5702
OK	Oklahoma City	874	664	532	180	36	—	—	—	12	148	474	775	3695
OR	Portland	834	622	598	432	264	128	48	56	119	347	591	753	4792
PA	Philadelphia	1014	871	716	367	122	—	—	—	38	249	564	924	4865
SC	Columbia	608	493	360	83	12	—	—	—	—	112	341	589	2598
SD	Sioux Falls	1575	1277	1085	567	259	65	10	18	165	465	957	1395	7838
TN	Nashville	828	672	524	176	45	—	—	—	10	180	498	763	3696
TX	Dallas–Fort Worth	626	456	335	88	—	—	—	—	—	60	287	530	2382
	El Paso	663	465	328	89	—	—	—	—	—	92	402	639	2678
	Houston	416	294	189	23	—	—	—	—	—	24	155	333	1434
UT	Salt Lake City	1147	885	787	474	237	88	—	5	105	402	777	1076	5983
VA	Richmond	853	717	569	226	64	—	—	—	21	203	480	806	3939
WA	Seattle–Tacoma	831	636	648	489	313	167	80	82	170	397	612	760	5185
	Spokane	1228	918	853	567	327	144	21	47	196	533	885	1116	6835
WV	Charleston	946	798	642	287	113	10	—	—	46	267	588	893	4590
WI	Milwaukee	1414	1190	1042	609	348	90	15	36	140	440	855	1265	7444
WY	Cheyenne	1190	1088	1035	669	394	156	22	31	225	530	885	1110	7255

Source: U.S. National Oceanic and Atmospheric Administration, *Climatic Data*. Adapted with permission.

monthly heating load calculated from Eq. (5-6) for solar energy systems, is based on the degree-day temperature difference (65°F minus the daily average temperature). The degree-day approach of Eq. (5-6) is probably the more accurate for determining a monthly average, since design temperatures are chosen at levels near the 20-year lows for the locality, and therefore they are not very indicative of a typical winter or of a typical month.

Illustrative Problem 5-6

A small elementary school in Kansas City, Missouri, has been designed for solar space heating. The winter outside design temperature is 2°F. A design heating load calculation was made with the following results:

$$Q_{transmission} = 350{,}000 \text{ Btu/hr}$$
$$\underline{Q_{infil\text{-}ventil} = 210{,}000 \text{ Btu/hr}}$$
$$\text{Total design load} = 560{,}000 \text{ Btu/hr}$$

Find the average monthly heating load for January.

Solution

The design heating load (560,000 Btu/hr) is based on an indoor–outdoor (design) temperature difference $\Delta t = (70 - 2)°F = 68°F$. For use in Eq. (5-6), this load must be converted to a Btu/hr-F° basis by dividing by Δt, the design temperature difference.

$$\frac{560{,}000 \text{ Btu/hr}}{68°F} = 8{,}235 \frac{\text{Btu}}{\text{hr-F}°}$$

From Table 5-5, the monthly degree-days for January for Kansas City are:

$$DD_{Jan} = 1153$$

Substituting in Eq. (5-6),

$$Q_{Jan} = 8235 \frac{\text{Btu}}{\text{hr-F}°} \times 1153 \frac{\text{F}°\text{-day}}{\text{month}} \times 24 \frac{\text{hr}}{\text{day}}$$
$$= 2.28 \times 10^8 \text{ Btu/month} \qquad Ans.$$

Illustrative Problem 5-7

A residence in Bakersfield, California has a calculated building transmission loss rate of 62,500 Btu/hr based on the winter design temperature of 20°F. Its infiltration–ventilation loss rate is estimated at 26,000 Btu/hr. Determine the average monthly heating load for December, given that the December heating degree-days are 502.

Solution

The design heating load (based on $\Delta t = 70 - 20 = 50°F$) is 62,500 Btu/hr plus 26,000 Btu/hr, or

$$Q_{total} = 88{,}500 \text{ Btu/hr}$$

Dividing by Δt gives

$$\frac{88{,}500 \text{ Btu/hr}}{50°F} = 1770 \frac{\text{Btu}}{\text{hr-F}°}$$

The monthly degree-days for December (given) are:

$$DD_{Dec} = 502$$

Substituting in Eq. (5-6),

$$Q_{Dec} = 1770 \frac{\text{Btu}}{\text{hr-F}°} \times 502 \frac{\text{F}°\text{-day}}{\text{month}} \times 24 \frac{\text{hr}}{\text{day}}$$
$$= 2.13 \times 10^7 \text{ Btu/month} \qquad Ans.$$

5-10   Additional Heating Load Coefficients Used in Solar Design

The unique requirements of solar energy system design center around the dominant feature of solar energy availability—the fact that, although over time a great amount of energy is available, the instantaneous availability is often not sufficient to meet energy demands of the moment. Consequently, as noted above, a different approach to determining energy needs (heating loads) is necessary in solar design, from that developed over the years for conventional winter air conditioning. This section, in concluding the chapter, lists and briefly discusses some heating load coefficients and parameters that will be used extensively in later chapters.

The Building Load Coefficient (BLC).[3]   This coefficient is the *per-degree Fahrenheit load* (or steady-state heat loss) of an entire room or building, including all glazings and solar collection elements. It is expressed in Btu per day (not per hour) per degree Fahrenheit of indoor–outdoor temperature difference (65°F minus the mean outdoor temperature). Note that the temperature difference here is the same as that associated with the degree-day, not with Δt, the design temperature difference.

Units of BLC are Btu/day-F° or Btu/DD

The Net Load Coefficient (NLC).   This coefficient is similar to the building load coefficient (BLC) except that the NLC is computed *excluding the solar elements* of the building. It is the per Fahrenheit degree steady-state heat loss, less the transmission loss through the total solar collection area. In computing the NLC, all solar collection areas are considered to be replaced by a thermally neutral (adiabatic) wall, which, for the purposes of the calculation, does not allow either a loss from or a gain of heat to the structure. The units of NLC are also Btu/day-F° or Btu/DD.

The Net Reference Thermal Load (NRTL).   This is the degree-day heating load (or steady-state heat loss) of the nonsolar elements of a solar building. The use of the word *reference* emphasizes that the load (or heat loss) is based on a constant indoor reference temperature of 65°F. The NRTL is, by definition, the NLC times the degree-days, for a stipulated time period such as a month or a year. The units of NRTL are Btu per the stipulated time period, as Btu per month.

These coefficients, along with others to be introduced later, are often involved in solar design applications, both passive and active. The next chapter begins the in-depth discussion of passive solar systems.

Illustrative Problem 5-8

An elementary school in Pendleton, Oregon, has an all-purpose room 70 ft by 40 ft with a sloping beam ceiling and roof (Fig. 5-6). It is built on a concrete slab, with 1-in. edge insulation all around the perimeter and extending 24 in. under the slab. The south (70 ft) wall has 700 ft² of solar windows (most of it fixed glass). There is also a total of 200 ft² of glass in the north wall (all fixed glass) for daylighting purposes. All glass is triple-pane insulating glass.

The exterior walls are 12-in. hollow-core concrete block with 4-in. face brick (total thickness 16 in.), finished on the inside with plaster on gypsum lath on furring strips, with rigid-board insulation between the furring strips.

The beam ceiling is sloped from a 14-ft height at the south wall to a 10-ft

---

[3] This term and a number of other coefficients and parameters defined here and in Chapter 6 are those used by the Los Alamos Laboratories research group. See J. D. Balcomb, et al., *Passive Solar Design Handbook*, vol. II (Washington: U.S. Dept. of Energy, January 1980), DOE/CS-0127/2.

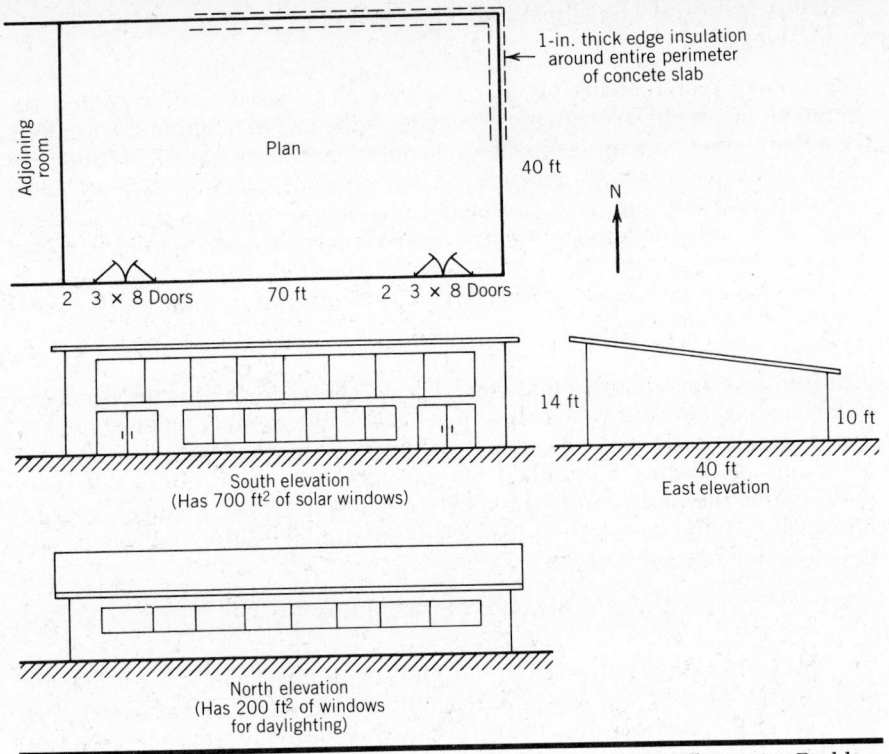

1-in. thick edge insulation
around entire perimeter
of concete slab

Adjoining room

Plan

40 ft

N

2 3 × 8 Doors   70 ft   2 3 × 8 Doors

South elevation
(Has 700 ft² of solar windows)

14 ft

10 ft

40 ft
East elevation

North elevation
(Has 200 ft² of windows
for daylighting)

Figure 5-6 Floor plan and elevations of elementary school for Illustrative Problem 5-8. (Not to scale.)

height at the north wall. The ceiling–roof structure from inside out is 2 × 4 tongue-and-groove fir nailed to the beams, covered with 2-in. rigid-board insulation. On top of this is 0.625 in. plywood sheathing covered with rolled felt and heavy cedar shakes.

There are four 3 ft by 8 ft doors (2-in.-thick solid hardwood) all on the south side. One of the 40-ft end walls is a common wall with an adjoining classroom. This wall is 8-in. concrete block, painted on both sides. This adjoining room will be held at a temperature no lower than 50°F.

The all-purpose room ventilation system has a capacity of 800 ft³/min (cfm), and it operates on all school days. Exhaust fans with a capacity of 200 ft³/min (cfm) also operate continuously on school days.

The design indoor temperature has been set by the school board at 70°F. The outside design temperature (winter) is −2°F at Pendleton.

Consideration is currently being given to the possibility of installing an active solar heating system for this space, and you have been asked to determine all of the following:

A. The winter design heating load, Btu per hour. (Solar gains through the windows and internal heat gains are not to be included in this calculation. These gains would enter into later calculations but are not involved in the design heating load.)
B. The average monthly heating load for January, Btu per month, based on the January degree-days—$DD_{Jan} = 1017$ F°-days for Pendleton.
C. The building load coefficient (BLC), Btu/DD.
D. The net load coefficient, (NLC), Btu/DD.
E. The net reference thermal load (NRTL), Btu_{Jan}.

Solution

The problem has many parts and subparts, so care is required in planning the calculations. The following layout is suggested.

A. The total design heating load, Q

Given, the outside winter design temperature for Pendleton is $-2°F$.

A–1. The Building Transmission Load, Q_{trans}

1a. Outside Walls
Area of outside walls

$$A_{OW} = \left[(70 \times 14) + (70 \times 10) + \left(40 \times \frac{14 + 10}{2}\right)\right] ft^2$$
$$- 900 \text{ ft}^2 \text{ window} - 96 \text{ ft}^2 \text{ doors}$$
$$= 1164 \text{ ft}^2$$

Wall U value (Table 5-3, section 2-D) = 0.12 Btu/hr-ft²-F°

$$Q_{outside \, walls} = 0.12 \text{ Btu/hr-ft}^2\text{-F}° \times 1164 \text{ ft}^2$$
$$\times [70 - (-2)]F°$$
$$= 10{,}100 \text{ Btu/hr} \qquad \qquad 10{,}100$$

1b. Common or Party Wall

Area $= \left(40 \times \dfrac{14 + 10}{2}\right) ft^2 = 480 \text{ ft}^2$

Wall U value (Table 5-3, section 2-C) = 0.56
$Q_{CW} = 0.56 \text{ Btu/hr-ft}^2\text{-F}° \times 480 \text{ ft}^2 \times (70 - 50) \text{ F}°$
$= 5400 \text{ Btu/hr} \qquad \qquad 5{,}400$

1c. Windows

Area of windows = 900 ft²
U value (triple-pane insul. glass) = 0.39
$Q_{windows} = 900 \text{ ft}^2 \times 0.39 \text{ Btu/hr-ft}^2\text{-F}° \times 72 \text{ F}°$
$= 25{,}300 \text{ Btu/hr} \qquad \qquad 25{,}300$

1d. Doors (with Storm Doors)

Area of doors = 4 × 24 ft² = 96 ft²
U value for doors as described = 0.24
$Q_{doors} = 96 \text{ ft}^2 \times 0.24 \text{ Btu/hr-ft}^2\text{-F}° \times 72 \text{ F}°$
$= 1700 \text{ Btu/hr} \qquad \qquad 1{,}700$

1e. Slab Floor

Use Eq. (5-1), $Q = PF \, \Delta t$
From Table 5-3, section 10, $F = 0.55$ Btu/hr-F°-lin. ft
$Q_{floor} = 220 \text{ lin. ft} \times 0.55 \text{ Btu/hr-F}°\text{-lin. ft} \times 72 \text{ F}°$
$= 8700 \text{ Btu/hr} \qquad \qquad 8{,}700$

1f. Ceiling–Roof

Area of sloping ceiling is calculated as 2815 ft².
From Table 5-3, section 6-B, for ceiling–roof as described,
$U = 0.08$ Btu/hr-ft²-F°

$$Q_{ceiling-roof} = 2815 \text{ ft}^2 \times 0.08 \text{ Btu/hr-ft}^2\text{-F}° \times 72 \text{ F}°$$
$$= 16{,}200 \text{ Btu/hr} \qquad \qquad \underline{16{,}200}$$

Total building transmission load, Q_{trans} $\qquad$ 67,400 Btu/hr

A–2. The outside air load, Q_{OA}

Estimating Infiltration Air: From Table 5-4 (windows and doors on three sides, but doors have storms, and much of the window area is fixed glass), estimate 1.5 air changes per hour.

$$V_{infil} = 1.5/hr \times (70 \times 40 \times 12) \text{ ft}^3 = 50{,}400 \text{ ft}^3/\text{hr}$$

$$\text{Ventilation air (given)} = 600 \text{ ft}^3/\text{min (cfm)}$$
$$= 36{,}000 \text{ ft}^3/\text{hr}$$

$$\text{Exhaust air (given)} = 200 \text{ ft}^3/\text{min (cfm)}$$
$$= 12{,}000 \text{ ft}^3/\text{hr}$$

Since infiltration air exceeds ventilation air by a considerable margin, the infiltration air volume is used. However (section 5-6), exhaust air must be added to infiltration air in this case to determine total outside air.

$$V_{OA} = (50{,}400 + 12{,}000) \text{ ft}^3/\text{hr} = 62{,}400 \text{ ft}^3/\text{hr}.$$

From Eq. (5-4), the outside air load is

$$Q_{OA} = 0.018 \, V(t_i - t_o)$$
$$= 0.018 \text{ Btu/ft}^3\text{-F}^\circ \times 62{,}400 \text{ ft}^3/\text{hr} \times 72 \text{ F}^\circ$$
$$= 80{,}900 \text{ Btu/hr} \qquad\qquad 80{,}900$$

Total Design Heating Load

$$Q_D = Q_{trans} + Q_{OA}$$
$$= 67{,}400 \text{ Btu/hr} + 80{,}900 \text{ Btu/hr} \qquad 148{,}300 \text{ Btu/hr}$$
$$\textit{Ans.}$$

B. The January Average Monthly Heating Load, Q_{Jan}

Convert the design heating load from A above to the January average monthly heating load by the method of section 5-9, using Eq. (5-6). Remember that the degree-day (DD) concept uses a reference indoor temperature of 65°F, not the indoor design temperature, t_i.

$$Q_{Jan} = \frac{Q_D}{[65 - (-2)] \text{ F}^\circ} \times DD_{Jan} \times 24$$

From given data for Pendleton, $DD_{Jan} = 1017$ F°-days
Substituting,

$$Q_{Jan} = \frac{148{,}300 \text{ Btu/hr}}{67 \text{ F}^\circ} \times 1017 \text{ F}^\circ\text{-day} \times 24 \text{ hr/day}$$
$$= 54.03 \times 10^6 \text{ Btu for the month of January} \qquad \textit{Ans.}$$

C. The Building Load Coefficient, BLC

The BLC is the per degree Fahrenheit load per day for the entire space, based on the 65° F reference base of the DD concept.

$$BLC = \frac{Q_D}{[65 - (-2)] \text{ F}^\circ} \times 24 \text{ hr}$$
$$= \frac{148{,}300 \text{ Btu/hr}}{67 \text{ F}^\circ} \times 24 \text{ hr/day}$$
$$= 53{,}100 \text{ Btu/day-F}^\circ \qquad \textit{Ans.}$$

D. The Net Load Coefficient, NLC

For this calculation, the entire solar collection area (that is, the south windows) is considered thermally neutral—no gains or losses. From A (1-c) above,

$$Q_{solar\,windows} = 700 \text{ ft}^2 \times 0.39 \text{ Btu/hr-ft}^2\text{-F}^\circ \times 72 \text{ F}^\circ$$
$$= 19{,}700 \text{ Btu/hr}$$

Subtracting this solar window transmission loss from the total transmission loss gives

$$Q_{D_{net}} = (148{,}300 - 19{,}700) \text{ Btu/hr} = 128{,}600 \text{ Btu/hr}$$

$$\text{NLC} = \frac{Q_{D_{net}}}{\text{Ref. temp.} - t_o} \times 24 \text{ hr}$$

$$= \frac{128{,}600 \text{ Btu/hr}}{67 \text{ F}^\circ} \times 24 \text{ hr/day}$$

$$= 46{,}000 \text{ Btu/day-F}^\circ \qquad\qquad Ans.$$

E. The Net Reference Thermal Load, *NRTL*

This is the total net load for January, based on the reference temperature of 65°F.

$$\text{NRTL}_{Jan} = \text{NLC} \times \text{DD}_{Jan.}$$
$$= 46{,}000 \text{ Btu/day-F}^\circ \times 1017 \text{ F}^\circ\text{-days}$$
$$= 46.78 \times 10^6 \text{ Btu} \qquad\qquad Ans.$$

5-11 Seasonal Heating and Cooling Loads

As suggested in the foregoing paragraphs, monthly heating loads are very useful measures for the solar designer. Equally useful are *total seasonal loads*—the total heating load for the entire heating season Q_{HS} (see Chapter 15), and the total cooling load for the cooling season Q_{CS} (see Chapter 16).

The "heating season" is usually defined as October through April except for extreme latitudes. The total heating season load Q_{HS} is determined by first calculating the monthly loads as above for October through April using Eq. (5-6), and then summing these up for the heating season.

The total cooling season load is determined by summing up the monthly cooling loads for May through September. The extra complexity of cooling load analysis often necessitates the use of computer methods for determining Q_{CS} (see Chapter 16).

Conclusion

This chapter has discussed the elements that are included in the winter heating load for buildings and living spaces. Methods of calculating each element of the load and of summing them up to obtain the total heating load have been explained. Modifications in conventional heating load analysis that are needed for application to solar energy systems have been introduced, and these analyses will be developed in more detail in later chapters.

Loads for domestic and industrial hot-water heating have not been treated in this chapter. Load analysis for these applications will be included in later chapters that are devoted in their entirety to these applications. Cooling loads will also be considered in detail in a later chapter.

PROBLEMS

1. Calculate the conductance C of an R-19 insulation blanket.

2. The thermal conductivity K of hardwood is 1.10 Btu/hr-ft²-F°/in. Calculate the R value of a solid hardwood door 1⅜ in. thick.

3. The R value per inch of expanded polyurethane insulating board is 6.25. Calculate the thermal conductivity K of this material.

4. Compute the U value of a wall made up of (from inside out) 0.625-in. gypsum plaster on metal lath on 0.75-in. furring strips on 8-in. brick. Include the effects of convection on inside and outside wall surfaces.

5. Determine the U value of the floor–ceiling described below. It is between the ground floor and a basement. The construction (from living space downward) is: Carpet on fiber pad on 0.625-in. plywood base, on 0.75-in. plywood subfloor on wooden joists. The basement ceiling is 0.625-in. drywall (gypsum wallboard), decorated, and nailed directly to the bottom of the joists. Include a factor for the air space between joists (no insulation) in your calculations.

6. A residential frame wall has the following construction, from inside out: Gypsum plaster board (drywall) 0.625 in., on 2 × 4 studs, spaces between studs filled with 3.5 in. of fiberglass blanket insulation. On the outside of the studs there is 0.625-in. plywood sheathing, then building paper, and finally, 1-in. board-and-batt cedar (vertical) siding. Disregard the effect of studs (that is, assume that the insulation is continuous), and calculate the U value of the composite wall. Remember convection effects inside and outside.

7. An outside wall of an office building has the following construction, from inside out: Gypsum plaster on metal lath on 1.0-in-thick wood furring strips on 12-in. brick. The spaces between the furring strips are filled with rigid-board insulation ($K = 0.16$ Btu/hr-ft^2-F°/in.). Neglect the effect of the wood furring strips (assume that the insulation is continuous), and calculate the U value of the wall, taking into account convection effects on both surfaces.

8. For the residential wall of problem 6 above, with studs (1.625 in. by 3.5 in.) on 16-in. centers, recalculate the U value taking into account the effect of the studs.

9. A small retail store is built on a concrete slab floor measuring 80 ft by 50 ft. The outdoor design temperature, $t_o = 0$°F; indoor design temperature, $t_i = 70$°F. Edge insulation 1 in. thick and extending 24 in. under the slab was installed around the entire perimeter. Calculate the steady-state heat loss from this slab floor for the conditions stated.

10. A general office space has three outside walls, with windows and an exterior door on two sides. Door usage is moderate. The dimensions are 60 ft × 40 ft × 12 ft. Calculate the estimated hourly volume of infiltration air V by the air-change method.

11. A medical clinic has overall dimensions of 100 ft × 80 ft × 12 ft. There are windows and/or doors on two sides. Construction is new and quite tight; door usage is moderate. Average occupancy is 100 persons during business hours. Local codes require 7 cfm (ft^3/min) per person of ventilation air. Exhaust fans operate continuously during business hours, at a capacity of 600 cfm. Design temperatures are: $t_i = 70$°F; $t_o = 20$°F. Calculate:
   (a) The volume V of outside air to use in the outside air load calculation, cubic feet per hour.
   (b) The total outside air load, Btu/hr.

12. The average temperature for a winter day in Cedar City, Utah, was 24°F. How many degree-days would this day have?

13. The winter design heating load for a residence in Spokane, Washington, has been calculated as $Q_D = 118,000$ Btu/hr, based on $t_i = 70$°F; $t_o = -7$°F. The January degree-days for Spokane are 1228. Calculate the average monthly structure heating load for January, Q_{Jan}, in Btu per month.

14. A school building in Reno, Nevada, has a building transmission loss of 440,000 Btu/hr, based on winter design temperatures of $t_i = 70$°F and $t_o = 5$°F. The outside air load has been computed as 210,000 Btu/hr. Calculate the average monthly heating load for February.

15. The total winter design heating load of a small office building in Charlotte, North Carolina, has been determined as 510,000 Btu/hr, based on an indoor design temperature of 70°F. Of this total, 108,000 Btu/hr is transmission loss through south-facing windows. Calculate all of the following:
   (a) The January average monthly heating load.
   (b) The building load coefficient (BLC).
   (c) The net load coefficient (NLC).
   (d) The net reference thermal load (NRTL).

PART TWO

Passive Solar Energy Systems

Passive Solar Design Concepts

The energy requirements for heating and cooling buildings can be reduced significantly if the architectural design takes advantage of natural heating and cooling. Over the centuries, many civilizations have evolved climate-responsive architecture. The Greeks came to realize the importance of allowing the winter sun to shine into the interiors of their homes, and over the centuries the Greek house gradually developed into a structure with an interior courtyard on the south side. The main rooms of the house were placed on the north side of the courtyard (usually in a two-story block), as shown in Fig. 6-1. The rooms on the south side of the courtyard were kept low and were ordinarily used only for storage. This design ensured that the rooms on the north would have unobstructed access to the winter sun. The Greek cities built after about 500 B.C. were laid out with street patterns that provided each building site with access to the sun.[1]

In arid regions of the Middle East, where keeping cool in summer was important, architectural styles developed that would take advantage of natural cooling. Buildings were constructed to capture cool breezes and to channel the moving air through the

[1] See K. Butti and J. Perlin, *A Golden Thread* (Palo Alto: Cheshire Books, 1960), p. 5.

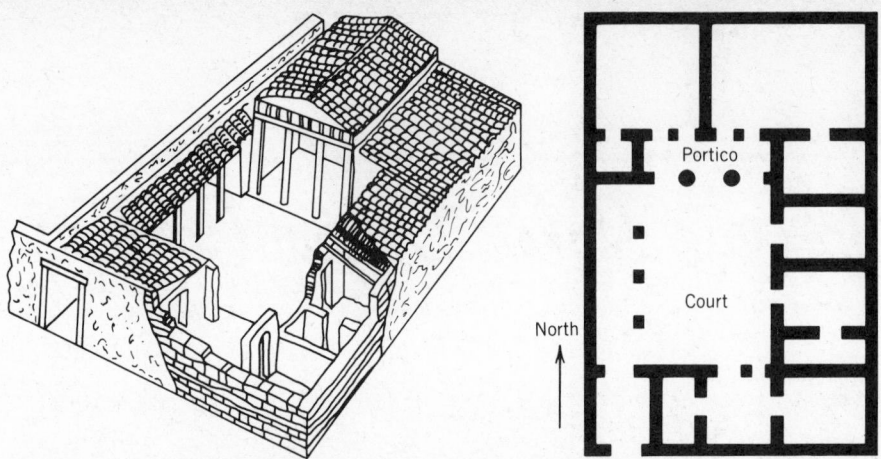

Figure 6-1 Representation of a classical Greek home from the ancient city of Priene. Rooms behind the portico faced into the south courtyard. Greek solar architecture was explained by Socrates, as quoted by Xenophon: "In houses that look toward the south, the sun penetrates the portico in winter, while in summer the path of the sun is right over our heads and above the roof so that there is shade."

Figure 6-2 Living with the sun in ancient cultures. Prehistoric Anasazi Indians of the American Southwest built their dwellings in the shelter of overhanging cliffs. This is Cliff Palace in Mesa Verde National Park, Colorado. It was occupied for about a century in the era A.D. 1200–1300 by the ancestors of present-day Native Americans of the Southwest. (U.S. National Park Service.)

interior rooms. Fountains and ponds of water were often used to help cool the air as it moved through the building.

In the southwestern United States, prehistoric Indians of the Anasazi culture (*Anasazi* is the Navajo name for "the ancient ones") built large residential complexes in south-facing caves formed by overhanging cliffs (Fig. 6-2). In the winter, when the sun is low in the sky, the houses built under the cliff overhang would receive full sun. The stone and mud walls were warmed by the sunlight, and the rooms behind the wall would then be heated. In the summer, when the sun is high in the sky, the cliff overhang would shade the buildings and help to keep them cool. The overhanging cliff also protected the building from rain, wind, and snow.

PASSIVE SOLAR ENERGY SYSTEMS

In the above examples, locally available building materials were used in an architectural style that enhanced the natural heating and/or cooling of the building. These same architectural styles can be used and improved upon today by using modern materials. For example, neither the Greeks nor the Anasazi Indians had glass to use for windows, but today glass and transparent plastics are very important building materials, since they allow solar radiation to enter the building while still providing a weather-resistant surface. These transparent materials are collectively termed glazings.

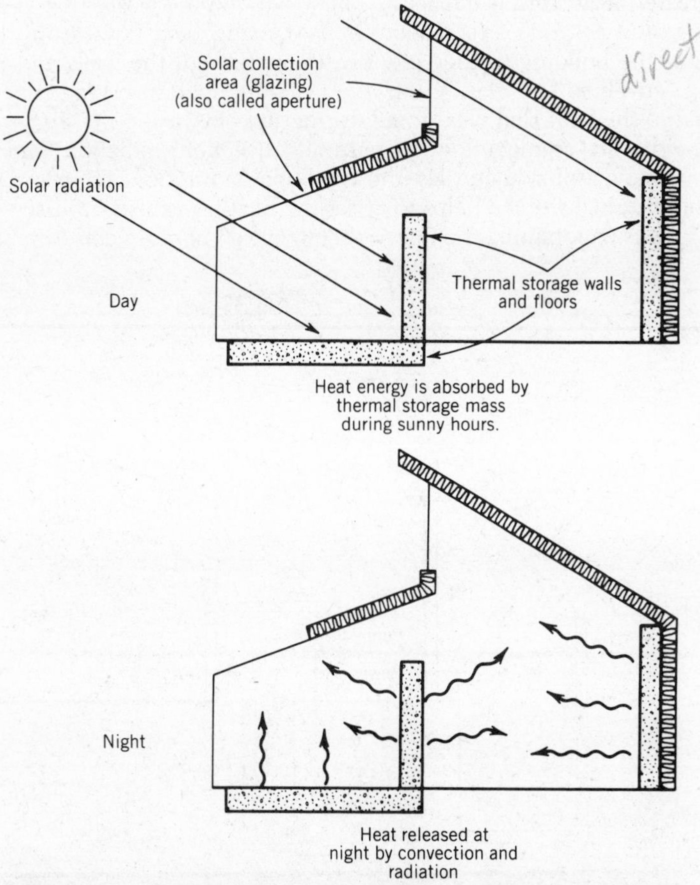

passive sol. en. sys.

direct-gain simplest draw on board

Figure 6-3 A passive solar system has two basic components: (1) the *glazing*, or *solar collection area* (also called the *solar aperture*), which admits solar radiation into the building; and (2) the *thermal storage*, which absorbs and stores heat energy, releasing it at night (or on cloudy days) into the building.

In Fig. 6-3, solar radiation enters the building through the glazed surfaces and strikes the interior building surfaces. These interior walls absorb the radiant energy, and their temperature rises as they store up heat. At night these interior thermal storage masses gradually release heat back into the building by natural convection and thermal radiation. The structure of Fig. 6-3 would be a naturally heated building if the solar radiation that is absorbed and stored inside it during the daylight hours were approximately equal to the energy lost from the building over a 24-hr period. Such a natural solar heating system is termed a passive solar energy system. To repeat, the glazings or openings through which solar radiation enters are collectively called the *solar collection area* (or *aperture*), and the materials that absorb and store heat energy are called *thermal storage* (or *thermal storage mass*).

6-1   Passive Systems versus Active Systems

There are two primary classifications of solar energy systems: *passive systems* and *active systems*. Passive systems collect and transport heat by non-mechanicial means, moving thermal energy by natural convection, conduction, and radiation. A passive solar system is integrated into the building in such a way that the building itself actually constitutes the system. On sunny days, solar radiation enters the building through glazed surfaces and strikes interior building surfaces, where it is either absorbed or reflected to strike other surfaces. The interior surfaces that are planned to receive solar radiation are typically constructed from heavy materials such as brick, stone, or concrete—materials that have greater heat-storing capacities on a per-volume basis than less dense materials such as wood (see section 3-2). At night, heat flows from the warm interior of the building to the colder exterior through the walls and roof (the building "envelope" or "skin"), and the inside temperature gradually falls. As this occurs, the heat that was stored during the day in the masonry walls and floor (the thermal storage) flows by natural convection, radiation, and conduction from the masonry materials into the building interior. This flow of stored heat from thermal storage helps to offset the heat transmission losses from the building, thus maintaining the interior temperature at more comfortable levels.

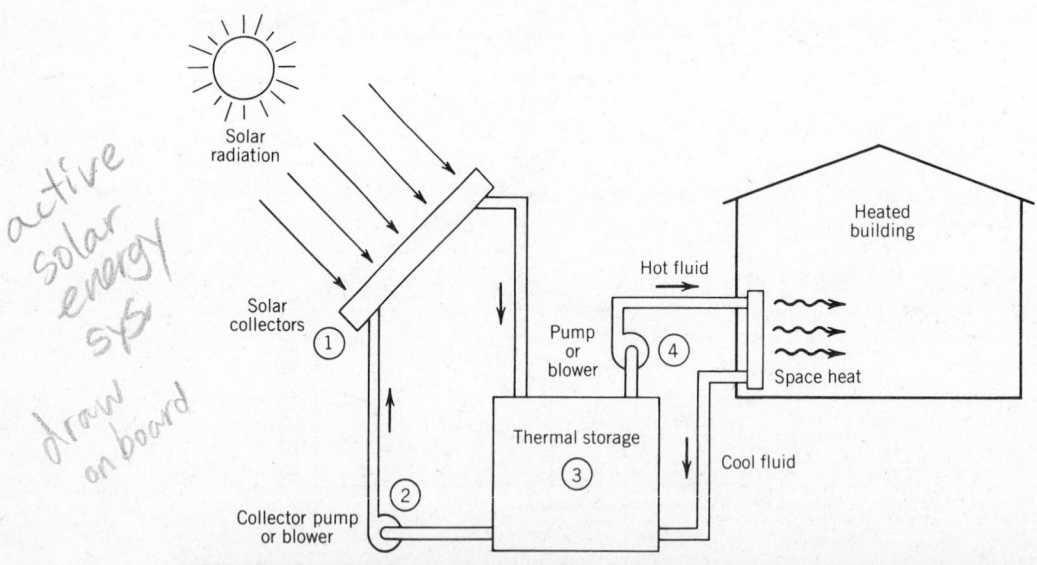

Figure 6-4 Schematic representation of the basic elements in an active solar energy system: (1) the solar collectors, which heat a fluid with the energy they absorb from the sun; (2) a pump or blower, which circulates the fluid from the collectors to thermal storage; (3) the thermal storage, which stores heat from the collectors; and (4) a pump or blower that distributes heat from the thermal storage to the heated space.

PASSIVE SOLAR DESIGN CONCEPTS

In contrast to a passive system, an active solar system mechanically moves heat from one component of the system to another. The components of an active solar system are the *solar collectors*, the *thermal storage*, and the *mechanical heat transport system*, as illustrated in Fig. 6-4. It is the function of the solar collectors to absorb solar radiation and transfer the heat absorbed to a fluid such as water or air. The thermal storage stores the heat from the solar collectors until needed. Mechanical systems are required to move heat from the solar collectors to thermal storage and from thermal storage to the heated space. Active solar systems will be the subject of several later chapters.

Ideally, a passive system should be integrated into the architecture of the building so that structural components serve a dual purpose. For example, windows with a southern orientation will collect solar radiation and at the same time provide visual access to the outside.[2] They can also provide natural daylighting and (when desired) ventilation for the interior space. The walls of a building, in addition to being essential load-bearing parts of the structure, may also provide thermal storage.

6-2   Classification of Passive Solar Systems

Passive solar systems have certain distinguishing characteristics that allow them to be classified under five major headings, as follows: *direct-gain systems*, *indirect-gain systems*, *isolated-gain systems*, *convective loop systems*, and *hybrid systems*. In some of these categories, variations exist that necessitate subcategories, as explained below.

Direct-Gain Systems.   The simplest of the passive systems is the direct-gain system, which was illustrated in Fig. 6-3. Solar radiation enters the interior of the building through glazed surfaces on the walls and/or roof. The sum total of all of these glazed surfaces makes up the solar collection area (solar aperture) for the building. The sunlight is converted to heat when it strikes absorbing surfaces within the building, causing the temperature of these absorbing materials (thermal storage mass) to rise. The heat is gradually dispersed throughout the space from the directly radiated absorbing surfaces to the surrounding surfaces (walls, floor, ceiling) and to the room contents by convection, conduction, and radiation. At night, heat is released by convection and radiation from the surfaces in the room that have stored up heat energy during the day. The essential feature of the direct-gain system is that once solar radiation enters through the solar collection area, it passes directly into the interior of the heated space.

Two variants of direct-gain systems are *sunspaces* (or sunrooms) and *solar greenhouses*. Figure 6-5 illustrates an attached sunspace. In its simplest form, the sunspace represents a direct-gain system as a part of a larger room (Fig. 6-5a). The solar energy collected by the sunspace heats air, which is carried by convection currents into the larger building space. In Fig. 6-5b, the sunspace is shown as a thermal zone separated from the rest of the building by a thermal storage wall. In this case, the sunspace itself remains as a direct-gain space, while the adjacent living space is heated indirectly from heat that passes through the thermal storage wall. Vents may be added through the thermal storage wall to allow warm air to move by convection into the adjacent space from the sunspace. Such a sunspace may act as a greenhouse or a sunroom. If the former, it is referred to as a solar greenhouse.

The various methods of incorporating direct-gain solar systems into buildings will be examined in Chapter 7.

Indirect-Gain Systems.   When thermal mass is placed between the glazed solar collection area and the interior space, solar radiation is blocked from entering directly into the interior spaces of the building. Heat contained in the

---

[2] It should be noted that this book deals with solar energy systems primarily from a northern-hemisphere perspective. All of the basic solar geometry would be reversed for the southern hemisphere.

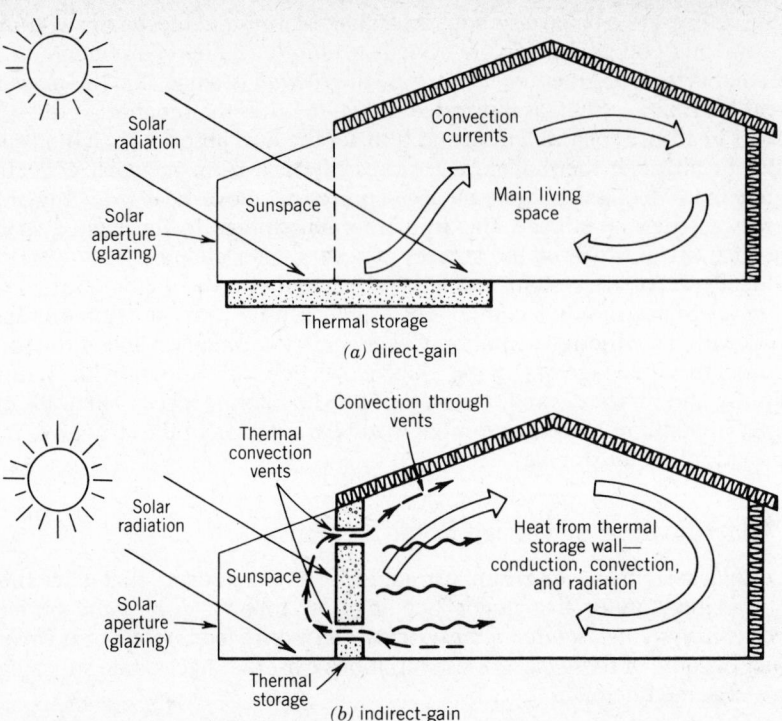

Figure 6-5 Sunspaces collect solar heat. (*a*) A sunspace can be attached to a larger room and act as a direct-gain system. (*b*) When thermal storage is placed between the sunspace and the rest of the building, two thermal zones are created—the sunspace, which is a direct-gain space, and the adjacent (indirect-gain) space, which receives heat from the thermal storage. Optional vents can be placed through the thermal storage to allow a convection loop to and from the sunspace.

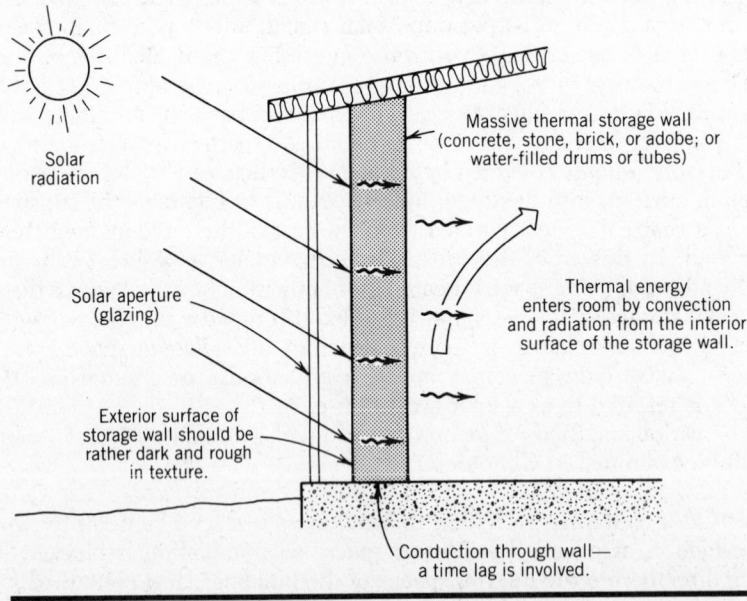

Figure 6-6 An indirect-gain system. A thermal storage wall absorbs solar radiation on its outer surface. Sunlight does not enter the living space behind the wall. The energy absorbed must be conducted through the thermal storage before it enters the building interior.

solar radiation that is absorbed by the thermal storage enters the building indirectly by natural convection, conduction, and radiation. These types of solar systems are termed indirect-gain systems. Indirect-gain systems may feature a thermal storage wall (Fig. 6-6) or a thermal storage roof (Fig. 6-7). The sunspace or greenhouse with an interior storage wall (Fig. 6-5*b*), although it is direct gain for the sunspace itself, heats the interior room by indirect gain. Heat transfer from the thermal storage is continuous as long as its temperature is above that of the interior. In passive systems, this heat transfer is not controlled by mechanical means. The rate at which heat is transferred is dependent upon the temperature of the thermal mass and the interior space temperature. Again, heat transfer is the result of conduction, convection, and radiation.

If a thermal storage wall is used, it is placed directly behind the glazing, between the glass and the interior spaces. The storage wall may be constructed of masonry or concrete materials or formed by the use of water-filled containers. As solar radiation passes through the glazing and strikes the wall, it is absorbed by the dark exterior surface of the wall. Heat flows into the thermal storage by conduction at a rate dependent upon the conductivity of the storage material. After a period of time, the temperature of the interior wall surface will start to rise also, and heat will begin to flow into the interior of the building. This time period is called the *time lag* of the wall. Time lag is dependent upon the thickness and the material of the thermal storage wall.

The *thermal storage roof*, illustrated in Fig. 6-7, represents another kind of indirect-gain system. It is similar to the thermal storage wall with the exception that the thermal mass is located in the plane of the roof. Water contained in bags or ponds on a flat roof is the thermal storage mass most often used. A movable insulation system is used to minimize heat losses at night and on cloudy days.

Isolated-Gain Systems. When thermal barriers are placed between the interior spaces and the solar-heated thermal storage, the heat flow from the storage into the building can be controlled or even shut off entirely. This type of passive system is an isolated-gain system. An isolated-gain solar system is basically an indirect-gain system except that a distinct separation (insulating or physical separation) exists between the thermal storage and the interior space. A thermal storage wall can become an isolated-gain system with the addition of an

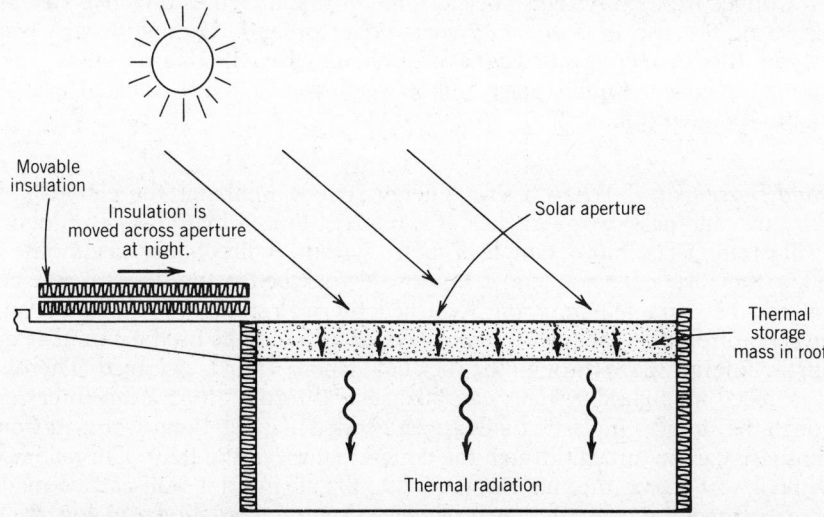

Figure 6-7 Roof thermal storage. The exterior surface of the thermal storage that stores and absorbs energy is exposed to solar radiation during the day. At night, the exterior surface of the storage is covered by movable insulation to reduce thermal losses. Heat continues to enter the interior spaces by thermal radiation from the ceiling.

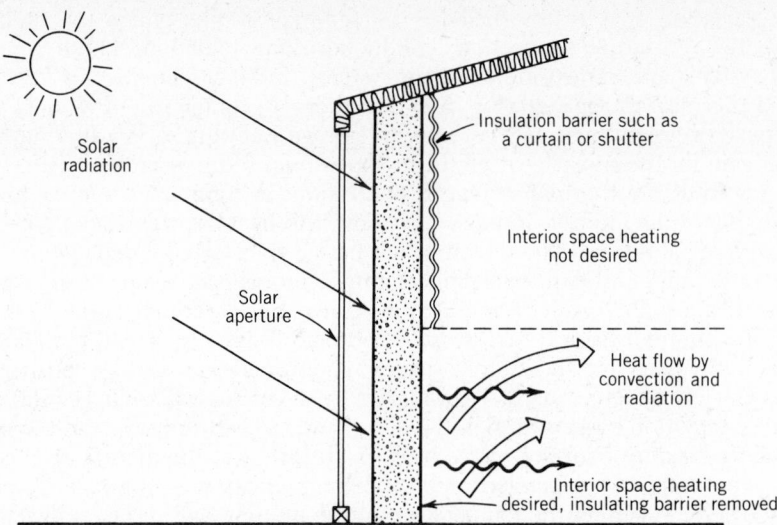

Solar radiation

Solar aperture

Insulation barrier such as a curtain or shutter

Interior space heating not desired

Heat flow by convection and radiation

Interior space heating desired, insulating barrier removed

Figure 6-8 Isolated-gain system. Thermal energy is stored during the day in walls or floors. The insulating barrier prevents heat from flowing into the building interior. When heating is required, the insulating barrier is removed, and heat flows from the thermal storage by convection and radiation.

insulating barrier, as shown in Fig. 6-8. The barrier is removed from the wall when heating is desired from the thermal storage. The insulating barrier may be in the form of an insulating curtain, or it could be a shutter or a sliding panel.

Convective Loop Systems. Although conforming to the general characteristics of an isolated-gain system, convective loop systems are sufficiently unique to justify a separate classification. In convective loop systems, the heat from solar radiation is ordinarily stored in an insulated rock bed and isolated from the interior spaces, as shown in Fig. 6-9a. When heat is required (Fig. 6-9b), vents are opened to the thermal storage to allow warm air to flow into the building by convection.

Convective loop systems possess some of the attributes of active systems because of the use of a separate solar collector and thermal storage bed. However, they can be classified as passive systems when the thermal energy flow between the collector and storage and between storage and the heated space is by natural convection.

Hybrid Systems. When a solar energy system combines the elements of both active and passive approaches, it is referred to as a hybrid system. Figure 6-10 illustrates a common example of such a system. A direct-gain greenhouse is used as the solar collector, and a fan-forced rock bed is the thermal storage. Heat can be extracted from the rock bed thermal storage by radiation and convection from the slab over the rock bed, or a fan can be used to circulate air from the interior space through the rock bed, where it will be warmed. Thermal storage used for hybrid systems can take several forms, since a fan forces air through the storage. If plastic tubes are embedded in a slab floor, warm air from a sunspace can be forced through the concrete mass of the floor. This allows a slab floor to become thermal storage mass. Or an interior wall can be made thermal storage by constructing it of heavy masonry, as shown in Fig. 6-11. Hollow cores or sections are left in the wall so that warm air can be circulated through it. Hybrid systems allow the thermal storage to be placed in locations that would not be feasible in pure passive systems.

Later chapters will deal at considerable length with the design of the systems just discussed.

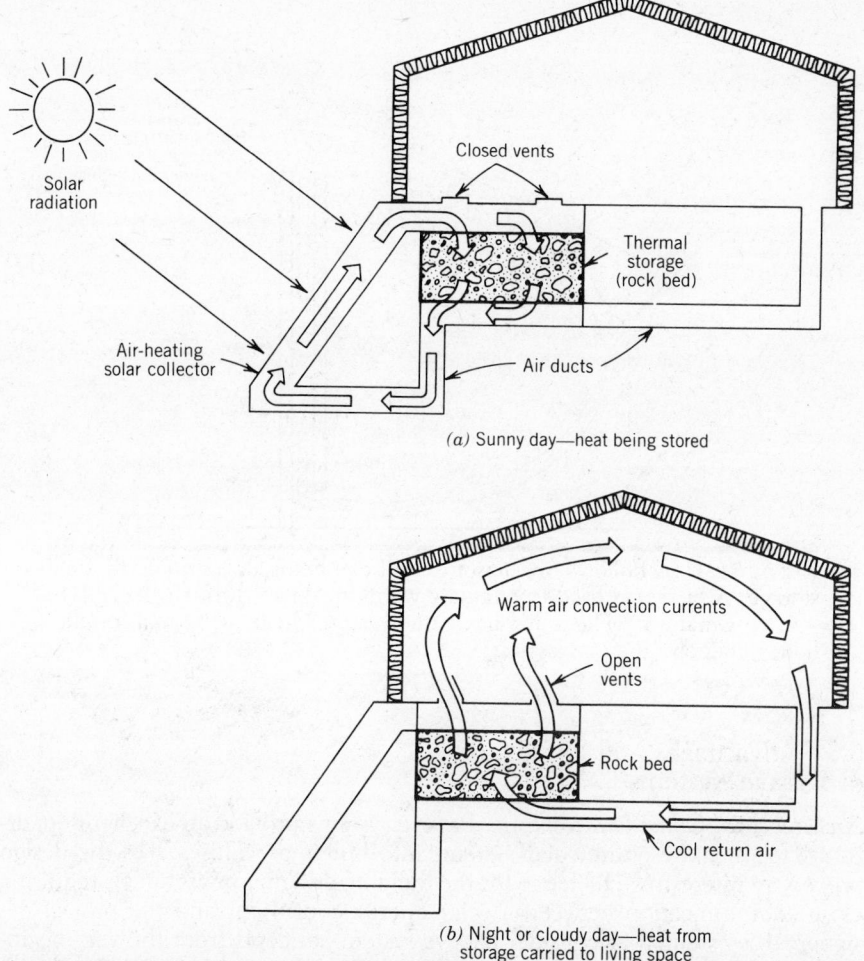

(a) Sunny day—heat being stored

(b) Night or cloudy day—heat from storage carried to living space

Figure 6-9 Convective loop solar energy system. (*a*) During the day, natural convection moves heat from the solar collector to thermal storage—a rock bed in this case. (*b*) When heat is needed at night or on cloudy days, vents are opened to the thermal storage. Heat moves from thermal storage into the building by natural convection.

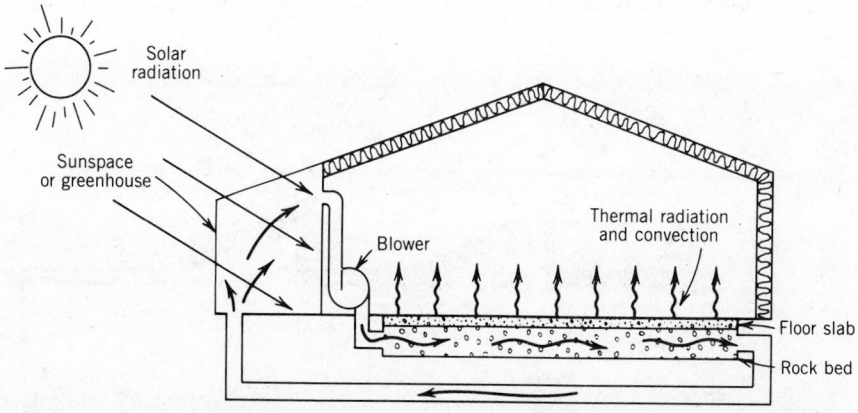

Figure 6-10 Hybrid solar energy system. A mechanical energizer, such as a blower, is used to move heat from the solar collection area to the thermal storage—in this case, an underslab rock bed. Heat can enter the building by radiation and convection from the slab, which is heated by the shallow rock bed, or room air can be blown through the rock bed to be heated and carry its heat back to the rooms.

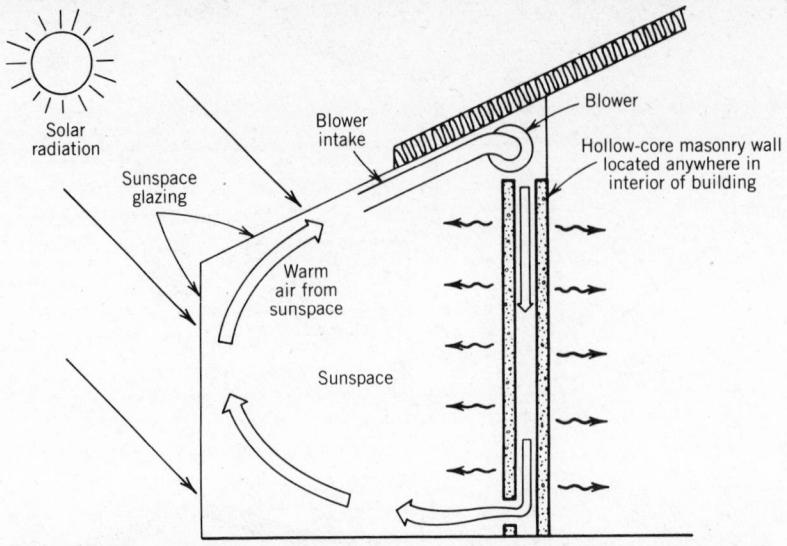

Figure 6-11 A hollow-core masonry wall can be used as remote thermal storage. A blower is used to circulate warm air down through the wall to warm the masonry. The wall will release the stored heat by thermal conduction, radiation, and convection.

6-3 Advantages and Disadvantages of Passive Systems

Architectural Considerations. Passive systems provide an excellent opportunity to integrate entire solar systems into buildings. This makes the design process an interesting challenge for the building designer or architect, requiring close communication between a solar energy consultant and the project designer. It is essential to consider passive system concepts from the very beginning of the design process. In a well-designed passive building, the solar elements are also able to perform other functions within the building. For example, it may be possible to have south-facing windows act as solar apertures for collecting energy while also providing views to relate the outside environment to the inside of the building. Skylights or clerestory windows, as shown in Fig. 6-12, can provide not only a source of heat but also interior lighting during the day. Thermal storage mass walls may serve a second function as load-bearing structural elements of buildings. By thus integrating passive compo-

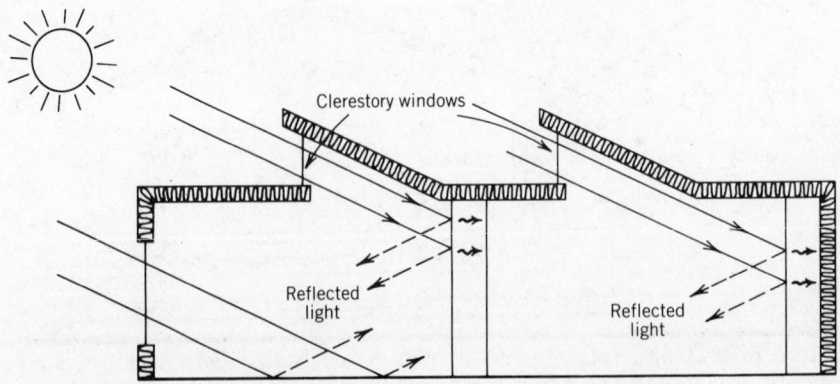

Figure 6-12 Clerestory windows may be used to bring sunlight to reflecting walls as a means of providing daylighting for the interiors of buildings, and also as a means of introducing radiant energy.

nents into the building structure, the additional cost of a passive solar system can be partially offset by the other functions its elements provide.

A well-designed passive solar system may also enhance a building space from a visual and artistic perspective. For example, a lobby area that is part of a passive solar system will be naturally lit with moving sunlight, resulting in constantly changing shadows and highlights in the space during the day. Furthermore, a variety of plants—even trees—can thrive in such a space. Such a passive-space lobby is far more interesting and livable than an artificially illuminated space.

Passive systems are not without problems, however. In systems where sunlight enters directly into the living space, the resulting glare is often unacceptable. Sunlight may fade or bleach out colors in fabrics and woods and can result in discomfort for occupants seated or working in direct sunlight. In the following chapters, control measures aimed at mitigating these effects will be discussed.

Building Site Considerations. The fundamental consideration for a solar project is solar access to the structure. The site should allow the building to have a southern orientation, or, as a minimum requirement, solar components should be oriented between 20° east of south and 20° west of south. Passive systems require that south walls have no obstruction to the south that would block the winter sun from windows and mass walls that are used as solar collection areas. When clerestory windows, skylights, or roof ponds are used, the roof must have unobstructed access to the winter sun.

There are many situations that can reduce solar access to a building. There might be shadowing by hills, buildings, or trees, for example. The lot orientation and/or zoning restrictions could prevent the structure from being oriented to the south. Solar access is so important that a thorough site analysis must be conducted by a solar designer as part of the preliminary consideration of any solar project (see section 6-6).

Simplicity of Passive Systems. A passive solar energy system, by nature, is simple in design. Passive systems have no pumps, pipes, fans, or ducts to move heat from one location to another, and, therefore, the operation of a passive building is quiet and uninterrupted. The day-to-day operation of a passive system may, however, necessitate opening and closing circulation vents in thermal mass walls, moving shading or insulating devices to control overheating during very clear days, seasonally adjusting shades or reflectors to enhance system performance, and opening and closing exhaust vents to remove overheated air. Some or all of the above operations can readily be automated in a passive system if desired, thus making it unnecessary for the occupants to adjust control devices manually on a daily or hourly basis.

The maintenance of a building with a passive system is not very much different from the maintenance of a standard building. Common construction materials are used with most passive systems, and they are easily maintained and serviced by personnel with only a limited technical background. Typical maintenance procedures might involve repairing caulking around glazing to prevent air infiltration, painting exposed parts of the structure, or keeping vents operable.

HUMAN COMFORT IN PASSIVE SOLAR BUILDINGS

Temperature control within a passive structure is of paramount importance, since the human body interacts with its surroundings in a complex way. For one thing, the body generates internal heat as a result of body metabolism, and in order for a more or less constant body temperature (about 98.6°F) to be maintained, heat must be dissipated at the same rate as it is gained. Heat is dissipated from the body by four mechanisms: *evaporation, respiration, convection,* and *thermal radiation.* The average rate of body heat generation is about 400 Btu/hr per person when people are seated at rest, and this heat is dissipated as follows: approximately 100 Btu/hr by evaporation of moisture from the skin and from the sensible and latent heat in respiration, about 100 Btu/hr by convection to the surrounding air, and about 200 Btu/hr by radiation from the body surface to surrounding surfaces which are cooler than the body surface temperature (about 88 to 92°F).

6-4 Human Comfort

Comfort is a complex subject that is very difficult to define. Psychological as well as physiological factors affect an individual's feeling of comfort, making it difficult to establish—in terms of objective physical measurements—what conditions actually provide comfort. However, there have been a number of at-

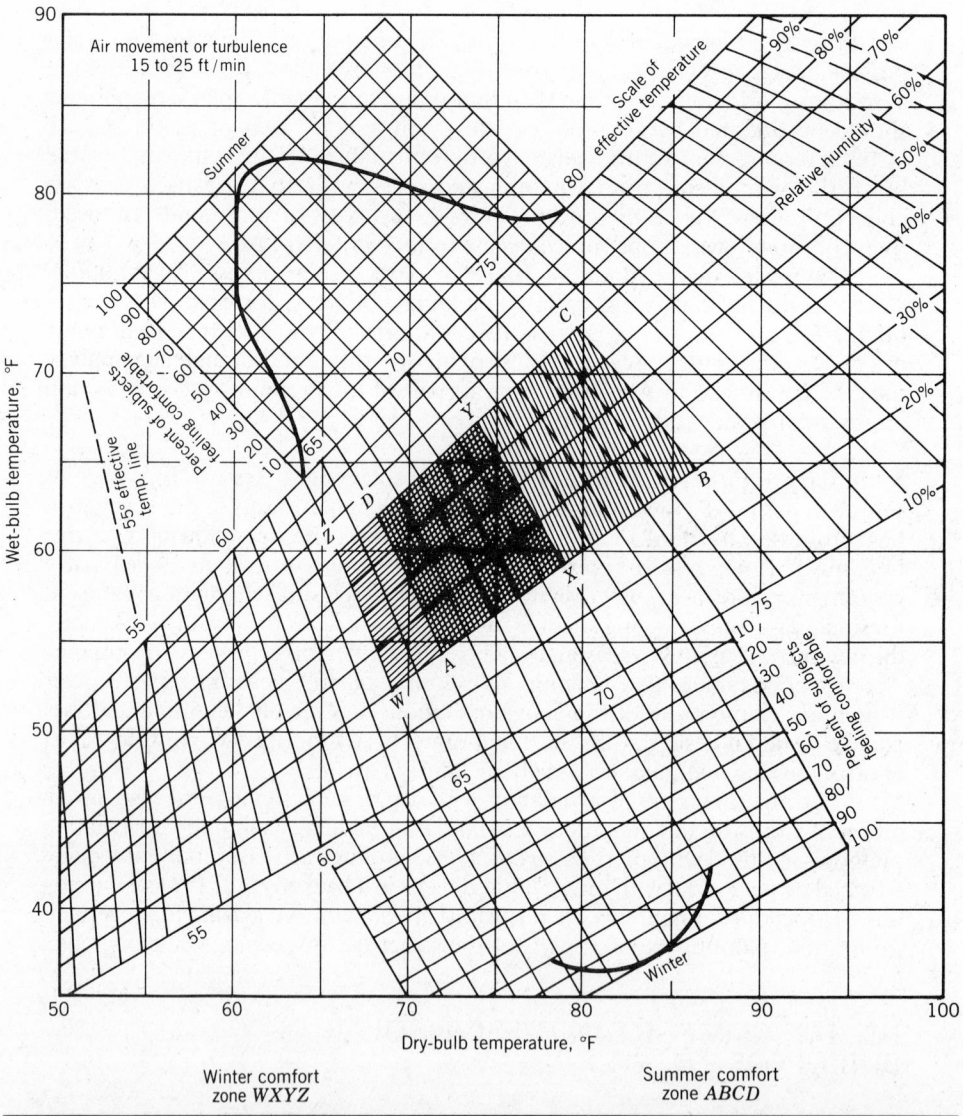

Winter comfort
zone *WXYZ*

Summer comfort
zone *ABCD*

Figure 6-13 ASHRAE *Comfort Chart*, relating dry-bulb temperature (DB), wet-bulb temeprature (WB), relative humidity (RH), and effective temperature (ET) to human comfort. The shaded section of the chart (the *Comfort Zone*) indicates the conditions desired by most persons for a feeling of comfort. Note the scales that denote the percentage of persons comfortable at various winter and summer conditions. *Source: ASHRAE Handbook of Fundamentals,* Atlanta: ASHRAE, 1967, p. 122. Adapted with permission.

tempts to establish physiological measurement scales of human comfort. In Great Britain, the *British Equivalent Temperature* has been used to compare the equivalent warmth of different environments. In the United States, research scientists working with air-conditioning engineers under the auspices of the American Society of Heating, Refrigerating, and Air Conditioning Engineers (ASHRAE), have developed the *ASHRAE Comfort Chart*, an essential element of which is the *Effective Temperature Scale* (Fig. 6-13). The Effective Temperature Scale combines the effects of temperature, relative humidity, and air movement on human comfort. The *Comfort Chart* is intended for use in spaces with reasonably still air, for situations where occupants are seated at rest or doing light work, and for spaces where the enclosing surfaces are at a mean temperature nearly equal to the dry-bulb temperature of the room air. Radiation gains to or losses from the human body are assumed to be zero.

Studies performed by different researchers with hundreds of human subjects under controlled indoor climate conditions have placed the *Comfort Zone* between 30 percent and 70 percent relative humidity. For winter (area WXYZ on the Chart), the optimum effective temperature was determined as 67° to 68°, with a range of 64° to 71° for men and women sitting at rest with average clothing. For summer (area ABCD), the optimum effective temperature was found to be 71°, with a range of 66° to 76°. These two areas make up the Comfort Zone, the shaded area on Fig. 6-13. Acclimatization probably accounts for the shift of the comfort zone between winter and summer. The body becomes seasonally adjusted to colder temperatures in winter and feels comfortable in a somewhat cooler environment, while in summer the adjustment is to hotter temperatures and the body soon feels comfortable in a warmer environment. Acclimatization according to geographical locations also shifts the zone of comfort. It has been found that the effective temperature increases approximately 1F° for every five degrees of latitude change (i.e., toward the equator). The zone of comfort has also been found to vary with individuals, with type of clothing worn, and with the kind and level of activity. There is some sex variation also, with women generally preferring an effective temperature for winter comfort about 1 F° higher than that preferred by men. Age also enters into the thermal requirements for comfort, with persons older than 40 years of age preferring an effective temperature 1 F° higher than men and women below this age.

Illustrative Problem 6-1

For winter conditions and a relative humidity of 50 percent, find the range of dry-bulb temperatures bounded by the Comfort Zone.

Solution

From Fig. 6-13, the lower boundary of the winter Comfort Zone on the 50 percent relative humidity line is at a dry-bulb temperature (horizontal scale) of approximately 67°F. The upper boundary of the winter Comfort Zone is approximately at 76.5°F. The range of dry-bulb temperatures in which at least some persons would report being comfortable is therefore 67°F to 76.5°F. At dry-bulb temperatures outside these limits, very few, if any, persons would report comfort.   *Ans.*

Illustrative Problem 6-2

The dry-bulb temperature of the air in a passive solar building is 68°F, and most of the occupants complain that they are too cold. The building engineer notes that the relative humidity is 20 percent. The engineer can switch on the backup furnace and provide heat to the room air until the occupants are warm enough at, say, 72°F air temperature. What alternative action could be taken if the proper equipment were available and operable?

Solution

If humidifying equipment is a part of the system, add moisture to the room air until the relative humidity is, say, 40 percent. Note from the Comfort Chart that a condition of 68°F dry-bulb, 40 percent RH (64° ET), results in comfort for about 70 percent of normally clothed persons. It might, however, require as much or more energy to raise the humidity as it would to increase the dry-bulb temperature to, say, 70°F. *Ans.*

The Bioclimatic Chart. A somewhat different approach to a comfort chart has been proposed by V. Olgyay. It is called the *Bioclimatic Chart*, and it is also based on a concept of human comfort related to dry-bulb temperature and relative humidity, but, in addition, it takes into account radiation effects, air velocity, and evaporation of moisture as they influence human comfort. (The ASHRAE Comfort Chart presumes negligible radiation effects.)

Figure 6-14*a* presents Olgyay's Bioclimatic Chart, indicating the comfort zones during winter and summer conditions for moderate climates in the United States (regions of 40° to 42° N Lat). For clarity, the Bioclimatic Chart has been redrawn and enlarged for winter conditions only in Fig. 6-14*b*. The lower boundaries of both comfort zones are called *shading lines*. Typically, persons wearing normal clothing and not in direct sun will feel comfortable both summer and winter when the values of humidity and dry-bulb temperatures are within the indicated comfort zones. At indoor conditions above the shading lines, a person must remain out of direct sun in order to be comfortable. At dry-bulb temperatures below the respective shading lines, a normally clothed person may feel too cool in either winter or summer. The feeling of winter comfort can be maintained below the shading line, however, if the body absorbs radiant energy from warm walls (floors or ceilings) or from direct solar radiation, to compensate for the increased thermal losses from the body by conduction and convection at the lower dry-bulb temperatures. Studies have shown that at dry-bulb temperatures below 70°F, a 1 F° drop in the air dry-bulb temperature can be compensated for by raising the *mean radiant temperature* (MRT) of the surfaces in the space by 0.8 F°.[3] A scale of mean radiant temperature is shown in Fig. 6-14*b* left and below the shading line, giving the MRT values necessary to compensate for air dry-bulb temperatures lower than 68°F. This compensation technique has practical limitations, however, since no more than about a 4 to 5 F° difference can be maintained between the room air and wall surface temperatures.

Another way to retain a feeling of winter comfort is for the body to absorb direct solar radiation. Studies have shown that a drop of 3.85 F° in air dry-bulb temperature can be compensated for by the body's absorption of 50 Btu/hr of direct solar radiation. The curves of Fig. 6-14*b* below the winter Comfort Zone (scale at right of the chart) indicate the level of body-absorbed radiation necessary to give a feeling of comfort at temperatures below 68°F dry-bulb.

Illustrative Problem 6-3

What steps can be taken to bring a condition of 60 percent relative humidity (RH) and 65°F dry-bulb temperature back into the winter Comfort Zone?

Solution

Enter the Bioclimatic Chart (winter, Fig. 6-14*b*) along the 60 percent RH line, and locate the room condition at R (60 percent RH, 65°F DB). Note that the 50

---

[3] Mean radiant temperature (MRT) is defined as the temperature at which all surfaces of an artificial room would have to be held in order to provide the same thermal radiation transfers between the body and the room walls (ceilings, floor, windows, etc.) that would occur in a real room with surfaces at many different temperatures. Thus, mean radiant temperature is an indication of the average of the surface temperatures of a space.

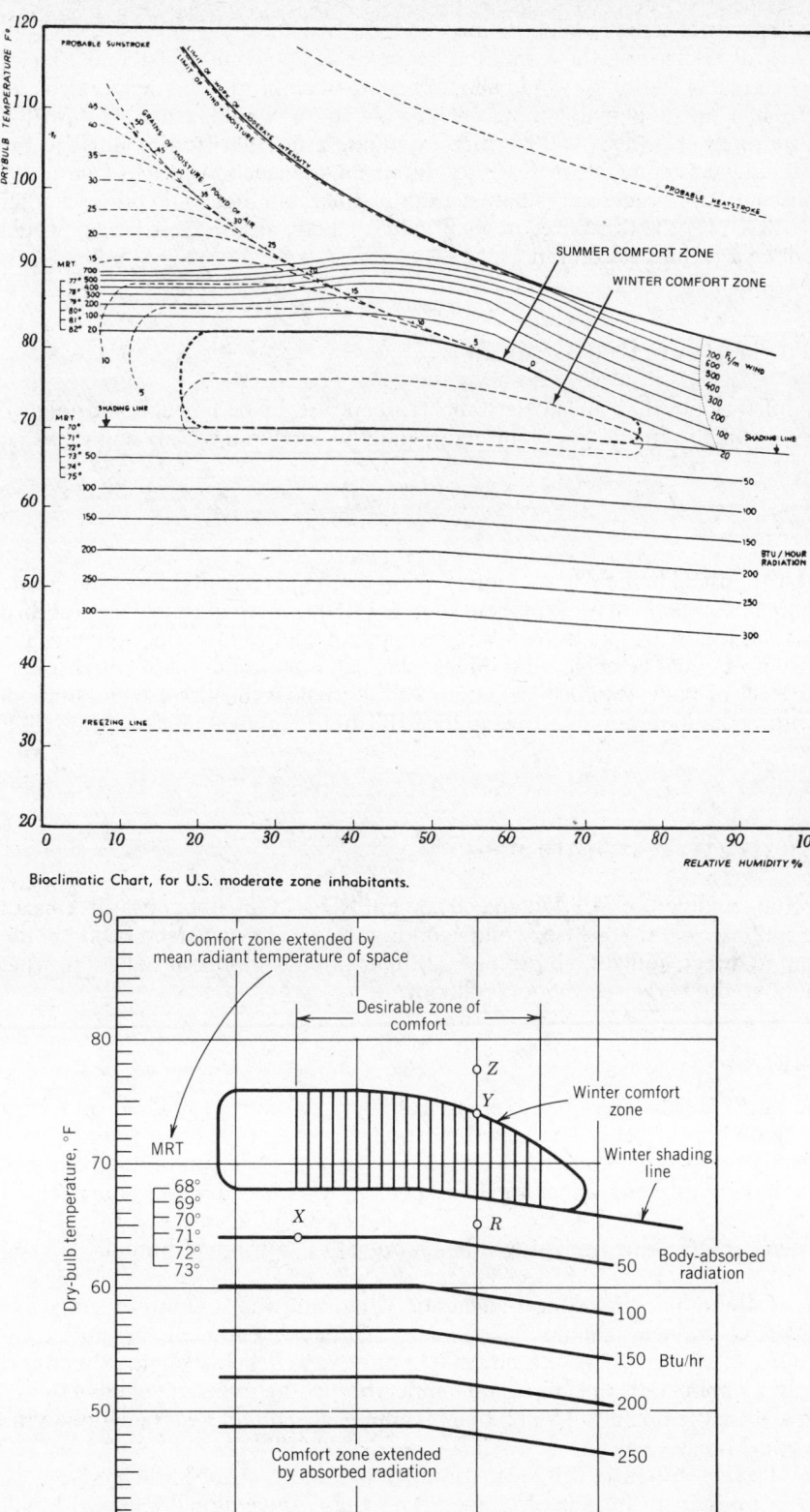

Bioclimatic Chart, for U.S. moderate zone inhabitants.

F i g u r e 6-14 The Bioclimatic Chart is another graphical representation of comfort conditions. The comfort zone can be extended in winter by countering low air (dry-bulb) temperatures with radiant energy absorbed by the human body. (*a*) The complete chart shows both the winter and the summer comfort zones. (Victor Olgyay with Aladar Olgyay, *Design with Climate: Bioclimatic Approach to Architectural Regionalism*, Princeton: Princeton University Press, 1963, p. 22. Reprinted with permission of Princeton University Press.) (*b*) The winter comfort zone is enlarged in this diagram with an expanded scale of *mean radiant temperature* (MRT) for easier reading.

Btu/hr body-absorbed radiation line will compensate for dry-bulb temperatures as low as 64°F. For the condition at point R, body-absorbed radiation of approximately 30 Btu/hr would allow a feeling of comfort to be maintained. An alternative means of maintaining a feeling of comfort would be to hold the MRT of the room at or above 70°F ..ince extending a line left from point R to the mean radiant temperature scale results in an intersection at MRT = 70°F. Consequently, direct solar radiation entering the room that could provide either an MRT of 70°F or above or 30 Btu/hr of body-absorbed radiation would produce a feeling of comfort.   Ans.

Illustrative Problem 6-4

An office lobby has an air dry-bulb temperature of 64°F when the relative humidity is 30 percent. Everyone agrees that the condition is "very chilly". What might be done?

Solution

Enter the Bioclimatic Chart (winter) along the 30 percent RH line, and locate the room condition at X (30 percent RH, 64°F DB). The required body-absorbed radiation that corresponds to point X is approximately 50 Btu/hr, and a feeling of comfort could be achieved if direct solar radiation in the room could supply this level of body-absorbed radiation. An alternative means of maintaining a comfort condition would be to hold the MRT of the room at or above the 71.3°F level.   Ans.

Illustrative Problem 6-5

A room condition of 74°F DB and 60 percent RH is at the upper (warm) edge of the winter comfort zone (see point Y on Fig. 6-14b). For a person sitting in the room in direct sunlight, absorbing 50 Btu/hr of body-absorbed radiation, what would be the perceived room conditions?

Solution

The winter Bioclimatic Chart is entered along the 60 percent RH line, and room condition Y is located and marked on the chart. Since 50 Btu/hr of absorbed radiation can compensate for a 3.85 F° dry-bulb temperature difference, proceed from point Y upward along the 60 percent RH line a distance equivalent to 3.85 F° to point Z. Note that the equivalent condition is about 78°F DB at 60 percent RH, which might well be described as "warm and humid."   Ans.

Careful analysis of the Bioclimatic Chart and the solution of the above illustrative problems should make it clear that people can really be quite comfortable in rooms where the air temperature is well below standard comfort levels, if either (1) the mean radiant temperature of the room surfaces is 4 to 5 F° above the air temperature, or (2) there is appreciable direct solar radiation in the occupied areas.

Passive structures with heat-storing floors, walls, or roofs will tend to have higher interior mean radiant temperatures than conventionally heated buildings. Thus, it is possible to maintain a sensation of comfort with much lower air temperatures in passive buildings than in forced-air-heated buildings. Passive buildings can also provide a more invigorating interior space, with natural lighting and sunlight penetration giving psychological as well as physical feelings of warmth and comfort.

For cooling and summer comfort (Fig. 6-14a) lowering the mean radiant temperature of the room surfaces can likewise counter the effect of fairly high air temperatures as indicated by the MRT scale to the left of and above the comfort zone. The cooling effects of passive structures will be discussed as part of natural cooling in later chapters.

6-5 Controlling Temperature in the Living Space

In contrast to the advantages of passive systems, the major disadvantage is one of control. A passive system has a large thermal storage capacity, which is an integral part of the building structure. This large mass has a great deal of "thermal inertia", or capacitance, which restricts the ability of the building to respond quickly to abrupt changes in either outdoor ambient temperature or received solar radiation. For a passive building to have good interior comfort control, the following conditions must exist:

1. The structure must be well insulated and have a low rate of air infiltration. This will ensure that abrupt changes in outdoor ambient air temperature will not result in sudden air temperature changes inside the building.
2. The solar collection area provided for the building must be adequate to collect the quantity of heat that the system designer has assigned to passive solar.
3. The thermal storage must be distributed in the building in such a way that it can absorb the energy of the solar radiation entering the building without an uncomfortable air temperature rise in the building.
4. The building should be provided with shading devices for the solar collection area, for use when energy collection is not needed, and with a method to exhaust excess energy already collected as warm air.

The daily operation of a passive system can best be described as a *thermal flywheel*. Just as a mechanical flywheel operates to moderate the speed fluctuations of an engine, so does a passive system with thermal storage operate to moderate the temperature fluctuations of a building. During the day, as solar radiation is transmitted into the building, sunlight strikes and is absorbed by the interior surfaces and thermal masses of the building, and their temperatures rise as a result. When the surfaces are those of good thermal storage materials, heat flows readily into the thermal storage, and the interior temperature of the thermal storage increases along with the surface temperature. This is the energizing process of the thermal flywheel—the absorption of heat into the thermal storage.

During this energizing or charging of the thermal storage, however, not all of the energy in the absorbed solar radiation enters storage. Since the surface of the storage is warmer than the surrounding air temperature, some heat is transferred directly to the room air from the warm surface (see section 3-10). Sunlight may also strike some surfaces that are not good thermal storage materials, such as rugs, furniture, and non-heat-storing walls. The surfaces of these objects will be warmed up, resulting in immediate heat loss to the room air by convection and to other surfaces by radiation. It is important to minimize air heating in a passive solar building so that the interior air temperature increase during a sunny day is not more than 8 to 10 F° above the lower limit of the Comfort Zone (68°F).

At night, heat will be lost to the outdoors through the building skin, and the interior surfaces that are not thermal mass will cool down very quickly. Air temperature will also drop, and the thermal storage, which at night is warmer than the other building surfaces, will begin to provide energy to the interior of the building. Interior air will be heated by convection and surface conductance from the warm surfaces of the thermal storage, while radiation from the warm surfaces will transfer heat to cooler surfaces in the building. This is the releasing

---

of energy from the flywheel, and it opposes and moderates the tendency for the interior temperature to decrease markedly as heat is lost from the building at night.

DESIGN FACTORS FOR PASSIVE SYSTEMS

The first step in the passive solar design process is to determine the *solar access* or availability of solar radiation at the proposed building site. If ample sunlight does not reach the building site because of climate, site orientation, or obstructions, then there is limited solar access, and it would probably be impractical to consider a solar energy system.

6-6   Determining Solar Access

Solar Access Devices.   There are a variety of commercial (patented) devices available that provide a visual assessment of the solar access at a building site. These devices impose either a sunchart (see section 4-10) onto the visible skyline or the visible skyline onto a sunchart. It will be recalled that a sunchart is a graphical representation of the sun's apparent motion in the sky at a specific latitude (see Fig. 4-17).

One very simple and useful device for solar access determination is the Solar Card™. The card is held in front of the observer, as shown in Fig. 6-15, and the southern skyline is viewed through the chart from the proposed building site. When the skyline is above the sun's path, the point where the observer is standing will not receive sunlight during the times of the day and the periods of the year indicated.

Another device for essentially the same purpose is the Solar Site Selector™ (Fig. 6-16). It has a level and a compass to assist in the setup and orientation of the tool. A different sunchart must be used for each latitude at which an observation is made.

Still another such device is illustrated in Fig. 6-17. It projects the skyline

Figure 6-15 The Solar Card® is used to project sunpath lines onto the skyline. The card with sunpath curves on it is held in front of the viewer's eyes, and the southern skyline is viewed through it, thus superimposing the sunpaths on the skyline. (*Solar Age Magazine*, Solar Vision, Inc. Harrisville, N.H. Reprinted by permission.)

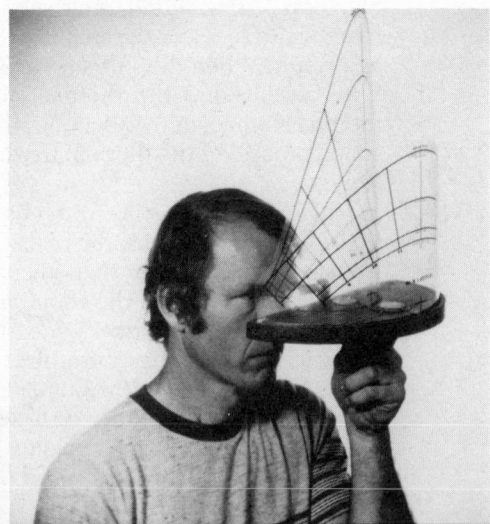

Figure 6-16 The Solar Site Selector® is another device used to project sunpath lines onto the horizon. (Lewis and Associates, Grass Valley, Calif.

Sunpaths are shown
on chart under
Pathfinder dome.

Reflected image of
skyline and of
shadow-casting
objects in Pathfinder
dome.

Figure 6-17 The Solar Pathfinder® is a device that projects the skyline down onto a circular sunpath diagram. (Solar Pathways, Inc., Glenwood Springs, Colo.).

down onto the sunchart by the use of an "optical dome." This handy tool is called the Solar Pathfinder™.

Use of Suncharts. The basic method of determining solar access is simply to measure the altitude of the skyline at various azimuths across the southern horizon and then plot these onto a sunchart for the latitude of the site. A compass is needed to determine azimuth angles along the skyline, and a transit or hand level with a protractor is required for measuring the altitude of the skyline. First, true south must be found using the hand compass. (Remember to correct the compass for magnetic variation.) Then, the altitude of the skyline at due south is measured, using the hand level and protractor. This altitude is then plotted on the sunchart above the azimuth angle of zero (due south). Other altitude measurements of the southern skyline should then be made at 15° increments in azimuth, up to 120° east and 120° west of south (Fig. 6-18*a*). This procedure yields a total of 17 altitude readings plotted on the sunchart (Fig. 6-18*b*). If large isolated objects are present on the skyline, such as buildings or evergreen trees, the outline of these objects should be measured and plotted (sketched in) on the sunchart. Deciduous trees should be indicated with dotted lines, as in Fig. 6-18*b*. This is to indicate that they lose their leaves in winter and will not effectively block winter sun. However, some kinds of trees, even when leafless, block out a good deal of winter sun—some trees as much as 50 percent.

Once the skyline is plotted on the sunchart, the open areas of the chart above the skyline indicate the periods when sunlight will reach the point from which the observations were made. If the horizon does not block the winter sun appreciably, as depicted in Fig. 6-18*b*, the site will have full solar access. If there are obstructions to the east or west, as shown in Fig. 6-19, they can be partially compensated for by orienting the building surfaces away from the obstructions so that the building faces into the open part of the plotted sunchart diagram. This procedure can be used as long as the solar collection areas of the proposed structure are not moved more than approximately 20° east or west of south. If a sizable obstruction occurs close to due south, the midday sun will be blocked during the winter. Should such a blockage last for only 30 to 40 min, it will not have a serious effect on the performance of a solar energy system, but in the

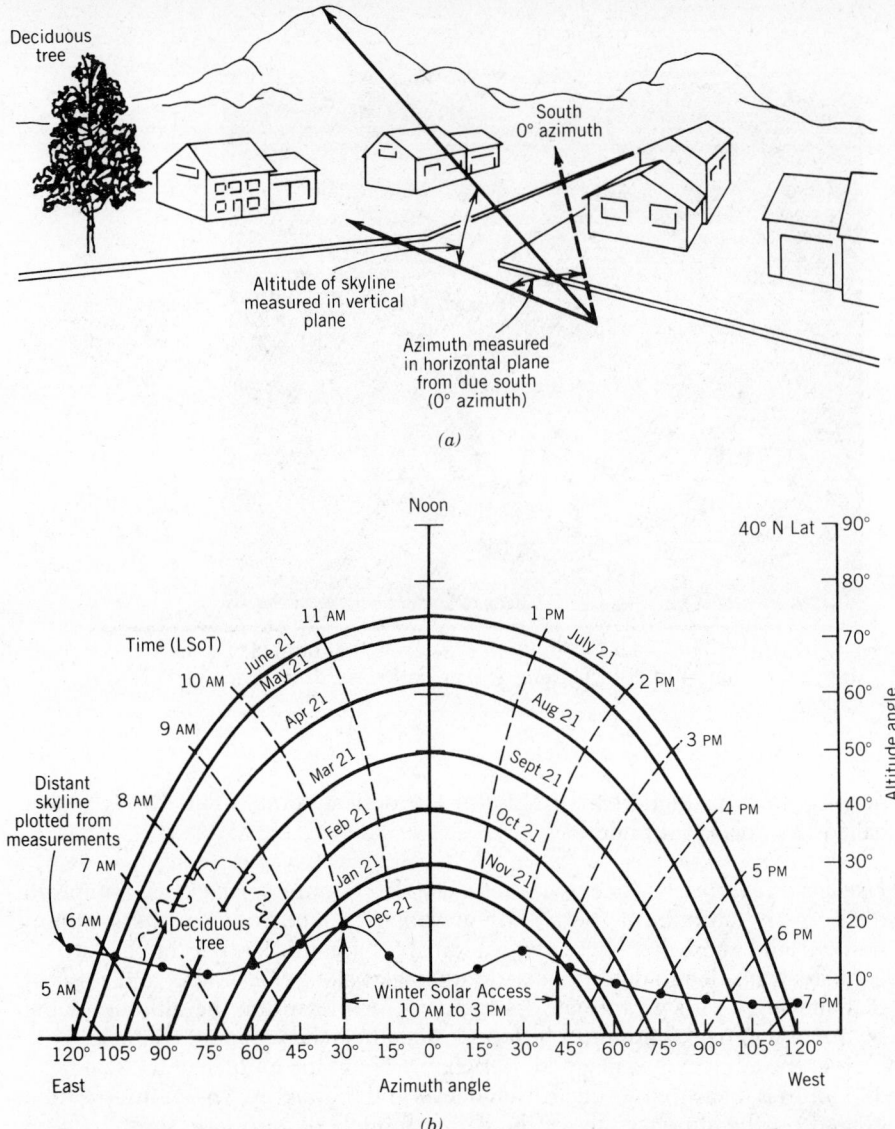

(a)

(b)

Figure 6-18 Determining solar access. (a) The altitude of the skyline is measured at 15° intervals in azimuth, east and west of due south. (b) The measured altitudes are plotted on a sunpath diagram (40° N Lat), which then shows at what time of the day (LSoT) the skyline blocks the sun for each month of the year. The deciduous tree is plotted as a dotted line on the sunchart to indicate that it does not completely block the winter sun. From this plot, it is apparent that winter sun (December 21) is available between the hours of 10 A.M. and 3 P.M.

case of a large obstruction resulting in a 2- or 3-hr blockage, as shown in Fig. 6-20, a significant portion—perhaps 35 to 40 percent—of the effective solar radiation may be blocked. In such cases, a careful analysis of the system performance (see section 6-9) will have to be made to determine whether or not a solar installation would be practicable.

Illustrative Problem 6-6

The skyline was viewed from two different locations, A and B, on a large building site at 40° N Lat, and the altitude and azimuth angles were recorded as follows:

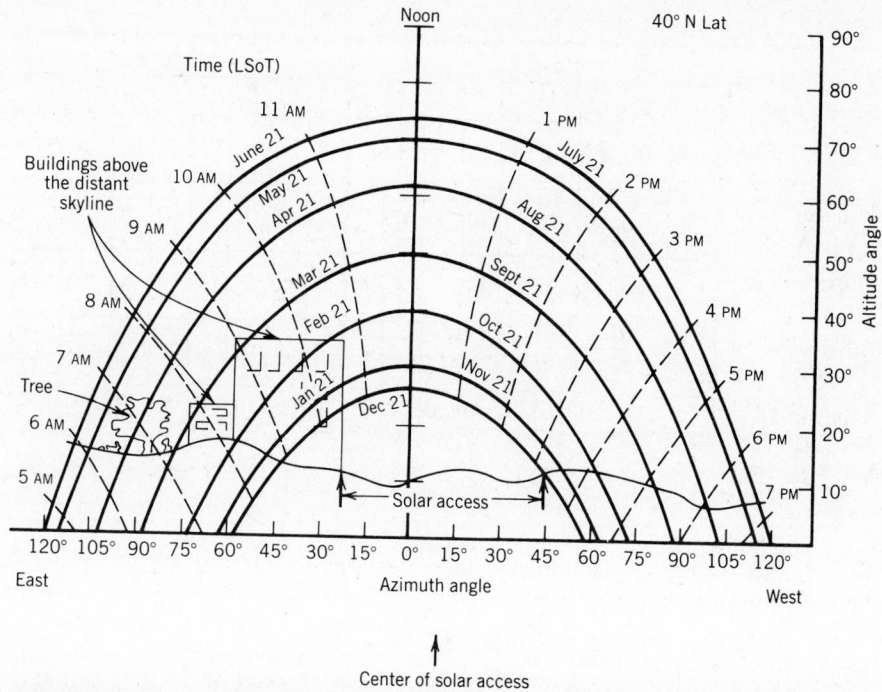

Figure 6-19 If an obstruction occurs in the east or west, blocking the winter sun, it can be partly compensated for by facing the solar elements toward the center of the solar access rather than due south. In this case, the proper orientation would be about 15° west of south.

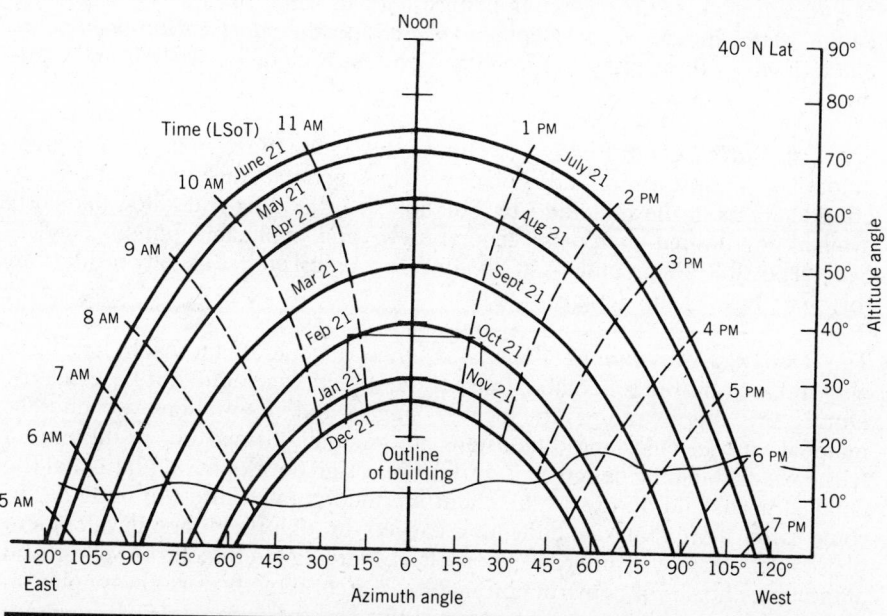

Figure 6-20 When a tree or building is directly to the south of a solar site, severe blockage of the winter sun can occur. In the case shown, three hours of the midday sun on December 21 will not reach the site.

DESIGN FACTORS FOR PASSIVE SYSTEMS

	East								Due south	West							
Azimuth angle	120	105	90	75	60	45	30	15	0	15	30	45	60	75	90	105	120
Horizon line, site A	9	10	11	14	15	15	7	6	7°	10	11	8	11	15	14	10	10
Building, site A	0	0	0	0	0	0	0	37	37	37	0	0	0	0	0	0	0
Horizon line, site B	9	10	11	14	15	15	7	6	7°	10	11	8	11	15	14	10	10
Building, site B	0	0	0	0	0	0	0	0	0	0	0	0	0	27	27	27	0
Tree, site B	0	0	0	26	25	0	0	0	0	0	0	0	0	0	0	0	0

Plot the two skylines on suncharts, and determine which location provides the best solar access.

Solution

The altitudes and elevations are plotted with the results shown in Figs. 6-21a and 6-21b. At site A, there is a large obstruction that blocks the winter sun from 11 A.M. to 1 P.M., severely limiting the solar access. At site B, the observer has moved away from the obstructing building so that its outline does not block the winter sunpath. There is now a deciduous tree in the east that blocks the morning sun, but this does not affect the solar access significantly.

It is readily apparent that location B is a much better site than location A, with respect to solar access.   *Ans.*

6-7   Basic Steps in Passive Solar Design

The design process for passive buildings involves three phases: the *schematic design phase*, the *design development phase*, and the *construction documents phase*. Each of these phases represents a process of detailed analysis and planning.

The Schematic Design Phase.   During the schematic design phase, a great amount of "brainstorming" takes place with only general constraints imposed by considerations of the building site, building size, building use, allowable cost, available or desired materials, general style, and applicable building codes. Several possible design options in the form of rough sketches usually result from this first phase of the design process.

The Design Development Phase.   Once a general approach has been chosen based on the schematic designs, the design development phase begins. During this phase, rooms are located, angles and orientations are chosen, materials are selected, and dimensions are gradually firmed up, resulting in a fairly specific building design. It is during the design development phase that the detailed needs and desires of the client are incorporated into the building. In some cases, more than one concept is carried through the design development phase—for example, different floor plans, varying architectural styles, and alternative choices of construction materials. Economic factors always play an important role in the decision processes at this phase. For example, the decision on whether to construct a thermal storage wall out of poured-in-place concrete or to use a brick wall, water containers, or stone would be based on economic and other considerations such as structural requirements and aesthetics. Tenta-

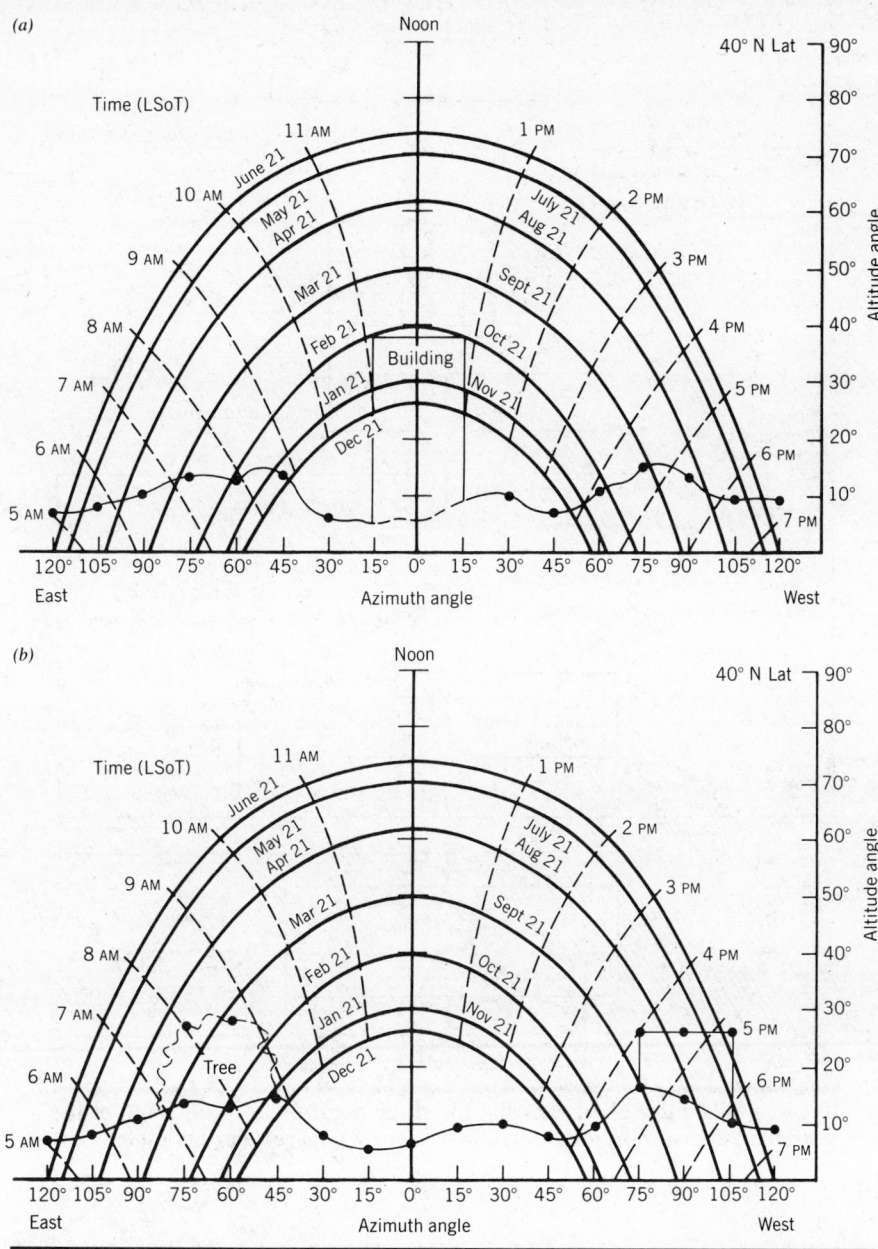

F i g u r e 6-21 Plots of horizon for Illustrative Problem 6-6. (*a*) For site A. (*b*) For site B.

tive approval by the client of a building design marks the end·of the design development phase. This is a major milestone in the passive solar design process, after which it is increasingly more difficult and expensive to make changes.

The Construction Documents Phase.   During this final phase, the exact dimensions, the detailing, and the specified material choices are made and committed to the drawings and specifications that will be used for the construction of the building.

The thermal design of a building must fit into the above design process as it evolves from the general to the specific. Traditionally, in conventional (i.e., nonsolar) buildings, the heating and air-conditioning design has been relegated to the final construction documents phase (except for space allocation for mechanical equipment, which is usually a part of the design development

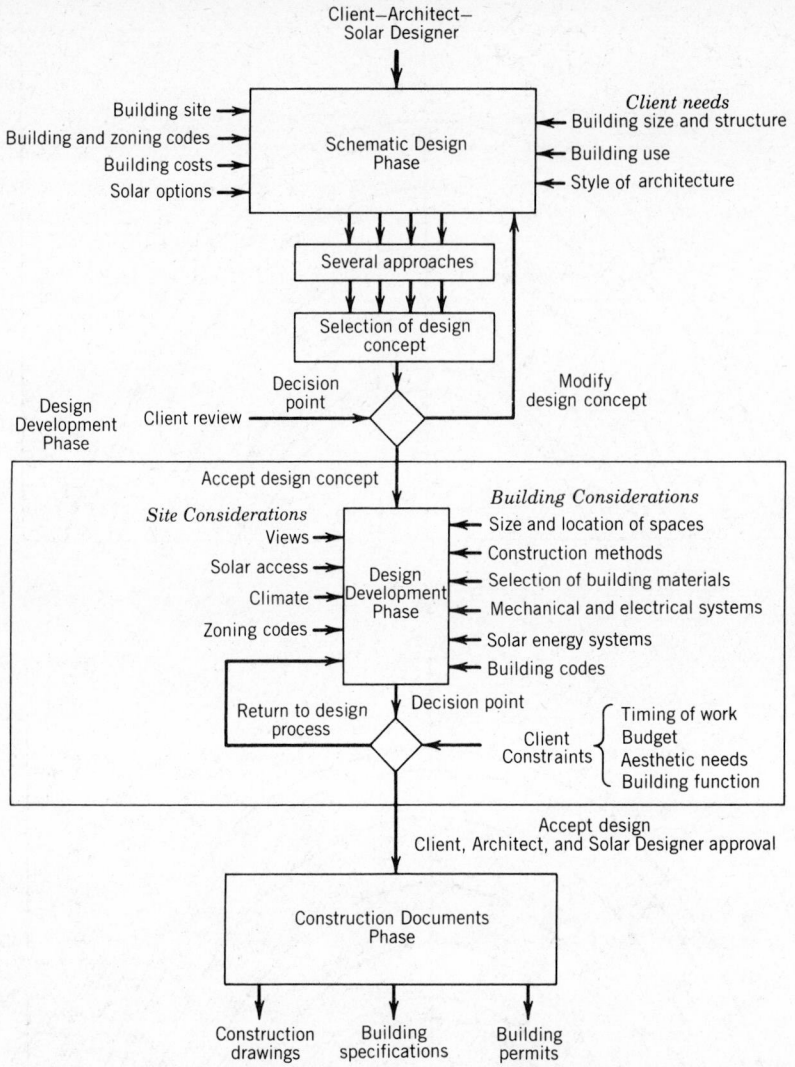

Client–Architect–
Solar Designer

Building site →
Building and zoning codes →
Building costs →
Solar options →

Schematic Design
Phase

Client needs
← Building size and structure
← Building use
← Style of architecture

Several approaches

Selection of design
concept

Decision
point

Design
Development
Phase

Client review →

Modify
design concept

Accept design concept

Site Considerations
Views →
Solar access →
Climate →
Zoning codes →

Design
Development
Phase

Building Considerations
← Size and location of spaces
← Construction methods
← Selection of building materials
← Mechanical and electrical systems
← Solar energy systems
← Building codes

Return to design
process

Decision point

Client
Constraints
{ Timing of work
Budget
Aesthetic needs
Building function

Accept design
Client, Architect, and Solar Designer approval

Construction Documents
Phase

Construction
drawings

Building
specifications

Building
permits

Figure 6-22 Flow chart showing the three main phases and some of the associated activities in the design process for a passive solar building.

phase). But for passive solar buildings it is too late to start the design of the solar elements at the construction documents phase. The commitment to a passive solar system must be made in the schematic design phase, with further development and refinement in the design development phase before construction documents are prepared.

Figure 6-22 illustrates in flow-diagram form the processes involved in designing a passive solar building.

6-8   Rules of Thumb for the Schematic Design Phase

In the schematic design phase, some general rules of thumb are needed to help in roughing out the sizes of the solar collection areas relative to other building dimensions, to help decide on the amount and location of heat-storing thermal mass, and to influence the building orientation.

Extensive research has been conducted at the Los Alamos National Laboratory in New Mexico on performance evaluation and analysis of passive solar heating systems.[4] This work has involved a great deal of technical experimentation and system simulation. The experimental program has consisted of data acquisition in 14 passive test rooms, a number of test cells, and 15 monitored passive solar buildings. The technical evaluation and analysis was based primarily on the PASOLE computer code developed at the Los Alamos Laboratory. Based on this work and on weather data for different locations in the United States, rules of thumb for the schematic design phase were developed for solar collection area, thermal storage mass, and building orientation. These will be referred to herein as *The Los Alamos Rules of Thumb*.

The Los Alamos Solar Collection Area Rule of Thumb. This rule can be summarized as follows:

> A solar collection area of $R1$ percent to $R2$ percent of the total floor area can be expected to reduce the annual heating load of a building in (location) by $S1$ percent to $S2$ percent; or if R-9 insulation is used at night in the solar collection area, by $S3$ percent to $S4$ percent.

The values of the parameters of $R1$ and $R2$ and of $S1$, $S2$, $S3$, and $S4$ are provided for selected locations in Table 6-1. A more complete table for many U.S. locations is given in the reference cited under the table. It is recommended that the larger of the two solar collection areas indicated by the percentages given not be exceeded, or else building overheating may result on clear winter days. The use of the Los Alamos solar collection area rule of thumb is illustrated in the following problem.

Illustrative Problem 6-7

A 1600-ft^2 residence is planned for Denver. Use the solar collection area rule of thumb to estimate the collection area for a 60 percent reduction in the annual heating load—that is, a 60 percent *Solar Savings*.[5]

Solution

From Table 6-1, for Denver, find:

$R1$ = 12%	$S1$ = 27%	$S3$ = 47%
$R2$ = 23%	$S2$ = 43%	$S4$ = 74%

The floor area is multiplied by $R1$ and $R2$ to find the solar collection area for the annual reductions indicated by $S1$, $S2$, $S3$, and $S4$. The results are given below.

Solar Collection Area (ft^2)	Reduction in Heating Load (No Night Insulation)	Reduction in Heating Load (with Night Insulation)
192	27%	47%
368	43%	74%

[4] The Los Alamos National Laboratory at Los Alamos, New Mexico, has done extensive research on the performance of both active and passive solar energy systems under grants from the Energy Research and Development Administration (ERDA) and now the Department of Energy (DOE). The work has included computer simulations and long-range study and analysis of the performance of solar buildings. Several important publications have come from the Los Alamos Laboratory, and some of them will be cited frequently throughout this book. It should be pointed out that there are a large number of other centers for solar energy research and study, some at major universities and some connected with state and federal agencies. We shall have occasion to cite their work in these pages also.

[5] The reduction in the annual heating load that results from the presence of a passive solar collection area is here referred to as the Solar Savings. The term *Solar Savings* will be more rigorously defined in section 6-9.

Table 6-1 Calculated Values of Parameters for Use with the Los Alamos Solar Collection Area Rule of Thumb

Locality	R1(%)	R2(%)	No Night Insulation		With R-9 Night Insulation	
			S1(%)	S2(%)	S3(%)	S4(%)
Albany, NY	21	41	13	15	43	66
Albuquerque, NM	11	22	29	47	46	73
Boston, MA	15	29	17	25	40	64
Charleston, SC	7	14	21	41	34	59
Chicago, IL	17	35	17	23	43	67
Denver, CO	12	23	27	43	47	74
Ely, NV	12	23	27	41	50	77
Fort Worth, TX	9	17	26	44	38	64
Fresno, CA	9	17	29	46	41	65
Hartford, CT	17	35	14	19	40	64
Houston, TX	6	11	25	43	34	59
Indianapolis, IND	14	28	15	21	37	60
Knoxville, TN	9	18	20	33	33	56
Lexington, KY	13	27	17	26	35	58
Little Rock, AR	10	19	23	38	37	62
Madison, WI	20	40	15	17	51	74
Medford, OR	12	24	21	32	38	60
Miami, FL	1	2	27	48	31	54
Midland, TX	9	18	32	52	44	72
New Orleans, LA	5	11	27	46	35	61
Phoenix, AZ	6	12	37	60	48	75
Salt Lake City, UT	13	26	27	39	48	72
San Francisco, CA	6	13	34	54	45	71
Santa Maria, CA	5	11	31	53	42	69

Source: Abstracted, with permission, from J. D. Balcomb, et al., *Passive Solar Design Handbook*, vol. II (Washington: U.S. Dept. of Energy, January 1980) DOE/CS-0127/2, pp. 22–23.

To obtain a 60 percent reduction, night insulation will have to be used, since only a 43 percent reduction can be obtained without the use of night insulation. The solar collection area can be interpolated for a 60 percent reduction as follows: A 60 percent annual reduction is located 0.48 of the interval between 47 percent and 74 percent. Thus, the solar aperture needed for a 60 percent reduction will be proportionally 0.48 of the interval between 192 and 368 sq ft. Thus, the solution:

$$\text{Solar collection area} = 192 \text{ ft}^2 + (368 - 192) \text{ ft}^2 \times 0.48 = 276 \text{ ft}^2 \text{ Ans.}$$

Table 6-2 Assumed (Normalized) Net Load Coefficients (Btu/DD-ft²) for the Los Alamos Solar Collection Area Rule of Thumb

Annual Heating Degree-Days Range	Normalized Net Load Coefficient (NLC) *per sq ft* (Exclusive of Solar Collection Area) (Btu/DD-ft²)
Less than 1000 DD	7.6
1000–3000	6.6
3000–5000	5.6
5000–7000	4.6
Greater than 7000	3.6

Source: Abstracted, with permission, from J. D. Balcomb, et al., *Passive Solar Design Handbook*, vol. II (Washington: U.S. Department of Energy, January 1980), p. 24.

The Los Alamos collection area rule of thumb assumes that the building will have certain stipulated thermal loss characteristics. Table 6-2, for example, gives assumed values of the Net Load Coefficient NLC (see section 5-10), normalized to the building floor area, in Btu per degree-day per square foot of building floor area for selected ranges of degree-days. From Table 5-5, it is noted that Denver has 6016 degree-days of heating. From Table 6-2, the Net Load Coefficient corresponding to 6016 DD is about 4.6 Btu/DD-ft². The house must be designed so that its thermal loss does not exceed 4.6 Btu/DD-ft², or the desired annual reduction of 60 percent will not be achieved.

The Thermal Storage Rule of Thumb.   A rule of thumb is also needed for making a rough estimate of the amount of thermal mass. The Los Alamos rule can be stated:

> A thermal storage mass of (0.6 × the percent annual reduction) lb of water, or (3 × the percent annual reduction) lb of concrete or masonry is recommended for each square foot of south-facing glazing.

This rule of thumb assumes that all of the thermal mass is in direct sun all day, as in a thermal storage wall. The mass predicted by the Los Alamos rule of thumb is also adequate thermal storage for a direct-gain application, provided the following conditions are met:

1. The mass is within the direct-gain space or encloses the direct-gain space.
2. The mass is not insulated from the direct-gain space.
3. The mass has an exposed surface area in the direct-gain space equal to three times the glazed area.

Masonry or concrete used in a direct-gain space is not considered effective beyond a depth of 4 to 6 in. from the surface. If thermal mass is located completely out of the sun, about three or four times the indicated amount would be required.

Table 6-3 lists recommended values of thermal storage mass to attain specified percentages of solar savings. It also shows that the amount of thermal storage depends upon the percentage of the building heat supplied by the solar system. Field experience reveals that for small values of heat load reduction (less

Table 6-3 Recommended Amounts of Effective Thermal Storage per Square Foot of Solar Collection Area for Stipulated Percentages of Expected Annual Reduction (Solar Savings)

Expected Annual Reduction (%)	Water $\left(D = 62.4 \frac{lb}{ft^3}\right)$		Concrete or Masonry $\left(D = 120 \frac{lb}{ft^3}\right)$	
	lb/ft^2 of Collection Area	ft^3/ft^2	lb/ft^2 of Collection Area	ft^3/ft^2
10	6	0.10	30	0.25
20	12	0.19	60	0.50
30	18	0.29	90	0.75
40	24	0.38	120	1.00
50	30	0.48	150	1.25
60	36	0.58	180	1.50
70	42	0.67	210	1.75
80	48	0.77	240	2.00
90	52	0.83	270	2.25

Source: Abstracted, with permission, from J. D. Balcomb, et al., *Passive Solar Design Handbook*, vol. II (Washington: U.S. Department of Energy, January 1980), p. 26.

than 30 percent), the solar heating will just about match the daytime heating requirements of the building, with little solar energy that can be or needs to be stored, and thus the amount of required storage is small. When the expected reduction is in the range of 30 to 70 percent, some of the solar energy collected during the day must be stored for use at night and possibly the next day. This daily storage and release of energy is the characteristic feature of passive applications. For expected reductions beyond 70 percent, some energy must be stored for several days, making the required volume of thermal mass quite large.

Illustrative Problem 6-8

For a 1600-ft² house with an expected solar savings of 60 percent, estimate the amount of thermal storage needed for 260 ft² of solar collection area. Assume that 100 ft² will be direct-gain windows with concrete or masonry storage, and the other 160 ft² is fixed glass with a water storage wall.

Solution

From Table 6-3, for a 60 percent solar savings, 36 lb of water (0.58 ft³) is needed per square foot of collection area, and 180 lb (1.5 ft³) of masonry per square foot of collection area. For the water wall, the amount of water needed is:

$$160 \text{ ft}^2 \times 36 \text{ lb/ft}^2 = 5760 \text{ lb water, or } 92.3 \text{ ft}^3 \text{ of water}$$

This is approximately 0.58 ft³ of water per square foot of solar collection area.

$$\left(\frac{92.3 \text{ ft}^3}{160 \text{ ft}^2} = 0.577 \frac{\text{ft}^3}{\text{ft}^2} \right)$$

For the direct-gain windows, the amount of concrete or masonry needed is

$$100 \text{ ft}^2 \times 180 \text{ lb/ft}^2 = 18,000 \text{ lb, or approximately } 150 \text{ ft}^3$$

This thermal storage mass can be obtained by using a 6-in.-thick slab or brick floor with an area equal to 300 ft² (three times the area of the direct-gain windows). The floor would be placed immediately behind the direct-gain windows so that the thermal storage would receive solar radiation for most of the day. For direct-gain systems, the thermal storage area should have a surface area of three times the aperture area. Hence, the 150 ft³ of masonry (or concrete) was made 6 in. thick with a surface area of 300 ft². *Ans.*

The Building Orientation Rule of Thumb.   It may not always be possible to orient the solar collection area toward due south, hence the following Los Alamos rule of thumb for orientation:

> For optimum solar gain, the orientation of the solar collection area should lie between 20° east and 32° west of true south.

If this rule is obeyed, the decrease in solar performance compared to that for a due south orientation will nearly always be less than 10 percent and typically will be less than 6 percent. In any case, the decrease in solar savings associated with orientations other than due south will be about as follows:

1. A 5 percent decrease at about 18° east or 30° west.
2. A 10 percent decrease at about 28° east or 40° west.
3. A 20 percent decrease at about 42° east or 54° west.

These values are the results of investigations conducted at the Los Alamos National Laboratory using weather data for the five cities of Albuquerque, New Mexico; Madison, Wisconsin; Medford, Oregon; Boston, Massachusetts; and Nashville, Tennessee.

The three rules of thumb that have just been discussed represent broad guidelines to be followed during the schematic design phase of a building project. Once the building has taken on a number of specific characteristics, an accurate quantitative estimate of solar performance should be made as the building evolves during the design development phase.

ESTIMATING SOLAR PERFORMANCE

Predicting the performance of a passive solar building is an important part of the design process. The rules of thumb just discussed are simple tools that are useful during the schematic design phase. However, as the project progresses into design development and on into the construction documents, more accurate methods are required. These methods must be able to predict with considerable precision the performance of different types of passive solar systems.

There are two approaches to estimating the performance of a passively heated building. First, the energy flows within the passive building can be simulated using existing computer programs available for purchase from a variety of sources.[6] Such analyses normally make use of hourly solar and weather data for a typical year and yield monthly and yearly performance estimates. Interior room temperatures, auxiliary heating requirements, thermal storage, and predictions of overheating are examples of information that can be obtained from such simulations. However, for a yearly analysis of a passive building, the time required to set up the building parameters,[7] the cost of computer time, and access to computer hardware and software are all barriers to the practical, everyday use of mainframe computer-simulated performance analyses. On most projects, solar designers need simpler techniques such as those involving handheld calculators or desktop computers, which can yield results in a short time at low cost.

Such methods have been developed, and they represent a second approach to evaluating passive solar heating performance. These methods are based on correlation techniques where the desired result (solar performance) is related to one or more derived parameters or coefficients, which are generally dimensionless.[8] Two of these correlation techniques are described in the remaining sections of this chapter.

6-9 The Solar Load Ratio Method

Researchers at the Los Alamos National Laboratory have recently developed the Solar Load Ratio (SLR) method for determining solar performance. This method utilizes a coefficient called the Solar Load Ratio (SLR), which relates the monthly net solar energy absorbed by the building to the monthly net building heating load. Solar Load Ratio methods have been applied extensively to passive systems by the Los Alamos National Laboratory research group.[9]

[6] These are generally programs that run on large mainframe computers. Some programs are available on time-sharing computer systems, which can be used from remote terminals. See "Analysis Methods for Solar Heating and Cooling Applications, Passive and Active Systems," Solar Energy Research Institute (SERI), Golden, Colo. SERI/SP-35-232R, August 1980. Other computer programs can be run on desktop or personal computers.

[7] Building parameters are mathematical coefficients used to represent the characteristics of the building. Thermal loss coefficients for different elements of a building (roof, wall, windows, doors, etc.) are some examples of thermal parameters.

[8] The *f-Chart* technique, developed at the University of Wisconsin, is an example of a performance analysis technique that uses two correlation parameters. The f-Chart analysis system was developed primarily for active solar systems.

[9] For more detailed information on SLR methods, see J. D. Balcomb, et al., *Passive Solar Design Handbook*, vol. II; and W. R. Jones, ed., *Passive Solar Design Handbook*, vol III (Washington: U.S. Department of Energy, January 1980 and July 1982, respectively).

6-10 Definitions of Terms Used in the Solar Load Ratio (SLR) Method

Before proceeding with an explanation of the Los Alamos Laboratory's SLR method, it is necessary to define with precision a number of measures related to solar performance. These definitions are collected and presented here for ready reference. They need not be memorized at this point, but with frequent use in the discussions and the design problems to follow, their meaning will become clear. (Some of the definitions are, with only slight adaptation, those proposed by Balcomb and Jones in *Passive Solar Design Handbook*, vols. II and III. See also section 5-10 of this book.)

1. *Building Load Coefficient (BLC)*—the building heat loss in Btu/F°-day or W-hr/C°-day caused by heat transmission through the building envelope and by infiltration of outside air into the building. (See Chapter 5 for methods of determining the BLC.)

2. *Net Load Coefficient (NLC)*—a term that has replaced Building Load Coefficient (BLC) in recent research to emphasize that the load is a net load, calculated by excluding gains and losses through the solar elements of the building. The units of NLC are also Btu/F°-day or W-hr/C°-day. The solar elements are assumed to be replaced by a thermally neutral wall.

3. *Auxiliary Heat (AH)*—heat delivered to the building by conventional space-heating equipment as a supplement to the solar heat received by the building. Same as backup heat.

4. *Net Reference Thermal Load (NRTL)*—the degree-day heating load of the nonsolar elements of the building, assuming a constant indoor temperature. This load is defined by the equation

$$NRTL = NLC \times DD \qquad (6\text{-}1)$$

   The units of NRTL are Btu, summed up for a stipulated time period. Expressed in another way, the NRTL is the conventional energy heat requirement for a comparable nonsolar building for a stipulated time period such as a month or a year. The NRTL concept (a) is based on the desired indoor temperature, or thermostat setting (the *reference* temperature); (b) assumes that the solar collection area (aperture) is replaced with a thermally neutral (adiabatic) wall, thus providing for a net load determination; and (c) does not take into account any internally generated heat. Figure 6-23 illustrates the exact meaning of NRTL.

5. *Degree-Days (DD)*—the differences between a fixed base temperature (usually 65°F) and a daily mean outdoor temperature, summed up for a stipulated period such as a month or a year. For example, if the mean outdoor temperature is 45°F for every day in a 30-day month, the monthly degree-days would be

$$DD_{month} = 30 \text{ days} \times 20 \text{ F}° = 600 \text{ DD}$$

6. *Solar Savings (SS)*—the reduction in heating requirements, in Btu per month (or Btu per year), resulting from the presence of passive solar collection elements. The term only has meaning when referred to the net reference thermal load (NRTL), as in this defining equation:

$$SS = NRTL - AH \qquad (6\text{-}2)$$

7. *Solar Load Ratio (SLR)*—

$$SLR = \frac{\text{Monthly net solar energy absorbed}}{\text{Monthly net reference thermal load}} \qquad (6\text{-}3)$$

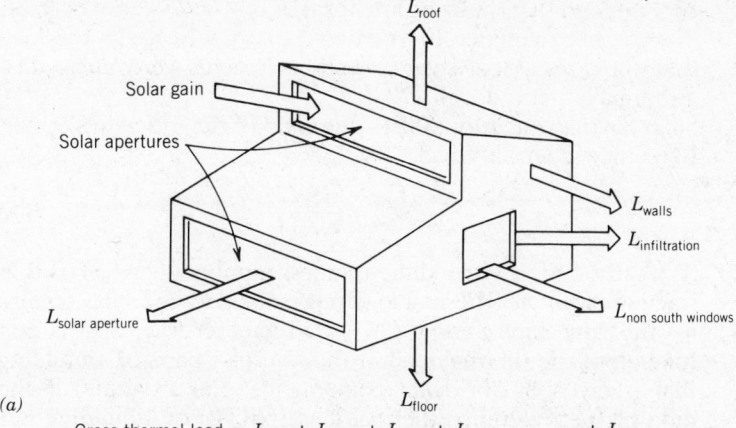

$$\text{Gross thermal load} = L_{\text{roof}} + L_{\text{walls}} + L_{\text{floor}} + L_{\text{non south windows}} + L_{\text{solar aperture}} + L_{\text{infiltration}} - \text{solar gain}$$

(a)

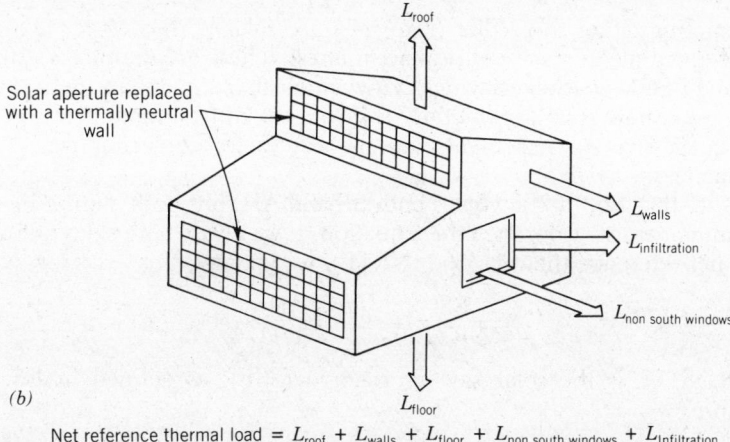

(b)

$$\text{Net reference thermal load} = L_{\text{roof}} + L_{\text{walls}} + L_{\text{floor}} + L_{\text{non south windows}} + L_{\text{Infiltration}}$$

Figure 6-23 A comparison of the Gross Thermal Load of a solar building with the Net Reference Thermal Load. (a) Thermal load elements for a typical solar house. Note that the heat loss and heat gain of the solar collection area enter into Gross Thermal Load. (b) The same solar house, but with the entire solar collection area replaced by a thermally neutral (adiabatic) wall. There is no heat gain and no heat loss through the solar collection area. The total heat loss under these conditions is termed the Net Reference Thermal Load (NRTL).

where:

Monthly net solar energy absorbed (Q_s) is the total solar radiation energy absorbed by the interior surfaces of the passive building for a particular month. It is the energy stored and available to provide part of the monthly heat required for space heating.
Monthly net reference thermal load (MNRTL) is the heating load summed up for a particular month and is calculated as follows:

$$\text{Monthly net reference thermal load} = \text{NLC} \times \text{DD}_{\text{month}} \quad (6\text{-}4)$$

Equation (6-3) can also be written

$$\text{SLR} = \frac{Q_s}{\text{MNRTL}} \quad (6\text{-}5)$$

Since both the numerator and the denominator of the ratio are in energy units, SLR itself is a dimensionless number. It must be kept in mind that, in using any form of Eqs. (6-3), (6-4), or (6-5), it is the net reference thermal load that is involved—the load that would result if all solar collection elements were replaced by a thermally neutral wall.

8. *Solar Savings Fraction* (SSF)—the ratio of Solar Savings to Net Reference Thermal Load,

$$\text{SSF} = \frac{\text{SS}}{\text{NRTL}} \qquad (6\text{-}6)$$

As a ratio, SSF is also a dimensionless number. The SSF is that fraction of the net thermal load represented by the solar savings for the same time period. Or, stated another way, SSF is that fraction of the thermal load of the nonsolar parts of a building that is saved by the solar components. For example, if the monthly net reference thermal load (NRTL) of a building is 8 million Btu and the solar savings (SS) is 4 million Btu for the month, then the solar savings fraction (SSF) is 1/2 or 0.50.

In attempting either to predict or to measure solar performance, two of the above-defined measures are of primary interest. These are number 3 (Auxiliary Heat) and number 6 (Solar Savings). How much auxiliary or backup heat will be required and how great a reduction or savings of conventional heat will result from the passive solar elements are the key factors of economics in solar performance.

Note that Eq. (6-2) involves both SS and AH, but both cannot be evaluated simultaneously from the same equation. If we divide both sides of Eq. (6-2) by the net reference thermal load (NRTL), we obtain

$$\frac{\text{SS}}{\text{NRTL}} = 1 - \frac{\text{AH}}{\text{NRTL}}$$

But SS/NRTL is the solar savings fraction (SSF), as defined in Eq. (6-6). Therefore,

$$\text{SSF} = 1 - \frac{\text{AH}}{\text{NRTL}}$$

And, from Eq. (6-1),

$$\text{SSF} = 1 - \frac{\text{AH}}{(\text{NLC})(\text{DD})} \qquad (6\text{-}7)$$

Now, solving Eq. (6-7) for AH gives

$$\text{AH} = \text{NLC} \times \text{DD} \times (1 - \text{SSF}) \qquad (6\text{-}8)$$

which expresses auxiliary heat (AH) in terms of structure-related (NLC) and weather-dictated (DD) data, and also in terms of the parameter SSF.

In like manner, from Eq. (6-6),

$$\text{SS} = \text{NRTL} \times \text{SSF}$$

And substituting from Eq. (6-1),

$$\text{SS} = \text{NLC} \times \text{DD} \times \text{SSF} \qquad (6\text{-}9)$$

With Eqs. (6-8) and (6-9), we now have both Solar Savings (SS) and Auxiliary Heat (AH) expressed in terms of readily determinable data (NLC and DD) for any building and location, and in terms of the parameter SSF.

6-11 Estimating Solar Performance with the Solar Load Ratio Method

Based on the definitions and equations presented above, this section discusses the monthly solar load ratio method as a tool for estimating the solar performance of passive solar buildings. The monthly SLR method is most applicable during the final stages of the building design, when an assessment of the building performance is required to finalize the design by comparing the solar performance of different options under consideration for the passive solar system. The SLR method does not replace the rules of thumb that have been discussed in section 6-8, but it works with them as a tool to help the solar designer to resolve the several possible approaches considered during the schematic design phase and to arrive at a final decision for the passive solar system.

The monthly SLR method measures the performance of a passive solar building in terms of Solar Savings for the building, given by Eq. (6-2). The Solar Savings is referenced to the Net Reference Thermal Load [from Eq. (6-2)] of the building. The NRTL is the degree-day heating load of the nonsolar building elements and is the reference base for which values of Solar Savings and Auxiliary Heat are derived for different types of passive solar systems. A solar designer can compare solar performances of the same building when different passive elements are used in the solar collection area, such as a direct-gain system, a thermal storage mass wall, or an attached greenhouse. Also, the performance of mixed passive solar sytems can be evaluated, as in the case in which part of the solar collection area is a thermal storage mass wall and part is an attached greenhouse (see Chapter 9).

The SLR method is a correlation method, as previously mentioned. The SLR has been correlated to the solar savings fraction based on extensive research and computer analysis done at the Los Alamos National Laboratory. The SLR is the variable that has climate information and the thermal characteristics of the building contained within it.

The solar load ratio was given by Eq. (6-5) as

$$\text{SLR} = \frac{Q_s}{\text{MNRTL}}$$

The monthly net solar energy absorbed (Q_s) can be defined as

$$Q_s = S \times A_p \qquad (6\text{-}10)$$

where

S = the monthly net solar radiation absorbed in the passive
    solar building per unit of projected collection area, Btu/month-ft^2
A_p = the projected vertical plane area of the solar collection
    area, square feet

If the collection area is not vertical, but tilted at an angle less than 90° to the horizontal plane, such as a sloping glass roof or skylight, A_p will be less than the actual collection area, as shown in Fig. 6-24.

The value of A_p is calculated from

$$A_p = \sin\theta \times \text{solar collection area} \qquad (6\text{-}11)$$

where

θ is the tilt angle of the solar aperture plane measured
    from the horizontal, and solar collection area is the actual
    area of the solar glazing, square feet

Using Eqs. (6-4) and (6-10), the solar load ratio can be written as

$$\text{SLR} = \frac{Q_s}{\text{MNRTL}} = \frac{S \times A_p}{\text{DD}_{\text{month}} \times \text{NLC}} = \frac{S}{\text{DD}_{\text{month}}} \times \frac{A_p}{\text{NLC}} \qquad (6\text{-}12)$$

The solar load ratio is shown as the product of two fractions $S/\text{DD}_{\text{month}}$ and A_p/NLC. The first fraction is the ratio of the monthly net solar radiation absorbed in the building per unit of projected collection area to the monthly heating degree-days. This ratio contains the information relating to the local climate. For each building location, these values will be different because they depend on the local climate conditions.

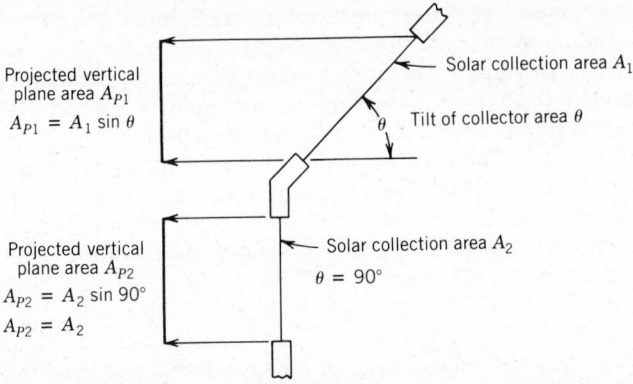

Projected vertical plane area A_{P1}

$A_{P1} = A_1 \sin\theta$

Solar collection area A_1

Tilt of collector area θ

Projected vertical plane area A_{P2}

$A_{P2} = A_2 \sin 90°$

$A_{P2} = A_2$

Solar collection area A_2

$\theta = 90°$

Figure 6-24 The relationship between net glazing area (solar aperture) and projected area, A_p. $A_p = A \sin\theta$, where θ is the angle between the horizontal and the plane of the glazing.

The second fraction is the ratio of the projected vertical plane area of the solar collection area to the net load coefficient of the passive solar building. This fraction is dependent on the design and construction of the building, because the Net Load Coefficient is determined by the thermal losses from the building excluding the solar collection area. The projected vertical plane area of the solar collection area is determined by the size of the passive solar collection area and by the tilt angle.

The Solar Load Ratio correlations prepared from research conducted at Los Alamos make use of the two fractional parts of the Solar Load Ratio. The solar savings fraction (SSF) is plotted along the vertical axis, and the ratio (S/DD_{month}) is plotted along the horizontal axis as shown in Fig. 6-25. A group of curves results, one for each of several different values of the ratio NLC/A_p, which is the reciprocal of the second fractional part of the Solar Load Ratio. The ratio NLC/A_p has been given the name *Load Collector Ratio* (LCR) because it is the ratio of the Net Load Coefficient to the projected vertical plane area of the solar collection area, A_p. The Load Collector Ratio can then be written as

$$LCR = \frac{NLC}{A_p} \qquad (6\text{-}13)$$

where

LCR = the Load Collector Ratio, Btu/DD-ft^2 of projected collection area
NLC = the Net Load Coefficient for the building, Btu/DD
A_p = the projected vertical plane area of the solar collection area, square feet

The Solar Load Ratio can now be written in the following form, from Eq. (6-12)

$$SLR = \frac{S}{DD_{month}} \times \frac{1}{LCR} \qquad (6\text{-}14)$$

Each of the curves plotted in Fig. 6-25 corresponds to a specific value of the ratio NLC/A_p (i.e., LCR, see LCR scale at left of curves). These curves allow the monthly solar savings fraction to be determined by using the two factors that comprise the Solar Load Ratio, S/DD_{month} and LCR. The curves shown in Fig. 6-25 are for a particular direct-gain solar system with the following characteristics:

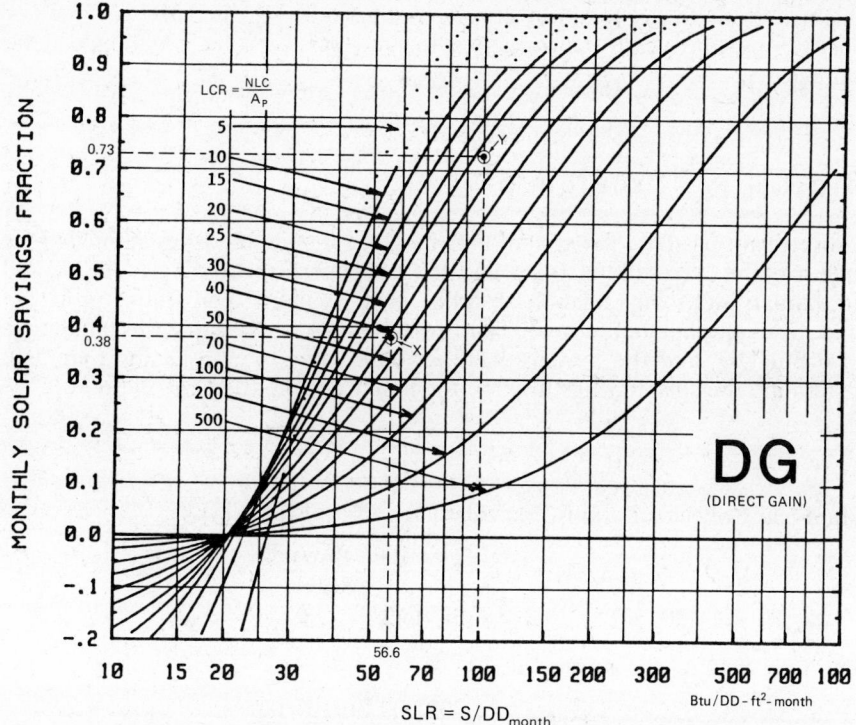

Figure 6-25 Correlation curves for one type of passive solar element or system. The correlation parameters are the Solar Load Ratio (SLR) (plotted along the horizontal axis), and the Solar Savings Fraction (SSF) (plotted along the vertical axis). The curves themselves are related to different values of the Load Collector Ratio (LCR). These curves are for a direct-gain system. *Source:* Reprinted by permission from Balcomb, J. D., et al, *Passive Solar Design Handbook*, vol. II, U.S. Dept. of Energy, Jan. 1980, p. 148, Fig. F1.

1. The thermal mass is either concrete or masonry with a thickness of about 6 in.
2. The exposed surface area of the thermal mass within the building is about three times the projected aperture area.
3. The aperture area is assumed to be double-glazed.

This set of curves should not be used if conditions vary appreciably from those stated here.[10] These curves are independent of climate and can be used for any location where monthly weather data exist from which values of monthly net solar radiation absorbed per unit of projected area (S) and monthly degree-days can be determined. To make use of the SLR correlation curves for determining the monthly solar savings fraction (SSF_M), the ratio NLC/A_p (Load Collector Ratio) must be calculated for the particular passive system. Then the ratio S/DD_{month} must be calculated for the specific month in question. The monthly solar savings fraction (SSF_M) is then found using the calculated values of LCR and S/DD_{month} from SLR correlation curves such as those in Fig. 6-25.

Then the monthly auxiliary heat (AH) and monthly solar savings (SS) can be calculated from Eqs. (6-8) and (6-9) respectively, using the value of SSF

[10] Correlation plots like Fig. 6-25 are provided for many passive solar configurations in J. D. Balcomb, et al., *Passive Solar Design Handbook*, vol. II (Washington: U.S. Department of Energy, January 1980); and R. W. Jones, ed., *Passive Solar Design Handbook*, vol. III (Washington: U.S. Department of Energy, July 1982). Plots are provided for different configurations of direct-gain systems, thermal storage walls, and sunspaces. The utilization of these correlation plots will be discussed further in Chapter 9.

obtained from the correlation curves of Fig. 6-25. By making 12 monthly calculations, the yearly performance for the passive solar building can be estimated.

Illustrative Problem 6-9

A direct-gain passive solar system has the general characteristics stipulated for the use of the curves in Fig. 6-25. The projected aperture area (A_p) is 200 ft^2, and the January net solar radiation absorbed in the building per unit of projected area (S) was calculated to be 30,000 Btu/ft^2-Jan. The NLC for the building is 8000 Btu/DD, and the January heating degree-days for the location are 530. Determine the solar savings fraction for January using the SLR method.

Solution

The Load Collector Ratio is first calculated using Eq. (6-13):

$$\text{LCR} = \frac{\text{NLC}}{A_p} = \frac{8000 \text{ Btu/DD}}{200 \text{ ft}^2}$$
$$= 40 \text{ Btu/DD-ft}^2$$

The ratio $\text{S/DD}_{\text{month}}$ is calculated as

$$\text{S/DD}_{\text{month}} = \frac{30,000 \text{ Btu/ft}^2\text{-Jan}}{530 \text{ DD/Jan}}$$
$$= 56.6 \text{ Btu/DD-ft}^2$$

Fig. 6-25 is entered at $\text{S/DD}_{\text{month}}$ equal to 56.6 Btu/DD-ft^2. A vertical line is drawn upward until it intersects the LCR = 40 curve at X. A horizontal line from X to the SSF (vertical) axis indicates a monthly solar savings fraction,

$$\text{SSF}_{\text{Jan}} = 0.38 \hspace{3cm} \textit{Ans.}$$

Illustrative Problem 6-10

The October net solar radiation absorbed per unit of projected area (S) is calculated to be 25,000 Btu/ft^2 for a direct-gain passive building. The projected aperture area (A_p) is 500 ft^2. The October heating degree-days are 250. The building has a calculated *NLC* of 17,000 Btu/DD. Determine the solar savings fraction for October using the SLR method.

Solution

The Load Collector Ratio is calculated using Eq. (6-13) as

$$\text{LCR} = \frac{\text{NLC}}{A_p} = \frac{17,000 \text{ Btu/DD}}{500 \text{ ft}^2}$$
$$= 34 \text{ Btu/DD-ft}^2$$

The ratio $\text{S/DD}_{\text{month}}$ is calculated as

$$\text{S/DD}_{\text{month}} = \frac{25,000 \text{ Btu/ft}^2\text{-Oct}}{250 \text{ DD/Oct}}$$
$$= 100 \text{ Btu/DD-ft}^2$$

Fig. 6-25 is entered at $\text{S/DD}_{\text{month}}$ = 100 Btu/DD-ft^2. Then point Y is located on the curve corresponding to an LCR of 34 Btu/DD-ft^2. From point Y, the indicated monthly Solar Savings Fraction,

$$\text{SSF}_{\text{Oct}} = 0.73 \hspace{3cm} \textit{Ans.}$$

Illustrative Problem 6-11

Determine the February Solar Savings (SS) and Auxiliary Heat (AH) for a building that has a February Solar Savings Fraction of 0.63. The NLC for the building is 25,000 Btu/DD, and the February heating degree-days are 650.

Solution

The Solar Savings can be calculated using Eq. (6-9) as:

$$
\begin{aligned}
SS &= NLC \times DD \times SSF \\
&= 25{,}000 \text{ Btu/DD} \times 650 \text{ DD} \times 0.63 \\
&= 10.2 \times 10^6 \text{ Btu} \qquad\qquad\qquad \textit{Ans.}
\end{aligned}
$$

The Auxiliary Heat can be calculated using Eq. (6-8) as:

$$
\begin{aligned}
AH &= NLC \times DD \times (1 - SSF) \\
&= 25{,}000 \text{ Btu/DD} \times 650 \text{ DD} \times (1 - 0.63) \\
&= 6.0 \times 10^6 \text{ Btu} \qquad\qquad\qquad \textit{Ans.}
\end{aligned}
$$

Illustrative Problem 6-12

The January Solar Savings Fraction (SSF_{Jan}) was calculated to be 0.58 for a direct-gain passive solar building. The January NRTL was calculated as 11.2×10^6 Btu. Determine the Solar Savings (SS) and the Auxiliary Heat (AH) for January.

Solution

By rearranging Eq. (6-6), we obtain

$$
\begin{aligned}
SS_{Jan} &= SSF \times NRTL \\
&= 0.58 \times 11.2 \times 10^6 \text{ Btu} \\
&= 6.5 \times 10^6 \text{ Btu} \qquad\qquad\qquad \textit{Ans.}
\end{aligned}
$$

From Eqs. (6-8) and (6-1),

$$
\begin{aligned}
AH_{Jan} &= NLC \times DD \times (1 - SSF) \\
&= NRTL \times (1 - SSF) \\
AH_{Jan} &= 11.2 \times 10^6 \text{ Btu} \times (1 - 0.58) \\
&= 4.7 \times 10^6 \text{ Btu} \qquad\qquad\qquad \textit{Ans.}
\end{aligned}
$$

To complete a monthly SLR calculation, the SSF, the monthly net reference thermal load MNRTL, SS, and AH must be calculated for each month of the heating season. An example of a typical layout is shown in Table 6-4. The *annual* Solar Savings and *annual* Auxiliary Heat are obtained by adding the monthly values.

In summary, then, there are six steps to the monthly SLR method:

1. Calculation of the parameters that are related to the building—the net load coefficient (NLC), the projected aperture area (A_p), and the Load Collector Ratio (LCR).
2. Determination of local climate information, the monthly degree-days (DD_{month}), and the monthly insolation on the collection area (Btu/ft²-month).
3. Determination of the monthly net solar radiation absorbed in the passive solar building per unit of projected collection area (S). (Methods for this calculation will be given in Chapter 9.)
4. Determination of the monthly solar savings fraction for the

Table 6-4 Results of a Sample Monthly SLR Calculation—Illustrating a Suggested Format

Heating Month	SSF	NRTL	SS	AH
Oct	1.0	1.5×10^6 Btu	1.5×10^6 Btu	0.0×10^6 Btu
Nov	.83	6.8	5.64	1.16
Dec	.65	10.2	6.63	3.57
Jan	.53	11.0	5.83	5.17
Feb	.69	9.8	6.76	3.04
Mar	.81	5.9	4.78	1.12
Apr	.93	1.1	1.02	0.08
Aggregate annual		46.3×10^6	32.16×10^6	14.14×10^6

Annual solar savings = 32.16×10^6 Btu
Annual auxiliary heat = 14.14×10^6 Btu.

---

      particular passive solar system type from curves like those of Fig. 6-25.

5. Evaluation of the Solar Savings (SS) and Auxiliary Heat (AH) for each month from Eqs. (6-8) and (6-9).

6. Summing up the monthly values to obtain annual values of Solar Savings and Auxiliary Heat.

The monthly SLR method allows a considerable degree of flexibility in evaluating passive system performance. Structures can have several solar collection units, each with its own area, orientation, passive system type, and night insulation. The shading effect of overhangs can also be included. A disadvantage of the monthly SLR method is that all of the calculations must be completed for each month of the heating season. Further details of the monthly SLR method will be discussed in Chapter 9 after a thorough discussion of the most common types of passive solar systems is presented in Chapters 7 and 8.

6-12 Annual Load Collector Ratio Method

The monthly SLR method requires that the monthly Solar Savings for each month of the heating season be calculated before an annual value for the Solar Savings or Auxiliary Heat can be obtained. This requires several sets of calculations for each month. The researchers at the Los Alamos National Laboratory have therefore proposed the annual Load Collector Ratio (LCR) method. This method involves having the monthly calculations of the SLR method precomputed for a given geographic location. The annual solar savings fraction can then be estimated from the Load Collector Ratio that has been calculated for the passive solar building. Extensive tables have been prepared which provide directly the annual values of the solar savings fraction for many different types of passive solar systems and for more than 200 U.S. and Canadian locations.[11] Table 6-5 is an excerpt from these tables, and it gives annual values of the solar savings fraction (SSF) for various values of the LCR for six different types of passive systems. Only three locations are provided here—Knoxville, Tennessee; Albany, New York; and Salt Lake City, Utah. For other locations, the complete Los Alamos tables must be consulted. The annual SSF is determined from the tables by using the LCR as a parameter.

    When the annual value of the SSF for a passive solar system is determined from prepared tables such as Table 6-5, using the LCR as a parameter, the

---

[11] W. R. Jones, ed., *Passive Solar Design Handbook*, vol. III (Washington: U.S. Department of Energy, July 1982), Appendix F, provides data for 94 different passive solar systems and 219 locations in the United States and Canada.

Table 6-5 Solar Savings Fraction (SSF) Versus Load Collector Ratio (LCR)

Knoxville, Tennessee				*Degree-Days—3478*					
Solar Savings Fraction:	0.1	0.2	0.3	0.4	0.5	0.6	0.7	0.8	0.9
Selected Types of Passive Elements	*Load Collector Ratios*								
Thermal waterwall[a]	195	90	53	35	24	17	12	9	6
Thermal waterwall, night insulation[b]	190	114	77	55	40	30	23	16	11
Masonry or concrete thermal wall[c]	140	65	38	25	18	12	9	6	4
Masonry or concrete thermal wall, night insulation	162	92	60	42	31	23	17	12	8
Direct-gain[d]	127	50	23	—	—	—	—	—	—
Direct-gain, night insulation	198	91	55	38	27	20	14	9	4

Albany, New York				*Degree-Days—6888*					
Solar Savings Fraction:	0.1	0.2	0.3	0.4	0.5	0.6	0.7	0.8	0.9
Selected Types of Passive Elements	*Load Collector Ratios*								
Thermal waterwall[a]	61	22	9	—	—	—	—	—	—
Thermal waterwall, night insulation[b]	89	52	34	24	17	12	9	6	4
Masonry or concrete thermal wall[c]	43	15	6	—	—	—	—	—	—
Masonry or concrete thermal wall, night insulation	75	41	26	18	13	9	7	5	3
Direct-gain[d]	—	—	—	—	—	—	—	—	—
Direct-gain, night insulation	96	42	24	16	11	7	4	—	—

Salt Lake City, Utah				*Degree-Days—6137*					
Solar Savings Fraction:	0.1	0.2	0.3	0.4	0.5	0.6	0.7	0.8	0.9
Selected Types of Passive Elements	*Load Collector Ratios*								
Thermal waterwall[a]	198	92	54	35	24	17	12	8	5
Thermal waterwall, night insulation[b]	193	116	78	55	41	30	22	16	11
Masonry or concrete thermal wall[c]	143	66	39	26	18	12	9	6	4
Masonry or concrete thermal wall, night insulation	164	94	61	43	31	23	17	12	8
Direct-gain[d]	131	52	24	—	—	—	—	—	—
Direct-gain, night insulation	200	92	56	39	28	20	14	9	4

Source: Adapted with permission from R. W. Jones, ed., *Passive Solar Design Handbook*, vol. III (Washington: U.S. Department of Energy, July 1982), pp. 440, 475, 494.

[a] Thermal waterwall has a thermal storage capacity of 46.8 Btu/ft² glazing-F° and is double-glazed.

[b] The night insulation used has an insulating value of R-9.

[c] Masonry or concrete thermal wall has a thermal storage capacity of 30 Btu/ft² glazing-F° and is directly behind a double-glazed solar aperture.

[d] Direct-gain has a thermal storage capacity of 30 Btu/ft² glazing-F°, and the solar aperture is double-glazed. The thermal mass has an exposed surface area six times the glazed area.

technique is referred to as the annual LCR method. Once the annual value of SSF is obtained from the tables, then the annual Solar Savings and the annual Auxiliary Heat can be calculated using Eqs. (6-8) and (6-9).

In summary, these are the three steps in estimating the annual solar performance of a passive system using the annual LCR method:

1. Determine the net load coefficient (NLC) of the building from the methods explained in Chapter 5.
2. Calculate the Load Collector Ratio (LCR) using Eqs. (6-11) and (6-13).
3. From an LCR-versus-SSF table (like Table 6-5), find the solar savings fraction (SSF) for the type of passive solar system and geographic location that corresponds to the calculated value of the LCR. The annual Auxiliary Heat (AH) can then be calculated using Eq. (6-8), or the annual Solar Savings (SS), using Eq. (6-9).

ESTIMATING SOLAR PERFORMANCE

Illustrative Problem 6-13

An architect would like to know the annual Auxiliary Heat (AH) and the Solar Savings (SS) for a proposed building in Knoxville, Tennessee. The architect has estimated the net load coefficient (NLC) for the nonsolar elements of the building at 9000 Btu/DD. The thermal storage wall is 12-in.-thick concrete behind double glazing, with no night insulation used. The vertical projected solar collection area A_p is 360 ft².

Solution

The Load Collector Ratio is first calculated, using Eq. (6-13):

$$LCR = \frac{NLC}{A_p}$$
$$= \frac{9000 \text{ Btu/DD}}{360 \text{ ft}^2}$$
$$= 25 \text{ Btu/DD-ft}^2$$

This value of the LCR is located in the Knoxville section of Table 6-5 for the masonry or concrete thermal wall. The corresponding value of the solar savings fraction (SSF) is found to be 0.4. The annual degree-days for Knoxville are given in Table 6-5 as 3478. The Auxiliary Heat is then calculated from Eq. (6-8) as

$$AH = NLC \times DD \times (1 - SSF)$$
$$= 9000 \text{ Btu/DD} \times 3478 \text{ DD/yr} \times (1 - 0.4)$$
$$= 18.8 \times 10^6 \text{ Btu/yr} \qquad \qquad Ans.$$

The Solar Savings is calculated from Eq. (6-9) as

$$SS = NLC \times DD \times SSF$$
$$= 9000 \text{ Btu/DD} \times 3478 \text{ DD/yr} \times 0.4$$
$$= 12.5 \times 10^6 \text{ Btu/yr} \qquad \qquad Ans.$$

The annual LCR method gives the solar designer a relatively quick method of estimating the annual performance of a passive solar system. It is a very useful tool during the design development phase, since the annual performance of different design variations can be estimated with a fair degree of precision. However, there are limitations to the annual LCR method that should be pointed out. For instance, the tables assume that the solar collection area is oriented due south and has unobstructed solar access. Thus, values obtained for orientations other than south or for cases of limited solar access would have to be used with caution. Calculations are also limited to the geographic locations and the types of direct-gain systems provided in the tables. Additional examples of SLR and LCR performance analysis will be presented in later chapters.

6-13   A Design Exercise

An architect has requested assistance in the development of a design for a solar house for a location near Albany, New York.

Schematic Design Phase   The property has two potential building sites that must be evaluated for solar access. Figure 6-26 indicates the results of a solar access study. The trees at building site A will limit the solar access throughout the winter. At site B, the building to the south will cause some afternoon shading during a few days close to the winter solstice, but after that time there is almost unobstructed solar access except for a tree in the east, the shading effects of which occur only in early morning hours. Site B will have the most hours of winter sun available and is recommended as the building site as far as solar access is concerned. The solar collection area will face due south.

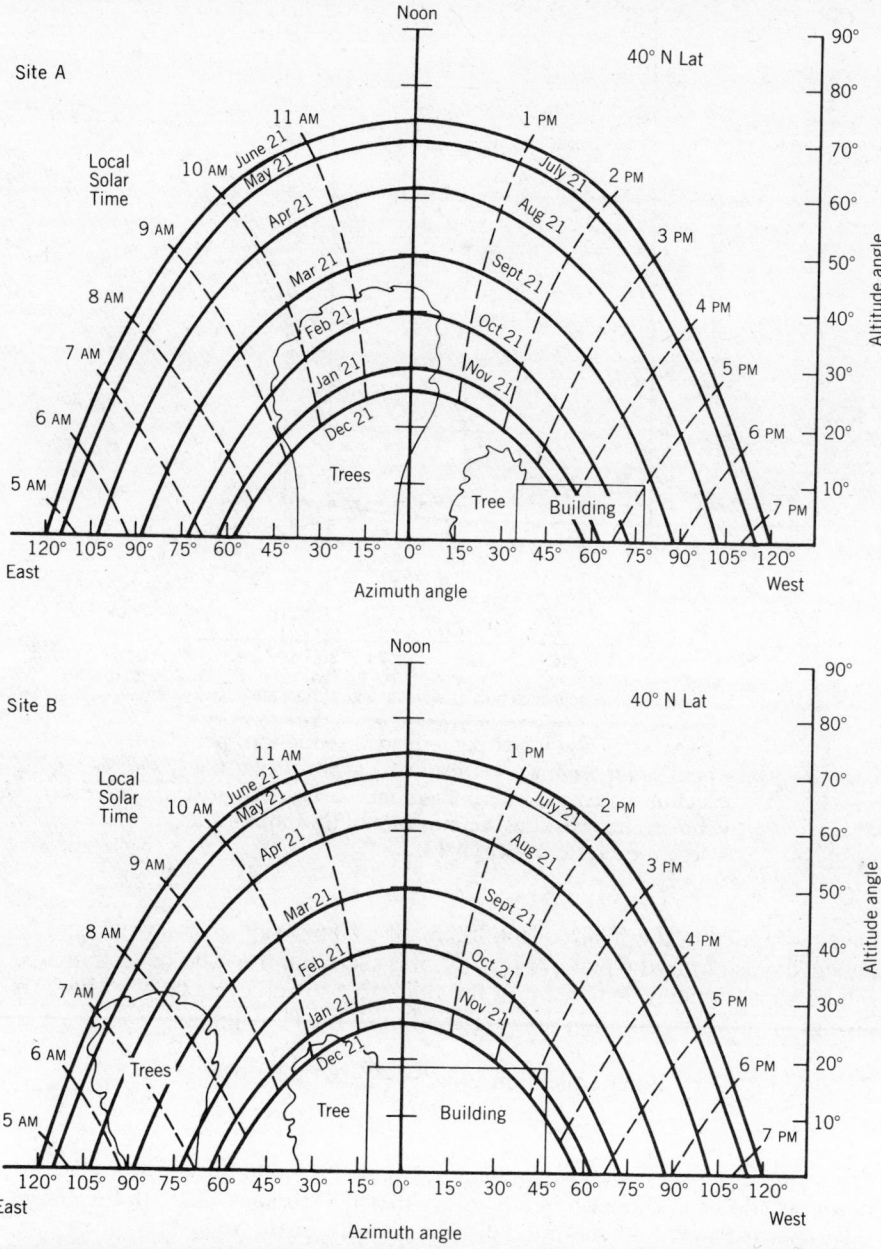

Figure 6-26 Plotted results for determining solar access for the building site in the design exercise, section 6-13.

Applying the Collection Area Rule of Thumb   The designer has decided on an approximate floor area of 1800 ft² for the house. The required solar collection area for a 60 percent solar savings will be determined.

The solar collection area rule of thumb will be used to estimate the collection area. From Table 6-1, the values of $R1$ and $R2$ and $S1$, $S2$, $S3$, and $S4$ are found for Albany.

$R1 = 21\%$ $S1 = 13\%$
$R2 = 41\%$ $S2 = 15\%$
              $S3 = 43\%$
              $S4 = 66\%$

It is often helpful to plot such values as shown in Fig. 6-27 as an aid to interpreting the information. From the plot, it is clear that night insulation will

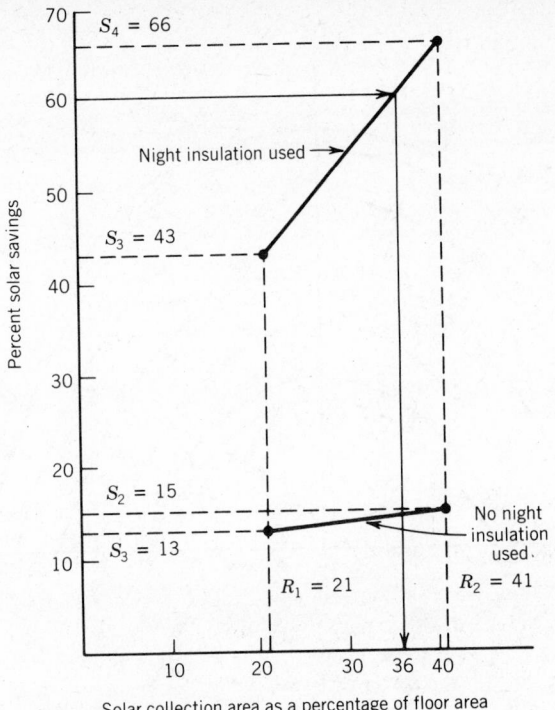

Figure 6-27 Plot of percent load reduction (percent Solar Savings, vertical axis) against solar collection area as a percentage of floor area (horizontal axis, Los Alamos rule of thumb) for the design exercise of Sec. 6-13.

have to be used in the solar collection area if a 60 percent solar savings is to be obtained. Also from the plot of Fig. 6-27, it is seen that the solar collection area will be approximately 36 percent of the floor area to achieve a 60 percent solar savings. The solar collection area for the house is calculated as

$$\text{Solar collection area} = \frac{36}{100} \times 1800 \text{ ft}^2$$
$$= 648 \text{ ft}^2$$

After some discussion with the designer, it is decided that the south wall of the house should be in the form of a masonry thermal storage wall, with the glazed solar aperture of 648 ft² mounted directly in front of the wall.

Design Development Phase   The designer has developed a set of preliminary floor plans and elevations for the house and now needs an estimate of the solar performance. The NLC for the nonsolar elements of the building has been calculated to be 8280 Btu/DD. The masonry thermal wall is a 12-in.-thick concrete storage wall with double glass in front. There will be night insulation with a thermal resistance of R-9. The actual area of the solar aperture as shown on the drawings is 650 ft². It is all in a vertical plane.

Calculating the SSF   The Load Collector Ratio is calculated from Eq. (6-13) as

$$\text{LCR} = \frac{\text{NLC}}{A_p} = \frac{8280 \text{ Btu/DD}}{650 \text{ ft}^2} = 12.7 \text{ Btu/ft}^2\text{-DD}$$

From Table 6-5 for Albany for a masonry thermal storage wall with night insulation, we find that an LCR of 13 gives an SSF of 0.5, and an LCR of 9 gives an SSF of 0.6. These values must be interpolated as follows: 12.7 represents a value that is 8 percent of the interval from 13 to 9, so the solar savings fraction (SSF) corresponding to 12.7 will be 8 percent of the interval from 0.5 to 0.6, or 0.508. The degree-days for Albany are 6888 DD/year (Table 6-5). The Solar Savings is then calculated, using Eq. (6-9):

$$
\begin{aligned}
SS &= NLC \times DD \times SSF \\
&= 8280 \text{ Btu/DD} \times 6888 \text{ DD/yr} \times 0.508 \\
&= 29.0 \times 10^6 \text{ Btu/yr}
\end{aligned}
$$
Ans.

The Auxiliary Heat is calculated using Eq. (6-8):

$$
\begin{aligned}
AH &= NLC \times DD \times (1 - SSF) \\
&= 8280 \text{ Btu/DD} \times 6888 \text{ DD/yr} \times (1 - 0.508) \\
&= 28.0 \times 10^6 \text{ Btu/yr}
\end{aligned}
$$
Ans.

These calculations indicate that approximately one half of the annual building heating load requirement will be met by the passive solar design.

At this point, the architect presents the plans and solar calculations to the client and it is decided that working drawings will be started after a few minor changes are made. The owners want more south windows and a small attached greenhouse on the south side of the house. These changes will be incorporated into a design example in Chapter 9.

PROBLEMS

1. Determine the relative humidity (RH) and the effective temperature (ET) of the air if measurements give a dry-bulb temperature of 75°F and a wet-bulb temperature of 50°F. (Use Fig. 6-13.)

2. What percentage of building occupants would probably feel comfortable with an effective temperature of 65°F (a) in winter? (b) in summer?

3. The Bioclimatic Chart includes the cooling effects of air velocity and evaporation of moisture. For a dry-bulb (DB) temperature of 85°F and a relative humidity (RH) of 50 percent, determine the necessary air velocity to produce a feeling of summer comfort. (Use Fig. 6-14a.)

4. For an indoor condition of 65°F DB, RH 40 percent, suggest the steps that could be taken to achieve a feeling of winter comfort. (Consult Fig. 6-14b.)

5. For a relative humidity of 40 percent, what is the DB temperature at which about 90 percent of people present would feel comfortable in winter?

6. For a person sitting in a sunspace with the relative humidity at 60 percent and the DB temperature at 60°F, what is the rate of radiation that should be absorbed for a feeling of winter comfort?

7. From Fig. 6-19, determine the total hours of available solar access at the site on January 21 and April 21.

8. For a residence being designed for Little Rock, Arkansas, determine the approximate percentage of floor area required in the solar collection area to provide a 62 percent reduction in the annual heating load (Solar Savings). Use Table 6-1, and assume that night insulation will be used.

9. For a Chicago residence of 1800 ft² to achieve an approximate reduction in the annual heating load (SS) of 43 percent, what must be the approximate solar collection area, if night insulation is used?

10. A passive solar system is designed to provide a 30 percent Solar Savings. What total quantity of water (gallons) should be used if the water is contained in a thermal-storage wall directly behind the glazing?

11. A passive solar system is designed to provide an 80 percent reduction in the annual heating load. What quantity of concrete or masonry should be used per square foot of collection area for the thermal storage?

12. If a solar building has an orientation that is between 30° and 40° west of south, what will be the expected decrease in the system performance compared to a due south orientation?

13. A solar collection area has an area of 250 ft² and is tilted at an angle of 60° from the horizontal. Determine the vertical projected area, A_p.

14. A passive solar building has a projected area A_p of 300 ft² and a net load coefficient NLC of 7000 Btu/DD. Determine the Load Collector Ratio (LCR) for the building.

15. Calculate the Auxiliary Heat (AH) and the Solar Savings (SS) for a passive solar building if the November SSF is 0.65, the NLC is 11,500 Btu/DD, and the November degree-days are 650.

16. Determine the April SSF for a direct-gain passive system with characteristics that correspond to those assumed for Fig. (6-25). The aperture area is 350 ft², the April net solar radiation absorbed in the building per unit of projected area (S) has been calculated as 35,000 Btu/ft²$_{\text{April}}$, the building NLC is 14,000 Btu/DD, and the April heating degree-days for the location are 420.

17. Estimate the annual auxiliary heat (AH) and the annual solar savings (SS) by the Annual Load Collector Ratio method for a passive solar building using a thermal water wall. The location is Albany, New York. The solar collection area has a projected area of 450 ft², and the NLC of the building is 9500 Btu/DD. The water wall is fitted with R-9 night insulation.

18. A passive solar building has an attached solar greenhouse as a passive heating element. The greenhouse has 300 ft² of vertical glazing and 200 ft² of glazing tilted at an angle of 40° from the horizontal, all facing south. Calculate the total projected aperture area for the greenhouse.

Direct-Gain Passive Systems

*I*n the preceding chapter, passive solar energy systems were defined as those systems in which the structure itself is the solar collector, and thermal energy flows involving radiation, conduction, and convection take place naturally without mechanical energy input. Further, it was indicated that all passive solar energy systems have two basic elements: the solar collection area (south-facing glazing) to allow solar energy to enter the space, and thermal mass for absorption, storage, and distribution of heat energy. The arrangement of these two elements and their relationship to the living space determines the type or classification of the passive system. Five system classifications were identified: direct gain, indirect gain, isolated gain, convective loop, and hybrid systems. This chapter will deal with direct-gain systems, the simplest approach to passive solar heating.

7-1 Essential Features of Direct-Gain Systems

In a direct-gain system, the living space itself acts as the solar collector, and the walls, floor, and contents of the living space provide the thermal storage mass. Sunlight is admitted into the space through south-facing glazing, and concrete, masonry, or water thermal storage walls, along with the floor of the living space itself, all intercept solar radiation directly and also absorb reflected or reradiated energy. Simply stated, the living space in a direct-gain system is the heat collector, thermal storage, and distribution system all in one.

Figure 7-1 illustrates an interior water wall that stores heat during the day and releases heat by convection and radiation into the living space at night.

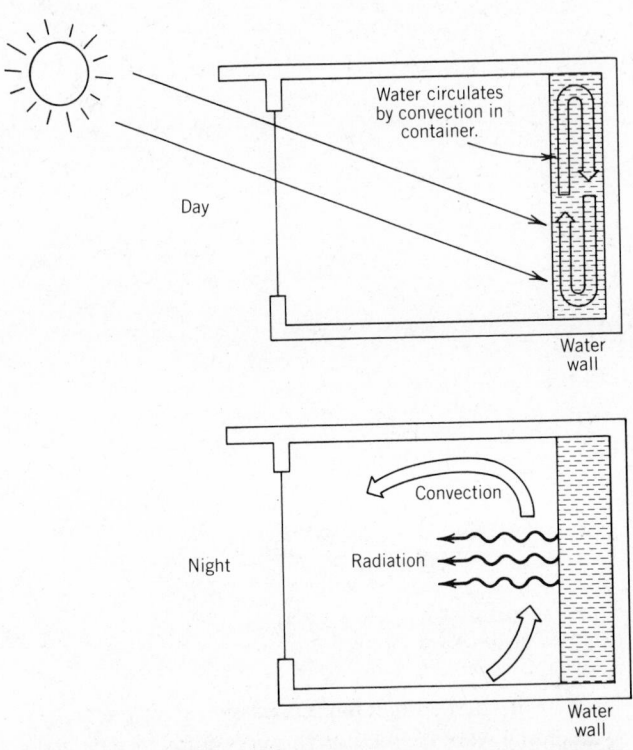

Figure 7-1 A simple direct-gain design with an interior water wall for heat storage. Heat stored in the water wall during the day is released into the living space at night by natural convection and thermal radiation.

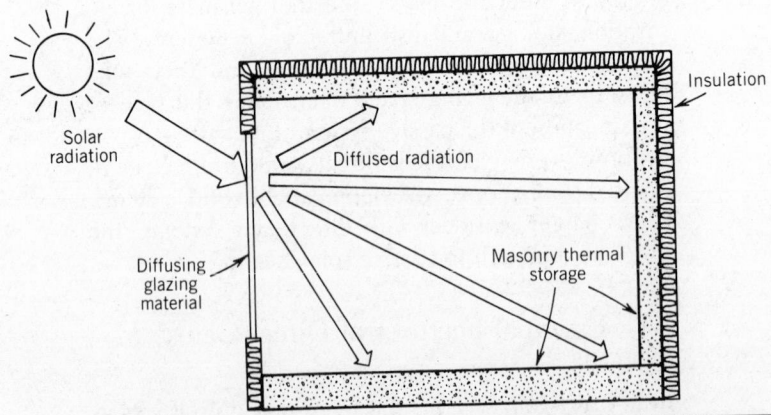

Figure 7-2 Diffusing glazing material. By using translucent glazing in windows where a view to the outdoors is not important, entering sunlight is diffused and scattered to all storage surfaces in the living space.

DIRECT-GAIN PASSIVE SYSTEMS

Figure 7-2 shows how solar radiation can be scattered or diffused by glazing onto masonry walls or onto a concrete or brick floor for absorption and storage.

7-2 Advantages and Disadvantages of Direct-Gain Passive Systems

For new construction, direct gain can be the least expensive solar heating option available. It can ordinarily be included at only modest additional cost if the building designer or architect is experienced with passive solar. However, solar houses must usually be somewhat unconventional, and it should be determined in advance whether or not the prospective owners will accept the necessary design innovations. Passive systems can be constructed of common industrial materials, and they usually have a long life and low maintenance costs. Retrofitting existing buildings for direct-gain solar heating is, however, usually quite difficult. Design considerations and cost–benefit analyses frequently rule out a direct-gain approach for a retrofit application.

Direct-gain systems also have the advantage of providing daylighting along with heat gain. However, this could become a disadvantage if excessive south glazing results in either overheating or unpleasant glare.

On the one hand, south glazing must be extensive enough to allow ample solar radiation to reach the thermal storage mass, wherever these masses are located in the space. But the designer must avoid the glare problem and the problem of overheating in areas where people congregate, such as at tables or in family rooms or living rooms. Given such additional considerations as furniture and art placement, privacy, and aesthetic preference, the extensive use of skylights and clerestory windows (see section 7-6 and Fig. 7-12) may be preferable to vertical south-facing glass. Such a solution allows more options in the placement of thermal mass, reduces the possibility of unacceptable glare, provides increased privacy, and avoids or minimizes shading problems from external and internal obstructions.

Two very important factors are comfort and control. The air temperature required for human comfort in a passively heated space tends to be lower than that required for a conventionally heated space (recall section 6-4 and the Bioclimatic Chart). Conventional (warm-air) heating systems rely on moving air around in the space—air that is significantly warmer than the steady-state ambient air temperature in the room. Generally speaking, the building masses—walls, floors, and so on—remain at temperatures well below that of the ambient air, and the body radiates to these cold surfaces. In order for people to feel warm, the air temperature has to be quite warm. In contrast, a passive solar structure radiates heat from thermal storage masses to the human body, and the air temperature can remain well below 70°F with most people still feeling comfortable.

Perhaps the greatest disadvantage of a direct-gain system is the problem of control. Large thermal masses possess a great deal of thermal inertia. They are slow to heat up and slow to cool down. This property (thermal inertia) dictates an appreciable time lag in the response of the system to temperature control measures. Proper sizing and location of thermal masses can help alleviate control problems, and so can a variety of control devices like shades, shutters, and automatic operation of vents and clerestory windows.

PROVIDING FOR DIRECT SOLAR GAIN

Windows and fixed glass on the southern face of a building constitute the standard method of providing direct solar gain for space heating. The extent of the solar collection area will be determined by the size of the structure to be heated and by its heating load. An absolute due south orientation is not required, since windows 20° east or west of due south will receive at least 90 percent of the solar radiation that a due-south-facing window would receive. (See section 6-8 for the building orientation rule of thumb.)

7-3   Sizing the Solar Collection Area

South-facing glass becomes an energy-producing element for a building because it will transmit more energy into the building, on the average, than it will lose from heat transmission out through the window.

Using the Los Alamos Collection Area Rule of Thumb.   Estimating the solar collection area is an important part of the schematic design phase (see section 6-7). By making initial estimates for the solar collection area in the early stages of the building design, adequate building surface area can be allocated for the collection of solar radiation. Recall from section 6-8 the Los Alamos solar collection area rule of thumb:

> A solar collection area of (R1) percent to (R2) percent of the floor area can be expected to reduce the annual heating load of a building in (location) by (S1) percent to (S2) percent; or, if R-9 insulation is used at night in the solar collection area, by (S3) percent to (S4) percent.

The values of $R1$ and $R2$ and of $S1$, $S2$, $S3$, and $S4$ are selected from Table 6-1. It will be recalled that the Rs represent the ratio of solar collection area to floor area, and the Ss represent the expected solar savings (SS). The following problems illustrate the usefulness of this rule of thumb for direct-gain applications.

Illustrative Problem 7-1

Estimate the solar collection area that will provide approximately 60 percent of the space heating for an 1800 ft^2 house in Chicago by direct gain. The building site has a full southern solar access.

Solution

Since the building site has full southern solar access, the solar collection area will face due south. From Table 6-1 for Chicago, we find the following values for use with the solar collection area rule of thumb:

$R1$ = 17%, $R2$ = 35%, $S1$ = 17%, $S2$ = 23%, $S3$ = 43%, $S4$ = 67%.

It is helpful to plot the above values as shown in Fig. 7-3 as an aid to interpreting the information. The percentage reduction in annual load (Solar Savings) is plotted on the vertical axis, and the percentage that the solar collection area is of the floor area is plotted along the horizontal axis. For the case of no night insulation in the solar collection area, two points, A_1 and A_2, are plotted. For the case where R-9 night insulation is used, B_1 and B_2 are plotted. Lines are drawn connecting the plotted points for the two cases. To determine the solar collection area as a percentage of floor area corresponding to a 60 percent reduction in the annual space-heating load, a horizontal line is drawn to the right from the 60 percent point on the vertical axis. The results are as follows:

With no night insulation:

The 60 percent line does not intersect the no-night-insulation curve at all, and it is apparent that a 60 percent reduction cannot be achieved without night insulation.

With night insulation:

The horizontal line from 60 percent solar savings intersects the R-9 night insulation line (point C), and a vertical line to the scale at bottom indicates that the solar collection area should be 30 percent of the building floor area.

Solar collection area = 0.30 × 1800 ft^2 = 540 ft^2 *Ans.*

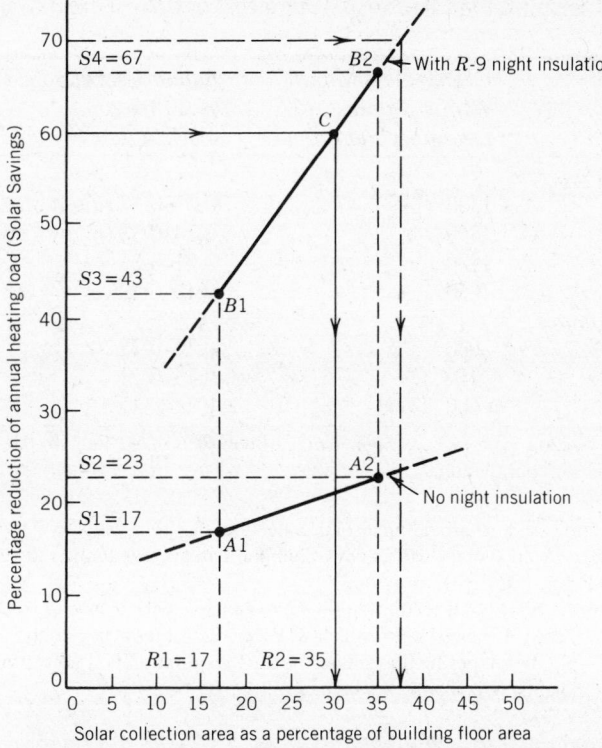

Figure 7-3 A plot of Solar Savings as a function of the percent that the solar collection area is of the living-space floor area.

Illustrative Problem 7-2

Suppose the house in Illustrative Problem 7-1 could not be given a due south orientation. Assume instead that the solar collection area will face 30° east of south. Estimate the solar collection area for a reduction in annual heating load (Solar Savings) of 60 percent.

Solution

Recall the Los Alamos orientation rule of thumb from section 6-8. When the variation from true south is 28° east of south, the solar savings will be reduced by about 10 percent. Because the solar savings is decreased, the solar collection area will have to be increased to compensate for the orientation away from due south. Figure 7-3 is the plot for a due south orientation, so to obtain an actual 60 percent solar savings, a 70 percent indicated solar savings will have to be used for the due south case, to compensate for the 30° east of south orientation. The dashed lines added to Fig. 7-3 indicate that the required solar collection area would be approximately 37 percent of the building floor area with R-9 insulation.

$$\text{Solar collection area} = 0.37 \times 1800 \text{ ft}^2 = 666 \text{ ft}^2 \qquad Ans.$$

The solar collection area rule of thumb indicates only the total area of glazing for the direct-gain structure. It does not provide information on how to divide the glazing among particular rooms or spaces in the building. It also does not provide information on daily or hourly performances. To determine the daily performance, different methods must be used. These will be discussed later in this chapter and in subsequent chapters.

Table 7-1 Sizing Guide for Solar Windows Used for Direct-Gain Solar Systems

Average Winter Outdoor Temperature (°F)	Degree-Days/Month Winter Months December and January	Square Feet of Solar Window Required for Each Square Foot of Floor Area
Cold Climates		
15°	1500	0.27–0.42 (night insulation used)
20°	1350	0.24–0.38 (night insulation used)
25°	1200	0.21–0.33
30°	1050	0.19–0.29
Temperate Climates		
35°	900	0.16–0.25
40°	750	0.13–0.21
45°	600	0.11–0.17

Source: Edward Mazria, *The Passive Solar Energy Book* (Emmaus, Pa.: Rodale Press, 1979), p. 122. Reprinted with permission.

Notes:
1. These rules are based on providing a solar collection area that will allow enough solar energy to be collected on an average winter day to maintain an average temperature in the direct-gain space of 70°F during the 24-hour period.
2. The ratios of the right-hand column apply to a residence with a space heat loss of 8 to 10 Btu/DD-ft^2 of floor area. If space heat loss is less, lower the values accordingly. The ratios can be used for other building types having similar heating requirements if adjustments are made for heat gains from lights, people, and appliances.

Mazria's Rule.   Another approach to estimating the required solar collection area has been proposed by Edward Mazria.[1] Rules for sizing the solar collection area were developed by Mazria based on computer simulations for different climates, which are categorized by average winter temperature and degree-days of heating. Table 7-1 gives Mazria's collection area sizing guide for direct-gain applications. The first column of the table gives the average winter outdoor temperature (°F) that characterizes the climate. The second column indicates the winter degree-days per month for December and January. The third column gives the square feet of window area needed for solar collection for each square foot of floor area. It is assumed that the solar collection area is double glazed. For each of the two climate zones, a range is given for the size of the solar collection area. For more southern latitudes (35° N or less), the smaller window-to-floor-area ratio should be used. For northern latitudes (48° N or greater), use the larger window-to-floor-area ratio. For latitudes in between, interpolation may be used.

It should be noted that these rules consider only the local temperature range and not the locally available solar radiation (insolation). They should be used with caution in areas where there is extended cloud cover that reduces winter insolation. In any case, they may serve as another guideline for sizing collection areas. A complete system evaluation should be made near the end of the design development phase to check the rules of thumb and guidelines used earlier in the design process.

Illustrative Problem 7-3

A house has 1800 ft^2 of floor space. How many square feet of due south glazing should the house have for a direct-gain solar energy system? The house is located in Seattle, Washington, with an average winter design temperature of 39°F. Use Mazria's guidelines.

[1] Edward Mazria, *The Passive Solar Energy Book* (Emmaus, Pa.: Rodale Press, 1979). Other rules and guidelines developed by Mazria will be cited in later sections of this book.

Solution

From Table 7-1, interpolate between 35°F and 40°F for the 39°F values. For 35°F the ratio is 0.25, and for 40°F it is 0.21 (using the larger values of the ranges given, since Seattle is a northern location). Interpolation suggests a value of 0.22 ft² of south glazing per square foot of floor space. As a result,

$$\text{Area of south glazing} = 0.22 \text{ ft}^2 \text{ glazing/ft}^2 \text{ floor} \times 1800 \text{ ft}^2 \text{ floor}$$
$$= 396 \text{ ft}^2 \text{ glazing} \qquad \textit{Ans.}$$

Illustrative Problem 7-4

An addition is to be made on a house in Little Rock, Arkansas, which will be heated with direct solar gain. The addition will have 800 ft² of floor area. Using Mazria's guidelines, determine the number of square feet of south glazing that will be needed. Average January temperature is 40°F.

Solution

From Table 7-1 for 40°F, a range of glazing ratios from 0.13 to 0.21 is found. Since Little Rock is at about 35° N Lat, the smaller ratio will be used. The indicated area of south-facing glazing is

$$0.13 \times 800 \text{ ft}^2 = 104 \text{ ft}^2 \qquad \textit{Ans.}$$

7-4   Estimating the Solar Gain

Solar Heat Gain Factors (SHGF).   Once the solar collection area has been estimated for a building, it is good practice to estimate also the amount of solar energy that will actually enter the building through the glazed window areas. This information can be used to determine how the solar collection area is to be distributed in the rooms of the building, and it can also serve as a check on the sizing of the solar collection area. The solar gains through the windows can then be compared to the heating load of the building on a daily basis to check the fraction of the total heating requirement that the solar gains will provide.

First, the quantity of solar energy transmitted by a window in one hour or one day depends on the intensity of the solar radiation striking the window. The latter, in turn, depends on the factors listed below, most of which were discussed in Chapters 4 and 6:

1. The latitude of the building site.
2. The time of year.
3. The azimuth angle of the window surface (measured from due south) and the tilt of the window (usually 90° or vertical).
4. The amount of reflected radiation incident on the window, I_r.
5. The percentage of the actual clear-day insolation incident on the window.
6. The amount of shading of the window by nearby trees, buildings, mountains, window overhangs, trellises, or other obstructions.

The percentage of the incident radiation energy that is transmitted through the window and becomes heat energy in the space depends on all of the following:

1. The type of glazing used in the window and the shading effect of curtains or other sun controls.
2. The reflection losses off the glass, which vary with the angle of incidence of the solar radiation.
3. The effectiveness of the building structure and contents in absorbing, storing, and using the transmitted energy.

The heat energy entering a structure through glass is made up of three flows: (1) radiation transmitted directly through the glass, (2) heat flow from radiant energy absorbed by the glass, and (3) conduction heat flow caused by the outdoor–indoor temperature difference ($\Delta t = t_o - t_i$). The total instantaneous rate of heat gain into the space through the glass is, then, the sum of the radiation energy transmitted directly through the glass plus the inward heat flow by conduction, which is ultimately transferred to the space by convection and radiation from the inner glass surface.

Solar heat gains through glass into interior spaces have been designated as *solar heat gain factors* (SHGF) by the American Society of Heating, Refrigerating, and Air Conditioning Engineers (ASHRAE). Tables of SHGF values have been compiled for various latitudes, for daylight hours of the twenty-first day of each month and for each of 16 compass orientations.[2] Portions of the ASHRAE solar heat gain tables are included as Table 7-A in the Chapter Appendix. Solar heat gain factors (SHGF) provide one means for estimating the solar energy entering a space through windows.

Table 7-A gives clear-day solar heat gain factors (SHGF) in Btu per hour and half-day radiant heat gain totals through single-pane, double-strength glass. This table is tabulated for the twenty-first day of each month, for different wall azimuths, and for a ground reflectance of 0.2. The SHGF values take into consideration such site factors as time of year, local solar time, window orientation, and ground reflectance. When other types of glazing such as double or triple pane are used, or when windows are shaded inside by drapes, blinds, or screens, a correction factor called the *shading coefficient* (SC) must be applied to the tabular SHGF values (see Table 7-2).

The shading coefficient (SC) is defined as a ratio converted to a percentage:

$$SC = \frac{\text{Solar heat gain of specified window}}{\text{Solar heat gain for clear double-strength glass (unobstructed)}} \quad (7\text{-}1)$$

The total instantaneous heat gain through glass is the sum of solar heat gain (SHGF $\times$ SC) and conduction heat gain (or loss) ($U\Delta t$) where $\Delta t = t_o - t_i$ for heat gain, or $t_i - t_o$ for heat loss.

As an equation, for 1 ft^2 of glazing area,

$$H = \text{SHGF} \times \text{SC} + U\Delta t \qquad \text{in Btu/hr-ft}^2 \quad (7\text{-}2)$$

[2] See *ASHRAE Handbook, 1977 Fundamentals* (Atlanta: ASHRAE, 1977), pp. 26.17–26.25.

Table 7-2 Shading Coefficients for Selected Window Situations

Glass Type	Shading Coefficient (SC)			
	No Shade	Light Venetian Blinds	Drapery	
			Light	Dark
Clear glass 1/8″	1.00	0.55	0.55	0.70
Clear glass 1/4″	0.94	0.55	0.55	0.70
Clear glass 1/2″	0.87	0.55	0.52	0.66
Gray glass 1/4″	0.69	0.53	0.44	0.52
Thermopane 1/8″ glass with a 1/4″ airspace (double-glazing)	0.88	0.51	0.48	0.58

Source: Abstracted from *ASHRAE Handbook 1977 Fundamentals* (Atlanta: ASHRAE, 1977), Chapter 26, Tables 28, 35, and 38.

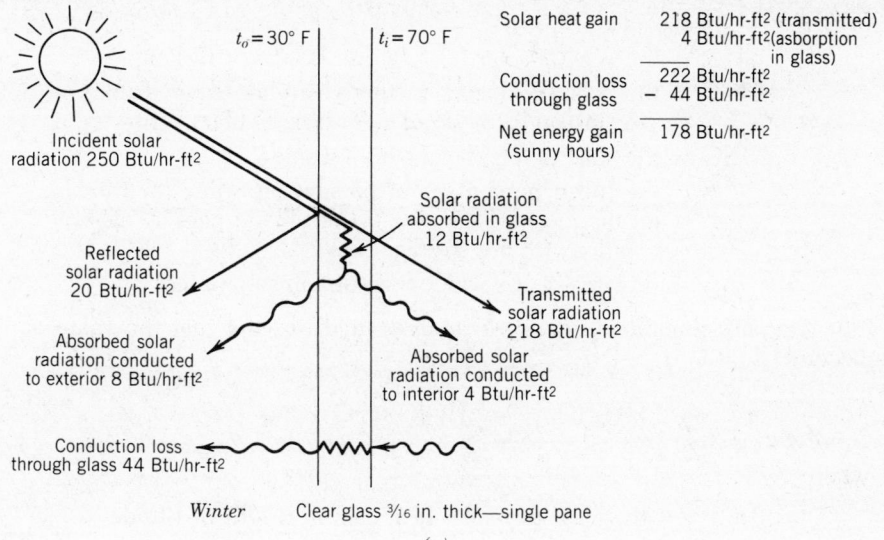

Solar heat gain   218 Btu/hr-ft² (transmitted)
                 4 Btu/hr-ft²(absorption in glass)
                     222 Btu/hr-ft²
Conduction loss through glass   − 44 Btu/hr-ft²
Net energy gain (sunny hours)   178 Btu/hr-ft²

$t_o = 30°$ F $t_i = 70°$ F

Incident solar radiation 250 Btu/hr-ft²

Solar radiation absorbed in glass 12 Btu/hr-ft²

Reflected solar radiation 20 Btu/hr-ft²

Transmitted solar radiation 218 Btu/hr-ft²

Absorbed solar radiation conducted to exterior 8 Btu/hr-ft²

Absorbed solar radiation conducted to interior 4 Btu/hr-ft²

Conduction loss through glass 44 Btu/hr-ft²

Winter   Clear glass ³⁄₁₆ in. thick—single pane

(a)

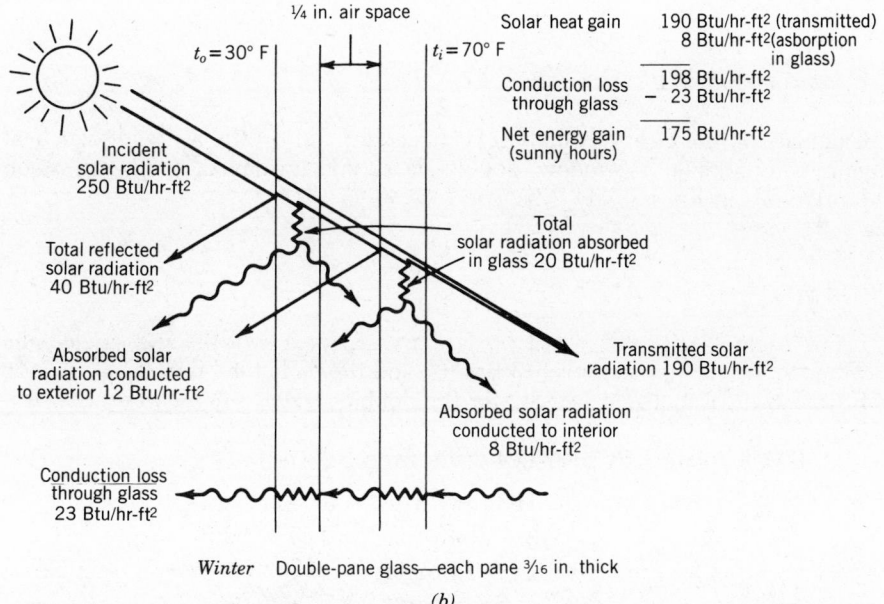

¼ in. air space

Solar heat gain   190 Btu/hr-ft² (transmitted)
                 8 Btu/hr-ft²(absorption in glass)
                     198 Btu/hr-ft²
Conduction loss through glass   − 23 Btu/hr-ft²
Net energy gain (sunny hours)   175 Btu/hr-ft²

$t_o = 30°$ F $t_i = 70°$ F

Incident solar radiation 250 Btu/hr-ft²

Total solar radiation absorbed in glass 20 Btu/hr-ft²

Total reflected solar radiation 40 Btu/hr-ft²

Absorbed solar radiation conducted to exterior 12 Btu/hr-ft²

Transmitted solar radiation 190 Btu/hr-ft²

Absorbed solar radiation conducted to interior 8 Btu/hr-ft²

Conduction loss through glass 23 Btu/hr-ft²

Winter   Double-pane glass—each pane ³⁄₁₆ in. thick

(b)

Figure 7-4 Net energy gain through (*a*) single-pane glass and (*b*) double-pane glass. The net gain for double-pane glass is slightly less than that for single-pane glass because of the greater absorption and reflection losses of the double-pane glass. The advantage of double-pane glass in cold climates is its far lower conduction loss for heat generated inside the building.

Figure 7-4 shows in diagram form the approximate values of heat gains and losses through two different types of glazings—single-pane glass (*a*) and double-pane glass (*b*) for winter conditions. Both diagrams assume an incident solar radiation intensity of 250 Btu/hr-ft² for comparison purposes. Shading coefficients for different glazings and different types of interior shading are given in Table 7-2.

    When considering the solar heat gain only (through glass), the following equation can be used to calculate average solar heat gains:

$$\text{SHG} = \text{SHGF} \times \text{SC} \times P \times A_W \qquad (7\text{-}3)$$

where

SHG = the *average* solar heat gain in Btu/hr or Btu/day

$SHGF$ = the solar heat gain factor (clear-day) from Table 7-A, in Btu/hr or Btu/day for a particular month

SC = the shading coefficient from Table 7-2

P = the mean percentage of possible sunshine for the locality, from Table 4-C (applicable to per-day calculations for an "average" day)

A_W = the area of the window glazing in square feet

For *clear-day* conditions, the same analysis holds, except that the value of P becomes 1.00.

$$SHG_{CD} = SHGF \times SC \times A_W \qquad (7\text{-}4)$$

where

SHG_{CD} = the clear-day solar heat gain in Btu/hr or Btu/day

In calculations where daily totals are desired, double the half-day totals in Table 7-A.

Illustrative Problem 7-5

Calculate (a) the clear-day daily solar heat gain and (b) the average solar heat gain, through a 90 ft² window, double-glazed and unshaded, south-facing, on January 21, in Kansas City, Missouri (use 40° N Lat).

Solution

From Table 7-A for 40° N Lat on January 21, for a south-facing single-pane window, the half-day total is 813 Btu/ft², and the daily total is SHG_{CD} = 1626 Btu/ft²-day. From Table 7-2, SC for the double-glazed window as described is 0.88.

(a) Clear-day daily solar heat gain, from Eq. (7-4),

$$
\begin{aligned}
SHG_{CD} &= SHGF \times SC \times A_W \\
&= 1626 \text{ Btu/ft}^2\text{-day} \times 0.88 \times 90 \text{ ft}^2 \\
&= 128{,}780 \text{ Btu/day} \qquad\qquad\qquad \textit{Ans.}
\end{aligned}
$$

(b) Average daily solar heat gain for January (Eq. 7-3), (P = 0.55, from Table 4-C),

$$
\begin{aligned}
SHG &= 1626 \text{ Btu/ft}^2\text{-day} \times 0.88 \times 0.55 \times 90 \text{ ft}^2 \\
&= 70{,}830 \text{ Btu/day} \qquad\qquad\qquad\qquad \textit{Ans.}
\end{aligned}
$$

In winter, the daylight hours are relatively few, and the amount of time that sunlight enters south windows is in the range of only six to eight hours. Part of the transmitted solar energy will flow into the floor, walls, ceiling, and furnishings, becoming stored energy; the rest flows into the room air by convection and conduction off the surfaces receiving the solar radiation. There is always the possibility of uncomfortably high room temperatures during the afternoon hours, unless the design has provided for adequate thermal storage in the structure (see section 7-11 below). Direct-gain structures with large glass areas may experience overheating problems during clear winter days if thermal mass is inadequate and if some measure of sunlight control is not provided. In other words, during sunlight hours, the instantaneous rate of solar heat gain through glass will ordinarily be much greater than the structural heat loss rate of the space, allowing stored heat to offset structural heat losses at night.

7-5 Daily and Seasonal Sunlight Cycles

Daily Changes in the Sun's Rays in a Room.   Sunlight will enter a south-facing window at different angles during the day as the sun moves across the sky. To demonstrate this shifting pattern, the sunlit portions of a simple room are shown in Fig. 7-5 for three times during the day in winter, at 40° N Lat.

During the morning hours (Fig. 7-5a), the sun will be low in the east, and the sunlight shining through a window will strike part of the west and north walls and some of the floor, as shown by the shaded areas. As the sun moves higher and toward the south, the transmitted sunlight will move from the west wall to the north wall of the room and also farther out onto the floor. At noon, when the sun is highest and directly south, sunlight entering the window will strike the floor and the north wall. At noon, sunlight enters the space at its steepest angle, and the floor receives a great deal of the transmitted solar energy. The walls receive relatively little of this high-angle radiation, as is shown in Fig. 7-5b. In the afternoon, as the sun moves to the west and to a lower altitude (Fig. 7-5c), the east wall and the floor are the areas that receive solar energy. This movement around the room of the sun's radiant energy must be considered when planning the location of thermal storage masses. Needless to say, thermal storage is most effective when it receives direct solar radiation for most of the sunny hours of the day.

Seasonal Changes in Sunlight Patterns.   The altitude of the sun changes during the year, shifting the pattern of penetration of radiation into a room, as shown in Fig. 7-6. During the winter months, the sun is lowest in the sky and solar radiation will penetrate deepest into the room (Fig. 7-6a). At the spring and fall equinoxes, the sun is 23.5° higher in the sky than it is at the winter solstice, and during these months (March and September) the higher sun does not shine as far into the room (Fig. 7-6b). At the summer solstice, the sun is very high in the sky and very little sunlight penetrates the window, as shown in Fig. 7-6c.

Effect of Latitude on Sunlight Patterns.   The latitude of the building location also has a marked effect on the amount of sunlight that comes through windows. Consider the effect of moving the room to a location on or very near the equator but still in the northern hemisphere and still facing south. For the same three times of the year as above, there are very different results. At the

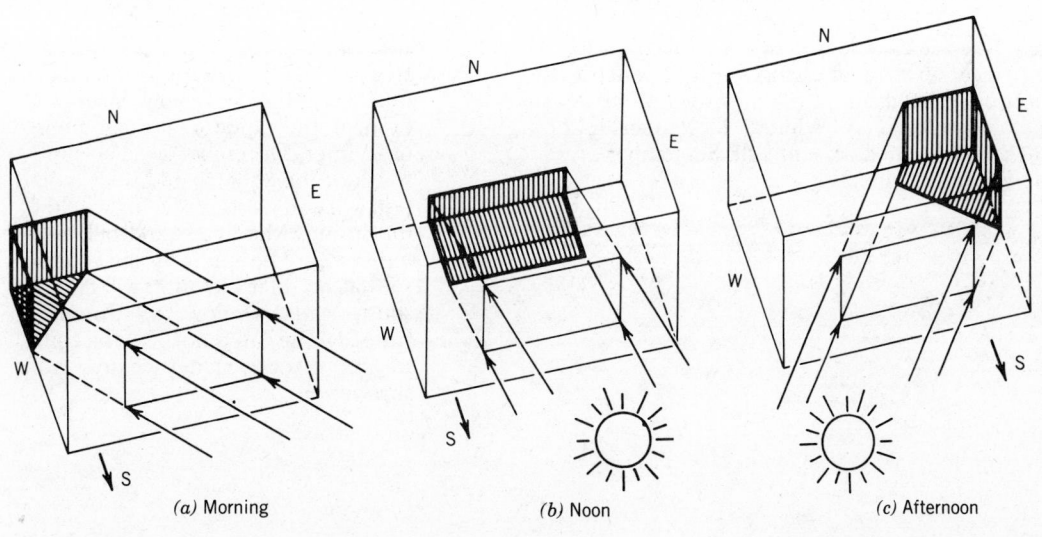

(a) Morning   *(b)* Noon   *(c)* Afternoon

Figure 7-5 Daily changes in sunlight patterns in a room. (*a*) Morning. (*b*) noon. (*c*) Afternoon.

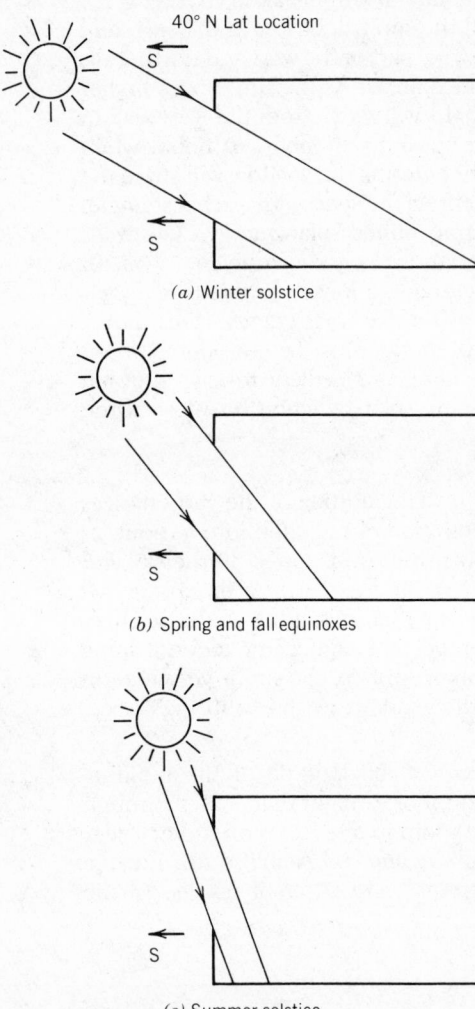

(a) Winter solstice

(b) Spring and fall equinoxes

(c) Summer solstice

Figure 7-6 Annual changes in the penetration of sunlight into a room located at 40° N Lat at noon. (*a*) Winter, December 21. (*b*) Spring, March 21, and Fall, September 21. (*c*) Summer, June 21.

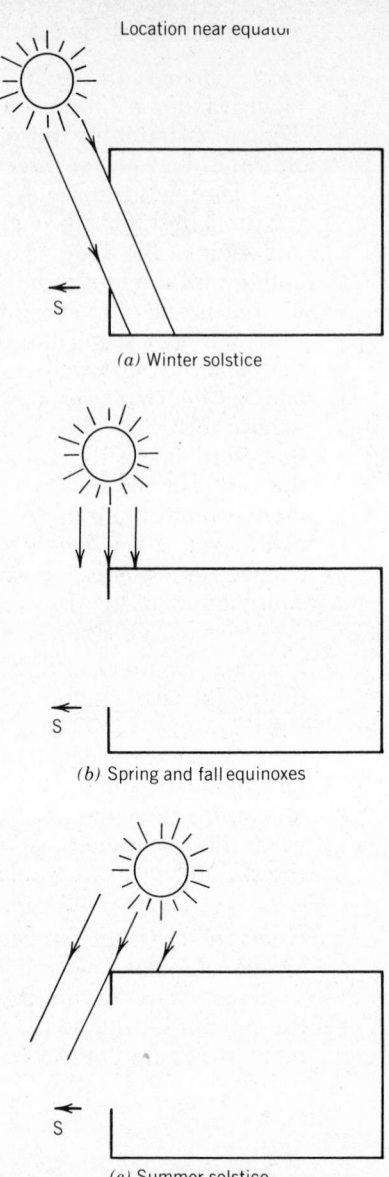

Location near equator

(*a*) Winter solstice

(*b*) Spring and fall equinoxes

(*c*) Summer solstice

Figure 7-7 Penetration of sunlight into a south-facing room located at or near the equator at solar noon. (*a*) Winter, December 21. The sun is south of the equator and solar radiation will enter a south-facing window. (*b*) Spring, March 21, and fall, September 21. The sun is on the equator so that no solar radiation will enter a south-facing window. (*c*) Summer, June 21. The sun is north of the equator and a south-facing window is in shadow.

winter solstice (sun at 23.5° S declination), sunlight penetrates into the room to some extent as shown in Fig. 7-7a. However, at the spring equinox (Fig. 7-7b), the sun is directly overhead, and the window is shaded by its own building from then on through summer solstice (Fig. 7-7c) and until the fall equinox. This suggests that at locations near the equator roof glazing would be more effective than vertical windows if, for any reason, solar heat gain were desired in winter, as it might be in certain high-altitude locations.

Now consider the case for a far northern latitude—for example, a location near Edmonton, Alberta, Canada, at about 54° N Lat. Figure 7-8 illustrates this situation. During the winter, the sun has a very small altitude angle, and solar radiation strikes the north wall more than it strikes the floor (Fig. 7-8a). During the spring and fall months, the sun is considerably higher and transmitted radiation will strike mostly on the floor (Fig. 7-8b). In the summer, the sun is at its maximum altitude (Fig. 7-8c) and much less sunlight enters the room. Almost all of it will be incident on the floor.

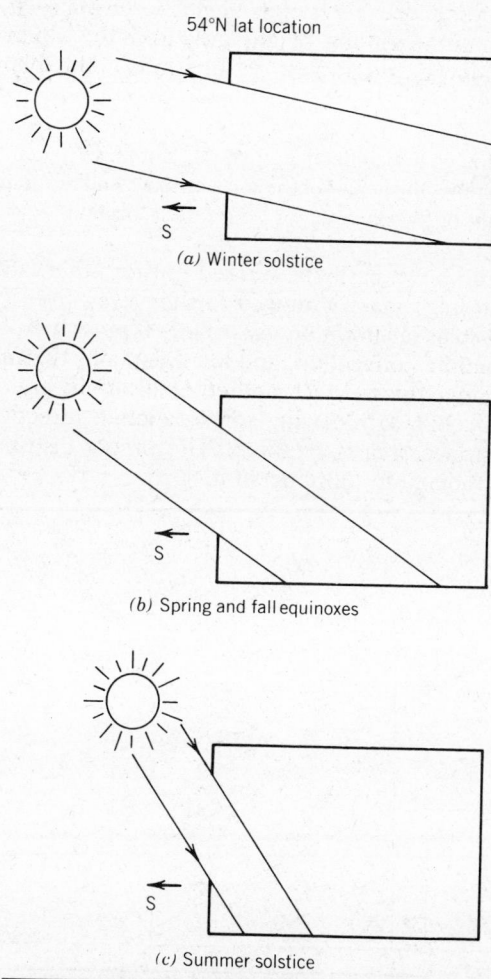

Figure 7-8 Penetration of sunlight into a south-facing room located at 54° N Lat at solar noon. (a) Winter, December 21. The sun is at a low altitude so that much of the north wall is sunlit. (b) Spring, March 21; and fall, September 21. Solar radiation strikes most of the floor. (c) Summer, June 21. The high-angle summer sun strikes only the portions of the floor near the glazing.

7-6 Factors Influencing Solar Collection Area Design

Limitations on Use of South-Facing Glass. In some situations, there may be considerations that preclude the use of a large expanse of south-facing windows for a passive solar heating system. One inhibiting factor is that room layout and living arrangements in a solar-window, direct-gain passive structure are somewhat limited. This limitation is partly caused by the fact that thermal storage cannot be set back more than a fixed distance from the windows, or the solar radiation will not strike the thermal storage.

Consider a 10-ft-high south-facing window in a structure located at 40° N Lat, as shown in Fig. 7-9. At the winter solstice, December 21, the noon solar altitude β (from Table 4-A) is 27°. On January 21, the solar altitude is 30°, as shown in Fig. 7-9. If the thermal storage wall is located too far back in the room, it will not receive any direct solar radiation, as in location A. For the thermal storage to be effective, it must receive solar radiation as indicated in locations B and C. From the dimensions and geometry of the sketch, it can be determined that thermal storage at location B has 2 ft of its height receiving solar radiation, and storage at location C has 3½ ft of its height receiving solar radiation. Since location C is 1¼ times the window height away from the window and location B is 1½ times the window height away from the window, the following generalization can be made:

> For thermal storage to be effective, it should not be set much farther back than about 1¼ times the height of the solar window, and never more than 1½ times the height of the window, for a 40° N Lat location.

Similar rules of thumb can easily be worked out for other latitudes.

As thermal storage mass is moved farther away from solar windows, it receives less direct solar radiation on its surface. It must then receive energy by reflected solar radiation, convection, and low-frequency thermal radiation from warmer surfaces. Since this indirect method of absorbing energy in the interior space is not as effective as receiving solar radiation directly, a much larger thermal storage surface area is required. The proper distribution of thermal storage will be considered in more detail in section 7-12.

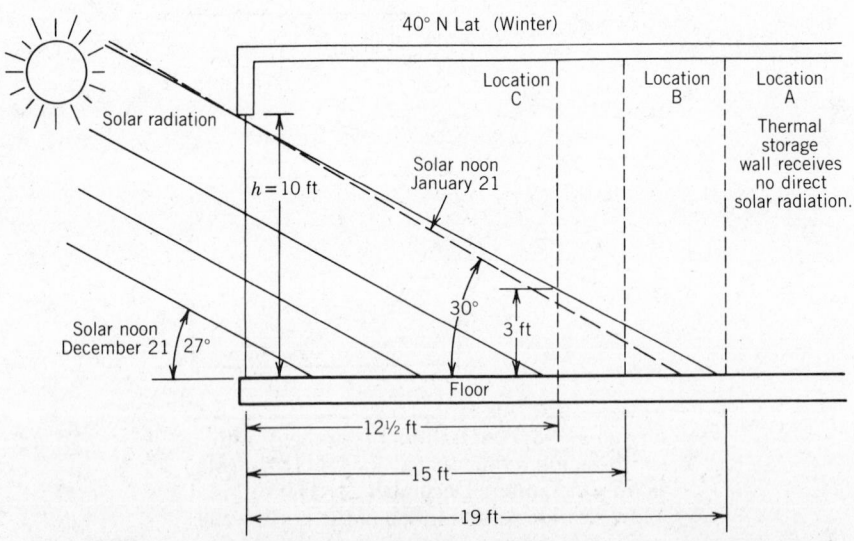

Figure 7-9 The depth to which sunlight will penetrate a room through a 10-ft-high window for a location at 40° N Lat on December 21 and January 21. The sunlit portion of a north wall is shown for location C, a distance 12.5 ft (1¼ times the height of the window) from the south wall.

The solar energy system designer must consider and discuss with the prospective owner the effects of direct sunlight on room furnishings and on wood and paint, since sunlight will fade fabrics and dry out and discolor wood finishes and painted surfaces. Materials such as brick, stone, and ceramic tile are not damaged by direct sunlight.

A final important consideration is the presence or possible future presence of nearby trees, shrubbery, hills, or new structures that may reduce direct solar gain on south walls sufficiently to render south-facing solar gain windows impractical. For example, if a large building were to be constructed on an adjoining lot, sited in such a manner that it would cast a shadow over a significant portion of the south-facing wall, the use of that wall for solar gain windows would be precluded. Some localities already have laws and ordinances protecting the sun rights of properties, but solar designers must always be alert to the possibility of future solar gain reduction from shading.

Clerestory Windows and Skylights. If the use of south-facing glazing is impractical because of one or more of the above conditions, there are alternative approaches to the use of direct gain, involving skylights and clerestory windows. A *clerestory window* is a window that projects up from the roof, as shown in Fig. 7-10. A clerestory window located on the south side of a roof allows winter solar radiation to enter through the roof structure of the building and strike interior walls (usually north walls) that would not normally have access to sunlight. The roof projection above a clerestory window provides shading in the summer months when the altitude of the sun is much higher than it is in the winter. A clerestory can also be used to admit solar radiation into rooms on the north side of buildings that have no southern exposure to solar radiation. Reflective surfaces may be used with clerestory windows to enhance their solar performance and to direct solar radiation to different parts of the room where thermal storage masses are placed.

The height of a clerestory window, *h*, as indicated in Fig. 7-10, is determined by the area of window needed for solar gain. The clerestory should be positioned so that the solar radiation entering through the glass will strike the thermal storage wall throughout the entire winter. The length *L* is the distance from the clerestory to the thermal storage wall, as shown in Fig. 7-10. The

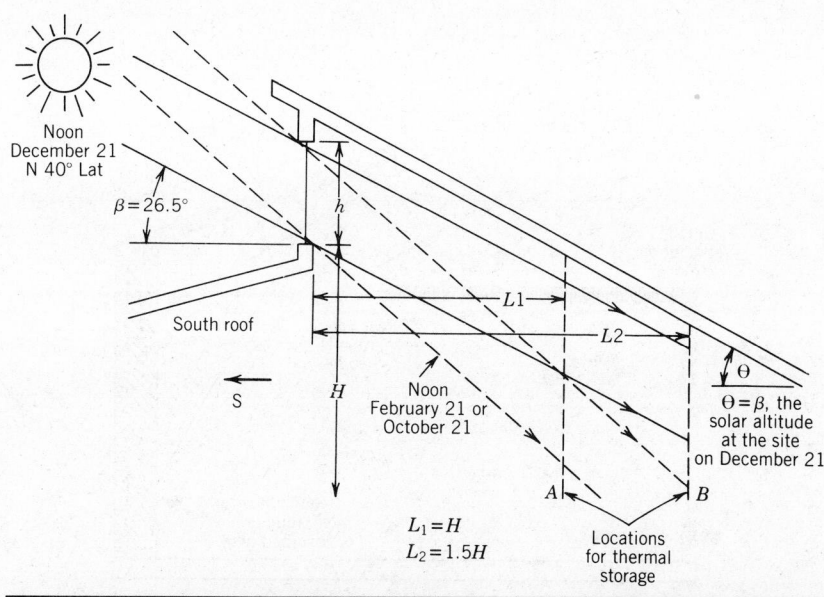

Figure 7-10 Clerestory windows can be used to provide solar radiation to interior walls that would normally not be exposed to direct winter sunlight. The placement of the thermal storage in relation to the clerestory window is shown.

PROVIDING FOR DIRECT SOLAR GAIN

height of the base of the clerestory above the floor is indicated as H. The angle of the roof behind the clerestory, θ, is generally designed to be the same as the solar altitude at the winter solstice, December 21, for the building location (in this case 40° N Lat).

When a thermal storage wall is located to the north of a clerestory window (Fig. 7-10), the solar radiation should strike the upper half of the wall during the months of December and January. The winter sun is shown for solar noon December 21 at 40° N Lat. Two locations for thermal storage are shown: A at $L_1 = H$, and B at $L_2 = 1.5 H$. If the thermal storage is placed farther away from the clerestory than location B, the winter solar radiation will strike only the upper part of the wall. During spring and fall, solar radiation will strike the floor rather than the wall. To make the thermal storage most effective, the winter solar radiation should strike the upper half of the wall, and the fall and spring solar radiation should strike the lower half of the wall. From Fig. 7-10 (for winter), it can be seen that the thermal storage should be located at a distance of $1.0 H$ to $1.5 H$ away from the clerestory, where H is the height or distance from the floor to the base of the clerestory.

The use of reflecting surfaces to direct sunlight to a masonry thermal storage wall is shown in Fig. 7-11. The low winter sun reflects off the outdoor reflector and on through the clerestory window, increasing the transmitted radiation. The ceiling reflector directs the radiation from the outdoor reflector onto the thermal storage wall.

For large rooms, clerestories can be arranged in a sawtooth pattern across the roof, as indicated in Fig. 7-12. With this type of roof, care must be taken to ensure that the clerestories do not shade each other.

A *skylight* is simply an opening in the roof that is covered with glazing. It can be either horizontal on a flat roof or pitched at the same angle as a sloping roof. Many skylights use an adjustable reflector/shutter to reflect winter sunlight into the space and to block the summer sun, as shown in Fig. 7-13.

Most large skylights require a shading device to block out summer solar radiation and thus prevent overheating. These shading devices may be on the outside of the skylight as with the reflector/shutter of Fig. 7-13, or they may be inside under the glazing. Sometimes the design calls for rigid insulation board incorporated into the shading device, in order to reduce heat loss through the skylight on winter nights and to minimize heat gain in the summer.

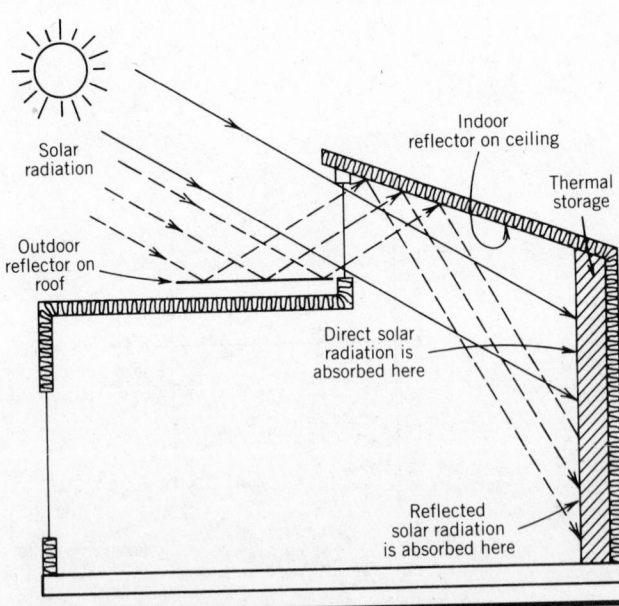

Figure 7-11 Reflecting surfaces can increase the solar radiation transmitted through a clerestory window and can be used to direct the radiation to thermal storage surfaces.

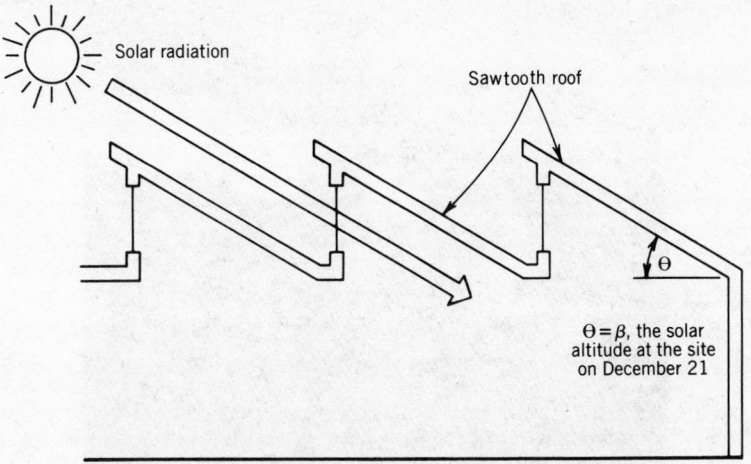

Figure 7-12 Clerestory windows may be arranged in a sawtooth pattern to provide solar gain and sunlight in a large enclosed space.

The Bateson building in Sacramento, California, is an example of a large building that uses a sawtooth skylight above a four-story central atrium (Fig. 7-14). There are adjustable louvers above the skylight glazing to control the amount of solar radiation entering the atrium. The atrium is also the core of the ventilating system in the building for providing heating and cooling. Solar radiation collected through the skylights provides a majority of the energy for heating the building during the winter. The offices around the atrium receive diffuse natural light, and the atrium itself provides an indoor garden area for people to enjoy all year.

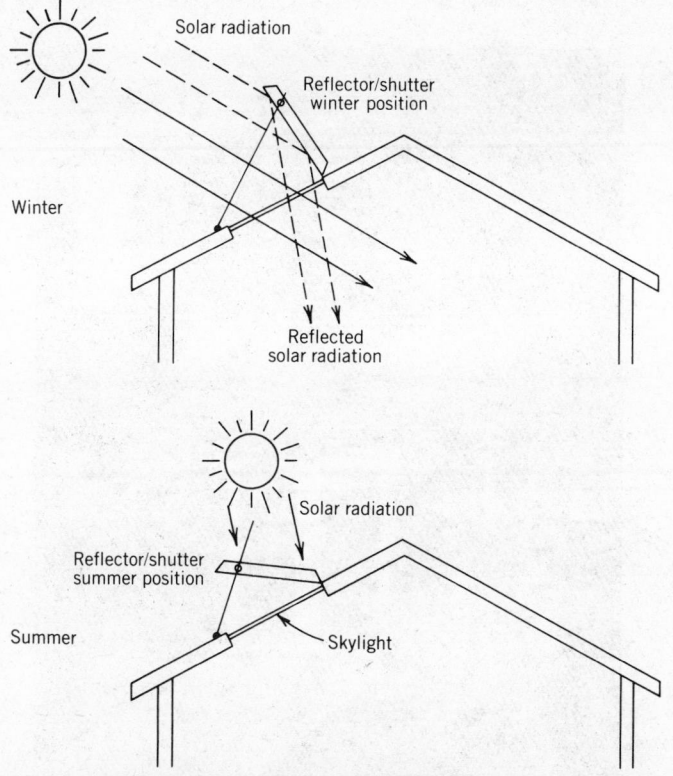

Figure 7-13 Skylights provide an alternative for direct solar gains. Shading devices must be included to prevent overheating during summer months.

PROVIDING FOR DIRECT SOLAR GAIN

Figure 7-14 Atrium of the Bateson building in Sacramento, California, showing the use of direct-gain skylights.

Figure 7-15 The entrance to a home developed as a sunspace.

The Use of Sunspaces. A sunspace is a room or part of a room that is, by design, very sunny. It usually has a south-glazed wall and roof. The primary purpose of the sunspace is to absorb solar radiation and distribute the collected heat to the rest of the room or building. Thermal mass may be included in a sunspace to store some of the solar heat collected (see also section 6-2).

Some examples of the use of a sunspace are an entrance into a house (see Fig. 7-15), a large bay window off a kitchen to provide a sunny breakfast nook, or a sunny section of a living room or family room.

The major purpose of a sunspace is to provide a sunny, bright space for use during the winter, but sunspaces may also serve as solar collectors for storage and later release of thermal energy. They may or may not be used at night, and can be isolated from the house by doors or shutters to eliminate heat loss through the sunspace glazing. From a heating standpoint, their purpose is to provide some heat during the day to adjacent spaces. They are by no means designed as a major source of direct-gain heating, but rather to be used as one element of a total design.

Sunspaces are distinguished from similar structures like solar greenhouses and atriums by their relatively small size and relatively small heating capability. Greenhouses and atriums (see Chapters 8 and 9) are structures that usually involve several hundred square feet of floor area and perhaps several hundred square feet of glazed surface, and they are intended to provide a significant fraction of the space-heating load to the house.

In summary, a number of methods of bringing sunlight into buildings have been discussed. They range from simple windows and glazed openings to sawtooth clerestories, skylight roof structures, and sunspaces, atriums, and greenhouses. All of these provide ways for the sun to heat the interiors of buildings, but they vary a great deal in the effectiveness with which controls can be added for governing indoor temperature on a daily and seasonal basis.

CONTROLLING DIRECT SOLAR GAIN

It has been mentioned that overheating can occur with direct-gain systems at two different times of the year. During the summer, solar gains through windows are not needed and, if allowed, will cause overheating. Also, on unusually clear winter days, the transmitted solar radiation through south glazing may overheat the interior of a building during the sunny hours of the day. A number of shading devices can be used to reduce the amount of solar radiation entering through glazing.

7-7 Roof Overhang

Roof overhang is the most common device for controlling summer solar gains on south-facing vertical glazing. Figure 7-16 illustrates the effect of a simple overhang above a south window for the winter and summer solstices.

A roof overhang will provide a considerable reduction in solar heat gain by providing shade on the window surfaces. It is applicable to south, southeast, and southwest exposures during the late spring, summer, and early fall. The solar altitude is generally so low on east and west exposures during the entire year that horizontal overhangs would have to be excessively long in order to provide effective shading for these exposures. For southern exposures during the winter, however, a horizontal overhang is a very useful device for passive solar systems, since it allows the winter sun to reach the solar collection area (windows) when solar gains are needed and then shades the solar collection area in the summer when solar gains are not needed.

Table 7-3 gives the length of horizontal overhang required to produce a shadow with a length of at least 10 ft down the wall from the overhang during the period from April 11 through September 1. The shadow will be longer than 10 ft during the summer months when the solar altitude is at its maximum. (Note that for P.M. times Table 7-3 is entered at bottom right.)

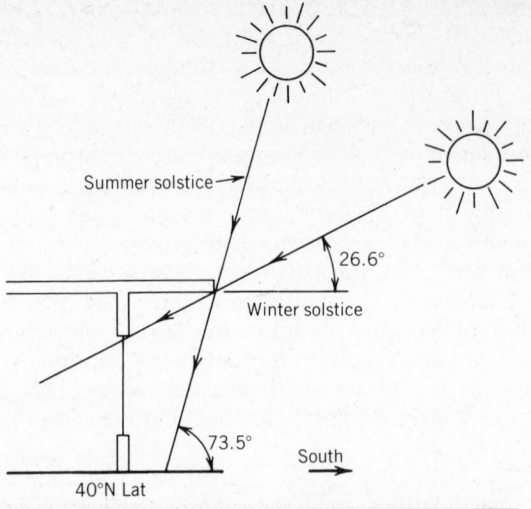

Figure 7-16 A roof overhang is a device that allows winter solar radiation to enter the window but obstructs summer solar radiation, thus shading the window.

Illustrative Problem 7-6

Determine the length of horizontal overhang required to shade a solar collection area with a 10-ft shadow from April 11 through September 1 on a southwest wall located at 48° N Lat. The wall should be in full shade through 3:00 P.M. sun time.

Solution

From Table 7-3, find for 48° N Lat that a 13.7-ft horizontal overhang is required to keep the wall in shade through 3:00 P.M. sun time for a southwest exposure. *Ans.*

The above method is very useful for determining shading during the summer, but no information is provided on how much shading occurs during the winter. Ordinarily, solar collection areas are not to be shaded during the winter when solar energy is being collected, but there are some instances where winter shading is desirable.

Profile Angle and Solar Altitude. In order to design an overhang for winter shading, it is necessary to know the angular relation between the glazed surface and the sun. The *profile angle* (as defined in Chapter 4) gives the angle from the horizon at which sunlight will enter through a window under an overhang. Recall (section 4-11) that the profile angle is defined as the angle through which a horizontal plane must be rotated about a horizontal line located in the plane of the window or wall in order to include the position of the sun (Fig. 4-19). It is again stressed that the profile angle is not the same as the solar altitude. The difference between these angles was shown in Fig. 4-19.

The geometry of glazing shaded by an overhang is detailed in Fig. 7-17. The overhang is defined by the projection P, and the distance between the top of the glazing and the overhang is the gap G. These parameters define the summer and winter shading lines. The glazing is completely shaded when the profile angle (*FDG* of Fig. 4-19) equals or exceeds angle A_1, as defined by the summer shading line (Fig. 7-17). When the profile angle is equal to or less than angle A_2, as defined by the winter shading line, the glazing will be completely

Table 7-3 Length of Horizontal Projection Required for Shading Windows and Walls
(For shading 10 ft down from projection for April 11 through September 1)

Latitude	Sun Time A.M. ↓	Projection (ft)						Sun Time
		N	NE	E	SE	S	SW	
24° N	6 A.M.	17.3	a	a	a	—	—	6 P.M.
	7	—	a	a	a	—	—	5
	8	—	10.8	16.5	15.8	1.6	—	4
	9	—	5.3	10.0	10.7	2.3	—	3
	10	—	2.3	5.8	7.5	2.7	—	2
	11	—	—	2.8	3.8	2.7	—	1
	12 noon	—	—	—	2.2	2.8	2.2	12 noon
32° N	6 A.M.	15.8	a	a	a	—	—	6 P.M.
	7	—	a	a	a	1.7	—	5
	8	—	10.0	17.3	14.2	3.0	—	4
	9	—	4.6	10.3	10.0	3.8	—	3
	10	—	1.4	6.0	7.3	4.2	—	2
	11	—	—	2.8	5.1	4.2	1.2	1
	12 N	—	—	—	3.0	4.2	3.0	12 N
40° N	6 A.M.	12.0	a	a	a	—	—	6 P.M.
	7	—	a	a	a	1.8	—	5
	8	—	9.7	18.9	16.1	4.3	—	4
	9	—	4.7	11.2	11.6	5.4	—	3
	10	—	1.3	6.5	9.1	5.8	—	2
	11	—	—	3.1	6.5	6.1	2.2	1
	12 noon	—	—	—	4.3	6.3	4.3	12 noon
48° N	6 A.M.	7.3	a	a	a	—	—	6 P.M.
	7	—	a	a	a	3.6	—	5
	8	—	9.3	19.6	18.7	6.5	—	4
	9	—	3.3	12.0	13.7	7.5	—	3
	10	—	—	7.3	10.8	8.4	—	2
	11	—	—	3.2	8.4	8.4	3.2	1
	12 noon	—	—	—	6.0	8.4	6.0	12 noon
56° N	6 A.M.	6.2	a	a	a	—	—	6 P.M.
	7	—	a	a	a	4.9	—	5
	8	—	9.6	a	a	7.8	—	4
	9	—	2.9	13.5	16.3	9.3	—	3
	10	—	—	7.9	13.0	10.4	1.4	2
	11	—	—	3.9	10.2	10.6	4.7	1
	12 noon	—	—	—	7.5	10.7	7.5	12 noon
		N	NW	W	SW	S	SE	← ↑ P.M.

a Projection greater than 20 ft required.

Source: ASHRAE Handbook of Fundamentals, 1971 (Atlanta: ASHRAE, 1971), p. 409.

unshaded. Using the tabulated values of profile angles given in Table 4-A, a geometric construction for an overhang design can be made as indicated in the following example.

Illustrative Problem 7-7

Design an overhang for a 7-ft high window facing due south at 40° N Lat. It is desired to have the window shaded from April 21 to August 21 during the hours of 10:00 A.M. to 2:00 P.M. and fully sunlit from October 21 to February 21 during the hours of 10:00 A.M. to 2:00 P.M. Determine the projection and gap of the overhang from a geometric construction.

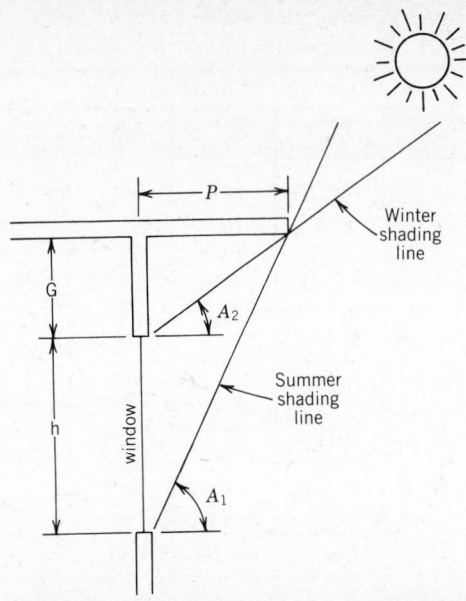

Figure 7-17 The geometry of a shading overhang. Angle A_1 is the profile of the summer shading line. The window is shaded whenever the profile angle exceeds A_1. The window is fully sunlit whenever the profile angle is less than A_2, which angle defines the winter shading line.

Solution

From Table 4-A, find the profile angle on April 21 and August 21 to be 63° at 10:00 A.M. and 2:00 P.M. and 62° at noon. The profile angle on October 21 and February 21 at 10:00 and 2:00 is 38°, and at noon it is 39°. The *summer shading line* is defined by the lowest position of the sun relative to the window between April 21 and August 21. This corresponds to the smallest profile angle on April 21 and August 21, which would be the noon value of 62°. The summer shading line is drawn from the bottom of the window (because the window is to be fully shaded at this date) at an angle equal to the profile angle of 62° as shown in Fig. 7-18. The *winter shading line* must be drawn from the top of the window, because from October 21 to February 21 the window is to receive full sun. The winter shading line is determined by the highest position of the sun on October 21 and February 21, which corresponds to the noon value of 39°. Thus, the winter shading line is drawn at a 39° angle from the top of the window, as shown in Fig. 7-18.

The point where the summer shading line and the winter shading line intersect determines the location of the overhang. The lengths of the projection P and the gap G are scaled from a geometric construction, as shown in Fig. 7-18:

Length of projection P = 6.5 ft

Length of gap G = 5.5 ft   *Ans.*

When a south-glazed solar collection area (windows) is designed to be unshaded at the winter solstice and shaded by an overhang at the summer solstice, the glazing will receive approximately 95 percent of the winter solar radiation that an unshaded collection area would receive. In summer, the shaded collection area will receive approximately 40 percent to 60 percent of the solar radiation that an unshaded collection area would receive. The diffuse sky and ground-reflected radiation account for this high level of solar radiation striking the shaded aperture in the summer.

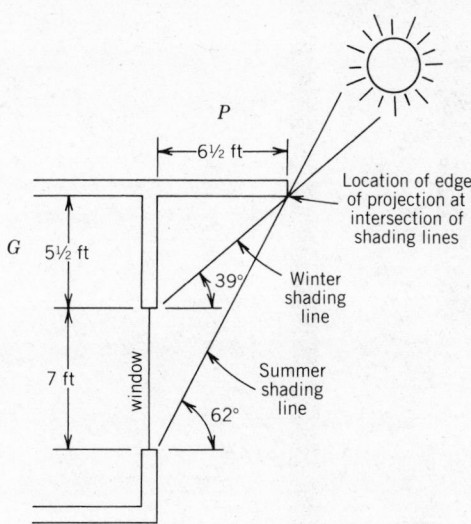

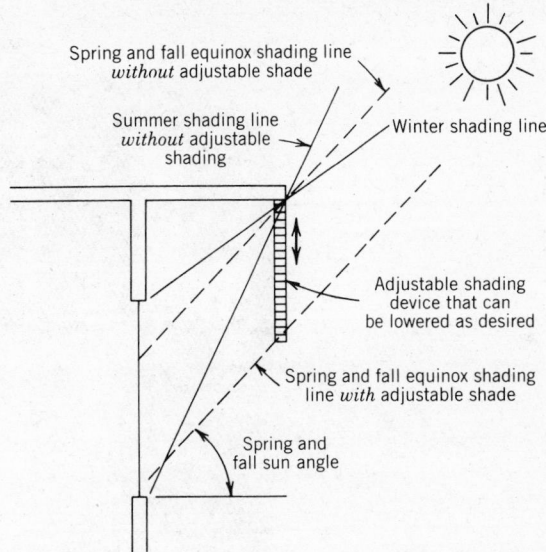

Figure 7-18 Geometric construction to determine the overprojection and gap for an overhang (Illustrative Problem 7-7). The winter and summer shading lines are shown. The point of intersection of the winter and summer shading lines determines the edge of the overhang.

Figure 7-19 Exterior adjustable shading device used to provide shading when the sun has a profile angle less than the profile angle of the summer shading line.

The shading provided by a fixed overhang is symmetric around the summer solstice; there are equal numbers of days when the window is completely shaded before and after the summer solstice. The number of days the overhang does not provide shade is also symmetric around the winter solstice; there are equal numbers of days when the window receives no shade before and after the winter solstice. This was the case for the overhang designed for illustrative problem 7-7.

The maximum heating and cooling requirements of a typical residence occur, in general, approximately 30 to 45 days after the respective solstices. Planning the shading on the basis of the solar cycle rather than with the heating and cooling cycle in mind can lead to problems in the spring and fall. For example, a particular design may require shading in September and solar gains during March. The solution to this problem is to provide summer shading that can be lowered, as shown in Fig. 7-19. For skylights, interior shades or sliding shutters can be mounted underneath to provide sun control. In many cases, the interior shading device is also an insulating barrier, which will insulate the aperture when closed.

7-8   Shading and Insulating Techniques

Many shading and insulating devices are available on the market for vertical glazings, windows, and doors. Figure 7-20 shows a roller shade for windows. Horizontal and vertical blinds can be used as shown in Figs. 7-21 and 7-22 to control solar gains. *Skylid*® is another solar control device which is also an insulating barrier when closed. Shown in Fig. 7-23, Skylid® insulating louvers consist of one to three insulation-filled louvers which fit beneath skylights or behind windows. The louvers are precisely balanced on smooth bearings and have pairs of refrigerant-filled canisters connected by tubing attached to them. One canister of each pair is on the outside near the edge of the louver, and the other is on the inside near the opposite edge of the louver; thus, the canisters sense outside and inside temperatures. The two canisters for each louver are

Figure 7-20 *Window Showcase,*® a roller shade type of screen which acts as an insulating cover for windows at night and blocks unwanted solar gains during the day. (Appropriate Technology Corporation.)

Figure 7-21 Horizontal blinds are a traditional method of providing shading.

Figure 7-22 Vertical blinds can be used to control the amount of solar energy entering a room and to direct it to chosen areas of the room.

Skylid®
insulating
louvers

Refrigerant
filled canister

Figure 7-23 *Skylid®* insulating louvers can be used under skylights to provide control of solar radiation and to act as an insulating barrier between the skylight and the room. (Zomeworks Corporation.)

supplied with just enough Freon to fill one canister.[3] The difference in vapor pressure that results from the different temperatures of the two canisters drives the volatile refrigerant to the cooler of the two canisters, resulting in a weight imbalance that either opens or closes the louver.

THERMAL STORAGE MASS

As emphasized earlier, an important aspect of dealing with a direct-gain passive solar design is the control of the interior temperature of the structure. All passive structures have some interior temperature fluctuation throughout the day–night cycle. This is the result of the thermal flywheel effect (section 6-5) inherent in the structure. The magnitude of the temperature fluctuation is determined by the ratio of thermal storage heat-storing capacity to the solar collection area (aperture), and also by the distribution of the thermal storage within the structure. Heat energy can be stored in thermal storage as either sensible heat or latent heat (see section 3-4).

7-9   Sensible Heat Storage

Selected properties of water, several masonry thermal storage materials, and some standard building materials are compared in Table 7-4. Included are the standard material properties such as conductivity K, specific heat c, and density D. A fourth property, *volumetric thermal capacity* C_V, is also included. It is derived from the product of the specific heat capacity c and the density D of a material ($C_V = cD$). The resulting units of volumetric thermal capacity are, then, Btu/lb-F° × lb/ft³ or Btu/ft³-F°.

Consider, for example, concrete with a density of 140 lb/ft³ and a specific heat capacity of 0.20 Btu/lb-F°.

$$\text{Volumetric thermal capacity } C_V = c \times D$$
$$= 0.20 \text{ Btu/lb-F°} \times 140 \text{ lb/ft}^3$$
$$= 28 \text{ Btu/ft}^3\text{-F°}$$

The materials listed in Table 7-4 are all sensible heat storage materials. This

[3] *Freon* is the trade name for the common refrigerant dichlorodifluoromethane, as supplied by the E. I. DuPont Company.

Table 7-4   Properties of Selected Sensible Heat Storage Materials

Masonry Materials	Thermal Conductivity K (Btu/hr-ft²-F°/ft)	Specific Heat Capacity c (Btu/lb-F°)	Density D (lb/ft³)	Volumetric Thermal Capacity C_V (Btu/ft³-F°)
Adobe	0.30	0.24	106	25.4
Brick (common)	0.40	0.20	123	24.6
Brick (magnesium additive)	2.20	0.22	158	35.1
Concrete (dense)	1.0	0.20	140	28
Stone (average)	—	0.20	90	18
Water (for comparison)	—	1.00	62.4	62.4
Building materials **Woods**				
Oak	0.102	0.57	47	26.8
Fir or pine	0.068	0.33	27	8.9
Plywood (Douglas fir)	0.068	0.29	34	9.9
Gypsum or plaster board	0.09	0.26	50	13.0

Source: ASHRAE *Handbook of Fundamentals,* 1977. Atlanta: ASHRAE, 1977, pp. 22.14–22.17, pp. 37.3–37.4; and manufacturers' data.

means that the gain or loss of energy by the material can be detected by a temperature change. Thus, their temperature increases as they absorb heat, and it falls as they lose heat.

7-10 Thermal Storage in the Form of Latent Heat

Another form of heat storage is found in materials that exhibit a phase change (change of state) during the absorption or loss of energy. This type of thermal storage is termed latent heat storage, because the temperature of the materials remains constant during the change of phase and thus cannot be sensed by a temperature change. The most common example of latent heat storage is the melting of ice into water. As the ice absorbs heat, it changes its physical state into water. The temperature of the ice and water remains at 32°F during the process despite the fact that heat is being added (see Fig. 3-2). Certain chemicals, often called *eutectic salts* (see section 3-5), exhibit a marked capacity for latent heat storage during phase change, in the range of 80°F to 90°F, when used in solar applications. These are often packaged in containers that are flat or tubular, to facilitate installation, as shown in Fig. 7-24. Table 7-5 gives the physical properties of some latent heat storage materials. They have relatively large thermal storage capacities when compared to sensible heat storage materials.

The volumetric thermal capacity C_V of latent heat storage materials is the product of the latent heat of phase change and the density of the material: $C_V = L_f \times D$, where L_f is the latent heat of phase change in Btu per pound.

Consider calcium chloride hexahydrate with a latent heat of phase change of 82 Btu/lb and a density of 97 lb/ft³:

$$
\begin{aligned}
\text{Volumetric thermal capacity } C_V &= L_f \times D \\
&= 82 \text{ Btu/lb} \times 97 \text{ lb/ft}^3 \\
&= 7954 \text{ Btu/ft}^3
\end{aligned}
$$

The thermal capacity of latent heat storage (eutectic) materials is not dependent upon temperature because the heat is stored at the temperature of phase

Figure 7-24 Latent heat storage packaged in a flat container. The containers can be mounted flat on the surface of a sunlit wall. (Dow Chemical Company.)

Table 7-5 Properties of Selected Latent Heat Storage Materials

Phase-Change Material	Approximate Temperature at Phase Change (°F)	Latent Heat of Phase Change (Btu/lb)	Density of Liquid (lb/ft³)	Volumetric Thermal Capacity C_V (Btu/ft³)
Calcium chloride hexahydrate $CaCl_2 \cdot 6H_2O$	81	82	97	7950
Sodium sulfate decahydrate $Na_2SO_4 \cdot 10H_2O$ (Glauber's salt)	89	97	83	8050

change, which for those eutectic materials most used with solar applications is in the range 80°F to 95°F. After completing the phase change, eutectic materials provide sensible heat storage in accord with their mass and specific heat.

7-11 Selection and Placement of Thermal Storage Mass

Absorption of Solar Radiation by Thermal Storage.   After solar radiation is transmitted through glazing and enters a building, it falls on some part of the building interior. The radiation striking the building material will either be absorbed or reflected. Any reflected radiation will usually strike another interior surface and be absorbed there. Energy absorbed at the surface of a material will flow into the material by conduction, and for sensible heat storage materials the absorption of energy will cause an increase in the temperature of the material as given by the equation:

$$\Delta t = \frac{H}{V \times C_V} \text{ (a form of the basic heat equation)} \qquad (7-5)$$

where

H = heat absorbed, Btu

V = volume of material, ft³

C_V = volumetric thermal capacity of the material in Btu/ft³-F°

Δt = change in temperature of material, F°

Consider the thermal effect on the three materials shown in Fig. 7-25 when they absorb 100 Btu of solar radiation during one hour (assuming for simplicity that there are no other energy gains or losses). For the square foot of 1-in. plywood (*a*), the volume of material is 1/12 ft³. From Table 7-4, C_V for fir plywood is 9.9 Btu/ft³-F°. From Eq. 7-5, solving for Δt,

$$\Delta t = \frac{100 \text{ Btu}}{1/12 \text{ ft}^3 \times 9.9 \text{ Btu/F°-ft}^3} = 121 \text{ F°}$$

Likewise, for 1 ft² of the 6-in. brick wall (*b*),

$$\Delta t = \frac{100 \text{ Btu}}{0.5 \text{ ft}^3 \times 24.6 \text{ Btu/F°-ft}^3} = 8.13 \text{ F°}$$

And for 1 ft² of the 6 in. of water (*c*),

$$\Delta t = \frac{100 \text{ Btu}}{0.5 \text{ ft}^3 \times 62.4 \text{ Btu/F°-ft}^3} = 3.2 \text{ F°}$$

From this example, it can be seen that building surfaces made from light construction, such as 2 × 4 stud walls with plywood or plaster board surfaces, do not store energy very well. When exposed to solar radiation, a small quantity of solar energy increases their temperature quite rapidly because of their low volumetric thermal capacity. This effect is very undesirable in a direct-gain building, because surfaces of materials with a low thermal capacity will transfer

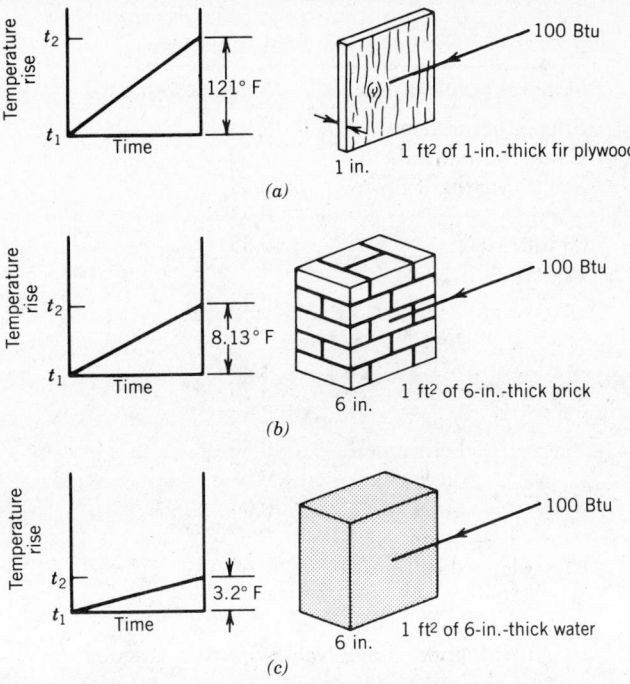

Figure 7-25 When solar radiation strikes building materials, they heat up, with the temperature rise dependent upon the thermal storage capacity of the particular material. Materials used in frame walls, such as plywood and gypsum, store small amounts of energy and experience large temperature rises. Masonry and water store larger quantities of heat and have smaller temperature rises.

the solar energy they absorb to the room air by convection from their relatively hot surfaces, resulting in a rapid rise of space temperature, perhaps well beyond the upper limit of the comfort zone. *Materials with a low thermal capacity must not be located where solar radiation will fall directly on them.* Heavier materials like water and brick ordinarily have larger thermal capacities and absorb greater amounts of heat for a given temperature rise. These materials should be placed in the direct sunlight so that they will be able to absorb the maximum amount of solar radiation. Their relatively small temperature rise will result in a comfortable space air temperature and minimum temperature variation during the day–night cycle. A well-designed direct-gain space will experience an interior temperature fluctuation of only 10 to 15 F° over a 24-hour period, while a poorly designed one may experience a 30 to 50 F° temperature fluctuation. The selection and location of thermal storage is critical to the performance of any direct-gain solar system.

Surface Colors. The surface layer of a material and its color will affect the amount of solar energy absorbed or reflected by that material.

The *absorptance* of a surface was defined in section 3-17 as the ratio of the absorbed solar energy to the incident solar energy. As an equation, absorptance

$$\alpha = \frac{\text{Absorbed solar energy}}{\text{Incident solar energy}} \tag{7-6}$$

Table 7-6 gives values of the absorptance of solar radiation for selected surfaces.

For direct-gain spaces, the following guidelines should be used for selecting surface colors. Masonry floors should have a medium dark color. All lightweight construction, such as woodframe partitions with low thermal capacity, should be light-colored to reflect solar radiation to more effective thermal storage surfaces. Masonry walls may have a range of colors, since solar radiation

Table 7-6 Absorptance of Solar Radiation by Various Surfaces (Normal Incidence)

Material or Surface	Solar Absorptance α
Brick, common red	0.68
Concrete (natural)	0.60
Sandstone, light fawn	0.54
red	0.73
Granite	0.55
Paints	
white	0.18
yellow	0.33
dark red	0.57
brown	0.79
gray	0.75
light green	0.50
dark green	0.88
black	0.94
Flat black	0.97
Clay tile, reddish brown	0.69
Wood, pine	0.60

In general, approximate values may be assigned using the following ranges:

White, smooth surfaces	0.25–0.40
Gray to dark gray	0.40–0.50
Green, red, and brown	0.50–0.70
Dark brown to blue	0.70–0.80
Blue to black	0.80–0.90

reflected from bright-colored masonry walls with an absorption of 0.20 to 0.30 will in turn be absorbed by other, darker, masonry surfaces. Water storage containers should be a dark color, black being best, but dark blue, green, or red show only a 5 to 10 percent decrease in storage performance. Latent heat storage containers should also be a very dark color in order to have maximum absorption of solar radiation.

Illustrative Problem 7-8

Calculate the difference in absorbed solar radiation for a cellulose white-painted surface and a cellulose black-painted surface. Assume that the incident solar energy (insolation) is 200 Btu/hr-ft^2 for both surfaces.

Solution

For the cellulose white surface (Table 7-6):

$$\text{Absorbed solar energy} = \alpha \times \text{incident solar energy}$$
$$= 0.18 \times 200 \text{ Btu/hr-ft}^2$$
$$= 36 \text{ Btu/hr-ft}^2$$

For the cellulose black surface:

$$\text{Absorbed solar energy} = 0.94 \times 200 \text{ Btu/hr-ft}^2$$
$$= 188 \text{ Btu/hr-ft}^2$$

The difference in absorbed solar radiation is

$$188 \text{ Btu/hr-ft}^2 - 36 \text{ Btu/hr-ft}^2 = 152 \text{ Btu/hr-ft}^2 \qquad \textit{Ans.}$$

Thermal Performance of Construction Materials.   Values of volumetric thermal capacity C_V and heat conductivity K for selected building materials were given in Table 7-4. The volumetric thermal capacity for adobe, brick, and concrete are approximately the same, ranging from 24 to 28 Btu/ft³-F°. However, there is a wide range in thermal *conductivity* for these materials. The thermal conductivity determines how rapidly heat is conducted from the sunlit surface to the interior portions of the material, and this rate of heat transfer controls to some extent the surface temperature of the sunlit surface.

To understand how the thermal conductivity of a material affects its performance as thermal storage, consider the simple direct-gain room in Fig. 7-26. There is a 12-in.-thick thermal storage wall on the north wall of the room and 50 ft² of south-facing clerestory window that is the solar collection area. The actual solar radiation absorbed on the sunlit surface of the thermal storage mass is shown hour by hour by the plot of Fig. 7-27. A computer model of the room was then used to predict the temperature of the thermal storage at the four locations T_1, T_2, T_3, and T_4, as indicated in Fig. 7-26b. The predicted temperatures are shown hour by hour by plots for both a concrete wall and an adobe

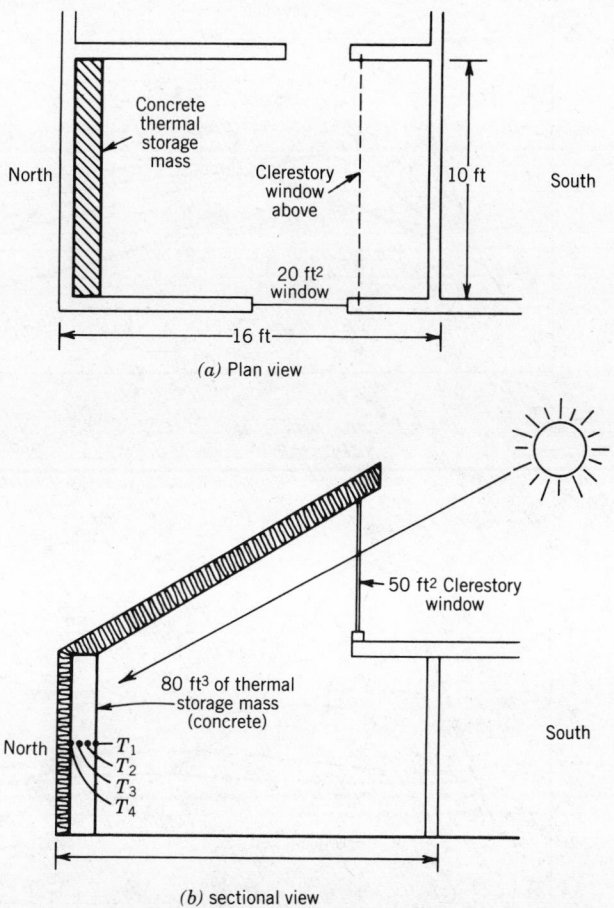

Figure 7-26 (*a*) Floor plan of a direct-gain room showing the locations of thermal storage mass and a clerestory window. (*b*) Section through room showing the clerestory window and thermal storage. The locations in the thermal storage wall for temperatures T_1, T_2, T_3, and T_4 are also indicated. T_1 is the surface temperature of the sunlit thermal storage; T_2 is the temperature 4 in. into the thermal storage; T_3 is the temperature 8 in. into the thermal storage; and T_4 is the temperature of the back surface of the thermal storage 12 in. from the absorbing surface.

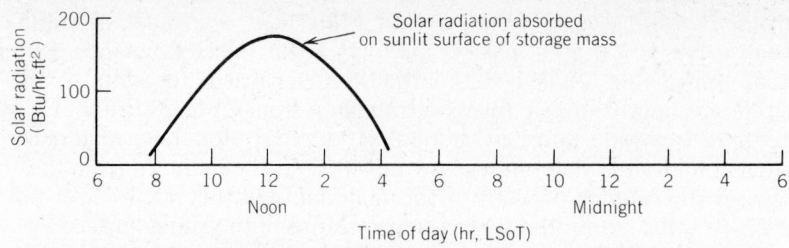

Figure 7-27 The solar radiation absorbed by the thermal storage wall of Fig. 7–26 on an hour-by-hour time scale.

wall, respectively, in Figs. 7-28a and 7-28b. These temperatures are based on the absorbed solar radiation indicated in Fig. 7-27.

The sunlit surface of the concrete wall heats up quickly, as shown in Fig. 7-28a, and then the interior parts of the concrete increase in temperature as heat is conducted into the concrete. It is interesting to note that there is approx-

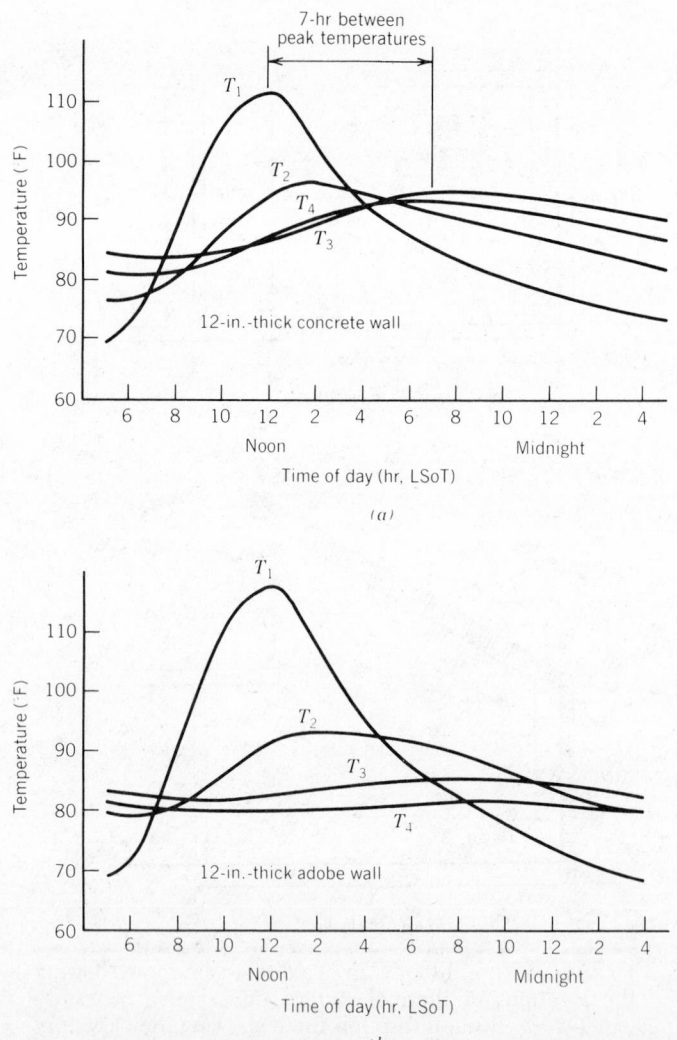

Figure 7-28 (a) Computed values of T_1, T_2, T_3, and T_4 on an hour-by-hour basis for the thermal storage temperatures of the 12-in.-thick concrete wall shown in Fig. 7–26. (b) Computed thermal storage temperatures for a 12-in.-thick adobe wall used as thermal storage as shown in Fig. 7–26.

imately a 7-hr time period after T_1 reaches its peak until T_4 reaches its peak. This is the length of time required for the heat that was absorbed on the surface to reach the far side of the thermal storage.

From Fig. 7-28b, it is noted that the sunlit surface of the adobe wall heats up quickly, as did the concrete wall. But the adobe wall attains a somewhat higher surface temperature than does the concrete wall. Since the conductivity of the adobe is lower, heat flows from the surface more slowly, and the surface therefore reaches a higher temperature. The temperature T_2 is lower for the adobe than for the concrete, since heat is not conducted through adobe as readily as it is through concrete. Less heat is conducted into the interior of the adobe wall, with the result that the interior temperature rise is lower. The temperature changes for T_3 and T_4 for the adobe wall are small compared to the temperature changes for the concrete wall, indicating that very little heat is being stored in the last third of the adobe wall because of its low conductivity. The indication here is that materials with low heat conductivities should be kept thin when used as thermal storage in a direct-gain situation, while materials with high conductivities can be made much thicker.

Water has a volumetric thermal capacity that is twice as great as concrete or masonry. Water is very effective as thermal storage because the entire volume of water will heat up uniformly, thus using all of its volumetric thermal storage capacity. The water is kept at a uniform temperature by thermally induced convection currents as shown in Fig. 7-29a. The energy absorbed by the sunlit

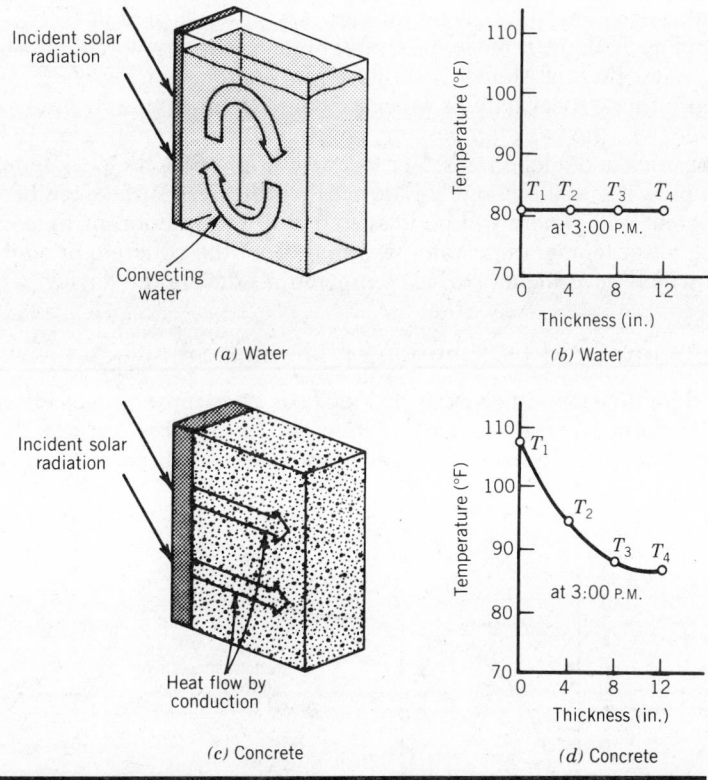

Figure 7-29 Water and concrete compared as thermal storage. (a) As heat flows into water from a surface absorbing solar radiation, convection currents are established in the water, mixing the water and heating it uniformly. (b) A thickness–temperature graph for water shows a uniform temperature throughout the water mass. (c) Solar radiation heats the surface of concrete, and heat flows from the surface into the concrete by conduction. The heat flow is, however, a function of time, and the surface remains hotter than the interior. (d) A thickness–temperature graph for concrete shows the decrease in temperature at points of increasing depth into the concrete.

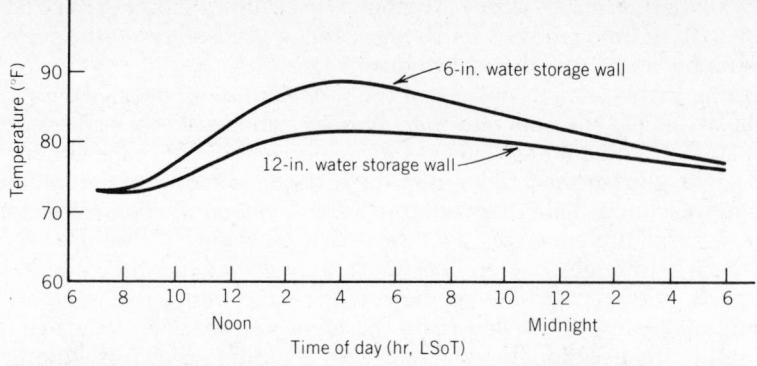

Figure 7-30 Time–temperature graph for a 6-in. and a 12-in. water wall placed in the room of Fig. 7–26. Compare these curves with those of Fig. 7-28 for concrete and adobe.

surface of the water container heats the water in contact with this surface. The heated water is less dense and rises, creating a convection loop in the container that distributes the heat throughout the container. Because the water container uses all of its volumetric thermal capacity, it attains a uniform temperature throughout (Fig. 7-29b), while concrete or masonry thermal storage has a temperature gradient through the storage mass, as indicated in Figs. 7-29c and 7-29d. In Fig. 7-30, the hour-by-hour water temperature is plotted for a 6-in. and a 12-in. water thermal storage wall placed in the room of Fig. 7-26. Since the volumetric thermal capacity of water is about 2.5 times that of brick or adobe (see Table 7-4), the peak temperatures of water walls are not nearly as great as those for brick and adobe walls. For example (Fig. 7-30), the peak temperature of the 6-in. water wall is about 20F° lower than the peak surface temperature of the concrete wall. There will be less heating of the interior air by convection from the lower temperature water wall than from the concrete or adobe walls, both of which have higher surface temperatures during the day.

7-12   Amount and Distribution of Thermal Storage

The total thermal storage capacity provided has a very pronounced effect on the overall performance of a passive system, as Fig. 7-31 illustrates. Based on research data obtained from the study of many typical direct-gain passive solar

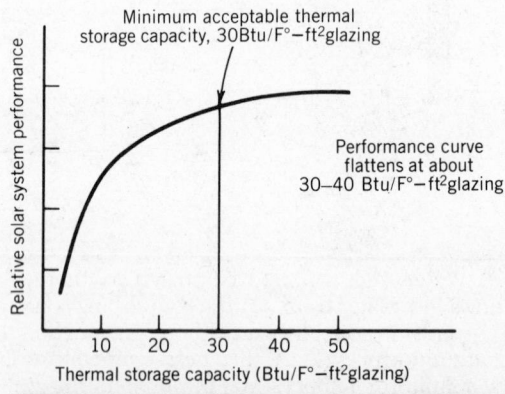

Figure 7-31 The performance of a passive system is affected by the thermal storage. Optimum storage capacity is in the range of 30 to 40 Btu/F°-ft² glazing. When storage capacity is increased above that, the increase in system performance is slight.

DIRECT-GAIN PASSIVE SYSTEMS

energy systems, the relative solar system performance can be plotted as a function of thermal storage capacity in Btu/F°-ft² of solar aperture. From the plot of Fig. 7-31, it can readily be seen that in the range of 30–40 Btu/F°-ft² glazing, the performance curve begins to flatten out, indicating that additional thermal storage capacity will result in diminishing returns in performance.

Guideline for Minimal Thermal Storage Capacity.   The minimum acceptable thermal storage capacity can be set at about 30 Btu/F°-ft² glazing for direct-gain passive systems. If less than 30 Btu/F° of thermal storage capacity is used per square foot of solar aperture, system performance will be reduced significantly because less energy is stored and more overheating will probably occur. Energy that is not stored and that contributes to overheating is usually vented from the building as warm air. Another method of preventing overheating is to reduce the solar collection area by a shading or shutter system. When inadequate thermal storage is used, the daytime space temperatures typically will be too high, while the night space temperatures are too low. An unsatisfactory space-heating system is the result.

The upper limit of thermal storage capacity is ordinarily determined by cost–benefit analysis based on performance curves like that of Fig. 7-31.

Illustrative Problem 7-9

Calculate the volume of (a) concrete and (b) water to provide 40 Btu/F°-ft² glazing of thermal storage capacity.

Solution

From Table 7-4, note that the thermal capacity for concrete is 28 Btu/F°-ft³, and that for water is 62.4 Btu/F°-ft³.

(a) The required volume of concrete is

$$\frac{40 \text{ Btu/F°-ft}^2 \text{ glazing}}{28 \text{ Btu/F°-ft}^3}$$

$$= 1.43 \text{ ft}^3/\text{ft}^2 \text{ glazing} \qquad\qquad Ans.$$

(b) The required volume of water is

$$\frac{40 \text{ Btu/F°-ft}^2 \text{ glazing}}{62.4 \text{ Btu/F°-ft}^3}$$

$$= 0.64 \text{ ft}^3/\text{ft}^2 \text{ glazing} \qquad\qquad Ans.$$

Distribution of Thermal Storage.   The placement of thermal storage in the direct-gain space influences the thickness and exposed surface area required. With sensible heat storage materials as interior surfaces, three possible situations exist. First, these materials may be exposed to direct solar radiation as in Fig. 7-32a. The absorbing surfaces are floors and walls of the room with one side receiving direct solar radiation and the other side (outside) being insulated. This storage geometry will be termed *solar-irradiated storage*. Figure 7-32b shows curves for several thermal storage materials obtained by plotting material thickness (horizontal axis) against the ratio of thermal storage surface area to solar collection area (vertical axis). If a 4-in.-thick concrete wall is used for thermal storage, Fig. 7-32b indicates that the ratio of storage surface area to solar aperture area is approximately 4. This means that 4 ft² of 4-in.-thick concrete wall must be placed in the building for each square foot of solar collection area. The concrete should be placed so that it receives direct solar radiation for at least four or more hours during the day.

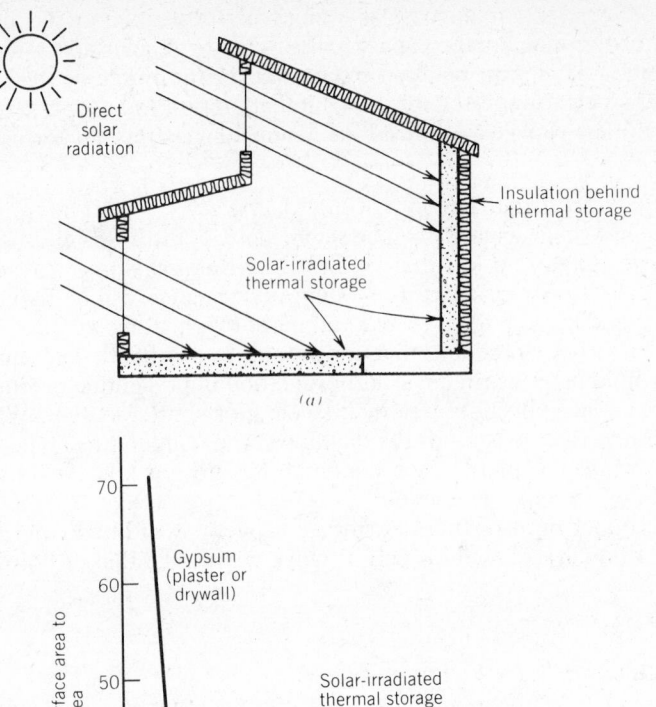

Figure 7-32 (*a*) *Solar-irradiated thermal storage* in a direct-gain solar application. The thermal storage is placed where solar radiation will strike it for 4 to 6 hours during the day. (*b*) Solar-irradiated thermal storage—the thickness of thermal storage material for different ratios of storage surface area to solar collection area. (Adapted with permission from *The Thermal Mass Pattern Book*, Total Environmental Action, Inc., Harrisville, N.H., 1980, pattern 1, p. 3.)

Secondary Irradiated Storage.   The second situation occurs when the thermal storage is located in positions where it does not receive direct solar radiation during the day (Fig. 7-33*a*). This thermal storage, usually ceiling surfaces and walls not in the path of direct solar radiation, receives heat from reflected solar radiation, convection of warm air, and thermal radiation from warmer surfaces

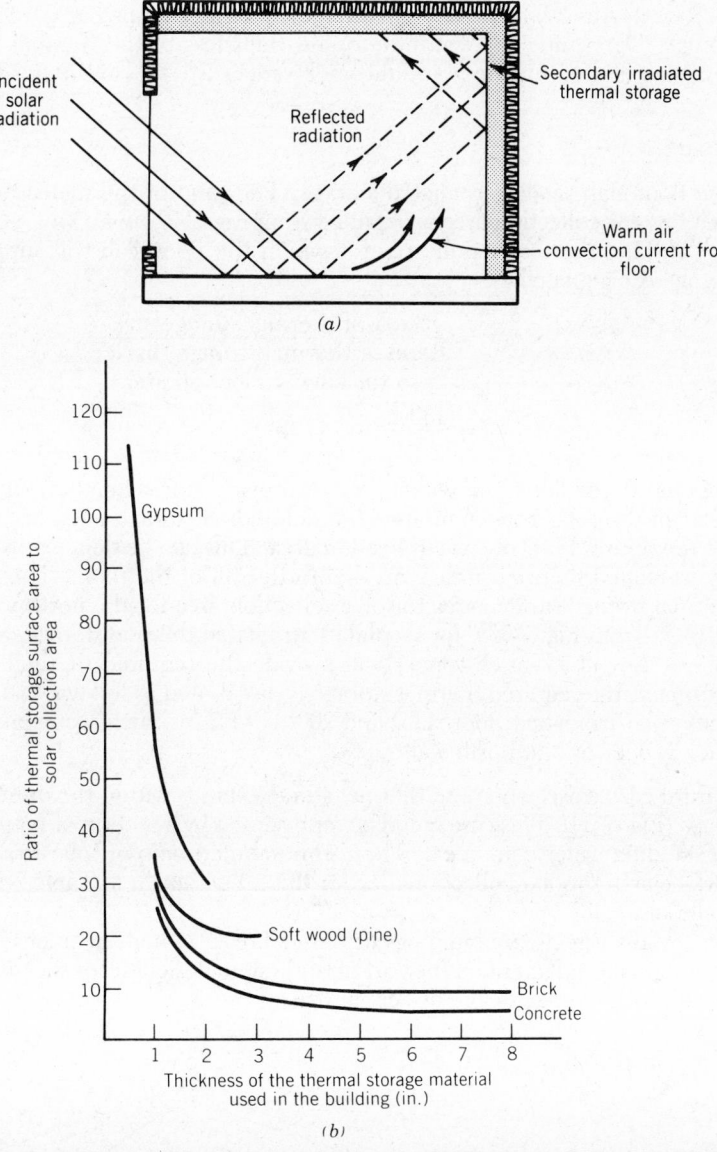

Figure 7-33 (a) Thermal storage located where sunlight does not reach it during the day is termed secondary irradiated thermal storage. This thermal storage receives energy from reflected solar radiation and convection of warm air. (b) The thickness of thermal storage material for different ratios of storage surface area to solar collection area, in the case of secondary irradiated storage. (Adapted with permission from *The Thermal Mass Pattern Book*.)

in the room. Such thermal storage geometry will be termed *secondary irradiated storage*. Again, material thickness is plotted against the ratio of thermal storage surface area to solar collection area, and the results for secondary irradiated storage are shown in Fig. 7-33b. The required surface area and volume for secondary irradiated storage are almost twice the values for direct solar-irradiated storage for a similar thermal performance.

Illustrative Problem 7-10

A room has a solar collection area of 60 ft², consisting of south-facing glass. It is estimated that 180 ft² of the floor will receive direct solar radiation during the day and can be used as thermal storage. The total floor area of the room is 250 ft². The north wall has 200 ft² of surface area that can also be used for thermal storage. Determine the distribution and thickness of the thermal storage if a 4-in. slab is used for the floor and brick veneer is used on the north wall.

Solution

The floor slab is solar-irradiated storage. The ratio of thermal storage surface area to solar collection area is obtained from Fig. 7-32b for a 4-in. concrete slab as 4.0. The solar collection area for which the floor slab will supply thermal storage is calculated as:

$$A_W = \frac{\text{Area of thermal storage slab}}{\text{Ratio of thermal storage surface to the solar collection area}}$$

$$= \frac{180 \text{ ft}^2}{4} = 45 \text{ ft}^2$$

The 180 ft² of floor slab will therefore provide sufficient thermal storage for 45 ft² of the solar collection area, and additional storage must be provided for the remaining 15 ft² of solar collection area. This can be done by using secondary irradiated thermal storage on the north wall of the room. The ratio of the thermal storage surface area to solar collection area for the north wall is 200/15 = 13.3. From Fig. 7-33b, for secondary irradiated thermal storage, for a ratio of 13.3, a 2-in.-thick brick veneer will provide the required thermal storage. In summary, the required thermal storage is distributed as follows: 180 ft² of 4-in.-thick solar-irradiated floor slab, and 200 ft² of 2-in.-thick secondary irradiated brick veneer on the north wall.   *Ans.*

A third case exists when the thermal storage stands within the direct solar gain space (Fig. 7-34). This case could be represented by a section of masonry wall or by standing water containers. The recommended ratio of solar-irradiated mass surface area to solar collection area for this case is given in Table 7-7 for several materials.

Many direct-gain buildings will use all three storage geometries, requiring discretionary judgment on the part of the system designer, for the distribution of the thermal storage in the direct-gain space.

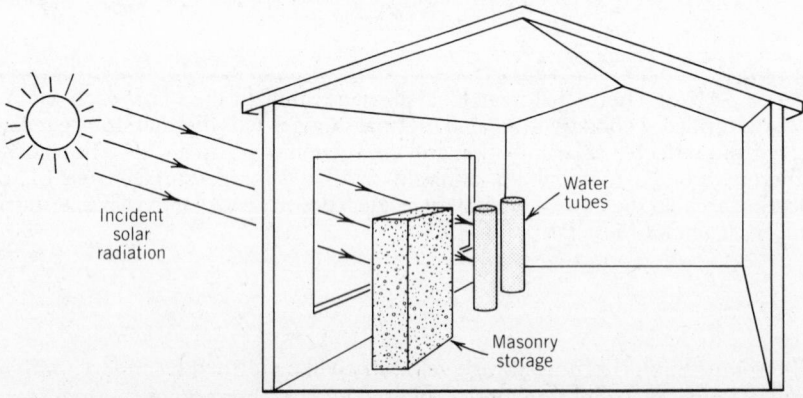

Figure 7-34 Two kinds of free-standing thermal mass. In this case, the thermal mass has all surfaces exposed to the air in the building space. Air is free to circulate around the thermal storage, and the space is heated by all the surfaces of the storage.

Table 7-7 Free-Standing Thermal Storage in Direct Solar Gain Space

Material	Ratio of Irradiated Mass Surface Area to Solar Collection Area
Brick, 8 in. thick	2
Concrete, 6 in. thick	2
Water tubes	At least 7 gal of water per square foot of solar collection area. This is approximately 1 ft³ of water per square foot of solar collection area

Source: Adapted with permission from *The Thermal Mass Pattern Book* (Harrisville, N.H.: Total Environmental Action, Inc., 1980), p. 7.

Problems on Providing Thermal Storage.   The following examples illustrate methods used to size and locate thermal storage for direct-gain applications. The *total thermal storage capacity* S_T is calculated by:

$$S_T = A_W \times C_A \tag{7-7}$$

where

S_T = total thermal storage capacity in Btu/F° for the entire room or space

A_W = the solar collection area, or glazed area, in square feet

C_A = the thermal storage capacity per square foot of solar collection area in Btu/F° per square foot of glazing

The *volume of a given storage material* to be used is determined by:

$$V = \frac{S_T}{C_V} \tag{7-8}$$

where

V = the volume of storage material, ft³

S_T = the total thermal storage capacity, Btu/F°, as defined for Eq. (7-7)

C_V = the volumetric thermal storage capacity of the specified material, Btu/F°-ft³, as previously defined

Illustrative Problem 7-11

A room is designed to be heated by direct solar gain. The room is 15 ft by 10 ft and has 40 ft² of window on the south (15-ft) wall. Using the minimum thermal storage guideline, determine the minimum volume of concrete to be used for thermal storage.

Solution

The minimum acceptable thermal storage capacity per square foot of aperture C_A is 30 Btu/F°-ft² glazing (Fig. 7-31). The total thermal storage S_T is obtained from Eq. (7-7):

$$S_T = A_W \times C_A$$
$$S_T = 40 \text{ ft}^2 \text{ glazing} \times 30 \text{ Btu/F}°\text{-ft}^2 \text{ glazing}$$
$$= 1200 \text{ Btu/F}°$$

The volume of concrete is found from Eq. (7-8):

$$V = \frac{S_T}{C_V} \text{ where } C_V = 28 \text{ Btu/F}° \text{ for concrete (from Table 7-4)}$$
$$V = \frac{1200 \text{ Btu/F}°}{28 \text{ Btu/F}°\text{-ft}^3}$$
$$= 42.8 \text{ ft}^3, \text{ the minimum volume of concrete thermal storage} \quad Ans.$$

The distribution of the thermal mass must be determined after the minimum value of the mass is determined. Several methods exist for doing this. The sunlit surfaces of the room can be ascertained from a sketch using the profile angle and the solar azimuth angle, and then the thermal mass can be assigned to these surfaces. The curves of Fig. 7-32b and Fig. 7-33b can be used to obtain surface areas of thermal mass for a given thickness of material. These curves will usually provide a thermal storage capacity larger than the minimum required value. The decision to use the minimum thermal capacity or a larger value will be influenced by cost considerations, and by the desired space–temperature swing (a larger thermal capacity will reduce the space–temperature swing).

Illustrative Problem 7-12

From the previous problem, assume that one half of the floor is sunlit during the day and can be used effectively as thermal storage. Determine the thickness of the floor for the minimum storage capacity.

Solution

The floor area is 10 ft by 15 ft = 150 ft^2, making the sunlit area that can be used for storage 75 ft^2. The thickness of the floor is then calculated as:

$$\text{Thickness} = \frac{\text{Minimum allowable volume of thermal storage}}{\text{Area used for storage}}$$
$$= \frac{42.8 \text{ ft}^3}{75 \text{ ft}^2}$$
$$= 0.57 \text{ ft or } 6.9 \text{ in.} \hspace{2cm} Ans.$$

Illustrative Problem 7-13

Determine the distribution of thermal mass for illustrative problem 7-11, using Figs. 7-32b and 7-33b. Assume that the building will have a standard 4-in. slab floor and that only one half of the floor will be solar irradiated during sunny hours. The solar collection area is 40 ft^2.

Solution

From Fig. 7-32b, note that for 4-in. concrete the ratio of solar-irradiated thermal storage surface area to solar collection area is approximately 4.0. The surface area of thermal storage required will therefore be:

$$\text{Storage surface area} = 4 \times 40 \text{ ft}^2$$
$$= 160 \text{ ft}^2$$

However, the floor area is only 150 ft^2, and only part of the floor area is effective as thermal storage. Therefore, additional thermal storage must be placed in the room. Since only one half of the floor is effective as thermal storage, then it alone will provide storage for a solar collection area given by:

$$A_W = \frac{0.5 \times 150 \text{ ft}^2}{4}$$
$$= 18.8 \text{ ft}^2$$

The total solar collection area for the room is 40 ft^2, and thermal storage must be provided for the remaining 21.2 ft^2. The walls of the room can be used to provide this storage. The north wall is 15 ft long and 8 ft high, yielding a surface of 120 ft^2. The end walls are 10 ft long and 8 ft high, giving a total area of 160 ft^2. The total wall area in the room is then 280 ft^2.

The walls of the room will not receive direct solar radiation, so they will be considered secondary irradiated thermal storage. They must provide thermal storage for the remaining 21.2 ft² of solar collection area.

The ratio of wall area to the remaining solar collection area is then 280/21.2 = 13.2. From Fig. 7-33b, for secondary irradiated thermal storage, for a ratio of 13, it is noted that the walls can be covered with about 2.5 to 3.0 in. of brick to provide the needed thermal storage. A standard 2½-in.-thick brick veneer can be used on the wall to provide the needed thermal storage.   *Ans.*

7-13   A Design Problem for a Direct-Gain Application

In previous sections, we have examined individual aspects of direct-gain systems. Now a complete analysis of a simple direct-gain system will be presented, using principles developed in this chapter and in Chapter 6.

Design Problem

A client has brought you a basic floor plan and sectional view of a two-room solar house to be built in Salt Lake City (Fig. 7-35). The interior walls are not shown, for simplicity, but in an actual application even conventional stud frame walls would affect the distribution and storage of thermal energy in the building to some extent, and their effect would have to be considered. The house is to be heated by direct solar gain.

The net load coefficient (NLC) for the house (*net* means that the solar aperture is excluded) has been determined as 3120 Btu/DD (4.6 Btu/DD-ft² of floor area, see Table 6-2). Room A will have one sliding glass door and three windows on the south wall. Room B will have five clerestory windows. Thermal storage can be placed along the walls or in the floor. Location of the house is 42.7° N Lat, with an average winter temperature of 30°F. Floors will be 4-in.-thick concrete slab.

You are to determine the total glazing area required and then size and locate the thermal storage.

Solution

Problem Approach.   Several approaches could be used to solve this problem. The chosen approach involves, first, estimating the required solar collection area; second, locating the collection area glazing on the walls of the building; third, determining the surfaces to receive solar radiation in the building; fourth, determining the amount and distribution of thermal storage mass; fifth, checking thermal storage against minimum requirements; and, finally, determining the solar savings fraction (SSF) and the annual Auxiliary Heat (AH) required.

Estimating the Solar Collection Area.   Since the client did not specify a specific percentage of the total load to be assigned to solar heating, the sizing criteria (ratios) of Table 7-1 (Mazria's Rule) will be used to estimate the window area based on the floor area. For an average winter temperature of 30°F, Table 7-1 gives the ratio of solar window area to floor area as a range from 0.19 to 0.29. Since the 42° N Lat of the site is about midway in the range of 35° to 48°, a midrange value of 0.24 will be used to determine the window area. Using this value, a reasonable window (solar collection) area for the house is calculated as:

Room A. From Fig. 7-35, room A has 384 ft² of floor space, and

$$\text{Solar window area} = 384 \text{ ft}^2 \times 0.24$$
$$= 92.2 \text{ ft}^2$$

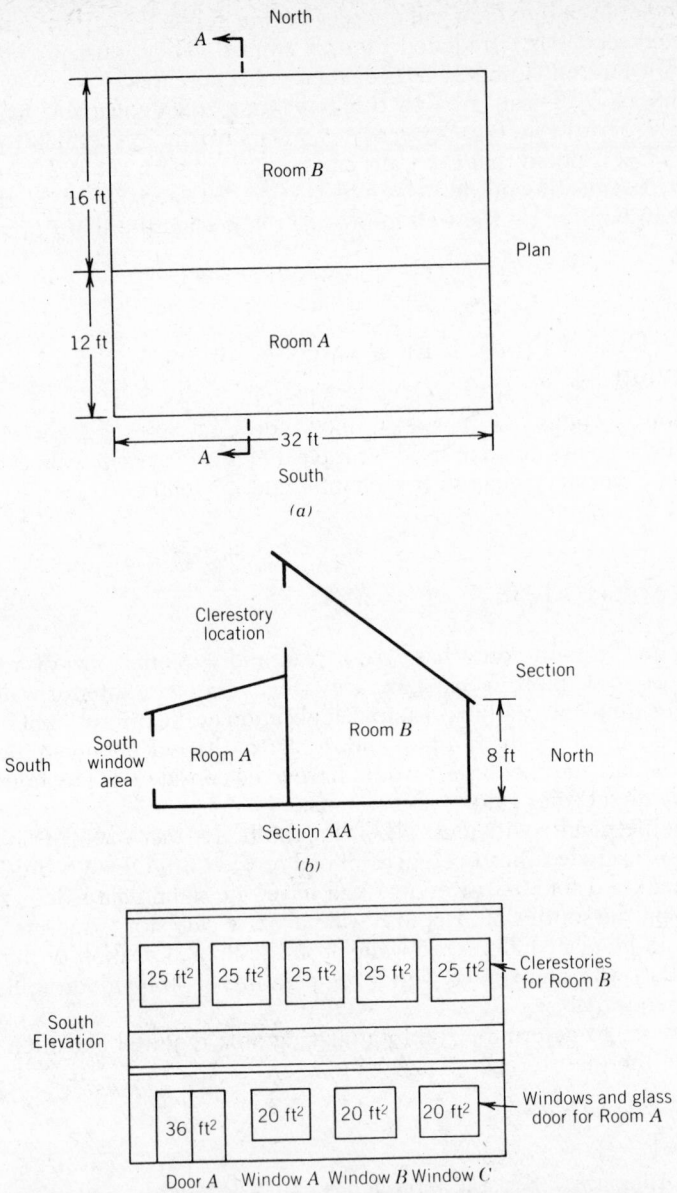

Figure 7-35 (*a*) Floor plan and (*b*) section of a simple two-room solar house to be used in connection with the Design Problem. (*c*) South elevation of the solar house showing location and size of windows (room A) and clerestories (room B).

Room B. Also from Fig. 7-35, room B has 512 ft² of floor space, and

$$\text{Solar window area} = 512 \text{ ft}^2 \times 0.24$$
$$= 122.9 \text{ ft}^2$$

Total solar collection area

$$A_W = 92.2 \text{ ft}^2 + 122.9 \text{ ft}^2$$
$$= 215 \text{ ft}^2$$

Location of Windows and Other Glazed Areas. The location of solar windows will, of course, be on south walls. For room A, the windows could be placed as shown in Fig. 7-35c. One 36-ft² sliding glass door and three 20-ft² windows will be used to give 96 ft² of south-facing double-glazed solar collection area for room A. For room B, on the north of room A, clerestory windows will be used to allow solar radiation to enter. A set of five windows, each 5 ft by 5 ft, will give 125 ft² of solar collection area for room B, as shown in Fig. 7-35c. The total solar collection area will then be 221 ft².

Determining Sunlit Portions of the Rooms. To determine the sunlit surfaces inside the rooms, a sketch showing a section of the house is made (Fig. 7-36a). The profile angles for this application are taken from Table 4-B for December 21 at a latitude of 40° N. The day can be characterized by three (local solar) times: 10:00 A.M., 12:00 noon, and 2:00 P.M. The house is oriented due south (recall that solar geometry is symmetric about local solar noon for a south-facing orientation). The profile angle at 10:00 A.M. will have the same numerical value as the profile angle at 2:00 P.M. From Table 4-B, the profile angle for a south wall

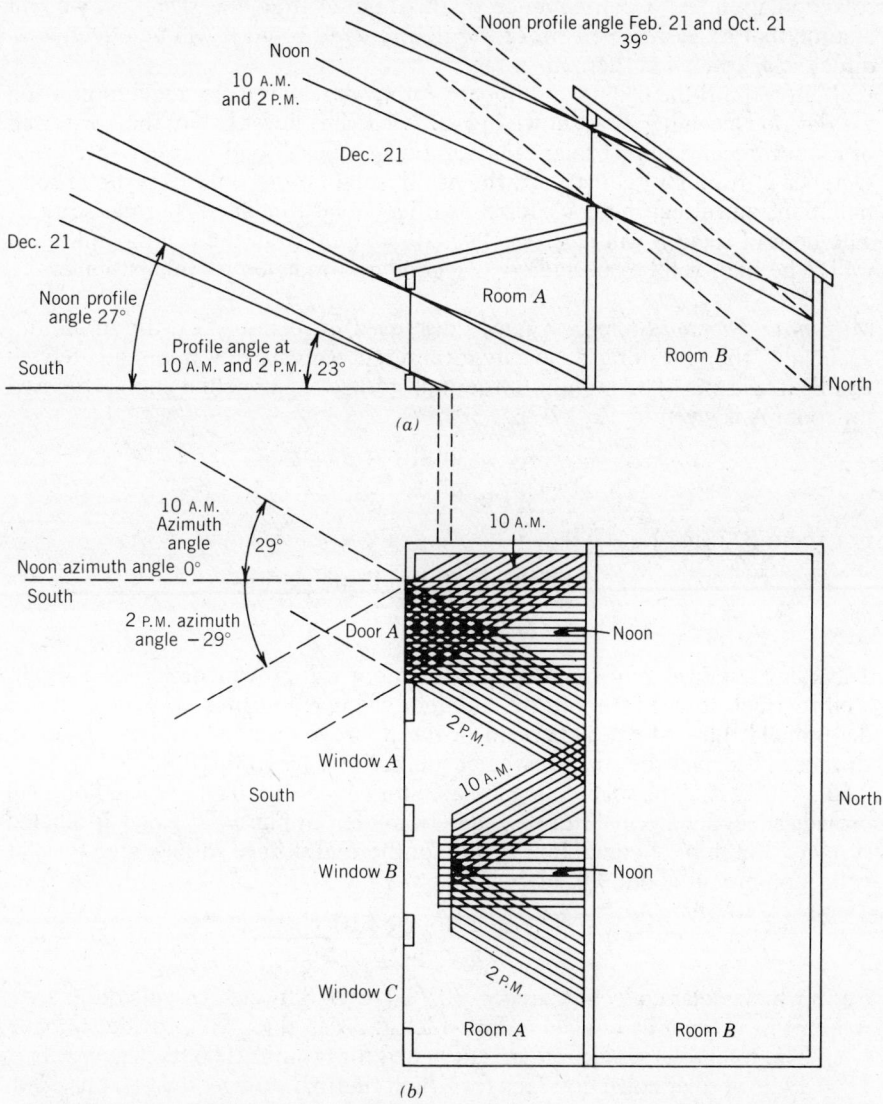

Figure 7-36 Sketches of solar house for the Design Problem, showing sun patterns. (*a*) Cross-section of house showing the sunlit portions of the profile angles for the building at 10:00 A.M., 12:00 noon, and 2:00 P.M. on December 21. (*b*) Floor plan of the house showing the sunlit floor of room A at 10:00 A.M., 12:00 noon, and 2:00 P.M. for door A and window B. The sunlit floor pattern for window A will be the same as for window B. The sunlit floor pattern for window C will be the same as for door A.

at noon is 27°, and the profile angle at 10:00 A.M. and 2:00 P.M. is 23°. These profile angles are used to sketch the sunlit sections of floor and walls in the building cross-section, as shown in Fig. 7-36a. The February 21 and October 21 profile angles are also shown for the clerestory windows. The winter sun on December 21 will strike the ceiling and upper part of the north wall of room B. The sun will strike the lower half of the north wall of room B on October 21 and February 21. Thus, major portions of the north wall of room B will receive direct solar radiation from October 21 through February 21.

The solar azimuth angles corresponding to the above profile angles are also obtained from Table 4-B. They are 0° for noon, −29° for 10:00 A.M., and 29° for 2:00 P.M. These azimuth angles along with the sunlit surfaces in room A (as shown in Fig. 7-36a) are used to project the sunlit sections of the floor. The sunlit parts of the floor are shown only for door A and window B in Fig. 7-36b. If the sunlit parts of the floor for windows A and C were drawn in also, it would be readily apparent that most of the floor surface in room A would receive direct solar radiation. As a consequence of this finding from the sketches, we will assume that at least 80 percent of the floor area of room A will be effective as direct solar-irradiated thermal storage.

In room B the ceiling and upper north wall receive solar radiation during the day on December 21 as shown in Fig. 7-36a. The lower half of the north wall of room B receives direct solar radiation on October 21 and February 21. From October 21 through February 21, the north wall of room B receives direct solar radiation, and it can be considered as direct solar-irradiated thermal storage. The floor of room B will receive reflected radiation from the ceiling and north wall. The floor will be considered as secondary irradiated thermal storage.

Minimum Thermal Storage. As was discussed in section 7-12, the minimum acceptable thermal storage capacity for a direct-gain space is approximately 30 Btu/F°-ft² glazing. The minimum thermal storage capacity that should be used for room A is given by Eq. (7-7):

$$S_T = 96 \text{ ft}^2 \times 30 \text{ Btu/F°-ft}^2 \text{ glazing}$$
$$= 2880 \text{ Btu/F°}$$

For room B, using Eq. (7-7):

$$S_T = 125 \text{ ft}^2 \times 30 \text{ Btu/F°-ft}^2 \text{ glazing}$$
$$= 3750 \text{ Btu/F°}$$

Thermal Mass in Room A. The concrete floor will provide direct solar-irradiated thermal storage for room A. The floor will be 4 in. thick, and above it was determined that at least 80 percent of the floor would be effective as thermal storage. This gives the surface area of thermal storage in the floor as 384 ft² × 0.80 = 307.2 ft². The solar collection area for which the 307 ft² of slab floor will provide adequate thermal storage is determined from Fig. 7-32b (solar-irradiated storage). For 4 in. of concrete, the ratio of thermal storage surface area to solar collection area is about 4.0, or

$$4.0 = \frac{\text{Thermal storage surface area}}{\text{Solar collection area}}$$

From which, solar collection area = 307/4.0 = 76.8 ft² of solar collection area. Again, this is the area of glazing for which the slab floor will provide storage.

The total solar collection area designed for room A is 96 ft², leaving 96 − 77 = 19 ft² of solar collection area for which thermal storage must be provided. The north wall of room A, which is secondary irradiated storage, will be used. The surface area of this wall is 320 ft², and 19 ft² of solar collection area must be provided for. Thus, the ratio of thermal storage surface area to solar collection area is 320 ft²/19 ft² = 16.8. From Fig. 7-33b, for secondary irradiated storage, a ratio of 16.8 gives a material thickness of approximately 2 in. for brick (manufactured half-brick) and 1½ in. for concrete. The wall between rooms A and B could be faced with 2 in. of brick or plastered with an extra thick plaster.

Thermal Mass in Room B.   The north wall of room *B* will be the location of the direct solar-irradiated storage (from clerestories). This wall has 256 ft² of surface (32 ft by 8 ft) and will be constructed of 6-in. concrete. The solar collection area for which the 256 ft² of wall will provide thermal storage is determined from Fig. 7-32*b*. For 6 in. of concrete, the ratio of thermal storage surface area to solar collection area is about 3.0, or

$$3.0 = \frac{\text{Thermal storage surface area}}{\text{Solar collection area}}$$

From which, solar collection area = 256/3.0 = 85.3 ft² solar collection area served by the north wall.

   But the solar collection area provided by the design for room *B* is 125 ft², leaving 125 ft² − 85 ft² = 40 ft² of solar collection area for which additional thermal storage must be provided. The floor of room *B* is secondary irradiated storage. It is a 4-in. concrete slab with an area of 512 ft². From Fig. 7-33*b* (secondary irradiated storage), the ratio of thermal storage surface area (4-in. concrete) to solar collection area is about 7.0, or

$$7.0 = \frac{\text{Thermal storage surface area}}{\text{Solar collection area}}$$

Therefore, solar collection area = 512 ft²/7.0 = 73.1 ft². It is apparent that the concrete slab will provide more thermal storage than is required for the remaining 40 ft² of clerestory windows in room *B*.

Thermal Storage Check.   The sizing and location of thermal mass is now completed. At this point, it is always a good idea to check back and make sure that the storage quantities arrived at are equal to or greater than the minimums as shown below. This process is illustrated for both rooms.

	Vol. ×	Thermal capacity =	Thermal storage capacity

Room A
Slab—0.8 × 384 ft² × ⅓ ft = 102 ft³ × 28 Btu/ft³-F° = 2856 Btu/F°
Brick Wall—320 ft² × ⅙ ft = 53 ft³ × 24 Btu/ft³-F° = 1280 Btu/F°
                                              4136 Btu/F°

Room B
Concrete wall—256 ft² × ½ ft = 128 ft³ × 28 Btu/ft³-F° = 3584 Btu/F°
       Slab—512 ft² × ⅓ ft = 171 ft³ × 28 Btu/ft³-F° = 4788 Btu/F°
                                              8372 Btu/F°

The required minimum thermal storage for room *A* was 2880 Btu/F°, and for room *B* 3750 Btu/F°. The calculated thermal storage is considerably greater than the required minimum for both rooms.

   The thermal storage calculated for room *B* is more than twice the minimum required thermal storage. This is somewhat excessive, and the thermal storage for this room should be recalculated. Start with the concrete slab (secondary irradiated storage) and then work from there to determine how thick the north wall facing material must be to provide the remainder of the required thermal storage. This recalculation will be presented as a problem at the end of the chapter.

Annual Auxiliary Heat.   The final step is to estimate the annual Auxiliary Heat (AH) for the building design using the annual Load Collector Ratio (LCR) method (as discussed in Chapter 6). The net load coefficient, NLC, for the house (excluding the solar aperture) has been determined as 3120 Btu/DD. From Table 5-5, the annual degree-days for Salt Lake City are found to be 5983.

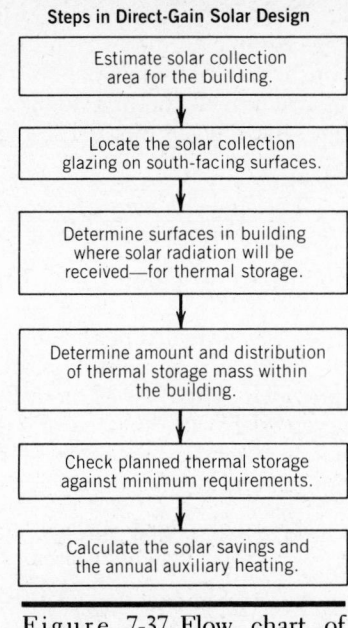

Steps in Direct-Gain Solar Design

Estimate solar collection area for the building.

Locate the solar collection glazing on south-facing surfaces.

Determine surfaces in building where solar radiation will be received—for thermal storage.

Determine amount and distribution of thermal storage mass within the building.

Check planned thermal storage against minimum requirements.

Calculate the solar savings and the annual auxiliary heating.

Figure 7-37 Flow chart of the six essential steps in the solution of a direct-gain solar energy design problem.

The Load Collector Ratio is calculated from Eq. (6-13) as:

$$\begin{aligned} \text{LCR} &= \frac{\text{NLC}}{A_p} \\ &= \frac{3120 \text{ Btu/DD}}{221 \text{ ft}^2} \\ &= 14.2 \text{ Btu/DD-ft}^2 \end{aligned}$$

From Table 6-5, the value of the solar savings fraction SSF for Salt Lake City is found to be 0.36 for double-glazed glass and 0.70 if night insulation with a thermal conductivity of 0.11 Btu/hr-F° (R-9) is used on the windows. The annual Auxiliary Heat required for the building is calculated by Eq. (6-8):

$$\begin{aligned} \text{Annual AH} &= \text{NLC} \times \text{annual DD} \times (1 - \text{SSF}) \\ &= 3120 \text{ Btu/DD} \times 5983 \text{ DD} \times (1 - 0.36) \\ &= 11.9 \times 10^6 \text{ Btu} \qquad\qquad\qquad\qquad\qquad Ans. \end{aligned}$$

For the case where R-9 night insulation is used:

$$\begin{aligned} \text{Annual AH} &= 3120 \text{ Btu/DD} \times 5983 \text{ DD} \times (1 - 0.70) \\ &= 5.6 \times 10^6 \text{ Btu} \qquad\qquad\qquad\qquad\qquad Ans. \end{aligned}$$

In this problem, the essential steps in the solution of a direct-gain solar design problem have been developed. As a review of the process and as an aid in solar energy design practice, the six basic steps are diagrammed in the flow chart of Fig. 7-37. Each solar design problem will have different details to consider, but basic steps and procedures will be found to be similar to those just illustrated. The auxiliary heat calculated for the building illustrates the influence that the building occupants can have on a solar design. If the occupants choose not to use night insulation, the energy usage for auxiliary heating will be twice as great as it would be with consistent use of night insulation.

CONCLUSION

This chapter has considered design methods for direct-gain solar systems and techniques for locating and sizing solar collection areas. Methods for controlling solar gain and the sizing and placement of thermal storage mass in the direct-gain space have also been discussed. In the next chapter, similar considerations will be covered for indirect-gain solar systems.

PROBLEMS

1. A new residence is to be designed with a direct-gain passive solar system for a location in Indianapolis, Indiana. The house floor area will be approximately 2500 ft². Estimate the solar collection area for a 65 percent reduction in the space-heating load (Solar Savings). (Assume a full southern exposure.)

2. An addition is to be added to a residence in Medford, Oregon. The addition will have 1000 ft² of floor area. The average winter temperature for December and January is 37°F. Estimate the solar collection area for a direct-gain solar system using Mazria's guidelines.

3. A residence is planned for Knoxville, Tennessee, with 2000 ft² of floor space. Estimate the solar collection area for a solar savings of 70 percent.

4. A room has 84 ft² of double-glazed south–southeast glass. Determine the solar heat gain in Btu per day on December 21 for a location in Fort Worth, Texas.

5. A large room has 150 ft² of south–southwest double-glazed glass. For a location in Kansas City, Missouri, estimate the daily solar heat gain on March 21.

6. A direct-gain room has south-facing windows with a height of 10 ft. The north wall of the room will provide part of the thermal storage. At what distance should the wall be located behind the window for a location at 40° N Lat?

7. A large room has clerestory windows that allow direct solar gain to enter the room through the ceiling. The windows are 12 ft above the floor. What should be the distance between the clerestory windows and the thermal storage wall north of the clerestory windows?

8. For a location at 48° N Lat, determine the length of a horizontal projection that will provide shading 10 ft down from the projection between April 11 and September 1, for an east orientation, from 10:00 A.M. to 2:00 P.M. local solar time.

9. For a location at 24° N Lat, it is necessary to shade a 9-ft-tall window from April 11 through September 1 on a south wall from 9:00 A.M. to 3:00 P.M. local solar time. Determine the length of the required overhang.

10. Design an overhang to shade a 6-ft-high south window at 48° N Lat from April 21 to August 21 between the hours of 10:00 A.M. and 2:00 P.M. (LSoT). The window should be fully sunlit from October 21 to February 21 during the hours of 10:00 A.M. to 2:00 P.M. Determine the projection and gap of the overhang.

11. A southeast window at 40° N Lat must be shaded between April 21 and August 21 between 9:00 A.M. and 2:00 P.M. The window must be sunlit from October 21 to February 21 between 9:00 A.M. and 2:00 P.M. The height of the window is 8 ft. Determine the projection and gap of the overhang.

12. A thermal storage capacity of 10,000 Btu/F° is required. What volume of concrete will provide this capacity?

13. What thermal storage capacity is provided by 100 ft³ of calcium chloride hexahydrate as it changes state from a solid to liquid?

14. A direct-gain window has a daily solar heat gain of 85,000 Btu. What volume of water is required to store this energy with a 12 F° temperature change?

15. A thermal storage mass receives 45,000 Btu/day of incident solar radiation. How much energy will the thermal storage absorb if it is painted light green? Black? Yellow?

16. A room has 80 ft² of window area. For a typical direct-gain system, estimate the minimum volume of concrete that will provide acceptable thermal storage. (See Fig. 7-31 and Table 7-4.)

17. A direct-gain room will have 4-in. brick walls that provide thermal storage mass. If the brick is direct solar-irradiated storage, what ratio of storage surface area to solar collection area must be used?

18. A building is to receive part of its space heating from direct solar gains. The solar collection area along the south wall has 650 ft² of double glazing. Using the minimum thermal storage guideline, determine the minimum volume of concrete that should be used to provide thermal storage.

19. In the design problem of section 7-13, the thermal storage capacity from room B was calculated as 8372 Btu/F°. This is somewhat excessive. Recalculate the thermal storage for room B by first determining the solar collection area for which the secondary irradiated floor slab will provide adequate thermal storage, and then calculate the thickness of solar-irradiated storage along the north wall required for the remaining solar collection area.

20. A direct-gain solar system has a south-facing double-glazed solar collection area of 380 ft². The building is located near Knoxville, Tennessee, and has a net load coefficient (NLC) of 13,000 Btu/DD. Determine the Solar Savings and Auxiliary Heat using the annual LCR method for the building. Solve for both cases—(a) when no night insulation is used, and (b) when night insulation is used.

21. A house with a direct-gain solar system is being designed for a location near Salt Lake City, Utah. The house will have 1800 ft² of floor area and will be designed to have a normalized net load coefficient of 5.0 Btu/DD-ft² floor area. Determine the solar collection area to provide a solar savings of 50 percent when no night insulation is used and when R-9 night insulation is used (use the annual LCR method).

22. For a building in Albany, New York, with a net load coefficient of 25,000 Btu/DD, determine the solar collection area required to provide a solar savings of 40 percent. Use the annual LCR method, and assume that R-9 night insulation will be used on the solar collection area.

23. For the basic house shown in Fig. 7-35, repeat the six steps in the design problem of section 7-13 for a building location in Knoxville, Tennessee.

24. For the basic house shown in Fig. 7-35, repeat the six steps in the design problem of section 7-13 for a building location in Albany, New York.

CHAPTER APPENDIX

Table 7-A Solar Intensity (Insolation) and Solar Heat Gain Factors (SGHF) for Selected Northern Latitudes

24°N

Date	Solar Time am	Direct Normal Btuh/ft²	N	NNE	NE	ENE	E	ESE	SE	SSE	S	SSW	SW	WSW	W	WNW	NW	NNW	HOR	Solar Time pm
Jan 21	7	71	2	3	21	45	62	67	63	49	25	3	2	2	2	2	2	2	5	5
	8	239	12	12	41	128	190	221	218	181	114	28	12	12	12	12	12	12	55	4
	9	288	18	18	23	106	190	240	253	227	166	73	19	18	18	18	18	18	121	3
	10	308	23	23	24	53	144	211	245	241	200	125	38	24	23	23	23	23	172	2
	11	317	26	26	26	27	73	156	211	234	220	173	95	29	26	26	26	26	204	1
	12	320	27	27	27	27	29	82	160	210	227	210	160	81	29	27	27	27	214	12
	HALF DAY TOTALS		95	96	148	372	671	942	1076	1039	840	505	241	120	96	95	95	95	664	
Feb 21	7	153	6	12	67	114	141	145	128	90	33	6	6	6	6	6	6	6	17	5
	8	262	15	16	80	165	220	240	224	172	89	17	15	15	15	15	15	15	83	4
	9	297	21	22	46	138	208	244	243	205	133	42	22	21	21	21	21	21	153	3
	10	314	26	26	28	76	157	209	228	213	165	87	28	26	26	26	26	26	205	2
	11	321	29	29	29	31	80	148	191	203	185	137	68	31	29	29	29	29	238	1
	12	323	30	30	30	30	32	70	134	177	192	177	133	70	32	30	30	30	249	12
	HALF DAY TOTALS		113	119	257	527	806	1011	1072	965	699	374	200	127	113	113	113	113	820	
Mar 21	7	194	11	45	115	164	186	180	145	86	17	10	10	10	10	10	10	10	36	5
	8	267	18	35	124	195	234	237	204	138	48	19	18	18	18	18	18	18	112	4
	9	295	25	27	85	165	215	232	214	163	82	27	25	25	25	25	25	25	180	3
	10	309	30	31	41	103	162	194	195	168	112	47	31	30	30	30	30	30	232	2
	11	315	33	33	34	42	85	129	154	155	139	86	43	34	33	33	33	33	264	1
	12	317	34	34	34	34	35	56	96	126	137	126	95	56	35	34	34	34	275	12
	HALF DAY TOTALS		133	189	422	693	906	1011	970	778	458	249	169	139	133	133	133	133	962	
Apr 21	6	40	6	21	33	39	39	33	22	7	2	2	2	2	2	2	2	2	4	6
	7	203	20	88	151	189	197	176	127	55	15	14	14	14	14	14	14	14	58	5
	8	256	24	80	159	209	228	212	164	88	24	22	22	22	22	22	22	22	132	4
	9	280	30	54	126	181	208	203	169	105	39	29	28	28	28	28	28	28	195	3
	10	292	34	37	75	125	157	165	148	107	56	35	33	33	33	33	33	33	244	2
	11	298	36	37	40	59	85	103	106	94	70	45	38	37	36	36	36	36	274	1
	12	299	37	37	38	38	39	46	59	70	75	70	58	45	39	38	38	37	283	12
	HALF DAY TOTALS		168	339	607	826	940	924	773	494	244	180	163	157	155	155	154	154	1048	
May 21	6	86	25	57	79	87	84	66	38	8	6	6	6	6	6	6	6	6	13	6
	7	203	43	117	171	199	196	163	105	32	17	17	17	17	17	17	17	18	73	5
	8	248	38	114	178	214	218	190	132	54	26	25	25	25	25	25	25	26	142	4
	9	269	35	88	150	188	198	179	132	66	33	31	31	31	31	31	31	31	201	3
	10	280	38	59	103	137	150	141	111	67	39	36	35	35	35	35	35	36	247	2
	11	286	40	43	55	72	83	84	75	58	44	40	39	38	38	38	38	39	274	1
	12	288	41	41	41	41	42	43	44	46	46	46	44	43	42	41	41	41	282	12
	HALF DAY TOTALS		238	492	749	909	943	840	614	308	187	176	174	173	172	172	172	175	1089	
Jun 21	6	97	36	70	93	101	94	73	39	8	7	7	7	7	7	7	7	8	17	6
	7	201	55	127	177	199	192	155	94	26	18	18	18	18	18	18	18	20	77	5
	8	242	50	126	184	214	212	179	117	43	27	26	26	26	26	26	26	27	145	4
	9	263	43	102	158	189	192	168	116	53	34	32	32	32	32	32	32	33	201	3
	10	274	41	72	113	140	146	131	96	55	39	36	36	36	36	36	36	38	245	2
	11	279	42	50	65	77	82	77	64	49	42	41	40	39	39	39	40	41	271	1
	12	281	43	43	43	43	43	43	43	43	43	43	43	43	43	43	43	43	279	12
	HALF DAY TOTALS		284	562	802	933	932	797	544	255	187	181	180	179	179	179	180	187	1096	
Jul 21	6	81	26	56	76	84	80	63	36	8	6	6	6	6	6	6	6	7	13	6
	7	195	45	116	168	194	190	158	101	31	18	18	18	18	18	18	18	19	73	5
	8	239	41	115	176	210	213	185	128	52	27	26	26	26	26	26	26	26	141	4
	9	261	37	90	150	186	195	175	129	64	34	32	32	32	32	32	32	32	198	3
	10	272	39	62	104	137	149	139	108	65	39	37	36	36	36	36	36	37	243	2
	11	278	41	44	58	73	83	83	73	57	44	41	40	39	39	39	39	40	270	1
	12	280	42	42	42	43	43	44	45	46	46	46	45	43	42	42	42	42	278	12
	HALF DAY TOTALS		247	498	746	897	925	820	595	300	191	181	178	177	177	177	177	181	1076	
Aug 21	6	35	6	20	30	35	35	30	19	6	2	2	2	2	2	2	2	2	4	6
	7	186	22	87	144	179	186	165	119	51	16	15	15	15	15	15	15	15	58	5
	8	241	26	82	156	203	220	204	157	84	26	24	24	24	24	24	24	24	130	4
	9	265	32	57	126	178	202	197	162	101	39	31	30	30	30	30	30	30	191	3
	10	278	36	40	78	125	155	161	143	103	55	37	35	35	35	35	35	35	239	2
	11	284	38	39	42	61	85	101	104	91	68	46	40	38	37	37	37	37	268	1
	12	286	38	39	40	40	41	47	58	69	72	68	58	47	41	40	40	39	277	12
	HALF DAY TOTALS		179	347	601	806	910	889	740	473	243	186	171	165	164	163	163	162	1028	
Sep 21	8	248	19	36	119	185	222	225	194	132	48	20	19	19	19	19	19	19	108	4
	9	278	26	28	84	160	207	223	206	158	81	28	26	26	26	26	26	26	174	3
	10	292	31	32	42	101	158	188	190	163	110	48	32	31	31	31	31	31	224	2
	11	299	34	34	35	43	84	127	151	151	128	86	44	35	34	34	34	34	256	1
	12	301	35	35	35	36	37	57	95	124	134	124	94	57	37	36	35	35	266	12
	HALF DAY TOTALS		139	190	406	661	863	964	927	749	451	251	174	145	139	138	138	138	930	
Oct 21	7	138	6	12	62	104	129	133	117	82	31	7	6	6	6	6	6	6	17	5
	8	247	16	17	79	159	211	230	214	164	85	17	16	16	16	16	16	16	82	4
	9	284	22	23	47	135	202	237	235	198	128	41	23	22	22	22	22	22	150	3
	10	301	27	27	29	77	154	204	222	207	160	85	29	27	27	27	27	27	201	2
	11	309	30	30	30	33	80	145	186	198	180	133	67	32	30	30	30	30	233	1
	12	311	31	31	31	31	33	70	131	173	187	172	130	69	33	31	31	31	244	12
	HALF DAY TOTALS		116	123	255	512	778	974	1032	929	675	367	200	131	117	116	116	116	804	
Nov 21	7	67	2	3	20	43	59	64	60	46	24	3	2	2	2	2	2	2	5	5
	8	232	12	13	42	126	186	216	213	177	111	28	12	12	12	12	12	12	55	4
	9	282	19	19	23	106	187	236	249	223	163	71	20	19	19	19	19	19	120	3
	10	303	23	23	24	53	143	209	241	237	197	123	37	24	23	23	23	23	171	2
	11	312	26	26	26	28	73	154	209	230	217	171	93	29	26	26	26	26	202	1
	12	315	27	27	27	27	29	81	158	207	224	207	158	80	29	27	27	27	213	12
	HALF DAY TOTALS		97	97	149	368	661	926	1056	1020	825	497	239	121	98	97	97	97	659	
Dec 21	7	30	1	1	7	18	25	28	27	21	12	2	1	1	1	1	1	1	2	5
	8	225	10	10	29	112	174	208	209	178	118	35	11	10	10	10	10	10	44	4
	9	281	17	17	19	93	180	234	252	231	174	84	18	17	17	17	17	17	107	3
	10	304	22	22	22	44	137	209	247	247	209	137	44	22	22	22	22	22	157	2
	11	314	25	25	25	26	69	156	216	241	230	183	104	29	25	25	25	25	188	1
	12	317	26	26	26	26	27	85	167	219	237	219	167	84	27	26	26	26	188	12
	HALF DAY TOTALS		88	88	118	313	611	899	1054	1042	868	550	257	117	89	88	88	88	598	
			N	NNW	NW	WNW	W	WSW	SW	SSW	S	SSE	SE	ESE	E	ENE	NE	NNE	HOR	PM

Half Day Totals computed by Simpson's Rule with time interval equal to 10 minutes.
*Total Solar Heat Gains for DS (0.125 in.) sheet glass.
Based on a ground reflectance of 0.20 and values in Tables 1 and 26.

Table 7-A Solar Intensity (Insolation) and Solar Heat Gain Factors (SGHF) for Selected Northern Latitudes

32°N

Date	Solar Time am	Direct Normal Btuh/ft²	N	NNE	NE	ENE	E	ESE	SE	SSE	S	SSW	SW	WSW	W	WNW	NW	NNW	HOR	Solar Time pm
Jan 21	7	1	0	0	0	1	1	1	1	1	1	0	0	0	0	0	0	0	32	5
	8	203	9	9	29	105	160	189	189	159	103	28	9	9	9	9	9	9	88	4
	9	269	15	15	17	91	175	229	246	225	169	82	17	15	15	15	15	15	136	3
	10	295	20	20	20	41	135	209	249	250	212	141	46	20	20	20	20	20	166	2
	11	306	23	23	23	24	68	159	221	249	238	191	110	29	23	23	23	23	176	1
	12	310	24	24	24	24	25	88	174	228	246	228	174	88	25	24	24	24	512	12
	HALF DAY TOTALS		79	79	107	284	570	856	1015	1014	853	553	264	112	80	79	79	79	9	
Feb 21	7	112	4	7	47	82	102	106	95	67	26	4	4	4	4	4	4	4	64	5
	8	245	13	14	65	149	205	228	216	170	95	17	13	13	13	13	13	13	127	4
	9	287	19	19	32	122	199	242	248	216	149	55	20	19	19	19	19	19	176	3
	10	305	24	24	25	62	151	213	241	232	189	112	31	24	24	24	24	24	207	2
	11	314	26	26	26	28	76	156	208	227	212	165	87	28	26	26	26	26	217	1
	12	316	27	27	27	27	29	79	155	204	221	204	155	79	29	27	27	27	691	12
	HALF DAY TOTALS		100	103	201	445	735	978	1080	1010	780	452	228	122	100	100	100	100	691	
Mar 21	7	185	10	37	105	153	176	173	142	88	20	9	9	9	9	9	9	9	100	5
	8	260	17	25	107	183	227	237	209	150	62	18	17	17	17	17	17	17	164	4
	9	290	23	25	64	151	210	237	227	183	107	30	23	23	23	23	23	23	211	3
	10	304	28	28	30	87	158	202	215	195	144	70	29	28	28	28	28	28	242	2
	11	311	31	31	31	34	82	142	179	188	168	120	59	32	31	31	31	31	252	1
	12	313	32	32	32	32	33	66	122	162	176	162	122	66	33	32	32	32	874	12
	HALF DAY TOTALS		124	162	359	629	875	1033	1041	888	589	326	193	136	125	124	124	124	874	
Apr 21	6	66	9	35	54	65	66	56	38	12	4	3	3	3	3	3	3	3	7	6
	7	206	17	80	146	188	200	182	136	65	16	14	14	14	14	14	14	14	61	5
	8	255	23	61	144	200	227	219	177	107	30	22	22	22	22	22	22	22	129	4
	9	278	28	36	103	168	206	212	187	133	58	29	28	28	28	28	28	28	188	3
	10	290	32	34	52	108	155	177	172	141	87	39	33	32	32	32	32	32	233	2
	11	295	35	35	36	47	83	118	135	132	108	70	40	36	35	35	35	35	262	1
	12	297	36	36	36	37	38	53	82	106	115	106	82	53	38	37	36	36	1015	12
	HALF DAY TOTALS		161	296	550	792	952	992	889	645	360	228	177	157	153	152	152	152	1015	
May 21	6	119	33	77	108	121	116	94	56	13	8	8	8	8	8	8	8	8	28	6
	7	211	36	111	170	202	204	174	118	42	19	18	18	18	18	18	18	19	81	5
	8	250	29	94	165	208	220	199	149	73	27	25	25	25	25	25	25	25	146	4
	9	269	33	61	128	177	198	190	155	93	37	32	31	31	31	31	31	35	201	3
	10	280	36	40	76	121	150	156	138	99	54	37	35	35	35	35	35	35	243	2
	11	285	38	39	42	59	83	99	102	90	68	47	40	39	37	37	37	37	269	1
	12	286	38	39	40	40	41	47	59	70	74	70	59	47	41	40	39	39	277	12
	HALF DAY TOTALS		222	438	702	900	985	933	747	447	250	199	183	177	175	174	174	175	1098	
Jun 21	6	131	44	92	123	135	127	99	55	12	10	10	10	10	10	10	10	11	28	6
	7	210	47	122	176	204	201	168	108	35	20	20	20	20	20	20	20	21	88	5
	8	245	36	106	171	208	214	189	135	60	28	27	27	27	27	27	27	27	151	4
	9	264	35	74	137	178	193	180	139	77	35	32	32	32	32	32	32	32	204	3
	10	274	38	47	86	125	146	145	123	83	45	38	36	36	36	36	36	36	244	2
	11	279	40	41	47	64	82	91	89	75	56	43	41	40	39	39	39	39	269	1
	12	280	41	41	41	42	42	46	52	58	60	58	52	46	42	42	41	41	276	12
	HALF DAY TOTALS		261	504	762	935	985	897	678	372	225	197	189	185	184	184	183	186	1122	
Jul 21	6	113	34	76	105	117	113	90	53	12	9	9	9	9	9	9	9	9	22	6
	7	203	38	111	167	198	198	169	114	41	20	19	19	19	19	19	19	19	81	5
	8	241	31	95	163	204	215	194	145	70	28	26	26	26	26	26	26	32	198	4
	9	261	34	64	129	175	195	186	150	90	37	32	32	32	32	32	32	32	240	3
	10	271	37	42	78	121	148	153	134	96	53	38	36	36	36	36	36	36	265	2
	11	277	39	40	43	60	83	98	99	88	66	47	41	40	38	38	38	38	273	1
	12	279	40	40	41	41	42	48	58	68	72	68	58	48	42	41	41	40	1088	12
	HALF DAY TOTALS		231	444	701	890	967	912	726	433	248	202	187	182	180	179	179	180	1088	
Aug 21	6	59	10	33	50	60	60	51	34	11	4	4	4	4	4	4	4	4	8	6
	7	190	19	79	141	179	190	172	128	61	17	15	15	15	15	15	15	15	61	5
	8	240	25	63	141	195	219	210	170	102	31	23	23	23	23	23	23	23	128	4
	9	263	30	39	104	166	200	206	181	127	57	31	29	29	29	29	29	29	185	3
	10	276	34	36	55	109	153	173	167	136	84	40	35	34	34	34	34	36	256	2
	11	282	36	37	39	50	84	116	131	127	104	69	41	38	36	36	37	37	265	1
	12	284	37	37	37	39	40	54	81	103	111	103	81	54	40	39	37	37	999	12
	HALF DAY TOTALS		171	303	546	774	922	955	854	618	352	231	184	166	162	161	160	160	999	
Sep 21	7	163	10	35	96	139	159	156	128	80	20	10	10	10	10	10	10	10	31	5
	8	240	18	26	103	173	215	224	198	143	60	19	18	18	18	18	18	18	96	4
	9	272	24	26	64	146	202	227	218	177	105	31	24	24	24	24	24	24	158	3
	10	287	29	29	32	86	154	196	208	189	141	70	31	29	29	29	29	29	204	2
	11	294	32	32	32	36	81	139	174	182	163	118	59	34	32	32	32	32	234	1
	12	296	33	33	33	33	35	66	120	158	171	158	120	66	35	33	33	33	845	12
	HALF DAY TOTALS		130	164	345	598	831	982	993	852	574	325	197	142	130	129	129	129	845	
Oct 21	7	99	4	7	43	74	92	96	85	60	24	5	4	4	4	4	4	4	10	5
	8	229	13	15	63	143	195	217	206	162	90	17	13	13	13	13	13	13	63	4
	9	273	20	20	33	120	193	234	239	208	144	54	21	20	20	20	20	20	125	3
	10	293	24	24	26	62	147	207	234	225	183	109	32	24	24	24	24	24	173	2
	11	302	27	27	27	29	76	152	203	221	207	160	85	29	27	27	27	27	203	1
	12	304	28	28	28	28	30	78	151	199	215	199	151	78	30	28	28	28	679	12
	HALF DAY TOTALS		103	106	200	433	708	941	1038	972	753	441	226	125	104	103	103	103	679	
Nov 21	7	2	0	0	0	1	1	1	1	1	1	0	0	0	0	0	0	0	32	5
	8	196	9	9	29	103	156	184	184	155	100	27	9	9	9	9	9	9	88	4
	9	263	16	16	17	90	173	225	241	221	166	80	17	16	16	16	16	16	136	3
	10	289	20	20	21	41	134	206	245	246	209	138	45	21	20	20	20	20	165	2
	11	301	23	23	23	24	67	157	218	245	234	188	109	29	23	23	23	23	175	1
	12	304	24	24	24	24	25	87	171	224	243	224	171	87	25	24	24	24	509	12
	HALF DAY TOTALS		80	81	108	282	561	841	996	995	838	544	261	113	81	80	80	80	509	
Dec 21	8	176	7	7	19	84	135	163	166	143	97	31	7	7	7	7	7	7	22	4
	9	257	14	14	15	77	162	218	238	222	171	89	15	14	14	14	14	14	72	3
	10	288	18	18	18	34	127	204	246	251	216	148	52	19	18	18	18	18	119	2
	11	301	21	21	21	22	63	157	222	252	243	197	116	29	21	21	21	21	148	1
	12	304	22	22	22	22	23	89	177	232	252	232	177	89	23	22	22	22	440	12
	HALF DAY TOTALS		71	71	84	227	500	792	965	986	852	578	275	107	71	71	71	71	440	
			N	NNW	NW	WNW	W	WSW	SW	SSW	S	SSE	SE	ESE	E	ENE	NE	NNE	HOR	PM

Half Day Totals computed by Simpson's Rule with time interval equal to 10 minutes.   Based on a ground reflectance of 0.20 and values in Tables 1 and 26.
*Total Solar Heat Gains for DS (0.125 in.) sheet glass.

Table 7-A Solar Intensity (Insolation) and Solar Heat Gain Factors (SGHF) for Selected Northern Latitudes

40°N

Date	Solar Time am	Direct Normal Btuh/ft²	N	NNE	NE	ENE	E	ESE	SE	SSE	S	SSW	SW	WSW	W	WNW	NW	NNW	HOR	Solar Time pm
Jan 21	8	142	5	5	17	71	111	132	133	114	75	22	6	5	5	5	5	5	14	4
	9	239	12	12	13	74	154	205	224	209	160	82	13	12	12	12	12	12	55	3
	10	274	16	16	16	31	124	199	241	246	213	146	51	17	16	16	16	16	96	2
	11	289	19	19	19	20	61	156	222	252	244	198	118	28	19	19	19	19	124	1
	12	294	20	20	20	20	21	90	179	234	254	234	179	90	21	20	20	20	133	12
	HALF DAY TOTALS		61	61	73	199	452	734	904	932	813	561	273	101	62	61	61	61	354	
Feb 21	7	55	2	3	23	40	51	53	47	34	14	2	2	2	2	2	2	2	4	5
	8	219	10	11	50	129	183	206	199	160	94	18	10	10	10	10	10	10	43	4
	9	271	16	16	22	107	186	234	245	218	157	66	17	16	16	16	16	16	98	3
	10	294	21	21	21	49	143	211	246	243	203	129	38	21	21	21	21	21	143	2
	11	304	23	23	23	24	71	160	219	244	231	184	103	27	23	23	23	23	171	1
	12	307	24	24	24	24	25	86	170	222	241	222	170	86	25	24	24	24	180	12
	HALF DAY TOTALS		84	86	152	361	648	916	1049	1015	821	508	250	114	85	84	84	84	548	
Mar 21	7	171	9	29	93	140	163	161	135	86	22	8	8	8	8	8	8	8	26	5
	8	250	16	18	91	169	218	232	211	157	74	17	16	16	16	16	16	16	85	4
	9	282	21	22	47	136	203	238	236	198	128	40	22	21	21	21	21	21	143	3
	10	297	25	25	27	72	153	207	229	216	171	95	29	25	25	25	25	25	186	2
	11	305	28	28	30	78	151	198	213	197	150	77	30	28	28	28	28	28	213	1
	12	307	29	29	29	29	31	75	145	191	206	191	145	75	31	29	29	29	223	12
	HALF DAY TOTALS		114	139	302	563	832	1035	1087	968	694	403	220	132	114	113	113	113	764	
Apr 21	6	89	11	46	72	87	88	76	52	18	5	5	5	5	5	5	5	5	11	6
	7	206	16	71	140	185	201	186	143	75	16	14	14	14	14	14	14	14	61	5
	8	252	22	44	128	190	224	223	188	124	41	22	21	21	21	21	21	21	123	4
	9	274	27	29	80	155	202	219	203	156	83	29	27	27	27	27	27	27	177	3
	10	286	31	31	37	92	152	187	193	170	121	56	32	31	31	31	31	41	217	2
	11	292	33	33	34	39	81	130	160	166	146	102	52	35	33	33	33	33	243	1
	12	293	34	34	34	34	36	62	108	142	154	142	108	62	36	34	34	34	252	12
	HALF DAY TOTALS		154	265	501	758	957	1051	994	782	488	296	199	157	148	147	147	147	957	
May 21	5	1	0	1	1	1	1	1	0	0	0	0	0	0	0	0	0	0	0	7
	6	144	36	90	128	145	141	115	71	18	10	10	10	10	10	10	10	11	31	6
	7	216	28	102	165	202	209	184	131	54	20	19	19	19	19	19	19	19	87	5
	8	250	27	73	149	199	220	208	164	93	29	25	25	25	25	25	25	25	146	4
	9	267	31	42	105	164	197	200	175	121	53	32	30	30	30	30	30	30	195	3
	10	277	34	36	54	105	148	168	163	133	83	40	35	34	34	34	34	34	234	2
	11	283	36	36	38	48	81	113	130	127	105	70	42	38	36	36	36	36	257	1
	12	284	37	37	37	38	40	54	82	104	113	104	82	54	40	38	37	37	265	12
	HALF DAY TOTALS		215	404	666	893	1024	1025	881	601	358	247	200	180	176	175	174	175	1083	
Jun 21	5	22	10	17	21	22	20	14	6	2	1	1	1	1	1	1	1	2	3	7
	6	155	48	104	143	159	151	121	70	17	13	13	13	13	13	13	13	14	40	6
	7	216	37	113	172	205	207	178	122	46	22	21	21	21	21	21	21	21	97	5
	8	246	30	85	156	201	216	199	152	80	29	27	27	27	27	27	27	27	153	4
	9	263	33	51	114	166	192	190	161	105	45	33	32	32	32	32	32	32	201	3
	10	272	35	38	63	109	145	158	148	116	69	39	36	35	35	35	35	35	238	2
	11	277	38	39	40	52	81	105	116	110	88	60	41	39	38	38	38	38	260	1
	12	279	38	38	38	40	41	52	72	89	95	89	72	52	41	40	38	38	267	12
	HALF DAY TOTALS		253	470	734	941	1038	999	818	523	315	236	204	191	188	187	186	188	1126	
Jul 21	5	2	1	2	2	2	2	1	1	0	0	0	0	0	0	0	0	0	0	7
	6	138	37	89	125	142	137	112	68	18	11	11	11	11	11	11	11	12	32	6
	7	208	30	102	163	198	204	179	127	53	21	20	20	20	20	20	20	20	88	5
	8	241	28	75	148	196	216	203	160	90	30	26	26	26	26	26	26	26	145	4
	9	259	32	44	106	163	193	196	170	118	52	33	31	31	31	31	31	31	194	3
	10	269	35	37	56	106	146	165	159	129	81	41	36	35	35	35	35	35	231	2
	11	275	37	38	40	50	81	111	127	123	102	69	43	39	37	37	37	37	254	1
	12	276	38	38	38	40	41	55	80	101	109	101	80	55	41	40	38	38	262	12
	HALF DAY TOTALS		223	411	666	885	1008	1003	858	584	352	248	204	186	181	180	180	181	1076	
Aug 21	6	81	12	44	68	81	82	71	48	17	6	5	5	5	5	5	5	5	12	6
	7	191	17	71	135	177	191	177	135	70	17	16	16	16	16	16	16	16	62	5
	8	237	24	47	126	185	216	214	180	118	41	23	23	23	23	23	23	23	122	4
	9	260	28	31	82	153	197	212	196	151	80	31	28	28	28	28	28	28	174	3
	10	272	32	33	40	93	150	182	187	165	116	56	34	32	32	32	32	32	214	2
	11	278	35	35	36	41	81	128	156	160	141	99	52	37	35	35	35	35	239	1
	12	280	35	35	35	36	38	63	106	138	149	138	106	63	38	36	35	35	247	12
	HALF DAY TOTALS		164	273	498	741	928	1013	956	751	474	296	205	166	157	156	156	156	946	
Sep 21	7	149	9	27	84	125	146	144	121	77	21	9	9	9	9	9	9	9	25	5
	8	230	17	19	87	160	205	218	199	148	71	18	17	17	17	17	17	17	82	4
	9	263	22	23	47	131	194	227	226	190	124	41	23	22	22	22	22	22	138	3
	10	280	27	27	28	71	148	200	221	209	165	93	30	27	27	27	27	27	180	2
	11	287	29	29	29	31	78	147	192	207	191	146	77	31	29	29	29	29	206	1
	12	290	30	30	30	30	32	75	142	185	200	185	142	75	32	30	30	30	215	12
	HALF DAY TOTALS		119	142	291	534	787	980	1033	925	672	396	222	137	119	118	118	118	738	
Oct 21	7	48	2	3	20	36	45	47	42	30	12	2	2	2	2	2	2	2	4	5
	8	204	11	12	49	123	173	195	188	151	89	18	11	11	11	11	11	11	43	4
	9	257	17	17	23	104	180	225	235	209	151	64	18	17	17	17	17	17	97	3
	10	280	21	21	22	50	139	205	238	235	196	125	38	22	21	21	21	21	140	2
	11	291	24	24	24	25	71	156	212	236	224	178	101	28	24	24	24	24	168	1
	12	294	25	25	25	25	27	85	165	216	234	216	165	85	27	25	25	25	177	12
	HALF DAY TOTALS		88	89	152	351	623	878	1006	974	791	493	247	117	89	88	88	88	540	
Nov 21	8	136	5	5	18	69	108	128	129	110	72	21	6	5	5	5	5	5	14	4
	9	232	12	12	13	73	151	201	219	204	156	80	13	12	12	12	12	12	55	3
	10	268	16	16	16	31	122	196	237	242	209	143	50	17	16	16	16	16	96	2
	11	283	19	19	19	20	61	154	218	248	240	194	116	28	19	19	19	19	123	1
	12	288	20	20	20	20	21	89	176	231	250	231	176	89	21	20	20	20	132	12
	HALF DAY TOTALS		63	63	75	198	445	721	887	914	798	551	269	101	63	63	63	63	354	
Dec 21	8	89	3	3	8	41	67	82	84	73	50	17	3	3	3	3	3	3	6	4
	9	217	10	10	11	60	135	185	205	194	151	83	13	10	10	10	10	10	39	3
	10	261	14	14	14	25	113	188	232	239	210	146	55	15	14	14	14	14	77	2
	11	280	17	17	17	17	56	151	217	249	242	198	120	28	17	17	17	17	104	1
	12	285	18	18	18	18	19	89	178	233	253	233	178	89	19	18	18	18	113	12
	HALF DAY TOTALS		52	52	56	146	374	649	822	867	775	557	276	94	53	52	52	52	282	
			N	NNW	NW	WNW	W	WSW	SW	SSW	S	SSE	SE	ESE	E	ENE	NE	NNE	HOR	PM

Half Day Totals computed by Simpson's Rule with time interval equal to 10 minutes.

*Total Solar Heat Gains for DS (0.125 in.) sheet glass.

Based on a ground reflectance of 0.20 and values in Tables 1 and 26.

Table 7-A   Solar Intensity (Insolation) and Solar Heat Gain Factors (SGHF) for Selected Northern Latitudes

48°N

Date	Solar Time am	Direct Normal Btuh/ft^2	N	NNE	NE	ENE	E	ESE	SE	SSE	S	SSW	SW	WSW	W	WNW	NW	NNW	HOR	Solar Time pm
Jan 21	8	37	1	1	4	18	29	34	35	30	20	6	1	1	1	1	1	1	2	4
	9	185	8	8	8	53	118	160	176	166	129	69	10	8	8	8	8	8	25	3
	10	239	12	12	12	22	106	175	216	223	195	136	50	12	12	12	12	12	55	2
	11	261	14	14	14	15	53	144	208	239	233	190	116	26	14	14	14	14	77	1
	12	267	15	15	15	15	16	86	171	226	245	226	171	86	16	15	15	15	85	12
	HALF DAY TOTALS		43	43	46	117	316	567	729	776	701	512	259	85	43	43	43	43	203	
Feb 21	7	4	0	0	1	3	3	3	3	2	1	0	0	0	0	0	0	0	0	5
	8	180	8	8	36	103	149	170	166	136	82	17	8	8	8	8	8	8	25	4
	9	247	13	13	16	90	168	216	230	209	155	71	14	13	13	13	13	13	66	3
	10	275	17	17	17	38	131	203	242	244	207	138	44	18	17	17	17	17	105	2
	11	288	19	19	19	20	65	158	221	249	239	192	113	27	19	19	19	19	130	1
	12	292	20	20	20	20	22	89	176	231	250	231	176	89	22	20	20	20	138	12
	HALF DAY TOTALS		68	68	107	274	541	816	968	967	813	531	261	104	68	68	68	68	395	
Mar 21	7	153	7	22	80	123	145	145	123	80	23	7	7	7	7	7	7	7	20	5
	8	236	14	15	76	154	204	222	206	158	82	15	14	14	14	14	14	14	68	4
	9	270	19	19	3	121	193	234	239	207	142	52	20	19	19	19	19	19	118	3
	10	287	23	23	24	58	146	208	237	231	189	115	33	23	23	23	23	23	156	2
	11	295	25	25	25	26	74	156	210	232	218	172	94	28	25	25	25	25	180	1
	12	298	26	26	26	26	27	83	161	211	228	211	161	83	27	26	26	26	188	12
	HALF DAY TOTALS		100	118	250	494	775	1012	1100	1014	767	465	244	126	101	100	100	100	636	
Apr 21	6	108	12	53	86	105	107	93	64	23	6	6	6	6	6	6	6	6	15	6
	7	205	15	61	132	180	199	189	148	84	18	14	14	14	14	14	14	14	60	5
	8	247	20	32	111	179	219	225	196	138	55	21	20	20	20	20	20	20	114	4
	9	268	25	26	60	141	197	223	215	176	106	33	25	25	25	25	25	25	161	3
	10	280	28	28	31	77	148	193	209	194	150	80	31	28	28	28	28	28	196	2
	11	286	31	31	31	33	78	140	181	193	177	133	69	33	31	31	31	31	218	1
	12	288	31	31	31	31	34	71	131	172	186	172	131	71	34	31	31	31	226	12
	HALF DAY TOTALS		147	242	461	724	957	1098	1081	895	605	370	226	156	141	140	140	140	875	
May 21	5	41	17	31	40	42	39	29	14	3	3	3	3	3	3	3	3	3	5	7
	6	162	35	97	141	162	160	133	85	24	12	12	12	12	12	12	12	13	40	6
	7	219	23	90	158	200	212	191	142	68	21	19	19	19	19	19	19	19	91	5
	8	248	26	54	132	190	218	214	178	113	38	25	25	25	25	25	25	25	142	4
	9	264	29	32	82	151	194	208	192	147	77	32	29	29	29	29	29	29	185	3
	10	274	33	34	39	90	145	178	184	163	116	57	35	33	33	33	33	33	219	2
	11	279	35	35	36	40	79	126	155	160	142	101	54	37	35	35	35	35	240	1
	12	280	35	35	35	36	38	63	107	139	150	139	107	63	38	36	35	35	247	12
	HALF DAY TOTALS		215	388	645	893	1065	1114	1007	749	483	316	225	184	174	173	173	174	1045	
Jun 21	5	77	35	61	76	80	72	53	2	6	5	5	5	5	5	5	5	8	12	7
	6	172	46	110	155	175	169	138	84	22	14	14	14	14	14	14	14	16	51	6
	7	220	29	101	165	204	211	187	135	60	23	21	21	21	21	21	21	21	103	5
	8	246	29	64	139	191	215	206	168	101	34	27	27	27	27	27	27	27	152	4
	9	261	31	36	91	153	190	199	180	133	66	33	31	31	31	31	31	31	193	3
	10	269	34	36	45	94	143	169	171	148	101	50	36	34	34	34	34	34	225	2
	11	274	36	36	38	44	79	118	142	145	126	88	49	38	36	36	36	36	246	1
	12	275	37	37	37	38	40	60	96	124	134	124	96	60	40	38	37	37	252	12
	HALF DAY TOTALS		257	459	722	955	1095	1102	955	678	436	299	228	197	189	188	188	191	1108	
Jul 21	5	43	18	33	42	45	41	30	15	3	3	3	3	3	3	3	3	4	6	7
	6	156	37	96	138	159	156	129	82	24	13	13	13	13	13	13	13	14	41	6
	7	211	25	90	156	196	207	186	138	66	22	20	20	20	20	20	20	20	92	5
	8	240	27	56	132	187	214	209	174	110	38	26	26	26	26	26	26	26	142	4
	9	256	30	34	83	149	191	204	187	143	75	33	30	30	30	30	30	30	184	3
	10	266	34	35	41	90	143	174	180	158	113	56	36	34	34	34	34	34	217	2
	11	271	36	36	37	42	79	124	151	156	138	99	54	38	36	36	36	36	237	1
	12	272	36	36	36	37	39	63	104	136	146	136	104	63	39	37	36	36	244	12
	HALF DAY TOTALS		223	395	646	886	1050	1092	983	730	474	315	229	190	181	179	179	180	1042	
Aug 21	6	99	13	51	81	98	100	87	60	22	7	7	7	7	7	7	7	7	16	6
	7	190	17	61	128	172	190	179	141	79	19	15	15	15	15	15	15	15	61	5
	8	232	22	34	110	174	211	216	188	132	53	23	22	22	22	22	22	22	114	4
	9	154	27	28	63	139	192	216	108	169	102	34	27	27	27	27	27	27	159	3
	10	266	30	30	33	78	145	188	203	188	144	78	33	30	30	30	30	30	193	2
	11	272	32	32	32	36	78	137	175	187	171	129	68	35	32	32	32	32	215	1
	12	274	33	33	33	33	36	71	128	167	189	167	128	71	36	33	33	33	223	12
	HALF DAY TOTALS		157	251	459	709	929	1060	1040	862	587	366	231	165	151	149	149	149	869	
Sep 21	7	131	8	21	71	108	128	128	108	71	21	8	7	7	7	7	7	7	20	5
	8	215	15	16	72	144	191	207	193	148	77	16	15	15	15	15	15	15	65	4
	9	251	20	20	34	116	184	223	227	197	136	52	21	20	20	20	20	20	114	3
	10	269	24	24	25	58	141	200	228	221	182	112	34	24	24	24	24	24	151	2
	11	278	26	26	26	28	73	151	203	223	210	166	92	29	26	26	26	26	174	1
	12	280	27	27	27	27	29	82	156	204	220	204	156	82	29	27	27	27	182	12
	HALF DAY TOTALS		105	121	240	465	729	953	1040	963	737	453	243	131	106	105	105	105	614	
Oct 21	7	4	0	0	2	3	4	4	3	2	1	0	0	0	0	0	0	0	0	5
	8	165	8	9	35	96	139	159	155	126	77	16	8	8	8	8	8	8	25	4
	9	233	14	14	16	88	161	207	220	199	148	68	15	14	14	14	14	14	66	3
	10	262	18	18	18	39	128	196	233	234	199	133	43	18	18	18	18	18	104	2
	11	274	20	20	20	21	64	153	213	241	231	186	109	27	20	20	20	20	128	1
	12	278	21	21	21	21	23	87	171	223	242	223	171	87	23	21	21	21	136	12
	HALF DAY TOTALS		71	71	108	266	519	780	925	925	779	513	256	106	72	71	71	71	391	
Nov 21	8	36	1	1	4	18	29	34	35	30	20	6	1	1	1	1	1	1	2	4
	9	179	8	8	9	52	115	156	171	161	125	67	10	8	8	8	8	8	26	3
	10	233	12	12	12	22	104	172	212	218	191	133	49	13	12	12	12	12	55	2
	11	255	15	15	15	15	52	142	204	234	228	186	114	26	15	15	15	15	77	1
	12	261	15	15	15	15	17	85	168	222	240	222	168	85	17	15	15	15	85	12
	HALF DAY TOTALS		44	44	47	117	310	555	713	760	686	502	255	85	44	44	44	44	204	
Dec 21	9	140	5	5	6	36	86	120	133	127	100	56	8	5	5	5	5	5	13	3
	10	214	10	10	10	16	91	156	194	201	179	126	49	10	10	10	10	10	38	2
	11	242	12	12	12	13	46	134	195	225	220	180	111	25	12	12	12	12	57	1
	12	250	13	13	13	13	14	81	163	215	233	215	168	81	14	13	13	13	65	12
	HALF DAY TOTALS		33	33	34	73	233	458	610	665	616	468	247	76	34	33	33	33	141	
			N	NNW	NW	WNW	W	WSW	SW	SSW	S	SSE	SE	ESE	E	ENE	NE	NNE	HOR	PM

Half Day Totals computed by Simpson's Rule with time interval equal to 10 minutes.
*Total Solar Heat Gains for DS (0.125 in.) sheet glass.
Based on a ground reflectance of 0.20 and values in Tables 1 and 26.

Table 7-A Solar Intensity (Insolation) and Solar Heat Gain Factors (SGHF) for Selected Northern Latitudes

56°N

Solar Heat Gain Factors, Btuh/ft²

Date	Solar Time am	Direct Normal Btuh/ft²	N	NNE	NE	ENE	E	ESE	SE	SSE	S	SSW	SW	WSW	W	WNW	NW	NNW	HOR	Solar Time pm
Jan 21	9	78	3	3	3	21	49	67	74	70	55	30	4	3	3	3	3	3	5	3
	10	170	7	7	7	13	74	126	156	162	143	100	38	7	7	7	7	7	21	2
	11	207	9	9	9	10	40	116	169	194	190	156	96	21	9	9	9	9	34	1
	12	217	10	10	10	10	11	71	144	190	205	190	144	71	11	10	10	10	40	12
	HALF DAY TOTALS		23	23	24	46	163	343	468	517	487	378	206	61	24	23	23	23	80	
Feb 21	8	115	4	4	21	64	95	109	107	88	55	12	4	4	4	4	4	4	10	4
	9	203	10	10	11	71	139	183	197	182	136	66	10	10	10	10	10	10	36	3
	10	246	13	13	13	28	115	184	223	227	196	133	45	14	13	13	13	13	65	2
	11	262	15	15	15	16	57	148	210	239	232	188	112	25	15	15	15	15	84	1
	12	267	16	16	16	16	17	86	171	225	244	225	171	86	17	16	16	16	91	12
	HALF DAY TOTALS		49	50	66	182	409	666	821	846	737	509	253	89	50	49	49	49	241	
Mar 21	7	128	6	16	65	101	121	122	105	70	21	6	6	6	6	6	6	6	14	5
	8	215	12	13	61	136	185	205	194	152	84	15	12	12	12	12	12	12	49	4
	9	253	16	16	23	105	179	224	233	207	148	61	17	16	16	16	16	16	89	3
	10	272	19	19	20	46	136	203	238	236	198	128	39	20	19	19	19	19	122	2
	11	282	21	21	21	22	68	156	215	241	230	184	106	27	21	21	21	21	142	1
	12	284	22	22	22	22	24	86	170	222	241	222	170	86	24	22	22	22	149	12
	HALF DAY TOTALS		85	97	200	419	699	956	1071	1016	800	502	258	118	86	85	85	85	491	
Apr 21	6	122	13	58	95	118	121	107	75	29	7	7	7	7	7	7	7	7	18	6
	7	201	15	51	123	173	195	188	152	91	21	14	14	14	14	14	14	14	56	5
	8	239	19	23	95	167	211	223	201	148	68	20	19	19	19	19	19	19	101	4
	9	260	23	24	44	126	190	223	223	189	126	44	24	23	23	23	23	23	140	3
	10	272	26	26	27	63	142	196	220	212	171	102	33	26	26	26	26	26	170	2
	11	278	28	28	28	30	74	147	195	213	200	156	86	31	28	28	28	28	189	1
	12	280	28	28	28	28	28	79	149	194	210	194	149	79	31	28	28	28	195	12
	HALF DAY TOTALS		139	226	430	694	951	1132	1147	982	699	437	252	154	132	131	131	131	772	
May 21	5	93	36	68	89	95	88	66	33	7	6	6	6	6	6	6	6	7	14	7
	6	175	33	99	148	174	173	147	97	31	14	14	14	14	14	14	14	14	48	6
	7	219	21	77	149	195	212	197	152	81	22	19	19	19	19	19	19	19	92	5
	8	244	25	38	115	179	215	218	189	131	52	25	24	24	24	24	24	24	135	4
	9	259	28	30	62	136	189	213	206	168	102	36	28	28	28	28	28	28	171	3
	10	268	31	31	33	75	141	185	200	187	145	80	33	31	31	31	31	31	199	2
	11	273	32	32	32	35	76	135	174	187	172	131	71	35	32	32	32	32	216	1
	12	275	33	33	33	33	36	71	129	168	181	168	129	71	36	33	33	33	222	12
	HALF DAY TOTALS		222	391	644	906	1112	1202	1120	878	604	392	256	187	172	170	170	173	986	
Jun 21	4	21	13	19	22	21	18	11	3	1	1	1	1	1	1	1	2	5	3	8
	5	122	53	94	119	126	115	85	40	10	9	9	9	9	9	9	12	25	7
	6	185	42	111	160	185	182	152	97	30	16	16	16	16	16	16	17	62	6
	7	222	25	86	156	199	213	195	147	74	24	22	22	22	22	22	22	105	5
	8	243	27	46	122	181	213	213	181	122	46	27	26	26	26	26	26	146	4
	9	257	30	32	69	139	187	206	196	156	91	34	30	30	30	30	30	181	3
	10	265	33	33	36	79	139	178	190	174	132	71	35	33	33	33	33	208	2
	11	269	34	34	35	38	76	129	164	174	159	119	65	37	34	34	34	225	1
	12	271	35	35	35	35	38	68	119	155	168	155	119	68	38	35	35	35	231	12
	HALF DAY TOTALS		275	473	738	989	1162	1207	1082	822	562	376	160	103	190	189	189	196	1070	
Jul 21	5	91	37	69	89	95	88	66	33	8	7	7	7	7	7	7	7	8	16	7
	6	169	34	98	145	170	170	143	95	31	15	14	14	14	14	14	14	15	50	6
	7	212	23	77	147	192	208	193	148	79	23	20	20	20	20	20	20	20	93	5
	8	237	26	40	115	177	211	214	185	128	51	26	25	25	25	25	25	25	135	4
	9	252	29	31	63	135	186	209	201	164	99	36	29	29	29	29	29	29	171	3
	10	261	32	32	34	76	139	181	196	182	142	78	35	32	32	32	32	32	198	2
	11	265	33	33	33	37	76	133	171	183	168	128	70	36	33	33	33	33	215	1
	12	267	34	34	34	34	37	71	126	164	177	164	126	71	37	34	34	34	221	12
	HALF DAY TOTALS		231	398	646	901	1097	1180	1096	859	593	390	259	193	179	177	177	180	987	
Aug 21	5	1	0	1	1	1	1	1	1	0	0	0	0	0	0	0	0	0	0	7
	6	112	14	56	91	111	114	101	71	28	8	8	8	8	8	8	8	8	20	6
	7	187	16	51	119	165	186	179	144	86	22	15	15	15	15	15	15	15	58	5
	8	225	20	25	94	162	203	214	192	142	66	22	20	20	20	20	20	20	101	4
	9	246	25	26	46	124	184	216	215	182	121	44	26	25	25	25	25	25	140	3
	10	258	28	28	30	65	139	191	213	204	165	99	34	28	28	28	28	28	169	2
	11	264	30	30	30	32	74	143	189	206	193	152	84	33	30	30	30	30	187	1
	12	266	30	30	30	30	30	78	145	188	203	188	145	78	33	30	30	30	198	12
	HALF DAY TOTALS		149	235	429	680	923	1092	1104	946	678	431	256	163	142	140	140	141	771	
Sep 21	7	107	6	15	56	87	104	105	90	60	19	6	6	6	6	6	6	6	14	5
	8	194	12	14	58	126	171	189	179	140	78	16	12	12	12	12	12	12	48	4
	9	233	17	17	24	100	170	211	220	195	140	59	18	17	17	17	17	17	86	3
	10	253	20	20	21	46	131	194	227	225	189	123	39	21	20	20	20	20	118	2
	11	263	22	22	22	24	67	150	206	230	220	176	103	28	22	22	22	22	137	1
	12	266	23	23	23	23	25	85	163	213	231	213	163	85	25	23	23	23	144	12
	HALF DAY TOTALS		89	99	191	391	652	893	1004	958	761	484	255	121	90	89	89	89	474	
Oct 21	8	104	4	5	20	59	87	100	98	81	50	11	4	4	4	4	4	4	10	4
	9	193	10	10	11	68	132	173	186	171	129	63	11	10	10	10	10	10	37	3
	10	231	14	14	14	28	111	176	213	216	186	127	44	14	14	14	14	14	64	2
	11	248	16	16	16	17	56	142	202	229	222	180	108	25	16	16	16	16	84	1
	12	253	16	16	16	16	18	83	164	216	234	216	164	83	18	16	16	16	91	12
	HALF DAY TOTALS		52	52	68	177	390	633	779	804	702	487	246	90	53	52	52	52	240	
Nov 21	9	76	3	3	3	21	48	66	72	69	54	29	4	3	3	3	3	3	6	3
	10	165	7	7	7	13	72	122	152	157	139	98	37	7	7	7	7	7	21	2
	11	201	9	9	9	10	39	113	165	190	186	152	94	21	9	9	9	9	35	1
	12	211	10	10	10	10	11	70	140	186	200	186	140	70	11	10	10	10	40	12
	HALF DAY TOTALS		24	24	24	47	161	336	457	505	475	369	202	61	24	24	24	24	81	
Dec 21	9	5	0	0	0	1	3	4	5	5	4	2	0	0	0	0	0	0	0	3
	10	113	4	4	4	7	47	82	103	107	96	68	27	4	4	4	4	4	9	2
	11	166	6	6	6	7	30	92	135	156	154	127	78	17	6	6	6	6	19	1
	12	180	7	7	7	7	8	59	120	159	171	159	120	59	8	7	7	7	23	12
	HALF DAY TOTALS		14	14	14	20	88	217	311	354	343	277	163	47	15	14	14	14	40	
			N	NNW	NW	WNW	W	WSW	SW	SSW	S	SSE	SE	ESE	E	ENE	NE	NNE	HOR	PM

Half Day Totals computed by Simpson's Rule with time interval equal to 10 minutes.
*Total Solar Heat Gains for DS (0.125 in.) sheet glass.
Based on a ground reflectance of 0.20 and values in Tables 1 and 26.

Source: *ASHRAE Handbook, 1977 Fundamentals*. Atlanta: ASHRAE, 1977, pp. 26-20–26-24. Reprinted by permission.

Indirect-Gain and Isolated-Gain
Passive Systems

*A*llowing sunlight to enter directly into a building is a very simple method of providing a source of heat. Such use of direct solar gain can provide a brightly lit space with a feeling of warmth both from the visual presence of sunlight and because of the thermal energy contained in sunlight. There are many situations, however, when it is not desirable to have sunlight shining into living spaces, even though heat gain from the sun is desired. Systems that provide solar heat without the actual presence of the sun's rays are called *indirect-gain* solar energy systems.

One example of indirect gain is shown in Fig. 8-1. The thermal storage mass has been moved from the interior surfaces of the room and placed directly behind the windows in the wall. The mass absorbs solar radiation and stores energy, but the sunlight does not enter the living space. In a direct-gain system, energy is released into the room by convection and radiation from the warm surfaces of the thermal storage mass, which is located in the room. For indirect gain, the surface of the thermal storage mass that receives solar radiation is not in the room, and for heat to enter the living space it must be conducted all the way through the thermal storage from its sunlit (outer) surface to its interior surface.

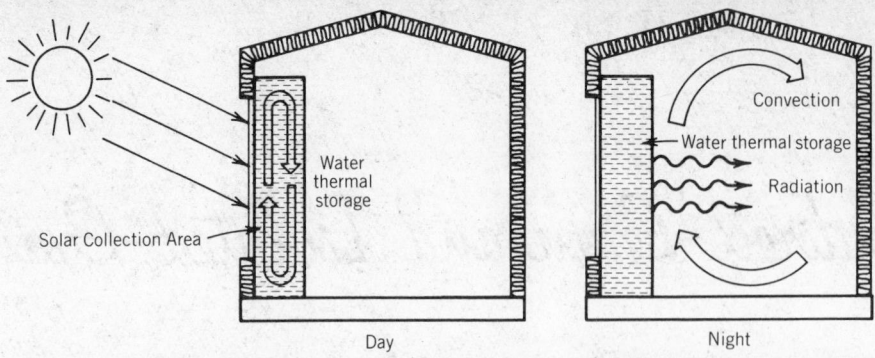

Figure 8-1 An indirect-gain system using a water wall for thermal storage. The diagram shows the water thermal storage placed between the solar collection glazing and the interior of the building.

INDIRECT-GAIN SOLAR SYSTEMS

There are several configurations for locating thermal storage mass between the glazing of the solar aperture and the interior of the building. The glazing and thermal storage can be incorporated into a wall in which the thermal storage mass can be a structural element of the building, or thermal storage can be designed into the roof structure of the building. In other designs, the thermal storage may not be a structural part of the building at all, taking the form of water cylinders or drums, for example.

8-1 Walls for the Collection and Storage of Solar Energy

There are three components in a solar collection and thermal storage wall, as illustrated in Fig. 8-2. The outer surface is covered with a material, such as glass, that transmits solar radiation. In most applications, this outer surface will be double-glazed—that is, two pieces of glazing material separated by an insulating air space. The second component is the glazing support system. This is a framework constructed from wood or metal that holds the glazing to the building. Next to the second (or inner) glazed surface there is an air space, the thickness of which depends on the wall design. Finally, there is the thermal storage material itself, with a front surface that is a good absorber of solar radiation.

Solar radiation transmitted by the glazing is absorbed on the surface of the thermal storage and converted into heat. Some of the absorbed energy heats the air between the glazing and the thermal storage, and some is lost back out through the glazing during both day and night. However, the greater part of the absorbed solar radiation is conducted into the thermal storage and increases the thermal energy stored in the wall. Subsequently, heat is released from the interior surface of the thermal storage into the building. The final transfer of heat to the space is by radiation and convection from the warm interior surface of the thermal storage.

The entire assembly illustrated in Fig. 8-2—the glazing, the glazing support system, and the thermal storage—is termed a *thermal storage wall*. All thermal storage walls have these three basic components and may have other features besides. In very cold climates, additional sheets of glazing or movable insulation may be provided to reduce thermal losses from the wall, as illustrated in Fig. 8-3 where roll-down insulation is shown on the inside of the glazing between the glazing and thermal storage. Another glazing system is the Beadwall™ developed by Zomeworks Corporation. In the Beadwall system, the air space between the two pieces of glazing is filled with tiny foam pellets by a

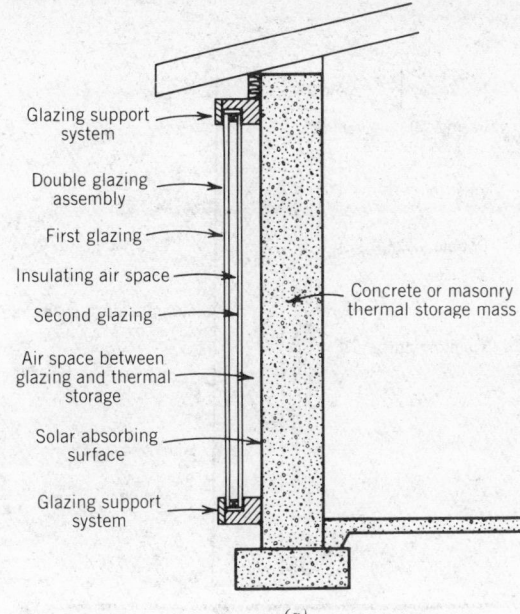

Glazing support
system

Double glazing
assembly

First glazing

Insulating air space

Second glazing

Air space between
glazing and thermal
storage

Solar absorbing
surface

Glazing support
system

Concrete or masonry
thermal storage mass

(a)

(b)

Figure 8-2 *(a)* Three basic components of a thermal storage wall: the glazing, the glazing support system, and the thermal mass. *(b)* Exterior view of a Trombe thermal storage wall—fiberglass glazing and wood frame glazing support.

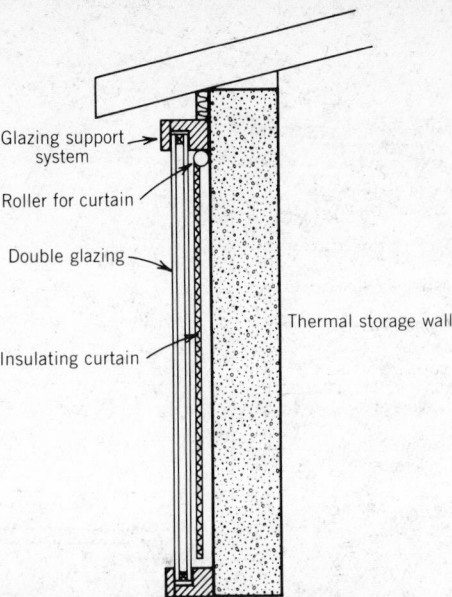

Glazing support system

Roller for curtain

Double glazing

Insulating curtain

Thermal storage wall

Figure 8-3 Reducing thermal losses through the glazing of a thermal storage wall. An insulating curtain on a roller system may be placed between the glazing and the thermal storage. The curtain is rolled up when solar radiation is available and desired and rolled down when radiation is not available or desired.

blower/vacuum system, when the wall requires insulation. Then, when the wall is to collect and store solar energy during the day, the pellets are vacuumed out to make the glazing transparent. Another location for insulation is on the exterior of the glazing. This may be in the form of sliding panels that slide across the front of the thermal wall to cover the glazing, or hinged panels that fold down in front of the mass wall. The installation of an insulating system for a thermal storage wall can increase the cost of the wall by 25 to 50 percent, and therefore a careful cost–benefit analysis should be made before specifying any particular system. There are several choices for the thermal storage mass to be used in a wall system, with the four most common types being masonry, poured concrete, water, and phase-change materials (eutectics). Thermal storage walls will be discussed under three headings, based on the above types of thermal storage.

8-2 Masonry and Concrete Thermal Storage Walls

An interesting property of masonry and concrete materials becomes apparent when they are used in thermal storage walls. In Fig. 8-4, the temperatures in a 12-in. concrete thermal storage wall are indicated and plotted for a 24-hr day. The sunlit surface (temperature T_4) undergoes an immediate temperature rise in the morning (Fig. 8-4c). The temperature of the air space between the thermal storage and the glazing (temperature T_5) follows closely the temperature of the sunlit surface of the thermal storage. As heat flows into the thermal storage from the sunlit surface, the temperature of the wall interior gradually increases (temperatures T_3 and T_2). From the time the outer surface of the wall first receives morning sunlight (say, 7:00 A.M.) until the time when a point 4 in. into the wall (point T_3) begins to incur a temperature increase (9:00 A.M.), about two hours will have elapsed (Fig. 8-4c). A six-hour time period elapses between the time the outer surface of the wall reaches its peak tem-

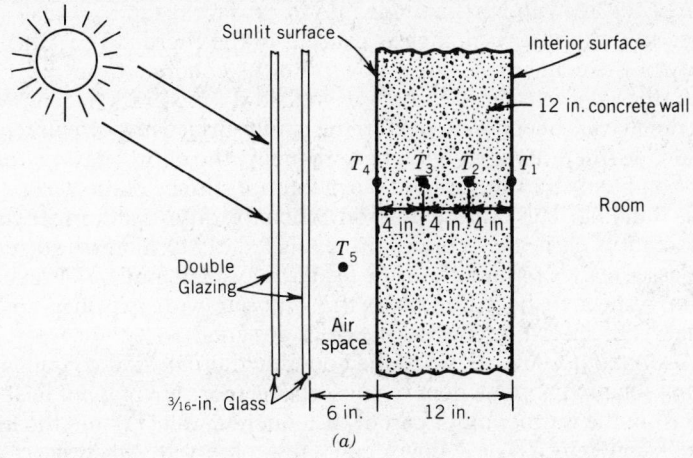

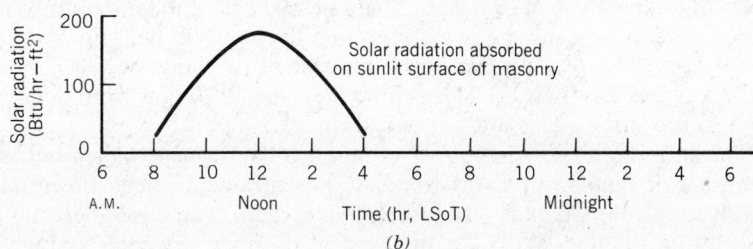

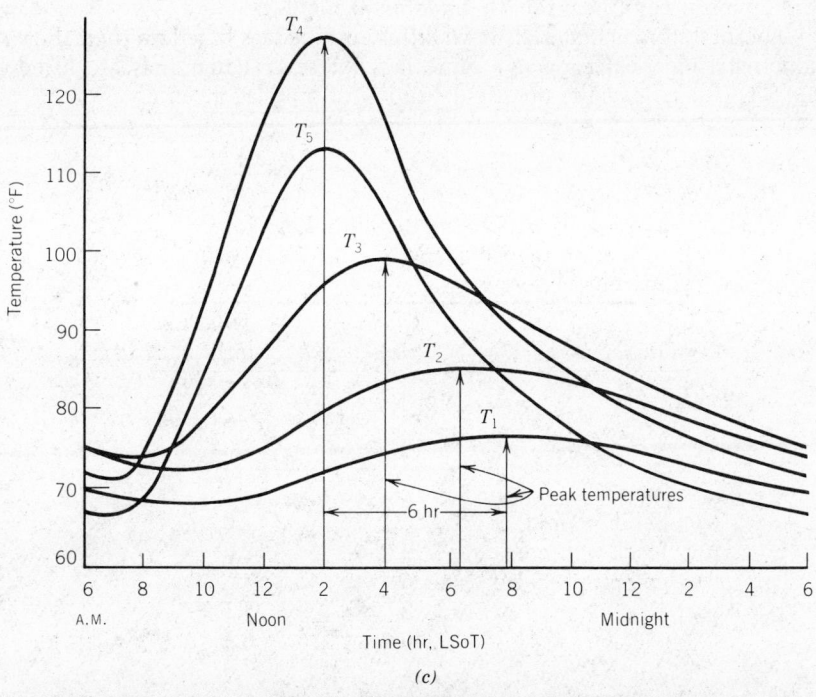

Figure 8-4 Time lag of heat transmission through a concrete wall. (a) Section through a 12-in. concrete thermal storage wall indicating the locations where temperatures T_1 through T_5 are calculated. (b)Plot of solar radiation absorbed on outer surface of storage wall. (c) Plot of wall temperatures for a 24-hr period showing the time at which each part of the wall reaches maximum temperature.

perature (2:00 P.M.) and the time the interior surface reaches its peak at 8:00 P.M. This lag in temperature change is caused by the thermal resistance and heat capacity of the concrete. As heat flows into the concrete from the warm sunlit surface, the concrete mass absorbs it and slowly increases in temperature. As succeeding layers of concrete next to the sunlit surface increase in temperature, heat flows further into the concrete. Gradually, the entire mass of concrete has increased in temperature, and the warm interior surface of the wall releases heat into the building. This time delay, from when the outer wall surface first absorbs solar radiation until the interior surface just starts to increase in temperature and release heat, is termed the *time lag* of the wall. All materials exhibit a time lag with respect to heat flow, and the time lag with masonry and concrete materials is of special interest to the solar designer.

After about 5:00 P.M., since the concrete thermal storage wall is no longer receiving solar radiation, it does not absorb any more energy, but heat continues to flow from the warmer outer part of the concrete wall (T_4) into the inner layers of the concrete. This flow can be observed in Fig. 8-4c from 6:00 P.M. until about 12:00 midnight. During this time, the outer part of the wall decreases in temperature because heat is flowing away from it to the cooler interior part of the wall. The curves show that the outer part of the wall begins to decrease in temperature at 2:00 P.M. At about 5:00 P.M., the temperature in the center of the concrete storage peaks and then decreases as heat flows from the center of the concrete wall toward the interior of the building. At about 7:00 P.M., T_4 becomes less than T_3. At this time, heat begins to flow from the interior (T_3) out to the surface (T_4) and on out to the glazing.

The time lag of a masonry or concrete solar storage wall provides the passive solar designer with a way to delay the entrance of heat into inhabited spaces. Recall from Chapter 7 that direct-gain rooms can easily become over-heated during the day. With the time lag of the masonry solar storage wall, entrance of absorbed solar radiation can be delayed for time periods as long as 10 to 12 hours. In some cases, direct-gain and indirect-gain systems are combined, direct gain providing light and heat during the daylight hours and indirect gain providing a source of heat for the building at night.

Thermal storage materials have different time lags based on their thermal conductivity and heat capacity. Time lags for several materials are listed in Table 8-1. (Remember the precise definition of time lag given above.)

Table 8-1 Approximate Time Lags for Thermal Storage Walls Made of Masonry and Concrete Materials

Material	Thickness (in.)	Time Lag (hr)
Brick	4	2.3
	8	5.5
	12	8.0
Concrete		
Solid (poured) or	2	1.0
block with cores	4	2.6
completely filled	6	3.8
	8	5.1
	10	6.4
	12	7.6
Stone	8	5.4
	12	8.0

Source: C. Strock, *Handbook of Air Conditioning, Heating, and Ventilating:* (New York: Industrial Press, 1965), pp. 1–60.

INDIRECT-GAIN AND ISOLATED-GAIN PASSIVE SYSTEMS

Illustrative Problem 8-1

A six-hour time lag is desired for a solar thermal storage wall. Suggest some masonry or concrete materials and material thicknesses that will yield this time lag.

Solution

From Table 8-1, we find that a 10-in. concrete block with cores filled, or a 10-in. poured concrete wall, will yield a time lag of 6.4 hr. An 8-in. concrete wall will yield a 5.1-hr time lag. By interpolating between these two values, it is found that a wall thickness of either material of 9.4 in. will yield approximately a 6-hr time lag.   *Ans.*

Using values from Table 8-1 for brick, one finds by interpolation that an 8.8-in. brick wall will have approximately a 6-hr time lag.   *Ans.*

8-3 Thermocirculation Vents for Masonry and Concrete Thermal Storage Walls—Trombe Walls

One disadvantage of the thermal storage wall is that heat absorbed by the wall cannot be moved directly into the building when and if heat is needed. In other words, time lag is often a problem, especially early in the day. This problem can be solved by modifying the masonry or concrete storage, as shown in Fig. 8-5. The placement of openings at the top and bottom of the storage wall allows the warm air between the glazing and the exterior surface of the thermal storage to circulate by convection to the building interior. Warm air rises up in the air space and flows into the building interior through the upper opening. Cool air is drawn from the room near the floor, through the lower opening into the air space to be heated. These openings are called *thermocirculation vents* and were first used by Felix Trombe and Jacques Michel in experimental houses built at Odeillo, France. When thermal storage walls are provided with thermocirculation vents, they are referred to as *Trombe walls*. The thermocirculation vents provide a means of bringing directly into the interior space some of the solar energy that is absorbed by the thermal storage walls, thus appreciably bypassing the time-lag property of the wall.

There are certain aspects of thermocirculation vents that must be taken into consideration by the solar designer. Most thermocirculation vents are manually operated, making the building occupants responsible for their proper use. If the occupants are not interested in operating the vents correctly, several conditions will result that negatively affect the performance of the Trombe wall. If the vents are never opened, the Trombe wall will simply perform as a thermal storage wall with a time lag. If the vents are left open at all times, the interior comfort of the building may be adversely affected. If they are open all day, heat will be transferred by convection from the outer surface of the masonry storage to the interior of the building without much time lag, which may result in overheating of the interior spaces, although doors and windows could be used to vent the warm air. When this happens, the maximum amount of heat is not stored in the masonry thermal storage, since much of it is lost by venting. This results in a reduction of the overall heating performance of the Trombe wall. If the vents are opened for a few hours during the day, they will provide a heat source to increase the interior air temperature. Ideally, the vents should be opened in the morning to provide a warmup of the interior of the building, and then they should be closed to allow the wall to store up heat for night release. When used in this manner, the thermocirculation vents provide a means of control for the interior comfort of the building, both day and night. A reverse circulation can occur in a Trombe wall under certain conditions. During cold, sunless days, the energy stored in the masonry material will be totally extracted, and the air between the glazing and masonry storage will become very cold, becoming more dense (i.e., heavier) than the warm interior air. If the vents are open, the heavier cold air will flow out of the lower vent into the room, drawing

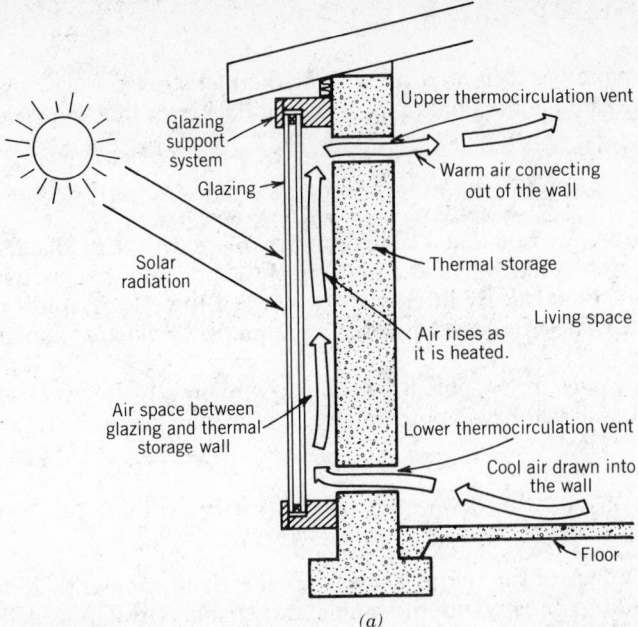

Figure 8-5 A Trombe thermal storage wall. (*a*) The placement of thermocirculation vents is shown. The vents allow warm air to circulate into the room. (*b*) Interior view of a Trombe wall showing thermocirculation vents.

warm room air into the upper vent. The warm room air will become chilled as it loses its heat out through the glazing, and it will then flow back into the room through the lower vent. This undesirable reverse circulation can be prevented by the use of a back-draft damper in the lower vent. The back-draft damper is constructed from a very thin (0.0005 in.) flap of plastic film, as shown in Fig. 8-6. The thin flap has very little weight, and the cold air between the glazing and masonry storage (being heavy) will push the flap up against the damper frame, making the flap seal against the frame. This will prevent the cold air from flowing out of the lower thermocirculation vent. When warm air starts to flow out of the upper vent, the flap will be drawn forward so cool air from the room can enter the lower vent.

INDIRECT-GAIN AND ISOLATED-GAIN PASSIVE SYSTEMS

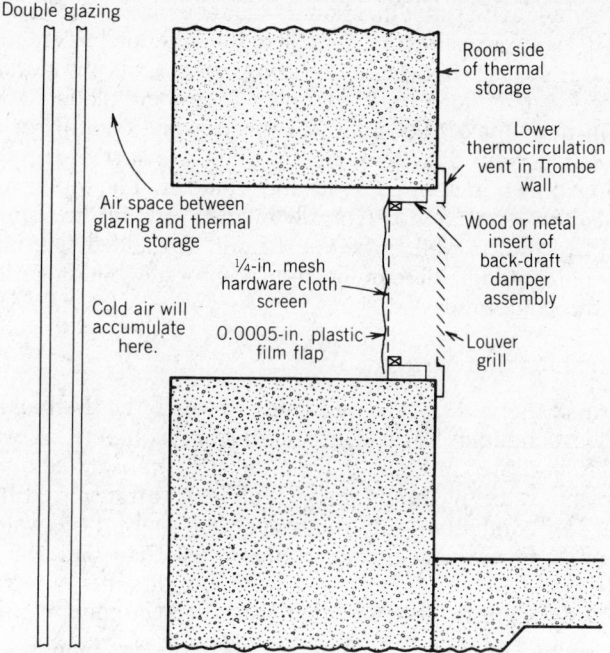

Double glazing

Room side of thermal storage

Lower thermocirculation vent in Trombe wall

Air space between glazing and thermal storage

Wood or metal insert of back-draft damper assembly

¼-in. mesh hardware cloth screen

Cold air will accumulate here.

0.0005-in. plastic film flap

Louver grill

Figure 8-6 Construction details of a back-draft damper for the lower thermocirculation vents of a Trombe wall. The back-draft damper prevents a reverse circulation of air in the wall when the damper is closed, stopping cold air from flowing out of the lower vent. The damper flap is made from a very thin plastic film.

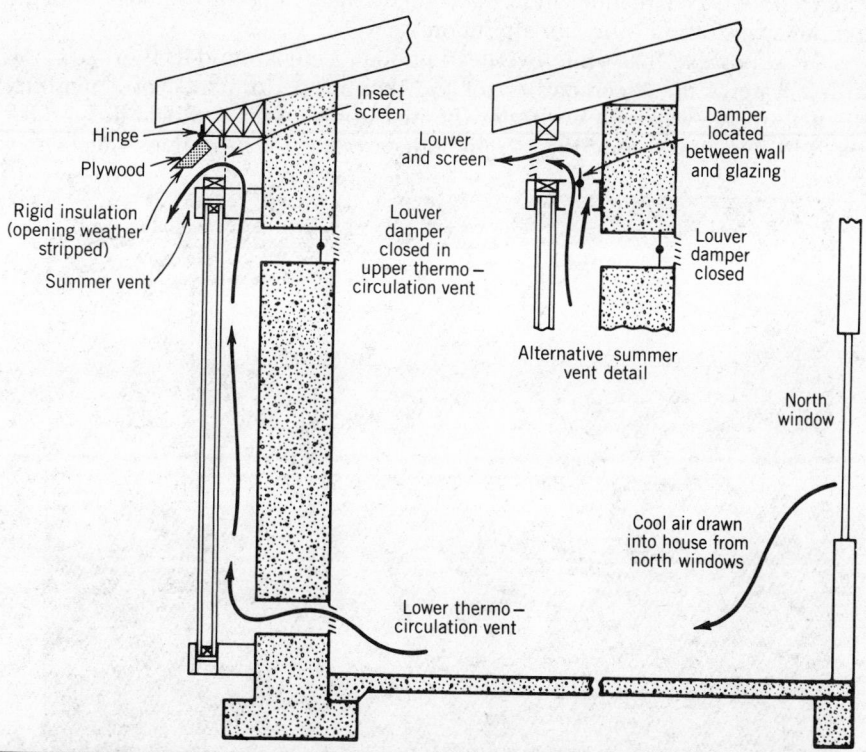

Insect screen

Hinge

Plywood

Rigid insulation (opening weather stripped)

Summer vent

Louver and screen

Louver damper closed in upper thermo–circulation vent

Damper located between wall and glazing

Louver damper closed

Alternative summer vent detail

North window

Cool air drawn into house from north windows

Lower thermo–circulation vent

Figure 8-7 Vents or dampers can be located along the top of the Trombe wall to allow warm air to be vented in the summer. Cool air is drawn into the air space from the lower thermocirculation vents.

INDIRECT-GAIN SOLAR SYSTEMS

There is one further modification that can be made to a Trombe wall. If it is necessary to vent warm air from the wall to the outdoors, a vent may be placed at the top of the air space between the glazing and the thermal storage (Fig. 8-7). When the top vent is open and the upper thermocirculation vent is closed, warm air will flow out of the outdoor vent, drawing cool air in at the lower thermocirculation vent. This flow creates a low pressure in the building, and cool air will be drawn in at open doors and windows. The top (outdoor) vent is termed the *summer vent* and is typically opened only in the summer to allow energy absorbed by the wall to be vented out. An added side benefit is that (relatively) cool outdoor air can be drawn into the building through open windows at the same time.

8-4   Water Thermal Storage Walls

Another form of thermal storage wall is one in which the thermal mass consists of water-filled containers instead of masonry or concrete. A water thermal storage wall is illustrated in Fig. 8-8. Solar radiation transmitted by the glazing is absorbed by the water-filled containers. The heat is transferred from the container surface by conduction into the water. The walls of the water containers are usually of plastic or metal and are ordinarily less than ¼ in. thick. They offer very little thermal resistance to heat flow from the container outer surface to the water within. Recall from Chapter 7 that water-filled containers will have only a small temperature variation throughout the volume of the water. There will be temperature differences of 5 to 10 F° from the bottom to the top of the container, but this is a relatively small difference compared to the 50 to 80F° temperature difference between the sunlit surface and the interior surface of a masonry or concrete thermal storage wall. The water thermal storage wall does not exhibit a pronounced time lag like the masonry wall does because the water is in a convection mode at all times. The entire volume of water heats up at nearly the same rate as it absorbs heat. A water thermal storage wall will reach a peak temperature in the late afternoon, resulting in the inner surface of the water containers radiating and convecting a peak amount of heat into the interior space, also in the late afternoon.

It is possible that some overheating may occur around 4:00 or 5:00 P.M. with a water wall. A comparison of the thermal performance of a masonry thermal storage wall with two water thermal storage walls is shown in Figs. 8-9 and 8-10. The basic room (Fig. 8-9) in both cases has 160 ft² of floor area. There

Figure 8-8 A water solar thermal storage wall is made up of water-filled containers positioned directly behind the solar collection area. (Kalwall–Solar Components.)

is a 20 ft² double-glazed window that does not receive solar radiation. The walls are insulated with 3½-in. fiberglass and the ceiling with 6-in. fiberglass insulation. The infiltration is assumed to be one-half air change per hour. The masonry wall is 12 in. thick, and the water walls are 6 in. and 12 in. thick. The solar aperture in each case is 60 ft². The temperatures were calculated using a computer program that models the heat flow in the room. The incident solar radiation and the ambient temperature are the same for all three cases, and these are plotted in Fig. 8-10a and b respectively. In Fig. 8-10c, the exterior and interior temperatures of the masonry wall are plotted, along with the average temperatures of the two water-storage walls. In Fig. 8-10d, the room air temperature is plotted for the three different thermal storage walls.

Several interesting points regarding the performance of solar thermal storage walls can be deduced from Fig. 8-10. First, the solar energy absorbed by the thermal storage is the same for the 12-in. masonry, the 6-in. water, and the 12-in. water walls. Average room temperatures $(T_{max} + T_{min})/2$ were approximately 66°F for the 12-in. masonry wall, 72°F for the 6-in. water wall, and 70°F for the 12-in. water wall. The room air temperature swings for the three walls were approximately 6 F° for the 12-in. masonry wall, 9.5 F° for the 12-in. water wall, and 14 F° for the 6-in. water wall.

The average room temperature gives a relative comparison of the thermal efficiency of each thermal wall system. The masonry wall has the lowest average room temperature, which indicates that less absorbed solar energy enters the room through the masonry wall than through water walls. This can be explained by the high temperature that the absorbing surface of the masonry wall reaches during the day, which increases its heat loss to outdoors through the glazing.

The masonry wall gives the greatest moderating effect since it has a room air temperature swing of only 6 F°, while the 12-in. water wall has a swing of 9.5 F° and the 6-in. water wall one of 14 F°. The peak room temperatures for the water walls occurred late in the afternoon between 4:00 and 5:00 P.M. However, in the case of the masonry wall, the peak room air temperature occurred between 7:00 and 8:00 P.M., illustrating the greater lag of a masonry thermal storage wall.

8-5 Thermal Storage Walls with Latent Heat Storage

Thermal storage walls using masonry (or concrete) and water as thermal storage materials increase in temperature as the thermal storage absorbs heat. As the thermal mass releases heat, their temperature decreases. Water and masonry, then, are sensible heat storage materials, materials that change in temperature as they absorb or lose thermal energy. Latent heat storage materials, on the other hand, store or release energy because a change of phase is occurring in the

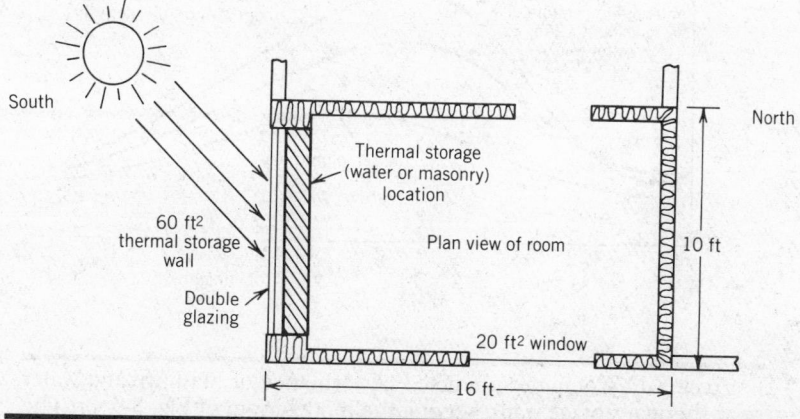

Figure 8-9 Floor plan of a model room showing location of thermal storage that will be used to compare the performance of a masonry and a water thermal storage wall.

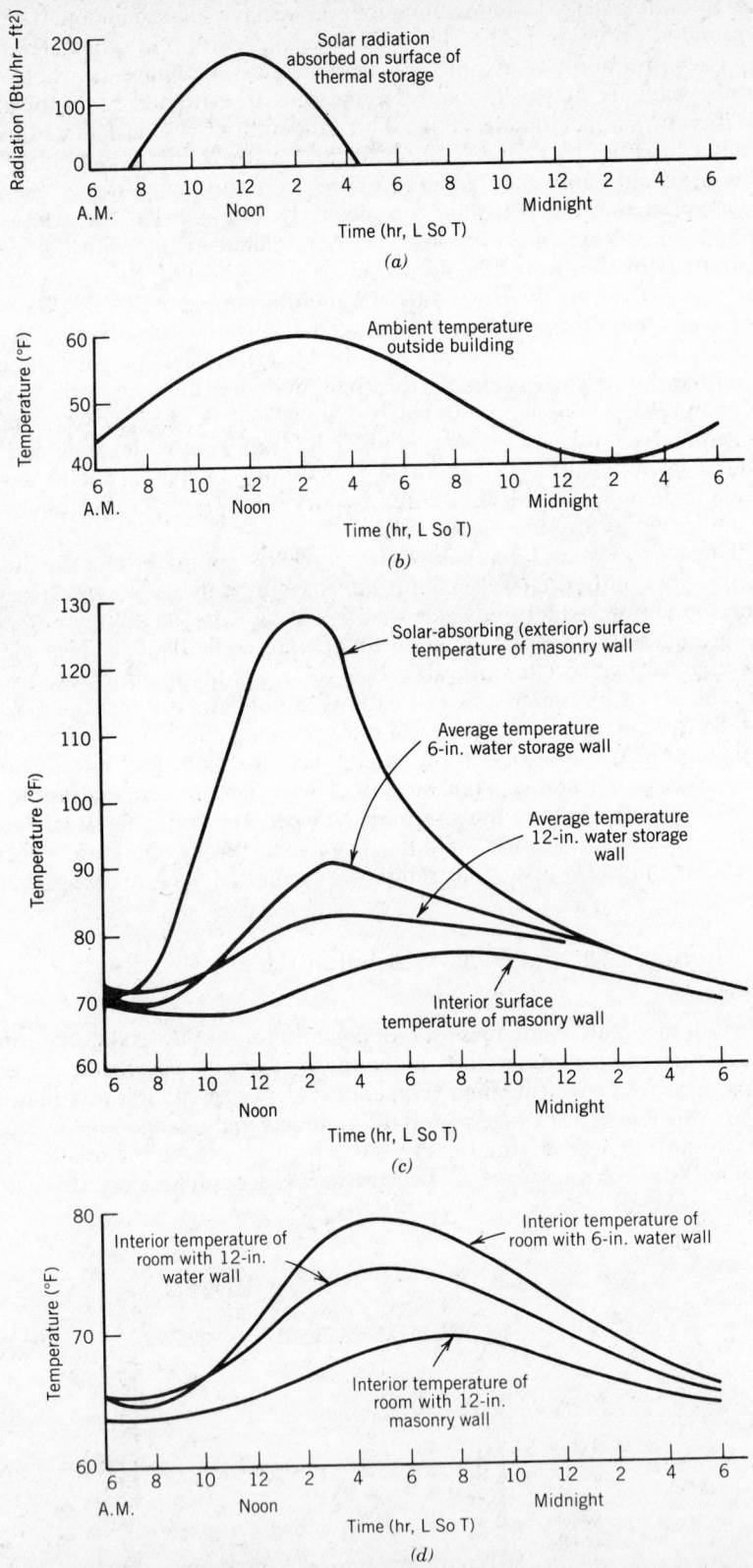

Figure 8-10 Computer-predicted performance of masonry and water solar thermal storage walls, serving the model room of Fig. 8-9. (a) The solar radiation assumed to be absorbed by the thermal storage. (b) The ambient (outdoor) temperature during the 24-hr day. (c) Temperature plots for the 24-hour day for the exterior and interior surfaces of the masonry wall, and the average temperature of the 6-in. and 12-in. water walls. (d) Interior room temperatures for the masonry wall and for the two thicknesses of water walls.

Figure 8-11 Thermal storage walls may also be constructed from phase-change heat storage materials in containers. These walls are made up using panels filled with eutectic materials. (Dow Chemical Company.)

material. (See Fig. 3-2 and section 7-10.) Materials that have phase-change temperatures in the range of 80°F to 120°F are suitable for thermal storage. These materials can be packaged in cylinders and flat cans and are commercially available for use in thermal storage walls. The phase-change materials can be placed behind the solar glazing as shown in Fig. 8-11. The phase-change material remains at a constant temperature while it is absorbing or releasing energy (that is, while it is undergoing a phase change) and does not exhibit a cyclical temperature swing as masonry and water thermal storage do. (See section 7-10.)

SIZING AND CONSTRUCTION OF SOLAR THERMAL STORAGE WALLS

In every solar design there is an initial point from which the design develops. This is usually a schematic design covering the general concepts of the building as was outlined in section 6-7. There must be an allocation of solar collection area for the building during this initial design phase.

8-6 Guidelines and Rules of Thumb for Storage Wall Sizing

The Los Alamos Laboratory's solar collection area rule of thumb (section 6-8) is a useful tool for obtaining preliminary estimates of solar collection areas. This rule can also be used to estimate a thermal storage wall area rather than a direct-gain window area.

Illustrative Problem 8-1

A client has brought you a rough floor plan for a house with 1500 ft² of floor area and would like to know the approximate solar collection area for the thermal storage wall needed to provide (a) 50 percent of the heating, and (b) 70 percent of the heating. The house is 15 mi from Knoxville, Tennessee.

Solution

From Table 6-1, the following values are obtained for the Los Alamos solar collection area rule of thumb: $R1 = 0.09$; $R2 = 0.18$; $S1 = 20$; $S2 = 33$; $S3 = 33$; $S4 = 56$. The rule of thumb states: A solar collection area of $R1$ percent to $R2$ percent of the floor area can be expected to reduce the annual heating load of a building in (location) by $S1$ percent to $S2$ percent or, if R-9 insulation is used at night in the collection area, by $S3$ percent to $S4$ percent. It is useful to plot the values, as shown in Fig. 8-12. From the plot, with a Solar Savings of 50 percent, the ratio of solar collection area to floor area is found to be approximately:

For no night insulation = 0.295, or 29.5 percent

Solar collection area = 0.295 × 1500 ft²
$$= 442.5 \text{ ft}^2 \qquad \qquad Ans.$$

For night insulation = 0.16, or 16 percent

Solar collection area = 0.16 × 1500 ft²
$$= 240 \text{ ft}^2 \qquad \qquad Ans.$$

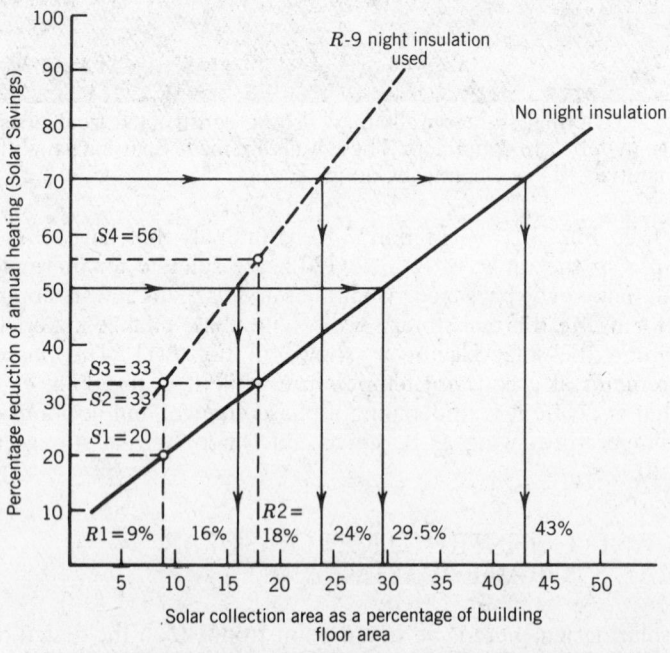

Figure 8-12 Plot of values for the Los Alamos solar collection area rule of thumb, for a location near Knoxville, Tennessee, for Illustrative Problem 8-1.

For a Solar Savings of 70 percent, the ratio of solar collection area to floor area is found to be approximately:

For no night insulation = 0.43, or 43 percent

$$\text{Solar collection area} = 0.43 \times 1500$$
$$= 645 \text{ ft}^2 \qquad \textit{Ans.}$$

For night insulation = 0.24, or 24 percent

$$\text{Solar collection area} = 0.24 \times 1500$$
$$= 360 \text{ ft}^2 \qquad \textit{Ans.}$$

It must be stressed that all rules of thumb make certain assumptions. In the case of the solar collection area rule of thumb, it is assumed that the building under consideration will have a net load coefficient (NLC) that satisfies the baseline conditions for the rule of thumb. If the NLC exceeds the baseline conditions, the actual Solar Savings will be less than the projected Solar Savings.

Mazria's Guidelines. The sizing guidelines developed by Mazria represent another approach to sizing thermal storage walls. Table 8-2 is a sizing guide (from Mazria's research) that gives the square feet of thermal storage wall recommended for each square foot of living-space floor area. Sizing is indicated for different climates based on the average winter outdoor temperature and on winter degree-days per month. The winter temperatures and degree-days represent typical December and January conditions. The wall areas are given for both masonry (or concrete) and water thermal walls. The ratio of thermal storage wall area to floor area is given as a range for each climate. For southern latitudes of 35° N or less, the smaller wall-to-floor-area ratio should be used. For northern latitudes of 48° N or greater, the largest ratio of wall to floor area should be used. These ratios assume that the normalized building load coefficient (BLC) is in the range of 8 to 10 Btu/DD-ft^2 of floor area. When night insulation is used with an R value of 8.0 or 9.0, the ratio of wall to floor area can be reduced to 85 percent of the values given in Table 8-2. These sizing guidelines are intended to provide enough energy on an average sunny day to supply all of the room's

Table 8-2   Mazria's Sizing Guide for Thermal Storage Walls

Average Winter Outdoor Temperature (°F)	Degree-Days/Month Winter Months December and January	Square Feet of Solar Wall Recommended for Each Square Foot of Floor Area	
		Masonry or Concrete Wall	Water Wall
Cold Climates			
15°	1500	0.72–1.00	0.55–1.00
20°	1350	0.60–1.00	0.45–0.85
25°	1200	0.51–0.93	0.38–0.70
30°	1050	0.43–0.78	0.31–0.55
Temperate Climates			
35°	900	0.35–0.60	0.25–0.43
40°	750	0.28–0.46	0.20–0.34
45°	600	0.22–0.35	0.16–0.25

Source: Adapted with permission from Edward Mazria, *The Passive Solar Energy Book* (Emmaus, Pa: Rodale Press, 1979), p. 156.

Notes: 1.   The wall-sizing ratios apply to a residence with a normalized Building Loss Coefficient (BLC) of 7 to 9 Btu/DD-ft^2 floor area.
2.   Temperatures and degree-days are listed for December and January, usually the coldest months.
3.   For southern latitudes (e.g., 35° N Lat), use the lower wall-to-floor ratios, and for northern latitudes (48° N Lat) the higher ratios. Use higher values for poorly insulated buildings. For thermal walls with night insulation (R-8 or R-9), use 85 percent of the listed ratios.

SIZING AND CONSTRUCTION OF SOLAR THERMAL STORAGE WALLS

space-heating needs, based on the climate data stipulated in the table. The average room temperature maintained over a 24-hour period should be in the range 65 to 75°F.

Illustrative Problem 8-2

Determine the solar thermal storage wall area for the 1500 ft² house located in Knoxville, Tennessee, that was used in Illustrative Problem 8-1. The December and January degree-days for Knoxville are 744 and 760, respectively. Use Table 8-2 to determine the wall areas for (a) a masonry wall and (b) a water wall.

Solution

(a) Masonry wall

From Table 8-2, for 750 DD/month, the ratio of wall area to floor area for a masonry wall is 0.28–0.46. The latitude of Knoxville is 35.8° N, so the smaller ratio should be used; or the range of ratios can be interpolated to find the ratio that corresponds to 35.8°, which is 0.29. The required area of the masonry thermal storage wall is found by multiplying the wall-to-floor ratio by the house floor area:

$$\text{Masonry wall area} = 0.29 \times 1500 \text{ ft}^2$$
$$= 435 \text{ ft}^2 \qquad \qquad Ans.$$

If night insulation is used, the area should be reduced to 85 percent of this value:

$$\text{Masonry wall area with insulation} = 0.85 \times 435 \text{ ft}^2$$
$$= 370 \text{ ft}^2 \qquad \qquad Ans.$$

(b) Water wall

From Table 8-2, the ratio of wall area to floor area for a water thermal storage wall is 0.20–0.34. The ratios can be interpolated to find a ratio that corresponds to 35.8°, which is 0.21. The required area of the water thermal storage wall is found by multiplying the wall-to-floor ratio by the house floor area:

$$\text{Water thermal storage wall area} = 0.21 \times 1500 \text{ ft}^2$$
$$= 315 \text{ ft}^2 \qquad \qquad Ans.$$

If night insulation is used, the area is reduced to 85 percent of this value:

$$\text{Water thermal storage wall area with insulation} = 0.85 \times 315 \text{ ft}^2 = 268 \text{ ft}^2$$
$$\qquad \qquad Ans.$$

The areas obtained by different rules of thumb will vary somewhat as illustrated by the answers obtained in Illustrative Problems 8-1 and 8-2. These estimated areas can be used as a basis for early planning in the solar design, but there are many considerations that must be dealt with during the continuing design process before the final size of the thermal storage wall is determined. These considerations include window area for views and natural lighting, solar access to the wall area, and the cost of the wall. When the solar design is concluded, the actual area of the thermal storage wall may be somewhat larger or smaller than the area initially determined by rules of thumb.

Rules of thumb are handy shortcuts, and they are widely used in the industry, but they should be used well within the assumptions made for their derivation and then only as starting points in the solar energy system design. More accurate methods of determining system performance must be used later in the design process to ensure that design goals are satisfied. These other methods include the Solar Load Ratio (SLR) method, the annual Load Collector Ratio (LCR) method, and/or computer simulations.

INDIRECT-GAIN AND ISOLATED-GAIN PASSIVE SYSTEMS

8-7 Capacity Requirements for Thermal Storage Walls

In the design of thermal storage walls, it is important to understand how the thermal storage capacity affects the overall performance of solar systems. Recall from section 7-9 that the volumetric thermal capacity C_V of a material is the product of the specific heat capacity c and the density D of the material: $C_V = cD$. The units of volumetric thermal capacity C_V are Btu/ft^3-F$^\circ$. Some properties of sensible heat storage materials were given in Table 7-4.

The thermal storage is placed directly behind the glazing in a thermal storage wall solar system. This placement ensures that it will receive solar radiation during the entire day as opposed to the direct-gain solar system where the thermal storage is distributed throughout the direct-gain space and may receive solar radiation for only a part of the day. Recall that the minimum thermal storage capacity for direct-gain solar systems has been established at about 30 Btu/F$^\circ$-ft^2 glazing (section 7-12), which is also a useful minimum for indirect-gain solar systems.

The performance of a thermal storage wall is determined to a great extent by the thermal conductivity of the thermal storage material. In Table 8-3, the thermal conductivity, thermal capacity, and recommended thickness are given for some typical materials used to construct thermal storage walls. The conductivity K affects the performance because solar radiation absorbed by the thermal storage wall must be conducted through it before heat can reach the interior of the building. A masonry material with a high conductivity will transfer the heat absorbed on its collecting surface to its interior surface quite rapidly. Therefore, walls made of high-conductivity materials must have a greater thickness so that heat is not provided on the interior surface at the wrong time. Conversely, walls constructed of low-conductivity materials such as adobe must be thinner so that heat can be transferred through them within a suitable time period. Traditional construction practice for adobe structures has been to build with 18- to 24-in. adobe walls, and when adobe thermal storage walls began to be built designers tended to follow this same practice. This is a mistake, however, since an adobe wall of 24-in. thickness will transfer roughly only one half the heat during the daily cycle that a 12-in. thick adobe wall will because of the low conductivity of adobe.

Each material exhibits a thickness range where the wall performance is optimal for thermal storage walls. These optimal thicknesses are listed in Table 8-3. The optimal thickness is determined by the thermal conductivity K and the

Table 8-3 Conductivity of Selected Storage Wall Materials and Approximate Space Air Temperature Fluctuations as a Function of Wall Thickness

Material	Thermal Conductivity (Btu/hr-ft^2-F$^\circ$)	Recommended Thickness (in.)	Approximate Indoor Temperature Fluctuation (F$^\circ$) as a Function of Wall Thickness[a]					
			4 in.	8 in.	12 in.	16 in.	20 in.	24 in.
Adobe	0.30	8–12	...	18°	7°	7°	8°	...
Brick (common)	0.42	10–14	...	24°	11°	7°	...	...
Concrete (dense)	1.00	12–18	...	28°	16°	10°	6°	5°
Brick (magnesium additive)[b]	2.20	16–24	...	35°	24°	17°	12°	9°
Water[c]	—	6 or more	31°	18°	13°	11°	10°	9°

[a]Assumes a double-glazed thermal wall. If additional mass is located in the space, such as masonry walls and/or floors, then temperature fluctuations will be less than those listed. Values given are for winter-clear days.

[b]Magnesium is commonly used as an additive to brick to darken its color. It also greatly increases the thermal conductivity of the material.

[c]When using water in tubes, cylinders or other types of circular containers, use *at least* a 9½-inch-diameter container or ½ cubic foot (31.2 lb or 3.74 gal) of water for each one square foot of glazing.

Source: Reprinted with permission from Edward Mazria, *The Passive Solar Energy Book* (Emmaus, Pa: Rodale Press, 1979), p. 163.

SIZING AND CONSTRUCTION OF SOLAR THERMAL STORAGE WALLS

thermal capacity C_V of the masonry (or concrete) material. In general, the greater the thermal conductivity, the thicker the optimal thickness of the thermal storage. The specified thickness of a thermal storage wall is influenced by three factors during the design: the thermal capacity required, the thermal time lag required, and the thickness for optimal performance. Any one of these three factors may become dominant and finally dictate the design thickness of the wall.

When water is used as the thermal storage medium, the thermal performance differs somewhat from that of masonry or concrete thermal storage walls, as was discussed in section 8-4. The thermal capacity of water is 62.4 Btu/ft³-F°, and the minimum thickness of a water wall required to provide a thermal storage capacity of 30 Btu/F°-ft² glazing is only approximately 6 in. Water thermal storage walls do not exhibit as large a thermal time lag as walls constructed of masonry materials, because the water heats up uniformly as a result of convection currents in the water containers. (In Fig. 8-10d, the room air temperature reached a peak about three to four hours earlier with the water wall than with the masonry wall.) When designing water solar thermal storage walls, their short thermal time lag must always be considered. Often the designer will increase the thermal storage capacity of the water wall by making its average thickness greater than the minimum thickness of 6 in. By increasing the thermal storage capacity in making a water wall 12 in. thick, room air temperature swing is reduced, as is illustrated in Fig. 8-10d. There is really no one optimum thickness for a water thermal storage wall, because the water is heated up uniformly by convection, but there are practical limits to the thickness, dictated by economic and perhaps aesthetic considerations. Generally, water walls will have an average thickness of between 6 and 12 in.

8-8   Construction Details of Thermal Storage Walls

The overall performance of a thermal storage wall is dependent to a considerable extent on the care with which it is constructed.[1] For example, if the wall loses heat from infiltration around the glazing or from conduction to the ground, the performance of the wall as a solar heating system will be greatly reduced. Figure 8-13 illustrates a poured concrete thermal storage wall with typical building construction details. The size and mass of the wall footings will vary depending upon local soil conditions. Note that there should be rigid insulation along the foundation varying in thickness from 1 to 2 in., with the thicker insulation used for colder climates. The wall must have steel reinforcing bars to strengthen it, the size and placement of which will vary for walls of different heights and thicknesses. The glazing is mounted in a wooden frame, which is bolted to the concrete wall using ramset or expansion bolts. A redwood or treated Douglas fir 2 × 4 is first bolted to the wall as shown in the details; then a glazing bar is screwed to the 2 × 4 as shown in the detail of the glazing mullion. A 1 × 4 wooden stop is screwed to the glazing mullion with brass screws to allow the removal of the glazing, if necessary. A high-quality glazing caulking or foam rubber sealing strip must be used to seal the glass to the wood glazing bar. All wood glazing channels and spacer strips should be painted with a high-temperature paint to protect the wood from direct exposure to solar radiation. The glazing channel and 2 × 4 spacer should have 1 in. of rigid insulation located on the inside of the air space along the top, bottom, and sides of the glazed wall to reduce heat loss through the framing.

A fully grouted concrete block wall can also be used as thermal storage. The block wall is easier to construct and usually less expensive than a poured concrete wall because of the elimination of the form work. The blocks used must be heavyweight masonry units, not lightweight concrete. Figure 8-14

---

[1] The construction details suggested here and on other pages are intended as general guidelines only. Local conditions can vary greatly. All applicable codes must be followed, and all work should be done by skilled personnel.

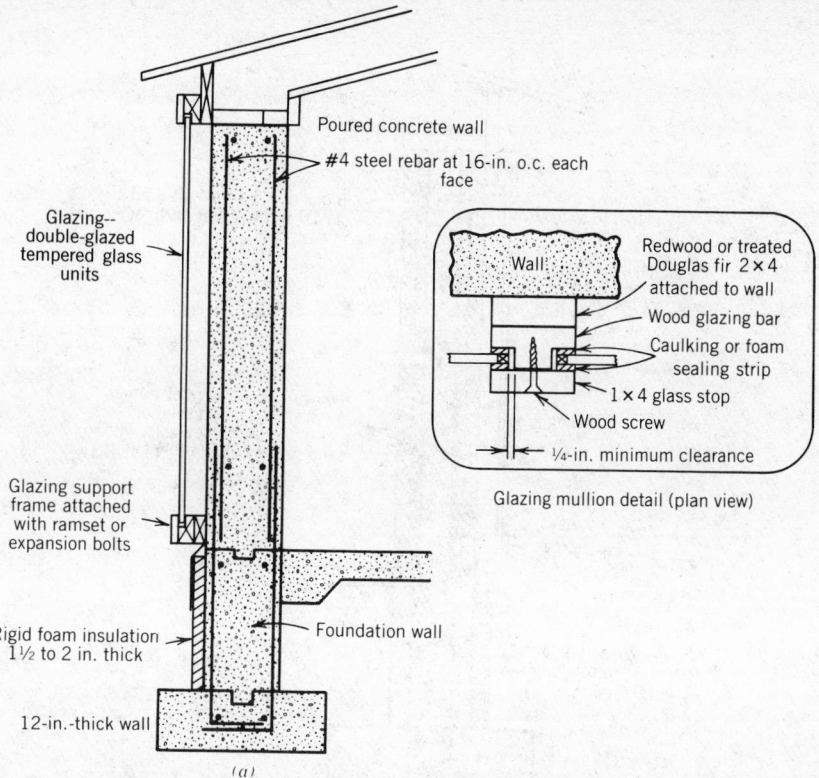

Poured concrete wall

#4 steel rebar at 16-in. o.c. each face

Glazing--double-glazed tempered glass units

Wall

Redwood or treated Douglas fir 2 × 4 attached to wall

Wood glazing bar

Caulking or foam sealing strip

1 × 4 glass stop

Wood screw

¼-in. minimum clearance

Glazing mullion detail (plan view)

Glazing support frame attached with ramset or expansion bolts

Rigid foam insulation 1½ to 2 in. thick

Foundation wall

12-in.-thick wall

(a)

(b)

Figure 8-13 Construction details of a poured concrete thermal storage wall. (a) A wooden glazing mullion is used to support the glazing on the concrete wall. (b) A two-story solar house with a concrete thermal storage wall on the south side behind the glazing.

shows a section through a masonry block thermal storage wall. In this case, the glazing is supported on a metal support system rather than a wood system. First, a treated 2 × 4 is bolted to the concrete block wall around the perimeter of the glazed wall. Then, a column is bolted to the perimeter wood frame as shown in

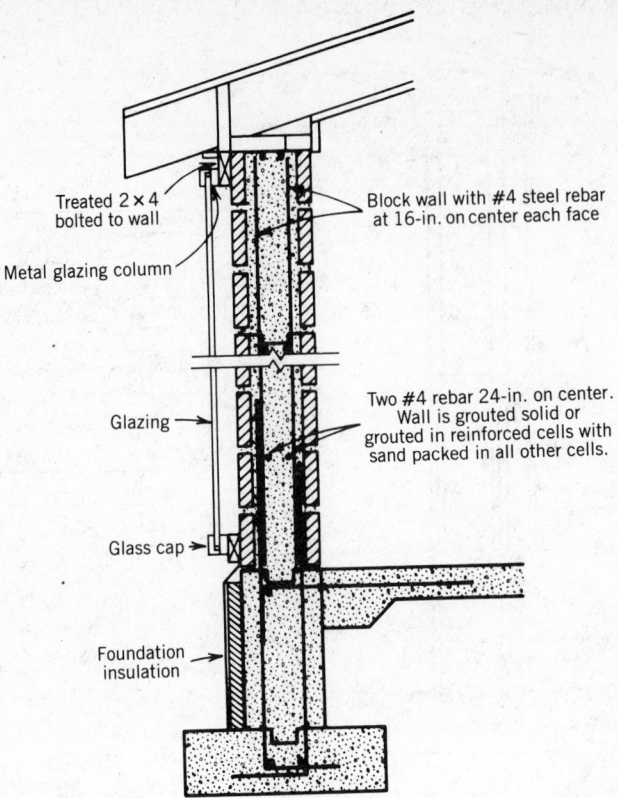

Treated 2 × 4
bolted to wall

Block wall with #4 steel rebar
at 16-in. on center each face

Metal glazing column

Glazing

Two #4 rebar 24-in. on center.
Wall is grouted solid or
grouted in reinforced cells with
sand packed in all other cells.

Glass cap

Foundation
insulation

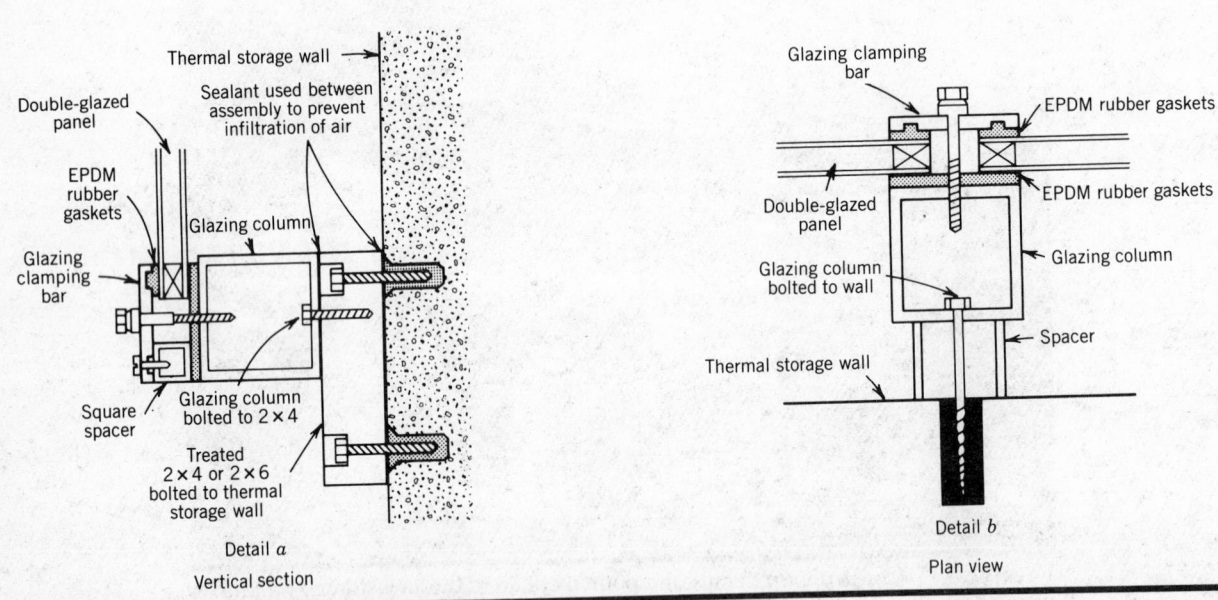

Thermal storage wall

Sealant used between
assembly to prevent
infiltration of air

Double-glazed
panel

EPDM
rubber
gaskets

Glazing column

Glazing
clamping
bar

Square
spacer

Glazing column
bolted to 2 × 4

Treated
2 × 4 or 2 × 6
bolted to thermal
storage wall

Detail *a*

Vertical section

Glazing clamping
bar

EPDM rubber gaskets

Double-glazed
panel

EPDM rubber gaskets

Glazing column

Glazing column
bolted to wall

Spacer

Thermal storage wall

Detail *b*

Plan view

Figure 8-14 Concrete block solar thermal storage wall. In this case, a metal system
is used to support the glazing as shown in detail *a* and *b*.

detail *a*, Fig. 8-14. Vertical columns are used where glazing panels meet, as
shown in detail *b*, Fig. 8-14. The vertical columns are attached to the wall with
ramset or expansion bolts on spacers between the wall and the glazing. The
glazing is mounted using EPDM rubber gaskets. A gasket is run along the

surface of the column against which the glazing is placed, and then a clamping bar with a gasket is placed against the glazing. The clamping bar is held in place by screws with a weather-seal washer. At the edges of the wall where the glazing ends, a square spacer is used between the clamping bar and column as shown in detail *a*, Fig. 8-14. A metal frame glazing system requires less labor to install than a wood frame glazing system and also needs less maintenance, such as painting. Additionally, the metal frame will not warp or deteriorate after long exposure to the sun and high air temperatures. Either a wood or metal frame glazing system, however, will give long-term satisfactory performance when properly designed and installed.

Thermal storage walls using water as the storage medium are constructed differently from walls using masonry. The glazing is supported on a structural building frame with water storage containers set behind the glazing as illustrated in Fig. 8-15*a*, or the glazing can be supported on a metal or wooden glazing frame that is attached to the structural building frame. There are many different containers that can be used to hold water, such as the vertical tubes indicated in Fig. 8-15*a*, which are made from metal or fiberglass. The tubes should be securely fastened to the structural frame of the building using steel support bands. Metal tubes may have a bottom plate that can be bolted to the slab as shown in Fig. 8-15*b*. Figure 8-16 shows steel drums being used for water thermal storage. These drums can be stacked horizontally by using a clip that clamps to the chime ring of the drums or a support rack made from angle iron, as shown in Figure 8-16*a*. The drums can also be stacked vertically, as shown in Fig. 8-16*b*, using a plywood platform and clamping blocks.

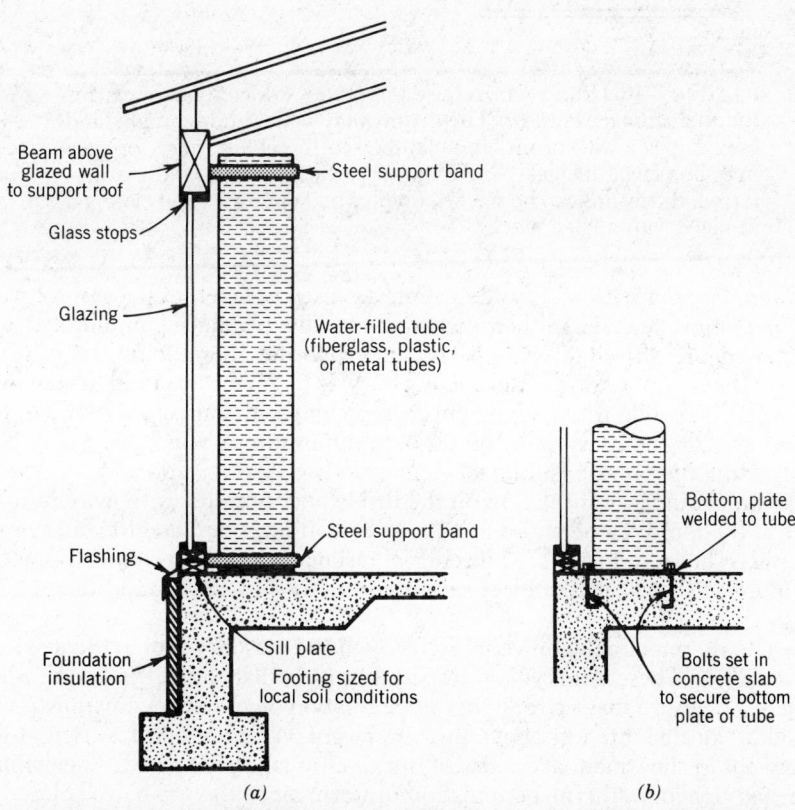

(a) (b)

Figure 8-15 Water tubes used for a thermal storage wall must be securely supported. (*a*) An extra thick slab should be used beneath the tubes to support their weight, and metal bands are used to hold the tubes in place at the top and bottom. (*b*) When steel tubes are used to contain thermal storage water, a steel plate can be welded to the bottom of the tubes as shown. This plate is then used to bolt the tubes to the floor slab.

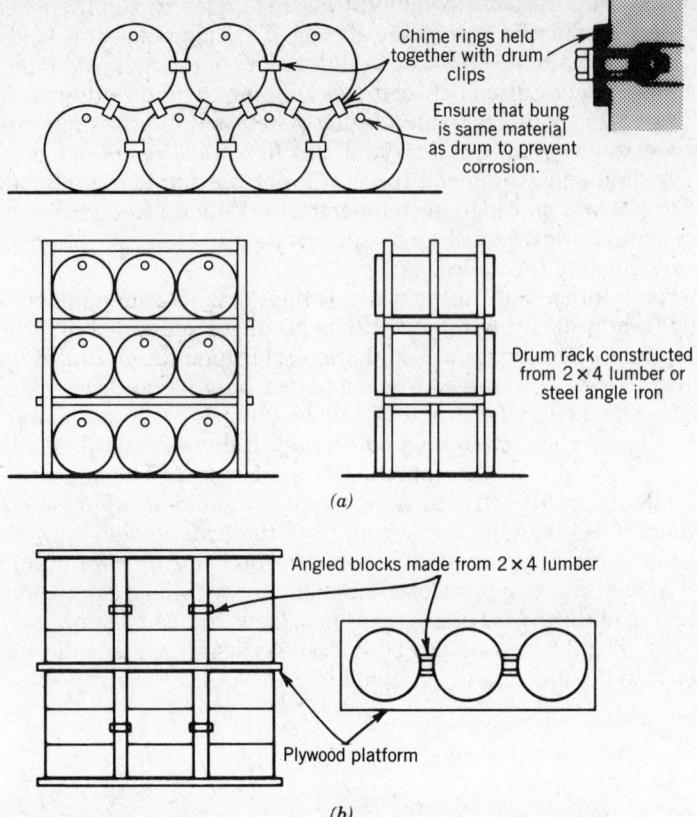

Chime rings held
together with drum
clips

Ensure that bung
is same material
as drum to prevent
corrosion.

Drum rack constructed
from 2 × 4 lumber or
steel angle iron

(a)

Angled blocks made from 2 × 4 lumber

Plywood platform

(b)

Figure 8-16 Drums can be used as water storage containers for thermal storage walls. (*a*) The drums may be laid horizontally and held in place with drum clips clamped to the chime rings, or they may be placed in racks. (*b*) When a vertical configuration is preferred, the drums can be stacked on plywood platforms and clamped in place with angled blocks.

Vents in Trombe Walls.   When vents are used with a thermal storage wall, great care must be taken in their construction. The two most common types of vents used are shown in Fig. 8-17. They are the thermocirculation vents through the thermal storage wall and the *outside thermal vents* used to vent heat in the summer. When any such vents are provided, the air space between the glazing and the masonry wall should be a minimum of 3 in. One method of constructing the outside (summer) vents is to use ¾-in. exterior plywood insulated with 1½-in. rigid insulation on the inside and mounted with a continuous hinge at the top. Weatherstripping across the bottom, sides, and top of the vent opening is shown in Fig. 8-17. The correct installation of the weatherstripping is important so that heat is not lost from the wall by air leaks around the outside vents.

The thermocirculation vents go through the masonry wall to the inside of the building. These vents allow warm air to circulate into the room when needed. To ensure that a strong thermocirculation of air occurs, the upper vent should be located 5 to 6 ft above the lower vent. A damper grill is used at the upper vent to allow manual control of the air circulation if desired. The damper is opened to allow a thermal circulation of warm air to the room, and closed to stop the thermocirculation. The bottom vent is covered with a grille and has a back-draft damper installed. The back-draft damper prevents cold air from flowing out of the lower vent into the room at night or on cold, sunless days. See Fig. 8-6 for back-draft damper construction details. The back-draft damper is mounted into the vent opening with screws so that it can be easily removed for servicing.

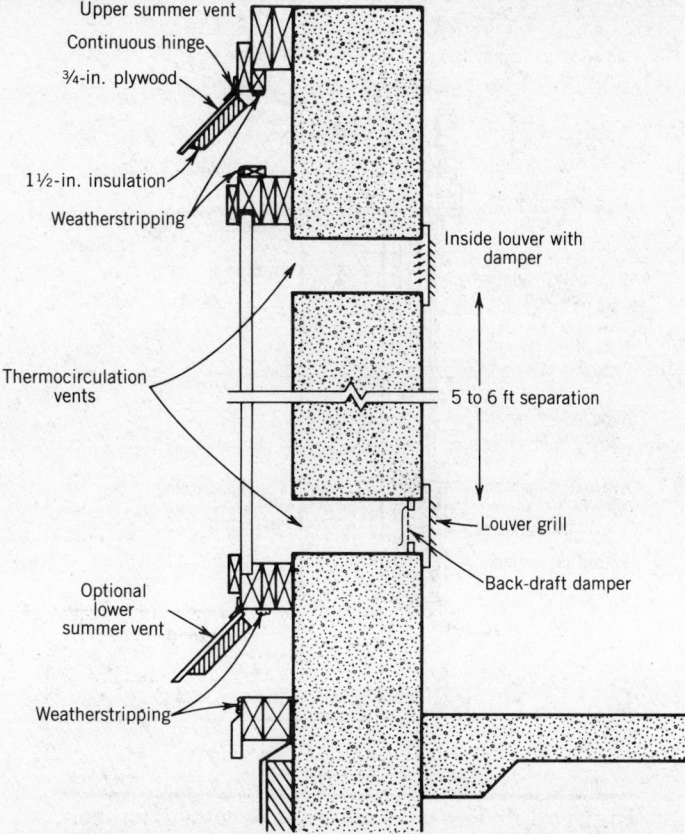

Upper summer vent

Continuous hinge

¾-in. plywood

1½-in. insulation

Weatherstripping

Inside louver with damper

Thermocirculation vents

5 to 6 ft separation

Louver grill

Back-draft damper

Optional lower summer vent

Weatherstripping

Figure 8-17 Construction details of thermocirculation vents. Exterior (summer) vents are shown made from 3/4-in. plywood and hinged at the top. The weatherstripping around the exterior vents is important to prevent loss of heat in winter. The upper inside vent is shown with a louver and damper to control the flow of warm air into the living space. The lower inside vent is shown with a back-draft damper.

In cold climates, the performance of a thermal storage wall will be increased if movable insulation is used. The insulation can be located between the glazing and the thermal storage, or it can be placed on the outside of the glazing. Two methods of installing an insulating curtain between the glazing and the thermal storage are shown in Fig. 8-18. The top half of the figure shows an inflatable insulating curtain—a commercially available product that is automated. A curtain assembly housing must be provided at the top of the wall to contain the curtain roller. There is a small motor and gear system in the curtain roller which allows the curtain to be rolled up or down. It is important that an access panel be provided in the outside cover of the curtain assembly housing, since the curtain roller is a mechanical device and may require maintenance at times. When insulating curtains are used, the air space should be increased to provide adequate space for the curtain. The glazing system should have a smooth inside surface so that there are no protrusions to interfere with the operation of the curtain.

The lower half of Fig. 8-18 shows an alternative method of storing an insulating curtain. In this case, the curtain folds up in a storage space at the bottom of the wall. Cords are used to raise and lower the curtain. This type of curtain allows the lower wall to be shaded during warm autumn weather when the roof overhang does not yet provide adequate shading for the bottom half of the glazed surface.

Exterior insulation can be installed as a fold-down shutter or as a set of horizontally sliding panels supported on tracks so that the panels stack up to one

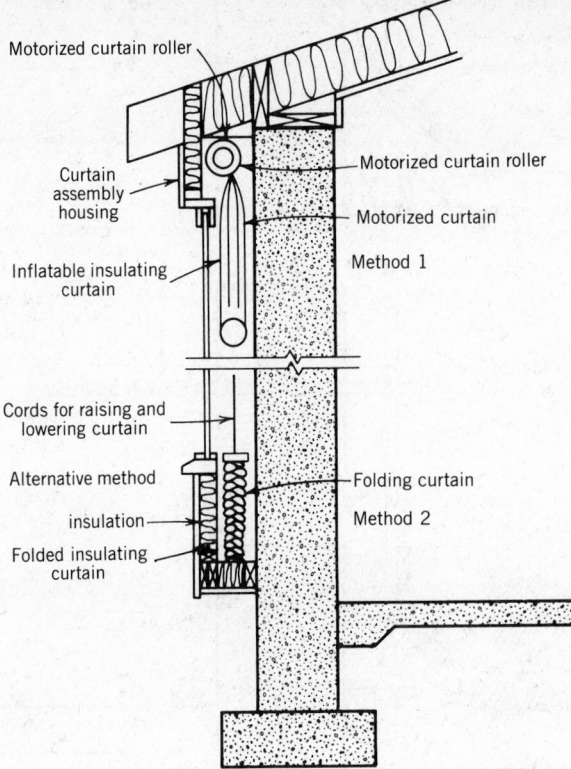

Motorized curtain roller

Curtain
assembly
housing

Inflatable insulating
curtain

Motorized curtain roller

Motorized curtain

Method 1

Cords for raising and
lowering curtain

Alternative method

insulation

Folded insulating
curtain

Folding curtain

Method 2

Figure 8-18 Two types of insulating curtains for thermal storage walls. One is an inflatable type of curtain that rolls up on a motorized roller. The other is a folding insulation curtain that is raised by pull cords.

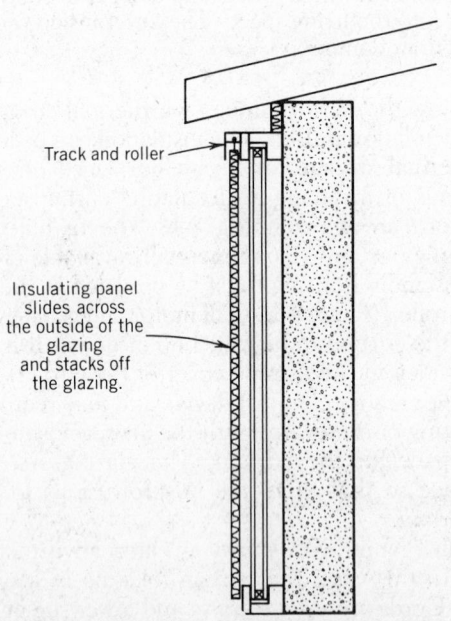

Track and roller

Insulating panel
slides across
the outside of the
glazing
and stacks off
the glazing.

Figure 8-19 Exterior movable insulation can be mounted on a track. The insulating panels can be slid across the solar collection area to insulate the thermal storage wall.

side of the glazing, as shown in Fig. 8-19. The seals should be snug fitting, or else the insulating value of the panels will be greatly reduced by infiltration.

A third type of movable insulation is the Beadwall™ system, referred to earlier in section 8-1.

PERFORMANCE OF SOLAR THERMAL WALL SYSTEMS

As the solar design process proceeds from schematic concepts into the design development phase, the expected performance of the thermal walls must be estimated in order to determine the solar savings provided by the thermal wall design as opposed to savings from other possible solar options, and also to evaluate the energy savings obtained from the use of a solar thermal wall.

There are two levels at which the performance of thermal wall systems can be estimated. First, daily and hourly performance levels give useful information about the interaction of the solar thermal wall with the interior spaces of the building. From an hourly performance determination, the following can be determined: (1) possibility of overheating or underheating of the interior spaces for different daily conditions, (2) the adequacy of the thermal storage capacity, and (3) the time lag of the wall and whether this time lag is consistent with optimum performance. The second level of performance is a long-term average performance prediction which gives the monthly or yearly energy savings for the building resulting from the passive system with solar thermal storage walls. This energy-savings information is used to determine cost savings resulting from the passive solar system compared with conventional heating and/or other solar options.

8-9   Estimating Hourly and Daily Performance of Thermal-Storage Walls

The sunlit surface of a masonry wall has a large daily temperature swing, as was indicated in Fig. 8-4c. There is no simple and accurate method to estimate what the interior temperature of the thermal storage wall will be at any hour and what quantity of heat will enter the interior of the building through the wall in any given hour. There are, however, computer and calculator programs that will model a thermal storage wall and give daily performance, as shown in Figs. 8-4 and 8-10. Computer-generated estimates of the daily performance of a thermal storage wall system are very useful because they will give information on daily overheating or underheating of the building.

Table 8-3 gives approximate indoor-air temperature fluctuations as a function of solar wall material and thickness, to give a general idea of the daily performance of a thermal storage wall. These daily temperature fluctuations are only approximate values and assume that the thermal wall has a double glazing and that it has been sized to maintain an indoor space temperature in the range of 65 to 75°F on clear winter days. Values from this table should be used only as general guidelines. For an actual design problem, a computer or calculator program should be used.

To estimate the long-term solar performance of thermal storage walls, the Solar Load Ratio (SLR) method can be used first to evaluate the monthly performance of the wall. These monthly values are then summed up to obtain the yearly performance. An example of the monthly SLR method will be given in Chapter 9.

The annual solar performance of solar thermal storage walls can also be quickly estimated by the annual Load Collector Ratio (LCR) method (as developed in Chapter 6). It should be recalled that the LCR method is a precomputed variant of the SLR method. The *Passive Solar Design Handbook, vol. III*, provides tables with tabulated values of the annual solar savings fraction (SSF) versus the annual Load Collector Ratio (LCR) for water thermal storage walls and masonry or concrete thermal storage walls without thermocirculation vents and with thermocirculation vents.[2] The *Handbook* tables are in a format similar to Table 6-5 of this book. The tables also provide 15 different types of water wall designs, which include thermal storage capacities from 15.6 Btu/ft^2-F° (3-in.-thick waterwall) to 124.8 Btu/ft^2-F° (24-in.-thick waterwall). Variation in the number of glazings from one to three glazings is provided, along with the use of night insulation. For masonry and concrete thermal storage walls, 21 different types are included in the tables for different ranges of thermal storage capacity and conductivities of the thermal wall material. Variations for different numbers of glazings and the use of night insulation are also included. These 21 different types of thermal storage wall configurations are also repeated for the case of thermal circulation vents being included in the thermal storage wall. The *Handbook* tables provide the passive solar designer with a wealth of design performance information, but their inclusion here is impossible because of space limitations.

8-10   Some Examples of Performance Estimates for Solar Thermal Storage Walls Using the Annual Load Collector Ratio

Illustrative Problem 8-3

A homeowner has started a design for a new house to be built on a lot with full solar access near Knoxville, Tennessee. The client has given you the rough drawings and has asked you to determine the expected solar performance. The house has been planned with 400 ft^2 of 12-in. concrete thermal wall. The net load coefficient (NLC) for the proposed residence is estimated at 16,000 Btu/DD. Determine the Solar Savings (SS) and the Auxiliary Heat (AH) for the solar building and the improvement in performance if night insulation is used.

Solution

The Load Collector Ratio is first calculated using Eq. (6-13):

$$\text{LCR} = \frac{\text{NLC}}{A_p} = \frac{16000 \text{ Btu/DD}}{400 \text{ ft}^2} = 40 \text{ Btu/DD-ft}^2 \text{ of collector}$$

The value of the solar savings fraction (SSF) is obtained from Table 6-5 for Knoxville and found by interpolation to be 0.29 for a masonry or concrete thermal wall with an LCR of 40. The degree days for Knoxville are listed in Table 6-5 as 3478 DD/year. The Auxiliary Heat (AH) is calculated from Eq. (6-8) as:

$$\begin{aligned}
\text{AH} &= \text{NLC} \times \text{DD} \times (1 - \text{SSF}) \\
&= 16000 \text{ Btu/DD} \times 3478 \text{ DD/year} \times (1 - 0.29) \\
&= 39.5 \times 10^6 \text{ Btu/year} \qquad\qquad\qquad\qquad\qquad\textit{Ans.}
\end{aligned}$$

The Solar Savings (SS) is calculated from Eq. (6-9) as:

$$\begin{aligned}
\text{SS} &= \text{NLC} \times \text{DD} \times \text{SSF} \\
&= 16000 \text{ Btu/DD} \times 3478 \text{ DD/year} \times 0.29 \\
&= 16.1 \times 10^6 \text{ Btu/year} \qquad\qquad\qquad\qquad\qquad\textit{Ans.}
\end{aligned}$$

When night insulation is used on the thermal storage wall, the solar savings fraction is obtained by interpolation from Table 6-5 for Knoxville as 0.418. The Auxiliary Heat is calculated from Eq. (6-8) as:

$$\begin{aligned}
\text{AH} &= \text{NLC} \times \text{DD} \times (1 - \text{SSF}) \\
&= 16000 \text{ Btu/DD} \times 3478 \text{ DD/year} (1 - 0.418) \\
&= 32.3 \times 10^6 \text{ Btu/year} \qquad\qquad\qquad\qquad\qquad\textit{Ans.}
\end{aligned}$$

[2] R. W. Jones, ed., *Passive Solar Design Handbook*, vol. III (Washington: U.S. Department of Energy, July 1982), pp. 344–516.

The Solar Savings is calculated from Eq. (6-9) as:

$$SS = NLC \times DD \times SSF$$
$$= 16000 \ Btu/DD \times 3478 \ DD/year \times 0.418$$
$$= 23.3 \times 10^6 \ Btu/year \qquad \qquad \text{Ans.}$$

Illustrative Problem 8-4

The client with the house plans described in Illustrative Problem 8-3 would like to know the required area of masonry thermal wall with night insulation for providing a solar savings fraction of 0.75. Also, the Auxiliary Heat and Solar Savings values are requested.

Solution

From Table 6-5, the Load Collector Ratio (LCR) for a solar savings fraction (SSF) of 0.75 for a masonry wall with night insulation is found by interpolation as 14.5 Btu/DD-ft^2. By rearranging Eq. (6-13) and solving for the solar collection area A_p, one obtains:

$$A_p = \frac{NLC}{LCR} = \frac{16000 \ Btu/DD}{14.5 \ Btu/ft^2\text{-}DD}$$
$$= 1103 \ ft^2 \ \text{of solar collection area} \qquad \text{Ans.}$$

The Auxiliary Heat is calculated using Eq. (6-8) as:

$$AH = NLC \times DD \times (1 - SSF)$$
$$= 16000 \ Btu/DD \times 3478 \ DD/year \times (1 - 0.75)$$
$$= 13.9 \times 10^6 \ Btu/year \qquad \qquad \text{Ans.}$$

The Solar Savings is calculated using Eq. (6-9) as:

$$SS = NLC \times DD \times SSF$$
$$= 16000 \ Btu/DD \times 3478 \ DD/year \times 0.75$$
$$= 41.7 \times 10^6 \ Btu/year \qquad \qquad \text{Ans.}$$

Illustrative Problem 8-5

An architect is in the process of designing a 2200 ft^2 solar house that will be built near Albany, New York. The house will be designed with a 500 ft^2 water thermal storage wall system with night insulation. The building envelope will be designed for a normalized net load coefficient (NLC) of 5 Btu/DD-ft^2 floor area. The architect wants estimates of the thermal performance of the building.

Solution

First, the NLC must be calculated. The house has 2200 ft^2 of heated floor area, and the net normalized heat loss is 5 Btu/DD-ft^2 floor area. The NLC can be obtained from the product of these values:

$$NLC = 2200 \ ft^2 \times 5 \ Btu/DD\text{-}ft^2$$
$$= 11,000 \ Btu/DD$$

The LCR is calculated using Eq. (6-13):

$$LCR = \frac{NLC}{A_p} = \frac{11,000 \ Btu/DD}{500 \ ft^2} = 22.0 \ Btu/ft^2\text{-}DD$$

The value of the solar savings fraction (SSF) is obtained from Table 6-5 for Albany and found by interpolation to be 0.429 for a thermal water wall with night insulation. The degree-days for Albany are listed in Table 6-5 as 6888 DD/year. The Auxiliary Heat is calculated from Eq. (6-8) as:

$$AH = NLC \times DD \times (1 - SSF)$$
$$= 11{,}000 \text{ Btu/DD} \times 6888 \text{ DD/year } (1 - 0.429)$$
$$= 43.3 \times 10^6 \text{ Btu/year} \qquad \qquad Ans.$$

The Solar Savings is calculated from Eq. (6-9) as:

$$SS = NLC \times DD \times SSF$$
$$= 11{,}000 \text{ Btu/DD} \times 6888 \text{ DD/year} \times 0.429$$
$$= 32.5 \times 10^6 \text{ Btu/year} \qquad \qquad Ans.$$

THERMAL STORAGE ROOF PONDS

The roof or ceiling of a building can also be used as a location for thermal storage. This type of indirect solar system is typically called a roof pond, since the thermal storage is usually a 6- to 10-in. layer of water contained in plastic bags or shallow tanks. Harold R. Hay pioneered and developed the original roof pond equipment, which is marketed under the Skytherm™ trademark.

8-11   Operation of Roof Ponds

The operation of a roof pond is illustrated in Fig. 8-20. One advantage of this type of storage is that it acts as a natural cooling system as well as a passive solar heating system. During the heating season, the movable insulating panels are opened in the day so that the water containers absorb solar radiation. At night, the insulating panels are closed to prevent heat losses from the thermal storage. The interior of the building is heated by thermal radiation from the ceiling, which in turn is heated by the water thermal storage.

During the summer, the roof pond system can provide a considerable amount of cooling if the insulating panels are opened at night to allow thermal radiation to radiate from the thermal storage to the cold night sky. During the summer day, the insulating panels are closed so that the thermal storage does not absorb any solar radiation. The cooled thermal storage absorbs heat from the building's interior during the day, providing a significant cooling effect. (See Chapter 16.)

Roof pond solar systems have a number of limitations. They must be used on flat roofs and perform best on single-story buildings. Additionally, buildings must have some added structural strength to support the roof pond. The cooling aspect of the roof pond requires clear night sky conditions to allow for the cooling of the water storage by thermal radiation. Such conditions are found most often in arid regions, as in the U.S. Southwest. Roof pond cooling by thermal radiation is not effective in climates with high humidity levels in the summer months, since water vapor in the atmosphere acts as a barrier to the long-wavelength radiation from the roof pond.

Three basic configurations are commonly used for incorporating roof ponds into buildings. The first is a horizontal roof structure holding the water containers, with a set of insulating panels that slide in a horizontal plane either to expose the roof ponds or to cover them. This system is illustrated in Fig. 8-20.

A second method of construction is to have the insulating panels hinged along the north side so that they will open as shown in Fig. 8-21. The undersides of the insulating panels are lined with a reflecting surface. This reflective lining increases the solar radiation absorbed by the roof pond since the slanted reflective surface has a better angle for the low winter sun. One disadvantage of this design is that the insulating panels, when open, are susceptible to wind damage. They must be designed to take the potential wind loads that may occur at specific building sites.

A third method of construction is to provide a roof and glazing arrangement over the pond as shown in Fig. 8-22. There is south-facing glazing on one side of the roof. A movable reflector/insulating panel is hinged so that the panel can either cover the glazing or swing back against the opaque roof and act as a

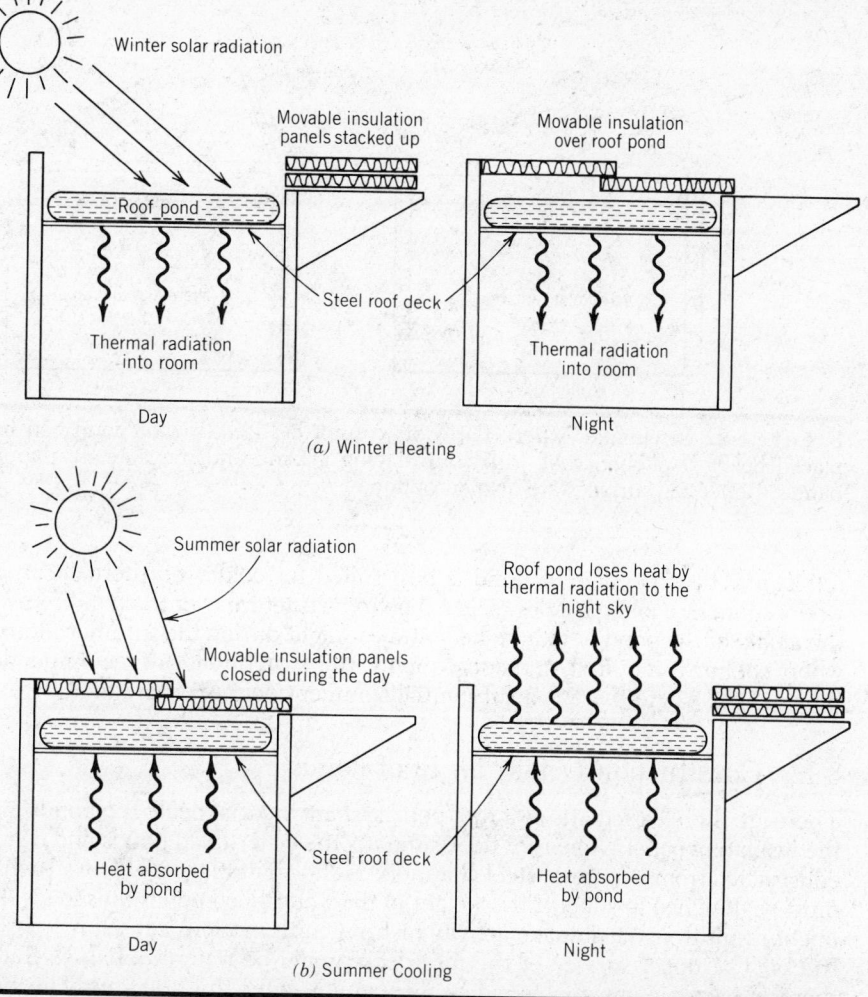

Figure 8-20 A solar roof pond system—four modes of operation. (*a*) The winter heating operation. The roof pond is exposed to solar radiation during the day by removing the insulation panels from above the pond. At night, the ponds are covered by the insulation, and thermal radiation from the ceiling heats the space. (*b*) The roof pond can also be used as a cooling system in the summer by cooling the pond at night by radiation to the night sky.

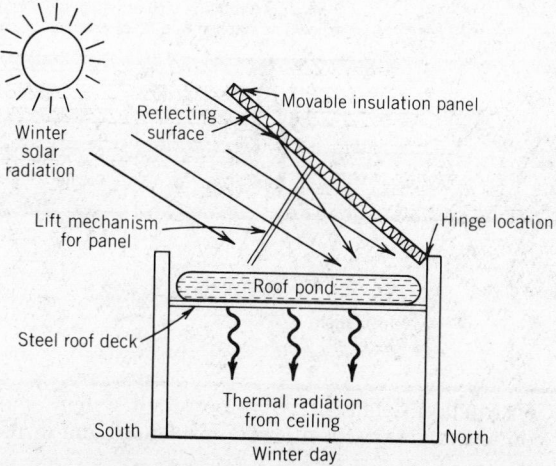

Figure 8-21 Roof ponds can be constructed with a tilt-up insulating panel rather than sliding panels. Reflectors are also often used, especially in northern latitudes where sun angles are low.

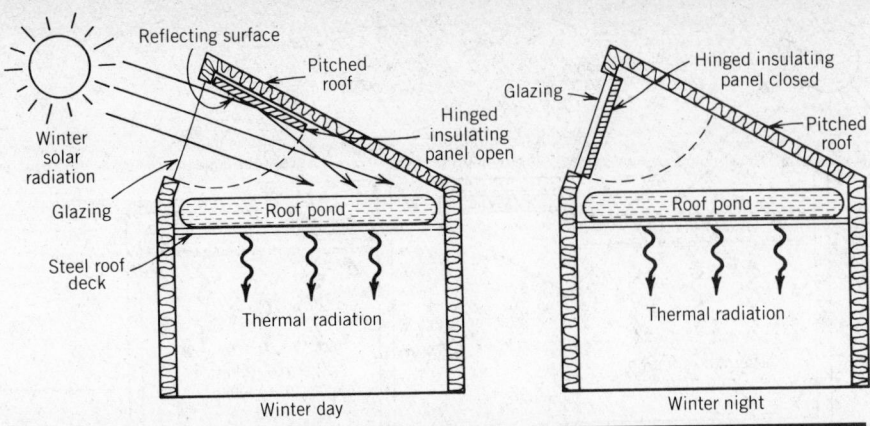

Figure 8-22 In climates where flat roofs cannot be used, a roof pond can be placed below a pitched roof with south-facing glazing and movable insulating panels. Reflecting surfaces are also an option.

reflector. This style of roof pond is best suited for colder northern locations where summer cooling is not needed. The roof structure over the pond reduces the ability of the pond to radiate heat away at night during the summer months where cooling is required, and consequently this structure is not recommended for localities where there is a substantial summer cooling requirement.

8-12 Construction Details for Roof Ponds

There are three basic parts of a roof pond system: the roof deck, the ponds, and the insulating panels. The roof deck supports the water-filled ponds and can be constructed from structural steel channels, as shown in Fig. 8-23. The structure must be designed to support the weight of the water-filled ponds plus loads that might result from earthquake activity or from earth motions caused by passing trucks. The upper surface of the roof deck should be waterproofed, so that if there is a leak in the roof pond, water cannot enter the building. The heat transfer from the roof pond to the roof deck should be as direct as possible. A hot-mopped asphalt roof with layers of felt presents an insulating barrier between the ponds and roof deck and is therefore not recommended. The roof deck can be sealed with a 2-ply rubber membrane such as Hypalon®, manufactured by Burke Rubber Company, or it can be coated with a thin layer of asphalt emulsion, then a layer of fiberglass cloth, and then a second layer of asphalt emulsion.

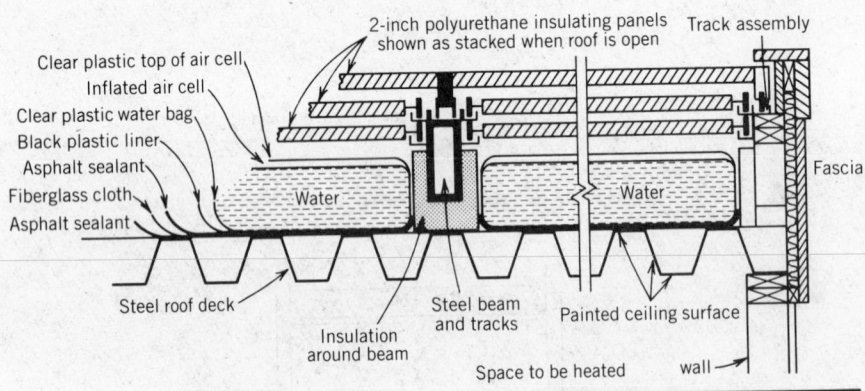

Figure 8-23 Construction details of a flat-roof pond system, showing the roof decking, water containers, tracks for movable insulation, and insulation panels.

The underside of the steel roof deck should be left exposed to the room to maximize thermal radiation from the deck into the room. It can be painted almost any color, but it should not be left as bare metal because bright metallic surfaces are poor radiators of thermal energy. These thermal requirements of the underside of the roof deck obviously will limit ceiling designs and room decor.

The roof ponds themselves can be constructed by forming shallow bags 6 to 12 in. in depth from polyethylene, polyvinylchloride, or another type of plastic. These bags would be somewhat similar to a water bed and would have a useful life of 5 to 10 years, depending on the type and thickness of plastic used. The ponds can also be constructed from metal or fiberglass tanks, using a rigid plastic cover such as an acrylic glazing.

As the water in the ponds is heated by the sun, the warm water will tend to rise to the top by natural convection. The cooler water will tend to remain at the bottom of the pond and inhibit the transfer of heat from the warm water through the roof to the interior of the building. It is important to minimize this temperature stratification in the ponds, and this can be achieved by constructing the pond with a clear cover and a black bottom, allowing solar radiation to be transmitted through the water and absorbed by the dark bottom of the pond. Thus, the water on the bottom of the pond is warmed first, and since warm water will rise by convection, there is a continual stirring of the water in the pond. A properly constructed roof pond should have only a 1 to 2 F° temperature difference between the top and bottom during the day when it is being heated.

The construction of the ponds should allow any accumulation of rainwater to drain away. This is quite important because the evaporation of rainwater will use up heat that otherwise would be absorbed by the water thermal storage.

One method of constructing a roof pond using plastic water bags is shown in Fig. 8-23. An inflatable air cell can also be added to the top of the pond to reduce heat losses during the winter days when the pond is absorbing solar radiation. The air cell can be deflated, removing the insulating air space during summer months when the pond must lose heat by night radiation.

Insulating Panels. The insulating panels are constructed from 2-in.-thick polyurethane insulation held in a metal frame with an aluminum or fiberglass skin on the top and bottom surfaces of the panel to protect the foam insulation. The panels are supported by rollers that run in tracks above the roof ponds, as shown in Fig. 8-23. The edges of the panels have weather seals to prevent air infiltration under them when the roof system is closed. The sealing of the roof panels is quite important, because a large fraction of the energy collected could potentially be lost to cold infiltrating air if sealing is inadequate.

8-13 Sizing Roof Ponds

The design of a roof pond system involves many specific details including waterproofing, water containers, insulation panels, tracks and rollers, and automatic opening and closing of the system.

The total area of roof pond required depends on whether the purpose of the system is heating or cooling or both, and on the share of the total load assigned to solar. For winter heating in latitudes from 28° to 36°N, the winter sun has an altitude high enough that a horizontal collection surface is adequate for energy collection. For such latitudes, the horizontal roof panel system, as shown in Fig. 8-20, will perform effectively. In more northern latitudes (40° to 56°N), the sun is lower in the sky, and the solar radiation incident on the roof pond needs to be increased with a reflector as shown in Figs. 8-21 and 8-22. Where snow loading is a problem, the roofed-over pond in Fig. 8-22 would have to be used. Table 8-4 gives some generalized sizing specifications for roof pond space heating as developed by Mazria.

Table 8-4 Ratio of Roof Pond Collection Area to Structure Floor Area for Solar Space Heating (suggested approximate ratios)

Average Winter Temperature	15°–25°F	25°–35°F	35°–45°F
Type of System			
Double-glazed roof ponds with night insulation (Fig. 8-20)		0.85–1.0	0.60–0.90
Single-glazed ponds with night insulation and reflector (Fig. 8–21)			0.33–0.60
Double-glazed ponds with night insulation and reflector (Fig. 8–21)		0.50–1.0	0.25–0.45
South-sloping roof glazing with night insulation (Fig. 8-22)	0.60–1.0	0.40–0.60	0.20–0.40

Source: Edward Mazria, *The Passive Solar Energy Book* (Emmaus, Pa: Rodale Press, 1979), p. 187. Used with permission.

Illustrative Problem 8-6

Determine the approximate size of double-glazed roof pond with night insulation for a 1600 ft² house located in Griffen, Georgia, using Mazria's ratios from Table 8-4. The average winter temperature is in the range of 49° to 51°F.

Solution

Check the latitude as being about 33°N. From Table 8-4, for an average winter temperature of 35° to 45°F, the roof-pond-to-floor-area ratio is in the range of 0.60 to 0.90. Since the average winter temperature for Griffen is higher than 35° to 45°F, the smaller ratio of 0.60 will be used. The roof pond area is then calculated as:

$$\text{Roof pond area} = 1600 \text{ ft}^2 \times 0.60$$
$$= 960 \text{ ft}^2 \qquad \textit{Ans.}$$

The optimum configuration for cooling is a flat pond that is exposed entirely to the night sky to maximize the cooling effect of thermal radiation from the pond. Under ideal conditions of clear sky with low humidity and cool summer night temperatures, a heat dissipation rate of 20 to 30 Btu/hr-ft² of pond can be achieved. If climate conditions are not optimal, the roof ponds can be sprayed or flooded with water to increase the cooling by evaporation. Table 8-5 gives some generalized sizing for cooling roof ponds as developed by Mazria.

Table 8-5 Ratio of Roof Pond Area to Structure Floor Area for Solar Space Cooling

	Ratio of Pond Area to Structure Floor Area	
Type of System	Hot, Humid Climate	Hot, Dry Climate
Single-glazed roof pond	1.0	0.75–1.0
Single-glazed roof pond augmented by evaporative cooling	0.75–1.0	0.33–0.50

Source: Edward Mazria, *The Passive Solar Energy Book* (Emmaus, Pa.: Rodale Press, 1979), p. 188. Used with permission.

ATTACHED SUNSPACES

Sunspaces were first discussed in Chapter 7 as direct-gain solar energy system types. In the direct-gain case, the sunspace area opens directly onto the adjacent building spaces. Thus, the sunspace is both physically and thermally a part of the adjacent living space. When the sunspace overheats or underheats—that is, reaches a temperature outside the comfort zone—this condition is coupled directly to the adjacent spaces by natural convection of air between the spaces and by thermal radiation. Thus, the sunspace could have a large influence on the comfort of the occupants of the building. This direct influence of a sunspace on adjacent building spaces can be tempered by placing a thermal storage mass between the sunspace and adjacent building spaces, as shown in Fig. 8-24. Such a configuration makes the sunspace a part of an indirect or isolated passive solar system.

8-14 The Interaction of a Sunspace with a Building

There are two thermal zones that result when thermal storage is placed between a sunspace and the interior spaces of a building. The first zone is the sunspace itself (see Fig. 8-25a), which receives the direct solar gain. The thermal capacitance of the storage provides thermal inertia, or acts like a thermal flywheel between the sunspace (zone 1) and the other building spaces (zone 2). During the day, when the sunspace is absorbing solar radiation, the thermal storage will be absorbing energy also. The sunspace air temperature may experience a large temperature rise during the day, but the thermal storage will undergo a relatively small temperature rise because of its large thermal capacity.

The second thermal zone represents the interior building spaces. Zone 2 does not interact directly with the sunspace when there is a thermal storage wall between, and therefore overheated air from the sunspace does not directly affect the comfort level in zone 2. Zone 2 is thermally affected by convection and radiation from the thermal storage surface.

There are several ways in which sunspaces and thermal storage can be used to construct indirect-gain solar systems. For example, a greenhouse structure can be used as shown in Fig. 8-25a and b. The greenhouse (or sun parlor) is placed on the south side of the building, and a thermal storage wall is built between the greenhouse and the rest of the building. Vents or doors may be placed in the thermal storage wall so that warm air can be brought directly into the rest of the building if desired. A greenhouse is (typically) a structure that has

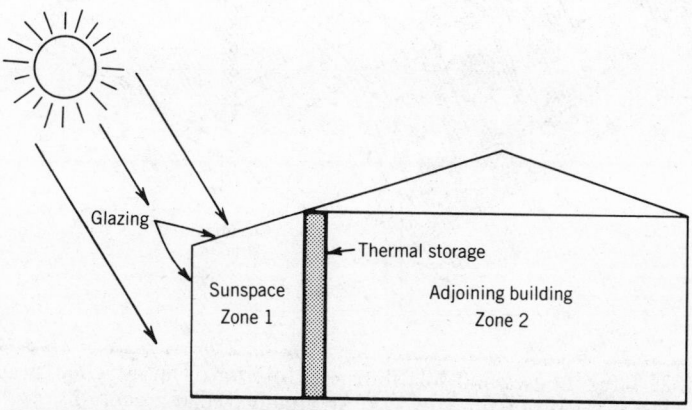

Figure 8-24 Sunspace attached to a building with thermal storage between the sunspace and the building resulting in a form of indirect solar gain. The sunspace forms one thermal zone, and the building is the second thermal zone. The thermal storage is a buffer between the two thermal zones.

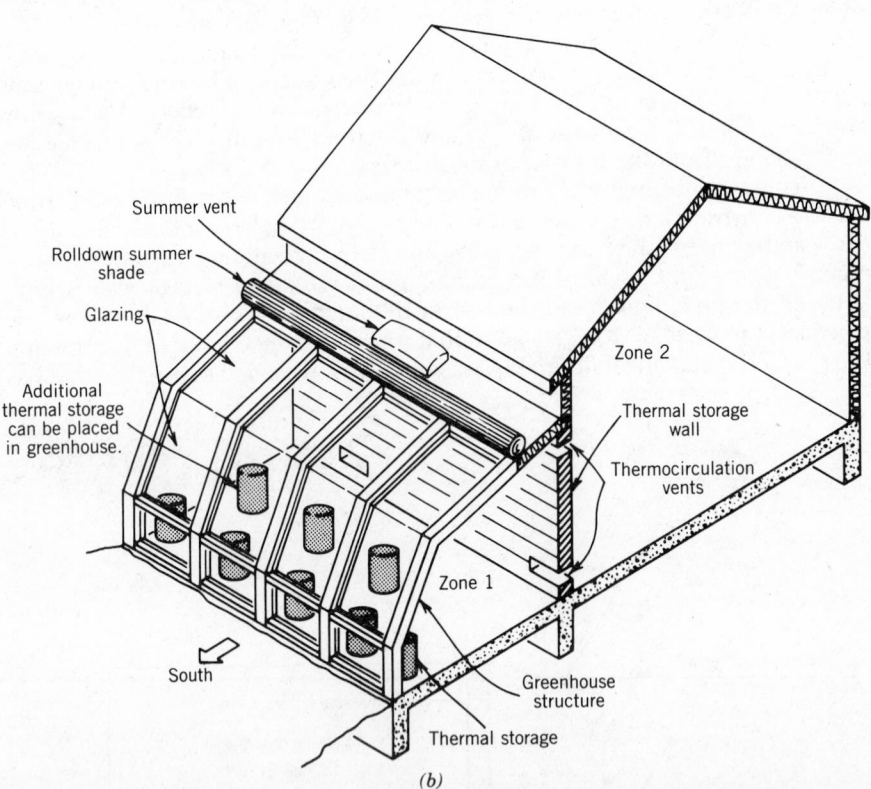

Summer vent

Rolldown summer shade

Glazing

Additional thermal storage can be placed in greenhouse.

Zone 2

Thermal storage wall

Thermocirculation vents

Zone 1

South

Greenhouse structure

Thermal storage

(b)

Figure 8-25 Greenhouses and solariums. (a) Interior of an attached sunroom or solarium. (The Rollscreen Company.) (b) An attached solar greenhouse with a glass south wall and a glass roof. Thermal storage must be available, and vents are necessary to allow warm air to circulate to the adjacent living spaces and to exhaust excess heat from the greenhouse in the summer. (c) A solarium sunspace will typically have a vertical south wall but only a small amount of roof glass. This type of sunspace depends on direct solar gain through the south-facing glazing.

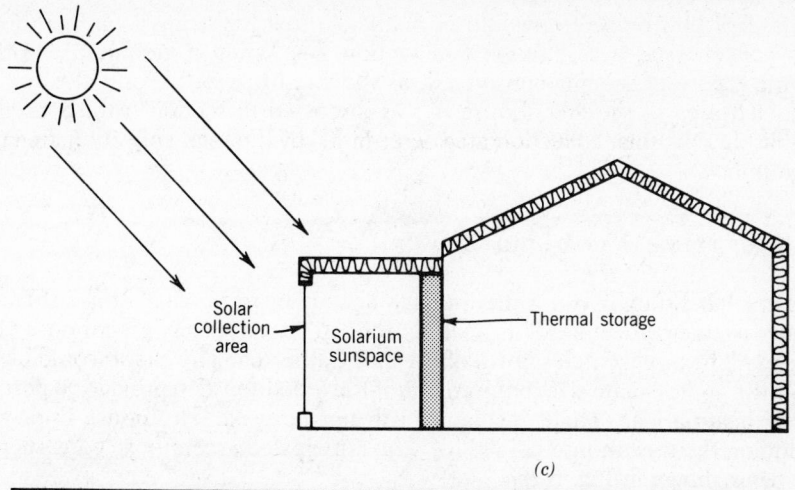

Figure 8-25 *(continued)*

at least one glazed wall and a glazed roof. A *greenhouse sunspace* will be one in which the south wall and at least 50 percent of the roof is glazed and oriented south. A *solarium sunspace* is characterized by a large south-glazed wall as shown in Fig. 8-25c. Part of the roof may have glazing, but the glass typically makes up less than 50 percent of the roof area.

The sunspace may also be coupled to a remote thermal storage mass. Warm air from the sunspace is removed by a duct and mechanical blower system to a thermal storage mass such as a rock bed. The storage mass absorbs heat from the air, and the cooled air is then returned to the sunspace to be heated again. Recall from Chapter 6 that this would be considered a hybrid solar energy system because it utilizes a sunspace as the solar collection unit, which is a *passive* element, and a mechanical (or *active*) system to move the thermal energy from the sunspace to the thermal storage. The design of hybrid solar systems will be considered further in Chapter 9.

The energy flows caused by natural convection, conduction, and radiation between the attached sunspace and the rest of the building are quite complex. This makes predicting the daily performance of attached sunspaces difficult, which in turn makes difficult the accurate sizing of sunspaces to provide a specific contribution of space heating to a building on a daily basis. Some guidelines will, however, be provided in the next section.

Attached sunspaces are often used as a solar retrofit for existing (nonsolar) buildings. A greenhouse or solarium can be constructed on the south side of an existing building without major modification of the existing structure. The quantity of heat that a sunspace provides is dependent upon many variables, such as latitude, local climate, type and quantity of thermal storage, size and insulating properties of the sunspace, and size and insulating properties of the building spaces to be heated. In the case of retrofits, it is usually the size and insulating properties of the spaces to be heated that are of prime importance. These must be known in order to size the sunspace properly and predict its performance with any degree of accuracy.

8-15 Guidelines for Sizing Attached Sunspaces

As with other passive systems, there are some basic guidelines for providing initial sizing estimates for greenhouses when they are used as solar heating elements. The Los Alamos solar collection area rule of thumb discussed in section 6-8 is basically a guideline for sizing vertical solar collection areas such as windows and thermal storage walls, but it can also be used to provide approximations for sunspaces when the solar collection area is vertical, as shown in

Fig. 8-25*c*. This procedure would be similar to that for sizing a vertical direct-gain collection area as discussed in section 7-3. When a greenhouse with a sloping glass roof is being considered, as shown in Fig. 8-25*b*, only the vertical projected area of the greenhouse A_p, as discussed in section 6-10, is used in establishing the solar collection area determined by the solar collection area rule of thumb.

Illustrative Problem 8-7

A client has brought you a preliminary floor plan for a solar house that will receive a large percentage of its heating from an attached solar greenhouse. The house will be built 20 miles north of Fresno, California. The client would like to know the approximate size requirement of the greenhouse to provide 50 percent of the heating load. Night insulation will not be used. The house floor area excluding the greenhouse is 1800 ft². The anticipated greenhouse construction will be as shown in Fig. 8-26*a*.

Solution

From Table 6-1, the following values are obtained for the solar collection area rule of thumb: R1 = 9; R2 = 17; S1 = 29; S2 = 46; S3 = 41; S4 = 65. These values are plotted in Fig. 8-26*b*. The ratio of solar collection area to floor area is

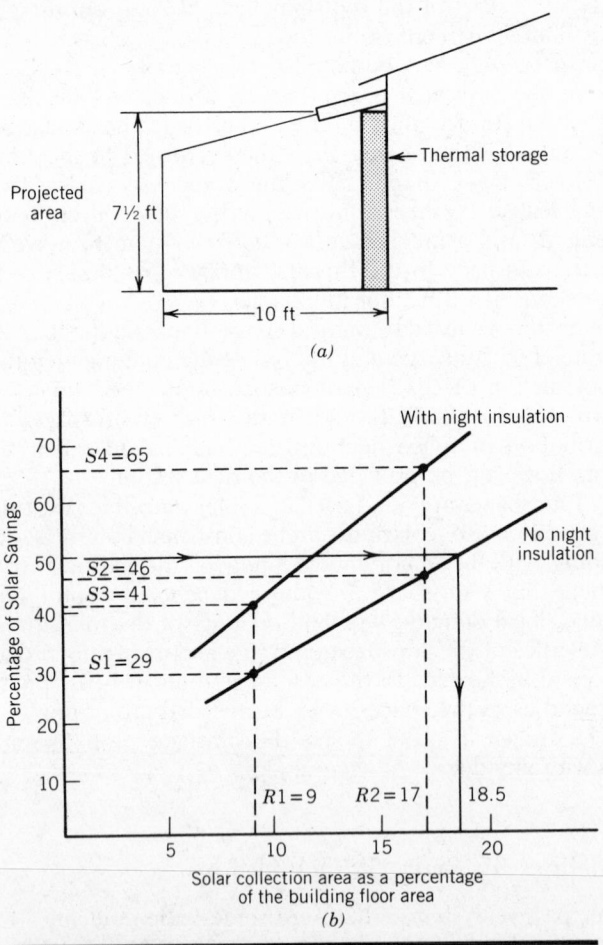

Figure 8-26 (*a*) Section through attached greenhouse described in Illustrative Problem 8-7. (*b*) Plot of Solar Savings versus the solar collection area as a percentage of the building floor area, from the Los Alamos solar collection area rule of thumb, for Illustrative Problem 8-7.

found to be approximately 0.185 (18.5 percent) (no night insulation) for a Solar Savings of 50 percent.

The projected collection area of the greenhouse per foot of greenhouse length is the sum of the vertical glass plus the projected area of the roof glass. The vertical projected height of the greenhouse glazing is shown in Fig. 8-26a as 7½ ft. The projected collection area of the greenhouse is then 7½ ft² per foot of linear length of the greenhouse.

The required solar collection area is:

$$\text{Solar collection area} = 0.185 \times 1800 \text{ ft}^2$$
$$= 333 \text{ ft}^2$$

The length of greenhouse needed is then determined by dividing the required solar collection area by the projected vertical greenhouse area per linear foot of greenhouse:

$$\text{Greenhouse length} = \frac{333 \text{ ft}^2}{7.5 \text{ ft}^2/\text{linear ft}}$$
$$= 44.4 \text{ linear ft} \qquad Ans.$$

The sizing guidelines developed by Mazria provide another approach to the preliminary sizing of attached greenhouses or sunspaces.[3] Table 8-6 gives the actual total square feet of greenhouse glazing required (rather than projected vertical area) to heat an adjacent living space of 1 ft² of floor area. The values in this table are based on computer modeling of buildings with attached greenhouses. The areas of south-facing glazing are assumed to be double-glazed and are also assumed to collect sufficient solar energy on a clear winter day to maintain the greenhouse and adjoining building space at an average temperature of 65° to 70°F. The computer model also assumed that the common wall between the greenhouse and building incorporated either a masonry, concrete or water storage wall, as shown in Fig. 8-25b.

Illustrative Problem 8-8

Determine the total greenhouse glazed area for heating a 900 ft² adjacent space for a house located in Bakersfield, California, where the average winter tem-

[3] Edward Mazria, *The Passive Solar Energy Book* (Emmaus, Pa.: Rodale Press, 1979), pp. 173–179.

Table 8-6   Sizing Guide for Attached Greenhouses

Average Winter Outdoor Temperature (°F)	Degree-Days/Month Winter Months December and January	Square Feet of Actual Total Greenhouse Glazing Required for Each Square Foot of Living Space Floor Area	
		Masonry/Concrete Common Storage Wall	Water Common Storage Wall
Cold Climates			
20°	1350	0.90–1.50	0.68–1.27
25°	1200	0.78–1.30	0.57–1.05
30°	1050	0.65–1.17	0.47–0.82
Temperate Climates			
35°	900	0.53–0.90	0.38–0.65
40°	750	0.42–0.69	0.30–0.51
45°	600	0.33–0.53	0.24–0.38

Source: Adapted with permission from Edward Mazria, *The Passive Solar Energy Book* (Emmaus, Pa.: Rodale Press, 1979), p. 175.

Notes: 1. It is assumed that the adjacent building has a building loss coefficient (BLC) of 7 to 9 Btu/DD-ft². The temperatures and degree-days are for December and January (usually the coldest months).

2. For more southern latitudes (e.g., 35° N Lat), use the lower glass-to-floor-area ratios; and for more northern latitudes (48° N Lat), use the higher ratios. If greenhouse or adjoining building is poorly insulated, use the higher ratios of glass to floor area.

perature is 45°F. (This is one half the size of the house used in the previous example.) Assume a water thermal storage wall between the greenhouse and the living space.

Solution

Check the latitude of Bakersfield as about 36° N. From Table 8-6 (water storage wall), the ratio of greenhouse glazing to floor area is in a range of 0.24 to 0.38. Recall from section 7-3 how latitude compensation for Mazria's rules is made. Since the latitude at 36° N is a lower latitude, the value of 0.25 is arrived at by interpolation. The greenhouse glazed area is then calculated as:

$$\text{Greenhouse glazed area} = 0.25 \times 900 \text{ ft}^2$$
$$= 225 \text{ ft}^2 \qquad Ans.$$

8-16   Sunspace Construction Details

The construction details of a sunspace are important if it is to provide the thermal performance called for by the design. A typical attached greenhouse sunspace was illustrated in Fig. 8-25b. The basic elements of the sunspace are (1) the structure, (2) the glazing, (3) the thermal storage, (4) the venting system, and (5) the shading system.

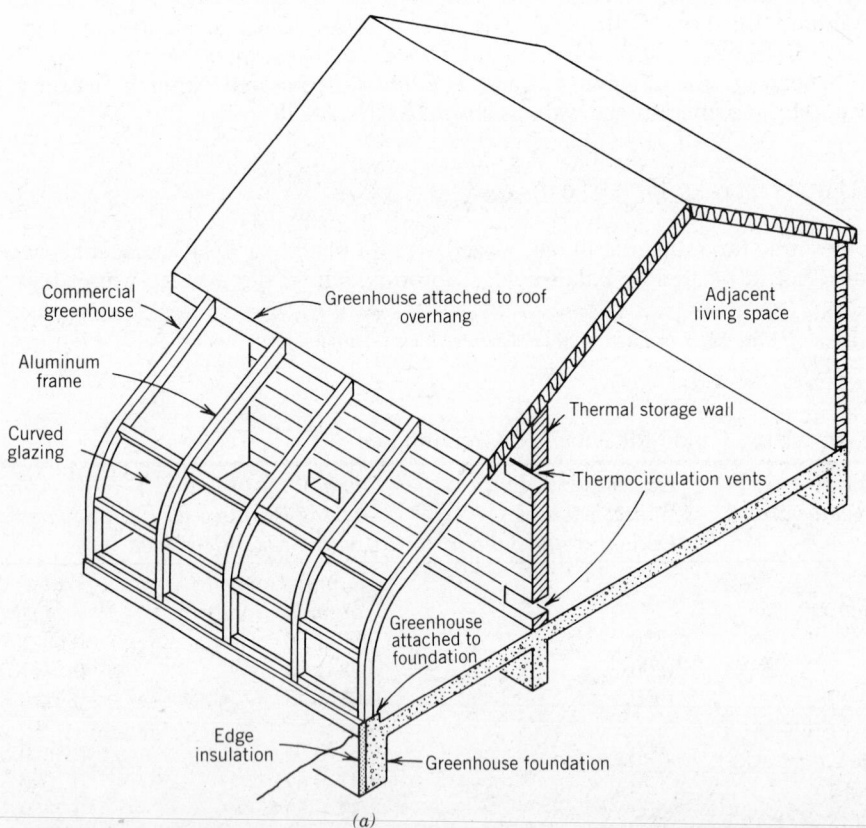

Figure 8-27 Construction of greenhouse sunspaces. (a) A commercially manufactured greenhouse may be preferred by the client. These units are shipped knocked down and can be quickly assembled once the footings and slab (or brick) floor have been prepared. (b) A sunspace can be constructed on the site, using standard construction materials, as shown. (c) Details of mounting and sealing the sunspace glazing. If not properly done, there will be potential for leaks at the seams and joints.

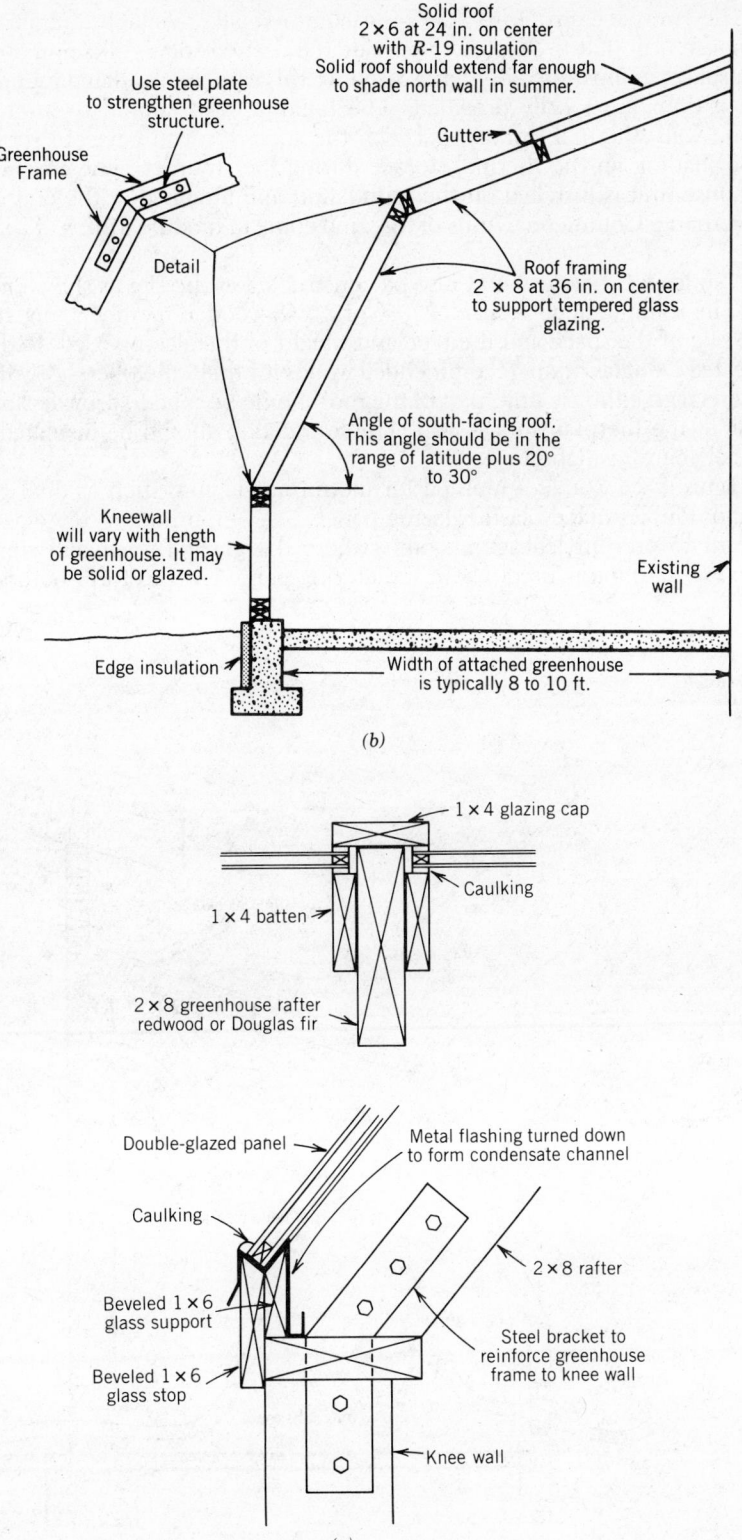

Solid roof
2 × 6 at 24 in. on center
with *R*-19 insulation
Solid roof should extend far enough
to shade north wall in summer.

Use steel plate
to strengthen greenhouse
structure.

Greenhouse
Frame

Detail

Gutter

Roof framing
2 × 8 at 36 in. on center
to support tempered glass
glazing.

Angle of south-facing roof.
This angle should be in the
range of latitude plus 20°
to 30°

Existing
wall

Kneewall
will vary with length
of greenhouse. It may
be solid or glazed.

Edge insulation

Width of attached greenhouse
is typically 8 to 10 ft.

(b)

1 × 4 glazing cap

Caulking

1 × 4 batten

2 × 8 greenhouse rafter
redwood or Douglas fir

Double-glazed panel

Metal flashing turned down
to form condensate channel

Caulking

2 × 8 rafter

Beveled 1 × 6
glass support

Steel bracket to
reinforce greenhouse
frame to knee wall

Beveled 1 × 6
glass stop

Knee wall

(c)

Figure 8-27 *(continued)*

The sunspace structure may be a commercially available prefabricated greenhouse unit that is adapted to provide the desired solar collection area for the sunspace, as illustrated in Figure 8-27a. In this case, an insulated foundation and slab floor are locally provided. The building roof should overhang the common wall (thermal storage) between the sunspace and adjacent rooms to provide shading on the thermal storage during the summer. The commercial greenhouse unit is installed on the foundation and attached at the end of the roof overhang. Commercial units of this kind come in module widths of 24 in. or 30 in.

A sunspace structure may also be constructed at the site as shown in Fig. 8-27b. The framing may be either 2 × 6 or 2 × 8 wooden members, depending on the size of the space and the type and weight of the glazing used. Redwood and treated Douglas fir are recommended woods for these members. As with the commercial greenhouse unit, part of the roof should be solid to provide summer shading of the thermal storage wall. The foundation should be insulated with perimeter foam insulation.

Figure 8-27c shows a method for mounting glazing, such as dual-glazed glass and double-walled plastic glazing panels. The mounting of the glazing is important to prevent leakage at joints where the glazing stops and starts. A sheet-metal flashing is used where the glazing panel ends to form both a drip

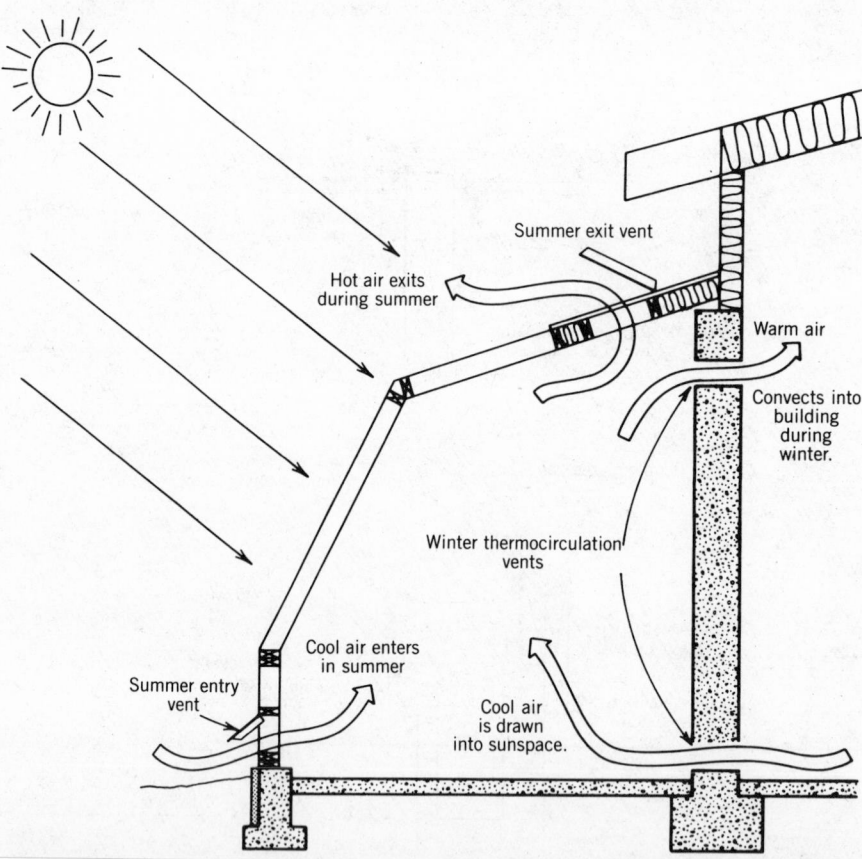

Figure 8-28 The venting system for a sunspace has two parts. The winter thermocirculation vents allow the warm air from the sunspace to enter the adjacent building during the day. The summer vents allow excess heat to be exhausted from the sunspace when the building does not require heating. Shading systems represent a second control option for sunspaces. They provide a screen of the solar radiation to reduce the energy absorbed by the sunspace. The shading may be either an exterior system or an interior curtain. The latter may also serve as an insulating system at night. (See Fig. 8-25.)

edge on the outside of the roof and a condensate channel on the inside of the roof. The caulking used must be a high-grade acrylic latex or silicone to ensure that leaks do not occur at the seams. It is also important to have a water diverter strip or gutter on the house roof just above the sunspace roof to divert water away from the glazed roof. This will help to minimize the possibility of leaks also. Steel brackets or angles should be used as needed to strengthen the structure (Fig. 8-27*b* and *c*).

Venting is very important in a sunspace. Both winter thermocirculation vents and summer cooling vents should be provided, as shown in Fig. 8-28. The winter thermocirculation vents allow a circulation of warm air from the sunspace to the adjacent building. The total area of the thermocirculation vents should be approximately 10 to 16 percent of the floor area of the sunspace, with upper vents and lower vents having nearly equal areas. Alternatives to the upper and lower vents would be to replace them with a sliding glass door between the sunspace and the adjacent living space, or with operable windows. A combination of vents and a door can also be used. The summer vents provide a means of exhausting to outdoors the excess solar gains, as hot air from the sunspace. Total area of summer vents should also be about 10 to 16 percent of the sunspace floor area. The summer entry vent should be low on the south side of the space and should have an area of approximately 8 to 10 percent of the sunspace floor area. The summer exit vent should be located at the highest point of the sunspace and have an area of approximately 6 percent of the floor area of the sunspace. Summer vents can also be located high up in the end walls of the greenhouse or sunspace.

Shading control for the sunspace is necessary to allow solar radiation to be screened out when heat collection is not desired. Figure 8-25*b* shows one method of exterior shading using a roll-down shade. Exterior shading is most effective when a roll-down system of slats is used, the most durable of which are those made from aluminum or plastic. These are available from commercial greenhouse manufacturers. Interior shading systems typically operate with a track support system attached under the greenhouse rafters. These may also act as thermal shades to reduce the heat loss from the sunspace at night. In milder climates, single glazing may be used in conjunction with an inside thermal curtain/shade system.

8-17   Thermal Storage for Sunspaces

The amount and placement of thermal storage for a sunspace is critically important to its overall performance, since the sunspace has a large glazed surface area and will intercept large quantities of solar radiation. The thermal storage must be properly distributed in the sunspace so that it can effectively absorb and store the incoming solar radiation. Otherwise, the air temperature in the sunspace will rapidly increase to unacceptable levels. Figure 8-25*b* indicates a possible placement of water thermal storage in a greenhouse sunspace.

The common wall between the sunspace and the adjoining building is a favored location for thermal storage. This wall absorbs solar radiation during the day and then releases the energy to both the sunspace and the adjacent room on the north of the sunspace. When a thermal storage wall is placed between the sunspace and the adjacent living space, it should have an approximate thickness as recommended in Table 8-7.

When a common masonry or concrete wall is used as the only thermal storage for the sunspace, large daily temperature fluctuations will result in the sunspace. This temperature swing may be as much as 40 to 60 F° on clear winter days and is caused by the fact that masonry and concrete, as was discussed in section 7-11, must have a high surface temperature in order for heat to flow into it and be stored. The sunspace air will be heated by the hot surface of the masonry or concrete thermal storage, resulting in the wide air temperature fluctuations referred to. The air is also heated from the floor and from other objects in the sunspace that absorb solar radiation. To reduce the magnitude of

Material	Thickness (in.)
Adobe	8–12
Brick (common)	10–14
Concrete (dense)	12–18
Water	8 or more

Source: Edward Mazria, *The Passive Solar Energy Book* (Emmaus, Pa.: Rodale Press, 1979), p. 181. *Used with permission.*

the air temperature swing, additional thermal storage, such as water-filled containers, should be placed in the sunspace, as shown in Fig. 8-25b. The water containers will absorb a great deal of the radiation, preventing the high masonry surface temperature and moderating the interior space temperature fluctuations. This additional thermal storage will also release heat to the sunspace at night to help reduce the night temperature drop in the sunspace.

If the common thermal storage wall is constructed from water-filled containers, the need for additional thermal storage is considerably reduced. If the water storage wall has a volume of 0.70 to 1.00 ft³ of water for each square foot of sunspace glazing, additional thermal storage distributed in the sunspace is usually not necessary. The water thermal storage should be placed so that it has a maximum surface exposed to the sunspace and to the adjacent space, for satisfactory heat absorption during the day and heat release at night.

An alternative to having a common thermal storage wall between the sunspace and building is to have remote thermal storage located in the building. This arrangement requires the use of a mechanical system to circulate warm air from the sunspace to the storage. Hybrid systems of this type will be discussed in detail in the next chapter.

8-18 Estimating the Performance of Attached Sunspaces

The *Passive Solar Design Handbook*, vol. III, appendix F, provides tabular data for determining the annual performance of five different configurations of attached sunspaces. Sketches of these five sunspaces are shown in Fig. 8-29. As was discussed in section 6-12 in connection with the annual Load Collector Ratio method, the use of these tabulated values from the *Handbook* allows the solar savings fraction (SSF) to be determined. There are two types of sunspaces shown. First (sketches A and B of Fig. 8-29) is an attached sunspace that has a common wall with the rest of the building and a width of 30 ft. The case A sunspace has a depth of 12 ft and case B a depth of 5.2 ft. The second type is a semienclosed sunspace that has three walls common to the adjacent building as shown in cases C, D, and E. All of these three sunspaces are 24 ft wide with a depth of 12 ft. In all five cases, the common wall has a height of 9 ft.

In order for precomputed tables to be used with the annual LCR method, the sunspace being designed must be quite comparable to one of these reference cases, as will be explained later. This limitation places some restrictions on the geometric configuration of the sunspace. The more detailed monthly SLR method must be used to evaluate the performance of a sunspace when the design deviates appreciably from these five reference cases.

Two types of thermal storage are used in the reference cases. One is a masonry or concrete common wall 12 in. thick, as shown in Fig. 8-30a. This wall is assumed to have a thermal conductivity K of 1.0 Btu/hr-ft²-F°/ft, and a volumetric thermal capacity C_V of 30 Btu/ft³-F°. The other thermal storage system (Fig. 8-30b) uses a row of water containers in front of an insulated frame

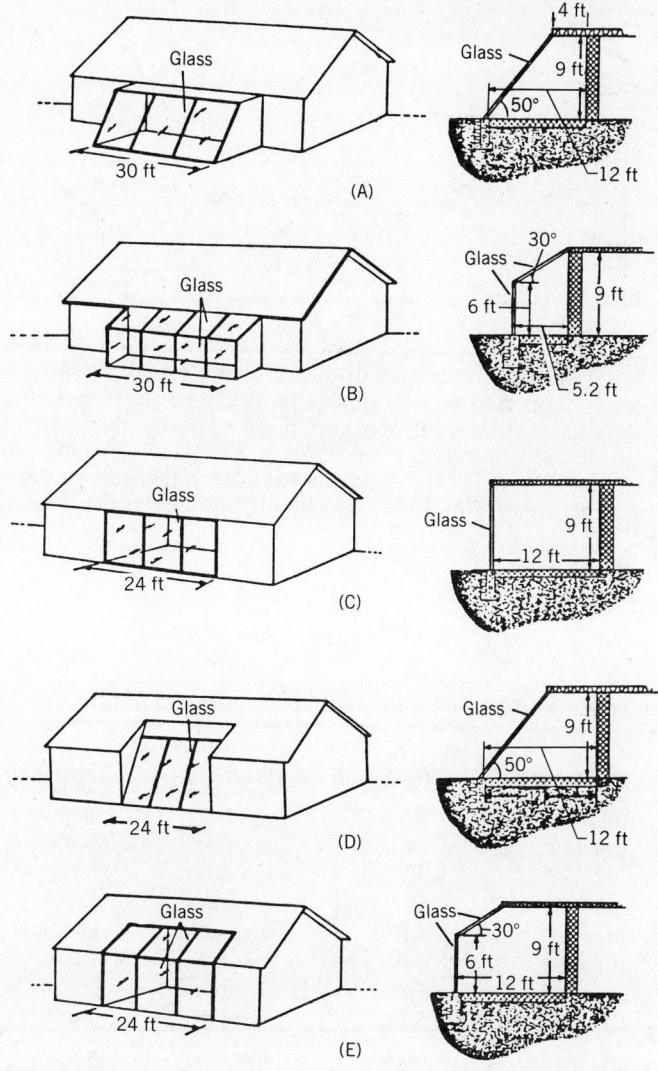

Figure 8-29 There are five basic sunspace configurations proposed for use with the Los Alamos Lab data. These tabulated data are for use with the annual Load Collector Ratio (LCR) method. The five reference cases are sketched here. With several variations for each case, a total of 28 reference designs or configurations are listed in Table 8-8. *Source:* R. W. Jones, ed., *Passive Solar Design Handbook*, vol. III, Washington: U.S. Dept. of Energy, July 1983, p. 88.

wall (for thermal isolation). The containers are twice as high as they are deep, and the water volume is assumed to be 1 ft³ per square foot of common wall area. It is assumed that the common wall has thermocirculation vents with the upper vent area equal to 3 percent of the common wall area and the lower vent area also equal to 3 percent of the common wall area. The sunspaces of Fig. 8-29 may have end walls that are either glazed or opaque. If opaque, they should have R-20 insulation. For the semienclosed sunspaces, the end walls are assumed to be common walls with the living space. The glazing is assumed to be equivalent to two panes of ⅛-in. window glass with a ½-in. air gap between. The floor of the sunspace is assumed to be a 6-in.-thick slab of concrete with a thermal conductivity of 0.5 Btu/hr-ft²-F°/ft, and a volumetric heat capacity of 30 Btu/ft³-F°. There is assumed conduction of heat through the floor to the soil below. Only the floor and the masonry (or concrete) wall (or water containers)

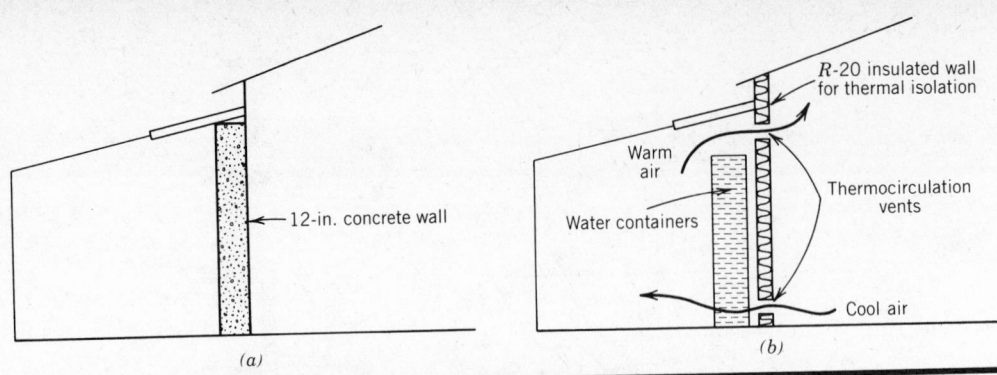

Figure 8-30 Standard construction details for the reference designs of Fig. 8-29.
(a) The masonry or concrete thermal storage for the five sunspace configurations is
assumed to be placed as shown. (b) When water is used as the thermal storage for the
five basic cases, it will be located as suggested in the sketch. An insulating wall is
assumed to separate thermally the sunspace from the adjacent living space. Thermocirculation vents are used in this case to allow heat to enter the living space from
the sunspace.

Table 8-8 Sunspace Systems—Twenty-Eight Reference Designs

Designation	Type	Tilt (degrees)	Common Wall	End Walls	Night Insulation
A1	attached	50	masonry	opaque	no
A2	attached	50	masonry	opaque	yes
A3	attached	50	masonry	glazed	no
A4	attached	50	masonry	glazed	yes
A5	attached	50	insulated	opaque	no
A6	attached	50	insulated	opaque	yes
A7	attached	50	insulated	glazed	no
A8	attached	50	insulated	glazed	yes
B1	attached	90/30	masonry	opaque	no
B2	attached	90/30	masonry	opaque	yes
B3	attached	90/30	masonry	glazed	no
B4	attached	90/30	masonry	glazed	yes
B5	attached	90/30	insulated	opaque	no
B6	attached	90/30	insulated	opaque	yes
B7	attached	90/30	insulated	glazed	no
B8	attached	90/30	insulated	glazed	yes
C1	semienclosed	90	masonry	common	no
C2	semienclosed	90	masonry	common	yes
C3	semienclosed	90	insulated	common	no
C4	semienclosed	90	insulated	common	yes
D1	semienclosed	50	masonry	common	no
D2	semienclosed	50	masonry	common	yes
D3	semienclosed	50	insulated	common	no
D4	semienclosed	50	insulated	common	yes
E1	semienclosed	90/30	masonry	common	no
E2	semienclosed	90/30	masonry	common	yes
E3	semienclosed	90/30	insulated	common	no
E4	semienclosed	90/30	insulated	common	yes

Source: Robert W. Jones, ed., *Passive Solar Design Handbook*, vol. III (Washington: U.S. Department of
Energy, 1982), p. 86. Adapted with permission.

Note: All of the walls designated "insulated" in the *Common Wall* column have water thermal storage mass.

INDIRECT-GAIN AND ISOLATED-GAIN PASSIVE SYSTEMS

Table 8-9   Solar Savings Fraction (SSF) versus Load Collector Ratio (LCR) for the Sunspace Systems of Table 8-8.

Richmond, Virginia

Solar Savings Fraction →	0.1	0.2	0.3	Degree-Days 3939		0.6	0.7	0.8	0.9
				0.4	0.5				
Greenhouse Designation				LCR					
A1	381	112	57	35	23	16	11	7	5
A2	354	149	85	55	38	27	19	14	9
A3	359	98	49	29	18	12	8	5	3
A4	358	146	83	53	37	26	18	13	8
A5	634	111	51	29	19	12	8	6	3
A6	351	146	84	54	37	27	19	13	9
A7	703	90	39	21	13	8	5	3	2
A8	357	142	80	51	35	24	17	12	8
B1	258	84	44	27	18	12	9	6	4
B2	279	121	70	46	32	23	16	12	8
B3	234	75	39	23	15	10	7	5	3
B4	272	117	68	44	30	22	16	11	7
B5	333	74	35	21	13	9	6	4	2
B6	262	115	67	44	31	22	16	11	7
B7	288	60	28	16	10	6	4	3	2
B8	253	109	63	41	28	20	14	10	6
C1	162	71	41	27	18	13	9	6	4
C2	181	93	58	39	28	20	15	11	7
C3	169	57	31	19	13	9	6	4	3
C4	179	83	50	33	23	17	12	9	6
D1	348	134	74	47	31	22	15	10	7
D2	313	162	101	69	49	36	26	18	12
D3	440	127	64	39	25	17	12	8	5
D4	320	159	97	65	46	33	24	17	11
E1	254	102	57	36	24	17	12	8	5
E2	256	131	81	55	39	28	21	15	10
E3	310	88	44	27	17	12	8	5	3
E4	264	123	74	49	34	24	18	12	8

Source: Robert W. Jones, ed., *Passive Solar Design Handbook*, vol. III (Washington: U.S. Department of Energy, 1982), p. 496. Reprinted with permission.

are thermal storage. All other surfaces are assumed to have negligible thermal storage mass. The use of night insulation is also an option for the different configurations. Table 8-8 gives a summary of the 28 possible variations of the sunspace reference cases, and Table 8-9 gives the Solar Savings (SS) as a function of the Load Collector Ratio (LCR) for a location near Richmond, Virginia.

The design analysis is not, of course, restricted to the exact dimensions shown in Fig. 8-29. The dimensions specified establish the relative sunspace configuration, and it is the configuration and not the absolute dimensions that determine performance per unit of projected area, A_p. For example, configuration A is normally 30 ft wide by 9 ft high with a 4-ft-wide ceiling and a glazing tilt of 50°, which gives an A_p of 270 ft². If the dimensions are increased by one third, the sunspace would be 40 ft wide by 12 ft high with a 5-ft-4-in.-wide ceiling and a glazing tilt of 50°. The projected area, A_p, is now 480 ft². The effect of the absolute size of the sunspace is accounted for by the value of the projected area, A_p, which is calculated on a per-linear-foot basis. Thus, the absolute dimensions of the sunspace can be proportionally increased or decreased as long as essentially the same geometric configuration is maintained.

The following problems illustrate the methods of calculating design data from these reference cases.

Illustrative Problem 8-9

A 2000 ft² home has been designed for a location 20 mi south of Richmond, Virginia. The house has an estimated net load coefficient (NLC) of 10,000 Btu/DD. An attached solar greenhouse corresponding to the designation B2 of Tables 8-8 and 8-9, with a masonry or concrete common wall and night insulation, is planned for the south side of the house. (a) Determine the Solar Savings and the Auxiliary Heat required. (b) Determine the effect on the performance of not using night insulation.

Solution

(a) The attached greenhouse is the case illustrated in Fig. 8-29B. The projected area A_p is equal to 9 ft × 30 ft = 270 ft². The Load Collector Ratio is calculated using Eq. (6-13):

$$LCR = \frac{NLC}{A_p} = \frac{10,000 \text{ Btu/DD}}{270 \text{ ft}^2} = 37 \text{ Btu/DD-ft}^2$$

The value of the solar savings fraction (SSF) is obtained from the B2 values of Table 8-9 for Richmond, Virginia, and found (by interpolation) to be 0.46. The degree-days for Richmond are listed in the table as 3939 DD/year. The Solar Savings is calculated from Eq. (6-9) as:

$$\begin{aligned} SS &= NLC \times DD \times SSF \\ &= 10,000 \text{ Btu/DD} \times 3939 \text{ DD/year} \times 0.46 \\ &= 18.1 \times 10^6 \text{ Btu/year} \end{aligned}$$ *Ans.*

The Auxiliary Heat is calculated from Eq. (6-8) as:

$$\begin{aligned} AH &= NLC \times DD \times (1 - SSF) \\ &= 10,000 \text{ Btu/DD} \times 3939 \text{ DD/year} \times (1 - 0.46) \\ &= 21.3 \times 10^6 \text{ Btu/year} \end{aligned}$$ *Ans.*

(b) If night insulation is not used on the attached greenhouse, then it would correspond to designation B1 listed in Tables 8-8 and 8-9. The value of the solar savings fraction (SSF) is found from Table 8-9 for designation B1 by interpolation to be 0.34. The Solar Savings is calculated from Eq. (6-9) as:

$$\begin{aligned} SS &= NLC \times DD \times SSF \\ &= 10,000 \text{ Btu/DD} \times 3939 \text{ DD/year} \times 0.34 \\ &= 13.4 \times 10^6 \text{ Btu/year} \end{aligned}$$ *Ans.*

The Auxiliary Heat is calculated from Eq. (6-8) as:

$$\begin{aligned} AH &= NLC \times DD \times (1 - SSF) \\ &= 10,000 \text{ Btu/DD} \times 3939 \text{ DD/year} (1 - 0.34) \\ &= 26.0 \times 10^6 \text{ Btu/year} \end{aligned}$$ *Ans.*

Illustrative Problem 8-10

An existing home in the Richmond, Virginia area has an estimated NLC of 13,000 Btu/DD. The owners are interested in adding an attached greenhouse for passive solar space heating. They would like to provide 70 percent of their space-heating needs with the passive solar system. Since this will be a retrofit greenhouse, it will fall into the designation of an attached greenhouse as shown in Fig. 8-29, designs A and B.

Determine the projected solar collection area (aperture) required for a Solar Savings of 70 percent. Assume that water thermal storage will be used in the greenhouse, that the common wall between the greenhouse and house will be insulated, and that night insulation will be used on the glazing. The end walls will be opaque with no glazing, insulated to R-20.

INDIRECT-GAIN AND ISOLATED-GAIN PASSIVE SYSTEMS

Solution

The reference cases include two types of attached greenhouses—class A with a 50° glazing tilt and class B with a 90°/30° glazing tilt.

(a) For the 50° glazing tilt with water thermal storage, night insulation, and opaque end walls, the designation from Table 8-8 is A6. To provide a 70 percent reduction in the space heating, a solar savings fraction of 0.70 would be needed. From Table 8-9, the LCR for an SSF of 0.7 for designation A6 is 19 Btu/DD-ft².

By solving Eq. (6-13) for A_p,

$$A_p = \frac{NLC}{LCR}$$
$$= \frac{13,000 \text{ Btu/DD}}{19 \text{ Btu/DD-ft}^2}$$
$$= 684 \text{ ft}^2 \qquad\qquad Ans.$$

(b) For the 90°/30° glazing tilt with water thermal storage, night insulation, and opaque end walls, the designation from Table 8-8 is B6. From Table 8-9, the LCR for an SSF of 0.7 is 16 Btu/DD-ft²:

$$A_p = \frac{NLC}{LCR}$$
$$= \frac{13,000 \text{ Btu/DD}}{16 \text{ Btu/DD-ft}^2}$$
$$= 812 \text{ ft}^2 \qquad\qquad Ans.$$

CONCLUSION

Indirect-gain passive solar systems and their design have been treated in this chapter. These systems were here considered as single or stand alone systems, supplying heat into a building. In the following chapter, the more typical configuration—use of several passive elements or systems to supply heat to a single building load—will be dealt with. Hybrid or combination solar systems and natural daylighting of buildings will also be considered. Also, the Solar Load Ratio (SLR) and Load Collector Ratio (LCR) methods of performance analysis will be applied in somewhat more advanced problems of the type often encountered in solar design practice.

PROBLEMS

1. If an 8-in.-thick brick wall is used as thermal storage, what is the approximate time lag that will result?

2. It is desired to have an 8-hr time lag for a thermal storage wall. Specify a material and thickness that will provide this time lag.

3. A 2000 ft² house is being planned for Boston, Massachusetts. The normalized BLC will be 9 Btu/DD-ft² of floor area. What should be the approximate size of the solar collection area and Trombe wall to provide 65 percent of the space heating? Assume that night insulation will be used.

4. If a solar collection area and thermal storage wall of 300 ft² is used to provide solar heat to a 1500 ft² house, what would the approximate solar savings be? The house is located in Medford, Oregon.

5. A 1600 ft² home is planned for Salt Lake City, Utah. The monthly winter degree-days are 1120 DD/month, and the latitude is N 40.8°. From Table 8-2, determine the solar collection and storage wall area for a masonry or concrete wall and for a water wall.

6. A 2000 ft² home is under construction near Indianapolis, Indiana. The monthly degree-days are 990 DD/month, and the latitude is N 39.7°. From Table 8-2, determine the solar collection and storage wall area for a water thermal storage wall.

7.   A 1400 ft^2 house is to have a solar roof pond system. The location is Ely, Nevada, at 39.3° N Lat. Average winter temperature is 25°F. The roof pond will be double-glazed with night insulation and a reflector. Determine the approximate roof pond area for a 60 percent share of the space-heating load.

8.   A single-glazed roof pond is to be sized for space cooling. The location is Albuquerque, New Mexico, where the daytime climate is hot and dry and the nights are cool. The building has 3000 ft^2 of floor space. How large should the surface area of the roof pond be if it is to be augmented by evaporative cooling? See Table 8-5.

9.   The thickness of a concrete or masonry wall not only influences the time lag of the wall but also affects the interior space temperature fluctuation. How thick should a concrete solar thermal storage wall be if an interior temperature fluctuation of about 10 F° is desired?

10.   A new home near Albany, New York, has an estimated net load coefficient (NLC) of 14,000 Btu/DD. It is desired by the owners that 70 percent of the space heating be provided by a masonry or concrete solar thermal storage wall and collection area. Determine the required area of the wall if night insulation of R-9 is used.

11.   Calculate the solar savings and auxiliary heat for a house located near Knoxville, Tennessee, if the house has a 425 ft^2 brick thermal storage wall. The estimated net load coefficient is 12,000 Btu/DD.

12.   An addition is planned for an existing home in Albany, New York. It will have a 600 ft^2 water thermal storage wall. The NLC for the house and the addition is estimated to be 15,000 Btu/DD. Determine the solar savings and the auxiliary heat if the wall uses R-9 night insulation.

13.   A greenhouse retrofit is planned for an existing house. The house has 1600 ft^2 of space to be heated. The house is located at 40° N Lat, and the average winter temperature is 45°F. How many square feet of south-facing glazing should be specified for the greenhouse, to provide 60 percent of the heating requirement? The (common) thermal storage wall is to be brick.

14.   An attached greenhouse is to provide space heating for a new residence near Seattle, Washington. The house will have 2400 ft^2 of floor space, and the common wall for thermal storage is to be brick. The monthly winter degree-days are 800 DD/month, and the latitude is 47.5° N. Estimate the required south-facing glazed area required for the greenhouse if the system is to provide 70 percent of the heating requirement.

15.   A new residence in Richmond, Virginia, is to have an attached greenhouse. The net load coefficient (NLC) for the house is 15,000 Btu/DD. The greenhouse will have a projected area (A_p) of glazing of 300 ft^2, and it will have a 90°/30° tilt on the glazing (designation type B of Fig. 8-29). The common wall is masonry, and the end walls are glazed. Night insulation will be used. Determine the solar savings and the auxiliary heat.

16.   A semienclosed sunspace with vertical glazing is being designed for an existing home, also near Richmond, Virginia. The sunspace is to provide 60 percent of the annual heating load. The NLC for the house is estimated at 12,000 Btu/DD after the house was weatherized. Determine the projected area of the solar aperture and the values of solar savings and auxiliary heat in Btu per year.

17.   An existing home near Cincinnati, Ohio, has an estimated net load coefficient of 11,000 Btu/DD. Determine the required projected area (A_p) of the solar aperture for an attached sunspace of designation B (Fig. 8-29). Assume that an insulated common wall is used, with water as the thermal storage mass, and that night insulation is used. Also determine the dimensions of the sunspace to provide the required solar aperture.

18.   The projected area (A_p) for a semienclosed sunspace with a glazing tilt of 90°/30° was determined to be 600 ft^2. Determine the sunspace dimensions for sunspace designation E (Fig. 8-29).

Advanced Passive Methods – Selected Applications

*I*n previous chapters we have treated passive solar systems as separate isolated systems. In this chapter we examine them in a larger context, considering among related topics the following: (1) the effect of passive solar systems on cooling loads, (2) a method of determining the annual Load Collector Ratio for evaluating buildings with a mixture of passive systems, (3) the use of remote storage rock beds, (4) design procedures for passive solar retrofits for existing structures, (5) the use of the monthly Solar Load Ratio method as a means of providing a final performance evaluation on the completed design of passive solar buildings, and (6) the use of daylighting in large commercial buildings as a major contribution to energy savings.

9-1   The Effects of Passive Solar Elements on Total Annual (Cooling and Heating) Load

When designing passive solar systems, it is very important to evaluate the solar apertures during the summer and early fall to assess the impact of the energy gains they contribute to the cooling load of the building. (See also Chapter 16.) For the purposes of this discussion, four types of generalized climates will be discussed. First, there is the heating-dominated climate, where the degree-days of heating are significantly greater than the degree-days of cooling. Such climates are found in New England and the northern Pacific Coast regions. The second climate type has approximately equal heating and cooling requirements during the year. Such climates are characteristic of the midwestern United States, where the winters are cold and cloudy while the summers are warm and humid. The third climate is one in which the need for cooling and dehumidification is dominant, with degree-days of cooling far exceeding heating degree-days—a climate typical of the south-central and southeastern United States. A fourth climate type is the dry climate, with hot summers and mild to cool winters, characteristic of the southwestern United States. There are many local climates that will fall between these generalized climates, and each one must be considered as an individual case by solar designers.

In a passive solar system, the use of large, south-facing glazed apertures dramatically increases the energy that enters the structure during the summer over that which an opaque, insulated wall would allow. Energy enters the building by direct solar gains, reflected radiation, diffuse radiation, radiative energy gains from heated surfaces, and conductive energy gains, as illustrated in Fig. 9-1. The effects of these energy gains will vary, depending upon the cooling load of the building and the summer climate.

Another factor that has a significant effect on the amount of heat gain through an aperture is the nature of the controls used with the aperture itself. For example, studies of thermal control of solar apertures during the cooling season have been made by M. J. Siminovitch on a residence in the Indianapolis area. An 1800 ft^2 residence was modeled using the BLAST computer program, with various strategies of control over the amount of heat introduced into the house by south-facing glazing in summer.

The effects of the following different aperture control strategies were considered (Fig. 9-2);

1. The use of ventilation to reduce the interior temperature during the cooling season.
2. The use of horizontal shading devices to block direct solar radiation.

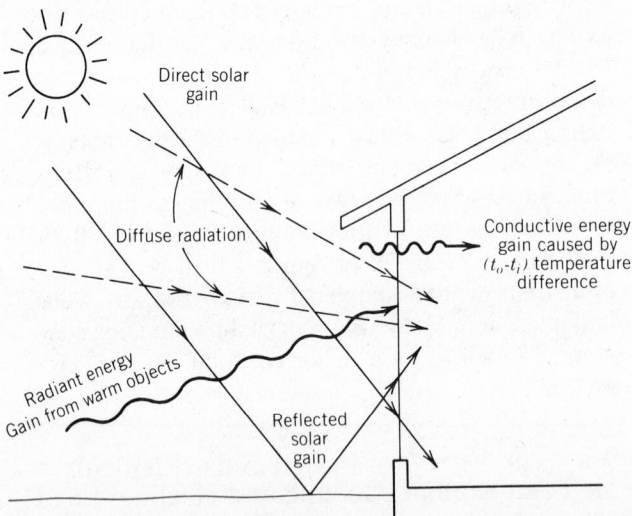

Figure 9-1 Heat energy enters a building through a glazed area by several methods. Direct and reflected solar radiation is transmitted through the glazing. Diffuse radiation is also transmitted through the glazing. Radiation from other objects is absorbed by the glazing, causing some heat to flow through the glazing by conduction. When the outside temperature is greater than the indoor temperature, heat will also flow through the glazing by conduction.

ADVANCED PASSIVE METHODS—SELECTED APPLICATIONS

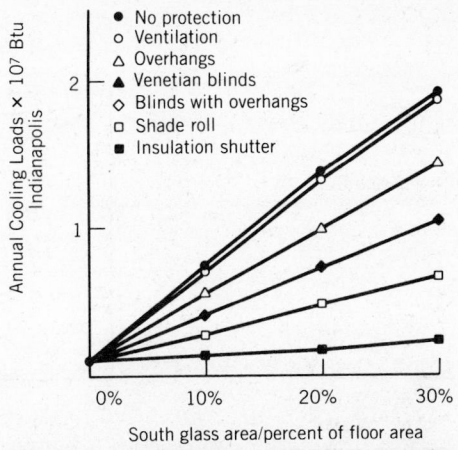

Figure 9-2 Effects on building cooling loads of various solar apertures protected by different screening devices. Annual cooling loads for an 1800 ft² house in Indianapolis are plotted on the vertical axis, and the area of the solar aperture (south glass area) is plotted along the horizontal axis as a percentage of the floor area of the house. Curves are plotted for different solar aperture protection strategies. The reference case is the plot corresponding to no protection. *Source:* M. J. Siminovitch, "Thermal Protection of the Solar Aperture in the Cooling Season—A Quantitative Performance Study," School of Architecture and Small Homes Council—Building Research Council, University of Illinois at Urbana-Champaign, 7th National Passive Solar Conference, Knoxville, Tenn., September 1982, pp. 883–88.

3. The use of horizontal blinds to screen and reflect solar radiation.
4. The use of roll-type shades for selective screening of glazed areas.
5. The use of insulated shutters for selective reduction of the solar aperture area.

The results of each of these cooling reduction concepts as applied to the annual cooling load are shown by the plots of Fig. 9-2. It is apparent that the ventilation strategy was not at all successful, probably because of the hot, humid summer climate. There is not sufficient outside–inside temperature difference (Δt) to provide appreciable cooling when ventilation is used. The roof overhangs provided protection from direct solar gains and reduced the cooling load by about 25 percent as compared to the unprotected control case. The use of venetian blinds had almost exactly the same effect. The combined use of overhang and venetian blinds provided approximately a 45 percent reduction in the cooling load compared to the unprotected control case. Interior roller shades reduced the cooling load by about 65 percent compared to the control case. Roller shades allow some diffuse light to enter through the shade, but they block the entrance of all other radiation. Roller shades also create a semidead air space between the shade and the glazing, with good insulating effect. The disadvantage of the roller shade is that it blocks the view during the day, whereas the overhang and venetian blinds provide both daylighting and a view. Exterior insulation shutters were the most effective in reducing daytime heat gains through solar apertures. Their use reduced the cooling load by about 85 percent compared to the unprotected reference case. The disadvantage of a shutter is

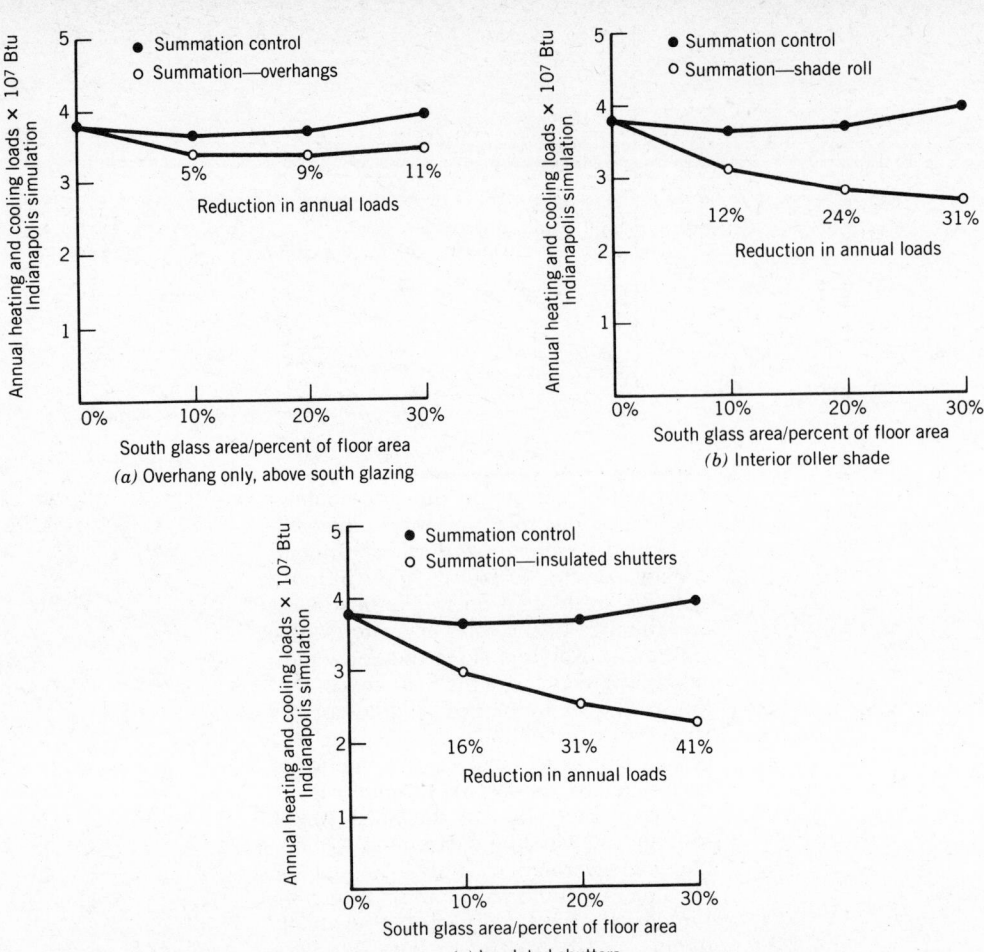

Figure 9-3 The effects of different shading systems for solar apertures on the total annual combined heating and cooling load for an 1800 ft² house located in Indianapolis. The "summation control" curve indicates what the combined heating and cooling load would be without the use of any of the aperture controls. *Source:* M. J. Siminovitch, "Thermal Protection of the Solar Aperture in the Cooling Season—A Quantitative Performance Study," School of Architecture and Small Homes Council—Building Research Council, University of Illinois at Urbana-Champaign, 7th National Passive Solar Conference, Knoxville, Tenn., September 1982, pp. 883–88.

that, like the roller shade, it is opaque and eliminates any view or daylighting through the solar aperture.

The effects of solar aperture and shading systems on annual heating and cooling loads in Indianapolis are shown in Fig. 9-3. In the case of an overhang only (chart *a*), the energy savings for the winter heating load from the radiation gains through the solar aperture are just about canceled out by an increase in the cooling load caused by the solar aperture. Only a small decrease in annual energy load results. However, when roller shades (chart *b*), or insulated shutters, (chart *c*) are used, a significant net decrease in the annual energy load results with increased area of solar aperture.

The effects of an overhang on a passive solar building using a Trombe wall are shown in Fig. 9-4*a*. This study was done for a building in Los Alamos, New Mexico. The overhang configuration is shown in Fig. 9-4*b*. The midwinter months (November through March) show very little change in the heating performance; that is, the heating load with and without the overhang remains about the same. This is because the overhang causes no shading on the solar aperture during these months. However, the heating performance is greatly

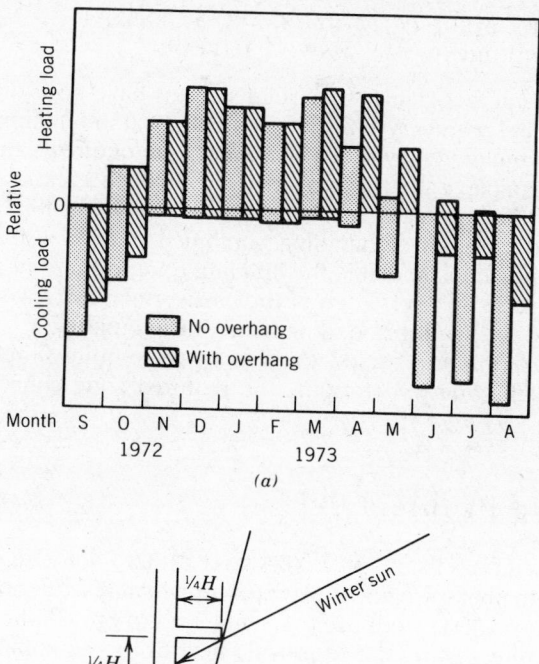

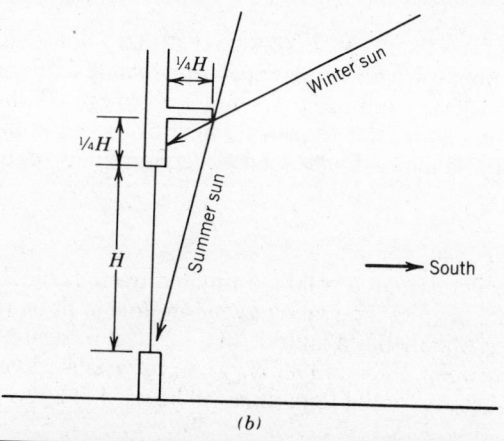

Figure 9-4 The effect of overhang on total cooling and heating loads. (a) The relative cooling and heating loads, month by month, for a passive building located in Los Alamos, New Mexico, with and without an overhang for the south-facing glazing. (b) The configuration of the overhang for the building whose loads are charted in a. Source: J. D. Balcomb, et al., *Passive Solar Design Handbook*, vol. II, Washington: U.S. Department of Energy, January 1980, p. 111.

affected in the spring during April and May, and the greatest benefit of the overhang is in the reduction of the summer cooling loads from June through August. During the fall months, the overhang reduces the cooling load by about a quarter.

These examples illustrate the importance of considering the net yearly effect on heating and cooling when solar apertures are used for space heating and/or daylighting. In climates where heating is the dominant factor and cooling loads are minimal, simple shading and ventilation techniques will offset much of the summer gain through the solar aperture. In cooling-dominated climates, the use of southern apertures for winter space heating and daylighting should be carefully analyzed for the effect on the summer cooling load.

In areas where the heating and cooling loads are approximately equal, solar control methods in addition to a fixed overhang are recommended. These methods could include an adjustable overhang or exterior shade, a movable insulation system either on the interior or exterior of the solar aperture, and the use of roller shades, horizontal blinds, or curtains to screen out heat gains.

9-2 Evaluating Solar Performance of Buildings Having a Combination of Passive Elements

Direct-gain and indirect-gain passive solar systems have been discussed in preceding chapters as separate systems for providing heat in buildings. In actuality, a typical passive building may utilize a combination of direct- and indirect-gain systems. For example, a house may have an attached greenhouse to provide heat to a part of the house and a Trombe wall to provide heat in another part of the house. Two methods of handling combined passive systems will be discussed. One approach is to divide the building up into thermal zones and treat each zone as a separate load. Each of the passive solar systems is then sized to the heating load of the corresponding zone of the building.

With Mazria's sizing guides, this can be done quite simply by using the floor area of each zone to determine the required zone collection area from Mazria's tables.

Illustrative Problem 9-1

A house designed for Albany, New York, has a total floor area of 2000 ft². The average winter temperature for Albany is approximately 25°F, and the latitude is 42.7°N. The house design will use a Trombe wall to provide heat to 1100 ft² of the house and direct-gain solar to provide heat to the remaining 900 ft² of the house. Determine the approximate solar collection areas required.

Solution

The direct-gain collection area will be estimated from Table 7-1 for an average winter temperature of 25°F. The ratio of window area to floor area has a range of 0.21 to 0.33. Since Albany has a latitude of 42.7°N, it is located about six tenths of the distance between 35°N and 48°N, so an interpolated value of 0.28 ft² of solar window per square foot of floor area will be used. The solar window area is then:

$$A_{window} = 900 \text{ ft}^2_{floor} \times 0.28 \text{ ft}^2_{window/ft^2 floor}$$
$$= 252 \text{ ft}^2 \qquad\qquad Ans.$$

The Trombe wall solar collection area will be estimated from Table 8-2 for an average winter temperature of 25°F. The ratio of wall area to floor area has a range of 0.51 to 0.91 for masonry or concrete. By interpolation, a value of 0.75 ft² of solar-irradiated wall area per square foot of floor area will be used. The Trombe wall area is then:

$$A_{wall} = 1100 \text{ ft}^2_{floor} \times 0.75 \text{ft}^2_{wall/ft^2 floor}$$
$$= 825 \text{ ft}^2 \qquad\qquad Ans.$$

When evaluating passive system performance using the annual LCR, combinations of different system types can be handled by the following procedure:

1. A single LCR is calculated based on the total building load and the total combined collection areas of the different passive systems.
2. The corresponding values of the solar savings fraction (SSF) are obtained from the LCR tables for each of the passive system types using the above LCR.
3. The solar savings fraction for each passive system type is mathematically weighted, based on the relative proportion of that system's glazed area compared to the total combined collection area.
4. The weighted values of the solar savings fractions of the several systems are then summed up to obtain the net solar savings fraction for the combination.

ADVANCED PASSIVE METHODS—SELECTED APPLICATIONS

5. The Solar Savings and annual Auxiliary Heat can then be calculated from the net solar savings fraction obtained in step 4. The following problem illustrates the method.

Illustrative Problem 9-2

Calculate the annual Solar Savings and Auxiliary Heat for the house described in Illustrative Problem 9-1. Assume that all of the collection area faces due south and that the direct-gain and the masonry concrete thermal storage (Trombe) wall will correspond to the design case for the data that were given in Table 6-5. Make the calculations for the case where R-9 insulation is used at night. The house has an estimated net load coefficient (NLC) of 16,000 Btu/DD.

Solution

From Illustrative Problem 9-1, the glazed area of the direct-gain system is 252 ft^2, and the area of the solar-irradiated thermal storage (Trombe) wall is 825 ft^2.

1. The net solar collection area A_p is then 1077 ft^2. The LCR is calculated using Eq. (6-13);

$$\text{LCR} = \frac{\text{NLC}}{A_p} = \frac{16{,}000 \text{ Btu/DD}}{1077 \text{ ft}^2}$$
$$= 14.86 \text{ Btu/DD-ft}^2$$

2. The solar savings fractions for the direct-gain system and the thermal storage wall are obtained by interpolation from Table 6-5 (direct gain with night insulation; and masonry or concrete thermal storage wall with night insulation):

$$\text{SSF}_{\text{DG}} = 0.423$$
$$\text{SSF}_{\text{TW}} = 0.463$$

3. The relative proportions of the passive systems are calculated as:

$$\text{Percent}_{\text{DG}} = \frac{252 \text{ ft}^2}{1077 \text{ ft}^2} = 0.23$$

$$\text{Percent}_{\text{TW}} = \frac{825 \text{ ft}^2}{1077 \text{ ft}^2} = 0.77$$

The weighted values of the solar savings fractions are then calculated as:

$$\text{SSF}_{\text{WDG}} = 0.23 \times 0.423$$
$$= 0.10$$

$$\text{SSF}_{\text{WTW}} = 0.77 \times 0.463$$
$$= 0.36$$

4. The value of the solar savings fraction for the combination system is then found to be:

$$\text{SSF}_{\text{Comb.}} = 0.10 + 0.36$$
$$= 0.46$$

5. The degree-days for Albany are listed as 6888 in Table 6-5. The Auxiliary Heat is calculated from Eq. (6-8) as:

$$\text{AH} = \text{NLC} \times \text{DD} \times (1 - \text{SSF})$$
$$= 16{,}000 \text{ Btu/DD} \times 6888 \text{ DD/year} \times (1 - 0.46)$$
$$= 59.5 \times 10^6 \text{ Btu/year} \qquad \textit{Ans.}$$

The Solar Savings is calculated from Eq. (6-9):

$$SS = NLC \times DD \times SSF$$
$$= 16{,}000 \text{ Btu/DD} \times 6888 \text{ DD/year} \times 0.46$$
$$= 50.7 \times 10^6 \text{ Btu/year} \qquad \qquad \text{Ans.}$$

9-3 Continuation of Design Exercise from Section 6-13

The solar house design problem of section 6-13 was concluded with a thermal storage wall solar aperture area of 648 ft², and a solar savings fraction of 0.508. The owners later decided that they would like to add an attached sunspace that would provide a winter greenhouse for growing plants and also provide some solar heat gain to the house. They also wanted windows to be used in conjunction with the thermal storage wall to allow direct solar gains and natural daylighting. Additional insulation and energy conservation measures were added to the house to reduce heat losses.

The house as modified has an estimated net load coefficient of 7300 Btu/DD. The solar collection area of the masonry thermal storage wall was reduced, and it now has an area of 450 ft². South-facing, direct-gain solar windows and clerestory windows have been added, which have a solar collection area of 200 ft². An attached sunspace has been added, which has a 50° slope with projected aperture A_p of 270 ft² (corresponding to designation type A in Fig. 8-29). The sunspace will use water thermal storage with an insulated common wall between the sunspace and the residence. The end walls are opaque, and night insulation of R-9 is used on the solar aperture. This is designation A6 in Table 8-9.

The Solar Savings and Auxiliary Heat must be calculated for this new design, which is a mixture of direct-gain, masonry thermal storage wall, and attached sunspace.

Solution

The total combined solar collection area is 920 ft². The LCR based on the total load and total combined solar collection area is calculated using Eq. (6-13) as

$$LCR = \frac{NLC}{A_p} = \frac{7300 \text{ Btu/DD}}{920 \text{ ft}^2}$$
$$= 7.93 \text{ Btu/DD-ft}^2$$

The solar savings fraction for direct gain with night insulation is determined from Table 6-5 for Albany (interpolating) as

$$SSF_{DG} = 0.578$$

The solar savings fraction for the masonry thermal storage wall with night insulation is also determined from Table 6-5 as

$$SSF_{TW} = 0.655$$

Solar savings fractions for selected sunspace designations are given in Table 9-1 for Albany based on LCRs for six sunspace types. The solar savings fraction for the attached sunspace designation A6 is found (by interpolation) from Table 9-1 as

$$SSF_{AS} = 0.636$$

The relative proportions of the passive systems are calculated as

$$P_{DG} = \frac{200 \text{ ft}^2}{920 \text{ ft}^2} = 0.22$$

$$P_{TW} = \frac{450 \text{ ft}^2}{920 \text{ ft}^2} = 0.49$$

$$P_{AS} = \frac{270 \text{ ft}^2}{920 \text{ ft}^2} = 0.29$$

Table 9-1 Solar Savings Fraction (SSF) versus Load Collector Ratio (LCR) for Attached Sunspaces

Albany, New York Solar Savings Fraction	0.1	0.2	0.3	0.4	0.5	0.6	0.7	0.8	0.9
Sunspace Designation (from Table 8-9)					Load Collector Ratios				
A2 attached	177	69	37	23	15	10	7	4	3
A6 attached	171	66	35	21	14	9	6	4	2
B2 attached	140	57	31	20	13	9	6	4	2
B6 attached	128	52	29	18	12	8	5	3	2
D2 semienclosed	155	77	46	30	20	14	10	7	4
D4 semienclosed	159	75	43	28	19	13	9	6	4

Degree-Days 6888

Source: R. W. Jones, ed., *Passive Solar Design Handbook*, vol. III, Washington: U.S. Department of Energy, July 1982, p. 440. Adapted with permission.

The weighted values of the solar savings fractions are then calculated as

$$SSF_{WDG} = 0.22 \times 0.578$$
$$= 0.13$$

$$SSF_{WTW} = 0.49 \times 0.655$$
$$= 0.32$$

$$SSF_{WAS} = 0.29 \times 0.636$$
$$= 0.18$$

The value of the solar savings fraction for the mixed systems is found as

$$SSF_{MIX} = 0.13 + 0.32 + 0.18$$
$$= 0.63$$

The Solar Savings is calculated from Eq. (6-9) as

$$SS = NLC \times DD \times SSF$$
$$= 7300 \text{ Btu/DD} \times 6888 \text{ DD/year} \times 0.63$$
$$= 31.7 \times 10^6 \text{ Btu/year} \qquad \text{Ans.}$$

The Auxiliary Heat is calculated from Eq. (6-8) as

$$AH = NLC \times DD \times (1 - SSF)$$
$$= 7300 \text{ Btu/DD} \times 6888 \text{ DD/year} \times (1 - 0.63)$$
$$= 18.6 \times 10^6 \text{ Btu/year} \qquad \text{Ans.}$$

ROCK BEDS AND THEIR APPLICATION TO HYBRID SYSTEMS

In a pure passive solar system, the thermal storage absorbs solar energy and releases it to the building spaces that are heated by the passive system. Generally, this requires the thermal storage to be present in the building along the south side of the structure. However, in some cases it may be desirable to have the thermal storage located in another part of the building for rooms that do not have direct access to the south side. The hybrid system provides an opportunity to achieve such a result.

9-4 Fan-forced Rock Beds

A commonly used hybrid system utilizes a fan-forced rock bed as a heat-storing system, coupled with a source of hot air produced by a passive solar element such as an attached greenhouse or sunroom. Generally, the rock bed is located at a lower elevation than the source of warm air, and natural convection cannot be used to transfer heat to the rock storage. Under these circumstances, a fan (an active system element) is employed to circulate the air, and the term *hybrid system* is used. Rock beds located under the floor are generally used as the thermal storage, although other configurations such as a thermal storage wall in the building interior can be used (see Fig. 6-11).

Hybrid systems are very effective in situations where the air temperature in the direct-gain space has risen above the comfort level. The overheated air may result from a variety of situations, such as insufficient thermal storage volume, thermal storage not properly distributed in the sunspace, or an exceptionally clear day resulting in increased insolation. The hybrid system allows overheated air to be removed with the following results:

1. Air temperature in the solar collection space is reduced, resulting in improved thermal comfort.
2. Excess heat that otherwise would be vented is stored and is available for later use in the building.
3. Heat is redistributed from the upper southern part of the building, where hot air will typically accumulate, to colder parts of the building such as the lower northern spaces. This redistribution tends to correct temperature imbalances in the building.

In most passive solar heating systems, the heated air temperature will typically be in the range of 85° to 95°F. However, the air temperature between the glazing and thermal storage of a Trombe thermal storage wall may be in the range of 100° to 130°F. In any case, the air from passive solar systems is at a much lower temperature than air from (active) air-heating solar collectors, whose temperature is ordinarily 150° to 180°F. Therefore, the design and operation of a rock bed that serves a hybrid system will be considerably different from that of a rock bed for an active air-heating system that features solar collectors. (See Chapter 15.)

There are two methods of removing heat from a rock bed. First, the stored heat can be removed as warm air and circulated to the building spaces, as shown in Fig. 9-5. Such a system requires careful design because of the low air temperatures from the rock bed, typically 75° to 85°F. In order to transfer large quantities of heat at these relatively low air temperatures, large air-flow rates are required, resulting in high air velocities at the discharge registers. Unless special

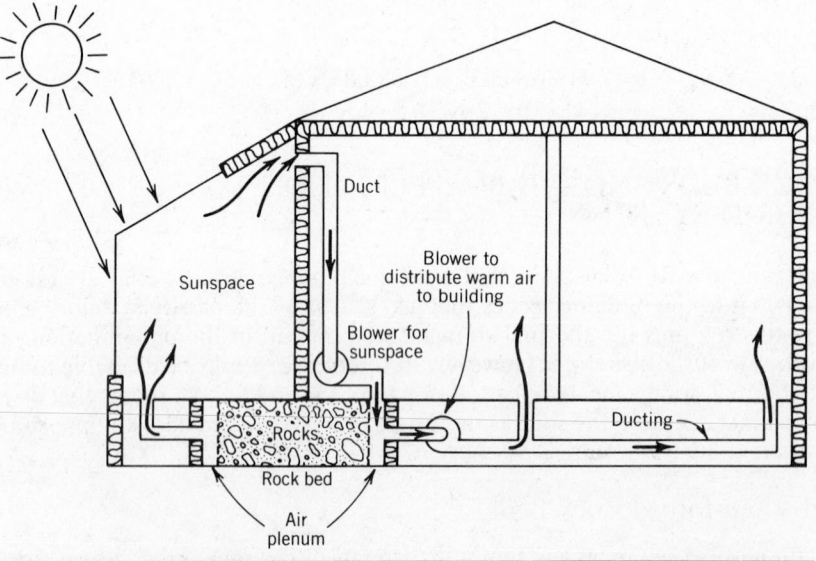

Figure 9-5 One method for removing heat from a rock bed is to use a fan and duct system to circulate warm air from the rock bed to the interior rooms.

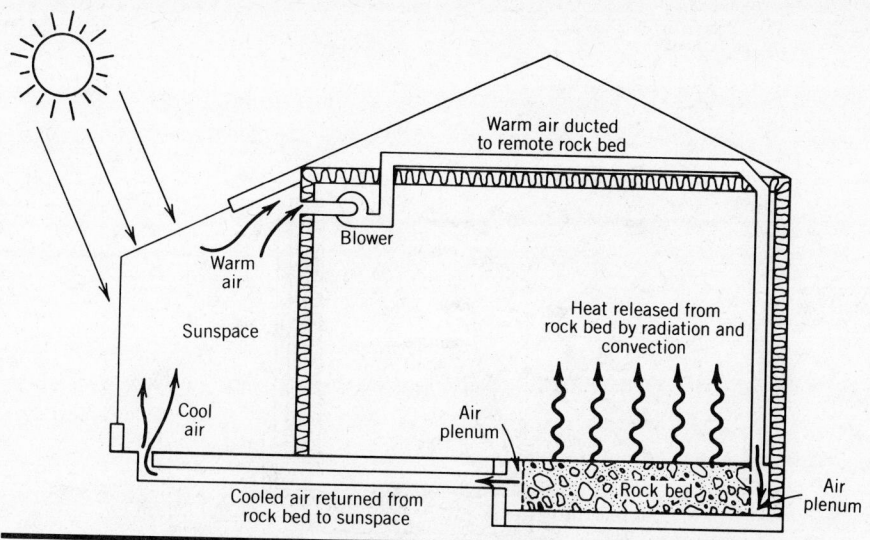

Figure 9-6 Heat can be released from a rock bed through a concrete slab into a room by radiation and convection from the slab, which is heated by the warm rocks beneath the slab.

attention is given to the size and location of the registers, drafts and an unsatisfactory noise level may result.

The second approach is to allow the heat to be released into the space by radiation and convection from the surface of the rock bed container. This method requies that the rock bed container be an integral part of the interior building surfaces. The rock bed can be placed beneath the building floor, as shown in Fig. 9-6, or it may be located behind walls or take other forms depending on the architectural design. The temperatures of these radiating floors or walls will be 5 to 10F° above the typical surface temperatures of a nonradiant heated room. Winter comfort in the room is enhanced because heating is provided from a large radiant surface. Since the mean radiant temperature of the space is increased, the interior air temperature can be lower than it is with a conventional (forced-air) heating system.

9-5   Space Configurations Used with Hybrid Rock Beds

There are two recommended configurations for hybrid rock bed systems: the *single-zone design* and the *dual-zone design*. The single-zone design (Fig. 9-7) has the rock bed under the space that collects the solar radiation. During the day, warm air is drawn from the high point of the solar collection space and circulated through the rock bed to store the heat. At night, heat is released from the rock bed into the space to maintain the space temperature. A single-zone system can be used in a greenhouse to store heat from the afternoon solar heat gains and release this heat at night. The fan can also be operated at night to force warm air from the rock bed back into the space. One disadvantage of the single-zone design is that large air volumes have to be circulated, and special attention must be given to the size and location of the air grilles.

The dual-zone approach allows the building to be divided into two parts, similar to the attached sunspaces as discussed in section 8-14. The difference is that part of the thermal storage mass has been removed from the sunspace and placed in other parts of the building in the form of a rock bed, as shown in Fig. 9-8. Zone 1 becomes a direct-gain space, and large daily temperature swings can be expected to occur because of the large solar collection area in this space. Zone 2 is buffered from the temperature swings in zone 1 by the thermal separation between the two zones. The thermal separation may be a thermal storage wall, an insulated wall, or a combination of thermal storage and insula-

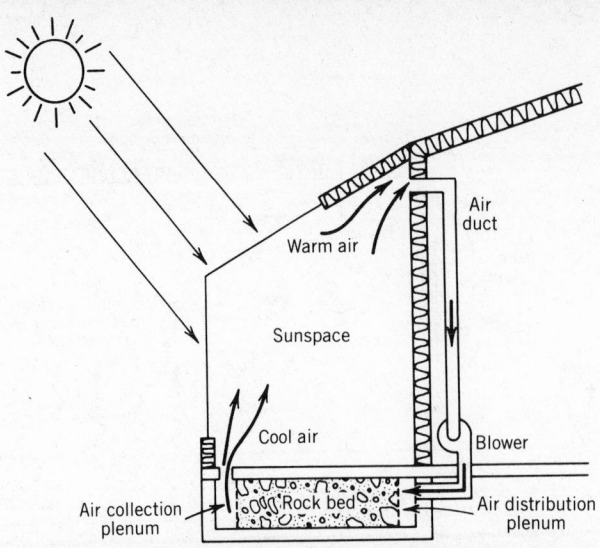

Figure 9-7 A single-zone hybrid rock bed design is one in which the rock bed is located within the sunspace (usually under a slab floor). The warm air is drawn from the sunspace and used to heat the rock bed, and the cool air, which has released its heat to the rock bed, is returned to the sunspace to be heated again. Heat can be released from the rock bed into the sunspace when the temperature in the sunspace falls below the rock bed temperature.

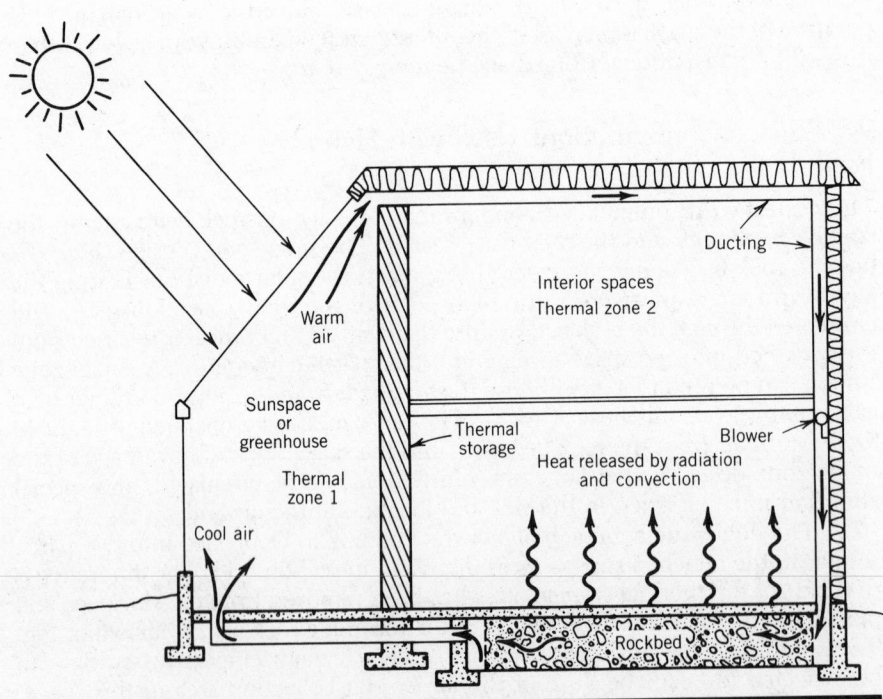

Figure 9-8 A two-zone hybrid system has the rock bed placed outside the sunspace. The typical location is under the floor slab of northern rooms. A fan is used to circulate warm air from the sunspace through the rock bed.

tion. The temperature swings in zone 1 may be in the range of 25° to 35F°, and these swings can be acceptable when zone 1 is used as a greenhouse, sunroom, atrium, conservatory, transit or circulation area, or a vestibule or airlock entry.

The fan-forced rockbed allows warm air to be drawn from the sunspace and stored in the stones of the rockbed. This heat removal reduces the overall daily temperature swings in zone 1.

A hybrid rock bed system works well with the two-zone approach because the temperatures in zone 1 are appreciably higher than the desired space temperature in zone 2. This allows heat to be moved from zone 1 to a rock bed in zone 2 with the temperature of the rock bed above the desired space temperature of zone 2. Heat can then be released from the rock bed either by natural means (radiation and convection) or by mechanical means (forced-air circulation).

9-6   Rock Bed Design Procedures

Rock bed size is dependent upon the desired amount of heat to be stored in the rock bed. As a general rule of thumb, not more than about one third of the net heat gain of a sunspace during daylight hours should be transferred from the sunspace to a rock bed. The net (daylight) heat gain is the solar radiation the sunspace absorbs during the day minus the daytime heat losses. Retaining most of the collected solar heat in thermal storage within the sunspace itself reduces dependence on external power and mechanical equipment. Transferring more than one third of the net heat from the sunspace would also require large air-flow rates, and this could result in pronounced space–temperature fluctuations (swings) since there would be so little stored heat in the sunspace. It is possible to design hybrid systems that will transfer 50 percent to 70 percent of the net heat to remote rock beds, but this makes the system design heavily dependent upon exterior energy sources and on the air circulation system.

Some Characteristics of Rock Beds.   To design rock beds, a thorough understanding of their performance is necessary. Figure 9-9 depicts air flow into and through a rock bed. The surface of the rock bed into which air flows is termed the *front face*. The *back face* is the surface from which air leaves the rock bed. The front face may be either a vertical or a horizontal surface, and air flow through the rock bed may be either horizontal or vertical. *Rock bed length* is the dimension of the rock bed measured in the direction of the air flow. The *face velocity* is the velocity of the air just *before entering* the front face of the rock bed, not the air velocity between and among the rocks, which will vary from place to place because of restricted air passages between the rocks. The volume of the rock bed is given by:

$$\text{Vol}_{\text{RB}} = A_{\text{face}} \times L_{\text{bed}} \tag{9-1}$$

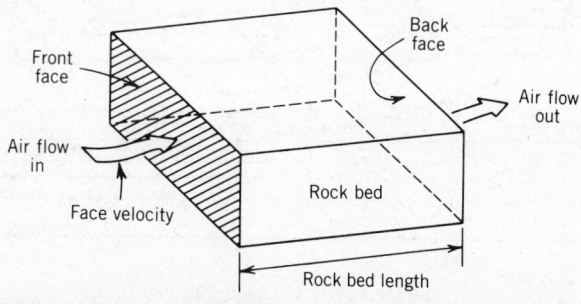

Figure 9-9 A rock bed can be characterized by the *front face*, which is the side into which the air flows, and the rock bed *length* in the direction of the air flow through the rock bed. The *back face* is the side from which air leaves the rock bed.

Where

$$Vol_{RB} = \text{the rock bed volume in cubic feet}$$
$$A_{face} = \text{the face area of the rock bed in square feet}$$
$$L_{bed} = \text{the length of the rock bed in feet}$$

The face velocity is given by:

$$V_{face} = \frac{Q}{A_{face}} \qquad (9\text{-}2)$$

Where

$$Q = \text{air flow, CFM}$$
$$V_{face} = \text{face velocity, ft/min}$$

The volumetric heat capacity C_V of a rock bed is given by:

$$C_V = D \times c \times (1 - \text{Void}) \qquad (9\text{-}3)$$

Where

C_V = the volumetric heat capacity, Btu/ft³-F°
D = the density of the rock, lb/ft³
c = the specific heat capacity of the rock, Btu/lb-F°
Void = the fraction of the rock bed that is void space between the rocks. The void comprises the air flow passageways in the rock bed.

The pressure drop, ΔP, and other heat transfer characteristics of a rock bed are shown in Fig. 9-10. Heat transfer in rock beds is very good because of the large surface area provided by the rocks as the air passes through the rock bed.

    The variable parameters on the chart of Fig. 9-10 that are dependent on the rock bed design are the rock diameter and the face air velocity. Based on these two parameters, it is possible to determine the pressure drop of the rock bed per unit of rock bed length (inches of water gauge per foot of rock bed). The length of the rock bed required to transfer 95 percent of the heat from the air is

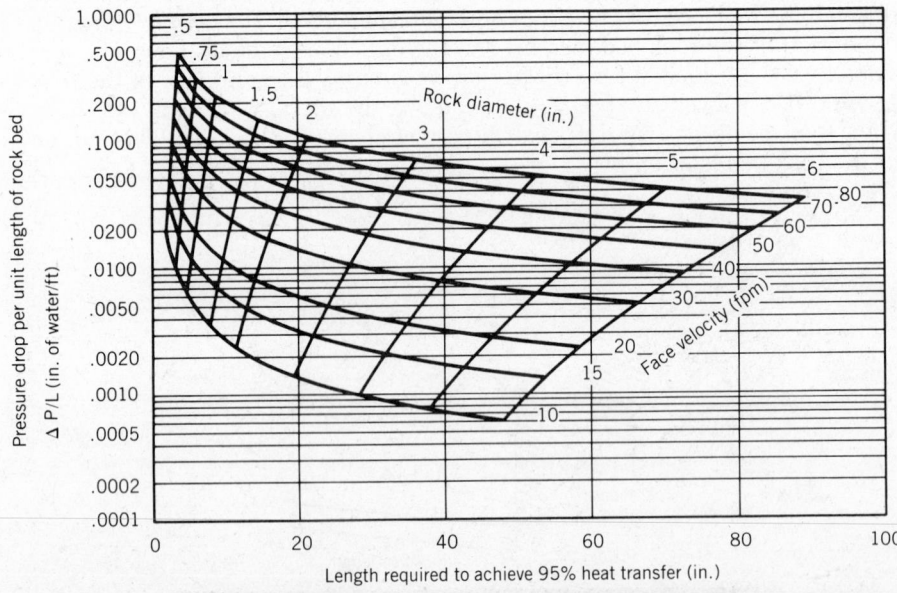

Figure 9-10 Pressure drop per unit length for air flow through a rock bed for different rock diameters and face air velocities. The length of rock bed to achieve a 95 percent transfer of heat from the air to the rocks is also indicated. *Source:* J. D. Balcomb, et al., *Passive Solar Design Handbook,* vol. II, Washington: U.S. Department of Energy, January 1980, p. 198.

ADVANCED PASSIVE METHODS—SELECTED APPLICATIONS

Table 9-2 Properties of Air at Various Altitudes

Altitude above Sea Level (ft)	Density (lb/ft³)	Heat Capacity (Btu/ft³-F°)	Heat-Transfer Capacity (Btu/hr-F°-CFM)	Air Density Ratio (ADR)
0	0.0750	0.0180	1.08	1.00
1,000	0.0724	0.0174	1.04	0.96
2,000	0.0698	0.0167	1.01	0.93
3,000	0.0672	0.0161	0.97	0.90
4,000	0.0648	0.0155	0.93	0.86
5,000	0.0625	0.0150	0.90	0.83
6,000	0.0601	0.0144	0.86	0.80
8,000	0.0559	0.0134	0.81	0.74
10,000	0.0516	0.0124	0.74	0.69

Source: J. D. Balcomb, *Passive Solar Design Handbook*, vol. II (Washington: U.S. Department of Energy, January 1980), p. 200. Used with permission.

also given on the chart. Generally, during sunlit hours, air from the collectors will be about 20F° hotter than the rock bed.

A rock bed length of 20 in. is typically sufficient to provide 95 percent of the transfer of heat from the air to the rock for rock diameters of 2 in. or less. The pressure drop of the rock bed is always a matter of concern in rock bed design since it will affect the size and power of the fan that circulates the air.

When designing a rock bed, the properties of air must be taken into account. Table 9-2 gives some important properties of air at different altitudes. The density of air is given and so is the heat capacity in Btu/ft³-F°. The heat-transfer capacity is given in Btu/hr-F°-CFM. The air density ratio, ADR, is also listed, and it is referenced to sea level where ADR = 1.00.

The pressure drop in a rock bed is proportional to the air density. Thus, for altitudes other than sea level, the pressure drop should be corrected as follows:

$$\Delta P' = \Delta P_{\text{sea level}} \times \text{ADR} \qquad (9\text{-}4)$$

The design procedure for a rock bed consists of finding a good match between the pressure drop characteristics of the rock bed and its associated ducting, and the operating characteristics of the fan or blower. Before designing a rock bed, it is necessary to determine two parameters based on the building design: the $\Delta T_{\text{working air}}$ and the $\Delta T_{\text{working bed}}$. The $\Delta T_{\text{working air}}$ is the air temperature difference between the air entering the rock bed and the air leaving the rock bed. The $\Delta T_{\text{working bed}}$ is the temperature variation of the rocks as they store and release heat. As a rule of thumb, the $\Delta T_{\text{working bed}}$ can be taken as approximately one half of the $\Delta T_{\text{working air}}$.

For a sunspace–rock-bed hybrid system, the $\Delta T_{\text{working air}}$ is determined by the air temperature in the sunspace and the average temperature at which the rock bed will tend to stabilize. In most attached sunspace residential situations, the sunspace temperature during midday hours will range from 80° to 90°F. The rock bed temperature will generally be in the range of 65° to 80°F. The temperature of the air exiting from the rock bed will be slightly lower than the average rock bed temperature by about 3 to 5°F. Ordinarily, the $\Delta T_{\text{working air}}$ will have a general range of from 15 to 20F°. The typical $\Delta T_{\text{working bed}}$ will have a general range of 7 to 10F°.

The seven major steps in the design of a rock bed are outlined below:

Step 1. The net energy collected by the sunspace must be estimated. This can be done by using Eq. (7-2) to calculate the instantaneous heat gain (H) of the sunspace. This calculation would be done for each of the daylight hours. The net daily energy collected would then be the sum of the hourly values of H, or

$$Q_n = H_8 + H_9 \ldots H_5 \qquad (9\text{-}5)$$

where

Q_n = the net daily energy collected

H = the hourly values of the instantaneous heat gain for the sunny hours of the day (8:00 A.M. to 5:00 P.M.).

Step 2. The rock bed size and air-flow requirements must be established for the overall design. Based on the net daily energy collected by the sunspace in step 1, it must be determined what portion of the net daily energy collected will be stored in the rock bed:

$$Q_R = Q_n \times f \qquad (9\text{-}6)$$

where

Q_R = the heat to be removed for storage in the rock bed, Btu/day

Q_n = the net daily energy collected, Btu/day

f = the fraction or portion of net daily energy to be stored in the rock bed.

After Q_R has been determined, the heat transport rate Q_t must be estimated. This is the average hourly rate at which heat is moved to and stored in the rock bed. This estimate is obtained by dividing Q_R by the midday hours during which the majority of the solar radiation is collected by the sunspace:

$$Q_t = Q_R/T \qquad (9\text{-}7)$$

where

Q_t = the heat transport rate to the rock bed, Btu/hr

Q_R = the heat to be stored in the rock bed, Btu/day

T = the number of hours during which heat is transported to the rock bed, generally taken as the number of hours during which the instantaneous heat gain (H) is significant (within 20 percent of the midday vales of H)

The air-flow rate to the rock bed is determined as:

$$\text{Air flow (CFM)} = \frac{Q_t}{\Delta T_{\text{working air}} \times c_{\text{air}} \times 60 \text{ min/hr}} \qquad (9\text{-}8)$$

Where

Q_t = the heat transport rate in Btu/hr

c_{air} = the heat capacity of air in Btu/ft³-F°, as listed in Table 9-2

The required rock bed volume is determined by:

$$Vol_{\text{RB}} = \frac{n \times Q_R}{C_V \times \Delta T_{\text{working bed}}} \qquad (9\text{-}9)$$

Where

n = a multiplier for the storage capacity of the rock bed

The multiplier n provides for a storage capacity greater than the daily energy to be stored in the rock bed. Typically, the value n is in the range of 1.5 to 2.5, but its actual value will depend on the local climate and how much extra storage capacity is desired for cloudy or overcast days.

Step 3. The dimensions of the rock bed must now be determined. The length of the rock bed, L_{bed}, is more or less fixed by the dimensions of the rooms under which it will be placed and by the pressure drop of the rock bed, ΔP, as will be discussed later. Once the rock bed length has been determined, the face area is calculated using Eq. (9-1).

Step 4. The face velocity of air into the rock bed is calculated using Eq. (9-2).

Step 5. Rock size selection is based on materials available in the local area and on the pressure drop desired for the rock bed. Once the size of rock to be used in the rock bed has been decided upon, the pressure drop per unit length of rock bed, $\Delta P/L$, is found from Fig. 9-10, using the rock diameter and face velocity, V_{face}. The total pressure drop for the rock bed is then found by multiplying by the length of the rock bed:

$$\Delta P_{RB} = \Delta P/L \times L_{bed} \qquad (9\text{-}10)$$

Where

ΔP_{RB} = the pressure drop for the entire rock bed, in. wg

$\Delta P/L$ = the pressure drop per unit length of rock bed, in. wg, from Fig. 9-10

L_{bed} = the length of the rock bed, ft

The selected rock bed length should be checked against the length required to provide a heat transfer of 95 percent as given in Fig. 9-10. If the selected rock bed length is shorter, then one should return to step 3 and select a longer rock bed length.

Step 6. The pressure drop caused by air friction in the ducts to and from the rock bed must be estimated and then added to the pressure loss of the rock

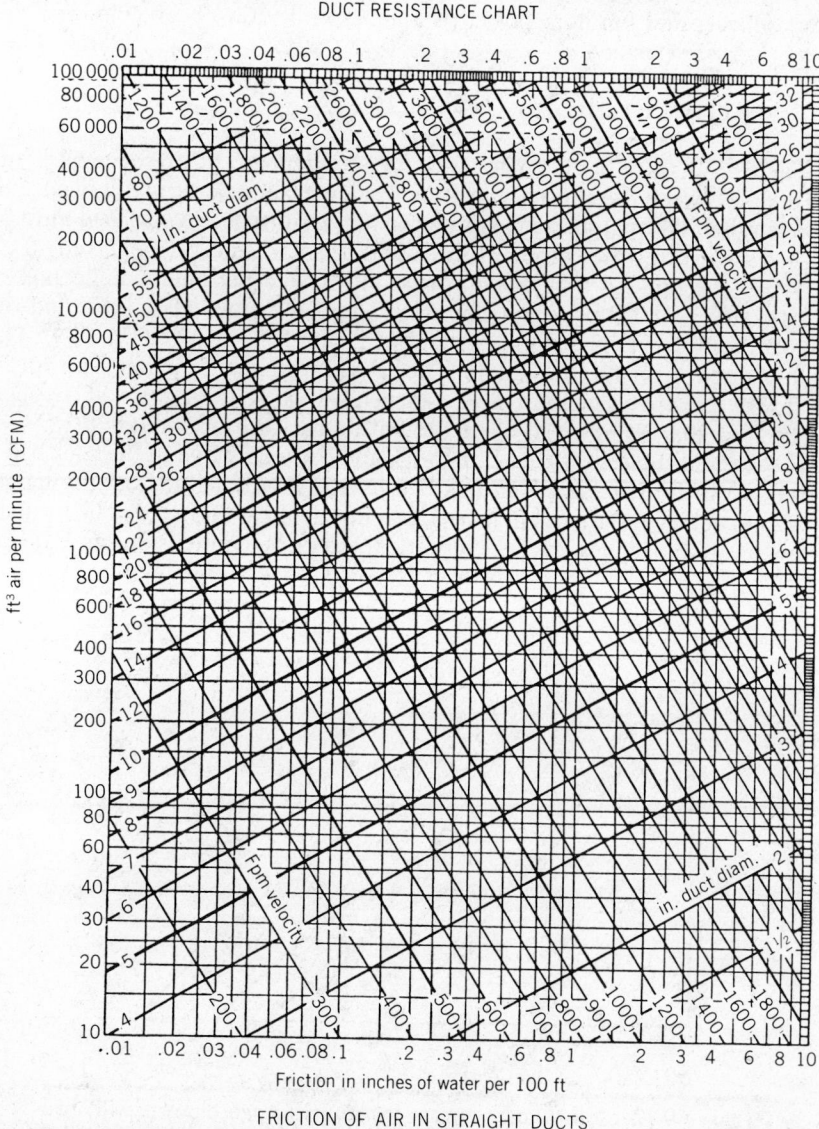

DUCT RESISTANCE CHART

FRICTION OF AIR IN STRAIGHT DUCTS

Figure 9-11 Duct resistance chart for diffferent duct diameters and air-flow rates.

ROCK BEDS AND THEIR APPLICATION TO HYBRID SYSTEMS

bed to give the total pressure loss for the rock bed system. The total equivalent length of the ducting is found approximately as:

$$\text{Equivalent length} = \text{Actual length} + \text{Number of elbows} \times 15 \text{ ft} \quad (9\text{-}11)$$

Pressure losses caused by friction in ducts (ΔP_D) are shown in Fig. 9-11 for different duct diameters and air-flow rates. Values are given in inches of water gauge per 100 ft of duct. The pressure drop for the ducting can be calculated from:

$$\Delta P_D = \frac{\text{Equivalent length (ft)}}{100 \text{ ft}} \times \Delta P_{D100} \quad (9\text{-}12)$$

The total pressure drop for the entire rock bed and ducting system will then be:

$$\Delta P_T = \Delta P_{RB} + \Delta P_D \quad (9\text{-}13)$$

Step 7.   The fan size and horsepower must be chosen so that the fan will supply the desired air flow at the static pressure given by ΔP_T. Figure 9-12 gives the air delivery and fan horsepower of some typical squirrel-cage blowers.

Illustrative Problem 9-3

An attached sunspace has been designed for a residence. The presence of sliding glass doors between the sunspace and the residence allows only about two thirds of the required thermal storage to be placed in the sunspace. A rock bed must be designed to store the other third of the solar energy collected (i.e., $f = \frac{1}{3}$). It is estimated that the sunspace will have a net energy collection of approximately 300,000 Btu/day. The $\Delta T_{\text{working air}}$ is estimated at 18F°, and the multiplier for the rock bed storage capacity (n) should be 1.5. See Eq. (9-9). The length of ducting to and from the rock bed is 80 ft with eight elbows. The duct is round with a diameter of 16 in. The rock bed will be 12 ft long. Local suppliers provide rock that is 1½ in. in diameter, with a rock density of 165 lb/ft³. The heat capacity of the rock is 0.21 Btu/lb-F°, and the void space in the rock bed is assumed to be 40 percent. The rock bed length should be at least the length required to achieve a 95 percent heat transfer of the heat in the air. The location has an elevation near sea level. Determine the rock bed dimensions and the fan size.

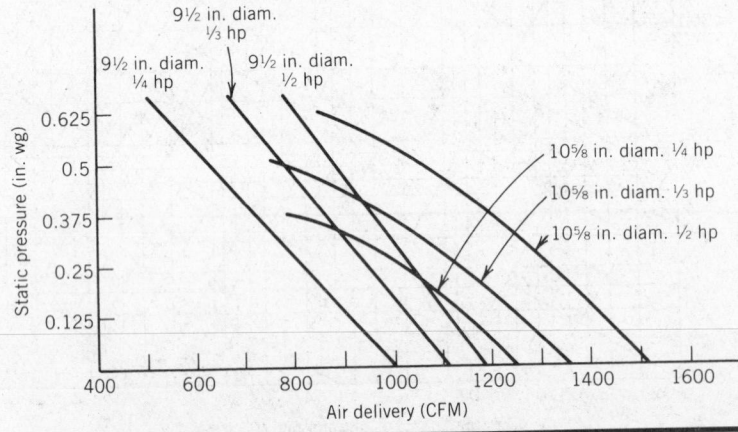

Figure 9-12 Performance curves for squirrel-cage blowers with two different diameters and three different motor sizes. Air delivery (CFM) is plotted against static pressure (in. wg). *Source:* Plotted from manufacturers' fan performance data.

Solution

The average daily energy stored in the rock bed is calculated using Eq. (9-6):

$$Q_R = Q_n \times f$$
$$= 300{,}000 \text{ Btu/day} \times 1/3$$
$$= 100{,}000 \text{ Btu/day}$$

The heat transport rate for the rock bed is found from Eq. (9-7) (T will be taken as 5 hr daily):

$$Q_t = \frac{Q_R}{T}$$
$$= \frac{100{,}000 \text{ Btu/day}}{5 \text{ hr/day}}$$
$$= 20{,}000 \text{ Btu/hr}$$

The air flow is calculated using Eq. (9-8):

$$\text{Air flow} = \frac{Q_t}{\Delta T_{\text{working air}} \times c_{\text{air}} \times 60 \text{ min/hr}}$$
$$= \frac{20{,}000 \text{ Btu/hr}}{18\text{F}° \times 0.018 \text{ Btu/ft}^3\text{-F}°}$$
$$= 1028 \text{ ft}^3/\text{min}$$

The rock bed volume is determined by Eq. (9-9):

$$Vol_{\text{RB}} = \frac{n \times Q_R}{C_V \times \Delta T_{\text{working bed}}}$$

Here, n will be taken as 1.5, and $\Delta T_{\text{working bed}}$ will be assumed to be half of $\Delta T_{\text{working air}}$. The volumetric heat capacity C_V for the rock bed is calculated from Eq. (9-3):

$$C_V = D \times c \times (1 - \text{void})$$
$$= 165 \text{ lb/ft}^3 \times 0.21 \text{ Btu/lb-F}° \times (1 - .40)$$
$$= 20.8 \text{ Btu/ft}^3\text{-F}°$$

The required rock bed volume is, from Eq. (9-9),

$$Vol_{\text{RB}} = \frac{1.5 \times 100{,}000 \text{ Btu/day}}{20.8 \text{ Btu/ft}^3\text{-F}° \times 9\text{F}°}$$
$$= 801 \text{ ft}^3$$

The face area of the rock bed is calculated from Eq. (9-1) as:

$$A_{\text{face}} = \frac{Vol_{\text{RB}}}{L_{\text{bed}}}$$
$$= \frac{801 \text{ ft}^3}{12 \text{ ft}}$$
$$= 66.8 \text{ ft}^2$$

Rock bed dimensions then (given the 12-ft length) could be 12 by 10 by 7 ft depth.   *Ans.*

The face air velocity is calculated using Eq. (9-2):

$$V_{\text{face}} = \frac{801 \text{ ft}^3 \text{ min}}{66.8 \text{ ft}^2}$$
$$= 12 \text{ ft/min}$$

From Fig. 9-10 for a rock diameter of 1.5 in. and a face velocity of 12 ft/min, the pressure drop for the rock bed is 0.006 in. wg/ft length. The total pressure drop for the bed is found by Eq. (9-10):

$$\Delta P_{RB} = \Delta P/L \times L_{bed}$$
$$= 0.006 \text{ in. wg/ft} \times 12 \text{ ft}$$
$$= 0.072 \text{ in. wg}$$

The equivalent length of the ducting is calculated using Eq. (9-11):

$$\text{Equivalent length} = \text{Actual length} + \text{Number of elbows} \times 15 \text{ ft}$$
$$= 80 \text{ ft} + (8 \times 15) \text{ ft}$$
$$= 200 \text{ ft}$$

The pressure loss for the ducts is found from Eq. (9-12):

$$\Delta P_D = \frac{\text{Eq. length}}{100 \text{ ft}} \times \Delta P_{D100}$$

ΔP_{D100} is found from Fig. 9-11 as 0.05 in. wg/100 ft (1028 CFM in a 16-in. round duct)

$$\Delta P_D = 200 \text{ ft} \times 0.05 \text{ in. wg/100 ft}$$
$$= 0.10 \text{ in. wg}$$

The total pressure loss for the system, from Eq. (9-12), is

$$\Delta P_T = \Delta P_{RB} + \Delta P_D$$
$$= 0.07 \text{ in. wg} + 0.10 \text{ in. wg}$$
$$= 0.17 \text{ in. wg}$$

The fan size is found from Fig. 9-12. For an air-flow rate of 1028 cfm and a pressure loss of 0.17 in. wg, a 9½ in. fan with a ⅓ hp motor could almost meet the requirement, but the best choice would be the 9½ in. fan with a ½ hp motor.

Ans.

9-7 Rock Bed Options

The design of a rock bed can be very flexible, with the configuration amenable to change to reduce the pressure drop as necessary. For example, consider the rock bed in Fig. 9-13, with air flow entering at left. If the pressure drop with this configuration were too great, the air flow could be put into a side with larger area, as shown. Another alternative would be to split the rock bed up into several segments with air plenums. This would increase the face area and reduce the bed length, thus reducing the pressure loss. There are several methods of constructing air plenums for rock beds. The most common is to use the foundation walls of the building as one side of the plenum and construct the front and rear faces of the rock bed using wire or steel mesh, as shown in Fig. 9-14*a*. Concrete blocks can be used to form flow channels to distribute warm air

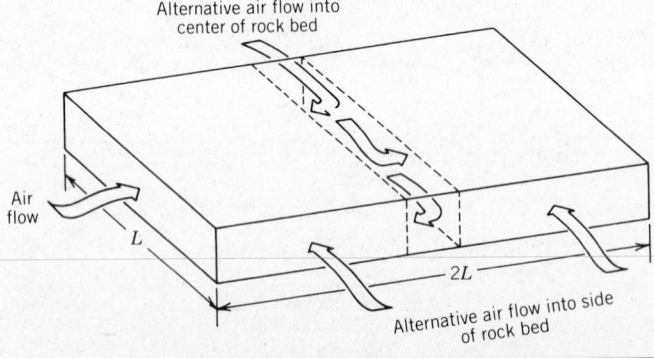

Figure 9-13 Alternative methods of circulating air through a rock bed. When the pressure drop is too large because of bed length, the rock bed can be divided into two parts, or the longer side can be used as the face area.

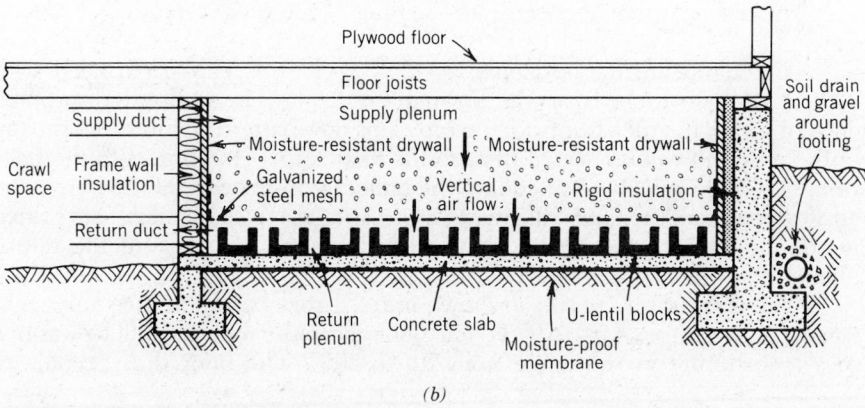

Figure 9-14 Details of rock bed construction. (a) A rock bed under construction. The concrete walls for containing the rocks are shown. Air plenums are constructed by using steel angle and mesh to hold the rocks back and form a space at each end of the rock bed. (b) An air plenum can be constructed using U-lentil concrete blocks to support a galvanized steel mesh at the bottom of a rock bed. A plenum is also present at the top of the rock bed between the floor and the rocks. The air flow is downward in this case as the rocks are heated. The upper plenum is the supply plenum, and the lower plenum is the return plenum.

into a rock bed, as shown in Fig. 9-14b. Perforated pipe may be used as a distribution system for a rock bed. When using perforated pipe, the distance between the supply pipe and the return pipe should be at least the length required to achieve 95 percent heat transfer, as was indicated by Fig. 9-10.

When rock beds are used under concrete slabs, the depth of the rocks should not be more than 2 to 3 ft. This depth allows the heat to migrate up through the rocks and warm the concrete slab above it, which then radiates the heat into the interior space. If the depth of the rock bed is greater than 2 to 3 ft, the heat stored in the lower portion of the rock bed will not reach the floor during the night and thus will not be effective in providing heat to the interior space.

When a fan is used to circulate air back through the rock bed to remove heat, there is no such limitation on the depth of the rock bed, since the heat is removed by the air flow.

9-8 Construction Details for Rock Beds

Rock beds should be carefully constructed if they are to operate satisfactorily for the life of the building. Figure 9-15 shows a cross-section of a typical rock bed construction, which features horizontal air flow through the rock bed. A soil drainage system is recommended around the rock bed to remove ground water. Rigid insulation and a moisture-proof membrane should be run around the exterior walls of the rock bed to discourage soil moisture from seeping through the concrete walls. A moisture-proof membrane should also be placed under the rock bed. It is recommended that a thin concrete slab be used in the bottom of the rock bed to enclose it. Rock beds can be constructed with just a moisture-proof membrane on the ground and the rocks placed directly on top of the membrane, but care must be taken not to damage the membrane. Galvanized steel mesh should be used at each end of the bed to contain the rocks and form a supply-and-return air plenum. A baffle is recommended down the center of the rock bed parallel to the air plenums. This baffle can be a rigid piece of insulation that extends down into the rocks several inches and up into the slab ½ to 1 in. Its purpose is to prevent air from flowing between the slab and the rocks if the rocks settle away from the bottom of the radiant slab. A plastic film must be placed over the filled rock bed before the concrete slab is poured over the rock bed. This prevents concrete from leaking into the bed. The slab over the rock bed should be designed to be self-supporting between footings. This will ensure that the slab does not crack if there is any settling of the rock in the rock bed with time.

Air channeling in horizontal air-flow rock beds can be avoided with vertical air-flow rock beds, as was shown in Fig. 9-14b. The air flow is distributed down through the rock bed from the top. This flow pattern ensures that the top of the rock bed will be heated first, so the heat is first stored next to the floor where it will be quickly released into the building. In the case of the horizontal air-flow rock bed, the floor will gradually be warmed as rocks beneath are heated in the direction of the air flow from the supply plenum toward the return plenum (see Fig. 9-15).

The temperature of the air drawn into the rock bed from the sunspace is usually in the range of 75° to 95°F. The floor above the rock bed will be warmed to a temperature in the range from 70° to 80°F. The floor thus becomes a

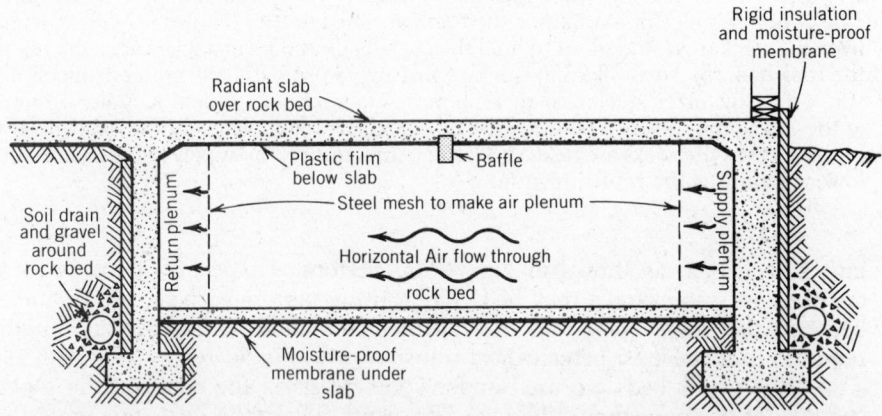

Figure 9-15 The construction details of a rock bed are important if the rock bed is to function correctly. A moisture-proof membrane should extend around the rock bed to prevent moisture from entering. A soil drain should be placed around the perimeter to ensure that ground water cannot flood the rock bed. The exterior walls should be insulated to prevent heat loss to the surrounding ground. A baffle should be installed in the middle of the rock bed to prevent air channeling between the floor slab and the rocks. The supply and return plenums should be constructed of galvanized steel mesh and angle, of thickness strong enough to hold back the weight of the rock.

radiating surface that releases heat to the interior space by infrared radiation and convection.

The warm air from a sunspace usually has a relative humidity below 50 percent. This relatively dry air circulated through the rock bed will generally produce no problems with algae or mold that could produce musty odors. If, however, the sunspace is used as a greenhouse, heavily watered and crowded with plants, the humidity level could be quite high, and the use of a rock bed under these conditions might well be accompanied by dank and musty conditions. Other types of storage than rock beds are better for use with humid air.

9-9 Other Storage Configurations for Hybrid Systems

Vertical walls may be used for thermal storage. Ungrouted cells in concrete block walls may be filled with rock, and air can be circulated down through these cells, as shown in Fig. 6-11. Tubes may be laid under slabs through which air is circulated, as shown in Fig. 9-16. Whenever these other thermal storage configurations are used, care must be taken to provide sufficient storage surface area so that heat may be readily absorbed into the thermal storage mass. If there is not adequate surface area, the heat will not be effectively transferred from the air, and the performance of the hybrid system will not meet the design expectations.

PASSIVE SOLAR RETROFITTING

While the incorporation of passive solar systems into new construction represents a challenging and innovative design process for professionals and technicians in the solar field and in related building professions, the successful

Figure 9-16 A concrete slab may be made into a heat storage system by placing tubes through the concrete slab to provide air passages. The plastic tubes shown will first be covered with concrete. A finished slab of concrete will then be poured on the top of the concrete-imbedded tubes. Warm air in a direct-gain space will be circulated through the tubes to heat the concrete slab mass.

addition of solar elements to *existing buildings* presents an even greater challenge. While many new buildings will be built between now and 2000, representing opportunities for solar design work, a great many buildings already in existence are prime prospects for some passive solar energy applications. Many of these buildings, both residential and commercial, were constructed when energy was relatively inexpensive, and they lack adequate insulation, effective heating and cooling systems, and efficient lighting systems. These older buildings have become or are becoming increasingly expensive to operate.

The term *retrofit* is used to describe the process of adding a solar energy system to an existing building. A methodology for approaching a retrofit situation will be developed along with some examples of passive solar retrofits.

9-10 Retrofit Design Procedures for Passive Solar Applications

The following steps are necessary in considering most retrofit jobs:

1. Determine existing energy loads for heating, cooling, and lighting.
2. Examine the building location and the surrounding environment for solar access, natural cooling potentials, and the potential for natural lighting (daylighting).
3. Develop a sequence of energy conservation measures for the building's envelope and the existing heating, cooling, and electrical systems.
4. Develop a series of options for the use of passive solar heating, natural cooling, and daylighting.

Each of the options for reducing the building's energy usage will have a cost of implementation and also a cost benefit (the energy savings resulting from the improvement). The building owner will usually compare the cost benefit to the cost of the energy-saving option and decide which options are most appropriate in his or her judgment. Frequently other factors may also enter in, such as improved environment and increased comfort within the building.

Determining Existing Energy Loads.　A good starting point for determining existing energy loads is an analysis of the utility bills for the building. A comparison of the energy used during different months may allow reasonably accurate estimates of the energy use for space heating and space cooling, as well as for domestic water heating. If it is not possible to determine energy usage from the utility bills directly, then the building loads can be calculated using the methods given in Chapter 5.

After energy conservation techniques have been considered and evaluated, passive solar retrofit options can be considered. There are almost unlimited options for passive solar retrofits, and of these the most appropriate ones for use must be determined for each situation.

Window and Skylight Retrofits.　These are probably the simplest form of retrofit. South-facing windows can easily be added to most frame walls to increase direct solar gains into rooms. Thermal storage must also be added to rooms when direct-gain windows are added, to prevent overheating. Skylights or clerestory windows can be added to allow natural light and solar gains into north-oriented rooms or rooms that have no south exposure. For a single room, one or more small, prefabricated greenhouse units may be attached to a wall or window with a southern exposure, as shown in Fig. 9-17.

Thermal Storage Walls with Glazing.　The next least complicated retrofit is probably a thermal storage wall. A large glazed area can be added to south-facing framed walls in the same manner that a window or a sliding glass door is added. Thermal storage must be provided behind the glazing, of course. Water-

Figure 9-17 One method of providing or increasing the direct solar gain for a room is to add a prefabricated greenhouse window. This type of unit can be used to replace an existing window, or it can be added as a new window area.

filled containers are often used in these retrofit situations, since these units are modular and can be set up behind the glazing with much less work than is involved in the construction of a masonry or concrete wall. In cases where a masonry or concrete wall already exists with a south-facing orientation, this wall can be used as the thermal storage. Such a wall might be a brick wall of a house or the exposed concrete basement wall of a house on a sloping lot. To convert the wall to a solar system, a glazing system is attached to the existing masonry or concrete wall in the same manner as was discussed in section 8-8 for new construction. The existing windows may be used as thermocirculation vents, or new openings can be cut through the wall for these vents. Modular thermal storage units designed to fit into wall stud spaces are also available. These units are made to appear like windows (Fig. 9-18) and will allow diffuse light into the room through the translucent water containers.

South-Oriented Porches. These provide an excellent opportunity for adding a passive solar system. A south-facing porch may be enclosed with glass or other glazing to create a winter sunspace. The glazing may be mounted in removable frames so that the porch can be opened during the summer months for air circulation. One option is to have a double set of removable frames, one with glazing for the winter and a second set with window screen for summer use. Depending on the size of the porch–sunspace retrofit, thermal storage may or may not be required. For small sunspaces with less than 150 ft^2 of south glazing, small duct fans may be used to circulate the warm air from the sunspace throughout the house. There is usually sufficient thermal capacity within the existing house to absorb the heat produced by a small sunspace without overheating occurring. The porch–sunspace conversion can also be made permanent by completely converting the porch into a sunspace, as shown in Fig. 9-19a. A utility porch at the back of the house shown in Fig. 9-19 was expanded and converted into a sunspace to collect winter sun for space heating and to use as a plant-growing area. The porch-sunspace is adjacent to the master bedroom and can supply heat to that room through a common French door (Figs. 9-19b

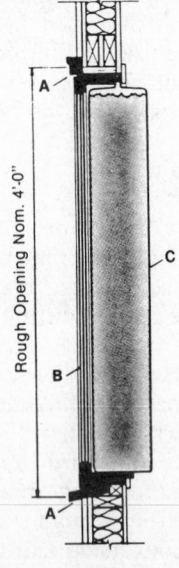

A Stud Space Module™ frame, of conventional jamb, sill and head construction.

B Triple layer of insulating glass with high solar transmittance.

C Stud Space Module™ by One Design

Rough Opening Nom. 4'-0"

(c)

Figure 9-18 Modular wall units are available which can be used to retrofit a thermal wall system. (*a*) The wood-framed module is installed in the usual manner, the same as a window. (*b*) The interior of the translucent, ribbed water container can be left exposed as shown, or it can be covered with a sheer curtain to let light pass into the room. (*c*) Cross-section of the preglazed Stud Space Module™. (All three illustrations courtesy One Design, Inc.)

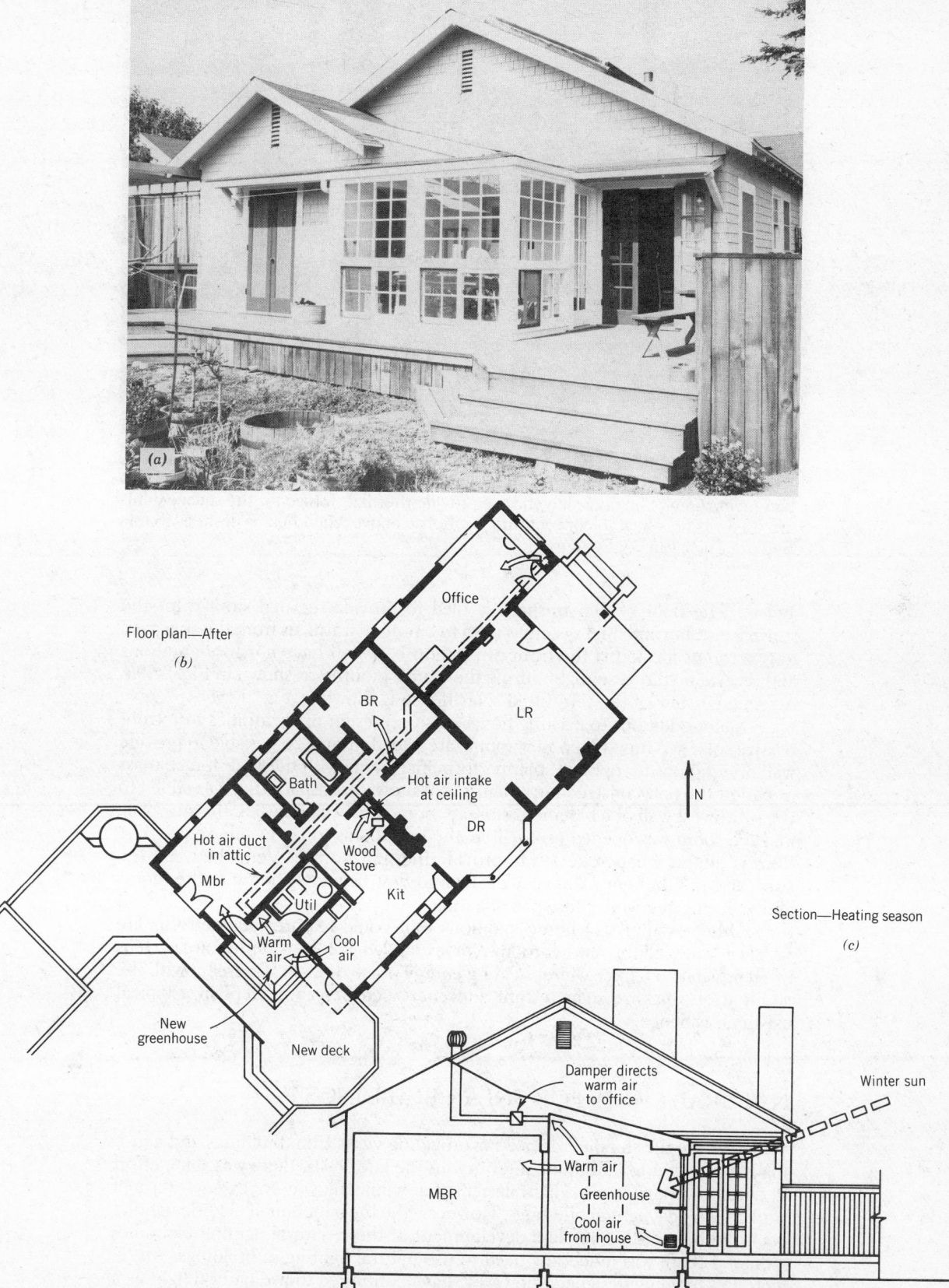

Floor plan—After
(b)

Office

BR

LR

Bath

Hot air intake
at ceiling

Hot air duct
in attic

Mbr

Wood
stove

DR

Util

Kit

Warm
air

Cool
air

New
greenhouse

New deck

N

Section—Heating season
(c)

Damper directs
warm air
to office

Winter sun

MBR

Warm air

Greenhouse

Cool air
from house

Figure 9-19 Sunspaces as passive solar retrofits. (a) A back porch was expanded
into a sunspace with southwest and southeast glazing. (b) The sunspace supplies heat
to the adjacent bedroom, and a duct system is used to circulate heat to rooms on the
north side of the building. (c) A section through the sunspace shows how sunlight
enters the space and where heat is drawn off for distribution to the north rooms. A
wind turbine provides a summer exhaust for the sunspace. (Bob Easton Design
Associates—L. Dennis Thompson, Architect.)

Figure 9-20 A 16 ft by 5 ft greenhouse addition was adddded to the south wall of a basement room. The floor slab and tile provide thermal storage for the space. Vents are used to allow heat to convect into the house above when heat is desired. (Santa Barbara Solar Energy Systems.)

and *c*). The floor of the sunspace is tiled to provide thermal storage for the sunspace. A fan and duct system is used to circulate warm air from the sunspace to two rooms located at the front of the house, which have north orientations and receive no direct sunlight during the winter months, as shown in Fig. 9-19*b*. A sectional view of the retrofitted solar house is shown in Fig. 9-19*c*.

Major additions to existing houses also represent opportunities to retrofit passive solar systems. When new rooms are added, it may be possible to provide wall orientations for optimal solar performance along with thermal storage mass as part of the new construction. Figure 9-20 shows a sunspace that was added to the southwest wall of a basement storage room. The wall between the sunspace and the room was opened up to allow the two rooms to act as one space. The floor is tile on a concrete slab to provide thermal storage. Vents between the sunspace and the rooms above allow heat to flow by convection into the rooms above during the winter for space heating.

Other examples of home additions that could feature solar retrofits are family rooms, additional bedrooms, or expanded living or dining areas. It is often possible to reduce winter heating energy usage by 25 to 50 percent with the addition of a passive solar retrofit and energy conservation steps in a typical existing residence.

NATURAL LIGHT FOR SOLAR BUILDINGS

Windows are the standard device for providing views from buildings and allowing natural light to illuminate interiors. In the late 1800s, there was some effort to develop window and skylight designs that would allow more extensive use of sunlight and daylight in buildings. However, the introduction of electric lighting occurred before the complete development of these natural lighting technologies, and there was much less need to use natural lighting in buildings. Architects and building designers, however, have continued to use natural light as a supplement to electric lighting design to enhance the quality of light in a building and the experience of working or living in the building. The architectural creation of a space means, among other things, the control of light by the use of openings, planes, textures, and colors within the building. Control of interior lighting provides a means of controlling the mood and ambience of the interior space.

9-11 Daylighting in Residences and Commercial Buildings

In most residential buildings, such as single-family residences and low-rise apartments, daytime lighting is provided almost entirely by natural lighting because there is sufficient wall area for windows to allow ample daylight to enter the building. In the case of large commercial buildings where the volume within the structure is very large in relation to the exterior surface area, the opportunity to allow natural light to enter the interior becomes more limited.

In many modern commercial buildings, the energy consumption for artificial lighting is frequently the largest single use of energy. The use of artificial lighting also produces heat, which adds to the summer cooling load of a building. As energy costs escalate, natural lighting can become not only an aesthetic element in buildings but also part of a strategy for reducing electric energy consumption and electric loads.

The use of natural lighting is a somewhat complex and evolving technology of which only the basic concepts will be covered in this discussion. It should be recognized that the use of daylight as a lighting design tool may not always result in big energy savings. For example, the use of daylight might result in an increased cooling load from solar heat gains, that could conceivably cancel out the energy savings realized from the daylighting. It is important to consider the goals and objectives of natural lighting as a combined architectural and energy strategy in order to be able to evaluate the success or failure of a design concept in fulfilling these goals.

When daylighting is used as a design criterion, there are four primary factors that should be considered:

1. *Lighting quality.*   The quality and quantity of natural light and electric light must meet the requirements for the tasks that are to be performed. Occupant productivity is an important consideration in most commercial buildings, and the lighting quality should encourage and enhance productivity.

2. *Energy savings.*   The use of electricity or other purchased energy for lighting should be minimized. The costs associated with the use of daylighting should also be minimized by optimizing the energy-related functions of windows and skylights.

3. *Load management.*   The electric lighting loads should be controlled within the building in accordance with the natural lighting levels available. This will minimize operating costs and maximize benefits for the local utility by contributing to a reduction in the daytime peak load.

4. *Design and aesthetics.*   The use of natural lighting should improve the experience of working in a space by contributing to a pleasant and healthful working environment.

It is difficult to integrate successfully all of the above concerns. The danger in emphasizing energy savings and load management is that the lighting quality and aesthetics may suffer. Conversely, achievement of lighting quality does not always guarantee the most energy-efficient building. Since it is much easier to quantify energy usage than to quantify ambience, there is the ever-present danger that lighting quality will be neglected in an attempt to minimize energy usage. It should be remembered that the interior of a building is lighted for the needs of the occupants and that the effectiveness of a lighting system should be evaluated not only on the basis of energy consumption but also with regard to its impact on human energy resources.

Concern for natural lighting will influence design decisions at all levels, from urban planning down to such details as the colors of carpets and walls in a single room. Urban planning influences the daylight available at a building site and places constraints on the length-to-width ratio and the height of the building. The size of a building is an important factor, since the total task area to be illuminated will be proportional to the total floor area, while the potential to admit natural light is proportional to the exterior surface area (or exposed skin) of the building.

Diffuse light from windows on one wall of a room has an illumination intensity that decreases rapidly as a person moves away from the window. In general, it can be assumed that the diffuse light from a window will provide adequate working illumination to a room depth of approximately one to three times the height of the window opening. This basic rule of thumb had a marked effect on the buildings built during the first part of this century. Buildings of that era were designed with articulated shapes using wings, light courts, and other design layouts so that windows could admit natural light to most of the interior spaces. The floor-to-ceiling height was also generous to enhance the penetration of light into the space. Today, buildings that are designed to maximize daylighting tend to utilize design techniques similar to those used in these early twentieth century buildings, to increase building skin area and window area. Figure 9-21 shows two views of the State Energy Commission Building in Sacramento, California. A central courtyard allows light to penetrate down four stories to provide natural lighting to the interior of the building. A translucent awning blocks out direct sunlight, thus minimizing the effects of glare.

For any daylighting system, three basic criteria should be met:

1. Adequate illumination for the occupants and their tasks.
2. Glare control.
3. Responsive control systems for (supplemental) electric lighting.

A good daylighting system should provide illumination at adequate levels over the widest possible area and in addition provide for visual comfort and glare control.

The amount of light reaching any location in a room depends on the amount of light available at the surface of the window. There are two sources for the light reaching the window: the *sky* and the *surround*.

The sky is the source of direct sunlight and of diffuse light, which is sunlight scattered by the atmosphere. Different regions of the sky have different brightness levels. A uniform overcast sky will have the bright region at the zenith with the dimmest region near the horizon, as shown in Fig. 9-22a. The relative brightness levels will remain the same during the day for the overcast sky, being independent of the sun's position. The clear sky condition, however, has a very different brightness distribution, as shown in Fig. 9-22b. The brightest region is near the position of the sun, while the dimmest region is directly opposite the sun at an angle of 90° from the sun's position. Clear-sky brightness also varies during the day as the sun traverses the sky from east to west. To ensure the greatest amount of available daylight at a window surface, the window must have access to as much of the sky as possible. In clear-sky conditions, this would be a vertical south-facing glazing. In locations where the skies are always predominantly overcast, a skylight may be most effective in maximizing the diffuse light from the sky. The least desirable option for a building that requires a good deal of heating is north-facing glazing. Since north-facing glass views a dim source of sky illumination, the result is a rather low level of natural lighting accompanied by the large heat loss associated with north glazing. In cases where space heating is not a major load, north-facing glass may be desirable, since northern light is relatively free from glare.

The *surround* consists of all of the non-sky objects outside the window that reflect light toward it. Included in the surround are the ground, nearby buildings, trees, ponds or lakes, and any sunlight reflectors that may be intentionally provided. For overcast skies, the surround usually exhibits an illumination brightness of only 10 percent of the overcast-sky illumination. For clear-sky conditions, however, the surround may become a very effective source of illumination—the ground (with snow), water surfaces, or the walls of adjacent

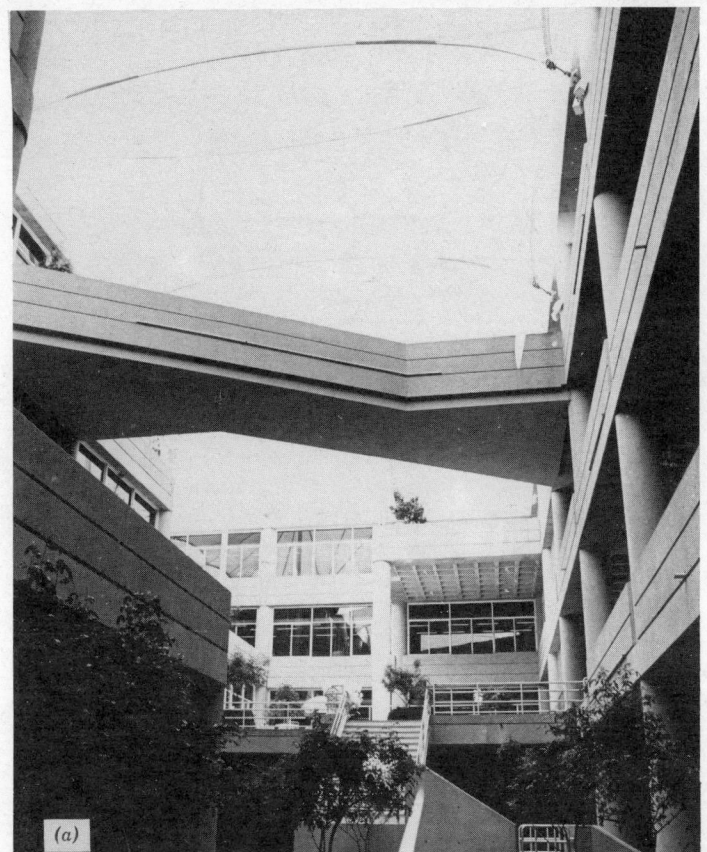

(a)

(b)

Figure 9-21 The California State Energy Commission Building in Sacramento, California. (*a*) A view up from the courtyard, showing the translucent awning above the courtyard. (*b*) A view down into a central courtyard. The sun-bathed courtyard allows natural light into the interior of the building.

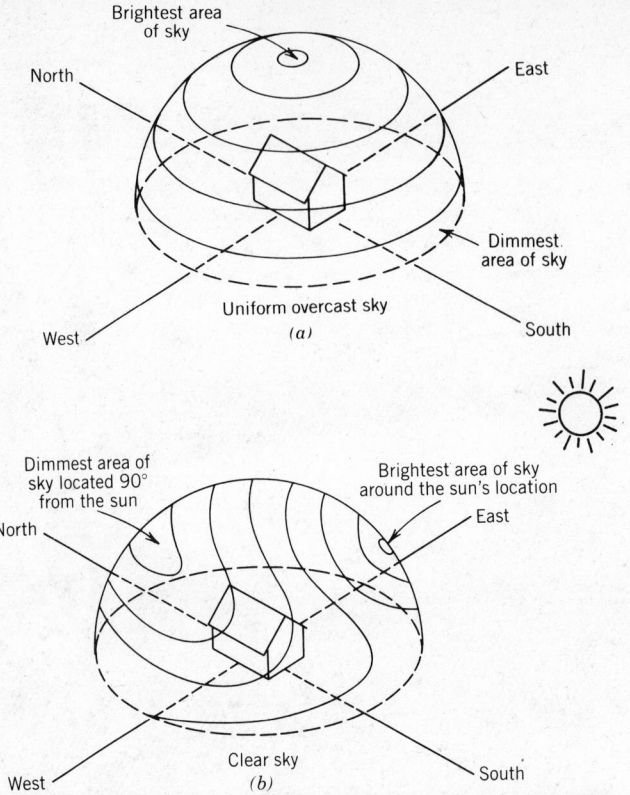

Figure 9-22 Brightness of regions in the sky. (a) The bright area of a uniform overcast sky is at the zenith, the point directly above the observer. The brightness decreases from the zenith toward the horizon. The dimmest area is located at the horizon. (b) For a clear sky, the brightest area is located around the sun. The dimmest region is in the northern sky at 90° from the sun's position. The brightest and dimmest regions will move across the sky as the sun moves during the day.

buildings being at times brighter than the sky. Such sources should not be allowed to produce imbalance or glare in interior spaces, since the occupants may resort to the use of electric lights rather than natural daylighting.

Inside the building, the walls and ceilings become secondary light sources when illuminated with daylight. Therefore, the shape of a room and the color of

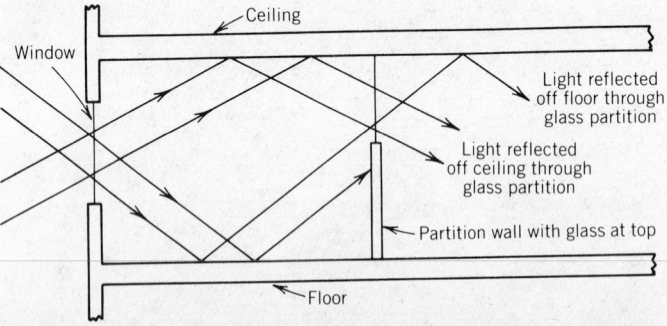

Figure 9-23 When partitions are used to divide a large room into smaller areas (as in office situations), the partitions should have glass in the top half to allow light reflected off the ceiling and other surfaces to pass through the partition into adjacent spaces.

its walls will influence the amount and quality of the light available away from the area near the window. Ceilings should always be white to reflect as much light as possible. Walls that are perpendicular to the plane of the window are reflectors and should be light colored. Partitions can be glazed in the upper portion so that daylight can reach adjacent spaces, as shown in Fig. 9-23.

Problems of Glare.   The control of glare is important if a natural lighting system is to be successful. Visual comfort seems to be best when the visual task is a bit brighter than the surrounding room. Bright sources of light in the immediate surround cause unacceptable glare, and measures must be taken to screen out bright spots of high light intensity such as sunlit areas on nearby (white or light-colored) buildings. Such screening may diminish overall light levels, but this can be offset by illumination of interior surfaces to provide the proper interior light balance. This additional illumination should ideally be from a natural source, such as a clerestory window or a light shelf (as discussed below), rather than resorting to electric lighting.

Direct sunlight entering the space usually results in some degree of glare, and it should be controlled by shades, louvers, or screens. The control of direct sunlight is also necessary for thermal reasons. In the case where thermal gain is desired, direct sunlight is admitted, but it should be diffused in ways that will not cause glare problems. If thermal gain is not desired, direct sunlight is screened out while allowing diffuse daylight to enter the window. Figure 9-24 shows fixed fins and louvers on the south side of a building that will block direct solar gains but allow diffuse light to enter. Movable fins along the east wall are also shown. These allow direct solar gains to be screened out as the sun moves across the sky. The movable fins are controlled by a computerized building-control system that governs the operation of all of the building systems.

The *light shelf* is a variation on clerestory lighting, as shown in Fig. 9-25. Direct sunlight is reflected off a horizontal reflector onto the ceiling, which then acts as a diffusing surface to distribute the light into the space. The light shelf can reflect daylighting deep into the room, thus providing illumination levels that exceed the capability of a window.

Figure 9-24 Louvers and fins are used to block direct solar gains and glare through windows. Fixed louvers and fins are shown on the south wall of this building. Adjustable vertical fins along the east wall provide a means of blocking direct solar gains and maximizing natural lighting during the day.

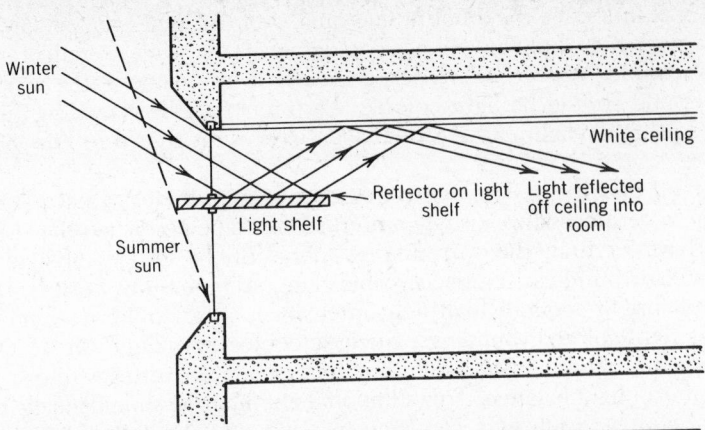

Figure 9-25 A light shelf can be used to reflect sunlight to the ceiling, thus providing a source of natural lighting deep into a room. The front of the light shelf can be extended to provide summer shading of the window below the light shelf.

Lighting controls are an essential part of a natural lighting system, because the electric lights must be turned off or dimmed in response to available daylight levels if energy savings are to be achieved. Simple on–off switches constitute the most direct method, but such a system is apt to result in an excessive use of electricity. Lighting fixtures should be wired in alternating banks so that the lighting level can be varied in steps rather than by turning on and off complete areas of lights, which may result in sharp and perhaps annoying changes in light levels. Photosensitive, automatic on–off controls constitute a good approach, but they are rather expensive. The ultimate in technical sophistication (and cost) is to use multilevel step dimming (in steps small enough to be perceived as continuous), controlled by a photosensitive control system that maintains the indoor illumination within a preset range of levels.

In summary, then, daylighting is a means of using solar energy for interior lighting in buildings. Such use of natural light can reduce energy consumption during daylight hours when coupled with minimization and load management of artificial lighting. Additionally, the use of natural light can be aesthetically pleasing and healthful. When designing for daylighting, a fundamental consideration is that the quality and quantity of the natural light should meet the requirements of the tasks to be performed in it.

ADDITIONAL METHODS OF PERFORMANCE EVALUATION

To provide additional flexibility in evaluating passive solar systems, researchers at the Los Alamos National Laboratory have taken four basic passive systems and analyzed the system performance as selected elements were varied, such as thermal storage, the number of glazings, and the uses of night insulation. Each of the passive systems analyzed is called a *reference design case*. Each reference design case has a set of specified requirements. These are:

1. The thermal storage per unit of projected area of the solar collection area, Btu/F°-ft² glazing.
2. The ratio of the thermal storage surface area to the projected area of the solar collection area (direct-gain systems).
3. The number of glazings used in the solar collection area.
4. The use or nonuse of night insulation on the solar collection area.
5. The type of absorbing surface used on the thermal storage.

Table 9-3 Reference Design Cases for Direct-Gain Solar Systems

Designation	Thermal Storage Capacity[a] (Btu/F°-ft²)	Mass Thickness[b] (in.)	Surface Area to Glazing Area Ratio	No. of Glazings	Night Insulation
A1	30	2	6	2	no
A2	30	2	6	3	no
A3	30	2	6	2	yes
B1	45	6	3	2	no
B2	45	6	3	3	no
B3	45	6	3	2	yes
C1	60	4	6	2	no
C2	60	4	6	3	no
C3	60	4	6	2	yes

Source: R. W. Jones, ed., *Passive Solar Design Handbook*, vol. III (Washington: U.S. Department of Energy, July 1982), p. 236. Adapted with permission.

[a] The thermal storage is per unit of projected area, Btu/F°-ft² glazing.

[b] The mass thickness is listed only as an appropriate guide, assuming that the volumetric heat capacity C_V is 30 Btu/ft³-F°.

The Los Alamos group has prepared solar performance data for 94 different reference design cases. These reference design cases are briefly categorized below.

Direct-Gain Passive Systems.   Nine reference designs for direct-gain systems have been analyzed. They are listed in Table 9-3 and are grouped by three designations—A, B, and C—which correspond to thermal storage capacities of 30, 45, and 60 Btu/F°-ft² glazing. The ratio of mass surface area to projected area is also given in the table. The number of glazings used varies from two to three. The use of night insulation is also indicated.

Water Thermal Wall Systems.   Fifteen reference designs are given in Table 9-4 for water thermal wall systems. They are grouped into three designa-

Table 9-4 Reference Design Cases for Water Wall Solar Systems

Designation	Thermal Storage Capacity[a] (Btu/ft²-F°)	Wall Thickness[b] (in.)	No. of Glazings	Wall Surface	Night Insulation
A1	15.6	3	2	normal	no
A2	31.2	6	2	normal	no
A3	46.8	9	2	normal	no
A4	62.4	12	2	normal	no
A5	93.6	18	2	normal	no
A6	124.8	24	2	normal	no
B1	46.8	9	1	normal	no
B2	46.8	9	3	normal	no
B3	46.8	9	1	normal	yes
B4	46.8	9	2	normal	yes
B5	46.8	9	3	normal	yes
C1	46.8	9	1	selective	no
C2	46.8	9	2	selective	no
C3	46.8	9	1	selective	yes
C4	46.8	9	2	selective	yes

Source: R. W. Jones, ed., *Passive Solar Design Handbook*, vol. III (Washington: U.S. Department of Energy, July 1982), p. 233. Adapted with permission.

[a] The thermal storage capacity is per unit of projected area of glazing.

[b] The wall thickness is based on the volumetric heat capacity C_V of water, 62.4 Btu/ft³-F°.

ADDITIONAL METHODS OF PERFORMANCE EVALUATION

tions—A, B, and C. The thermal storage capacity ranges from 15.6 to 124.8 Btu/F°-ft² glazing. The number of glazings used ranges from one to three. The use of night insulation is also indicated. Additionally, the type of absorbing surface used is indicated as either a normal surface or a selective surface. A normal surface is assumed to have an infrared emittance of 0.9, while the selective surface is assumed to have an infrared emittance of 0.1, thus significantly reducing thermal losses from the surface.

Masonry or Concrete Thermal Storage Walls.   Twenty-one design cases for unvented and 21 design cases for vented masonry or concrete thermal storage walls are indicated in Table 9-5. The thermal storage capacity has a range from 15 to 45 Btu/F°-ft² glazing. The number of glazings used varies from one to three for the different cases. The use of night insulation and type of wall surface are also indicated.

Sunspaces.   The 28 design cases for sunspaces were listed in Table 8-8, and details of these various sunspaces were shown in Fig. 8-29. These design cases were discussed at some length in section 8-18.

All of the reference design cases also have other assumed characteristics, which are listed in Table 9-6. The values of thermal conductivity, density, and specific heat capacity for concrete or masonry thermal storage assumed for the reference design cases are listed. The solar absorptances for the surfaces in the different passive systems are listed along with the infrared emittance assumed for the thermal storage surfaces. The assumed glazing properties are also given for the solar aperture. Interior temperature ranges for sunspaces and rooms are indicated. A description of the assumptions for thermocirculation vents, night insulation, and solar radiation is included.

Table 9-5   Reference Design Cases for Masonry or Concrete Thermal Wall Solar Systems

Designation Vented	Designation Unvented	Thermal Storage Capacity[a] (Btu/ft²-F°)	Wall Thickness[b] (in.)	Wall Conductivity (Btu/hr-ft-F°)	No. of Glazings	Wall Surface	Night Insulation
A1	F1	15.0	6	1.00	2	normal	no
A2	F2	22.5	9	1.00	2	normal	no
A3	F3	30.0	12	1.00	2	normal	no
A4	F4	45.0	18	1.00	2	normal	no
B1	G1	15.0	6	0.50	2	normal	no
B2	G2	22.5	9	0.50	2	normal	no
B3	G3	30.0	12	0.50	2	normal	no
B4	G4	45.0	18	0.50	2	normal	no
C1	H1	15.0	6	0.25	2	normal	no
C2	H2	22.5	9	0.25	2	normal	no
C3	H3	30.0	12	0.25	2	normal	no
C4	H4	45.0	18	0.25	2	normal	no
D1	I1	30.0	12	1.00	1	normal	no
D2	I2	30.0	12	1.00	3	normal	no
D3	I3	30.0	12	1.00	1	normal	yes
D4	I4	30.0	12	1.00	2	normal	yes
D5	I5	30.0	12	1.00	3	normal	yes
E1	J1	30.0	12	1.00	1	selective	no
E2	J2	30.0	12	1.00	2	selective	no
E3	J3	30.0	12	1.00	1	selective	yes
E4	J4	30.0	12	1.00	2	selective	yes

Source: R. W. Jones, ed., *Passive Solar Design Handbook*, vol. III (Washington: U.S. Department of Energy, July 1982), pp. 234, 235. Adapted with permission.

a The thermal storage capacity is per unit of projected area of glazing.

b The mass thickness is listed only as an appropriate guide, assuming that the volumetric heat capacity C_V is 30 Btu/F°-ft³.

ADVANCED PASSIVE METHODS—SELECTED APPLICATIONS

Table 9-6 System Characteristics for Reference Design Cases Used for Solar Load Ratio Correlations

Masonry Properties	
Thermal conductivity (K)	
Sunspace floor	0.5 Btu/hr-ft²-F°
All other masonry	1.0 Btu/hr-ft²-F°
Density (D)	150 lb/ft³
Specific heat (c)	0.2 Btu/lb-F°
Solar Absorptances	
Water wall	1.0
Masonry, Trombe wall	1.0
Direct gain and sunspace	0.8
Sunspace, Water containers	0.9
Lightweight common wall	0.7
Other lightweight surfaces	0.3
Infrared Emittance	
Normal surface	0.9
Selective surface	0.1
Glazing properties	
Transmission characteristics	diffuse
Orientation	due south
Index of refraction	1.526
Extinction coefficient	0.5 in.⁻¹
Thickness of each pane	⅛ in.
Air gap between panes	½ in.
Infrared emittance	0.9
Control Range	
Room temperature	65°F to 75°F
Sunspace temperature	45°F to 95°F
Internal heat generation	0
Thermocirculation Vents (when used)	
Vent area/projected area (sum of both upper and lower vents)	0.06
Height between vents	8 ft
Reverse flow	none
Night Insulation (when used)	
Thermal resistance	R9
In place, solar time	5:30 P.M. to 7:30 A.M.
Solar Radiation Assumptions	
Shading	none
Ground diffuse reflectance	0.3

Source: R. W. Jones, ed., *Passive Solar Design Handbook*, vol. III (Washington: U.S. Department of Energy, July 1982), p. 232. Adapted with permission.

Solar Load Ratio—Solar Savings Correlations. Correlations have been prepared that relate the Solar Savings (SS) to the Solar Load Ratio (SLR) for all of the 94 reference design cases. These correlations provide a vast amount of data that the solar designer can utilize. Extensive tables have also been prepared for 200 U.S. and Canadian locations, based on the 94 reference design cases, for use with the annual Load Collector Ratio (LCR) method as discussed in section 6-11. These tables are contained in *The Passive Solar Design Handbook*, vol. III, and have a format similar to Table 6-5 and Table 8-10. These tables, it will be recalled, allow a very quick evaluation of a passive system using the annual LCR method. The evaluations are restricted to a due south solar aperture and the given locations in the tables. For nonsouth orientations and locations not tabulated, a more general approach to evaluating passive system performance must be used.

9-12 The Monthly Solar Load Ratio (SLR) Method—Further Discussion

The monthly Solar Load Ratio (SLR) method was introduced in Chapter 6, sections 6-9 to 6-11. This method allows the monthly performance of a passive solar system to be calculated and provides the solar designer with an effective tool for evaluating passive system designs. The SLR method provides a means of accounting for:

1. Any effect that modifies the monthly energy absorbed by the passive system, such as orientation, shading, and the use of reflectors.
2. The effects of interior space thermostat setting and internal heat generation within the building.

In the following section, the procedure for making the above calculations will be discussed, but first a brief review of the SLR method is given here. The Solar Load Ratio was defined as

$$SLR = \frac{\text{Monthly net solar energy absorbed}}{\text{Monthly net reference thermal load}} \qquad (6\text{-}3)$$

The SLR can then be written as the product of two fractions:

$$SLR = \frac{S}{DD_{month}} \times \frac{A_P}{NLC} \qquad (6\text{-}12)$$

It will be recalled that the Load Collector Ratio (LCR) is defined as

$$LCR = \frac{NLC}{A_p} \qquad (6\text{-}13)$$

The SLR can therefore be written with the LCR as a factor:

$$SLR = \frac{S}{DD_{month}} \times \frac{1}{LCR} \qquad (6\text{-}14)$$

Correlations that relate the monthly solar savings fraction to the LCR and to the ratio S/DD_{month} have been plotted for many of the reference design cases (see Figs. 6-25 and 9-28). The six basic steps in the monthly Solar Load Ratio method are repeated here:

1. Calculation of the parameters that are related to the building: the net load coefficient (NLC), the projected aperture area (A_p), and the Load Collector Ratio (LCR).
2. Determination of local climate information: the monthly degree-days (DD_{month}) and the monthly insolation on the collection area (Btu/ft^2-month).
3. Determination of the monthly net solar radiation absorbed in the passive solar building per unit of projected collection area S. See Eq. (6-10).
4. Determination of the monthly solar savings fraction from the parameters LCR and the S/DD_{month} from prepared correlation curves.
5. Evaluation of the Solar Savings (SS) and Auxiliary Heat (AH) for each month from Eqs. (6-8) and (6-9).
6. Summing up the monthly values to obtain annual values of Solar Savings and Auxiliary Heat.

Each of these six steps will now be elaborated in detail. When making a monthly SLR calculation, it is helpful to use worksheets for recording the calculations as the analysis progresses.

Building Characteristics. The passive building is characterized by the net load coefficient (NLC) of the building and by the projected area (A_p) of the solar collection area. The net load coefficient was defined in section 5-10 along with methods for determining and calculating the NLC. The projected area (A_p) was defined in section 6-11 along with the method for calculating A_p. When more than one passive system is used on the same building, the analysis may be done by treating the NLC as if it were divided into portions in the same ratios as the relative projected areas of the various passive system types, similar to the procedure discussed in section 9-2 for the annual load collector ratio. Then each passive system is analyzed independently and the results are combined.

Local Climate Information for the Building Site. For SLR calculations, it is necessary to have average monthly insolation values. The clear-day solar radiation values in Table 4-E can be used to obtain approximate average monthly solar radiation values by using the percent possible sunshine listed in Table 4-C and the following equation:

$$\bar{I}_M = I \times P \times N \qquad (9\text{-}14)$$

Where

$\bar{I}_M$ = the average monthly insolation on a surface, either horizontal or tilted, Btu/ft²-month

I = the clear-day insolation on a surface, either horizontal or tilted, Btu/ft²-day, for the twenty-first day of the month, from Table 4-E.

P = the percent possible sunshine for the month, Table 4-C

N = the number of days in the month

Illustrative Problem 9-4

Determine the average monthly insolation in November for a location at 48°N Lat which has 44 percent possible sunshine. The surface is tilted at 48° from the horizontal and faces due south.

Solution

From Table 4-E, the clear-day insolation for November 21 for a south-facing surface tilted at 48° is found to be 1449 Btu/ft²-day by summing up the hourly values. The number of days in November is 30. Using Eq. (9-14) and substituting values,

$$\bar{I}_M = I \times P \times N$$
$$= 1449 \text{ Btu/ft}^2\text{-day} \times 0.44 \times 30 \text{ day/month}$$
$$= 19{,}100 \text{ Btu/ft}^2\text{-month} \qquad \textit{Ans.}$$

Another source of insolation values is from tables that give average insolation for different locations. Table 9-7 gives insolation values as well as certain other data for four selected U.S. locations. Two values of insolation are provided in these tables, as follows:

HS = the mean monthly value of the total hemispheric radiation incident on a horizontal surface, Btu/ft²-day

VS = the mean value of daily total radiation incident on a vertical, south-facing surface, Btu/ft²-day

To obtain monthly values of insolation, the values of HS or VS must be multiplied by the number of days per month.

Monthly heating degree-days must also be obtained for the building location. Heating degree-days are usually based on a 65°F base temperature and can be obtained from published tables. Degree-days and average daily temperatures (TA) are also given for the four locations listed in Table 9-7. They are given for

Table 9-7 Climate Data for Use with Monthly Solar Load Ratio Calculations—Four Selected U.S. Locations

Albany, New York   Elev 292   Lat 42.7

	HS	VS	TA	D50	D55	D60	D65	D70	KT	LD
JAN	456	674	22	884	1039	1194	1349	1504	.39	64
FEB	688	871	24	742	882	1022	1162	1302	.43	56
MAR	986	933	33	515	670	825	980	1135	.43	45
APR	1335	922	47	136	253	395	543	693	.45	33
MAY	1570	873	58	12	49	125	253	384	.46	24
JUN	1730	882	68	1	3	12	39	125	.47	19
JUL	1725	908	72	0	1	3	9	58	.49	21
AUG	1499	947	70	0	1	7	22	93	.48	29
SEP	1170	1009	62	3	15	58	135	253	.47	41
OCT	817	961	51	66	150	276	422	577	.45	53
NOV	457	623	40	317	463	612	762	912	.36	62
DEC	356	521	26	747	902	1057	1212	1367	.34	66
YR	1068	843	48	3424	4428	5586	6888	8403	.45	

Knoxville, Tennessee   Elev 981   Lat 35.8

	HS	VS	TA	D50	D55	D60	D65	D70	KT	LD
JAN	621	785	41	302	449	602	756	911	.41	57
FEB	863	923	43	219	346	483	630	762	.44	49
MAR	1191	953	50	98	191	322	484	624	.47	38
APR	1599	932	60	8	30	89	173	300	.51	26
MAY	1803	863	68	1	3	14	47	123	.52	17
JUN	1902	856	76	0	0	0	0	29	.52	12
JUL	1804	839	78	0	0	0	0	15	.51	14
AUG	1666	889	77	0	0	0	0	19	.51	22
SEP	1383	998	72	0	1	6	10	71	.51	34
OCT	1121	1152	61	7	27	83	175	294	.53	46
NOV	759	969	49	106	201	331	474	624	.47	55
DEC	569	757	42	277	422	574	729	884	.40	59
YR	1275	909	60	1018	1671	2504	3478	4654	.49	

five different base temperatures of 50°, 55°, 60°, 65°, and 70°F, which correspond to the column headings of D50, D55, D60, D65, and D70. For most residential calculations, the 65°F base degree-days are used. When it is necessary to make compensation for internal heat gains, such as in commercial buildings, a new base temperature can be calculated to compensate for the internal heat gains, as follows.

The internal heat sources in a building will produce an increase in the interior space temperature. This increase in temperature can be calculated using the following equation:

$$\Delta T_I = \frac{Q_{int}}{BLC} \tag{9-15}$$

Where

ΔT_I = the temperature rise associated with the internal heat gains, F°

Q_{int} = the daily internal heat generation by all heat sources other than the auxiliary heating sources or the sun, Btu/day

BLC = the building load coefficient including the heat losses for the solar aperture, Btu/DD

An effective base temperature can now be determined, which will take into consideration the effects of internal heat sources for the building:

$$T_b = T_{set} - \Delta T_I \tag{9-16}$$

$$= T_{set} - \frac{Q_{int}}{BLC} \tag{9-17}$$

Table 9-7 (*continued*)

Richmond, Virginia

	HS	VS	TA	D50	D55	D60	D65	D70	KT	LD
JAN	632	863	38	390	543	698	853	1008	.44	59
FEB	877	1004	39	301	438	577	717	857	.47	51
MAR	1210	1025	47	140	262	408	569	716	.49	40
APR	1566	953	58	11	46	119	226	369	.51	28
MAY	1762	872	67	1	4	17	64	148	.51	18
JUN	1872	864	74	0	0	0	0	32	.52	14
JUL	1774	849	78	0	0	0	0	11	.50	16
AUG	1601	891	76	0	0	0	0	18	.50	24
SEP	1348	1020	70	0	1	6	21	84	.50	36
OCT	1033	1097	59	7	32	98	203	336	.50	47
NOV	733	988	49	100	199	334	480	630	.48	57
DEC	567	810	39	345	497	651	806	961	.43	61
YR	1250	936	58	1296	2021	2909	3939	5170	.49	

Elev 164   *Lat 37.5*

Salt Lake City, Utah

	HS	VS	TA	D50	D55	D60	D65	D70	KT	LD
JAN	639	1017	28	683	837	992	1147	1302	.51	62
FEB	989	1363	33	467	605	745	885	1025	.58	54
MAR	1454	1463	40	334	481	633	787	942	.62	43
APR	1894	1284	49	116	208	334	474	625	.63	31
MAY	2362	1184	58	20	61	135	237	371	.68	22
JUN	2561	1122	66	3	9	31	88	167	.70	17
JUL	2590	1186	77	0	0	0	0	29	.73	19
AUG	2254	1356	75	0	1	4	5	47	.71	27
SEP	1843	1693	65	4	13	43	105	195	.72	39
OCT	1293	1724	52	72	150	259	402	548	.68	51
NOV	788	1276	39	337	480	628	777	927	.58	60
DEC	570	953	30	612	766	921	1076	1231	.50	64
YR	1606	1301	51	2648	3612	4725	5983	7409	.66	

Elev 4226   *Lat 40.8*

Source: R. W. Jones, ed., *The Passive Solar Design Handbook*, vol. III (Washington: U.S. Department of Energy, July 1982), Appendix D. Reprinted with permission.

Note: A complete set of these tables for 209 U.S. and 10 Canadian locations is in the source cited.

Where

$$T_b = \text{the effective base temperature for the building, } °F$$
$$T_{set} = \text{the themostat set point, } °F$$
$$Q_{int} = \text{the daily internal heat generation, Btu/day}$$
$$BLC = \text{the building load coefficient, Btu/DD}$$

Once the new base temperature has been calculated, the monthly heating degree-days can be determined from tables like Table 9-7, by interpolation from the tabulated degree-days.

Illustrative Problem 9-5

Calculate the degree-day base temperature for a building that has a BLC of 10,000 Btu/DD and an internal heat generation of 80,000 Btu/day. The thermostat setting is 68°F.

Solution

The degree-day base temperature is calculated using Eq. (9-17):

$$T_b = T_{set} - \frac{Q_{int}}{BLC}$$

ADDITIONAL METHODS OF PERFORMANCE EVALUATION

Substituting values,

$$T_b = 68°F - \frac{80,000 \text{ Btu/day}}{10,000 \text{ Btu/DD}}$$

$$= 68°F - 8F°$$

$$= 60°F \qquad \qquad \textit{Ans.}$$

Calculation of the Monthly Net Solar Radiation Absorbed in the Building per Unit of Projected Area (S). The value of S must be calculated using a set of factors that correct for the solar aperture orientation, tilt, glazing transmittance, and the solar absorptance of the thermal storage surface. Other corrections may be made for a ground reflectance that is different from the assumed 0.3, for overhang shadowing, for site shadowing, and for the effect of reflectors.

For each component of the solar aperture that has a different orientation or tilt, the net monthly solar radiation is calculated using the following equation:

$$Q_S = I_M \times \left(\frac{\text{Incident}}{\text{Horizontal}}\right) \times \left(\frac{\text{Transmitted}}{\text{Incident}}\right) \times \left(\frac{\text{Absorbed}}{\text{Transmitted}}\right) \times A \quad (9\text{-}18)$$

Where

Q_S = the net monthly solar radiation absorbed through the solar aperture

I_M = the average monthly insolation on a horizontal surface

A = the area of the solar aperture, ft²

Incident/horizontal is the ratio of the monthly solar radiation incident on the solar aperture to the monthly radiation on a horizontal surface. This factor can be obtained from plots like those of Fig. 9-26, using the parameters of surface tilt, surface azimuth, Lat–Dec (LD), and K_T.

Transmitted/incident is the ratio of the monthly solar radiation transmitted through the solar aperture to the monthly solar radiation incident on the solar aperture. This factor can be obtained from plots like those of Fig. 9-27, using the parameters of surface tilt, surface azimuth, Lat–Dec (LD), and K_T.

Absorbed/transmitted is the ratio of the monthly solar radiation absorbed within the building to the monthly solar radiation transmitted through the solar aperture. This factor can be obtained from Table 9-8.

The two parameters Lat–Dec (LD) and K_T are utilized to determine the above ratios and are defined as follows:

Lat–Dec is the site latitude minus the midmonth solar declination. It can be obtained from Table 9-7 under the LD column. K_T is the average monthly clearness ratio, the ratio of the total hemispheric radiation incident on a horizontal surface to the extraterrestrial radiation that would be incident on the horizontal surface if it were above the earth's atmosphere. The values of K_T can be obtained from Table 9-7 under the KT column.

The total monthly solar energy absorbed in the building is the sum of the radiation absorbed by each component of the solar collection area. The monthly solar radiation absorbed in the building per unit of projected area S is then,

$$S = \frac{Q_S \text{ (total)}}{A_p \text{ (total)}} \qquad (9\text{-}19)$$

Where

S = the monthly net solar radiation absorbed in the building per unit of projected area, Btu/ft²-month

Q_S (total) = the total net monthly solar radiation absorbed for all components of the solar aperture, Btu/month

A_p (total) = the projected area of the total solar aperture, ft²

ADVANCED PASSIVE METHODS—SELECTED APPLICATIONS

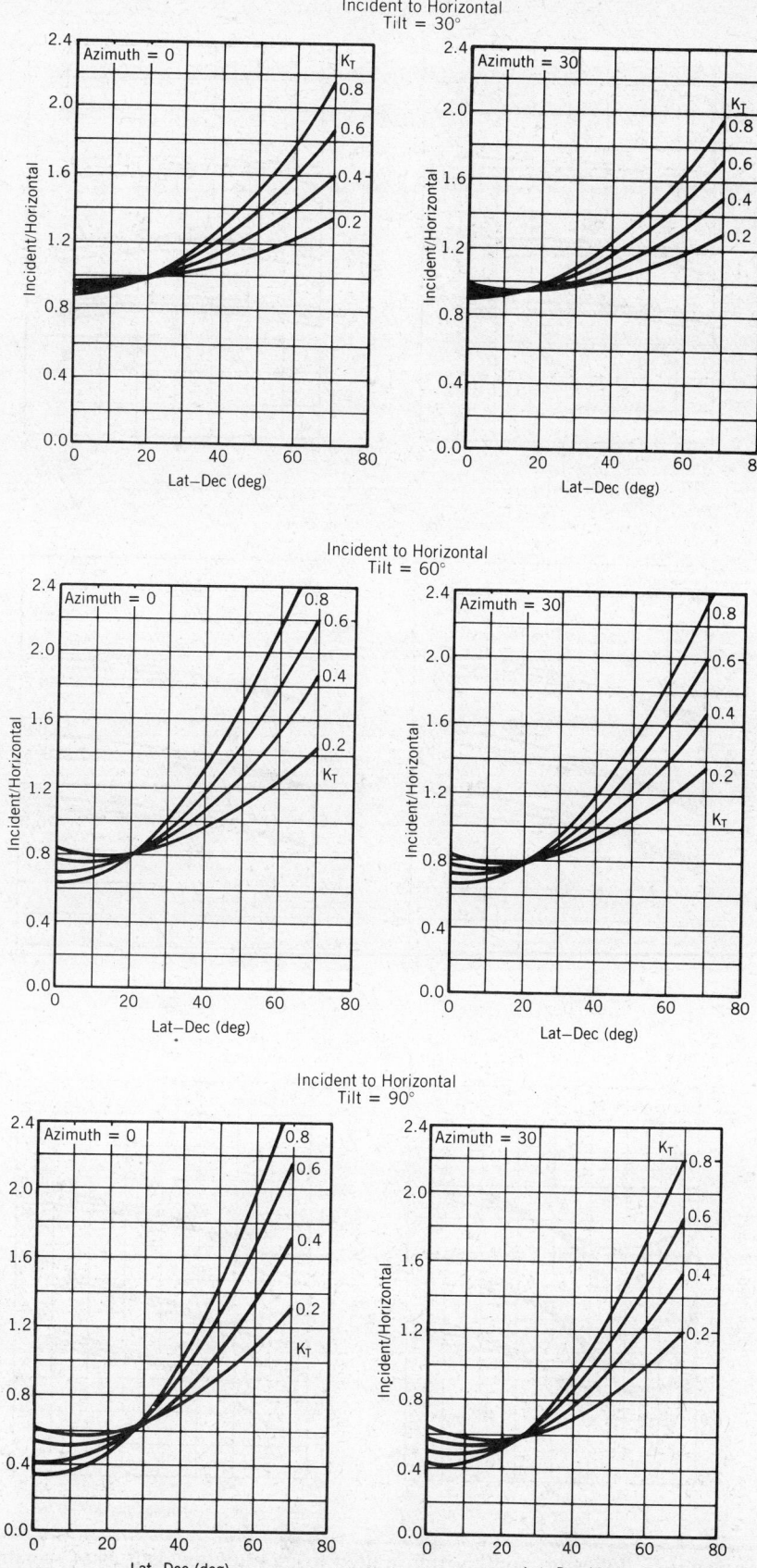

Figure 9-26 Ratios of monthly incident to monthly horizontal solar radiation. The latitude minus the solar declination is plotted along the horizontal axis. The clearness ratio K_T is shown as a parameter for the curves. Plots are shown for azimuths of 0° and 30° and tilts of 30°, 60°, and 90°. *Source:* R. W. Jones, ed., *The Passive Solar Design Handbook*, vol. III, Washington: U.S. Department of Energy, July 1982, Appendix E, Figs. E-1, E-2, E-5, E-6, E-9, E-10. Reprinted with permission.

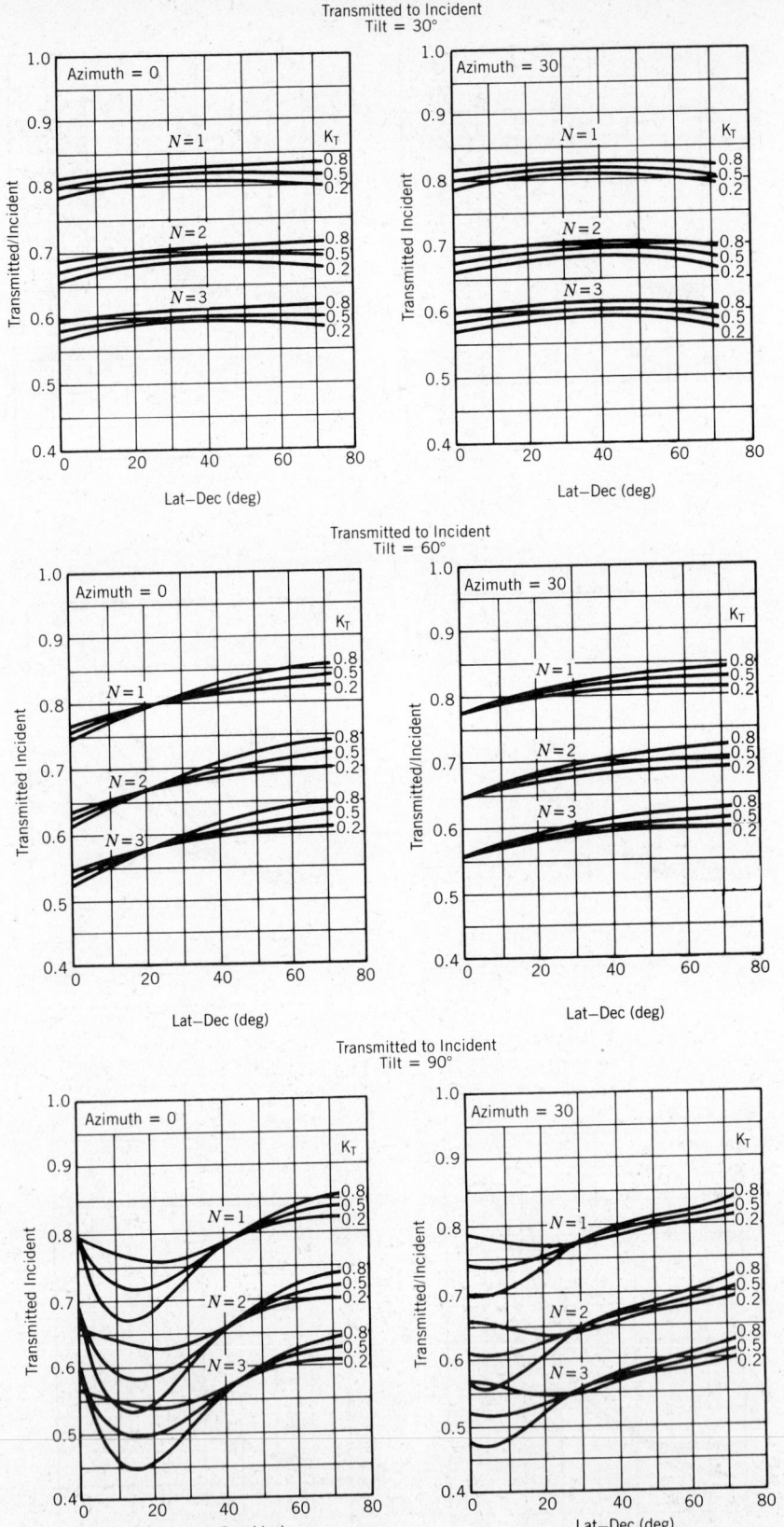

Figure 9-27 Ratios of monthly transmitted to monthly incident solar radiation. Curves are shown for single glazing ($N = 1$), double glazing ($N = 2$), and triple glazing $N = 3$). The latitude minus the solar declination is plotted along the horizontal axis. The clearness ratio K_T is shown as a parameter for the curves. Plots are shown for azimuths of 0° and 30° and tilts of 30°, 60°, and 90°. *Source:* R. W. Jones, ed., *The Passive Solar Design Handbook*, vol. III, Washington: U. S. Department of Energy, July 1982, Appendix E, Figs. E-13, E-14, E-17, E-18, E-21, E-22. Reprinted with permission.

Table 9-8 Ratio of Absorbed to Transmitted Solar Radiation for Thermal Storage Wall Solar Systems

Wall Surface Absorption	Effective Absorption of Wall		
	No. of Glazings		
	1	2	3
1.00	1.000	1.000	1.000
.95	.957	.962	.964
.90	.914	.922	.928
.85	.869	.882	.890
.80	.824	.840	.851
.75	.779	.798	.811
.70	.733	.754	.769
.65	.686	.710	.726
.60	.638	.664	.682
.55	.589	.617	.636
.50	.540	.568	.588
.45	.490	.519	.539
.40	.439	.468	.488
.35	.387	.415	.435
.30	.335	.361	.380

Source: R. W. Jones, ed., *The Passive Solar Design Handbook*, vol. III (Washington: U.S. Department of Energy, July 1982), p. 310. Adapted with permission.

Calculation of the Monthly Solar Savings Fraction. The ratio S/DD_{month} is calculated for each month. Then the monthly solar savings fraction is obtained from a correlation chart like those of Fig. 9-28, using the ratio S/DD_{month} and the LCR. Figure 9-28 is a correlation chart of the solar savings fraction for a vented masonry or concrete thermal wall of designation A4 (Table 9-5). Figure 6-25 gave a similar correlation chart of the solar savings fraction for a direct-gain system.

Evaluation of the Solar Savings and Auxiliary Heat. The monthly Solar Savings is calculated, using a modified form of Eq. (6-9), as

$$SS_M = DD_{month} \times NLC \times SSF_M \qquad (9\text{-}20)$$

Where

$$SS_M = \text{the monthly solar savings, Btu/month}$$
$$DD_{month} = \text{the degree-days per month}$$
$$NLC = \text{the net load coefficient, Btu/DD}$$
$$SSF_M = \text{the monthly solar savings fraction}$$

The monthly Auxiliary Heat is calculated, using a modified form of Eq. (6-8), as

$$AH_M = DD_{month} \times NLC \times (1 - SSF_M) \qquad (9\text{-}21)$$

Annual Solar Savings and Auxiliary Heat. The annual values of the Solar Savings and Auxiliary Heat are found by summing up the monthly values.

Illustrative Problem 9-6

A house has been designed for Knoxville, Tennessee using a masonry thermal wall with thermocirculation vents. There are two components to the solar collection area: 450 ft² of vertical south-facing wall and 200 ft² of vertical wall facing 30° east of south. The house has a calculated net load coefficient of 10,000 Btu/DD. The thermal storage wall will be painted flat black with an absorptance of 0.95 and will have double glazing in front. No night insulation

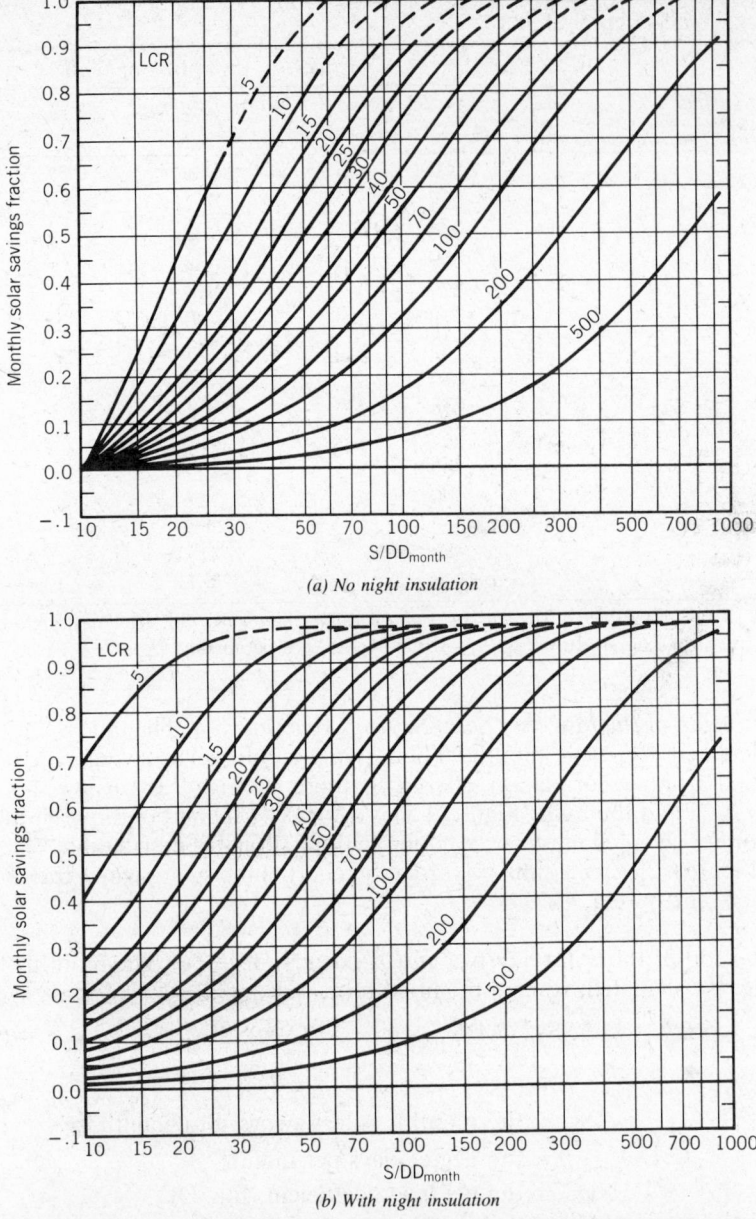

(a) No night insulation

(b) With night insulation

Figure 9-28 (*a*) A correlation chart of the solar savings fraction for a vented masonry or concrete thermal wall of designation A4, without night insulation. The ratio of the monthly absorbed radiation (*S*) to the monthly degree-days for the solar building (S/DD_{month}) is plotted along the horizontal axis. The solar savings fraction (SSF) is plotted along the vertical axis. Curves are given for different values of the Load Collector Ratio (LCR). The SSF is found for a specific value of LCR and S/DD_{month} from the chart. (*b*) Monthly solar savings fraction (SSF) plotted as a function of S/DD_{month} for various values of the Load Collector Ratio (LCR) for a vented masonry or concrete thermal wall of designation A4 with night insulation used on the solar collection area. *Source:* J. D. Balcomb, et al., *The Passive Solar Design Handbook*, vol. II, Washington: U. S. Department of Energy, January 1980, p. 130. Reprinted with permission.

will be used, and the internal heat gains are assumed to be negligible. Determine the annual Solar Savings and the annual Auxiliary Heat for this building using the monthly SLR method and the charts and tables provided above.

Solution

There are no internal heat gains, so the degree-day base temperature used will be 65°F. The six steps for the monthly solar load ratio calculation are as follows, using Table 9-9 as a worksheet.

1. The net load coefficient is given as 10,000 Btu/DD. The solar collection area is a vertical wall and the projected area A_p will then be the same as the actual solar collection area of 650 ft². The Load Collector Ratio is then calculated using Eq. (6-13):

$$LCR = \frac{NLC}{A_p}$$
$$= \frac{10,000 \text{ Btu/DD}}{650 \text{ ft}^2}$$
$$= 15.4 \text{ Btu/DD-ft}^2$$

2. For the Knoxville location, the monthly degree-days for a base temperature of 65°F can be obtained from Table 9-7. The monthly solar radiation on the solar collection area is also obtained from Table 9-7. The solar collection area has two parts: a due south vertical wall and a vertical wall facing 30° east of south. Tables 9-9 and 9-10 will be used to calculate the net monthly solar radiation absorbed through the solar collection area. For the vertical south wall, the daily total radiation incident on a vertical south-facing wall, VS, can be used. The values of VS for Knoxville from Table 9-7 are then entered into the average daily insolation column of the Table 9-9 worksheet. The value for September (for example) is 988 Btu/ft²-day. The average monthly insolation on a vertical surface is then found by multiplying by the number of days in the month. For September:

$$\bar{I}_M = 988 \text{ Btu/ft}^2\text{-day} \times 30 \text{ days/month}$$
$$= 29,900 \text{ Btu/ft}^2\text{-month}$$

This value is entered in the worksheet (Table 9-9). From Table 9-7, the September value of Lat–Dec is found from the LD column as 34°. The value of K_T is found from the KT column of Table 9-7 as 0.51. For the rest of the months, $\bar{I}_M$ is calculated in a similar manner and the values of Lat–Dec and K_T are also obtained from Table 9-7.

The Table 9-10 Worksheet is for the calculation of Q_S—Eq. (9-18)—for the wall facing 30° east of south. In this case, the average monthly insolation on a horizontal surface will have to be used and then corrected for the off-south orientation. The average daily insolation on a horizontal surface, HS (Table 9-7), is used to fill in the average daily radiation column of Table 9-10. $\bar{I}_M$ for a horizontal surface is then calculated by multiplying the average daily radiation by the number of days in each month. The values of Lat–Dec and K_T are not shown in Table 9-10, since they are the same as in Table 9-9.

3. The monthly net solar radiation absorbed in the building per unit of projected area is now calculated. First, the net monthly solar radiation absorbed through the two solar collection areas is calculated as Q_{S1} and Q_{S2} using Eq. (9-18). For the due south orientation, $\bar{I}_M$ in Table 9-9 is the incident radiation on a vertical, south-facing surface, since VS was used. Therefore, the incident/horizontal factor is equal to 1.0. The transmitted/incident factor is found from Fig. 9-27 for a 90° tilt and an azimuth of 0° (due south). For September, this factor is found to be approximately 0.64 for $K_T = 0.51$ and Lat–Dec = 34°. For the other months, the transmitted/incident factor is found in a similar manner. The absorbed/transmitted factor is obtained from Table 9-8. For a wall absorption of 0.95 and two glazings, this factor is found to be 0.962 and is

ADDITIONAL METHODS OF PERFORMANCE EVALUATION

Table 9-9 Worksheet for the Calculation of Net Solar Radiation Absorbed Q_{S1} for a Vertical South-Facing Wall with 450 ft² Area (Illustrative Problem 9-6)

Month	Days per Month	Lat–Dec (degrees)	K_T	Average Daily Radiation Vertical (Btu/ft²-day)	I_M Vertical Surface (Btu/ft²-month)	Incident Horizontal	Transmitted Incident	Absorbed Transmitted	Q_{S1} (Btu/month)
Sept	30	34	.51	998	29,900	1.0	.64	.962	8.28 × 10⁶
Oct	30	46	.53	1152	34,600	1.0	.67	.962	10.0 × 10⁶
Nov	30	55	.47	969	29,100	1.0	.70	.962	8.82 × 10⁶
Dec	31	59	.40	757	23,500	1.0	.70	.962	7.12 × 10⁶
Jan	31	57	.41	785	24,300	1.0	.70	.962	7.36 × 10⁶
Feb	28	49	.44	923	25,800	1.0	.68	.962	7.59 × 10⁶
Mar	31	38	.47	953	29,500	1.0	.65	.962	8.30 × 10⁶
Apr	30	26	.51	932	28,000	1.0	.60	.962	7.26 × 10⁶
May	31	17	.52	863	26,800	1.0	.58	.962	6.73 × 10⁶
June	30	12	.52	856	25,700	1.0	.59	.962	6.56 × 10⁶

Table 9-10 Worksheet for the Calculation of Net Solar Radiation Absorbed Q_{S2} for the Vertical Wall Facing 30° East of South, with 200 ft² Area (Illustrative Problem 9-6)

Month	Days per Month	Average Daily Radiation Horizontal (Btu/ft²-day)	$\bar{I}_M$ Vertical Surface (Btu/ft²-month)	Incident Horizontal	Transmitted Incident	Absorbed Transmitted	Q_{S2} (Btu/month)
Sept	30	1383	41,500	.70	.66	.962	3.69 × 10⁶
Oct	30	1121	33,600	.91	.67	.962	3.94 × 10⁶
Nov	30	759	22,800	1.21	.68	.962	3.61 × 10⁶
Dec	31	569	17,600	1.22	.68	.962	2.81 × 10⁶
Jan	31	621	19,300	1.20	.68	.962	3.03 × 10⁶
Feb	28	863	24,200	1.02	.67	.962	3.18 × 10⁶
Mar	31	1191	36,900	.79	.66	.962	3.70 × 10⁶
Apr	30	1599	48,000	.59	.64	.962	3.49 × 10⁶
May	31	1803	55,900	.53	.62	.962	3.53 × 10⁶
June	30	1902	57,100	.53	.61	.962	3.55 × 10⁶

ADVANCED PASSIVE METHODS—SELECTED APPLICATIONS

entered in the Table 9-9 worksheet. The value of Q_{S1} for September is obtained, using Eq. (9-18), as

$$Q_{S1} = \bar{I}_M \times \frac{\text{Incident}}{\text{Horizontal}} \times \frac{\text{Transmitted}}{\text{Incident}} \times \frac{\text{Absorbed}}{\text{Transmitted}} \times A$$
$$= 29{,}900 \text{ Btu/ft}^2\text{-month} \times 1.0 \times 0.64 \times 0.962 \times 450 \text{ ft}^2$$
$$= 8.28 \times 10^6 \text{ Btu/month}$$

The values of Q_{S1} for the other months of the year are calculated in a similar manner and are shown in the Table 9-9 worksheet.

In Table 9-10, $\bar{I}_M$ is the incident radiation on a horizontal surface. The incident/horizontal factor must be used to correct for the 30° azimuth of the wall. These factors can be obtained from Fig. 9-26 for a 90° tilt and an azimuth of 30°. For September, this incident/horizontal factor is found as approximately 0.70 for $K_T = 0.51$ and Lat–Dec = 34°. The transmitted/incident factor is found from Fig. 9-27 for a 90° tilt and an azimuth of 30°. For September, this factor is found as approximately 0.66 for $K_T = 0.51$ and Lat–Dec = 34°. The factors for the other months are found in a similar manner. The absorbed/transmitted factor will be 0.962 as before. For September, Q_{S2} is then found, using Eq. (9-18), as

$$Q_{S2} = 41{,}500 \text{ Btu/ft}^2\text{-month} \times 0.70 \times 0.66 \times 0.962 \times 200 \text{ ft}^2$$
$$= 3.69 \times 10^6 \text{ Btu/month}$$

The values for the other months of the year are calculated in the same manner.

The monthly net solar radiation absorbed in the building per unit of projected area is then calculated using Eq. (9-19), as follows:

$$S = \frac{Q_S \text{ (total)}}{A_p \text{ (total)}}$$
$$= \frac{Q_{S1} + Q_{S2}}{A_p \text{ (total)}}$$

Substituting values for September,

$$S = \frac{8.28 \times 10^6 \text{ Btu/month} + 3.69 \times 10^6 \text{ Btu/month}}{650 \text{ ft}^2}$$
$$= 18{,}400 \text{ Btu/ft}^2\text{-month}$$

The values for S for the other months are calculated in a similar manner and entered in the Table 9-11 worksheet.

Table 9-11 Worksheet for the Calculation of SSF_M, SS_M, and AH_M (Illustrative Problem 9-6)

Month	DD_{month}	S (Btu/month)	$\dfrac{S}{DD_{month}}$ (Btu/ft²-F°)	SSF_M	SS_M ($\times 10^6$ Btu)	AH_M ($\times 10^6$ Btu)
Sept	10	18,400	1840	1.0	0.10	0.0
Oct	175	21,400	122	1.0	1.75	0.0
Nov	474	19,100	40	0.55	2.61	2.13
Dec	729	15,300	21	0.25	1.82	5.47
Jan	756	16,000	21	0.25	1.89	5.67
Feb	630	16,600	26	0.35	2.21	4.10
Mar	484	18,500	38	0.55	2.66	2.18
Apr	173	16,600	96	0.93	1.61	0.12
May	47	15,800	336	1.0	0.47	0.0
June	0	15,600	—	—	—	—
Annual totals					15.1	19.7

4. The degree-days for a base temperature of 65°F are entered in the Table 9-11 worksheet for Knoxville from Table 9-7. The ratio S/DD_{month} is then calculated for each month. For September,

$$\frac{S}{DD_{month}} = \frac{18,400 \text{ Btu/ft}^2\text{-month}}{10 \text{ DD/month}}$$
$$= 1840 \text{ Btu/ft}^2\text{-F}°$$

The values of S/DD_{month} are calculated in a similar manner for the other months.

The monthly solar savings fraction is then found from a correlation chart, using the ratio S/DD_{month} and the LCR. The correlation curves in Fig. 9-28*a* are for a thermal storage wall with vents and no night insulation. For September, with S/DD_{month} = 1840 Btu/ft²-F° and calculated LCR = 15.4 Btu/ft²-F°, the SSF_M (from Fig. 9-28*a*) will be 1.0. For October, the SSF_M will also be 1.0. In November, with S/DD_{month} = 40 Btu/ft²-F°, the SSF_M is found to be approximately 0.55. The values of SSF_M for the other months are found in a similar manner.

5. The Solar Savings (SS) values are calculated using Eq. (9-20) and entered in the Table 9-11 worksheet. The Auxiliary Heat (AH) values are calculated using Eq. (9-21) and entered in the Table 9-11 worksheet.

6. The monthly values are summed up to obtain

$$SS_{annual} = 15.1 \times 10^6 \text{ Btu} \qquad\qquad Ans.$$

$$AH_{annual} = 19.7 \times 10^6 \text{ Btu} \qquad\qquad Ans.$$

CONCLUSION

This chapter concludes the section on passive solar systems. Many aspects of system design, construction, and performance for passive solar systems have been considered. The treatment has included direct-gain and indirect-gain systems, thermal storage wall systems, sunspaces, natural lighting, and hybrid systems. Chapter 9 has gone beyond basic principles and standard practices to explore some advanced passive methods and examine the results of recent research and correlation techniques for calculating Solar Savings and Auxiliary Heat for passive solar buildings with a mix of solar elements.

In the following chapters, active solar energy systems will be examined. These will include solar water-heating systems, swimming pool and spa systems, space-heating systems, space-cooling systems, solar energy systems in industry and commerce, and systems for the production of electricity from solar energy.

PROBLEMS

1. A direct-gain sunspace will use a rock bed to store one half of the daily energy collected by the sunspace, which is 275,000 Btu/day. Determine Q_R, the energy stored in the rock bed, and Q_t, the heat transport rate for the rock bed. (Assume that heat will be removed from the sunspace for 4 hr daily.)

2. A hybrid solar system has a heat transport rate of 30,000 Btu/hr and a $\Delta T_{working\ air}$ of 20 F°. Determine the required air flow to transport the heat to the rock bed.

3. A rock bed has a face area of 50 ft². The air flow into the rock bed is 1000 CFM. Determine the face velocity and the total pressure drop for the rock bed if it has a length of 20 ft and rock diameter of 2 in.

4. The ducting for a hybrid system has an overall length of 120 ft with 10 elbows. The duct diameter is 14 in. and the air flow is 950 CFM. Determine the pressure loss for the ducts.

5. The pressure loss for a rock bed is 0.11 in. water, and the pressure loss for

the ducting is 0.12 in. water. Determine a fan size and motor horsepower that will provide an air-flow rate of 1300 CFM to the rock bed.

6. A rock bed has a length of 15 ft, a width of 8 ft, and a depth of 2 ft. The air flow into the rock bed is 600 CFM. Determine the pressure loss of the rock bed if the 2 ft by 8 ft side is used as the face area and the rock diameter is 1½ in.

7. A rock bed will use vertical air flow with a plenum at the top and bottom. The rock bed is 20 ft by 15 ft with a depth of 2½ ft. Determine the pressure loss for an air flow of 2000 CFM for a rock diameter of 2 in.

8. It is necessary to remove 30,000 Btu/hr of heat from a sunspace. If the working air temperature difference is assumed to be 25 F°, determine the required air-flow rate.

9. A house is being designed with three passive components: a water thermal storage wall, an attached greenhouse, and a direct-gain solar element. The house has a floor area of 2100 ft². Assume that each of the three passive components will provide approximately one third of the space heating. Determine the solar aperture for each component for a climate where the monthly winter heating degree-days are 600. (Use Mazria's guidelines, Tables 8-2, 8-6, and 7-1.)

10. A house has been designed for Salt Lake City, Utah. The house has a net load coefficient of 9500 Btu/DD and uses two passive solar heating components: 250 ft² of direct-gain solar aperture (with night insulation) and 400 ft² of water thermal storage wall (with night insulation). Determine the Solar Savings (SS) and Auxiliary Heat (AH) for the house using the annual LCR method. (Use data in Table 6-5.)

11. For the house in Illustrative Problem 9-6, calculate the monthly solar savings fraction, the Solar Savings, and the Auxiliary Heat if night insulation is used on the solar collection area. (Use the correlation curves in Fig. 9-28b.)

12. For the passive solar house in Illustrative Problem 9-6, determine the Solar Savings (SS) and Auxiliary Heat (AH) if the house were to be located in Richmond, Virginia. (Use the monthly SLR method.)

13. A passive solar building has a building load coefficient (BLC) of 23,000 Btu/DD (including the solar collection area). The net load coefficient (NLC) is 20,000 Btu/DD. The building uses two passive solar components: a concrete thermal storage wall and a direct-gain element. The thermal storage wall will have 500 ft² of vertical south-facing area, and the direct-gain element will have 500 ft² of vertical south-facing area. The solar collection area will be double-glazed. The building will have a daily internal heat gain of 250,000 Btu/day and a thermostat setting of 68°F. Use the monthly (SLR) method to determine the Solar Savings (SS) and the Auxiliary Heat (AH) for a location in Salt Lake City, Utah. (Use the correlation curves in Fig. 6-25 for the direct-gain calculations and the correlation curves in Fig. 9-28a for the thermal storage wall calculations. Determine the degree-day base temperature to compensate for the internal heat gains.)

PART THREE

Active Solar Energy Systems

CHAPTER 10

Introduction to Active Solar Energy Systems

*I*n the preceding pages, passive solar systems involving the nonmechanical collection and transport of heat have been discussed. The basic components of passive systems—south-facing glazing and thermal mass—were examined in relation to the concepts of direct and indirect solar gain. Passive systems involve the building, or elements of it, in natural thermal energy flows such as convection, conduction, and radiation. *Active solar systems*, on the other hand, involve the mechanical collection, transport, and storage of heat by elements separate from the building structure itself.

An active solar system is composed of five basic elements:

1. Solar collectors for absorbing solar radiation.
2. Thermal storage for storing collected energy.
3. The heat transfer medium for moving heat between and among components of the active system.
4. A mechanical distribution system for circulating the heat transfer medium, such as a pump for liquids or a blower for air.
5. Controls that actuate the different parts of the mechanical system to ensure integrated operation.

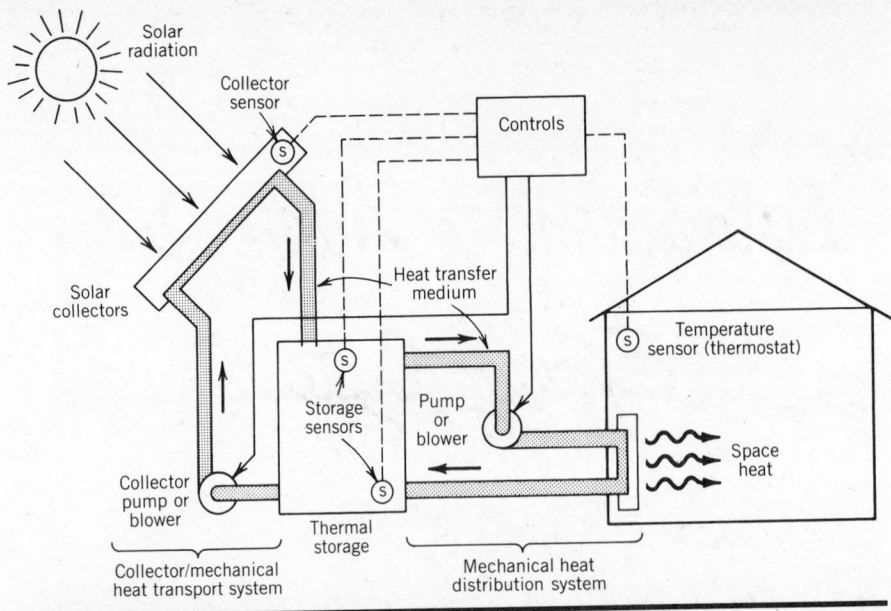

Figure 10-1 A diagrammatic representation of the arrangement of the five basic elements that comprise an active solar system.

Figure 10-1 illustrates a typical arrangement of active system elements. The illustration is schematic only, intended to portray general concepts, not actual installation practices. It is the interaction of the above elements (with mechanical assistance) that distinguishes active systems from passive systems.

The primary individual element of an active solar system is the *solar collector*. Solar collectors can be grouped in two general classifications: flat-plate (low to medium temperature) collectors and focusing (high temperature) collectors (Figs. 10-2 and 10-3).

Figure 10-2 Flat-plate medium-temperature solar collectors used to heat domestic hot water for five apartment units in Lompoc, California. The orientation of these collectors is due south with the collectors tilted at an angle of 30° from the horizontal. (Santa Barbara Solar Energy Systems.)

INTRODUCTION TO ACTIVE SOLAR ENERGY SYSTEMS

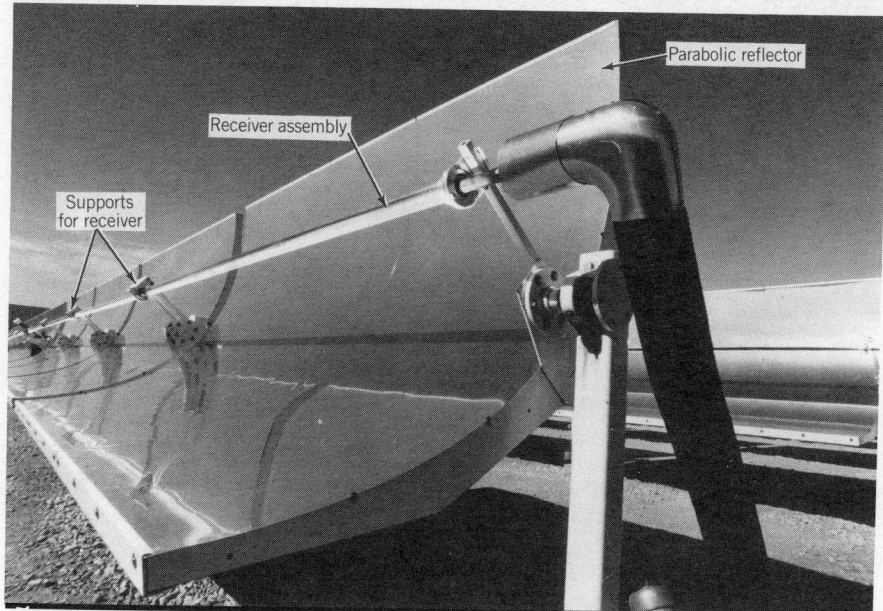

Figure 10-3 Focusing parabolic-trough collectors used to heat water for high-temperature, industrial-process applications. The parabolic mirror reflects concentrated solar radiation onto the metal receiver tube, which absorbs the solar energy and heats the fluid flowing within it. (Acurex Corporation, Mountain View, California.)

10-1   Flat-Plate Collectors

The flat-plate collector derives its name from its construction. The part of the collector that absorbs solar radiation typically is a flat metal surface with a black coating. To understand how a flat-plate collector operates, the basic principles of radiation (see sections 3-12 to 3-18) should be recalled. The amount of energy that an object exposed to direct sunlight absorbs depends on the color and texture of its surface, darker-colored surfaces typically absorbing more radiant energy than lighter-colored surfaces. As a surface warms up, it too becomes a radiator and starts to lose heat to its surroundings in the form of low-frequency (infrared) wave energy. It may also lose heat by means of conduction and convection, unless conditions are such as to inhibit or prevent these losses.

Radiation on a Black Metal Plate.   The diagram of Fig. 10-4 indicates a flat black metal plate placed in direct sunlight. The painted surface has an absorptance α of 0.95, which means that 95 percent of the sun's energy that falls on it is absorbed and converted into heat at the surface of the plate. The remaining 5 percent is reflected from the painted surface and is lost to the sky or the surroundings.

As the plate temperature increases above the temperature of the air and surrounding objects, heat will flow from the warm plate to other objects. (It will be recalled that heat always flows from hot objects to cooler objects as a result of temperature difference.) Heat flows through to the back of the plate by conduction and on into the material on which the plate rests. The air next to the top surface of the plate will become heated, and a convection current will be set up that will carry heat away from the plate. The plate will also radiate some electromagnetic energy in the far-infrared wavelengths, thus losing heat by radiation. As solar energy is absorbed by the plate, its temperature will continue to rise until an equilibrium is reached between the radiant energy being absorbed from the sun and the total heat energy being lost from the plate. The maximum plate temperature is reached when the rate of energy absorption is exactly matched by the total rate at which heat is being lost. This temperature is termed the *stagnation temperature* of the plate.

INTRODUCTION TO ACTIVE SOLAR ENERGY SYSTEMS

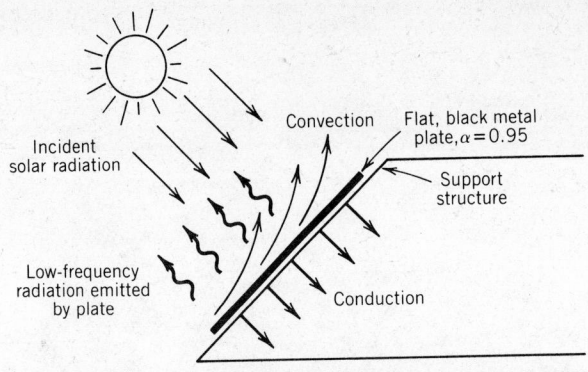

Figure 10-4 A flat-black plate exposed to solar radiation will increase in temperature as it absorbs solar energy. Heat will then be lost from the plate by convection, conduction, and radiation because it is at a higher temperature than its surroundings.

Illustrative Problem 10-1

The black metal plate shown in Fig. 10-5 is exposed to direct solar radiation at normal incidence. Its back surface is insulated so that its stagnation temperature will be influenced only by convection losses and radiation losses. Calculate the intensity of solar radiation (insolation) I necessary to maintain the black plate at a temperature of 120°F, when the ambient air temperature is 70°F and the wind speed is 5 mph. Assume a clear sky and a black-surface absorptance of 0.96.

Solution

Since conduction losses are assumed to be zero, it remains only to calculate the convection and radiation losses from the front of the plate.

Recall from section 3-10 that the rate of heat loss from a surface caused by *convection* is given by

$$Q_c = h \, A \, \Delta t \qquad (3\text{-}14)$$

where h is the convection coefficient. Values of h for different wind speeds are given in the chart of Fig. 10-6. Note that for a wind speed of 5 mph, an

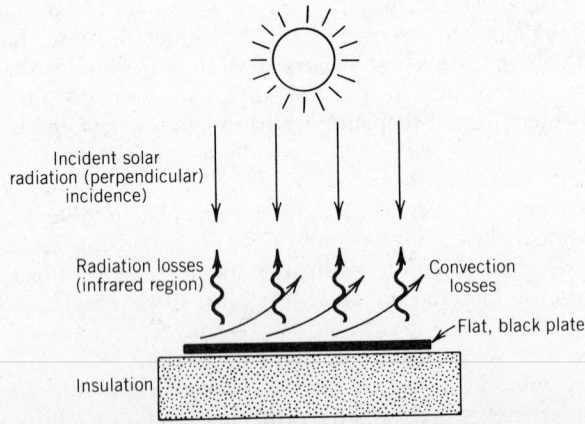

Figure 10-5 When the back of a flat-black plate is insulated, heat will still be lost from the front surface by convection and radiation. The conduction loss through the insulation will be very small compared to the loss from the front of the black plate.

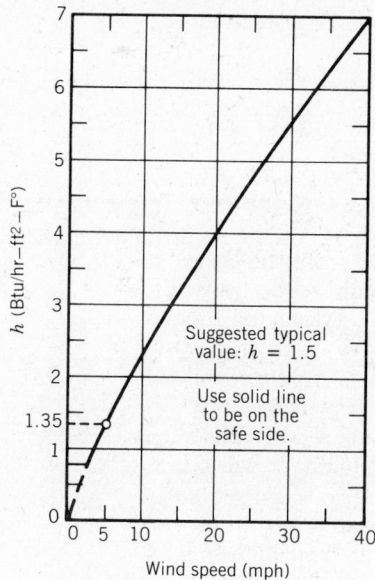

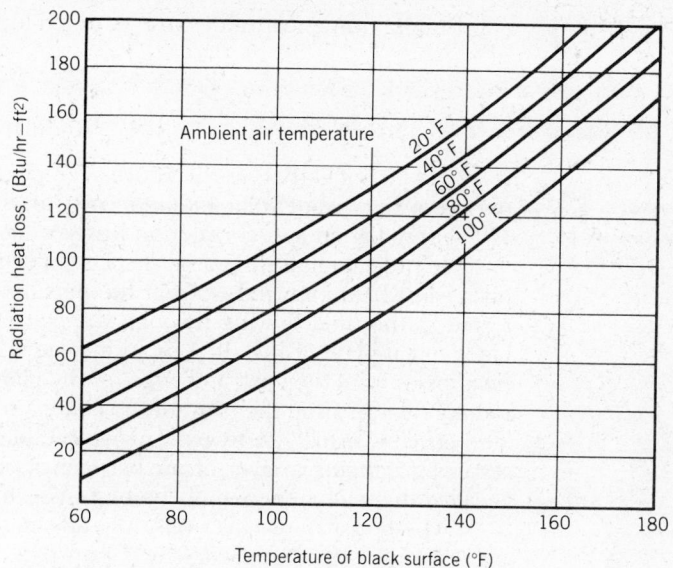

Figure 10-6 The convection coefficient h is the rate of convection
heat loss from a flat surface. Here,
values of h are plotted as a function of wind speed. *Source:*
F. deWinter, *How to Design and
Build a Solar Swimming Pool
Heater*, New York: Copper Development Association, 1974, p. 4.

Figure 10-7 Radiation heat loss, Btu/hr-ft², from a horizontal black surface to a clear sky, for the ambient air temperatures
shown on the curves.

approximate value of 1.35 Btu/hr-ft²-F° is obtained for h. We will use unit area A = 1.0 ft². For the given temperatures, $t_2 - t_1 = \Delta t = 50$ F°, and the convection heat loss,

$$Q_c = 1.35 \text{ Btu/hr-ft}^2\text{-F°} \times (1.0 \text{ ft}^2) \times (50 \text{ F°})$$
$$= 67.5 \text{ Btu/hr for the 1 ft}^2 \text{ of area.}$$

The *radiation* heat loss can be determined with the aid of Fig. 10-7, which is a plot of radiation heat loss from a black surface as a function of the black surface temperature and the ambient air temperature. For a plate temperature of 120°F and an ambient air temperature of 70°F, the radiation heat loss is read as (approximately)

$$Q_r = 100 \text{ Btu/hr-ft}^2$$

The total heat loss per unit area is the sum of the convection and radiation losses,

$$Q_T = Q_c + Q_r = 67.5 \text{ Btu/hr-ft}^2 + 100 \text{ Btu/hr-ft}^2$$
$$= 167.5 \text{ Btu/hr-ft}^2$$

Now, the *absorbed solar radiation* must be exactly matched by the total energy loss Q_T, at the steady-state or stagnation temperature. But the absorbed solar radiation is equal to the incident solar radiation multiplied by the absorptance of the black surface. Consequently,

Q_T = Absorbed solar radiation = Absorptance × Incident solar radiation

Or, rearranging,

$$I = \text{Incident solar radiation} = \frac{Q_T}{\alpha}$$

INTRODUCTION TO ACTIVE SOLAR ENERGY SYSTEMS

Substituting numerical values, the required incident solar radiation (insolation)

$$I = \frac{167.5 \text{ Btu/hr-ft}^2}{0.96}$$

$$= 174.5 \text{ Btu/hr-ft}^2 \qquad \qquad \textit{Ans.}$$

It should be noted that, as the above example shows, a black metal plate would require a significant solar radiation just to attain and hold a temperature of 120°F, if radiation and convection losses were uncontrolled.

In order to make a practical solar collector, thermal losses from the basic plate must be minimized, so that the maximum amount of heat can be transferred to the fluid flowing through the collector. This is done by placing the black metal plate of Fig. 10-4 in an insulated box, thus markedly reducing the heat losses from the back and edges of the plate. To reduce the convection and radiation losses from the front of the plate, a transparent cover is placed over it. The cover is usually ½ to 1 in. above the plate, and it minimizes convection losses by trapping a layer of air between itself and the black plate. It will be recalled that still air is one of the best insulators known.

The transparent cover (glazing) also minimizes radiation losses from the plate surface, since, although it is transparent to the medium- and high-frequency radiation from the sun, it effectively stops the lower-frequency radiation that is being reradiated from the black plate (the greenhouse effect discussed in Chapter 3). A typical flat-plate collector is constructed as shown in Fig. 10-8. A detailed discussion of the construction and operation of flat-plate collectors will be provided in Chapter 11.

10-2 Focusing Solar Collectors

Focusing collectors represent a second class of active solar collectors. Focusing collectors are often called *concentrating collectors*, because solar radiation is concentrated on the absorber by mirrored parabolic surfaces or by lenses. This

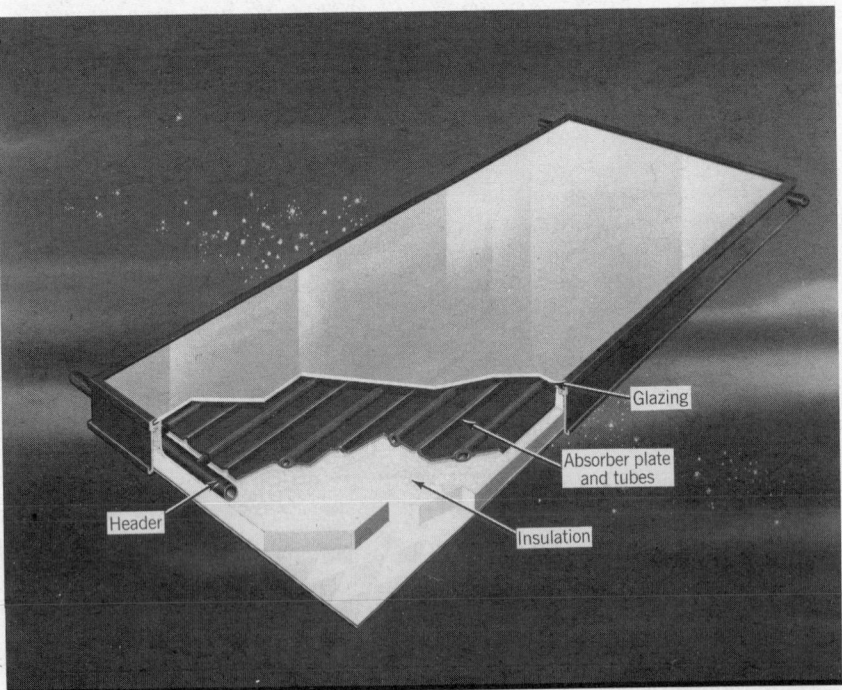

Figure 10-8 A typical flat-plate solar collector for heating liquids. To reduce convection and radiation heat losses, one or more glazings are used above the absorber plate. The collector case is insulated all across the back and up the sides. The frame is constructed of extruded aluminum. (American Solar King Corporation.)

INTRODUCTION TO ACTIVE SOLAR ENERGY SYSTEMS

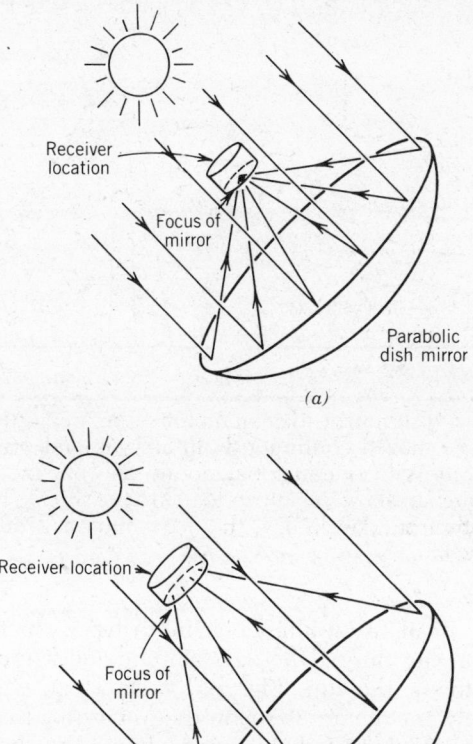

Figure 10.9 (a) A parabolic dish mirror will focus sunlight onto a very small area, thus concentrating the solar radiation to a very high intensity and producing high temperatures. (b) A spherical dish mirror focuses sunlight into a somewhat larger area than does a parabolic mirror and, therefore, the concentration is not quite as intense.

allows the collectors to operate more efficiently at higher temperatures than flat-plate collectors, because the area of the absorber is much smaller in the concentrating collector than in a flat-plate collector of the same collection area and, thus, the thermal losses from the absorber (assuming the same temperature) for the concentrating collector are much less. The basic disadvantages of focusing collectors are that they must track the sun as it moves across the sky during the day and that they can utilize only direct solar radiation—radiation that comes directly from the sun unscattered by clouds or haze.

Two basic configurations of circular focusing collectors are illustrated in Fig. 10-9. The parabolic dish mirror (Fig. 10-9a) focuses solar radiation into a small circular area which is then absorbed by the *receiver*. The receiver is the device that absorbs the focused solar radiation and transfers the absorbed energy to a heat exchange fluid. With high-quality optical surfaces, a parabolic mirror can obtain temperatures as high as 6000°F. Solar concentrators of this type are used as solar furnaces to produce very high temperatures. A spherical reflector may be used if a precise focus is not required but concentration is needed. The spherical mirror, Fig. 10-9b, produces a broad focus that is not as concentrated as that produced by a parabolic mirror.

Parabolic and spherical reflectors must track the sun across the sky, and this requires a special mounting as shown in Figure. 10-10. The mounting

INTRODUCTION TO ACTIVE SOLAR ENERGY SYSTEMS

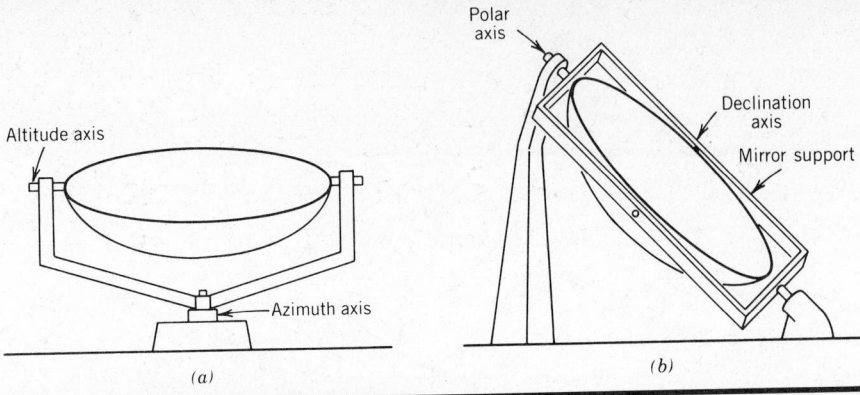

Figure 10-10 (*a*) A dish mirror may be mounted on an altitude-azimuth support. The mirror must be moved continuously in altitude and azimuth, to follow the motion of the sun. (*b*) With an equatorial mount, only one motion—rotation of the polar axis—is required to allow the mirror to follow the sun. As the declination of the sun changes slightly from day to day, the declination axis of the mirror must be adjusted every few days.

illustrated in Fig. 10-10*a* is an altitude-azimuth type, which requires two motions to follow the sun during the day, altitude and azimuth being changed continuously to follow the sun. Fig. 10-10*b* shows an equatorial mounting, which requires only one motion to follow the sun during the day. The mirror is rotated on the polar axis to follow the sun across the sky. The polar axis is mounted parallel to the earth's axis and can be used to counter the effect of the earth's rotation, keeping the mirror pointed toward the sun. The second axis is the declination axis. The sun's motion in declination is very small during any one day, and the mirror must be adjusted for declination only about once a week.

Focusing collectors can also be constructed in a cylindrical or trough shape, as shown in Fig. 10-3. The cross-section of the cylinder may be in the form of a parabola or a circle. The cylindrical mirror focuses the solar radiation on an axis running the entire length of the mirror. The focused radiation is absorbed by a linear structure, such as a pipe, which is placed along the focal axis. This structure is termed the receiver because it receives the concentrated solar radiation and absorbs it. The cylindrical collector may be mounted on a north—south axis, as shown in Fig. 10-11*a*, in which case the cylinder is rotated from east to west during the day to track the sun. An alternative mounting is shown in Fig. 10-11*b*, with the cylindrical collector mounted on an east—west axis. The east—west collector is left stationary during the day and moved only periodically as the sun's declination slowly changes during the year.

10-3   Thermal Storage for Active Systems

The second component of active solar energy systems is the thermal storage. The type of thermal storage provided is ordinarily determined by the heat-transfer medium used in the solar collector. When water or another liquid is used to transfer heat from the solar collector, large tanks of water are typically used to store the heat. If water is used as the heat-transfer medium, it is often circulated directly from the thermal storage to the collectors to be heated and then back to the thermal storage, as shown schematically in Fig. 10-12*a*. In the case of an antifreeze solution's being used in the solar collectors as the heat transfer fluid, a heat exchanger is placed in the thermal storage tank as shown in Fig. 10-12*b*.

The heat-transfer fluid is circulated between the collector and the heat exchanger, carrying heat from the collector to the heat exchanger. As the hot heat-transfer fluid circulates through the heat exchanger, heat flows through the heat exchanger walls to the surrounding water.

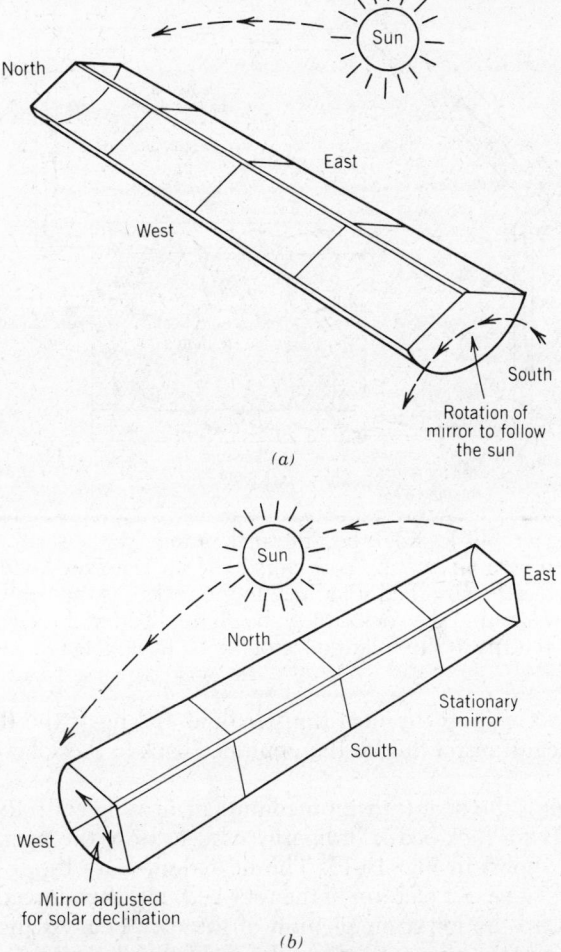

Figure 10-11 Mounting cylindrical or parabolic trough collectors. (*a*) A cylindrical mirror may be mounted in a north–south direction and rotated from east to west to follow the sun's motion during the day. (*b*) An alternative mounting for a cylindrical mirror is along an east-west axis. The mirror is adjusted periodically for the changing solar declination.

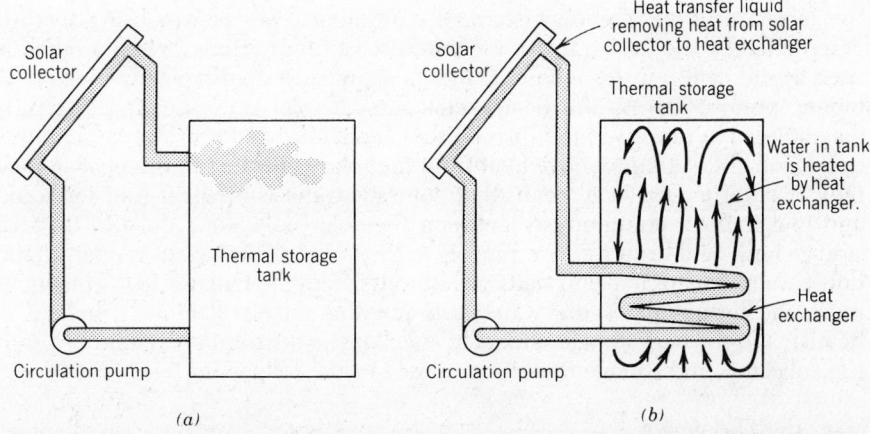

Figure 10-12 Water as thermal storage. (*a*) When water is used as the thermal storage, it may also be the heat transfer fluid circulating through the collectors. The heat transferred from the absorber plate to the water in the collectors is then stored as hot water in the thermal storage tank. (*b*) In some active solar systems, a separate heat exchange fluid is used (e.g., an antifreeze solution) to carry heat from the solar collectors to the storage tank. A heat exchanger is placed in the storage tank to transfer the heat from the collector fluid to the water in the storage tank.

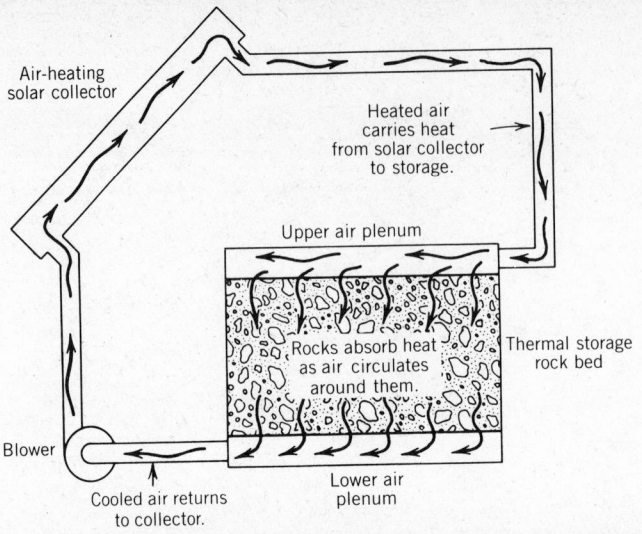

Figure 10-13 Rock bed thermal storage systems are often used with air-heating collectors. The air from the collectors carries the absorbed solar heat to the rocks. As the heated air flows around the rocks, they absorb the heat, and the cooled air returns to the solar collectors to be heated again.

This process cools the heat-transfer fluid and heats the thermal storage. The cooled heat-transfer fluid is then pumped back to the solar collectors to be heated again.

When air is the heat-transfer medium (air-heating collectors are common in the industry), a rock bed is ordinarily used to store the heat from the solar collectors, as shown in Fig. 10-13. The heated air from the solar collectors is blown into the upper air plenum of the rock bed. It is then forced down through the rocks toward the lower air plenum of the rock bed. As the hot air passes around the rocks, the air is cooled and the rocks absorb the heat. There is a large surface-area-to-volume ratio in a rock bed, and this ensures an efficient transfer of heat from the air to the rocks. The cooled air collects in the lower air plenum of the rock bed, and is circulated back to the solar collectors for reheating.

10-4 The Heat Transfer Medium and the Mechanical Distribution System

The movement of thermal energy in an active solar system is dependent upon the mechanical distribution system, the different types of which are in turn determined by the heat-transfer medium used in the system. When a liquid is used as the heat-transfer medium, the mechanical system typically consists of piping, pumps, valves, and heat exchangers. A solar water-heating system is diagrammed in Fig. 10-14 to illustrate the mechanical system. This system uses open-loop circulation to move heat from the solar collectors to the heat storage tank. The term *open-loop* means that domestic water is the heat-transfer liquid and that it flows continuously between the solar collectors and the thermal storage tank, and from there to faucets in the house. The system is open to the house water line for makeup water as hot water from the tank is used. The pump mechanically circulates the water between the solar collectors, where it is heated, and the heat storage tank. The mechanical distribution system between the collectors and storage tank is composed of the following:

1. The piping.
2. The pump.
3. The valves and air vents.

INTRODUCTION TO ACTIVE SOLAR ENERGY SYSTEMS

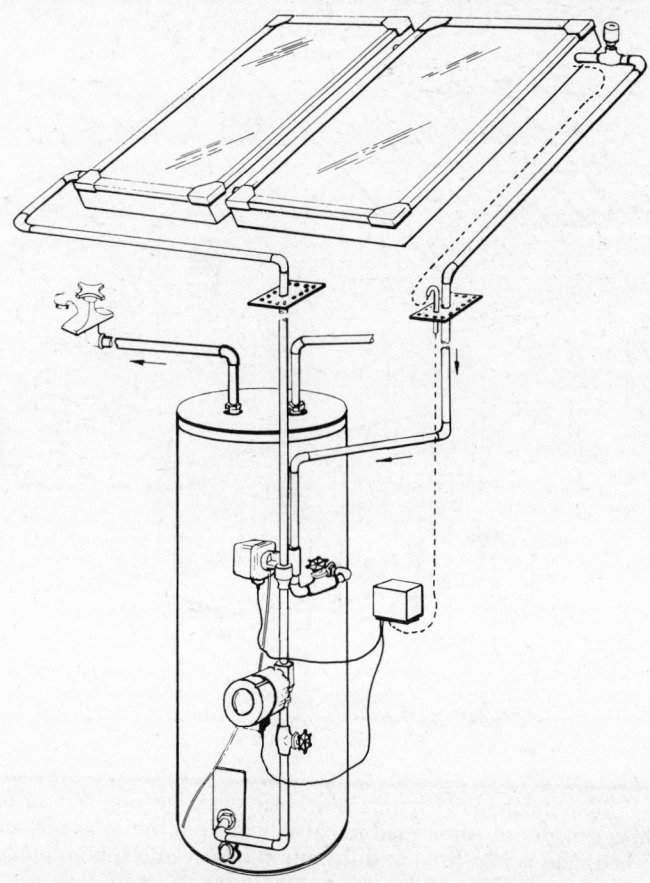

Figure 10-14 Flow diagram of an open-loop domestic solar water-heating system. This active solar system uses water as the heat transfer fluid and heats the domestic water in the thermal storage tank. (Western Solar Development, Inc.)

Domestic hot water from the storage tank is distributed to the house fixtures by the hot-water piping in the house. (Hot-water solar-heating systems will be treated in detail in Chapter 13.)

One possible design for an active solar space-heating system is illustrated by the sketch of Fig. 10-15. The heat transfer medium in this case is air. A blower circulates air through ducting to the air-heating collectors. The heated air then enters the solar air director from the collector return duct. The solar air director contains dampers which control and direct the air flow from the collectors either into the rock bed thermal storage or directly into the spaces to be heated. Another air director allows the backup furnace blower to use heat either directly from the collectors or from the thermal storage. If there is insufficient solar heat, the furnace itself will supply heat to the building. (Space heating by active solar systems will be dealt with in detail in Chapter 15.)

10-5 Controls for Active Solar Systems

The control system of an active solar energy system determines which parts of the system operate at any given time. First, the control system determines when it is effective to remove heat from the solar collectors, either for storage or for some other use. The second function of the control system is to determine when and where the heat from storage or the collectors is to be used. The control system may also have a number of secondary functions, such as:

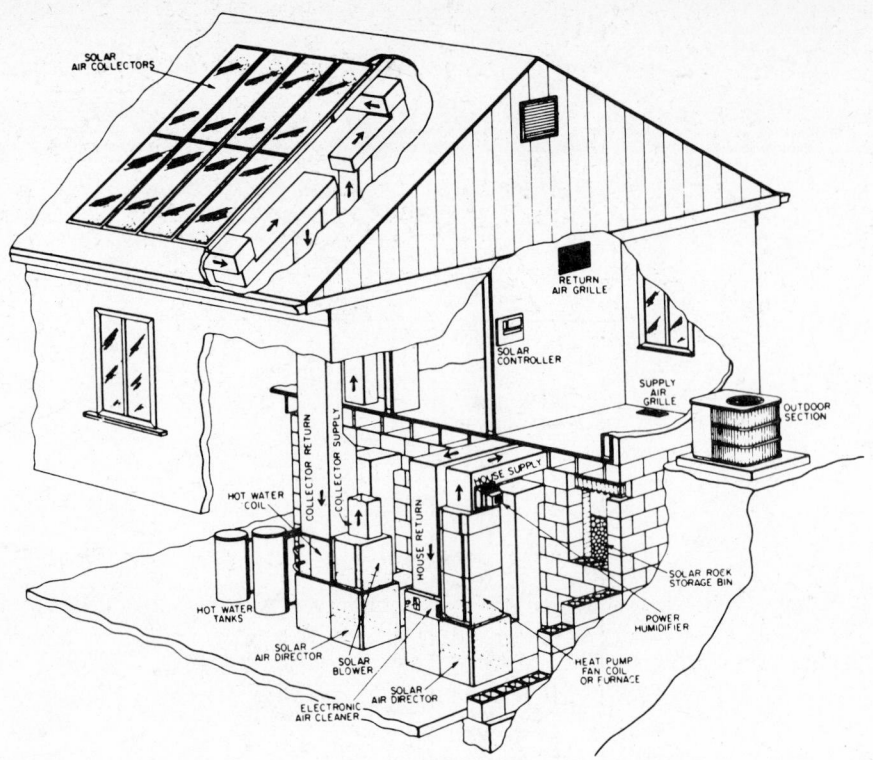

Figure 10-15 The mechanical system of a solar space-heating system using air as the heat transfer medium. Air is used to carry heat from the solar collectors to the rock storage bed, and is also used to distribute the heat into the building.

1. A protection mode for freeze protection in systems using water as the heat-transfer medium.
2. "Priority logic," which directs solar heat to the most critical heating load.
3. Built-in safety features that will protect equipment in the case of component failure and give a warning of system problems.

The *differential-temperature thermostat* (or, more simply, the *differential thermostat*) is the basic control around which most other solar system controls are designed. A differential thermostat is typically an electronic device that compares the temperature at two different locations—for example, at the solar collector outlet and at a point near the *lower portion* of the thermal storage. The purpose of a differential thermostat is to turn on the mechanical system when there is ample heat available for removal from the solar collectors, and to turn off the mechanical system when the collectors are not producing heat in significant amounts.

The electronic circuits in the differential thermostat produce a voltage that is proportional to the temperature difference between the collector outlet temperature and the lower thermal storage temperature. This temperature difference is termed the temperature differential. The voltage representing the temperature difference is compared to two reference voltages, one of which represents an On state and the other an Off state. A typical differential thermostat for solar use is shown (with cover removed) in Fig. 10-16a, and its operating principles are illustrated in Fig. 10-16b.

As the temperature of the collector sensor increases (as the collectors start to warm up), the temperature difference ($T_{coll} - T_{storage}$) will change from a negative value to a positive value, as illustrated in Fig. 10-16b. The control will remain in the Off state, as shown along line FEA. When the temperature

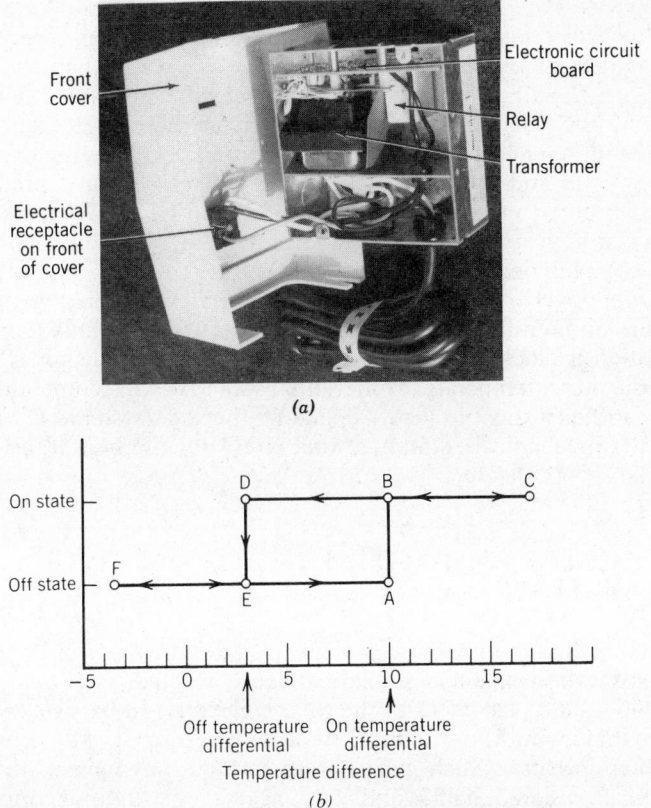

Front cover

Electronic circuit board

Relay

Transformer

Electrical receptacle on front of cover

(a)

On state

Off state

D   B   C

F

E   A

−5   0   5   10   15

Off temperature   On temperature
differential   differential

Temperature difference

(b)

Figure 10-16 The differential thermostat is the central control of many solar energy systems. (*a*) Delta-T® differential thermostat with cover removed, showing internal components. (*b*) In the operation of a differential thermostat, the collector temperature is compared to the storage temperature along the temperature-difference scale. When the temperature difference is greater than the On temperature differential (at *A*), the control will switch to the On state (at *B*). When the temperature difference is less than the Off temperature differential (at *D*), the control will switch back to the Off state (at *E*). (Delta-T is a registered trademark of Heliotrope General Corporation.)

difference becomes equal to the On temperature differential (see point A), the control will switch from the Off state to the On state—that is, from point A to point B in the diagram. The control will remain in the On state as long as the temperature difference is greater than the Off temperature differential. While the control is in the On state, it operates along line *CBD*. When the temperature difference decreases to the Off differential (point *D*), the control will switch from the On state to the Off state (from point *D* to point *E*). It will remain in the Off state (operating along line *FEA*) as long as the temperature difference is less than the On temperature differential (point A).

Manufacturers supply these controls with the option of selecting differential On and differential Off settings. The differential On settings typically range from 5°F to 25°F with 5 F° steps. Some manufacturers supply a control with an adjustable On differential calibrated from 5°F to 25°F. The Off differential is a fixed value in the range of 2 F° to 5 F°.

The differential thermostat controls the system in the following manner. The solar collectors start to heat up in the morning as they absorb the early-morning solar radiation. When the collector outlet temperature exceeds the lower thermal storage temperature by an amount equal to the preset On temperature differential, the control will switch to the On state and turn on the

INTRODUCTION TO ACTIVE SOLAR ENERGY SYSTEMS

pump or blower that circulates the heat transfer medium through the collector. The control will remain On as long as the temperature difference between collector outlet and the lower thermal storage is greater than the preset Off temperature differential. During this time, heat will be removed from the solar collectors by the circulating fluid and placed in thermal storage. When the temperature difference between the collector outlet and the lower thermal storage decreases and becomes equal to the Off temperature differential, the control will switch to the Off state. This condition typically occurs in the late afternoon (on a clear day), when the level of solar radiation absorbed by the solar collectors has decreased to the point that the collector outlet temperature is not high enough to maintain a temperature difference greater than the Off temperature differential. This condition also occurs on cloudy days when the solar radiation is blocked periodically by clouds. If the collector is not able to maintain the preset temperature differential above the Off temperature, it is not producing sufficient heat to justify operating the solar system, and the control turns off the mechanical equipment that circulates the heat transfer medium through the solar collectors.

SOLAR ACCESS AND ACTIVE SOLAR SYSTEMS

An important consideration in maximizing overall system performance of active solar systems is the availability of solar radiation, which dictates to a great extent the orientation and placement of the solar collectors. In passive systems, glass surfaces that face south, or not more than 45° east or west of due south, can be used as *solar apertures*. Such glazed areas—which may have as their primary purpose the provision of light, ventilation, access, or structural continuity—are also able to perform as elements of a passive solar space-heating system. In contrast, the "hardware" of active solar systems is strictly for the purpose of providing a source of heat, and only rarely does it perform as a part of the building skin (i.e., roof or wall). For optimal performance, solar collectors must also be oriented as close to due south as possible, as the following elementary discussion indicates.

10-6 The Geometry of Collector Orientation

It will be recalled that the incident solar energy on any surface is given by the equation

$$I = I_{DN} \cos \Theta + I_d + I_r \qquad (4\text{-}3)$$

The terms of this equation were defined and discussed at length in section 4-4, but brief definitions are repeated here, for convenience:

I is the total solar radiation incident on the collector surface (Btu/hr-ft^2).
I_{DN} is the intensity of direct solar radiation on a surface normal to the direction of the sun's rays (Btu/hr-ft^2).
Θ is the angle of incidence of solar radiation on the collector surface, in degrees.
I_d is the diffuse component of solar insolation (Btu/hr-ft^2).
I_r is the reflected component—the radiant energy of sunlight reflected by the ground and all the surrounding objects, including buildings (Btu/hr-ft^2).

The basic insolation equation (4-3) applies directly to the absorber plate surfaces of solar collectors.
      The angle of incidence Θ is a function of the sun's position in the sky and the orientation of the collector surface. The orientation of the collector is described by two angles: the azimuth angle and the collector tilt angle (Figs. 10-17 and 4-14). The azimuth angle is an angle measured relative to compass points of the horizon. In the northern hemisphere, the reference point is the

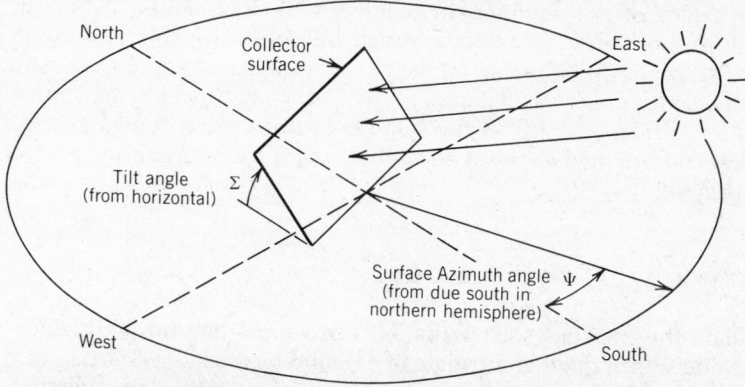

Figure 10-17 Collector orientation angles. The *tilt angle* of a collector is the angle at which the collector surface is tilted from the horizontal plane. The *azimuth angle* is measured from due south in the northern hemisphere.

due south point, and in the southern hemisphere it is the due north point. The collector tilt angle is the angle measured from the horizontal to the plane of the collector surface. (Review also the discussion of section 4-11.)

The collector tilt angle determines how the collector will face the sun at different times of the year. Some general rules can be given for collector orientation, but it must be kept in mind that these may require corrections for local weather patterns in specific locations. One rule is to incline the collector at an angle with the horizontal that is between the latitude and the latitude plus 10°. This collector inclination will give approximately equal insolation on the collector during the summer and winter months—an optimal angle of tilt for a solar system expected to serve a constant load throughout the year, such as a domestic hot-water system. However, for a solar energy system designed to provide space heat for winter only, the collectors should be tilted up to face the (lower) average solar altitude during the winter. This tilt angle would be in the range of latitude plus 15° to latitude plus 20°.

It may not always be practical to face a solar collector directly toward the equator, for a variety of reasons, including such factors as existing roof orientation, lot lines, zoning ordinances, or simply customer preference. Figure 10-18

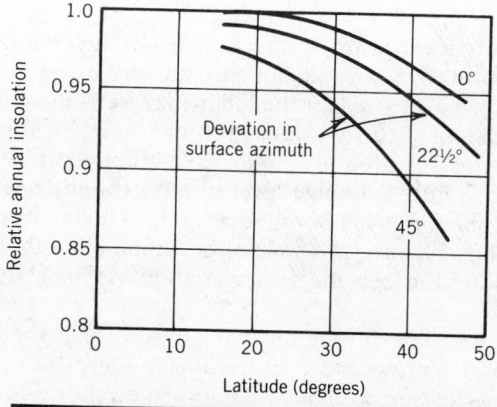

Figure 10-18 Variation in relative annual insolation as a function of deviation in surface azimuth for selected latitudes. *Source:* R. C. Jordan, *Low-Temperature Engineering Applications of Solar Energy*, Atlanta: ASHRAE, p. 69.

SOLAR ACCESS AND ACTIVE SOLAR SYSTEMS

shows a plot of the variation in relative annual insolation as a function of latitude and deviation in azimuth (from due south) for three different surface azimuth deviations. Note that on the curve labeled zero azimuth deviation (that is, a due south collector facing) a 15° N Lat location yields maximum annual insolation, designated 1.0. If the same collector were installed at a 40° N Lat location and oriented 45° away from due south, the yearly insolation would be (relatively) only 0.89.

Illustrative Problem 10-2

Calculate the incident solar radiation for a clear day on a collector surface facing due south, tilted at an angle of 45°, and located at 40° N Lat, on January 21 at 11:00 A.M. LSoT. Assume conditions of no reflected radiation on the surface and full solar access. Use data provided in tables, as indicated.

Solution

From Table 4-E (Chapter 4 Appendix) for 40° N Lat, the insolation I on a 40° tilted surface facing due south at 11:00 A.M. LSoT on January 21 is found to be 277 Btu/hr-ft^2. For a surface tilted at 50°, I is 290 Btu/hr-ft^2. The 45° tilted surface lies midway between the two table values.

Interpolating,

$$I_{45} = 277 \text{ Btu/hr-ft}^2 + \frac{(290 - 277)}{2} \text{ Btu/hr-ft}^2$$

$$= 283.5 \text{ Btu/hr-ft}^2 \qquad \qquad \textit{Ans.}$$

10-7 Solar Access Surveys for Active Systems

The basic considerations of solar access and several commercially available instruments for evaluating solar access were discussed in section 6-6. At this point, the actual procedure for making a solar access survey or audit will be explained.

Before an active solar system is even considered, the availability of solar radiation must be determined. If collectors are installed where they are shaded for much of the day, the performance of the system will be seriously hindered. If, during a solar survey or audit, obstructions are noted that materially reduce the solar access, these facts can be presented to the owners for consideration as part of an overall proposal.

The procedure for measuring the elevation of the skyline and transferring the measured information onto a sunchart was discussed in section 6-6. The Solar Site Selector® will be used for the following discussion. Figure 10-19 shows the Solar Site Selector with its winter grid and summer grid. The Solar Site Selector has a compass (with a magnetic declination chart included) for determining geographic south, a bubble level to level the instrument, and a wide-angle eyepiece to allow viewing of the solar grids. The instrument is set up and leveled on a tripod and oriented toward geographic south by correcting the magnetic compass for the local magnetic variation. A typical field setup is shown in Fig. 10-20a.

A photograph taken through the Solar Site Selector is shown in Fig. 10-20b. This view is representative of the image seen through the wide-angle eyepiece, in which the sunpath lines are imposed on the skyline. The same kind of representation can be obtained by plotting the skyline on a sunchart, as discussed in section 6-6.

In making a solar site audit, a parameter called the *Prime Solar Fraction* is determined. From the solar grids of the Solar Site Selector, a fairly accurate estimate of the percent of total insolation that is blocked by shading obstacles can be made. This is termed *percent of occlusion*. The *percent of insolation* is

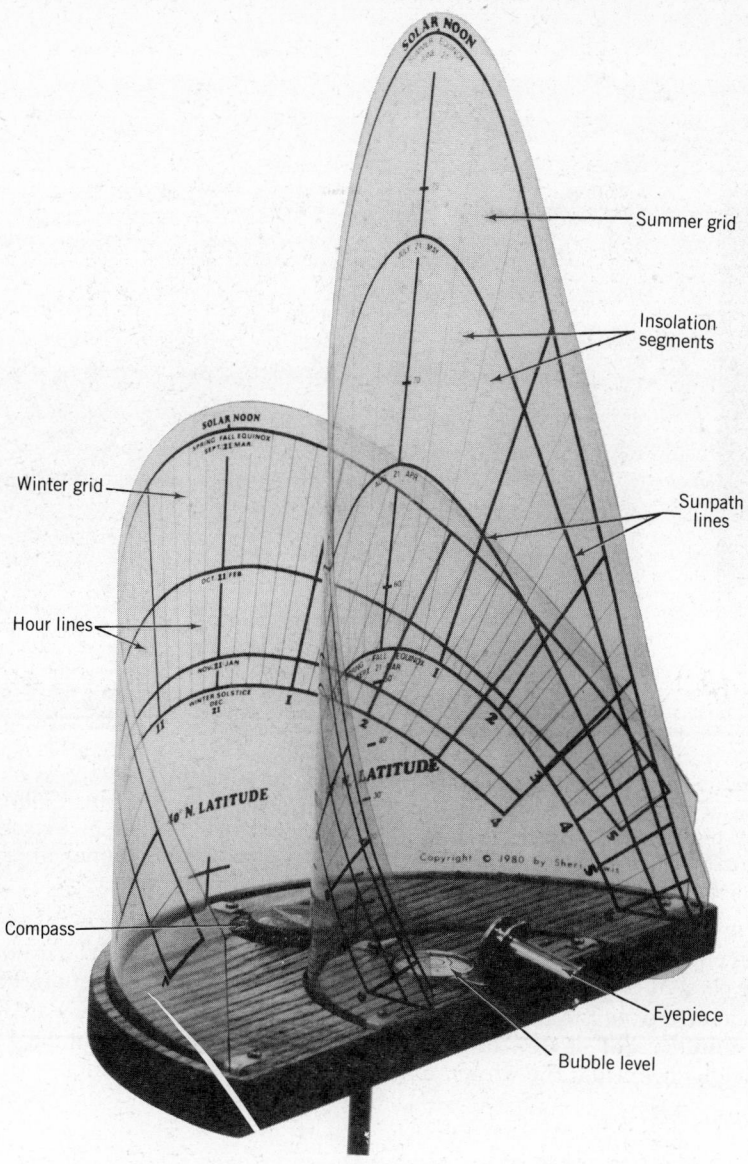

Summer grid

Insolation segments

Sunpath lines

Winter grid

Hour lines

Eyepiece

Compass

Bubble level

Figure 10-19 The Solar Site Selector® with 40° N Lat grid in place. The instrument uses two transparent sun charts: the winter solar grid for evaluating solar access in winter and the summer solar grid for evaluations of solar access in summer. (Lewis and Associates, Grass Valley, California.)

100 minus the percent of occlusion. The prime solar fraction (PSF) is then defined as the percent of insolation divided by 100 percent:

$$\text{Prime solar fraction} = \frac{\text{Percent of insolation}}{100 \text{ percent}} \qquad (10\text{-}1)$$

The percent of occlusion is determined in two steps using the Solar Site Selector. First, a judgment must be made as to whether the insolation segments of the solar grid are solidly covered (70 percent to 100 percent shaded) or partially covered (less than 70 percent shaded). A grid segment may be partially covered in two ways: a solid object may cover less than 70 percent of the grid segment, or a nonsolid object, such as a leafless tree, may cover most or all of a grid segment. The number of solidly covered grid segments is added to half of the total number of partially covered grid segments. This number of covered

Figure 10-20 Field use of the Solar Site Selector. (*a*) The instrument is leveled on a tripod, and the skyline is viewed through it to determine the solar access for that location and time of the year. (Lewis and Associates, Grass Valley, California.) (*b*) View photographed through a Solar Site Selector showing buildings and trees on the horizon, all superimposed on the pattern of sunpath lines of a sunchart. (Photo courtesy Santa Barbara Solar Energy Systems.)

grid segments is multiplied by a correction factor given in Table 10-1 to obtain the percent of occlusion. The percent of insolation is then obtained by subtracting the percent of occlusion from 100 percent. The prime solar fraction (PSF) is then obtained from Eq. (10-1). A winter PSF is determined from the winter grid, and a summer PSF is determined from the summer grid. An annual PSF is obtained by averaging the winter and summer prime solar fractions.

Illustrative Problem 10-3

A simulated view through a Solar Site Selector at 34°N Lat is shown in Fig. 10-21. Determine the winter, summer, and annual prime solar fractions.

Table 10-1 Correction Factors for the Solar Grids Used with the Solar Site Selector

Winter Grid	
Grid Latitude	Correction Factor for Obtaining the Percent of Occlusion
20°–22°	0.88
24°–32°	0.91
34°–42°	0.94
44°–56°	1.00
58°–60°	1.09

Summer Grid
   All summer grids have 100 segments; thus, the correction factor is 1.0 for all latitudes.

Source: Adapted with permission from *Solar Site Selector Instructions,* Lewis and Associates, Grass Valley, California.

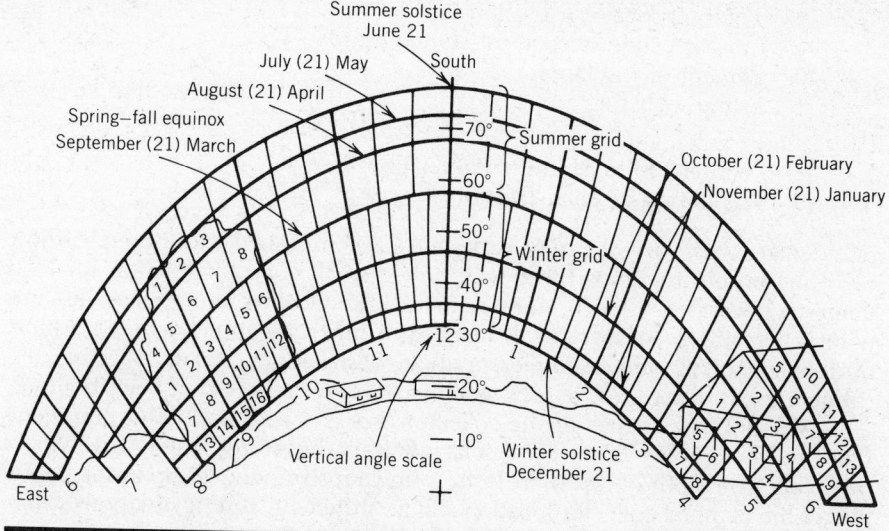

Figure 10-21 Diagram of a simulated view through a Solar Site Selector, for Illustrative Problem 10-3. A building to the west obstructs the late afternoon sun. A deciduous tree in the east results in partial winter shading and full summer shading.

Solution

The winter PSF is calculated as follows. The building in the west solidly covers (70 percent to 100 percent shaded) 8 winter grid segments and partially covers 2 winter grid segments. The deciduous tree in the east without its leaves (winter) partially covers 16 winter segments. The total of covered winter segments is then $8 + (2 \div 2) + (16 \div 2) = 17$ segments. For 34° N Lat, the correction factor from Table 10-1 is 0.94. The percent of occlusion is then $17 \times 0.94 = 15.9$, or 16 percent. The winter percent of insolation is 100 percent $-$ 16 = 84 percent, and from Eq. (10-1) the winter PSF is 0.84.   *Ans.*

The summer PSF is calculated as follows. The deciduous tree (east) will have full foliage in the summer and will provide solid shading. The tree covers 8 summer grid segments. The building (west) solidly covers another 13 summer grid segments and partially covers 4 additional segments. A total of 23 insolation segments $(8 + 13 + 4 \div 2)$ are blocked. From Table 10-1, the correction factor is 1.0 for the summer months, and the percent of occlusion will be 23 percent. The summer percent of insolation is therefore 77 percent, and the summer PSF is 0.77.   *Ans.*

The annual PSF is found by averaging the winter and summer PSFs:

$$\frac{(0.84 + 0.77)}{2} = 0.81$$

Ans.

The above problem is summarized in the following sample audit worksheet.

Solar Site Audit Worksheet for Illustrative Problem 10-3

I. Winter calculations
   1. Solidly shaded segments   8
   2. Partially shaded segments $18 \div 2 =$ 9
   3. Total uncorrected percent of occlusion $\overline{17}$
   4. Corrected percent of occlusion $.94 \times 17 =$ 16
   5. Percent of insolation $100\% - 16 =$ 84
   6. Prime solar fraction (PSF) $84 \div 100 =$ $\underline{0.84}$

II. Summer calculations
   7. Solidly shaded segments   21
   8. Partially shaded segments $4 \div 2 =$ 2

9. Total percent of occlusion $\underline{23}$
   (no correction needed for summer grids)
10. Percent of insolation $100\% - 23 = \underline{77}$
11. Prime solar fraction (PSF) $77\% \div 100 = \underline{0.77}$

III. Annual calculations
12. Winter PSF 0.84 plus summer PSF 0.77 $= \underline{1.61}$
13. Annual prime solar fraction $1.61 \div 2 = \underline{0.81}$

Some solar access surveys can be very simply accomplished—for example, those involving no obstructions, those with only a small collector area (as for solar domestic hot-water systems), or those involving collectors mounted at the same point where the solar survey was made. But other situations demand a more thorough analysis—for example, multiple obstructions, solar collectors distributed over a large area, or jobs with complex elevation difference situations.

The anticipated size of the collector area is a very important consideration. To check solar access for a solar domestic hot-water system involving a total collector width of less than 15 ft, a single instrument set up at the center line of the planned collector location will be sufficient. But in situations where the collector area will have a width greater than 15 ft, at least two instrument setups are required—one at the east end and one at the west end of the planned collector bank. The reason for this is that the solar survey instrument views the skydome or solar window from a single point. A single survey set up at the center of a long collection area, as shown in Fig. 10-22 at point B, would indicate no obstructions between the hours of 9:00 A.M. and 3:00 P.M. However, a glance at the field of view from points A and C clearly shows that if the obstructions indicated in the diagram are tall enough, they would produce both morning and afternoon shading. In such cases it is necessary to break the collection area up into two or more segments and calculate the PSF for each segment. Then the PSF obtained can be averaged by segment area to obtain a total PSF.

The Problem of Elevation Difference.   If the collectors will be at a different elevation from that at which the solar survey is made, an elevation correction is in order. The solar access from an elevated position, for example, will not be the same as that viewed from ground level. Many survey situations will show significant solar obstructions from ground level while showing little or no obstruction from rooftop or collector level. In some cases it may not be possible to make a rooftop sighting, or, indeed, the building that the solar energy installation will serve may not yet exist. In these situations an extrapolation of a ground-level sighting to a rooftop level will have to be made.

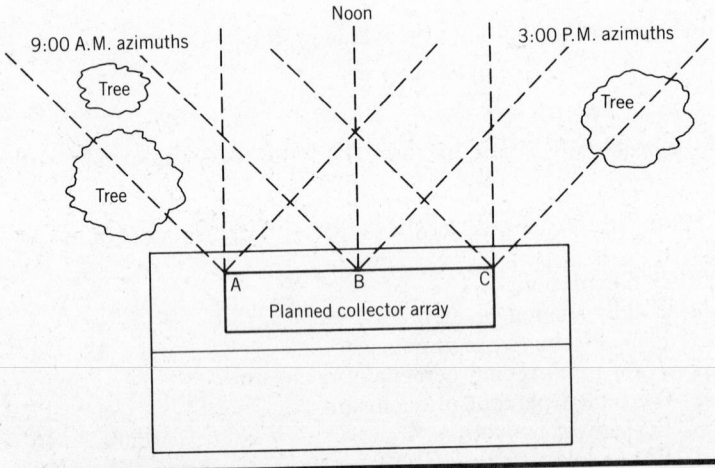

Figure 10-22 Solar access as viewed from three points along an extensive solar collector array. From point B, there appear to be no obstructions. From points A and C, the trees will interfere with solar access during part of the day.

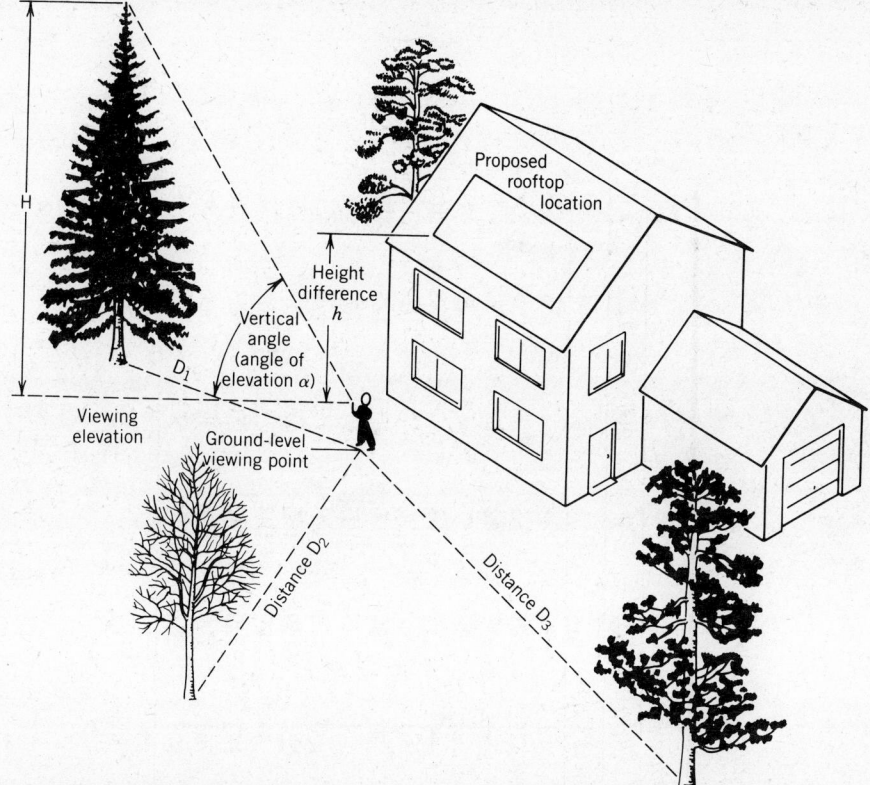

Figure 10-23 Extrapolating a ground-level solar survey to a rooftop location. The height difference h is measured from the (planned) collector location on the roof to the actual point of the observations. For any potential obstruction, the distance to the obstruction and its vertical angle (angle of elevation) must be noted.

To extrapolate shading patterns from ground level to any elevated position, the following procedure is recommended:

1. Select a ground-level viewing point closely adjacent to the proposed rooftop location, as shown in Fig. 10-23. Then measure the height difference h between the viewing elevation and the rooftop location.
2. Place a piece of clear acetate or vinyl in front of the solar grid of the Solar Site Selector.
3. Use a grease pencil or felt-tip transparency pen to trace the skyline or shading profile on the clear acetate.
4. Measure the horizontal distance D from the Solar Site Selector to each object that might appear on the rooftop profile (trees, buildings, etc., that may shade the roof).
5. Determine the angle of elevation of each object using the angle scale on the noon line of the solar grid. (Note that the Solar Site Selector uses the term *vertical angle* for the angle commonly referred to as the angle of elevation.)
6. From Table 10-2, determine the height H of each object above the viewing elevation, using the vertical angle α and distance D to the object.
7. Determine the height of each object above the roof H_R, from

$$H_R = H - h$$

8. For each object, a rooftop vertical angle is now determined from Table 10-2 using the distance D and the height H_R. With the

Table 10-2 Height Calculation Table for Solar Access Surveys

Vertical Angle of Object α	Distance to Object D (ft)											
	10	20	30	40	50	60	70	80	90	100	125	150
	Height of Object H (ft)											
15°	3	5	8	11	13	16	19	21	24	27	33	40
20°	4	7	11	15	18	22	26	29	33	36	45	55
25°	5	9	14	19	23	28	33	37	42	47	58	70
30°	6	12	17	23	29	35	40	46	52	58	72	87
35°	7	14	21	28	35	42	49	56	63	70	88	105
40°	8	17	25	34	42	50	59	67	76	84	104	126
45°	10	20	30	40	50	60	70	80	90	100	125	150
50°	12	24	36	48	60	72	83	95	107	119	149	179
55°	14	29	43	57	71	86	100	114	129	143	179	214
60°	17	35	52	69	87	104	121	139	156	173	217	260
65°	21	43	64	86	107	129	150	172	193	214	268	322
70°	27	55	82	110	137	165	192	220	247	275	343	412
75°	37	75	112	149	187	224	261	299	336	373	466	560
80°	57	113	170	227	284	340	397	454	510	567	709	851

$H = D \tan \alpha$

clear acetate laid flat over the solar grid, plot the new vertical angles for each object and then draw a new skyline profile.

9. The prime solar factor (PSF) can now be determined using the extrapolated rooftop skyline profile and the methods elaborated above.

Although the above procedure may seem cumbersome, after two or three trials it will become quite routine.

Illustrative Problem 10-4

A ground-level skyline profile is depicted in Fig. 10-24, with obstructing objects sketched in and with their angles and horizontal distances indicated. Extrapolate to a rooftop location that is 20 ft above the ground-level point from which the solar access survey was made.

Solution

The skyline is traced on a clear piece of acetate in front of the solar grid, as illustrated in Fig. 10-25a.

The horizontal distance D to each obstruction is measured and noted, as in Fig. 10-25a.

Obstruction	Distance D
Building (east)	100 ft
House (south)	60 ft
Tree (west)	20 ft

The vertical angle α of each obstruction is read from the vertical angle scale of the grid.

Obstruction	Vertical Angle α
Building	32°
House	35°
Tree	50°

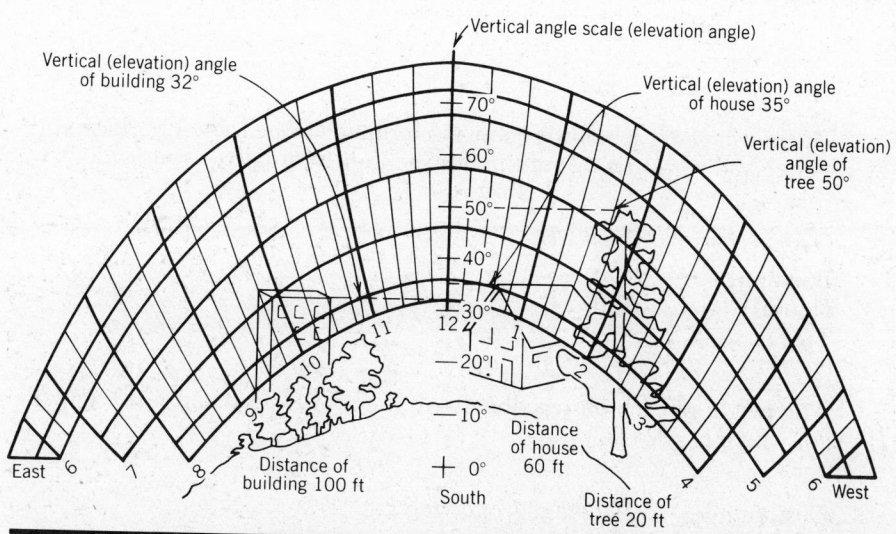

Figure 10-24 Simulated skyline for Illustrative Problem 10-4. There are three obstructions: a building, a house, and a tree. Distances from the observation point to each obstruction are shown.

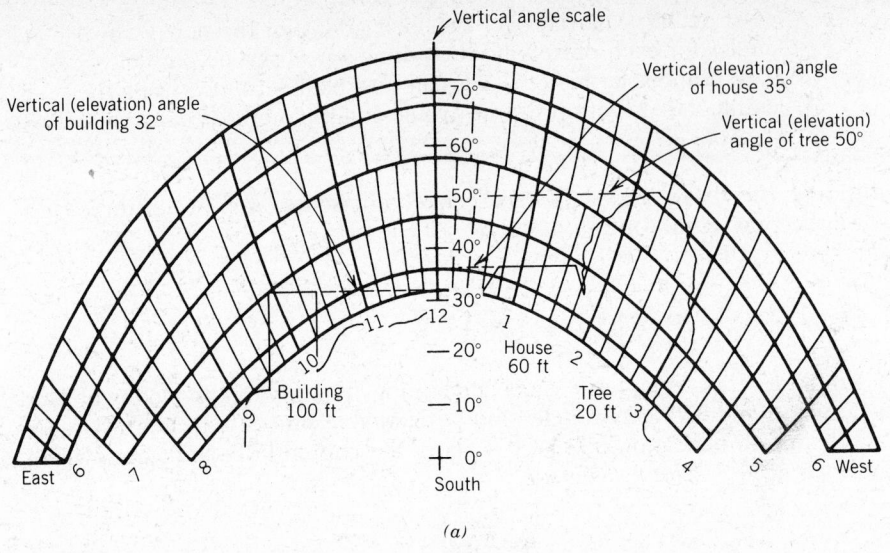

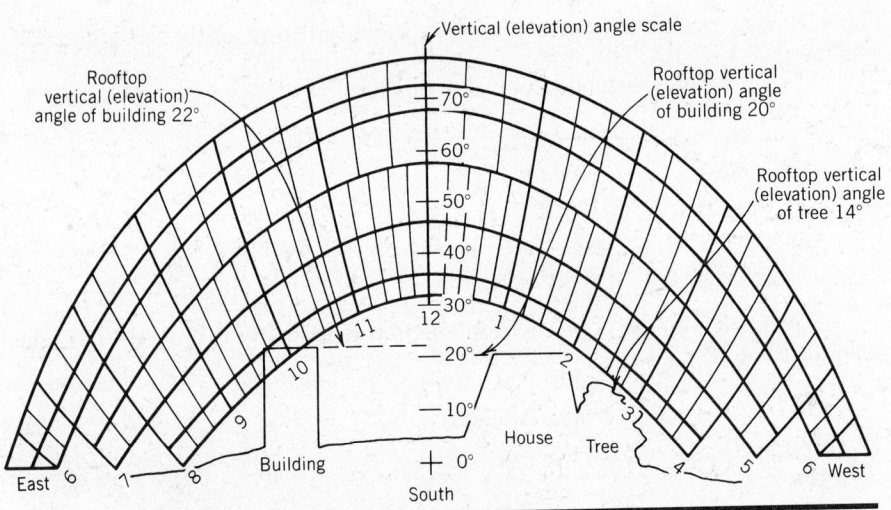

Figure 10-25 Skyline profiles for Illustrative Problem 10-4. (*a*) Profile as viewed from the ground location. (*b*) The skyline profile as extrapolated to the rooftop location 20 ft above the actual viewing point.

The height H of each obstruction is determined from Table 10-2, based on the distance and vertical angle of the obstruction. Interpolate as needed.

Obstruction	Height H
Building	61 ft
House	42 ft
Tree	24 ft

The height of each obstruction above the roof H_R is determined. ($H_R = H - h$), where $h = 20$ ft.

Obstruction	Height above rooftop H_R
Building	41 ft
House	22 ft
Tree	4 ft

A new (rooftop) vertical angle for each obstruction is now determined from Table 10-2, interpolating, using the distance D and the height H_R (as determined above).

Obstruction	Rooftop vertical angle
Building	22°
House	20°
Tree	14°

A new skyline profile is now drawn using these rooftop vertical angles, as shown in Fig. 10-25b.

The skyline profile, as extrapolated to the planned location of the collectors on the roof, shows that there is almost 100 percent solar access from the rooftop location.

CONCLUSION

This chapter has provided a brief introduction to active solar system design and has identified the essential elements that comprise an active system: solar collectors, thermal storage, heat transfer media, mechanical systems for heat distribution, and controls. Most of the topics treated briefly in this chapter as an introduction will be subjects for detailed study in later chapters. We turn next to the subject of solar collectors—their construction and their performance on selected applications.

PROBLEMS

1. A black metal swimming pool collector (unglazed) has an operating temperature of 95°F. The ambient temperature is 75°F. Determine the convection heat loss per square foot when the wind speed is 2 mph.

2. Estimate the radiation loss per square foot surface area for an uncovered black surface at 130°F with an ambient temperature of 80°F.

3. An unglazed black metal surface is exposed to solar radiation at normal incidence. Calculate the intensity of solar radiation (insolation) necessary to maintain the plate at a temperature of 95°F when the ambient air temperature is 65°F and the wind speed is 10 mph. Assume a black surface absorptance of 0.92.

4. Calculate the incident solar radiation for a clear day on a collector surface facing southwest, tilted at an angle of 34°, and located at 24° N Lat, on March 21 at 2:00 P.M. LSoT. Assume no ground reflection.

5. What is the total clear-day solar radiation incident on a southeast-facing collector, tilted at an angle of 42° and located at 32° N Lat, on November 21? Assume no ground reflection.

6. Estimate the prime solar fraction for the ground-level location shown in Fig. 10-25a. Location 34° N Lat.

7. Estimate the prime solar fraction for the rooftop location shown in Fig. 10-25b. Location 34° N Lat.

8. For the ground-level skyline shown in Fig. 10-24, make a rough sketch of the skyline profile as seen from a rooftop location 10 ft above the viewing point.

9. For the ground-level skyline shown in Fig. 10-21b, make a sketch of the skyline profile as seen from a rooftop location 15 ft above the viewing point. Assume that the tree in the east has a horizontal distance from the viewing point of 30 ft, and the house in the west a horizontal distance of 60 ft.

10. Determine the prime solar fraction for the rooftop skyline obtained in the solution of problem 10-9. Assume a 34° N Lat location.

Flat-Plate Solar Collectors

*I*n Chapter 10, solar collectors were considered as one element of an active solar system. As the energy collecting element, the solar collector is the single most important component of an active system. Without the solar collector, the other elements of the system would serve no function. Therefore, it is important to understand how collectors are constructed and how they function as energy collection devices.

Of the several types of solar collectors, the most common is the *flat-plate solar collector*. The purpose of this chapter is to explain in detail the construction of liquid-heating and air-heating flat-plate solar collectors.

Flat-plate collectors can be divided into two basic categories:

1. Those that heat a liquid, such as water, and have fluid passages in the collector (Fig. 11-1).
2. Those that heat a gas, such as air, and have large air-flow passages (Fig. 11-2).

Flat-plate collectors are most commonly used for space heating and for domestic hot-water heating, as well as for pool and spa heating.

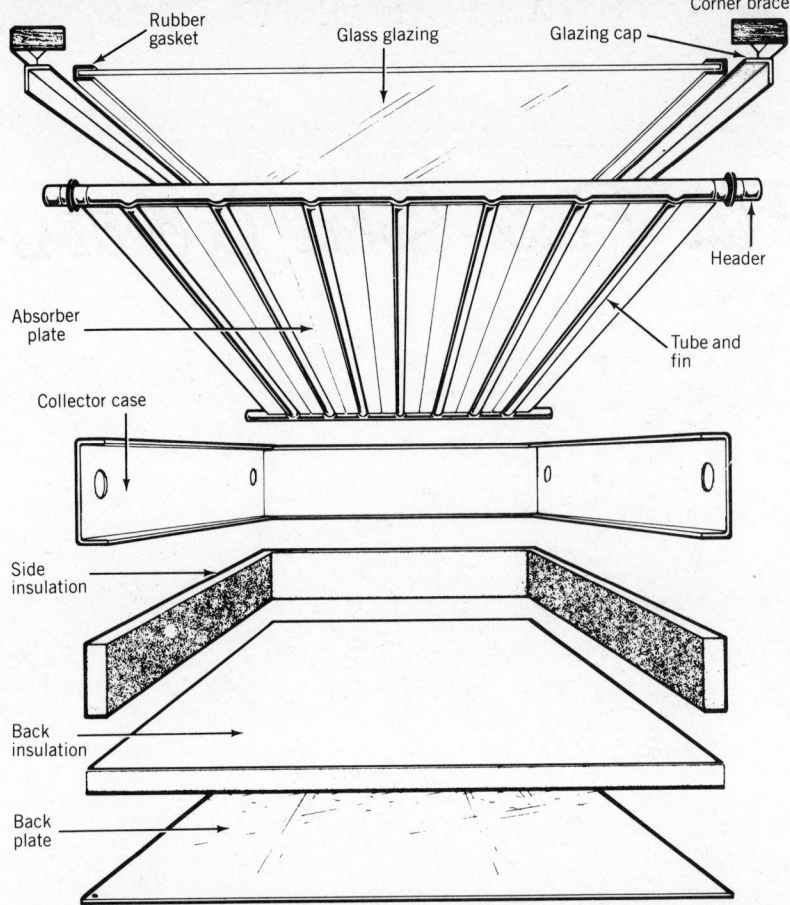

Figure 11-1 Flat-plate liquid-heating solar collector in exploded view, showing glazing, absorber plate with tubes, insulation on bottom and sides, and metal case. This is representative of a factory-assembled collector. (Western Solar Development, Incorporated.)

11-1   Components of a Flat-Plate Collector

Figure 11-3 shows a cross-section cutaway view of a typical flat-plate collector. The absorber plate inside the collector converts sunlight into heat energy and transfers this thermal energy to a fluid such as water or air. This fluid, in turn, removes the thermal energy from the collector and transfers it either to the space or product being heated or to thermal storage. Various types of absorber plate designs will be discussed later.

The front surface of the absorber plate has a black coating that absorbs solar radiation. The coating can be either flat-black paint or any one of a number of chemically deposited films. The back and sides of the absorber are insulated to prevent heat losses from the edges and back of the collector, as was pointed out in Chapter 10. Figure 11-1 shows the side insulation.

The boxlike enclosure containing the absorber plate and its insulation is termed the *collector case*. It provides a rigid frame for supporting the collector assembly and a weathertight enclosure for protection of the absorber plate and insulation. Collector cases can be made of aluminum, steel, or wood.

The collector assembly is completed by placing a transparent covering (called the *glazing*) over the absorber plate, secured by the collector case. This glazing transmits solar radiation to the absorber plate and minimizes heat losses from the hot absorber plate. It also provides a weatherproof front surface for the collector. Some collectors have more than one layer of glazing material, the

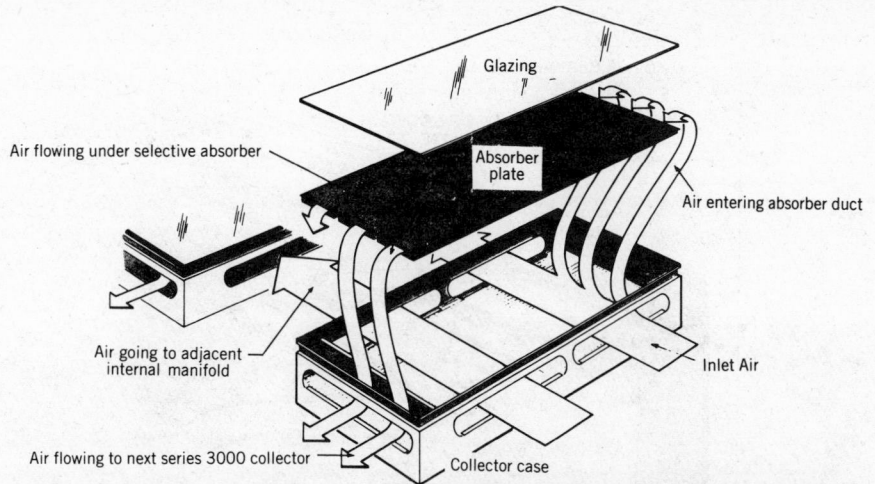

Glazing

Air flowing under selective absorber

Absorber plate

Air entering absorber duct

Air going to adjacent internal manifold

Inlet Air

Air flowing to next series 3000 collector

Collector case

(b)

Figure 11-2 Flat-plate air-heating solar collector. (*a*) Exploded view of an air-heating collector. The air flow through the collector is shown. There are internal air manifolds which allow adjacent collectors to be connected together. (*b*) A single roof-mounted air heating collector. *Source:* Solaron Corporation, Englewood, Colorado.

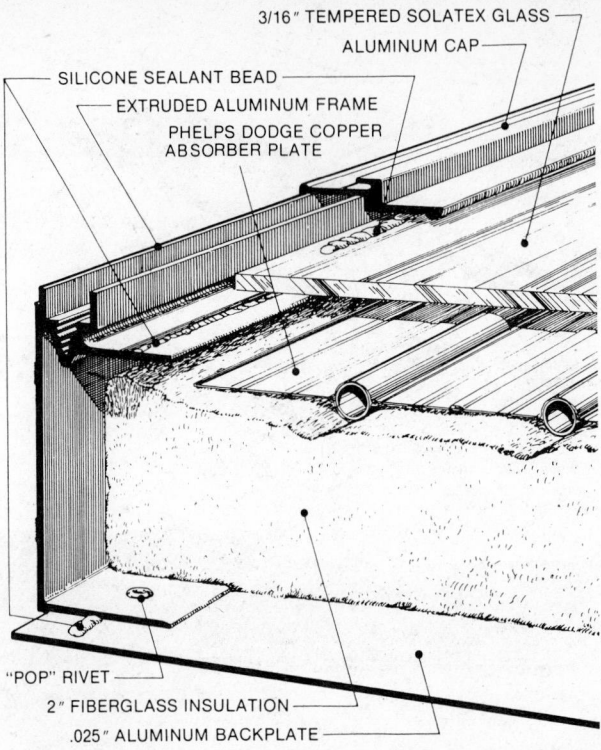

3/16" TEMPERED SOLATEX GLASS
ALUMINUM CAP
SILICONE SEALANT BEAD
EXTRUDED ALUMINUM FRAME
PHELPS DODGE COPPER
ABSORBER PLATE
"POP" RIVET
2" FIBERGLASS INSULATION
.025" ALUMINUM BACKPLATE

Figure 11-3 Section through a flat-plate liquid-heating solar collector, showing the absorber plate, fluid passages, absorber plate fins, the collector case, collector insulation, glazing, glazing gasket, and glazing cap. (Archer Industries, Solar Systems Division.)

type and number of glazings depending on the operating temperature of the collector and the type of coating on the absorber plate. A later section will deal with glazings in detail.

11-2 Liquid-Heating Absorber Plates

The absorber plate is the essential component of the solar collector. It is generally metallic, although flat-plate collectors can be made of a variety of materials, including rubber and plastic. The use to which the plate is to be put—heating water or water with antifreeze additives, heating air or other gases—will often determine the material of which it is made. The basic function of the absorber plate is to absorb as much of the solar radiation reaching it as possible. Other elements of the collector are provided to minimize losses and to assist the absorber plate in transferring the absorbed solar energy to the heat transport fluid flowing through the collector.

Collector Tubes and Fins.   When the collector is to heat liquids, the tubes or fluid passages must be intimately attached to the absorber plate with a good thermal bond. One of the most common kinds of absorber plates is the tube-and-fin type illustrated in Fig. 11-4. As solar radiation heats the fin, its temperature increases above that of the fluid in the tube. This causes heat to flow from the ends of the fin toward the tube. A temperature gradient (Fig. 11-5) is established along the fin, with the highest temperature at the ends of the fin and the lowest temperature where the fin is attached to the tube. The fin temperature never drops as low as the fluid temperature in the tube because of the thermal resistance of the bond between the fin and the tube and the thermal resistance of the metal of the tube itself. Fluid temperature in the tube is less

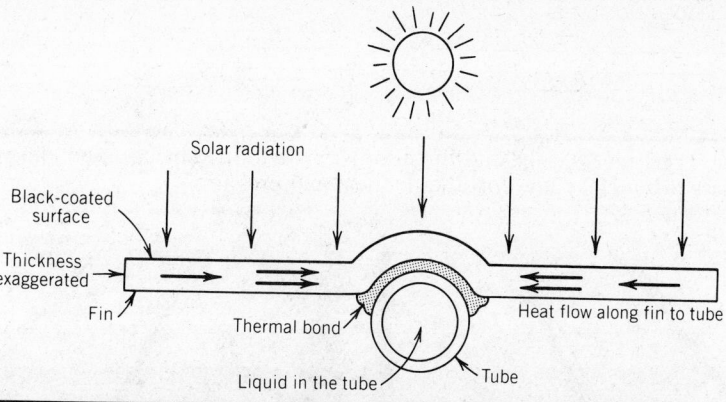

Figure 11-4 Heat flow in a section of a tube-and-fin type of absorber plate. In order for the absorbed solar radiation to reach the liquid flowing in the tube, heat must flow from the fin across the thermal bond between the fin and tube.

than the tube wall temperature, since cool fluid is continually being supplied to the tubes by a circulating pump or by natural convection. Fin temperature is affected by the thermal bond, the thickness of the fin, the length of the fin, and the conductivity of the fin material. Fin temperature will be high if the fin is too thin to provide good conduction or if the fin is too long; and it will be high if the thermal bonding between fins and tubes is poor. It is important to have the fin temperature fairly close to the fluid temperature, since excessively high fin temperatures result in substantial heat losses from the collector.

The major challenge with the tube-and-fin type of absorber plate has been to obtain a good thermal bond between the absorber plate and the fluid passages without incurring excessive production costs.

Materials most frequently used for absorber plates are copper, aluminum, steel, and stainless steel. In addition, many types of plastics and rubber have been used for low-temperature absorber plates for collectors used in swimming pool and spa heating.

Integral-Tube Absorber Plates. An absorber plate with integral tubes formed in the plate itself is shown in the diagram of Fig. 11-6. This design ensures a good thermal connection between the plate and the tubes and there-

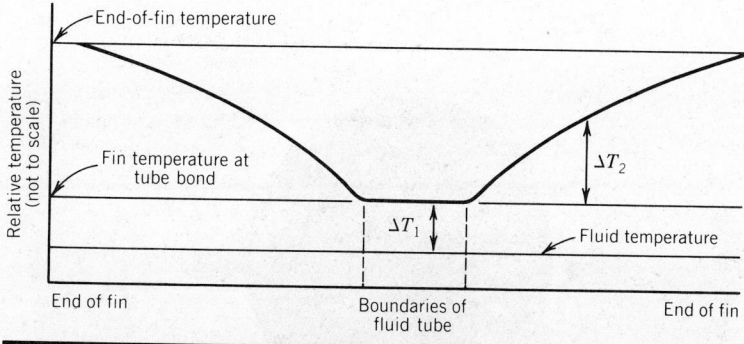

Figure 11-5 Typical fin-temperature profile for a finned-tube absorber plate. Where the fin is bonded to the tube, the fin will have a relatively constant temperature slightly higher than the fluid temperature. The temperature difference (ΔT_1) between fluid and fin is caused by the thermal resistance of the metal tube wall plus the thermal resistance of the thermal bond. The fin temperature increases as one moves from the fluid tube toward the end of the fin. The temperature difference ΔT_2 is caused by the thermal resistance of the fin material.

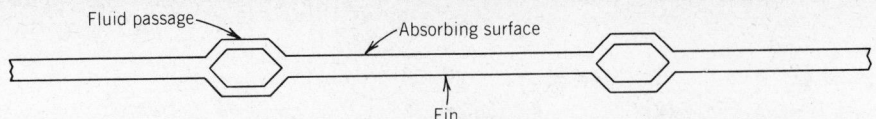

Figure 11-6 Integral tube-and-fin absorber plate formed by a roll-bonding process that fuses two metal plates together by heat and pressure.

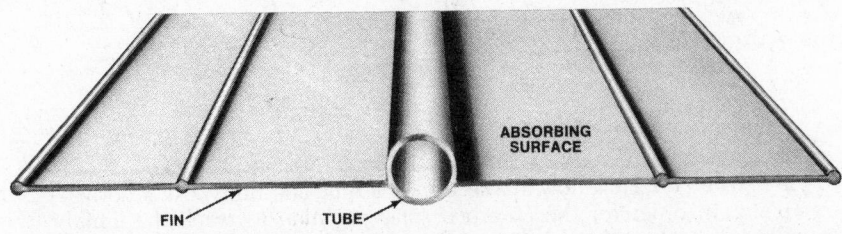

Figure 11-7 An integral tube-and-fin section can be extruded from a die. The sections are then combined to form the complete absorber plate. (Reynolds Aluminum Company.)

fore provides excellent heat transfer. These absorbers can be fabricated of aluminum or copper plates by means of a roll-bonding process similar to that used in the manufacture of radiators, refrigerant coils, condensers, and other types of heat exchangers. Another method of fabrication of the integral-tube plate is to produce an extrusion of the form desired (Fig. 11-7). With this design, a parallel grouping of several tube–fin units is used to form the complete absorber plate.

Brazed or Soldered Tube-and-Fin Construction. Some absorber plates are fabricated by brazing or soldering copper or steel tubes to the top or bottom of the absorber plate. Figure 11-8 shows how copper tubes are soldered or brazed

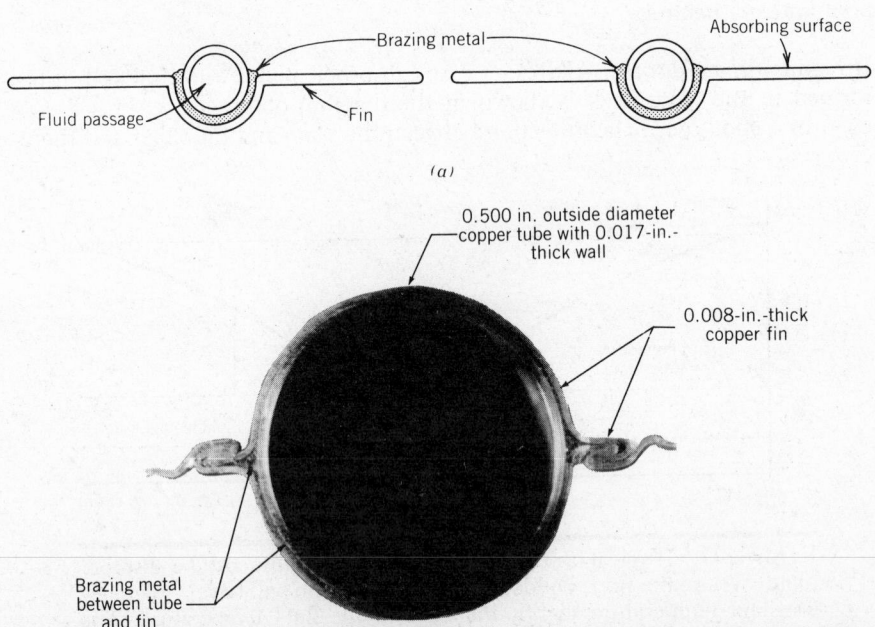

Figure 11-8 Brazed copper tube-and-fin assembly. (*a*) Copper fins are first formed to fit around the outside diameter of the tube. Then the tube and fin are mechanically and thermally attached by brazing them together. (*b*) Photo of section cut through a tube and fin. (Phelps Dodge Solar Enterprises.)

to copper fins. The bonding is important for reasons of mechanical strength as well as for low thermal resistance. Laboratory studies have shown that there is a wide variation in the efficiency with which the bonds will conduct heat. Values as high as 1000 Btu/hr-ft²-F° are reported for securely soldered or brazed tubes, and values as low as 3.2 Btu/hr-ft²-F° have been noted for poorly crimped or badly soldered tube–fin bonds.

Mechanically Fastened Tubes.   Figure 11-9 shows two methods of mechanically fastening fins to a copper tube. In Fig. 11-9*a*, an aluminum fin extrusion is clamped onto a copper tube. A filler of thermally conducting material is often placed in the extrusion before it is clamped over the tube, to fill any voids that might occur between the tube and fin surfaces. Figure 11-9*b* illustrates another approach to mechanically fastened tubes, with the tube on the back of the fin extrusion.

    Both the integral-tube design and the tube-and-fin design have the advantage of being able to accommodate a wide range of fluid pressures. Operating pressures for these designs are determined by the tube wall thickness in the same manner as with standard tubing used in any other application.

Connecting the Tubes.   There are two basic patterns for connecting the fluid passages in a flat-plate absorber. Both are illustrated in Fig. 11-10. To create the *sinusoidal* or *serpentine* absorber (Fig. 11-10*a*), the finned tube is bent back and forth so that the finned-tube sections form a series flow path in the collector case. The liquid to be heated enters a tube at the lower end of the sinusoidal tube and flows upward, absorbing heat from the finned-tube structure as it moves toward the exit at the top.

    In the *parallel-tube* absorber plate (Fig. 11-10*b*), the finned tubes run parallel to each other and are connected to a *header* tube at each end of the absorber plate. Liquid to be heated enters the lower header and is broken into parallel fluid streams passing up the parallel tubes. The streams are combined again at the upper header, and the liquid flows out. Generally, the liquid enters the lower header on one side of the absorber and leaves the upper header on the opposite side so that flow paths of equal resistance will be provided, ensuring equal flow rates in all branches.

Non-Tube-and-Fin Absorber Plates.   There are other methods of fabricating absorber plates that do not use the tube-and-fin design. Figure 11-11 illustrates a method of fastening two sheets of metal together to form a watertight flow channel. This is called *parallel-plate* construction, with the two plates ordinarily spot-welded or riveted together. Copper, steel, or stainless steel can be used, and the fluid flows under the upper plate, in direct contact with the surface that is absorbing solar radiation. Excellent heat transfer results, and the

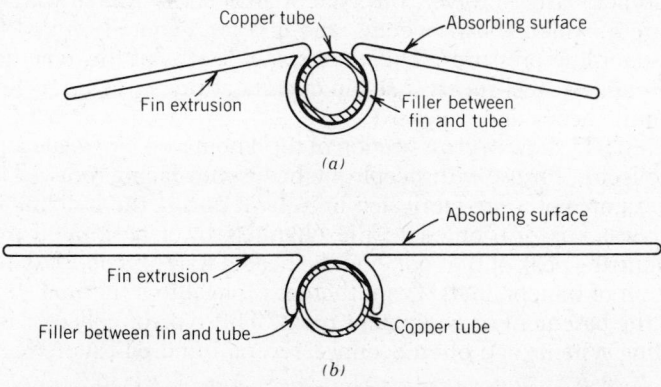

Figure 11-9 Clamped aluminum fin and copper tube assemblies. (*a*) This fin is clamped onto the tube from the bottom. (*b*) Fins may also be clamped over the top of the tube.

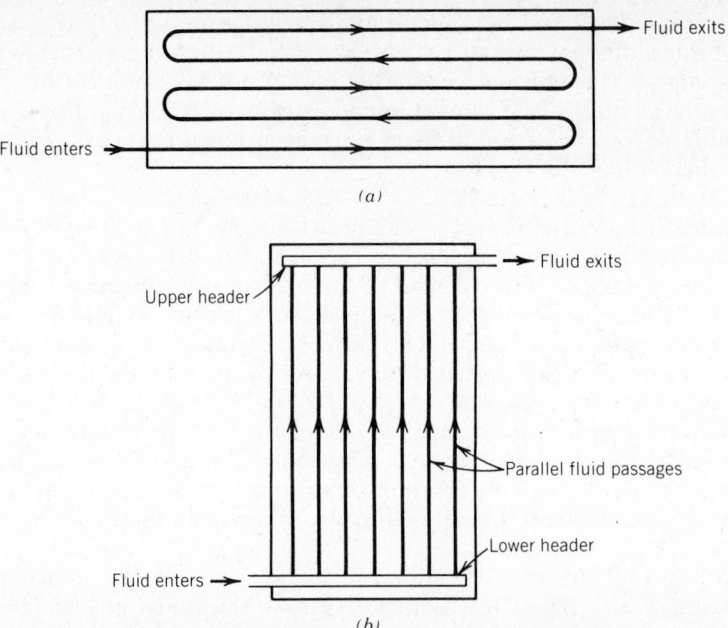

(a)

(b)

Figure 11-10 Two commonly used fluid flow patterns for finned-tube absorber plates. (*a*) A sinusoidal or serpentine configuration with fluid entering at bottom left and undergoing *series flow* through the coils of tubing, exiting top right. (*b*) A *parallel flow* configuration with lower and upper tube headers, allowing fluid flow through several parallel paths (tubes).

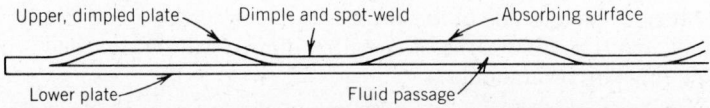

Figure 11-11 A parallel-plate absorber is constructed from a dimpled plate and a flat plate spot-welded together to form fluid-flow passages. Recommended only for low pressure applications.

problems of bonding fins to tubes—thermal resistance and mechanical lesions—do not exist. However, this type of absorber is limited with respect to the pressure at which it can operate, and it is not generally suited for use at domestic water-line pressures. There are some designs of this type using thick copper or stainless steel sheets that will operate at pressures up to 100 psi, but these are quite heavy and expensive.

Figure 11-12 shows a cross-section of the Thomason corrugated aluminum channel collector, for use with steeply pitched, south-facing roofs.[1] This collector is one example of a site-fabricated unit, built during the building construction. Cold water is distributed to the channels by a perforated metal pipe running along the peak of the roof, and the heated water is collected in a trough at the bottom of the channels. From there it is piped to a thermal storage tank located in the basement or equipment room. This type of collector is used for space-heating systems and often occupies several hundred square feet of area.

[1] See Harry E. Thomason and Harry J. L. Thomason Jr., *Solar House Plans* (Barrington, N.J.: Edmund Scientific Co., 1972) p. 2.

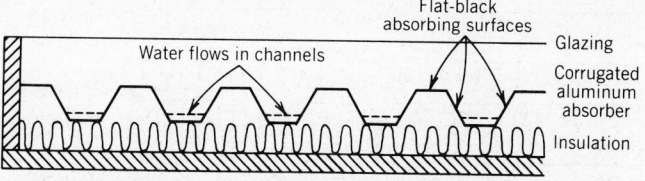

Figure 11-12 Thomason-type trickle collector constructed from a corrugated aluminum sheet. Water is not contained in tubes, but flows down the channels in the corrugated sheet in a shallow trickle.

11-3   Air-Heating Absorber Plates

Air or other gases can be heated with flat-plate absorbers if the absorber plate is designed with an extended surface to overcome the poor heat transfer between metal surfaces and air. A finned-plate absorber for heating air is shown in Fig. 11-13a. Note the fins in the air stream behind the absorber plate. These fins increase the surface area of the plate in the air stream and maximize the heat transfer from metal to air. Laboratory studies indicate that the surface area of the absorber in contact with the air flow should be two to four times the surface area exposed to the sunlight. The actual ratio will also be a function of the air velocity and the turbulence of the air flow in the collector.

Illustrative Problem 11-1

An air-heating finned-plate absorber is to be designed with a fin length of 2 in. Calculate the number and the spacing of fins for the maximum and minimum ratios cited above (i.e., ratios of 2:1 and 4:1).

Solution

1. The minimum case—2 ft² of surface for each square foot of absorber face exposed to sunlight.

    The back of the absorber provides 1 ft² of surface, and the fins must provide the other square foot of surface area. The surface area of a 2 in. × 12 in. fin for both sides is $2 \times 2 \times 12 = 48$ in.².

    The number of fins for 1 ft² of surface is, then,

    $$\text{Fins/ft}^2 = \frac{144 \text{ in.}^2/\text{ft}^2}{48 \text{ in.}^2/\text{fin}} = 3 \text{ fins/ft}^2$$

    Three 2-in. fins would have to be placed across each square foot of collector absorber, at a spacing of 4 in.   *Ans.*

2. The maximum case—4 ft² of surface for each square foot of absorber face exposed to sunlight.

    Again, the back surface of the absorber provides 1 ft² of surface, leaving 3 ft² of fin surface needed.

    $$\text{Fins/ft}^2 = \frac{3 \times 144 \text{ in.}^2/\text{ft}^2}{48 \text{ in.}^2/\text{fin}} = 9 \text{ fins/ft}^2$$

    Nine fins would have to be placed across each square foot of collector absorber, spaced 1.33 in. apart.   *Ans.*

A second way to increase the heat-transfer surface in an air-heating collector is to use a corrugated absorber plate, as shown in Fig. 11-13b.

Another way to increase the heat-transfer surface is to use a boxed-type absorber plate, as shown in Fig. 11-13c. This design features boxed air-flow channels enclosed by an upper absorbing surface, a back surface, and metal

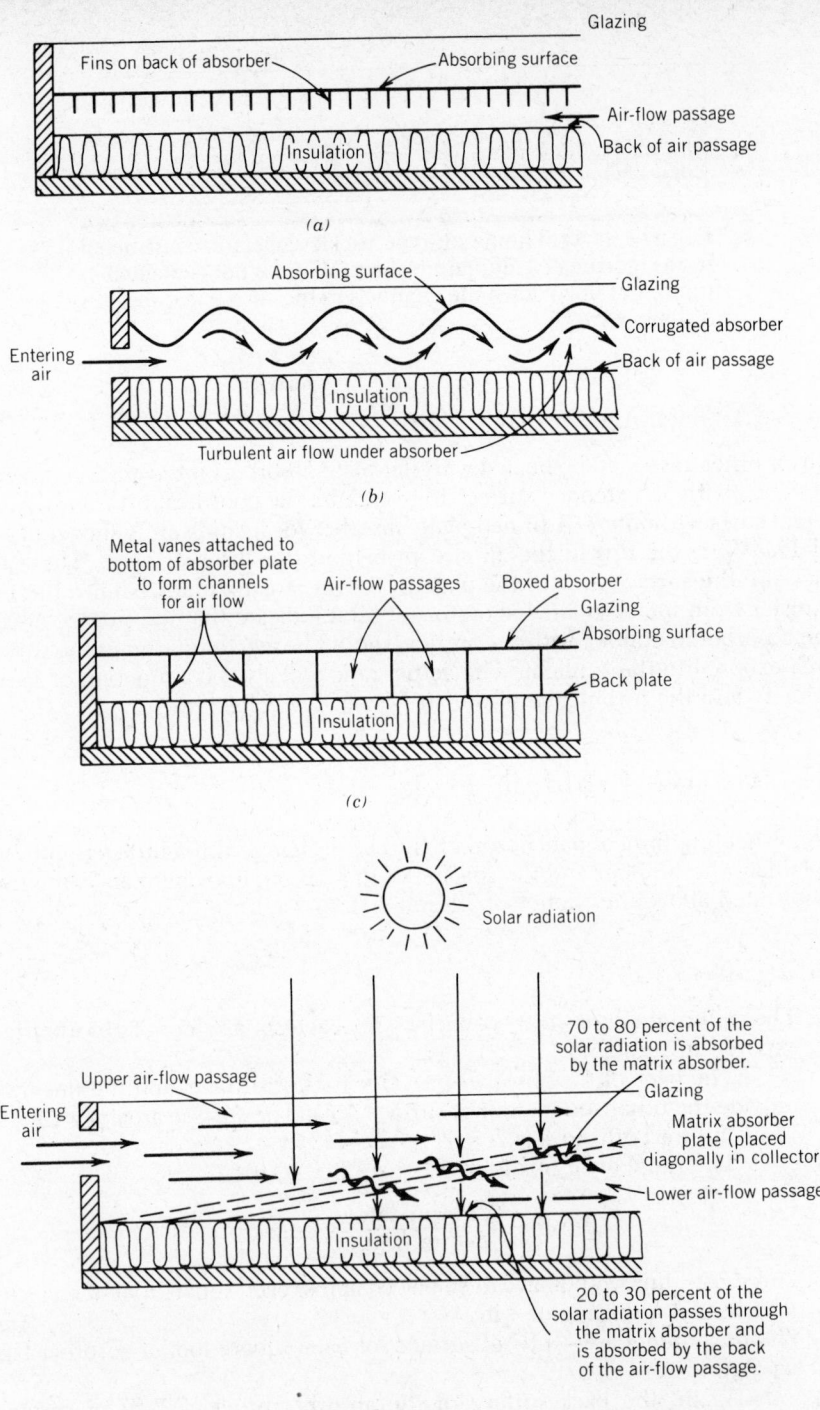

Figure 11-13 Air-heating absorber plates. (*a*) Finned-plate absorber for heating air. The fins on the back of the plate increase the surface area of the plate exposed to the air for more efficient heat transfer from plate to air. (*b*) Corrugated air-heating absorber plate. Air flowing across the corrugated plate creates turbulence along the plate, which increases the heat transfer from the plate to the air. (*c*) Boxed-frame absorber plate for heating air. The box frame creates air-flow passages between the vanes. The vanes conduct from the absorbing surface plate to the back plate. Heat is transferred to the air by all of the surfaces of each boxed air-flow channel. (*d*) Matrix absorber plate for heating air. The matrix is formed by stacking several sheets of metal mesh such as expanded metal plastering lath. The matrix is placed diagonally in the collector so that the air will flow through the matrix.

vanes between these two surfaces. The surface area exposed to the air is increased by the presence of the vanes and the back plate, both of which help to transfer heat to the air flowing in the collector. Thus, the entire box structure acts as a heat-transfer surface.

Matrix- or perforated-type absorbers constitute yet another kind of air-heating absorber. Figure 11-13*d* shows a matrix type of absorber made up of layers of expanded and blackened metal mesh placed diagonally in the collector case. The mesh will absorb about 70 to 80 percent of the solar radiation entering the collector. This energy will be transferred to the air as it flows through the matrix of metal mesh. The matrix provides a large surface area to the air and thus is very effective in transferring the heat absorbed to the air flow. Some solar radiation will pass through the matrix (about 20 to 30 percent). This radiation is absorbed by the back of the lower air-flow passage, and the heat is transferred to the air flowing across this surface.

11-4   Absorber Surface Coatings

In earlier chapters it was pointed out that dark-colored objects become warmer in the sun than light-colored objects. For instance, the surface temperature of black asphalt will be much warmer than that of white concrete when both are exposed to the same solar radiation. This principle also applies to solar collectors. Since the surface coating on a collector absorber plate should absorb as much solar radiation as possible, dark-colored coatings that absorb 90 to 96 percent of the solar radiation falling on them are used. (See Chapter 10.)

When sunlight falls on a surface, part of it is absorbed and part of it is reflected, as shown in Fig. 11-14. The radiation falling on the surface is termed the *incident energy*. *The angle of incidence* is the angle the sun's rays make with a line perpendicular (normal) to the surface. The *reflectance r* of a surface is the ratio of the reflected energy to the incident energy (see also section 3-17):

$$r = \frac{\text{Reflected solar energy}}{\text{Incident solar energy}} \qquad (11\text{-}1)$$

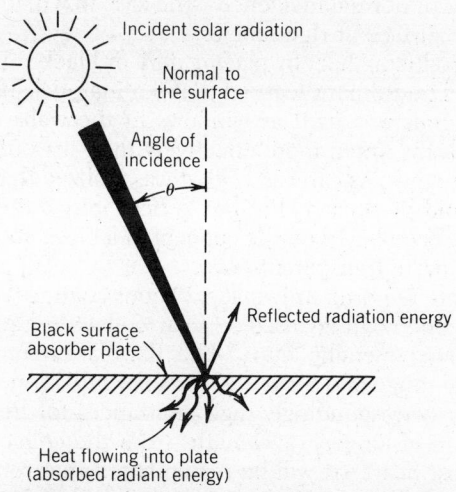

Figure 11-14 Absorption and reflection of solar radiation at a black surface. The radiant energy beam is shown incident on a flat-black absorber plate at an angle with the normal to the absorber plate surface. A small amount of the total energy is reflected (not more than 4 to 8 percent for a good flat-black surface), but most of it is absorbed and converted into heat energy, which raises the temperature of the metal plate.

Recall that, from Chapter 7, section 7-11, the absorptance α of a surface is the ratio of the absorbed energy to the incident energy:

$$\alpha = \frac{\text{Absorbed solar energy}}{\text{Incident solar energy}} \qquad (7\text{-}6)$$

Since the incident energy is either absorbed or reflected at an opaque surface, it follows (see section 3-17) that

$$\alpha + r = 1.00 \qquad \text{(for any opaque surface)} \qquad (11\text{-}2)$$

Illustrative Problem 11-2

From measurements made at a flat-black surface, it is found that the incident solar radiation is 200 Btu/hr-ft^2 and the reflected energy is 30 Btu/hr-ft^2. Find the absorptance and reflectance for this surface.

Solution

$$r = \frac{\text{Reflected energy}}{\text{Incident energy}} = \frac{30 \text{ Btu/hr-ft}^2}{200 \text{ Btu/hr-ft}^2} = 0.15 \qquad \textit{Ans.}$$
$$\alpha = 1 - r = 1 - 0.15 = 0.85 \qquad \textit{Ans.}$$

The absorptance and reflectance at a surface are dependent on the following factors:

1. The angle of incidence of the incoming radiation.
2. The texture and color of the surface.
3. The wavelength of the solar radiation.
4. The optical properties of the surface film or coating.

The following discussions deal with the properties of solar absorber surface coatings for the case of normal incidence—the case in which incident radiation strikes the absorber surface at right angles.

The idea of producing heat by placing dark or black surfaces in sunlight is as old as antiquity. The Romans learned that if a material such as thin sections of transparent minerals was used as windows in the walls of rooms, the sun would heat these rooms much more effectively than it would heat rooms with open (unglazed) windows. As early as 1770, it was realized that a glass cover over a black surface would produce a "hot box" effect. Since then, almost all solar heat collectors have been based on the concept of a black absorbing surface in a "box," with one or more transparent covers.

Since metals are not ordinarily black, various coatings are used to increase their absorptance. The coatings used in solar collectors range from flat-black paints to electroplated metallic films. The flat-black paints have very high absorptances, in the range of 0.95 to 0.97 for solar radiation. The problem with these paints is their correspondingly high emittances for thermal radiation (in this case, reradiation of longer wavelengths from the collector absorber plate, which represents lost heat). (It will be remembered that reradiation from low-temperature sources such as an absorber plate will be long wavelength radiation in the infrared region.)

Better than black paint would be a coating of some kind that would combine *high absorptance* in the wavelength region of the total solar spectrum with *low emittance* in the infrared region. Such a coating would be *selective* in its action. Figure 11-15 illustrates the attributes of an idealized selective coating—idealized in the sense that it does not actually exist. Shown at the left of the figure is a rough approximation of a solar spectrum chart with wavelength plotted against radiation intensity. It can be seen that most of the sun's radiant energy occurs in the region from 0.4 microns to 2.0 microns in wavelength (see also Fig. 4-6). In this wavelength range, the reflectance (see scale at right of

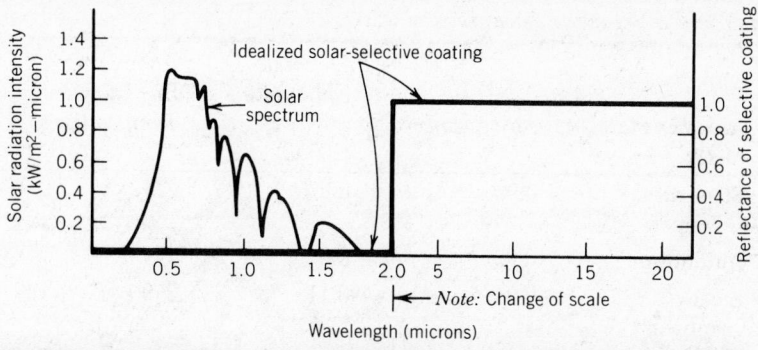

Figure 11-15 Idealized wavelength-selective coating. This coating would have 100 percent absorptance for solar radiation of wavelengths up to about 2.0 microns, at which point the coating would become 100 percent reflective, emitting zero radiation energy. No such coating exists. (Adapted with permission from R. C. Jordan, ed., *Low Temperature Engineering Applications of Solar Energy*, Atlanta: ASHRAE, 1976, p. 45.)

diagram) of such an idealized coating would be near zero, which means that the absorptance would be near its maximum value of 1.00 from Eq. (11-2). At wavelengths longer than most solar radiations (say, 2.0 microns), an idealized selective coating would have its reflectance increase suddenly to a perfect 1.00, which means (see section 3-17) that its emittance would be zero (good reflectors are poor emitters) in the wavelength region of the low-temperature radiation that is emitted from the hot absorber plates of solar collectors. Radiation losses from a hot absorber plate coated with such an idealized selective coating would thus be nearly zero.

During the past 30 years, a great deal of research has gone into the development of selective coatings. Figure 11-16 compares the actual reflectances of some coatings currently used for solar collectors. Note that, although their selectivity does not approach that of the idealized situation of Fig. 11-15, these coatings nevertheless exhibit high absorptance (low reflectance) in the range of the solar spectrum and low emittance (high reflectance) in the far infrared region. Table 11-1 compares the properties of a number of selective coatings. Note that some of them are adversely affected by high humidity. The

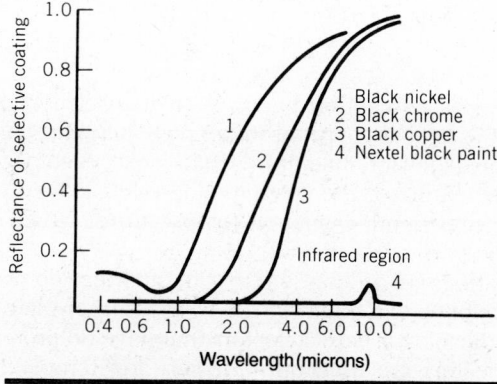

Figure 11-16 Reflectance of some actual selective coatings. These coatings have low reflectance (high absorptance) in the solar spectrum region from 0.4 to 2.0 microns, and high reflectance (low emittance) at wavelengths in the infrared region (wavelengths greater than 2.0 microns). (Adapted with permission from *Solar Engineering Magazine*, August 1978, p. 32.)

FLAT-PLATE SOLAR COLLECTORS

Table 11-1 Properties of Selective Absorber Coatings

Coating	Absorber Plate Metal	Absorptance α	Thermal Emittance e (100°C)	Approximate Breakdown Temperature[a] (°C)	Humidity Degradation (10 da MIL-STD 810B)[b]
Black nickel on nickel	Steel or copper	0.96	0.07	275	Variable
	Aluminum	0.96	0.07	275	Variable
Black chrome on nickel	Steel or copper	0.95	0.11	350	No effect
	Aluminum	0.96	0.12	350	No effect
Black chrome	Steel	0.91	0.07	350	Completely rusted
	Copper	0.95	0.14	300	Little effect
	Galvanized steel	0.95	0.16	350	Complete removal
Black copper	Copper	0.88	0.15	300	Complete removal
Iron oxide	Iron	0.85	0.08	350	Rust pinholes
Honeywell selective paint	Any	0.90	0.30		
Permanganate treatment	Aluminum	0.70	0.08		
Organic overcoat on iron oxide	Iron	0.90	0.20		Little effect
Organic overcoat on black chrome	Steel	0.94			Little effect

Source: Reprinted with permission from Y. B. Mar, "Selective Black Paints Have Potential Advantages," *Solar Engineering Magazine*, August 1978, p. 31.

[a] At the breakdown temperature, the chemical structure changes and the coating loses its selective properties.

[b] A 10-day test under Military Standard 810-B.

most durable (i.e., least susceptible to corrosion) is black chrome on an original nickel coating on the steel or copper absorber plate. Organic paints and selective coatings all have breakdown temperatures that, if exceeded, will result in deterioration and shortened life of the coating. These breakdown temperatures are usually above the maximum stagnation temperatures, which ordinarily occur with flat-plate collectors at about 350°F to 400°F.

Selective coatings are ordinarily very thin, and they do not provide protection against atmospheric corrosion for the base absorber-plate metal. They are therefore not practicable for bare collectors that have no protective covering or glazing. Bare collectors are used only for very low-temperature applications such as swimming pool heating, however, and selective coatings are not appreciably more effective at these temperatures (80°F to 100°F) than are flat-black paints.

The decision about whether to use a selective coating or a flat-black paint is based on cost–benefit analysis. Selective coatings cost a good deal more than paint, but in cases where collector performance is sufficiently increased to reduce system costs, selective coatings would be indicated. At collector operating temperatures above 180°F, selective coatings have a definite advantage over flat-black paint in lower thermal losses and should probably be used. At operating temperatures of 150°F and below, only a careful cost–benefit analysis will tell whether their higher cost is justified.

11-5   Collector Glazings

The transparent covering on the front of a solar collector is called the collector glazing. It usually consists of one or two sheets of glass or transparent plastic. The collector glazing performs two functions:

1.   It reduces thermal losses from the collector absorber plate.
2.   It provides a weather-resistant surface for the absorber plate and the collector insulation.

The physical mounting used to secure the glazing to the collector case and the sealants that are used are very important. Sealants must be able to withstand atmospheric weathering as well as the stagnation temperature the collector will attain, which may be as high as 350°F to 400°F when it is not being cooled by fluid flow.

Surface reflection and internal absorption are the two most important optical properties of the glazing. Figure 11-17 shows what happens as sunlight strikes and passes through a single glazing of glass or clear plastic. The sum of the reflected energy, the transmitted energy, and the absorbed energy must equal the incident energy, or

$$I_{incident} = I_{trans} + I_{refl} + I_{absorb} \qquad (11\text{-}3)$$

The *transmittance t* of a material is defined as the ratio of transmitted energy to incident energy as given by the equation

$$t = \frac{I_{trans}}{I_{incident}} \qquad (11\text{-}4)$$

The reflectance *r* of a material was defined (section 3-17 and section 11-4) as the ratio of the reflected energy to the incident energy, Eq. (11-1).

The absorptance α of a material was also defined in sections 3-17, 7-11, and 11-4 as the ratio of the absorbed energy to the incident energy, Eq. (7-6). Eqs. (11-1) and (7-6) are repeated here for convenience:

$$r = \frac{I_{refl}}{I_{incident}} \qquad (11\text{-}1)$$

$$\alpha = \frac{I_{absorb}}{I_{incident}} \qquad (7\text{-}6)$$

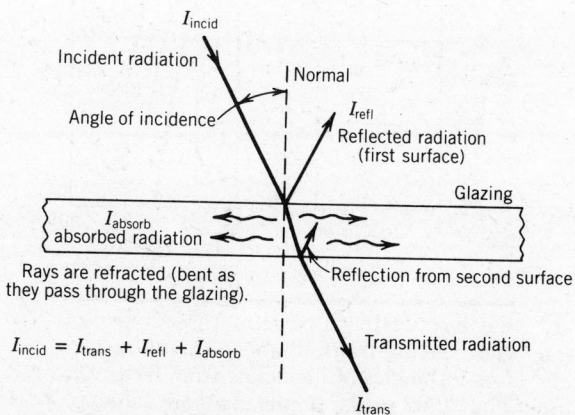

Figure 11-17 Transmission of light through a glazing, showing reflected radiation I_{refl}, absorbed radiation I_{absorb}, and transmitted radiation I_{trans}.

By adding Eqs. (11-1), (7-6), and (11-4), and then making use of Eq. (11-3), a highly useful relationship is obtained (for transparent and translucent materials):

$$\alpha + r + t = 1 \qquad (11\text{-}5)$$

Reflectance and absorptance vary as the angle of incidence changes and a corresponding change in the transmittance occurs. The reflectance is at a minimum when the angle of incidence at the surface of the glazing is zero—that is, when sunlight strikes the glazing perpendicularly. As the angle of incidence increases, the reflectance increases, causing a decrease in the transmitted light through the glazing.

The absorptance of a glazing increases gradually as the angle of incidence increases until an angle of about 60° to 70° is reached, at which point it decreases for angles of incidence up to 90°. As the angle of incidence approaches 90°, the reflectance approaches unity, the transmittance approaches zero, and the absorptance approaches zero.

In practice, the transmittance, reflectance, and absorptance are measured for normal incidence, and the values are then corrected for different angles of incidence. Figure 11-18 gives (experimentally determined) correction factors for the transmittance of three types of glazings. In this plot, the correction factor is unity at a zero angle of incidence. The correction factor decreases as the angle of incidence increases, in accord with the reduced transmittance at increasing angles of incidence, as predicted by the equation

$$t = k\, t_{\text{normal}} \qquad (11\text{-}6)$$

where t is the corrected transmittance, t_{normal} is the transmittance at normal incidence, and k is the correction factor from Fig. 11-18. To find k for a specific angle of incidence on a specified type of glass, consider the correction for single glass with an angle of incidence of 50°. A line is drawn up from 50°, as indicated by the dotted line of Fig 11-18, until it meets the single-glass curve. Then a horizontal line is drawn left, where the transmittance correction factor k is read as 0.94.

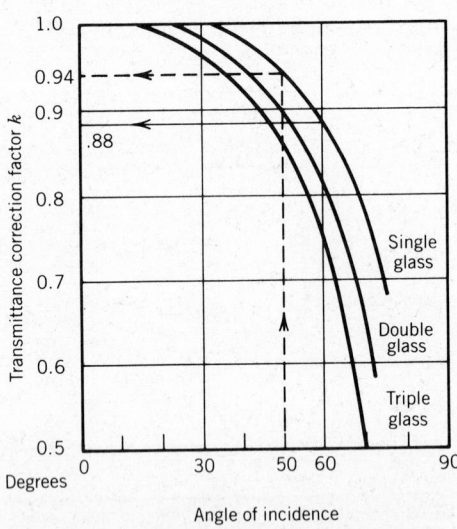

Figure 11-18 Curves for three types of glass, giving transmittance correction factors for angles of incidence from 0° to 90°. The values read from the chart are values of k for use in Eq. (11-6). (Adapted with permission from R. C. Jordan, ed., *Low Temperature Engineering Applications of Solar Energy*, Atlanta; ASHRAE, 1976, p. 32.)

Table 11-2 Selected Properties of Collector Glazings

Glazing Material	Transmittance (normal)	Absorptance (normal)	Reflectance (normal)	Maximum Operating Temperature (°F)	Nominal Thickness (in.)
Polyvinyl fluoride[a]	0.92–0.94	0.02	0.04	227	0.004
Polyethylene terephthalate or polyester[b]	0.85	—	—	220	0.001
Polycarbonate[c]	0.82–0.89	—	—	250–270	0.125
Fiberglass-reinforced plastics[d]	0.77–0.90	—	—	200° produces 10% transmission loss	0.040
Methyl methacrylate[e]	0.89	—	—	180–190	0.125
Fluorinated ethylene-propylene[f]	0.97	—	—	248	0.002
Ordinary clear lime glass (float) (0.10–0.13% iron) (window glass)	0.85	0.06	0.09	400	0.125
Sheet lime glass (0.05–0.06% iron)	0.87	0.04	0.09	400	0.125
Crystal glass (0.01% iron)	0.90	0.02	0.08	400	0.125

[a] Tedlar

[b] Mylar

[c] Lexan, Merlon

[d] Kalwall's Sunlight

[e] Lucite, Plexiglas, Acrylite

[f] Teflon

Source: Abstracted with permission from "Solar Heating and Domestic Hot Water Systems," *Intermediate Minimum Property Standards Supplement*, U.S. Department of Housing and Urban Development, 1977, p. B-1.

Table 11-2 lists some common materials used for collector glazings and provides values of transmittance, reflectance, and absorptance, along with other physical characteristics.

Illustrative Problem 11-3

Determine the transmittance of a glazed collector cover using common single-pane window glass at an angle of incidence of 60°.

Solution

From Table 11-2, the normal transmittance for ordinary (lime) window glass is 0.85. From Fig. 11-18 for 60° incidence, the correction factor k is 0.88. The corrected transmittance is then found from Eq. (11-6):

$$t = k\, t_{\text{normal}} = 0.88 \times 0.85 = 0.75 \qquad \textit{Ans.}$$

Illustrative Problem 11-4

Measurements made on a solar collector glazing yielded the following data:

Incident energy	250 Btu/hr-ft^2
Reflected energy	20 Btu/hr-ft^2
Transmitted energy	190 Btu/hr-ft^2

Determine the absorptance of the glazing material.

Solution

From Eq. (11-3)

$$I_{incident} = I_{trans} + I_{refl} + I_{absorb}$$

from which

$$I_{absorb} = I_{incident} - I_{trans} - I_{refl}$$

substituting,

$$I_{absorb} = (250 - 190 - 20) \text{ Btu/hr-ft}^2$$
$$= 40 \text{ Btu/hr-ft}^2$$

From Eq. (7-6), the definition of absorptance,

$$\alpha = \frac{I_{abs}}{I_{incident}} = \frac{40 \text{ Btu/hr-ft}^2}{250 \text{ Btu/hr-ft}^2}$$
$$= 0.16 \qquad \text{Ans.}$$

The reduction of incident solar radiation on the absorber plate surface caused by the collector glazing is a necessary tradeoff in order to have an insulating and weather-resistant covering on the collector. However, as pointed out on an earlier page, the glazing creates a greenhouse effect by minimizing thermal losses from the collector absorber plate because of reradiation in the infrared region. Glass is a natural wavelength-selective material (see sections 3-16 and 4-2), and it acts as a trap for solar energy, as illustrated in Fig. 11-19a, where a sheet of glass has been placed over a black absorber. Figure 11-19b shows the spectral transmittance (i.e., the transmittance at various wavelengths of the spectrum) for glass, superimposed on a curve of the solar spectrum. Note that glass is quite transparent over almost the entire range of wavelengths in the solar spectrum but becomes opaque to radiation with a wavelength longer than about 2.5 microns. Beyond 2.5 microns, glass "looks like" a black body to infrared radiation, stopping the long-wavelength reradiation from the black absorber. Part of this radiation is reflected back to the absorber, and the rest is absorbed by the glazing, with a consequent rise in its temperature. However, if a strong wind is blowing across the glass, it will be cooled by convection to nearly the ambient temperature of the air, thus setting up a conduction temperature gradient that materially reduces the heat trap properties of the glazing. The main effect of the glazing in this case is to protect the absorber surface from convective cooling by providing a layer of still air between the glass and the absorber surface.

Multiple Glazings. When two or more glazings are used, the inner glazing is protected from convective cooling to the ambient air, as shown in Fig. 11-20. Radiation from the absorber does not reach the outer glazing directly; however, the inner glazing will be heated by absorber radiation and then lose some heat to the next glazing by both conduction and thermal radiation. The net effect is that heat loss through a double-glazed cover is much less than that through a single-glazed cover. Figure 11-21 provides information on the effect of different numbers of glass glazings on collector heat loss.

The air spaces between glazings and between the absorber and the inner glazing have an effect on the heat transmission of the glazing. These air spaces

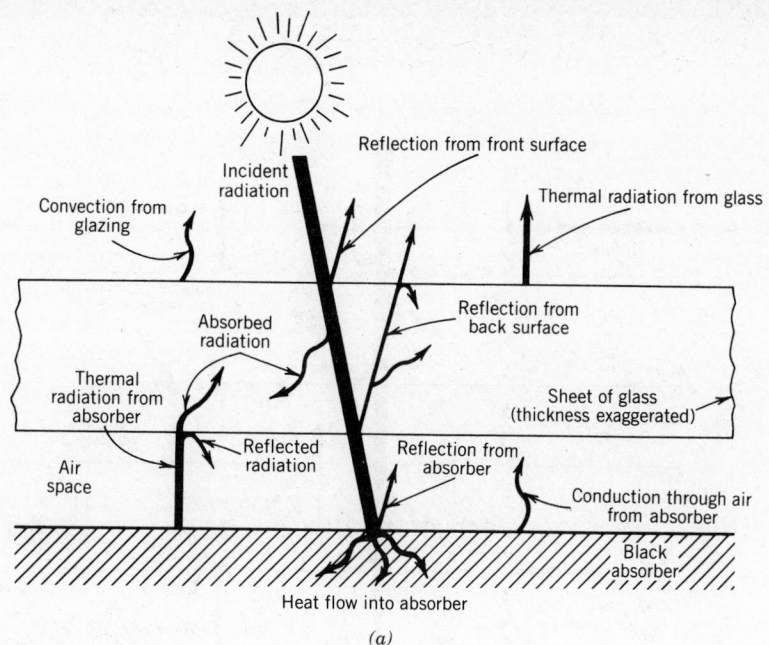

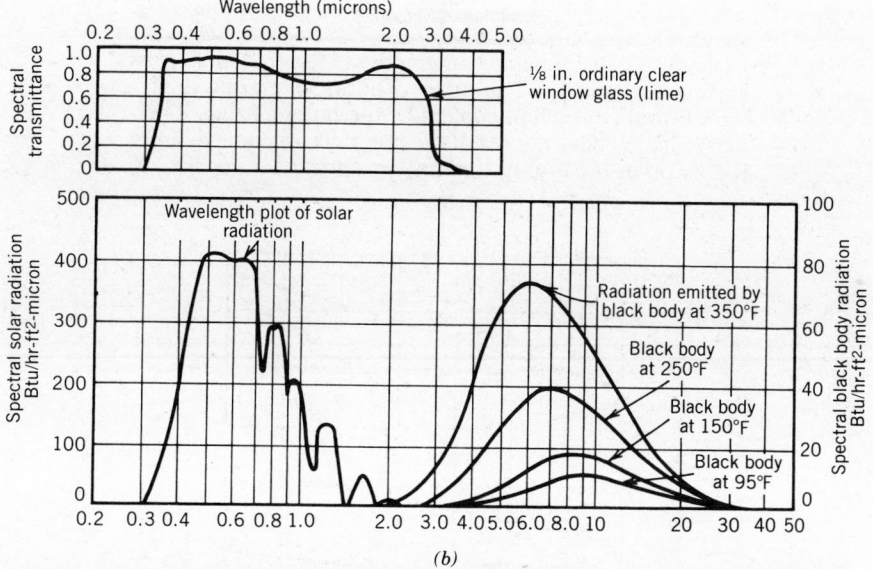

Figure 11-19 Radiation interactions for a glass glazing above a black absorber plate. (*a*) The flow of energy from incident radiation. Part of the incident radiation is reflected from the surfaces of the glazing, and part is absorbed by the glazing. The energy transmitted by the glazing strikes the absorber, where most of it is converted to heat. Some of the energy is reflected from the absorber. The heated absorber loses heat by thermal radiation and by conduction through the air space between the absorber and the glazing. (*b*) Spectral transmittance as a function of wavelength. Spectral transmittances for normal window glass are shown, plotted as a function of wavelength. Also given are the wavelength plots of solar radiation, and those of a black surface at 350°F, a black surface at 250°F, a black surface at 150°F, and a black surface at 95°F. (Adapted by permission from *ASHRAE Handbook of Fundamentals*, Atlanta: ASHRAE, 1972, p. 387.)

FLAT-PLATE SOLAR COLLECTORS

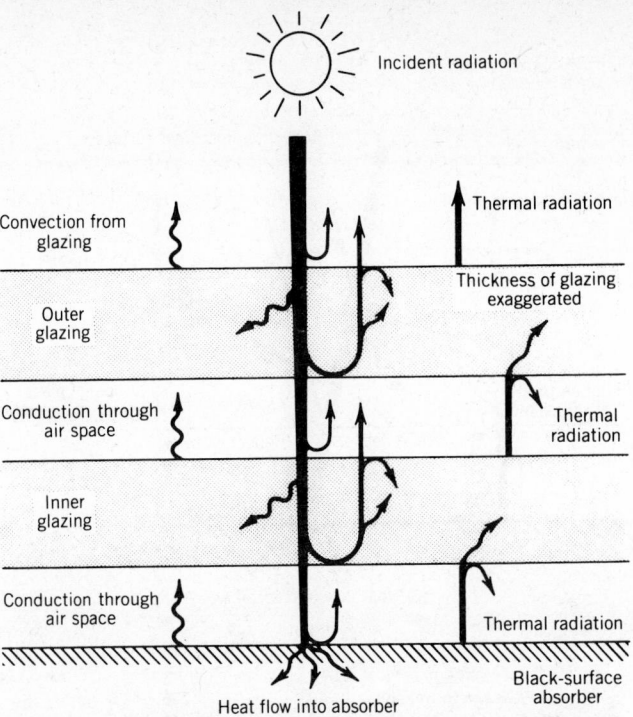

Figure 11-20 Radiation interactions for a double-glass glazing above a black absorber. The inner glazing is protected from convection cooling by the outer glazing. Thermal radiation does not reach the outer glazing directly from the absorber because of the inner glazing.

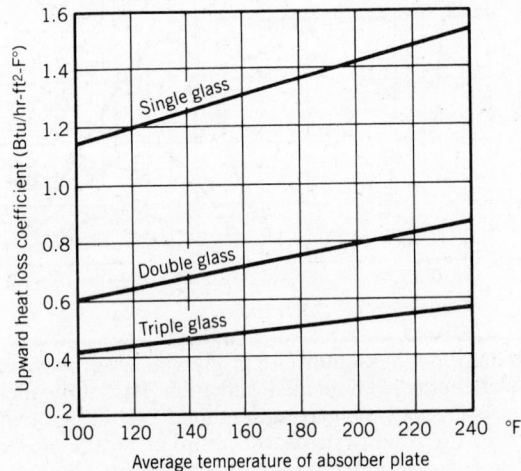

Note: Ambient air temperature assumed to be 50° F.

Figure 11-21 Chart showing how the upward heat loss coefficient varies with the average temperature of a black absorber plate for a typical flat-plate solar collector. Curves for three different glazings are shown. Note that double glazing produces a marked improvement over single glazing, while triple glazing results in only a small additional improvement. (Adapted with permission from R. C. Jordan, ed., *Low Temperature Engineering Applications of Solar Energy*, Atlanta: ASHRAE, 1976, p. 32.)

Table 11-3 Conductances of Double-Glazed Glass Covers with Varying Air-Space Thicknesses

Air-Space Thickness	Conductance of Double-Glass Glazing Btu/hr-ft²-F°
⅛ in.	0.65
¼ in.	0.61
½ in.	0.56

will range from ⅜ in. to 1 in. thick for different types of collectors. Air spaces 1 in. or less in thickness effectively prevent convection currents from occurring in the air space, but when the air space is much larger than 1 in., small convection currents can begin to occur. If the air space is too small, on the other hand, the thermal conductance of the air space itself will become a significant problem. Table 11-3 gives thermal conductances for double-glazed glass covers with air spaces of different thicknesses.

Single glazing is normally used on collectors in regions where the average ambient temperature during daylight hours is in the range above 40°F. Where temperatures are less than 40°F, and especially in windy locations, double glazing may be necessary to improve collector performance. Many other factors also enter into the decision about whether to use single or double glazing. Some of these are the operating temperature of the collector, the intensity of solar radiation at the locality, the type of coating on the absorber plate, and cost–benefit considerations.

Figure 11-22 shows the upward heat loss in Btu/hr-ft² from large flat-plate collectors plotted for different absorber coatings and glazings, for a range of operating temperatures. The upward heat loss is the sum of conduction losses q_c and radiation losses q_r through the collector glazing. Heat loss is plotted as a function of the average absorber plate temperature rise above the ambient temperature. An important inference from this plot is that the use of a selective coating and a single-glass glazing has about the same result as a flat-black coating and a double-glass glazing.

Illustrative Problem 11-5

Find the upward heat loss from a 24 ft² collector operating at an average absorber temperature of 100°F above ambient, with single glazing and a flat-black absorber coating.[2]

Solution

From Fig. 11-22 (curve BG), the upward heat loss is about 120 Btu/hr-ft². The total heat loss is then

$$Q = 120 \text{ Btu/hr-ft}^2 \times 24 \text{ ft}^2$$
$$= 2880 \text{ Btu/hr}$$

Ans.

Illustrative Problem 11-6

Calculate the reduction in upward heat loss if a selective coating is used instead of flat-black paint for the absorber surface of the collector described in Illustrative Problem 11-5.

---

[2] The average operating temperature of a collector can be taken as the average of the fluid-in temperature and the fluid-out temperature.

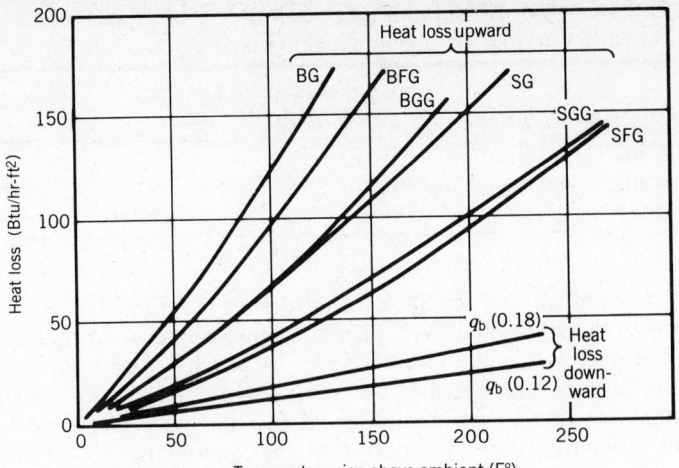

Computed heat losses from large flat-plate collectors.
  30° tilt, 1 in. air gaps, ambient temperature 70° F
  External heat loss coefficient 4 Btu/hr-ft²-F° (10 mph wind)
Heat loss upward ($q_r + q_c$)

Key:			
BG	normal black	e = 0.95	one glass pane
BGG	normal black	e = 0.95	two glass panes
BFG	normal black	e = 0.95	one glass, one film (plastic)
SG	selective black	e = 0.12	one glass pane
SGG	selective black	e = 0.12	two glass panes
SFG	selective black	e = 0.12	one glass, one film (plastic)

Figure 11-22 The upward heat loss from large flat-plate collectors, for various combinations of absorber coatings and glazings. Two absorber coatings are used: a flat-black coating with an emittance of 0.95 and a selective coating with an emittance of 0.12. Two types of glazing are used. One is glass, and the other is a thin plastic film. The upward heat loss is composed of two components: the conduction heat loss q_c and the radiation loss q_r. The heat loss downward is $q_b = 0.18$ Btu/hr-ft²-F° for 2-in. back insulation; and $q_b = 0.12$ Btu/hr-ft²-F° for 3-in. back insulation. (Adapted with permission from R. C. Jordan, ed., *Low Temperature Engineering Applications of Solar Energy*, Atlanta: ASHRAE, 1976, p. 46.)

Solution

From Fig. 11-22 (curve SG), the upward heat loss is about 68 Btu/hr-ft². The total heat loss is

$$Q = 68 \text{ Btu/hr-ft}^2 \times 24 \text{ ft}^2$$
$$= 1630 \text{ Btu/hr}$$

The net reduction in upward heat loss is

$$2880 - 1630 = 1250 \text{ Btu/hr} \qquad Ans.$$

11-6 Installing Glazings on the Collector Case

Installation procedures for glazings depend on the type and thickness of the glazing material. Collector glazings can be classified in three groups: class one—rigid, clear, self-supporting sheets such as glass, Lexan, and Acrylite; class two—fiberglass-reinforced plastics, not self-supporting, but mechanically supported or stretched taut to prevent sagging; class three—thin plastic films such as Teflon and Tedlar.

    Collector frames are constructed to provide for mountings and seals for glazings. The seals and mounting must support the glazing, allow for thermal expansion and contraction, and prevent atmospheric materials from entering the collector. Moisture and dust must be kept out of the collector, since

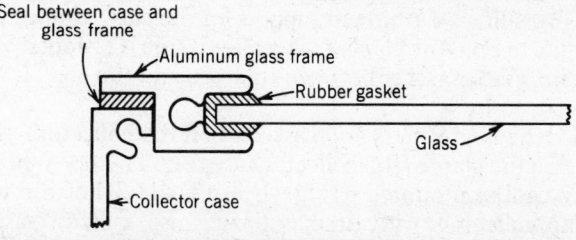

Figure 11-23 Extruded aluminum *glass frame*. The collector glazing is supported by a rubber gasket, which is held in place by the aluminum glass frame. The seal between collector case and glass frame is usually a foam rubber strip. It prevents moisture and dust from entering the collector.

moisture will result in damage to the absorber coating and to the insulation, and dust, if a layer of it collects on the absorber coating—either selective or flat black—will reduce the plate absorptance a great deal.

There are a number of methods of mounting, sealing, and securing glass glazings. One of the common techniques is to fit the glass with a rubber, U-shaped gasket, which is then slipped into a channel in an aluminum extrusion (the "glass frame"), as shown in Fig. 11-23. This type of seal allows the glass and aluminum frame to expand and contract without stressing the glass, with the

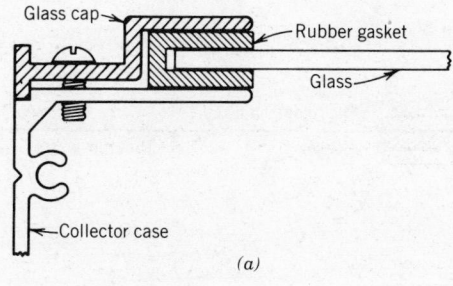

(a)

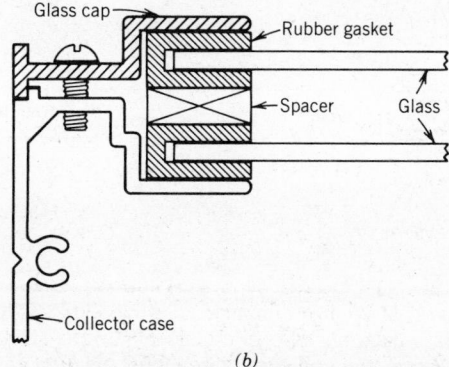

(b)

Figure 11-24 Glass cap assemblies. (*a*) Assembly for single glazing. The glass cap clamps the rubber gasket to the collector case. This provides a weathertight seal between the glass and collector case. (*b*) Assembly for double glazing using separate sheets of glass. The two sheets of glass are supported on rubber gaskets. A spacer is used to separate the two sheets of glass. The glass cap securely clamps the assembly to the collector case.

rubber gasket providing a cushioned support for the glass. Some collectors use a metal glass frame to hold the glazing. The glass frame is fastened to the collector case with a foam rubber gasket between the case and the glass frame, as shown in Fig. 11-23.

In other designs, the glass and gasket sit on the collector case, and a glass cap is used to seal the glass to the collector case (Fig. 11-24a). When two glazings are used, they can be mounted as shown in Fig. 11-24b if the two pieces are separate. Some collectors with double glazing use a factory-assembled unit, similar in arrangement to Fig. 11-24b.

Plastic glazings such as Lexan have a thickness similar to glass and can be mounted in the same manner as glass. Fiberglass-reinforced resin sheets have a thickness of from 0.025 in. to 0.060 in. and must be mounted differently. They are not self-supporting like a sheet of glass and therefore must be stretched across the collector and clamped to prevent sag. One such method is shown in Fig. 11-25a. Support struts may also be run under the sheets at approximately 2-ft intervals to insure against sag.

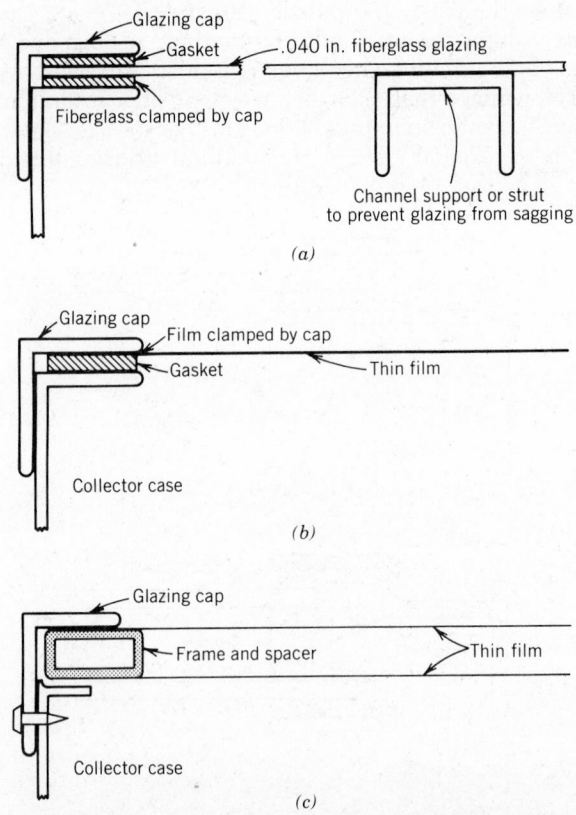

Figure 11-25 (a) Glazing cap assembly for a fiberglass glazing. The fiberglass glazing is stretched taut across the collector case. The glazing cap securely clamps the fiberglass glazing to the collector case. A channel or strut may be used to prevent the glazing from sagging. (b) Thin-film glazing assembly. The thin plastic film is stretched taut over the collector case. The glazing cap securely clamps the film to the collector case. (c) Double-glazed plastic film assembly. The two thin films are stretched across a frame (spacer) and fastened together. This forms a double-glazed assembly with the two films separated by the thickness of the frame (spacer). The double-glazed assembly is secured to the collector case with a glazing cap.

FLAT-PLATE SOLAR COLLECTORS

Thin plastic film glazings, such as Teflon and Tedlar, must be stretched taut on a frame. Figure 11-25*b* shows one method for securing a thin plastic film where an aluminum glazing cap is used, and Fig. 11-25*c* illustrates a double-glazed mounting where the film is wrapped around a frame, which also serves as a spacer for the two films. This unit is then mounted in the collector case with a gasket and glazing cap.

11-7 Collector Insulation

Collector glazings provide the insulating cover for the top of the solar collector above the absorber plate. The sides and back of the collector must also be insulated, or else untenable thermal losses will occur (Fig. 11-26). Insulation used inside the collector must be stable under the maximum temperature the absorber plate may reach during stagnation—the temperature that occurs when the collector is not being cooled by fluid flow. Stagnation temperatures depend on collector design and on the heat loss from the collector. A well-designed flat-plate collector can have absorber plate temperatures at stagnation in excess of 400°F, and, consequently, in the immediate area of the absorber plate, along the back and edges of the plate, the insulation can be subject to a very high temperature.

Collector insulation must not "outgas" or emit vapors that might coat the underside of the glazing and reduce its transmittance; and it must not melt, expand, or shrink at the maximum temperature to which it may be subjected.

Fiberglass insulation that is free of resinous binders is one type of collector insulation. A pink, yellow, or orange color indicates a resinous binder, while binder-free fiberglass is white. Table 11-4 gives some properties of selected glass fiber insulations, including the maximum service temperature recommended for their use. Those with low-temperature resin binders should be avoided entirely for solar collector use, and those with high-temperature resin binders should be used only if the stagnation temperature does not exceed 400°F. Manufacturers' literature indicates that high-temperature resins are stable at temperatures up to 400°F to 450°F. At temperatures in excess of 450°F, these organic binders will oxidize and emit gases.

The thermal conductivity of fiberglass varies with its mean temperature, as illustrated in Fig. 11-27. The mean temperature of insulation is defined as the temperature midway through the slab or blanket of insulation as depicted in Fig. 11-28. Note (Fig. 11-27) that at 400°F the thermal conductivity of fiberglass insulation is about 0.38 Btu/hr-ft²-F°/in., instead of the K value of about 0.25 associated with building heat loss calculations in the temperature range 0°F to 90°F.

Foam insulations are also used in solar collectors, but care must be used in their selection since many foam insulations melt or outgas at temperatures of 300°F to 400°F. Table 11-5 gives some properties of selected foam insulations. The isocyanurate and urethane foams are often used in solar collectors. A considerable advantage that foam insulation has over fiberglass insulation is a lower thermal conductivity, typically half that for fiberglass. On the negative side,

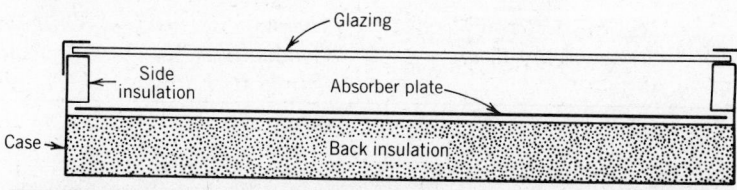

Figure 11-26 Placement of insulation in a flat-plate solar collector. The insulation across the back of the absorber plate is termed *back insulation*. The insulation along the side of the collector case between the absorber plate and glazing is termed *side insulation*.

Table 11-4 Physical Properties of Selected Glass Fiber Insulations

Type of Insulation	Maximum Recommended Service Temperature (°F)	Coefficient of Thermal Conductivity K (Btu/hr-ft^2-F°/in.	Density (lb/ft^3)
Glass fiber with low-temperature resin binder	250	0.25–0.50	0.5–1.5
Glass fiber with high-temperature resin binder	450 500°F–1000°F with outgassing (not recommended at this temperature range)	0.3–0.4	3
Glass fiber with mechanical bonding of fibers—no resin	1200	0.35–0.50	9–11

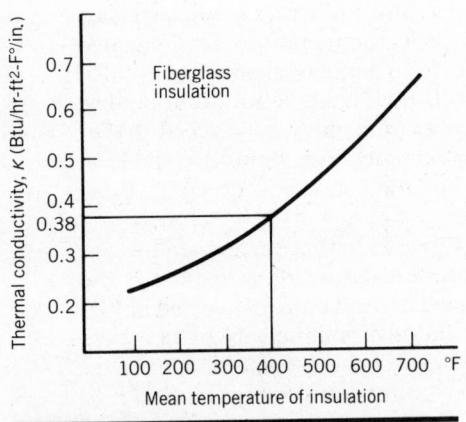

Figure 11-27 Variation in thermal conductivity with mean temperature for fiberglass insulation.

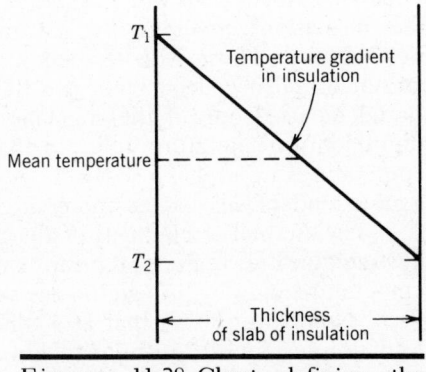

Figure 11-28 Chart defining the mean temperature of insulation. The temperature midway through the insulation is the mean temperature of the slab of insulation.

Table 11-5 Physical Properties of Selected Foam Insulations

Type of Insulation	Maximum Recommended Service Temperature (°F)	Coefficient of Thermal Conductivity K, (Btu/hr-ft^2-F°/in.)	Outgassing Temperature (°F)	Density (lb/ft^3)
Polystyrene foam	185	0.24		1–2
Polyurethane foam	300	0.14–0.17	350	0.5–3.0
Isocyanurate	250	0.13		2
Isophenal formaldehyde	350	0.2		2.5

however, foam deterioration temperatures are lower than those of glass fibers. In any event, collector design must ensure that stagnation temperatures do not exceed the maximum service temperature, no matter what insulation is used.

Illustrative Problem 11-7

The insulation for a certain solar collector must have a thermal resistance (R value) of 9.00. Calculate the thickness of (a) isocyanurate insulation and (b) resin-free fiberglass needed to provide this thermal resistance. Assume that the mean temperature of the fiberglass insulation will be 100°F.

Solution (Review section 3-9.)

Thermal resistance

$$R = \frac{1}{C}$$

And

$$C = \frac{1}{R} = \frac{1}{9.0} = 0.111 \text{ Btu/hr-ft}^2\text{-F}°$$

Let x = the thickness of the insulation:

$$C = \frac{K}{x}$$

$$x = \frac{K}{C}$$

(a) For isocyanurate (from Table 11-5),

$$K = 0.13 \text{ Btu/hr-ft}^2\text{-F}°/\text{in.}$$

$$x = \frac{0.13 \text{ Btu/hr-ft}^2\text{-F}°/\text{in.}}{0.111 \text{ Btu/hr-ft}^2\text{-F}°}$$

$$= 1.17 \text{ in.} \qquad \qquad \textit{Ans.}$$

(b) For resin-free fiberglass (from Table 11-4),

$$K = 0.25 \text{ Btu/hr-ft}^2\text{-F}°/\text{in. at } 100°\text{F}$$

$$x = \frac{0.25 \text{ Btu/hr-ft}^2\text{-F}°/\text{in.}}{0.111 \text{ Btu/hr-ft}^2\text{-F}°}$$

$$= 2.25 \text{ in.} \qquad \qquad \textit{Ans.}$$

As the above example illustrates, a solar collector can be made somewhat thinner with foam insulation than with fiberglass insulation and still be just as well insulated against heat conduction. This can be an advantage since less material is required to fabricate the collector case. Foam insulation also weighs less than fiberglass insulation of equal volume. One method of using foam insulation is to protect it with a thin layer of fiberglass insulation, as illustrated in Fig. 11-29. The foam insulation provides the greater part of the thermal resistance for the collector, and the fiberglass provides thermal isolation for the foam insulation during the high-temperature conditions of stagnation in the collector.

All glazed collectors have insulation behind the absorber plate, called back insulation, ranging from 0.5 in. of foam to as much as several inches of fiberglass. Some manufacturers insulate the sides of their collectors and some do not. As might be predicted, other factors being equal, side-insulated collectors will generally give better thermal performance than those without side insulation.

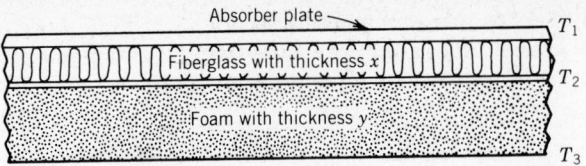

Figure 11-29 Composite collector insulation. A foam insulation can be protected from high stagnation temperatures by a layer of fiberglass insulation positioned between the absorber plate and the foam insulation.

11-8   Assembling the Collector

In the preceding sections, the various components of flat-plate solar collectors have been described and evaluated. This concluding section will explain how all of the components fit together to make a finished collector.

Basically, there are two methods of fabricating a flat-plate collector: site fabrication and factory assembly. Site-fabricated collectors are built by the building contractor or a subcontractor as an integral part of the building, while factory-assembled units are finished products ready to install in multiples, to provide an array or bank large enough to meet the design load. The factory-assembled unit is more commonly encountered in industrial practice.

The collector case provides the structural frame for the collector and a weather-resistant enclosure for the insulation and absorber plate. The case is usually made of steel or aluminum in a manner similar to that illustrated in Fig. 11-30. The collector frame is an aluminum extrusion that forms the sides and ends of the collector. A support bracket is used to carry the absorber assembly. Thermal isolation is provided between the absorber plate and the support bracket to prevent conduction and resulting heat loss to the collector case. At the top of the collector frame, a ledge is provided for the glazing gasket. In the type of collector shown, the glazing consists of two pieces of glass held to the collector case by an aluminum-angle glazing cap. The back and sides of the collector case are insulated with closed-cell foam. The complete side of the frame is insulated up to the glazing, thus minimizing heat loss from the edges of the case. The back of the case is covered with an aluminum sheet that protects the insulation.

Figure 11-31 shows the details of assembly of a collector of different design, called a through-header collector. In this design, the absorber plate is not fastened to the collector case. Instead, it is held in position by four large

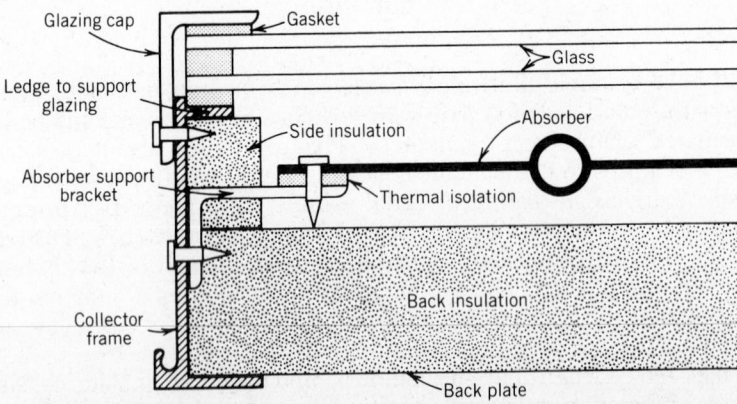

Figure 11-30 A typical collector case section. The mechanical assembly of the collector is shown, illustrating how the glazing is secured, how the absorber plate is mounted to the case, and how the insulation is installed in the case.

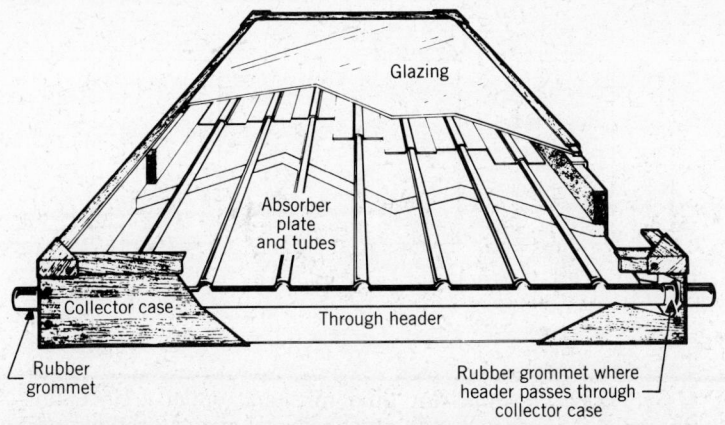

Figure 11-31 Design of a through-header collector. Note that the absorber plate is held in by rubber grommets.

rubber grommets that provide seals where the header passes through the case. The absorber plate is free-floating within the collector case and has no mechanical contact to the case except at the four rubber grommets.

Some collectors are designed to mount directly into the roof of the building, where they are sealed to the roof membrane with flashing. In this case, the collectors must be waterproof because they are actually a part of the roof of the building, in the same manner that a skylight is. This type of collector system is often feasible on new construction but would rarely be used on existing buildings. Figure 11-32 shows some of the details of this kind of collector design.

11-9   Connections to and Mounting of Collectors

Most collectors are mounted on top of the waterproof roof membrane. They may be mounted on short brackets, allowing an air space between the back of the collectors and the roof membrane. Collector-mounting instructions are given in Chapter 13.

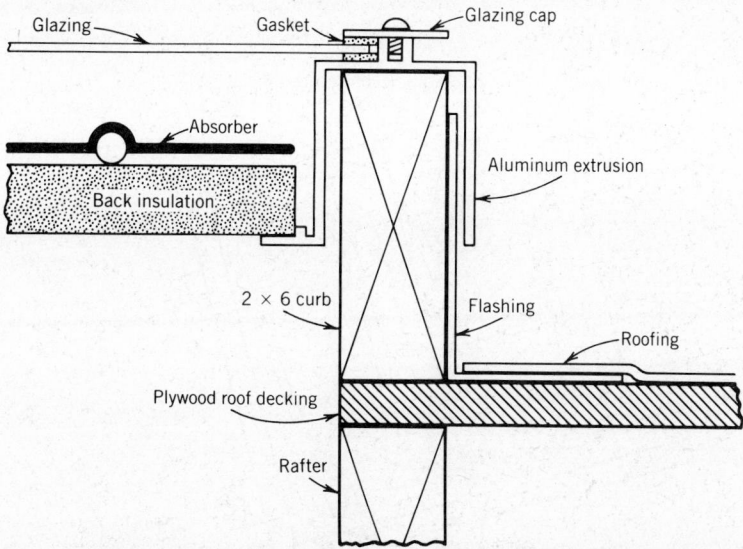

Figure 11-32 Integral roof collector details. This type of collector has a waterproof glazing and a weather-resistant case. The case is mounted in and sealed to the roof structure. The sides and back of the collector are inside the roof structure and they need not be completely weathertight.

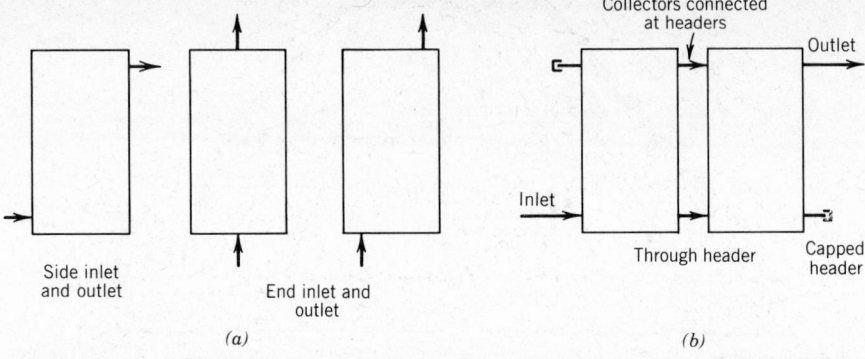

Figure 11-33 Typical locations for fluid inlet and outlet connections—liquid-heating collectors. (*a*) Single collectors may have side or end locations for inlets and outlets. (*b*) Collectors with through headers are designed so that several collectors may be connected by coupling the headers together.

Liquid-heating collectors must be provided with an inlet and an outlet for fluid flow. These connections can be located on the ends, the sides, or the back of the collector case, as indicated in Fig. 11-33*a*. Through-header collectors are connected by simply connecting the internal headers together (Fig. 11-33*b*), eliminating the requirement of an external header for the collector array. (Chapters 13, 14, and 15 deal with collector plumbing.)

Air-heating collector cases are in some ways similar to liquid-heating cases, as can be noted from Fig. 11-2. The case frame for air-heating collectors is deeper, however, because air-flow passages must be larger than tubes for liquid flow. The glazing assembly is similar in both types, but the absorber plates are very different (see sections 11-2 and 11-3), and the locations of air duct inlets and outlets differ from those of liquid inlets and outlets. Most air collectors have the duct connections on the back of the collector, with air manifolds running under the collectors as shown in Fig. 11-34. Some air collectors are designed with internal manifolds built into the collector so that collectors can be connected side to side, without the need of an external manifold (Fig. 11-35).

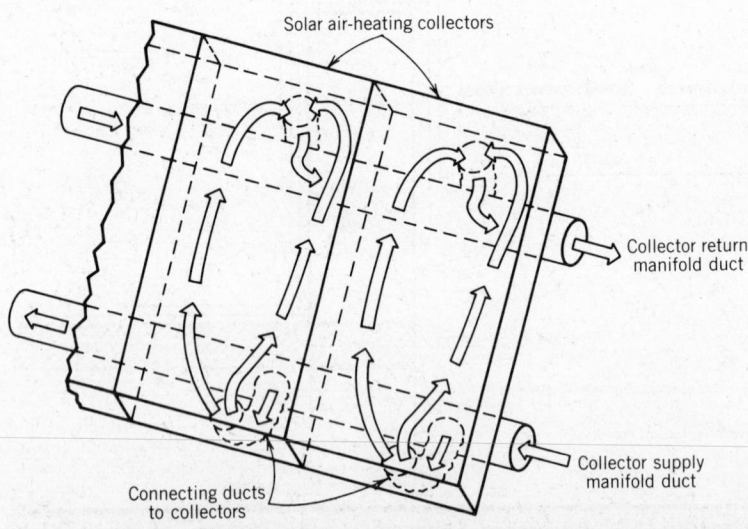

Figure 11-34 Manifold ducts for air-heating solar collectors. To supply air to the collectors, a main air-supply manifold duct has small connecting ducts attached to each collector. A return manifold duct collects the heated air from the collectors.

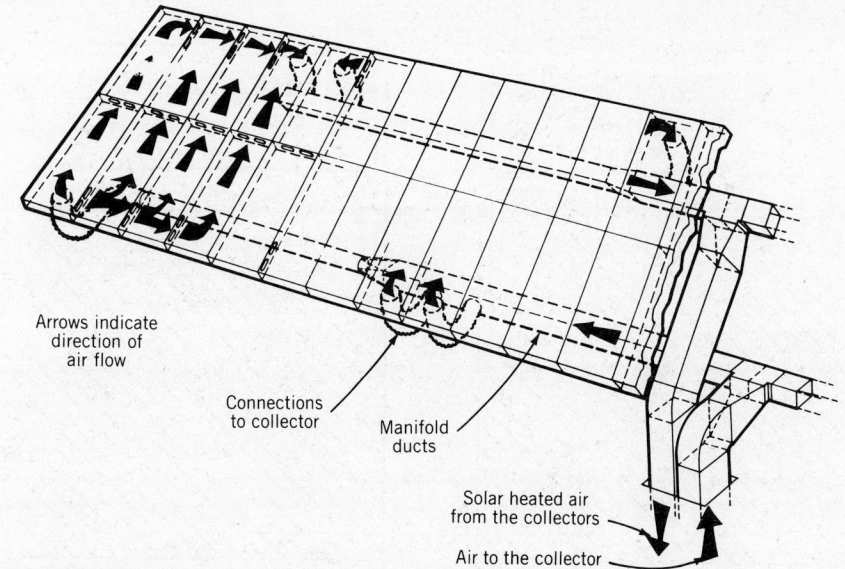

Arrows indicate
direction of
air flow

Connections
to collector

Manifold
ducts

Solar heated air
from the collectors

Air to the collector

Figure 11-35 Air collectors with internal manifolds. These collectors have internal air ducts that allow air to circulate through several collectors internally. This construction reduces the number of connections that are required to the collectors. (Solaron® Solar Energy Systems.)

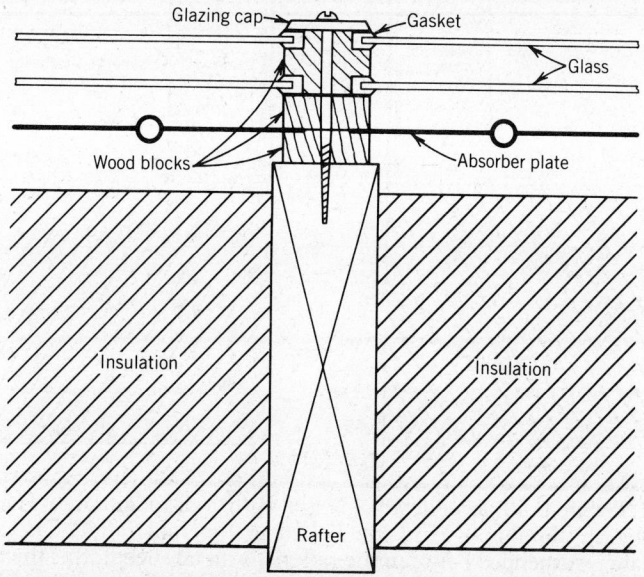

Glazing cap

Gasket

Glass

Wood blocks

Absorber plate

Insulation

Insulation

Rafter

Figure 11-36 Site-built integral roof collector. This collector is assembled on the roof rafters and replaces part of the roof. In this case, the wood collector parts are not thermally isolated from the absorber plate, so it is not a recommended construction.

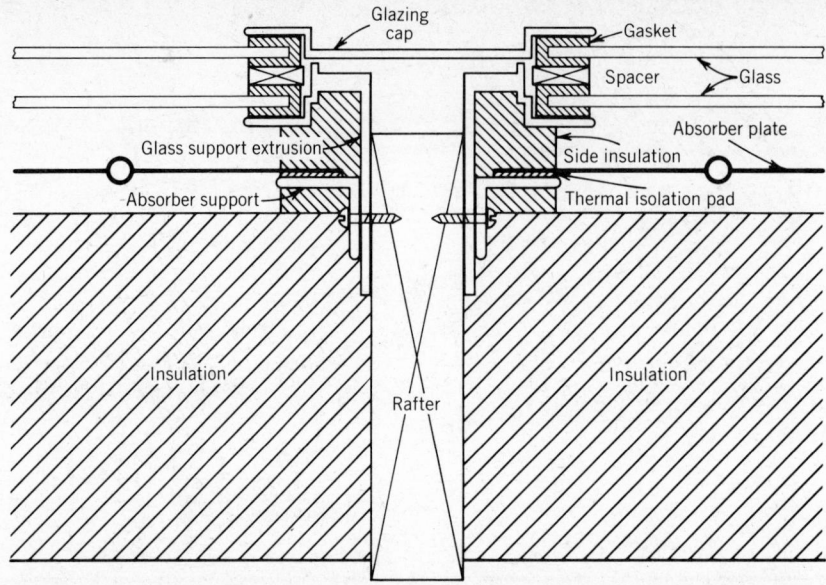

Figure 11-37 Site-built integral liquid collector with thermal isolation for wood rafters. The tube-and-fin absorber plate is thermally isolated from the support structure. Side insulation is used to isolate the rafter from internal collector temperatures.

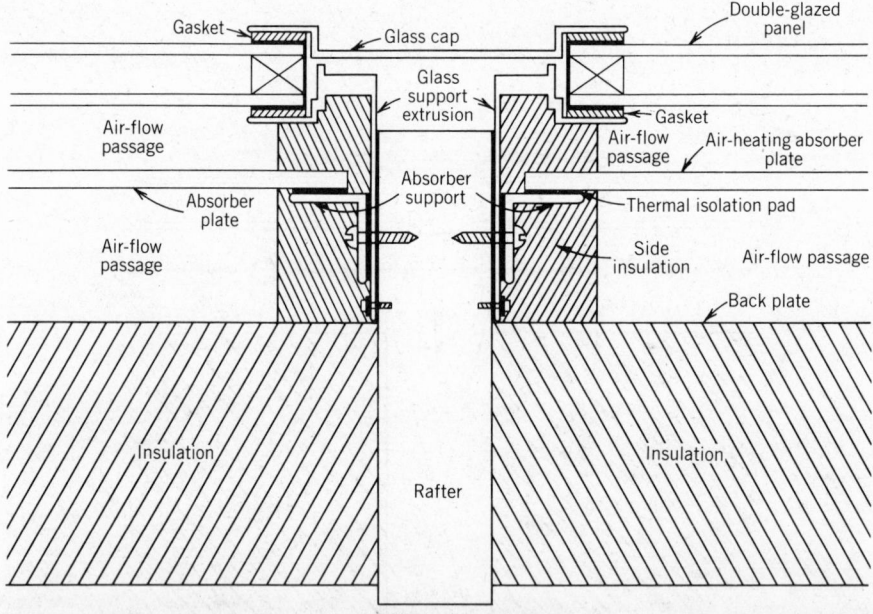

Figure 11-38 Site-built air collector integral with roof membrane. The glazing is supported by an aluminum extrusion bolted to the roof rafter. The air-heating absorber plate is suspended between the rafters. A metal sheet forms the back plate, which is attached to the rafter. The back plate and glazing form a reasonably airtight chamber through which air flows on both sides of the absorber plate. The absorber plate could be a finned plate (Fig. 11-13*a*), a corrugated plate (Fig. 11-13*b*), or a matrix absorber (Fig. 11-13*d*). Insulation is placed behind the back plate and side insulation is placed along the rafter to protect it from the high temperatures in the collector.

FLAT-PLATE SOLAR COLLECTORS

Air collectors are often mounted on the roof structure and flashed in place like a skylight. The ducts from the building below can then be connected directly into the back of the collector in the attic space.

At hillside locations, air collectors are sometimes located on the downhill slope just below the building to be heated, so that hot air from the collectors will flow upward by gravity into a thermal storage rock bed located in the basement or under the house. (See Fig. 6-9.)

Site-built collectors are usually part of a very large collector installation. In some cases, system cost can be reduced when the solar collector system is assembled as a part of the building construction. The cost of a collector case is saved, and the building roof insulation does double duty and acts as the collector insulation. These savings must, however, be evaluated against the additional labor costs of fabricating the collectors on the job. Figure 11-36 shows one method (not recommended) of mounting collector absorbers between roof rafters, with the glazing mounted on top of the rafters. In this type of collector, the structural frame is wood, and it is exposed directly to solar radiation and to the internal temperatures of the collector. Wood will gradually deteriorate over time when exposed to direct sunlight and to the high temperatures in solar collectors. With wood-frame collectors, the wood should be thermally isolated from the absorber plate and from the air space between the absorber plate and the glazing; also, protection from direct sunlight should be provided. Figure 11-37 illustrates a design that provides these requirements.

Air-heating solar collectors are often site-built. These collectors typically use absorber plates of the corrugated type (Fig. 11-13b) or the matrix type (Fig. 11-13d). Absorber plates like these can be constructed from standard building materials. For example, aluminum roofing can be used for the corrugated absorber plate, and four to five layers of expanded metal or plastering lath can be used to make a matrix absorber plate. The absorber plates can be installed between roof rafters, as shown in Fig. 11-38.

CONCLUSION

This chapter has described in detail the construction of flat-plate solar collectors. Their modes of operation, their efficiencies, and their application to a wide range of solar energy systems will be dealt with in subsequent chapters.

Concentrating (focusing) collectors will be given a more detailed treatment in Chapter 17.

PROBLEMS

1. An air-heating absorber plate is being designed with the fins on its air-side surface being 1 in. in depth. The surface exposed to the air stream must be 3 ft^2 of surface for each square foot of absorber plate. Calculate the fin spacing.

2. A new absorber-plate coating is being tested under conditions where the insolation (normal incidence) is 250 Btu/hr-ft^2 on its surface. The radiation reflected from its surface is measured as 25 Btu/hr-ft^2. Calculate the values of the absorptance and reflectance.

3. A flat-black paint has a measured reflectance of 0.06. What is the absorptance for this paint?

4. The energy reflected from an absorber-plate coating is 35 Btu/hr-ft^2. The insolation (normal incidence) was measured at 275 Btu/hr-ft^2. Calculate the values of the absorptance and reflectance.

5. Determine the solar reflectance of a black chrome selective coating on a copper absorber plate. (See Table 11-1.)

6. The transmittance at normal incidence for a sample of ⅛-in.-thick glass is 0.89. What is the transmittance when the angle of incidence is 70°F?

7. The measured energy transmitted through a glazing is 200 Btu/hr-ft². The measured reflected radiation is 20 Btu/hr-ft². The normal incident energy is measured at 250 Btu/hr-ft². Determine the transmittance, reflectance, and absorptance for this glazing.

8. Tests made on a sample of double-glass glazing gave these data: incident energy, 265 Btu/hr-ft²; reflected energy, 24 Btu/hr-ft²; transmitted energy, 190 Btu/hr-ft². What are the values of the absorptance, the reflectance, and the transmittance?

9. A flat-plate solar collector with a single-glass glazing and black absorber plate is operating at an average temperature of 150°F. The collector has an area of 32 ft². If the ambient temperature is 40°F, (a) calculate the upward heat loss, and (b) compute how much heat could be saved per hour if a double glazing were used.

10. A solar collector has a flat-black coating and a single-glass glazing. The collector area is 32 ft². The average collector operating temperature is 140°F. If the ambient temperature is 30°F, (a) calculate the upward heat loss, and (b) compute how much heat could be saved per hour if a selective coating were used on the same metal plate.

11. A solar collector must have insulation with a thermal resistance (R value) of 12.00. Calculate the thickness of (a) isocyanurate insulation and (b) resin-free fiberglass needed to provide this thermal resistance. Assume that the mean temperature of the fiberglass insulation will be 110°F.

12. A solar collector has 2½ in. of polyurethane foam insulation. What is the thermal resistance and conductance for this thickness of insulation?

Performance of Flat-Plate Collectors

*I*n Chapter 11, the various components of flat-plate solar collectors were explained, and so was their assembly as a complete unit. We now turn to an examination of the performance of the solar collector, as affected by absorber plate design, absorber coating, collector glazing, and collector insulation. Other equally important factors that influence the useful heat production of a flat-plate collector are the orientation of the collector, the ambient air temperature, and, of course, the intensity of the solar radiation (insolation) incident on the collector.

A great amount of scientific and technical information exists about collector design and performance. From such information the solar design engineer or technician must decide which collector to use on a specific job. Many of the factors that should influence collector selection were discussed in the preceding chapter. One additional and extremely important consideration is the efficiency of the collector.

12-1 Efficiency of Flat-Plate Collectors— Definitions and Equations

Collector efficiency is defined as the ratio of the usable energy output from the collector to the solar energy input (i.e., the incident solar radiation). As an equation,

$$\eta = \frac{E_o}{I} \qquad (12\text{-}1)$$

where

$$\eta = \text{the (instantaneous) collector efficiency}$$
$$E_o = \text{usable energy output rate of the collector, Btu/hr-ft}^2, \text{ or W/m}^2$$
$$I = \text{total incident solar energy input rate to the collector, Btu/hr-ft}^2, \text{ or W/m}^2$$

The symbol η is the Greek letter eta.

A solar energy system has other components whose efficiencies are also of critical importance. *Thermal storage efficiency* is defined as the ratio of the energy that can be recovered from storage to the energy that was put into storage. *System efficiency* is the ratio of usable energy output from the entire system to the total energy input into the system, including not only solar energy input but also any electrical or mechanical energy input. A simplified flow diagram of a domestic solar hot-water system is shown in Fig. 12-1 to aid in clarifying these terms. For this diagram, the system efficiency is the net energy added to the water (hot-water energy output in the diagram), divided by the sum total of the solar energy input and the electrical energy input over the same period of time.

As was pointed out in Chapter 3, all energy conversions are accompanied by losses of usable energy, and solar collectors are no exception to this First Law of Thermodynamics. Consider Fig. 12-2, illustrating a simple collector without glazing. The absorber plate has a temperature greater than the surroundings, and it will therefore lose some energy (convection losses) to the ambient air. It will also radiate infrared energy to the sky and other objects. But, since all objects emit radiation, the sky and surrounding objects will also radiate energy back to the plate. A net energy flow from the plate to the cold sky and surrounding objects will be the end result, however, since the absorber plate is at a quite high temperature. This net loss in exchanged thermal radiation (excluding direct solar gain) is termed *radiation loss*.

As shown in Fig. 12-2, there are two ways in which absorbed solar energy leaves the plate. One is by the radiation and convection losses just discussed, and the other is by the heat-transfer fluid in the collector whose specific purpose is to extract usable energy from the plate. As an equation,

$$\text{Usable energy output} = \text{Absorbed solar energy} - \text{Energy losses} \qquad (12\text{-}2)$$

If Eq. (12-2) is divided on both sides by the incident solar energy, the left side becomes the collector efficiency—Eq. (12-1)—and we obtain

$$\eta = \frac{E_o}{I} = \frac{\text{Absorbed solar energy}}{\text{Incident solar energy}} - \frac{\text{Energy losses}}{\text{Incident solar energy}} \qquad (12\text{-}3)$$

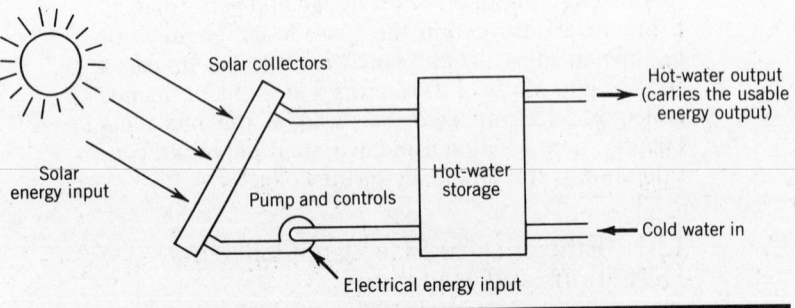

Figure 12-1 Flow diagram of a domestic solar hot-water heating system showing input energy flows and hot-water energy output.

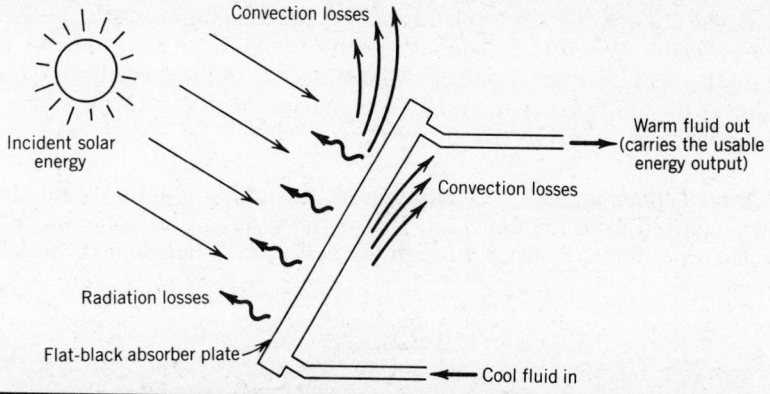

Figure 12-2 Simple (unglazed) flat-plate collector showing energy flows. Usable energy output is in the warm fluid. Heat transfers by convection and radiation represent losses.

In section 7-11, Eq. (7-6), the absorptance of an absorber surface was defined as

$$\alpha = \frac{\text{Absorbed solar energy}}{\text{Incident solar energy}}$$

The first term on the right side of Eq. (12-3) can therefore be replaced by the absorptance of the absorber plate, with the result that

$$\eta = \alpha - \frac{\text{Energy losses}}{\text{Incident solar energy}} \qquad (12\text{-}4)$$

The second term on the right side of Eq. (12-3) is the ratio of heat energy losses to incident solar radiation. This ratio cannot be expressed in a simple form, so each part will be evaluated separately. The thermal losses can be approximately represented by assuming that they are proportional to the difference between the average temperature of the upper surface of the absorber plate T_p and the ambient air temperature T_a. As an equation, this becomes

$$\text{Energy losses} = k\,(T_p - T_a) \qquad (12\text{-}5)$$

The proportionality constant k combines the heat losses resulting from conduction, convection, and thermal radiation into one term. From a strict heat-transfer approach this is only an approximation, since it is valid only for a small range of temperatures and for a specific value of k. We shall use this approximation, however, for simplicity and rename the constant U_L, the total *thermal loss coefficient per unit area of the collector*, in Btu/hr-ft²-F°, or W/m²-C°. The thermal loss coefficient U_L is a critical factor in evaluating flat-plate collector performance, and the smaller it is numerically, the better. Equation (12-5) can now be written

$$\text{Energy losses} = U_L\,(T_p - T_a) \qquad (12\text{-}6)$$

As explained in Chapter 4, the total incident solar energy on a surface is represented by the symbol I, and it is actually an energy flow *rate*, expressed in Btu/hr-ft² or in kW/m². If both sides of Eq. (12-6) are now divided by the total incident solar energy I, the result is

$$\frac{\text{Energy losses}}{I} = \frac{U_L\,(T_p - T_a)}{I} \qquad (12\text{-}7)$$

Eq. (12-7) may now be combined with Eq. (12-4) to obtain an expression for the *efficiency of an unglazed collector*:

$$\eta = \alpha - \frac{U_L\,(T_p - T_a)}{I} \qquad (12\text{-}8)$$

In this equation, the factors related to the physical characteristics of the collector are α, the absorptance of the plate, and U_L, the total thermal loss coefficient for the collector. When the plate temperature is close to the ambient temperature, the heat losses are small, and the efficiency of the collector is close to the value of the absorptance α. In fact, the maximum efficiency for a bare (no glazing) collector is equal to the absorptance α when $T_p = T_a$.

Stagnation Temperature.   Next, consider the case where no useful energy is being removed from the collector, that is, η = zero. This condition would occur when no fluid is flowing through the collector. When $\eta = 0$, Eq. (12-8) becomes

$$\alpha = \frac{U_L (T_p - T_a)}{I} \qquad (12\text{-}9)$$

If both sides of Eq. (12-9) are now multiplied by I, the result is

$$\alpha I = U_L (T_p - T_a) \qquad (12\text{-}10)$$

Note that αI is the solar energy absorbed by the collector—from Eq. (7-6)—and $U_L (T_p - T_a)$ is the energy loss from the collector to the surroundings—Eq. (12-6). Consequently, when no useful energy is being removed from the collector (when $\eta = 0$), all of the absorbed energy must be lost to the surroundings. The only quantity in Eq. (12-10) that can vary is T_p since α, U_L, T_a, and I are fixed at any given time, either by the environment or by the materials of construction. As solar energy is absorbed by a collector from which no energy is being withdrawn, the plate temperature will increase until Eq. (12-10) is satisfied, at which time T_p is at its maximum. The maximum temperature reached by the plate under the condition that $\eta = 0$ is called the *stagnation temperature* (see also section 10-1).

When collector efficiency η is plotted (on a vertical axis) as a function of $(T_p - T_a)/I$ (on a horizontal axis), a sloping, straight-line curve results. Four such curves are shown in Fig. 12-3. The variable $(T_p - T_a)/I$ is often referred to as a *parameter* (meaning a characteristic element or constant factor). Its units are quite unusual and are derived as follows. The units of $(T_p - T_a)$ are F°, while the units of I are Btu/hr-ft². Combining these to obtain the units of the parameter gives F°/(Btu/hr-ft²), or F°-hr-ft²/Btu (or, in SI units, C°-m²/W).

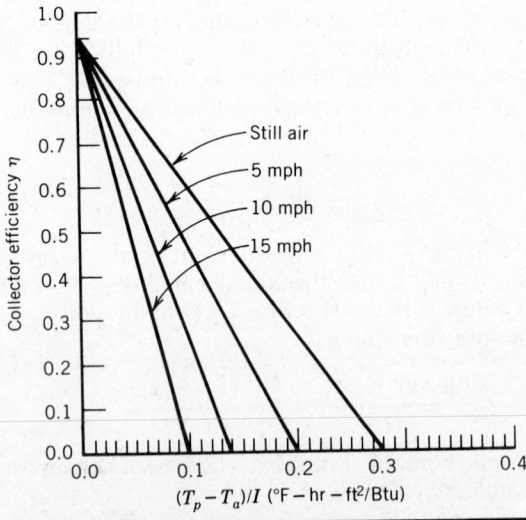

Figure 12-3 Efficiency plots for an unglazed swimming pool collector, for still air and for three wind velocities. Efficiency η is plotted as a function of the parameter $(T_p$-$T_a)/I$.

PERFORMANCE OF FLAT-PLATE COLLECTORS

12-2 Efficiency of an Unglazed Collector

Unglazed collectors are low-temperature collectors operating at temperatures fairly close to the ambient air temperature, usually not more than 10F° to 20F° above ambient. They find frequent application for swimming pool heating where pool temperatures may range from 75°F to 85°F. The collector efficiency plots shown in Fig. 12-3 are for an unglazed swimming pool collector for still air and for three different wind velocities. Wind velocity is an appreciable factor for unglazed collectors, since the absorber is exposed to the air, and convection is enhanced, with consequent effects on the value of U_L.

Illustrative Problem 12-1

Using the collector efficiency plots shown in Fig. 12-3, determine the absorptance of the collector surface and also the stagnation temperature of the collector absorber plate, for still air and for a wind velocity of 15 mph, when the ambient temperature is 75°F and the insolation I on the absorber is 200 Btu/hr-ft².

Solution

From Fig. 12-3, the plots of all of the efficiency lines intercept the vertical axis at an efficiency of 0.94. Thus, the maximum efficiency is 0.94 when the value of the parameter $(T_p - T_a)/I$ is zero. Under this condition (see Eq. 12-8), $\alpha = \eta$ and

$$\text{Absorptance} = \text{Collector efficiency} = 0.94 \qquad \textit{Ans.}$$

To calculate the stagnation temperature, the value of the parameter $(T_p - T_a)/I$ must be determined for the case $\eta = 0$.

Stagnation temperature for still air: When $\eta = 0$, $(T_p - T_a)/I = 0.28$ F°-hr-ft²/Btu (read from horizontal axis of Fig. 12-3). Now calculate T_p when $T_a = 75$°F and $I = 200$ Btu/hr-ft².

$$T_p = \left(0.28 \frac{\text{F°-hr-ft}^2}{\text{Btu}}\right)\left(200 \frac{\text{Btu}}{\text{hr-ft}^2}\right) + 75\text{°F}$$

$$= 56\text{°F} + 75\text{°F} = 131\text{°F} \qquad \textit{Ans.}$$

Stagnation temperature for 15 mph wind velocity:

$$(T_p - T_a)/I = 0.10 \text{ F°-hr-ft}^2/\text{Btu when } \eta = 0.$$

$$T_p = \left(0.10 \frac{\text{F°-hr-ft}^2}{\text{Btu}}\right)\left(200 \frac{\text{Btu}}{\text{hr-ft}^2}\right) + 75\text{°F}$$

$$= 95\text{°F} \qquad \textit{Ans.}$$

12-3 Efficiency of Glazed Collectors

Although unglazed collectors work well for producing *warm* water, they are entirely unsuitable for domestic hot-water and space-heating systems, since the temperature of the collector absorber plate for these latter systems must be in the 120°F to 180°F range. As indicated above, the efficiency of an unglazed collector will be quite low when the collector absorber plate operates at temperatures well above ambient air temperature; that is, when $(T_p - T_a)$ is large. This low efficiency results because the unglazed collector is then operating close to its stagnation temperature, where the collector losses are very large and almost equal to the solar radiation that the collector is absorbing.

In Eq. (12-8), the two factors that depend directly on the physical construction of the collector are α and U_L. To increase the efficiency of a collector and still maintain a given plate temperature T_p, either α must be increased or U_L reduced. Since typical values of α are in the 0.90 to 0.96 range, it is impossible (with a maximum value of 1.00) to increase α appreciably. The most effective

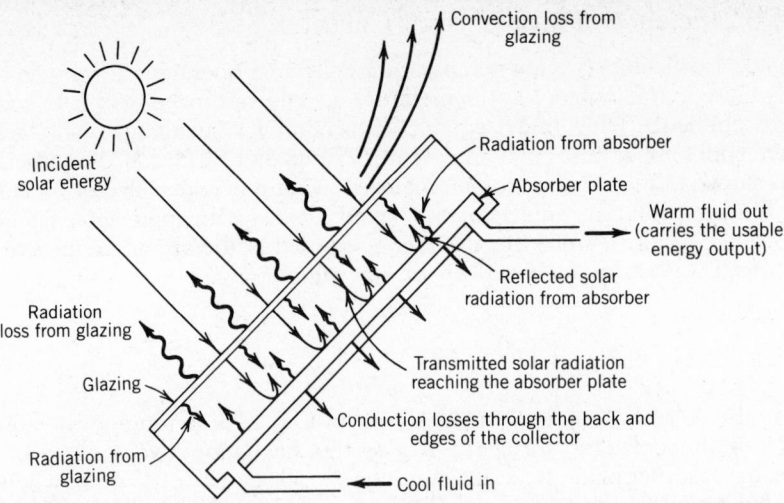

Figure 12-4 Energy flows for a glazed collector. Convection losses, radiation losses, conduction losses, and transmitted and absorbed solar energy are shown.

approach, therefore, is to reduce U_L by placing glazing over the absorber plate and installing insulation along the back and sides of the collector frame.

Glazing reduces collector energy losses from both convection and reradiation, but it also reduces somewhat the solar radiation incident on the absorber plate. Energy flows for a glazed collector are shown in the diagram of Fig. 12-4. The radiant energy transmitted through the glazing and reaching the absorber plate is:

$$\text{Transmitted solar energy} = \tau \times \text{incident solar energy}$$

where τ is the *transmittance* of the glazing. The absorbed energy is then:

$$\text{Absorbed solar energy} = I\,\tau\,\alpha \qquad (12\text{-}11)$$

The ratio of absorbed solar energy to incident solar energy for glazed collectors is an important parameter, and it follows directly from Eq. (12-11) that

$$\frac{\text{Absorbed solar energy}}{\text{Incident solar energy}} = \tau\,\alpha \qquad (12\text{-}12)$$

Eq. (12-12) is actually only approximate, since the solar radiation not absorbed by the absorber plate is reflected back to the glazing. A portion of this reflected solar radiation is transmitted out through the glazing, but some of it is reflected by the glazing back onto the absorber plate, where it is reabsorbed. These internal reflections that occur between glazing and absorber plate result in slightly more energy being absorbed than is indicated by the product $I\,\tau\,\alpha$. The difference is small and depends on the incident angle of the direct-beam radiation striking the glazing, and on τ and α. An effective *transmittance-absorptance product* is used to compensate for this difference. This product is designated $\overline{\tau\alpha}$, and it is the fraction of the incident total solar radiation that is actually absorbed by the collector absorber plate. For a glazed collector, then, the efficiency equation—see Eq. (12-8)—can be written

$$\eta = \overline{\tau\alpha} - U_L \frac{(T_p - T_a)}{I} \qquad (12\text{-}13)$$

So far, we have related efficiency discussions to the plate surface temperature T_p. However, the plate surface temperature is a variable along the absorber plate because the fluid temperature increases as the fluid moves through the absorber plate, entering the collector as a cool fluid and leaving it as a much warmer fluid. The plate will therefore have a lower temperature where the fluid is cool and a higher temperature where the fluid is hot. In practice, it is much more convenient to measure the inlet fluid temperature T_i than the plate temperature T_p. This is the approach used in the ASHRAE Standard 93-77, *Method of Testing to Determine the Thermal Performance of Solar Collectors*.[1] A factor is inserted into Eq. (12-13), which allows the substitution of inlet fluid temperature T_i for absorber plate temperature T_p. The test information can thus be used directly by the solar system designer or installer, since the temperature of the fluid coming from storage to the collector is easily determined, whereas collector plate surface temperature is very difficult to determine and at best would have to be an average value.

The factor that allows the substitution of the inlet fluid temperature T_i, for the absorber plate surface temperature T_p is called the *collector heat removal overall efficiency factor*, F_R. The mathematical derivation of F_R must consider the fluid, the fluid flow rate, the collector–fluid interface, and the interface geometry in order to relate T_i and T_p for a particular collector. Such factors can be derived mathematically but are beyond the scope of consideration here. The F_R factor compensates for the situation that the absorber plate must be warmer than the entering fluid (i.e., $T_p > T_i$) in order for heat to flow from the plate to the fluid. The use of the factor F_R allows Eq. (12-13) to be written in terms of T_i:

$$\eta = F_R \left\{ \overline{\tau\alpha} - U_L \frac{(T_i - T_a)}{I} \right\} \tag{12-14}$$

or

$$\eta = F_R \overline{\tau\alpha} - F_R U_L \frac{(T_i - T_a)}{I} \tag{12-15}$$

Note that Eq. (12-15)[2] is in the form of an equation whose type form is $y =$ constant $+ mx$. Collector efficiency curves are ordinarily plotted with η (collector efficiency) on the vertical axis, as a function of $(T_i - T_a)/I$ plotted along the horizontal axis (Fig. 12-5). Maximum efficiency occurs when $T_i = T_a$, since at that condition there are no heat losses from the collector. This condition occurs where the efficiency curve intersects the vertical (y) axis of the plot at $\eta_{max} = F_R \overline{\tau\alpha}$. The plotted efficiency slopes downward to the right (decreases) as $(T_i - T_a)/I$ increases at a rate of $-F_R U_L$ (the slope) until it intersects the horizontal axis and $\eta = 0$. This point (the x-axis intercept) represents stagnation conditions when no useful heat is being removed from the collector.

[1] The American Society of Heating, Refrigerating, and Air Conditioning Engineers (ASHRAE) has been responsible for the development of most of the heating, ventilating, and air-conditioning testing standards. Their standard for testing solar collectors was approved in 1977 as ASHRAE Standard 93-77.

[2] The terms in Eq. (12-15) fit the equation of a straight line as follows: y is η; the constant is $F_R \overline{\tau\alpha}$; the slope m is $- F_R U_L$; and the variable x is $(T_i - T_a)/I$.

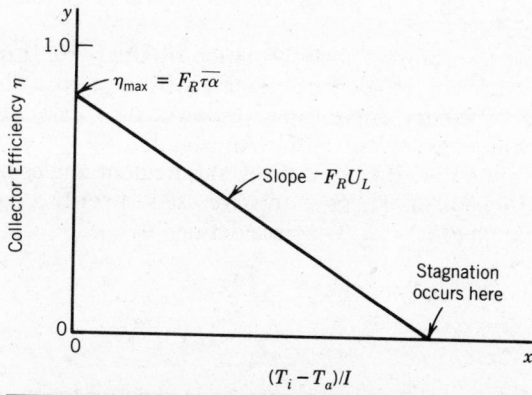

Figure 12-5 Collector efficiency plot for a glazed collector, showing y intercept at η_{max}, the slope ($-F_R U_L$), and the stagnation point at the x intercept. (Not to scale.)

PERFORMANCE OF FLAT-PLATE COLLECTORS

Collector efficiency curves represent the standardized means of comparing the thermal performance of manufactured solar collectors. Efficiency curves are determined from measurements made by independent laboratories, and the curves, along with the specified collector operating conditions, become a part of the manufacturer's thermal specifications for the collector. It is important that the solar designer or installer understand how to obtain $F_R \overline{\tau \alpha}$ and $F_R U_L$ from the efficiency curve, because these two parameters are used in calculations for seasonal or yearly system performance.

Solar Collector Energy Production Rate.   The hourly rate at which a collector produces heat energy is calculated from the relation

$$Q = A I \eta \qquad\qquad (12\text{-}16)$$

where

Q = the rate of heat production, Btu/hr
A = the gross area of the collector, ft^2
I = the incident solar radiation (insolation), Btu/hr-ft^2
η = the collector efficiency, determined (for glazed collectors) from Eq. (12-15), or from the manufacturer's specifications or test laboratory reports

Illustrative Problem 12-2

A manufacturer's specification stipulates that a certain 32 ft^2 liquid collector has an efficiency of 0.61 under the midday thermal conditions for which it is designed. Ten of these collectors will be used to heat water for an apartment house. Calculate the hourly energy output at noon for the array of 10 collectors when the solar insolation rate is 200 Btu/hr-ft^2.

Solution

$$A = 32 \text{ ft}^2 \times 10 = 320 \text{ ft}^2$$
$$I = 200 \text{ Btu/hr-ft}^2$$
$$\eta = 0.61$$

Using Eq. (12-16), $Q = A I \eta$, and substituting values,

$$Q = 320 \text{ ft}^2 \times 200 \text{ Btu/hr-ft}^2 \times 0.61$$
$$= 39{,}000 \text{ Btu/hr} \qquad\qquad Ans.$$

Collector efficiency curves are plotted from data obtained with normal solar incidence—that is, with the collector facing directly into the sun. In actuality, the incident angle changes continually for a fixed flat-plate collector as the sun moves across the sky.

The effective transmittance–absorptance product, $\overline{\tau \alpha}$ [Eq. (12-13)], decreases as the angle of incidence increases from zero. This causes the *y*-intercept of the efficiency curve to move down the *y* axis, thus shifting the whole efficiency curve downward. ASHRAE Standard 93-77 requires that a test be made to determine the efficiency at several incident angles other than zero degrees (normal incidence). These results are used to verify a parameter called the *incident angle modifier* $K_{\tau\alpha}$, which is defined as:

$$K_{\tau\alpha} = \frac{\overline{\tau\alpha}_\theta}{\overline{\tau\alpha}_N}$$

where

$\overline{\tau\alpha}_N$ = the effective transmittance–absorptance product for normal incidence
$\overline{\tau\alpha}_\theta$ = the effective transmittance–absorptance product for an incident angle θ

The variation of the incident angle modifier with the incident angle can be approximated by the following equation:

$$K_{\tau\alpha} = 1 - b_0\left(\frac{1}{\cos\theta} - 1\right) \qquad (12\text{-}18)$$

where

$K_{\tau\alpha}$ = the incident angle modifier
b_0 = the incident angle modifier constant
θ = the incident angle

The incident angle modifier constant b_0 is determined by calculations fitted to experiment. The value of b_0 for a collector is calculated by a first-order least squares method so that the values of $K_{\tau\alpha}$ obtained from Eq. (12-18) match as closely as possible the experimentally measured values. Consequently, the value of b_0 will usually be stated on the manufacturer's specifications or on the laboratory test report for each model of collector. Figure 12-6 is a plot of the incident angle modifier $K_{\tau\alpha}$ as a function of the incident angle θ for a collector where b_0 was determined to be 0.11.

The incident angle modifier $K_{\tau\alpha}$ is used to correct the glazed collector efficiency equation—Eq. (12-15)—for angles of incidence other than normal incidence, as follows:

$$\eta_\theta = K_{\tau\alpha}F_R\overline{\tau\alpha} - F_R U_L \frac{(T_i - T_a)}{I} \qquad (12\text{-}19)$$

where

η_θ = the instantaneous collector efficiency for an incident angle θ
$K_{\tau\alpha}$ = the incident angle modifier for an angle of incidence θ
$F_R\overline{\tau\alpha}$ = the maximum efficiency for the collector, where the collector efficiency plot curve intercepts the y axis
$-F_R U_L$ = the slope of the efficiency plot curve

It should be noted (Eq. 12-19) that the incident angle modifier $K_{\tau\alpha}$ affects only the y-axis intercept $F_R\overline{\tau\alpha}$ and not the slope of the efficiency equation $-F_R U_L$.

Illustrative Problem 12-3

A solar collector efficiency curve has a y-axis intercept of 0.70 and a slope of -0.74 Btu/hr-ft²-F°. Determine the collector efficiency for angles of incidence of (a) 35° and (b) 60° when the direct solar radiation I_{DN} is 200 Btu/hr-ft², the diffuse component I_d is 15 Btu/hr-ft², and the reflected component I_r is 0.0

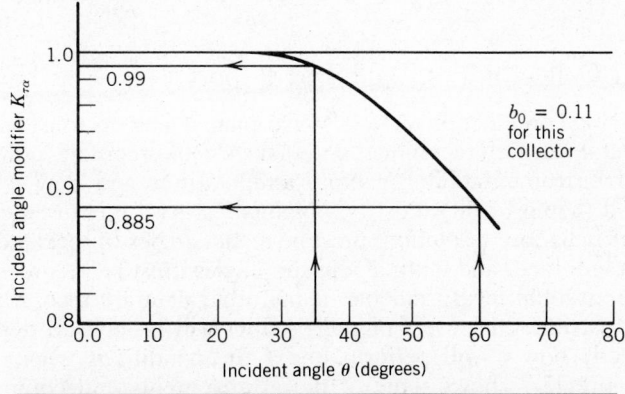

Figure 12-6 Incident angle modifier $K_{\tau\alpha}$ plotted as a function of the incident angle θ, for a collector with a calculated value of $b_0 = 0.11$.

Btu/hr-ft². The ambient temperature T_a is 45°F, and the collector inlet fluid temperature T_i is 100°F. The collector's incident angle modifier may be estimated from Fig. 12-6 ($b_0 = 0.11$ for this collector).

Solution

(a) For 35° incidence

From Fig. 12-6, $K_{\tau\alpha}$ is approximately 0.99.

From Eq. (4-3), calculate the incident total solar radiation.

$$
\begin{aligned}
I &= I_{DN}\cos\theta + I_d + I_r \\
&= 200\ \text{Btu/hr-ft}^2 \times \cos 35° + 15\ \text{Btu/hr-ft}^2 + 0.0\ \text{Btu/hr-ft}^2 \\
&= 178.8\ \text{Btu/hr-ft}^2
\end{aligned}
$$

From Eq. (12-19),

$$
\begin{aligned}
\eta_{35} &= 0.99 \times 0.70 - 0.74\ \text{Btu/hr-ft}^2\text{-F}° \times \frac{100°F - 45°F}{178.8\ \text{Btu/hr-ft}^2} \\
&= 0.69 - 0.23 \\
&= 0.46\ \text{or 46 percent} \qquad\qquad\qquad\qquad\qquad\qquad\qquad\qquad Ans.
\end{aligned}
$$

(b) For 60° incidence

From Fig. 12-6, $K_{\tau\alpha}$ is approximately 0.885.

The incident total solar radiation is

$$
\begin{aligned}
I &= 200\ \text{Btu/hr-ft}^2 \times \cos 60° + 15\ \text{Btu/hr-ft}^2 + 0.0\ \text{Btu/hr-ft}^2 \\
&= 115\ \text{Btu/hr-ft}^2
\end{aligned}
$$

From Eq. (12-19),

$$
\begin{aligned}
\eta_{60} &= 0.885 \times 0.70 - 0.74\ \text{Btu/hr-ft}^2\text{-F}° \times \frac{100°F - 45°F}{115\ \text{Btu/hr-ft}^2} \\
&= 0.62 - 0.35 \\
&= 0.27\ \text{or 27 percent} \qquad\qquad\qquad\qquad\qquad\qquad\qquad\qquad Ans.
\end{aligned}
$$

Although we have developed the above calculations in some detail to show how various parameters, constants, and chart values are actually obtained, it should be noted that these detailed computations are not often necessary. In actual field practice, the required design data on collector performance can ordinarily be obtained directly from the summary sheets supplied by solar collector manufacturers.

12-4   Solar Collector Tests and Test Reports

Laboratories that conduct official tests on flat-plate collectors must operate with the most careful research techniques and make use of precision equipment and sophisticated instrumentation. Accurate temperatures and flow rates of the collector fluid (liquid or air) must be obtained, precise measurements of the incident solar radiation (insolation) must be made, angles of incidence must be accurately determined, and latitude and sun angles must be recorded. In addition to these essential measurements, many other data are recorded during a test run so that potential users of the product will be able to determine in advance exactly how it will perform under the conditions where it will be installed. Figure 12-7 shows some of the arrangements and equipment at a collector testing laboratory.

In day-to-day design practice, the solar designer will ordinarily use collector data in summary form, as displayed in Figs. 12-8 and 12-9.

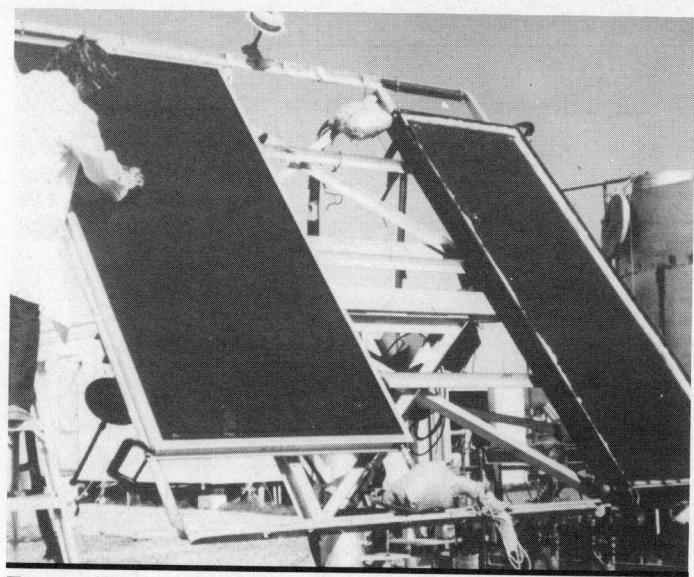

Figure 12-7 Two collectors installed on a test stand for measuring collector performance. The technician is cleaning the glazing prior to the start of the test. The pyranometer is mounted between the collectors in the same plane as the collectors, in order to measure the incident solar radiation on the collectors. The test stand can be adjusted in both altitude and azimuth. (Courtesy DSET Laboratories, Inc., Phoenix, Arizona.)

The collector data summary example in Fig. 12-8 is from the California Energy Commission's Testing and Inspection Program for Solar Equipment (TIPSE). It lists the manufacturer, collector model, and date of certification. The Energy Rating section gives average energy output and efficiency for typical operating conditions. The warranty on the collector is summarized. The collector construction materials are listed, along with the physical properties of the collector. The dry weight is the weight of the collector without fluid; the gross area is the overall surface area of the collector; the heat transfer fluid is the fluid used in the collector for the test run; the fluid capacity is the volume of fluid the collector holds; and the number of covers is the number of glazings. Note that the Technical Information section gives all of the following: the slope of the efficiency plot $(-F_R U_L)$; the y intercept $(F_R \overline{\tau\alpha})$; the second-order thermal performance equation, which is used only when Eq. (12-15) is not accurate enough to describe collector performance; the incident angle modifier $(K_{\tau\alpha})$; the empirically determined value of b_0 (0.215); the maximum rated pressure for the collector fluid passages; the collector time constant, which is the time needed for the collector to respond to changes in solar radiation level or inlet fluid temperature; the estimated maximum stagnation temperature, which is the temperature that would be reached under a no-fluid-flow condition; and the manufacturer's maximum recommended fluid flow rate. An approximate performance curve is also provided.

A sample Summary Information Sheet from the Florida Solar Energy Center is shown in Fig. 12-9. It provides essentially the same information as the report of Fig. 12-8, but it is included as an example of a test report using SI–metric units. As a third example, technical data for the Grumman Sunstream Solar Collector 300 series are given in Fig. 12-10. The sheet provides data on physical characteristics, collector performance, dimensions, and pressure drop. Collector performance data are given for the gross collector area (overall size). The y-axis intercept $F_r(T)$ and the slope of the efficiency curve $-F_r(UL)$ are given, and so is the incident angle modifier constant (B_0). Note that the notation used by this manufacturer differs somewhat from that used in this chapter. $F_r(T)$ is the efficiency η_{max}, for example.

PERFORMANCE OF FLAT-PLATE COLLECTORS

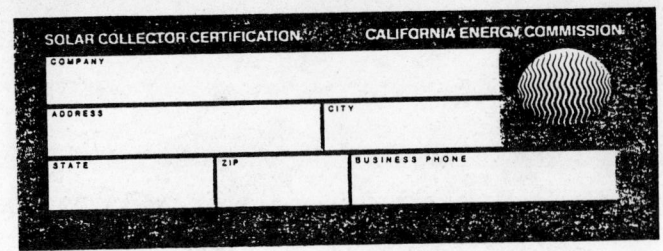

Model No. ___WSD7-CR___   Trade Name _____

Date of Certification ___January 26, 1980___

Energy Rating		BTU's/hr/ft²	Efficiency, %
	Low temperature	171	74
	Medium temperature	118	51

Warranty: ___3 years: parts and labor___
___1 year: labor for corrosion___
___Excludes freezing damages and glass breakage.___

Dry Weight: ___76___ lbs
Gross Area: ___18.38___ ft²
Heat Transfer Fluid: ___water___
Fluid Capacity: ___.7___ gal
Number of Covers: ___one___

Materials:
Cover ___Glass_____
Absorber:
   material ___Copper_____
   coating ___Black chrome_____
Frame ___Aluminum_____
Insulation ___Isocyanurate foam_____

Remarks: _____

TECHNICAL INFORMATION

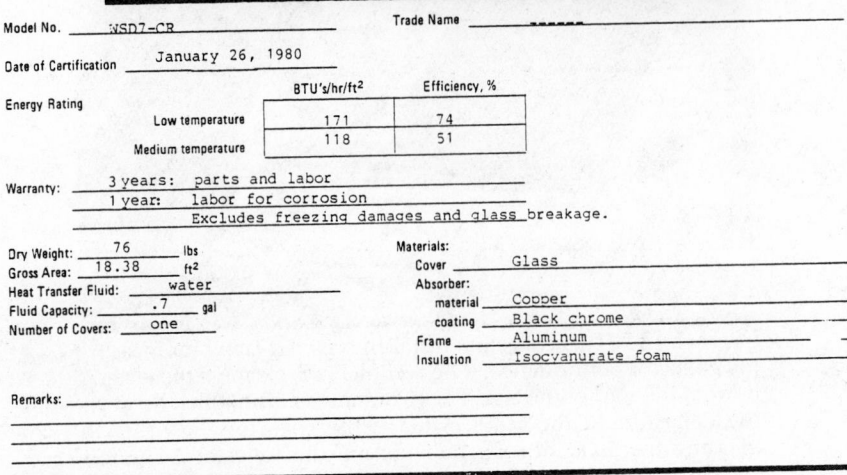

Slope ___-.803___ BTU/hr/ft² °F

Y intercept ___74___ %

Thermal Performance Equation

$$\eta = \underline{.739} - \underline{.641}\left(\frac{T_i \cdot T_a}{I}\right) - \underline{.564}\left(\frac{T_i \cdot T_a}{I}\right)^2$$

Incident Angle Modifier

$$K_{\alpha\tau} = \underline{1} - \underline{.215}\left(\frac{1}{Cos\,\theta \cdot 1}\right)$$

Time Constant ___1.1___ Minutes

Maximum Rated Pressure ___120___ psi

Estimated Maximum Stagnation Temperature ___300___ °F

Manufacturer's Maximum Recommended Flow Rate ___10___ gpm

APPROXIMATE
PERFORMANCE CURVE

Figure 12-8 A solar collector certification from the California Energy Commission. A test report on the performance of a liquid-heating collector. (Reprinted from *TIPSE* by permission of the California Energy Commission.)

Illustrative Problem 12-4

From the test data given in Fig. 12-8, and assuming normal incidence, determine the collector efficiency equation—Eq. (12-15)—and the collector efficiency for the following operating conditions:

$$I = 150 \text{ Btu/hr-ft}^2$$
$$T_a = 30°F$$
$$T_i = 120°F$$

Solution

From Fig. 12-8, the slope is -0.803 Btu/hr-ft²-F°, and the y intercept is 0.74. Eq. (12-15) then becomes

$$\eta = 0.74 - 0.803 \text{ Btu/hr-ft}^2\text{-F}° \times \frac{(T_i - T_a)}{I} \qquad \text{Ans.}$$

PERFORMANCE OF FLAT-PLATE COLLECTORS

SUMMARY INFORMATION SHEET

FLORIDA SOLAR ENERGY CENTER
300 STATE ROAD 401, CAPE CANAVERAL, FLORIDA 32920, (305) 783-0300

FSEC #79081S

MANUFACTURER

Western Solar Development, Inc.
1236 Callen Street
Vacaville, CA 95688

Collector Model
WSD - 7

This solar collector was evaluated by the Florida Solar Energy Center (FSEC) in accordance with prescribed methods and was found to meet the minimum standards established by FSEC. This evaluation was based on solar collector tests performed at Solar Energy Analysis Laboratory, San Diego, California. The purpose of the tests is to verify initial performance conditions and quality of construction only. The resulting certification is not a guarantee of long term performance or durability.

DESCRIPTION

Gross Length	1.948	meters	6.39	feet
Gross Width	0.876	meters	2.88	feet
Gross Depth	0.096	meters	0.32	feet
Gross Area	1.707	square meters	18.38	square feet
Transparent Frontal Area	1.566	square meters	16.86	square feet
Volumetric Capacity	2.6	liters	0.7	gallons
Weight (empty)	34.5	kilograms	76.0	pounds
Number of Cover Plates	One			
Flow Pattern	Parallel			

Incident Angle Modifier $K_{\tau a} = 1.0 - 0.165\ (1/\cos\theta - 1)$

Efficiency Equations   First Order $\eta = 74.2 - 687.6\ (T_i - T_a)/I$

Second Order $\eta = 73.7 - 625.5\ (T_i - T_a)/I - 993.9\ [(T_i - T_a)/I]^2$

Tested per ASHRAE 93-77   Units of $T_i - T_a/I$ are $^\circ C/Watts/m^2$

MATERIALS

Enclosure	Aluminum frame
Glazing	ASG Sunadex (Tempered water white glass)
Absorber	Copper tube soldered to copper sheet, black paint coating
Insulation	Foil faced polyisocyanurete 2.54 cm thick

RATING

The collector has been rated for energy output on measured performance and an assumed standard day. Total solar energy available for the standard day is 5045 watt-hour/m^2 (1600 BTU/ft^2) distributed over a 10 hour period.

Output energy ratings for this collector based on the second-order efficiency curve are:

Collector Temperature		Energy Output		
Low Temperature, 35°C (95°F)	20,300	Kilojoules/day	19,300	BTU/day
Intermediate Temperature, 50°C (122°F)	14,900	Kilojoules/day	14,100	BTU/day
High Temperature, 100°C (212°F)	2,500	Kilojoules/day	2,400	BTU/day

Figure 12-9 Florida Solar Energy Center *Summary Information Sheet*. A test report in SI–metric units on the performance of a liquid-heating collector. (Western Solar Development, Inc.)

For the above operating conditions,

$$\eta = 0.74 - 0.803\ \text{Btu/hr-ft}^2\text{-F}^\circ \times \frac{120°F - 30°F}{150\ \text{Btu/hr-ft}}$$

$$= 0.74 - 0.803 \times 0.60$$

$$= 0.26\ \text{or}\ 26\ \text{percent} \qquad\qquad\qquad Ans.$$

Illustrative Problem 12-5 (SI–Metric)

From the collector test data presented in Fig. 12-9, determine the collector efficiency for an angle of incidence of 45°, when the direct solar radiation I_{DN} is

ITEM	STANDARD MODELS		
	321A	332A	340A
Length L, in. (cm)	84 (213.4)	96 (243.8)	121 (307.3)
Width W, in. (cm)	35 (88.9)	47.75 (121.3)	47.75 (121.3)
Depth D, in. (cm)	2.75 (7.0)	2.75 (7.0)	2.75 (7.0)
Header, center-to-center C, in. (cm)	81.25 (206.4)	93.25 (236.9)	118.25 (300)
Gross area Ag, ft.^{2} (M^2)	20.4 (1.89)	31.8 (2.95)	40.12 (3.7)
Aperture area A_a, ft.^{2} (M^2)	18.9 (1.76)	30 (2.79)	38.08 (3.5)
Weight wet/dry lbs. (kg)	83/79 (37.7/35.9)	132/126 (60.0/57.3)	183/175 (83.0/79.4)
ABSORBER PLATE			
Material	Cu- Plank™ copper absorber/fluid passages		
Coating	Black chrome		
No. parallel fluid passages	8	11	11
Risers, in. (cm)	3/8 (0.95) OD		
Inlets/outlets, in. (cm)	1 1/8 (2.86)		
Heat transfer fluid	Any fluid compatible with copper		
Max. Operating/Test Pressure psig (kp)	150/200 (1034/1379)		
Flow rate¹, gpm (1pm) Note (1)	0.5 to 0.7 (1.9 to 2.6)	0.7 to 1.0 (2.6 to 3.8)	1.0 to 1.5 (3.8 to 5.7)
GLAZING CHARACTERISTICS			
Type	Low Iron Tempered		
Thickness, in. (mm) Transmissivity, % (per lite)	1/8 (3.175) 90	3/16 (4.763) 89	3/16 (4.763) 89
Design wind² mph (km/hr)	181 (291)		

Dimensions: inches (metric)
(1) For water or water/glycol systems. For other fluids contact Grumman Energy Systems Company
(2) Factor of safety = 1.5; collector mounted flat.

PRESSURE DROP

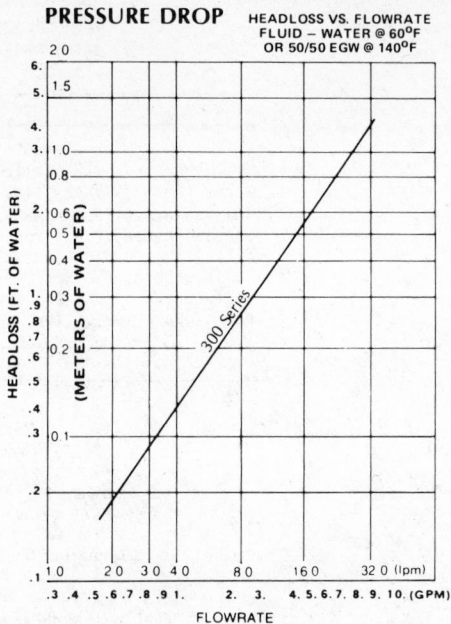

HEADLOSS VS. FLOWRATE
FLUID – WATER @ 60°F
OR 50/50 EGW @ 140°F

DIMENSIONS

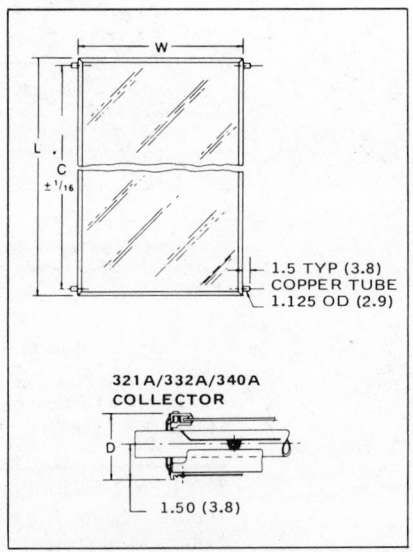

321A/332A/340A COLLECTOR

Dimensions: (see Specifications)

PERFORMANCE EFFICIENCY *

COLLECTOR MODEL (GROSS AREA)	F_r(T) INTERCEPT	F_r (UL) LOSS CO-EFFICIENT BTU/HR FT² °F	INCIDENT ANGLE MODIFIER, B_0
321A — (21 Ft.²)	.736	.861	−.125
332A — (32 Ft.²)	.730	.844	−.086
340A — (40 Ft.²)	.734	.738	−.142

*BASED ON INDEPENDENT TEST DATA USING ASHRAE 93-77 TEST PROCEDURES

Figure 12-10 Manufacturer's technical information sheet for the Grumman Sunstream Solar Collector—300 Series. Included is information on the physical characteristics of the collectors, performance efficiency, dimensions, and pressure drop. Note that this manufacturer's notation for various parameters is somewhat different from that used in the text. (Grumman Energy Systems Company.)

850 W/m² , the diffuse component I_d is 45 W/m², and the reflected component is 0.0 W/m². Assume that T_a is 10°C and T_i is 38°C.

Solution

From Fig. 12-9, the collector efficiency equation in the form of Eq. (12-15) is the first-order equation:

$$\eta = 74.2 - 687.6 \text{ W/m}^2\text{-C}° \times \frac{(T_i - T_a)}{I} \text{ (converted to percent at the outset)}$$

The incident angle modifier for 45° is calculated as

$$K_{\tau\alpha} = 1.0 - 0.165 \left(\frac{1}{\cos 45} - 1\right)$$
$$= 1.0 - 0.165 (1.41 - 1)$$
$$= 0.93$$

The incident solar radiation is calculated from Eq. (4-3) as

$$I = 850 \cos 45 \text{ W/m}^2 + 45 \text{ W/m}^2 + 0.0 \text{ W/m}^2$$
$$= 646 \text{ W/m}^2$$

From Eq. (12-19),

$$\eta_{45} = 0.93 \times 74.2 - 687.6 \text{ W/m}^2\text{-C}° \times \frac{38°C - 10°C}{646 \text{ W/m}^2}$$
$$= 69.0 - 29.8$$
$$= 39.2 \text{ percent}$$

Ans.

Illustrative Problem 12-6

For Grumman Collector Model 332A, as described in Fig. 12-10, determine the collector efficiency based on the gross collector area. Assume that the insolation rate on the collector is 225 Btu/hr, the incident angle is 40°, T_a is 45°F, and T_i is 95°F.

Solution

From Eq. (12-15), the efficiency of this collector using the performance efficiency parameters from Fig. 12-10 [$F_r(T) = \eta_{max}$] is:

$$\eta = .730 - 0.844 \text{ Btu/hr-ft}^2\text{-F}° \frac{(T_i - T_a)}{I}$$

The incident angle modifier for 40° is calculated from Eq. (12-18) as

$$K_{\tau\alpha} = 1.0 - 0.086 \left(\frac{1}{\cos 40} - 1\right)$$
$$= 1.0 - 0.086 (1.31 - 1)$$
$$= 0.97$$

From Eq. (12-19),

$$\eta_{40} = 0.97 \times 0.730 - 0.844 \text{ Btu/hr-ft}^2\text{-F}° \left(\frac{95°F - 45°F}{225 \text{ Btu/hr-ft}^2}\right)$$
$$= 0.71 - 0.19$$
$$= 0.52$$

Ans.

12-5 Efficiency of Air-Heating Collectors

Efficiency curves for a typical air-heating collector are shown in Fig. 12-11. Three plots are provided, for three different air-quantity flow rates through the collector. Air velocity has a strong influence on the heat transfer that takes place from the absorber plate to the airstream, turbulent air flow providing better heat transfer than laminar flow.

Air-heating collector performance generally is not quite as good as liquid-heating collector performance; that is, air collector efficiencies are ordinarily lower, as indicated in Fig. 12-11. The dashed line in Fig. 12-11 is the efficiency curve for the liquid-heating collector described in Fig. 12-8, for comparison purposes. Both the air and the liquid collectors have single glazing and black-painted absorbers. It will be noted that the air-heating collector efficiency curves are lower than the liquid collector curve, except for the 5 CFM/ft² curve. In part, this is because it is more difficult to transfer heat from a metal plate to an airstream than to a liquid. It can be noted from Fig. 12-11 that, in the range of air flows plotted, the highest efficiency occurs with the largest of the three air flows.

A test data summary sheet for an air-heating collector from a different manufacturer is shown in Fig. 12-12. Note that test data reports for air collectors provide about the same information as reports for liquid collectors.

Illustrative Problem 12-7

From the air collector efficiency plots of Fig. 12-11, determine the useful heat energy output rate from a 32 ft² collector with an air flow of 3.3 CFM/ft² of collector area. The insolation rate on the absorber plate surface is 200 Btu/hr-ft² at normal incidence, the inlet air temperature is 85°F, and the ambient air temperature is 50°F.

Solution

First, calculate the value of the parameter $(T_i - T_a)/I$:

$$(T_i - T_a)/I = \frac{85°F - 50°F}{200 \text{ Btu/hr-ft}^2} = 0.18 \text{ F°-hr-ft}^2/\text{Btu}$$

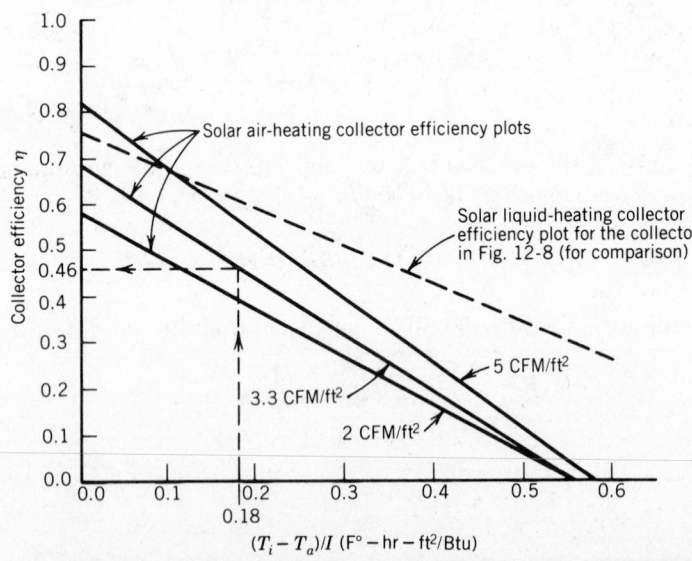

Figure 12-11 Collector efficiency plots for a typical air-heating collector. There are three efficiency plots for different air-flow rates in the collector. A plot of the collector efficiency for the liquid-heating collector of Fig. 12-8 has been added for comparison.

PERFORMANCE DATA

The industry-accepted standard for showing collector performance is the efficiency curve as shown. The two parameters that characterize the curve are the intercept and the slope. These change as a function of the collector flow rate and number of panels connected in series. Therefore, the parameters are presented in this table. The performance data was determined by an independent testing agency. i.e., Desert Sunshine Exposure Test, Incorporated, using the ASHRAE 93-77 test method.

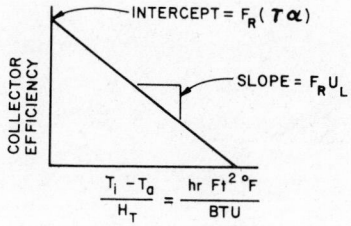

	FR =	Heat Removal Factor
INTERCEPT = $F_R(\tau\alpha)$	τ =	Cover Transmittance
SLOPE = $F_R U_L$	α =	Plate Absorptance
	U_L =	Overall loss coeff.
	T_i =	Coll. inlet temp.
	T_a =	Ambient Air Temp.
	H_T =	Solar Energy Available

$$\frac{T_i - T_a}{H_T} = \frac{hr\ Ft^2\ {}^\circ F}{BTU}$$

NOTE: Because of the unique nature of the measured incident angle modifier for the Solaron Series 3000 collector, conventional methods of calculating long-term solar system performance (e.g., f-chart) will yield a conservative estimate of system performance.

SOLARON COLLECTOR EFFICIENCY - Series 3000

FLOW RATE (SCFM/FT²)	No. of PANELS IN SERIES	$F_R(\tau\alpha)$	$F_R U_L$
2	2	0.522	0.846
	3	0.568	0.921
3	2	0.618	1.001
	3	0.623	1.010
4	2	0.653	1.059

SPACE HEATING OF A BUILDING: EFFICIENCY CURVES
Note: Even though the AIR collector may have a lower instantaneous efficiency curve, the system operating efficiency range is higher for AIR than LIQUID systems when the air systems are used for space heating with pebble bed storage.

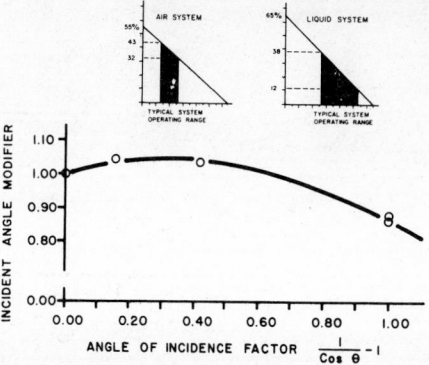

Figure 12-12 A test report on the performance of an air-heating collector. (Solaron® Solar Energy Systems.)

From Fig. 12-11, for an air flow of 3.3 CFM/ft² of collector area and a value of 0.18 F°-hr-ft²/Btu for the parameter, the efficiency η is seen to be about 0.46.

From Eq. (12-16), the useful energy output is:

$$Q = A\,I\,\eta = 32\ ft^2 \times 200\ Btu/hr\text{-}ft^2 \times 0.46$$
$$= 2940\ Btu/hr \qquad\qquad\qquad Ans.$$

12-6 Daily Performance of Solar Collectors

It must be remembered that collector test data are determined *for normal incidence* of solar radiation on the collector, and consequently heat energy yields as calculated above are actually *instantaneous* values—that is, *rates* of heat energy flow for normal incidence. For a stationary collector (most flat-plate collector installations are fixed in one position), the angle of incidence changes throughout the day, with marked effects on collector performance. The incident angle modifier, discussed in section 12-3 and defined by Eq. (12-17), can be used to plot collector performance curves for incident angles other than normal. The following example, together with Fig. 12-13, illustrates this process.

Illustrative Problem 12-8

For the collector test data of Fig. 12-9, determine the incident angle modifier for angles of 15°, 30°, 45°, 60°, and 75°.

Solution

From the data on the Summary Information Sheet (Fig. 12-9), the equation for the incident angle modifier is

$$K_{\tau\alpha} = 1.0 - 0.165\,(1/\cos\theta - 1)$$

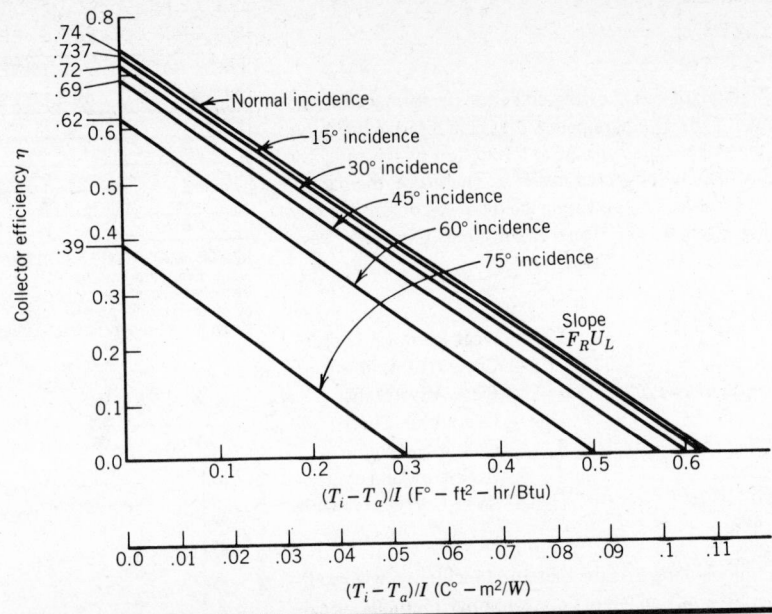

Figure 12-13 Collector efficiency plots for the different angles of incidence involved in illustrative problem 12-8.

For an angle of 15°,

$$K_{\tau\alpha} = 1.0 - 0.165 \ (1/\cos 15° - 1)$$
$$= 1.0 - 0.165 \ (1.04 - 1)$$
$$= 0.994 \qquad \qquad Ans.$$

With similar calculations, the values for the required angles are:

θ	$K_{\tau\alpha}$
15	0.994
30	0.974
45	0.932
60	0.835
75	0.527

Using the values of $K_{\tau\alpha}$ just calculated, a plot of the collector performance for different incident angles can be prepared, as shown in Fig. 12-13. These efficiency curves all have the same slope $(-F_R \, U_L)$, but the values of the y intercept, $\eta_{max} = F_R \, \overline{\tau\alpha}$—see Eq. (12-15) and Fig. 12-5—have been reduced for all angles except normal incidence, by multiplying by the proper value of $K_{\tau\alpha}$, as indicated in Eq. (12-19).

Plots like Fig. 12-13 can be used to determine the daily performance of a solar collector. Tabulations like those shown in Table 12-1 (English units) and Table 12-2 (SI units) can be prepared by choosing the typical day to be analyzed and inserting the expected values of T_a in the proper column. A column of expected fluid storage temperatures should also be prepared for use as values of T_i. Then the values of $(T_i - T_a)$ can be calculated. The insolation values (I column) are obtained from insolation tables for the specific location and date involved. Total solar radiation is used, and curves like those of Fig. 12-13 can be used to correct for different incident angles. The values of $(T_i - T_a)/I$ are then calculated and entered in the proper column. Incident angles for each hour are then obtained from solar tables for the locality, taking into account the tilt angle, latitude, and azimuth of the collector installation. Efficiencies can then be read from a plot like that of Fig. 12-13 for each value of $(T_i - T_a)/I$, by interpolating between the curves for the different incident angles. If a certain

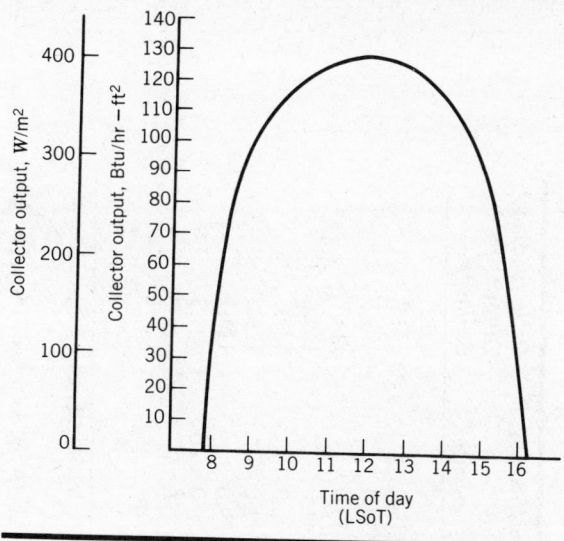

Figure 12-14 A plot of hourly collector energy output, from data obtained from Tables 12-1 and 12-2.

value of $(T_i - T_a)/I$ has no intercept on any of the incident angle curves of Fig. 12-13, the value of the efficiency η is zero. After the values of η are obtained (corrected for angle of incidence), the incident solar radiation is multiplied by η—see Eq. (12-16)—to give collector energy output. After the completion of a table like either Table 12-1 or 12-2, a plot of the day-long performance of the collector can be made, as illustrated in Fig. 12-14.

A set of tables like Table 12-1 or Table 12-2 can be made for typical days to give the solar system designer or contractor a tool to determine the daily collector energy output. Different collectors can then be evaluated for a particular application, from an energy-output standpoint. At this point, other considerations such as collector cost, availability, and installation costs can be taken into account. Finally, a decision can be reached on the most appropriate collector to use for the system being designed. Further attention to these design problems will come in later chapters.

CONCLUSION

We have presented in this chapter the general method of calculating hourly and/or daily collector performance from the collector test data that appear on laboratory test reports or on manufacturers' specification sheets. In many cases, however, a long-term performance estimate of the solar energy system is needed for an economic feasibility study. These long-term calculations may be for a season or for an entire year. Different methods such as F-chart, SOLCOST, and MC-2 programs have been developed to calculate long-term performance. Some of these involve tabular calculations, but others are full-fledged computer programs for either mainframe or personal (desk-top) computer analysis. All such calculations require that the values of $F_R \overline{\tau\alpha}$ and $F_R U_L$ be obtained from the test data to be used as inputs for the calculation.[3]

---

[3] See *Solar Age*, May, 1984, pp. 26–34, for a good analysis of currently available computer programs and a listing of firms that market microcomputer software for solar energy system design.

Table 12-1 Determination of Hourly Collector Performance Based on the Efficiency Plots of Fig. 12-13. Collector is at 40° N Lat, South-Facing, Tilt 41°, Date February 21 (English units)

Hour (LSoT)	T_a (°F)	T_i (°F)	$(T_i - T_a)$ (F°)	Insolation I (Btu/hr-ft²)	$\dfrac{(T_i - T_a)}{I}$	Incident Angle θ (degrees)	Efficiency η (percent)	Collector Energy Output, Btu/hr-ft²
7	41.0	131.0	90.0	68.8	1.31	75.0	0	0
8	42.8	116.6	73.8	206.0	0.36	60.5	15.0	30.9
9	44.6	109.4	64.8	275.0	0.24	45.9	34.0	93.5
10	46.4	113.0	66.6	294.0	0.23	31.5	39.0	114.7
11	51.8	113.0	61.2	304.0	0.20	18.0	41.0	124.6
12	53.6	113.0	59.4	308.0	0.19	10.0	42.0	129.4
13	59.0	111.2	52.2	304.0	0.17	18.0	41.0	124.6
14	57.2	111.2	54.0	294.0	0.18	31.5	39.0	114.7
15	55.4	116.6	61.2	275.0	0.22	45.9	31.0	85.3
16	50.0	131.0	81.0	206.0	0.39	60.5	12.0	24.7
17	44.6	125.6	81.0	68.8	1.18	75.2	0	0
								$\overline{842.4}$

Total Btu/ft²-day

Source: This data first appeared in the November 1978 issue of Solar Age. © 1978 Solar Vision, Inc., Harrisville, N.H. 03450. All rights reserved. Adapted and published with permission.

Note: Typical hourly collector performance data adapted from, Alwin B. Newton, "Getting the Most Out of ASHRAE 93-77," Solar Age, November 1978, pp. 26-29.

Table 12-2 Determination of Performance of Collector Based on the Efficiency Plots of Figure 12-13. Collector at 40° N Lat, South-Facing, Tilt 41°, February 21 (metric units)

Hour (LSoT)	T_a (°C)	T_i (°C)	$(T_i - T_a)$ (C°)	Insolation I (W/m²)	$\dfrac{(T_i - T_a)}{I}$	Incident Angle θ (degrees)	Efficiency η (percent)	Collector Energy Output, (W/m²)
7	5	55	50	217	0.23	75.0	0	0
8	6	53	47	706	0.067	60.5	15	106
9	7	50	43	867	0.050	45.9	34	295
10	8	53	45	929	0.048	31.5	39	362
11	11	56	45	961	0.047	18.0	41	394
12	13	58	45	971	0.046	10.0	42	408
13	15	59	44	961	0.046	18.0	41	394
14	16	60	44	929	0.047	31.5	39	362
15	13	60	47	867	0.054	45.9	31	269
16	10	60	50	706	0.071	60.5	12	85
17	7	59	52	217	0.240	75.2	0	0

Total watt-hours per day   2675

Note: Typical hourly collector performance data adapted from, Alwin B. Newton, "Getting the Most Out of ASHRAE 93-77," Solar Age, November 1978, pp. 26–29.

PROBLEMS

1. Determine the stagnation temperature for an unglazed collector, using the collector efficiency plots shown in Fig. 12-3, for an ambient temperature of 80°F and a wind velocity of 10 mph.

2. An unglazed swimming pool collector is operating under the following conditions: ambient temperature 75°F, average collector plate temperature 77° F, wind velocity 5 mph, and insolation on the collector 195 Btu/hr-ft². Determine the collector efficiency from Fig. 12-3.

3. A solar collector efficiency curve has a y-axis intercept ($F_R \overline{\tau\alpha}$) of 0.74 and a slope ($F_R U_L$) of -0.93 Btu/hr-ft²-F°. Determine the collector efficiency for normal incidence when the insolation is 215 Btu/hr-ft², T_a is 50°F, and T_i is 95°F.

4. A solar collector has an incident angle modifier constant (b_0) of 0.11. Determine the incident angle modifier ($K_{\tau\alpha}$) for incident angles of 45°, 55°, and 65°. (Use Fig 12-6.)

5. A solar collector efficiency curve has a y-axis intercept ($F_R \overline{\tau\alpha}$) of 0.68 and a slope ($F_R U_L$) of -1.06 Btu/hr-ft²-F°. Determine the collector efficiency for (a) normal incidence and (b) an incident angle of 55°, when the direct solar radiation I_{DN} is 225 Btu/hr-ft², the diffuse component I_d is 25 Btu/hr-ft², and the reflected component I_r is 25 Btu/hr-ft². The ambient temperature is 30°F, and the collector inlet fluid temperature T_i is 95°F. (Use Fig. 12-6 for $K_{\tau\alpha}$)

6. Ten of the solar collectors described in Fig. 12-8 are to supply hot water for an apartment house. Determine (a) the collector efficiency at incident angles of 0°, 45°, and 60°; and (b) the hourly energy output Q (Btu/hr) of the 10 collectors at the three incident angles. Assume that the direct solar radiation I_{DN} is 210 Btu/hr-ft², the diffuse component I_d is 25 Btu/hr-ft², the reflected component I_r is 20 Btu/hr-ft², the ambient temperature is 35°F, and the collector inlet fluid temperature is 98°F.

7. Calculate the energy output of a Grumman collector model 321A (use Fig. 12-10) at normal incidence. Assume that the normal insolation is 230 Btu/hr-ft², the fluid inlet temperature is 110°F, and the ambient temperature is 53°F.

8. For the solar collector described in Fig. 12-8, determine the collector hourly energy output for (a) normal incidence and (b) an incident angle of 50°. Assume the following operating conditions:

$$I_{DN} = 175 \text{ Btu/hr-ft}^2$$
$$I_d = 40 \text{ Btu/hr-ft}^2$$
$$I_r = 15 \text{ Btu/hr-ft}^2$$
$$T_a = 39°F$$
$$T_i = 110°F$$

9. An array of 10 Grumman model 332A collectors is located at 48° N Lat, facing due south. The collectors are tilted at 58° from the horizontal. Determine clear-day hourly energy output at noon, 2:00 P.M., and 4:00 P.M. LSoT, on April 21. Remember to correct for the angle of incidence. Use Table 4-E for the incident solar radiation and Table 4-A for the angle of incidence.

10. A solar water heater uses three of the collectors described in Fig. 12-8. They are located at 40° N Lat and tilted at 50° from the horizon. Calculate the clear-day energy output for noon, 2:00 P.M., and 4:00 P.M. LSoT on December 21. Use Table 4-E for the incident solar radiation and Table 4-A for the angle of incidence.

11. Air-heating solar collectors will be used for a space-heating system. It is necessary to know the hourly heat output per square foot of collector at solar noon. Use the data in Fig. 12-11 at an air-flow rate of 3.3 CFM/ft². The incident solar radiation is 265 Btu/hr-ft², the incident angle is 15°, the ambient temperature is 40°F, and the inlet air temperature is 95°F. Use the plot in Fig. 12-6 to determine the incident angle modifier.

12. For the air-heating collector described in Fig. 12-12, determine the collector efficiency equation for the following incident angles: 0°, 40°, 50°, 60°. Assume three panels in series with an air-flow rate of 3 CFM/ft$_c^2$.

13. For the collector data given in Fig. 12-8, prepare a plot of the collector efficiencies for the incident angles of 0°, 30°, 45°, 55°, 65°, and 75°. This plot should be similar to Fig. 12-13.

14. For the Grumman Collector Model 321A, prepare a plot of the collector efficiencies for the incident angles of 0°, 30°, 45°, 55°, 65°, and 75°. (See Fig. 12-10.)

15. A space-heating system is being designed for a house using the collectors described in Fig. 12-12. Prepare collector efficiency plots for the incident angles of 0°, 30°, 45°, 55°, 65°, and 75°. Assume two panels in series with an air flow of 4 CFM/ft$_c^2$.

16. It is necessary to estimate the total daily energy output of the collector described in Fig. 12-8. Construct a table similar to Table 12-1, using the efficiency plots calculated in problem 13 above. Assume the same values of T_a, T_i, I, and incident angle θ as given in Table 12-1. Calculate the collector efficiency for each hour, the collector energy output (Btu/hr-ft²), and the total output (Btu/ft²-day) for the collector.

17. Twenty Model 321A collectors, as described in Fig. 12-10, will be used in a solar space-heating system. Construct a table similar to Table 12-1, using the efficiency plots calculated in problem 14. Assume the same values of T_a, T_i, I, and incident angle θ as in Table 12-1. Calculate the collector efficiency for each hour of the daylight hours, the energy output in Btu/hr-ft², the energy output for the 20 collectors (Btu/hr), and the total daily energy output in Btu per day (7:00 A.M. to 5:00 P.M.).

C H A P T E R   13

Solar Heating of Domestic Hot Water

*T*he use of solar radiation for heating water has been established during the past 80 years in many regions of the world. In the United States, experimentation with solar water heating around the turn of the century led to an extensive solar water-heating industry in Florida and southern California in the 1920s and 1930s. Solar water heating was often the least expensive option for heating water in these areas before the introduction of low-priced natural gas and electricity in the 1930s. Many of the basic principles for heating water by solar energy were well established at this time. Solar water heating has also undergone extensive development in Australia, Japan, and Israel.

SYSTEM TYPES AND OPERATING MODES

13-1   Broad Classification of Solar Water Heaters

Two distinctly different configurations of solar water heaters are in current use. One type combines the solar collection and the thermal storage into a single unit, as illustrated in Fig. 13-1a. This type of unit is called an *integral collector storage* (ICS) system or *batch heater*.

The second type of solar water heater utilizes ordinary solar collectors to absorb solar radiation and

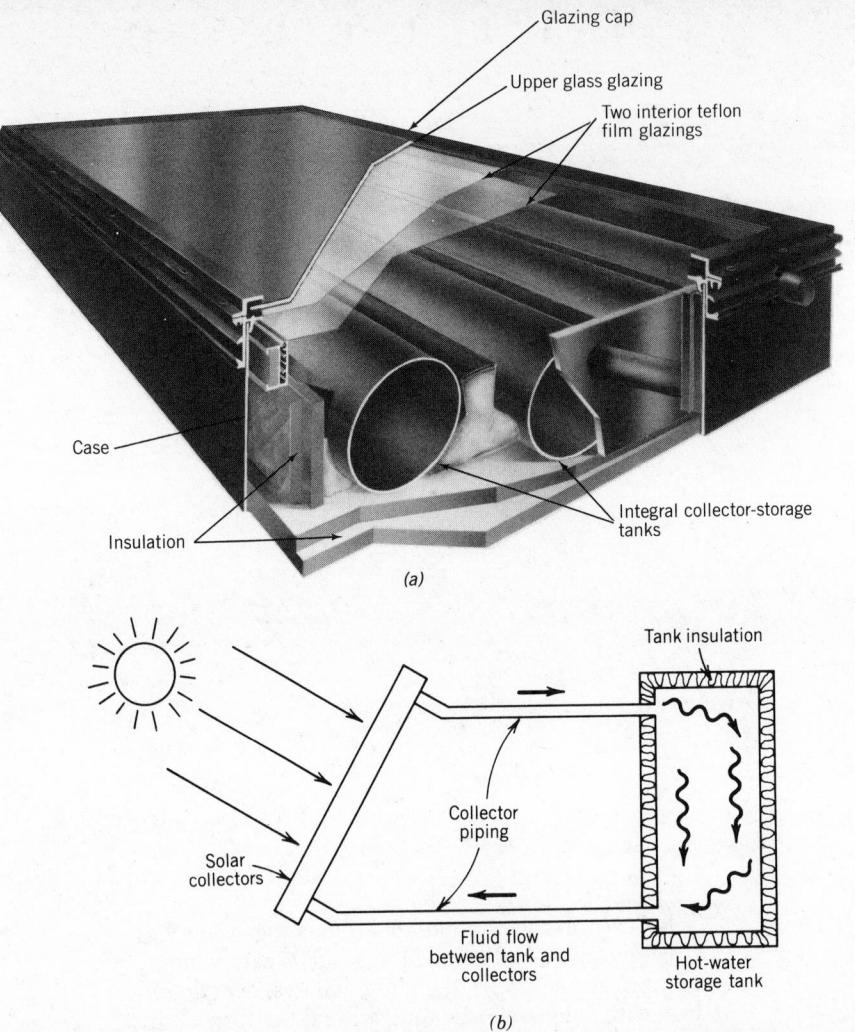

Glazing cap

Upper glass glazing

Two interior teflon
film glazings

Case

Insulation

Integral collector-storage
tanks

(a)

Tank insulation

Collector
piping

Solar
collectors

Fluid flow
between tank and
collectors

Hot-water
storage tank

(b)

Figure 13-1 Two basic types of solar water heaters. (a) Construction of a batch-type solar water heater. Also called integral collector storage (ICS) water heaters, these units typically consist of black-painted tanks mounted in an insulated box with a glass cover. Solar radiation is absorbed by the black tanks, and this energy heats the water. (Gulf Thermal Corporation.) (b) Flow diagram for a collector-and-storage-tank solar water heater. This type of heater utilizes separate solar collectors to heat the water and an insulated tank to store the hot water. The water may circulate between the collector and the storage tank by natural convection, in which case the arrangement is called a thermosiphon system. If the water is circulated by a pump, the arrangement is called an active system.

heat domestic water, which is then stored in an insulated tank, as shown in Fig. 13-1b. These systems may have forced circulation or operate on a thermosiphon principle.

13-2 Integral Collector Storage or Batch Water Heaters

Many of the early solar water heaters built during the first part of the twentieth century were batch water heaters. They were basically tanks that were painted black and installed in glass-covered boxes, as shown in Fig. 13-2. The water gradually heated up as solar radiation was absorbed by the black-painted tank. By the end of a typical sunny day, the water in the tank could reach temperatures of from 95°F to 115°F. A disadvantage of the batch water heater is that

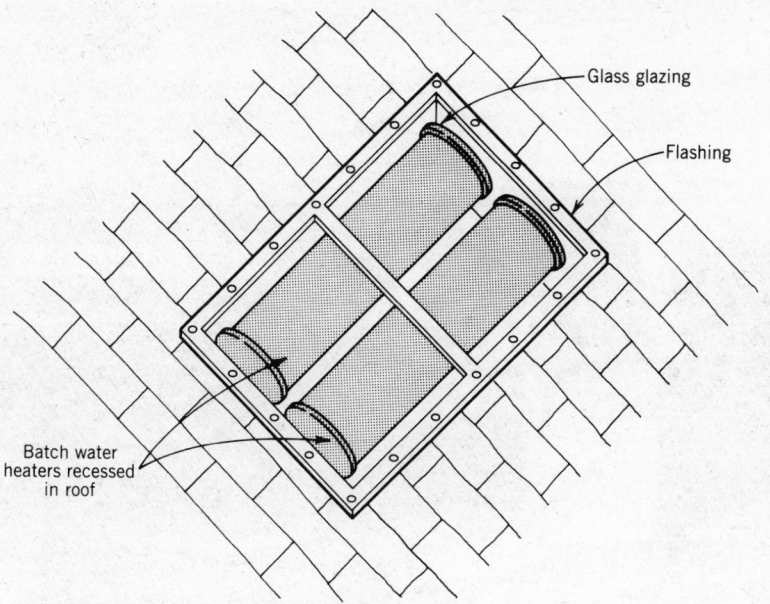

Figure 13-2 Early batch solar water heaters were simply black-painted tanks placed in an insulated cavity in the roof. A glass covering was placed over the tank to create a heat trap. (Sketch adapted from U.S. Patent No. 705.167, July 22, 1902.)

heat will be lost from the stored hot water through the glazing, thus gradually cooling the heated water during the night. This heat loss can be reduced somewhat by using multiple glazings.

A batch water heater that minimizes heat losses is shown in Fig. 13-3. It is called a "breadbox" water heater because of its likeness to an old-fashioned breadbox. It is double-glazed to reduce heat losses. The insulating panels are hinged so that they can be opened during the day and closed at night to insulate the water tanks. The inside of the insulating panels is reflective, to direct solar radiation onto the black tanks in the breadbox heater. The term *breadbox heater* is often applied to batch heaters even when hinged insulating panels are not provided.

The batch water heater can be built directly into the roof of a building, as shown in Fig. 13-4, where black-painted tanks are mounted in an insulated box with a double-glazed cover. This type of installation acts as a preheater for a standard gas-fired or electric water heater and will provide a large percentage of the energy for heating domestic hot water.

Many breadbox-type water heaters have been constructed by homeowners or by persons in community improvement organizations. Commercial batch water heaters are readily available from several manufacturers. An example is the one shown in Fig. 13-5a, manufactured by the Gulf Thermal Corporation. The combined absorber–storage tank assembly is manufactured from stainless steel and holds 39.8 gallons of water. The absorber has a selective coating for absorbing solar radiation and reducing heat losses. The case is made of sheet and extruded aluminum insulated with R-12 insulation. Figure 13-5b gives the manufacturer's specifications and system performance data, based on ASHRAE Standard 95-1981.[1] Such a unit will provide 14,500 Btu/day as heated water. This energy output would heat 40 gallons of water from 65° to 108°F during a typical day, as specified in the ASHRAE standard. Water heaters of this kind are mounted on a roof as units or in multiples, based on the volume of water that is to be heated.

[1] ASHRAE Standard 95-1981 is a test standard for comparing the performance of solar water-heating systems. It will be discussed in section 13-18.

Movable insulating panels

Figure 13-3 A breadbox batch heater for a mountain cabin. This unit utilizes movable insulating panels to insulate the tank(s) at night. During the day, the panels are opened and they reflect additional solar radiation onto the black storage tank(s).

There are several precautions that should be observed when installing a batch water heater. First, the weight of the installation must be considered, since the empty water heater itself may weigh several hundred pounds. It is not unusual for a small batch water heater to weigh 500 pounds or more when filled with water. Therefore, when they are to be mounted on a roof, it should be ascertained in advance that the roof is capable of supporting such a load. If there is any doubt about the strength of the roof, additional support must be provided.

Batch water heater tanks

Figure 13-4 A solar house with batch water heaters. Note that the heaters are built directly into the roof of the building, appearing like skylights from a distance.

Possible freezing of the water is another consideration. In moderate climates (3500 heating degree-days or less), this will not be a problem because of the large volume of warm water in the unit; however, the water pipes to and from the unit should be insulated to protect them from freezing. In colder climates (3500 to 4500 DD), batch heaters must be used with caution; and in climates with more than 4500 DD, their use is not recommended.

13-3   Forced-Circulation Solar Water Heaters

Forced-circulation solar water heaters can be divided into two basic types: open-loop (or direct) systems and closed-loop (or indirect) systems. The open-loop system is illustrated in Fig. 13-6a. The term *open-loop* is derived from the fact that the domestic water is circulated through the solar collectors as the heat transfer fluid and is directly heated by the solar collector. A closed-loop system is illustrated in Fig. 13-6b. The term *closed-loop* is derived from the fact that a heat-exchange fluid is circulated through the solar collectors and also through a heat exchanger, where the heat collected by the heat-exchange fluid is transferred to the domestic (potable) water. Thus, the domestic hot water is indirectly heated.

Open-Loop Domestic Water Heaters.   A typical residential open-loop solar water heater is shown in Fig. 13-7. The 4 ft by 8 ft collectors are mounted on the roof, as shown in Fig. 13-7a. The solar storage tank, circulation pump, differential thermostat control, and conventional gas (auxiliary) water heater are shown in Fig. 13-7b. A schematic representation of this type of system is shown in Fig. 13-8. A solar domestic water heater for a typical residence will have one or more solar collectors (SC), and the storage tank (ST) will range in volume from 66 to

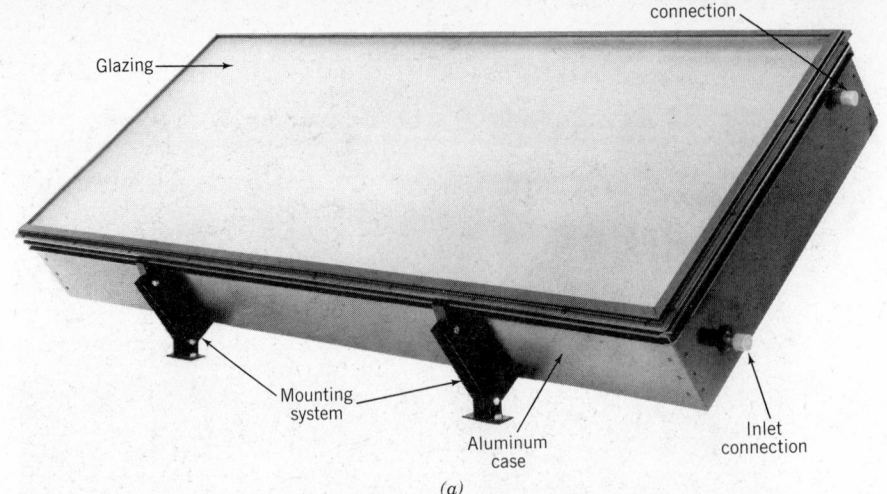

SPECIFICATIONS

Volumetric capacity:	39.84 gallons
Gross length:	86.25 inches
Gross width:	32.5 inches
Gross depth:	9.625 inches
Gross area:	19.46 FT2
Transparent frontal area:	17.4 FT2
Weight:	Dry-240 lbs.; Wet-559 lbs.
Flow pattern:	Series
Test pressure:	225 psi.
Design pressure:	150 psi.
Maximum Design temperature:	350°F
Normal operating temperature range:	40°F to 200°F

The following three tests were performed on the PT-400. The results vary because of different system sizes, piping configuration, or absorber coating. Everyday use performance also depends on water use patterns, hot water demand, and local weather conditions.

PERFORMANCE TEST RESULTS

FLORIDA SOLAR ENERGY CENTER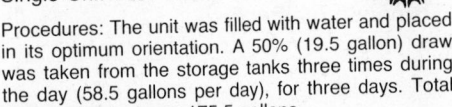
FSEC Certification # 82136
Single Unit Test

Procedures: The unit was filled with water and placed in its optimum orientation. A 50% (19.5 gallon) draw was taken from the storage tanks three times during the day (58.5 gallons per day), for three days. Total draw during test was 175.5 gallons.

Average air temperature	80.3° F
Average inlet temperature	81.7° F
Average outlet temperature	113.9° F
Average daily collection:	15,522 Btu
Average daily insolation:	1886 Btu/ft2
Average daily efficiency:	47.3 based on transparent area of front face
Absorber coating	Selective paint

TESTS PERFORMED BY DSET LABORATORIES, INC. CERTIFIED IN ACCORDANCE WITH SRCC STANDARD 200-82 BY THE SOLAR RATING AND CERTIFICATION CORP. (S.R.C.C.) Washington D.C.

 **DSET** LABORATORIES, INC.

SRCC NCC M82A

SRCC Test Parameters

Daily Insolation:	1500Btu/ft2
Water Supply Temperature:	71.6°F

Ambient Temperature:	71.6°F
Standard Test Load:	40,199 Btu/DAY

SRCC System Test Results "A"

DSET test #26565SYS
Model #PT-40

System Type:	Integral Collector/Storage
Transfer Fluid:	Water
System Configuration:	Two units connected in series. 76.5 gallons
Absorber surface:	selective paint
Array gross area:	39.20 ft2
Array Aperture area:	35.08 ft2
Number of collectors:	Two
Single solar collector tank volume:	38.25 gallons

System Performance "A"

$Q(NET)* = 7.22$ Kwh (24,600 Btu)
% Load Delivered* = 61
$Q(RES)* = 3.57$ Kwh (12,200 Btu)
$L* = 14.15$ w/°C (26.83 Btu/(hr.°F))

SRCC System Test Results "B"

DSET test #26884SYS
Model #PT-40

System Type:	Integral Collector/Storage
Transfer Fluid:	Water
System Configuration	Single unit. 39.84 gallons
Absorber Surface:	Selective Black Nickel
Array Gross Area:	19.5 ft2
Array Aperture Area:	17.4 ft2
Number of Collectors:	One
Single solar collector tank volume:	39.84 gallons

System Performance "B"

$Q(NET)* = 4.26$ Kwh (14,500 Btu)
% Load Delivered* = 36
$Q(RES)* = .26$ Kwh (900 Btu)
$L* = 6.28$ w/°C (11.91 Btu/(hr.°F))

---

*Q(NET) is the net energy delivered by solar in a DHW system at the specified test load.
*% Load Delivered is the percentage of the specified desired load that is delivered by the solar DHW system.
*Q(RES) is the reserve capacity of the solar DHW system after the specified daily use has been withdrawn. The reserve capacity is determined by a continuous drawdown of the system at the end of the test sequence.
*L is the rate of heat loss from the storage unit determined by a stagnant heat loss test and a mathematical model based on an exponential function.

Figure 13-5 Commercial batch water heaters. (*a*) A typical mass-produced batch (integral storage) water heater. This PT-40 unit is built by Gulf Thermal Corporation. (*b*) Specifications, certifications, and test results on the PT-40 batch water heater. The specifications give the size, volumetric capacity, and operating temperatures, while the test data give the performance of the unit under specified laboratory standard conditions. (Gulf Thermal Corporation.)

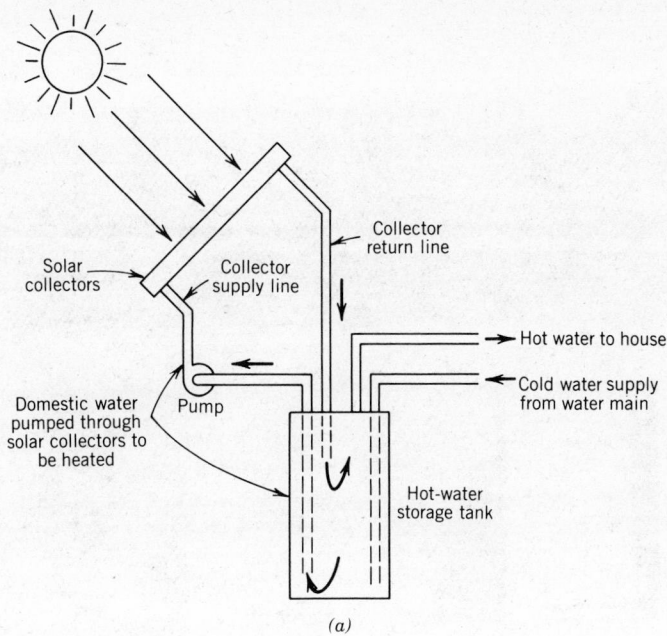

Solar collectors

Collector supply line

Collector return line

Hot water to house

Cold water supply from water main

Pump

Domestic water pumped through solar collectors to be heated

Hot-water storage tank

(a)

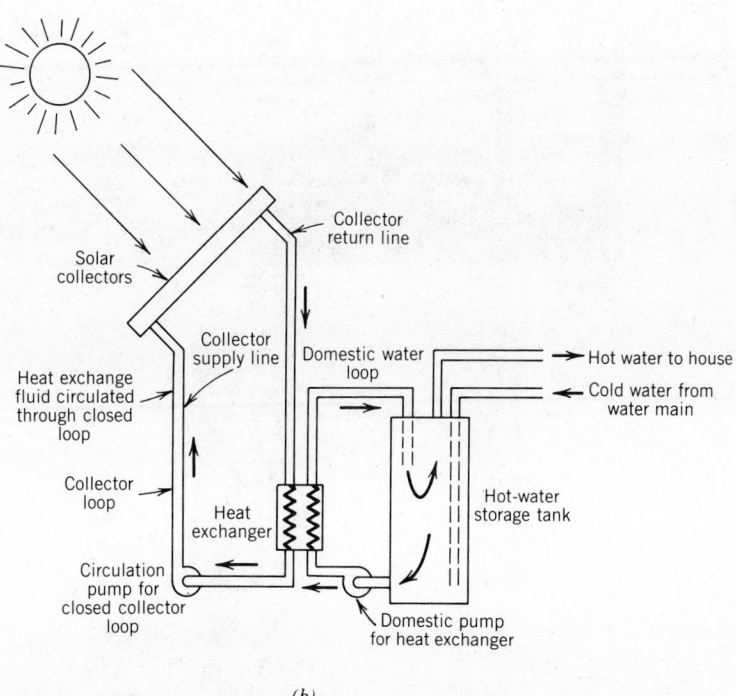

Solar collectors

Collector return line

Collector supply line

Domestic water loop

Hot water to house

Cold water from water main

Heat exchange fluid circulated through closed loop

Collector loop

Heat exchanger

Hot-water storage tank

Circulation pump for closed collector loop

Domestic pump for heat exchanger

(b)

Figure 13-6 Open-loop and closed-loop active solar water heater designs. (*a*) Flow diagram of an open-loop solar water heater. The domestic water to be heated is circulated directly through the collectors. As hot water is drawn out of the system, cold water enters to replace it. The collector–storage tank loop is open to a continual supply of cold water. (*b*) A closed-loop active water heater system. In this design, the domestic water does not flow through the collectors but is heated instead in a heat exchanger by the hot fluid in the collector–heat exchanger loop. Hence the name *closed-loop*.

Figure 13-7 An open-loop water heater installation. (*a*) Two of these solar collectors on a south-facing roof provide the heating elements for domestic hot water. The other two heat water for a spa or hot tub. (Santa Barbara Solar Energy Systems.) (*b*) Other essential components of an open-loop system. The hot-water storage tank, pump, and controller are mounted in a utility enclosure. The auxiliary gas heater is next to the storage tank.

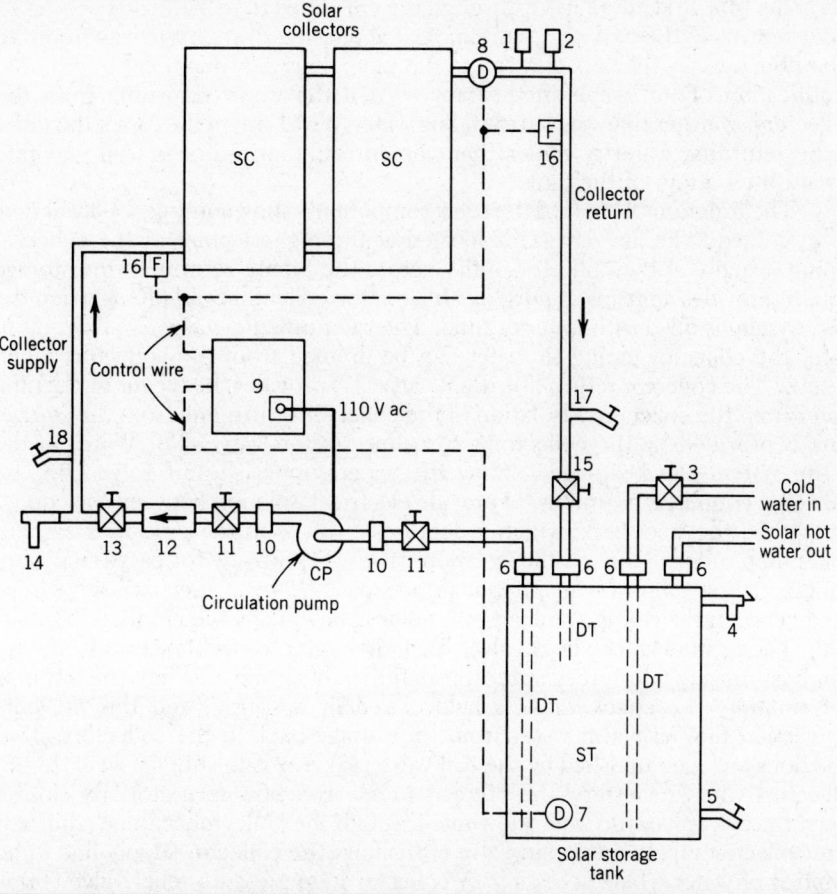

Figure 13-8 Diagrammatic representation of an open-loop solar water heater. The system components are identified by the code numbers and the legend.

SC   Solar water heating collectors
ST   Solar storage tank
CP   Circulation pump
DT   Dip tube
1   Air vent
2   Vacuum relief valve
3   Water supply shutoff valve
4   Solar storage tank temperature and pressure relief valve
5   Solar storage tank drain
6   Dielectric coupling or union
7   Storage differential sensor
8   Collector differential sensor
9   Differential thermostat controller
10   Pump union
11   Isolation valves
12   Spring-loaded check valve
13   Ball valve (flow control)
14   Collector-loop pressure relief valve
15   Collector return line shutoff valve
16   Freeze snap switch
17   Collector return line drain valve
18   Collector supply line drain valve

120 gallons. The tank shown in Fig. 13-8 shows three dip tubes (DT), plastic tubes that run from the tank fittings down into the tank. The cold-water dip tube feeds cold water to the bottom of the storage tank, thus preventing the cold water from mixing with the hot water at the top of the tank. The circulation pump (CP) draws cool water from the bottom of the storage tank through the

pump dip tube and circulates it through the collectors to be heated. The heated water returns to the tank and is discharged about one third of the way from the top of the tank by the collector return dip tube. This dip tube helps to promote stratification of hot water in the tank. When the water returning from the collector is warmer than the surrounding water, it will rise to the top of the tank. If the returning water is cooler than the surrounding water, it will gravitate toward the bottom of the tank.

The function of each of the solar components shown in Fig. 13-8 will now be examined. The air vent (1) is located at the highest point of the collector piping, usually at the collector outlet before the piping returns to the storage tank. It provides a means of purging air from the collectors and piping when the solar system is filled with water or fluid. The vacuum relief valve (2) allows air to enter the collector piping so water can be drained from the collectors when desired. The collector return line drain valve (17) provides the means of draining water from the collectors. Isolation of the collector return line from the storage tank is provided by the collector return line shutoff valve (15). Water to the entire system can be turned off by the water supply shutoff valve (3). The dielectric couplings or unions (6) provide electrical isolation between the copper piping and the steel tank, which reduces electrolysis in the pipe–tank system. The pump unions (10) and the isolation valves (11) provide for easy removal of the circulation pump for maintenance or repair. Water is mechanically circulated between the storage tank and the solar collectors by the circulation pump (CP). The spring-loaded check valve (12) allows water to circulate only when the pump is running and forces water through the check valve. When the pump is not running, the check valve is held closed by a spring, and this prevents convection flow of warm water from the storage back to the collectors. Two functions are accomplished by the ball valve (13): one is to shut off water to the collectors when necessary; the other is to act as a flow restrictor. By closing down the ball valve, the water flow rate through the collectors can be adjusted. The collector supply line drain valve (18) allows the collector supply line to be emptied of water when necessary. A collector loop pressure relief valve (14) is necessary because the collectors can be isolated from the tank when valves 13 and 15 are closed. If these valves are closed and water is not drained from the collectors, pressure from the expansion of the water heated in the collectors may become high enough to burst the piping. The pressure relief valve (14) prevents a high pressure from building up in the collectors by releasing water at a preset pressure in the range of 100 to 125 psi. The solar storage tank temperature and pressure relief valve (4) protects the storage tank from excessive temperatures and pressures by releasing water from the tank when either its temperature or its pressure set point is exceeded. The differential thermostat controller (9) turns the circulation pump on when the solar collectors are able to provide heat to the storage tank. The freeze snap switches (16) protect the solar collectors from freezing by closing at 38° to 40°F, which actuates the controller (9) to turn on the circulation pump. The pump circulates warm water from the storage tank up to the collectors until the freeze snap switches open as they are warmed up. This type of freeze protection is termed *recirculation freeze protection*. The collector differential sensor (8) senses the collector outlet temperature and provides an input to the differential thermostat controller. The storage differential sensor (7) senses the storage tank temperature and also provides an input to the controller. A means of draining the storage tank is provided by the storage tank drain (5). Control wiring between the controller and the sensors is usually 18-gauge low-voltage wire. The piping between the collectors and storage tank will ordinarily be either ½-in. or ¾-in. type *L* or *M* copper tubing.

Freeze Protection for Open-Loop Systems. The open-loop system just described has freeze protection provided by recirculation. The freeze snap switches attached to the lower collector header and the collector return piping have contacts that close when the temperature reaches 38° to 40°F. This will cause the control to turn the pump on, and warm water will be circulated

SOLAR HEATING OF DOMESTIC HOT WATER

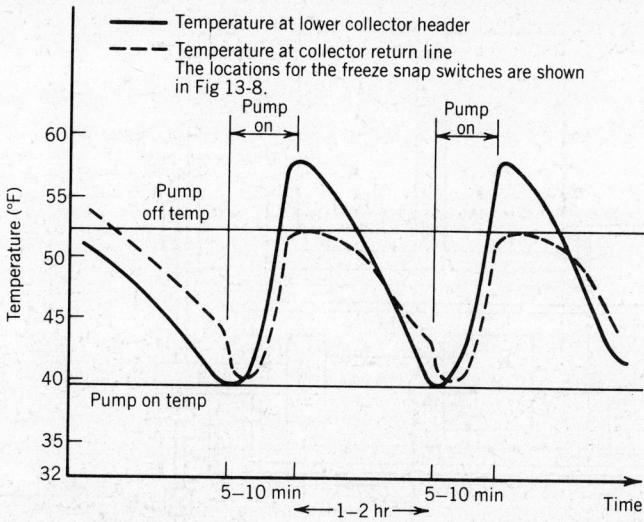

Figure 13-9 A temperature–time diagram for a recirculation system. In these systems, the cold water settles to the lowest part of the collectors. When the temperature at the lower collector header (solid line curve) reaches 38° to 40° F, the freeze snap switch (Fig. 13–8) will close, causing the controller to turn on the pump to circulate warm water from storage into the collector. The collector then warms up rapidly. The temperature at the outlet of the collector (dashed line curve) will drop as the cold water in the collector is pumped out. The pump will turn off when the collector freeze sensors are warmed above 52° F.

through the collector until both freeze snap switches (16 in Fig. 13-8) are warmed above 52°F.

The operation of a recirculation system is illustrated in Fig. 13-9. The night temperature at the lower collector header and the collector return line temperature are shown plotted against time, with an indication of the periods when recirculation occurs to warm the collectors. Note that recirculation occurs for only a few minutes each time.

The main advantage of recirculation freeze protection is that it is simple and requires only the freeze snap switches as extra components. A disadvantage is that collected solar heat is used from the storage tank to warm the solar collectors. In geographic locations where freezing conditions occur for only 15 to 20 days a year, the recirculation system will provide satisfactory and economical protection for the solar collectors. However, in locations where freezing conditions occur often, a large percentage of the collected solar heat could be used up at night in warming the solar collectors by means of recirculation. Another disadvantage of the recirculation system is that it will not function if the electric power is off for any reason. Consequently, in locations with frequent freezing conditions or where power outages might occur during a cold spell, the recirculation system should not be used.

Draindown freeze protection is another option for open-loop solar systems. A typical draindown system is diagrammed in Fig. 13-10. This system has all of the basic components of the recirculation system with the addition of the electric draindown valve. The differential thermostat controller must have the draindown option in order to control the draindown valve (DDV).

The draindown system will operate like the recirculation system during the day as the controller maintains the draindown valve in the normal position so that water can circulate to the collectors. When freezing conditions occur, the freeze snap switches (16 in Fig. 13-10) will operate at 50°F, and the controller

SYSTEM TYPES AND OPERATING MODES

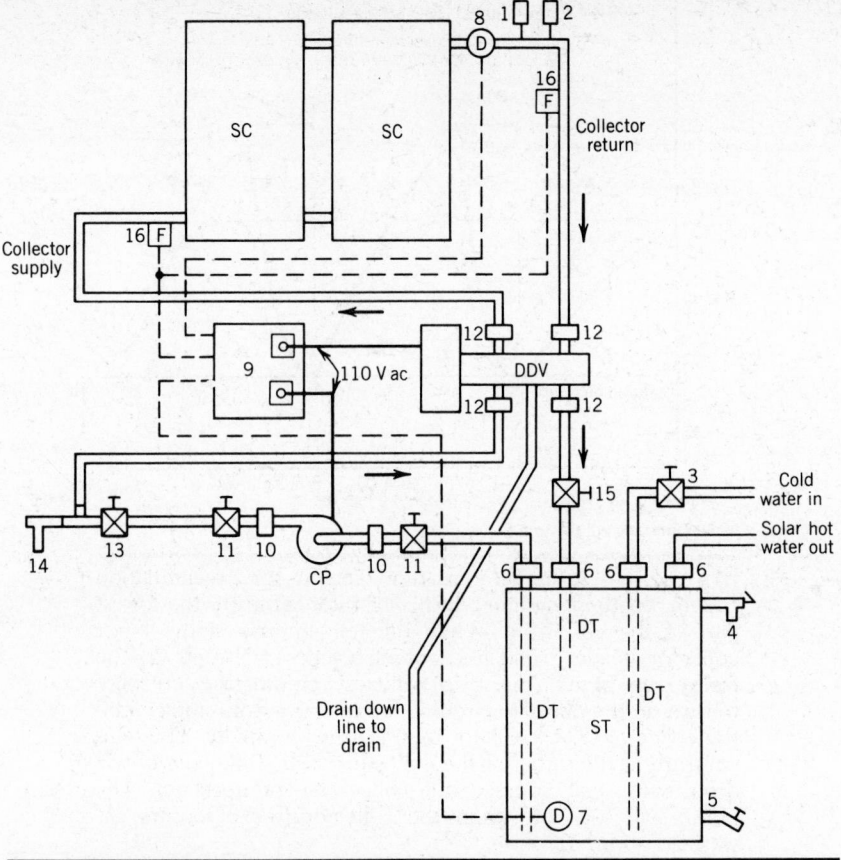

Figure 13-10 Basic components of an open-loop solar water-heating system utilizing a draindown freeze protection method. Components are identified by code numbers to the legend.

SC	Solar water-heating collectors
ST	Solar storage tank
CP	Circulation pump
DDV	Draindown valve
DT	Dip tube
1	Air vent
2	Vacuum-relief valve
3	Water-supply shutoff valve
4	Solar storage tank temperature and pressure-relief valve
5	Solar storage tank drain
6	Dielectric coupling or union
7	Storage differential sensor
8	Collector differential sensor
9	Differential thermostat controller with draindown option
10	Pump union
11	Isolation valves
12	Draindown valve unions
13	Ball valve (flow control)
14	Collector-loop pressure-relief valve
15	Collector return line shutoff valve
16	Freeze snap switch

will turn off electric power to the draindown valve. The DDV automatically changes from the normal position to the drain position whenever it is not electrically energized. In the drain position, the collector supply and return lines are closed off from the storage tank and opened to the drain port of the draindown valve. This allows the water in the collectors and the piping to drain out, thus eliminating the possibility of water freezing in the collectors. The

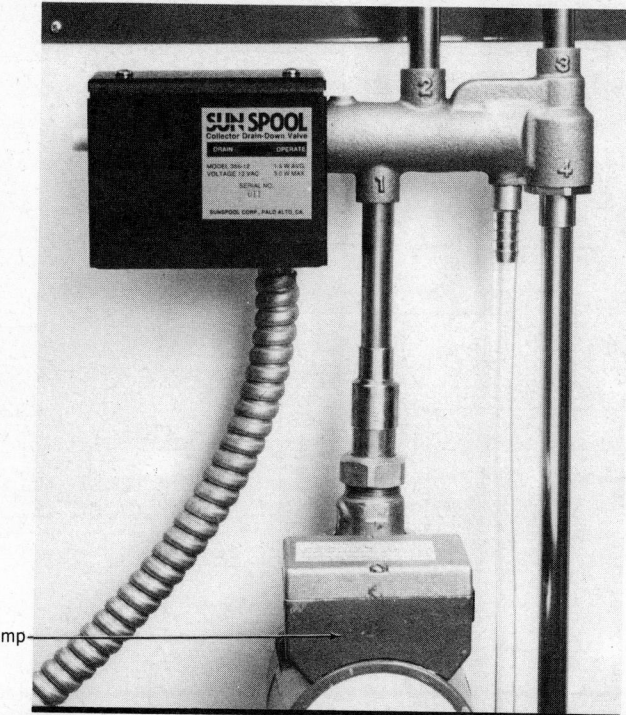

Pump

Figure 13-11 A draindown valve has four ports, as shown. The circulation pump is connected to port 1; port 2 goes to the collector; port 3 is connected to the collector return; and port 4 is connected to the storage tank. The serrated hose connector is the draindown port. (Sunspool Corporation.)

collectors will also drain if the power fails, since the DDV reverts to the drain position when it is not electrically energized. The collectors are therefore protected from freezing during power outages. A typical draindown valve is shown in Fig. 13-11.

Closed-Loop Domestic Water Heaters. Closed-loop domestic hot-water systems are used when open-loop systems are not appropriate. Examples are in locations where the climate is too cold to use water as a heat-transfer fluid, or where the water quality precludes circulating the domestic water through the solar collectors. A closed-loop water-heating system differs from an open-loop system by the addition of a heat exchanger. The heat exchanger transfers heat from the fluid flowing through the solar collectors to the domestic (potable) water that is to be heated. The advantage of the closed-loop system is that the same fluid is always circulated through the solar collectors. This minimizes the possibility of corrosion or the buildup of mineral deposits from the local water. The fluid circulated through the collectors can be a solution of water and antifreeze such as a 50–50 solution of water and propylene glycol, which has a freezing point of approximately −28°F, or a heat-transfer fluid such as a synthetic oil.[2] The use of an antifreeze solution in the collector loop also eliminates the possibility of damage to the collectors or collector piping from freezing.

The heat exchanger used in a closed-loop system can either be a separate component or it can be integral with the hot-water storage tank. In Fig. 13-12, the components of a closed-loop domestic hot-water system using a double-walled heat exchanger are shown. Figure 13-13 shows the construction of a

[2] Propylene glycol is recommended because ethylene glycol (the most common antifreeze) is quite toxic and would be dangerous to use in any system handling domestic water.

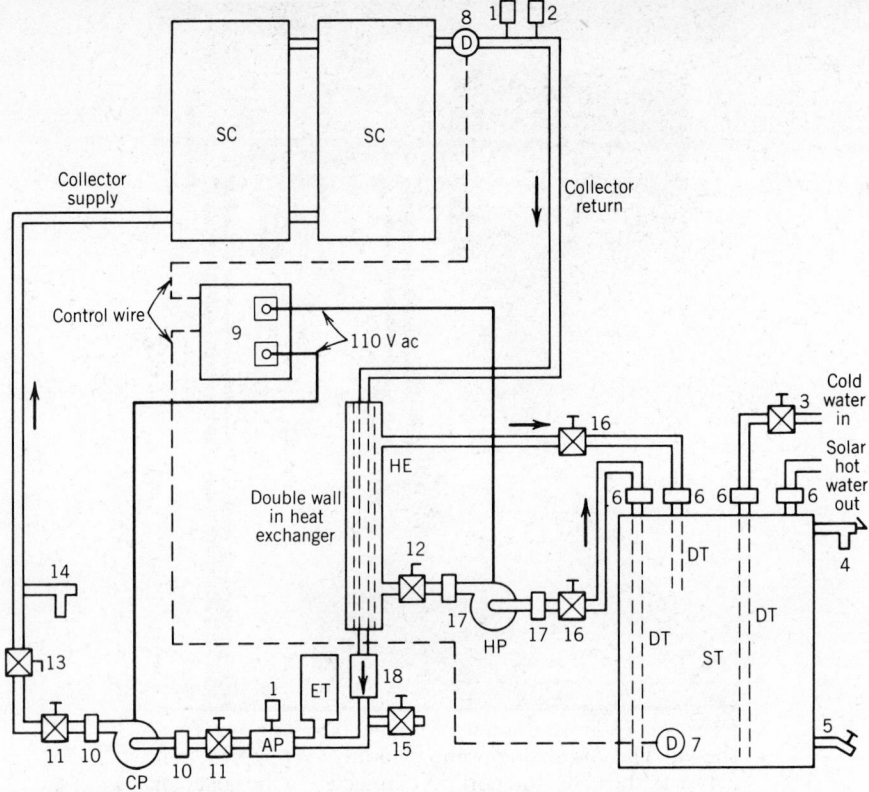

Figure 13-12 The essential components of a closed-loop solar water-heating system using a heat exchanger external to the storage tank. Code numbers identify the components.

SC	Solar collectors
ST	Solar storage tank
CP	Closed-loop circulation pump
HE	Double-walled heat exchanger
ET	Expansion tank
AP	Air purger
HP	Heat exchanger circulation pump
DT	Dip tube
1	Air vent
2	Vacuum breaker
3	Water supply shutoff valve
4	Solar storage tank temperature and pressure relief valve
5	Solar storage tank drain
6	Dielectric coupling or union
7	Storage differential sensor
8	Collector differential sensor
9	Differential thermostat controller
10	Pump union
11	Isolation valve
12	Ball valve (flow control)
13	Ball valve (flow control)
14	Collector-loop pressure relief valve
15	Fill-and-drain valve for closed loop
16	Isolation valves to heat exchanger
17	Heat exchanger pump union
18	Check valve

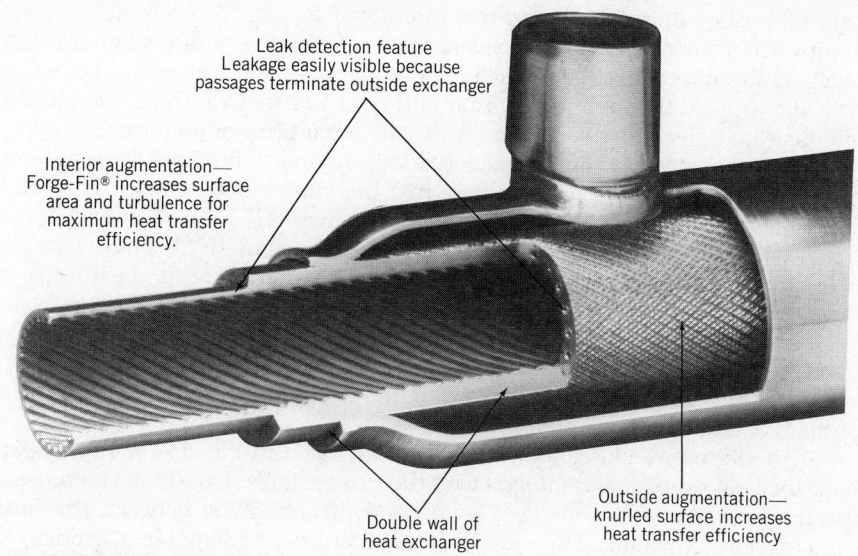

Leak detection feature
Leakage easily visible because
passages terminate outside exchanger

Interior augmentation—
Forge-Fin® increases surface
area and turbulence for
maximum heat transfer
efficiency.

Double wall of
heat exchanger

Outside augmentation—
knurled surface increases
heat transfer efficiency

Figure 13-13 A double-walled heat exchanger for a solar hot-water system. (Noranda Metal Industries.)

double-walled heat exchanger. The double wall prevents the contamination of the domestic water by the heat-exchange fluid if a wall of the heat exchanger develops a leak. The space between the double walls allows any leaking fluid to exit the heat exchanger body so that the leak will be visible and the heat exchanger can be replaced. Many governmental jurisdictions require by strict codes that a double-walled heat exchanger be used in closed-loop solar systems to protect against the possible contamination of the domestic water supply.

Each component of the closed-loop system shown in Fig. 13-12 performs a definite function. The solar collectors (SC) heat the closed-loop fluid flowing through them. The heat exchanger (HE) transfers the heat in the heat-transfer fluid of the closed loop to the domestic water circulated from the heat storage tank. The solar storage tank (ST) stores the heated domestic hot water. Air can be released from the closed-loop system by air vents (1) as the system is being filled. The vacuum breaker (2) allows air to enter the closed loop when it is being drained. In order that the heat-transfer fluid can expand and contract as it is heated and cooled each day, an air cushion is provided by the expansion tank (ET). An air purger (AP) separates out any air trapped in the closed-loop fluid and vents it through an air vent (1). A fitting for filling or draining the closed loop is provided by the fill-and-drain valve (15). Pump unions (10) and isolation valves (11) allow the circulation pump to be removed from the system without draining the closed loop. The closed-loop circulation pump (CP) provides for the forced circulation of the heat-exchange fluid in the closed loop. The spring-loaded check valve (18) allows fluid to circulate only when the pump (CP) is on. Flow can be adjusted in the closed loop for optimum system performance by the flow adjustment valve (13). In the event of excessive pressure buildup in the closed loop, the pressure relief valve (14) allows the release of fluid. The water supply shutoff valve (3) turns off the water supply to the solar storage tank. Dielectric unions (6) are used between the solar storage tank and the copper piping to reduce corrosion and electrolysis. The tank temperature and pressure relief valve (4) allows water to be released from the storage tank in the event of excessive pressure or temperature. If the heat exchanger or heat exchanger pump must be repaired, the isolation valves to the heat exchanger (16) allow the domestic water to be shut off. The pump unions (17) allow the heat exchanger pump to be removed for maintenance. The heat exchanger pump (HP) circulates cool water from the bottom of the storage tank through the heat exchanger to be heated. The flow of water through the heat exchanger can be adjusted for optimum system performance by the flow adjustment valve (12). The differential thermostat controller (9) and the differential sensors (7 and 8) turn on the pumps when solar heat can be collected and turn off the pumps when it is not practical to collect solar heat. In the closed-loop system, the controller (9) is typically just a differential thermostat and does not include freeze protection options, since the closed-loop fluid is usually an antifreeze solution.

Comparing Fig. 13-12 of the closed-loop system to the open-loop systems of Figs. 13-8 and 13-10, it can be seen that the closed-loop system has several additional components, namely the heat exchanger (HE), the heat-exchanger pump (HP), the air purger (AP), and the expansion tank (ET). These components increase the cost of the system as does the addition of the heat-transfer fluid. A closed-loop system typically will cost 20 to 30 percent more than an open-loop system of the same capacity. The net system performance of the closed-loop system will be about 5 to 10 percent less than that of an open-loop system of the same capacity because of the extra losses caused by the heat exchanger and additional piping.

An alternative closed-loop system is illustrated in Fig. 13-14. In this system, the heat exchanger is integral with the storage tank. This design eliminates the need for the heat-exchanger pump and the plumbing between the heat exchanger and the tank. The use of a storage tank with integral heat exchanger simplifies the installation because there is less plumbing and fewer components to install. The heat exchanger heats the water in the storage tank, which then rises away from the heat exchanger by convection. This sets up a natural convection in the tank, which keeps the domestic water moving across the heat exchanger surfaces in the tank as long as solar heat is collected by the solar collectors.

Another type of closed-loop system is one that combines the features of a closed-loop heat exchanger system with draindown freeze protection. This type of system is often given the name of *drainback*, and it will be discussed in more detail in a later section.

The disadvantage of the tank–heat exchanger combination is that (with some models), if either the tank or heat exchanger fails, the entire unit may have to be replaced.

Some equipment manufacturers provide preassembled control modules such as those illustrated in Fig. 13-15. These control modules or packages are designed for use with residential-sized solar domestic hot-water systems. An open-loop control module is shown in Fig. 13-15a. It consists of a "package" that has the circulation pump, the differential thermostat controller, and the draindown valve. In Fig. 13-15b, a closed-loop control module is shown. This package consists of a heat exchanger, collector loop circulation pump, expansion tank, check valve, pressure relief valve, closed-loop filling valve, pressure gauge, differential thermostat control, and heat-exchanger circulation pump.

13-4 Auxiliary Heat for Solar Hot-Water Systems

Solar water heaters are typically not designed to provide 100 percent of the hot-water load, and a source of auxiliary or backup heating for the water must be provided. One method of providing auxiliary heat is to introduce it into the same tank that is providing the solar storage volume, as shown in the flow diagrams of Fig. 13-16. This type of system is termed a *single-tank system*, because the solar storage and the auxiliary heating are combined in a single tank. Any of the collector loops that have been discussed above can be used to provide solar heat to the water in the storage tank. A single-tank open-loop system with electric auxiliary heating is shown in Fig. 13-16a. The electric heat elements are installed in the upper part of the tank, and the lower part of the tank provides storage volume for the solar-heated water.

A single-tank system using a conventional gas water heater is diagrammed in Fig. 13-16b. This type of installation is not recommended because the source of auxiliary heat is at the bottom of the tank and the entire water volume will be heated to the auxiliary temperature setting, with the result that solar energy will provide heat only at temperature levels above the auxiliary temperature setting.

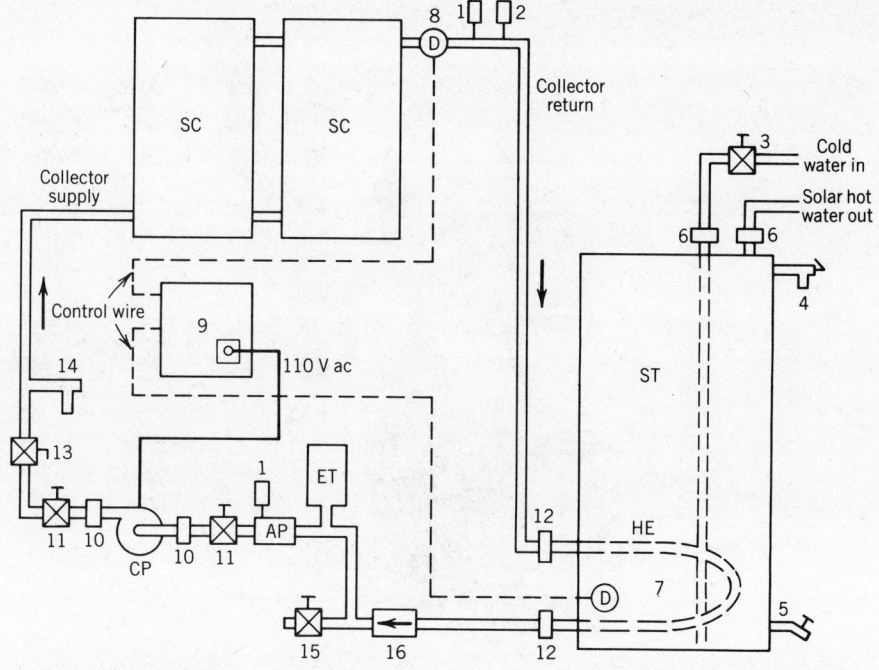

Figure 13-14 Flow diagram of a closed-loop system utilizing a solar storage tank with an internal heat exchanger. This arrangement eliminates the need for a second pump to circulate domestic water through the heat exchanger.

SC	Solar collectors
ST	Solar storage tank
CP	Closed-loop circulation pump
HE	Double-walled heat exchanger in storage tank
ET	Expansion tank
AP	Air purger
1	Air vent
2	Vacuum breaker
3	Water supply shutoff valve
4	Solar storage tank temperature and pressure relief valve
5	Solar storage tank drain
6	Dielectric coupling or union
7	Storage differential sensor
8	Collector differential sensor
9	Differential thermostat controller
10	Pump union
11	Pump isolation valve
12	Heat-exchanger unions
13	Ball valve (flow control)
14	Collector-loop pressure relief valve
15	Collector-loop fill-and-drain valve
16	Check valve

Such an arrangement is quite inefficient—first, because solar collectors are not very efficient in the higher temperature ranges, and second, because gas heat may often be used when solar heat is actually available. Solar collectors should provide the low-temperature heat up to the maximum of their output for the day and time, and conventional energy sources should provide the "topping-off" heat. The overall contribution by the solar system will be much lower in the case of Fig. 13-16*b* than in the case previously discussed (Fig. 13-16*a*), where the electric strip heater is in the top of the tank.

Specially designed single-tank gas heaters are commercially available, which have a submerged combustion chamber in the upper part of the tank, as shown in Fig. 13-16*c*. The upper part of the tank is heated by gas to provide

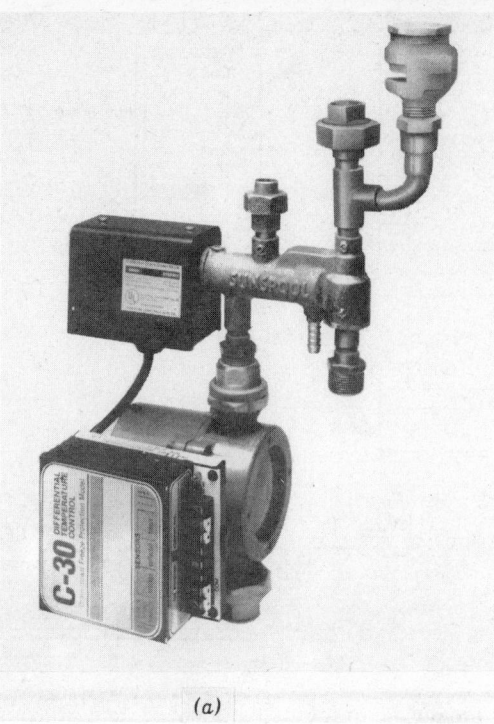

(a)

(b)

Figure 13-15 Controls for active solar water heaters. (a) A control package for an open-loop system. This unit provides the pump, draindown valve, and differential controller. (b) Control package for a closed-loop system. Provided are two pumps, a heat exchanger, expansion tank, valves, and the differential controller. (Both photos from American Appliance Manufacturing Corporation.)

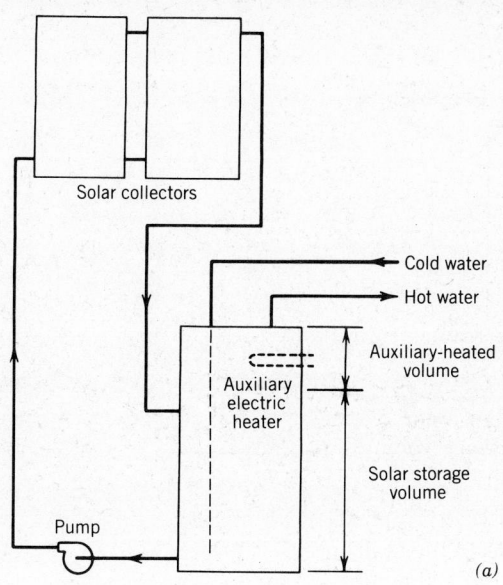

Solar collectors

← Cold water

→ Hot water

Auxiliary-heated volume

Auxiliary electric heater

Solar storage volume

Pump

(a)

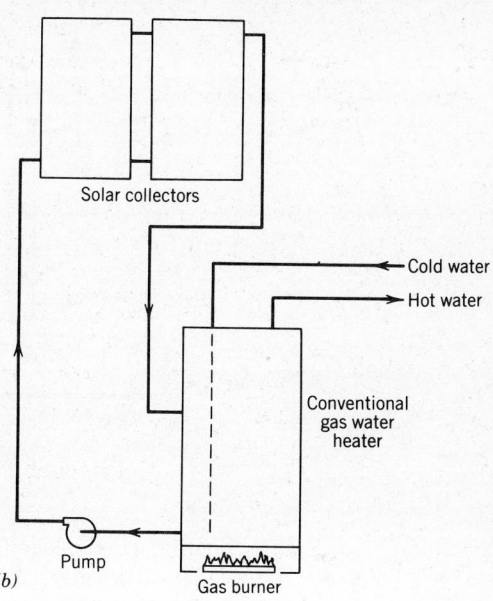

Solar collectors

← Cold water

→ Hot water

Conventional gas water heater

Pump

Gas burner

(b)

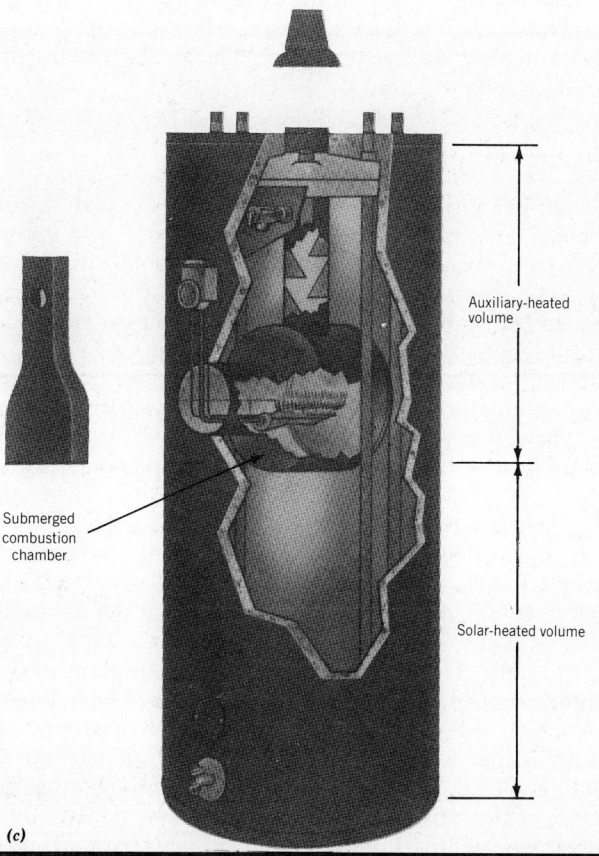

Auxiliary-heated volume

Submerged combustion chamber

Solar-heated volume

(c)

Figure 13-16 Flow diagrams for solar water heaters with different kinds of Auxiliary Heat. (*a*) A single-tank system using an electric auxiliary heater. The tank provides two storage volumes—one at the top for the auxiliary heater, and the lower part of the tank for the solar collectors. The collector return line enters the tank just below the auxiliary heater storage volume. If the solar-heated water is cooler than the auxiliary-heated water, the solar-heated water will diffuse downward and not mix appreciably with the auxiliary-heated water. When the solar-heated water is very hot, it will rise to the top of the tank by convection. (*b*) A single-tank solar water-heating system utilizing a standard gas water heater. In this system, there can be no separation of auxiliary and solar storage volumes because the entire tank is heated from the bottom by the gas burner. (*c*) Single-tank solar–gas water heaters are commercially available to provide separate auxiliary and solar-storage volumes. The gas burner is placed in a combustion chamber in the upper part of the tank. (American Appliance Manufacturing Corporation.)

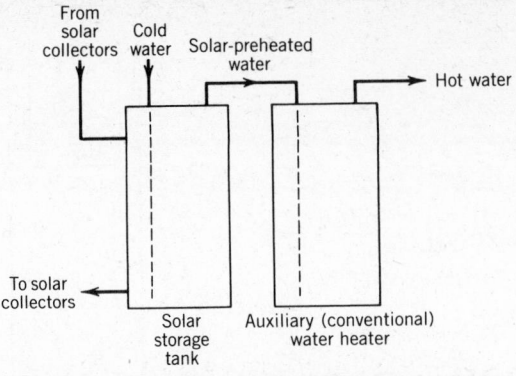

From solar collectors

Cold water

Solar-preheated water

Hot water

To solar collectors

Solar storage tank

Auxiliary (conventional) water heater

Figure 13-17 A dual-tank solar hot-water system uses one tank for the solar-storage volume and another tank for the auxiliary water heater.

auxiliary heating when needed, while the lower part of the tank provides storage for the solar-heated water, allowing for much greater efficiency of and energy input from the solar collectors.

A second arrangement for providing auxiliary heat is to use a separate solar storage tank and auxiliary heater, as shown in Fig. 13-17. This type of system is termed a *dual-tank* or *preheat system*, since the solar storage tank acts as a preheater for the auxiliary hot-water heater. The auxiliary tank can be heated by either gas or electricity.

13-5 Thermosiphon Systems

Some solar water-heating systems do not use forced circulation but make use of the *thermosiphon* principle. This type of system uses natural convection to circulate the heat-exchange fluid from the collectors to the storage tank, and there is no need for a pump and control system. Many of the solar water-heating systems installed before 1970 were the thermosiphon variety.

A thermosiphon system may be either an open-loop or a closed-loop system, as shown in Fig. 13-18a and b. In either case, the collectors must be located at a lower elevation than the storage tank to allow a thermal convection loop between the collectors and the tank to occur. As the fluid in the collectors is heated by the sun, it becomes less dense than the cooler water in the tank above. The warmer, less dense water will rise from the collectors and flow into the tank (the highest part of the system), while the cooler, heavier water will flow down to the collectors (the lowest part of the system). This thermal convection will cause water to circulate from the collectors to the tank as long as the collectors have a higher temperature than the water in the bottom of the tank. This convection stops if the collectors are at the same temperature as or colder than the tank. To prevent a reverse thermosiphon from occurring at night, the bottom of the tank should be a minimum of 18 in. above the top of the collectors, as shown in Fig. 13-18a. Otherwise, the thermosiphon will reverse at night, resulting in hot water being drawn from the tank into the collectors, where it will be cooled and flow back into the tank if the bottom of the collector and the bottom of the tank are at the same level.

The open-loop thermosiphon system presents a freeze protection problem. During cold weather, the coldest water will tend to settle in the bottom of the collectors and will eventually freeze, damaging the collector tubing. There are two methods of protecting against freeze damage. One method is to use either a manual or an automatic draindown system during cold weather. The other is to use a thermostatically controlled valve that bleeds the cold water out of the bottom of the collectors before the water can freeze.

The closed-loop thermosiphon system (Fig. 13-18b) is automatically protected by using an antifreeze solution in the closed loop. This type of system

SOLAR HEATING OF DOMESTIC HOT WATER

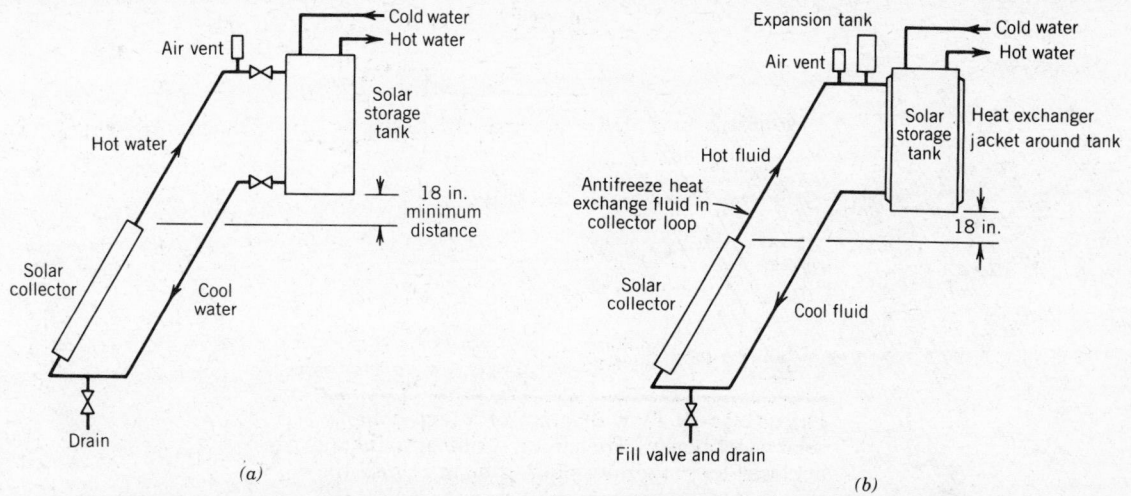

Figure 13-18 Thermosiphon solar hot-water heaters. (a) Flow diagram of an open-loop thermosiphon system. The tank must be placed above the collectors so that a natural convection current of hot water can take place between the collectors and the tank. (b) Freeze protection for thermsiphon water heaters can be provided by a closed-loop system, with the closed loop being filled with antifreeze fluid.

requires a fill valve, expansion tank, and air vent. A special type of heat exchanger, such as a jacket around the storage tank, is also required.

A modular open-loop thermosiphon system mounted on a roof is shown in Fig. 13-19.

13-6   Phase-Change Water Heaters

Figure 13-20 shows a flow diagram of a phase-change solar water heater. The solar collector is connected to the heat exchanger in the hot-water storage tank

Figure 13-19 A roof-mounted thermosiphon water heater. This unit consists of two solar water-heating collectors and a hot-water storage tank, which is mounted above the solar collectors. (Courtesy Santa Barbara Solar Energy Systems.)

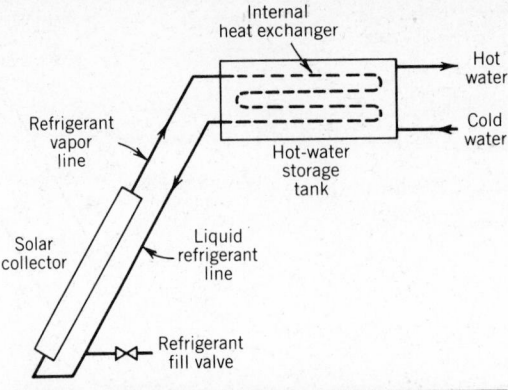

Figure 13-20 Flow diagram of a phase-change solar water heater. The circuit is similar to that of a closed-loop thermosiphon system, since the storage tank is at a higher elevation than the collectors. The essential difference is that the collector loop is filled with a refrigerant, which serves as the heat-exchange fluid. The transfer of heat energy from the collectors to hot-water storage is effected by latent heat transfer in and through the heat exchanger.

by a refrigerant vapor line and a liquid refrigerant line, the latter leaving the heat exchanger and returning to the bottom of the collector. This loop is filled with a suitable refrigerant.

When solar radiation is absorbed by the collector, the refrigerant is heated until it reaches its boiling point and vaporizes into a gas, a phase change in which it absorbs its latent heat of vaporization. The vapor circulates from the collector to the heat exchanger by natural convection, in a fashion similar to the natural convection of water in a thermosiphon solar water heater. When the vapor enters the heat exchanger, it releases heat to the water in the storage tank through the heat exchanger. This loss of heat allows the vapor to condense back to a liquid in the heat exchanger (phase change from vapor to liquid), and the liquid refrigerant then flows by gravity from the heat exchanger back to the collector through the liquid line.

The vapor pressure of the refrigerant in the collector loop is determined by the water temperature in the storage tank, because the water temperature determines the temperature at which the refrigerant vapor condenses back into a liquid. As the water storage temperature increases, the vapor pressure correspondingly increases. A refrigerant commonly used in phase-change solar water heaters is dichlorotetrafluoroethane ($CClF_2CClF_2$), whose refrigerant number is R-114. R-114 has the following properties:

1. Freezing point of $-137°F$.
2. Atmospheric pressure boiling point of $38.8°F$.
3. Vapor at $140°F$ condenses at 83.76 psia.
4. Nonflammable.
5. Low toxicity.
6. Good chemical stability.

These properties of R-114 make phase-change water heaters inherently freeze-proof, noncorroding in the collector loop, and nontoxic as compared to other closed-loop solar water-heating systems that use heat exchange fluids.

The ability of the refrigerant to undergo a phase change gives a refrigerant-charged system certain advantages in performance. Typically, a solar domestic hot-water system uses the sensible heat transfer of a liquid such as water, water and glycol, or other heat-transfer fluid. The refrigerant-charged

SOLAR HEATING OF DOMESTIC HOT WATER

system utilizes the latent heat of vaporization and condensation to transfer heat from the collector to the solar storage tank, allowing the heat to be removed at a lower collector temperature than with sensible heat-transfer fluids, and resulting in an increase in the system performance for phase-change solar water heaters. Manufacturers presently are indicating that the phase-change systems are 30 to 40 percent more efficient than other heat-exchanger systems on a lifetime basis.

The advantages of the phase-change system are that it has no moving parts, it is a passive system and requires no energy sources other than the sun, it is freezeproof, and it requires very little maintenance. The main disadvantage is that the storage tank–heat exchanger must be located above the collectors, and the horizontal distance between the collectors and the heat exchanger should not be more than 15 to 45 ft, depending on the collector area and vapor line diameter. Some manufacturers provide modular units where the collector and storage are shipped as a package. Others provide components that can be assembled on the job.

SOLAR WATER HEATER SYSTEM DESIGN

The solar designer or installer must be able to apply the essential techniques in designing, sizing, and predicting the performance of a solar water-heating system. Fortunately, there are a variety of design and prediction methods, ranging from general rules of thumb and hand calculations to computer simulations. It is important to be able to predict the effect of different collector tilt angles and azimuth angles and compensate for nonoptimal orientations. It is also important to be able to analyze the effect of shading when there is limited solar access. Potential purchasers of solar water-heating systems should be given as many facts as possible in order for them to make decisions about the purchase of a solar water heater. The client should be made aware of the expected performance of the system and should be apprised of factors that can limit performance, such as site orientation and solar access.

The design and sizing of a solar water-heating system can readily be broken down into the following steps:

1. The determination of the average domestic hot-water load on a monthly or yearly basis.
2. The type of solar energy system to be used—integral collector storage (batch heater), open-loop forced circulation, closed-loop forced circulation, or thermosiphon.
3. Sizing of the solar collection area to provide the desired portion of the hot-water load.
4. Determining the solar hot-water storage tank size, based on the collector area and hot-water load.
5. Determining pump sizes and pipe diameters, based on the distance between the collectors and the storage tank.
6. Selecting the controls for the system.

13-7 Determining the Domestic Hot-Water Load

The hot-water load is determined by a number of factors, among which are the number of occupants in the building, the type of hot-water usage, the incoming water temperature, and the hot-water design temperature. The average monthly domestic water-heating load, in Btu per month, can be calculated using the following equation:

$$Q_{WM} = c_W \times D \times WC_h \times (T_s - T_m) \times N \qquad (13\text{-}1)$$

where

c_W = the specific heat of water = 1.0 Btu/lb-F°
D = the density of water = 8.34 lb/gal
WC_h = the hot water consumption, gal/day
T_s = the temperature of the hot-water supply, °F (usually taken as 135° to 140°F)
T_m = the temperature in the cold water main, °F
N = the number of days per month
Q_{WM} = the average monthly water-heating load, Btu

Table 13-1 Daily Hot-Water Usage (140°F) for Typical U.S. Locations

Category	One- and Two-Family Units and Apartments up to 20 Units					Apartments of 20 to 200 Units (Central Hot Water)
Number of occupants	2	3	4	5	6	
Number of bedrooms	1	2	3	4	5	
Hot-water usage (gal/day-living unit)	40	55	70	85	100	40

Source: *HUD Intermediate Minimum Property Standards Supplement, Solar Heating and Domestic Hot Water Systems*, 1977, Table S-615-1.

Table 13-2 Daily Average Hot-Water Usage for Selected Appliances

Appliance	Average Daily Hot-Water Consumption (gal/person-day)
Clothes washer	4.5
Dishwasher	3.5
Showerhead or bathtub faucet	4.5
Faucet, kitchen	5.0
Other faucets	1.0

Source: *The Appliance Efficiency Program, Revised Staff Report, September 1977, Relating to Space Heaters, Storage Type Water Heaters, and Plumbing Fixtures*, California Energy Resources Conservation and Development Commission.

Determining the daily hot-water consumption for individual households can be difficult. Typically, average daily hot-water usage will be estimated unless the customer or owner has more specific data. Table 13-1 gives daily hot-water usage for typical U.S. locations, according to building type, number of occupants, and number of bedrooms. Table 13-2 gives average daily hot-water usage for various types of appliances.

Illustrative Problem 13-1

Estimate the average daily hot-water usage (U.S.) for a three-bedroom home with four regular occupants.

Solution

From Table 13-1, the hot-water usage for a typical U.S. location is 70 gal/day.
Ans.

Illustrative Problem 13-2

A house has two baths, a dishwasher, and a clothes-washing machine. There are two bedrooms, and three occupants live in the house. Estimate the daily hot-water usage.

Solution

From Table 13-2, the following estimated water usage is obtained:

Dishwasher	3.5 gal/person-day
Clothes washer	4.5
Shower or bath	4.5
Kitchen sink faucet	5.0
Bathroom lavatory faucet	1.0
	18.5 gal/person-day

Total daily hot-water usage will then be

$$18.5 \text{ gal/person-day} \times 3 \text{ persons} = 55.5 \text{ gal/day} \qquad Ans.$$

Note the rather close agreement with the 55 gal/day estimate given by Table 13-1 for a two-bedroom, three-occupant home.

The monthly hot-water heating load can be calculated using Eq. (13-1) once the daily hot-water usage is established. Typically, the design temperature for a domestic hot-water heater is in the range of 120°F to 140°F. The local water supply temperature may vary from 45°F to 75°F, depending on the climatic region and season of the year. An average water supply temperature of 55° F is often used for design purposes if local water temperatures are unknown. However, whenever possible, determine the local cold-water temperature, averaged on a monthly basis.

Illustrative Problem 13-3

Calulate the January hot-water load for a residence where the hot-water usage is 60 gal/day. The incoming water temperature is 50°F, and the hot-water system is set to deliver 130°F water.

Solution

From Eq. (13-1),

$$
\begin{aligned}
Q_{W_{Jan}} &= 1 \text{ Btu/lb-F°} \times 8.34 \text{ lb/gal} \times 60 \text{ gal/day} \\
&\quad \times (130 - 50) \text{ F°} \times 31 \text{ days} \\
&= 1.24 \times 10^6 \text{ Btu} \qquad\qquad Ans.
\end{aligned}
$$

To estimate the yearly hot-water load, Eq. (13-1) is used to calculate all 12 monthly loads, which are then summed up to give the yearly load.

An alternative to the use of Eq. (13-1) is the chart shown in Fig. 13-21. This chart allows the monthly or yearly hot-water load to be quickly estimated. The inputs to the chart are:

1. Number of persons occupying the structure on a regular basis.
2. Hot-water usage in gallons per person per day.
3. Year-round average temperature difference between the water supply main and the hot-water supply ($T_s - T_m$).

The use of this chart may not give results that are as accurate as those obtained from Eq. (13-1), but it is convenient for initial estimate purposes. The following problem illustrates its use.

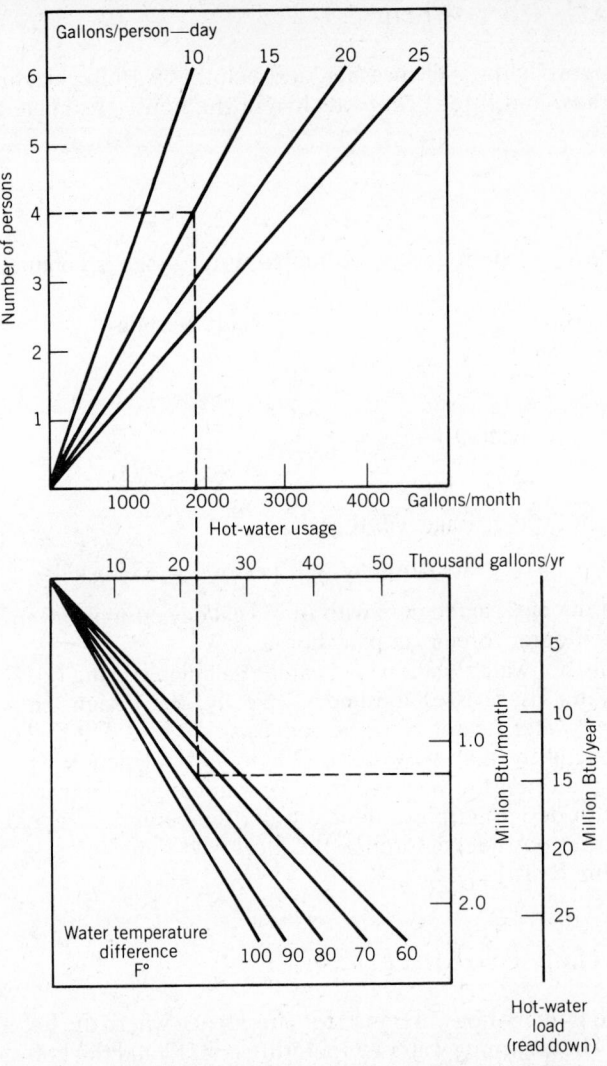

Figure 13-21 A load-sizing chart for determining monthly
and annual domestic hot-water usage. (W. A. Wright, "Sol-
graph/DHW," *Solar Age*, March, 1980; © 1980, Solar Vision,
Inc., Harrisville, N.H. All rights reserved. Adapted and published by permission.)

Illustrative Problem 13-4

The daily hot-water use in a house is estimated to be 15 gal/person-day. There
are four occupants. The water supply temperature is 55°F, and the hot-water
system is set to deliver 135°F water. Use Fig. 13-21 to estimate the monthly and
average yearly hot-water loads.

Solution

The temperature difference $(T_s - T_m)$ between water supply main and hot-
water supply is 135°F − 55°F = 80°F. Figure 13-21 is entered at four persons,
the number of occupants in the house. A horizontal line is projected until it
intersects the 15 gal/person-day line. A vertical line is projected down until it
intersects the 80°F water temperature difference line. A horizontal line is
projected until it meets the energy axes of the chart. The monthly energy load is
approximately 1.2 million Btu/month, and the yearly energy load is about 14.5
million Btu.   *Ans.*

13-8 Deciding on the Type of System

The type of solar system to be specified for a particular job will depend on several factors, such as the hot-water load, roof orientation, roof strength, location of existing hot-water heater, available space for additional hot-water storage tanks, and customer desires.

If an integral collector storage water heater is chosen, the test and performance data supplied by the manufacturer must be considered in sizing the system. Figure 13-5*b* gives some performance data for a PT-40 unit. The Florida Solar Energy Center indicates that the average energy collected by the unit on a day when the insolation averages 1886 Btu/ft² is 15,522 Btu/day, and the monthly energy collected would be

$$Q_{WM} = 30 \text{ days} \times 15,522 \text{ Btu/day}$$
$$= 465,660 \text{ Btu/month.}$$

If the hot-water load is 1.24×10^6 Btu/month, as is the case in Illustrative Problem 13-3, the use of two PT-40s would provide approximately 931,000 Btu/month, or about 75 percent of the hot-water load. The exact size of an integral collector storage system for a hot-water load is difficult to determine because the system performance is dependent upon the heat lost from storage through the glazing.

When a thermosiphon system or active system is used, the system performance and energy collected are based on the collector area and collector efficiency. The energy losses from storage are small because the storage tank will typically be completely insulated with R-12 to R-20 insulation. The sizing of these systems will be discussed in the following sections.

13-9 Determining the Size of the Solar Collection Area

Once the domestic hot-water load has been determined, the fraction of the domestic hot water that will be supplied by the solar energy system must be decided upon, which in turn determines the size of the solar collector area. The solar contribution of the domestic hot-water load is termed the *solar heating fraction* and is defined as:

$$\text{SHF}_{DHW} = \frac{Q_{SM}}{Q_{WM}} \tag{13-2a}$$

where

$$\text{SHF}_{DHW} = \text{the solar heating fraction of the}$$
$$\text{domestic hot-water heating load}$$
$$Q_{SM} = \text{the total } monthly \text{ solar heat input}$$
$$\text{to the system}$$
$$Q_{WM} = \text{the total monthly water-heating}$$
$$\text{load}$$

Now, let $Q_{Aux\,DHWM}$ be the auxiliary heat supplied to the system per month:

$$Q_{WM} = Q_{SM} + Q_{Aux\,DHWM}$$
$$Q_{SM} = Q_{WM} - Q_{Aux\,DHWM}$$

Substituting for Q_{SM} in Eq. (13-2a),

$$\text{SHF}_{DHW} = \frac{Q_{WM} - Q_{Aux\,DHWM}}{Q_{WM}}$$

or,

$$\text{SHF}_{DHW} = 1 - \frac{Q_{Aux\,DHWM}}{Q_{WM}} \tag{13-3a}$$

The solar heat fraction may also be written in terms of average *daily* values, with Q_{SD} and Q_{WD} replacing Q_{SM} and Q_{WM}, as follows:

$$SHF_{DHW} = \frac{Q_{SD}}{Q_{WD}} \qquad (13\text{-}2b)$$

Eq. (13-3a) then becomes

$$SHF_{DHW} = 1 - \frac{Q_{Aux\ DHWD}}{Q_{WD}} \qquad (13\text{-}3b)$$

where $Q_{Aux\ DHWD}$ is the average daily auxiliary heat required.

Illustrative Problem 13-5

The hot-water load for a residence is estimated to be 1.8×10^6 Btu in January. If a solar water-heating system provides a solar heat fraction of 70 percent, what will the January auxiliary heating requirement be?

Solution

From Eq. (13-3a),

$$SHF_{DHW} = 1 - \frac{Q_{Aux\,DHWM}}{Q_{WM}}$$

Solving for $Q_{Aux\,DHW_{Jan}}$,

$$
\begin{aligned}
Q_{Aux\,DHW_{Jan}} &= (1 - SHF_{DHW}) \times Q_{WM} \\
&= (1 - 0.70) \times 1.8 \times 10^6\ \text{Btu} \\
&= 0.54 \times 10^6\ \text{Btu} = 540{,}000\ \text{Btu} \qquad Ans.
\end{aligned}
$$

Illustrative Problem 13-6

A solar water-heating system was designed to provide 1.3×10^6 Btu in March. The average monthly hot-water load is 1.5×10^6 Btu. Determine the solar heating fraction for the system.

Solution

From Eq. (13-2a),

$$
\begin{aligned}
SHF_{DHW_{March}} &= \frac{Q_{SM}}{Q_{WM}} = \frac{1.3 \times 10^6\ \text{Btu}}{1.5 \times 10^6\ \text{Btu}} \\
&= 0.87 \qquad Ans.
\end{aligned}
$$

In sizing domestic solar hot-water systems, there are different levels of design tools. First, there are very general rules of thumb that apply to most cases; second, there are guidelines for specific geographic locations based on the local climate; and, third, there are performance predictors based on system simulation or system correlation methods.

The general rules of thumb deal with the solar heat fraction (SHF), the solar collector area, and the solar hot-water storage volume.

The rule of thumb for the solar heat fraction can be stated as:

A solar heat fraction of between 50 and 80 percent of the domestic hot-water load is generally found to be optimum, when all factors influencing system performance and cost–benefit considerations have been taken into account.

This rule of thumb implies that a solar hot-water system will usually be sized to

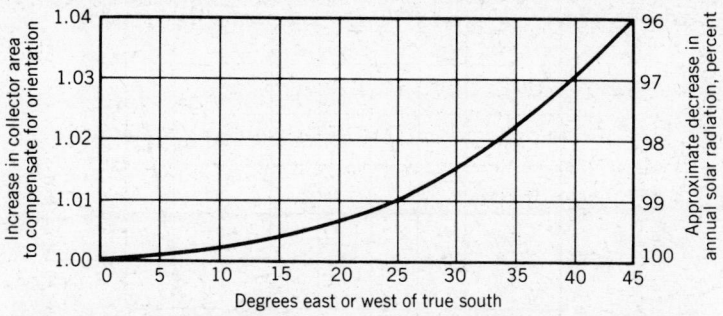

Figure 13-22 Chart showing the variation in the annual solar radiation available to solar collectors as a function of collector orientation, and the increase in collector area needed to compensate for orientations that are not due south. (Adapted from *National Solar Water Heater Workshop Handbook*, "Pumped Recirculation" Freeze Protection, Arizona State University, DOE Grant No. DE-FG-03-805F11444, 1981.)

provide between 50 and 80 percent of the hot-water load. There will be exceptions to this rule, such as a hot-water system in a remote location where there are no energy-related utilities available. In this situation, it may well be desirable for solar heat to provide almost 100 percent of the hot-water load.

The required solar collector area can be estimated from the following rule of thumb:

> During sunny conditions, an efficient commercial solar collector can be expected to heat between one and two gallons of water from 65°F to domestic hot-water temperatures (120°F to 140°F) per square foot of solar collector per day when the collector is mounted with an optimal orientation for the latitude at the site.

In some locations, this rule of thumb would not be applicable, as in areas with frequent heavy cloud cover.

The optimum collector orientation is generally due south with a collector tilt angle ranging from the latitude to the latitude plus 10°. If for any reason the solar collector cannot be given the optimum orientation, its performance may decrease appreciably. Figure 13-22 gives the approximate decrease in annual solar radiation collected as a collector is oriented away from south. Also included is a scale by which the collector area can be increased to compensate for the variation in azimuth.

Figure 13-23 indicates the approximate decrease in annual solar radiation collected as the collector tilt is changed from an optimum tilt equal to the local latitude. The chart also includes a scale by which the collector area can be increased to compensate for the departure from optimum tilt angle.

13-10 The Hot-Water Storage Tank Size

Storage tank size can be estimated using the following rule of thumb:

> The volume of the hot-water storage tank should be in the range of one to two times the daily hot-water demand. In industry practice, the optimum storage volume in most cases is in the range of 1.25 to 1.60 times the daily hot-water demand.

The following example illustrates the use of the three rules of thumb just discussed in obtaining preliminary estimates of collector area and water storage tank size.

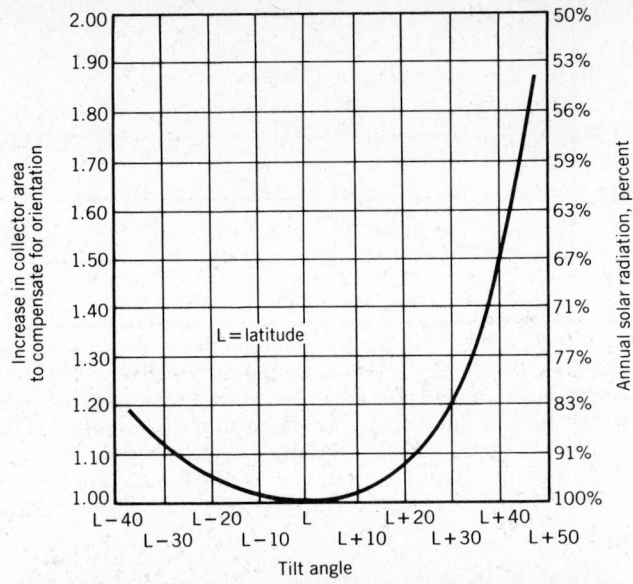

Figure 13-23 Chart showing the variation in the annual solar radiation available to solar collectors as a function of collector tilt, and the increase in collector area needed to compensate for tilt angles not equal to the latitude. (Adapted from *National Solar Water Heater Workshop Handbook*, "Pumped Recirculation" Freeze Protection, Arizona State University, DOE Grant No. DE-FG-03-805F11444, 1981.)

Illustrative Problem 13-7

Estimate the size of a solar domestic hot-water system for a family of four with a hot-water load of 80 gal/day.

Solution

First, the solar heat fraction must be chosen. In this case, 0.75 will be the assumed SHF. The solar heating system will then have to provide the following volume of hot water:

$$V_W = 0.75 \times 80 \text{ gal/day} = 60 \text{ gal/day}$$

Based on the collector area rule of thumb, the solar collector area will be in the range of 30 ft^2 to 60 ft^2. The storage tank volume will have a range of 1.25 to 1.60 times the daily demand of 60 gal/day. Thus, the tank size should be in the range from 75 gal to 96 gal.   *Ans.*

These rules of thumb are adequate to give preliminary estimates for a solar domestic hot-water system, but more often than not further specific information is desired. Local or regional sizing information is often useful or necessary. Sources of regional sizing information are available from state and federal agencies. Manufacturers will often provide system-sizing recommendations for their solar energy system components, and test data reports will usually give the average energy output of a collector or the average collector efficiency.

The Energy Research and Development Administration (ERDA) published a *Pacific Regional Solar Heating Handbook* in 1976, which contains data for the sizing of active solar water-heating systems. Figure 13-24 shows the ERDA collector sizing requirements for domestic hot-water systems using one- and two-tank solar systems. This chart represents an example of a regional sizing

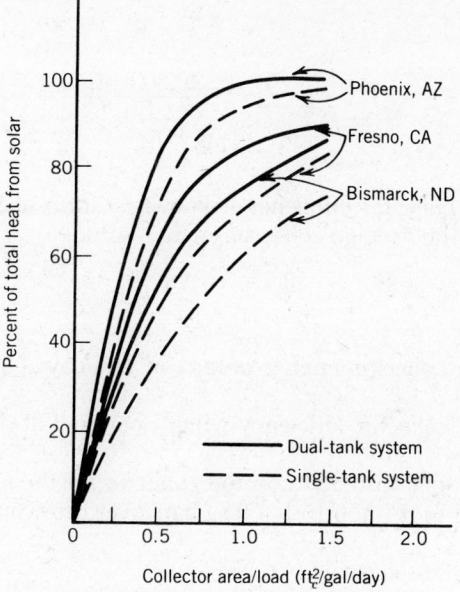

Figure 13-24 Solar hot-water system sizing chart for three selected locations in the western United States. (*Pacific Regional Solar Heating Handbook*, ERDA San Francisco Operations Office, 1976, p. 68.)

guideline. Another example is a series of design notes published by the Northeast Solar Energy Center for that region (not shown here).

When regional sizing guidelines are not adequate or not available, the solar designer or contractor will have to establish local design guidelines for determining the solar collector area. The daily energy output of a solar collector in Btu per day per square foot may be obtained as discussed in section 12-6 for a clear-day case or for an average-day case, depending on which values of insolation are used in the calculations.

Illustrative Problem 13-8

A domestic hot-water load has been estimated to be 1.75×10^6 Btu/month. Use the collector whose performance is delineated in Table 12-1 to estimate the area of solar collector needed for a solar heat fraction of 0.70.

Solution

The daily heat load can be estimated to be

$$Q_{WD} = \frac{1.75 \times 10^6 \text{Btu/month}}{30 \text{ days/month}}$$
$$= 58,300 \text{ Btu/day}$$

The solar contribution is calculated as:

$$Q_S = 0.70 \times Q_{WD}$$
$$= 0.70 \times 58,300 \text{ Btu/day}$$
$$= 40,800 \text{ Btu/day}$$

From Table 12-1, the daily energy output of the solar collector, Q_C, is 842 Btu/ft²-day. The required collector area is calculated as:

$$\text{Collector area} = A_C = \frac{Q_S}{Q_C}$$

$$= \frac{40,800 \ \text{Btu/day}}{842 \ \text{Btu/ft}^2\text{-day}}$$

$$= 48.5 \ \text{ft}^2 \qquad\qquad Ans.$$

When the average collector efficiency is known, a form of Eq. (12-16) can be used to determine the average collector energy output:

$$\overline{Q}_C = A_C \overline{I}\overline{\eta} \qquad\qquad (13\text{-}4)$$

where

$\overline{Q}_C$ = the average collector energy output in Btu/day (Q_{CD}) or Btu/month (Q_{CM})

$\overline{\eta}$ = the average collector efficiency when operating at domestic hot-water temperatures

$\overline{I}$ = the average solar insolation on the collector for the location, in Btu/ft²-day ($\overline{I}_D$) or Btu/ft²-month ($\overline{I}_M$) (the bars over the symbols mean average values)

A_C = the area of the solar collector, ft²

Average daily values $\overline{I}_D$ can be approximated by using the values for the twenty-first of each month in Table 4-E for clear-day solar radiation, and Table 4-C for percent possible sunshine, with the following equation:

$$\overline{I}_D = I \times P \qquad\qquad (13\text{-}5)$$

where

$\overline{I}_D$ = the average daily solar insolation on the collector, Btu/ft²-day

I = the clear-day solar insolation from Table 4-E, Btu/ft²-day, for the twenty-first day of the month

P = the percent possible sunshine for the given month and location from Table 4-C

Test data for solar collectors usually provide an approximate value for the collector efficiency, as was shown in Fig. 12-8. When the collector efficiency is not known, a calculation similar to those made in section 12-6 (Tables 12-1 and 12-2) will give hourly collector efficiencies during the day.

The average collector efficiency can then be approximated as:

$$\overline{\eta} = \frac{\overline{Q}_{CD}}{\overline{I}_D} \qquad\qquad (13\text{-}6)$$

where, again,

$\overline{\eta}$ = the average collector efficiency

$\overline{Q}_{CD}$ = the calculated average daily collector energy output, Btu/day

$\overline{I}_D$ = the average daily insolation on the collector, Btu/ft²-day

Illustrative Problem 13-9

Calculate the average efficiency of the collector whose performance is indicated by the calculations given in Table 12-1.

Solution

Summing up the I column of Table 12-1, the total daily insolation I_D is 2504 Btu/ft²-day (not average, but actual total for the given day). The calculated daily collector energy output is 842 Btu/ft²-day.

Using Eq. (13-6), the average collector efficiency for the day is calculated as:

$$\bar{\eta} = \frac{842}{2504} = 0.34 \qquad \textit{Ans.}$$

A general equation for collector area can be developed by solving Eq. (13-4) for the collector area and using $\bar{I}_D$ for the average daily insolation on the collector:

$$A_C = \frac{\overline{Q}_{CD}}{\bar{\eta}\bar{I}_D} \qquad (13\text{-}7)$$

Then rearrange Eq. (13-2b) as

$$Q_{SD} = \mathrm{SHF}_{DHW}\, Q_{WD}$$

Now substitute this value of Q_{SD} for $\overline{Q}_{CD}$ in Eq. (13-7), to obtain:

$$A_C = \frac{\mathrm{SHF}_{DHW}\, Q_{WD}}{\bar{\eta}\bar{I}_D} = \frac{\mathrm{SHF}_{DHW}\, Q_{WD}}{\bar{\eta}\mathrm{PI}_D} \qquad (13\text{-}8a)$$

On a monthly basis,

$$A_C = \frac{\mathrm{SHF}_{DHW}\, Q_{WM}}{\bar{\eta}\bar{I}_M} = \frac{\mathrm{SHF}_{DHW}\, Q_{WM}}{\bar{\eta}\mathrm{PI}_M}, \qquad (13\text{-}8b)$$

Eq. (13-8a) is a basic formula for estimating the collector area required for an average daily hot-water load, Q_{WD}; and Eq. (13-8b) yields the collector area for the monthly hot-water load, Q_{WM}. These equations account for different collector orientations in tilt and azimuth, because the values of I used will be those for the specified collector orientation.

Illustrative Problem 13-10

A new house has four bedrooms, and the size of the domestic hot-water system needs to be estimated. The house is located at 40° N Lat near Kansas City, Missouri. The desired SHF for the hot-water system is 0.80. Assume that the solar collectors have an average collector efficiency of 0.35 in the winter and 0.45 in the summer. Determine the required collector area for December and for June, for a due south orientation with collector tilts of 30° and 50°.

Solution

The domestic hot-water load (Q_{DHW}) is estimated by using Table 13-1. For four bedrooms, the hot-water usage is 85 gal/day. The hot-water load is calculated using Eq. (13-1), with $T_m = 55°F$ and $T_s = 140°F$.

$$\begin{aligned}
Q_{DHW} &= 1.0\ \mathrm{Btu/lb\text{-}F°} \times 8.34\ \mathrm{lb/gal} \times 85\ \mathrm{gal/day} \\
&\quad \times (140 - 55)\mathrm{F°} \\
&= 60{,}200\ \mathrm{Btu/day}.
\end{aligned}$$

For December 21, 40° N Lat, 30° tilt, due-south-facing, from Table 4-E, $I_D = 1479$ Btu/ft²-day by summing the hourly values. From Table 4-C, the percent possible sunshine in December is 54.

Using Eq. (13-8b) and substituting values,

$$\begin{aligned}
A_C &= \frac{(0.8)(60{,}200\ \mathrm{Btu/day})}{(0.35)(0.54)(1479\ \mathrm{Btu/ft^2\text{-}day})} \\
&= 172\ \mathrm{ft^2} \qquad \textit{Ans.}
\end{aligned}$$

For December 21, 50° tilt, from Table 4-E, $I_D = 1740$ Btu/ft²-day, by summing the hourly values.
Using Eq. (13-8b) and substituting values,

$$\begin{aligned}
A_C &= \frac{(0.8)(60{,}200\ \mathrm{Btu/day})}{(0.35)(0.54)(1740\ \mathrm{Btu/ft^2\text{-}day})} \\
&= 146\ \mathrm{ft^2} \qquad \textit{Ans.}
\end{aligned}$$

SOLAR WATER HEATER SYSTEM DESIGN

For June 21, 30° tilt, from Table 4-E, I_D (summed up) = 2430 Btu/ft²-day. The percent possible sunshine for June is 72.

Using Eq. (13-8b) and substituting values,

$$A_C = \frac{(0.8)(60,200 \text{ Btu/day})}{(0.45)(0.72)(2430 \text{ Btu/ft}^2\text{-day})}$$

$$= 61 \text{ ft}^2 \qquad\qquad\qquad Ans.$$

For June 21, 50° tilt, from Table 4-E, I_D (summed up) = 1969 Btu/ft²-day.

Using Eq. (13-8b) and substituting,

$$A_C = \frac{(0.8)(60,200 \text{ Btu/day})}{(0.45)(0.72)(1969 \text{ Btu/ft}^2\text{-day})}$$

$$= 75.5 \text{ ft}^2 \qquad\qquad\qquad Ans.$$

The above calculations for the collector area do not consider heat losses from the hot-water storage or the piping. These losses may be as high as 10 percent of the total energy collected by the collectors. The estimated collector area should be increased by enough area to compensate for the estimated system heat losses.

The above example shows that the calculated winter collector areas are much larger than those for summer. Generally, it is not practical to provide the total winter collector area, and the collector area actually used will be determined in part by the collector size (model) chosen. Consider the areas (in Illustrative Problem 13-10) for a collector tilt of 50°. The December A_C = 146 ft², and the June A_C = 75.5 ft². If, for example, collectors with an area of 32 ft² were to be used on the system, the number of collectors for June would be 2.4. The number of collectors for December would be 4.6. Three collectors could be used, which would provide more than adequate collector area for summer but a somewhat undersized area for the winter months. Four solar collectors could be used, of course, but the economics of such oversizing (for most of the year) should be subjected to careful analysis.

13-11 Pipe and Pump Sizing—The Collector Loop

After the type and number of solar collectors to be used has been determined, the collector-loop plumbing must be specified. When more than one solar collector is used in a system, the collectors must be interconnected in such a way that equal flow through each collector will be ensured. Basically, there are three patterns in which solar collectors can be interconnected; *parallel flow with direct return, parallel flow with reverse return,* and *series–parallel flow.*

A parallel flow plan with direct return is diagrammed in Fig. 13-25a. In a direct-return plumbing circuit, the heat-transfer fluid circulates from a supply header through the collectors and on to a return header at the top of the collectors. However, this arrangement may cause operating problems resulting from flow imbalance in the collectors. Each collector will ordinarily present the same resistance to fluid flow, and this resistance is termed the *pressure drop* for the collector. The pressure drop is usually given in feet of water head pressure for different flow rates in gallons per minute. Note that 1 ft of water head pressure is equal to the pressure that a 1-ft high column of water exerts at its base (1 ft of water head pressure equals 0.434 psi).

The supply and return headers also offer resistance to fluid flow, and the flow to the furthest collector will encounter more resistance than the flow to the first collector. As a result, there will be a larger fluid flow volume through the first collector than through succeeding collectors. Consequently, the temperature at which each collector operates will be different because of their differing fluid flow volumes. In extreme cases, there could be almost no flow at all through the end collectors, resulting in very little collection of heat from them. To equalize the fluid flow through the collectors, balancing valves or orifices are installed at each collector to allow the flow therein to be adjusted.

Manual adjustment of these balancing valves is a difficult and tedious task, and whenever possible the collectors should be plumbed in a manner that is self-

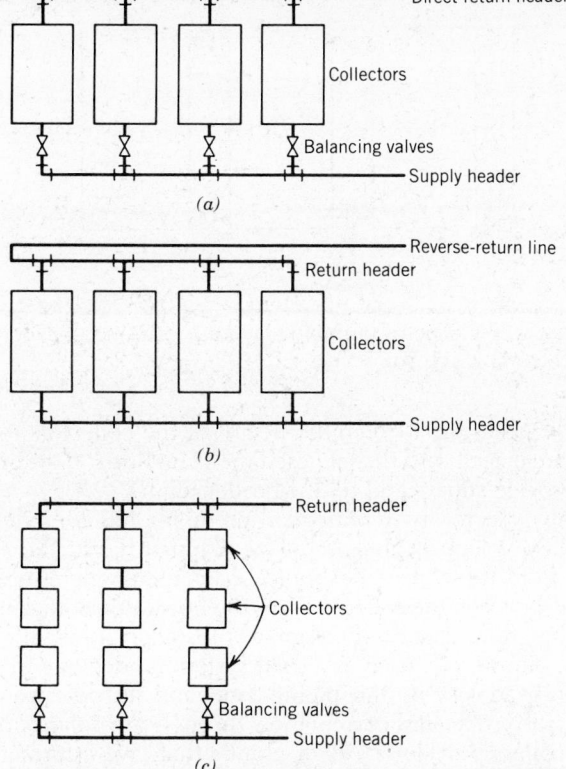

Figure 13-25 Flow arrangements through collector arrays. (*a*) A collector array using parallel flow with a direct-return header. With this piping arrangement, balancing valves must be used to adjust the flow through each collector. (*b*) An array using parallel flow with a reverse-return header. This arrangement is self-equalizing, and balancing valves are not needed. (*c*) A collector array using series–parallel flow.

equalizing, as shown in Fig. 13-25*b*. This arrangement is termed parallel flow with reverse return. The total length of supply piping and return piping to each collector is the same with this arrangement, and the pressure drop through each collector path is theoretically the same, since the pressure drop through each collector is the same and the path length of manifold piping for each collector is the same.

A third method for interconnecting collectors is the series–parallel method, as shown in Fig. 13-25*c*. The series–parallel flow arrangement is often used in large arrays of collectors to reduce the amount of piping required by allowing several collectors (connected in series) to be served by the same supply and return headers. The series flow is also employed to increase the output temperature of the collector system. Either a direct-return or a reverse-return distribution circuit can be used. Since a typical single-family residential domestic hot-water system will rarely have more than six solar collectors, the series–parallel flow is not often encountered on residential installations. Series–parallel arrangements result in greater flow resistance than parallel arrangements of the same number of collectors.

The *through-header collector* with internal parallel flow is commonly used. Through-headers make the interconnection of several collectors to form an array a very simple task. The headers are connected together, as shown in Fig. 13-26, using couplings or unions. These collectors have a parallel-flow pattern and should be connected in a parallel flow with reverse-return piping circuit as shown.

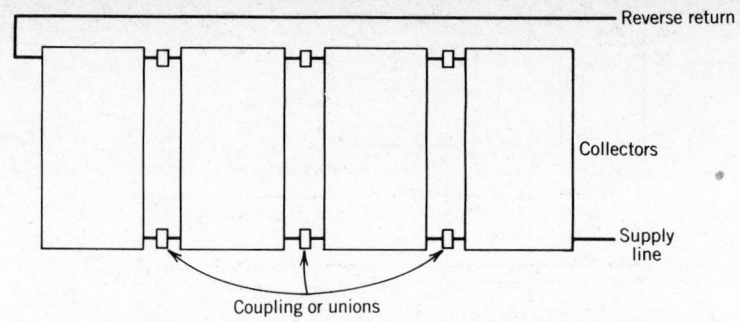

Figure 13-26 Collector plumbing layout for through-header collectors with reverse return.

When it is necessary to keep the height of the collectors to a minimum, the collectors are placed with the longest dimension horizontal. In this case, it is necessary to provide supply and return header piping.

When the collector interconnection plumbing has been determined, the rest of the collector loop piping should be planned. First, the total length of piping to and from the collectors should be accurately determined, since the length of piping between the collectors and the hot-water storage tank presents a resistance to fluid flow. The solar designer or installer should always verify that the circulation pump to be used in a solar system is adequate to overcome the system friction-head loss in the piping loop and also to provide sufficient pressure head for proper fluid circulation through the solar collectors. Many equipment suppliers provide charts or graphs that give information on pump size and pipe diameter based on the collector area used and the total piping length. An example of such a chart is given in Fig. 13-27.

Illustrative Problem 13-11

A domestic hot-water system will have three solar collectors connected in parallel with reverse return. The distance from the collectors to the storage tank

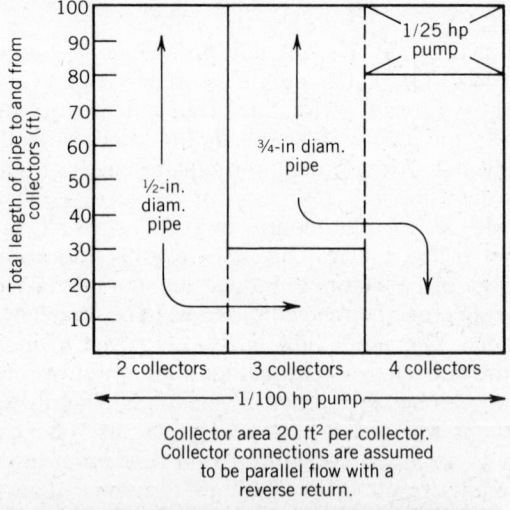

Figure 13-27 Pump and pipe sizing chart for solar hot-water systems, using a parallel-connected, reverse-return piping circuit. (Adapted from *National Solar Water Heater Workshop Handbook*, "Pumped Recirculation" Freeze Protection, Arizona State University, DOE Grant No. DE-FG-03-805F11444, 1981.)

is 40 ft in one direction. From Fig. 13-27, determine the pipe diameter and pump size that should be used for this system. Assume that the reverse return adds a total of 14 ft to the piping length.

Solution

The total piping length will be approximately 94 ft. From Fig. 13-27, for three collectors and a 94-ft piping length, ¾-in. copper tubing should be used with a ¹⁄₁₀₀ hp pump.   *Ans.*

13-12   Fluid Flow Calculations in More Complex Systems

The designer will invariably encounter situations that depart from the typical or the norm. To satisfactorily handle the unusual situation, a basic understanding of the fluid hydraulics of a collector loop is necessary. The purpose of the fluid is to carry heat away from the collector and deliver it to storage. The efficiency of the collector is quite dependent upon the fluid flow rate. With high fluid flow rates, the temperature rise of the fluid will be less, making the average temperature of the collector lower, which makes the thermal losses from the collector smaller—or, stated another way, making the collector efficiency higher.

Consider two collectors with 70°F water entering them. One collector has a high fluid-flow rate, with the water leaving the collector at 75°F. The average collector temperature would then be approximately 72.5°F. The second collector has a very low flow rate, with the water leaving the collector at 100°F. The average collector temperature in this case would be approximately 85°F. The collector with the high flow rate will have a higher efficiency because the average collector temperature is lower, and thus heat losses from the collector are lower.

There is a practical range of flow rates for solar collectors, with many manufacturers recommending a flow of 0.02 to 0.03 gal/ft$_c^2$-min.[3] (The notation ft$_c^2$ means square foot of solar collector.) The temperature rise of a fluid flowing through a collector is given by the following relation:

$$\Delta t = \frac{AI\eta}{V \times D \times c_W \times 60 \text{ min/hr}} \tag{13-9}$$

where

Δt = the temperature rise of the fluid in passing through the collector, F°
A = the collector area, ft^2
I = the solar insolation on the collector surface, Btu/hr-ft^2
η = the collector efficiency, a decimal fraction
V = the fluid flow rate, gal/min
D = the density of the fluid, lb/gal
c_W = the specific heat of the fluid (1.0 Btu/lb-F° for water)

Illustrative Problem 13-12

Calculate the fluid temperature rise for a 32 ft^2 solar collector with a water flow rate of 0.02 gal/ft$_c^2$-min. Assume that the insolation is 225 Btu/hr-ft^2, and the efficiency of the collector is 0.45 under these conditions.

Solution

The total fluid flow rate through the collector V is 0.02 gal/ft$_c^2$-min × 32 ft^2 = 0.64 gal/min.

[3] This should be checked, of course, for any collector model being specified in the design.

Substituting in Eq. (13-9),

$$\Delta t = \frac{32 \text{ ft}^2 \times 225 \text{ Btu/hr-ft}^2 \times 0.45}{0.64 \text{ gal/min} \times 8.34 \text{ lb/gal} \times 1.0 \text{ Btu/lb-F}^\circ \times 60 \text{ min/hr}}$$

$$= 10.1 \text{ F}^\circ \qquad\qquad\qquad\qquad\qquad \textit{Ans.}$$

For domestic solar hot-water systems, the typical fluid temperature rise for a single pass through the collectors is in the range from 10 to 20 F°.

Pressure losses in the collector loop must be overcome by the circulating pump in order to maintain the proper flow through the collectors. All of the components in the collector loop contribute to the fluid friction pressure loss, and this pressure loss varies as the square of the flow rate. The components contributing to the pressure head loss are given in Table 13-3. Figure 13-28 gives the friction loss for hot water per 100-ft run of copper pipe. The pipe sizes correspond to plumbers' copper tube because of its general use in new domestic construction. The water flow rate is given on the vertical axis in gallons per minute; and the friction loss (pressure drop) is given in feet of water, or pounds per square inch per 100-ft length of pipe, along the horizontal axis. Table 13-4

Table 13-3 Collector Loop Pressure Losses for a Typical Parallel-Flow, Reverse-Return Collector Array

Component	Typical Percentage of Total System Head Loss
Piping runs	74
Fittings	20
Valves	5
Collectors	1

Source: Charles Hill, "Gauging Hydraulic Performance for Solar Hot Water Systems," *Solar Engineering Magazine,* March 1978, p. 31.

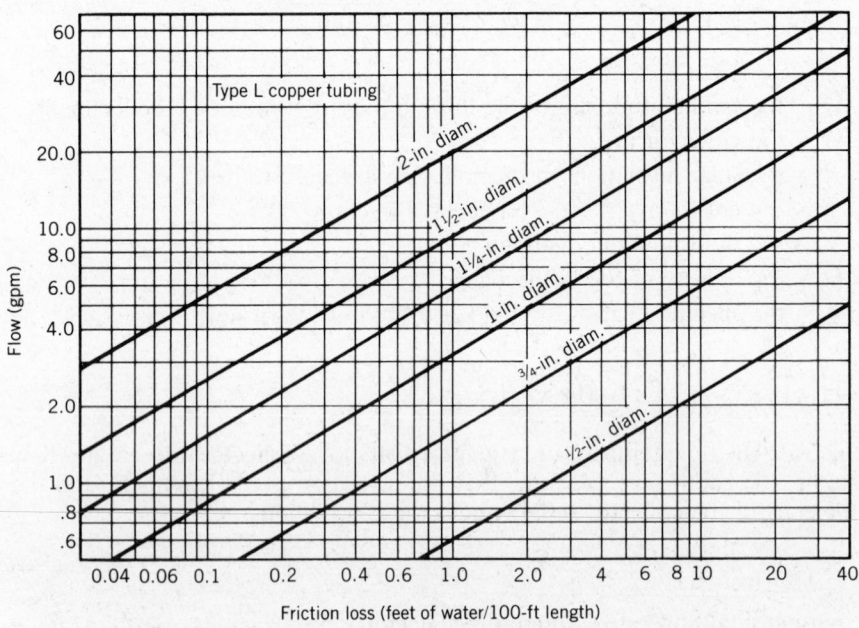

Figure 13-28 Friction loss chart for various diameters of copper pipe. The losses for various flow rates can be read from the chart. (Adapted from *ASHRAE Handbook of Fundamentals,* 1972, p. 501.)

SOLAR HEATING OF DOMESTIC HOT WATER

Table 13-4 Friction Loss of Fittings and Valves Expressed in Equivalent Length of Straight Copper Pipe (ft)

Diameter of Fitting (in.)	90° Elbow	45° Elbow	90° Branch of Tee	Coupling or Straight Run of Tee	Ball Valve	Gate - Valve	Poppet Check Valve
½	2.0	1.2	3.0	0.6	0.1	0.4	18
¾	2.5	1.5	4.0	0.8	0.2	0.5	35
1	3.0	1.8	5.0	0.9	0.3	0.6	36
1¼	4.0	2.4	6.0	1.2	0.35	0.8	37
1½	5.0	3.0	7.0	1.5	0.4	1.0	39
2	7.0	4.0	10	2.0	0.5	1.3	40

Sources: Handbook of Fundamentals, ASHRAE, 1972, p. 510; Charles Hill, "Gauging Hydraulic Performance for Solar Hot Water Systems," Solar Engineering Magazine, March 1978, p. 31.

gives the friction loss caused by pipe fittings, in equivalent length of straight pipe. The equivalent length of a piping run is calculated by adding the actual length of straight pipe and the equivalent length of all of the fittings. The pressure loss for the piping run is then found from Fig. 13-28 when the design flow rate is known.

Illustrative Problem 13-13

Determine the pressure loss for a collector loop that has 125 ft of straight pipe, 25 90° elbows, a check valve, two ball valves, and 90 ft^2 of collector area. The manufacturer recommends a flow rate of 0.02 gal/ft^2_c-min. Assume that ½-in. copper tubing will be used.

Solution

The flow rate for the collector array is calculated as follows:

$$V = 0.02 \text{ gal/ft}^2_c\text{-min} \times 90 \text{ ft}^2$$
$$= 1.8 \text{ gal/min}$$

The equivalent pipe length for all fittings can be calculated with the aid of Table 13-4, as follows:

Fitting	No. of Fittings × Equivalent Pipe Length		= Total Equivalent Length
90° elbows	25 × 2.0	=	50.0 ft
Check valve	1 × 18	=	18.0 ft
Ball valve	2 × 0.1	=	0.2 ft
			68.2 ft

The total effective length of piping is then the sum of the total equivalent length of the fittings and the actual length of the piping run:

$$\text{Total effective length} = 125 \text{ ft} + 68.2 \text{ ft}$$
$$= 193.2 \text{ ft}$$

From Fig. 13-28 (½-in. tubing), the pressure loss for a flow rate of 1.8 gpm is approximately 7 ft of head or 3.0 psi per 100 ft of pipe. For 193 ft of pipe, the pressure loss will be:

$$\text{Pressure loss} = 7 \text{ ft head/100ft} \times 1.93 \text{ (100 ft)}$$
$$= 13.5 \text{ ft of (water) head, or 5.85 psi} \qquad Ans.$$

The pressure loss through the collectors themselves is usually quite small. Fig.

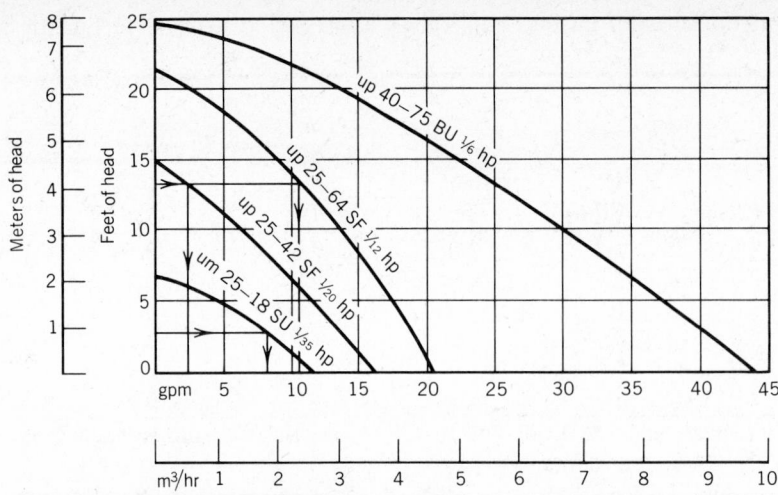

Figure 13-29 Performance chart for various sizes of hot-water circulators (pumps) used in domestic hot-water systems. (Grundfos Pump Corporation.)

12-10 for example, shows the pressure loss as a function of collector flow rate. In the previous example, the flow rate through each of the three collectors is 1.8/3 gpm = 0.6 gpm. From Fig. 12-10, the pressure loss for each collector would be approximately 0.2 ft of head, which is very small compared to the other pressure losses in the loop.

The pump size must be determined based on the total collector loop pressure loss. Figure 13-29 gives performance curves for some hot-water circulating pumps. The flow rate is given along the horizontal axis, and the head is given along the vertical axis.

Illustrative Problem 13-14

A pump for the collector system in Illustrative Problem 13-13 must be selected. Determine a pump size suitable for ½-in. piping and one for ¾-in. piping.

Solution

(a) ½-in. piping

From Illustrative Problem 13-13, the total head loss for ½-in. piping was calculated as 13.5 ft. On Fig. 13-29, a horizontal line is drawn across the graph from a head pressure of 13.5 ft. The point where this line intercepts the pump curves gives the flow rate for the system. The UP25–42SF pump has a flow rate of approximately 2 gpm, and the UP25–64SF pump has a flow rate of approximately 10.5 gpm. If ½-in. tubing is used, the UP25–42SF pump should be used because the UP25–64SF provides too great a flow. A flow-control valve would be needed with the UP25–64SF to adjust the collector loop flow rate to the design rate of 1.8 gpm, and this would waste pump energy. *Ans.*

(b) ¾-in. piping

The pressure loss for the ¾-in. piping run is calculated as follows, using values from Table 13-4:

Fitting	No. of Fittings × Equivalent Pipe Length			= Total Equivalent Length
90° elbows	25	×	2.5	= 62.5 ft
Check valve	1	×	35	= 35 ft
Ball valve	2	×	0.2	= 0.4 ft
				97.9 ft

The total effective length of piping is, then,

$$125 \text{ ft} + 98 \text{ ft} = 223 \text{ ft}$$

From Fig. 13-28, the pressure loss for a flow rate of 1.8 gpm is approximately 1.3 ft of head per 100 ft of ¾-in. pipe. For 210 ft of pipe, the pressure loss will be:

$$\text{Pressure loss} = 1.3 \text{ ft of head/100ft} \times 210\text{ft}$$
$$= 2.73\text{ft of head}$$

From Fig. 13-29, the UM25–18SU pump will provide a flow rate of approximately 7.8 gpm at a head of 2.7 ft. This pump would provide adequate circulation when ¾-in. piping is used.   *Ans.*

13-13 Drainback Systems

The plumbing requirements for a drainback system are quite different from those for closed- and open-loop systems discussed above. The drainback system is a special case of a closed-loop solar system which uses a small fixed volume of water as the heat-transfer fluid in the closed loop. This water is circulated through the collectors, and a heat exchanger transfers the heat from the collector loop to the domestic water. When heat is not being collected, the heat-transfer water drains from the collectors and collector loop piping into a small reservoir, thus eliminating the possibility of water freezing in the collectors or piping runs. There are two major types of drainback systems: a *siphon return* and an oversized return line or *downcomer* return.

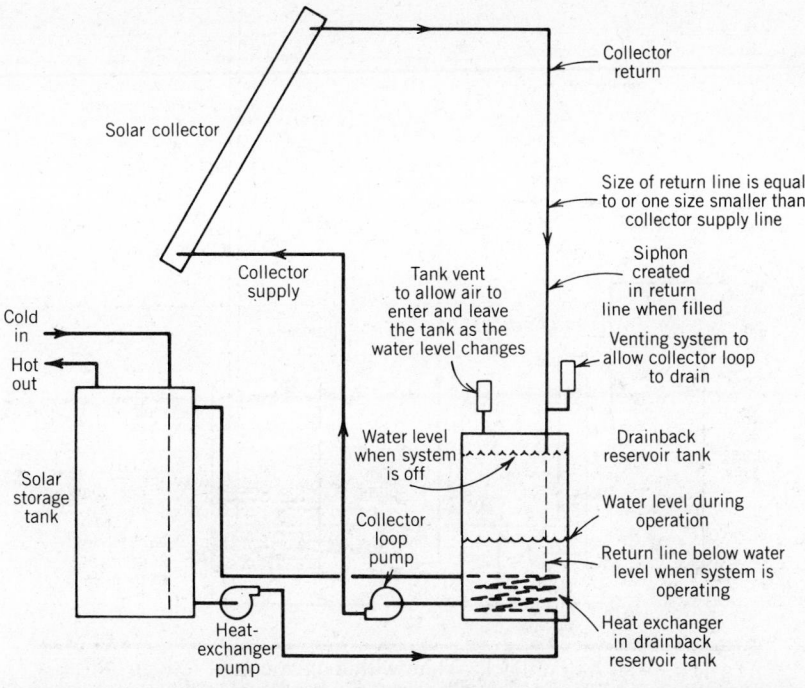

Figure 13-30 Flow diagram of a siphon-return drainback system. This system requires an air vent to break the siphon when the pump turns off so that the collectors will drain back into the reservoir.

The siphon return system operates with completely filled supply and return piping and has a venting system that allows air to enter the collector loop (Fig. 13-30) when the pump stops. The entering air then allows the water to drain out of the collectors and piping runs. A siphon develops when the pump fills the collector and the return line with water. The advantage of the siphon return is that the pumping horsepower requirements are minimized for the system; however, there are two disadvantages to the siphon return. First, the venting system must be properly located, as shown in Fig. 13-30, or else air will not enter the return line and allow the collectors to drain. Second, air is introduced through the exterior vent each time the system fills and drains. This creates a potential for corrosion in the system as well as for contamination of the water with organic material, which could result in the growth of algae and bacteria in the water holding tank. For these reasons, the siphon return system is used only when water has to be pumped up several stories and it is necessary or desirable to decrease pumping energy.

The downcomer return system is shown in Fig. 13-31. Water is forced by the collector loop pump through the collectors to the top of the collector array and then flows down the return piping in the oversized downcomer. Because the downcomer never fills with water, air is constantly available at the top of the system to vent the collectors and allow them to drain. The advantage of the downcomer is that the system is completely closed at all times, avoiding the introduction of new air into the system. The disadvantage is that higher pumping horsepower may be required to lift the water to the top of the collectors, and larger diameter piping is required for the return line.

A section through a typical drainback–tank heat exchanger is shown in Fig. 13-32. The drainback system requires two pumps (see Fig. 13-31)—one to circulate domestic water from the solar storage tank through the heat exchanger and another to circulate the closed-loop water through the collectors. The domestic water (heat-exchanger) pump is typically a small hot-water circulator in the range of $\frac{1}{100}$ to $\frac{1}{25}$ hp. The collector-loop pump must be capable of

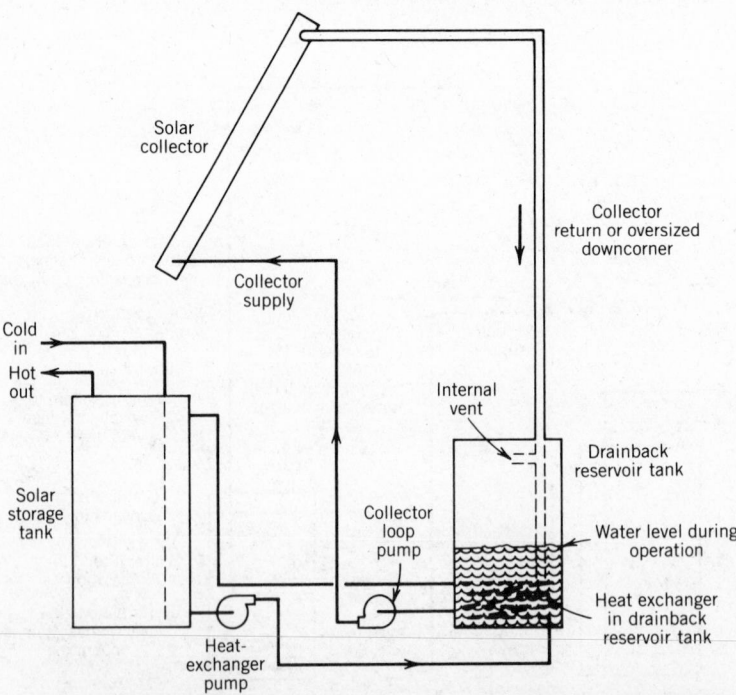

Figure 13-31 A drainback system with a downcomer return. This is a closed-loop system with a heat exchanger in the drainback reservoir. When the collector loop pump turns off, the water in the collectors drains down into the reservoir tank, because the downcomer allows air from the reservoir to enter the collectors.

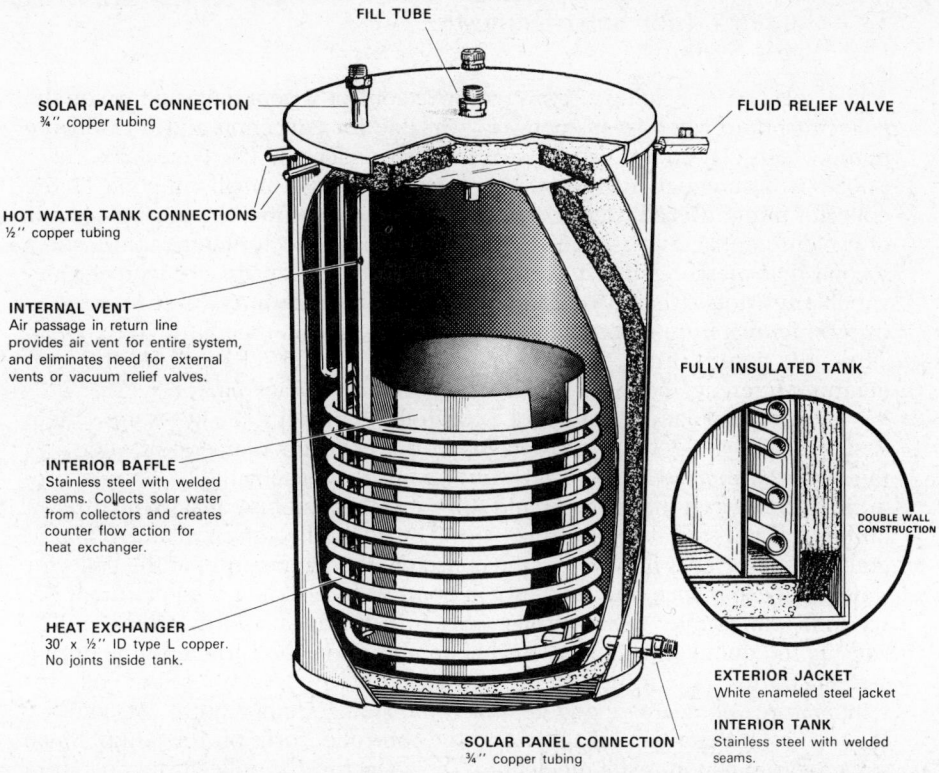

FILL TUBE

SOLAR PANEL CONNECTION
¾" copper tubing

FLUID RELIEF VALVE

HOT WATER TANK CONNECTIONS
½" copper tubing

INTERNAL VENT
Air passage in return line
provides air vent for entire system,
and eliminates need for external
vents or vacuum relief valves.

FULLY INSULATED TANK

INTERIOR BAFFLE
Stainless steel with welded
seams. Collects solar water
from collectors and creates
counter flow action for
heat exchanger.

DOUBLE WALL
CONSTRUCTION

HEAT EXCHANGER
30' x ½" ID type L copper.
No joints inside tank.

EXTERIOR JACKET
White enameled steel jacket

INTERIOR TANK
Stainless steel with welded
seams.

SOLAR PANEL CONNECTION
¾" copper tubing

Figure 13-32 Drainback tank and heat exchanger. The drainback tank must have a reservoir large enough to fill the collectors and collector piping with water and still have the heat exchanger covered with water. (Sunburst Solar Energy Division, ACRO Energy Corporation.)

pumping water from the drainback reservoir to the top of the collector array. The height of the system, the pipe and collector pressure losses, the net positive suction-head requirements for the pump, and the volume of water that can be drawn from the reservoir must all be considered when sizing this pump and the piping.

First, the volume of water required to fill the supply piping and collectors must not exceed available water in the drainback reservoir. Second, the circulator pumps typically used in solar systems have a minimum net positive suction-head requirement. This simply means that the water level in the drainback reservoir must be above the pump center line. This suction-head requirement is typically 2 to 4 ft, which means that the water level in the drainback reservoir must be 2 to 4 ft above the pump. A drainback system will have a head requirement that is equal to the height of the top of the system above the water level in the drainback reservoir plus the pressure drop in the collector supply line resulting from friction losses. There is no calculation for the friction loss in the downcomer return, since the water is flowing because of gravity, and the pump is not providing any energy for this portion of the system. For example, if the collectors are mounted on a second-story roof 20 ft above the drainback reservoir, and if the pressure loss in the supply piping and collectors totals 3 ft of head, the system head will be 23 ft. Assume that a collector-flow rate of 3.0 gpm is desired. From Fig. 13-29, the UP40–75BU pump (⅙ hp) will provide this head at 6 gpm flow rate. A flow restrictor valve would be needed to set the collector flow at 3 gpm. As an alternative, two UP25–42SF pumps could be used in series to provide a flow of about 4.5 gpm. A flow restrictor valve would still be necessary to adjust the collector flow rate to 3 gpm. The use of two small pumps in series can often result in providing the required head with lower energy usage than would be required by a single larger pump.

13-14 Controls for Solar Domestic Hot-Water Systems

Active solar water-heating systems require a control system that will turn on the collector pump when solar energy is available for collection and turn off the pump when solar energy cannot be effectively collected. The typical controller utilizes a differential thermostat to control the basic on–off function of the collector pump. Recall from section 10-5 that a differential thermostat consists of two differential sensors—one that measures the collector temperature, and a second that measures the storage temperature. The sensors are temperature-variable resistors. Electronic circuits compare these temperatures to a preset turn-on temperature differential and a preset turn-off temperature differential. Thus, the control turns the circulation pump on and off based on the temperature difference between the collector and the storage tank.

The typical daily operation of a controller for a domestic hot-water system is shown in Fig. 13-33. As the sun rises, the collector temperature increases rapidly since there is no fluid flow until point A is reached. The temperature difference between the collector and storage at this point is equal to the turn-on differential (ΔT_{ON}). The controller turns on the collector circulator pump at point A. As the cool fluid enters the collector, the temperature at the collector sensor will experience a dip as shown at point B. If the ΔT_{ON} temperature is not set sufficiently large, the controller may turn off again shortly after point B, causing the pump and control to cycle on and off in the early morning during system startup. Late in the day, as the solar intensity decreases, the collector temperature will decrease and approach the storage temperature. At point C, the turn-off differential is reached and the controller turns off the pump. Since some solar radiation is still incident on the collector, the collector temperature will rise as indicated between points C and D until the collector heat loss rate equals the decreasing rate of insolation. The control and pump may also cycle on and off at the end of the day if the ΔT_{ON} setting is not chosen correctly.

It is apparent from the above discussion and the plot of Fig. 13-33 that the key to proper controller function is the correct selection of ΔT_{ON} and ΔT_{OFF}. When ΔT_{ON} is made very high, the system will be stable and will not cycle on and off during starting and stopping. However, if ΔT_{ON} is set too high, some collectible energy will be lost while the collector is warming up to the starting condition. It has been found that, for home water-heating systems, values of 10 to 15 F° for ΔT_{ON} and 2 to 3 F° for ΔT_{OFF} provide stable operation. Some controllers have provisions for adjusting ΔT_{ON} in the field, and this allows the system to be optimizied by the installer or service technician if a cycling condition develops.

Controllers for active systems have other options in addition to the basic differential thermostat function. Some of these functions are discussed below.

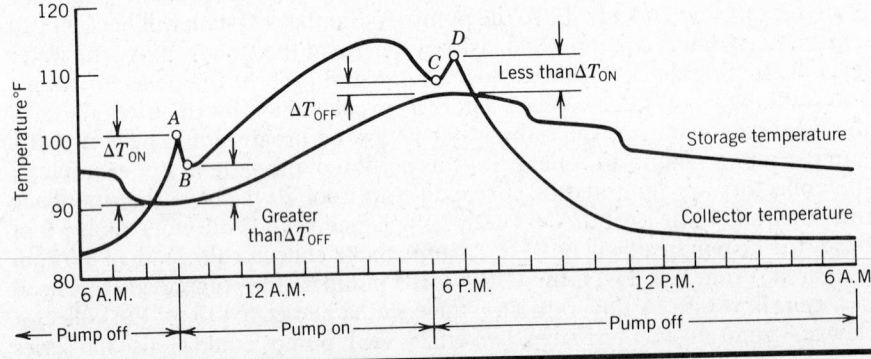

Figure 13-33 The on-off cycle of a differential thermostat. Point A represents a morning condition where the collector/storage temperature differential ΔT becomes equal to ΔT_{ON}. The controller turns on the system and it remains on to Point C, which represents the late afternoon condition where ΔT becomes equal to ΔT_{OFF}.

Freeze Protection. A recirculation freeze-protection option is available for many controllers by the addition of *freeze sensors*, which are thermostatically activated snap switches. These sensors are usually wired in parallel with the collector sensor and located at the collector array inlet and outlet. They should also be placed at locations where cold water can settle in the collector piping. A typical sensor wiring circuit for a recirculation freeze-protection controller is shown in Fig. 13-34*a*. The freeze-recirculation snap sensors will ordinarily make contact at 42°F, turning on the recirculation, and will open at 52°F, turning off the recirculation.

When a controller has a draindown valve option (as discussed in section 13-3), a separate control circuit is ordinarily required for the freeze-protection sensors. This is necessary because the draindown function must control the draindown valve independently of the collector pump. Figure 13-34*b* shows a sensor wiring circuit for a draindown controller. (The freeze–snap-switch sensors are wired in series.) Sensors used for a draindown circuit ordinarily remain closed until a temperature of about 42°F is reached, at which point they open. When the continuity of the series circuit is broken by any one of the snap-switch sensors, the draindown function will be activated by the controller. The sensors will reclose at 52°F; thus, when all of the sensors are at 52°F, the draindown function will deactivate and the system will refill with water. The controller will not turn on the collector pump as long as the draindown function is activated.

High-Temperature Limits. Controllers usually provide an option for turning off the collector pump when the storage tank reaches a maximum temperature. For a differential thermostat or recirculation controller, the high-limit option requires adding a third temperature sensor to the storage tank, as shown in Fig. 13-35*a*. This third (high-limit) sensor measures the tank temperature. When the tank temperature exceeds 160°F, the controller will turn off the collector pump, thus stopping the collection of heat. When this option is used on open- and closed-loop systems, provision must be taken to handle the temperatures and pressures that will result when the fluid flow is stopped in the collectors.

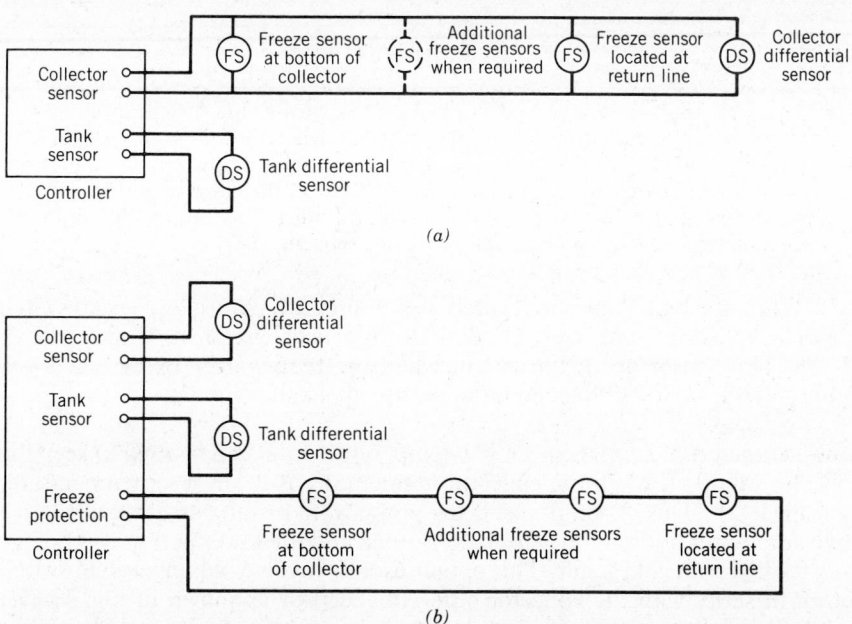

(a)

(b)

Figure 13-34 Wiring diagrams for sensors. (*a*) Typical sensor wiring circuit for a recirculation freeze-protection control system. (*b*) The sensor wiring circuit for a draindown freeze-protection control system. The freeze-protection circuit is separate from the differential sensors' circuits.

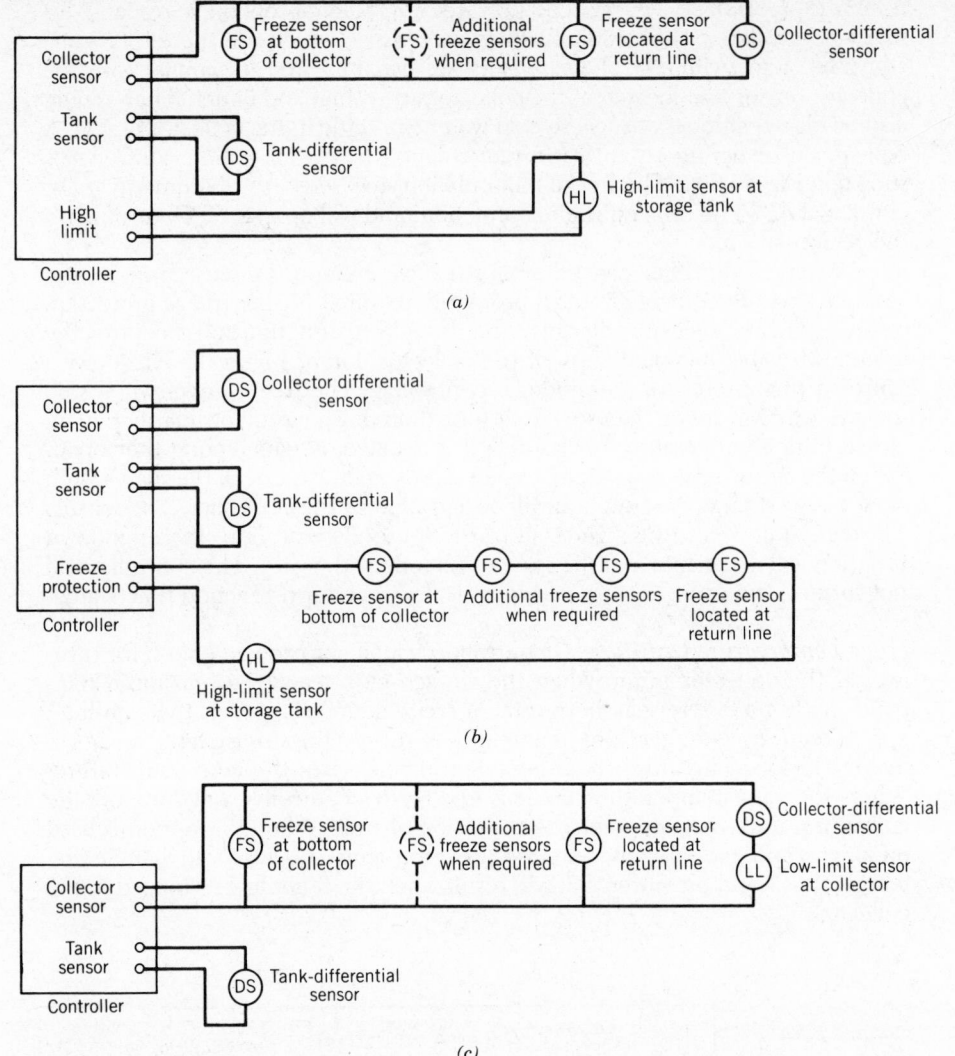

Figure 13-35 High- and low-limit control sensor wiring. (*a*) A high-limit control sensor is typically a separate sensor that is wired to the recirculation freeze-protection controller. (*b*) On a draindown controller, the high-limit sensor is wired in series with the freeze sensors. (*c*) A low-limit sensor can be installed in series with the collector differential sensor as shown. This will inhibit the action of the controller until the collector reaches a given temperature, typically 80° F.

When the high limit is to be used on a draindown controller, a snap-switch sensor is added in series with the draindown sensor circuit, as shown in Fig. 13-35*b*. This sensor opens the circuit when the temperature exceeds a preset figure—160°F to 200°F, depending upon the application.

Low-Temperature Limits.   A low-temperature limit can be used to keep the collector pump off when the collector temperature is below a set temperature, typically 80° F. This option prevents the pump from turning on at night when a large amount of hot water has been consumed and the lower part of the storage tank is filled with cold water. This option usually involves adding a snap-switch sensor in series with the collector differential sensor as shown in Fig. 13-35*c*. This sensor will open the differential sensor circuit at temperatures below 80°F, thus preventing the differential thermostat from turning on the pump.

The above functions are basic options available for most controllers. Other options, such as digital temperature readouts and adjustable settings, are also available from most controller manufacturers.

The installation of domestic hot-water heaters requires not only a good basic knowledge of solar theory and equipment but also a working knowledge of the related fields of plumbing, carpentry, circulating hot-water systems, roofing, and electrical work. In some instances, of course, the solar installer may install only the solar equipment such as collectors, tanks, pumps, and controls, as on new construction and large installations where many subcontractors are working on a building. On new construction, the plumbing contractor may install all of the piping including the solar piping, the electrician might make all electrical connections to electrical equipment including pumps and controls, and the carpenters and roofers may provide mounting pads or a support rack for the collectors. Even though solar installers might not perform the related trade work, they should have enough knowledge of the trade to determine if the work that was performed is satisfactory before installing the solar equipment. Invariably, if there is some difficulty experienced with the solar system, it will be the solar contractor or installer who is called back to the job to deal with the problems.

In the case of retrofitting solar domestic hot-water systems to existing homes or apartment buildings, the solar installer may perform all of the work and will thus require skills in all related building trades.

13-15   Location and Types of Collector Mountings

Solar collector location is the basic starting point for an installation, since the type of collector mounting and the piping runs required are determined by the location of the collectors. Solar collectors typically can be mounted in three locations: on a roof, on an awning rack or trellis-type rack, or on a ground-mounted rack. The simplest collector installation is one where the collector is mounted directly on a sloping roof, as shown in Fig. 13-36a. In cases where the collector tilt is to be greater than the roof pitch, a rack or tilted structure must be used to mount the collectors, as shown in Figs. 13-36b and 13-36c. In some cases, the collectors may have to be mounted above a north-facing roof, as shown in Figs. 13-36d and 13-36e. When collectors are to be placed on a flat roof, a collector support must be used, as shown in Fig. 13-37. The mounting of collectors on an east- or west-facing roof, skewed to the roof surface, requires a more elaborate support system in order to face the collectors south, as shown in Fig. 13-38.

Collectors may also be mounted off the wall of the house using an awning-type rack, as shown in Fig. 13-39. When it is not possible or practical to mount collectors on the building structure, they can be ground-mounted, as shown in Fig. 13-40.

Pitched-Roof Attachments.   The attachment of the collectors or collector racks to the roof structure is a critical part of the installation. The roof membrane must be penetrated at each attachment point, and this creates a potential for roof leaks if not properly done. The attachment must also be securely tied to the roof structural elements such as rafters or the blocking between rafters. Anchoring just to the roof sheeting is absolutely unacceptable.

When collectors are mounted on shingled roofs (either wood or composition) it is common to use a lag-bolt mounting, as shown in Fig. 13-41. A pilot hole (typically ¼ in.) is drilled into a 2 × 4 or 2 × 6 nailed to the roof rafter and is then filled with silicone rubber caulking. An area around the pilot hole is coated with silicone caulking or roofing mastic. A spacer block is then placed on the roof over the pilot hole. A mounting angle, collector mounting clip, or collector rack is placed over the block, and a lag bolt and washer (typically ⅜ in.) is inserted into the hole and securely screwed into the 2 × 4 or 2 × 6 member. The head of the bolt and washer should be given a generous coating of roof cement or caulk. The mounting block should raise the back of the solar collec-

(a)

(b)

Figure 13-36 Methods of mounting collectors. (*a*) Collectors can be mounted directly on the existing roof when no pitch adjustment is required. This is the simplest and least expensive, and involves the least weight of any mounting method. In regions where ice damming is prevalent, the collectors should be mounted at the roof peak. (*b*) When collectors must be mounted at a different angle from the existing roof pitch, a rack can be used. If adequate space is left under the rack, reroofing may be done without removing the collectors or the rack. (*c*) A dormer can be used to provide a tilted surface for collector mounting. The weight can be distributed evenly over the roof surface. (*d*) When it is necessary to mount collectors on a north-facing roof, a rack can be used to provide the correct collector tilt angle. (*e*) Collectors facing south, but mounted against a tall chimney on the north slope of the roof. (All photos courtesy of Santa Barbara Solar Energy Systems.)

tor 1½ to 3 in. above the roof surface. This is to allow sufficient air space under the collector to prevent the buildup of debris under the collector and to allow water and ice to flow freely under the collector. The collector mounting should not dam the roof and block the flow of water. Local building codes should be checked for the minimum spacing between roof and collector.

When it is not possible to anchor directly to the roof rafters, threaded rods and 2 × 4 or 2 × 6 spanners can be used to provide a secure roof anchor, as shown in Fig. 13-42. A 7/16-in. clearance hole is drilled through the roof; then a spanner is securely nailed to two rafters, and a hole is drilled in the spanner. The hole in the roof is filled with silicone caulking after the threaded rod is inserted.

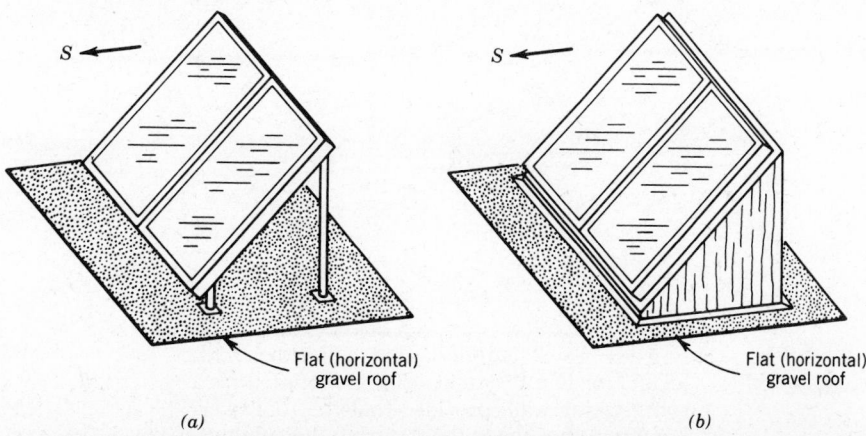

(c)

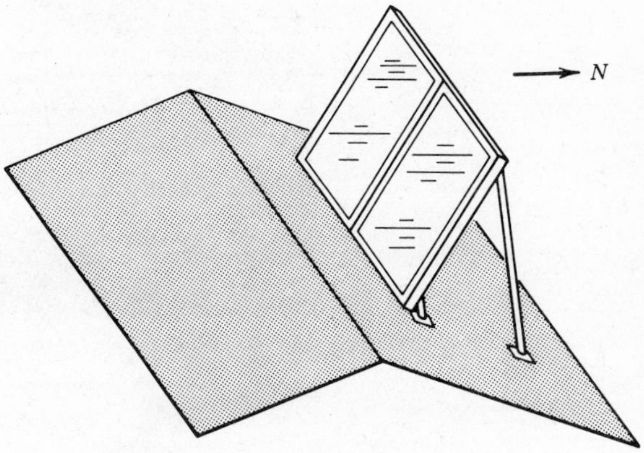

(d)

N

(e)

Figure 13-36 (*continued*)

S ←

S ←

Flat (horizontal)
gravel roof

Flat (horizontal)
gravel roof

(a)

(b)

Figure 13-37 Mounting collectors on flat roofs. (*a*) On flat roofs, collectors must
be mounted on a rack structure to give them the correct tilt. The rack itself can easily
be oriented due south, since there is no preexisting roof pitch or orientation to
contend with. (*b*) An alternative flat-roof mounting involves building a dormer
structure that is waterproofed and flashed to the existing roof.

Figure 13-38 (*a*) When collectors are mounted on a non-south-facing roof and it is desired to give them a south orientation, it is necessary to construct a rack or support. This results in the collector's being angled across the sloping roof. Greater wind loads may result on this type of mounting. (*b*) View of support structure back of collectors.

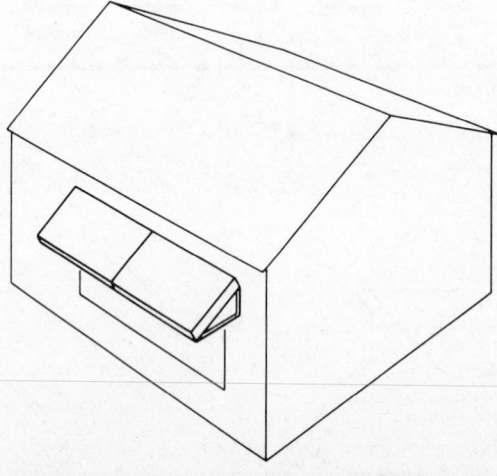

Figure 13-39 Wall-mounted collectors. Collectors for DHW systems can be mounted on a south-facing wall, provided that the roof overhang does not shade them during the summer.

Figure 13-40 Ground-mounted collectors. A rack with adjustable legs is required to support the collectors and provide the correct tilt angle.

On either side of the rod, 2 × 4 blocking is placed between the roof sheathing and the spanner. This prevents the roof from being deformed when the bolts are tightened. The roofing around the rod is coated with silicone caulk or roof cement, and a mounting block is set on the roof. Then the collector rack or mounting angle is placed on the mounting block. Nuts and washers are then placed on the threaded rod, and the entire assembly is securely tightened. Double nuts should be used on each end of the rod to prevent loosening of the nuts. The nuts and washers should be covered with roof cement.

A superior roof attachment is shown in Fig. 13-43a, where the spacer block has a flashing around it that extends back under the existing shingles so that the

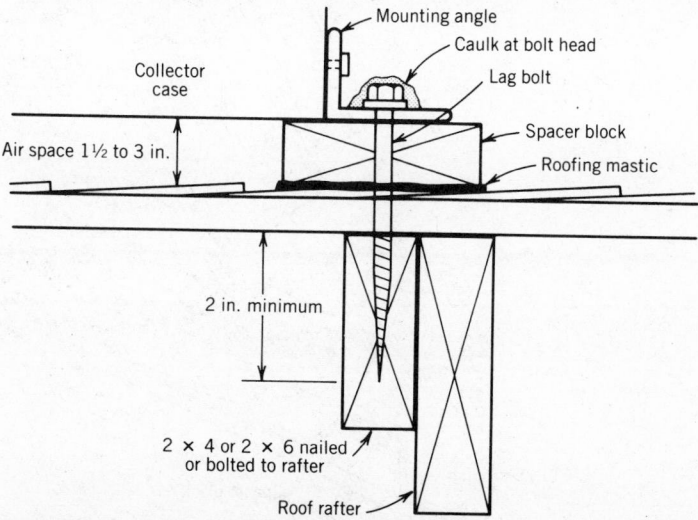

Figure 13-41 Lag-bolt collector mounting. A spacer block is used to raise the collector above the roofing so that proper drainage and air circulation can occur under the collector. The lag bolt should penetrate into the 2 × 4 or 2 × 6 a minimum of 2 in.

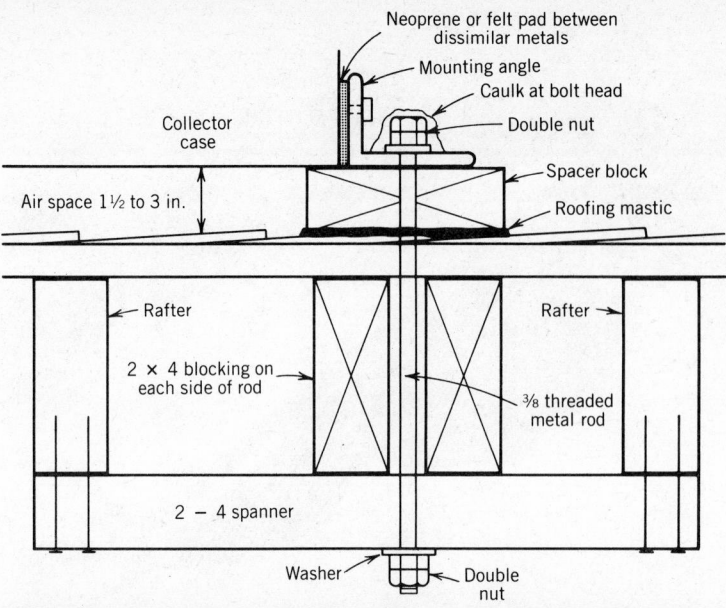

Figure 13-42 A spanner is used between roof rafters when it is not possible to bolt directly to a rafter. A threaded rod is then run from the mounting bracket through the roof and spanner. Double-locking nuts are used at each end of the rod to secure the assembly.

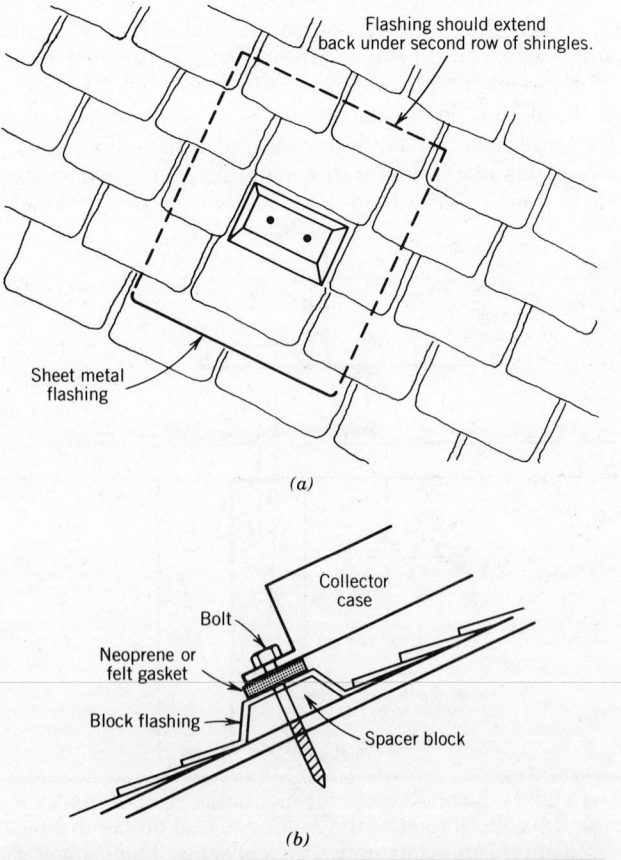

Figure 13-43 Collector attachment has good weather resistance when the spacer block is flashed and becomes an integral part of the roof membrane, with a gasket between the flashed top of the block and the collector mounting bracket. The disadvantage of this type of mounting is the added cost.

whole assembly will shed runoff. The spacer block may be fastened with either lag bolts or a spanner system. A neoprene or felt pad, as shown in Fig. 13-43*b*, should be used to separate dissimilar metals such as galvanized flashing and the aluminum collector frame.

Flat-Roof Attachments.   Flat roofs typically have pooling problems because of low places in the roof. When the roof membrane is pierced to make an anchor for a solar collector rack, the sealing at the anchor point must be exceptionally good or water may leak in. One method is to use a pitch (asphalt) pan, as shown in Fig. 13-44. Before a mounting is placed on an existing roof, the installer should consult the roof manufacturer's warranty for requirements regarding the installation of accessories on the roof. The roofing and rigid roof insulation will have to be cut out where the pitch pan will be placed. The collector rack leg must be securely attached to the roof deck, and solid blocking must be attached to the roof deck around the leg. The flange of the pitch pan is nailed securely to the blocking, after which the roof is patched, from the old roof membrane to the pitch pan, covering all of the pan flange. The pitch pan is then filled with asphalt.

An alternative to the pitch pan is to use a flashed mounting block, as shown in Fig. 13-45. It is important that the flashing up the sides of the block extend higher than the depth to which water may accumulate on the roof.

When the collectors are to be mounted on a new roof or on a roof that is to be reroofed, any of the roof attachment procedures that have been discussed can be used. Attachments are mounted on the roof with the proper flashing, and the new roof is then installed. The collector racks and collectors are then mounted using the roof attachments. Care should be taken when installing the collector to minimize any damage to the new roof. A roof mounting rack used

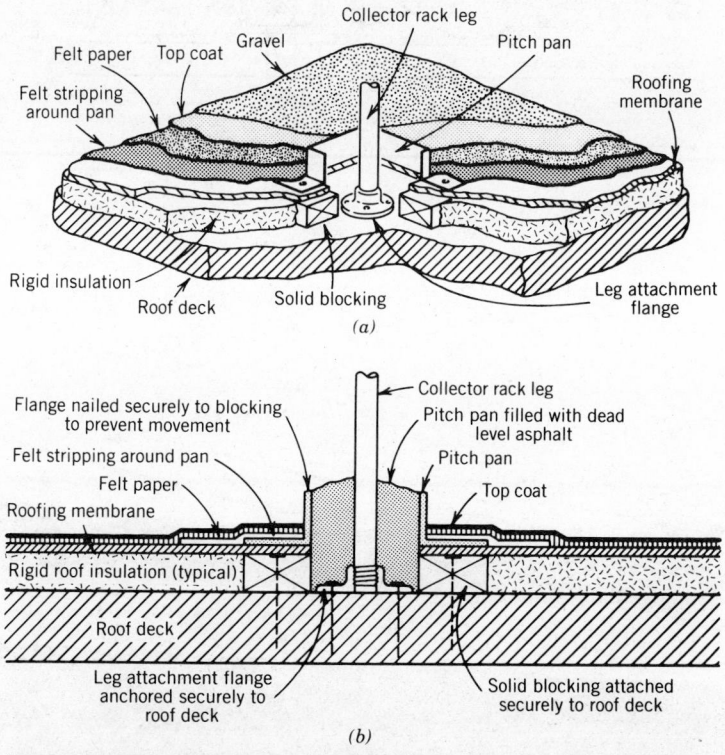

Figure 13-44 A pitch pan is typically used when a collector assembly must be anchored to a flat roof with a builtup roof membrane. (*a*) Cutaway view showing layers of roof construction. (*b*) Vertical section through pitch pan and collector rack leg.

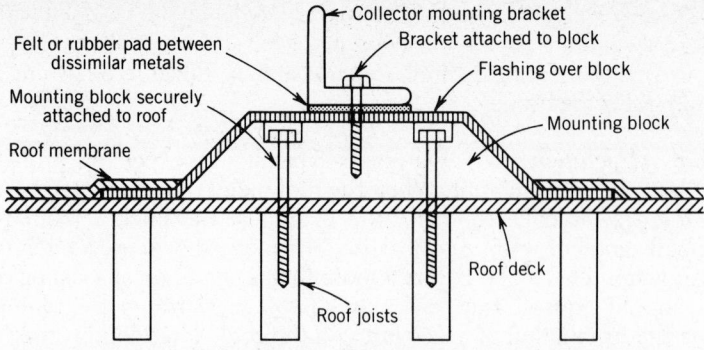

Figure 13-45 A flashed mounting block is an alternative to a pitch-pan assembly, but the top of the flashing should always be higher than the maximum level to which water can rise on the roof.

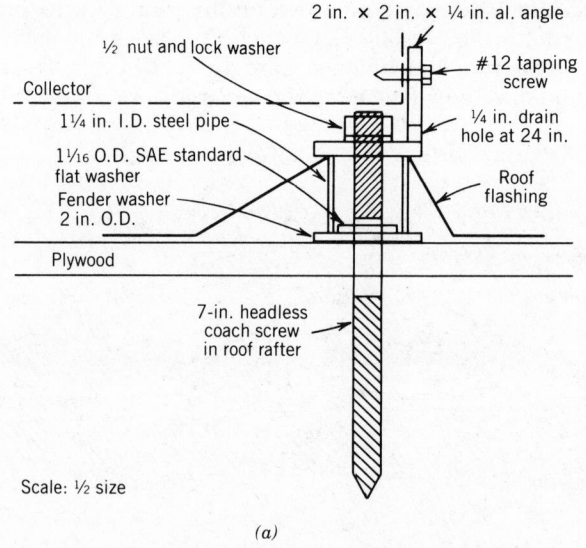

Scale: ½ size

(a)

Figure 13-46 Mounting collectors on new construction. (a) A flashed collector rack may be installed before the roofing is put on. (b) The photo shows standard pipe roof flashings around the rack pipe supports.

on new construction is shown in Fig. 13-46a. This mounting allows the shingles to be installed around the attachment points using a standard vent flashing. A portion of the rack construction is shown in Fig. 13-46b.

13-16   Completing the Collector Installation

Pipe Penetrations.   Another roof membrane penetration occurs where the collector piping and control sensor wiring pass through the roof to the collectors. Typically, there will be two points where the piping will pass through the roof—one for the collector supply line, and a second for the collector return line. On a shingled roof, a pipe flashing is worked under the shingles, as shown in Fig. 13-47. It will be necessary to lift some of the shingles above the pipe hole and remove any roofing nails where the flashing is to slip in. The pipe and insulation may be run through the flashing as a unit, as shown in Fig. 13-47a. In this case, the pipe insulation joints that are exposed to the weather must be waterproofed so that moisture cannot leak between the pipe and the insulation into the building. Alternatively, the pipe may be run through the flashing with the insulation stopping at the flashing, as shown in Fig. 13-47b. Figure 13-48 shows several ways in which a penetration for piping can be made through a wall.

Piping Runs.   The actual piping layout for a collector array will depend upon the location of the collectors and the particular type of collector. In general, the following guidelines will apply:

1.   All "horizontal" piping should be run at a constant pitch (typically ⅛ to ¼ in. per foot).

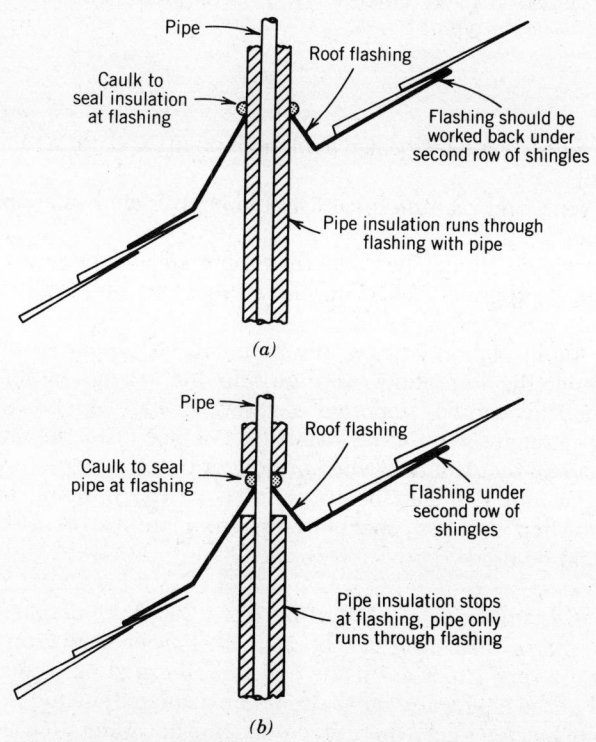

(a)

(b)

Figure 13-47 The pipe and insulation can be run through the flashing (*a*) as a single unit, or (*b*) only the pipe itself may be run through, with the insulation stopping at the flashing.

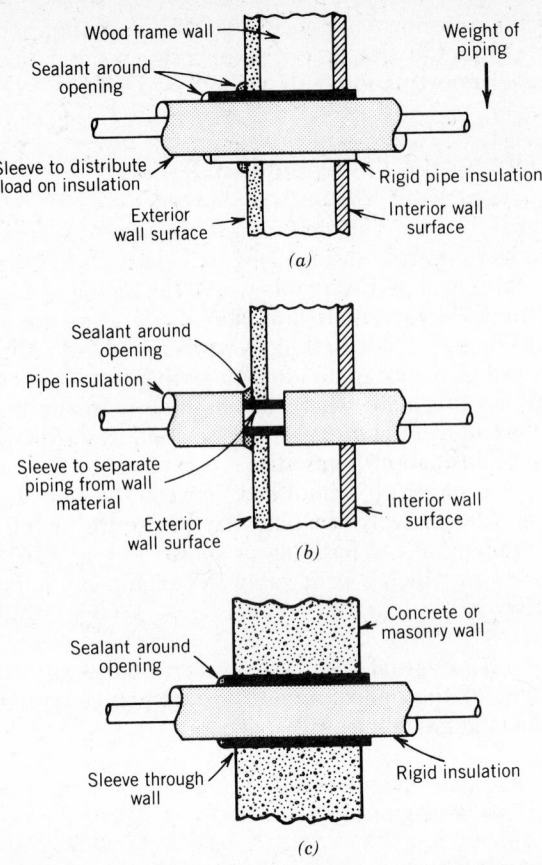

Figure 13-48 Three ways of making wall penetrations for piping.

2. Air vents should be located at all high points and drains at all low points.
3. Air vents and vacuum breaks should be insulated when exposed to the weather.
4. The piping should be properly supported and provisions provided for thermal expansion and contraction on long runs.

When rigid pipe insulation is used, the pipe can be supported by a clamped assembly around the insulation, as shown in Fig. 13-49. In the interior of buildings, pipe runs can be supported as shown in Fig. 13-50. Care should be taken to allow enough room for the installation of pipe insulation and to make a minimal impact on the structure when drilling or notching structural members. Remember that vibration can be transmitted to the building through pipe hangers, and when this is a possibility, as from pumps or motors, resilient supports should be used.

Application of Insulation. After the system has been pressure-tested successfully, the piping insulation can be applied. Proper insulation of the solar energy system is very important if the expected thermal performance is to be achieved. All of the piping and tanks should be insulated, including at least two feet of the cold-water inlet where it enters the hot-water storage tank. The interior pipe insulation should have a minimum insulation value of *R*-2.5, and the exterior pipe insulation should have a minimum insulation value of *R*-4.0. Thermal storage tanks should have a minimum of *R*-12 insulation (*R*-20 is recommended).

SOLAR HEATING OF DOMESTIC HOT WATER

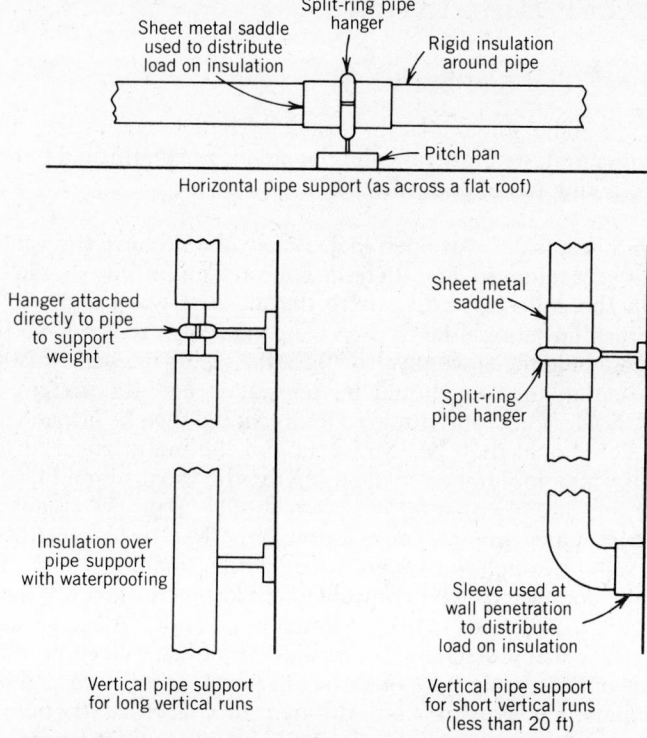

Figure 13-49 Three methods of supporting piping when it is exposed on the exterior of buildings.

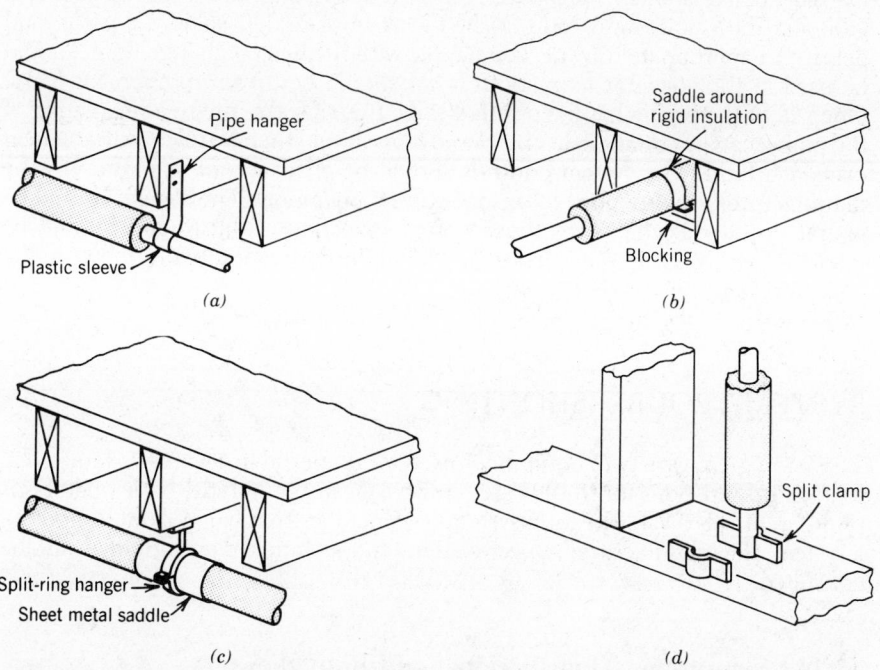

Figure 13-50 Interior pipe-support details. (*a*) Hanger attached directly to pipe. This type of support will transmit vibration and heat loss, but it is the necessary method when using nonrigid insulation. Hangers that have a plastic sleeve at the point of contact with the pipe should be used. (*b*) Pipes may be strapped to the blocking when running parallel with the joists. (*c*) The use of a saddle and split-ring hanger will minimize vibration and heat loss. (*d*) On long vertical pipe runs, it is necessary to use a split clamp to support the pipe.

STARTING UP THE SYSTEM

13-17   Check Test and Start

When the plumbing for a solar hot-water system is completed, the system should be pressure-tested for leaks and the piping system flushed to remove bits of solder, soldering paste, and other debris.

Open-Loop Systems.   An open-loop system should have the vacuum relief valve and air vent removed, and then the collector supply line should be flushed out through the collector array where the air vent was removed. Then the collector return line should be flushed. The air vent is removed so that debris will not be caught in the air vent valve. Once the piping has been flushed, the air vent and vacuum breaker should be reinstalled and the system filled and pressurized. Solder joints and threaded joints should then be inspected for leaks. The controller should then be turned on and the pump checked for correct operation. If water does not circulate properly, the valves should be checked to ensure that they are open, and the system should be purged of any air blocks that may exist. Once the system is found to be free of leaks with the pump circulating water through the system satisfactorily, final insulating of all of the piping can be completed. The controller should be checked out according to manufacturer's instructions to ensure correct operation for the solar collection mode and the freeze protection mode. With all of the collectors receiving full sunlight, the outlet temperature of each collector in an array should be checked to ensure that there is equal flow through all collectors. If there is a flow imbalance, it will be indicated by unequal collector outlet temperatures. Finally, any necessary adjustments of flow control valves should be made to obtain the desired temperature rise in the collectors.

Closed-Loop Systems.   Closed-loop systems should have both the collector loop and the storage tank loop flushed to remove solder, soldering paste, and debris. An appropriate solvent, compatible with the heat-exchange fluid, should be used in the collector loop. Both loops should be pressure-tested for leaks. Once all of the piping has been found to be free of leaks, the final insulating of the piping can be completed. The closed loop should then be filled with the heat exchange fluid. The system controls should be checked out according to the manufacturer's instructions to ensure correct operation. The system should be operated and the collectors checked for equal flow. Adjustments should be made on flow controls as necessary, to obtain the desired temperature rise in the collectors and the heat exchanger.

SYSTEM TROUBLESHOOTING

There is a great variety of equipment used for domestic hot-water systems. The following discussions for trouble shooting cover, in general, the basic operations of different solar water-heating systems. Remember that when troubleshooting a system, the manufacturer's specifications and technical data should always be consulted.

13-14   Symptoms, Malfunctions, and Remedies

Open-Loop DHW with Recirculation Freeze Protection

1. Pump does not run.
    (a) Verify that full power is available.
    (b) Check switches on controller.
    (c) Plug pump directly into power outlet.

SOLAR HEATING OF DOMESTIC HOT WATER

(d) Check power cord on pump.

(e) If pump still does not run, remove pump from system and check that impeller turns freely and is not clogged with foreign material.

2. Pump runs but is noisy.

(a) A gurgling or swishing noise is caused by air trapped in the pump. Steps should be taken to purge the system of trapped air.

(b) If pump makes a loud vibration or a squeaking noise, the motor bearings may be worn or need oiling, depending upon the type of pump.

3. Pump runs excessively at night.

(a) Freeze recirculation should occur during cold nights, and the pump should run at least two to four minutes per hour. The total time the pump runs would be 25 to 40 minutes nightly, depending on the ambient temperature.

(b) The freeze-protection time duration can be checked using an electric clock patched (use a power splitter) into the controller that governs the pump. The clock will run whenever the pump is on. The elapsed time the pump runs at night can be determined by reading the time indicated on the clock at night and in the morning.

(c) If the elapsed time shows that the pump is running excessively at night, this indicates a controller problem. The instructions of the manufacturer should be referred to and the following checks made:

(1) The resistance of the sensors should be determined using an ohmmeter.

(2) Freeze switches should also be checked.

4. Pump runs all the time, day and night.

(a) This is a controller problem.

(b) The control wiring to the sensors should be checked for a possible short-circuit.

(c) The sensors should be checked to verify proper installation.

(d) The manufacturers' instructions should be followed for field-testing the controller.

(e) The resistance of the sensors should be checked according to the manufacturer's instructions.

5. Pump cycles on and off during clear, sunny days.

(a) This indicates that the flow through the collectors is too large. The collector flow should be restricted by closing the flow adjustment valve(s).

6. Water leaks from air vent or vacuum breaker.

(a) The air vent or vacuum breaker should be cleaned of any mineral deposits or debris. Replace if the unit continues to leak.

7. Storage tank is cold by morning.

(a) This indicates that heat is being lost from the tank. This can be caused by excessive recirculation at night or by a thermosiphoning effect at night.

(b) The system should be checked for excessive recirculation.

(c) The piping should be checked at night to determine if a thermosiphon is occurring. If the piping to the collectors is warm several feet from the storage tank, a thermosiphon is occurring. The check valve should be cleaned.

Open-Loop DHW with Draindown Freeze Protection

1. Pump does not run.

(a) See item 1 under Open-Loop DHW above.

2. Pump runs but is noisy.
    (a) See item 2 under Open-Loop DHW above.
    (b) The draindown valve may be in the drain position and the pump is not able to circulate water to collectors. Check the draindown valve.
3. Pump runs at night or all the time.
    (a) This indicates a controller problem. With a draindown system, the pump should only run during the day when solar energy is being collected. The control wiring should be checked for shorts or open circuits. The differential sensors should be checked for correct resistance. The controller should be field tested according to the manufacturer's instruction.
4. Pump cycles on and off during clear, sunny days.
    (a) See item 5 under Open-Loop DHW above.
5. System will not fill.
    (a) Check to see if the controller is sending power to the draindown valve.
    (b) Check the draindown valve. It may be stuck in the draindown position or may be defective.
6. System will not drain.
    (a) Check the draindown valve. It may be stuck in the open position.
    (b) Check the vacuum breaker valve. It may be stuck shut and need opening to allow air to enter the collectors for draining them.
7. Draindown valve leaks constantly.
    (a) This indicates that the valve is defective. Check the seals in the valve and make repairs or replace the valve.

Closed-Loop Systems with Nonfreezing Heat Transfer Fluid

1. Collector loop pump and heat exchanger pump do not run.
    (a) See item 1 under Open-Loop DHW (Recirculation) above.
2. Collector loop pump runs, but is noisy.
    (a) A gurgling or swishing noise is caused by air trapped in the pump. The fluid level in the collector loop should be checked. If it is low, the collector loop piping should be checked for leaks, the system filled with additional fluid, and air purged from pump and collector loop.
    (b) See item 2 under Open-Loop DHW (Recirculation) above.
3. Pump runs at night.
    (a) This indicates a controller problem. The control wiring should be checked for shorts or open circuits. The differential sensor should be checked for correct resistance readings. The controller should be field-tested for correct operation according to the manufacturer's instructions.
4. Pumps cycle on and off during clear, sunny days.
    (a) This indicates that the fluid flow rate through the collectors is too large. The flow rate through the collector loop and heat exchanger loop should be adjusted to give a temperature rise in the collectors of 10 to $20F°$.
5. Heat exchange fluid leaks from air vents.
    (a) The closed loop should be depressurized, the air vent removed, the valves cleaned or air vent replaced.

13-18   Solar Hot-Water System Ratings

The American Society of Heating, Refrigerating, and Air Conditioning Engineers (ASHRAE) has developed a testing standard for domestic solar hot-water (DHW) systems. This standard is ASHRAE Standard 95-1981. It provides a means of comparing the performance of different solar hot-water systems under standard conditions. Up to 1981, the ASHRAE Standard 93-77 instantaneous efficiency test had been used, but it cannot be applied to certain kinds of solar water heaters, such as thermosiphon systems and integral collector storage (ICS or breadbox) systems, which are becoming increasingly popular. ASHRAE Standard 95-1981 will allow designers, contractors, and consumers to evaluate the different systems by comparing thermal performance against dollar cost. In the future, most DHW systems may have performance certifications based on this standard (or on other methods now being investigated for the testing of complete systems by such groups as the Air Conditioning and Refrigeration Institute, the Oregon Department of Energy, and the Florida Solar Energy Center). This means that salespeople and dealers will have accurate data on which to base their claims for system performance.

ASHRAE Standard 95-1981 defines three categories for domestic hot-water systems: *solar only*, *solar preheat*, and *solar plus supplemental*. A solar-only system is one in which all of the hot water must be supplied by solar energy, with no other source. A solar-preheat system is the type of system installed in a retrofit situation and includes all system components except the auxiliary water heater. The solar-plus-supplemental system includes the auxiliary heating system, which may be an electrical heating element in the solar storage tank or a separate hot-water (gas-fired) heating tank.

The solar system that is to be evaluated is placed in an environmental test chamber where instruments are installed to monitor the cold-water supply temperature, the ambient air temperature, the number of gallons of hot water drawn from the system, and the temperature of the water drawn from the system. These measurements allow the calculation of the net amount of daily energy provided by the sun to a solar hot-water system, Q_{NET}.[4]

For a solar-plus-supplemental system, the Q_{NET} is calculated as:

$$Q_{NET} = Q_{DEL} + Q_{LOSS} - Q_{AUX} - Q_{PAR} \qquad (13\text{-}10)$$

where:

Q_{DEL} = the total daily energy delivered by the solar hot-water system, including any energy supplied by an auxiliary heating system

Q_{LOSS} = the daily standby loss of the auxiliary tank, specified in Appendix C of ASHRAE Standard 95-1981 as 3.1 kW-hr for an electric auxiliary heater, and as 14.9 kW-hr for a gas auxiliary heater

Q_{AUX} = the auxiliary energy used (It is the daily energy, either electric or gas, used by the solar hot-water system in order to deliver the required amount of hot water.)

Q_{PAR} = the sum of parasitic energy losses (It is the daily energy used to supply power to pumps, controllers, and trackers to operate the solar hot-water system.)

For solar-only and solar-preheat systems, Q_{NET} is calculated as:

$$Q_{NET} = Q_{DEL} - Q_{PAR} \qquad (13\text{-}11)$$

In this case, there is no auxiliary heating system, so Q_{LOSS} and Q_{AUX} are not present.

ASHRAE Standard 95-1981 is a testing standard that gives a detailed procedure for performing the tests. It does not specify the environmental and domestic hot-water load conditions. Generally, industrial rating organizations specify these conditions. The Solar Rating and Certification Corporation (SRCC) has announced a certification program based on ASHRAE Standard 95-1981. The SRCC program requires that the test day provide 1500 Btu/ft² of simulated solar radiation to a solar aperture inclined at a 45° angle. This energy

[4] The terminology used here and the description of the test procedure are adapted from William J. Putnam, "The Solar Hot Water System Rating Test." This article first appeared in the August issue of *Solar Age*. © 1983 Solar Vision, Inc., Harrisville, N.H. 03450. All rights reserved. Adapted and published by permission.

is to be provided during a nine-hour period from 8:00 A.M. to 5:00 P.M. This simulated day is very similar to an equinox day (March 21 or September 21). The ambient air and supply water temperatures are held constant at 71.6°F. During the test run, the wind velocity across the collectors is maintained at 7.6 mph. The daily total energy load to be drawn from the solar system, Q_{DL}, is 40,119 Btu taken in three equal draws of 13,373 Btu at 8:00 A.M., noon, and 5:00 P.M. (Each draw of 13,373 Btu is approximately equal to 33 gal of water heated from 71.6°F to 120°F.)

Other parameters that are measured are Q_{RES}, the reserve capacity of the system (that is, the energy remaining in storage after the last day's load is drawn off at the end of a test); and Q_{CAP}, the energy storage capacity of the auxiliary heating system tank (this parameter is measured only for solar-plus-supplemental systems).

For systems that have the thermal storage located outdoors, such as integral collector storage (ICS) systems and some thermosiphon systems, an exponential heat loss coefficient L (Btu/hr-F°) is measured.

Figure 13-5b gave the ASHRAE Standard 95-1981 test performance factors for the PT-40 integral collector storage (ICS) system. System performance A test results were for two PT-40 units connected in series, providing 76.5 gal of storage and 39.2 ft² of gross collector aperture. The system performance was as follows (see Fig. 13-5b):

The net energy delivered by the two PT-40 units was 24,600 Btu/day. The standard test load placed on these units was 40,199 Btu/day. The two PT-40 units were able to provide 61 percent of the test load with solar heat. Auxiliary Heat would be required to supply the balance of the test load. The overall system performance can be determined by dividing the net energy delivered from the two PT-40 units by the total incident solar energy, or

$$\eta = \frac{Q_{NET}}{A_c \times I}$$

$$= \frac{24,600 \text{ Btu/Day}}{39.2 \text{ ft}^2 \times 1500 \text{ Btu/day-ft}^2}$$

$$= 0.45$$

The average temperature of the water off the heater at the end of the day can be estimated by using Eq. (13-1) as follows:

$$Q_{WM} = c_W \times D \times WC_h \times (T_s - T_m) \times N$$

where

Q_{WM} = the net energy delivered, 24,600 Btu
WC_h = the water capacity of the two units, 76.5 gal
c_W = the specific heat of water, 1.0 Btu/lb-F°
D = the density of water, 8.34 lb/gal
T_m = the temperature of the initial water, taken as 60°F
T_s = the temperature of the water at the end of the day, which will be estimated
N = the number of days, taken as 1

substituting

$$24,600 \text{ Btu} = 1.0 \text{ Btu/lb-F}° \times 8.34 \text{ lb/gal} \times 76.5 \text{ gal} \times (T_s - 60°F) \times 1$$

Solving for T_s, we obtain

$$T_s = 98.6°F$$

The two PT-40 units would provide an average water temperature of 98.6° F at the end of the day, assuming that no hot water has been drawn off during the day.

13-19 A Solar Hot-Water System Design Problem

A solar domestic hot-water system is assigned 75 percent of the hot-water load for a four-unit apartment building. Each unit has one bedroom, bath, and kitchen. Hot-water usage may be estimated from the 1977 HUD standards of Table 13-1. The collectors will be located on a south-facing roof with a tilt of 22°. Estimate the hot-water load assuming that the temperature in the water main is 60°F. The desired hot-water temperature is 140°F. The location is Los Angeles, California. Estimate the collector area for December and June and the size of the solar storage. Then decide on a final choice for the collector area. Use the collector data in Fig. 12-8 for a WSD7-CR-31 to estimate average collector performance. Finally, determine pressure relationships and size the collector-loop pump.

Solution

Determination of the hot-water load

From Table 13-1, the hot-water consumption is 40 gal/day for a one-bedroom unit. From Eq. (13-1), the daily hot-water load Q_{WD} is:

$$Q_{WD} = 1.0 \text{ Btu/lbF}° \times 8.34 \text{ lb/gal} \times 40 \text{ gal/day} \times$$
$$(140 - 60)\text{F}° \times 4 \text{ units}$$
$$= 106,800 \text{ Btu/day}$$

Estimating collector area

From Fig. 12-8, the medium-temperature efficiency of the WSD7-CR-31 collector is given as 51 percent. This can be used as the average collector efficiency $\bar{\eta}$, during average daily insolation conditions.

For December 21

The solar radiation tables for 32° N Lat will be used to estimate the solar radiation available at Los Angeles.

From Table 4-E, the total daily solar radiation on a 22°-tilted south surface is 1705 Btu/ft²-day. From Table 4-C, the average percentage of possible sunshine is 71 percent. Using Eq. (13-8), the collector area is calculated as:

$$A_{c_{Dec}} = \frac{SHF_{DHW} \, Q_{DHW}}{\bar{\eta}PI} = \frac{(0.75)(106,800 \text{ Btu/day})}{(0.51)(.71)(1705 \text{ Btu/ft}^2\text{-day})}$$
$$= 129.7 \text{ ft}^2 \qquad\qquad\qquad\qquad \text{Ans.}$$

For June 21

From Table 4-E, the total daily solar radiation on a 22°-tilted south surface is 2440 Btu/ft²-day. From Table 4-C, the average percentage of possible sunshine is 65 percent. Using Eq. (13-8), the collector area is calculated as:

$$A_{c_{Jun}} = \frac{(0.75)(106,800 \text{ Btu/day})}{(0.51)(.65)(2440 \text{ Btu/ft}^2\text{-day})}$$
$$= 99 \text{ ft}^2 \qquad\qquad\qquad\qquad \text{Ans.}$$

The WSD7-CR-31 collector in Fig. 12-8 has a gross area of 31.2 ft². The number of collectors required in winter would be 129.7 ft²/31.2 ft²/coll = 4.16 collectors. The number of collectors in summer would be: 99 ft²/31.2 ft²/coll = 3.2 collectors.

A reasonable compromise between the winter and summer requirements would be to use four solar collectors in parallel with a total area of 124 ft². This arrangement would provide a little less than the desired 75 percent SHF in winter and more than the assigned 75 percent of the load in summer. The system falls into the optimum-range SHF as discussed in section 13-5.

Sizing the solar storage

The rule of thumb for hot-water storage volume is 1.25 to 1.6 times the daily hot-water demand. In the present case, the hot-water demand is 160 gal/day, and the solar storage would have a volume of from 200 to 256 gal. The storage volume could be obtained by using two 100-gal solar tanks to give 200 gal or two 120-gal solar tanks to yield 240 gal. Alternatively, a single large storage tank with a volume of 250 gal could be used. The decision of whether to use two small tanks or a single large tank may depend upon the space available for the tanks and the size of openings through which the tanks will have to pass to be installed.

Determining pump size

The approximate pipe distance between the collectors and storage tank will be 40 ft one way. The total length of piping will then be 80 ft. There will be approximately 18 90° elbows, one check valve, and two ball valves in the collector loop. The collectors will be connected in parallel, with through headers and reverse return, as shown in Fig. 13-26. If a different collector plumbing arrangement were used, as one with branch tees, then the equivalent lengths for the tees would have to be calculated. The pipe size used should be ¾ in., since more than 100 ft² of collector is being used. The equivalent length of pipe for the fittings is calculated using this simple layout and data from Table 13-4:

Fitting	No. of Fittings	×	Equivalent Pipe Length	=	Total Equivalent Length
90° elbows	18	×	2.5	=	45.0 ft
Check valve	1	×	35.0	=	35.0
Ball valve	2	×	0.2	=	0.4
					80.4 ft

The total effective length of piping is then:

$$80.4 + 80 = 160.4 \text{ ft}$$

The flow rate through the collectors is adjusted to the recommended flow rate of 0.02 gal/ft^2_c-min. The flow rate for the four collectors is then:

$$0.02 \text{ gal/ft}^2_c\text{-min} \times 31.2 \text{ ft}^2/\text{coll} \times 4 \text{ coll} = 2.5 \text{ gal/min},$$
$$\text{or about 0.6 gal/min per collector}$$

From Fig. 13-28, the piping pressure loss for a flow rate of 2.5 gal/min is approximately 2.5 ft of head per 100 ft of ¾-in. pipe. For 160.4 ft of pipe, the pressure loss will be:

$$\text{Piping pressure loss} = 2.5 \text{ ft water/100 ft pipe} \times 160.4 \text{ ft pipe}$$
$$= 4 \text{ ft water}$$

The pressure drop in the collectors themselves can be estimated from Fig. 12-10. For a flow rate of 0.6 gal/min per collector, the pressure drop is approximately 0.2 ft of water.

The total collector loop pressure drop is then 4 ft + 0.2 ft = 4.2 ft of water, head. The pump size can now be obtained from Fig. 13-29. A UM25–18SU will provide 5.5 gpm with a head pressure of 4.2 ft. This pump will be more than adequate to provide a circulation of 4 gpm.

CONCLUSION

Most of the commonly used systems of heating domestic hot water with solar energy have been presented in this chapter. Passive systems requiring no outside energy, such as the integral-collector-storage (ICS) system and the thermosiphon system, were discussed. Active systems of both open-loop and closed-

SOLAR HEATING OF DOMESTIC HOT WATER

loop configurations were thoroughly described, and sample sizing and design problems were presented.

Heating water is perhaps the most universal use of solar energy, and in the next chapter two more applications of solar-heated hot-water technology will be considered—heating swimming pools and spas or hot tubs.

PROBLEMS

1. Make a preliminary estimate of the domestic hot-water usage in a 10-unit apartment house using the HUD standards in Table 13-1. Each unit has two bedrooms.

2. A four-bedroom house has two baths, a dishwasher, and a clothes washer. There are five occupants living in the house. (a) Estimate the monthly hot-water usage from the data in Table 13-2. (b) If the water main temperature is 50°F and the hot-water delivery temperature is 130° F, calculate the monthly hot-water heating load Q_{WM}, using Eq. (13-1).

3. For a family of four with a clothes washer, two baths, and a dishwasher, estimate the average monthly and yearly hot-water loads from Fig. 13-21. Assume that the cold-water supply temperature is 50°F and the hot-water system is set to deliver 140°F water.

4. A 30-unit apartment house is planned to have a central solar hot-water system. Estimate the average monthly hot-water load Q_{WM} if the water main temperature is 50°F and the delivered water temperature is 130°F. Use Table 13-1.

5. The hot-water heating load for a commercial building is estimated to be 2.5 $\times$ 10^6 Btu/month. It is desired that the solar heat fraction SHF_{DHW} be 0.75. Estimate the solar collector area required based on the collector performance data in Table 12-1.

6. An apartment house has an estimated hot-water load of 12.5 $\times$ 10^6 Btu/month. Determine the collector area using the performance data in Table 12-1 if a solar heat fraction of 0.70 is required.

7. A three-bedroom house is located at 40° N Lat in Kansas City, Missouri. The roof where the collectors will be mounted has a 30° slope. Determine the required collector area for December 21 and June 21 to provide a solar heat fraction of 0.70. Assume that the water main temperature is 55°F and the hot-water delivery temperature is 135°F. Use Table 13-1 to estimate hot-water usage. Use the medium-temperature efficiency rating for the WSD7-CR-31 collector shown in Fig. 12-8.

8. A solar hot-water system will use two solar collectors with parallel flow and reverse return. The distance from the collector location to the storage tank location is estimated to be 20 ft. Pipe diameter is ½ in. Determine the pump horsepower that should be used with the system.

9. Four solar collectors of the type shown in Fig. 12-8 will be used on a domestic hot-water system. The array is parallel-connected with reverse return. The distance from the collectors to the hot-water storage tank is 40 ft. Determine a combination of pipe diameter and pump horsepower that would be suitable.

10. Determine the fluid temperature rise for a 24 ft² solar collector with a flow rate of 0.015 gal/ft²_c-min. Assume that the insolation is 210 Btu/hr-ft² and the collector efficiency is 0.54.

11. A fluid temperature rise in a collector loop is specified to be 20F° at midday. The collector efficiency is 0.48; the insolation is 225 Btu/hr-ft². Determine the fluid flow rate to give the 20 F° temperature rise.

12. A solar system has a collector loop with 75 ft of pipe, 10 90° elbows, a check valve, and two ball valves. The flow rate through the loop will be 2 gal/min. Determine the pressure drop due to the piping and fittings for ½-in. copper tubing.

13.   An apartment house solar DHW system will have 200 ft² of solar collectors, parallel-connected with reverse return. The piping from the collector array to the solar storage will be 75 ft in one direction. There are 24 90° elbows used on the collector loop, two ball valves, and a check valve. The flow rate through the collectors will be 0.021 gal/ft²-min. Determine the pressure loss for the collector loop if ¾-in. pipe is used.

14.   A collector loop has a pressure loss of 10 ft of water, head. The flow rate is to be 5 gal/min. What size pump should be installed on this system? Use the data in Fig. 13-29.

15.   A collector loop has a total of 65 ft of ¾-in. piping, 18 90° elbows, two valves, and a check valve. The flow rate is 5 gal/min. There are four solar collectors, in a parallel-connected array, with reverse return. Determine the pressure drop for the collector loop and choose a pump for the system, using Fig. 13-29. The collector pressure loss can be determined from the pressure-drop chart of Fig. 12-10.

16.   A drainback solar hot-water system has the top of the collectors located 16 ft above the water level in the drainback reservoir. The length of piping from the pump to the collectors is 45 ft of ¾-in. copper tube, with 12 90° elbows. There are three collectors in parallel flow with reverse return, and the desired flow rate is 2 gal/min. Determine the system head requirement and choose a pump for the system. Use Figs. 13-29 and 12-10. (The system head is composed of the pressure drop in the collector supply piping, the collector pressure loss, and the elevation head pressure resulting from the height of the collectors.)

Solar Heating of Swimming Pools and Spas

Solar energy provides an ideal source of heat for swimming pools and spas or hot tubs. The typical temperatures for swimming pools are in the range of 75°F to 85°F, while spas are maintained at temperatures ranging from 95°F to 110°F. Both spas and swimming pools are open bodies of water exposed to ambient conditions when in use, unless they are enclosed in a building, and thermal losses from open bodies of heated water can be very high. A spa or swimming pool loses heat from its surface by thermal radiation, convection, and evaporation, and there is also additional heat loss by conduction through the sides and bottom. With pools that are below grade, conduction losses into the surrounding earth are minimal, and it is the surface heat losses that are of real consequence.

Figure 14-1 shows plots of the calculated average daily heat loss for an open swimming pool located in Santa Maria, California, with the contributions to the heat loss by radiation, convection, and evaporation indicated. The calculations are from the equations developed in section 14-3 below. Note how both the evaporative heat loss and the convection heat loss dramatically increase when a breeze of even 4 mph is present at the surface of the pool.

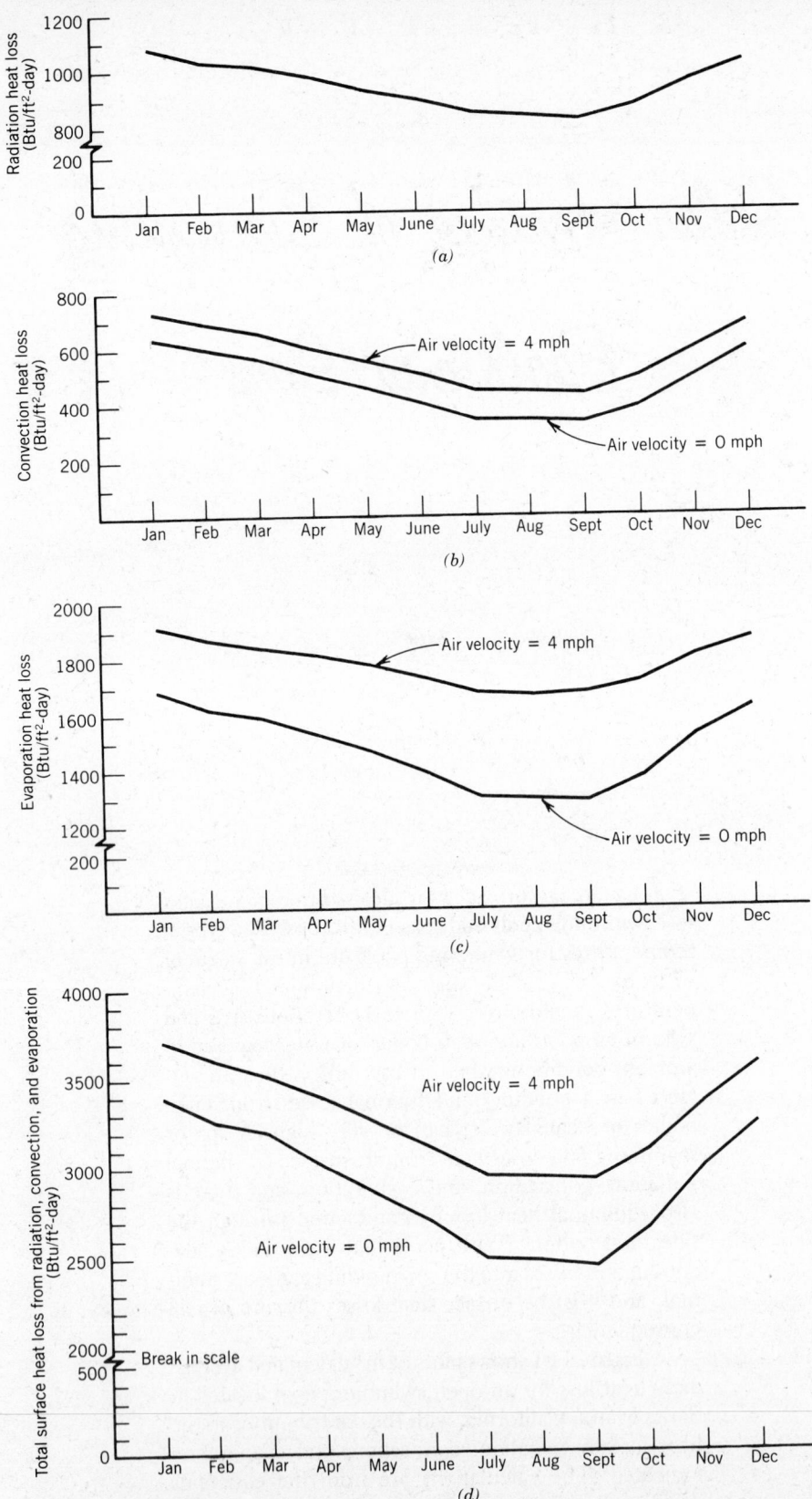

Figure 14-1 Average daily heat losses (Btu/day-ft²) from the surface of an outdoor swimming pool with a water temperature of 80° F, located in Santa Maria, California. (a) Radiation heat loss from the pool surface. (b) Convection loss from the pool surface for an air velocity (wind) of 0 mph and 4 mph. (c) Evaporation heat loss from the pool surface for an air velocity of 0 mph and 4 mph. (d) Total heat loss from the pool surface.

Illustrative Problem 14-1

Estimate the average total daily heat loss from a 500 ft² swimming pool located in Santa Maria, California, in the month of April for still air conditions (no breeze). Pool temperature is 80°F.

Solution

From Fig. 14-1d, the total heat loss for April is indicated as approximately 3000 Btu/ft²-day for a wind velocity of 0 mph. The total heat loss for the pool is then calculated as:

$$\text{Total daily heat loss} = 3000 \text{ Btu/ft}^2\text{-day} \times 500 \text{ ft}^2$$
$$= 1.5 \times 10^6 \text{ Btu/day} \qquad \textit{Ans.}$$

A spa or hot tub is operated at higher temperatures than a swimming pool, so the heat losses per square foot are much larger, as indicated in Fig. 14-2.

14-1 Solar Collectors for Use with Pools and Spas

There are two basic categories of pools and spas: outdoor and indoor. The outdoor category will be considered first.

The outdoor swimming season varies with the location and local climate, usually lasting from early June to mid-September in the northern United States. In the central tier of states, the season may extend from April through October, while in the South and Southwest it may extend through the entire year. During the swimming season, the daytime air temperature is seldom lower than 15 to 20 F° below the pool water temperature.

The performance of unglazed low-temperature collectors was discussed in section 12-2 and found to be satisfactory when the operating temperature is limited to the range of 10 to 20 F° above ambient. Since outdoor pools are rarely heated to temperatures greater than 20 F° above the average daytime ambient air temperature, unglazed low-temperature collectors are quite effective in supplying solar heat to pools during the normal swimming season.

Pool Collectors. Low-temperature pool collectors are typically manufactured from either plastic or metal. The plastic collectors are often made from extruded plastic, as shown in Fig. 14-3. These collectors usually have an integral

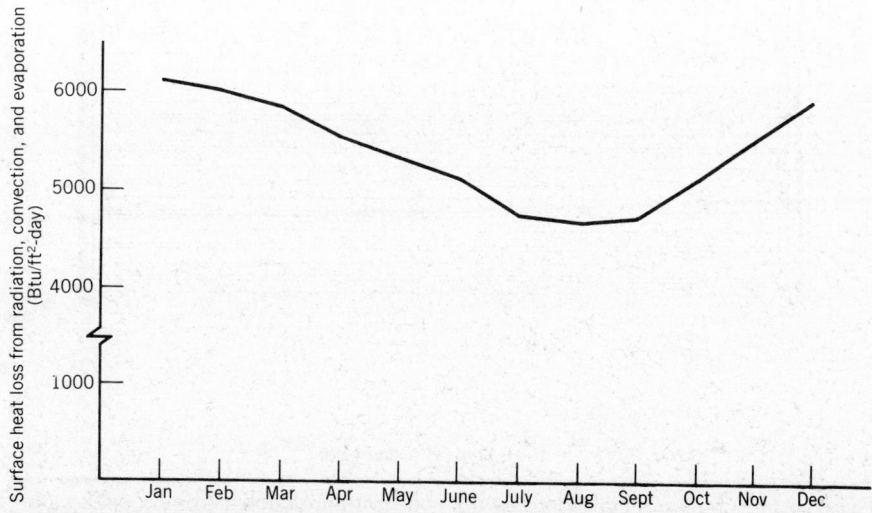

Figure 14-2 The average daily heat losses (Btu/day-ft²) from the surface of an exposed spa located in Santa Barbara, California. The spa water temperature was 102° F, and the wind velocity was 0 mph.

SOLAR HEATING OF SWIMMING POOLS AND SPAS

HCP® Absorber Panel

Manifold

Substrate

Frame

Back Brace

Figure 14-3 A typical plastic solar collector for swimming pool application. This collector has a molded header at the top and bottom. (Sealed Air Corporation.)

header at the top and bottom so that rubber couplings can be used to make interpanel connections. The plastic panels are held to the roof or mounting structure by straps that allow the collectors to expand and contract with temperature change. The length of a ten-foot plastic panel may change by more than an inch during the daily cycle of heating and cooling, and if the panels were rigidly fastened at the top and the bottom they would buckle as they heat up.

Metallic pool collectors usually have a parallel-flow pattern of ⅜- or ½-in. tubing, with headers at top and bottom. Either copper fins or aluminum fins are attached to the parallel-flow tubes. Figure 14-4 illustrates a common construction of metallic pool collectors. Copper waterways are used in metallic collectors because copper is more resistant to corrosion than most metals. Pools have a complex chemistry, which is the result of adding different chemicals such as chlorine, acids, and bases. The presence of the pool chemicals can have detrimental effects even on copper waterways if proper precautions are not observed.

Water-flow velocities in the collector waterways should be kept below 6 ft/sec. This is necessary because at greater water velocities the copper oxide coating that forms on the inside of the copper waterways begins to erode, resulting in reduced collector life and probable copper oxide stains on the bottom of the pool.

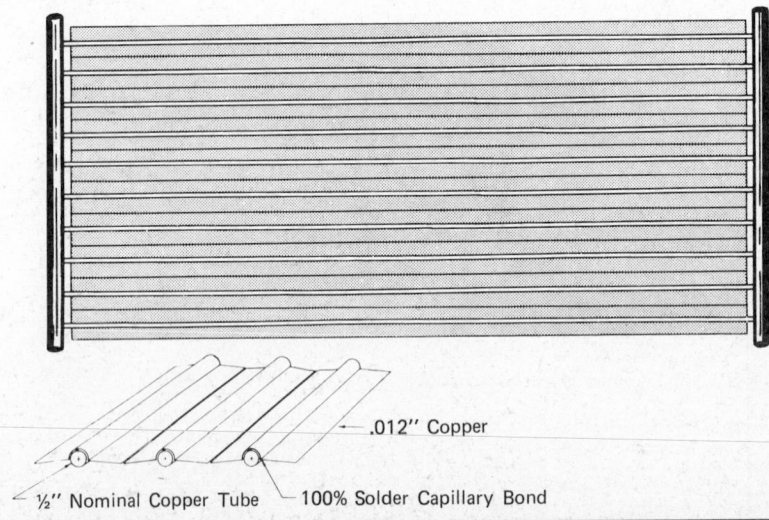

.012″ Copper

½″ Nominal Copper Tube — 100% Solder Capillary Bond

Figure 14-4 A bare metal pool-heating collector constructed from copper waterways and copper fins. These collectors come with a 1½-in. header, so they can be arranged in banks or arrays with couplings at the header connections. (Western Solar Development, Inc.)

Proper pH and chlorine levels must be consistently maintained in the pool. If the pool pH is allowed to become acidic, serious corrosion may occur in copper collectors. Most pool collector warranties have disclaimers against collector failures from corrosion caused by improper pool pH. The pool water in the collector loop should never be allowed to reach temperatures above 110°F, or the chlorine will begin to outgas from the water.

Metals other than copper and certain special grades of stainless steel are not recommended for pool systems where the pool water is circulated directly through the collectors, because excessive corrosion tends to occur from chloride attack.

Plastic collectors have the advantage of low initial cost, and they have good corrosion resistance to the chemicals present in pool water. However, they are more easily damaged than metal collectors, and they will gradually be weakened by the influence of the elements and by ultraviolet radiation, even though they are fabricated from ultraviolet-stabilized plastics. Plastic collectors will normally provide good performance over a useful lifetime of about 10 to 15 years. Warranties are usually for a 5- or 10-year period. Metal collectors, on the other hand, can be expected to have useful lifetimes in excess of 20 years provided that corrosion caused by improper pH balance of the pool is not allowed to occur.

Spa Collectors.   Spas operate in a much higher temperature range than pools. At a typical temperature of 104 to 105°F, they are 35 to 45 F° above ambient daytime temperatures in the spring and fall and perhaps 20 F° above the average summer daytime temperature in most locations. Low-temperature unglazed collectors might be useful for heating spas during the summer months in some locations, but spas are often used all year round, so medium-temperature collectors with glazing and insulation are usually specified. These are the same collectors as those used for heating domestic hot water.

SOLAR ENERGY SYSTEMS FOR OUTDOOR POOLS

14-2 Sizing and Design of Solar Energy Systems for Outdoor Pools

The major design considerations for swimming pool solar-heating systems are (1) selecting the type of solar collector, (2) determining the required collector area, (3) sizing the collector loop piping, and (4) selecting appropriate valves and controls.

Rules of Thumb for Sizing the Collector Area (Outdoor Pool).

> When solar collectors are to provide pool heating only during the summer swimming season, the solar collector area required ranges from one half to two thirds of the swimming pool surface area. The *minimum collector area* for south-facing collectors tilted at an angle equal to the site latitude minus 10° is 50 percent of the pool surface area. If the collector area is much less than 50 percent of the pool surface area, the solar heat contribution to the pool may be small when compared with the pool-heating requirements.

As in the case of domestic hot-water heaters, both orientation and tilt angle of pool collectors are very important for system performance. Although collectors installed at almost any inclination will collect some solar radiation and produce some useful heat, the most efficient collector performance occurs when the collector is perpendicular to the average daily position of the sun, which changes with both the season of the year and the latitude of a specific location.

Table 14-1 Percentage Increase in
Collector Area Needed to Compensate for
Collector Tilts or Slopes That Vary from
Summer or Winter Optimums
(outdoor pools)

Slope Misalignment	Extra Solar Collector Area Needed
20°	6%
30°	13%
40°	23%

Source: Manufacturers' literature.

As a general rule, the recommended collector tilt angle is:

1. Summer (May through September)—latitude of the site minus 10°.
2. Winter (October through April)—latitude of the site plus 10°.

It is usually not practical to change the collector tilt angle on a seasonal basis, and consequently the collectors should be installed for that season when pool heating is most needed. In latitudes where outdoor pools can be used all year, collectors should be tilted for the winter season (latitude plus 10°). In northern areas where outdoor pool use is limited to the summer, the summer tilt angle would be used (latitude minus 10°). Regardless of latitude, the final decision on collector tilt angle may be influenced by the owner or determined by the slope of the roof on which the collectors will be mounted. Table 14-1 gives some corrections for collector area as a function of variations in tilt angle from the optimums listed above.

Illustrative Problem 14-2

Determine the minimum recommended collector area for a 400 ft² swimming pool located at 38° N Lat. Assume a summer swimming season only. The roof where the collectors will be located has a south-facing slope of 45°. Use the rules of thumb stated above.

Solution

The minimum recommended collector area is 50 percent of the pool surface area, or 200 ft², if tilt angle is optimum.

For the summer swimming season, the recommended collector tilt angle is latitude minus 10° or, in this case,

$$\text{Tilt} = 38° - 10° = 28°$$

The existing roof slope is 45°, so there is a slope misalignment of 17°. From Table 14-1 for a 20° slope misalignment, the collector area should be increased by approximately 6 percent. The collector area should be increased by 12 ft², or a minimum total of 212 ft² of solar collector should be used.   *Ans.*

More Accurate Methods of Collector Area Sizing.   Rules of thumb give the solar designer or installer basic guidelines for estimation. A more accurate sizing of the pool collector array can be achieved if the actual pool energy requirements are known. For existing pools, the most accurate method of determining the energy requirements is to review the fuel usage records for the pool heater over a period of several years. The monthly pool-heating requirement can be expressed as follows:

$$Q_{pool} = E_{heater} \times Q_{fuel} \tag{14-1}$$

where
Q_{pool} = the heat delivered to the pool per month
E_{heater} = the efficiency rating of the pool heater
Q_{fuel} = the Btu equivalent of the fuel used by the pool heater per month

Illustrative Problem 14-3

The fuel used by a pool heater for the month of April is metered as 240 therms $(24 \times 10^6 \text{ Btu})$.[1] Determine the heat delivered to the pool in April if the pool heater has an efficiency rating of 70 percent.

Solution

$$Q_{pool} = E_{heater} \times Q_{fuel}$$
$$= 0.70 \times 24 \times 10^6 \text{ Btu}$$
$$= 16.8 \times 10^6 \text{ Btu} \qquad \textit{Ans.}$$

When the monthly heat load for the pool has been established, the collector area can be quite accurately estimated using the following equation, a variant of Eq. (13-8):

$$A_c = \frac{Q_{pool}}{\overline{\eta} \overline{I}_D N} = \frac{Q_{pool}}{\overline{\eta} I P N} \qquad (14\text{-}2)$$

where
A_c = the area of the solar collector array, ft^2
Q_{pool} = the daily (or monthly) (as stated) heat load of the pool
$\overline{\eta}$ = the average collector efficiency for the month
$\overline{I}_D$ = the average daily solar insolation
I = the clear-day insolation, from Table 4-E
P = the percent possible sunshine for the given location, from Table 4-C
N = the number of days in the month.

Illustrative Problem 14-4

The pool in Illustrative Problem 14-3 has an average monthly load for the swimming season of 16.8×10^6 Btu/month. Determine the collector area required if the clear-day insolation is 2000 Btu/ft^2-day, the percent possible sunshine for the location (in April) is 60 percent, and the average efficiency of the solar collectors is 0.40.

Solution

Eq. (14-2) can be used to estimate the collector area required:

$$A_c = \frac{Q_{pool\,month}}{\overline{\eta} I P N}$$
$$= \frac{16.8 \times 10^6 \text{ Btu/month}}{(0.40)(2000 \text{ Btu/ft}^2\text{-day})(0.60)(30 \text{ days/month})}$$
$$= 1167 \text{ ft}^2 \qquad \textit{Ans.}$$

[1] 1 therm = 10^5 Btu.

14-3 Engineering Methods in Pool-Heating Load Estimation

When it is not possible to estimate the pool-heating load from the fuel usage of the pool heater because the pool fuel use information cannot be separated from other fuel usage, or because the pool does not have a heater, or because it is a new pool to be designed, then the pool-heating load can be estimated on an average daily basis for each month, as follows.

The *daily energy balance* for an open pool can be expressed as follows:

$$Q_{SRD} - Q_{RD} - Q_{CD} - Q_{ED} - Q_{GD} + Q_{Heat} = 0$$

or,

$$Q_{Heat} = Q_{RD} + Q_{CD} + Q_{ED} + Q_{GD} - Q_{SRD} \qquad (14\text{-}3)$$

where the terms are defined as follows and illustrated in Fig. 14-5:

Q_{SRD} = the daily solar energy absorbed per square foot of pool from radiation, direct and diffuse, Btu/ft²-day

Q_{RD} = the daily energy lost from the pool by radiation, Btu/ft²-day

Q_{CD} = the daily energy lost from the pool by convection, Btu/ft²-day

Q_{ED} = the daily energy lost from the pool by evaporation of water, Btu/ft²-day

Q_{GD} = the daily energy lost from the pool by conduction into the ground surrounding the pool, Btu/ft²-day

Q_{Heat} = the pool heat load, the net total daily energy lost per square foot of pool, or the daily energy per square foot needed to maintain the pool at the design temperature, T_w

The ground energy loss, Q_{GD} usually cannot be economically prevented because of difficulty in insulating the sides of a pool. Fortunately, these losses tend to be small because of the poor heat transfer characteristics of the ground and the relatively small temperature difference between the ground and the pool water. We therefore neglect Q_{GD} in the rest of the calculation.

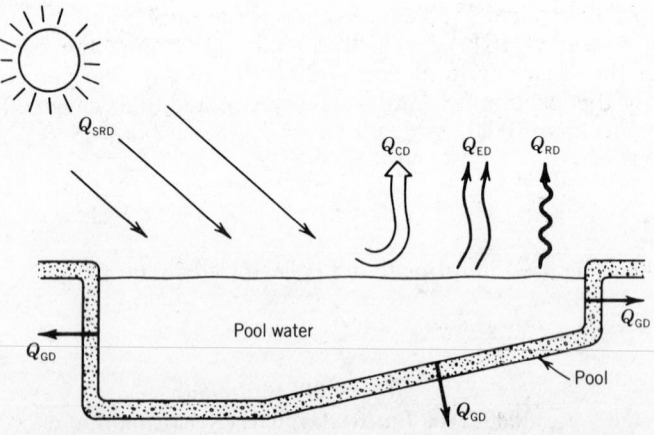

Figure 14-5 Schematic representation of the energy gains and losses for an outdoor swimming pool. Note that losses from radiation, convection, conduction, and evaporation are all indicated. The only gain in this case is from solar radiation.

The *daily radiation loss* Q_{RD} can be calculated as:[2]

$$Q_{RD} = 3.95 \times h_R \times [T_w - (T_a - 20)] \qquad (14\text{-}4)$$

where

Q_{RD} = the rate of heat loss by radiation, Btu/ft²-day

h_R = the radiation heat transfer coefficient, Btu/ft²-day-F°, which is obtained from Fig. 14-6 for values of T_w and T_a

T_w = the average (design) pool water temperature, °F

T_a = the average ambient air temperature, °F

The term $(T_a - 20)$ provides an approximation of the sky temperature.

The *daily convective loss* Q_{CD} can be calculated as:

$$Q_{CD} = (24 \text{ hr/day}) \times h_C \times (T_w - T_a) \qquad (14\text{-}5)$$

where

Q_{CD} = the rate of heat loss by convection, Btu/ft²-day

h_C = the convective heat transfer coefficient, Btu/hr-ft²-F°, which can be obtained from Fig. 10-6, as a function of wind velocity

T_w = the average (design) pool water temperature, as before

T_a = the average ambient air temperature

The *daily evaporative loss* Q_{ED} can be calculated as:

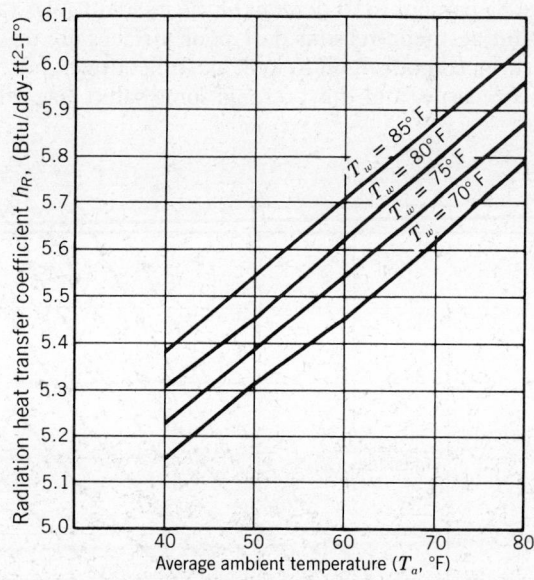

Figure 14-6 The radiation heat-transfer coefficient h_R for outdoor swimming pools, plotted as a function of the pool water temperature and the ambient air temperature. Values of h_R are plotted from the relation:

$$h_R = 4 \times 10^{-8} \times \left[\frac{T_w + (T_a - 20)}{2} \right]^3 \text{ Btu/day-ft²-F°}$$

[2] The general method outlined here, along with Eqs. (14-4), (14-5), and (14-6), is adapted from J. T. Czarnechi, "A Method of Heating Swimming Pools by Solar Energy," *Solar Energy*, vol. 7, no. 1, 1963, pp. 3–7. The empirical constants 3.95 and 4.8×10^3 hr-F°/day-psi are also from Czarnechi.

$$Q_{ED} = 4.8 \times 10^3 \, \frac{\text{hr-F}^\circ}{\text{day-psi}} \times h_C \times (P_w - P_a) \qquad (14\text{-}6)$$

where

Q_{ED} = the rate of heat loss by convection, Btu/ft²-day
h_C = the convective heat transfer coefficient
P_w = the vapor pressure (psi) of the swimming pool water at temperature T_w
P_a = the water vapor pressure (psi) in the air

Values for P_w can be obtained from Fig. 14-7 using the 100 percent relative humidity line and the pool water temperature T_w for the temperature. Values for P_a can also be obtained from Fig. 14-7 by using the mean maximum air temperature for the month and the mean minimum relative humidity for the month.

The *solar radiation absorbed* by the pool during the day Q_{SRD} will typically be at least 75 percent of the incident solar radiation, but not more than about 85 percent. These values assume a well-filtered pool with an average depth of 5 ft, and with the sides and bottom of the pool finished with smooth white plaster. Approximately 5 percent of the incident solar energy is reflected from the surface of the water and provides no heating. A significant amount of the solar radiation, particularly the short wavelength (blue light), is able to penetrate to the bottom of the pool and is reflected back up through the water and out of the pool. The white plaster typically used in pool construction is a poor absorber of short-wavelength radiation. Enough of the short wavelengths are able to pass through the water, reflect off the pool surfaces, and pass back out of the pool to give the illusion that the water in the pool is blue. If the pool had a dark absorbing surface, 85 percent to 95 percent of the incident solar radiation would be absorbed, but for aesthetic reasons dark pool surfaces are not very popular. Also, since dark-surfaced pools tend to look like reflecting ponds, objects in the pool cannot be seen easily, and this presents some safety hazards.

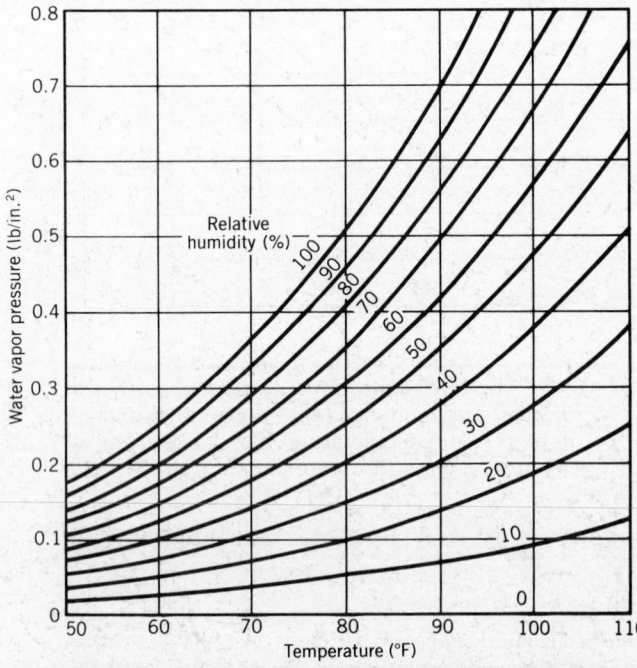

Figure 14-7 Relationships among water vapor pressure in the air, air dry-bulb temperature, and relative humidity.

Illustrative Problem 14-5

A pool is located at 32° N Lat. The desired pool temperature is 75°F, the average spring daily air temperature is 64°F, and the average minimum relative humidity is 50 percent. The average daily maximum temperature is 75°F. The collectors will be mounted on an existing south-facing roof at a tilt of 22°. The percent possible sunshine for the location is 55 percent. The average daily efficiency for the solar collectors is estimated at 0.67. The pool has a wall around it and wind velocity is estimated to be zero mph. The pool has a surface area of 450 ft². Size the solar collector array to provide 100 percent of the pool heating for the month of April.

Solution

The approach to solving this problem will be to calculate the pool heat load using Eqs. (14-3) through (14-6), and then to estimate the solar collector area using Eq. (14-2).

The radiation loss Q_{RD} is calculated using Eq. (14-4):

$$Q_{RD} = 3.95 \times h_R \times [T_w - (T_a - 20)]$$
$$= 3.95 \times h_R [75°F - (64 - 20)°F]$$
$$= 3.95 \times h_R [31°F]$$

From Fig. 14-6, the value of h_R is found as approximately 5.62 Btu/ft²-day-F°. Substituting values,

$$Q_{RD} = 3.95 (5.62 \text{ Btu/ft}^2\text{-day-F°})(31°F)$$
$$= 688 \text{ Btu/ft}^2\text{-day}$$

The convective loss Q_{CD} is calculated using Eq. (14-5):

$$Q_{CD} = (24 \text{ hr/day}) \times h_C \times (T_w - T_a)$$

From Fig. 10-6 for a wind velocity of zero mph, h_C is found as approximately 0.8 Btu/hr-ft²-F°. Substituting,

$$Q_{CD} = 24 \text{ hr/day} \times 0.8 \text{ Btu/hr-ft}^2\text{-F°} \times (75°F - 64°F)$$
$$= 211 \text{ Btu/ft}^2\text{-day}$$

The evaporative loss Q_{ED} is calculated using Eq. (14-6):

$$Q_{ED} = 4.8 \times 10^3 \frac{\text{hr-F°}}{\text{day-psi}} \times h_C \times (P_w - P_a)$$

From Fig. 14-7, P_w is found to be approximately 0.44 psi for a relative humidity of 100 percent and a temperature of 75°F. The value of P_a is found to be approximately 0.22 psi for a relative humidity of 50 percent and an average maximum temperature of 75°F. Substituting,

$$Q_{ED} = 4.8 \times 10^3 \frac{\text{hr-F°}}{\text{day-psi}} \times (0.8 \text{ Btu/hr-ft}^2\text{-F°})$$
$$\times (0.44 \text{ psi} - 0.22 \text{ psi})$$
$$= 845 \text{ Btu/ft}^2\text{-day}$$

The daily solar radiation absorbed by the pool can be estimated as:

$$Q_{SRD} = 0.75 \times \bar{I}_{horiz}$$

when the pool receives full sunshine during the day.

The clear-day solar radiation on a horizontal surface at the pool location is found from Table 4-E as $I_{horiz} = 2390$ Btu/ft²-day.

The value of $\bar{I}_{horiz}$ is then $0.55 \times 2390 = 1315$ Btu/ft²-day.

$$Q_{SRD} = 0.75 \times 1315 \text{ Btu/ft}^2\text{-day}$$
$$= 985 \text{ Btu/ft}^2\text{-day}$$

The pool heat load is then calculated using Eq. (14-3):

$$Q_{heat} = Q_{RD} + Q_{CD} + Q_{ED} + Q_{GD} - Q_{SRD}$$

Q_{GD} will be assumed to be negligible compared to the other losses and will be assigned a value of zero.

$$Q_{heat} = 688 \text{ Btu/ft}^2\text{-day} + 211 \text{ Btu/ft}^2\text{-day} +$$
$$845 \text{ Btu/ft}^2\text{-day} - 985 \text{ Btu/ft}^2\text{-day}$$
$$= 759 \text{ Btu/ft}^2\text{-day}$$

and the total pool heat requirement,

$$Q_{pool} = 450 \text{ ft}^2 \times 759 \text{ Btu/ft}^2\text{-day}$$
$$= 341,500 \text{ Btu/day}$$

The collector area is estimated using Eq. (14-2):

$$A_c = \frac{Q_{pool}}{\eta IPN}$$

From Table 4-E, I is found to be 2445 Btu/ft²-day for a surface tilt of 22° at 32° N Lat. N is taken as 1.0 since the pool load is in Btu/day. Substituting,

$$A_c = \frac{341,500 \text{ Btu/day}}{(0.67)(2445 \text{ Btu/ft}^2\text{-day})(0.55)(1.0)}$$
$$= 380 \text{ ft}^2 \qquad\qquad Ans.$$

14-4   Pool Covers as Energy-Saving Devices

A large percentage of outdoor pool heat loss is caused by evaporation. Figure 14-1 indicates the relative magnitude of the evaporative, convective, and radiative heat losses, and shows that during breezy conditions evaporative and convective losses can increase significantly. The use of a pool cover when the pool is not being used will reduce evaporative losses to a minimum. Pool covers also reduce convective losses and, to some extent, radiative losses as well. It is impossible to predict the total reduction in the pool-heating load resulting from the use of a pool cover, because the effectiveness of the cover is dependent upon the hours the pool is in use and on the consistency with which the pool owner uses the cover. In general, the consistent use of a good pool cover will reduce the yearly pool-heating load by 40 to 50 percent. A typical pool cover installation is shown in Fig. 14-8.

Some guidelines for estimating the effect of a semi-transparent floating pool cover with air cells are given here:

1. The hourly evaporative loss can be assumed to be reduced by 80 percent for each hour the cover is on the pool.
2. The hourly convective loss will be reduced by 40 percent for each hour the cover is on the pool.
3. Radiative losses are not appreciably reduced, since the transparent plastic will transmit a good deal of the infrared radiation from the warm pool.
4. If the cover is on the pool during daytime hours, the pool absorptance for solar radiation should be reduced by 5 percent to compensate for the reduced transmittance of the cover.

Illustrative Problem 14-6

Assume that the pool in Illustrative Problem 14-5 has a semi-transparent floating pool cover with air cells, in place for 12 hours during the night and off for 12 hours during the day. Recalculate the pool-heating load taking the cover into account, and redetermine the solar collector area.

Figure 14-8 A semi-transparent floating pool cover reduces the heat losses from a pool. The roller shown in the background is used to remove the cover when the pool is in use. As a general guide, a pool with a good cover will stay about 10 F° warmer during the summer swimming season than an uncovered pool in the same location. (Santa Barbara Solar Energy Systems.)

Solution

In Illustrative Problem 14-5, the following pool losses were obtained:

$$Q_{RD} = 688 \text{ Btu/ft}^2\text{-day}$$
$$Q_{CD} = 211 \text{ Btu/ft}^2\text{-day}$$
$$Q_{ED} = 845 \text{ Btu/ft}^2\text{-day}$$

The average hourly losses are obtained by dividing the above values by 24 hr/day:

$$Q_R = 29 \text{ Btu/ft}^2\text{-hr}$$
$$Q_C = 8.8 \text{ Btu/ft}^2\text{-hr}$$
$$Q_E = 35 \text{ Btu/ft}^2\text{-hr}$$

The above heat losses must be adjusted for the use of the pool cover—on for 12 hours and off for 12 hours. The hourly convective loss will be assumed to be reduced by 40 percent and the evaporative losses by 80 percent.

The daily losses will be calculated as:

$$\begin{aligned} Q_{CD} &= 8.8 \text{ Btu/ft}^2\text{-hr} \times 12 \text{ hr/day} + 8.8 \text{ Btu/ft}^2\text{-hr} \\ &\quad \times (1 - 0.40) \times 12 \text{ hr/day} \\ &= 106 \text{ Btu/ft}^2\text{-day} + 64 \text{ Btu/ft}^2\text{-day} \\ &= 170 \text{ Btu/ft}^2\text{-day} \\ Q_{ED} &= 35 \text{ Btu/ft}^2\text{-hr} \times 12 \text{ hr/day} + 35 \text{ Btu/ft}^2\text{-hr} \\ &\quad \times (1 - 0.80) \times 12 \text{ hr/day} \\ &= 420 \text{ Btu/ft}^2\text{-day} + 84 \text{ Btu/ft}^2\text{-day} \\ &= 504 \text{ Btu/ft}^2\text{-day} \end{aligned}$$

The radiation loss Q_{RD} is assumed to remain the same since the plastic pool cover is transparent to infrared wavelengths.

The solar radiation absorbed by the pool will remain essentially the same since the pool is uncovered during the daytime for 12 hr.

From Eq. (14-3), assigning Q_{GD} = zero,

$$Q_{heat} = Q_{RD} + Q_{CD} + Q_{ED} + Q_{GD} - Q_{SRD}$$
$$= 688 \text{ Btu/ft}^2\text{-day} + 170 \text{ Btu/ft}^2\text{-day}$$
$$+ 504 \text{ Btu/ft}^2\text{-day} - 985 \text{ Btu/ft}^2\text{-day}$$
$$= 377 \text{ Btu/ft}^2\text{-day}$$

The total heat load for the pool is then:

$$Q_{pool} = 377 \text{ Btu/ft}^2\text{-day} \times 450 \text{ ft}^2$$
$$= 169,600 \text{ Btu/day.}$$

The collector area is estimated using Eq. (14-2), noting that for this case $Q_{heat} = Q_{pool}$:

$$A_c = \frac{Q_{pool}}{\overline{\eta} IPN}$$

I was found as 2445 Btu/ft^2-day (from Table 4-E), and N is taken as one day since the pool load is in Btu per day rather than Btu per month:

$$A_c = \frac{169,600 \text{ Btu/day}}{(0.67)(2445 \text{ Btu/ft}^2\text{-day})(0.55)(1.0)}$$
$$= 188 \text{ ft}^2 \qquad\qquad Ans.$$

SYSTEM DESIGN AND INSTALLATION

14-5   Swimming Pool System Layout

Once the required collector area has been established, the pipe sizing and layout, water flow rates, and system controls must be considered in order to complete the system design. There are two basic system designs for solar-heated swimming pools: the open-loop system, where the pool water is circulated directly through the solar collectors as illustrated in Fig. 14-9, and the closed-loop system, using a heat exchanger to transfer the collected heat from the collector loop to the pool water as illustrated in Fig. 14-10.

Referring to Fig. 14-9, note that the filter pump circulates water through the pool, the filter, and the pool heater. A time clock controls the running time of the filter pump. The solar energy system will ordinarily have an automatic valve that will direct water through the solar collectors when solar energy can be collected. When it is not practical to remove energy from the solar collectors, the automatic valve (actuated by the solar controller) bypasses the solar collectors. The solar controller uses sensors that measure the pool water temperature and either the collector temperature or the intensity of solar radiation. The solar controller determines if energy can be removed from the solar collectors and controls the position of the automatic valve either to direct water through the collectors or to bypass the solar collectors. The check valve after the filter prevents the pool pump from losing its prime. The check valve in the collector return line prevents water from backing up into the collectors when the valve is in the bypass position. The draindown bypass line allows water in the collectors and in the collector supply line to drain back into the pool through the collector return line. The vacuum relief valve on the collector return header allows air to enter the collectors so that water can drain from the collectors and piping; and the air vent allows air to escape from the collectors and piping as the system fills with water.

The components of a closed-loop pool solar system are shown in Fig. 14-10. The filter pump circulates pool water through the filter, heat exchanger, and pool heater. The closed-loop pump circulates the heat-exchange fluid through the collectors where it is heated and through the heat exchanger where

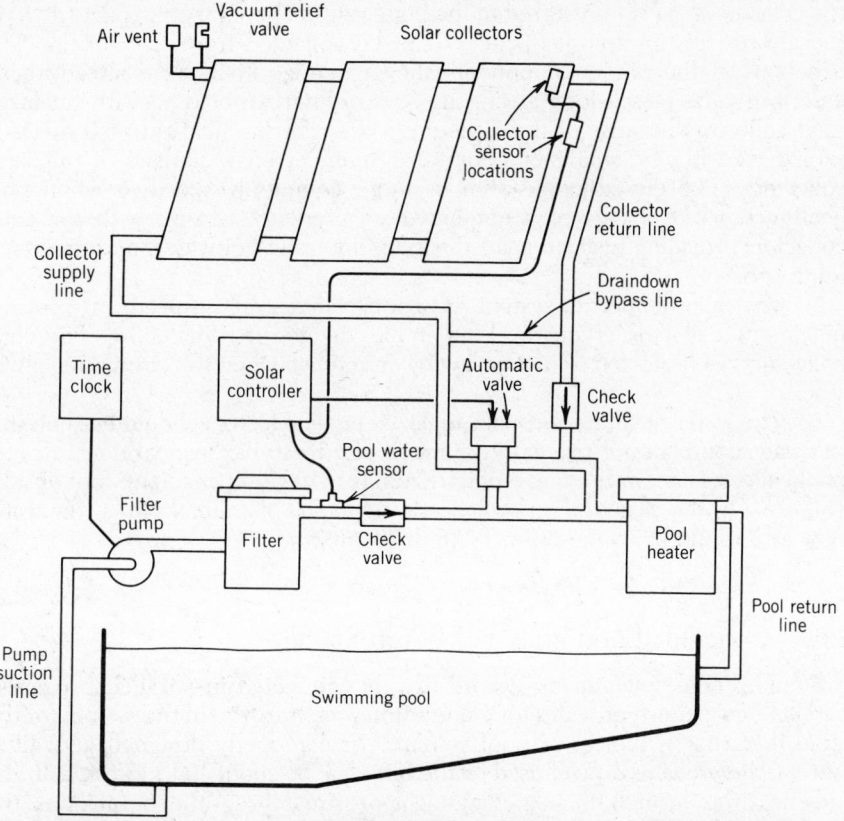

Figure 14-9 Flow diagram of an open-loop solar heating system for a swimming pool. The pool water is pumped directly through the solar collectors, and it serves as the heat-transfer fluid.

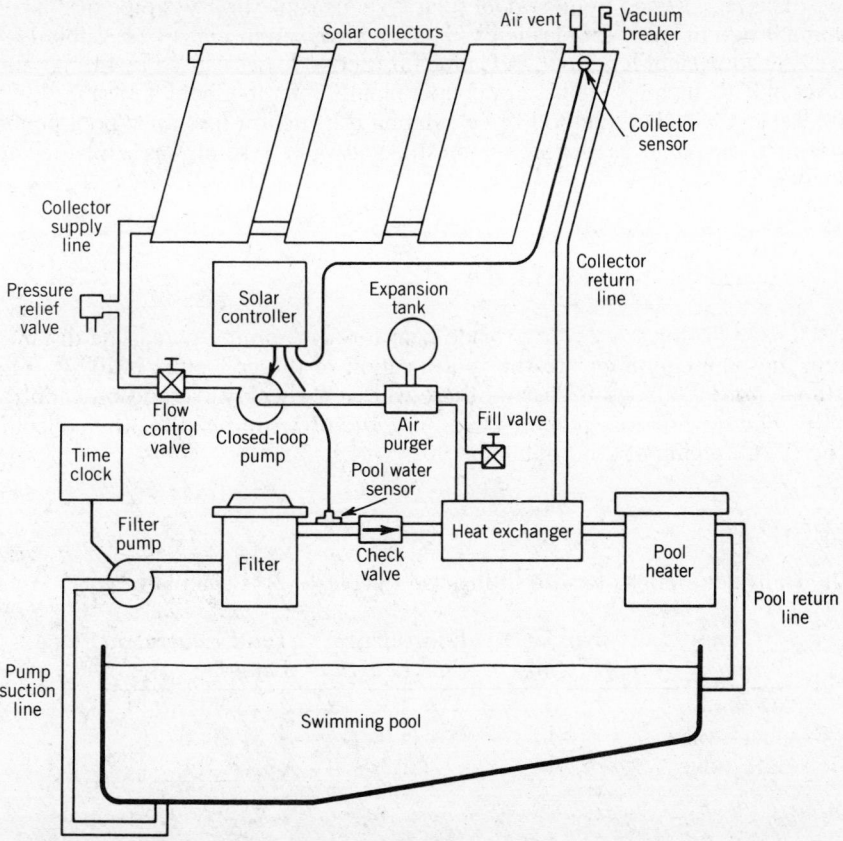

Figure 14-10 In a closed-loop pool solar-heating system, a heat-transfer fluid is circulated between the solar collectors and a heat exchanger. The heat exchanger transfers the collected energy in the heat-transfer fluid to the pool water.

the collected heat is transferred to the pool water. The solar controller turns on the closed-loop pump when energy can be collected from the system. The functions of the other components shown in Fig. 14-10 have already been described. The closed-loop arrangement is ordinarily not used with unglazed solar collectors because of the difficulty of transferring heat through the heat exchanger with the small temperature differences characteristic of unglazed collectors. The closed-loop system is more commonly specified when pool heating is to be a part of a combined solar energy system, with the solar collectors providing heat for space heating, domestic hot water, and the swimming pool.

The open-loop pool system has the advantage of simplicity, it provides maximum collector efficiency, and it has the lower initial cost. The solar collectors are protected from freezing by draindown when the circulating pump stops.

The water flow rate through unglazed pool collectors should be such that the water temperature rise is in the range of 5 to 10 F°. Typical flow rates for pool collectors are in the range of 0.04 gpm to 0.10 gpm per square foot of solar collector. In the case of an installation with 400 ft² of collector area, the water flow rate would be in the range of 16 to 40 gpm.

14-6 Collector Loop Pipe and Pump Sizing

Swimming pool collector arrays are usually connected in parallel, because a parallel-flow pattern provides for a minimum pressure drop in the system for the large flow that is typical of pool systems. In a properly designed pool filter system, the pressure drop caused by the filter will be about 10 to 15 psi, with the pressure drop through the heater adding approximately another 5 psi. Thus, the total pressure drop is in the range of 15 to 20 psi. When a solar-heating system is added to an existing pool filter system (see Fig. 14-9), the pressure head at the pump increases because of the aggregate friction head contributed by the collector-loop piping, the collectors, and the valves and fittings.

The piping used in most pool solar systems is plastic PVC pipe or copper tubing. Friction losses for plastic PVC piping are given in Fig. 14-11. Table 14-2 gives the equivalent length of PVC pipe for friction losses caused by fittings and valves in PVC piping systems. For copper piping, refer to Chapter 13 (Fig. 13-28 and Table 13-4). The method of calculating the friction loss for a pool piping system is the same as that for domestic hot-water systems, as explained in Chapter 13.

Illustrative Problem 14-7

A 400 ft² collector array is to provide heat for a swimming pool. The distance from the pool equipment to the roof location of the collectors is 100 ft. The estimate for 90° elbows is 20, and there will be a check valve and one control valve. The flow rate is 40 gpm. Determine the pressure head for the collector loop (PVC) piping of 2 in. diameter.

Solution

The equivalent length for the fittings (see Table 14-2) is calculated as:

Fitting	No. of Fittings	×	Equivalent Pipe Length	=	Total Equivalent Length
90° elbows	20	×	3.5	=	70 ft
Check valve	1	×	45.0	=	45.0
Gate valve	1	×	1.3	=	1.3
					116.3 ft.

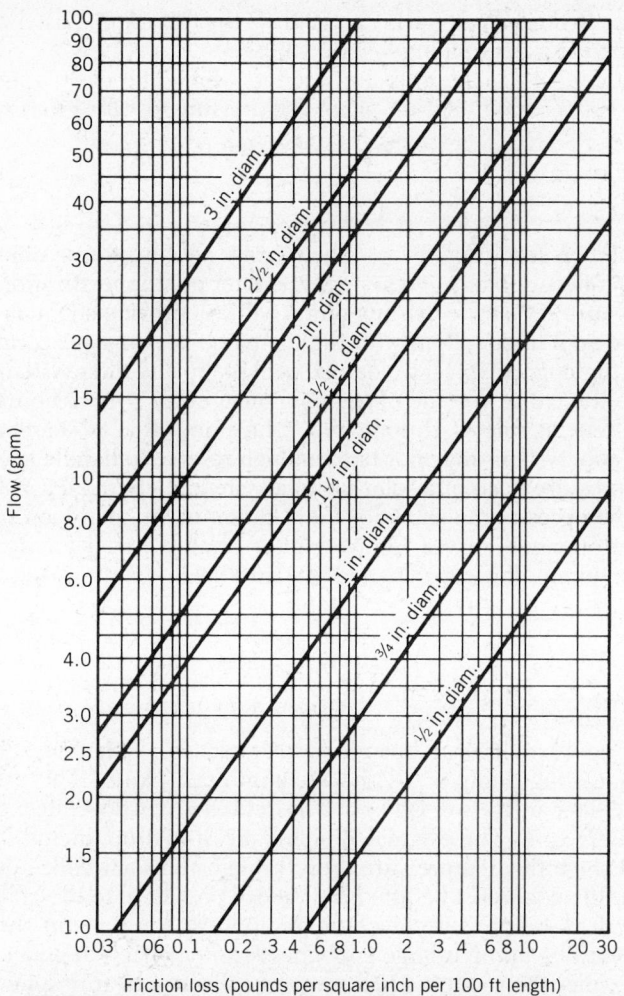

Figure 14-11 Pressure loss caused by flow friction in schedule-40 plastic (PVC) pipe. (Adapted from *ASHRAE Handbook of Fundamentals*, Chapter 26, Fig. 19.)

Table 14-2 Allowance in Equivalent Length of Pipe for Friction Loss in Valves and Fittings (for use with PVC pipe)

Fitting or Valve	Equivalent Feet of Pipe for Various Sizes		
	1 in.	1½ in.	2 in.
45 ° elbow	0.9	1.5	2.0
90° elbow	1.5	2.5	3.5
Tee, run	0.45	0.75	1.0
Tee, branch	2.5	3.5	5.0
Globe valve	25.0	45.0	55.0
Gate valve	0.6	1.0	1.3
Check valve	22.0	22.0	45.0

Source: Uniform Plumbing Code, 1982, p. 136, and ASHRAE Handbook of Fundamentals, 1972, Chapter 26. Adapted with permission.

The distance to the collector array is 100 ft, so the total piping used will be 200 ft and the total equivalent piping length will be 316 ft.

From Fig. 14-11, the pressure loss for a flow rate of 40 gpm is approximately 1.4 psi per 100 ft of 2-in. pipe. The pressure loss for 316 ft will then be:

$$\text{Pressure loss} = 1.4 \text{ psi/100 ft} \times 316 \text{ ft}$$
$$= 4.4 \text{ psi} \qquad\qquad Ans.$$

Retrofit Solar Heat for Pools. When a solar pool-heating system is added to a filter and conventional heater system, the filter pump specifications should be checked to ensure that the existing pump will still provide adequate water flow through the pool filter system with the increased pressure head caused by the solar collector loop. A typical residential pool pump and filter system is designed to circulate the entire volume of the pool once every 8 to 12 hours.

As a general rule of thumb, the ½-hp pump that is standard on small residential pool systems may not have enough reserve to handle adequately the additional pressure head of a solar-heating system. Generally, a ¾-hp or larger pump will be necessary to allow for the addition of an open-loop solar-heating system. Performance curves for pool filter pumps are given in Fig. 14-12, showing the pump flow capacity and pressure head.

Illustrative Problem 14-8

An existing pool is to have an open-loop solar collector system installed (see Fig. 14-9). The existing pump is a ½-hp medium-head pump. The pressure drop through the filter and heater is 15 psi. The desired water-flow rate after installing solar heat is 25 gpm. The calculated piping pressure drop (including fittings) is 4.0 psi and the estimated pressure drop through the solar collectors is 1.0 psi. Water from the pool must be lifted 20 ft above pool level to the collector array. Determine the total pressure head the pump must handle, and check whether or not the existing pump is large enough. Assume no siphon effect in collector return line, since the vacuum breaker opens collectors to atmospheric pressure.

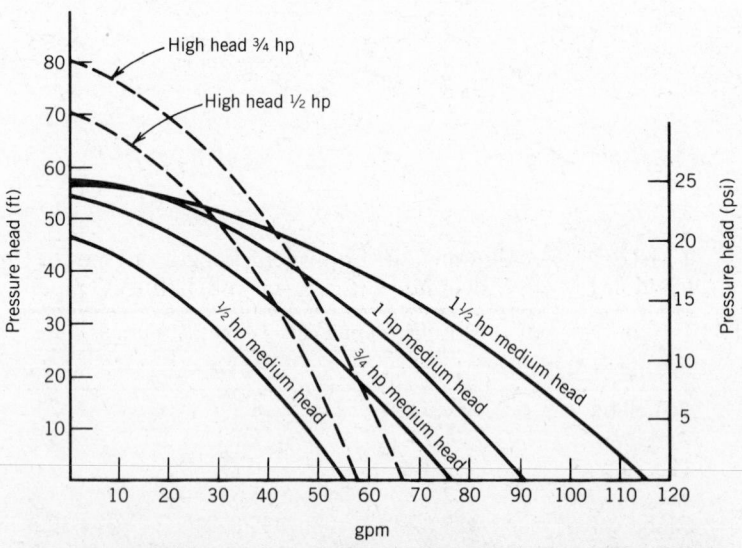

Figure 14-12 Flow capacity and pressure head for six different sizes of swimming pool pumps from one manufacturer. Pressure head is given in psi and in feet of water. (Data from pump manufacturer's literature.)

Solution

The static pressure (elevation head) of lifting water 20 ft is calculated as:

$$p = 20 \text{ ft water} \times 0.434 \text{ psi/ft water}$$
$$= 8.68 \text{ psi}$$

The total head pressure that the pump will have to work against is calculated as:

Filter and heater	15.0 psi
Collector piping	4.0
Collectors	1.0
Static head of 20 ft	8.7
Total head	28.7 psi, or 66.3 ft of head

From Fig. 14-12, the high-head, ½-hp pump will be able to provide only about 10 gpm against a 66-ft head, which is not adequate since the desired flow rate is 25 gpm. A high-head, ¾-hp pump will provide approximately 26 gpm at this total head pressure. The existing ½-hp, medium-head pump will have to be replaced with a ¾-hp high-head pump.   *Ans.*

In the case of a closed-loop pool system, the filter pump will encounter additional head pressure because of the heat exchanger in the filter loop, as was shown in Fig. 14-10. The closed-loop pump will have to cope with a head pressure that is the aggregate of the pressure drop of the collectors and the pressure drop in the closed-loop piping and fittings, but no static (elevation) head since it is a closed circuit. The design of a closed-loop pool system is quite similar to that of a closed-loop domestic hot-water system (Chapter 13), the main difference being that the pool system has a much larger collector area than a residential domestic hot-water system.

14-7   Pool Controls

It is necessary to control the flow of pool water through the collector array so that heat will be collected when solar radiation is available and to stop the flow when insolation is inadequate. This is accomplished with a controller and valve system similar to that shown in Fig. 14-13. The auto valve is a three-way motor-operated valve that directs the flow of water either to the solar collector array or directly to the pool heater, bypassing the collector array. The controller has a differential thermostat that compares the pool water temperature at the pump and filter to the collector array temperature. When the temperature difference between the pool water and the collector array water exceeds the On differential of the controller, the auto valve will direct water through the collector array to be heated. The On differential for a pool controller is typically 3 to 5 F°. The controller will continue to direct water through the collector array until the temperature difference between the pool water and the collector array falls below the Off differential for the control, which is typically 1.5 to 2.0 F°. When the controller switches to the Off state, the auto valve will direct the water flow to the pool heater, bypassing the solar collectors. (Conventional pool heaters are controlled by an internal thermostat, which can be set for the desired pool temperature.) Pool controllers usually have a second control function, which is an adjustable high-limit control. This function senses the pool water temperature and compares it to a high-limit temperature setting. When the pool water reaches the set high limit, the control will cause the water flow to bypass the solar collectors. A typical swimming pool controller and motorized valve is shown in Fig. 14-13, and a diagram of the controller functions is given in Fig. 14-14.

In most pool systems, the filter pump is controlled by a time clock that is programmed to turn the filter system on for a set number of hours during the day. When a solar-heating system is added to a pool and the filter pump also acts as the circulator pump for the solar collectors, it is important to inform the

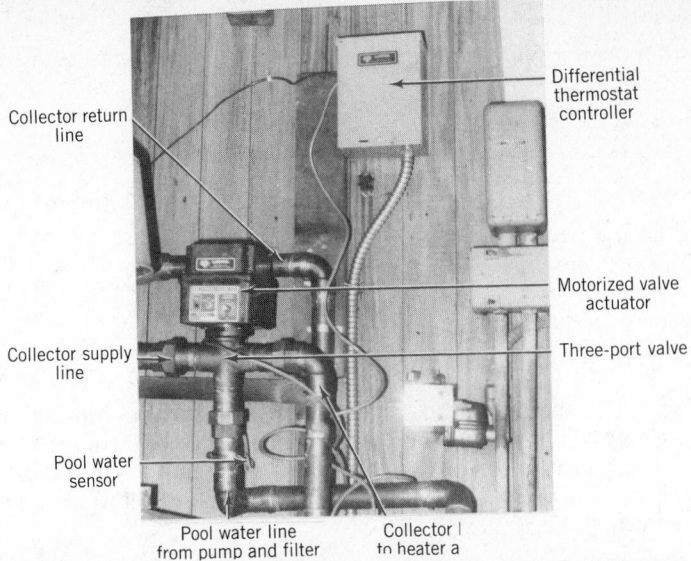

Figure 14-13 Solar energy controls for an open-loop solar pool-heating system. The controller with a differential thermostat operates a motorized (auto) valve, which either directs the flow of water through the collector array or bypasses the array and directs the flow to the pool heater. (Santa Barbara Solar Energy Systems.)

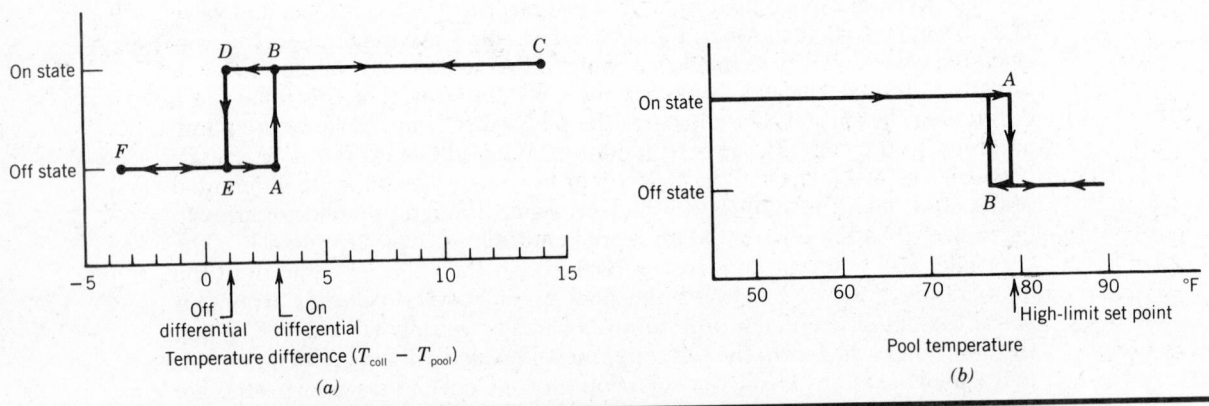

Figure 14-14 Pool and spa auto-control operation. (a) The operation of the differential thermostat. The temperature of the collector fluid is compared to the temperature of the pool or spa. When the temperature difference is greater than the On temperature differential (at A), the control will switch to the On state (at B). When the temperature difference is less than the Off temperature differential (at D), the control will switch back to the Off state (at E). For pools and spas, the On and Off differentials are typically smaller than they are for domestic hot-water systems. (b) The operation of the high-limit set point. The pool or spa controller usually has an adjustable high-limit control. In this case (a pool) it is set at approximately 79° F. When the water temperature at the pool sensor reaches this high-limit setting, the control will switch off the system (at A). When the pool water temperature falls below the set-point temperature at B, the controller turns the solar system back on, provided the differential thermostat is still in the On state.

SOLAR HEATING OF SWIMMING POOLS AND SPAS

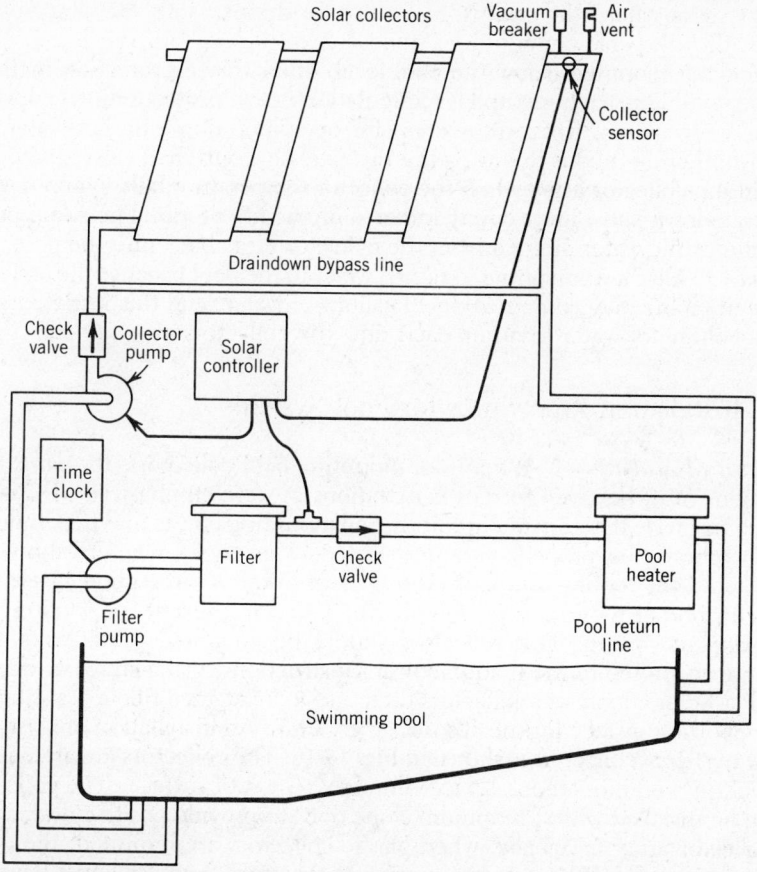

Figure 14-15 Pump circuit for large solar-heated pools. On large pool systems, a separate circulating pump is often used for the collectors. In this case, the pool controller will turn the collector pump on and off, and the time clock will govern the operation of the filter pump.

owner or the pool maintenance personnel that the filter pump time clock must be set to run the pump during the midday hours when the collection of solar energy is greatest.

For large pools with surface areas in excess of 1500 ft², it may be necessary to add a separate pump to circulate the water through the pool collector circuit. Such a system is illustrated in Fig. 14-15. In this system, the controller turns on the circulating pump whenever heat can be obtained from the solar collector array. The controller must have a motor starting relay, which is rated for the pump horsepower rating, or a separate motor relay must be used between the controller and the pump motor.

14-8   Freeze Protection for Pool Installations

In most residential solar pool systems, the collectors are protected from freezing simply by drainback into the pool. When the circulation pump stops, vacuum breakers at the top header of the collector array open to allow air to enter the collectors, and this allows the water to drain out of the collectors and collector piping into the pool. When check valves are installed to prevent the circulation pump from losing its prime, a ⅜-in. or ½-in. draindown bypass line must be installed between the collector supply line and the collector return line, as shown in Figs. 14-9 and 14-15. Such a bypass line provides for complete drainage of the collector array and piping when the pump is off.

In cases where the collectors cannot be drained into the pool, either because the allowable piping routes trap water in the collectors or because the collectors are mounted below the pool level, other freeze-protection methods must be used. One such method is recirculation freeze protection, which works like the recirculation freeze protection for open-loop domestic hot-water systems. Another method is to have a thermostatically controlled valve that opens to drain the collector array when the collector temperature falls to about 40°F. This draindown valve must empty into a sump with a pump. The sump pump then pumps the water drained from the collector array back into the pool. It is necessary to have a sump pump to return water to the pool because the collector array and piping may contain 20 to 40 gallons of water, and this represents too large a volume of water to dump each time the collectors are drained.

14-9   Installation Procedures for Pool Systems

Collector Mounting.   When roof mounting pool collectors, the same considerations must be given to roof penetrations as were emphasized in Chapter 13, section 13-11. It is very important that all mounting points that penetrate the roof membrane be properly weatherproofed. The pool collectors should be raised above the roofing by about 1½ to 3 in. as discussed in section 13-11. This allows for proper roof drainage and air circulation to prevent the roofing from mildewing or rotting. The collectors should be anchored to the mounting structure according to the manufacturer's instructions, using straps or clamps.

Plastic or copolymer collectors often use a corrugated fiberglass substrate as a supporting surface for the absorber (Fig. 14-3). An installation of plastic (or copolymer) pool collectors is shown in Fig. 14-16. The collectors are attached to the roof, in a position about 1½ to 3 in. above the roof surface. The collectors should be installed with a minimum slope of 8° to provide positive drainage of the collector array. Typically, when plastic collectors are mounted, the upper header is securely clamped to the mounting structure, and the lower header is free to move back and forth as the collector expands and contracts with tem-

Figure 14-16 An installation of swimming pool collectors on a roof. (Sealed Air Corporation.)

SOLAR HEATING OF SWIMMING POOLS AND SPAS

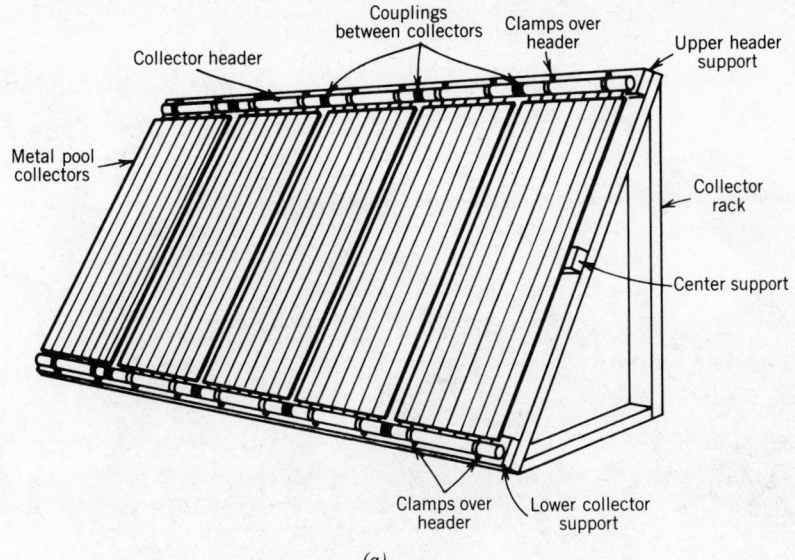

Figure 14-17 Bare metal collectors. (*a*) Bare metal collectors mounted on a wood frame. The collectors are supported at the top and bottom headers and also at the center of the panel. (*b*) Bare copper pool collectors used to heat a municipal swimming pool. (Western Solar Development, Inc.)

perature change. The straps or other mounting hardware hold the collectors down to the support structure.

With bare metal collectors, the copper header waterway usually provides a reasonably rigid structure. These collectors can be mounted by clamping the upper and lower header to the support structure with an additional support member at the center of the collector, as shown in Fig. 14-17*a*. Metal collectors should also have a minimum slope of 8° to provide positive drainage of the collector array. Figure 14-17*b* shows a large collector array for a municipal swimming pool.

When glazed collectors are used on the pool system, they should be mounted in the same manner as was discussed for domestic hot-water collectors in section 13-11. Figure 14-18 shows a group of glazed collectors mounted with a leg-support system on a flat roof.

Support leg for first collector

Solar collector piping

Pool equipment enclosure

Figure 14-18 Glazed collectors for pool heating mounted at an angle on a flat roof.

Collectors made from synthetic rubber tubes connected to each other laterally to form a 6-in. ribbon or mat are sometimes used for swimming pool heating. These can be glued or strapped to almost any support structure. The material (EPDM, ethylene propylene diene monomer) can safely withstand temperatures ranging from −50°F up to about 350°F. The individual tubes are separated and inserted into a predrilled manifold. Because the rubber tubes are not damaged by freezing, only the manifolds require draining for freeze protection.

Piping. The collector loop piping should be securely supported and fastened at approximately 5-ft intervals. When PVC plastic pipe is run horizontally, it is good practice to place a 2 × 4 or 2 × 6 support under it so that the pipe will not sag when it becomes warm, as shown in Fig. 14-19. When glazed metallic collectors are used, a 10-ft length of copper tubing should be placed between the collectors and the plastic (PVC) piping, as shown in Fig. 14-20. The 10-ft length of copper tubing acts as a heat dissipator between the plastic piping and the glazed collectors so that the plastic pipe is not deformed or melted if the collectors should reach a stagnation-temperature condition. The same precaution should be used when connecting PVC plastic piping between the filter and pool heater. There should be at least 4 to 6 ft of copper piping between the plastic pipe and the pool heater inlet connection. When connecting PVC plastic to copper or brass fittings, a female plastic threaded adapter should be used on a male threaded nipple or adapter, as shown in Fig. 14-21*a*. A large hose clamp can be clamped over the plastic female adapter to reinforce it. The plastic pipe tends to shrink and become soft when it is warmed by the heated water coming from the solar collectors, and by using a female plastic adapter on metal male threads there is less probability of leaks occurring. When a plastic male adapter is screwed into a metal valve or metal female adapter, the plastic adapter can shrink from being heated, and the threaded connection can develop a leak. Another method of connecting plastic to copper piping is to use a stainless steel and rubber clamp-type coupling, as shown in Fig. 14-21*b*.

Controls. Most pool equipment is powered with 220-V ac circuits. When working on power connections to controls and pump motors, care should be taken to ensure that all power has been turned off and that the circuits are dead,

Figure 14-19 Plastic (PVC) pipe should be well supported when it is not buried in the ground. Adequate support will prevent the pipe from deforming under the weight of water it carries.

by checking with a voltmeter or other indicator before starting work. All electrical equipment must be properly grounded to ensure that electric shocks cannot occur from the equipment.

Controllers for pool solar systems are generally of two types: (1) a differential thermostat that operates an automatic valve, as shown in Fig. 14-9; or (2) a differential thermostat that turns on a circulation pump for the collector loop, as shown in Figs. 14-10 and 14-15. When a controller and automatic valve are used, the valve and control unit usually are factory wired. The field connections that must be made are connections to the collector sensor and the pool water temperature sensor. Power connections must also be made to the controller. Most pool solar controllers will have an adjustable high-temperature cutoff, which can be set anywhere in the range of 70°F to 110°F, thus allowing a

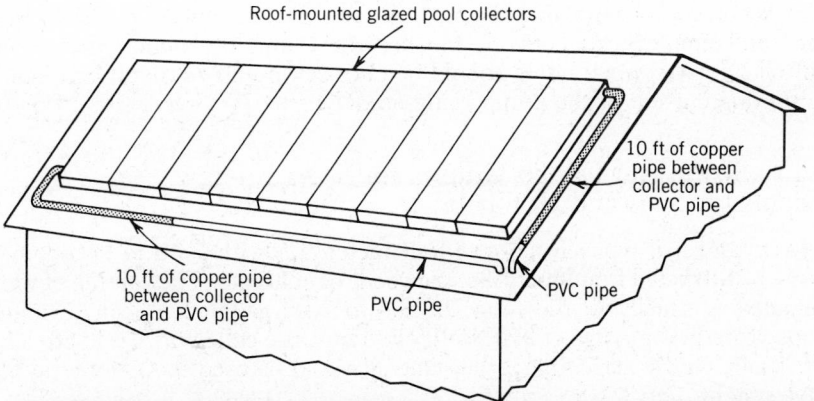

Figure 14-20 Plastic pipe should not be connected directly to a glazed collector because of the high temperatures involved. A length of copper tubing should be placed between the plastic pipe and the glazed collector to act as a heat dissipator, so that the plastic pipe will not be subjected to the stagnation temperature of the glazed collector.

Figure 14-21 Methods of connecting plastic to metal fittings for solar pool systems. (*a*) A female threaded plastic adapter is used with a male threaded metal fitting. A hose clamp is used over the plastic female adapter. (*b*) An alternative method of connecting plastic and metal piping in pool systems is to use a clamp-type coupling. (Santa Barbara Solar Energy Systems.)

maximum temperature to be set for the pool. The extended cutoff range to 110°F allows the controller to be used also with spas.

When the controller turns on a circulating pump for the collector loop, the field connections that must be made are sensor connections, power connections, and connections between the controller and the pump motor. The controller relay contact rating should be checked to ensure that it can handle the starting current of the pump being used.

14-10   Sizing and Design Criteria for Solar Energy Systems for Spas and Hot Tubs

A spa or hot tub is typically a 5- to 8-ft diameter tank with a 3½- to 4-ft depth. It may be constructed from fiberglass, redwood, or concrete. The volume of water contained is usually in the range of 500 to 1000 gallons with an operating temperature in the range of 95°F to 105°F. Thus, the collectors used with a spa will usually be the glazed type, the same as would be used for a domestic hot-water system.

In order to use solar heating effectively for a spa, heat losses from the spa must be minimized. Figure 14-2 shows the heat losses from an open spa at 102°F. Note that the range is from 6000 Btu/day-ft² in winter to only about 4700 Btu/day-ft² in summer. A spa cover is essential to reduce heat losses when the spa is not in use. There are a number of types of covers for use with spas, which

Figure 14-22 Spa and hot tub covers. (*a*) A floating bubble cover can be used to reduce the heat losses from a spa when it is not in use. (*b*) A foam spa cover may have shaped wooden supports to make it self-supporting when it is rolled out on top of the spa. The cover is removed by rolling it up and off the spa. (Both photos courtesy Santa Barbara Solar Energy Systems.)

range from simple floating covers to rollout foam and wooden lids. Two typical spa covers are shown in Fig. 14-22. These covers reduce the surface losses from radiation, convection, and evaporation significantly. A maximum insulating effect is achieved when a floating cover is used on the surface of the water in combination with a tight-fitting insulating lid. Also, some spas have insulated sides and bottoms that further reduce heat losses from the unit.

A general rule of thumb for sizing the collector area for a residential spa or hot tub with an insulated cover is to use a collector area that is between 1 and 1½ times the spa surface area. This rule applies generally to latitudes of 35° to 40°. In warmer climates, the collector area may only need to be equal to the spa surface area, and in colder climates it may be necessary to have a collector area as large as 2 times the spa surface area.

A more accurate method for sizing the collector area for existing spas is to measure the spa fuel usage, if that is possible. For new installations, the spa heat losses will have to be determined by methods elaborated below.

The procedure for measuring heat losses from an existing spa or hot tub is as follows:

1. The spa should be heated to the temperature at which it is typically used. Record this temperature and turn off the heater. The insulating cover should be placed over the heated spa.
2. After a period of time, such as 12 or 24 hours the filter pump should be run for several minutes to thoroughly mix the spa water, and then the temperature of the spa should be measured.
3. The average ambient air temperature for the test period should be estimated by using local weather data, a maximum–minimum thermometer, or a recording thermometer.
4. The heat loss from the spa can then be calculated, using the following equation:

$$Q_{Spa} = c_w(T_1 - T_2)(8.34 \text{ lb/gal})(V_{Spa}) \qquad (14\text{-}7)$$

where

c_w = the specific heat of water = 1.0 Btu/lb-F°

Q_{Spa} = the heat loss from the spa for the time period chosen, Btu

T_1 = the starting temperature of the spa

T_2 = the ending temperature of the spa

V_{Spa} = the volume of the spa in gallons

5. The average *heat loss coefficient* for the spa can then be estimated using the following equation:

$$U_{Spa} = \frac{Q_{Spa}}{\Delta t\left[\dfrac{T_1 + T_2}{2} - T_a\right]} \qquad (14\text{-}8)$$

where:

U_{Spa} = the average heat loss coefficient of the spa, Btu/hr-F°

Q_{Spa} = the heat loss from the spa for a specific time period, as above, Btu

Δt = the specific time period in hours

T_1 = the starting temperature of the spa, °F

T_2 = the ending temperature of the spa, °F

T_a = the average ambient air temperature, °F

$\dfrac{T_1 + T_2}{2}$ = the average spa temperature for the time period Δt

Illustrative Problem 14-9

A 500-gal spa was heated to 104°F and then covered with its insulating cover. Ten hours later the average temperature of the spa was measured to be 96°F. During this period, the average ambient temperature was estimated to be 50°F. Determine the heat loss from the spa and the average heat loss coefficient U_{Spa}.

Solution

The heat loss from the spa is determined by using Eq. (14-7):

$$Q_{Spa} = 1.0 \text{ Btu/lb-F}° \times (104°F - 96°F) \times 8.34 \text{ lb/gal} \times 500 \text{ gal}$$
$$= 33,360 \text{ Btu loss in the 10-hour period} \qquad \textit{Ans.}$$

The average heat loss coefficient is determined from Eq. (14-8). Substituting values,

$$U_{Spa} = \frac{33,360 \text{ Btu}}{(10 \text{ hrs})\left[\dfrac{(104 + 96)}{2} - 50°\right]°F}$$

$$= \frac{33,360 \text{ Btu}}{(10 \text{ hrs})(100 - 50)°F}$$

$$= 66.7 \text{ Btu/hr-F}°. \qquad \textit{Ans.}$$

An alternative method is to actually calculate the conduction loss from the sides, bottom, and cover of the spa using the basic conduction heat-flow equations that were presented in Chapter 5.

Once the spa heat loss is determined on a daily or monthly basis, the collector area can be determined using a form of Eq. (13-8):

$$A_c = \frac{SHF_{Spa}Q_{Spa}}{\overline{\eta}PI} \qquad (14\text{-}9)$$

where

$$A_c = \text{the collector area}$$
$$SHF_{Spa} = \text{the desired solar heat fraction}$$
$$Q_{Spa} = \text{the spa heat loss, as before, Btu/day}$$
$$\overline{\eta} = \text{the average collector efficiency}$$
$$P = \text{the percent possible sunshine}$$
$$I = \text{the clear-day solar radiation on the tilted collector surface, Btu/ft}^2\text{-day}$$

Illustrative Problem 14-10

The average heat loss from a spa was determined to be 41,000 Btu/day in April. The collector efficiency is approximately 0.45, the location is 32° N Lat, and the percent possible sunshine is 55 percent for the location. The collector tilt is 42°, orientation due south. The desired solar fraction is 90 percent. Determine the collector area needed for April.

Solution

Use Eq. (14-9). From Table 4-E, the value of I is found to be 2207 Btu/day-ft^2:

$$A_c = \frac{SHF_{Spa}Q_{Spa}}{\overline{\eta}PI}$$

Substituting,

$$A_c = \frac{(41,000 \text{ Btu/day}) \times 0.90}{(0.45)(0.55)(2207 \text{ Btu/day})}$$

$$= 67.6 \text{ ft}^2 \qquad \textit{Ans.}$$

In general, the solar collector area for a spa-heating system will be in the range of 30 ft^2 to 80 ft^2. The water-flow rate should be similar to that of a pool system, as discussed in section 14-4, with a flow in the range of 0.04 to 0.10 gpm per square foot of solar collector. The maximum water-flow rates through the spa solar-heating system will be in the range of 3 to 8 gpm. Typical filter pumps used with spas and hot tubs are in the 1 to 2 hp range—a much larger capacity than is needed to circulate 3 to 8 gpm to solar collectors.

It is not recommended that a solar-heating system be retrofitted to a spa using an existing large-capacity pump. This practice will result in excessive flow through the collectors. The recommended method is to use a separate pump and controller to circulate water through the spa collectors, as shown in Fig. 14-23. In this case, the controller will be the same as a pool controller, and it will operate the collector loop pump. The high-limit control will be set to the upper desired temperature of the spa. A filter or strainer is essential in the line before the pump to catch dirt, hair, leaves, or other contaminants. The collectors are normally protected from freezing by drainback into the spa when the circulating pump is off. If for some reason the collectors cannot be drained back into the spa, then recirculation freeze protection can be used in mild climates.

Pump capacity for a spa system is determined based on the flow rate required, the friction loss for the piping loop and collectors, and the static head based on the height to which water must be pumped up to the collectors. These calculations have already been discussed for drainback solar domestic hot-water heaters in Chapter 13 and for swimming pools in section 14-6. The collector pump size for the collector–spa loop will be in the range of 1/20 to 1/4 hp.

If the spa is installed in the ground and is essentially inaccessible, the collector supply and return lines may have to be installed in the existing piping, as shown in Fig. 14-24. In this case, the collector supply line is "tee'd" into the spa filter pump suction line, as shown. When this method is used, it is important to use a time clock or relay so that the spa filter pump does not operate when the solar collector pump is operating. If this should happen, the large spa pump's suction will prevent the small solar pump from operating correctly. It is also important to use check valves to prevent the pumps from losing their prime if they are mounted above the water level of the spa. When the pumps are mounted below the water level and given sufficient suction head, the filter check valve is not needed, but the check valve on the solar loop should always be present to prevent air from being drawn through the collector vacuum breaker and into the filter pump. A draindown bypass line is required to allow the collectors to drain, since there is a check valve in the collector pump circuit.

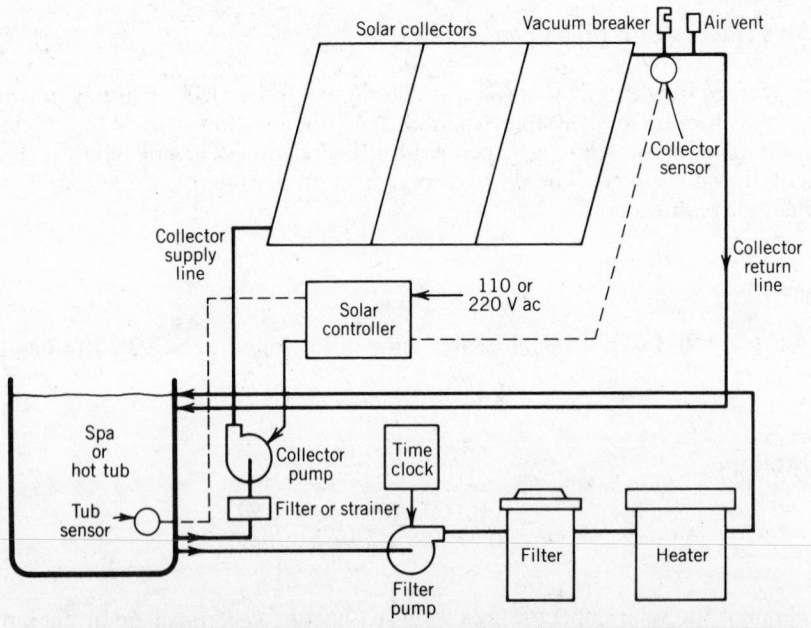

Figure 14-23 Flow diagram of a solar-heating system for a spa with independent connections to the spa. Note the separate collector pump. The solar collector loop is controlled by the controller, and it can operate independently of the filter system.

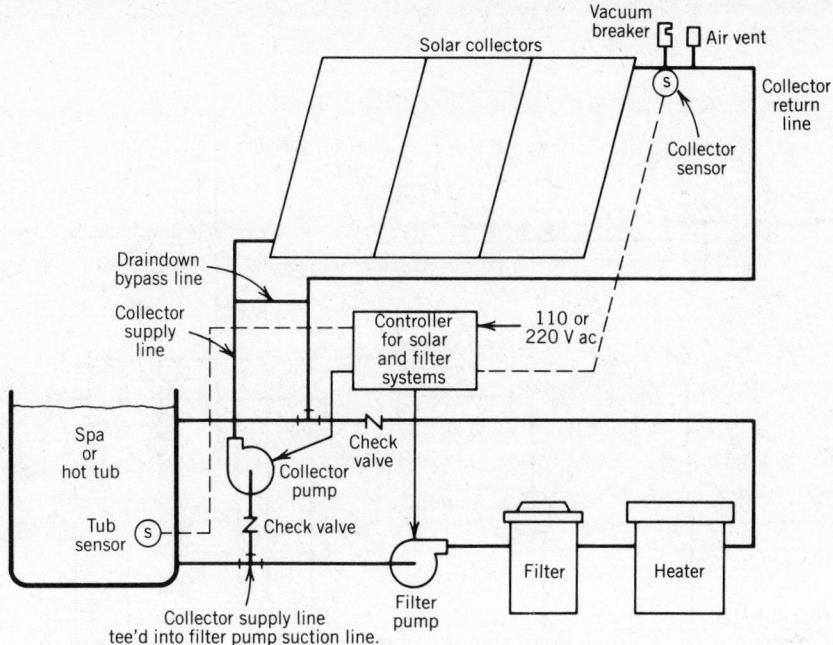

Figure 14-24 A retrofitted spa solar-heating system. In this case, the solar-heating piping must be connected to the existing filter system piping. Check valves must be installed to prevent the pumps from losing their prime. The controller circuit must be designed so that the spa filter pump cannot operate when the collector pump is running.

14-11 Combined Systems

It is possible to have pool and spa solar-heating systems combined with other types of solar energy systems, such as domestic hot-water systems or space-heating systems. These combined systems make for good year-round operating economies for the owner, and they represent good business for the solar energy contractor because of their maximized use of solar hardware. In Fig. 14-25, a combined spa and domestic hot-water system is shown in schematic representation. A heat exchanger is provided in the spa through which hot water from the domestic hot-water system is circulated to provide heat for the spa. (Check local codes on this.) An aquastat controls the circulator pump for the spa heat exchanger. The water returning from the spa heat exchanger is directed either back to the solar storage tank or to the auxiliary heater. A controller compares the temperature of the returning water from the spa heat exchanger with the temperature of the solar storage tank. When the heat exchanger water is warmer than the storage tank water, it is returned to the auxiliary heater by the three-way valve. When the storage tank water is warmer than the heat-exchanger water, the heat-exchanger water is directed to the solar storage tank. This control valve system prevents the returning heat-exchanger water from supplying heat into the storage tank when the tank itself is colder than the returning heat-exchanger water. The controller, pump, and solar collectors are typical of a domestic hot-water system, as discussed in Chapter 13. Obviously, extra collector area must be provided for the heat load of the spa beyond that required for the domestic hot-water load.

Another possibility is to combine swimming pool and spa heating, as shown in Fig. 14-26. In this case, the same filter system is used for both the pool and the spa. Either manual or automatic three-way valves are used to direct the filter system and the solar-heating system to either the pool or the spa. The control system should have a priority controller that will provide for heating the spa to a preset temperature first and then provide heat to the pool for the rest of the day.

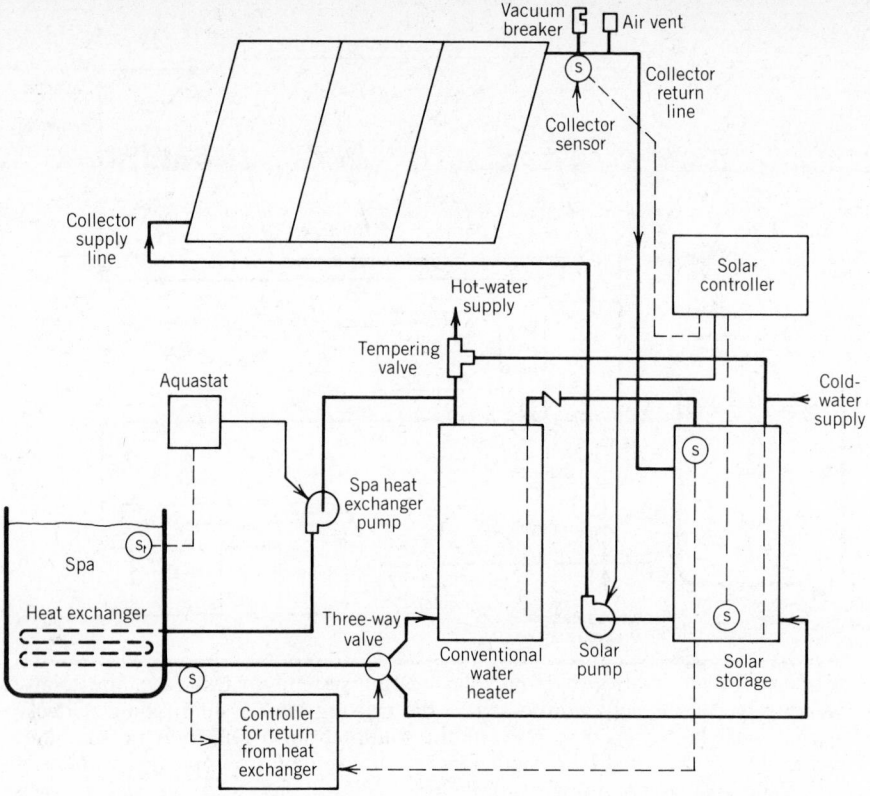

Figure 14-25 Flow diagram for a combined solar spa and domestic hot-water system. This type of system allows the use of the solar collectors to be maximized and eliminates the gas heater for the spa. Controls for such a combined system are more complex than those for a single-purpose system.

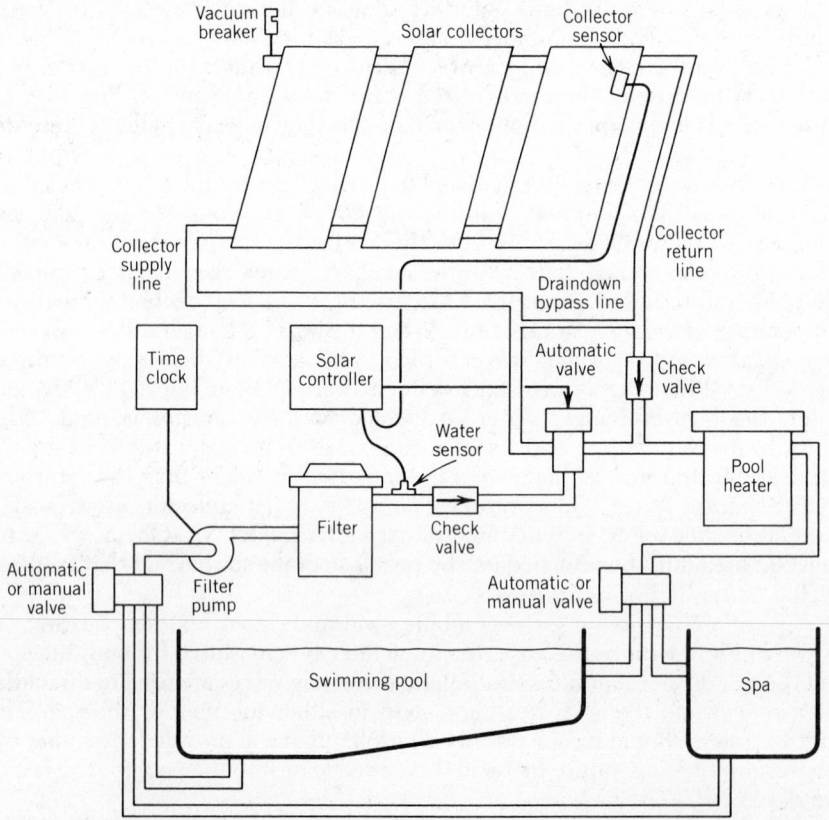

Figure 14-26 The same solar-heating system may be used to heat both a pool and a spa when they share a common filter and heater system. This type of system requires valves to divert the water circulation to either the pool or the spa.

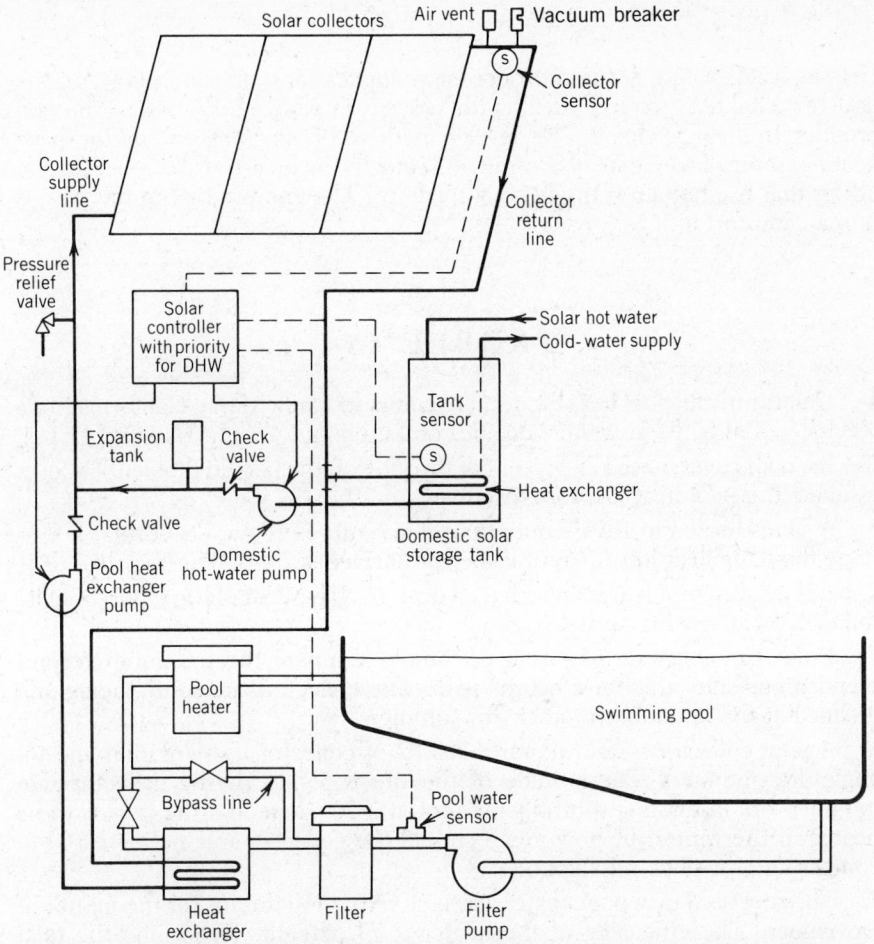

Figure 14-27 Pool heating and domestic hot-water heating can be combined. This combination requires a closed-loop system with heat exchangers to transfer the heat either to the domestic hot-water tank or to the pool water. The controller exerts priority for the domestic hot-water system.

A combined pool and domestic hot-water heating system is diagrammed in Fig. 14-27. In this case, a closed-loop system is used on the collector loop, and heat exchangers are used to heat both the domestic hot water and the pool water. This type of system would be used, for example, in cold climates for an indoor pool. Glazed collectors are necessary to provide reasonable collection efficiency and sufficiently high temperatures during winter conditions. A priority controller is used to provide control functions for domestic hot-water heating and pool heating.

It is also possible to have a combined space-heating and pool-heating system where the collectors provide space heating in winter and pool heating at other times. (See Chapter 15.)

One advantage of combined systems is economic, in that the amortization of the combined-system cost is more effective because the solar equipment supplies heat to more loads during the year and its use is maximized. The disadvantage is that these systems are more complex than single-function solar-heating systems because of the multiple function controls, extra valves, and pumps required.

Care should be taken when choosing materials for the heat-exchanger tubing for pools and spas. Copper is the recommended material for heat-exchanger tubing. When designing heat exchangers, it is important to avoid unduly high velocities in the tubes because of the erosive effects of pool and spa water that might have a high count of suspended solids.

SYSTEM DESIGN AND INSTALLATION

CONCLUSION

Heating water is one of the most common applications of solar energy. In the past two chapters, we have dealt with systems in which hot water is the end product. In the next chapter, we develop methods of designing systems for space heating, some of which involve using hot water as the heat-transfer medium and others that use hot air as the heating medium. The end product in this case is human comfort in living spaces.

PROBLEMS

1. Determine the pool heat loss for November in Santa Maria, California, for a 400 ft^2 pool at 80°F, for wind velocities of 0 mph and 4 mph. Use Fig. 14-1.

2. If a pool cover is used 75 percent of the time on the pool in problem 1 above, estimate the pool heat loss for November.

3. A spa is heated to 102°F continuously in Santa Barbara, California. Determine the daily heat loss in April if the spa surface is 25 ft^2. Use Fig. 14-2.

4. If the spa in problem 3 above has a cover on 80 percent of the time, estimate the daily heat loss for April.

5. From the collector area rule of thumb, estimate the minimum recommended collector area for a 600 ft^2 pool. The collectors are south-facing and inclined at the optimum tilt angle for summer.

6. A pool collector system requires 500 ft^2 of collector if orientation and tilt angle are optimum. The latitude of the site is 35° N. If the collectors are mounted on an existing south-facing roof at a 20° slope and the pool is to be heated in the winter, by how much should the collector area be increased to compensate for slope misalignment?

7. The fuel used by a pool heater was metered at 150 therms for the month of November. The efficiency of the heater is 73 percent. Determine the heat delivered to the pool in Btu.

8. A public swimming pool has a heating load of 55×10^6 Btu for the month of March. The location is 32° N Lat with a percent possible sunshine of 65 percent. The solar collectors to be used for the pool-heating system have an average efficiency of 0.53, and they will be mounted facing due south with a tilt angle of 42°. Determine the collector area required to meet this heating load.

9. A residential pool has a heating load of 16×10^6 Btu for the month of May. The location is 40° N Lat with a percent possible sunshine of 63 percent. If the collectors are mounted at a 30° tilt facing due south, determine the required collector area. Assume that the collectors have an average efficiency of 0.55.

10. A pool collector array has 600 ft^2 and is located 15 ft above the pool equipment. The piping run from the pool equipment to the collector array is estimated to total 60 ft with seven 90° elbows. There will be a check valve and a control valve in the collector loop. The pipe diameter is 1½ in. PVC. Determine the collector loop piping pressure drop for a flow rate of 40 gpm. (Include the static head.)

11. A pool collector and piping system has an estimated pressure drop of 5.0 psi. The collectors are mounted at 15 ft above the pool water level. The existing pressure head of the filter system is 13 psi. Determine the pump size from Fig. 14-12 to deliver a 30 gpm flow rate through the solar system. (Include static pressure head.)

12. A residential spa has 25 ft^2 of surface. Make a preliminary estimate of the collector area that should be used to provide solar heating for the spa.

13. A 600-gal hot tub showed a 15 F° temperature drop over a 24-hour period. Determine the daily heat loss from the spa.

14. The average daily heat loss from a spa was measured as 32,000 Btu/day in

October. The location is N 40° Lat with a percent possible sunshine of 60 percent. The collector tilt will be 50°, and the estimated average collector efficiency is 0.60. Determine the collector area for a solar fraction of 85 percent.

15. An outdoor pool is located at N 40° Lat. Determine the pool heat losses (Btu/ft^2-day) from radiation, convection, and evaporation, and the total net energy loss Q_{pool} (Btu/ft^2-day) for a pool temperature of 80°F. The average daily ambient air temperature is 60°F, and the average relative humidity is 50 percent. Assume a 2-mph wind velocity.

Solar Energy Systems for Space Heating

*I*n this chapter, we will examine system components for liquid- and air-based space-heating systems. The five basic and essential elements of an active solar system were introduced in Chapter 10. In the following sections, the requirements of these elements for space-heating systems will be explored.

The first step in designing an active solar space-heating system is to determine the heating load of the building. The size of the solar system components will be influenced by the magnitude of the heating load that they are designed to balance.

INITIAL DESIGN CONSIDERATIONS

In the design process for an active solar space-heating system, *energy conservation* must be a major consideration in order to reduce the total space-heating load. Generally, it is more cost-effective to reduce the space-heating load by energy conservation than it is to supply additional energy by an active solar system. Active solar system components are more expensive than insulation, weatherstripping, or even double glazing of windows and fixed glass. The heat losses for a building with active solar space heating should be the same as or even less than the heat losses for a

Table 15-1 Recommended Building Load Coefficients (BLC) for Active Solar Space-Heating Systems, Normalized to the Building Floor Area

Annual Heating Degree-Days (range)	Normalized Building Load Coefficient (BLC) (Btu/DD-ft^2)
Less than 1000	10
1000–3000	8
3000–5000	7
5000–7000	6
Greater than 7000	5

Sources: Abstracted with permission from *Pacific Regional Solar Heating Handbook*, U.S. Energy Research and Development Administration, San Francisco Operations Office, March 1976; and from J. D. Balcomb, *Passive Solar Design Handbook*, vol. II (Washington: U.S. Department of Energy, January 1980).

passive solar building. The normalized building load coefficient (BLC) values given in Table 15-1 provide a reference point for the architect or designer to work from during the early design stages of a building that is to have active solar heat.

The overall energy usage in the building is another important consideration, because the energy efficiency of lighting and appliances will determine the quantity of internal heat generated within the building. The internal heat generated in a single-family residence typically is quite small compared to the heating load and has minimal effect on the overall heating design. In commercial buildings, however, the internal heat generated can be very large, making a significant contribution to balancing the space-heating load. In very large commercial buildings (multistory office buildings or department stores), the internal heat generated often exceeds heat losses, and the building may require cooling during both winter and summer. It is important in all buildings to have a well-designed HVAC system that can maximize the use of internally generated heat to offset winter space-heating loads and also to minimize the need to use mechanical cooling in the summer months. In buildings with active solar heating, it is even more important to have an efficient HVAC system in order to maximize the use of the collected solar energy. The design procedures for HVAC systems are outside the scope of this book, and the reader is referred to any standard air-conditioning text. It must be kept in mind that the HVAC system will influence the performance of the active solar system, and both should be designed to provide optimum performance as they interact with the building and with each other.

15-1 Initial Sizing Guidelines

The designer of active solar buildings needs some guidelines that provide approximate sizes for the solar collector area and the thermal storage, so that these requirements can be incorporated into the schematic design and design development phases. Table 15-2 gives some general sizing criteria in terms of the ratio of collector area to building floor area, based on the average winter degree-days per month. The ratios of collector area to building floor area are given as a range of values for selected levels of winter degree-days per month. Climates that have clear winter sky conditions can ordinarily use the smaller ratio of the range given.

There are some assumptions with regard to Table 15-2 that must be observed. First, it is assumed that the solar collectors face due south with a tilt equaling latitude plus 10° to 15°. Second, it is assumed that the solar collectors will have an average efficiency of 0.45 or greater. Third, it is assumed that the building has a heat loss rate not greater than 8 to 10 Btu/DD per square foot of

Table 15-2 Approximate Ratio of Collector Area to Building Floor Area Required for Active Solar Space-Heating Systems, for Selected Levels of Winter Degree-Days per Month

Average Winter Degree-Days/Month	Ratio of Collector Area to Building Floor Area
1600	0.52–0.65
1400	0.50–0.62
1200	0.46–0.58
1000	0.42–0.55
800	0.32–0.48
600	0.18–0.35
400	0.08–0.20

The above values are based on the following assumptions:
1. The heat loss from the building is 8 to 10 Btu/DD per square foot of floor area.
2. The solar collectors are oriented due south with a tilt of latitude plus 10° to 15°.
Abstracted with permission from *Pacific Regional Solar Heating Handbook*, U.S. Research and Development Administration, San Francisco Operations Office, March 1976.

floor area or in the range of values given in Table 15-1. Given these conditions, the collector areas calculated from the given ratios will provide approximately 60 to 70 percent of the space-heating load.

Illustrative Problem 15-1

A residence of approximately 1800 ft² is being designed in a location where the winter degree-days per month are approximately 800. What area should be allocated for active solar collectors for space heating?

Solution

From Table 15-2, the collector area to building floor area ratio has a range of 0.32 to 0.48. The required collector area will then be in the range from 0.32 × 1800 ft² to 0.48 × 1800 ft², or 576 ft² to 864 ft².   *Ans.*

In the above example, the house should be designed to provide a sloping roof area facing south. The roof must have an area suited for the installation of from 580 to 870 ft² of solar collectors at a tilt angle of latitude plus 10° to 15°. It is emphasized that the above rules of thumb are for preliminary design estimates only.

15-2 Thermal Storage Requirements for Active Systems

In order to allocate space for thermal storage within a building during preliminary design work, a rule-of-thumb guide for sizing the thermal storage is useful. The thermal storage is usually sized based on the square footage of the collector area. For active space-heating systems, it has been the practice to provide a thermal storage capacity, C_c, of 10 to 18 Btu/F° for each square foot of solar collector.[1] The temperature rise of the thermal storage can then be determined as:

---

[1] The designation C_c means heat storage capacity per unit area of solar collector, Btu/F°-ft².

$$\Delta T_s = \frac{Q_c}{C_c} \qquad\qquad (15\text{-}1)$$

where

ΔT_s = the rise in temperature of the thermal storage, $F°/day$

Q_c = the heat energy actually collected by the collector, $Btu/ft_c^2\text{-}day$

C_c = the heat storage capacity of the thermal storage, $Btu/F°\text{-}ft_c^2$

For example, consider the collector performance shown in Table 12-1. The daily energy collected by this collector is 842 $Btu/ft_c^2\text{-}day$. For a thermal storage capacity of 10 $Btu/F°\text{-}ft^2$ of collector, the temperature rise of thermal storage would be:

$$\Delta T_s = \frac{Q_c}{C_c}$$
$$= \frac{842 \ Btu/ft_c^2\text{-}day}{10 \ Btu/F°\text{-}ft_c^2}$$
$$= 84.2 \ F°/day$$

If the thermal storage capacity were (for example) 18 $Btu/F°\text{-}ft_c^2$, the temperature rise would be 46.8F°/day. Theoretical studies and practical experience have shown that if the thermal storage capacity is reduced to a value below 10 $Btu/F°\text{-}ft_c^2$, the temperature rise of the storage becomes so large as to unduly increase the operating temperature of the solar collectors, resulting in a lower system efficiency because of larger heat losses. If the thermal storage is increased above about 18 $Btu/F°\text{-}ft_c^2$, there is a small increase in the system performance, but the increased cost of the thermal storage is usually not justifiable. In most applications, the thermal storage capacity should be in the range of 10 to 18 $Btu/F°\text{-}ft_c^2$.

Illustrative Problem 15-2

If a thermal storage capacity of 10 $Btu/F°\text{-}ft_c^2$ is needed, calculate (a) the volume of water required, and (b) the volume of rock required, for thermal storage. Assume that the rock has a density D of 100 lb/ft^3, a specific heat c of 0.19 $Btu/lb\text{-}F°$, and a void of 0.40.

Solution

The volume of required storage can be determined using a modified form of Eq. (7-8):

$$V = \frac{C_c}{C_V}$$

where

V = the required volume of thermal storage material, ft^3

C_c = the thermal capacity per unit of collector area, $Btu/F°\text{-}ft_c^2$

C_V = The volumetric thermal storage capacity, $Btu/F°\text{-}ft^3$

(a) For water, C_V is 62.4 $Btu/ft^3\text{-}F°$, or 8.34 $Btu/gal\text{-}F°$. Substituting values:

$$V = \frac{10 \ Btu/F°\text{-}ft_c^2}{8.34 \ Btu/gal\text{-}F°}$$
$$= 1.2 \ gal/ft_c^2 \qquad\qquad Ans.$$

(b) For a rock bed, C_V is found from Eq. (9-3). Substituting,

$$C_V = D \times c \times (1 - void)$$
$$= 110 \ lb/ft^3 \times 0.20 \ Btu/lb\text{-}F° \times (1 - 0.40)$$
$$= 13.2 \ Btu/F°\text{-}ft^3$$

The required volume is then found as:

$$V = \frac{10 \text{ Btu/F°-ft}_c^2}{13.2 \text{ Btu/F°-ft}^3}$$
$$= 0.76 \text{ ft}^3/\text{ft}_c^2 \qquad \textit{Ans.}$$

Rules of Thumb for Early Design Stage. When water is used for thermal storage, the volume required will range from 1.2 to 2.0 gal per square foot of solar collector. For an air system using rock bed thermal storage, the volume of rock required will range from 0.5 to 1.0 ft^3 of rock per square foot of collector.

Illustrative Problem 15-3

For a house that will have a solar collector area of 500 ft^2, determine the thermal storage requirements for water and rock storage.

Solution

For water storage, the recommended storage volume is 1.2 to 2.0 gal per square foot of collector. Therefore, the required volume of water will be in the range from 1.2 × 500 to 2.0 × 500, or 600 to 1000 gal of water. *Ans.*
For rock, the recommended storage volume is 0.5 to 1.0 ft^3 per square foot of collector. Therefore, the required volume of rock will be in the range from 0.5 × 500 to 1.0 × 500, or 250 to 500 ft^3 of rock. *Ans.*

ACTIVE SPACE-HEATING SYSTEMS AND THEIR COMPONENTS

It is important to understand the interaction of all of the components in active solar energy systems. In the following sections, liquid- and air-based active systems will be discussed in detail with emphasis on how the system components function and interact in a complete system.

15-3 System Configurations for Liquid-Based Systems

A liquid-based space-heating system has a collector loop configuration similar to that of a domestic hot-water system. The basic difference is that a space-heating system will generally have a much larger collector area and storage volume than the typical residential domestic hot-water system. The most commonly used collector configurations are a closed-loop system and a drainback system. (See pages 473–475 and 509 for definitions of these terms.)

Closed-Loop System. The closed collector loop and the heat-exchanger–storage loop for an active space-heating system are shown in Fig. 15-1. An active space-heating system for a residence may have several hundred square feet of collector area, while one for a commercial building may have thousands of square feet. The solar collectors will typically be grouped in banks to facilitate plumbing and to allow the adjustment of flow rates.

In Fig. 15-1, two banks of solar collectors are shown: SCB1 and SCB2. Each collector bank is shown with an inlet valve (10) and an outlet valve (4). These valves allow each bank to be isolated from the rest of the collector loop, if necessary, for maintenance. Each bank is also shown with a drain (11), vacuum breaker (2), and air vent (1). Pressure relief valves (3) are provided to protect each collector bank from unduly high pressure conditions. Thermometer T_1 indicates the temperature of the collector-loop fluid returning from the collector array. The heat exchanger (HE) transfers heat from the collector loop to the

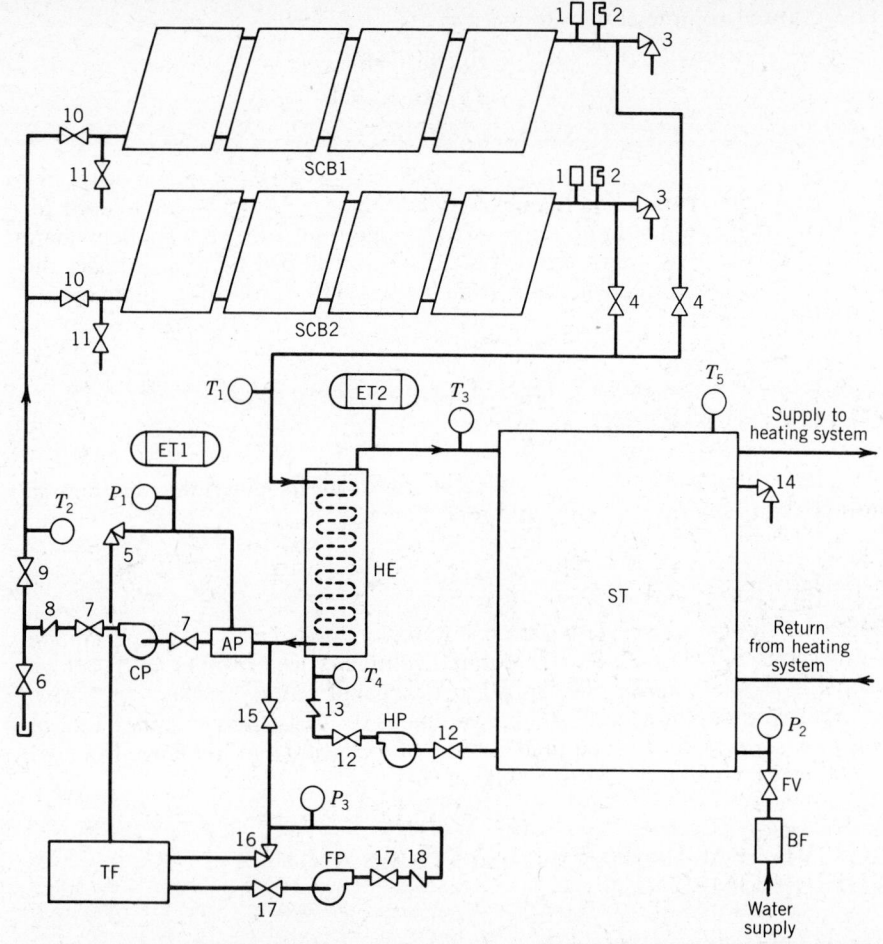

Figure 15-1 Closed-loop space-heating system flow diagram, showing the solar collectors, heat exchanger, thermal storage, collector-loop storage, pumps, and valves. This system uses the heat exchanger (HE) to transfer the heat from the collector heat-exchange fluid to the water in the thermal storage tank (ST). There are three subsections to this system: the collector loop, the thermal storage loop, and the collector-loop fluid reservoir.

Collector Loop

SCB	Solar collector banks in collector array
HE	Heat exchanger
AP	Air purger
ET1	Expansion tank, closed-loop system
T_1	Collector return thermometer
T_2	Collector supply thermometer
CP	Pump for collector loop
P_1	Collector-loop pressure gauge
1	Air vents
2	Vacuum breakers
3	Collector relief valves
4	Collector bank outlet valves
5	Collector-loop pressure relief valve
6	Collector-loop drain
7	Isolation valves for collector-loop pump
8	Check valve for collector loop
9	Flow control valve for collector loop
10	Collector bank inlet valves
11	Drain valve for collector banks

Heat Exchanger–Storage Loop

HP	Pump for heat exchanger–storage loop
ST	Thermal storage tank
P_2	Thermal storage loop pressure gauge
T_3	Heat exchanger outlet thermometer
T_4	Heat exchanger inlet thermometer
T_5	Tank thermometer on thermal storage tank
ET2	Expansion tank for thermal storage
FV	Fill valve for thermal storage
BF	Back-flow device
12	Isolation valves for collector-loop pump
13	Check valve for heat exchanger–storage loop
14	Pressure relief valve for thermal storage system

Collector-Loop Fluid Storage Loop

FP	Collector-loop fill pump
TF	Collector-loop fluid storage tank
P_3	Closed-loop fill system pressure gauge
15	Fill valve for collector loop
16	Pressure relief valve
17	Isolation valves for pump
18	Check valve

thermal storage loop. The air purger (AP) removes any air from the collector-loop system. The expansion tank (ET1) provides an air cushion for the expansion and contraction of the collector-loop fluid as it heats and cools. This expansion tank (ET1) should be sized based on the type of collector-loop fluid used and the expected operating temperature range in the collector loop. A pressure gauge (P_1) indicates the pressure in the closed-loop system. The pressure relief valve (5) allows the heat exchange fluid to escape from the closed-loop system if excessive pressure builds up. The pressure setting on this valve should be 10 to 15 psi lower than the setting on the relief valves (3) located on the collector banks, because if fluid must be released it should be done in the mechanical equipment room so that the fluid can be saved in an overflow tank (TF). A drain valve (6) allows all of the closed-loop fluid to be drained from the system. It should be located in the lowest part of the piping loop. The collector-loop pump (CP) circulates the heat transfer fluid through the collector loop. Isolation valves (7) allow the circulation pump to be removed for maintenance without completely draining the system. A flow control valve (9) allows for adjusting the flow rate through the collectors. In very large systems this valve would not be used and the collector pump would have multiple speeds or a variable speed motor to allow for controlling the circulation rate. Thermometer (T_2) indicates the temperature of the heat-exchange fluid returning to the collector array.

The heat-exchanger–storage-tank loop is shown as a pressurized system. The thermal storage tank (ST) will generally hold from 500 to 2000 gal for residential-size systems, and many thousands of gallons for commercial systems. The heat exchanger pump (HP) circulates water from the storage tank through the heat exchanger and back to the storage tank. The isolation valves (12) allow the heat exchanger pump to be removed for maintenance without draining the thermal storage system. The water temperature into and out of the heat exchanger is indicated by thermometers T_3 and T_4. The check valve (13) allows only the forced circulation by the pump to go through the heat exchanger. The storage tank temperature is indicated by T_5. An expansion tank (ET2) is provided for the heat-exchanger storage loop. The volume of the expansion tank should be approximately 8 percent of the total volume of water in the thermal storage and heat distribution systems. The actual size of the expansion tank should be calculated based on the operating temperature and volume of the storage system. A method of providing water to the storage subsystem is provided by the fill valve (FV) and the back-flow device (BF). The back-flow device prevents water from the storage tank from backing into the water supply main and causing possible contamination. Pressure in the storage subsystem is indicated by P_2. A pressure relief valve (14) prevents overpressure conditions from occurring in the storage system.

A collector-loop fluid reservoir (TF) is often provided on very large systems. This is a tank that has a capacity larger than the volume of fluid in the closed-loop system. It provides storage for extra collector-loop fluid and also overflow storage in the event that fluid is released from the collector loop. A pump (FP) is shown for adding heat-exchange fluid into the closed loop when necessary. Isolation valves (17) are shown for the pump, and so is a check valve (18). A pressure gauge (P_3) shows the pressure of the fluid-filling system. A valve (15) allows fluid to be added to the closed-loop system, and a relief valve (16) allows fluid to return to the reservoir when the pressure in the filling system reaches the pressure setting on the relief valve.

Drainback System. A drainback collector-loop storage system with downcomer is shown in Fig. 15-2a. It is a sealed system; that is, air cannot enter the system. Since, in a drainback system, water is both the collector fluid and the thermal storage, it is quite important that the drainback feature be carefully designed and installed in order to avoid freeze-ups in cold climates.

The solar collectors in Fig. 15-2 are shown in banks (SCB1, SCB2) as before. The collector bank outlet valves (1) and inlet valves (4) allow the collector banks to be isolated, if required, for maintenance. The pressure relief

ACTIVE SPACE-HEATING SYSTEMS AND THEIR COMPONENTS

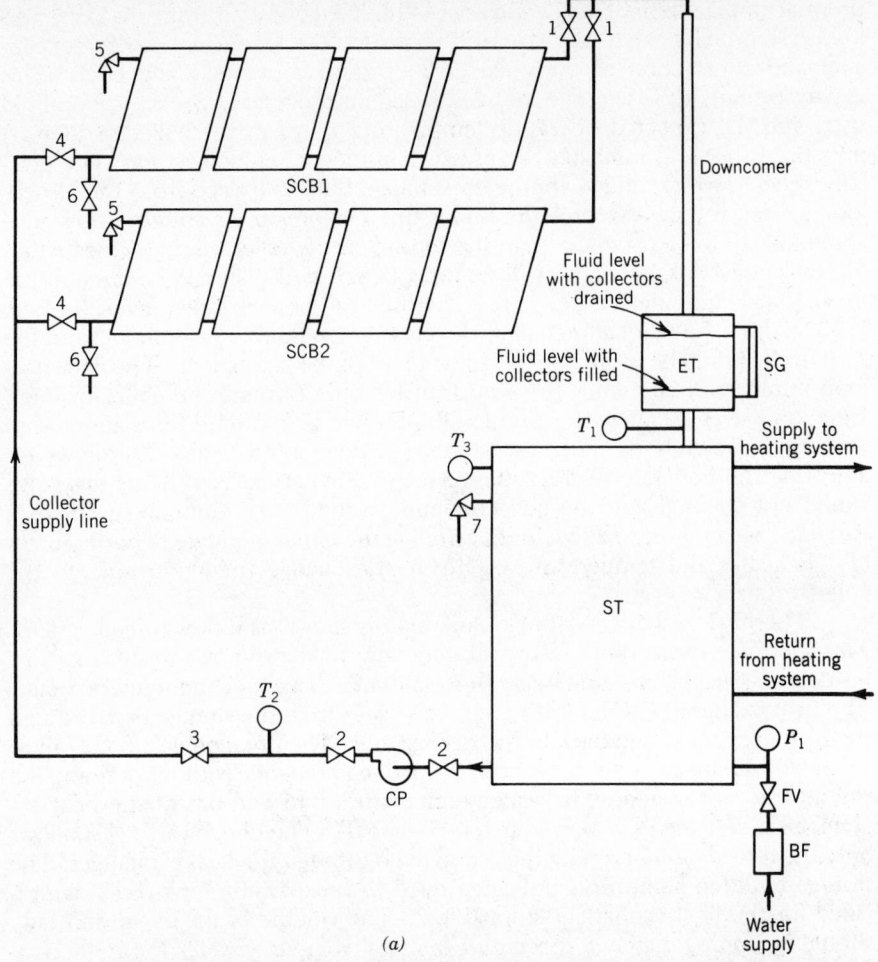

Figure 15-2 Drainback collector loop system for active space heating. (*a*) The expansion tank (ET) provides a volume larger than the fluid capacity of the collectors and the collector piping. The sight gauge (SG) indicates the fluid level in the expansion tank. The downcomer allows air from the expansion tank to enter the collectors to allow the water to drain out when the pump turns off. (*b*) This configuration—a siphon drainback system—uses a booster pump to fill the collectors and the collector piping with water. This pump then turns off. The collector-loop pump is then able to sustain the fluid circulation through the collectors. The air is displaced into the expansion tank (ET). The collector return line enters the expansion tank below the water level when the collectors are filled. A separate air line is run to the collector return line. Air entering the return line breaks the siphon so that water can drain from the collectors when the pump turns off.

SCB	Solar collector banks in collector array
ET	Expansion tank
SG	Sight gauge for expansion tank
T_1	Collector return thermometer
T_2	Collector supply thermometer
T_3	Storage tank thermometer
ST	Thermal storage tank
CP	Pump for collector–thermal storage loop
P_1	Pressure gauge for storage and collector loop
FV	Fill valve
BF	Back-flow device
1	Collector bank outlet valve
2	Isolation valves for collector-loop pump
3	Flow control valve for collector loop
4	Collector bank inlet valve
5	Collector bank relief valve
6	Drain valve for collector bank
7	Storage tank pressure-relief valve

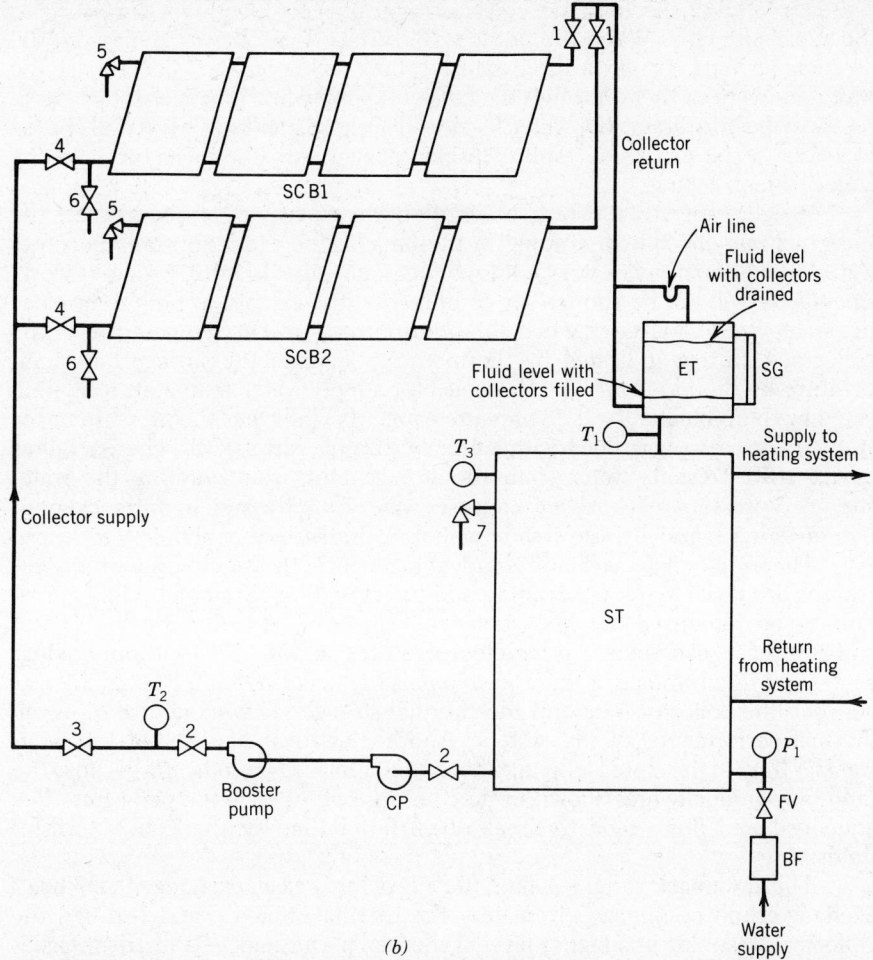

Figure 15.2 (*continued*)

valve (5) protects the collector banks from overpressure conditions if they are isolated and not drained. A drain (6) for each collector bank is provided for times when maintenance work must be done. The collector pump (CP) circulates cool water from the bottom of the thermal storage (ST) through the collector banks. Isolation valves (2) provide for the removal of the pump without draining the system. A flow control valve (3) allows for adjustment of the flow rate through the collector array. The downcomer from the collector array is one size larger than the collector supply line to ensure that the water will drain from it into the expansion tank (ET) when the pump turns off. This will allow air to enter the top of the collectors from the expansion tank (ET) and allow water to drain from them back through the collector pump into the bottom of the storage tank. As the water drains from the collectors, the water level will rise in the expansion tank as the air moves into the collectors. The expansion tank (ET) must be large enough to provide for the expansion of the water volume in the system and also to provide a volume equal to the fluid volume of the collectors and collector piping.

When the pump for the collector–thermal-storage loop (CP) is off, the collectors will be filled with air and the expansion tank (ET) will be almost filled with water. When the pump (CP) turns on, the water level in the expansion tank will drop as the collectors are filled with water and the air that was in the collectors is displaced into the expansion tank. When the collectors are filled with water, the water level in the expansion tank will be low, and all of the air

ACTIVE SPACE-HEATING SYSTEMS AND THEIR COMPONENTS

that was in the collectors and collector-loop piping will now be displaced to the expansion tank. When the pump (CP) turns off, water will drain from the downcomer, and the air in the expansion tank will enter the collectors as the water drains from them through the collector supply line into the storage tank. As the collectors drain, the water level in the expansion tank rises, and the air that was in the expansion tank is displaced back into the collectors and the collector-loop piping.

A sight gauge (SG) should be used on the expansion tank to indicate the water volume and the air volume in the tank. If the expansion tank becomes waterlogged (too much water and not enough air), the drainback will not work, since there will not be enough air contained in the expansion tank to displace the water in the collectors when the pump turns off. The temperature of the collector return is indicated by thermometer T_1, and the storage tank temperature by thermometer T_3. The collector supply water temperature is indicated by thermometer T_2. The pressure gauge (P_1) indicates the pressure in the system. The system is filled with water by the fill valve (FV). The back-flow device (BF) prevents water from the storage tank from entering the water supply. A pressure relief valve (7) releases water from the system in the event of overpressure. Normally, the system would be pressurized at about 10 to 15 psi.

There are advantages and disadvantages to both the closed-loop system and the drainback system. Advantages of the closed-loop system are that corrosion can be minimized and the system can be pressurized so that the fluid (water and antifreeze) can operate at high temperatures (above 212°F) without boiling. The major disadvantage of the closed loop is the necessity for a heat exchanger between the collector loop and the thermal storage. This reduces the overall thermal performance of the system. Another disadvantage is that the heat-transfer fluid in the closed loop may have to be replaced periodically because the fluid may gradually break down in time because of the high temperatures. The aqueous-based fluids tend to break down faster than synthetic heat-transfer fluids.

The drainback system avoids the need for a heat exchanger and heat-exchange fluid by directly circulating the thermal storage water through the collectors, resulting in a higher level of thermal performance. A disadvantage is that the circulation pump must be able to lift the water to the top of the collector array, which increases the energy required for pumping. Also, there are additional requirements in piping design to ensure that the system will drain properly. When water must be lifted several stories, a *siphon drainback* can be used to reduce the pumping energy. In this siphon drainback system (Fig. 15-2*b*), two pumps are used, one being the collector pump (CP) and the second a booster pump. When the system first comes on, the collector pump (with the help of the booster pump) fills the collector supply line, the collectors, and the collector return line with water. The air in the system is pushed into the expansion tank. Once flow is established, the booster pump is turned off, and the collector pump continues to circulate water through the collectors. The hydrostatic head in the collector supply line will be partially balanced by the gravity pull of the hydrostatic column of water in the collector return line. Thus, the collector pump will have to cope only with the friction head once the collector loop has been filled with water and flow begins. When the collector loop pump is turned off, the air line from the expansion tank allows air to enter the collector return line. This will allow air into the collector array from the expansion tank. As air enters the collector array, water will drain from the system. The height between the top of the collector array and the bottom of the expansion tank should not exceed 30 feet for unpressurized systems.

For large collector arrays, it is often advantageous to have the systems pressurized at a level above atmospheric pressure. This higher pressure will reduce the possibility that the collector fluid will flash boil when it is pumped into dry collectors that are at stagnation temperatures. Heat transfer fluids (water and antifreeze solutions and oils) should not be allowed to remain for more than a few hours in the collectors at stagnation temperatures, since these temperatures may be high enough to cause chemical decomposition of the

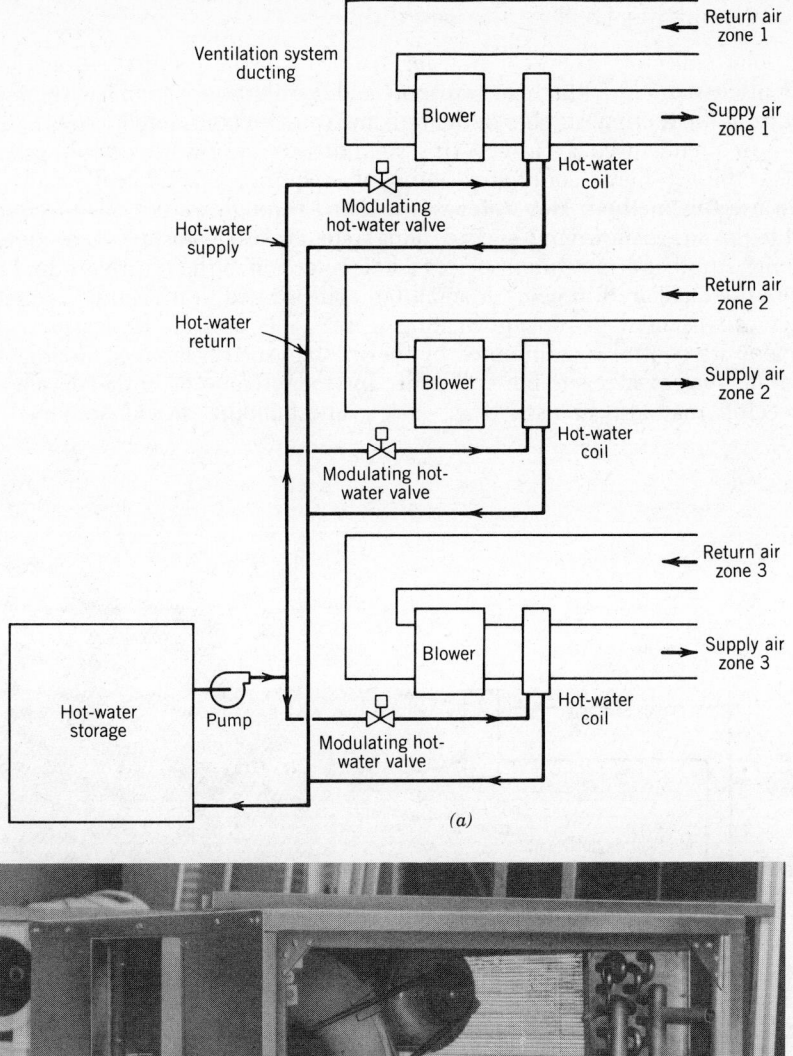

Figure 15-3 Hot-water coils in a ducted air distribution system may be used as a means of transferring stored solar heat into the building. A fan or blower circulates air through the fins and past the tubing of the hot water coils. (*a*) The diagram illustrates a three-zone system. When heating is required in a building zone, hot water is circulated to the coil for that zone, which then heats the air in that particular building zone. Thermostats and motorized (modulating) hot-water valves control system operation. (*b*) A hot-water coil mounted in a zone air handling unit. (The Trane Company.)

fluids and subsequent corrosion. High system temperatures and stagnation conditions can be avoided by using an air-cooled or water-cooled heat rejector to remove excess heat from the system when necessary. Auxiliary pumps are often provided in case the main pumps have a failure. For very large systems, an emergency generator may be necessary to keep the solar system operating in the event of a power failure.

15-4 Supplying Heat to the Building

The solar collectors, the heat transfer system, and the thermal storage have been discussed. Next, the heat transport and distribution system for supplying heat from the thermal storage to the building must be considered. One method that is frequently used is a forced-air system that uses hot-water coils to provide heat to the air. Large commercial buildings requiring mechanical ventilation often use this method. Hot water is circulated through water coils to provide heat to the air, as shown in Fig. 15-3. This system is also applicable to residential systems. Another arrangement places a hot-water coil in the return air duct just before the auxiliary furnace. Or small fan coils located in individual rooms or zones can be used to provide heating (Figs. 15-4a and b). Room (or zone) temperature control is maintained by thermostats and automatic (modulating) hot-water valves at each fan-coil unit. Instead of fan-coil units, baseboard convectors may also be used (Fig. 15-4c). The building should be zoned for

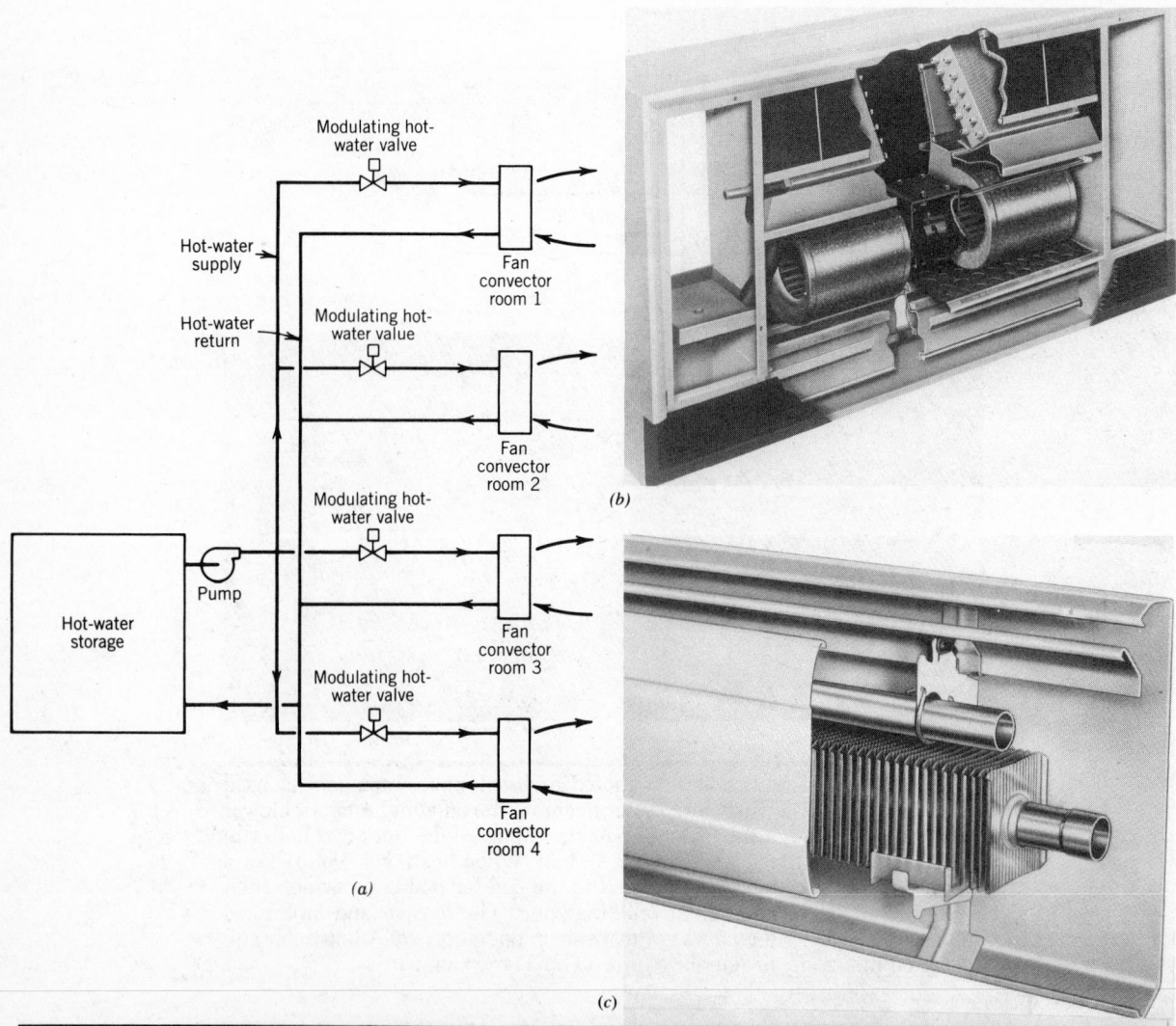

Figure 15-4 (a) Individual fan-forced convectors can be located in each room or zone to be heated. This allows zoning of the heating, since the areas not requiring heating can have the fan convectors turned off. (b) Cutaway view of individual room fan-forced convector. (Modine Manufacturing Company.) (c) Section of typical baseboard radiator–convector unit, as used in hydronic heat systems in residences or small commercial buildings. (Sterling Radiator.)

effective temperature control, and hot water from thermal storage is pumped through the baseboard convector units. Again, modulating hot-water valves are used to control heat flow to the several zones.

Radiant warm-water heating can also be used. In this case, warm water is circulated through coils in a concrete slab floor, as diagrammed in Fig. 15-5. The warm water heats the floor to 85 to 90°F, and heat is released to the room by radiation and convection from the floor. Coils may also be placed in plastered walls and ceilings.

The details of designing different heat distribution systems are outside the scope of this book. Readers involved in this work may consult any standard text on heating and air conditioning.

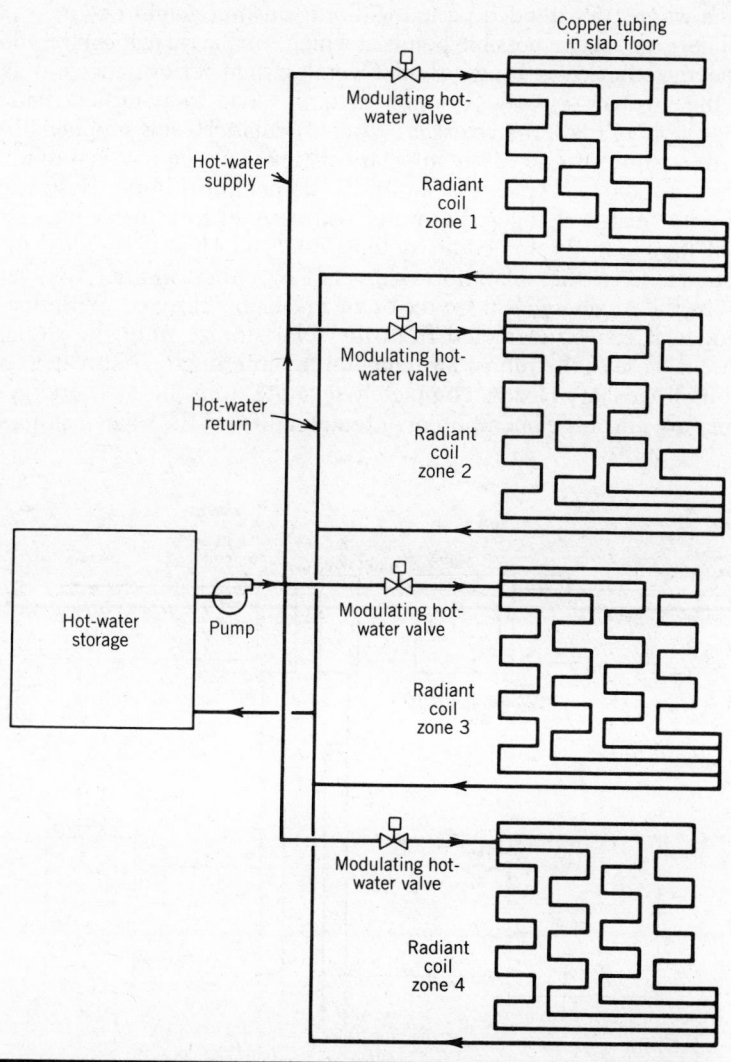

Figure 15-5 Radiant heating panels provide a very good method of distributing heat into a building from a solar energy system. With the proper design, the mean water temperature for a radiant floor system can be kept low, in the range of 105° to 115° F. Heat can be effectively used from the thermal storage until the storage reaches the radiant floor design water temperature. Thus, a radiant panel system with a design water temperature of 110° F will effectively use more of the stored solar heat than will some other type of system with a design temperature of, say, 130° F.

ACTIVE SPACE-HEATING SYSTEMS AND THEIR COMPONENTS

15-5   Backup and Auxiliary Heat Sources for Liquid Systems

As with solar hot-water heating systems, it is usually not practical or cost-effective to design a solar space-heating system to supply 100 percent of the annual space-heating requirement. Solar space-heating systems are usually designed to supply between 50 and 80 percent of the annual heating energy requirement, although in certain circumstances the design may aim at a figure above 80 percent.

There are many methods of providing backup or auxiliary heating. They include gas-fired or oil-fired furnaces or boilers, electric resistance heat, wood or coal furnaces or stoves, central steam or hot water generated offsite, and electric heat pumps. The backup system supplies auxiliary energy when the solar storage is depleted. It is approved industry practice to provide a backup system that will meet the total design heating load of the building, to meet the worst-case situation where an extended period without sunshine might occur.

There are several possible points at which Auxiliary Heat can be added to a solar-energized space-heating system. Overall system performance is maximized when the auxiliary energy is supplied to the load loop rather than to the collector loop, or to thermal storage. When Auxiliary Heat is supplied directly to the load, as shown in Fig. 15-6, auxiliary energy consumption is minimized. In this case, a forced-air system is shown as the heat distribution system. A hot-water-to-air heat exchanger (hot-water coil) is used to transfer heat from the thermal storage to the forced-air system. Auxiliary Heat is supplied to the air downstream from the solar hot-water coil, by either another hot-water coil heated by the Auxiliary Heat source or an air-heating furnace. With this configuration, heat can be extracted from the solar storage until the storage temperature is close to the return air temperature entering the hot-water coil. The addition of Auxiliary Heat is completely separate from the solar energy system and does not affect the operation of or temperatures in the solar-heating system.

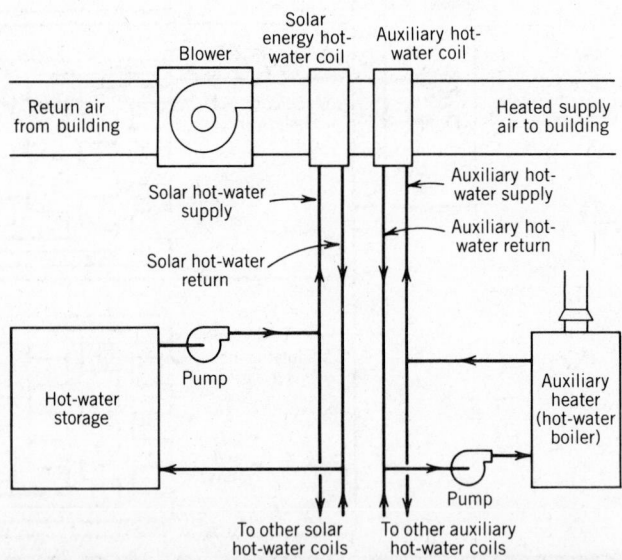

Figure 15-6 The most effective way of providing Auxiliary Heat in an active solar-heating system is to introduce it independently of the solar system. In the configuration shown, the solar hot-water coil heats the building air first. If the solar system water temperature is not high enough to provide all of the heat needed, the auxiliary hot-water coil adds heat to the air. *Note:* The auxiliary system shown (hot-water boiler and coil) could be replaced by a hot-air furnace.

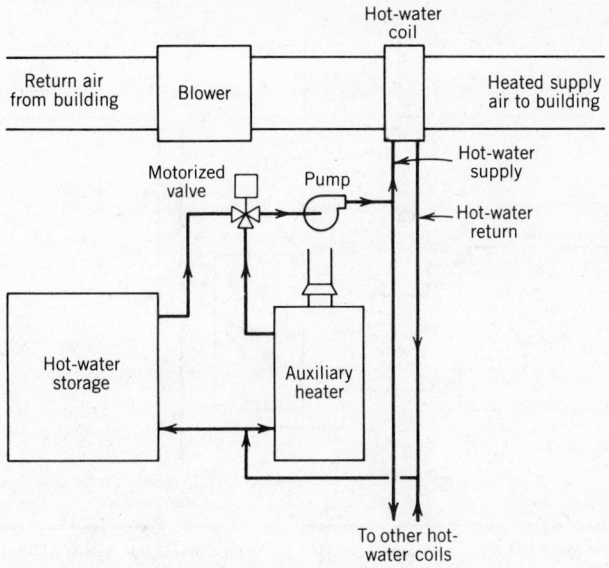

Figure 15-7 Auxiliary Heat can be added in a parallel configuration with solar heat. In this case, heat from either the solar storage or the auxiliary can be supplied to the heat distribution system, but heat cannot be supplied simultaneously from both heat sources. A motorized valve is used to switch from the solar storage to the auxiliary (hot-water boiler) when the solar storage temperature drops below the hot-water supply design temperature for the heat distribution system.

The disadvantage of this arrangement is that it requires a completely separate auxiliary heating system.

Another solar-heating system configuration is shown in Fig. 15-7. It places the auxiliary heater in parallel with the solar thermal storage. In this arrangement, solar-heated water from the thermal storage is used as long as its temperature is equal to or greater than the required design temperature of the hot-water distribution system (heating coil, fan convector, or radiant panel). When the thermal storage temperature is too low to provide adequate solar heat, circulation of solar-heated water is discontinued, and water heated by the Auxiliary Heat source is used exclusively. With this type of arrangement, the same hot-water supply and return system is used for both solar heat and Auxiliary Heat; thus, there is some economic advantage related to equipment over the system portrayed in Fig. 15-6. The disadvantage, however, is that heat can only be extracted from the solar thermal storage when its temperature is greater than the minimum operating temperature of the heat distribution system.

Auxiliary heating may also be supplied in series with the solar heat, as shown in Fig. 15-8. A conventional hot-water boiler is placed between the solar storage and the heat distribution system. Solar-heated water from storage is pumped through the auxiliary heater on its way to the heat distribution system. If required, the auxiliary heater will switch on and boost the temperature of the water. The disadvantage of this configuration is that the water returning to storage may have a higher temperature than the storage itself, which will result in adding part of the auxiliary energy to storage. This effect can be minimized in part by directing the returning water to the upper part of the thermal storage whenever the auxiliary heater is on. Another configuration for supplying heat in series with the solar energy system is to have the auxiliary heater heat the upper portion of the thermal storage tank. This configuration is shown in a later section.

ACTIVE SPACE-HEATING SYSTEMS AND THEIR COMPONENTS

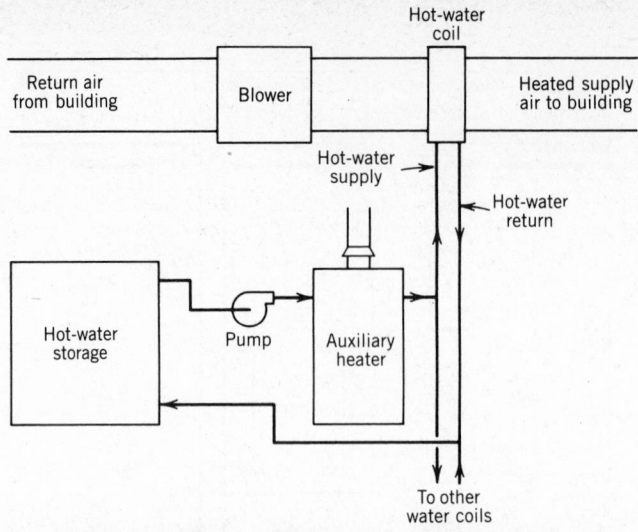

Figure 15-8 Auxiliary heater in series with the solar storage. The auxiliary heater will boost the water temperature when solar heat is inadequate to cope with the load.

15-6   Air-Based Active Space-Heating Systems

In Chapter 11, the construction and manifolding of air-heating solar collectors were discussed. Now the other components of a space-heating system energized from air-heating collectors will be examined. Figure 15-9 is a diagram of a typical air-based space-heating system. A blower (BC) provides for circulation of air through the solar collectors and thermal storage rock bed (the collector loop). Dampers D1 through D5 control the air flow in the system. Air filters should be provided to entrap dust that otherwise would enter the system. Filter 1 is located in the return air duct to filter out dust from the house return air entering the system. Filter 2 is located at the intake of the collector blower to remove dust from the air before it is circulated through the collectors. A third filter is shown in the auxiliary heater to filter the air before it is returned to the house. A water-heating coil is often placed in the collector return duct to preheat domestic hot water. Auxiliary Heat is typically supplied by a conventional air heater such as a standard gas furnace. Oil furnaces, electric resistance heaters, and conventional air-to-air electric heat pumps may also be used. The only modification that is required with a conventional furnace system is that the furnace blower motor should be capable of operating in temperatures of 150°F if it is located in the airstream. (The air supplied to the furnace is coming from the air-heating collectors or rock bed and will typically have a temperature in the range of 110°F to 150°F.) With this system, the furnace (or heat pump) is located in the load loop, which is the preferred location as discussed in section 15-5. Auxiliary energy will be added to the airstream only when needed, and it is added after the air has been heated by the solar system (either from the collectors or from thermal storage). Adding auxiliary energy in the load loop does not have any effect on the thermal performance of the solar collectors or the thermal storage.

Heat Pumps as an Auxiliary Heat Source.   When an electric heat pump is used as the Auxiliary Heat source, there are two possible configurations the system may take: the *solar-augmented heat pump* and the *solar-assisted heat pump*. The solar-augmented heat pump would have the configuration shown in Fig. 15-10. The heat pump is in parallel with the solar system, and either the solar system or the heat pump can supply heat to the building. A split heat-pump system (air to air) is most convenient; that is, the indoor unit is located with the solar equipment, and the compressor is located in the outdoor unit, the

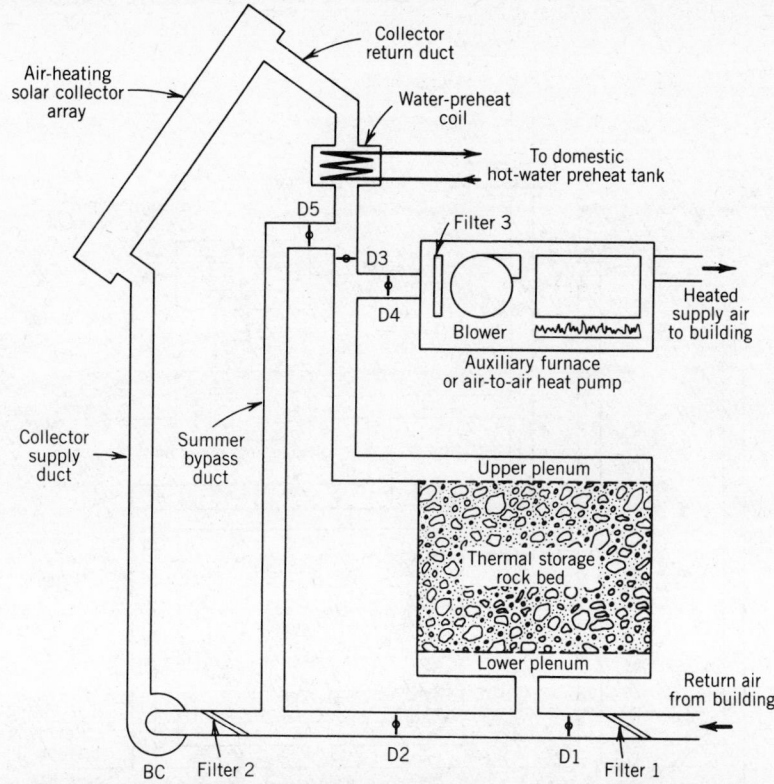

Figure 15-9 The basic components of an air-based space-heating system are the air-heating collector array, the rock bed thermal storage, the collector-loop blower, the auxiliary furnace (or heat pump), and the control dampers. A water-heating coil in the collector return duct may be added to preheat domestic hot water.

BC	Collector-loop blower
Filter 1	Return-air filter
Filter 2	Collector-loop filter
Filter 3	Auxiliary furnace filter
D1	Return-air damper
D2 and D3	Collector-loop dampers
D4	Supply-air damper
D5	Summer bypass damper

two units being connected with refrigerant lines. The indoor unit will generally have electric resistance strip heaters for auxiliary heat to meet peak heating loads. The solar-augmented heat pump is not used to boost the temperature of solar-heated air, so a heat-pump bypass duct is added between the heat pump and the return air ducting.

With a solar-augmented heat-pump system, the building load is first supplied with heat from the solar system. If solar heat is insufficient to maintain the comfort level, the solar mode is turned off (dampers D1 and D4 closed), and the damper (D6) in the heat-pump bypass duct opens, so that return air from the building is heated directly by the heat-pump system.

The disadvantages of the solar-augmented heat pump are twofold. First, all of the available heat in the solar storage may not be utilized because the heat pump may be required to operate to maintain the interior temperature of the building during the the time the solar system is bypassed. Second, the system would require peak electrical consumption during cold, sunless days, the same as a nonsolar system would.

The solar-assisted heat-pump system is one in which the solar system and the heat pump are in series. The solar system supplies heat to the evaporator of

ACTIVE SPACE-HEATING SYSTEMS AND THEIR COMPONENTS

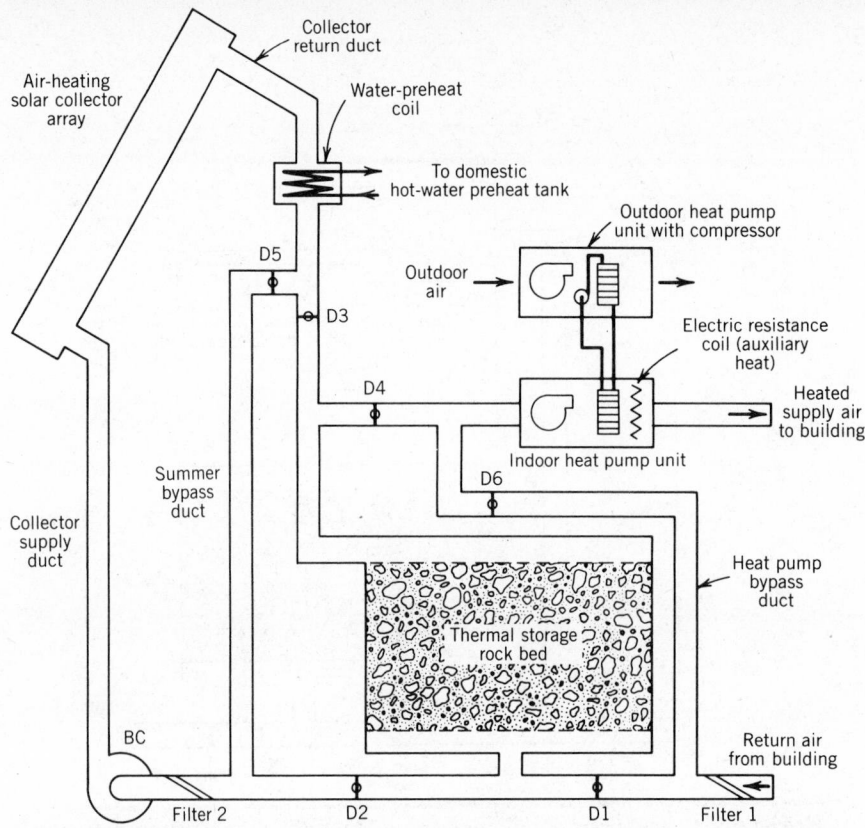

Figure 15-10 In an air-based solar-heating system with electric heat-pump auxil-iary, a heat-pump bypass duct should be used. The heat pump will operate in a parallel mode with the solar system (solar-augmented heat pump), where either the solar system or the heat pump will supply heat to the building. When the heat pump is operating, damper D6 opens and allows the heat pump to heat the return air from the building, bypassing the solar-heating system.

the heat pump, and the condenser of the heat pump supplies heat to the building. The advantage of this system is that a reduction in the peak load can be realized when the solar collector area and thermal storage are sized to provide heat to the heat pump through worst-case conditions, which minimizes the need for operating auxiliary electric strip heaters. Solar-assisted heat pumps will be discussed further in section 17-10.

Operating Modes for an Air-Based System. An air-based system has four operating modes, which are determined by the positions of the dampers in the ducting, as shown in Fig. 15-11. Solar heat is stored when the collectors are warmer than the thermal storage and the building does not require heat, as shown in Fig. 15-11a. Dampers D3 and D2 are open, and the blower (BC) draws cool air from the bottom of the rock bed and circulates it through the solar collectors to be heated. The heated air returns to the rock bed and flows down through the rocks and heats them progressively from top to bottom during sunny hours of the day. The very large heat-transfer surface of a rock bed effectively removes the heat from the air. When the differential sensor in the system controller (not shown) senses that the collectors are no longer providing a sufficient air temperature rise, the collector blower (BC) is turned off, and damper D3 is closed to prevent air circulation through the collector loop. Damper D3 must be a motorized damper that is operated by the system control-ler. Damper D2 may be either a motorized damper or a gravity backdraft damper that allows air to flow in only one direction, from the bottom of the rock

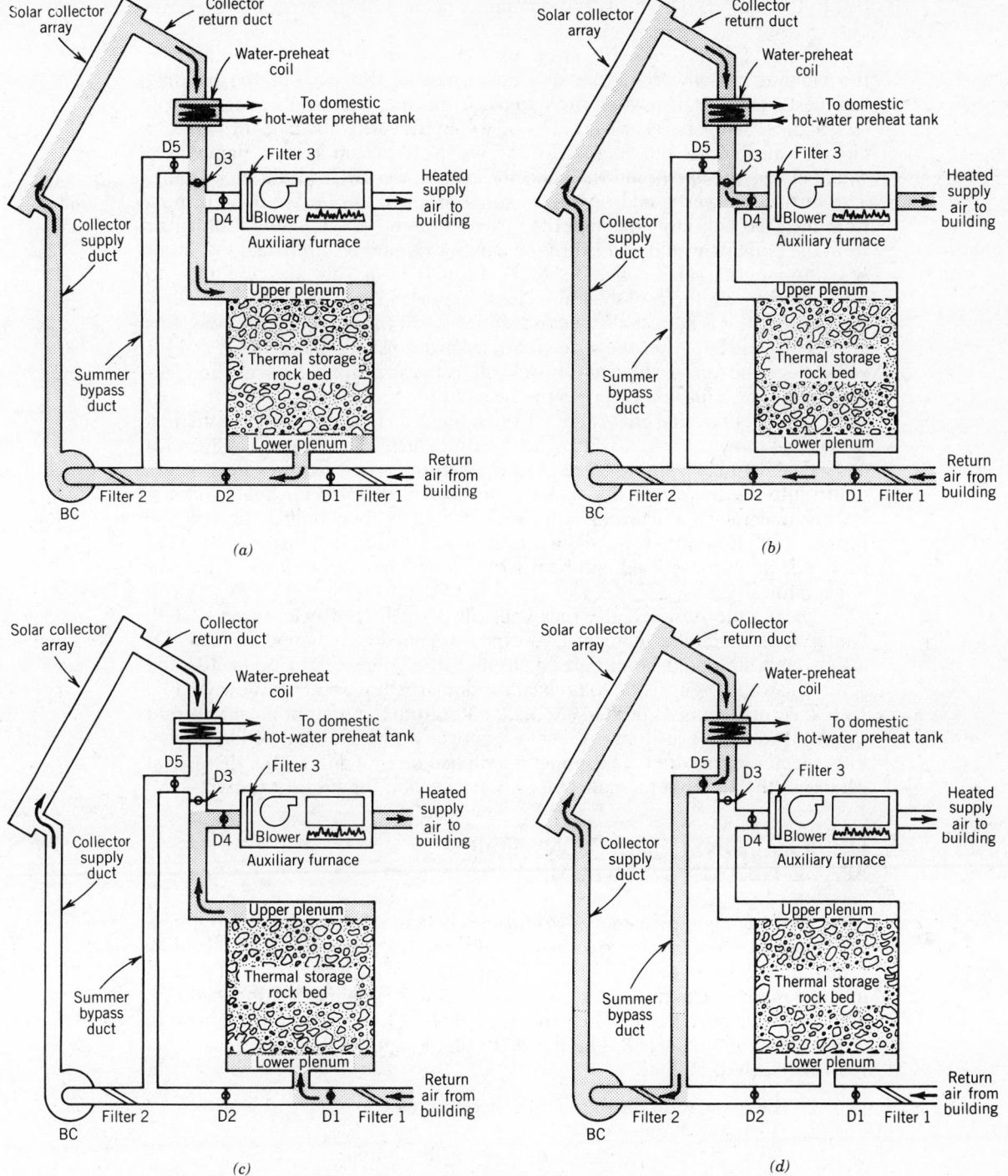

Figure 15-11 There are four basic operating modes for an air-based space-heating system: (*a*) Solar heat supplied to thermal storage by the solar collectors. (*b*) Solar heat supplied directly to the building by the solar collectors. (*c*) Heat removed from the solar storage to be supplied to the building. (*d*) Summer heating of domestic hot water.

ACTIVE SPACE-HEATING SYSTEMS AND THEIR COMPONENTS

bed to the blower. The backdraft damper is opened by air pressure when the blower is on.

Figure 15-11b shows the air circulation pattern when heat is supplied to the building directly from the solar collectors. In this case, if the building thermostat calls for heat when the system is collecting heat, the system controller will open dampers D4 and D1. The blower in the furnace will be turned on to circulate air into the building. Since the system is collecting heat, damper D3 would already be open, and the collector loop blower (BC) would be running. The furnace blower would now draw some or all of the heated air returning from the solar collectors through the furnace and into the building. Return air from the building would enter through damper D1 and be circulated by blower BC through the solar collectors to be heated. Some air may continue to circulate through the rock bed also. Heat may also be added by the auxiliary furnace, but this is not usually necessary during a sunny day when the collectors are heating the air. After the space thermostat is satisfied, dampers D4 and D1 will close as the furnace blower is turned off by the system controller. The solar-heating system will then revert to the heat-storing mode.

At night or during cloudy days, the heating load must be met from thermal storage, as shown in Fig. 15-11c. The system controller will open dampers D4 and D1 and turn on the furnace blower when the space requires heat. Air is drawn into the bottom of the rock bed through the return air duct, and it is heated as it moves up through the rock bed. It is then pulled through the furnace and blown into the building. If there is not sufficient heat stored in the rock bed, the furnace itself can be used as an auxiliary to boost the supply air temperature.

During the summer, the rock bed will normally not need to be heated. Under these conditions, damper D5 would be opened with dampers D2 and D4 closed, and air would be circulated through the collectors to be heated and across the water preheat coil to preheat the domestic hot water, as shown in Fig. 15-11d. A small pump is used to circulate water from the solar preheat tank (not shown) through the air-to-water heat-exchanger (water-preheat) coil located in the collector return duct. This pump is switched on by a differential thermostat whenever the air from the collectors is warmer than the preheat tank.

DESIGN CONSIDERATIONS FOR SPACE-HEATING SYSTEMS

After the building design moves from the schematic design phase into design development, more specific information with respect to the active solar-heating system must be considered. The building exterior skin will be fairly well defined in the design development phase, and an accurate building heating-load estimate will have been made. This estimate will allow the design of the system to proceed and be optimized. The details in the design development phase that must be resolved include:

1. Final decision on the exact collector area.
2. Thermal storage volume.
3. Type of heat distribution system for the building.
4. Types of controls required.
5. Nature of the auxiliary heat source.
6. Locations for the equipment and space allocations for piping and air ducts.

The first step in finalizing the system design is to estimate accurately the building-heating requirements.

15-7 Estimating the Building-Heating Load

The detailed method for calculating building-heating loads was discussed in Chapter 5. The two heating loads that are of interest in designing a space-heating system are the *design heating load* (Btu per hour) and the *average monthly heating load* (Btu per month) (See Sections 5-7 to 5-9). The space-heating system must be designed to meet the design heating load of the building. This means that the auxiliary heating system must be designed so that it alone is capable of supplying the total design heating load of the building. This is necessary because at times there may be no heat available from the solar system for several days because of inclement weather. The design heating load usually represents the worst-case design point. The winter design temperature may actually occur for only 2 or 3 percent of the days in a heating season for a typical year, since it is chosen near the 20-year low for a given locality. During a very severe winter however, the design temperature may occur for several days in sequence, and a building's heating system must be capable of supplying the design heating load in order to maintain the interior-space temperatures at a comfortable level. The solar collector area itself is not ordinarily sized for the design heating load, because this would grossly oversize the solar collection area for all but the worst-case years. Collector areas are typically sized based on the average monthly heating load as calculated from Eq. (5-6):

$$Q_{\text{month}} = (UA + 0.018V)\frac{\text{Btu}}{\text{hr-F}^\circ} \times DD\frac{\text{F}^\circ\text{-day}}{\text{month}} \times 24\frac{\text{hr}}{\text{day}} \qquad (5\text{-}6)$$

The degree-day approach is based on the degree-day temperature difference, which is the difference between a base temperature (generally 65°F) and the daily average temperature.

For single-family residences and small multiple-family residences such as two- and three-story apartment houses, the degree-day base temperature of 65°F is ordinarily used. However, when it is necessary to take into consideration internal heat gains, as in commercial buildings where there is a large generation of internal heat, the method used in Chapter 9, section 9-12, establishes a new degree-day base temperature based on the internal heat gain. This new degree-day base temperature can be calculated from Eq. (9-17):

$$T_b = T_{\text{set}} - \frac{Q_{\text{int}}}{\text{BLC}} \qquad (9\text{-}17)$$

where

T_b = the new base temperature, °F
T_{set} = the desired interior temperature, °F
Q_{int} = the rate of internal heat generation, Btu/day
BLC = the building load coefficient, Btu/degree-day

The above equation can also be used to compensate for internal gains inside well-insulated residences when the internal gains are significant compared to the total load.

The average monthly space-heating load for the building can then be calculated as:

$$Q_{\text{month}} = \text{BLC} \times DD \qquad (15\text{-}2)$$

where:

Q_{month} = the average monthly space-heating load, Btu/month
BLC = the building load coefficient, Btu/DD
DD = the monthly degree-days for a base temperature of 65°F or for a base temperature T_b when internal heat gains are considered

The average monthly space-heating load should be calculated using Eq. (15-2) for each month of the heating season. It is the average monthly heating load that provides the basis for sizing the solar collector area and the thermal storage.

15-8 Determining the Type of System

A decision on the type of system (liquid-based or air-based) must be made before considering detailed design. This decision will be influenced by the type of heat distribution system desired by the architect, builder, or owner. If, for instance, a

DESIGN CONSIDERATIONS FOR SPACE-HEATING SYSTEMS

Table 15-3 System Parameters for Standard Reference Liquid-Based Solar Space-Heating System

Values of parameters used for the standard solar-heating system using liquid-heating solar collectors, a heat exchanger, water tank thermal storage, and forced-air heat distribution system to the building. The values are normalized to 1 ft² of collector (ft²$_c$).

Parameter	Nominal Value
Solar Collector	
1. Orientation	Due South
2. Tilt (from horizontal)	Latitude + 10°
3. Number of glazings	One
4. Glass transmittance (at normal incidence)	0.86 (6% absorption, 8% reflection)
5. Surface absorptance (solar)	0.98
6. Surface emittance (infrared)	0.89
7. Coolant flow rate × specific heat of coolant	20 Btu/hr-F°-ft²$_c$
8. Heat transfer coefficient to liquid coolant	30 Btu/hr-F°-ft²$_c$
9. Back insulation *U* value	0.083 Btu/hr-F°-ft²$_c$
10. Heat capacity	1 Btu/F°-ft²$_c$
Collector Plumbing	
11. Heat loss coefficient (to ambient)	0.04 Btu/hr-F°-ft²$_c$
Heat Exchanger	
12. Heat transfer coefficient	10 Btu/hr-F°-ft²$_c$
Thermal Storage	
13. Heat capacity	15 Btu/F°-ft²$_c$
14. Heat loss coefficient[a]	0 Btu/hr-F°-ft²$_c$
Heat Distribution System	
15. Design water distribution temperature[b]	133°F
Controls	
16. Building maintained at	68°F
17. Collectors on when energy can be collected	

Source: *Pacific Regional Solar Heating Handbook*, U.S. Energy Research and Development Administration, San Francisco Operations Office, November 1976, p. 17. Reprinted with permission.

[a] It is assumed that all heat loss from the thermal storage is into the heated space.

[b] The hot-water coil and air circulation are sized to meet the building load with an outside temperature of −2°F with 133°F water and an air-flow rate adequate to make up the space heat losses at an air discharge temperature of 120°F.

forced-air system is desired, then an air-based solar system would interface directly with the forced-air heating system. On the other hand, if a hydronic space-heating system is selected, then a liquid-based solar system would be indicated. If the systems are carefully designed and installed, the overall system performance of a liquid-based system and an air-based system of similar collector areas will be quite comparable. System studies done at the Los Alamos National Laboratory and published in the *Pacific Regional Solar Heating Handbook* show this to be the case.[2] Table 15-3 gives the parameters that describe the reference liquid-based solar-heating system studied at Los Alamos, and Table 15-4 gives the parameters that describe the reference air-based solar system.

[2] *Pacific Regional Solar Heating Handbook*, U.S. Energy Research and Development Administration, San Francisco Operations Office, March 1976, p. 15.

Table 15-4   System Parameters for Standard Reference Air-Based Solar Space-Heating System

Values of parameters used for the standard solar-heating system using air-heating solar collectors, rock bed thermal storage, and a forced-air heat distribution system to the building. The values are normalized to 1 ft² of collector (ft²$_c$).

Parameter	Nominal Value
Solar Collector	
1. Orientation	Due South
2. Tilt (from horizontal)	Latitude + 10°
3. Number of glazings	One
4. Glass transmittance (at normal incidence)	0.68 (6% absorption, 8% reflection)
5. Surface absorptance (solar)	0.98
6. Surface emittance (infrared)	0.89
7. Coolant flow rate	2 ft³/min-ft²$_c$
8. Heat transfer effectiveness	4 Btu/hr-F°-ft²$_c$
9. Back insulation U value	0.083 Btu/hr-F°-ft²$_c$
10. Heat capacity	0.5 Btu/F°-ft²$_c$
Collector Duct Work	
11. Heat loss coefficient	0.1 Btu/hr-F°-ft²$_c$
Thermal Storage	
12. Heat capacity	15 Btu/F°-ft²$_c$
Heat Distribution System	
13. Heat loss coefficient	0 Btu/F°-ft²$_c$
Controls	
14. Building maintained at	68°F
15. Collectors on when energy can be collected	

Source: Pacific Regional Solar Heating Handbook, U.S. Energy Research and Development Administration, San Francisco Operations Office, November 1976, p. 45. Reprinted with permission.

a The heat transfer effectiveness is the product of the effective heat transfer coefficient of the collector times the effective heat transfer area that the air flows across in the collector. The value is normalized to the area of the collector.

The reference active solar-heating systems listed in Tables 15-3 and 15-4 were analyzed for system performance in a number of different climates. System performance was also studied as system parameters were varied—for example, collector tilt, the number of collector glazings, the collector area, and so on. Figure 15-12 shows the performance of liquid-based and air-based space-heating systems at Fresno, California. The performances of the two systems are indeed very nearly identical, and the choice between the two is ordinarily determined by considerations other than just system performance. Some of these other considerations will now be discussed.

Air systems often have certain cost advantages in new installations. Air collectors have a simpler construction in that their absorbers do not have to be liquid-tight. The air collector absorber plate has only to transfer heat into the airstream flowing across the absorber. Also, the air collector operates at a pressure only slightly above atmospheric pressure. Because of the simplicity of air collectors, they can be fabricated on the building site with reasonable success and economy. Considerable savings can sometimes be realized with a site-fabricated air collector system when compared to the cost of liquid-based collectors. Another advantage of an air system is that an air leak in the system does not have the disastrous results that a leak in a liquid system can have. While air leaks are certainly to be avoided, generally the only negative result is that the performance of the system is reduced somewhat. Piping system corro-

DESIGN CONSIDERATIONS FOR SPACE-HEATING SYSTEMS

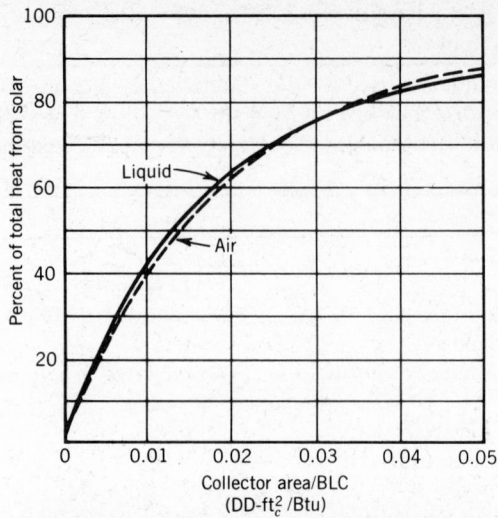

Figure 15-12 A comparison of air-based and liquid-based space-heating systems for a location in Fresno, California. The solar collector area has been normalized to the building load coefficient, and this ratio is plotted along the horizontal axis. The percent of total heat supplied by the solar system is plotted along the vertical axis. (From *Pacific Regional Solar Heating Handbook*, U.S. Energy Research and Development Administration, San Francisco Operations Office, November 1976, p. 15. Reprinted with permission.)

sion—common in liquid-based systems—is not a problem with air-based systems.

One disadvantage of air systems is that air ducts and rock beds are much bulkier than the pipes and tanks used in liquid-based systems. For example, a 1500-gal water tank will take up about 224 ft³ of volume. This tank will provide the same capacity of heat storage that a 625-ft³ rock bed does. It is therefore more difficult to retrofit an air-based system into an existing building than it is a liquid system because of the physical size of such components.

15-9   Effects of System Components on System Performance

The effects of varying different system parameters on active heating system performance has been intensively studied at the Los Alamos National Laboratory for the two reference systems listed in Tables 15-3 and 15-4. The results of these studies are summarized here to provide an understanding of how the varying of different aspects of an active system may affect its overall performance.

Collector Considerations.   The optimum collector orientation is generally due south for both liquid-based and air-based systems. It was found that a variation of up to 30° east or west of due south reduced the system performance from 2.4 to 5.0 percent. Variations greater than 30° from due south reduced the performance substantially. In locations with morning fog or cloud cover, westerly orientations affected the performance less than an equal easterly orientation. The optimum collector tilt has been found to be somewhat dependent on the local climate. The optimum tilt has a range from latitude plus 10° up to latitude plus 25°. Common practice is to use a collector tilt of latitude plus 10° to 15°.

The monthly distribution of the heating load also has an effect on the optimum collector tilt. If the heating load is concentrated in a two- or three-month period, the collectors should be tilted to maximize the solar energy collection during this period. Conversely, if the heating load is distributed over a six-month or longer period, then the collector tilt should be chosen to provide balanced collector performance over this longer period.

The collector glazing provides a cover transparent to solar radiation and reduces the energy losses from the absorber plate caused by convection and radiation. Solar collectors, both liquid and air, are ordinarily available with either single glazing or double glazing, as was discussed in Chapter 11. The decision to use single- or double-glazed collectors depends on the added cost of the second glazing compared to the performance gained. Table 15-5 gives decimal fraction multipliers by which the collector area may be reduced if single glazing is replaced by double glazing. The decimal fractions are ratios of the double-glazed collector area to single-glazed collector area. The improvement in system performance as reflected by reduced collector area is basically dependent on how cold the location is, as shown by the January degree-days in Table 15-5. The table shows that for the reference systems of Tables 15-3 and 15-4, with a 75 percent solar heating fraction and a flat-black absorber coating in Phoenix, Arizona, for example, about 94 percent of the single-glazed collector area is needed when double glazing is used. In Bismarck, North Dakota, only 74 percent of the single-glazed collector area is needed when double glazing is used. The use of double glazing will provide a significant improvement in cold climates for both liquid and air systems, but no great improvement in warmer climates. Whether double glazing is used, then, will depend on a detailed cost–benefit analysis.

The effect of the glazing transmittance for both liquid and air systems was found to have a relatively small effect on overall system performance. When the glazing transmittance was varied by 1 percent, the yearly system performance varied by only 0.36 to 0.94 of 1 percent, depending on the site location. For standard float glass, the absorptance is 0.06 and the reflectance is 0.09, which gives a transmittance of 0.85 (see Table 11-2). By using a low-iron glass, the transmittance can be increased to 0.90 (see Table 11-2). If low-iron glass is used to replace standard glass, then the system performance would increase by approximately 1.8 to 4.7 percent for a single-glazed collector, depending upon the location and size of the solar energy system.

The absorptance of the collector absorber plate is another important factor in system performance. Most collectors have either a black-painted or a

Table 15-5 Decimal Fraction Ratios of Collector Area for Double-Glazed Collectors, Compared to Area for Single-Glazed Collectors, for Selected Locations with a Variation in Heating Degree-Days

Location	January Degree-Days	System Solar Heating Fraction	
		75%	40%
Phoenix, AZ	425	0.94	0.97
Santa Maria, CA	453	0.94	0.94
Fresno, CA	629	0.91	0.95
Medford, OR	862	0.86	0.91
Seattle, WA	753	0.84	0.88
Bismarck, ND	1730	0.74	0.85

Source: Pacific Regional Solar Heating Handbook, U.S. Energy Research and Development Administration, March 1976, p. 24. Adapted with permission.

Note: Collector area ratio (flat-black absorber coating): $\dfrac{\text{Double-glazed}}{\text{Single-glazed}}$

Table 15-6 Decimal Fraction Ratios of Collector Area for a
Selective Surface Collector Absorber Plate Compared to a Flat-Black
Surface Collector Absorber Plate (collectors are assumed to be
single-glazed)

| Location | January Degree-Days | System Solar Heating Fraction | |
		75%	40%
Phoenix, AZ	425	0.84	0.85
Santa Maria, CA	453	0.82	0.82
Fresno, CA	629	0.79	0.83
Medford, OR	862	0.70	0.87
Seattle, WA	753	0.69	0.75
Bismarck, ND	1730	0.62	0.73

Source: Pacific Regional Solar Heating Handbook, U.S. Energy Research and
Development Administration, March 1976, p. 29. Adapted with permission.

Note: Collector area ratio: $\dfrac{\text{Selective surface area}}{\text{Flat-black surface area}}$

selective-coated absorber surface. These surfaces will typically have absorptances in the range of 0.92 to 0.98. The overall system performance for both liquid and air systems was found to decrease by 0.44 to 0.67 of 1 percent as the surface absorptance was decreased by 1 percent. Thus, the difference in performance with an absorptance of 0.92 compared to an absorptance of 0.95 is quite small. In any case, an absorptance of less than 0.90 should not be used. The absorber coating should have good durability for the temperature at which the solar collector will be operating.

It will be recalled that the infrared emittance of the solar collector absorber has a significant effect on the performance of a solar collector (section 11-4). The *overall system* performance will also be affected by the infrared emittance of the solar collector absorber surface. The two absorber surfaces used for solar collectors are a flat-black surface and a selective surface. The flat-black surface will have an infrared emittance in the range of 0.85 to 0.95, while the selective surface will have an infrared emittance in the range of 0.10 to 0.15.

Table 15-6 shows the improvement in system performance as indicated by collector area ratios for single-glazed collectors. Ratios of selective-surface collector area to flat-black (matte) surface collector area are given for two different SHFs. For example, if a solar space-heating system is designed for an SHF of 0.75 in Medford, Oregon, with 500 ft² of solar collector using a flat-black absorber surface, then only 0.70 × 500 ft² = 350 ft² of solar collector with a selective coating absorber surface would be required to provide similar solar system performance. The use of a selective surface with a single-glazed collector reduces the radiation losses from the collector appreciably, allowing a smaller collector area to be used. There is usually little need to use double glazing with a selective surface for normal space-heating systems. If very high temperatures are needed, such as temperatures over 180°F, or if the location has very cold daytime temperatures in winter, then there may be some benefit to using double glazing in combination with a selective surface.

Illustrative Problem 15-4

An active space-heating system is planned for Santa Maria, California. The collector area for collectors with single glazing and flat-black absorber is 500 ft² for a system with a 75 percent solar heating fraction. A collector manufacturer can provide a collector with single glazing and flat-black absorber for $10.50/ft², or a similar collector with a selective absorber for $12.73/ft². Determine if there is any advantage in using the selective absorber.

Solution

The cost of the collectors with flat-black absorber is:

$$Cost_1 = 500 \text{ ft}^2 \times \$10/\text{ft}^2$$
$$= \$5000.00$$

According to Table 15-6, a selective surface will reduce the collector area by a factor of 0.82. The selective surface collector area needed will be:

$$A_c = 0.82 \times 500$$
$$= 410 \text{ ft}^2$$

The cost of the collectors with the selective absorber is:

$$Cost_2 = 410 \text{ ft}^2 \times \$12.73/\text{ft}^2$$
$$= \$5219.30$$

The collector cost for the selective absorber is greater than the cost for the flat-black collector, so, based only on collector costs, there is no advantage to using a selective absorber. But, since less collector area is required with the selective absorber, fewer collectors can be installed. The cost of installing the 410 ft² of selective-absorber collectors will be less than the cost of installing the 500 ft² of flat-black absorber collectors. If the labor cost of installing the 410 ft² of selective-absorber collectors saves more than $200 compared to installing 500 ft² of the flat-black absorber collectors, then the use of selective-absorber collectors should be considered.   *Ans.*

Illustrative Problem 15-5

An active space-heating system planned for Bismarck, North Dakota requires 1000 ft² of solar collector for a 75 percent solar heating fraction. How much savings in collector cost can be realized by using a single-glazed collector with a selective absorber that costs $13.20 per square foot, compared to a single-glazed collector with a flat-black absorber that costs $10.50 per square foot?

Solution

The cost of the collectors with flat-black absorber is:

$$Cost_1 = 1000 \text{ ft}^2 \times \$10.50/\text{ft}^2$$
$$= \$10,500$$

The use of a selective absorber reduces the required collector area by a factor of 0.62. The required selective surface collector area is then:

$$A_c = 0.62 \times 1000 \text{ ft}^2$$
$$= 620 \text{ ft}^2$$

The cost of the collectors with a selective absorber is:

$$Cost_2 = 620 \text{ ft}^2 \times \$13.20/\text{ft}^2$$
$$= \$8184$$

There is a cost savings of $2316 realized by using selective-absorber collectors in this case.   *Ans.*

Flow Rate Through Collectors.   The flow rate of the heat-exchanger fluid through solar collectors is important because it determines the operating temperature of the collectors. This factor is one that the system designer can control. For liquid collectors, the flow rate should be in the range of 0.02 to 0.03 gal/min-ft²$_c$ when water is used. This provides a heat removal rate of 10 to 15 Btu / hr-F°-ft²$_c$. When fluids with a lower specific heat than water are used, the fluid flow rate should be increased to provide about the same heat removal rate.

For air collectors, the air-flow rate through the collector has a pronounced effect on system performance, because heat transfer to air in the air collector is dependent on the air-flow rate. The performance of an air system decreases severely when the air-flow rate drops below 1.0 CFM/ft$_c^2$. For most air systems, the optimum air-flow rate can be determined from the manufacturer's specification for a factory-manufactured unit. The range for air-heating collectors is recommended as 2.0 to 4.0 CFM/ft$_c^2$. In the case of a site-fabricated collector system, the collector absorber plate should be designed to maximize heat transfer from the plate to the air flowing through the collector, as was discussed in section 11-3.

Heat losses between collectors and thermal storage can be minimized by using an adequate level of insulation on the collector-loop piping or ducting. Heat losses from the piping or ducting can be kept reasonably low by using enough insulation to keep the heat loss rate at 0.04 Btu/hr-F°-ft$_c^2$ or less. In liquid systems, the collector loop piping is typically in the range of 1¼ to 2 in. in diameter for a residential-sized system, resulting in a relatively small surface area to insulate. In the case of air systems, where the duct diameter may be in the range of 12 to 16 in., there is a very large duct surface area to insulate.

Illustrative Problem 15-6

An active space-heating system has a collector array with 1000 ft^2 of area. The total length of collector-loop piping or ducting between the collector array and the thermal storage is 100 ft. It is desired to have a heat loss rate of 0.03 Btu/hr-F°-ft^2 collector.

(a) Determine the R value for the insulation required for a 2-in. diameter pipe, and for a 14-in.-diameter air duct.

(b) If fiberglass insulation is used with an R value of 3.5/in., determine the thickness of the insulation for the piping and ducting.

Solution

(a) The heat loss rate is normalized to the collector area, so to find the heat loss for the system, the collector area must be multiplied by the normalized heat loss rate:

$$UA = 1000 \text{ ft}^2 \times 0.03 \text{ Btu/hr-F}°\text{-ft}^2 \text{ collector}$$
$$= 30 \text{ Btu/hr-F}°$$

The surface area of the 2-in. piping is calculated as:

$$A = \pi DL \quad (D = \text{diameter of pipe}; L = \text{length})$$
$$= \pi \times \frac{2 \text{ in.}}{12 \text{ in./ft}} \times 100 \text{ ft}$$
$$= 52.4 \text{ ft}^2$$

The necessary heat loss coefficient for the piping is found by:

$$U = \frac{UA}{A} = \frac{30 \text{ Btu/hr-F}°}{52.4 \text{ ft}^2}$$
$$= 0.57 \text{ Btu/hr-ft}^2\text{-F}°$$

The R value of the insulation for the 2-in. pipe is then

$$R = \frac{1}{U} = \frac{1}{0.57}$$
$$= 1.75 \qquad\qquad Ans.$$

The surface area of the 14-in. duct is calculated as:

$$A = \pi DL$$
$$= \pi \times \frac{14 \text{ in.}}{12 \text{ in./ft}} \times 100 \text{ ft}$$
$$= 367 \text{ ft}^2$$

The necessary heat loss coefficient for the ducting is found by

$$U = \frac{UA}{A} = \frac{30 \text{ Btu/hr-F}°}{367 \text{ ft}^2}$$
$$= 0.082 \text{ Btu/hr-ft}^2\text{-F}°$$

The R value of the insulation for the duct is then:

$$R = \frac{1}{U} = \frac{1}{0.082}$$
$$= 12.2 \qquad \qquad Ans.$$

(b) The thickness t of the fiberglass insulation required for the pipe insulation is:

$$t = \frac{R \text{ value}}{R \text{ value/in.}}$$
$$= \frac{1.75}{3.5/\text{in.}}$$
$$= 0.5 \text{ in.} \qquad \qquad Ans.$$

The thickness of the fiberglass insulation required for the ducting is:

$$t = \frac{12.2}{3.5/\text{in.}}$$
$$= 3.5 \text{ in.} \qquad \qquad Ans.$$

The above example shows that duct insulation must be much thicker compared to pipe insulation in order to keep the heat losses from the collector loop for an air system equal to those for a liquid system.

15-10   Thermal Storage Considerations

The total performance of an active system can be severely reduced if sufficient thermal storage is not provided. When a condition of insufficient storage exists, the storage will be heated to a high temperature, resulting in the collectors operating at a higher temperature. The efficiency of the collectors will therefore be lowered, and the heat losses from the collector loop and thermal storage will be increased. Consequently, the overall system performance will be reduced. Operating experience has shown that the storage heat capacity (C_c) should be at least 10 Btu/F°-ft^2_c (1.2 gal of water/ft^2_c). Values less than this will generally result in a significant loss of system performance. Thermal storage capacities above 15 Btu/F°-ft^2_c do not adversely affect system performance, but they do not significantly improve it either. Furthermore, there will be increased thermal losses from the unnecessarily large thermal storage.

Storage capacity in the range of 10 to 15 Btu/F°-ft^2_c will generally provide about 24 hours of heat storage. For example, if a solar collector system heats up the thermal storage during a sunny day and then the following day is cloudy, at some point during the second day the heat stored will be used up and Auxiliary Heat will be required.

Some climates have a cyclical pattern of clear and cloudy days during the winter. In this case, it may be advantageous to design the system with enough thermal storage so that heat can be provided for several sunless days. This would also require that the solar collector area be large enough to collect sufficient energy during sunny periods to fill up the thermal storage with heat for the sunless periods. It generally is not economic to provide thermal storage capacity for more than three to five days. The following formula can be used to determine the carry-through time for a thermal storage system—that is, the elapsed time during which the thermal storage is able to maintain the interior space temperature at the desired room temperature.[3]

[3] *Pacific Regional Solar Heating Handbook*, U.S. Energy Research and Development Administration, San Francisco Operations Office, March 1976, p. 42. Adapted with permission.

$$T_{ct} = \frac{24\frac{hr}{day} \times S_T}{BLC} \times \left[\frac{t_s - t_{rm}}{t_{rm} - t_a} - \frac{t_{dw} - t_i}{t_i - t_o}\right] \qquad (15\text{-}3)$$

where

T_{ct} = carry-through time, hr

S_T = thermal capacity of the storage, Btu/F°

t_s = initial storage temperature, °F

t_{rm} = room thermostat setting, °F

t_{dw} = design water temperature for the space-heating system, °F

t_a = ambient temperature, °F (assumed a constant, which for cloudy winter weather is a reasonable assumption)

t_i = indoor design temperature, °F

t_o = outdoor design temperature, °F

BLC = degree-day building load, Btu/DD

After the carry-through time has been exceeded, heat would continue to be extracted from the thermal storage, but additional heat from the auxiliary would have to be used to maintain the desired room temperature.

Illustrative Problem 15-7

A space-heating system has been designed with a thermal storage capacity of 15,000 Btu/F°. The outdoor design temperature is 28°F, the degree-day building load is 18,000 Btu/DD, the design water temperature is 130°F, and the interior design temperature is 70°F. Determine the carry-through time if the thermal storage has an initial temperature of 150°F, the room thermostat setting is 68°F, and the ambient temperature is 40°F.

Solution

Using Eq. (15-3) and substituting values,

$$T_{ct} = \frac{24\frac{hr}{day} \times 15{,}000 \text{ Btu/F°}}{18{,}000 \text{ Btu/F°-day}} \times \left[\frac{150°F - 68°F}{68°F - 40°F} - \frac{130°F - 70°F}{70°F - 28°F}\right]$$

$$= 20 \text{ hr} \times \left[\frac{82}{28} - \frac{60}{42}\right]$$

$$= 20 \text{ hr} \times (2.93 - 1.43)$$

$$= 30 \text{ hr} \qquad \qquad Ans.$$

Heat losses from the thermal storage can have serious effects on system performance. To minimize losses, the best location for the thermal storage is either within or beneath the heated space. In this case, any heat that is lost from the thermal storage will enter the heated space, and there is no net thermal loss from the system. Even though the thermal storage is located within the heated space, it still must have insulation, because one would not want the thermal storage to lose heat so rapidly to the interior space that overheating occurs. When the thermal storage is located within the heated space, it should have at least R-12 insulation.

When the thermal storage is located outside the heated space, the insulation should have a minimum value of R-20. In cases where the thermal storage is a tank buried in the ground, the tank should have a moistureproof jacket around it to prevent moisture from entering the insulation. A gravel drainage system should also be used to prevent ground water from accumulating around the tank.

SYSTEM PERFORMANCE FOR ACTIVE SOLAR SPACE HEATING

The performance of the two reference space-heating systems cited in Tables 15-5 and 15-6 was extensively studied at the Los Alamos National Laboratory. One result of these studies is a monthly Solar Load Ratio method, which gives the monthly solar heat fraction and the annual solar heat fraction. It should be noted that, while the Solar Load Ratio method for active systems is similar in concept to the Solar Load Ratio method for passive systems, there is a difference in the terminology and in the calculation procedure, as will be explained.

15-11 Solar Load Ratio (SLR) Method for Active Solar Space Heating

The Solar Load Ratio (SLR) for active systems with the characteristics indicated in Tables 15-5 and 15-6 is defined as the ratio of the total solar energy incident on the collectors per month to the total energy required to heat the building. The Solar Load Ratio can be calculated as:

$$ \text{SLR} = \frac{A_c \times \bar{I}_M}{\text{BLC} \times \text{DD}_{t_b}} \qquad (15\text{-}4) $$

where

SLR = the Solar Load Ratio for a given month
A_c = the collector area, ft^2
$\bar{I}_M$ = the average monthly insolation on the collector surface, Btu/ft^2-month
BLC = the building load coefficient, Btu/DD
DD_{t_b} = the monthly degree-days for a base temperature of t_b [t_b is taken as 65°F unless one wishes to compensate for internal heat gains, in which case a different base temperature is calculated using Eq. (9-17)]

In previous chapters, two methods of estimating the average monthly insolation have been utilized. One method is to use the clear-day insolation from Table 4-E and the average percentage of possible sunshine from Table 4-C. The average monthly insolation can then be determined as:

$$ \bar{I}_M = I \times P \times N \qquad (15\text{-}5) $$

where

$\bar{I}_M$ = the average monthly insolation on the collector surface, Btu/ft^2-month
I = the clear-day insolation from Table 4-E for the collector tilt and orientation, Btu/ft^2-day
P = the percentage of possible sunshine for the particular month and location, from Table 4-C
N = the number of days per month

The second method of determining the average monthly insolation was developed in Chapter 9. The average monthly incident solar radiation on a horizontal surface can be corrected by using Fig. 9-26 to find the average incident solar radiation on the collector surface using the parameters of collector tilt, collector azimuth, the latitude minus the midmonth solar declination (LD), and the monthly clearness ratio for the location (K_T). The monthly incident solar radiation on the collector surface can be calculated as follows:

$$ \bar{I}_M = HS \times \frac{\text{Incident}}{\text{Horizontal}} \times N \qquad (15\text{-}6) $$

where

$\bar{I}_M$ = the average monthly insolation on the collector surface, Btu/ ft^2-month

HS = the average daily horizontal insolation for a specific location, Btu/ft^2- day

$\dfrac{\text{Incident}}{\text{Horizontal}}$ = the factor to convert the horizontal radiation to radiation for a specific surface tilt and orientation from Fig. 9-26 using K_T and Lat–Dec (LD)

N = the number of days per month

15-12   The Solar Heating Fraction (SHF)

The solar heating fraction (SHF) for an active space-heating system is defined as:

$$\text{SHF} = \frac{Q_S}{Q_L} \qquad (15\text{-}7)$$

where

SHF = the solar heating fraction, either monthly or yearly, as desired

Q_S = the total monthly or yearly solar heat input to the building, Btu/month or year

Q_L = the total monthly (or yearly) space-heating load, BLC $\times$ DD$_{t_b}$, Btu/month or year

The solar heating fraction is shown as a function of the Solar Load Ratio in Fig. 15-13. This curve is valid only for active solar space-heating systems that correspond rather closely to either the standard reference liquid-based system specified in Table 15-3 or the air based system specified in Table 15-4. The solar heating fraction for a particular month can be read from Fig. 15-13 once the Solar Load Ratio for that month has been determined. For example, if a Solar Load Ratio is calculated to be 2.0, then the corresponding solar heating fraction would be approximately 0.65. The curve of Fig. 15-13 flattens at a Solar Load Ratio of approximately 5.7 and a solar heating fraction of 1.0. This indicates that a solar space-heating system with a Solar Load Ratio of 5.7 or greater will have a solar heating fraction of 1.0 and will supply 100 percent of the space-heating

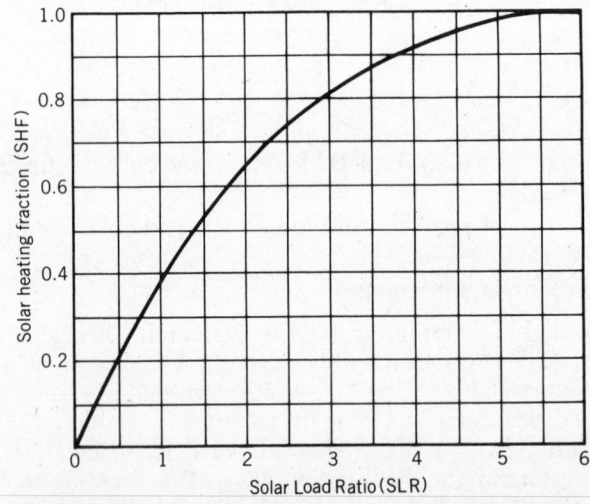

Figure 15-13 The solar heating fraction (SHF) plotted as a function of the Solar Load Ratio (SLR) for the reference active solar systems given in Tables 15-5 and 15-6. (From *Pacific Regional Solar Heating Handbook*, U.S. Energy Research and Development Administration. San Francisco Operations Office, November 1976, A62. Adapted with permission.)

SOLAR ENERGY SYSTEMS FOR SPACE HEATING

load for that month. The solar heating fraction can also be written in terms of the Auxiliary Heat by an equation similar to Eq. (13-3), as:

$$SHF = 1 - \frac{AH}{Q_L} = 1 - \frac{AH}{BLC \times DD_{t_b}} \qquad (15\text{-}8)$$

where

AH = the Auxiliary Heat supplied to the building, Btu/month

Q_L = the total monthly or yearly space-heating load, Btu/month, (BLC $\times$ DD_{t_b}/month) or Btu/year

Eq. (15-8) can be rearranged to give the Auxiliary Heat used as:

$$AH = BLC \times DD \times (1 - SHF) \qquad (15\text{-}9)$$

Seven steps are necessary to complete a monthly SLR calculation for an active system:

1. Calculation of the building load coefficient (BLC) for the building by the methods outlined in Chapter 5.
2. Calculation of the monthly solar energy incident on the solar collectors by the methods developed in Chapters 9 and 13; see Eqs. (15-5) and (15-6).
3. Determination of the degree-day base temperature to be used, and the calculation of the monthly heating degree-days. If there are no internal heat gains to compensate for, 65°F should be used as the degree-day base temperature. If internal heat gains are significant, then Eq. (9-17) should be used to calculate the degree-day base temperature. The monthly heating load (Q_L = BLC $\times$ DD/month) is then calculated.
4. Calculation of the monthly values of the SLR, from Eq. (15-4).
5. Determination of the monthly values of SHF from the SLR values, using Fig. 15-13.
6. Calculation of the Auxiliary Heat, using Eq. (15-9) for each month.
7. Calculation of the annual AH and SHF, by summing up monthly values.

Illustrative Problem 15-8

An active solar system has been designed for a residence located in Richmond, Virginia. The house has an estimated building load coefficient of 18,000 Btu/DD. The solar collector area is 800 ft², and the collector tilt is 60°. The degree-day temperature base is 65°F. Calculate the monthly solar heating fraction, the monthly Auxiliary Heat, and the annual solar heating fraction. Assume that the active system will have the characteristics of the standard reference liquid system cited in Table 15-3. Use the solar radiation and local climate information in Table 9-7 for Richmond, Virginia.

Solution

Since the proposed active system has the characteristics of the standard reference liquid system on which the Los Alamos SLR method for active systems is based, this method will be used to predict the system performance. The solution follows the sequence of steps listed above:

1. The building load coefficient is given as 18,000 Btu/DD.
2. The monthly solar energy on a horizontal surface can be obtained from Table 9-7 for Richmond. The incident solar energy on the collector surface with a 60° tilt can be obtained using Eq. (15-6) and the incident-to-horizontal correction factor in Fig.

Table 15-7 Worksheet for Solar Load Ratio Calculation for Liquid-Based Solar Space-Heating System with 800 ft²
Collector Area (Illustrative Problem 15-8)

Month	Days per Month	Lat- Dec (degrees)	K_T	HS (Btu/ft²-day)	Incident Horizontal	$\bar{I}_M$ (Btu/ft²-month)	DD_{month} ($t_b = 65°F$)	SLR	SHF	$Q_L \times 10^6$ Btu	$AH \times 10^6$ Btu
Sept	30	36	.50	1348	1.02	41,200	21	87.2	1.00	0.38	0.00
Oct	31	47	.50	1033	1.24	39,700	203	8.7	1.00	3.65	0.00
Nov	30	57	.48	733	1.60	35,200	480	3.3	0.85	8.64	1.30
Dec	31	61	.43	567	1.60	28,100	806	1.6	0.55	14.51	6.53
Jan	31	59	.44	632	1.58	31,000	853	1.6	0.55	15.35	6.91
Feb	28	51	.47	877	1.35	33,200	717	2.1	0.67	12.91	4.26
Mar	31	40	.49	1210	1.12	42,000	569	3.3	0.85	10.24	1.54
Apr	30	28	.51	1566	0.90	42,300	226	8.3	1.00	4.07	0.00
May	31	18	.51	1762	0.80	43,700	64	30.4	1.00	1.15	0.00
June	30	14	.52	1872	0.78	43,800	0	—	—	—	—
Annual totals								—	—	70.9	20.54

9-26, which gives the ratio of incident-to-horizontal solar radiation for a 60°-tilted, south-facing collector. It is helpful to use a worksheet to enter numerical values as the calculations for the SLR method are performed. The worksheet in Table 15-7 has the values of Lat–Dec (LD), K_T, and HS filled in for Richmond from Table 9-7. The incident-to-horizontal correction factor is found for September to be approximately 1.02 for the values of LD = 36 and K_T = 0.5 from Fig. 9-26. Values for the other months are also obtained from Fig. 9-26 and entered on the worksheet. The values of average monthly insolation $(\bar{I}_M)$ are calculated using Eq. (15-6). For September, we obtain:

$$\bar{I}_M = \text{HS} \times \frac{\text{Incident}}{\text{Horizontal}} \times N$$
$$= 1348 \text{ Btu/ft}^2\text{-day} \times 1.02 \times 30 \text{ days/month}$$
$$= 41,200 \text{ Btu/ft}^2\text{-month}$$

3. The degree-day base temperature is given as 65°F. The monthly heating degree-days can be obtained directly from Table 9-7 for a base temperature of 65°F and entered into Table 15-7 in the DD_{month} column. The monthly heating load Q_L is calculated using Eq. (15-2). For September,

$$Q_{\text{Sept}} = \text{BLC} \times \text{DD}_{t_b}$$
$$= 18,000 \times 21$$
$$= 0.38 \times 10^6 \text{ Btu}$$

The values of Q_L for the other months of the year are calculated in the same manner and entered on the worksheet.

4. The monthly Solar Load Ratio is then calculated using Eq. (15-4). For September,

$$\text{SLR}_{\text{Sept}} = \frac{A_c \times \bar{I}_M}{\text{BLC} \times \text{DD}}$$
$$= \frac{800 \text{ ft}^2 \times 41,200 \text{ Btu/ft}^2\text{-Sept}}{18,000 \text{ Btu/DD} \times 21 \text{ DD}}$$
$$= 87.2$$

The SLR values for the other months are calculated in a similar manner and entered on the worksheet.

5. From Fig. 15-13, the monthly values of the solar heating fraction are determined using the calculated values of Solar Load Ratio. For September, since the SLR is greater than 5.7, the SHF is assumed to be 1. The SHF for October will also be 1. The SLR for November is 3.3, and the SHF is found from Fig. 15-13 as approximately 0.85. The SHFs for the other months are found in a similar manner.

6. The Auxiliary Heat required is calculated using Eq. (15-9). For September and October, the AH will be zero since the SHF is 1.0. The AH for November is calculated as:

$$\text{AH} = \text{BLC} \times \text{DD} \times (1 - \text{SHF})$$
$$= Q_{L\text{Nov}} \times (1 - \text{SHF})$$
$$= 8.64 \times 10^6 \text{ Btu} \times (1 - 0.85)$$
$$= 1.3 \times 10^6 \text{ Btu}$$

The Auxiliary Heat values for the other months are calculated in a similar manner and entered on the worksheet.

7. The annual Auxiliary Heat is calculated by adding the monthly values to obtain:

$$\text{AH} = 20.5 \times 10^6 \text{ Btu/year} \qquad \textit{Ans.}$$

The annual SHF is calculated using Eq. (15-7):

$$SHF = 1 - \frac{AH}{Q_L}$$

$$= 1 - \frac{20.5 \times 10^6 \text{ Btu/year}}{70.9 \times 10^6 \text{ Btu/year}}$$

$$SHF = 0.71 \qquad\qquad Ans.$$

Illustrative Problem 15-9

In Illustrative Problem 15-8, the standard liquid reference system cited in Table 15-3 was used. This system uses a flat-black absorber plate in the solar collectors. Determine the collector area required if a selective absorber is used in the collectors.

Solution

The collector area with flat-black absorber is 800 ft^2, and the annual solar heating fraction was calculated to be 0.71. For Richmond, the January heating degree-days are known as 853. This corresponds approximately to Medford, Oregon in Table 15-6. From Table 15-6, the area of selective-surface collector for a system with a 75 percent solar heating fraction is 0.70 of the flat-black collector area. For a system with a 40 percent solar heating fraction, the selective-surface collector area is 0.87 of the flat-black collector area. By interpolation, for a solar heating fraction of 71 percent, the selective-surface collector area would be 0.72 of the flat-black collector area, or:

$$A_c = 0.72 \times 800 \text{ ft}^2$$

$$= 576 \text{ ft}^2 \qquad\qquad Ans.$$

When the horizontal solar radiation is not available for a given location, the clear-day insolation from Table 4-E can be used to estimate the average monthly insolation on the collector surface using Eq. (15-5).

Illustrative Problem 15-10

A small commercial building is located at 40° N Lat and an active solar space-heating system is being proposed. The owner would like to know the estimated performance of the solar system in terms of the annual Auxiliary Heat and the annual solar heating fraction. The proposed system will use 2000 ft^2 of single-glazed air-heating collectors with flat-black paint on the absorber. The collectors will be tilted at 60° facing due south. The building load coefficient has been estimated to be 40,000 Btu/DD. The internal heat gain averages 320,000 Btu/day. The building thermostat setting will be 68°F. The building location has a heating climate similar to that of Knoxville, Tennessee, so the monthly degree-day information shown in Table 9-11 can be utilized. The percentage of possible sunshine for Memphis, Tennessee will be used to estimate the monthly insolation. Assume that the system characteristics are the same as the reference air-based system of Table 15-4.

Solution

Since the proposed active system has the characteristics of the standard reference air system on which the Los Alamos SLR method for active systems is based, this method will be used to predict the system performance.

1. The building load coefficient is given as 40,000 Btu/DD.
2. The clear-day insolation from Table 4-E will be used to estimate the average monthly insolation on the collector. It is helpful to use a worksheet for these calculations, as shown in Table 15-8.

Table 15-8 Worksheet for Solar Load Ratio Calculation for Air-Based Solar Space-Heating System with 2000 ft² of Collector Area (Illustrated Prob. 15-10)

Month	Days per Month	I (Btu/ft²-day)	P	I_M (Btu/ft²-month)	DD_{month} (t_b = 60°F)	SLR	SHF	Q_L ×10⁶ Btu	AH ×10⁶ Btu
Sept	30	2076	69	42,970	6	358.0	1.00	0.24	0.00
Oct	31	2074	71	45,650	83	27.5	1.00	3.32	0.00
Nov	30	1950	58	33,930	331	5.1	0.98	13.24	0.26
Dec	31	1796	49	27,280	574	2.4	0.72	22.96	6.43
Jan	31	1944	48	28,930	602	2.4	0.72	24.08	6.74
Feb	28	2177	54	32,920	483	3.4	0.85	19.32	2.90
Mar	31	2175	57	38,430	322	5.9	1.00	12.88	0.00
Apr	30	1955	63	36,950	89	20.7	1.00	3.56	0.00
May	31	1763	69	37,710	14	134.0	1.00	0.56	0.00
June	30	1674	73	36,660	0	—	—	—	0.00
Annual totals						—	—	100.16	16.33

From Table 4-E, the clear-day insolation (I) is found for a south-facing surface tilted at 60° for a latitude of 40° N. For September 21, $I = 2076$ Btu/ft²-day by summing the hourly values for the day. The values for the other months are then entered into the worksheet (Table 15-8). The percentage of possible sunshine (P) is entered into the worksheet from Table 4-C for Memphis. The average monthly insolation is then calculated, using Eq. (15-5), and entered in the worksheet. For September,

$$\bar{I}_M = I \times P \times N$$
$$= 2076 \text{ Btu/ft}^2\text{-day} \times 0.69 \times 30 \text{ days/month}$$
$$= 42{,}970 \text{ Btu/ft}^2\text{-month}$$

3. The degree-day base temperature is calculated using Eq. (9-17) to compensate for the internal heat gain:

$$T_b = T_{set} - \frac{Q_{int}}{BLC}$$
$$= 68°F - \frac{320{,}000 \text{ Btu/day}}{40{,}000 \text{ Btu/DD}}$$
$$= 68°F - 8°F$$
$$= 60°F$$

The monthly heating degree-days can be obtained directly from Table 9-7 for a base temperature of 60°F for Knoxville. These values are entered in the DD_{tb} column of the worksheet. The monthly heating load Q_L is calculated using Eq. (15-3). For September,

$$Q_{Sept} = BLC \times DD_{tb}$$
$$= 40{,}000 \text{ Btu/DD} \times 6 \text{ DD}$$
$$= 0.24 \times 10^6 \text{ Btu}$$

The values of Q_L for the other months of the year are calculated in the same manner and entered on the worksheet.

4. The monthly Solar Load Ratio is then calculated, using Eq. (15-4), for September,

$$SLR_{Sept} = \frac{A_c \times \bar{I}_M}{BLC \times DD_{tb}}$$
$$= \frac{2000 \text{ ft}^2 \times 42{,}970 \text{ Btu/ft}^2\text{-month}}{40{,}000 \text{ Btu/DD} \times 6 \text{ DD}}$$
$$= 358$$

The SLR values for the other months are calculated and entered on the worksheet.

5. From Fig. 15-13, the monthly values of the solar heating fraction are determined using the calculated values of SLR and these are entered on the worksheet.

6. The Auxiliary Heat required is calculated using Eq. (15-9). The values are entered on the worksheet.

7. The annual Auxiliary Heat is calculated by summing the monthly values to obtain:

$$AH = 16.23 \times 10^6 \text{ Btu/yr} \qquad Ans.$$

The annual SHF is calculated using Eq. (15-7):

$$SHF = 1 - \frac{AH}{Q_L}$$
$$= 1 - \frac{16.23 \times 10^6 \text{ Btu}}{100.16 \times 10^6 \text{ Btu}}$$
$$= 0.84 \qquad Ans.$$

15-13   Specification of Mechanical System Components

After a solar collector area that will provide the required solar heating fraction has been determined, the mechanical specifications must be prepared for the collector loop. The controls for the solar energy system must also be specified.

Components for Liquid-Based Systems.   The sizing of the collector-loop piping and circulation pump is similar for space-heating systems and for domestic hot-water systems. The following steps will be discussed:

1. The collector flow rate is determined from the manufacturer's recommendations, or by Eq. (13-9), to give a desired temperature rise for midday conditions.
2. The pressure drop for the collector array and the collector-loop piping is calculated using the techniques discussed in section 13-12.
3. A pump is chosen that will provide the collector flow rate at the determined pressure drop, as discussed in section 13-12.

In an active, closed-loop space-heating system, a heat exchanger is used to transfer heat from the collector heat-exchange fluid to the thermal storage. Water is typically circulated from the bottom of the thermal storage tank through the heat exchanger, where it is heated by the collector heat-exchange fluid. The heated water is then returned to the upper part of the tank, as shown in Fig. 15-1. Another possible design is to use an immersed heat exchanger coil near the bottom of the storage tank, as shown in Fig. 15-14. Natural convection will cause the water in the tank to circulate around the heat exchanger coil with the heated water rising to the top of the tank.

   The heat exchanger can have a negative effect on system performance if it is not properly sized. Overall system performance may be decreased by 5 to 10 percent if the heat exchanger is undersized. The heat exchanger heat-transfer coefficient should be at least 10 Btu/hr-F$^\circ$-ft$_c^2$. This would mean that if a collector array were delivering energy at a rate of 150 Btu/hr-ft$_c^2$ the temperature difference between the collector-loop fluid and the storage water should be 15 F$^\circ$ or less. From a performance standpoint, it is desirable to have the heat exchanger heat-transfer coefficient as large as possible so that the temperature difference between the collector heat-exchange fluid and thermal storage is as small as possible. There are practical limitations, of course, dictated by size and

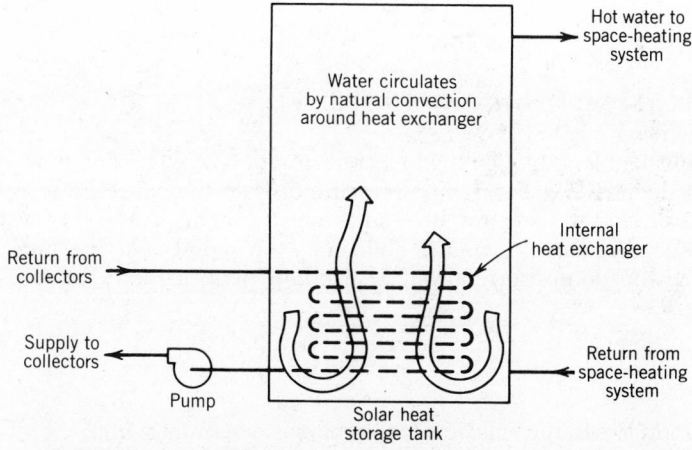

Figure 15-14 An immersed heat exchanger at the bottom of the thermal storage tank, used to transfer heat from the collector loop to the thermal storage. Water will circulate in the storage tank by natural convection.

cost of the heat exchanger. Generally, the average temperature difference between the collector-loop fluid and the storage tank will be in the range of 5 to 15 F°.

The heat exchanger will also add friction to the flow of fluid in the collector loop, increasing the overall pressure drop in the collector loop. The pressure drop caused by the heat exchanger should be in the range of 2 to 6 ft of water head.

The drainback collector-loop system uses no heat exchanger, since the water from thermal storage is circulated directly through the collectors. The thermal system performance with this system will always be somewhat better than it is with a comparable closed-loop system, because with the closed-loop system the collector loop must be 5 to 15 F° warmer than the thermal storage, due to the required temperature difference across the heat exchanger. The collector loop of the closed-loop system will therefore operate at 5 to 15 F° warmer than the collector loop of a comparable drainback system. Thus the closed-loop system, will have a somewhat larger heat loss from the collector loop than will the drainback system.

Illustrative Problem 15-11

An active space-heating system has a collector area of 900 ft². The efficiency of the collectors, from manufacturer's data, is 0.51. Calculate the water flow through the collectors to give a temperature rise of 15 F° when the insolation on the collectors is 240 Btu/hr-ft².

Solution

Eq. (13-9) can be rearranged to give the fluid flow rate as:

$$V = \frac{AI\eta}{\Delta t \times D \times c_w \times 60 \text{ min/hr}}$$

Substituting values:

$$V = \frac{900 \text{ ft}^2 \times 240 \text{ Btu/hr-ft}^2 \times 0.51}{15 \text{ F}° \times 8.34 \text{ lb/gal} \times 1.0 \text{ Btu/lb-F}° \times 60 \text{ min/hr}}$$
$$= 14.7 \text{ gal/min} \qquad\qquad Ans.$$

After the collector flow rate has been calculated, then the piping size and pump size must be determined.

Illustrative Problem 15-12

The equivalent length of piping in a collector loop is 120 ft. The flow rate in the collector loop is 15 gal/min. The pressure drop in the collector array is 2 ft of water head, and the pressure drop in the heat exchanger is 6 ft of water head. Determine the diameter of the collector loop piping and the pump size to minimize the pump horsepower. Use pump performance data as shown in Fig. 13-29.

Solution

The friction loss in the collector loop piping is determined from Fig. 13-28. The pressure drop for 1-in.-diameter pipe is found to be approximately 15.7 ft of head per 100 ft, which gives a pressure drop of

$$15.7 \times \frac{120}{100} = 18.8 \text{ ft of head}$$

For 1¼-in.-diameter pipe, the pressure drop is 5.8 ft of head per 100 ft, which gives a pressure drop of

$$5.8 \times \frac{120}{100} \text{ ft} = 7 \text{ ft of head}$$

The total collector-loop pressure drop with 1-in.-diameter pipe is:

$$p = 2 \text{ ft} + 6 \text{ ft} + 18.8 \text{ ft} = 26.8 \text{ ft}$$

The total collector-loop pressure drop with 1¼-in.-diameter pipe is:

$$p = 2 \text{ ft} + 6 \text{ ft} + 7 \text{ ft} = 15 \text{ ft}$$

The 1-in. pipe gives a much larger pressure drop, and the pump performance shown in Fig. 13-29 indicates that none of the pumps could provide 15 gal/min in a 1-in. pipe at a pressure head of 26 ft. The ⅙-hp pump, however, will provide 15 gal/min at a head pressure of 19 ft. Therefore, the ⅙-hp pump should be used with 1¼-in.-diameter collector-loop piping. *Ans.*

Components for Air-Based Systems.   These systems require that the pressure loss in the ducting, collectors, and rock bed be determined. Then the fan size and fan horsepower can be chosen. The methods for estimating duct and rock bed pressure losses were discussed in Chapter 9, section 9-6. The same methods apply to active space-heating systems. The basic difference is the configuration of the rock bed. For active space-heating systems, the rock bed typically has the air flow in a vertical direction, with the rock bed length (in the vertical direction) being 4 to 8 ft. The heated air flowing into the rock bed quickly gives up its heat as it flows through the multiple paths among the rocks. The rocks near the air-entry end of the rock bed can be at a much higher temperature than the rocks at the exit end of the rock bed.

During the day, when the solar collectors are providing heat, hot air is circulated from the top of the rock bed down through the rocks, as shown in Fig. 15-11*a*. The rocks at the top are heated first, and so the highest temperature zone in the rock bed is at the top. If, on the other hand, the heat were supplied at the bottom of the rock bed, the highest temperature zone would be at or near the bottom, and this would be undesirable for the following reasons.

When the collection of solar energy stops and air ceases to be circulated through the rock bed, the lower portion of the rock bed would be hot and the upper portion relatively cool. A sharp temperature gradient usually exists between the hot and cool portions of a rock bed. This temperature gradient is ordinarily about 15 to 20 in. long, the length of rock required for 95 percent heat transfer, as indicated in Fig. 9-10. The heat at the bottom of the rock bed will readily move upward by convection and conduction, the end result being that the high-temperature region at the bottom of the rock bed cools down while the cool region at the top warms up, as the temperature gradient in the rock bed decreases. When air is circulated back through the rock bed to be heated, it will not be heated to as high a temperature, because there is no remaining region of high temperature in the rock bed. This problem is avoided by heating the top of the rock bed first. Heat will not migrate downward in a rock bed to any great extent, and therefore the top of the rock bed will remain at a relatively high temperature, facilitating heat transfer to the air. Remember that it is *temperature difference*, not total heat quantity stored, that causes heat transfer.

Temperature profiles for a rock bed at different hours of the day and night are shown in Fig. 15-15, (*a*) when heat is being stored, and (*b*) when heat is being removed. The rock bed is assumed to be cold in the morning (8:00 A.M.), with the bottom of the storage at 70°F (room temperature) and the top at 80°F (Fig. 15-15*a*). The air temperature from the solar collectors increases during the day until a peak is reached at about midday, after which the temperature decreases. At midday, the highest temperature zone in the rock bed is the top layer. In the afternoon, the highest temperature zone moves gradually down into the rock bed, and the temperature of this zone decreases as the zone moves downward. The temperature of the bottom of the rock bed remains relatively constant

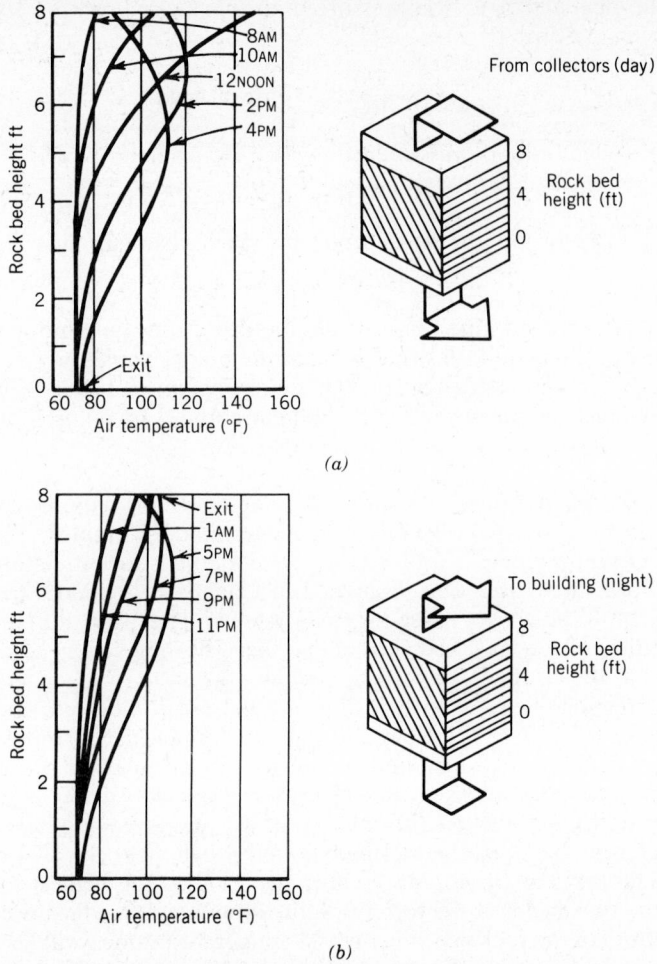

Figure 15-15 Temperature distribution within a rock bed used for storing heat. (*a*) Temperature profiles are shown for various times during the day as heat is supplied to the rock bed from the solar collectors. (*b*) The temperature profiles in a rock bed for nighttime hours as heat is removed from the rock bed at night to provide heat for the building. (From *Pacific Regional Solar Heating Handbook*, U.S. Energy Research and Development Administration, San Francisco Operations Office, November 1976, p. 50. Adapted with permission.)

during the day in the range of 70° to 75°F. This is because the rock bed is extremely effective in absorbing the heat in the air as it flows down through the rock.

When the collector loop is turned off and heat is required from the rock bed, the air flow is reversed through the rock bed as shown in Fig. 15-15*b*. This reversal allows the air to exit from the hottest part (top) of the rock bed. (If the air flow were not reversed, it would take several hours before the stored heat would move downward through the bed, enabling hot air to exit at the bottom.) The air temperature will rise for the first few hours of night operation as the highest temperature zone moves back up toward the top of the rock bed, after which the air temperature will gradually decrease as heat continues to be removed from the rock bed.

The design and specification of the mechanical components of an air-heating collector loop are very similar to the procedure described in section 9-4 for rock bed design. The same equations—(9-1) through (9-12)—are applicable for active air systems.

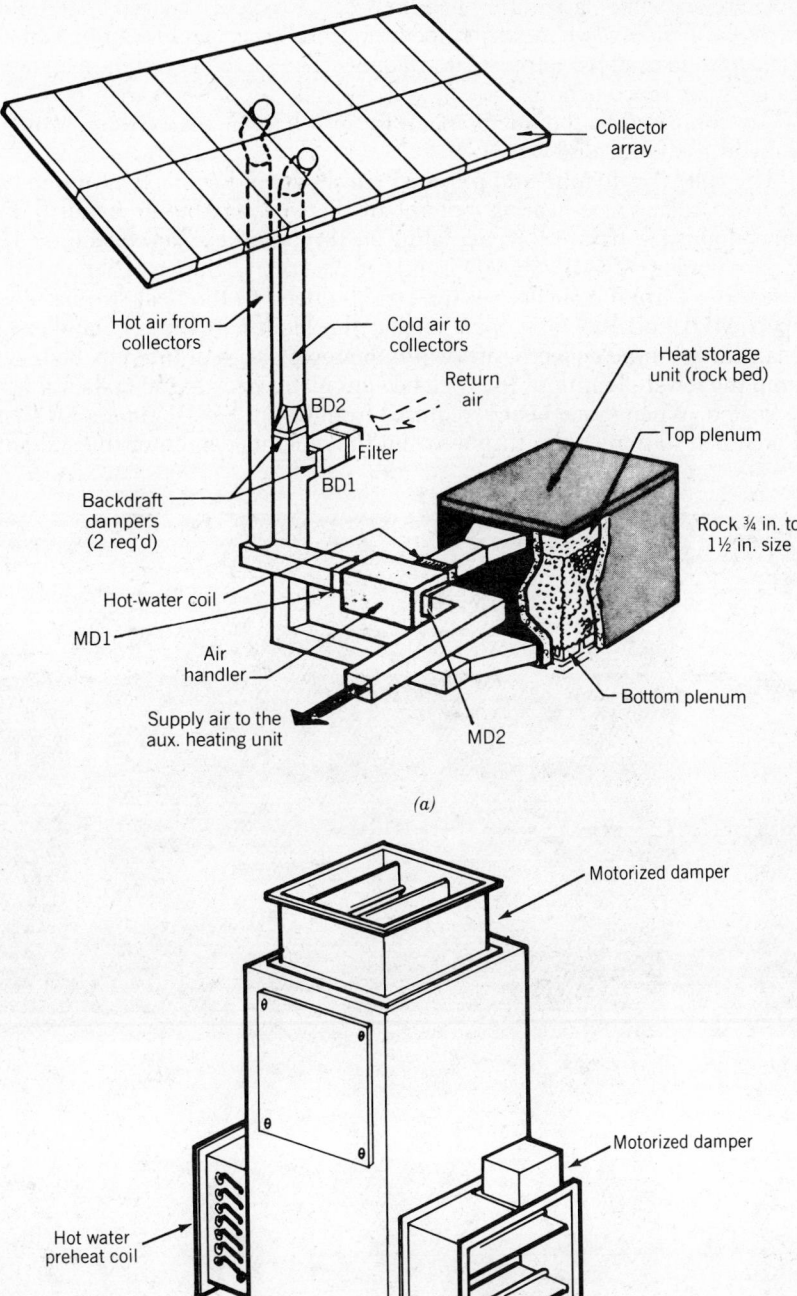

Figure 15-16 Components of an air-based solar space-heating system. (*a*) Isometric drawing of a system showing air ducts between air-heating solar collectors and rock bed. Dampers to control air flow in the system are also shown, along with the air handler unit. (Solaron Corporation.) (*b*) Air handler unit used with an air-based solar system. This unit contains the blower for circulating air between the air-heating solar collectors and the rock bed. It also has the motorized dampers on it which control the air flow in the system. (Solaron Corporation.)

There are some basic differences between a rock bed hybrid system for a passive solar building, where air temperatures are lower, and the rock bed and collector system used for active solar buildings. The air temperature in an active air-based solar space-heating system will typically be in the range of 120° to 160°F, as compared to the air temperature in a hybrid solar system, which is typically in the range of 80° to 95°F.

The collector ducting and rock bed are shown in Fig. 15-16a for one type of air-based solar space-heating system. In this system, the air handler (Fig. 15-16) contains the blower for circulating air through the solar collectors. Two motorized dampers (MD1 and MD2) and two backdraft dampers (BD1 and BD2) are used to control the air flow in the system ducts. In the heat-storing mode, dampers MD1 and BD2 are open so that the blower in the air handler can circulate air from the collector array into the top plenum of the rock bed. Cool air from the lower plenum of the rock bed circulates back to the collector array to be heated. When space heat is required from the system, dampers MD2 and BD1 are open so that cool return air from the building can enter the system to

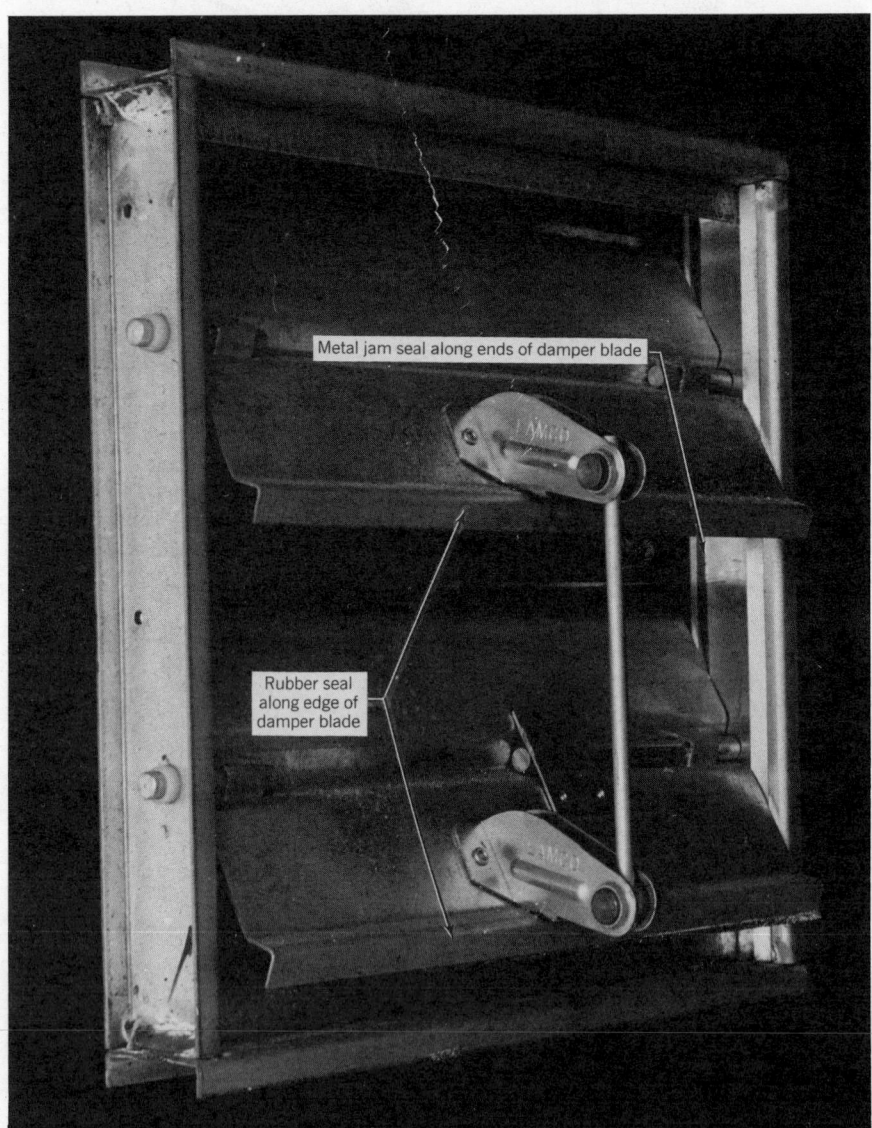

Figure 15-17 A control damper for air ducts. This type of damper is operated by a motor and is used to open or close an air duct. The damper blades have rubber seals along the edges and a metal jam seal along the ends to ensure that the damper is airtight when closed.

be heated, and heated air can be drawn either directly from the solar collectors or from the rock bed when the collectors are not collecting heat.

The dampers used in an air-based system should have tight-fitting seals so that air cannot leak past them when they are closed. A typical damper is shown in Fig. 15-17. It has a rubber seal along the edge of the damper blades and a jam seal at the ends of the blades in the damper frame. In Fig. 15-16a, damper MD2 closes to prevent warm air from the rock bed from convecting into the supply ducting. Damper MD1 closes to prevent cold air from convecting down from the collector array and hot air from convecting out of the rock bin into the collector array. Damper BD1 closes to prevent a reverse flow of air in the return air ducting. Damper BD2 closes to prevent cold air from convecting down into the ducting.

The rock bed may be constructed from either a concrete box or a wood frame box. Figure 15-18a shows a rock bed in cutaway with the hot air opening at the top entering the upper air plenum. The lower plenum is constructed by using U-shaped bond-beam concrete blocks and galvanized mesh across the concrete blocks to support the rocks. Construction details for a typical concrete rock bed are shown in Fig. 15-18b, and the construction details for a wood-framed rock bed are shown in Fig. 15-18c.

Illustrative Problem 15-13

A space-heating system will use 500 ft² of air-heating collectors. The rock bed will contain 375 ft³ of rock with an average rock diameter of 1.5 in. The vertical length of the rock bed is 5 ft. The air flow through the collectors will be 2.5 ft³/min per square foot of collector. The air pressure drop in the collectors is given by the manufacturer as 0.08 in. of water, gauge. The effective length of the collector loop round ducting is 80 ft, with a 14-in. diameter. Determine the pressure drop for the rock bed and ducting; then pick a blower size and motor horsepower to provide the required air flow in the collector loop.

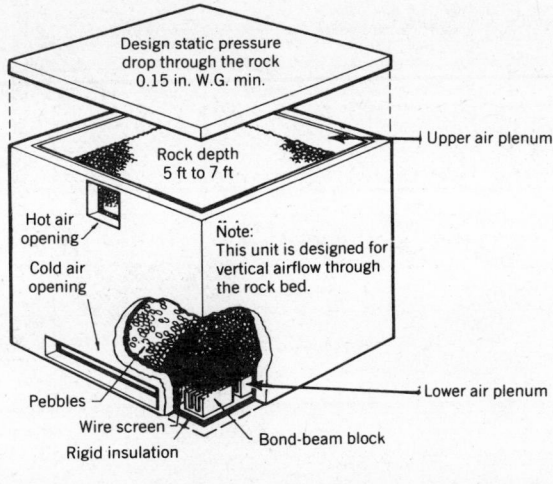

(a)

Figure 15-18 Rock bed heat storage units for air-based solar space-heating systems. (a) Cutaway of a rock bed showing the upper air plenum and the lower air plenum. The cold air and hot air duct openings are also shown. (See next page for Figs. 15–18b and c).

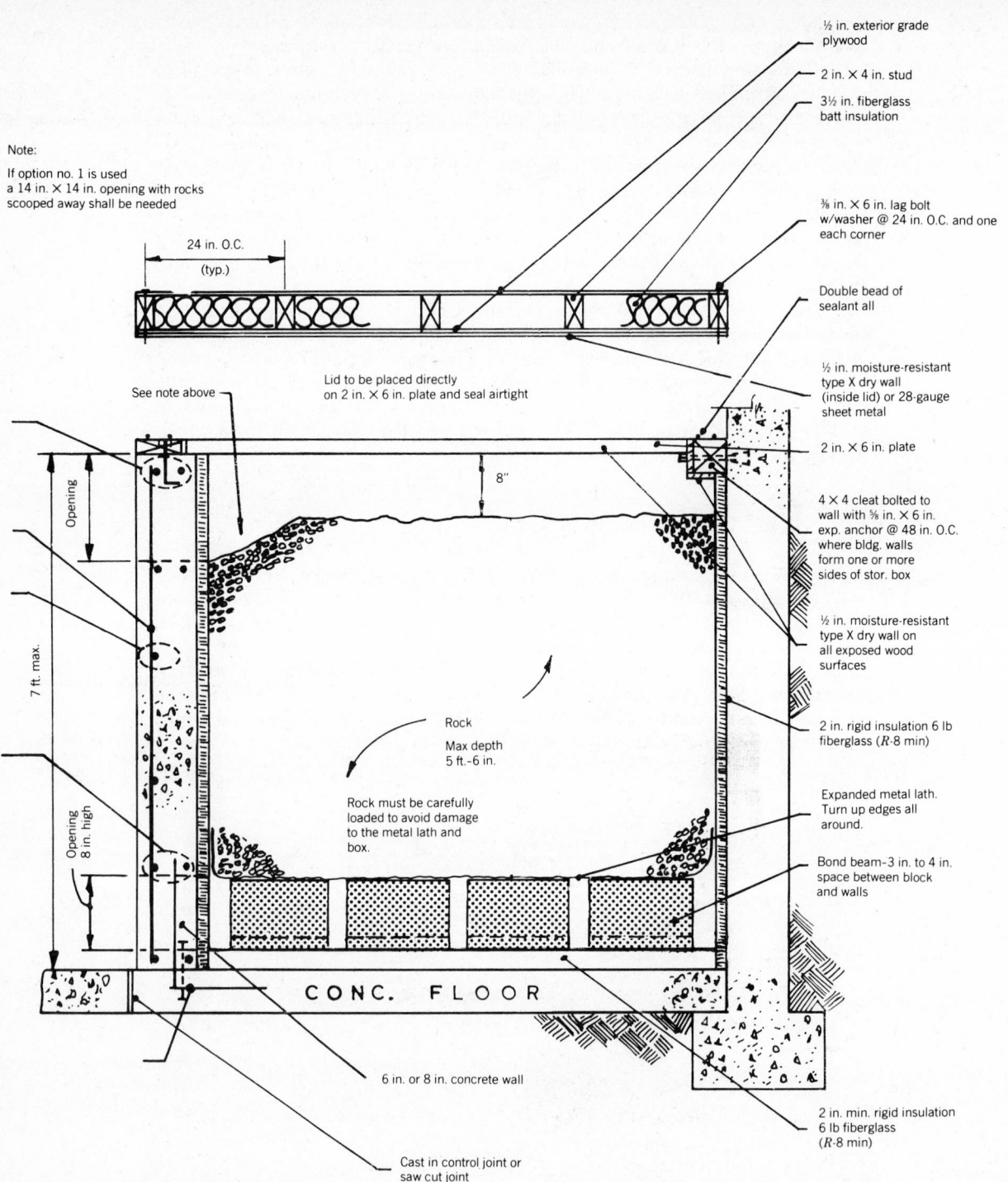

Note:

If option no. 1 is used
a 14 in. × 14 in. opening with rocks
scooped away shall be needed

24 in. O.C.
(typ.)

½ in. exterior grade
plywood

2 in. × 4 in. stud

3½ in. fiberglass
batt insulation

⅜ in. × 6 in. lag bolt
w/washer @ 24 in. O.C. and one
each corner

Double bead of
sealant all

See note above

Lid to be placed directly
on 2 in. × 6 in. plate and seal airtight

½ in. moisture-resistant
type X dry wall
(inside lid) or 28-gauge
sheet metal

2 in. × 6 in. plate

Opening

8"

4 × 4 cleat bolted to
wall with ⅜ in. × 6 in.
exp. anchor @ 48 in. O.C.
where bldg. walls
form one or more
sides of stor. box

7 ft. max.

Rock

Max depth
5 ft.-6 in.

½ in. moisture-resistant
type X dry wall on
all exposed wood
surfaces

2 in. rigid insulation 6 lb
fiberglass (R-8 min)

Rock must be carefully
loaded to avoid damage
to the metal lath and
box.

Expanded metal lath.
Turn up edges all
around.

Opening
8 in. high

Bond beam-3 in. to 4 in.
space between block
and walls

CONC. FLOOR

6 in. or 8 in. concrete wall

2 in. min. rigid insulation
6 lb fiberglass
(R-8 min)

Cast in control joint or
saw cut joint

Note: R-11 min. (R-30 in unheated area)

Note: Interior surfaces and insulation of rock box must be
noncombustible and suitable for temperatures up to 200° F.

(b)

Figure 15-18 (*continued*)
(*b*) A cross-section showing construction details for a rock bed constructed in a
concrete enclosure.

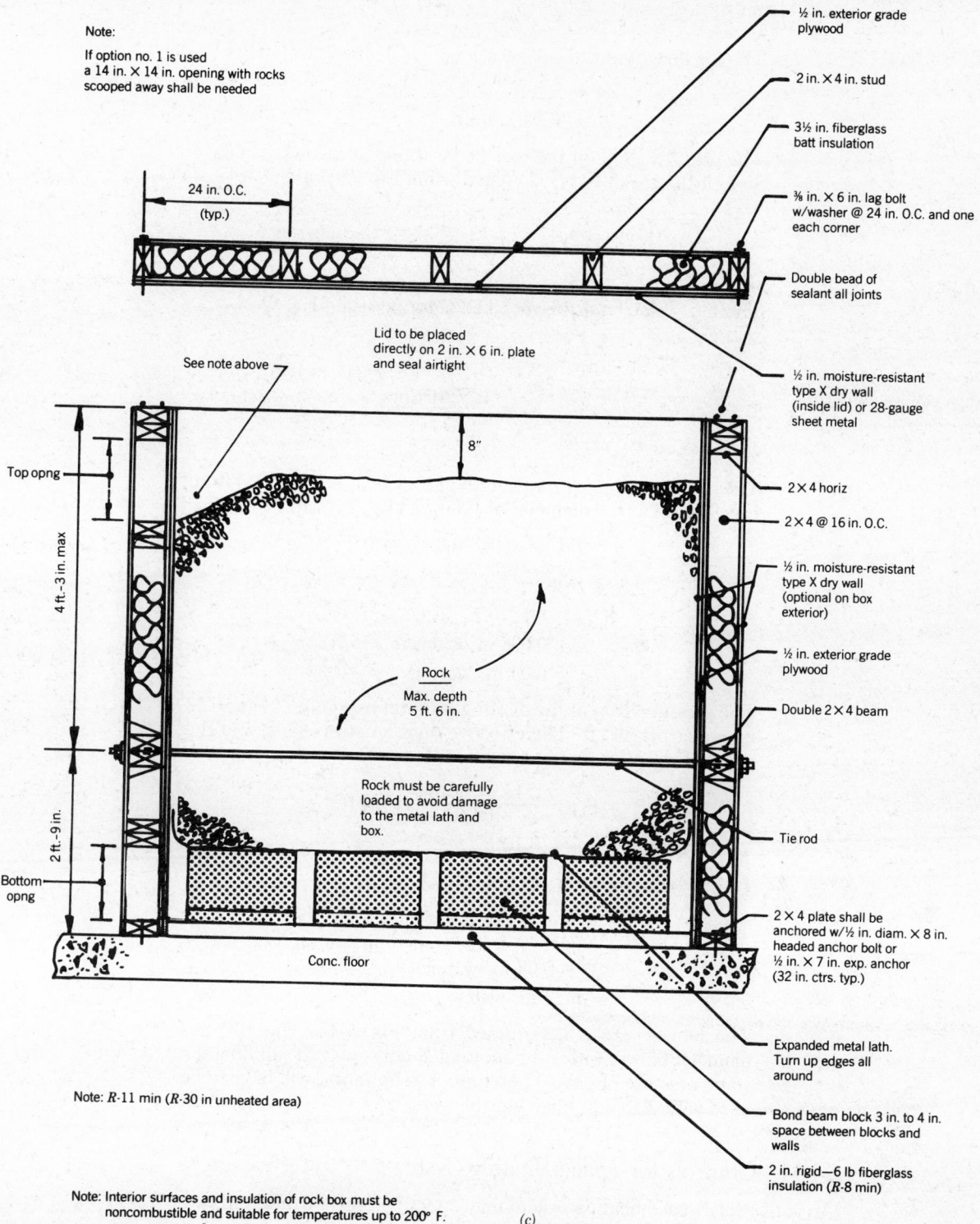

Note:

If option no. 1 is used a 14 in. × 14 in. opening with rocks scooped away shall be needed

24 in. O.C. (typ.)

½ in. exterior grade plywood

2 in. × 4 in. stud

3½ in. fiberglass batt insulation

⅜ in. × 6 in. lag bolt w/washer @ 24 in. O.C. and one each corner

Double bead of sealant all joints

Lid to be placed directly on 2 in. × 6 in. plate and seal airtight

See note above

8"

½ in. moisture-resistant type X dry wall (inside lid) or 28-gauge sheet metal

Top opng

2 × 4 horiz

2 × 4 @ 16 in. O.C.

4 ft.–3 in. max

½ in. moisture-resistant type X dry wall (optional on box exterior)

½ in. exterior grade plywood

Rock
Max. depth 5 ft. 6 in.

Double 2 × 4 beam

Rock must be carefully loaded to avoid damage to the metal lath and box.

2 ft.–9 in.

Tie rod

Bottom opng

Conc. floor

2 × 4 plate shall be anchored w/½ in. diam. × 8 in. headed anchor bolt or ½ in. × 7 in. exp. anchor (32 in. ctrs. typ.)

Note: R-11 min (R-30 in unheated area)

Expanded metal lath. Turn up edges all around

Bond beam block 3 in. to 4 in. space between blocks and walls

2 in. rigid—6 lb fiberglass insulation (R-8 min)

Note: Interior surfaces and insulation of rock box must be noncombustible and suitable for temperatures up to 200° F.

(c)

Figure 15-18 (continued)
(c) A cross-section showing construction details for a rock bed constructed in a wood-framed enclosure. (All three sketches courtesy Solaron Corporation.)

Solution

1. The air flow Q in the collector loop is calculated as follows:

$$Q = 500 \text{ ft}^2 \times 2.5 \text{ ft}^3/\text{min-ft}_c^2$$
$$= 1250 \text{ ft}^3/\text{min}$$

2. The pressure drop for the rock bed is found as follows. The face area of the rock bed is calculated using Eq. (9-1) as:

$$A_{face} = \frac{V_{RB}}{L_{Bed}} = \frac{375 \text{ ft}^3}{5 \text{ ft}}$$
$$= 75 \text{ ft}^2$$

The face velocity of the rock bed is found using Eq. (9-2):

$$V_{face} = \frac{Q}{A_{face}}$$
$$= \frac{1250 \text{ ft}^3/\text{min}}{75 \text{ ft}}$$
$$= 16.7 \text{ ft}/\text{min}$$

The pressure drop per unit length of rock bed, P/L is found from Fig. 9-10 for a rock diameter of 1.5 in. and V_{face} of 16.7 ft/min as:

$$P/L = 0.008 \text{ in. water/ft}$$

The total pressure drop for the rock bed, from Eq. (9-9), is:

$$P_{RB} = P/L \times L_{bed}$$
$$= 0.008 \text{ in. water/ft} \times 5 \text{ ft}$$
$$= 0.04 \text{ in. water}$$

3. The pressure drop in the ducting is found from Fig. 9-11 as 0.13 in. water per 100 ft. The effective duct length is 80 ft, and the pressure drop for the ducting is:

$$P_D = \frac{0.13 \text{ in. water}}{100 \text{ ft}} \times 80 \text{ ft}$$
$$= 0.10 \text{ in. water}$$

4. The pressure drop for the collector loop is then:

$$P_T = P_{RB} + P_D + P_C$$
$$= 0.04 \text{ in. water} + 0.10 \text{ in. water}$$
$$+ 0.08 \text{ in. water}$$
$$= 0.22 \text{ in. water}$$

5. The blower size is determined from Fig. 9-12. The 10⅝-in.-diameter blower with ⅓-hp motor will not supply the air flow at a static pressure of 0.22. Therefore, a ½-hp motor will have to be used with a 10⅝-in.-diameter blower.   *Ans.*

15-14 Controls for Space-Heating Systems

An active solar space-heating system may operate in several different ways. The most common operating modes are:

1. Heat is supplied from the solar energy system to the building load.
2. Heat from the solar collectors is placed in thermal storage.
3. Domestic hot water is preheated.
4. Auxiliary Heat is supplied to the building load.

The selection of the operating mode is dependent on temperature differences in the system. The collection of heat is usually controlled by a differential thermostat, which compares the collector outlet temperature to the thermal storage temperature. The collector loop is operated to collect heat only when the useful quantity of energy collected exceeds a minimum threshold value. For liquid-systems, the collector-to-storage On temperature differential is in the range of 10 to 15 F°, while the Off temperature differential is in the range of 3 to 5 F°. Air-based systems usually use slightly higher temperature differentials. The collector-to-storage On temperature differential for air-based systems is usually in the range of 15 to 25 F°, while the Off temperature differential is in the range of 5 to 10 F°. The collector-to-storage differential thermostat will turn on and off the pumps or blower used in the collector loop.

Figure 15-19 shows a block diagram of a space-heating control system. A second differential thermostat is shown, which controls the domestic hot-water preheating. This control has a sensor on the collector return and a sensor on the preheat storage tank. When the temperature differential between the collector fluid and the preheat tank is in the range of 10 to 15 F°, the control will turn on a pump that circulates domestic water through a heat exchanger to preheat the domestic hot water with heat from the collectors. The control will turn off the pump when the temperature differential falls below 3 to 5 F°.

Space heating is usually controlled with a two-stage wall thermostat. The first stage (T_1) will activate the building heat distribution system and remove heat from the solar energy system, provided the temperature in the thermal storage is above a preset low-limit setting controlled by a low-limit thermostat. This low-limit control prevents the heat distribution system from operating when there is not sufficient heat in storage. The auxiliary heat source is controlled by a second stage (T_2), which is set ½ to 1 F° below T_1. The second stage will turn the Auxiliary Heat source on when the thermal storage temperature is less than the preset low-limit setting, or if the temperature of the solar storage is not high enough to supply the required heat to the building to maintain the desired temperature.

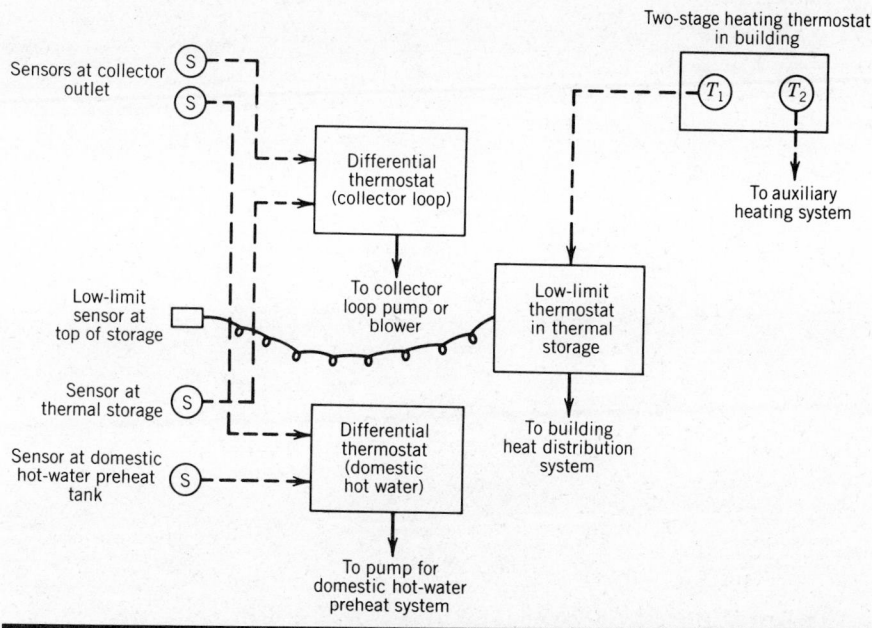

Figure 15-19 The controls for a space-heating system will typically contain a differential thermostat to control the collector loop. A differential thermostat may also be present for control of the domestic hot-water preheating. The building will ordinarily have a two-stage heating thermostat. The first stage will control the distribution of heat from the solar storage to the building, and the second stage will control the auxiliary heating system.

15-15 Examples of Space-Heating Systems

In this section, three active space-heating systems already designed and built will be examined. The Gerber residence is a solar air-based system with separate subsystems for the domestic hot water and pool, as shown in Fig. 15-20. The house consists of two sections that are double-story with roofs oriented south and pitched at 45° to optimize the solar exposure at latitude 35°. The larger double-storied section contains the living and dining rooms, kitchen, and a work loft. The active air-heating collectors are on the 45° roof surface above these rooms. Heated air from the collectors is either stored in a rock bed for later use or circulated into the rooms through a forced-air heating system. A schematic diagram of the solar space-heating system is shown in Fig. 15-21. The collector loop blower is controlled by a differential thermostat. A two-stage heating thermostat controls the forced-air system. The first stage opens the motorized dampers to the rock bed and turns on the furnace blower, allowing heated air to be drawn from the collectors if they are collecting energy or from the rock bed

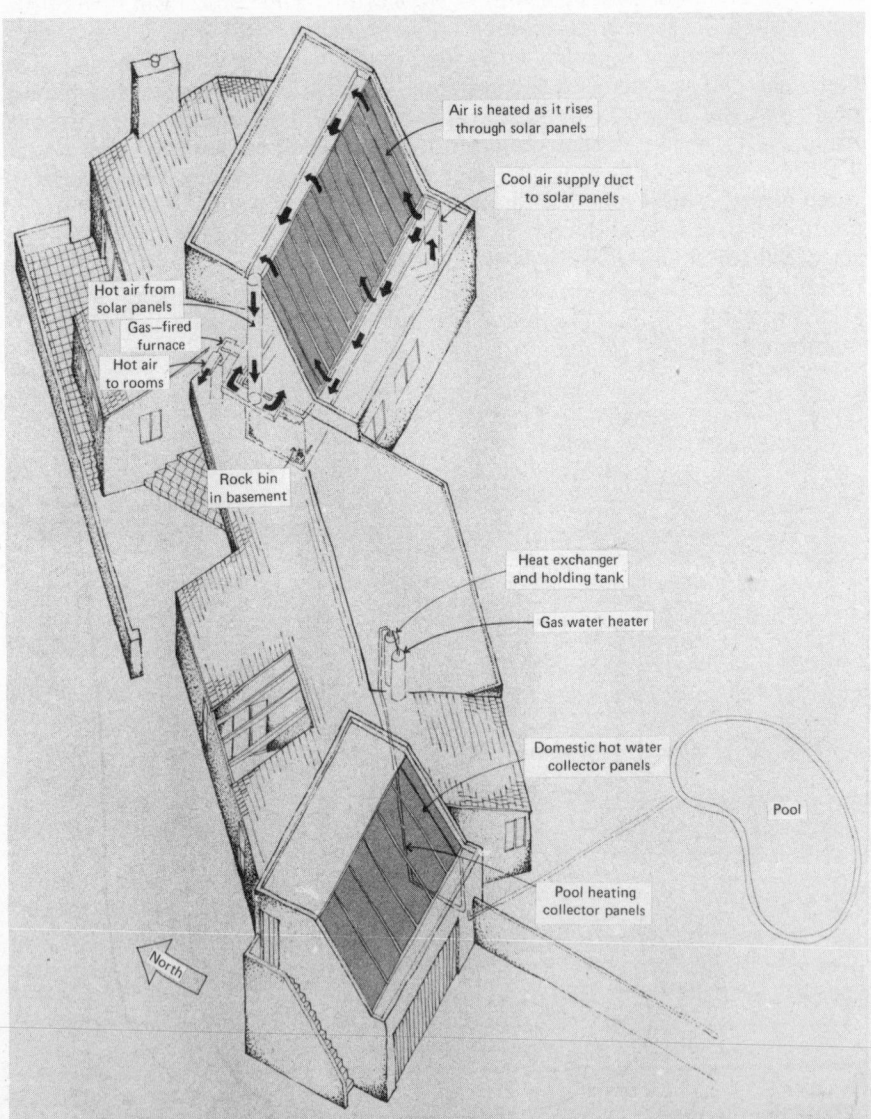

Figure 15-20 Isometric view of a residence with active solar systems. The space heating is provided by an air-heating collector system and rock bed storage. The domestic hot water is provided by a separate solar system. The swimming pool is heated by a solar pool-collector system. (Bob Easton Design.)

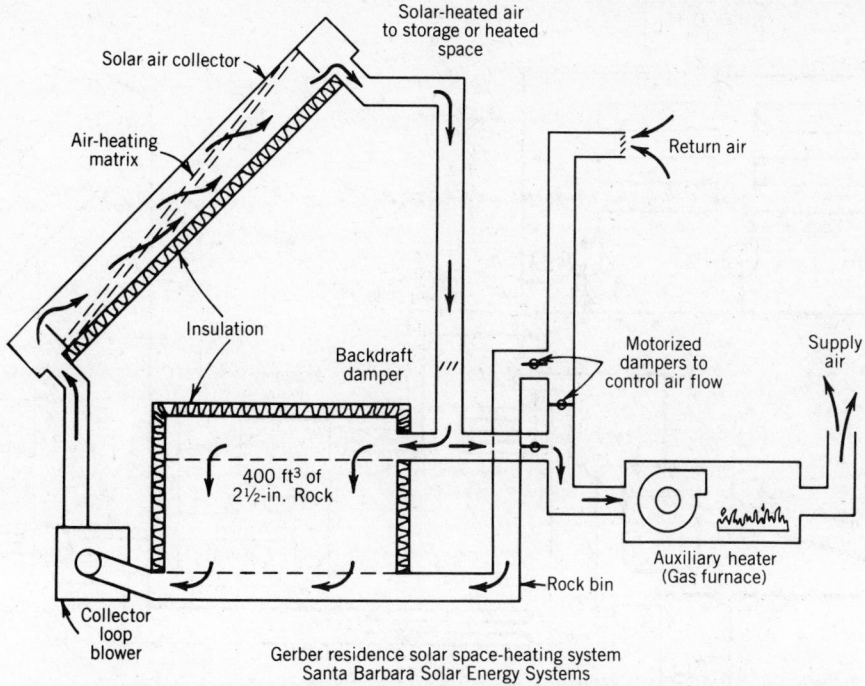

Solar-heated air
to storage or heated
space

Solar air collector

Air-heating
matrix

Return air

Insulation

Backdraft
damper

Motorized
dampers to
control air flow

Supply
air

400 ft³ of
2½-in. Rock

Auxiliary heater
(Gas furnace)

Rock bin

Collector
loop
blower

Gerber residence solar space-heating system
Santa Barbara Solar Energy Systems

Figure 15-21 Schematic of the air-based solar-heating system of the residence shown in Fig. 15-20. A blower is used to circulate air through the collector loop, the operation of which is controlled by a differential thermostat. The furnace blower and the motorized dampers are controlled by the first stage of the house thermostat to distribute heat from the solar system into the house. The second stage of the thermostat activates the Auxiliary Heat source.

when the collectors are off. If the house temperature cannot be maintained by heat from the solar energy system, the thermostat's second stage turns on the gas valve in the furnace and auxiliary heat is supplied to the house.

The smaller double-storied house section contains the garage and a guest room above. This 45° roof section holds the liquid-based collectors for the domestic hot-water system and the swimming pool. Each of the solar energy systems in this house is an independent system with its own controls. This approach allows each system to operate independently, and the controls for each system are relatively simple.

The functions of space heating, domestic hot-water heating, and pool heating are all combined in one system in the Chenovick house, as shown in the diagram of Fig. 15-22a. In this system, a collector array of 400 ft² (Fig. 15-22b) provides heat to a 1000-gal thermal storage tank located in the basement. Figure 15-23 is a photograph showing the thermal storage tank in the basement, along with associated valves, pumps, and controls. There are three heat exchangers in the storage tank (Fig. 15-22a). One is for the collector loop (HE1) to transfer heat from the collector to the storage. A second heat exchanger (HE2) transfers heat from the storage tank to the radiant heating panels. The third heat exchanger (HE3) provides heat to both the swimming pool and/or the spa. Motorized valves MV1 and MV2 allow the filter system to filter either the pool or the spa. Motorized valve MV3 allows the filter system to circulate water through heat exchanger HE3 when either the pool or the spa requires heating or to bypass the heat exchanger when heating is not required. The auxiliary heater provides heat to the upper quarter of the thermal storage tank. The controller for the system is shown in Fig. 15-24. It provides the following operating modes:

1. Collector pump speed 1 (differential thermostat 1).
2. Collector pump speed 2 (differential thermostat 2).

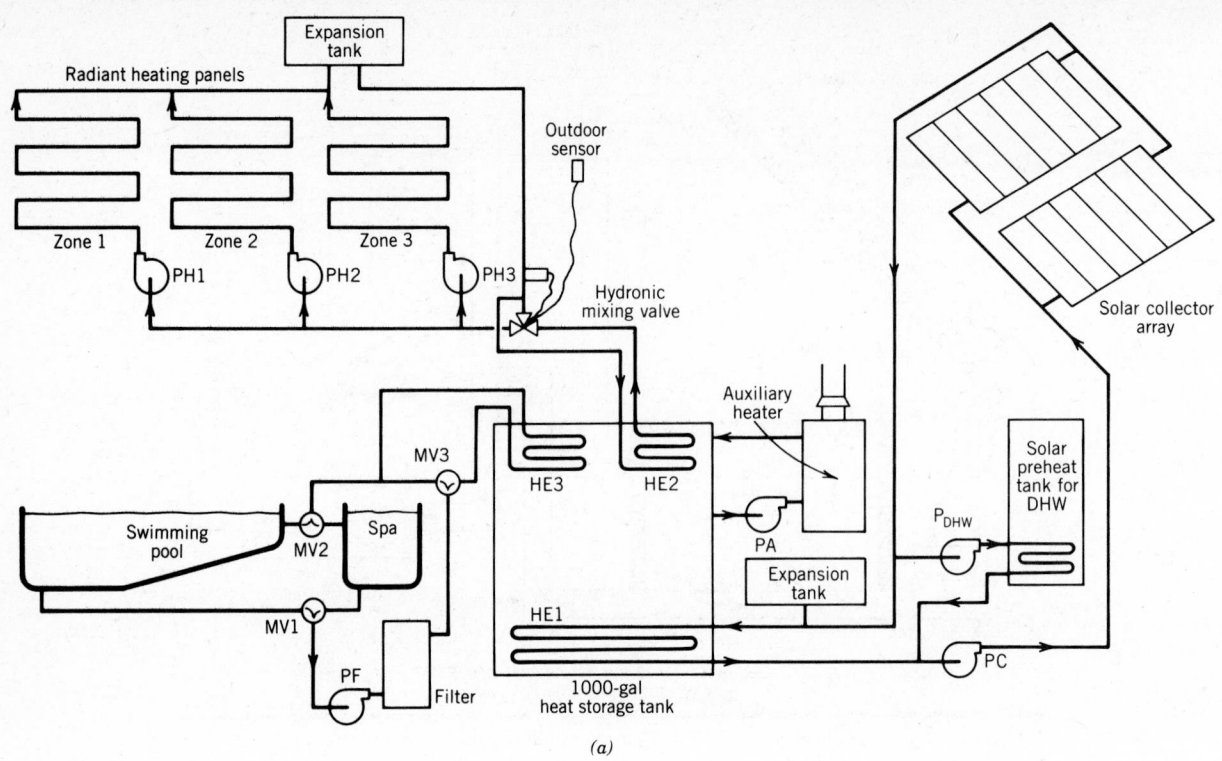

(a)

Figure 15-22 (a) System diagram for a liquid-based system that provides space heating, domestic hot water, pool heating, and spa heating. (b) Collector array for the Chenovick house. (Used with permission.)

Figure 15-23 Fiberglass heat storage tank. This tank holds 1000 gal and has insulation of value *R*-30. It is the thermal storage unit for the system in Fig. 15-22. (Chenovick house; photo used with permission.)

Figure 15-24 The radiant-heating loop pumps are shown along with the Heliologic unit that provides the control function for the active system of Fig. 15-22. (Chenovick house; photo used with permission.)

3. Domestic hot-water heating (differential thermostat 3).
4. Supply heat to heating zones 1, 2, or 3.
5. Filter pool.
6. Filter spa.
7. Heat pool.
8. Heat spa.
9. Auxiliary Heat to storage.
10. High-limit storage control.
11. High-limit on domestic hot water.

This liquid-based system is much more complex than the air-based system just described, because there are seven pumps, three motorized valves, and one hot-water loop control valve. The thermal storage tank is fiberglass and cannot be pressurized, so heat exchanger HE2 was required in order that the radiant heating (hydronic) system could be kept pressurized.

A retrofitted active space-heating system for the Mulac/Lundell residence is diagrammed in Fig. 15-25. This system provides space heating, domestic hot-water heating, and summer (evaporative) cooling. The system operating modes in winter are:

1. Heating of rock bed from the air-heating solar collectors.
2. Space heating from the collectors during the day.
3. Space heating from storage.
4. Heating domestic hot water while collectors are supplying heat to storage or to the house.

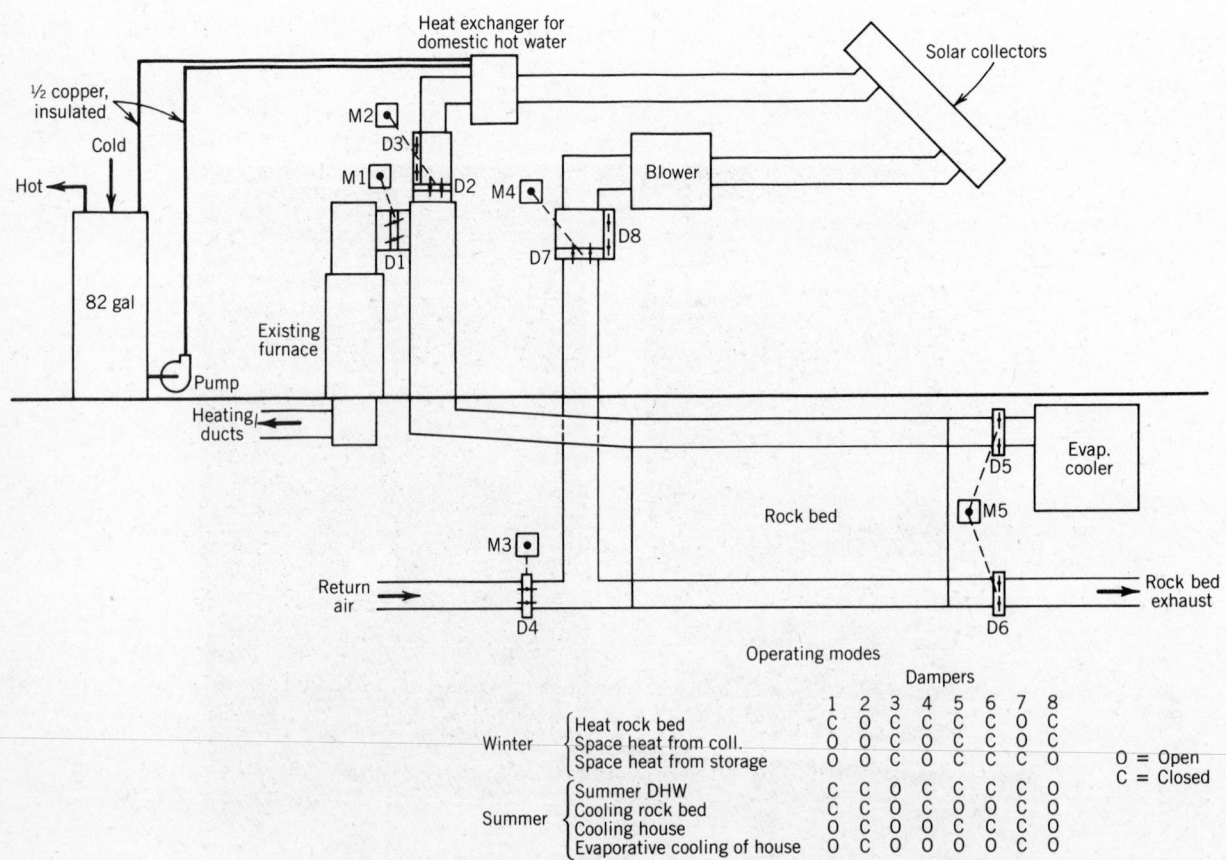

	Operating modes									
		Dampers								
		1	2	3	4	5	6	7	8	
Winter	Heat rock bed	C	O	C	C	C	C	O	C	
	Space heat from coll.	O	O	C	O	C	C	O	C	
	Space heat from storage	O	O	C	O	C	C	C	O	
Summer	Summer DHW	C	C	O	C	C	C	C	O	O = Open
	Cooling rock bed	O	C	O	C	O	O	C	O	C = Closed
	Cooling house	O	C	O	C	C	C	C	O	
	Evaporative cooling of house	O	C	O	O	O	O	C	O	

Figure 15-25 Air-based space-heating system that uses the rock bed as cool storage for summer cooling. This system has two sets of operating modes—one for winter space heating, and a second for summer cooling.

Figure 15-26 Interior view of a rock bed looking into the upper air plenum. The rock bed container is heavily insulated and lined with moisture-resistant drywall. All seams and joints are sealed to prevent air leaks.

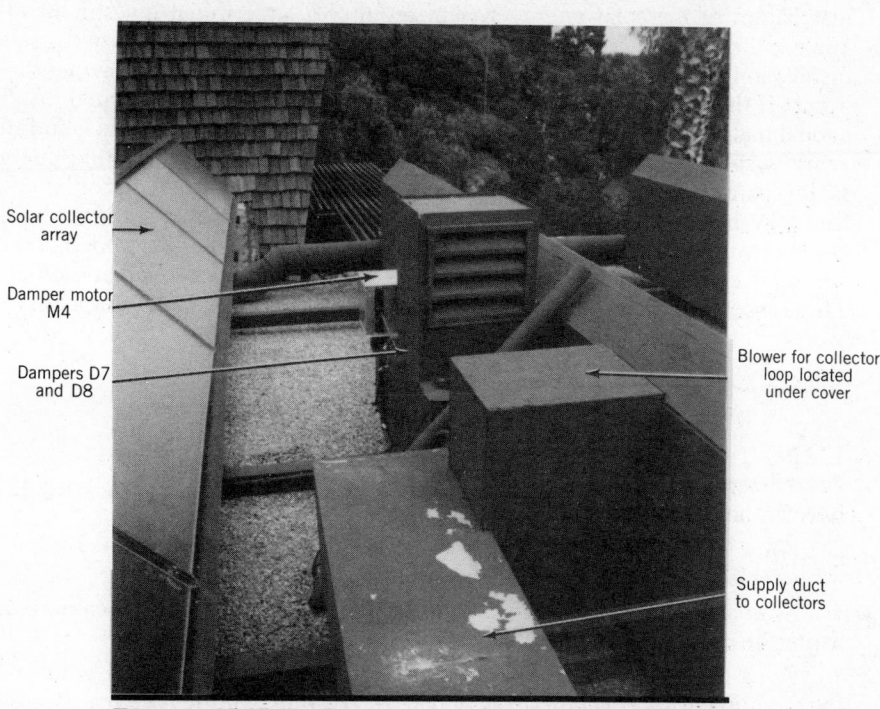

Figure 15-27 Rooftop view of damper motor M4 and dampers D7 and D8 of the system diagrammed in Fig. 15-25. The blower is in the foreground under the plywood cover. The front of one collector bank and the backs of two other collector banks can be seen. (Mulac/Lundell house; photo used with permission.)

SYSTEM PERFORMANCE FOR ACTIVE SOLAR SPACE HEATING

The summer operating modes are:

5. Summer heating of domestic hot water.
6. Night cooling of the rock bed using cool night air and the blower in the evaporative cooler.
7. Cooling the house using the cool rock bed.
8. Cooling the house directly with the evaporative cooler. (The rock bed is not cooled with evaporative cooling because of probable moisture buildup in the rock bed.)

This system has eight dampers and five damper motors to control the operations. There are three differential thermostats, four ordinary thermostats, and 14 relays in the control system. Figure 15-26 shows the inside of the rock bed.

The blower for the collector loop and damper motor M_4 are shown in Fig. 15-27. This system uses site-fabricated solar collectors for heating the air.

The three space-heating systems just described are merely representative of the ways in which active solar systems may be utilized for space heating. In actual practice, there are many possible variations in the design and operation of active space-heating systems.

CONCLUSION

This chapter has dealt specifically with the design and performance of active space-heating systems. Not covered in this chapter, but a matter of much interest, is the very important consideration of economics. The primary function of an active solar energy system is to collect solar energy to replace the use of conventional energy sources such as natural gas, oil, propane, or electricity. A solar energy system will save the owner money on an operating basis by reducing the amount of conventional energy that must be purchased, but the installation of the solar energy system requires a large initial investment. To analyze the economics of a solar system, one must consider the costs of the solar installation, which would include either the principal and interest payments on a loan if the system were financed or the return that would be earned on the capital had it been invested in a different project. System maintenance and the yearly system operating costs must also be considered. The operating energy savings must be related to the system costs, along with any tax credits or benefits that may be available. At some future time, the total energy savings should equal or surpass the total costs associated with the installation and operation of the solar system, including the returns foregone of the capital investment. These aspects of solar system design are presented in more detail in Appendix II.

PROBLEMS

1. For a 1500 ft² house located in an area with 3500 annual heating degree-days, determine what the building load coefficient should be if the house is to have an active solar space-heating system. (Use Table 15-1.)

2. For a 1500 ft² residence located in a climate with 600 DD/month during the winter, estimate the solar collector area. (Use Table 15-2.)

3. Estimate the solar collection area for a 2000 ft² house in a location with winter heating degree-days of 1200 DD/month.

4. For a thermal storage capacity of 15 Btu/F°-ft²$_c$, determine the volume of rock required for a collector array of 600 ft². Assume that the rock has a density of 100 lb/ft³, a specific heat of 0.18 Btu/lb-F°, and a void of 0.40.

5. An active solar space-heating system has a collector area of 450 ft². If the thermal storage capacity for the system is to be 12 Btu/F°-ft²$_c$, determine the volume of water required for a water thermal storage system.

6. If a collector array produces 900 Btu/ft²-day of heat, determine the rise in

temperature of the thermal storage during the day if it has a heat storage capacity of 15 Btu/F°-ft^2_c. Assume no losses or use of stored heat.

7. For an active solar space-heating system with 3000 ft^2 of solar collector, determine the approximate volume of water required for thermal storage. Use a rule of thumb of 15 Btu/F°-ft^2_c.

8. A building has a building load coefficient (BLC) of 50,000 Btu/DD. The estimated internal heat gain is 350,000 Btu/day. Determine a base temperature (T_b) for a thermostat setting of 68°F to take into account the internal heat gain.

9. An active solar space-heating system has been designed with 700 ft^2 of single-glazed solar collectors with flat-black absorbers. Determine the collector area required if a comparable double-glazed collector is used. Assume that the location has an average of 800 DD/month in winter.

10. A commercial building has an active solar space-heating system that uses 3000 ft^2 of single-glazed collectors with flat-black absorbers. The location has an average of 600 DD/month in winter. Determine the collector area if comparable collectors with selective absorber surfaces are used.

11. A liquid-based solar space-heating system uses 600 ft^2 of solar collector. The piping diameter is 1¼ in. and the total length is 80 ft in the collector loop. It is desired that the heat loss rate from the collector-loop piping be 0.04 Btu/hr-F°-ft^2_c. Determine the R value for the pipe insulation.

12. The collector-loop ducting used for an air-based solar space-heating system is 16 in. in diameter. The total length of the ducting is 150 ft. There are 8000 ft^2 of solar collectors in the system. Determine the R value for the duct insulation if the desired heat loss from the collector loop ducting is to be 0.06 Btu/hr-F°-ft^2_c.

13. An active space-heating system (liquid-based) has a total thermal storage capacity of 10,000 Btu/F°. The outdoor design temperature is 26°F, the degree-day building load is 15,000 Btu/DD, the hot-water design temperature is 125°F, and the interior-space design temperature is 70°F. Determine the carry-through time if the water thermal storage has an initial temperature of 140°F, the room thermostat setting is 68°F, and the ambient temperature is 32°F.

14. Estimate the average monthly insolation for January on a due-south-facing surface tilted at 50° and located at 40° N Lat. Assume that the percent possible sunshine is 48 percent. Use Eq. (15-5).

15. An active solar space-heating system is being designed for a residence in Salt Lake City, Utah. The house has a building load coefficient of 14,000 Btu/DD. The solar collector area is 600 ft^2 with a due-south orientation and a tilt of 60°. A 65°F temperature base will be used for the degree-days. The system will have the characteristics of the air-based system cited in Table 15-4. Use the Solar Load Ratio method to estimate the monthly and annual Auxiliary Heat and also the annual solar heating fraction.

16. A solar space-heating system with rock bed storage will use 600 ft^2 of air-heating solar collectors. The air flow through the collectors is 3.0 CFM per square foot of collector. The air pressure drop in the collectors is 0.09 in. of water, gauge. The rock bed will contain 400 ft^3 of rock with a rock diameter of 2.0 in. The vertical length of the rockbed is 4 ft. The effective length of the collector loop ducting is 120 ft, and the duct diameter is 16 in. (a) Determine the pressure drop for the rock bed and ducting. Then (b) determine the total collector loop pressure drop and a blower size and motor horsepower to provide the required air flow in the collector loop.

17. The equivalent length of piping in the collector loop for a liquid-based solar space-heating system is 150 ft. The pipe diameter is 1½ in. The flow rate is 20 gal/min in the collector loop. The pressure drop in the collector array is 1.5 ft of water head. The heat exchanger in the system has a pressure drop of 5 ft of water head. Determine the pressure drop for the collector loop.

18. Determine the annual Auxiliary Heat and the annual solar heating fraction for the small commercial building in Illustrative Problem 15-10 if it is located in Salt Lake City, Utah.

C H A P T E R   16

Summer Cooling of Solar Buildings

*I*n most temperate-zone locations, winter heating
of living spaces is essential for human comfort.
Conditions of the ambient air for human comfort in
winter are discussed in Chapter 6, and the concept of
the winter Comfort Zone was developed in section
6-4. The corresponding conditions for summer comfort are the subject of the first part of this chapter.

Although people generally exhibit somewhat
more tolerance for summer heat than they do for
winter cold, there is nevertheless a good measure of
agreement that the upper limit of the summer Comfort Zone is in the range of 76° to 80°F (24° to 27°C),
depending on the relative humidity of the air in the
space. In the United States, from 40° N Lat southward, except for locations at higher elevations, it is
now common practice to at least consider the provision of summer air conditioning as homes, apartments, hotels, and institutional and commercial
buildings are being designed.

The term *summer air conditioning* means much
more than merely cooling the air in a building. In
addition to cooling the air, it also implies controlling
the relative humidity, providing proper ventilation,
filtering out contaminants (air cleaning), and dis-

tributing the conditioned air to the lived-in spaces in proper amounts, without appreciable drafts or objectionable noise.

Summer cooling is a term that is often loosely used. Summer cooling may sometimes mean complete air conditioning, but it is often used in connection with systems or equipment that do only a part of the job, accomplishing cooling but neglecting some or all of the other four functions of complete air conditioning. For example, the common window air conditioner or room cooler cools, dehumidifies, and filters air, but blows the cool air into only one room. Other rooms of the structure receive very little of the cooled air.

This chapter deals with the summer cooling of solar buildings—both passive and active solar designs. The emphasis, for the most part, will be on air-conditioning systems that are themselves solar-energized, but there will also be some attention given to the use of electrically powered equipment, such as blowers and fans, water-cooling towers, and evaporative (desert) coolers for use with solar buildings.

HUMAN COMFORT IN SUMMER

16-1 Air and Water Vapor

Absolutely dry air is rarely encountered, and atmospheric air usually contains significant amounts of water vapor. *Humidity* is the general term for moisture (i.e., water vapor) in air. The amount of water vapor that air can hold depends on the temperature—the higher the temperature, the more moisture it can hold. When air contains all of the water vapor it can hold at a given temperature, we say that it is *saturated*. The actual amount of water vapor in air at a given temperature (pounds of water vapor per pound of air) is called *specific humidity* W. Since air at any temperature can conceivably be either bone dry or saturated—or for that matter at any condition in between these extremes—a third measure of humidity called *relative humidity* is necessary. Relative humidity RH is the ratio, at any given temperature, of the actual (i.e., specific) humidity W to the specific humidity required for saturation at that temperature, W_{sat}:

$$RH = \frac{W}{W_{sat}} \qquad (16\text{-}1)$$

Relative humidity RH is an expression of the percentage of saturation of air at the given conditions. The RH of saturated air is 100 percent.

Psychrometrics and the Psychrometric Chart.   The science of air–water vapor mixtures and the study of their properties is called *psychrometrics*. Psychrometrics is a complex subject, and its theoretical treatment involves extensive mathematical calculations. However, most of the relationships among air, water vapor, and heat energy that are needed in the practice of air conditioning are either contained in tables or are conveniently expressed in graphical form by means of the *Psychrometric Chart*, a common form of which is displayed in Fig. 16-1. For the brief treatment of psychrometrics that space allows here, we shall rely wholly on the chart, without reference to tables or detailed mathematical analyses.

Before explaining the use of the psychrometric chart in system design, it is necessary to review some of the terms used in psychrometrics:

> *Dry-bulb temperature* (DB) is the temperature of the ambient air as measured by an ordinary thermometer.
> *Wet-bulb temperature* (WB) is the lowest temperature attained by a thermometer whose bulb is surrounded by a water-saturated cloth or *wick*. The WB thermometer is placed in a moving current of air, or it is whirled in still air, thus producing relative air movement to cause the maximum rate of

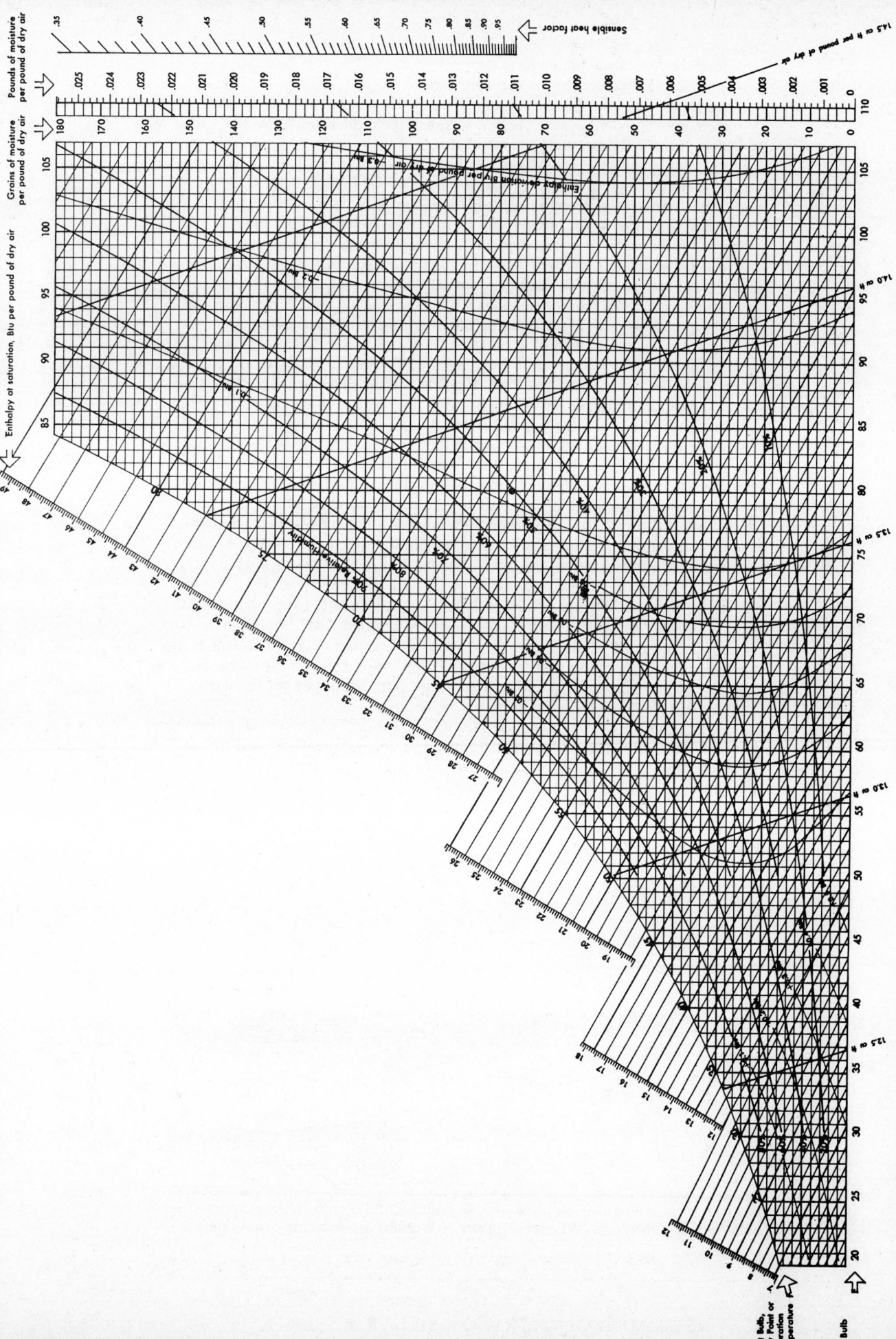

Figure 16-1 Psychrometric Chart for normal temperatures. (Carrier Air Conditioning Corporation.)

evaporation of the water from the wet wick. Evaporation causes a temperature drop at the wet bulb.

Wet-bulb depression is the difference in degrees between the DB temperature and the WB temperature. The drier the ambient air, the greater the rate of evaporation from the wet wick, and the greater the WB depression. If the WB depression is zero—that is, if WB temperature equals DB temperature—the air contains all of the water vapor it can hold at that temperature, and we say it is *saturated*.

Dew-point temperature (DP) is the temperature at which moisture begins to condense out of air as the air is cooled. Presence of dew on the grass in the morning means that during the night the air temperature has reached or fallen below the DP temperature.

A *hygrometer* is an instrument for determining the DB and WB temperatures. Perhaps the most common form of hygrometer, and the one used almost exclusively by air-conditioning engineers and technicians, is the *sling psychrometer*, illustrated in Fig. 16-2.

Now refer to the Psychrometric Chart (Fig. 16-1). Note that the DB temperature scale is the horizontal scale across the bottom. The vertical lines rising from this scale are lines of equal DB temperature.

Next, locate the line at the far left of the chart that curves to the right and sharply upward. This is the *saturation line* (the line of 100 percent relative humidity), and both the WB temperature and the DP temperature are plotted on a scale along this line. Note especially that when air is saturated, WB = DB = DP. At saturation, any increase in temperature would reduce the relative humidity; any drop in temperature would reduce the specific humidity, because moisture would start to condense out of the air; that is, dew would start to form.

The family of curves that more or less parallel the saturation line are relative humidity (RH) lines. They are labeled in percent RH from 10 percent to 90 percent. The saturation line is, again, the 100 percent RH line.

At the far right of the chart, the vertical scale is marked off with values of specific humidity W. Note that the values for W are given in pounds of water vapor per pound of dry air, and also in grains of water vapor per pound of dry air (1 lb = 7000 gr).

The Psychrometric Chart incorporates a great deal of additional information, but for our purposes the above-defined factors and measures will suffice.[1] Some exercises follow to illustrate the use of the Psychrometric Chart.

[1] Readers desiring a more extensive treatment of psychrometrics may consult *ASHRAE Handbook, Fundamentals,* 1977 or any air-conditioning text, such as Norman C. Harris, *Modern Air Conditioning Practice,* 3rd ed. (New York: McGraw-Hill Book Company, 1983), Chapters 5, 6, and 7.

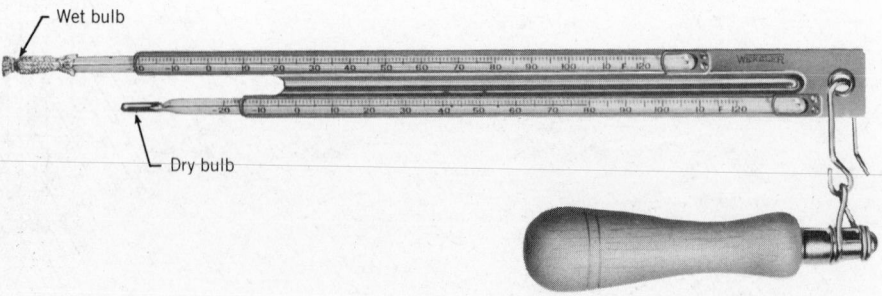

Figure 16-2 A sling psychrometer for obtaining simultaneous readings of the dry-bulb (DB) and wet-bulb (WB) temperatures. (Weksler Instruments Corporation.)

Illustrative Problem 16-1

A sling psychrometer, after being whirled in the room air, gives the following readings: DB = 80°F; WB = 66.8°F. What is the RH in the room?

Solution

Enter the chart of Fig. 16-1 along the DB temperature scale at bottom. Locate the 80°F dry-bulb line and follow it vertically upward until it intersects the 66.8°F wet-bulb line (interpolate as necessary) that comes to the right and down from the WB scale on the saturation line. Note that the intersection is on the 50 percent RH line. The relative humidity is 50 percent.   *Ans.*
(The condition 80°F, 50 percent RH is called the *reference point*, and it is circled on the Carrier chart used in this book.)

Illustrative Problem 16-2

On a hot, muggy day in summer, a sling psychrometer gives 92°F DB, 85°F WB when it is whirled in outdoor air. What is the relative humidity?

Solution

Enter the chart at the 92°F DB line, and follow it up vertically until it intersects the 85°F WB line as the latter comes to the right and down from the saturation line. Estimate (interpolating) the relative humidity,

$$RH = 75\%$$ *Ans.*

Illustrative Problem 16-3

A client tells you that the summer inside-design condition for a dental clinic being planned should be 78°F, 50 percent RH. At this condition, what is the specific humidity W of the room air? (Refer to Fig. 16-3.)

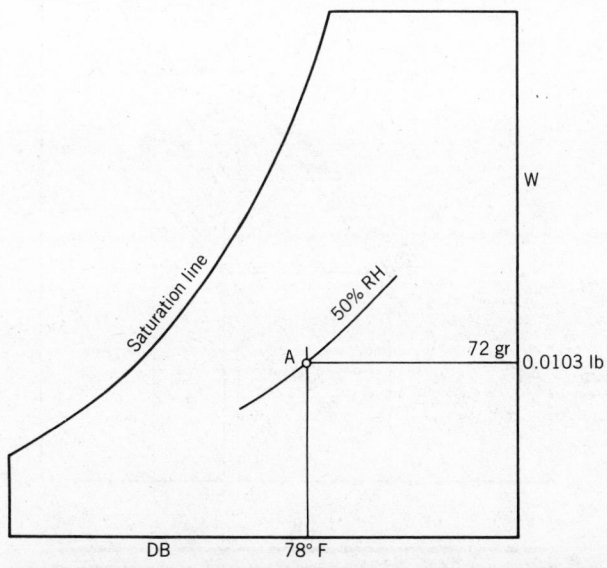

Figure 16-3 Skeleton section of a psychrometric chart to show the steps in the solution of Illustrative Problem 16-3.

Solution

Enter the Psychrometric Chart at 78°F DB, and follow this line vertically upward to its intersection with the (curving) 50 percent RH line (see plotted point A on Fig. 16-3). From point A, follow a horizontal (constant moisture) line right to the specific humidity scale at the far right of the chart. Read:

$$W = 72 \text{ gr water vapor per lb dry air}$$
$$= 0.0103 \text{ lb water vapor per lb dry air} \qquad Ans.$$

Illustrative Problem 16-4

Hot, humid, outdoor air at 90°F DB and 70 percent RH enters the cooling coil of a window air conditioner and is cooled (and dehumidified in the cooling process) so that it comes off the cooling coil at 60°F, saturated. It is discharged into the space being conditioned. Assuming that no other air (or moisture) enters the room and that the temperature (DB) in the room is 80°F, what is the RH in the room? (Refer to Fig. 16-4.)

Solution

Locate the condition point of the outside air on the Psychrometric Chart and label it in pencil, point O. (See point O on Fig. 16-4.) As this air is cooled by the cold coil, it first undergoes a drop in DB temperature (move horizontally left from point O to the saturation line). Its condition is now approximately DB = WB = DP = 79°F (point B, Fig. 16-4). As the air is cooled further, it also loses moisture (water will actually drip off the cold coil in the unit), and its condition moves down along the saturation line to the air-off-the-coil condition of 60°F, saturated (point C of Fig. 16-4). Here it is discharged into the room, where it soon warms up (all sensible heat, since we assumed that no additional moisture gets into the room) to the room DB temperature of 80°F. This final process is shown by a straight line drawn horizontally (along a line of constant specific humidity) to the right from point C to the room DB temperature line of 80°F.

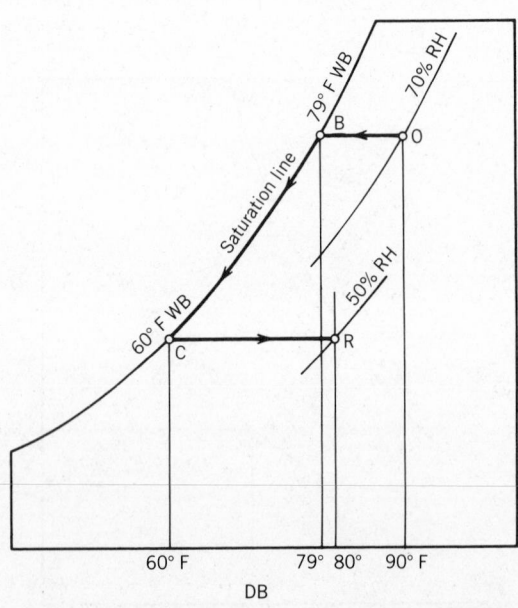

Figure 16-4 Skeleton section of a psychrometric chart for reference in the solution of Illustrative Problem 16-4.

Locate the room condition as point R on Fig. 16-4. Read the room relative humidity from the Psychrometric Chart as

$$RH = 50\% \text{ (approximately)} \qquad \textit{Ans.}$$

16-2 Design Conditions for Summer Cooling

The amount of cooling that has to be accomplished to keep buildings comfortable in hot summer weather depends on the desired condition indoors and on the outdoor conditions on a given day. These conditions are, respectively, termed the indoor design condition and the outdoor design condition.

Indoor Design and Human Comfort. It will be recalled from the discussions in Chapter 6 that human comfort is very difficult to define, since it is an individual feeling, not an objective measurement. Most persons agree, however, that comfort is determined by temperature, humidity, and air movement. The *Comfort Chart* (Fig. 6-13) has been devised on the basis of research conducted by ASHRAE over a period of many years, with hundreds of human subjects. A winter comfort zone (WXYZ) and a summer comfort zone (ABCD) are plotted on a basic grid of dry-bulb versus wet-bulb temperatures. Scales are superimposed on the Comfort Chart indicating the percent of subjects feeling comfortable under any set of conditions falling within the comfort zones. Relative humidity lines are prominently plotted on the chart, and so is an *effective temperature scale*. For human comfort in winter, it was noted in section 6-4 that an effective temperature (ET) of about 67°F was found to be optimum for subjects wearing average-weight clothing, sitting at rest indoors. This condition—67°F ET—is equivalent to 72°F DB and 40 percent RH.

Although winter comfort is affected to some extent by humidity, most people can tolerate a rather wide variation in winter relative humidity without any great sense of discomfort, provided the dry-bulb temperature stays in the range of 70° to 74°F. Not so in summer, however. The familiar saying, "It's not the heat, it's the humidity," has a strong basis in fact for summer comfort. Of the four mechanisms by which the body loses heat in order to maintain temperature equilibrium, one of them—evaporation of perspiration—depends on the relative humidity of the ambient air; if RH is low, rapid evaporation of perspiration occurs, the body surface is cooled, and people feel comfortable even with rather high dry-bulb temperatures. Conversely, if the relative humidity is high on a warm day, perspiration does not evaporate from the skin, body heat builds up, and people sweat profusely and complain about the muggy weather.

Reference to the *Comfort Chart* (Fig. 6-13) shows that for 50 percent of any indoor group of people to feel comfortable in summer, the effective temperature should be between 67°F and 75°F. The optimum summer ET is 72°F. Note that virtually no one feels comfortable if the relative humidity exceeds 70 percent, no matter what the air DB temperature is.

A good measure of agreement could be reached on the following conditions for indoor summer comfort, for lightly clothed persons sitting at rest or doing desk work:

1. For long periods of occupancy (several hours) for employees in a building or residents in houses: 75°F DB, 50 percent RH
2. For short periods of occupancy (up to one hour) for customers or clients in a commercial building: 78 to 80°F DB, 50 percent RH

In summary, indoor design conditions for both winter and summer are not fixed values. Owner or occupant preference and/or government energy-conservation standards may affect the actual design condition to be specified for a given building. As a general guideline for summer air-conditioning design, engineers for many years have used 78°F DB, 50 percent RH (73°F ET) as a desirable indoor design condition for comfort air conditioning. The high cost of energy today and the recognition of need for energy conservation have influenced some designers to go to 80°F DB, 50 percent RH (74°F ET). It is a common observation that many public buildings are overly refrigerated in summer. Persons coming into such indoor climates from the hot outdoor climate actually shiver as the dry, cold air chills the skin and evaporates perspiration. Investigation often reveals that such buildings are on thermostat settings of 74° to 76°F, even when the outdoor temperature is 95° to 100°F. This temperature difference is actually a shock to the body temperature control systems of short-occupancy customers and clients. A good rule for public buildings is not to exceed a 20° outdoor–indoor temperature difference.

Illustrative Problem 16-5

An air-conditioning engineer whose past practice has been to use a summer indoor design condition of 78°F DB, 50 percent RH is told by the client to design for 80°F DB, 60 percent RH in the interest of energy conservation. (a) Will this condition be in the summer comfort zone? (b) What will be the effective temperature (ET)? (c) What percent of persons (lightly clothed and at rest) will probably feel comfortable? Use the Comfort Chart (Fig. 6-13).

Solution

(a) On the Comfort Chart, find the 80°F DB line, and follow it vertically upward to its intersection with the 60 percent RH line. Note that this condition point is in the summer comfort zone. *Ans.*
(b) From the plotted condition point, follow an effective temperature (ET) line up and to the left to the scale of effective temperature and read:

$$ET = 75°F \qquad Ans.$$

(c) Continue following the 75°F ET line up to the left until it intersects the bell-shaped curve of "percent persons feeling comfortable." Read 50 percent. *Ans.* On days when the indoor condition is just being met by the cooling equipment, about half of the building's occupants might report some degree of discomfort. The problem then becomes not an engineering consideration, but a personnel problem for the owner or building manager.

Outdoor Design Conditions and Design Temperature Difference. Outdoor design conditions depend on the latitude, the locality, and the climate. The American Society of Heating, Refrigerating, and Air Conditioning Engineers (ASHRAE) has prepared, from climatic data, tables of recommended outdoor design conditions for both summer and winter. The dry-bulb and wet-bulb temperatures listed are average values recorded over many years, and they exclude extremes (high or low) that may have occurred on fewer than 10 days each year. Table 5-1 provides suggested outdoor design conditions for both summer and winter for many localities in the United States and Canada.

The design temperature difference Δt is merely the arithmetic difference between the outdoor and indoor design temperatures. For example, if a summer indoor design temperature of 80°F DB is specified for Tulsa, Oklahoma, the design temperature difference (see Table 5-1 for Tulsa) is $\Delta t = 100° - 80° = 20°F$.

The standard outdoor design dry-bulb temperature recommended by ASHRAE and by the American Refrigeration Institute (ARI), as a basis for testing and rating summer air-conditioning equipment, is 95°F.

ANALYSIS OF THE COOLING LOAD

16-3 Summer Load Classifications

Summer cooling-load calculations are similar in many ways to winter heating-load calculations (review Chapter 5). However, there are some very important differences that result from two climatological factors:

1. There is a significant moisture load at most locations in the summer because of the higher specific humidity of the summer air.
2. Solar gains become a part of the cooling load in summer, whereas in winter they constitute a desirable source of heat.

Cooling Loads Classified by Source.   Cooling loads fall into the following categories, based on their sources:

1. Heat transfer (gain) through the building skin by conduction, as a result of the outdoor–indoor temperature difference.
2. Solar heat gains (radiation) through glass or other transparent materials.
3. Heat gains from ventilation air and/or infiltration of outside air.
4. Internal heat gains generated by occupants, lights, appliances, and machinery.

Cooling Loads Contain Two Kinds of Heat.   A second classification of cooling loads is based on these two forms of heat:

1. *Sensible heat* gains—the form of heat that produces an increase in temperature (see section 3-4).
2. *Latent heat* gains—the heat contained in water vapor. This is heat that was required to vaporize the water that is in the humid air, and it is heat that must be removed to condense the moisture out of the air. Latent heat does not cause a temperature rise, but it constitutes a load on the cooling equipment.

In calculating cooling loads, it is necessary to sum up the sensible heat gain from all sources and the latent heat gain from all sources, and to keep these two totals separate up to the point of final analysis and equipment selection. It is also a good idea to keep the heat gains from *external* sources separate from the heat gains produced in the conditioned space (*internal* heat gains), since the relative magnitude of these two loads may also have a bearing on equipment selection.

Summer cooling loads are *instantaneous rates* of heat gain (Btu per hour or watts), just as winter heating loads are instantaneous rates of heat loss.

16-4 Building Heat Gain

The cooling load is the hourly rate at which heat must be removed from a building in order to hold the indoor air temperature at the design value. The *design cooling load* is closely related to the building heat gain but is not necessarily equal to it for reasons that will be elaborated below.

Summer load analysis is considerably more complex than winter load analysis, for several reasons. First, there is usually a much greater outdoor temperature variation over a 24-hour period in summer than there is in winter. Second, there is a time lag or thermal inertia present in building components, which results in heat absorbed now being released into the building interior perhaps hours later. Third, solar heat gain is, as pointed out earlier, a plus factor in winter heating, but it may be a major part of the load for summer cooling. Fourth, there is the matter of the moisture content of summer air—latent heat—which has a great deal to do with human comfort. Much of this moisture must be removed from the indoor air in order to attain a comfortable condition, and this moisture load is a load on the cooling equipment. And fifth, internal heat sources such as lights, machinery, appliances, and people constitute cooling loads in summer, whereas in winter the heat from these sources is a plus factor.

For these and other reasons, in summer a condition of steady-state heat flow through the building envelope is rarely attained. Heat-gain maximums from the different sources do not peak at the same time. For example, in a

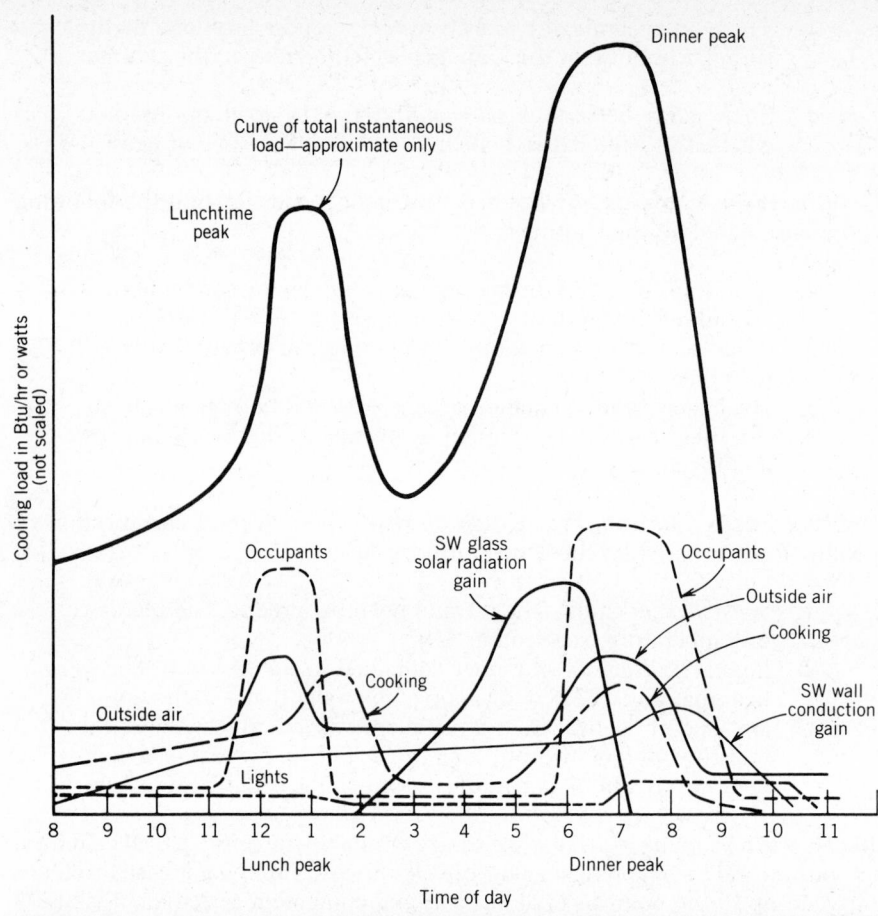

Figure 16-5 A plot showing the hourly variations in the several components of the cooling load for a restaurant. Note that the several loads do not peak at the same time, and therefore the design cooling load is not the sum of the several peak loads.

restaurant with southwest glass and wall exposures, there will be a lunchtime peak of the occupant and cooking loads, but the solar radiation load through glass may not peak until 5:00 or 6:00 P.M. (Fig. 16-5). By 7:00 P.M., when the internal loads peak again for the dinner hour, the solar radiation load through glass will have fallen to near zero, but the conduction load may still be increasing because of time lag in the building structure. The ventilation load might remain fairly constant throughout the day, since ventilation and exhaust fans would probably be operating all day long, but the infiltration load would peak sharply at the lunch and dinner hours, as diners arrive and leave, with frequent door usage.

As a consequence of the various loads peaking at different times, the design cooling load and the recommended cooling capacity for the installation are generally considerably lower than the sum total of all of the building heat-gain maximums. The structure thermal flywheel smooths out these peaks and allows engineers to design on a 12-hour or 24-hour basis, specifying equipment whose capacity may be appreciably less than the aggregate of the instantaneous heat-gain maximums. By operating the equipment during off-peak hours, with the thermostat set to allow a few degrees of subcooling during unoccupied periods, the building thermal mass can be precooled to some extent, in anticipation of the next day's or this afternoon's heat gains—using thermal mass in a sense to "store" cooling, just as in winter it is used to store heat. The temperature variation in the conditioned space that results from such operating schedules is referred to as *space-temperature swing*. Needless to say, in systems of significant size and/or complexity, control of all of the elements of the system is coordinated by programmed controllers or a computer.

As the above discussion indicates, the detailed analysis and calculation of summer cooling loads is a complex engineering problem—one that for large, multiuse buildings is well beyond the scope of this book. Readers interested in an in-depth treatment of cooling load analysis are referred to *ASHRAE Handbook, Fundamentals, 1977,* Chapter 25.[2]

Not all buildings exhibit the variability of the restaurant example above, however, and for simple buildings with low occupancy the problem of cooling load determination is not at all difficult, although it is time-consuming. In the next section, we provide an example of a simple solar building and illustrate one method by which a close approximation of its summer cooling load may be determined.

16-5 Calculating the Summer Cooling Load of a Simple Solar Building

The following example illustrates the general method of determining the design cooling load of a small passive solar building. Remember that the design cooling load may not equal the sum total of all of the load peaks but is usually a value somewhat less than that, taking into account the fact that all of the loads do not peak at the same time.

---

[2] See also Norman C. Harris, *Modern Air Conditioning Practice,* 3rd ed. (New York: McGraw-Hill Book Company, 1983), Chapters 8 and 9.

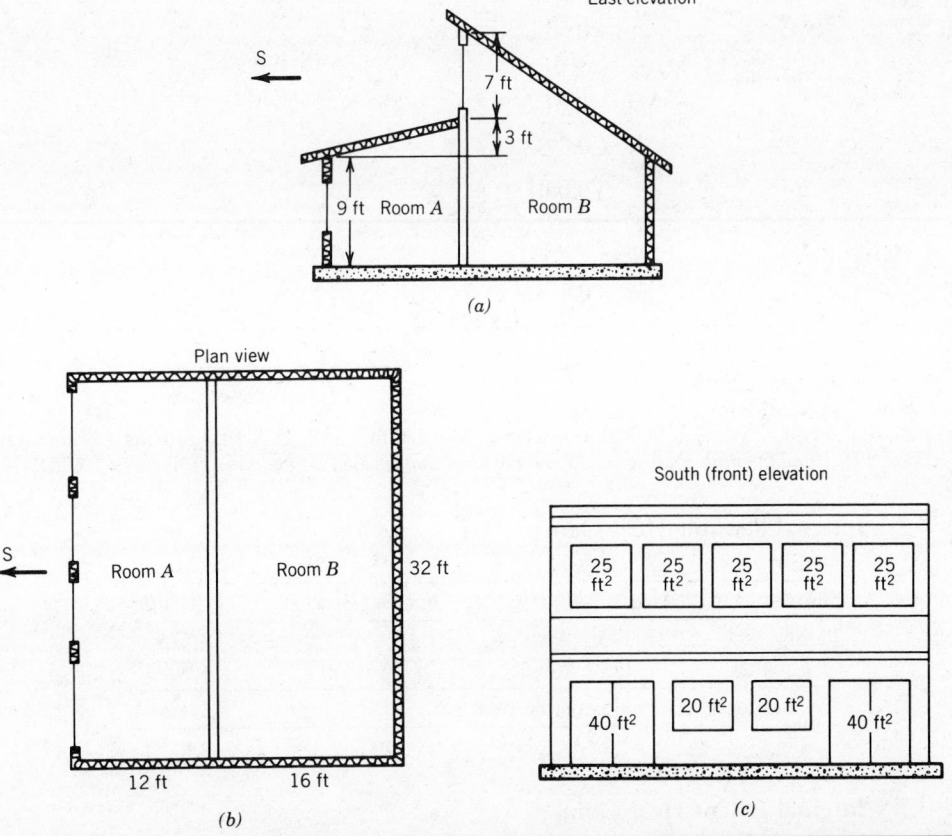

Figure 16-6 Simplified plan of a small passive-solar office building. (*a*) East elevation. (*b*) Floor plan. (*c*) South (front) elevation. (For use with Illustrative Problem 16-6.)

ANALYSIS OF THE COOLING LOAD

Illustrative Problem 16-6

Three views of a small passive solar-heated office building are shown in Fig. 16-6. A summer cooling installation is being planned for the building, which is located at 40° N Lat. Room A has 120 ft² of south-facing double-pane glass, and room B has 125 ft² of double-pane clerestory windows. All glass for both rooms is shaded in summer by roof overhangs and by indoor venetian blinds. All outside walls are insulated at R-11, and the cathedral ceilings of both rooms are insulated at R-19. It is assumed that there will be no heat gain through the floor.

For a building of this type, infiltration will normally provide 1.5 air changes per hour. Ventilation is per local code—10 CFM per person—and exhaust fans with a capacity of 100 CFM operate nearly all day. Internal heat sources are lights—2500 W; occupants—15, light activity. The heat gain from occupants is given as sensible—260 Btu/hr per occupant; latent—270 Btu/hr per occupant.

Outside design conditions are 90°F DB, 70 percent RH; and inside design is 78°F DB, 50 percent RH.

Calculate the design summer cooling load for the steady-state daytime conditions specified, using 3:00 P.M. July 21 as a base.

Solution

Lay out the work as follows. (See Tables 5-3, 7-2, and 7-A for tabular data.)

1. External Sensible Heat Gain

Heat Gain from	Area (ft²)	U	$\Delta t = (t_o - t_i)(F°)$	SHGF[a]	SC[b]	H(Btu/hr)
Windows and glass doors, south	245	0.58	22	52	0.55	10,130
	Eq. (7-2)	$H = UA$	$(t_o - t_i) + A\,(SHGF)(SC)$			
Walls (net)		$H = UA$	$(t_o - t_i)^c$			
East	350	$\frac{1}{11} = 0.091$	22			700
South	267	0.091	22			535
West	350	0.091	22			700
North	288	0.091	22			575
Beam ceilings	975	$\frac{1}{19} = 0.053$	22			1,135
Subtotal, external sensible heat gain						13,775

[a] See Table 7-A for solar heat gain factors.
[b] See Table 7-2 for shading coefficients.
[c] Solar heat gain on walls in direct sun is not taken into account when using $\Delta t = (t_o - t_i)$. For a more accurate treatment, see ASHRAE Handbook, 1977 Fundamentals for discussion and tables of cooling load temperature differences (CLTDs) for sunlit walls.

2. Internal Sensible Heat Gain

Source	Units	Factor (Btu/hr)	H(Btu/hr)
Lights	2500 W	3.41	8,525
Occupants	15, light activity	260	3,900
Subtotal, internal sensible heat gain			12,425

3. Internal Latent Heat Gain

Source	Units	Factor (Btu/hr)	H(Btu/hr)
Occupants	15, light activity	270	4,050

4. Heat Gains from Outside Air

Ventilation air: 10 CFM per person $\times$ 15 persons = 150 CFM

Infiltration: $\dfrac{1.5 \text{ air changes/hr} \times 10,000 \text{ ft}^3}{60 \text{ min/hr}}$ = 250 CFM

Since the infiltration volume is greater, it will be used.

Outside air volume = infiltration air plus exhaust air
= 250 CFM + 100 CFM = 350 CFM

or,

$$V = 350 \,\frac{\text{ft}^3}{\text{min}} \times 60 \,\frac{\text{min}}{\text{hr}} = 21,000 \,\frac{\text{ft}^3}{\text{hr}}$$

A. *Sensible Heat Gain from Outside Air*
From Eq. (5-4),

$$H_S = 0.018 \, V \times (t_o - t_i)$$
$$= 21,000 \,\frac{\text{ft}^3}{\text{hr}} \times 0.018 \,\frac{\text{Btu}}{\text{ft}^3\text{-F}^\circ} \times 22 \text{ F}^\circ$$
$$= 8315 \text{ Btu/hr}$$

B. *Latent Heat Gain from Outside Air*
Summer air often contains very large amounts of water vapor. Since the cooling system must remove (condense) much of this water vapor in order to produce the inside design RH condition, water vapor content in outside air constitutes a latent cooling load H_L. The expression for calculating this load is:

$$H_L = \text{CFM} \times 0.68 \, (W_o - W_i)$$

where W_o and W_i are the outdoor and indoor specific humidities, respectively, in grains of water per pound of air. (See the Psychrometric Chart, Fig. 16-1, for the values of W_i and W_o for the conditions given in the problem.) The calculation follows:

$$H_L = 350 \times 0.68 \times (151 - 72) = 18,800 \text{ Btu}$$

5. Summary of the Cooling Load

External sensible heat	13,775 Btu/hr
Internal sensible heat	12,425
Internal latent heat	4,050
Outdoor air sensible heat	8,315
Outdoor air latent heat	18,800
Total cooling load, 3:00 P.M., July 21	57,365 Btu/hr *Ans.*

In order to decide on a design cooling load and specify equipment, the designer would now analyze the peaks very carefully, perhaps preparing a time graph of them like that of Fig. 16-5, noting how much overlap there is among the peaks and determining the time lag associated with the several load components. An analysis of the building's thermal masses and their capacity to store cooling would then be made, and the final design load determined on a 12-hour or perhaps even a 24-hour basis. The decision on equipment and system capacity would have to reflect the owner–client's views on indoor comfort and on whether or not provision would be made for off-peak and night operation of the system for subcooling purposes. These matters are often made a part of the contract.

Short-Form Load Estimates. For small, simply constructed buildings and for summer cooling installations where a guarantee of performance to maintain a stipulated set of indoor conditions is not required, extended load calculations are sometimes omitted in favor of cooling load estimates made by using prepared estimating forms. These "short-form" estimate sheets use precalculated factors and coefficients to fit average construction and average climatic conditions. Although when properly used they produce reasonably good estimates (if the assumed conditions actually apply), they should be used with caution and with full knowledge of their limitations. Never use short-form estimating methods if the client wants a guaranteed indoor condition regardless of the outdoor weather.

Short-form cooling-load estimate sheets are available from the major manufacturers and marketers of air-conditioning equipment.

Total Season Cooling Load. For application to solar-cooling system design, the instantaneous cooling load expressed in Btu per hour is not a very suitable measure. Solar collectors and their associated equipment must be sized, taking into account average conditions of insolation and load demand over considerable periods of time, rather than for a peak hourly load on a given day. Furthermore, most solar-cooling installations are part of a year-round system, with the collectors being used for winter heating and perhaps domestic water heating as well.

Consequently, the methods of section 16-5 are usually incorporated into an expanded calculation that yields a total cooling load for the cooling season Q_{CS}, in Btu (see section 5-11). This seasonal load can then be used as a factor, along with the available seasonal insolation and various equipment parameters, to make a reasonably accurate estimate of the minimum solar collector requirements (see section 16-14). Such calculations are cumbersome and complex at best, involving the summing up of hourly loads (disregarding short-term peaks) to get daily loads; then summing these to obtain monthly loads for the cooling season (typically May through September); and finally aggregating the monthly loads to yield the total season cooling load Q_{CS}.

Computer programs for personal and desktop computers are often used to simplify these calculations.

PASSIVE SOLAR COOLING

The ideal initial step in the summer cooling of buildings is to prevent them from getting hot in the first place. In practice, of course, this ideal is not often attained, but every possible effort should be made in the design and construction of buildings to control the amount of summer heat that enters and the amount of heat generated inside.

Siting considerations are important—orientation of the building; presence of shade trees, vegetation, ponds, or streams; and the availability of daily breezes. South and west glass must be shaded in summer, by roof overhangs, awnings, deciduous trees, and interior blinds or drapes. The importance of preventing direct solar radiation from entering through glass can be appreciated from the calculation that 60 ft² of unshaded west-facing glass at 4:00 P.M. on a summer day (40° N Lat) will contribute a heat load of about 12,000 Btu/hr (1 ton of refrigeration) to the interior, requiring about 1.5 compressor horsepower for a conventional air-cooled air conditioner.

As emphasized repeatedly in earlier chapters, the building structure itself should be weatherstripped and extremely well insulated (R-11 to R-19 for walls, and R-19 to R-30 for ceilings). Double-pane insulating glass should be used wherever possible. Just as the south side is planned for maximum heat collection in winter, the north and east sides should be planned to maximize the effect of whatever cool air there is in summer.

Passive cooling means cooling without mechanical refrigeration and with little or no energy consumption involved, except for occasional use of fans or pumps. In locations without extreme summer temperatures or relative humidities—for example, not exceeding 90°F DB, 50 percent RH—passive cooling techniques, expertly employed, are often sufficient to maintain conditions either in or near the summer comfort zone. In the very hot and humid climates of the U.S. Southeast and Midwest, however (DB temperatures consistently over 85°F and RH in excess of 65 percent), or for buildings with large internal loads, passive cooling alone will not suffice, and mechanical refrigeration is usually necessary to create conditions for summer comfort. Mechanical cooling using solar heat as the energizer is called *active solar cooling*. It will be dealt with at length later in this chapter.

16-6   Passive Cooling and Human Comfort

The summer comfort zone of Fig. 6-13 identifies the two most important factors affecting summer comfort: the dry-bulb temperature (DB) and the relative humidity (RH) of the air. However, three other factors also influence summer comfort: (1) the rate at which heat is carried away from the body surface by air motion (convection), (2) the rate at which the body radiates heat to surrounding (colder) surfaces, and (3) the rate of evaporation of perspiration from the skin. Air motion, or ventilation, has a two-way effect: it carries away sensible heat from the skin and also speeds up evaporation of perspiration, thus removing latent heat. This double effect of air currents explains why we feel so much cooler in a breeze, even though the moving air itself is at the existing dry-bulb temperature. Ventilation air at temperatures above about 95°F, however, has very little perceived cooling effect, and at 100°F and above it feels like a hot wind.

Of these five factors affecting human comfort, four can be significantly influenced by passive cooling techniques. Only the relative humidity factor is not effectively influenced by passive measures. Some of the more common systems and techniques for passive cooling will now be described.

16-7   Ventilation as a Means of Passive Cooling

Ventilation of an interior space is achieved by removing air from the space and replacing it with outdoor air. When outdoor air conditions fall within the summer comfort zone, ample ventilation is all that is needed to air condition a building. Buildings in which the cooling load is largely internally generated make a great deal of use of ventilation air, the technique being called the *economizer cycle* in conventional air-conditioning practice. Ventilation also increases the flow velocity of air currents in a building, which increases the rate of heat removal from the body, both sensible and latent. Ventilation may be provided either by natural means (open windows, doors, and breezeways) or by fans and blowers, with only a modest energy input. Night ventilation with cool outdoor air can not only drop the air temperature in a building, but it will also cool the thermal masses in the building, storing "coolth" (as it is often called) for the next day. Cooling rockbeds by night ventilation is another method of storing cooling. Cooling by nocturnal ventilation is really effective only in climates where nighttime dry-bulb temperatures fall below 65°F.

Most passive solar designs make definite provision for summer ventilation modes. By their very nature, passive designs induce ventilation because warm air heated by solar gains will exhaust from vents in the building, causing a natural lowering of pressure in the structure. Replacement air from outdoors is then drawn through open windows, doors, or vents, and these should be on the cool (north) side of the building. Thermal storage (Trombe) walls can also operate in a summer ventilation mode, as was explained in section 8-3, with warm air being vented from the top of the wall, which draws indoor air into the lower vents and on out at the top. Cool outdoor air then flows into the building from the north side. High vents are provided in sunspaces for warm air to exit, thus inducing a flow of outdoor air into the building from openings on the cool (north) side. Cooling by ventilation can also be accomplished through the use of roof vents and scoops suitably positioned so that breezes will enhance natural ventilation. Roof vents take advantage of the low pressure that exists on the leeward side of the house to induce a positive air flow out of the living spaces or attic of the building.

When natural ventilation at the site is insufficient for a good cooling effect, fans can be used. Ceiling fans simply increase the air velocity inside the building spaces to promote convective and evaporative cooling from human body surfaces. Exhaust fans create a lowered pressure in the building, which can induce a flow of cooler outside air into the structure from the north side or from a garden or pond area. Large blowers may be used in windows or with a duct system to bring in large quantities of outdoor air, perhaps at night, to flush out all of the warm air in the building and replace it with cool air and at the same time cool down the thermal masses in the building for the next day.

Air Quantity for Night Ventilation.   There is no direct and accurate method of calculating the exact flow quantity of night air to meet the net (next day) cooling load of a building. However, based on actual experience with passive solar buildings in medium-dry to arid climates (southwestern U.S. and interior valleys of California), some guidelines can be suggested. These rules of thumb are applicable only in locations where the nighttime temperature falls below 65°F for most of the cooling season. Also, they apply only in buildings with significant total building mass, since the nighttime "coolth" must be stored to absorb the next day's heat.

Here are two guidelines.[3] They should be used with the above limitations clearly in mind:

1.   With 40 to 50 lb of building mass per square foot of floor area, 1 CFM/ft² floor area can yield 35 to 40 Btu cooling/day-ft².
2.   With 80 to 150 lb of building mass per square foot of floor area, 3 to 4 CFM/ft² floor area can yield 80 to 140 Btu cooling/day-ft².

Examples
1.   Under guideline 1, a 1500 ft² residence with a 1500 CFM blower operating all night with 65°F air could have about 60,000 Btu of cooling stored up for the next day.
2.   Under guideline 2, a 6000 ft² medical clinic building could be provided with 18,000 CFM of night ventilation with 65°F air and store up about 600,000 Btu of "coolth" for the next day.

Buildings with heavy occupancy loads at night, located in regions where the outdoor air is cool and dry, can often be adequately cooled by switching from the mechanical cooling system to ventilation with outdoor air—in other words, to the economizer cycle. The volume of outside air required depends on the magnitude of the sensible cooling load and on the outdoor–indoor temperature difference $\triangle t$. The equation for calculating the required air flow is a form of Eq. (5-4):

$$V_{\text{CFM}} = \frac{\text{Sensible heat to be removed, Btu/hr}}{1.08\ \triangle t}$$

The following problem is a typical situation.

Illustrative Problem 16-7

A restaurant–nightclub has 200 occupants engaged, on the average, in light activity. At 10:00 P.M., there is still a conduction heat load of 55,000 Btu/hr coming into the space because of the time lag of the walls and ceiling. This is all sensible heat. Cooking and served food contribute another 12,000 Btu/hr of sensible heat. The outdoor air temperature is 63°F DB, and the desired indoor

[3] Adapted from J. D. Balcomb, ed., *Passive Solar Design Handbook*, vol. I (Washington: U.S. Department of Energy, DOE CS-0127/1, March 1980), pp. D-275–276.

temperature is 74°F DB. What volume (CFM) of outside air must be introduced as ventilation to balance the sensible load and maintain the indoor temperature at 74°F?

Solution

The total (sensible) load is:

Conduction	55,000 Btu/hr
Cooking and meals	12,000
Occupants, light activity	
200 × 260	52,000
	119,000 Btu/hr

Volume of 63°F outdoor air required is:

$$V_{CFM} = \frac{119,000 \text{ Btu/hr}}{1.08 \times (74 - 63)} = 10,000 \text{ CFM} \qquad Ans.$$

Solar Thermal Chimneys. Solar thermal chimneys may be used to create an induced ventilation rate of significant volume. If they contain thermal storage, the heat stored will continue to induce ventilation long after sundown. For best results in inducing summer ventilation, thermal chimneys should be placed at the east or west ends of the building.

16-8   Evaporative Cooling

Evaporative cooling has been in use for centuries. The evaporation of water into hot, dry air from fountains, pools, and shrubbery in courtyards has been used to cool houses since ancient times. Water containers made of porous clay cool the water inside by evaporation of the water that seeps through to the outside. Dripping-wet burlap was the primary method of cooling milk and vegetables at the springhouse before the twentieth century. The following brief analysis of the evaporative cooling process makes use of the Psychrometric Chart (Fig. 16-1).

An evaporative cooler cools air without any heat transfer to or from the outside of the cooler itself. The heat required to evaporate the water comes from the sensible heat of the very air into which the water evaporates. Since the latent heat of evaporation (L_v) of water at 75°F is about 1050 Btu/lb and the specific heat of air is about 0.24 Btu/lb-F°, evaporating 1 lb of water into an airstream will cool 1000 lb of air by slightly more than 4 F°. Figure 16-7 illustrates the process in diagram form, as superimposed on a psychrometric chart. Hot, dry air at 95°F and 15 percent RH (point OA) is blown over wetted pads that supply sufficient water to result in 90 percent saturation of the air (that is, the RH of the air off the cooler is 90 percent). The psychrometric process occurs along a line of constant wet-bulb temperature (WB = 63.2°F). Note from the diagram that the air-off-the-cooler (point OC) has been cooled to 65°F DB. This air is very moist, but when it is blown into living spaces and mixes with the air there, the resulting relative humidity–dry-bulb temperature condition will be in or near the Comfort Zone. Note carefully that a dry climate is absolutely necessary for evaporative cooling to be feasible. For example, if the outside air is at 95°F DB and 60 percent RH (a condition often encountered in the U.S. Southeast and the Midwest), the air-cooling process would take place along the 83°F WB line, resulting in air off the cooler at a condition of 85°F DB, 90 percent RH—a condition of no improvement whatever from the outside air condition, as far as human comfort is concerned.

Two types of evaporative coolers are in common use: *direct evaporative coolers* and *indirect evaporative coolers*. A typical direct evaporative cooler is illustrated in Fig. 16-8. Direct coolers blow outside air across and/or through wetted pads or screens and then into a duct system or directly into the spaces to be cooled. Such coolers are manufactured and marketed by the thousands for

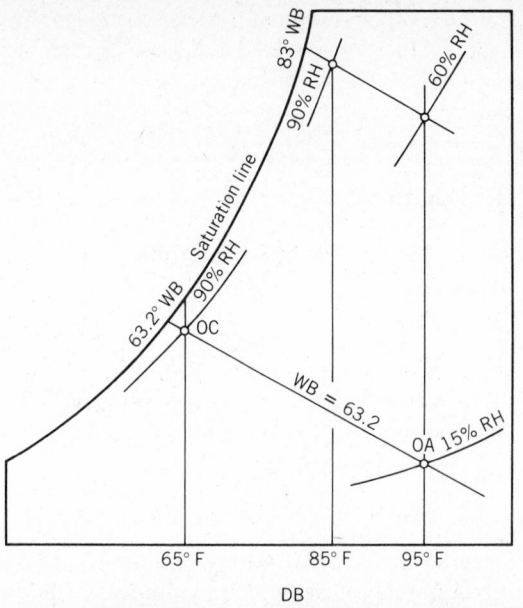

Figure 16-7 Skeleton section of a psychrometric chart showing the process of evaporative cooling of air along a constant wet-bulb temperature line.

Figure 16-8 A direct-type (or wet-pad) evaporative cooler. This model is intended for residential and small commercial use. It can be roof-mounted and attached to ductwork, or it can blow directly into the space to be cooled. The panel that has been removed also contains a wetted pad. (McGraw-Edison Company.)

use throughout the southwestern U.S. and other hot and arid regions of the world. They are regarded as a low-cost alternative to refrigerated air conditioning wherever the psychrometrics of the local climate permit their efficient operation. As a general rule of thumb, evaporative coolers can be used with considerable success in climates where dry-bulb temperatures of 90°F and above are accompanied by wet-bulb temperatures of 72°F and below. At higher wet-bulbs, their application ranges from marginal to useless.

Indirect evaporative coolers also evaporate water into a stream of hot, dry outside air, but the cooled air then goes through a heat exchanger instead of being blown directly into the living spaces. The other side ("dry side") of the heat exchanger is in the living-space air distribution circuit, and this air is cooled in the heat exchanger by means of heat transfer by conduction to the evaporatively cooled air passing through the "wet side" of the heat exchanger. Thus, no moisture is added to the air in the living-space air distribution loop.

With both types of evaporative coolers, significant amounts of electrical energy are required to operate fans or blowers. In a sense, then, systems employing such coolers are hybrid systems, not strictly passive cooling systems. However, in climates conducive to the operation of evaporative coolers, the energy requirements are so much less than those of refrigerated systems that their combination with passive cooling designs is economically attractive. Even in humid areas, the indirect type of evaporative cooler can be efficiently used as a precooler for air entering a refrigeration cooling coil. Studies show that such use can reduce the electric energy input to refrigeration equipment by as much as 50 percent if the system is properly operated with automatic controls. For effective operation, evaporative coolers of both types must be on a regular service and maintenance schedule.

Evaporative Cooler Rule of Thumb. Long experience with direct-type evaporative coolers operating in desert-type climates (90°F DB and above, 70°F WB and below) shows that when they are well maintained, with ample water wetting the pads, they will perform at about 80 percent adiabatic efficiency. Loosely defined, *adiabatic efficiency* is the percent reduction of the initial wet-bulb depression. For example, if the summer outdoor air is at 95°F DB, 70°F WB as it enters the cooler, the initial wet-bulb depression is 25 F°, and an adiabatic efficiency of 80 percent would give an air-off-the-cooler dry-bulb of 75°F (95 − 0.80 × 25 = 75).

A general rule of thumb for wetted-pad evaporative coolers in good condition operating in desert-type climates is:

A flow of 1000 CFM of air through the cooler will balance a 10,000 Btu/hr *sensible* cooling load for an inside design temperature of 80°F DB.

It is readily apparent that large volumes of air are required with evaporative cooling. Since the air off the cooler is usually no colder than 72° to 75°F, a very large volume of such air is needed to maintain 80°F or below in the living space. Air flows of 5000 CFM are quite common for medium-sized (1500 to 2000 ft²) residences in the hot climates of the U.S. Southwest. Consequently, appreciable drafts and air noise usually accompany evaporative cooling systems. There may be a problem also with dank, musty odors if the cooler and pads are not properly serviced.

One essential factor to remember about evaporative cooling of any type is that it *does not remove moisture* from the conditioned space. Air off an evaporative cooler is usually 90 percent saturated with moisture, so when it is blown into the living spaces it actually supplies more moisture. Indirect-type coolers do not send moisture into the space, but they do not take any out, either.

In addition to the operating modes suggested above—both of which (direct or indirect) feature cooling a stream of air and supplying it to living spaces in the daytime—two other methods of cooling using evaporative coolers are worthy of mention. Both depend on nighttime wet-bulb temperatures of 65°F and below. Nocturnal evaporative cooling is used not only to cool the air in the living space but also to cool down the thermal masses in the building and store up "coolth" for the ensuing day. Massive thermal storage walls, floors, and/or water walls can absorb large amounts of next-day heat if they can be cooled to, say, 65°F during the night. Evaporative coolers are also effective in cooling rock beds at night. In this case, the moist air is exhausted outside the house, and the rock mass is cooled to perhaps 66° to 70°F. Next day, using a house–rock bed circuit, house air is cooled without the addition of any moisture—a definite advantage in areas where humidity tends to increase as afternoon approaches. A heat exchanger (i.e., an indirect evaporative cooler) should be used, so that the rock bed will not take on a dank, musty odor.

16-9 Passive Cooling by Radiation

The night sky is an ideal *heat sink* for radiative cooling. Studies done at Trinity University in Texas have shown that, on the average, night sky temperatures are generally 15 to 20 F° lower than the dry-bulb temperature of the air near the ground. The surfaces of a building that are exposed to the night sky will lose energy in the form of long-wavelength radiation. The net radiant heat loss is greatest under low-humidity, clear-sky conditions. Surfaces that point directly to the sky lose more heat than do slanting or vertical surfaces, and therefore radiative cooling is most effective from horizontal roofs.

In night-sky radiant cooling, a massive body of water (roof pond) or masonry in or on the roof is cooled by radiation to the cold sky. Once cooled, the roof pond or thermal mass can act as a "cold storage" and draw heat away from the living spaces throughout the ensuing day to provide natural summer cooling (see Fig. 8-20b).

The Skytherm® system, developed and patented by Harold Hay, is one of the better-known types of roof-pond cooling. Typical construction and sizing details for roof ponds were discussed in section 8-12. In dry climates (WB temperatures below 72°F), it is possible to wet the surfaces of roof-pond water containers and significantly increase the total rate of cooling by adding the effect of evaporative cooling to the radiative cooling.

There are many variations in the use of radiative cooling systems. These may incorporate the use of active or thermosiphoning systems to move cooled air or water from the radiative cooling surfaces to the storage volume that is to be cooled. Roof radiators can be constructed that will cool either air or water that is circulated through them. Cooled air flows back into the building to cool the structure's thermal mass; or cooled water flows back to water thermal storage, where the "coolth" is stored for use next day.

Water may be trickled or sprayed on a roof radiator, and such an arrangement will provide radiative, evaporative, and convective cooling at night. The water that is cooled can then be stored and recirculated during the day through radiant cooling panels in the walls, floor, or ceiling. When using interior radiant cooling panels of any kind in humid climates, care must be taken not to allow the cooling surface to fall to the dew-point temperature of the indoor air, or the panels will condense moisture out of the air. The dew-point temperature itself can be controlled by dehumidifying the air in the building to the point where its wet-bulb temperature is lower than the temperature of the radiant cooling panel. Such a procedure would involve operating a mechanical dehumidifier, however, and might not be cost-effective.

The condensation factor points up a major problem of cold-panel cooling, since daytime dew-point temperatures in many regions of the United States are in excess of 70°F for extended periods of the summer. No matter how effectively the thermal storage water is chilled by nocturnal radiative cooling, its temperature must not be as low as the daytime wet-bulb as it is being circulated through the cooling panels. And, since the cooling water cannot be very cold, it follows that it will not absorb large amounts of heat from the interior spaces.

The cool earth can provide an important heat sink in certain climates. If the average summer ground temperature is lower than the upper limit of the summer comfort zone, a building can be cooled by direct conduction through its walls to the earth. If the natural terrain does not lend itself to the construction of a below-grade building, earth berms can be constructed along one or more sides of the structure and perhaps even extending over all or a portion of the roof (Fig. 16-9). In any case, the structure of any building in contact with the earth will have to have special treatment or construction materials to prevent leaks and water seepage.

Earth-contact cooling can take another form where the structure does not have to be underground. A network of tubes buried 4 to 5 ft deep in the earth can be used to circulate air or water. The tubes are called earth-cooling tubes, or simply *cool tubes*. The major advantage of earth-tube cooling is that the building can be thermally coupled to the earth when desired and isolated from it when conditions do not warrant its use. A network of small-diameter tubes (5 to 6 in.) has been found to be more effective than a single large-diameter tube. The total length of tubes required depends on the size of the building, with lengths of 100 to 200 ft being recommended for residences. Fans (for air) or pumps (for water) are required to circulate the cooling medium from the tubes to the building being cooled.

In the hot, arid climates of the U.S. Southwest, earth-cooling tubes have been used with considerable success. The tubes themselves are usually constructed from porous materials such as concrete or clay pipe. This construction allows soil moisture to seep into the tube, and, when hot, dry air is drawn through the tube, it is cooled sensibly by the earth and is further cooled evaporatively by the soil moisture that has penetrated into the tube. Some designs purposely place the cool tubes under watered landscaping so that there will be plenty of soil moisture in the tubes for evaporative cooling.

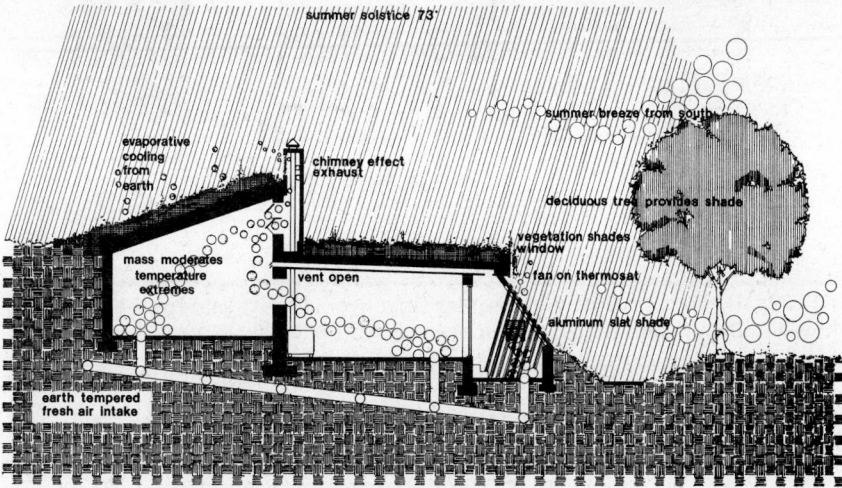

(a)

Figure 16-9 An earth-berm passive solar house. This residence is a HUD award-winning design, developed under a joint grant from the Department of Housing and Urban Development and the Department of Energy. The drawings illustrate many of the methods used in attaining year-round comfort in passive structures. (*a*) In summer, indoor comfort is enhanced by a variety of strategies—evaporative cooling from wet earth; earth-contact cooling; an earth-cooled fresh air intake ("cool tube"); open vents for convection currents and maximum use of breezes; and shade from shrubs and deciduous trees.

winter day

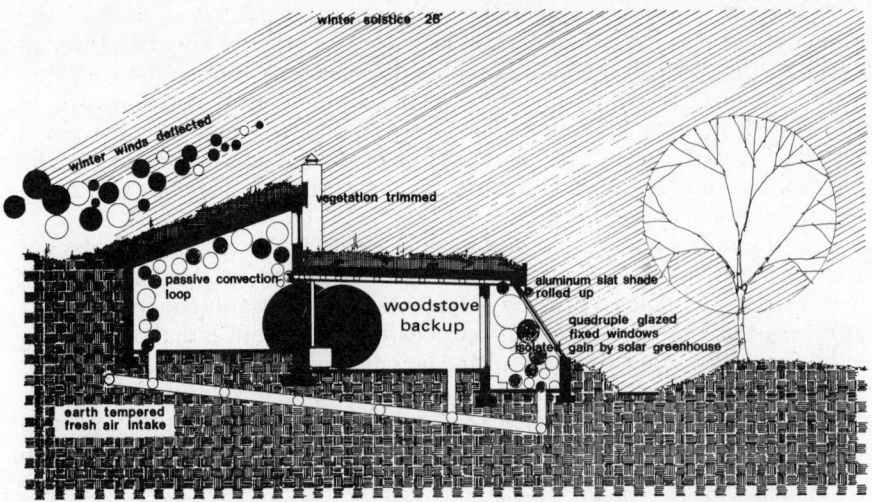

(b)

floor plan

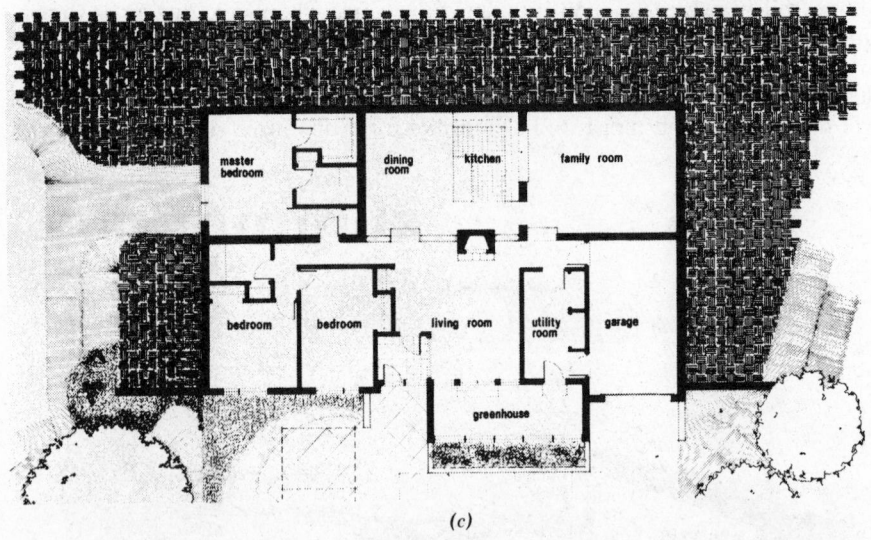

(c)

Figure 16-9 *(continued)* *(b)* In winter there is solar heat gain from the greenhouse and the clerestories, with a wood stove for auxiliary heat. Fresh air is tempered in the earth tube, and walls and ceilings are heavily insulated to minimize heat losses. Winter winds are deflected, and there is very little wind effect on the building envelope. *(c)* Floor plan of the house. (Design and construction by Milliner Construction, Inc., Frederick, Maryland).

In hot humid climates, however, night temperatures do not fall very far below day temperatures, and soil temperatures are also quite warm in the summer months. Earth-cooling tubes in these regions will provide a limited amount of sensible cooling to air passing through them but very little or no evaporative cooling. The air passing through the tubes is sensibly cooled, but at the same time its relative humidity rises, and what was warm muggy air now becomes cool muggy air, with little or no positive effect on human comfort.

In summary, and with regard to all of the passive cooling designs and techniques discussed in the preceding sections, passive cooling systems can be quite effective in warm, dry climates, and they require very low inputs of energy. Even if supplemented with evaporative cooling equipment, the energy input is still modest, and the resulting relative humidity will generally be within acceptable limits. By the opposite token, in hot, humid climates, passive cooling has severe limitations. Only a careful analysis of climatic data, soil temperatures, and energy costs can determine whether or not the necessary investment for a passive cooling system would be worthwhile. The major limitation of passive cooling is that it cannot effectively take moisture out of humid air. Solar-energized mechanical cooling can control both dry-bulb temperature and relative humidity, however, and we now turn to an analysis of active solar-cooling systems.

ACTIVE SOLAR-COOLING SYSTEMS

The sun is our primary source of energy, and since energy is required to operate refrigerating machinery, it is altogether logical that solar energy can be used to energize mechanical cooling equipment. The function of the refrigeration system in an air-conditioning application is to remove heat from the interior of a building, where the temperature may be in the 75° to 85°F range, and transfer it to warmer outdoor surroundings, where the temperature may be 95°F or more. In order to provide for effective heat transfer from a cooler to a warmer environment, the refrigeration condenser must operate at a much higher temperature than that of the warm outdoor environment. In other words, heat has to be forced to flow up a temperature gradient from the room temperature to the condenser temperature, and it cannot do this as a natural event. Energy must be supplied from some other source to move heat "up a temperature hill" (Second Law of Thermodynamics).

In conventional air conditioners, this energy input is supplied by electricity or steam or an internal combustion engine operating through a compressor. But solar energy can also be used, and this section will describe and analyze two types of solar-energized cooling systems: *solar Rankine-cycle cooling*, and *solar absorption cooling*.

At the outset, it should be noted that solar-energized mechanical cooling, of whatever type, is presently in the developmental phase. The technology is ready, but cost factors stand in the way of vigorous marketing programs. At present (1985), active solar cooling is not in a reasonably competitive position with respect to conventionally energized (electricity or fossil fuel) cooling systems. This situation could change quickly, however, within a decade, and all solar designers and technicians should understand the basics of active solar cooling in order to grow with the industry when its market potential begins to assert itself.

Two different refrigerating cycles are used with solar cooling: the *compression cycle* and the *absorption cycle*. Both will be briefly described.

16-11   The Compression Refrigeration Cycle

Most refrigeration systems depend on the evaporation of a *refrigerant* at low temperature and pressure. Various refrigerants are used, depending on the application, among them being ammonia, the Freon family of refrigerants (fluorocarbons), and a few others. The pressure–temperature relationships involved in a refrigeration cycle are critically important and are governed by various controls and flow valves. Refer to Fig. 16-10 for the following discussion of a simple compression refrigeration cycle. This cycle of operation is typical of an air-conditioning application.

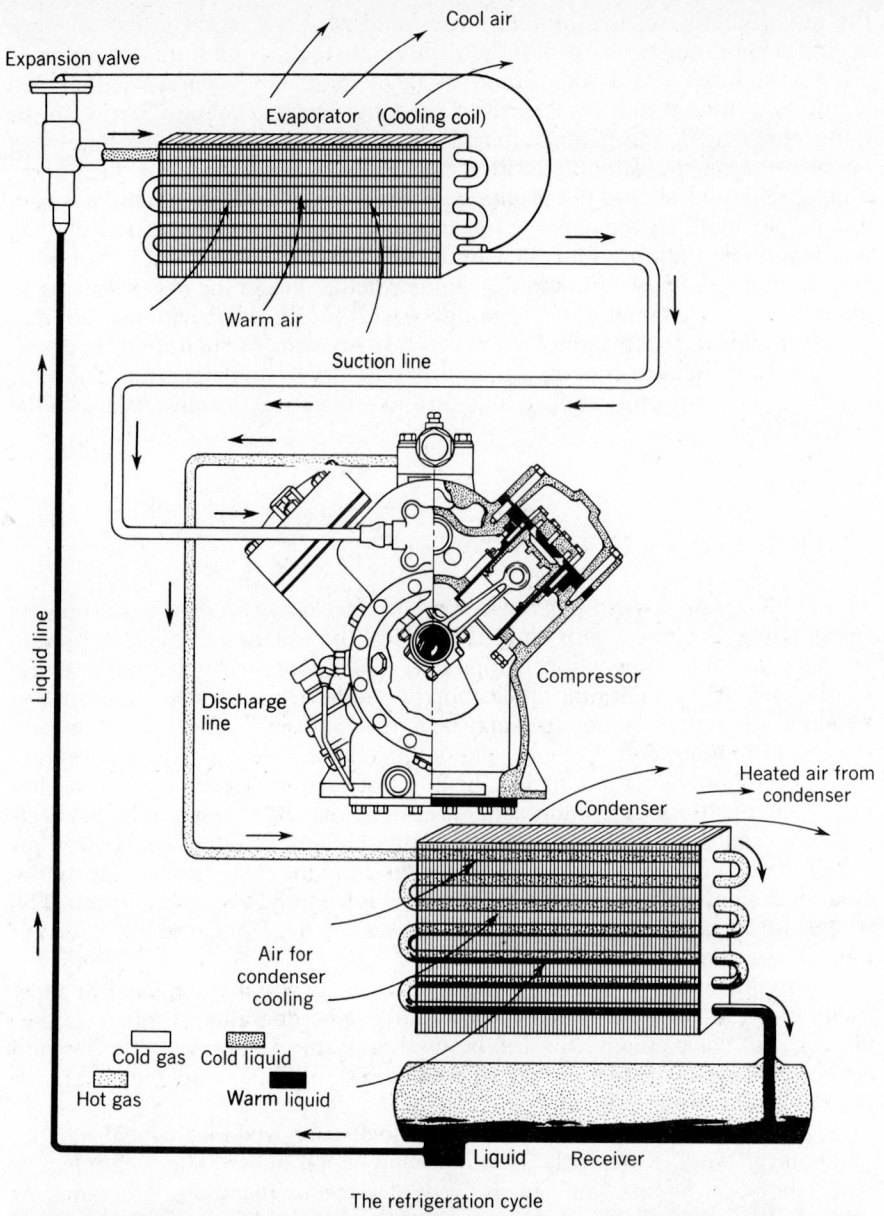

Expansion valve

Cool air

Evaporator   (Cooling coil)

Warm air

Suction line

Liquid line

Discharge
line

Compressor

Heated air from
condenser

Condenser

Air for
condenser
cooling

Cold gas Cold liquid

Hot gas   Warm liquid

Liquid   Receiver

The refrigeration cycle

F i g u r e 16-10 Flow diagram of a simple compression refrigeration cycle on air-conditioning service. Note that the evaporator is an air-cooling coil, and the condenser is also air-cooled. (Carrier Air Conditioning Corporation.)

    In the flow diagram, liquid refrigerant enters the *evaporator* and evaporates (boils) at a low temperature (about 40° to 45°F for the air-cooling coil shown) and low pressure. For air-conditioning purposes, the evaporator may be either a finned coil (Fig. 16-11) or a water chiller (Fig. 16-12). As the refrigerant evaporates, it absorbs heat from (i.e., it cools) warm air (or warm water) and changes phase from the liquid state to the vapor state, carrying with it its latent heat of vaporization.

    In order to reject this heat to the outdoor environment, the cold refrigerant vapor is compressed by the *compressor* so that its temperature becomes high enough to allow heat to flow down a descending temperature gradient from the hot refrigerant vapor to the external environment (which could be either outdoor air or water). The compressor also increases the pressure of the gaseous refrigerant.

Figure 16-11 A direct-expansion finned-tube cooling coil for air-conditioning service. Such coils are actually heat exchangers and are usually mounted in the main duct as it leaves the refrigerating equipment, or they can be a part of a zone air-handling unit. The operating temperature of these coils is typically in the range of 40° to 45°F. (Dunham-Bush, Inc.)

Heat rejection occurs in the *condenser*, and the high-pressure gas condenses, returning to the liquid state as a warm liquid. Condensers may be air-cooled or water-cooled. As a warm liquid under high pressure, the refrigerant is fed back to the evaporator through the *expansion valve*, and the cycle begins again. The purpose of the expansion (throttling) valve is to control the flow of

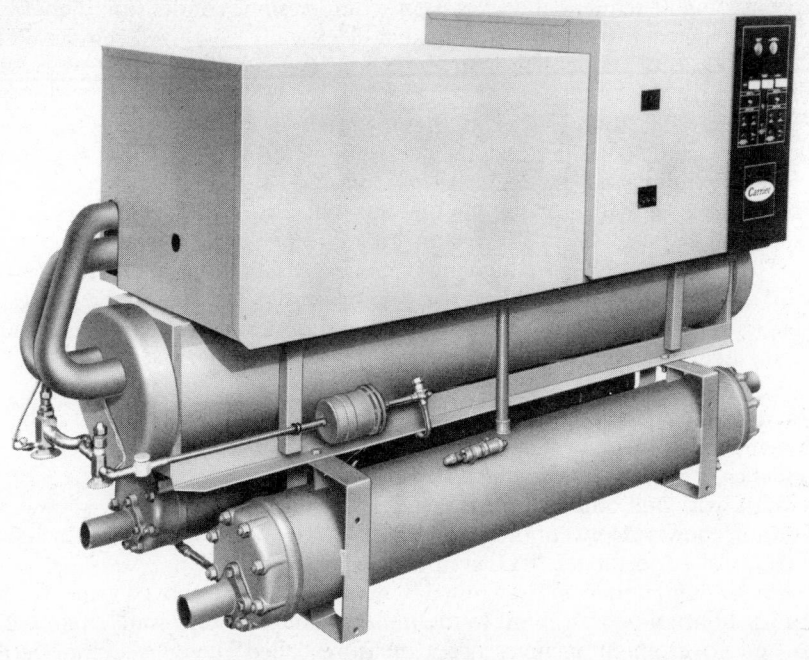

Figure 16-12 A water chiller of the direct-expansion type. Cold refrigerant absorbs heat from the input water, and the output is chilled water at about 55°F. (Carrier Air Conditioning Corporation.)

Condenser
fan

Condenser

Controls

Hermetic compressor

Figure 16-13 An electrically energized refrigeration condensing unit for residential and small commercial air-conditioning service, with components labeled. (Carrier Air Conditioning Corporation.)

refrigerant to the evaporator in response to the cooling demands of the system. The throttling action of the valve provides for a sudden expansion and pressure drop of the refrigerant, and some of it flashes to vapor immediately, dropping the temperature of the refrigerant liquid–vapor mixture as it enters the evaporator. The expansion valve separates the high-pressure side (*high side*) of the cycle from the low-pressure side (*low side*).

Summing up, the compression refrigeration cycle is a thermodynamic cycle consisting of sequential evaporation, compression, condensing (liquefaction), and expansion (throttling). A typical electrically energized refrigerating condensing unit for residential or small commercial air-conditioning service is shown in Fig. 16-13.

Conventional summer air-conditioning systems are most often energized by electricity, and their thermodynamic relationships are such that they can transfer several units of heat energy from indoors to outdoors for every unit of electrical energy provided to the compressor and fan motors. This statement may seem paradoxical—a violation of the First Law of Thermodynamics—but it is not. The First Law of Thermodynamics applies to the production of heat from work (or vice versa), and it says that there can be no loss or gain in the process, that energy must be conserved. But an air conditioner on cooling duty is not *producing* heat; it is merely *moving* it from indoors to outdoors.

Coefficient of Performance of a Refrigerating Machine. The ratio of heat units removed divided by energy units supplied is called the *coefficient of performance* of a refrigerating machine, abbreviated COP. To calculate COP, both input work and output heat must be in the same units—watts, kilowatts, Btu, or horsepower. Conventionally energized refrigerating units typically have coefficients of performance (COPs) in the range of 2.5 to 4.5.

For air conditioners in the United States, since heat removed is measured in Btu per hour and power input to the motors is measured in watts, manufacturers and government agencies use a measure called *energy efficiency ratio* (EER) or, more recently, the *seasonal energy efficiency ratio* (SEER), to express the effectiveness of so-called "packaged" air conditioners. SEERs in the range of 8.0 to 12.0 Btu/hr-W are typical of current production models (1984) of residential air conditioners.

Since 1 W = 3.41 Btu/hr, EER (Btu/hr-W) = 3.41 × COP (W/W)

The ratio of horsepower per ton of cooling (hp/ton) is also used as an efficiency measure in the refrigeration industry—usually for larger-capacity machines of 20 tons and up. The *ton* is the unit of refrigeration capacity, equal to a heat removal rate of 12,000 Btu/hr.

The compression of the refrigerant can be accomplished either by reciprocating compressors or by centrifugal compressors. Each type of compressor is suited to certain refrigerants and to different kinds of applications. Centrifugal (turbine) compressors are used with solar-energized systems, as briefly described in the next section.[4]

16-12   Rankine-Cycle Solar Cooling

The *Rankine cycle* is that used by turbine steam engines. A fluid (usually water, but see below) is evaporated in a boiler at high temperature and pressure (Fig. 16-14), and the vapor is allowed to expand from high pressure to low pressure doing work against the turbine blades, thus spinning the turbine. The exhaust vapor at low pressure and low temperature is condensed back to liquid in a condenser. An energy input then has to be provided to a pump to force the liquid back into the high-pressure boiler, where the cycle begins again. Rankine-cycle (steam) turbines typically have efficiencies of 25 to 35 percent; that is, of the heat energy supplied, usually less than one third shows up as output (shaft) power.

Rankine-cycle solar-cooling systems use a centrifugal compressor and the compression refrigeration cycle discussed above. The compressor is one component of a system designed for conditioning air. High-temperature solar heat (up to 550°F) is required to operate Rankine turbines on a steam cycle, and the use of concentrating solar collectors (see Chapters 10, 11, and 12) is essential in order to attain the high temperatures needed.

[4] Space limitations preclude a detailed presentation of refrigeration theory and practice. Readers desiring a more comprehensive treatment are referred to such sources as *ASHRAE Handbook, 1977 Fundamentals*, Chapter 1; Guy R. King, *Modern Refrigeration Practice* (New York: McGraw-Hill Book Company, 1971), Chapters 5 and 10; or Norman C. Harris, *Modern Air Conditioning Practice*, 3rd. ed. (New York: McGraw-Hill Book Company, 1983), Chapter 12.

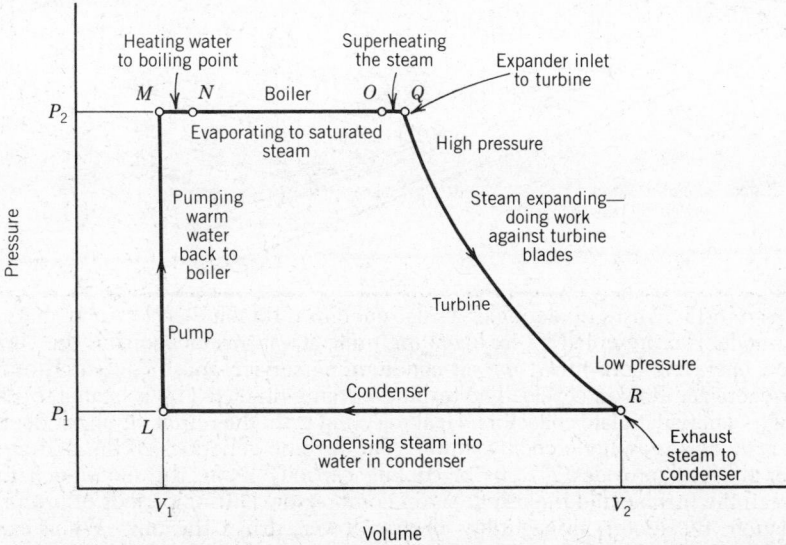

Figure 16-14 Pressure–volume diagram for a Rankine-cycle turbine on a steam cycle, showing the four major processes of the cycle (greatly simplified).

Solar-energized steam turbines operate near the lower end of the practical temperature–pressure range for steam turbines. The shaft power output can be direct-connected to a centrifugal compressor on air-conditioning service and/or to a motor–generator unit. Such an arrangement allows flexibility of operation under varying conditions of cooling load and available solar radiation, and provides for in-plant operating economies. When solar energy is in excess of building cooling needs, electric energy can be generated and fed into the electric power grid of the building or plant. On the other hand, when solar energy is limited or nonexistent, the air-conditioning system can be energized entirely by electric power input to the motor of the motor–generator unit.

Some Rankine-cycle equipment, especially for smaller units (up to about 100 tons of refrigeration capacity), is being designed for use with fluorocarbons as the working substance in the turbine, rather than steam. Fluorocarbon vapors have greater density than water vapor (steam), and therefore smaller volumes of vapor can be circulated, with consequent reduction in turbine size. Fluorocarbons also permit much lower turbine-operating temperatures and pressures, which opens up the possibility of using high-quality flat-plate solar collectors instead of the far more expensive concentrating (tracking) collectors. The two fluorocarbons that show promise so far are R-113 (Trichlorotrifluoroethane) and FC-88 (a fluorinert compound).[5] An R-113 machine, now in the

[5] See Ezzat Wali, "Optimum Working Fluids for Solar-Powered Rankine-Cycle Cooling of Buildings," *Solar Energy* (New York: Pergamon Press, 25:3, 1980).

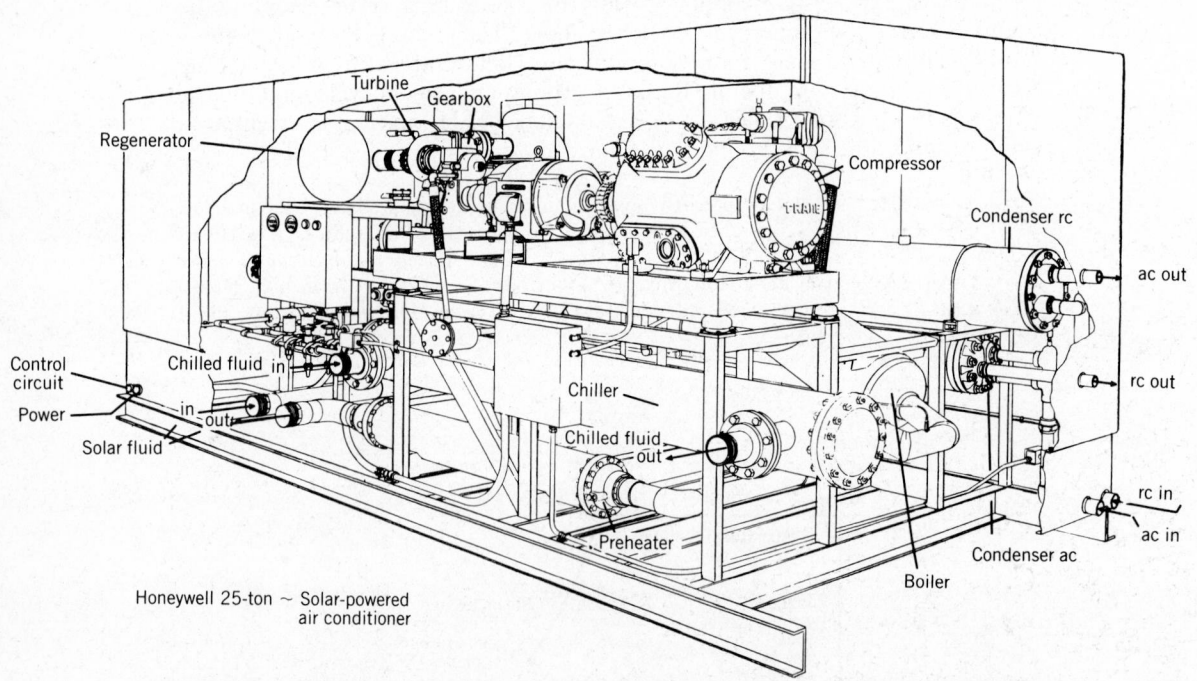

Honeywell 25-ton ~ Solar-powered air conditioner

Figure 16-15 Artist's rendering of a solar-energized Rankine-cycle air conditioner. This model is represented by six operating units at various locations in the United States, operating both for actual air-conditioning service and for research on the solar-energized Rankine cycle. The turbine working fluid (*R*-113) is heated to 195°F by high-quality flat-plate collectors. Heat rejection from the refrigerating condensing unit is to an atmospheric cooling tower. The turbine generates 20 hp, and at full power the unit provides 25 tons of cooling capacity. Note the motor–generator between the turbine and the refrigeration compressor. During periods of low or no insolation, the motor, using utility electric power, drives the unit. When excess insolation is available, the generator feeds electricity into the plant or the local power network. (A joint project of Honeywell, Inc., Barber-Nichols Engineering Company, and Lennox Industries, Inc. Illustration courtesy of Barber-Nichols Engineering Company.)

pilot-plant stage, is reported to have a turbine efficiency of 22 percent when operating at 118°F (46°C) condensing temperature and 391°F (200°C) expander inlet temperature (Fig. 16-15).

Rankine-cycle solar cooling is not a commonly encountered system. On the contrary, these machines are still pretty much in the research and development phase. Some pilot plants are in operation, and actual production models may appear in a few years. Based on current trends, the following observations about Rankine-cycle solar cooling can probably be substantiated:

1. It is best suited to systems of fairly large capacity—industrial or commercial buildings where other heavy machinery requires the presence of an operating engineer.
2. It would be impractical except in localities where ample solar radiation occurs along with heavy cooling loads.
3. The high technology of such systems brings with it very high capital costs. One way to amortize these costs in a shorter time is to generate and use steam or electrical energy from any excess solar energy available. If the building or plant has no need for internally generated steam or electric power, the economics of Rankine-cycle solar cooling become much less attractive.

Illustrative Problem 16-8

An electrically energized air conditioner is successfully coping with a cooling load of 65,000 Btu/hr. The electrical power input to all components of the unit (compressor motor and fan motors) is measured as 6.6 kW. Calculate the energy efficiency ratio (EER) and the coefficient of performance (COP).

Solution

$$\text{EER} = \frac{65,000 \text{ Btu/hr}}{6600 \text{ W}} = 9.85 \text{ Btu/hr-W} \qquad Ans.$$

$$\text{COP} = \frac{\text{EER}}{3.41} = \frac{9.85 \text{ Btu/hr-W}}{3.41 \text{ Btu/hr-W}} = 2.89 \qquad Ans.$$

Illustrative Problem 16-9

A Rankine-cycle fluorocarbon turbine drives a refrigerating plant whose overall COP is 2.6. The overall turbine and heat exchanger efficiency is 20 percent. The fluorocarbon vapor is heated by solar collectors with an average efficiency of 28 percent at the required operating temperature. The insolation (normal incidence) on the collectors averages 290 Btu/hr-ft^2 of collector during the midday hours on clear days in July. Find the minimum collector area required to cope with a cooling load of 1,000,000 Btu/hr.

Solution

The energy that must be supplied to the turbine by the hot fluorocarbon vapor is

$$H_T = \frac{1,000,000 \text{ Btu/hr}}{2.6 \times 0.20} = 1,923,000 \text{ Btu/hr}$$

The hourly heat output of the collector array Q_{coll} must equal or surpass this value. Therefore,

$$H_T = Q_{coll} = I \times A_c \times \eta$$

where

$$I = \text{the insolation on the collectors, Btu/hr-ft}^2$$
$$A_c = \text{the collector area, ft}^2$$
$$\eta = \text{the collector efficiency, percent}$$

From which,

$$A_c = \frac{H_T}{I\eta}$$
$$= \frac{1,923,000 \text{ Btu/hr}}{290 \text{ Btu/hr-ft}^2 \times 0.28}$$
$$= 23,680 \text{ ft}^2 \qquad\qquad\qquad Ans.$$

This array could be accommodated on a roof space with a usable area of about 80 ft by 300 ft.

16-13 Solar-Energized Absorption Cooling

Refrigeration by means of the compression cycle is the basis of the summer cooling methods discussed above. The essential processes are, in sequence, evaporation, compression, condensation (liquefaction), and expansion. The purpose of the compression sequence, it will be recalled, is to add enough energy to the refrigerant gas so that its temperature will be high enough to allow rejection of the heat absorbed in the evaporator to the outdoor environment. Also, a high pressure is needed to cause the hot gas to liquefy as it cools in the

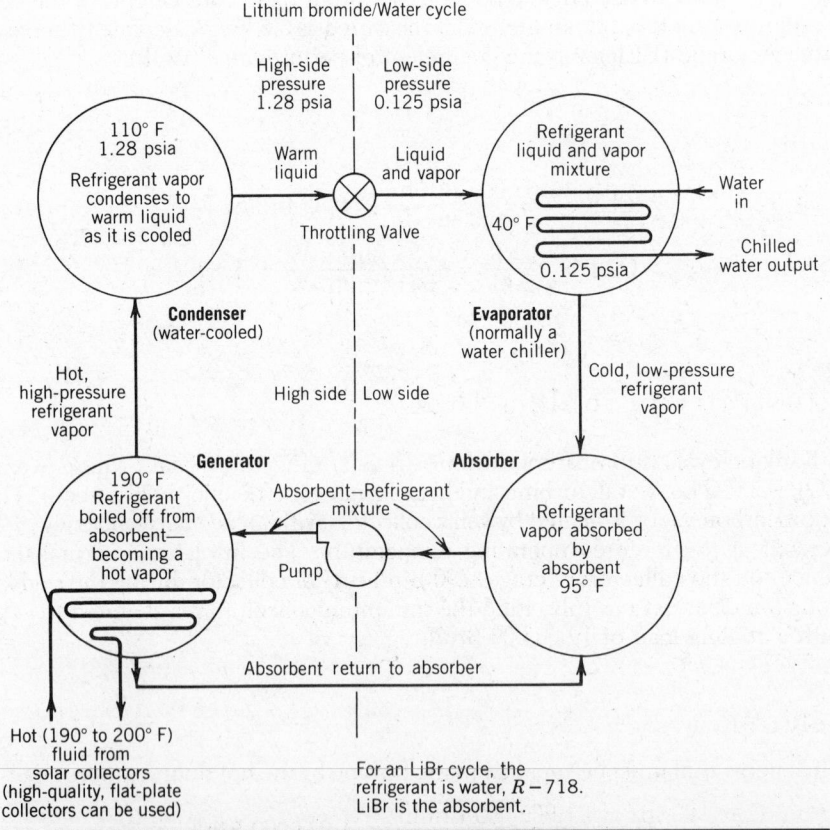

Figure 16-16 Block diagram of a solar-energized lithium bromide (LiBr) absorption-type refrigerating machine with its evaporator operating as a water chiller. Note the approximate temperatures and pressures at critical points in the cycle. These machines operate very satisfactorily when energized by a hot fluid (190° to 200°F) from solar collectors.

condenser. Compressing the cold refrigerant gas, however, is only one way to provide the requisite energy.

Another cycle that accomplishes the same result is based on adding energy directly in the form of heat instead of in the form of compressor work. The *absorption refrigeration cycle* is a heat-energized method of refrigeration. The heat input can be supplied by steam, by a gas or oil flame, by electricity, or by solar energy. A brief and much simplified discussion of solar-energized absorption-cycle cooling follows.

First, a block diagram (Fig. 16-16) will be used to clarify the basic principles. The *evaporator* is a water chiller in most absorption systems. Water is chilled to about 40° to 50°F and then pumped to one or more zone air-conditioning units (not shown) in the air distribution system of the building. The refrigerant vapor, having absorbed its latent heat of vaporization L_v as it evaporated and chilled the water, flows from the evaporator–water chiller to the *absorber*. The absorber contains an *absorbent*, a substance chosen for its ability to absorb large quantities of the refrigerant vapor immediately on contact. This sudden absorption of the vapor produces and maintains a very low pressure on the evaporator–absorber side (low side) of the cycle. The absorber also in a sense "compresses" the vapor, since a large volume of vapor becomes a very small volume as it goes into solution in the absorbent.

At this point of the cycle, the absorbent–refrigerant mixture is pumped to the *generator*, where energy is supplied from solar-heated hot water or steam. Refrigerant vapor is distilled (boiled) off from the absorbent in the generator, and the refrigerant vapor is raised to a high temperature and pressure (high side) so that it can reject the heat that was picked up in the evaporator to cooling water or to outdoor air.

From the generator, hot, high-pressure vapor flows to the *condenser*, where heat is removed, its temperature drops, and it condenses back to a warm liquid still at high pressure, ready to flow through the *expansion* (throttling) *valve* and begin another cycle. The process is continuous, refrigerant flow and heat input being controlled by automatic valves and controllers that constantly sense the needs of the air-conditioning system.

Refrigerants and Cycles for Absorption Systems.   At present, the two refrigerants of commercial importance for absorption refrigeration are ammonia (R-717) and water (R-718). Where ammonia is the refrigerant, water is used as the absorbent, since a very large quantity of ammonia vapor will immediately be absorbed in a very small volume of water.

Ammonia systems require rather high generator temperatures (220° to 280°F), and such systems are suitable only with high-temperature, concentrating solar collectors. For this and other reasons, the ammonia–water cycle absorption system is not currently being commercially produced for solar-energized air-conditioning.

The water–lithium bromide absorption cycle can operate quite well with solar-heated (190°F) water, and it is the cycle currently being used for solar cooling. This cycle uses ordinary water as the refrigerant and lithium bromide (LiBr) as the absorbent. Lithium bromide is a highly deliquescent salt (that is, it has great affinity for water), and it dissolves in the water that it absorbs. Water under a high degree of vacuum (say, 0.125 psia) boils at temperatures suited to an evaporator on air-conditioning service (35° to 45°F), and the thermodynamic properties of the water–LiBr cycle are such that generator temperatures in the 180° to 200°F range give reasonably good efficiencies.

Solar Absorption Refrigeration Rule of Thumb.   The cooling output from the evaporator of absorption equipment is ordinarily chilled water, which is circulated to air-cooling coils or to zone air-handling units at selected locations in the building air distribution system. The heat input to the generator is provided by hot water from either concentrating or flat-plate solar collectors. If flat-plate collectors are used, they must be of the highest quality (double- or triple-glazed and featuring the best of selective absorber coatings), capable of producing water temperatures in the 180° to 200°F range. At temperatures below 180°F, the efficiency (COP) of the water–LiBr cycle drops off sharply. Even at generator temperatures in the 190°F range, the COP of these machines is typically in the range of 0.72 to 0.80, compared to COPs of 2.5 to 4.0 for conventional electrically energized equipment. If collector efficiency can be maintained at 35 to 50 percent, and allowing for losses in other components of the system, a rule of thumb for the actual cooling output of a solar-energized absorption machine into the chilled water will be from 18 to 25 percent of the total insolation on the collectors during sunlit hours. In other words, overall COP is 0.18 to 0.25.

Since the efficiency (COP) of absorption refrigeration drops off sharply at lower generator temperatures, and the efficiency of solar collectors increases at lower temperatures, there is an optimum operating point (readily determinable by computer analysis) for the daily conditions encountered on any installation. For example, under certain climatological conditions, it might be advantageous to take water off the collectors at 150° to 160°F where collector efficiency would be high, put it into storage, and then boost its temperature to 200°F with fossil fuels or with a high-temperature heat pump, before supplying it to the generator of the absorption machine. As indicated, such analyses are as complex as the number of variables involved, and they are feasible only with computer analysis.

Water as a Refrigerant.   Water is not ordinarily thought of as a refrigerant, since it does not have a low boiling point at atmospheric pressure. However, under a *very low pressure* (0.125 psia, high degree of vacuum), its boiling point is

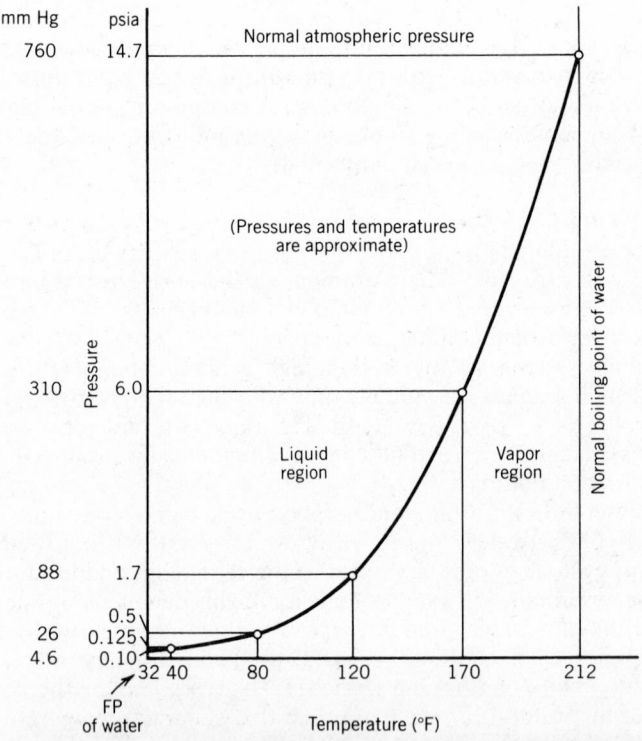

Figure 16-17 Pressure–temperature curve, or boiling point curve, for water. Note that the normally recognized boiling point of water (212°F) occurs only when the pressure is 1 atm (14.7 psia). In a vacuum at 0.125 psia, water boils at about 40°F, with the change of state picking up about 1070 Btu per lb of water evaporated. LiBr absorption refrigerators produce low-side pressures of about 0.125 psia, and the refrigerant water (R-718) chills the "product water" that is circulated to the air-conditioning units of the system.

sufficiently low for air-conditioning purposes (40°F) (Fig. 16-17). The advantages of water are that it is nontoxic, relatively noncorrosive, and relatively inexpensive. It is absorbed extremely rapidly by lithium bromide (a deliquescent salt), which then dissolves in the water it has absorbed, forming what is called the *weak absorbent* in the low-side shell of the equipment. The main advantage of water as a refrigerant, however, is its very high latent heat of vaporization, L_v (1070 Btu/lb at 40°F, as compared to about 64 Btu/lb for the common refrigerant, Freon-12). This very high value of L_v enables water, as a refrigerant, to accomplish a great deal of cooling per pound of water circulated.

The main disadvantage of water as a refrigerant is the very high degree of vacuum that must be maintained throughout the absorption system, even on the high side (Fig. 16-16). The cycle is completely sealed at the factory; it is a closed system and cannot be serviced in the field without high-tech equipment and factory-trained personnel.

Rejection of Indoor Heat to the Outdoor Environment.   The thermodynamics of LiBr absorption refrigerating equipment are such that condenser and absorber cooling are best accomplished by water rather than by outdoor air. Atmospheric water-cooling towers are the most frequently used means of transferring the heat picked up in the conditioned spaces and that produced in the solar-heated generator to the outdoors.

The cooling action of atmospheric water-cooling towers depends on the wet-bulb temperature of the ambient air. The lowest temperature to which water can be cooled by such equipment is the wet-bulb temperature at the time. In actual industrial practice, however, the cooled water leaving the tower is usually at least 5 F° warmer than the existing wet bulb. For optimum operating conditions with LiBr, solar-energized absorption machines, the cooling-tower water (CTW) temperature should be in the range of 68° to 78°F as it is supplied to the condenser. Figure 16-18 shows a water-cooling tower installation on air-conditioning service.

Figure 16-19 illustrates a current production model of a lithium bromide machine for air-conditioning duty, and Fig. 16-20 shows a solar-energized

Figure 16-18 A mechanical-draft cooling tower of the type used to cool condenser water for air-conditioning applications. (Marley Cooling Tower Company.)

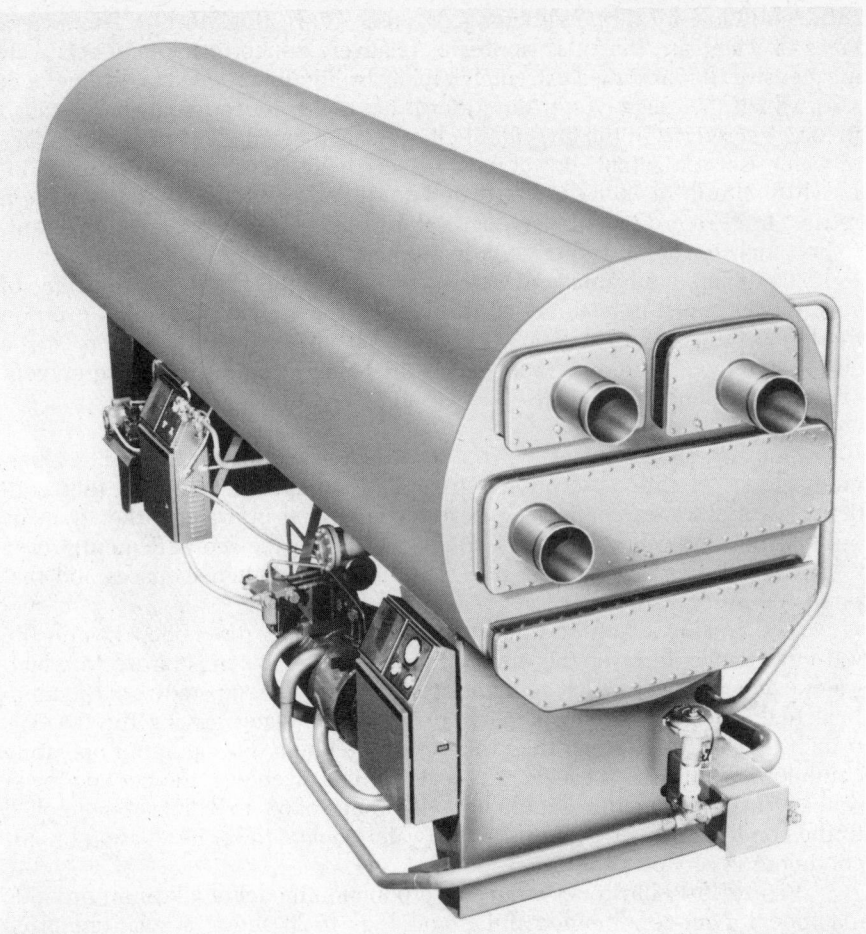

Figure 16-19 A LiBr–water absorption refrigerating machine for air-conditioning service. All of the functional elements described by the block diagram of Fig. 16-16 are factory-sealed in the cylindrical shell.

Figure 16-20 A 25-ton capacity LiBr-"packaged," solar-energized absorption machine for small to medium commercial air-conditioning application. (ARKLA Industries, Inc.)

machine designed for small to medium commercial applications. Figure 16-21 is a layout and flow diagram showing one possible arrangement of equipment for a combination solar-energized space-heating, space-cooling, and domestic hot-water system.

Solar-energized absorption cooling is well established for large-tonnage installations, especially in plants or buildings where steam is already available for auxiliary heat when needed, and where excess solar energy (when available) can be profitably used in the plant for other purposes. Concentrating (tracking) collectors are recommended for such systems.

In the residential and small commercial segments of the air-conditioning market, solar cooling is not moving ahead very rapidly at present. The economics of small-capacity solar-cooling systems is not yet very attractive, even though the costs of electric energy and fossil fuels have risen sharply in recent years. Based on a number of carefully documented engineering studies made in differing localities and climates, it appears that the pay back period for residential and small commercial solar-cooling systems is of the order of 20 to 25 years, when compared with the capital and operating costs of conventional summer

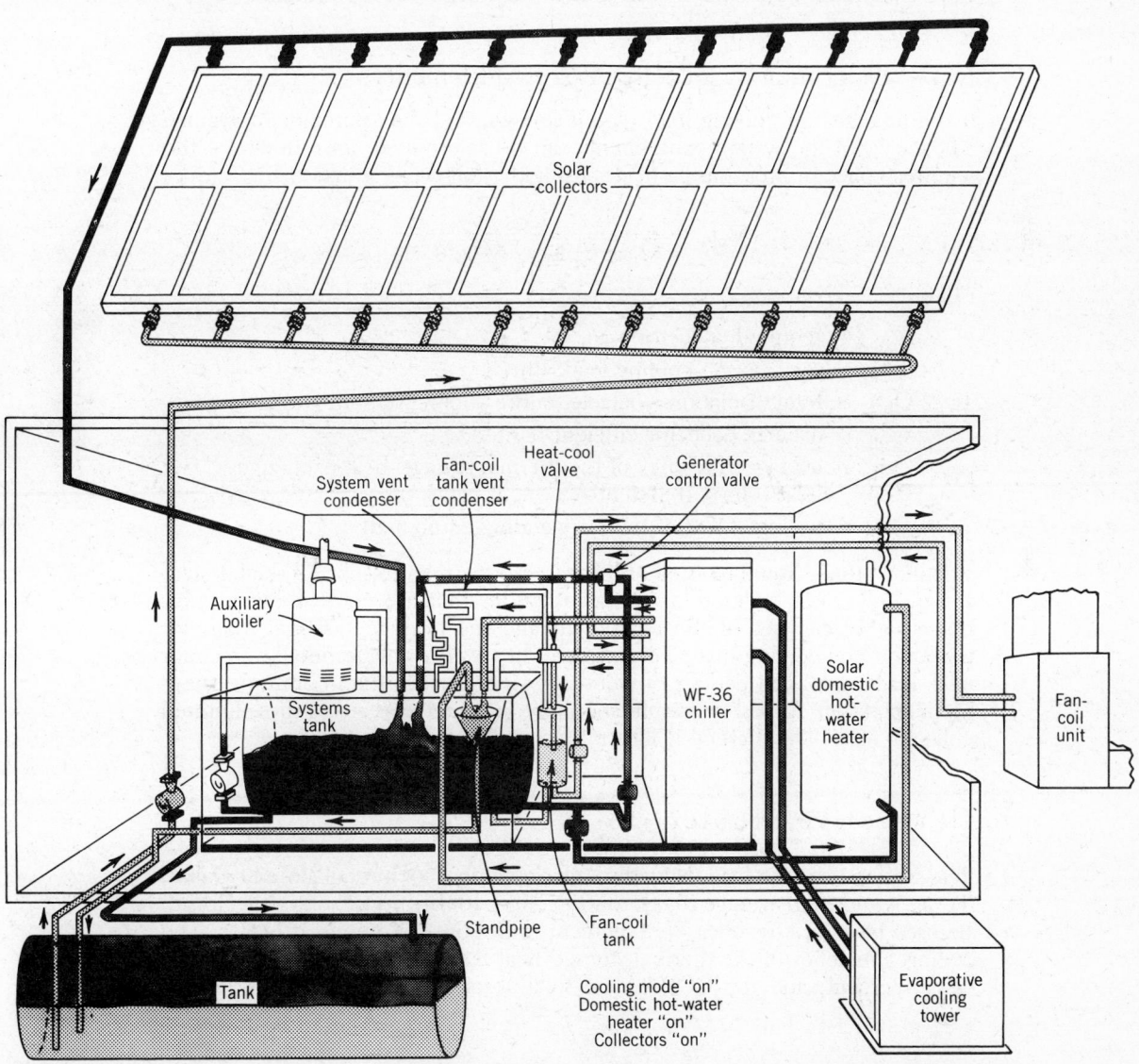

Figure 16-21 Schematic layout and flow diagram for a combination solar heating, cooling, and domestic water-heating system for a residence. The flow pattern shown is for the cooling mode. (ARKLA Industries, Inc.)

ACTIVE SOLAR-COOLING SYSTEMS

cooling equipment. There is some evidence that combined systems (solar-energized cooling, heating, and hot water) are in a much better competitive position, since the amortization of the capital investment is shortened by year-round operation.

There are many complex variables involved in solar-energized cooling—the variability of the cooling load, climatological and insolation factors, thermal storage problems, cost of installation and of backup facilities and fuels, and the optimization of solar collector efficiencies versus the COP of refrigeration cycles, to name just a few. Consequently, nothing short of a complete computer analysis will suffice in determining the feasibility of designing and installing solar-energized cooling systems. Such analyses often confirm the feasibility of large commercial and industrial systems, but more often than not cast doubt on the advisability of putting in residential and commercial systems of less than about 15 tons of refrigerating capacity.[6] Again, it is necessary to point out that the price of conventional energy sources could escalate beyond any currently imagined estimates within a decade. If this should happen, solar cooling would immediately become an attractive investment, and manufacturers that now have equipment development "on hold" would begin mass production.

16-14   Determining Collector Area Requirements

If the total season cooling load Q_{CS} is known, and the approximate average efficiencies of the system components can be determined, an estimate of the required collector area can be made, using the following equation:[7]

$$A_{coll.} = \frac{Q_{CS}}{Q_{ICS} \times \eta_{coll.} \times \eta_S \times \eta_R} \qquad (16\text{-}2)$$

where

$A_{coll.}$ = required collector area, ft^2
Q_{CS} = total season cooling load, Btu
Q_{ICS} = total insolation available, entire cooling season, Btu/ft^2
$\eta_{coll.}$ = average collector efficiency
η_S = average efficiency of the thermal storage–heat exchanger–heat transport system
η_R = average COP of the refrigerating equipment

Equation (16-2) should be used only for working estimates, not for final design. It can only be as accurate as the assumptions that are made about average operating efficiencies of the equipment and the average seasonal totals for insolation and cooling load. For example, its results provide no allowance for peak loads, extended periods of cloudy weather, or less than optimum equipment operation. At best, its results should be interpreted as giving the minimum collector area. The method is illustrated as follows.

Illustrative Problem 16-10

The seasonal cooling load (May through September) of a retail store in Dallas, Texas, is known to average 20,000 ton-hr (240×10^6 Btu). The solar collectors to be used have an efficiency of 40 percent under the conditions at Dallas. The average efficiency of the thermal storage–heat transport system is estimated to be 85 percent, and the COP of the solar-absorption refrigerating system is

[6] Students interested in a more complete analysis of trends in solar cooling are referred to Alwin B. Newton, "Solar Cooling," in Richard C. Jordan and B. Y. H. Liu, eds., *Applications of Solar Energy for Heating and Cooling of Buildings* (Atlanta: ASHRAE, 1977), Chapter XIII.
[7] *Ibid.*, p. XIII-3.

known to be 0.72 under the conditions imposed by the site. The collectors face south, with a tilt angle of 22° from the horizontal. Estimate the minimum total net collector area required in square feet.

Solution

First, determine the location of Dallas as being about 32° N Lat. Refer to tables of daily or monthly insolation (for example, Table 4-F) and, for 32° N Lat in the 22° tilt column, note the average daily total insolation for the months May through September. Multiply each of these daily totals by the number of days in each of the four months involved, and then sum up these monthly totals to obtain the total insolation available for the entire cooling season Q_{ICS}. For this problem,

$$Q_{ICS} \cong 367,000 \text{ Btu/ft}^2$$

Substituting numerical values in Eq. (16-2) gives

$$A_{coll.} = \frac{240 \times 10^6 \text{ Btu}}{3.67 \times 10^5 \text{ Btu/ft}^2 \times 0.40 \times 0.85 \times 0.72}$$
$$= 2670 \text{ ft}^2 \qquad\qquad\qquad\qquad \textit{Ans.}$$

16-15 Heat Pumps Combined with Solar Energy

In large buildings such as multistory office buildings, hospitals, and hotels, it is often the case that some spaces need cooling at the same time that other spaces need heat. Air-conditioning problems of this kind are often solved with a combination solar energy–electric heat-pump system. Solar energy is used to heat and cool those spaces where normal winter (heating) and summer (cooling) conditions prevail. Excess solar heat is, of course, put into storage, or it may be used to generate steam or electricity.

For those spaces that depart from the norm (that is, they may need cooling in winter or heat in summer), electric heat pumps are an ideal solution. They "pump" heat from (i.e., cool) spaces that are too warm and deliver it to spaces that are too cold. Serving this function, they operate in a cooling mode similar to that of an ordinary air conditioner except that they reject the heat to some other room instead of to the outdoor air. Or they may deliver the heat to the thermal storage system of the solar energy plant. Whether the season is winter or summer makes no difference to the operation of the heat pump; its heating and cooling modes of operation are cycled by controls within the spaces being air conditioned. On cooling cycle, heat pump COP is similar to that of an ordinary refrigeration cooling unit, but when operating as an air heater—using solar-heated water at 70° to 80°F, for example, as a heat source—the COP may be as high as 2.5 to 4.0, making this type of application economically very attractive.

Both air-to-air and water-to-air heat pumps are used in combination with solar energy systems.[8]

CONCLUSION

Passive solar cooling is a firmly established architectural concept, the techniques having been practiced in one form or another since ancient times. Passive cooling can create conditions of summer comfort even in rather severe summer climates, but two limitations must be remembered: It cannot be engineered to produce precisely controlled indoor climate, and it cannot cope successfully with moisture (latent) loads—that is, conditions of high humidity.

---

[8] For further treatment of heat-pump theory and operation, see *ASHRAE Handbook, 1979 Equipment* (Atlanta: ASHRAE, 1979), Chapter 43; and Norman C. Harris, *Modern Air Conditioning Practice*, 3rd ed. (New York: McGraw-Hill Book Company, 1983), Chapter 14.

Active solar cooling is a proven technology, capable of precise engineering design. However, although active solar equipment is ready and performs well, it is not yet competitive in cost. Within a decade, this may change, and when it does, active solar cooling will become an important industry.

PROBLEMS

1. On a day when the air feels warm and muggy, the sling psychrometer provides the following readings: DB = 88°F; WB = 80°F. What is the relative humidity (RH)?

2. At Santa Fe, New Mexico, on a summer day, the DB temperature is 90°F and the WB temperature is 60°F. What is the approximate RH?

3. On a summer day when DB = 95°F, WB = 85°F, what is the actual moisture content of the air in pounds of water per pound of air?

4. The air in a building is at a condition of 85°F DB, 70 percent RH. At what temperature would a cold coil have to operate to just begin condensing moisture out of this air?

5. A summer condition of 80°F DB, 50 percent RH is being maintained in an office building. Approximately what percentage of the occupants should report feeling comfortable?

6. An indoor condition in a small factory is 80°F DB, 70°F WB. What is the effective temperature (ET)? Does it fall within the Comfort Zone?

7. A solar building has the same design and construction as that described in Illustrative Problem 16-6 and Fig. 16-6. Make the following changes in the data: Location 32° N Lat, outside walls insulated R-9, cathedral ceilings R-30, lights 5000 W, occupants 40; outside design 95°F DB, 70 percent RH; inside design 80°F DB, 50 percent RH; date August 21. All other conditions are unchanged. Calculate the design cooling load at 4:00 P.M. solar time.

8. Night ventilation is being used for next-day cooling of a solar building. The structure has 130 pounds of thermal storage mass per square foot of floor area. The night air temperature is ordinarily about 62°F. About how much cooling (Btu) can be stored up for use the next day in a 2000 ft² building if 6000 CFM of cool air is circulated throughout the night?

9. An office building with a sizable internal sensible load requires cooling even when the outdoor ambient temperature is 60°F. The total sensible heat gain is 950,000 Btu/hr, and this heat is to be removed by a flow of outdoor air—the economizer cycle. What flow of 60°F air will be required to maintain the indoor temperature at 78°F DB?

10. A wetted-pad type of evaporative cooler operates in outdoor air whose condition is 95°F DB, 20 percent RH. Air off the cooler is at 90 percent of saturation. (a) What is its dry-bulb temperature? (b) If this air is blown into a living space where it maintains a temperature of 80°F DB, what is the RH in the space?

11. A homeowner unfamiliar with psychrometrics heard about evaporative coolers and decided to install one in a home which is in a location with extended periods of muggy weather in the summer. (a) If 90 percent saturation off the cooler is assumed, what is the temperature of the air off the cooler if the outdoor ambient condition is 90°F DB, 50 percent RH? (b) The owner discovers that 84°F DB is the lowest temperature attainable in the house with the cooler operating. What is the RH in the house at this temperature, assuming no other moisture added?

12. Using the rule of thumb for evaporative coolers in desert climates (relative humidity consistently lower than 30 percent), specify the size (air delivery volume) of a cooler for a 1500 ft² house with a sensible cooling load of 40,000 Btu/hr when the indoor temperature is 80°F? If the outdoor ambient condition is 100°F DB, 15 percent RH, and the cooler is well maintained, approximately what will be the indoor RH?

13. The refrigeration unit of a home air conditioner is known to have a COP of 3.3. Neglecting other energy inputs (such as to fan motors), what is the EER of this unit?

14. An air conditioner with a rated overall EER of 11.0 is successfully handling a cooling load of 90,000 Btu/hr. About what is the total electric power input in kilowatts?

15. The COP of a centrifugal refrigerating machine is 2.65. It is driven by a fluorocarbon turbine whose overall efficiency is 18 percent. The fluorocarbon vapor is heated by solar collectors whose efficiency under the operating conditions is 32 percent. The collectors track the sun for normal incidence all day. Specify the minimum collector area required for August 21 at solar noon for a location near Minneapolis, Minnesota.

16. A solar-energized absorption refrigerating machine has a COP of 0.78. Flat-plate collectors are used, supplying 190°F water to the generator at a collector efficiency of 40 percent. Estimate, from the solar absorption refrigeration rule of thumb, the approximate cooling output into chilled water when this machine is energized by a 5000 ft² collector array. The time is 3:00 P.M. solar time on July 21. Location is Pittsburgh, Pennsylvania.

17. The seasonal cooling load (May through September) of a large auto sales room and associated offices at Grand Junction, Colorado, has been computed as 32,000 ton-hr. Solar collectors with a rated efficiency of 38 percent will be used. The combined efficiency of the rest of the solar components is 82 percent, and the COP of the absorption refrigerating machine is 0.70. The collectors face due south, with a tilt of 22° from the horizontal. Estimate the minimum total net collector area needed, in square feet.

C H A P T E R   17

Commercial and Industrial Applications of Solar Energy

The industrial sector of the U.S. economy uses approximately 18 percent of the total annual fossil-fuel energy consumed in the nation. The combined energy consumption of both the commercial and the industrial sectors, including electrical energy, accounts for more than 50 percent of all energy consumed annually in the United States. A large portion of this energy is used as process heat in manufacturing, with millions of barrels of oil and trillions of cubic feet of natural gas being burned just to produce heat in the relatively low temperature range of 150° to 350°F. Studies have indicated that one third or more of the process heat used by industry involves temperatures below 350°F. Since the basic solar technology already exists to provide unlimited amounts of heat for this temperature range, it is a tragic error to continue the use of oil and natural gas for this purpose. In this chapter, techniques for utilizing solar heat for selected commercial and industrial processes will be presented. The coverage will be general in scope, since the design of specific industrial process-heat systems is very complex and well beyond the scope of a basic text.

17-1 Commercial Uses of Solar Energy

Many energy demands that occur in commercial and industrial buildings can be met, at least in part, by solar energy systems. Most commercial buildings require space air conditioning and lighting. Natural daylighting can be planned for new construction to minimize the use of electric energy during daytime. (See section 9-11.) Space heating for commercial buildings may be provided by either active or passive solar systems, as discussed in earlier chapters. It must be pointed out, however, that heating requirements of commercial buildings are often quite different from those of residential buildings. For example, office buildings tend to be occupied during daytime hours only, and heating energy needs may be minimal at night. The same can be said for most industrial buildings and manufacturing plants, unless two or more shifts are at work. Retail stores provide excellent opportunities to use natural daylighting and solar-heating techniques.

The demand for domestic hot water is generally quite low in office buildings and retail stores, being approximately 1 gal of hot water per occupant per day. Restaurants, on the other hand, require large amounts of hot water, with approximately 2.4 gal of hot water being used for each meal served.[1]

Institutional buildings such as hospitals, nursing homes, and rest homes require large amounts of hot water for food preparation, laundry, and personal hygiene, along with the energy requirements for space heating. Hotels also require ample hot water and heat for space air conditioning, with the temperature requirements for the hot water being generally in the range of 120° to 190°F, which is well within the operating range of high-quality flat-plate solar collectors, as discussed in Chapters 11 and 12. For certain commercial dishwashing requirements, water temperatures in the range of 180° to 210°F may be required for the sterilization standards set by health codes. These temperatures are somewhat high for standard flat-plate collectors, and a careful review of collector system performance should be made before deciding whether to use high-performance flat-plate collectors or one of the high-temperature collector systems to be discussed below. Commercial laundries are large users of hot water also, and solar water-heating systems are well suited to this application. A solar hot-water system for a large laundry might involve several thousand square feet of solar collector area.

Space-heating requirements for many commercial buildings can be provided by passive solar systems (if there is solar access) or by active solar systems using air or liquid flat-plate collectors. Space cooling for commercial, institutional, and industrial buildings can often be provided by such passive methods as ventilation with cool outside air (the economizer cycle), night-sky radiative cooling, evaporative cooling, and other passive cooling methods treated in sections 16-7 to 16-10. The same limitations of passive cooling that were discussed in Chapter 16 apply to commercial and industrial buildings—passive cooling is most effective when days are hot and nights are cool, when there is very little humidity, and when a good share of the load is an exterior load rather than an interior load. If a solar-energized active cooling system is planned for the building, the solar collector system must be capable of supplying fluid in the temperature range of 185°F to 210°F to energize refrigerated cooling equipment, as discussed in Chapter 16. The heat collector system in this case would have to be either a very high-performance flat-plate type or a system based on concentrating solar collectors.

17-2 Industrial Process-Heat Requirements

Temperatures used for different heat-dependent manufacturing processes in industry range from the relatively low temperatures for drying agricultural products such as grains and fruits to the very high temperatures found in the

[1] State of California Energy Resources Conservation and Development Commission, *Energy Conservation Design Manual for New Nonresidential Buildings*, Sacramento, Calif., October 1977, Appendix 5, Table 1.

Process Temperature	Heat Sources		
	Water	Steam	Direct Heat (combustion gases or hot air)
< 160°F	X		X
160–212°F	X	X	X
212–350°F	X	X	X
350–500°F	X (pressurized)	X	X

manufacture of steel and glass. Heat used in industrial operations, whether from solar or conventional sources, is called *process heat*. Table 17-1 shows the temperature ranges found in selected industrial processes at the low end of the industrial temperature range. Hot water, steam, or hot air (combustion gases) can be used to supply heat to such processes.

Low-temperature air (70°F to 160°F) can be supplied by standard flat-plate air-heating collectors, for such uses as heating makeup air for painting booths and for drying lumber or textiles.

Low-temperature water (100°F to 160°F) can be supplied from standard liquid-heating solar collectors. Water in this temperature range is used in the food-processing and beverage industry for washing and cleaning food before freezing, bottling, canning, or preserving (Fig. 17-1). The use of water at these temperatures also occurs in some chemical processes, in cleaning manufactured parts before painting or plating, and in clothes laundries and automatic car washers.

Water temperatures in the range of 160°F to 212°F must be supplied by either high-performance flat-plate collectors or concentrating collectors. This temperature range is used for cooking in the food-processing industry. It is also used in washing and cleaning manufactured parts.

Steam (or pressurized hot water) in the temperature range of 212°F to 350°F can be provided by concentrating solar collectors. This temperature range is used for heating chemical solutions, processing plastics and rubbers, cooking and concentrating food products, and finishing and drying textiles.

Figure 17-1 Solar collectors used to heat the wash water for the milking parlor at the USDA Animal Genetics and Management Laboratory in Beltsville, Maryland. (U.S. Department of Agriculture.)

17-3 Flat-Plate Collectors for Industrial Applications

High-performance flat-plate solar collectors can be used to heat water in the temperature range of 160°F to 212°F. These collectors will typically have a selective coating on the absorber plate, low-iron glass glazings (for high transmittance), and a well-insulated case, as shown in Fig. 17-2. The insulation and gaskets used in the collector case must be able to withstand the very high stagnation temperature (350°F to 400°F) that the collector can reach when there is no fluid flowing through it. These collectors usually have double glazing to minimize convection losses from the absorber plate.

The tilt angle of the solar collector array will be determined to some extent by the heating load the collector array is designed to meet. If it is a seasonal load, which might be the case for a processing plant for farm products, the collector array may be tilted to optimize for the late summer—that is, latitude minus 15°. This angle would also be used if the primary function of the collector array were to provide heat for a solar-energized cooling system used only during the summer months. To optimize a collector array for a load during winter months, the collector tilt would be latitude plus 15° to 20°, as was discussed in Chapter 15.

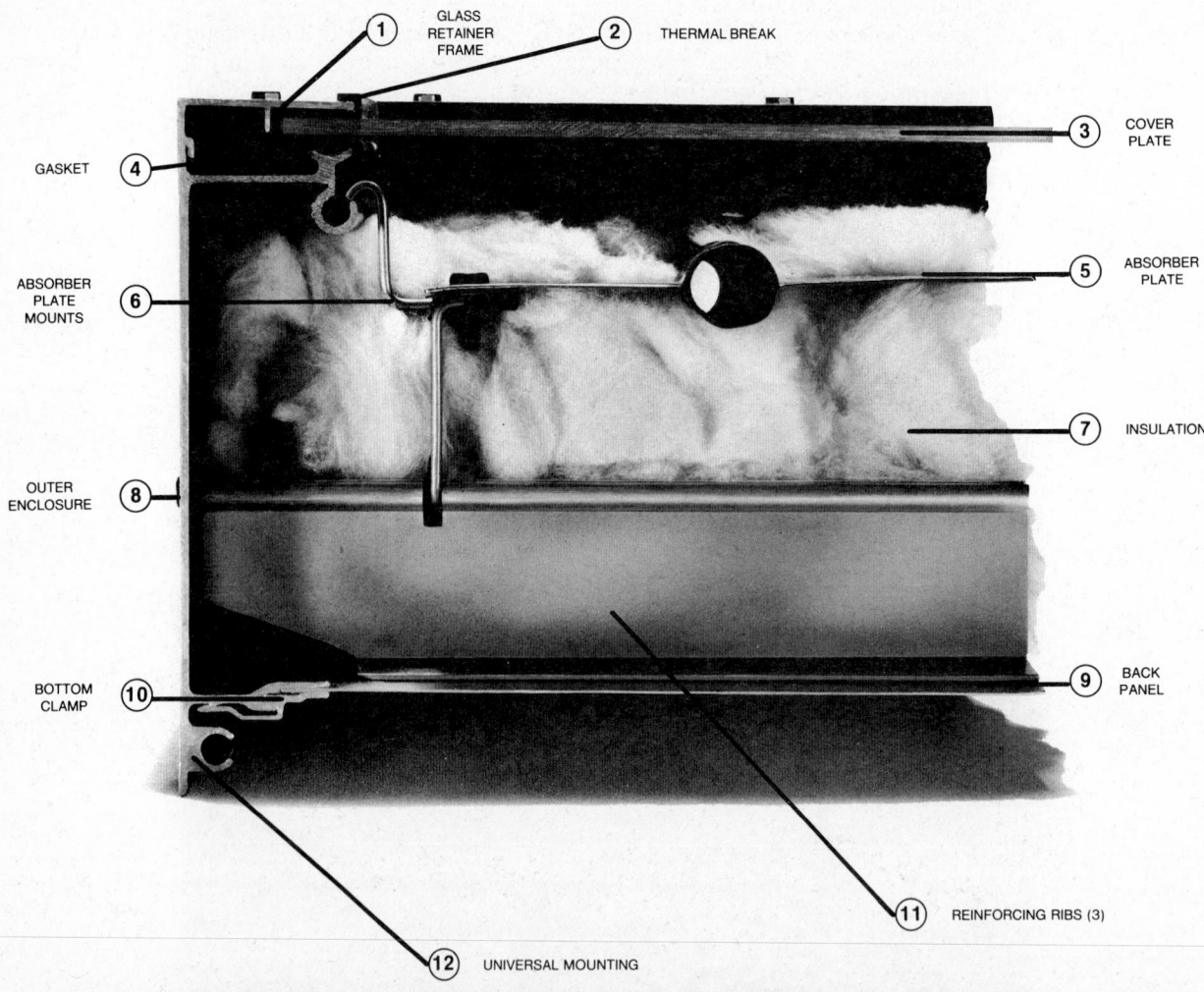

Figure 17-2 Cross-section of a high-performance flat-plate collector. These collectors operate in the range of 180°F to 212°F, and must have selectively coated absorber plates, insulation that will withstand high temperatures (such as non-resinous fiberglass), a high-transmittance glazing, and high-temperature glazing gaskets such as silicone rubber.

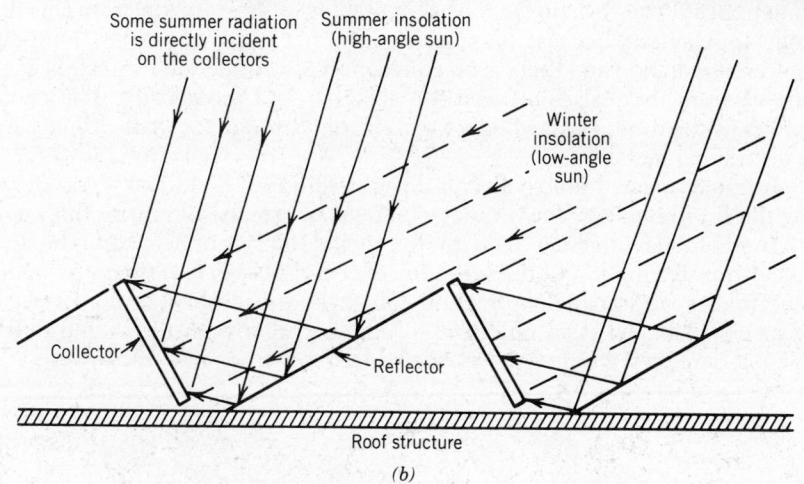

Figure 17-3 The summer performance of flat-plate collectors can be enhanced by using a sawtooth arrangement of collectors and reflectors on the roof. (a) A large solar heating and cooling system for an elementary school in Atlanta, Georgia. Reflectors on the back slope of each collector allow the collectors to absorb maximum radiation both winter and summer, even though the tilt angle favors winter operation. (U.S. Department of Energy, and Westinghouse Electric Corporation.) (b) Collectors tilted for optimum winter performance. The reflector is tilted so that the summer sun is reflected onto the collector, increasing the incident energy and compensating for the collector tilt, which is actually optimized for the winter sun.

In the case where the load is constant throughout the year, the collector should be tilted to provide a constant energy output year-round, a tilt angle of latitude plus 5° being a good compromise. The disadvantage of this tilt angle is that it favors the spring and fall over the summer and winter.

A sawtooth arrangement, as shown in Fig. 17-3, can be used to provide maximum collector performance during both summer and winter. The solar collectors are permanently mounted at an angle to optimize their winter performance (Fig. 17-3a). A sloping reflector is placed in back of each collector to reflect the high-angle solar radiation in summer onto the adjacent collector in the array.

17-4 Evacuated-Tube Collectors for High-Temperature Service

Collectors that operate above the boiling point of water (212°F) in the temperature range of 212°F to 350°F must be pressurized when water is used as the heat-transfer fluid. This is necessary to prevent the water from flashing into steam in the collectors at temperatures above 212°F. If a high-boiling-point heat-transfer fluid is used in the collectors, pressurization is not necessary. Solar collectors operating above 212°F will have significant thermal losses because of the large temperature differences between the collector and the ambient temperature.

One technique to reduce heat losses from collectors that operate above 212°F is to use evacuated glass covers. Design practicalities preclude constructing a flat-plate solar collector with an evacuated space under the glazing. Most evacuated collectors have tubular collector elements that are evacuated. The tubular geometry gives the glass tube strength so that it does not implode from atmospheric pressure after it has been evacuated. An evacuated-tube solar collector is shown in Fig. 17-4a. It consists of 16 separate collector elements. Each element or tube (Fig. 17-4b) consists of an evacuated enclosure between two concentric glass cylinders joined to form a long vacuum bottle. The outer glass cylinder serves as the glazing or "window" for the collector element. The inner glass cylinder is coated on its outer surface with a selective coating, and it is the solar absorber. The space between the two glass cylinders is evacuated to eliminate heat losses by convection. Thermal losses are very low from this type of collector element, because the selective absorber surface has a low emittance for infrared radiation and there is no convection or conduction from air because of the vacuum that exists between the absorber and the glazing. The energy absorbed by the inner glass cylinder is conducted through the glass cylinder wall to a cylindrical metal fin.

In the evacuated tube collector illustrated in Fig. 17-4, heat is transferred along the fin to one leg of a U tube, which carries the heat-transfer fluid. One leg of the U tube is attached to the fin to provide transfer of heat from the fin to the heat-transfer fluid, and the other leg of the U tube is free-floating to allow for thermal expansion and contraction. The cylindrical tube assembly is placed in a specially shaped cusp reflector, so designed that both direct and diffuse insolation is reflected onto the absorber tubes.

(a)

Figure 17-4 (a) An evacuated-tube solar collector has many individual absorber elements (tubes) enclosed by an evacuated glass cylinder. A reflector is used behind the absorber tubes to reflect sunlight back onto them. (General Electric Company.)

The circular-fin–U-tube assemblies inside each of the evacuated tubes are joined to headers, as shown in Fig. 17-4c. There are two sets of eight vacuum tube elements per collector. Eight of the U tubes are connected in series to form two continuous loops of eight U tubes in the collector. The two loops are connected in parallel to the supply and return headers that run through the center of the collector.

The performance of one type of evacuated-tube collector (GE Model TC-100) is shown in Fig. 17-5a, compared to that of a flat-plate collector with a selective coating. The evacuated-tube collector efficiency is shown as a function of collector inlet temperature for solar noon with an insolation of 300 Btu/hr-ft². The performance of the evacuated-tube collector is essentially unaffected by the ambient temperature, while there is a considerable decrease in the performance of the flat-plate collector when the ambient temperature drops from 80°F to 20°F.

In Fig. 17-5b, the energy collection of the TC-100 evacuated-tube collector, with a collector inlet temperature of 200°F, is shown and compared to that of a flat-plate collector. Note that the evacuated-tube collector collects useful energy over a longer period each day than does a flat-plate collector. This is because the heat loss from the evacuated-tube collector is very low, which allows a low insolation threshold level at which useful energy can be collected, making it possible to collect energy both early and late in the day, as shown in the illustration. An array of vacuum tube collectors is shown in Fig. 17-6, providing high-temperature water for a hotel. The solar energy produced is used to operate air conditioning equipment and to supply domestic hot water.

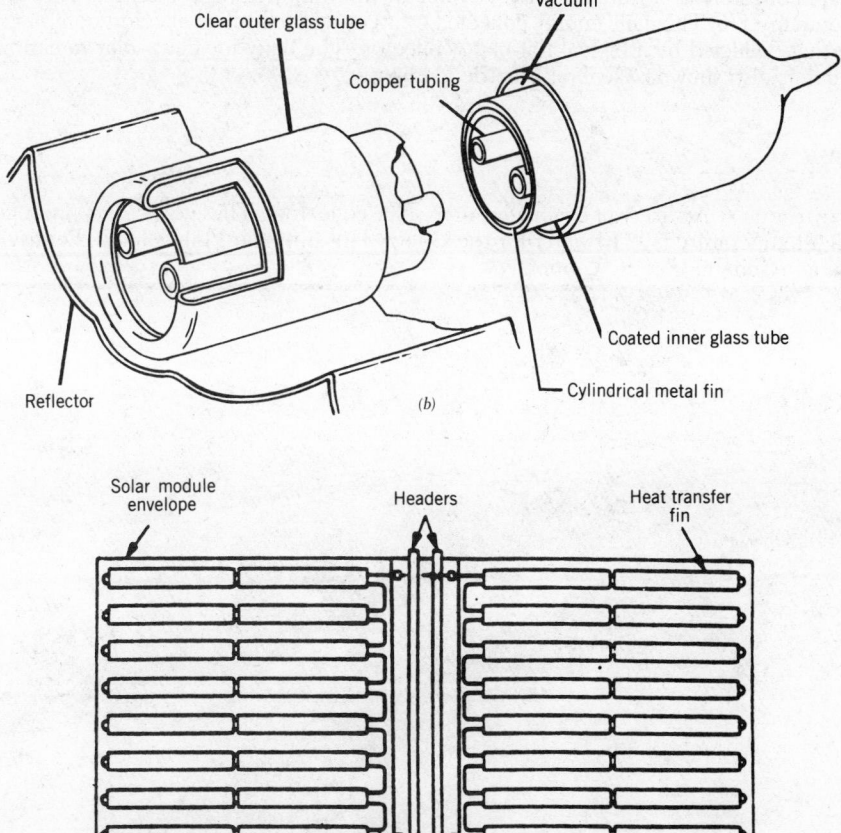

Figure 17-4 (continued) (b) Each absorber element is composed of an evacuated glass cylinder, a cylindrical heat-transfer fin, and a fluid tube for the heat-transfer fluid. (General Electric Company.) (c) The fluid tubes of all absorber elements of the collector are internally connected. Eight elements are connected in series and then connected to internal headers. The two groups of eight absorber elements are then connected in parallel. (General Electric Company.)

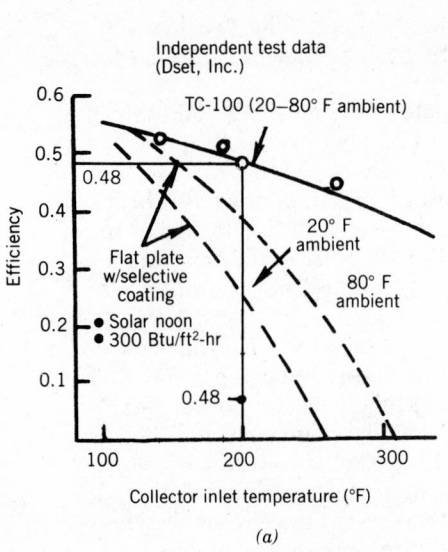

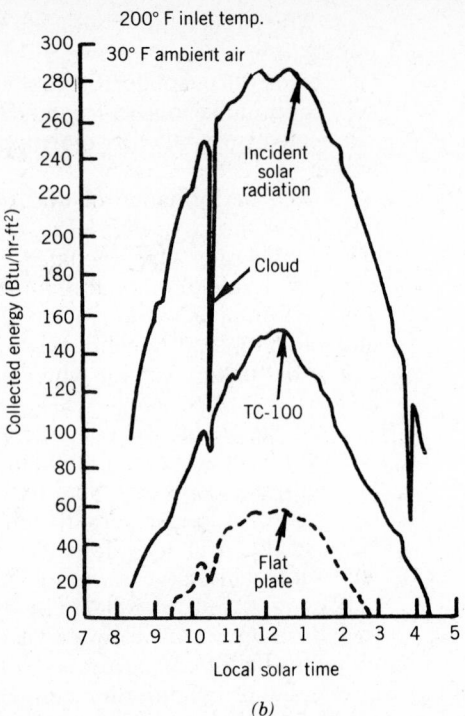

Figure 17-5 (a) The efficiency of an evacuated-tube collector compared to a flat-plate collector, as a function of the collector-inlet temperature. (General Electric Company.) (b) The daily energy collected by an evacuated-tube collector, compared to that collected by a typical flat-plate collector. The daily incident solar radiation curve is also shown. (General Electric Company).

Figure 17-6 An array of evacuated-tube solar collectors. These collectors produce high-temperature (190°F) water for the Cherry Hill Inn near Philadelphia, Pennsylvania. (General Electric Company.)

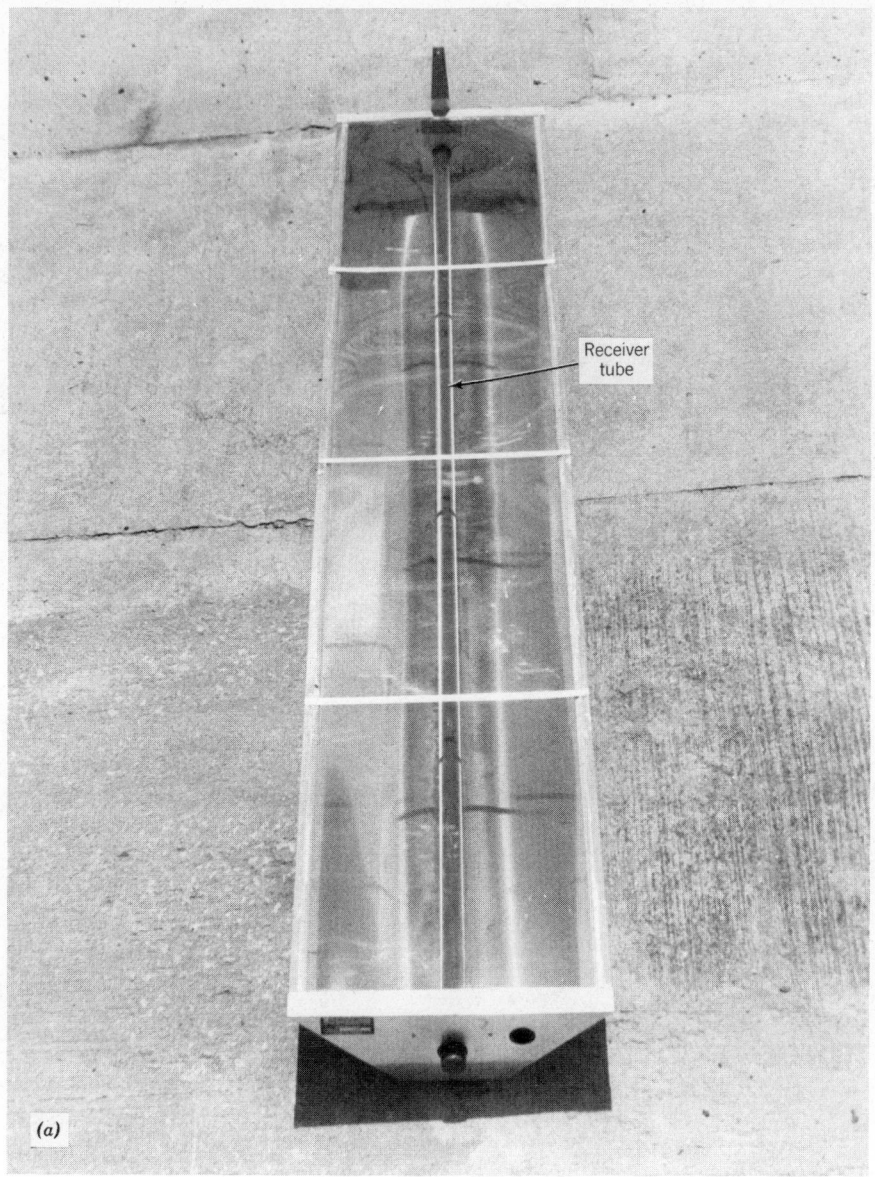

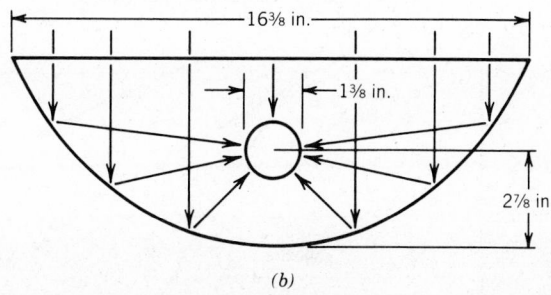

Figure 17-7 (*a*) A parabolic-trough concentrating solar collector with integral receiver tube along the focal axis. (Whiteline Concentrating Solar Collectors.) (*b*) This type of collector uses a very short-focus reflector to reflect sunlight onto the receiver tube. The concentration ratio in this case is 16⅜ in./1⅜ in. = 11.9. (Whiteline Concentrating Solar Collectors.)

17-5 Parabolic-Trough (Concentrating) Collectors

Another type of collector used for high-temperature applications is a parabolic reflector in the form of a trough. Figure 17-7*a* shows one type of parabolic concentrating solar collector, which has an operating temperature range of 160°F to 250°F. This collector has a *receiver tube* down the axis of the parabolic trough. The trough itself forms a very short-focus parabola in cross-section, and the receiver tube is positioned along the focal axis of the trough. Sunlight reflected from the parabolic reflector strikes a large portion of the receiver tube surface area, as shown in Fig. 17-7*b*. Short-focus collectors have a relatively low concentration ratio, a parameter obtained by dividing the width of the collector aperture by the width of the receiver. For the collector shown in the diagram, the concentration ratio is approximately 11.9 to 1 (16⅜ ÷ 1⅜). This particular collector has an acrylic glazing above the parabolic reflector to keep dust and dirt off the reflector and reduce convection from the receiver tube. These collectors, when mounted on an east–west axis, require a monthly seasonal (declination) adjustment to align the collector with the sun's path across the sky. The collectors can also be used with continuous tracking in elevation during the day, but this is not generally required for a short-focus concentrator. The performance of a parabolic-trough collector compared to a flat-plate collector and to an evacuated-tube collector is shown in Fig. 17-8. The parabolic-trough collector does not perform quite as well as the evacuated-tube collector even when the inlet temperature is as low as 100°F. As the inlet temperature increases, the parabolic collector's performance decreases somewhat more rapidly than that of the evacuated-tube collector. This is caused by heat loss from the receiver tube, which has no vacuum jacket around it. At any inlet temperature above about 160°F, the parabolic collector's performance becomes better than that of the flat-plate collector. Figure 17-9 shows an array of parabolic-trough collectors that provide space heat and domestic hot water for a residence.

A parabolic-trough collector with a high concentration ratio is shown in Fig. 17-10. The receiver is located above the reflector, indicating that the focal length is much longer than in the previous case. This type of collector operates well in the temperature range of 212°F to about 600°F. The parabolic trough has an aperture width of 7 ft and the receiver assembly has a 1.5-in. diameter

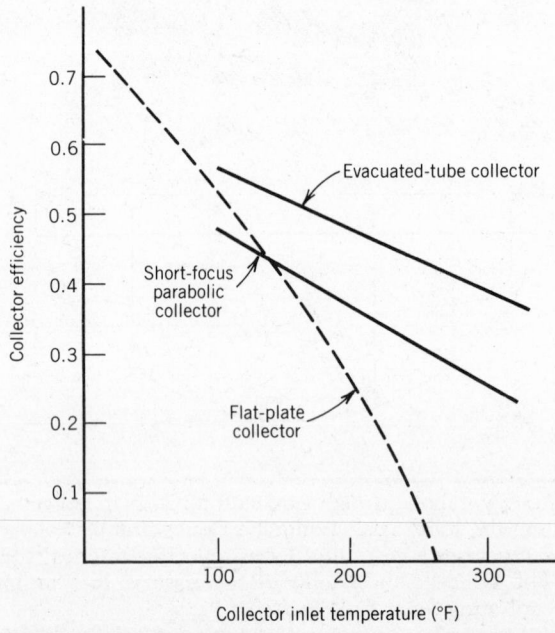

Figure 17-8 The performance of a short-focus parabolic-trough collector compared to a flat-plate collector and an evacuated-tube collector.

COMMERCIAL AND INDUSTRIAL APPLICATIONS

Figure 17-9 A 220 ft² array of concentrating parabolic solar collectors used for home heating and for domestic hot water. (Whiteline Concentrating Solar Collectors.)

receiver tube with a pyrex glass glazing. The concentration ratio for this collector is again the collector aperture width divided by the receiver width, approximately 56 to 1 (84 in. ÷ 1.5 in.). Concentrating collectors with such a high concentration ratio should continuously track the sun to ensure that the focused sunlight falls directly on the receiver assembly at all times of the day. The performance of a collector of this type is charted in Fig. 17-11. The

Figure 17-10 A high-concentrating parabolic reflector. Current models of this type of collector are capable of operating from 212°F to a temperature as high as 600°F, providing process heat over a wide range of temperatures. (Courtesy of Acurex Solar Corporation.)

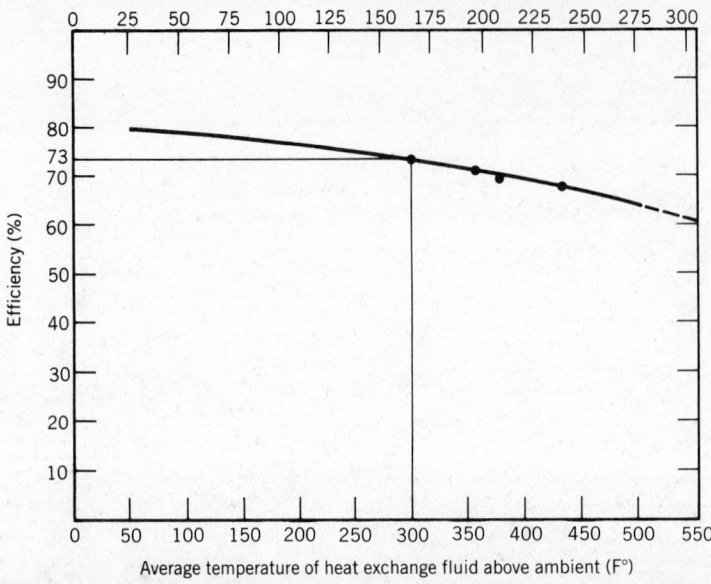

Average temperature of heat exchange fluid above ambient (F°)

Figure 17-11 The performance of a high-concentrating collector. The collector performance varies by about 18 percent as the average collector operating temperature above ambient increases from 50°F to 550°F. (Courtesy of Acurex Solar Corporation.)

collector efficiency remains very high, above 70 percent for operating temperatures up to 350°F. Maximum operating temperatures of over 500°F are typical of these collectors.

It should be remembered that concentrating collectors that use a mirror or a lens to focus sunlight onto a receiver can use only direct sunlight, because diffuse radiation will not be focused. Concentrating collectors should be used only in locations where direct-beam radiation makes up a large portion of the annual solar radiation, and only in applications where sun tracking (at least single-axis tracking) can be provided.

17-6   Solar Process-Heat Systems

When a process-heat system is designed many factors must be considered. Some general factors relate to the requirements of the process itself, such as the required temperature, the temperature tolerance, the seasonality of the process, or the continuous or intermittent nature of the process. A very important factor is the manner in which the auxiliary heating system is integrated into the system. The type of collectors used and the insolation available at the site also influence the system design.

When a system is designed for a specific process-heat application, the above factors will influence the design in a unique way. To some extent, each application will have a specific solution. We cannot, however, deal with all possible ramifications in a basic book and will instead discuss a few generalized solar process-heat systems to illustrate fundamental concepts. An industrial process may require that a liquid or gas be heated, that a substance be dried, or that steam be supplied to a turbine or other mechanical equipment. Three different systems will be considered—one that provides process hot water and/ or indirect heating of process fluids, one that provides indirect heating of process gases, and one that supplies process steam. These three are representative of the many types of process-heat systems used by industry.

17-7 Process Hot-Water and Fluid Heating

The system requirements for a process hot-water system are different from those of a domestic hot-water system or a space-heating system, because the process-heat system operates at much higher temperatures. High-temperature operation requires more exacting performance from system components, such as the capability of operating at the higher pressures that normally accompany high temperatures. A generalized process-heat system for heating fluids and hot water is shown in flow diagram form in Fig. 17-12.

The collector array consists of banks of high-temperature collectors (SCB) connected in series–parallel between the supply and return headers. Each bank has isolation valves, an air venting valve, and a pressure relief valve. The number of collectors connected in series will be determined by the desired or required temperature rise. The rest of the collector loop consists of high-pressure heat exchanger (HEP), air purger (AP), receiver tank for the closed-loop fluid (RT), collector loop pump (CP), pressure relief valve (7), back-pressure valve (6), vacuum-break valve (5), pump isolation valves (8), fill valve (10), shutoff valve (9), and purge valve (11).

The Collector Loop.   The collector-loop heat-exchange fluid will have a large daily temperature variation. Each day the fluid will start out near ambient temperature and be heated to a temperature in the range of 200°F to 350°F or more. The heat-exchange fluid will undergo expansion in the range of 5 to 12 percent when heated to an operating temperature in the range of 200°F to 350°F. To provide for the expansion of the heat exchange fluid, the receiver tank (RT) is provided. The capacity of the receiver tank must be large enough to store the heat-transfer fluid during periods of collector stagnation. The receiver tank is unpressurized and vented to the atmosphere. It is provided with a sight

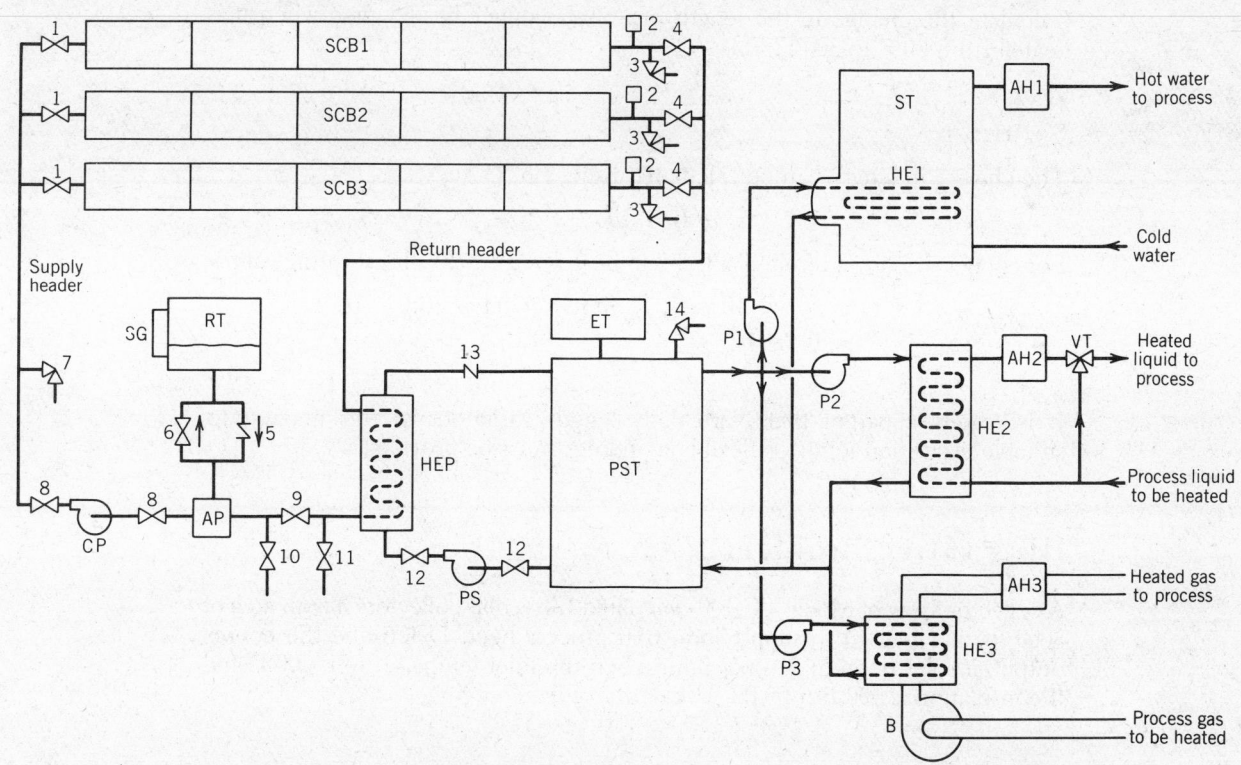

Figure 17-12 A generalized solar-energized process-heat system. The solar collectors heat a heat-exchange fluid, which in turn heats a pressurized hot-water storage system through heat exchanger HEP. The pressurized hot water is then used to heat process hot water, process liquids, or process gases.

COMMERCIAL AND INDUSTRIAL APPLICATIONS OF SOLAR ENERGY

gauge (SG). The pressurization of the collector loop is maintained with a pair of valves—a back-pressure valve (6) and a vacuum-break valve (5). The back-pressure valve allows expanding heat-exchange fluid to enter the receiver tank as the collector loop temperature increases. The vacuum-break valve is a one-way valve that allows fluid to flow from the receiver tank back into the collector loop. This flow occurs as the collector loop cools down and the heat-exchange fluid contracts, creating a vacuum in the collector loop. The vacuum-break valve then opens and allows heat-exchange fluid to reenter the collector loop, thus maintaining the fluid level.

The closed (collector–heat-exchanger) loop must be pressurized to keep the heat-transfer fluid from boiling. If a solution of either water and ethylene glycol or water and propylene glycol is used, the closed loop will generally be pressurized in the range of 50 to 75 psi dependent upon the fluid mixture and operating temperature range. The pressure setting on the back-pressure valve (6) should be set at the pressure equal to the desired closed-loop pressurization. The collector-loop pressure relief valve (7) should be set at a pressure of 15 to 20 psi greater than the desired closed-loop pressurization plus the collector-loop static pressure drop. When nonaqueous heat-transfer fluids are used, the closed-loop pressurization will depend on the fluid properties. The piping used for high-temperature systems is metal, usually copper or steel.

The piping connections between main headers and collector manifolds must allow for thermal expansion. When piping is heated and cooled over a temperature range of several hundred degrees, there can be a significant variation in length, and inadequate provision for expansion can have disastrous results (see section 3-6).

Illustrative Problem 17-1

Calculate the change in the length of a 40-ft copper header pipe when it is heated from 60°F to 300°F.

Solution

The change in length can be calculated using Eq. (3-8):

$$\triangle L = \alpha L_0 \triangle t$$

For copper, α is found from Table 3-2 to be $9.5 \times 10^{-6}/\text{F}°$. Substituting values,

$$L = 9.5 \times 10^{-6}/\text{F}° \times 40 \text{ ft} \times (300 - 60)\text{F}°$$
$$= 0.091 \text{ ft}$$
$$= 1.09 \text{ in.} \qquad \qquad Ans.$$

It is readily apparent that, with daily length variations of this magnitude, suitable expansion joints or flexible couplings must be provided.

Illustrative Problem 17-2

A collector array of Model TC-100 evacuated-tube solar collectors has an area of 3000 ft^2 and is used to supply industrial process heat. Determine the energy output at solar noon in Btu per hour when the inlet temperature is 200°F and the insolation is 300 Btu/hr-ft^2. Use data in Fig. 17-5.

Solution

From Fig. 17-5a, the TC-100 evacuated-tube collector has an efficiency of approximately 0.48 at an inlet temperature of 200°F and an insolation level of 300 Btu/hr-ft^2. Using Eq. (12-1), the energy output can be calculated as:

$$E_o = \eta \times I$$
$$= 0.48 \times 300 \text{ Btu/hr-ft}^2$$
$$= 144 \text{ Btu/hr-ft}^2$$

The total energy from the array is then

$$E = 144 \text{ Btu/hr-ft}^2 \times 3000 \text{ ft}^2$$
$$= 432,000 \text{ Btu/hr} \qquad\qquad Ans.$$

Illustrative Problem 17-3

A solar collector array uses 5000 ft² of high-concentrating parabolic-trough collectors to provide an average fluid temperature of 300°F above ambient to a low-pressure steam boiler for an industrial plant. Estimate the energy output for the collector array when the direct-beam insolation I_{DN} is 220 Btu/hr-ft². (Remember that focusing collectors only use the direct component of the solar radiation.) Use collector performance data in Fig. 17-11.

Solution

The collector efficiency is obtained from Fig. 17-11 as approximately 0.73. The energy output of the collector is calculated using Eq. (12-1):

$$E_o = \eta \times I$$
$$= 0.73 \times 220 \text{ Btu/hr-ft}^2$$
$$= 160.6 \text{ Btu/hr-ft}^2$$

The total energy output for the array is then

$$E = 161 \text{ Btu/hr-ft}^2 \times 5000 \text{ ft}^2$$
$$= 803,000 \text{ Btu/hr} \qquad\qquad Ans.$$

Expansion and contraction of the piping is usually compensated for by using flexible tubing between the piping connections. These flexible couplings may be in the form of high-temperature flexible braided steel hose or a bellows section of tubing, as shown in Fig. 17-13.

The Storage Loop. The thermal storage shown in Fig. 17-12 is the pressurized water tank PST. Table 17-2 gives the water vapor pressure for temperatures above 212°F. From this table, it is apparent that if a storage system were to operate at 250°F, the system would have to be pressurized above 15 psig to prevent boiling. At 350°F, a pressure of 120 psig would be required.

The storage loop must also provide for pressure relief and for expansion and contraction of the storage volume as it is heated and cooled. The expansion tank (ET) and the pressure-relief valve (14) serve these purposes. The expansion tank should be sized for approximately 8 to 10 percent of the total liquid volume in the storage loop.

Figure 17-12 also shows three ways in which heat from the pressurized thermal storage can be used. *Process hot water* can be heated using the heat exchanger (HE1). The process hot water must have a storage tank (ST) to provide a reserve of hot water. The process hot water might be consumed during the process (such as a washing process), or, if water conservation is mandatory, it would be reclaimed by processes not shown. Cold-water feed is supplied to the tank (ST) from the water supply. An auxiliary heat source (AH1) is provided to boost the water temperature when required.

The indirect heating of a process fluid can be accomplished by the use of the heat exchanger (HE2). A pump (P2) circulates the solar-heated fluid through this heat exchanger. Auxiliary heat may be added to the fluid when required (AH2). The temperature of the fluid is controlled by the tempering valve (VT) which mixes hot fluid and cool fluid to achieve the required temperature.

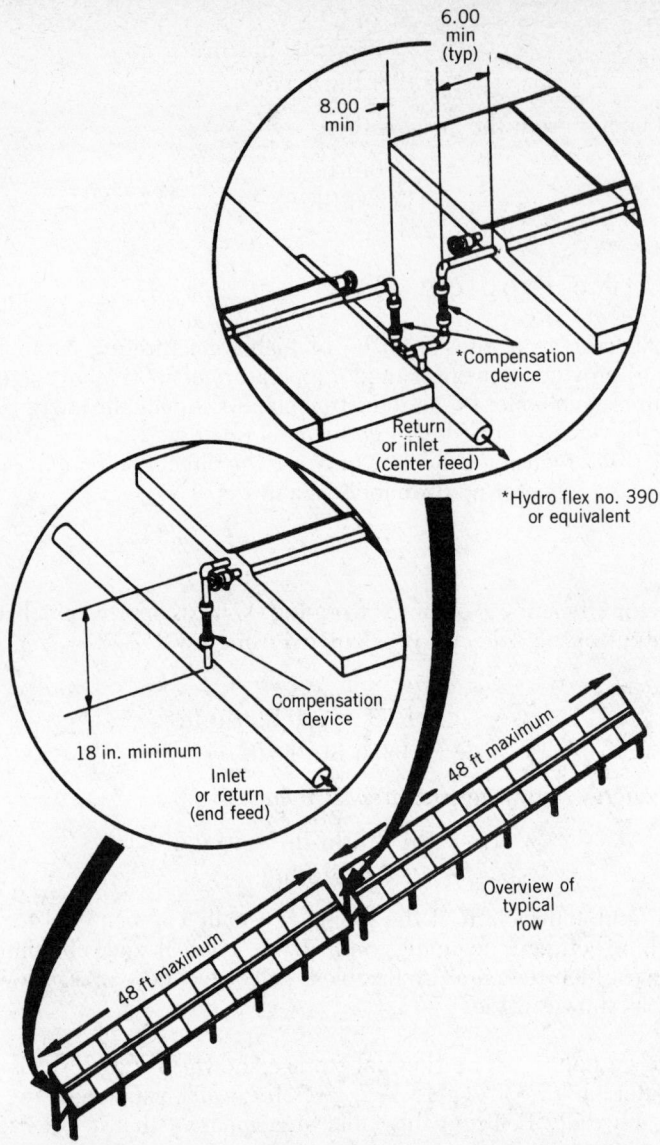

Figure 17-13 A compensation device must be used between the piping connections of high-temperature collectors to allow for the thermal expansion and contraction of the piping as it heats and cools. (General Electric Company.)

The indirect heating of process gases may be done by circulating hot water from the storage tank through a water-to-gas heat exchanger (HE3). Auxiliary heat may be added when required (AH3).

Process (low-pressure) steam can be produced by using a simple kettle evaporator or boiler, as shown in Fig. 17-14. High-temperature (e.g., 350°F) pressurized water is circulated through the heat exchanger in the boiler. The water contained in the boiler boils at a lower temperature than the pressurized water, generating low-pressure steam, which is sent to different processes. The condensate is returned to the boiler.

An alternative to pressurized hot water for a heat-storage system is to use a large tank filled with stones (similar to a rock bed for solar air-heating systems). A heat-transfer fluid with a low vapor pressure at the system operating temperature is used, such as a heat-transfer oil. This heat-transfer fluid is circulated

Temperature (°F)	Gauge Pressure (lb/in²)
212	0
220	2.5
230	6.1
240	10.3
250	15.1
260	20.7
270	27.2
280	34.5
290	42.9
300	52.3
310	63.0
320	75.0
330	88.3
340	103.3
350	119.9
360	138.3
370	158.7
380	181.1
390	205.7
400	232.6

through the collectors and then through the rock-filled tank, with hot rock providing the thermal storage. Solar-thermal electric generating plants use this technique (see section 18-19 and Fig. 18-30b).

17-8   Controls for Industrial Process-Heat Systems

Controls for high-temperature solar energy systems are often quite different from the control systems discussed previously. Instead of using a differential thermostat to compare the collector temperature to the storage tank temperature, a pyranometer sensor is used to measure the insolation at the collectors. When the insolation reaches a predetermined level each morning, the collector-loop pump and storage-loop pump are started. The collection of

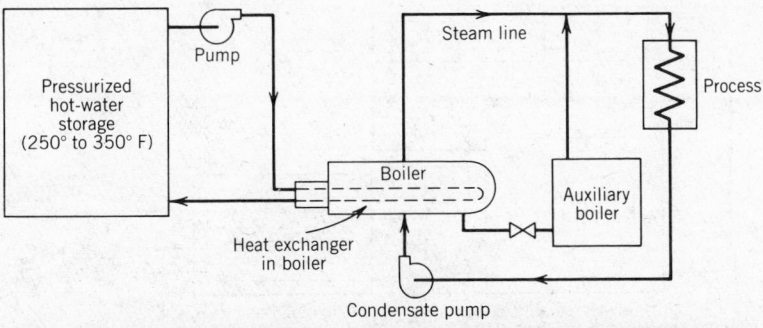

Figure 17-14 Process steam can be generated from high-temperature pressurized hot water (250° to 350°F). The high-temperature water is circulated through a heat exchanger in a low-pressure boiler, heating water to make steam. The steam can then be used to supply heat to different industrial processes. Typical conditions on the steam side of the boiler might be 240°F, 10 psi.

energy will continue until the insolation at the collector falls below a given threshold, at which point the collector-loop pump and storage-loop pump will be turned off. The control system must also protect the collectors from damage during stagnation conditions. With high-temperature collectors, stagnation temperatures may exceed 750°F. Controls that govern the operation of evacuated-tube collectors ordinarily have stagnation lockout sensors. If a stagnation condition results because of a pump failure, power failure, or some other problem, the stagnation lockout system prevents the heat-exchange fluid from being circulated into the stagnated collectors. If cool fluid is circulated into stagnated collectors, severe stresses will occur, which may damage the entire structure of the collector array.

Tracking collector controls should have stagnation sensors that override the tracking so that the concentrators will be positioned at angles where the receiver does not receive radiation, thus protecting the receiver from the high temperatures associated with stagnation.

17-10   Solar-Assisted High-Temperature Heat Pumps

An alternative method of providing process heat in the temperature range of 180°F to 220°F is to combine a low-temperature solar energy system with a high-temperature heat pump. High-temperature heat pumps take low-grade heat in the temperature range of 60°F to 120°F and boost it to a higher temperature in the range of 180°F to 220°F—a temperature range suitable for process-heat applications and space heating. The flow diagram of Fig. 17-15 shows a high-temperature heat pump operating with a low-temperature solar energy system. This type of system is called a *solar-assisted heat pump*. The solar collectors and solar heat storage ordinarily operate in the temperature range of 80°F to 120°F. The high-temperature heat pump uses the solar storage as a heat source and provides a temperature lift to the range of 180°F to 220°F. This type of system has several advantages over an all-solar-energized process-heat system. The solar storage operates at a relatively low average temperature of about 100°F, allowing standard flat-plate solar collectors to be used at optimum efficiency for collecting the heat, rather than the more expensive high-temperature solar collectors such as evacuated-tube units, which would be needed to provide process-heat temperatures directly. Operating efficiencies for a single-glazed flat-plate solar collector and for an evacuated-tube collector are compared in Fig. 17-16. With an assumed insolation of 250 Btu/hr-ft² and an ambient temperature of 50°F, the flat-plate solar collector will operate at an efficiency of

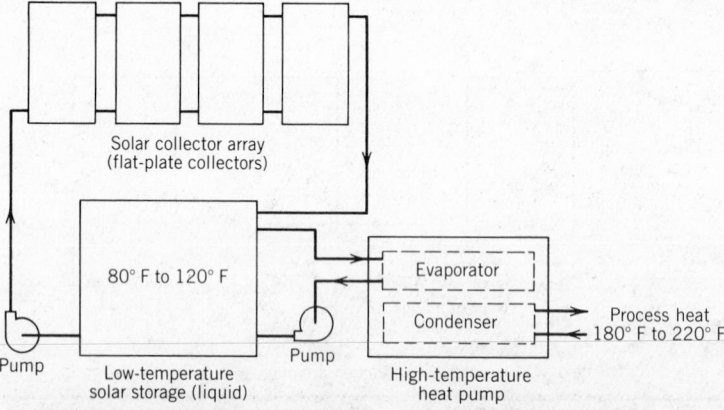

Figure 17-15 Flow chart of a solar-assisted high-temperature heat-pump system. Flat-plate solar collectors are used to heat the low-temperature solar energy storage. The high-temperature heat pump removes heat from the low-temperature storage (usually hot water) and provides a process-fluid output at a much higher temperature.

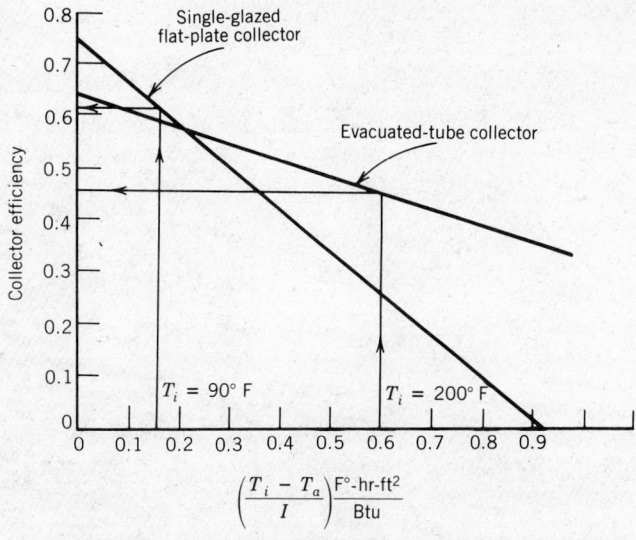

$$\left(\frac{T_i - T_a}{I}\right)\frac{F°\text{-hr-ft}^2}{Btu}$$

Figure 17-16 Comparison of the operating efficiencies of a single-glazed flat-plate solar collector and an evacuated-tube collector. Note that at temperatures under about 100°F, the flat-plate collector efficiency is as high as that of the evacuated-tube collector.

0.62 with an inlet temperature of 90°F. The evacuated-tube collector will operate at an efficiency of 0.46 with an inlet temperature of 200°F. Since the medium-temperature flat-plate collector is more efficient than an evacuated-tube collector at temperatures under 100°F, and also is much less expensive, teaming up with a heat pump may result in significant system economy because of the high COP of heat-pump systems.

The coefficient of performance (COP) of a high-temperature heat pump varies between 2.0 and 4.5, depending on the heat-source temperature and the final process-heat temperature. This means that for every unit of electric energy the heat pump consumes (expressed in Btu) it supplies 2 to 4.5 Btu of heat at the process-heat temperature. By using a solar-assisted heat pump, low-temperature solar collectors can be used, the initial investment in the solar energy system is less, and a heat source suited to heat-pump operation is provided. In some cases there can be as much as a 50 percent savings in the initial system investment. The disadvantage of the solar-assisted heat pump system is that there is an annual operating cost for the electric energy used by the heat pump. Only a life-cycle cost–benefit analysis (see Appendix II) could determine the economic advantage of one system over the other.

Illustrative Problem 17-4

A solar-assisted high-temperature heat pump is to provide 180°F water for a hotel. This water will be used for heating the domestic hot water and for space heating. The heat pump will operate with a COP of 3.0. If the average heat load in winter is 7×10^6 Btu/day, determine the energy that the solar collectors must provide to the low-temperature storage.

Solution

With a COP of 3.0, one third of the heat pump output will come from the electricity consumed and two thirds of the heat output will come from the solar-heated low-temperature storage. The energy that the solar collectors must supply is then:

Figure 17-17 Closeup view of the Templifier® high-temperature heat pump with front cover removed. Note the four hermetic compressors. This heat pump will operate from a heat source liquid at 90°F and provide output liquid at 190°F. (McQuay, Inc.)

$$Q_c = \frac{2}{3} \times 7 \times 10^6 \text{ Btu/day}$$
$$= 4.67 \times 10^6 \text{ Btu/day} \qquad Ans.$$

A high-temperature heat pump of current manufacture is shown in Fig. 17-17. This unit is to be installed in a solar-assisted high-temperature heat-pump system that supplies process heat for aluminum anodizing. The collector array

Figure 17-18 Flat-plate solar collectors for providing source heat to a Templifier® heat pump that furnishes 180°F hot water to an Ottawa, Canada apartment complex. (McQuay, Inc.)

for a different solar-assisted heat pump system is shown in Fig. 17-18. This system provides 180°F water to an apartment complex in Ottawa, Canada for space heating and domestic hot-water heating.

CONCLUSION

In this chapter we have discussed several solar-energy systems that are capable of providing process heat in the temperature range of 160°F to over 500°F, using hot water or steam as the heat-transfer fluid. These systems are much larger and more complex than residential or small commercial solar-energy systems.

    The presentation here has been descriptive and exploratory only, the major purpose being to indicate the range of industrial and commercial applications of energy from the sun. By 2020, up to one fourth of total U.S. industry needs for process heat (at temperatures up to 350°F) could be met from solar-energy installations.

PROBLEMS

1.   Determine the change in the length of a 150-ft-long copper pipe when it is heated from 50°F to 250°F.

2.   Determine the efficiency of an evacuated-tube solar collector when the inlet temperature is 300°F and the insolation level is 300 Btu/hr-ft^2. Use data from Fig. 17-5a.

3.   A solar collector array uses 7000 ft^2 (aperture area) of high-concentrating parabolic-trough collectors to provide heat to a chemical plant at a temperature of 425°F. Ambient temperature is 50°F. Determine the energy output for the collector array when the direct insolation is 240 Btu/hr-ft^2. Use data from Fig. 17-11.

4.   A solar-assisted heat pump supplies process hot water to a plating process at 250°F. The average temperature of the low-temperature storage is 90°F. The daily heat required is 2.5×10^6 Btu, and the heat pump operates with a COP of 2.75 under these temperature conditions. Determine (a) the heat removed daily from the low-temperature storage, and (b) the collector area required to supply the heat removed daily from the storage by the heat pump. Assume that the collectors have an average efficiency of 0.61 and the average daily insolation on the collectors is 1000 Btu/ft^2.

PART FOUR

Electricity from the Sun

CHAPTER 18

Solar Energized Electric Power

The sun is our greatest perpetually renewable source of energy. Solar power arrives at the surface of the earth with an intensity of about 900 to 1000 W/m², or 290 to 315 Btu/hr-ft² on clear days at local solar noon over much of the area of the temperate zones. Other chapters in this book have explained in detail how the sun's energy can be applied to passive and active solar heating and cooling systems. But electricity is the most convenient form of energy for almost any purpose, and it is therefore most fortunate that sunlight can readily be converted into electric current. Electricity from solar energy is generated by two very different methods: *photovoltaic generation* and *thermal generation*.

PHOTOVOLTAIC GENERATION OF ELECTRICITY

The conversion of light energy directly into electric energy is known as the photovoltaic effect, and the process occurs in *photovoltaic cells*, more commonly called PV cells by the solar-electric industry. The industry often uses the term *solar cells* as well.

A typical solar cell (Fig. 18-1) has as its primary

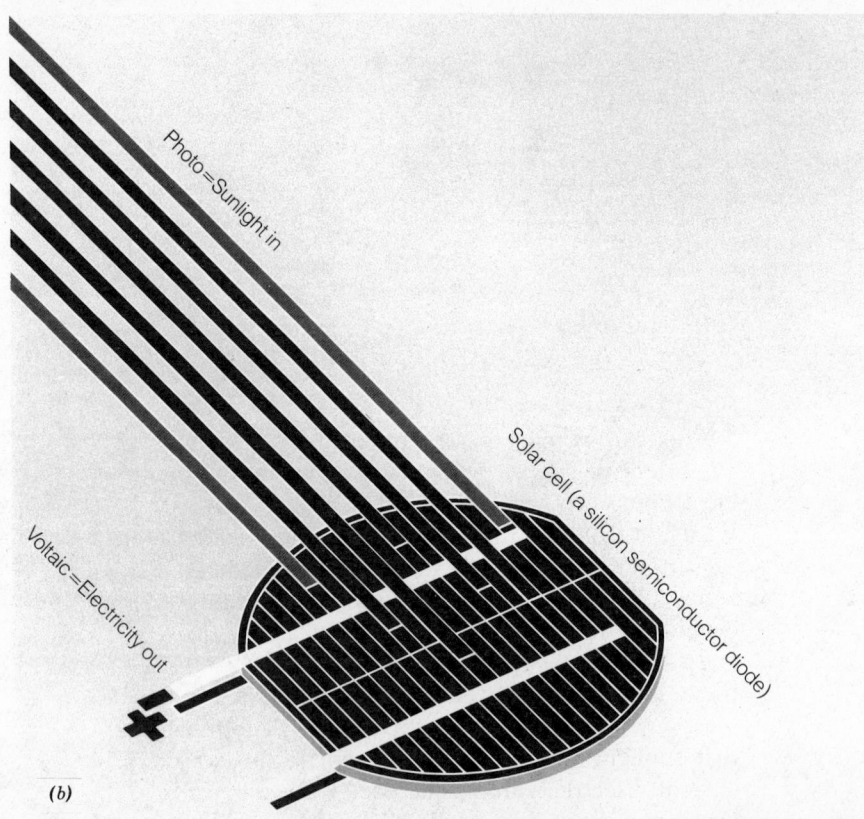

Figure 18-1 The silicon solar cell. (a) Typical 4-inch cells. (Left) New (1984) square-design cell. (Right) Older round cell. Note grids and metal ribbons for electron collection and flow. (b) Diagrammatic representation of a silicon solar cell. Streams of photons (sunlight) eject electrons from the outer orbits of silicon atoms. The metallic grids and connector ribbons indicated provide the means of collecting the freed electrons and establishing flow through an external load. (c) Four-inch round silicon solar cells assembled in a module. (Photos and diagram courtesy ARCO Solar, Inc.).

Figure 18-1 (*continued*)

photosensitive material a thin wafer of silicon (other semiconductors are also used) that has been treated ("doped") to create a permanent electric potential in the cell. When light energy in the form of particles (*photons*) strikes silicon atoms, electrons are ejected from outer orbits. These now-free electrons are swept all in the same direction by the permanent electric field in the cell, establishing an electron flow. The silicon wafer is coated with silver grids, which collect the electron flow. This electric current then flows from the wafer or cell by metal ribbons that connect the cell to the exterior load. Single cells can be connected in series or in parallel, or in series–parallel combinations to provide a direct current (dc) of the desired voltage and current. A more detailed treatment of the construction and operation of photovoltaic cells will be presented in a later section, after some of the attributes and potentialities of photovoltaic energy are explored.

18-1 History and Recent Developments in Photovoltaics

Although the photoelectric effect has been a well-known scientific phenomenon since at least 1905 when Albert Einstein proposed its theoretical base in the quantum theory, the practical uses of electricity from light were quite limited until the dawn of the Space Age in the late 1950s. Photoelectric cells (also called phototubes) were used in photography and motion-picture projection equipment, in automatic control devices, and in automated counters and building security systems from about 1930, but little if any use was made of such devices as actual sources of usable electric power because their output (from an evacuated tube) was so small. With the advent of solid-state devices in the 1940s and 1950s, however, photovoltaic cells were soon produced that could generate significant amounts of electric power from sunlight. The high technology of space vehicles and orbiting satellites demanded a lightweight, reliable, rugged, and continuous source of energy to operate onboard systems. (Skylab, for example, had an installation of about 150,000 solar cells generating power at a peak of 22 kW.)

In the 1960s, several federal agencies (NASA and the Department of Defense, for example) along with many universities and space- and energy-related private companies sponsored or actually conducted intensive research and development in the field of photovoltaics, and by 1970 the "spinoff" from space-related research was having a significant impact on industry and on our daily lives. Solar cells had come back to earth and were getting down to work.

In the past 15 years, photovoltaics has changed from a scientific, exotic, high-cost technology to a fast-growing industry serving homes, commercial and institutional buildings, irrigated farms, government agencies, and even public utilities with a renewable and nonpolluting source of electric power. Photovoltaic power installations range in capacity from a few watts for home, yacht, or hobby use to at least one generating plant with an output of 10 MW, feeding solar-generated electricity into the distribution network of a large public utility. Although large installations for industry and public utilities tend to claim the headlines, the photovoltaic industry is also heavily involved in providing electric power systems for remote locations and for unique applications—for example, operating signal lights (Fig. 18-2), navigational aids, remote microwave repeater stations, water pumping plants, and the electrical systems of yachts at sea or those of motor homes and house trailers. The application of photovoltaics to the power systems of homes and ranches in remote locations is currently a rapidly growing segment of the market for PV systems.

Unusual applications in recent years include a solar-powered aircraft (Fig. 18-3) and a solar-powered automobile. Such devices, though they suggest some

Figure 18-2 Solar power for remote signals. A 20W solar-electric module provides current to charge a Ni-Cad battery that energizes this wayside block signal on a railroad in Kansas. (ARCO Solar, Inc.)

SOLAR ENERGIZED ELECTRIC POWER

Figure 18-3 *Solar Challenger*, the first successful solar-powered aircraft, designed and built by Dr. Paul MacCready of Pasasdena, California. More than 16,000 solar cells arrayed on top of the wing and the horizontal stabilizer generate about 2.7 hp in flight. Its frame is constructed of lightweight, high-strength materials, and the skin is DuPont *Mylar*. Total weight of the aircraft is 217 lb. Its most famous flight to date took it across the English Channel in July 1981. It reached a height of 11,000 ft and a top speed of 47 mph. (E. I. duPont de Nemours and Co., Inc.)

interesting possibilities for the future, are not likely to become mainstays of the photovoltaic industry in the foreseeable future.

Cost Trends.   Improved designs, better semiconductor materials, and automated mass-production methods have all combined to bring down the cost of photovoltaic cells and systems over the past 20 years. For example, a photovoltaic system in the central region of the United States, installed on a residential roof, with its cost capitalized over a period of 20 years, can now (1984 estimates) produce electric power at a price within the range of 15 to 30 cents per kilowatt-hour, compared to the approximate 7 to 8 cents per kilowatt-hour charged by many electric utility companies. Continued production improvements and mass markets will undoubtedly bring much lower prices; and if fossil-fuel costs for conventional generation of electric power continue to rise, it may be only a few years until photovoltaic power will be price-competitive with other energy sources.

Some idea of the rapidity with which the cost of photovoltaic systems has been dropping in recent years is given by the chart of Fig. 18-4. In 1960, $200 per peak watt was common throughout the industry.[1] In 1984 the cost for a much-improved module system was in the vicinity of $9 per peak watt, and was decreasing.

The Solar-Electric Potential.   At this point another look at Fig. 1-10 is in order. Note that the 1985 energy gap between aggregate U.S. demand and total domestic production was estimated at more than 10 million barrels of oil equivalent per day (MB/DOE). This gap will grow wider by 1990, and at about that year it will begin to be alarming. By the year 2000 it could easily be 20 MB/DOE or more. Given the environmental and safety problems of the coal and nuclear options and the relatively small potentials of hydro, geothermal, ocean-thermal, and wind energy, the only truly viable option is solar energy. Although much of the solar contribution will be in the form of heat (see earlier chapters), an increasing amount will be solar-generated electricity, because electricity is the most convenient of all energy forms.

[1] The term *peak watt* refers to the operating mode in which the solar cell is exposed to the maximum solar radiation at perpendicular incidence.

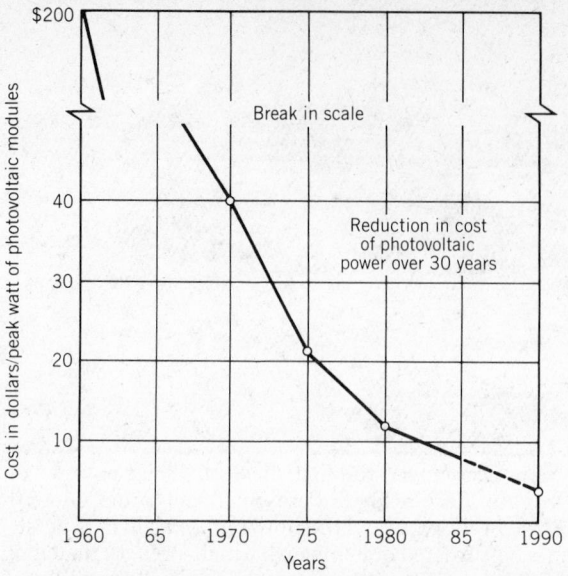

Figure 18-4 Chart showing the reduction in cost of photovoltaic modules over the past 25 years, with an estimate to 1990. (Data from various industry publications.)

Estimates of the solar-electric potential are as varied as the sources that publish them. Solar enthusiasts claim that up to 30 percent of the nation's electric energy needs should come from solar-electric generation by 2000. More conservative sources cite figures ranging from 3 percent to 10 percent.[2] No one knows for sure what the solar-electric contribution will be, but we do know that the technology is here, ready and waiting to meet the demand when the economic and politico-social conditions call for it.

Market Potential.   Total annual sales, U.S. and foreign, of photovoltaic cells and arrays amounted to about 4 MW of output in 1980. By 1990 the sales volume is expected to reach 25 MW, with a price tag of about $1 billion. The international market for photovoltaics is growing rapidly at present, and U.S. manufacturers currently have a large share of that business. Developing nations without centralized power networks and without petroleum or natural gas reserves are prime markets for photovoltaic power.

All predictions of the industry's future are, of course, subject to variables beyond the knowledge and control of the predictor. One statistic, however, does emerge with some clarity—when the cost of photovoltaic power gets down to $2 per installed watt, the industry will be truly competitive, and it will enter a sustained growth period lasting to the turn of the century and beyond. In the meantime, industry growth will depend to some extent on federal and state purchases and subventions and on the long-term capital investments made by private corporations with an eye to the future.

18-2 Photovoltaic System Components—
An Overview

The basic unit in photovoltaic power is the *solar cell*. A typical 4-in. diameter, single-crystal silicon solar cell weighs about 6 g, and the silicon "wafer" itself is about 0.015 in. thick. The wafer is provided with solder points and grid lines (thin strips of metal) to collect electrons and provide a conducting path for electron flow to and from the exterior circuit.

[2] For example, the President's Council on Environmental Quality (April 1978 estimates).

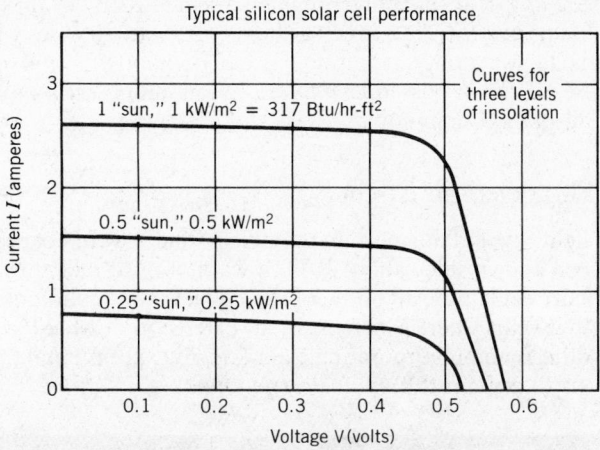

Figure 18-5 Curves of typical silicon solar cell performance for three levels of solar radiation intensity (insolation). For most areas of the United States and for clear-day
conditions at or near midday, an insolation of 1 kW/m²
(317 Btu/hr-ft²) is considered to be full sun. This value of
insolation is called "1 sun."

A 4-in. diameter round silicon solar cell in full sun (air mass $m = 1$), where
the insolation is of the order of 1 kW/m², will deliver about 2 amperes (amp) dc
at a potential difference of about 0.5 volts (V), for an approximate power output
of 1 W (Fig. 18-5).

Solar cells are laminated to glass and mounted in frames to form *solar
modules*, the number of cells in a module and the manner of connecting them
being determined by the current and voltage characteristics desired. The solar
module is the smallest, nondivisible, self-contained unit that incorporates interconnected PV cells to provide a dc electrical output. Standard modules from
one major manufacturer are illustrated in Fig. 18-6.

The module and its anodized aluminum frame rails are built to protect the
solar cells from impact damage and exposure to the elements. The cells are
incorporated into the module by being laminated into a "sandwich" of tempered
water-white glass, two layers of plastic, and a back cover of multiple layers of
polymers. Ten to 20 years of service is the standard expectation, even under
adverse conditions of high heat, intense cold, high humidity, rain, wind, and
hail.

Modules are combined to form *panels* (Fig. 18-6*c*) which are groupings of
two or more modules fastened together and electrically connected to form a unit
for ready installation in the field. Panels, in turn, are connected together to
form *solar arrays*, and arrays are the large movable units in huge *array fields* for
generating commercial amounts of electric power.

Balance-of-System Equipment.   Photovoltaic cells produce direct current
(dc), and they function only when the sun is shining. Since many users will want
alternating current (ac) and will also demand uninterrupted service day and
night and during clear or cloudy weather, photovoltaic systems must have
certain adjunct equipment, in addition to the photovoltaic modules or arrays
themselves. These adjunct or peripheral components are referred to in the
industry as *balance-of-system* (BOS) components. For a relatively simple system
such as that serving a residence, the BOS components might be as shown in the
block diagram of Fig. 18-7.

The solar array provides dc electricity to the storage batteries during sunny
hours. A charge regulator is provided for a multiple purpose—to guard against
overcharging or overdischarging of the batteries and to prevent the batteries
from discharging to the solar array at night. An alarm or meter with warning
light is provided to alert the user to any of several possible system malfunctions.

If there are any dc loads, power may be drawn directly from the batteries, but for ac loads an *inverter* is required to change dc to ac. In cases where a tie-in exists with an electric utility, additional BOS components and controls are required. These will be discussed in a later section, along with a more detailed treatment of all BOS equipment.

18-3 The Photoelectric Effect

The fact that light rays striking certain metallic surfaces would cause an electric current had been known since about 1839. It was not until 1905, however, that a satisfactory theory explaining all of the observed facts about photoelectricity was proposed. In that year Albert Einstein, in an extension of Max Planck's earlier quantum hypothesis and incorporating then-known experimental findings, proposed equations to explain the *photoelectric effect*.

Figure 18-6 Construction of solar modules. (*a*) Preparing the configuration of solar cells for heat lamination in glass. (*b*) A 1984 production solar module, consisting of 36 rectangular single-crystal silicon cells. This module is rated at 43 W under 1 sun, with an open-circuit voltage of 21.7 V. The overall efficiency is 11.5 percent under normal operating conditions. The dimensions of this module are 12 in. by 48 in. (*c*) Assembling modules into a photovoltaic panel. Note the connector posts on the back surface of each module, to be used in wiring the panel in whatever series–parallel arrangement is desired. (All photos, ARCO Solar, Inc.)

Figure 18-6 *(continued)*

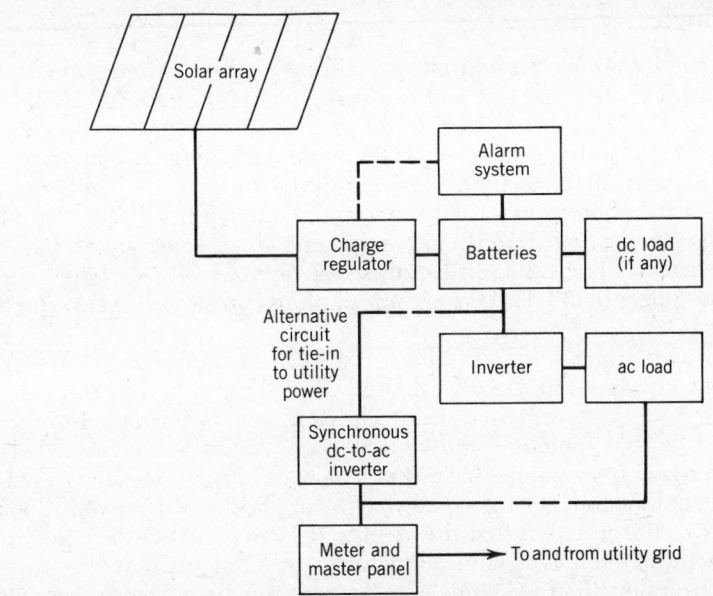

Figure 18-7 Block diagram showing the major components in a small photovoltaic system for a residence. The system could supply either direct current (dc) or alternating current (ac). An alternative circuit for tie-in to utility power is also indicated.

PHOTOVOLTAIC GENERATION OF ELECTRICITY

The following statements summarize a much-simplified concept of the photoelectric effect:

1. Light energy possesses dual characteristics: it sometimes acts as wave energy (as it travels) and it sometimes acts as if its energy is carried in the form of high-speed particles (on interaction with atoms).

2. In photoelectric phenomena, light is regarded as a stream of high-speed particles or packets of energy called *photons* (from the Greek, meaning "a particle of light").

3. Photons have differing energies, ranging from very high-energy photons to low-energy photons.

4. As a photon interacts with and imparts its energy to an electron in an orbit of a metallic atom, some of the photon energy is used up in doing the work of dislodging or ejecting the electron from its orbit, or, more precisely (in a semiconductor), the work of moving the electron from its *valence band* into a *conduction band*. This amount of energy is a characteristic of the particular metal, and it is called the *work function* of that metal—the minimum amount of energy required to move an electron out of its orbit in the metal atom.

5. Any photon energy over and above the work function energy will show up as kinetic energy of the emitted electron. If the photon's total energy upon arrival at the metallic atom is less than the metal's work function, no electron emission will occur.

6. Each electron ejected from a valence band acquires its energy from only one photon. The process is a one-on-one interaction.

7. High-intensity solar radiation (large values of insolation) means that more photons per second are striking the metal surface, resulting in more electrons being emitted from the metal.

8. High-energy photons are associated with light of high frequency and short wavelength, and low-energy photons with light of low frequency and long wavelength.

18-4 Electrical Properties of Silicon as a Semiconductor

In silicon, the minimum photon energy required to free an electron is 1.08 electron volts (eV). (An *electron volt* is a very small unit of energy, defined as that amount of energy gained or lost by an electron when it is accelerated through a potential difference of 1 V.) This amount of energy is that associated with infrared light of wavelength 11,550 angstroms (Å) or 1.15 microns (μ). About half of the photons in solar radiation have energies of 1.08 eV or more (wavelengths shorter than 1.15 μ) and are potentially effective in freeing electrons from silicon. The other half consists of photons of less than 1.08 eV (wavelengths longer than 1.15 μ), and these photons contribute no electron flow from silicon.

When light shines on pure silicon, valence electrons are ejected from silicon atoms to wander in the crystal lattice, but there is no electric field, or direction, or "sweep" to initiate a steady electron flow, or current. In order to provide such a field (potential difference), pure silicon is contaminated, or doped, with traces of elements that either supply or borrow electrons. There is an interface, or junction, between the two types of silicon created by the doping, and it is across this junction that the electric field sweeps the freed electrons. The silicon thus becomes a *semiconductor*, capable of maintaining a much greater electron flow than would be possible with the metal in its pure state.

The dopants most commonly used in making silicon solar cells are phosphorus, an element with five electrons in its outer orbit, and boron, with three electrons in its outer orbit. Silicon itself has four outer-orbit (valence) electrons (Fig. 18-8).

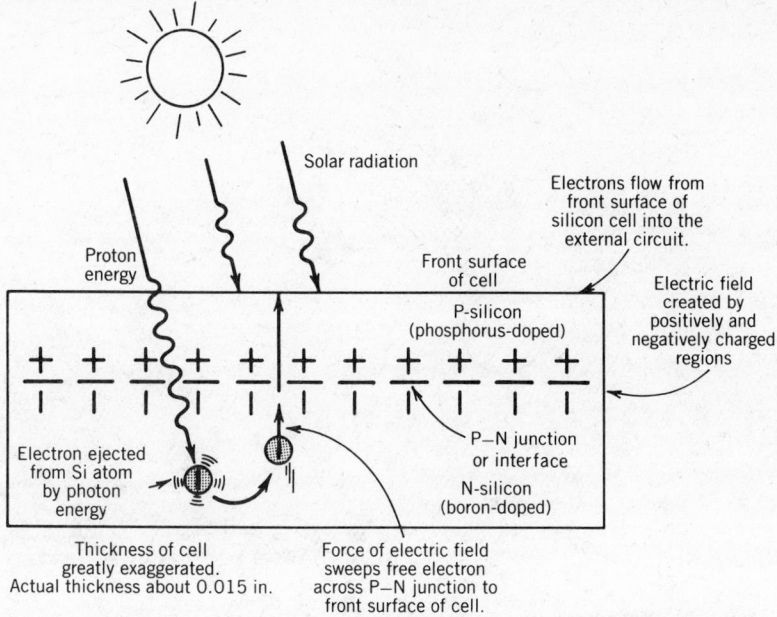

Figure 18-8 Diagrammatic representation of the action of the P–N junction of a silicon solar cell in "sweeping" freed electrons through the electric field at the junction to initiate electron flow, or current. *Source: ARCO Solar Training Manual*, vol. 1.

A phosphorus atom will share four of its outer-orbit electrons with four neighboring silicon atoms (we say it "bonds" with the four silicon atoms), but this leaves one electron left over, and it breaks free and becomes a conduction electron wandering in the silicon crystal lattice (Fig. 18-9a). Literally trillions upon trillions of these excess conduction electrons (negative charges) are present in phosphorus-doped silicon, and consequently it is called negative-type or N-type, or simply N-silicon.

When boron is the dopant, the boron atom also bonds with four adjacent silicon atoms, but now there is an electron lacking in one of the bonds. This lack of an electron acts just like a positive charge, so boron-doped silicon is called positive-type or P-type, or P-silicon. The position vacated by the electron is called a "hole". The concept of electrons and holes, then, is central to discussions of semiconductor theory. Holes not only act like positive charges, but they are also free to wander from atom to atom in the silicon crystal lattice (Fig. 18-9b).

At the interface where P-type and N-type silicon meet, the well-known P–N junction of semiconductor theory is created.[3] The junction, negative on one side and positive on the other, establishes the electric field that sweeps electrons in one direction through the cell and out and through the external load. The trillions of electrons that are freed as a result of photon interaction are pushed along into the external circuit, leaving positive charges or holes on the other side of the junction. If the silicon cell is now connected to an external circuit (load), when light strikes the cell, electron flow commences from the region of negative charge (high electron density) through the solder points and grid lines into the external circuit and then back again to the region of positive charge, where the holes are waiting to be filled with electrons. Electrons flowing through the external circuit can be put to work lighting lamps, energizing motors, operating communications equipment, or producing heat, or they can charge batteries for stored-up electric energy.

[3] The treatment of silicon as a semiconductor given here and in the following pages is elementary and greatly oversimplified. Space limitations do not permit a full discussion of semiconductor theory. Interested readers should consult any basic text on semiconductors.

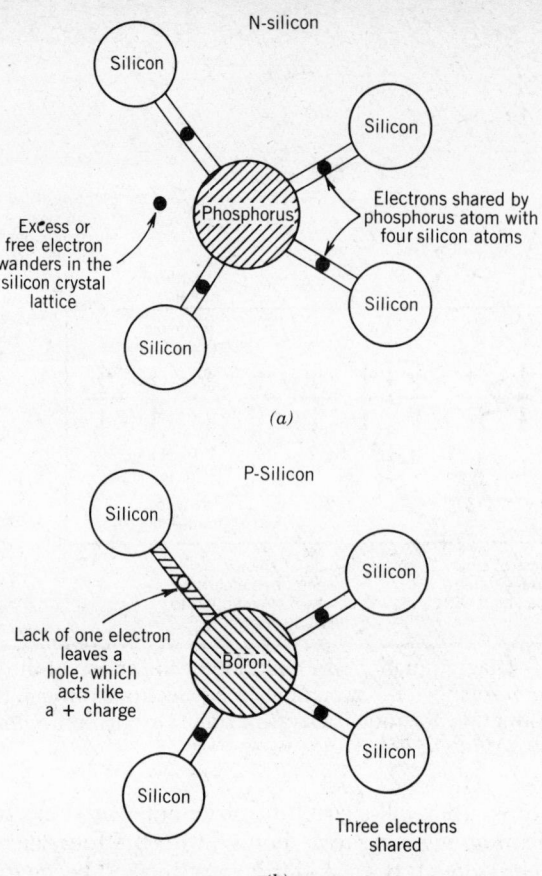

N-silicon

Silicon

Silicon

Electrons shared by
phosphorus atom with
four silicon atoms

Phosphorus

Excess or
free electron
wanders in the
silicon crystal
lattice

Silicon

Silicon

(a)

P-Silicon

Silicon

Silicon

Lack of one electron
leaves a
hole, which
acts like
a + charge

Boron

Silicon

Silicon

Silicon

Three electrons
shared

(b)

Figure 18-9 Schematic representation of the effects of doping silicon. (*a*) A phosphorus atom bonds with four nearby silicon atoms, and one free electron is left over to wander in the silicon crystal lattice. (*b*) Boron has only three valence electrons to share, so a vacancy, or "hole", is left. The hole acts like a positive charge. Holes are filled by returning electrons, and new holes are created by photon action, with the net effect that holes also seem to wander about in the crystal lattice. (Sketches courtesy of ARCO Solar, Inc.)

18-5   Single-Crystal Silicon

Silicon is the metallic element that, combined with oxygen, results in ordinary sand. In its purest natural form it is quartz or silica (SiO_2). Silicon is very plentiful in the earth's crust, but it is a costly process to obtain silicon of the requisite purity from ordinary silica (quartz), since the refining process has to result in 99.9999 percent purity. At this level of purity, the current (1984) price is about $100 per kilogram. Solar-grade silicon is purchased in rock form from metallurgical companies (Fig. 18-10), but in this form its crystals are not aligned. This form is called *polycrystalline silicon* or polysilicon. In order to improve the electrical properties of silicon, *single-crystal silicon* is produced. The polysilicon rocks are placed in an electric furnace with a trace of boron dopant. The batch is heated, melted, and maintained at a temperature of about 1400°C. At this point, a small seed crystal is lowered to the surface of the melt where the molten silicon begins to freeze on it. Under very precisely controlled temperature conditions, the growing seed is slowly pulled up out of the melt as it gradually grows to a diameter of 4 to 5 in. This ingot-growing method, known as the *Czochralski*

Figure 18-10 Solar-grade polycrystalline silicon "rocks", before being melted and doped with boron. (ARCO Solar, Inc.)

process, requires several hours, and when complete the ingot is a 4- to 5-in. diameter cylinder about 1 m long (Fig. 18-11*a*). This procedure results in a uniform crystalline structure throughout the ingot, with tiny pyramidal structures whose axes are all aligned with the axis of the cylindrical ingot itself. The ingot is now ready to be ground to a uniform diameter and sawed into "wafers".

A diamond-edged circular saw spinning at a high rate of speed saws one wafer after another from the ingot, under automatic control, cutting precisely in a plane perpendicular to the ingot axis (Fig. 18-11*b*). After being edge-ground to precise dimensions (round, hexagonal, or rectangular) the surfaces of the wafers are etched to remove saw damage and to delineate more sharply the small pyramids that help scatter and absorb the incident solar radiation.

The phosphorus dopant is now impregnated into one surface of the wafer by gaseous diffusion. This surface becomes the front (top) of the cell, where sunlight strikes. The wafer now has its P–N junction, and it is in effect an active solar cell, although there is as yet no means of collecting electrons and establishing electron flow.

The pattern or grid matrix of solder points and metallic grid lines is now bonded to the front and back surfaces, and the cell is ready to be assembled with many others into a photovoltaic module.

18-6   Electric Current from Solar Cells

In photovoltaics, we think of light as consisting of trillions and trillions of photons, or tiny packets of energy. Some photons have enough or more than enough energy to eject a valence electron out of a silicon atom (1.08 eV or more), and some do not. Remember that no matter how great a photon's energy is, it will eject only one electron—the one-on-one interaction that Einstein described as he explained the photoelectric effect. The excess energy from high-energy photons does not show up as usable electric energy; it is absorbed in the cell as heat, to be dissipated by conduction, convection, or reradiation.

Some light is reflected at the surface of the cell itself, and some is blocked by the solder points and grid lines and does not reach the silicon surface. Some photons pass completely through the cell and are absorbed on the back plate. As a result of these factors, there are a number of losses that occur in the light-to-

Figure 18-11 Preparation of "wafers" for 4-in. round silicon solar cells, (a) A silicon ingot as it is being slowly drawn from the melt. Note the nearly perfect cylindrical shape. When completely drawn, cooled, and polished it is a right circular cylinder about 5 in. in diameter and 30–36 in. long. (b) The cylinder is sawed into thin wafers about 0.015 in. thick. The stacks of wafers seen here have just emerged from the automatic saw and are ready for final shaping and surface smoothing, after which the phosphorus dopant will be added. (ARCO Solar, Inc.)

Table 18-1 Efficiency Losses in Mass-Produced Silicon Solar Cells (approximate)

Source of Energy Loss	Unconverted Solar Energy (%)
Photons with energies below 1.08 eV	25
Photons with excess energy	31
Reflection from front surface	4
Blocking by connectors on the front surface	4
Photons passing through cell	4
Other internal losses, inherent in the crystal structure	18
	86

Source: Solar industry technical data from several manufacturers.

electricity conversion process. The sources of these losses and their approximate percentages are given in Table 18-1.

As Table 18-1 indicates, mass-produced single-crystal silicon solar cells of current manufacture (1984) can operate at efficiencies as high as 14 or 15 percent, but actual modules and arrays convert to electricity only about 10 to 12 percent of the solar energy available on their total surfaces because of the unused areas between cells. Even with hexagonal and the newer rectangular cell shapes, the unused areas of an array account for losses of 2 to 3 percent.

The electrical response of a solar cell to the entire range of the solar spectrum is portrayed approximately in Fig. 18-12. Note that the electrical response begins near the blue end of the spectrum at a wavelength of about 4000 Å (0.4 μ), rises to a maximum at about 9000 Å, and drops back to near zero in the near infrared region at about 11,500 Å, or 1.15 μ.

The working efficiency of a solar module, panel, or array is determined by dividing the actual electrical power output by the measured insolation (solar power) on the total surface of the photovoltaic unit. The following problem illustrates the method.

Illustrative Problem 18-1

During a one-hour test when the insolation (perpendicular incidence) on a 12- by 36-ft solar array is constant at 310 Btu/hr-ft², the average electric power output of the array is measured as 4.5 kW. Calculate the overall efficiency of the array.

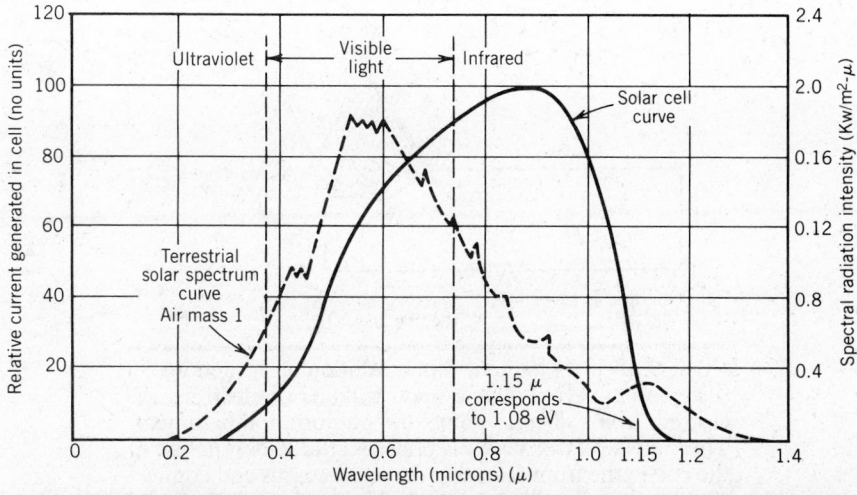

Figure 18-12 Electrical response of a silicon solar cell to the electromagnetic radiation of the solar spectrum. Note that a photon must have an energy of 1.08 eV (corresponding to a wavelength of 1.15 μ) in order to free electrons from silicon atoms. (Data on solar cell response from various industry specification sheets.)

Solution

From Table 4-3, 1 Btu/hr-ft² = 0.293 W/ft²

The insolation in watts is therefore

$$I_{watts} = 90.8 \text{ W/ft}^2 \times 12 \text{ ft} \times 36 \text{ ft} = 39,225 \text{ W}$$
$$= 39.23 \text{ kW}$$

$$\text{Eff} = \frac{4.5 \text{ kW}}{39.23 \text{ kW}} = 0.115$$
$$= 11.5 \text{ percent} \qquad\qquad Ans.$$

How the Silicon Solar Cell Works.   As photons enter the cell, most of them pass through the thin phosphorus-doped front layer (N-silicon) and are absorbed in the main body of the wafer, the boron-doped (P-type) silicon. For every free electron produced by photon absorption, one hole is created, and the hole acts as a positive charge, with trillions of them wandering and waiting in the P-silicon layer.

The free electrons also wander until they are swept into the electric field at the P–N junction. There they are accelerated through the field, losing some of their energy in collisions in the crystal lattice. Under usual conditions, however,they will have a remaining potential of about 0.5 V after passing through the P–N junction. Their kinetic energy and the electric field move them to the front face of the cell, where solder points and grid lines provide conducting paths to the external circuit (Fig. 18-13). The front of each cell has one or more interconnector strips or ribbons, which are connected in series to the back side of the next cell in the module, and thus the electron flow gains 0.5 V of electrical potential as it transits each cell. Series connection of the cells allows for module or array construction that will provide current to the load at any desired voltage.

Electrons return from the load to the back of the cells and fill the waiting holes in the P-silicon, thus completing the circuit. Nothing is used up, nothing

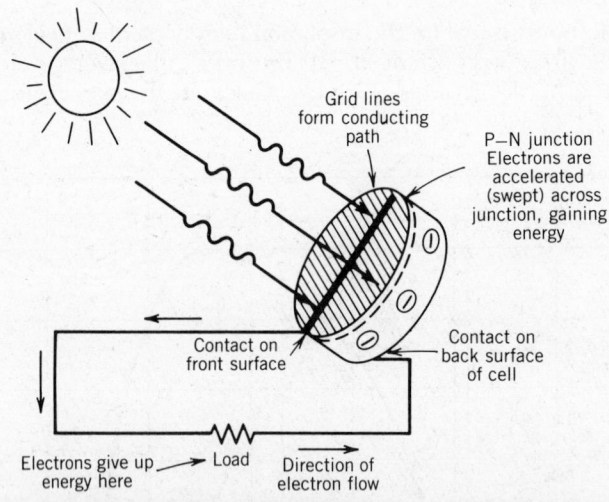

Figure 18-13 Schematic representation of the action of a silicon solar cell. Trillions upon trillions of electrons are ejected from silicon atoms by photon bombardment. These free electrons are swept across the P–N junction of the cell to the front face, where metallic grids and connector ribbons collect them and provide a path to the external circuit and the load. They return to the back surface of the cell and fill waiting holes in the boron-doped silicon. *Source:* Sketch idea from ARCO *Solar Training Manual,* vol. I.

wears out, there is no noise or pollution, and the only side effect or byproduct is a small amount of heat, which can be readily dissipated to the ambient air or, in some "total energy" systems, collected and used.

18-7   Other Types of Solar Cells

Although single-crystal silicon represents the current workhorse of the photovoltaics industry, other types of silicon cells and cells made of other semiconductors are also undergoing test and development, and some are actually on the market.

Current prices of single-crystal solar cell arrays are in the general range of $9 to $10 per peak watt (1984). Industry analysts estimate that this cost can be reduced to $2 to $3 per peak watt if a method of producing a thin continuous sheet of silicon by automated processes with no wasted material can be perfected. Other proposals for bringing costs down include using a less expensive form of silicon and a continuous dip process for laying a thin film of silicon on a ceramic base.

Polycrystalline Silicon.   The Czochralski process of drawing ingots of single-crystal silicon is both very costly and energy-intensive, and a great deal of research and development is currently taking place with a view to manufacturing acceptable solar cells from polycrystalline silicon. Silicon rocks can be melted and cast into a polycrystalline ingot at much lower cost than is required to produce a single-crystal ingot. However, to offset their lower cost, polycrystalline silicon solar cells are less efficient in converting solar radiation to electricity, typical efficiencies running about 6 to 10 percent, instead of 12 to 15 percent.

Ribbon-Growth Silicon.   Some photovoltaic firms are experimenting with a process that pulls a ribbon of silicon out of a graphite or ceramic die set in molten silicon. The ribbon (in one process) is about 12 cm wide and 0.25 mm thick. It is cut into rectangular sheets of a length determined by the desired cell size. These sheets are then doped and made into cells as before.

Thin-Film Deposition.   Another promising development being pursued is a method of depositing a thin film of silicon on a ceramic base or substrate. Under precise control, a thin film of silicon of uniform thickness is deposited on the substrate. Less than one fourth of the silicon material used for single-crystal wafers is required, and much less energy and time. So far, cells made from thin-film deposition have attained efficiencies of only 7 to 9 percent, but with the sharp decrease in production costs and encouraging signs of increased efficiency in the future, the method may be the industry standard in a few years. Proponents of thin-film deposition predict that solar cells of this kind will be on the market at less than $1 per peak watt by the end of the 1980s.

Other methods based on silicon are in various stages of development, but space limitations preclude their discussion here. It is probably safe to say that by 1990 photovoltaics will be well advanced into processes that are far more elegant and less costly than the single-crystal silicon wafer approach that dominates the industry today.

Solar Cells Using Other Metals.   A number of materials other than silicon are being used for solar cells. Some of them show a great deal of promise and are already in limited production, while others are in the research and development phase. Table 18-2 lists some of these materials, with a rough estimate of their efficiency in converting solar energy to electric energy.

These theoretical efficiency maximums are misleading in the sense that they give no indication of such factors as cost of manufacture, longevity and dependability of the cell, availability of raw materials, freedom from hazard in manufacture and use, and device stability. In noting that they all show promise

Table 18-2 Photovoltaic Materials and Conversion Efficiency

Material	Maximum Theoretical Efficiency
Cadmium sulfide	15–18
Cadmium telluride	20–22
Copper indium selenide— cadmium sulfide (thin film)	9–11
Gallium arsenide	25–27
Germanium	11–13
Indium phosphide	24–26
Silicon	20–25

Source: Industry technical data.

for the future (as do some other materials), it must at the same time be reemphasized that for now (mid-1980s) single-crystal silicon is the industry standard.

Concentrator Cells and Arrays.   Because of the high cost of solar cells and the rather low intensity of solar radiation, some manufacturers are producing *concentrating* photovoltaic arrays. The cells themselves are of a special design, capable of absorbing the solar radiation that would ordinarily fall on 20 or more cells. These 20-sun cells are now made of single-crystal silicon or gallium arsenide and are fitted with a very dense grid pattern and more efficient connectors to carry away the increased electron output.[4] The doping process is also much more carefully controlled. The potential difference produced is still about 0.5 V per cell, but the electric current produced is greatly increased. For example, one such cell of current manufacture can produce 2 W of power per square inch, compared with the 1 W produced by a standard 4-in. diameter (area 12.5 in.^{2}) cell. This 25 to 1 ratio is truly significant, even when increased cell costs and the cost of concentrators and sun trackers are taken into account.

The concentrators themselves take two forms: (1) parabolic reflectors or troughs that bring sunlight to a focus on a row of cells along the focal axis,

[4] Some manufacturers are marketing concentrator configurations that result in cells absorbing 100 suns.

Figure 18-14 A bank of high-efficiency silicon solar cells, with the sun's rays concentrated by Fresnel lenses. This installation provides solar-electric power for the joint United States–Saudi Arabia SOLERAS program. (Applied Solar Energy Corporation.)

and (2) Fresnel lenses, which can be either linear-type, focusing sunlight along a linear axis, or point-focus-type, focusing solar energy on a single cell (Fig. 18-14).

Under the intense radiation from the concentrator, the cells get hot, and the excess heat must be dissipated. Various heat-dissipation methods are used, including simple fins, forced air, or circulating water. Some methods collect and use the heat, thus improving the overall efficiency of the system. These are called total energy systems.

Sun-Tracking Systems. Concentrator arrays must track the sun in order to focus the sun's rays directly on the cells. Up to 50 percent more energy annually can be obtained from a given array by having it track the sun in both elevation and azimuth. This kind of daily and seasonal traverse is called *two-axis tracking.* It provides the maximum annual solar radiation to the array of cells, but the mounting and mechanism are very expensive, and cost–benefit ratios must be carefully evaluated. (Fig. 18-15).

Single-axis tracking often provides a good compromise between capital costs and energy delivered. The array is faced due south in the northern hemisphere, set at a permanent tilt angle, optimum for the site latitude, and daily tracking for azimuth only is provided (Fig. 18-16).

Concentrator systems that track the sun can result in much-improved efficiencies, smaller areas to be covered by arrays, and byproduct hot water or hot air for a total-energy system. On the negative side, they involve high capital costs, high-technology maintenance, and some safety hazards. Their principal market tends to be in industry for the commercial generation of electric power for manufacturing and for electric utilities.

PHOTOVOLTAIC POWER—SYSTEM COMPONENTS

Solar cells are offered by many manufacturers in a wide variety of shapes, sizes, and configurations. Some are round, some hexagonal, and some rectangular. All single-crystal silicon cells deliver about the same voltage (0.45 to 0.50 V), but

Figure 18-15 Bottom side of a large PV array mounted on a two-axis tracking mechanism. Tracking in both azimuth and elevation is provided on a daily basis. Frequent corrections are also made for seasonal changes in the sun's declination. The entire mechanism is controlled by a computer program. (ARCO Solar, Inc.)

Figure 18-16 A photovoltaic array built for single-axis tracking of the sun. This array is mounted on a fixed polar axis (pointed at Polaris, the North Star). Twenty modules are mounted in the array, and each is fitted with reflectors on both sides to increase the amount of solar radiation incident on the cells. The tracking mechanism moves the array in azimuth at one-minute intervals, keeping the solar cells pointed directly at the sun all day long. (Solarwest Electric.)

the current (amperage) depends on the size of the cell and on the intensity of solar radiation falling on its front surface. In full sunlight at solar noon, the 4-in. round or hexagonal single-silicon cells typical of recent production deliver about 2 amp at 0.5 V for a power output of 1 W. The operating characteristics of such a cell were diagrammed in Fig. 18-5. Newer rectangular cells have about the same output per cell, but they can be assembled into PV modules with almost no waste space between cells, and module outputs are much improved as a result (Fig. 18-6b).

18-8 Cells and Modules.

In order to increase voltage, solar cells are connected in series, the interconnecting leads or ribbons passing from the front surface of one cell (N) to the back surface of the next cell (P). The voltage from a series-connected module is then equal to the number of the cells in the module multiplied by (approximately) 0.5 V.

To increase the current or amperage, cells are connected in parallel, front surface to front surface and back surface to back surface. Various series–parallel combinations are used to provide modules with the desired voltage and current characteristics.

Cells are connected together and then heat-laminated into weathertight units called *modules*. With 30 to 40 cells connected in series, a module will deliver, under full sun, about 33 to 43 W at about 16 V, with a direct current of about 2 amp, suitable for charging 12-V batteries, where the charging voltage required is 13 to 15 V. If more current is needed at this voltage, several of these modules are connected in parallel until the power output meets the load demand, still at a voltage of 12 V. If higher voltages are needed, more modules or panels are connected in series.

Most manufacturers use multiple contact points on each cell, thus providing redundant or parallel current paths. Thus, if any cell is damaged or cracked, an open circuit in the module will not result. The multiple contact points provide a conducting path around the break or crack to bypass the open-circuit condition.

Modules are mounted in extruded aluminum frames (Fig. 18-6c) for protection and easy handling. The power output terminals are solder-plated and impact-tested to ensure long and trouble-free service. Junction boxes are provided on high-voltage modules to ensure safety and to provide a weatherproof center for interconnecting wiring.

Packaged solar modules are easy to handle and install. They are designed and built for outdoor installation on rooftops or hillsides or on free-standing mounts. All they need is a rugged supporting structure that will withstand wind and snow loads and that will last for the life of the module—at least 20 years. Wiring them to the load is a simple task, but it must conform to local codes.

18-9 Installing Arrays

Several modules are connected together to form panels and arrays. Arrays may be mounted on rooftops either directly, with the same slope as the roof, or on specially constructed racks or frames at any desired tilt angle. Very large arrays for commercial power production must be mounted on carefully engineered and precision-constructed mounts. These installations are ordinarily provided with sun-tracking mechanisms. If tracking is not used, the array should be oriented due south, and (for year-round use) a tilt angle equal to the latitude should be provided (Fig. 18-17). If the system is primarily for winter use, a tilt angle of latitude plus 10° to 15° is optimum; and for summer use, a tilt angle of latitude minus 10° to 15°.

Special hardware and fittings are required to install large arrays. Wind loads and snow loads on such large surfaces are tremendous at times. Hailstorms also may be a problem. Vandalism sometimes occurs, and precautions must be taken to prevent injury to anyone (including trespassers) who might come into contact with the system. On systems with sun tracking in elevation, provision may be made to stow the arrays in a vertical position at night, to allow for flow of condensed moisture or dew off the front surface of the array. This water flow carries with it accumulated dust, resulting in a self-cleaning array system.

The installation and wiring of large arrays is a job for trained riggers and electricians. Modules, panels, and small arrays, however, can be installed quite readily by "do-it-yourself" operators if manufacturers' instructions and local codes are carefully followed and if the installer has the necessary tools and the

Figure 18-17 A 10-kW nontracking array used to supply electric power for a water-pumping station at Sadat City, Egypt. The array faces due south and is mounted at a permanent tilt angle equal to the latitude—a compromise for year-round use. (ARCO Solar, Inc.)

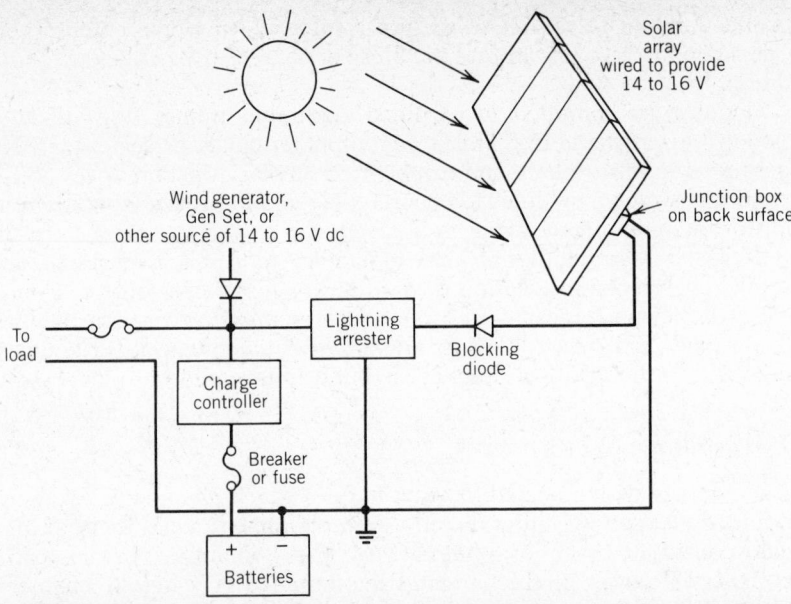

Solar array wired to provide 14 to 16 V

Junction box on back surface

Wind generator, Gen Set, or other source of 14 to 16 V dc

Lightning arrester

Blocking diode

To load

Charge controller

Breaker or fuse

+ − Batteries

Figure 18-18 Wiring diagram for a small dc photovoltaic system. Provision is made for input power from a wind generator or a diesel-generating set.

skills required to use them effectively. As a general rule, for the typical home-owner, it is probably best to engage the services of a licensed solar energy contractor, both for the design of the system and for its installation.

Wiring a Small Array. Electrical connections between junction-box-equipped modules and arrays should be made with number 10, multiple-conductor neoprene-jacketed cable. For open-terminal modules, number 10 single-conductor Teflon-coated wire is recommended. A simplified wiring diagram for connecting a small array to the balance-of-system (BOS) equipment is shown in Fig. 18-18. The *blocking diode* prevents discharge of the batteries into the solar array at night. A fuse as close to the battery terminals as possible is essential to protect the wiring from the heavy surge of current that would come from the batteries if a short-circuit developed.

18-10 Types of Photovoltaic Systems

Photovoltaic power systems can be classified on the basis of the source of energy into two categories:

1. *Stand-alone systems*, in which the only source of energy available to the system is photovoltaic energy.
2. *Combination or hybrid systems*, in which some other source of energy is tied in with the PV system, such as a wind generator, a diesel generator, a hydroelectric generator, or a power company's electric line. In the case of a PV system interfaced with an electric power line, the total system is called a *grid-connected* system. In this case, the system may use the power company as a source of backup power, or (with proper equipment) it may supply power to the electric company's grid when excess solar-generated power is available.

Markets for Photovoltaic Systems. There are five principal markets for photovoltaic equipment and systems, classified as follows: commercial, low-power remote (industrial), water pumping and delivery, home electrification, and central power stations (utility level).

Figure 18-19 Photovoltaic power for a mobile radio repeater station near Casper, Wyoming. Remote locations, where maintenance is difficult, are ideal applications for PV power. (ARCO Solar, Inc.)

The commercial market is mainly for very small solar cells or panels for photographic equipment, digital watches, electronic calculators, and audio–video equipment.

The industrial market encompasses applications that require a small amount of power in a remote location, such as telecommunications systems, space vehicle and satellite power systems, navigational aids and other warning signals, and systems for the cathodic protection of underground pipelines and structures (Fig. 18-19).

Pumping and delivering water in remote areas is an ideal application for PV systems. In many developing nations, there is no rural electrification, diesel fuel is scarce and costly, and maintenance personnel are not available. In such instances, PV systems are the most favored method of providing water for domestic and agricultural use (Fig. 18-17).

Home electrification is one of the major markets for photovoltaics. For now, the emphasis is on remote locations not served by an electric power line. When photovoltaics becomes economically competitive with grid-distributed electric power, however, this market will no doubt become a major segment of the industry. Included in this market are systems for schools and colleges, institutional and office buildings, and mobile living units such as recreational vehicles and yachts, as well as remote homes and small villages (Figs. 18-20 and 18-21).

Figure 18-20 Photovoltaic array providing electric power to apartment complex in Pearl City, Hawaii. Note also the flat-plate solar collectors for the domestic hot-water system.

Figure 18-21 Racing sloop equipped with a PV module designed for marine use. It provides power to the yacht's electrical systems and keeps the batteries charged for night and cloudy day operation. (Solarwest Electric.)

Central power stations (utility scale) constitute a relatively new field for photovoltaics. At \$8 to \$10 per peak watt, the PV electric generating systems now installed and operating are definitely not cost-competitive, but are regarded as pilot operations for research and development. With the possibility of a module cost of \$2 per peak watt within five years, a number of public utilities are already putting photovoltaic plants on-line, to gain valuable experience with an energy source that, in the future, may be cheaper and more readily available than oil or natural gas. To date, such systems have been heavily subsidized either by the electric utility itself or by some combination of corporate and government subsidy (see section 18-18).

18-11   Storing and Using Photovoltaic Energy

A photovoltaic module or array delivers electric energy only when the sun shines. But the need for electricity may exist around the clock and on cloudy days as well as sunny days. Consequently, electricity must be produced when solar radiation is plentiful and stored for later use when there is no sunlight (Fig. 18-22).

Several storage systems are possible, including mechanical storage systems, hydrogen gas storage, and storage battery systems. The first two are largely experimental at present, and only a word or two of explanation about them will be given before turning to a more detailed discussion of battery storage.

Mechanical Storage.   Two examples will be cited. (1) PV energy can be used to pump water to an elevated reservoir or tank, where the water and the potential energy gained are stored until either or both are needed. (2) Sophisticated flywheel systems have been proposed, and working models have been in use for years. PV energy drives an electric motor at 95-plus percent efficiency to spin a heavy flywheel at high rotative speeds. Energy is stored in the flywheel as rotational kinetic energy, to be drawn off as needed either as mechanical energy or through reconversion by means of a generator as electric energy. Mechanical storage ordinarily involves complex systems and ponderous equipment, not at all suited to homes or other small- to medium-capacity installations.

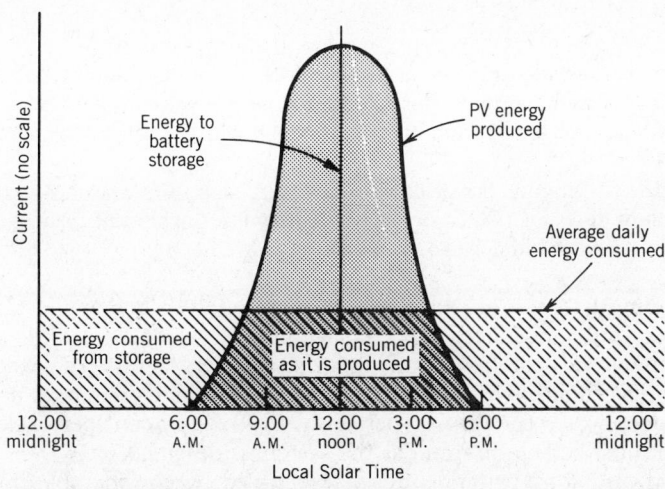

Figure 18-22 Daily photovoltaic energy production and daily energy consumption for a solar-energized residence, shown in chart form. Energy consumed is represented by an average current "curve". Solar-electric energy produced is represented by the bell-shaped curve, which is assumed to be symmetrical around solar noon.

Hydrogen Gas Storage. Water is readily separated into its constituent gases —hydrogen and oxygen—by passing low-voltage dc through weakly acidified water in a suitable cell. The hydrogen produced can be stored indefinitely (some energy will be required to compress it) and later used as needed. Hydrogen is an excellent fuel, and it can be reconverted to electricity by the use of fuel cells.

Hydrogen storage is quite inefficient if the desired end product is heat. For example, PV arrays are 10 to 12 percent efficient in converting solar energy to electricity; the electricity-to-hydrogen process represents another small energy loss; and when hydrogen is burned to heat water or air, the combustion loss is considerable. In contrast, heating water by flat-plate solar collectors is accomplished all in one operation (solar to hot water) at efficiencies of 35 to 50 percent.

18-12 Batteries for Electric Energy Storage

Most of the photovoltaic systems now in operation use storage batteries to store the energy generated. A storage (or secondary) battery changes electrical energy into chemical energy which is stored until a reversible chemical reaction yields a flow of electrons again.

Storage batteries can be charged and discharged again and again, and one of the important criteria for a PV-system battery is the *cycle life*—the number of charge–discharge cycles it can provide before it will no longer deliver energy at the system-design level. Cycle life is meaningful only when one also knows the depth of discharge (DOD) of each average daily cycle, say 15 or 20 percent for shallow-cycle batteries or 50 to 80 percent for deep-cycle batteries.

Among the types of batteries used for PV systems are lead–acid batteries, nickel–cadmium (Ni-Cad) batteries, and nickel–iron (Edison) batteries. The latter two have certain advantages related to long life and reduced maintenance, but their disadvantages—lower voltage per cell (1.2 and 1.1 V per cell, respectively), and high cost per ampere-hour—are such that they find limited application in the photovoltaic industry.

Lead–acid batteries (2 V per cell), despite their weight and the problems associated with acids and electrolytes, are the preferred storage for present-day PV systems. The term *lead–acid* is a generic classification. Actually, several types of lead are used—lead–calcium, lead–antimony, and pure lead. Among the manufacturers of batteries suited for photovoltaic system service are Delco, Exide, Globe, Willard, Varta, C. and D., and Trojan.

Rating criteria for batteries used in photovoltaic systems are summarized herewith:[5]

> *Self-discharge rate*—the rate at which a battery will discharge by itself, not supplying useful current. This rate should be as low as possible.
> *Maximum charge rate*—the maximum rate at which a battery will accept a recharge.
> *Cycle life*—the number of charge–discharge cycles to a specified percent depth of discharge (DOD) before the battery loses significant capacity.
> *Cost*—expressed in dollars per kilowatt-hour.

Lead–calcium batteries are adaptable to most photovoltaic applications. They bubble very little and rarely need replacement of water or electrolyte. Where maintenance-free operation is an important design factor, lead–calcium batteries are first choice. They do not accept as high a rate of charge, nor do they cycle as deeply as some other types (only 10 to 20 percent per cycle), so this limitation must be kept in mind as the system is designed.

Lead–antimony batteries are characterized by considerable bubbling of the electrolyte and will need occasional water replenishment and perhaps special venting precautions in the battery room. These limitations are offset by the ability to accept a high rate of charge and by the long cycle life at deep DOD (50

---

[5] These criteria and much of the accompanying discussion are drawn from *ARCO Solar Training Manual*, vol. I, pp. Bt-2 to Bt-5. Adapted with permission.

to 80 percent per cycle). These batteries are adaptable to many photovoltaic applications.

Pure-lead batteries are suited only to very shallow-cycling applications (5 to 10 percent per cycle). Their best use is in standby power systems that are discharged only for emergency power. They should not be regularly cycled below a DOD of about 10 percent. Their maintenance requirements are minimal, and they find application in remote locations for communications and cathodic protection, within the cycling limits cited above.

The final decision on battery selection for a particular PV application is made on the basis of optimizing system criteria and battery characteristics. Tradeoffs must be carefully analyzed. Detailed specifications on batteries, their suitability for various photovoltaic applications, their service needs, voltage and current characteristics, ease of transportation, warranties, and installation instructions can be obtained from the technical manuals of the battery manufacturers or from specifications prepared by the manufacturers of photovoltaic systems.

18-13   Home Electric Systems

More than one third of the world's people live without electric power. Even in the United States, the homes of many people are not served by an electric utility. Home electric systems constitute a major segment of the world market for photovoltaics. Home systems can be classified under four headings:

1. Stand-alone dc systems.
2. Stand-alone ac systems.
3. Hybrid (combination) systems—photovoltaic power combined with or tied into a diesel, wind, or hydro-generating system.
4. Grid-connected systems—photovoltaic power tied into an electric power grid. This kind of system may be one-way only, drawing power from the grid as needed; or two-way, with the capability of feeding power to the grid when there is excess solar electric power.

Each of these applications will be briefly discussed, along with the associated balance-of-system (BOS) equipment.

Stand-Alone Home dc Systems.   A typical small home electrification system might include the following equipment:

6	photovoltaic modules of 35 to 40 cells each, providing a total of about 220 peak watts of power at solar noon.
6	12-V batteries, probably lead–calcium type.
1	charge controller (voltage regulator).
1	meter–control panel. Connecting wire, clips, cable, mounting structures, and hardware.

A system with these components could, depending on the insolation at the site, provide up to 1 kW-hr of energy per day, enough to operate lights, radio or a music system, and intermittent use of some small dc-operated tools or appliances.

Stand-Alone Home ac Systems.   Most homeowners will need more energy than the above-described system will provide, and most users in the United States will want 110 to 120 V, 60-cycle power, since that is the standard in this country. For example, a system including the following equipment could provide about 700 peak watts of power and 3 kW-hr of energy per day in a location in the U.S. Southwest:

Figure 18-23 A portion of the photovoltaic array serving the electrical power requirements of the Natural Bridges National Park visitor center in southeastern Utah. At 100 kW, this installation was the largest stand-alone ac system in operation in the early 1980s. (Photo courtesy of ARCO Solar, Inc.)

24	photovoltaic modules of 35 to 40 cells each, providing a total of about 700 peak watts of power at solar noon.
24	12-V batteries, lead–calcium type.
1	charge controller.
1	meter–control panel.
1	dc-to-ac static inverter.
	Connecting wire, clips, cable, mounting structures, and hardware.

An ac system of this size could operate lights, television, stereo, and small ac appliances, within the 3 kW-hr/day capacity of the modules.

Remote ranches and government installations may require much larger stand-alone ac systems. A fully equipped and mechanized ranch in a remote region might require up to 50 kW-hr/day. An example of a really large stand-alone ac system is the installation at the Natural Bridges National Park in southeastern Utah, which has a capacity of 100 kW of peak power (Fig. 18-23).

Hybrid Systems. Either of the above types of systems can be combined with a wind generator, a diesel generator, or a small hydro generator. A solar array, charge controller, and batteries are provided, as before. In addition, a *battery charge controller* for the mechanical generator is needed so that it, too, can charge the batteries. If the system is to provide ac, an inverter is also needed. During sunny hours, the PV array charges the batteries. The main elements of a PV system to be tied in with an existing diesel generator are shown in Fig. 18-24.

If wind or hydro power is available, this energy can drive a generator to charge batteries. Or, if a diesel generator is the auxiliary, it will kick on as dictated by the controller to keep the batteries charged. The diesel generator (*gen-set* is the term used in the industry), when it is running, should be operated at or near full power, since it is much more efficient at that rating. A hybrid system with a diesel generator backup can give excellent service. The user gets quiet, pollution-free, fuel-free power during most of the year, and uses the gen-set only to carry through the severest months of winter. If a strictly photovoltaic system were designed for worst-case winter conditions, it would be significantly overpowered for most of the year, with poor amortization of capital costs.

Figure 18-24 A "package" of photovoltaic system equipment to tie in with an existing diesel generator. Included in the package are solar modules with a peak output of 720 W, deep-cycling storage batteries, a dc-to-ac inverter, a battery charger, metered control panel, and transfer switch. (Solarwest Electric.)

Grid-Connected Systems. Some homeowners, even though electric utility power is readily available, prefer to generate their own electricity, with the luxury of utility power as a backup. An application of this kind is also called a *line-tied system*. At current costs for photovoltaic systems, this option is not really cost-effective, but it may become so in the relatively near future.

With a line-tied system, a special line-tie synchronous inverter is used. It not only converts dc power from the PV array to ac power, but it also synchronizes its ac output (that is, brings it into phase) with the ac power on the utility company's lines. Batteries are not needed, and no energy is stored in the system. PV power is used to the extent that it is being generated (sunny hours), and any shortfall is made up by power drawn from the line tie to the utility grid. If more PV power is being generated than the consumer is using, it is synchronized, metered, and fed into the grid. The power company credits the owner's account for this energy.[6]

Line-tied or grid-connected systems are usually of medium to large capacity. Owners of these systems think in terms of 3 to 15 kW of peak power—enough to operate all lights, appliances, hobby shop and/or tools, perhaps a domestic water system, perhaps summer air conditioning, and some excess to sell to the power company on sunny days to partially offset power drawn from the grid at night and in cloudy weather.

A typical line-tie installation might consist of an array of 90 solar modules of 35 to 40 peak watts each, together with a synchronous inverter, a lightning protection and ground unit, a PV power output meter, and associated wiring

[6] The Public Utilities Regulatory Policy Act of 1978 (PURPA) requires that utility companies buy excess power supplied to their grids from renewable sources such as solar, wind, and hydro. The rates paid are set by the public utilities commissions of the states.

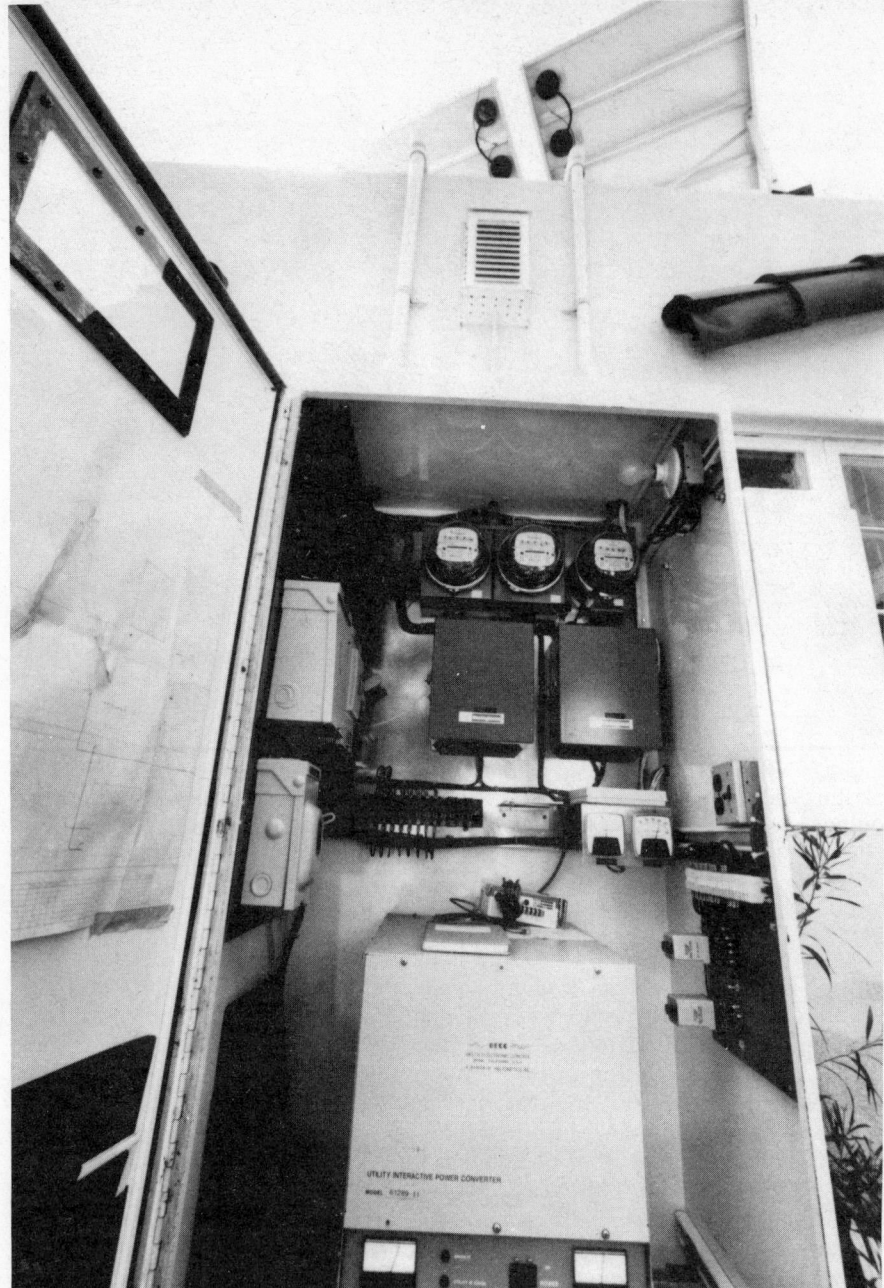

Figure 18-25 Control panel (cover opened) for a residential line-tie PV power installation. Note meters at top and utility-interactive power converter at bottom. (Solarwest Electric.)

and equipment for mounting the array. Larger systems of this type may feature a polar-axis tracker to ensure all-day maximum insolation on the solar array. Figure 18-25 illustrates some of the equipment needed in the control panel of a residential line-tie system.

PHOTOVOLTAIC SYSTEM DESIGN

It has been noted that PV cells and arrays are rated in terms of output in *peak watts*—that is, the energy rate produced on a clear day at solar noon on a south-facing surface tilted at the optimum angle for the site latitude. Obviously, peak-watt capacity is not a parameter to be used in designing a system that must deliver electric power day and night, on clear and cloudy days, throughout the year. A more meaningful figure is the expected energy output of the module or array, expressed in kilowatt-hours over a stipulated time interval, such as kilowatt-hours per day, per month, or per year.

We turn now to an analysis of electrical loads and the design of photovoltaic systems to match estimated loads. The discussion will, for the most part, center around residential and small to medium commercial or institutional systems. The methods suggested can readily be modified to apply to systems for other applications. The technical manuals and specification sheets of the photovoltaic equipment manufacturers contain detailed design criteria and instructions for many types of specialized systems.

18-14   Determining the Electrical Load

First, the energy requirement and nature of the electrical load must be determined. The client will decide whether ac is necessary or if dc will be satisfactory. Whether or not the system should be designed for 100 percent of the load depends on whether or not a power company service line is available, or whether supplemental power from wind, hydro, or a diesel engine is contemplated.

As an example, a typical day's load profile for a residence might be plotted as shown in Fig. 18-26. Such a plot is obtained by first listing on a worksheet all electric devices to be used (lights, TV, toaster, washing machine, iron, etc:), with their daily hours of use and their power requirements in watts. Each segment of the load profile can be converted to watt-hours, and the *average daily load* in watt-hours per day can be calculated. This daily load can be extrapolated to an average load for the month (kW-hr/month), and taking into account expected monthly load variations, an average annual load (kW-hr/year). The following problem shows how the calculations are made.

Illustrative Problem 18-2

The worksheet below was compiled from an analysis of the loads in a client's residence. The Average Daily Load column has been filled in and the sum of these loads has been obtained to get the Total Average Daily Load.

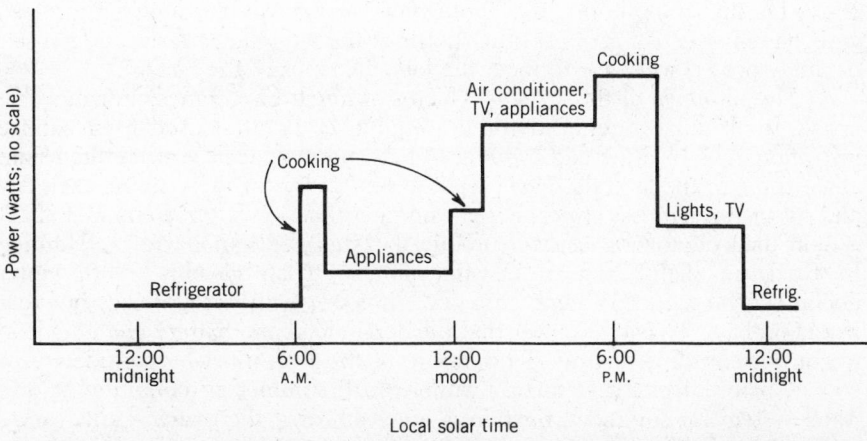

Figure 18-26 Graphical representation of a residential electric energy load for a typical summer day. An accurate profile of this kind can be plotted after a complete load analysis of the residence is made.

Load Analysis Worksheet

Residential Application (ac power)
(typical summer day)

Appliance	Rated Watts	Hours/Day	Days/Week	Average Daily Load (W-hr/day)
Lights	1000	3.0	7	3000
Toaster oven	1500	0.5	5	535
Vacuum cleaner	700	0.5	3	150
Iron	1100	1.0	2	315
Refrigerator	—	—	—	3000
Washing machine	500	2.0	2	285
Color TV	200	3.5	7	750
Stereo–Hi-Fi	50	2.0	7	100
Other small appliances	2000	0.25	4	285
Average daily net load				8420
Add 10% for inverter loss				840
Total average daily load				9260 W-hr/day
or				9.26 kW-hr/day

Ans.

A system may seldom operate at exactly the calculated average daily load level, but such a figure is a necessary starting point in sizing system components. A tabular calculation like that above should be done for each month of the year if any significant month-by-month variation is anticipated. The monthly and annual energy load demands can then be determined.

18-15 Estimating the PV Energy Available from Insolation

Methods of determining the insolation (solar power) available on horizontal and tilted surfaces at any hour of the day (solar time), for stipulated days of every month of the year, and for selected latitudes and regions of the world have been explained in detail in Chapter 4 and subsequent chapters. The reader should review these methods at this point.

On any given day, plots of average energy consumed and energy produced can be combined on the same chart with results similar to those plotted in Fig. 18-22. During sunny hours, the photovoltaic arrays will be supplying the load and charging the batteries. At other hours of the day, the batteries (or gen-set, or utility power) will have to meet the load demands of the system.

The tilt angle of the array is a factor of much importance. For movable arrays, a tilt angle optimized for the equinoxes is often used to maximize insolation in both winter and summer. A fixed-position array is often tilted at an angle equal to the latitude. However, the sun shines for more hours each day and passes through less atmosphere in summer than in winter, so the net effect is more daily photovoltaic energy produced in summer than in winter. By tilting the fixed array slightly to favor the winter sun, say to latitude plus 10°, the winter deficit can be reduced. Some summer PV power will be sacrificed, but that would probably be excess power that the design load and battery-charging load might not consume anyway. The nature of the load must be considered, of course, before deciding to favor a winter tilt. If summer air conditioning or a water system for summer irrigation is involved, or if the owner wants excess summer power to sell to a public utility, then the array design and tilt angle might be neutral or even favor the summer energy output.

Local weather (overcast days or percent possible sunshine) also affects the daily and seasonal power output from the PV array.

18-16 Sizing the Photovoltaic Array

When the tilt angle for the specific application is determined and when the average daily (monthly) load demand is known, the expected daily (monthly) output required from the PV array can be estimated. An allowance for system losses, usually 15 to 20 percent of the average energy load, is added to determine the required photovoltaic output. The following problem illustrates one method of sizing the photovoltaic array.

Illustrative Problem 18-3

A ranch home at 40° N Lat, in a location with 72 percent possible sunshine for August, has an average daily electric energy load demand for August of 16.25 kW-hr/day. The load is distributed much like that in the table of Illustrative Problem 18-2, except that a domestic water system (used also to water extensive landscaping in summer) and summer air conditioning are also involved. In view of these summer loads, the tilt angle of the array is selected equal to the latitude, or 40° with the horizontal.

The PV array is to be made up of single-crystal silicon solar cells, and the array will have an overall working efficiency of 8 percent. An allowance of 15 percent is made for system losses in batteries, inverter, and other BOS components.

Calculate the required total area of PV array to match this August load, assuming a clearness factor of 0.96.

Solution

From Table 4-F, 40° N Lat, note that with a tilt angle of 40° for August, the clear-day insolation available is 2560 Btu/day-ft² of surface. Converting this value to W-hr/day-ft² (from Table 4-3, 1 Btu = 0.293 W-hr), gives 750 W-hr/day-ft² for clear days and a clearness factor of 1.00. Applying the given percent possible sunshine and clearness number multipliers gives

$$750 \text{ W-hr/day-ft}^2 \times 0.72 \times 0.96 = 518 \text{ W-hr/day-ft}^2$$

which is the actual solar power input to the solar cell array, for a typical August day. The power to be delivered by the PV array must take into account the 15 percent BOS losses, or

$$\begin{aligned} \text{PV}_{\text{power}} &= [16{,}250 + (0.15 \times 16{,}250)] \text{ W-hr/day} \\ &= 18{,}700 \text{ W-hr/day} \end{aligned}$$

The array area (conversion efficiency 8 percent) is calculated from

$$\begin{aligned} A_{\text{array}} &= \frac{18{,}700 \text{ W-hr/day}}{518 \text{ W-hr/day-ft}^2 \times 0.08} \\ &= 452 \text{ ft}^2 \qquad\qquad \textit{Ans.} \end{aligned}$$

An array meeting these requirements could be made up of 113 48 in. by 12 in. photovoltaic modules, each with 35 to 40 single-crystal silicon cells. Such modules usually have an overall efficiency of 10 to 11 percent, and the array, over its total area, could easily show an efficiency of 8 percent.

The calculated array area of 452 ft² was obtained from August insolation data, but it would no doubt be suitable for all of the summer months. Typical fall, winter, and spring months would have to be analyzed to determine whether the load demands and the insolation available during those months would indicate more or fewer photovoltaic modules.

In most winter climates at 40° N Lat, an array sized to just balance the summer load will not be able to fully recharge the battery bank each winter day unless the winter load is much less than the summer load. Many overcast winter days will contribute to a seasonal deficit for the battery bank—a period of perhaps several months in which the batteries are never fully charged or "topped

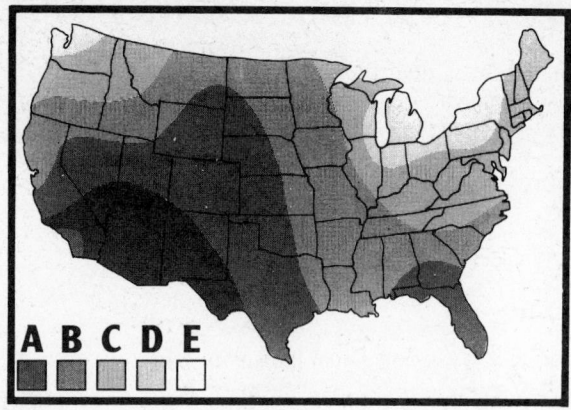

A B C D E

Table One: Power Budget for AC Power, DC Power, and Gen Set Plus Systems
Watt Hours per Day*

Map Code	Number of Modules in System									
	1	2	4	6	8	12	15	16	20	24
A	145	290	580	870	1160	1740	2175	2320	2900	3480
B	125	250	500	750	1000	1500	1875	2000	2500	3000
C	115	230	460	690	920	1380	1725	1840	2300	2760
D	100	200	400	600	800	1200	1500	1600	2000	2400
E	90	180	360	540	720	1080	1350	1440	1800	2160

*Figures are for M61 module only.
 For M51 module, multiply by 1.3. For ASI 16-2000 module, multiply by 1.17.

Power Budget for Line Tie Systems
Watt Hours per Day**

Map Code	Number of Modules in System				
	15	30	60	90	120
A	2,900	5,800	11,600	17,400	23,200
B	2,500	5,000	10,000	15,000	20,000
C	2,300	4,600	9,200	13,800	18,400
D	2,000	4,000	8,000	12,000	16,000
E	1,800	3,600	7,200	10,800	14,400

**Figures are for M51 module only.
 For M53 module, multiply by 1.43.

These figures are approximate; your actual output will vary depending on local weather conditions.

Table Two: Power Consumption of Various Appliances   *Watt hours per average use.

Lighting	Wattage (Average)
AC incandescent	60, 75, 100
AC fluorescent	20, 40, 60
DC incandescent	25, 50

Entertainment	
Color TV (AC)	200
9" Color TV (DC)	35

Tools	
Electric drill ¼"	350
Power saw	1,300
Table saw 10"	1,800

Household	
Refrigerator 16 cu. ft. manual defrost) 3000 watt hrs/day*	Varies
Refrigerator 16 cu. ft. (16 cu. ft. frost-free) 5000 watt hrs/day*	Varies
Electric range 2000 watt hrs/day*	12,000
Toaster	1,100
Microwave	1,400
Washing machine 300 watt hrs/load*	500

Household	
Electric dryer 3000 watt hrs/load*	4,850
Dishwasher 1000 watt hrs/load*	1,200
Gas dryer	350
Blender	300
Vacuum	600
Iron	1,100
Clock	2
Hair dryer	1,000
Electric shaver	14
Sewing machine	75

Heating/Air Conditioning	
Portable heater	1,500
Electric blanket 750 watt hrs/night*	150
Electric water heater 13,000 watt hrs/day*	4,475
Furnace blower ⅛ hp	300

Pumps	
Pool pump ¾ hp 500 watt hrs/day*	1,000
Water pump ½ hp	800
Circulation pump 1/25 hp	30

**Figures are for M51 module only.
 For M53 module, multiply by 1.43.

Figure 18-27 Chart and tables for one shortcut method of sizing residential and small commercial PV systems. Locate your site in one of the insolation regions (A through E) on the map. Then enter the tables for that region with the calculated energy budget for your application. Interpolation between vertical columns may be necessary to arrive at the recommended number of modules. Charts such as this are guidelines only. Values obtained from them should be carefully checked against local conditions and by an engineering analysis. Also, values in this table are for modules manufactured by ARCO Solar, Inc., and could not be expected to apply to equipment from other manufacturers. (Chart and tables courtesy of Solarwest Electric.)

off".[7] This seasonal deficit should be made up every summer, and the array should be sized with this topping off in mind. In the above problem, however, the extra capacity required by the water pump and the summer air conditioner could well be sufficient to prevent any winter battery deficit. A month-by-month analysis of loads and available insolation would be necessary to decide on the optimum number of modules for year-round service.

The type and number of batteries and the specific model and capacity of the BOS components will vary widely with every application. Manufacturers of photovoltaic arrays provide detailed design manuals for use by dealers and solar designers in selecting not only the photovoltaic modules and arrays themselves but also all items of BOS equipment. These manuals cover all of the common applications—homes, institutions, ranches, pumps and water systems, recreation vehicles and yachts, and communications applications.

Shortcut Methods of Sizing Arrays. If, for any reason, a detailed engineering analysis of a proposed application seems unnecessary, there are shortcut methods of matching modules and arrays to loads. One such method identifies codes for different insolation regions on outline maps that are provided. Tables are also provided that give for each of the coded map regions a recommended array size in numbers of photovoltaic modules of a given model and capacity, to match various values of the average daily load. Figure 18-27 provides a coded outline map and table for such a shortcut method. The following problems illustrate the use of the method.

Illustrative Problem 18-4

A remote mountain home in Idaho (code section C on the map) is to have a low-power dc photovoltaic system. The average daily energy load (power budget) has been carefully evaluated as 2400 W-hr/day. How many M61 modules should be recommended?

Solution

From the left-hand columns of Table 1 in Fig. 18-27, note that for the region coded C, a load of 2300 W-hr/day requires 20 M61 modules, and 2760 W-hr/day requires 24 M61 modules.

A good selection would be 21 M61 modules.   *Ans.*

Illustrative Problem 18-5

An elementary school in region B in Nebraska is planning a line-tie ac system with an estimated average daily load (power budget) of 17,500 W-hr/day. Excess PV-generated power to sell to the electric utility is not being planned for. How many M51 modules should be recommended?

Solution

From the right-hand columns of Table 1 in Fig. 18-27, the recommended number, by interpolation, is 105 M51 modules.   *Ans.*

---

[7] Prolonged periods at only a partial state of charge contribute to battery failure. Formation of lead sulfate crystals (sulfation) occurs. These crystals "lock up" battery capacity, and it is difficult to uform them under normal recharging procedures.

PHOTOVOLTAIC SYSTEM DESIGN

Illustrative Problem 18-6

A residence in upstate New York (region E of Fig. 18-27) has a calculated power budget of 7.8 kW-hr/day. How many M53 modules should be specified for this application?

Solution

From Table 1 in Fig. 18-27 (left-hand columns), note that the expected average electric energy output for one M61 module is 90 W-hr/day in region E. The M53 module, however, has an output 1.43 times greater, or 129 W-hr/day. The indicated number of M53 modules is given by the ratio:

$$\frac{7800 \text{ W-hr/day}}{129 \text{ W-hr/day}} = 60.5$$

Either 60 or 61 M53 modules could be installed.   *Ans.*

18-17 Current Costs of Photovoltaic Systems

The wide variety and special requirements of photovoltaic systems make general comments about costs inherently inaccurate and not very meaningful. Furthermore, the rapidly changing relative cost picture—PV module and array costs decreasing as utility electric rates continue to escalate—outdates cost comparisons and forecasts very rapidly.

We shall, however, present one rather common example—a home electric system—and give some comparative figures for the 1984–85 year.

The home is an 1800 ft^2 single-family residence in a U.S. Southwest location. The estimated average daily load is 40 kW-hr/day, 1200 kW-hr/month, or 14,600 kW-hr/year. The system will be line-tie, without storage batteries. It will be sized to provide 85 percent of the load on an annual basis, and when excess photovoltaic energy is available in summer, it will be fed to the power company lines for credit to the homeowner's account.

An array with a capacity of 1 peak kW (solar noon, summer) will produce about 6 to 8 kW-hr/day or about 2500 kW-hr/year in the U.S. Southwest. The system capacity for 85 percent of the annual load is

$$\text{Required system arrays} = \frac{0.85 \times 14,600 \text{ kW-hr/yr}}{2500 \text{ kW-hr/yr-array}} = 4.96 \text{ arrays}$$

Consequently, the installation will consist of 5 arrays of 1 peak kW capacity each.

Current (1984) installed costs for residential line-tie systems in the U.S. Southwest range from $10,000 to $15,000 per peak kilowatt-hour of capacity. Obviously, at this level of capital investment, such systems are far from competitive with public utility power. However, if the DOE goal of $1 per peak watt for PV arrays is met by 1990, installed system costs could drop to $2500 to $3000 per peak kilowatt-hour, making the system described above cost about $12,000 to $15,000.

Not counting any possible solar tax credits or other subsidies, the cost of amortizing the capital investment (say $14,000) on the five peak kW-hr system over a 20-year period, would be $154 per month or $1850 per year, based on a 12 percent interest rate (from mortgage loan payment tables). The interest portion of each annual payment is deductible on the owner's income tax return. The interest portion of the first year's payments will be $1670.

If the owner is, say, in a 35-percent federal income tax bracket, the federal income tax deduction for the first year will be $585. (In ensuing years, as the mortgage is paid off, the deduction for interest paid will decrease.)

The owner's first-year net cost for the system is then $1850 − $585 = $1265 per year, or $105 per month. This monthly cost buys 0.85 × 1200 kW-hr of electric energy, for a cost of $0.0971 per kilowatt-hour. For comparison purposes, a typical 1984 cost of power-company-supplied electric energy in the Southwest is $0.08 per kilowatt-hour.

In summary, the monthly net cost for the first year of operation of the photovoltaic line-tie system, under the assumptions made, is approximately $105, to produce 1020 kW-hr of electric energy. If this amount of energy were purchased from the power company at $0.08/kW-hr, the monthly cost would be $81.60, representing an advantage of some $23 per month for conventional energy sources. However, electric energy costs are increasing by 5 to 10 percent per year, and a breakeven point could be reached by the early 1990s.

Cost–benefit analysis for any investment in solar energy, including photovoltaics, is not a simple exercise. The installed cost of the system itself, the cost of money, the cost of power company electricity, the existence of possible subsidies for energy conservation, the owner's tax status, and the expected dependability of the system must all be considered. As ancillary considerations, the question about the system's effect on the resale value of the owner's property and the possibility of investing the same capital in some other venture with a greater rate of return on investment are factors to be evaluated.

POWER PRODUCTION FOR ELECTRIC UTILITIES

Some of the applications discussed on the preceding pages are, or soon may become, cost-effective. Others, as in the case of remote locations or in underdeveloped countries, are examples of the *only energy option available*, regardless of cost. We close this chapter with a brief treatment of an application that is not limited to the solar option and that is not, at present, very close to being cost-effective—solar electric power for public utilities. These installations are examples of investments in the future—research and development (and pilot plants) now, to ensure alternative energy sources in the 1990s and beyond.

Commercial electric power production in the United States is now tied closely to fossil fuels, nuclear fission, and hydro power. To be sure, geothermal power makes a limited contribution, and wind generators and ocean-thermal generators are in the test and development stage.

Solar-electric commercial power, though not significant as regards its total energy output at the present time, is undergoing vigorous development with two different technologies—*photovoltaic generation* and *solar-thermal generation*. Brief descriptions of recent developments in these technologies follow.

18-18 Commercial Electric Power from Photovoltaics

A good example of a commercial photovoltaic power plant that feeds its entire output to a public utility grid is the ARCO Solar Photovoltaic Power Plant near Hesperia, California (Fig. 18-28). It was put on line with the Southern California Edison Company in February, 1983. It is rated at 1 MW capacity and is capable of producing 3 million kW-hr of electric energy annually—three times the output of any other PV system in the world, at that time.

The photovoltaic array field covers 20 acres in the high desert, where the sun shines nearly every day of the year. There are 108 arrays in the field (Fig. 18-28*a*), each array consisting of 16 panels of 16 modules each. Each module contains 35 solar cells, giving a total for the entire field of 967,680 single-crystal silicon cells.

Each array is mounted on a two-axis sun tracker (see Fig. 18-15), which keeps the array pointed directly at the sun all day long and also makes adjustments for seasonal changes in solar declination.

The photovoltaic output is dc, and it has to be "power-conditioned" before it is fed to the utility grid. Synchronous inverters (Fig. 18-28*b*) change the dc power to ac power, and then this ac output has its voltage stepped up by transformers to the voltage required at the power company's substation.

Figure 18-28 Commercial photovoltaic power. (*a*) A portion of the array field at the ARCO Solar Photovoltaic Power Plant at Hesperia, California. Each of the arrays shown contains 8960 silicon solar cells. All 108 arrays are mounted on two-axis trackers to track the sun all day long for its traverse across the sky every day in the year. (*b*) Power-conditioning equipment (synchronous inverters) for converting dc power to ac power, and synchronizing it with the utility power on the regional network (Both photos ARCO Solar, Inc.)

A much larger plant—final project size 16 MW, with 10 MW already on line in 1984—is located on the Carrisa Plains in Central California. This is also a region of intense solar radiation (320 Btu/hr-ft^2 or 1 kW/m^2 at solar noon in summer), and the plant's full output will serve 6400 typical single-family homes. Pacific Gas and Electric Company is the utility involved in this project. Newly designed high-efficiency photovoltaic modules with square cells (Fig. 18-1*a*) are used on this installation, and reflector enhancement has been incorporated into the arrays (Fig. 18-29), producing the effect of a solar radiation concentrator.

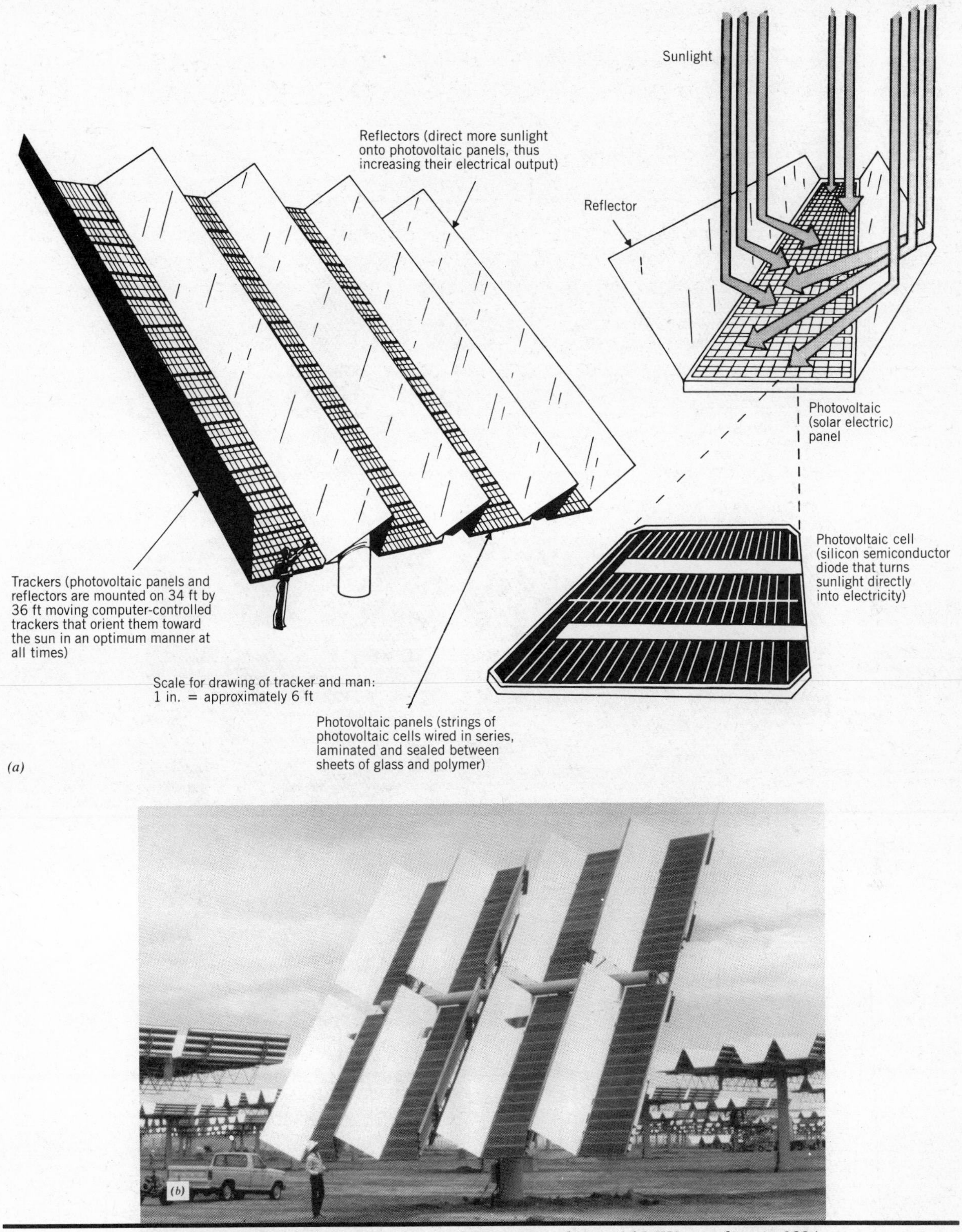

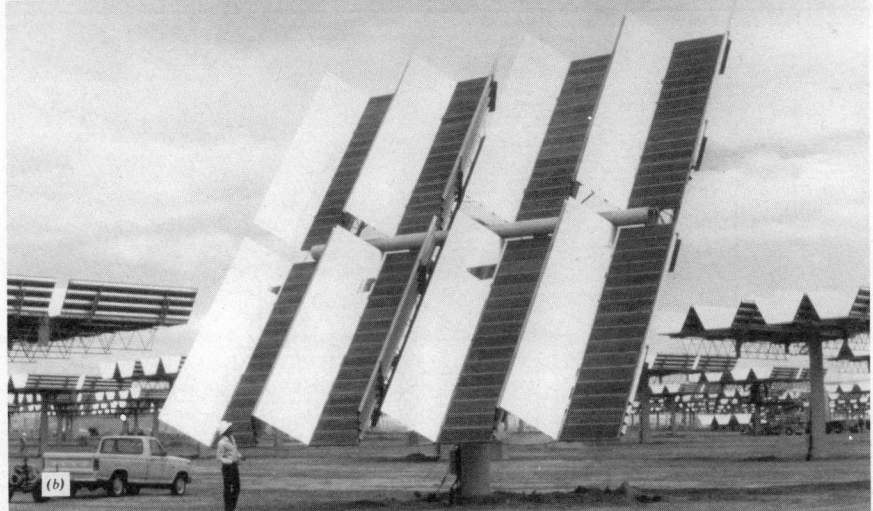

Reflectors (direct more sunlight
onto photovoltaic panels, thus
increasing their electrical output)

Sunlight

Reflector

Photovoltaic
(solar electric)
panel

Photovoltaic cell
(silicon semiconductor
diode that turns
sunlight directly
into electricity)

Trackers (photovoltaic panels and
reflectors are mounted on 34 ft by
36 ft moving computer-controlled
trackers that orient them toward
the sun in an optimum manner at
all times)

Scale for drawing of tracker and man:
1 in. = approximately 6 ft

Photovoltaic panels (strings of
photovoltaic cells wired in series,
laminated and sealed between
sheets of glass and polymer)

(a)

(b)

Figure 18-29 World's largest photovoltaic power plant—16 MW—on line in 1984.
(a) Diagram of the solar cell–module–panel–array configuration at the Carrisa Plains
PV plant. The panels are fitted with reflectors to direct more sunlight on the cells. (b)
Photo of a small portion of the array field taken during construction of the plant.
(ARCO Solar, Inc. and Pacific Gas and Electric Company.)

POWER PRODUCTION FOR ELECTRIC UTILITIES

(a)

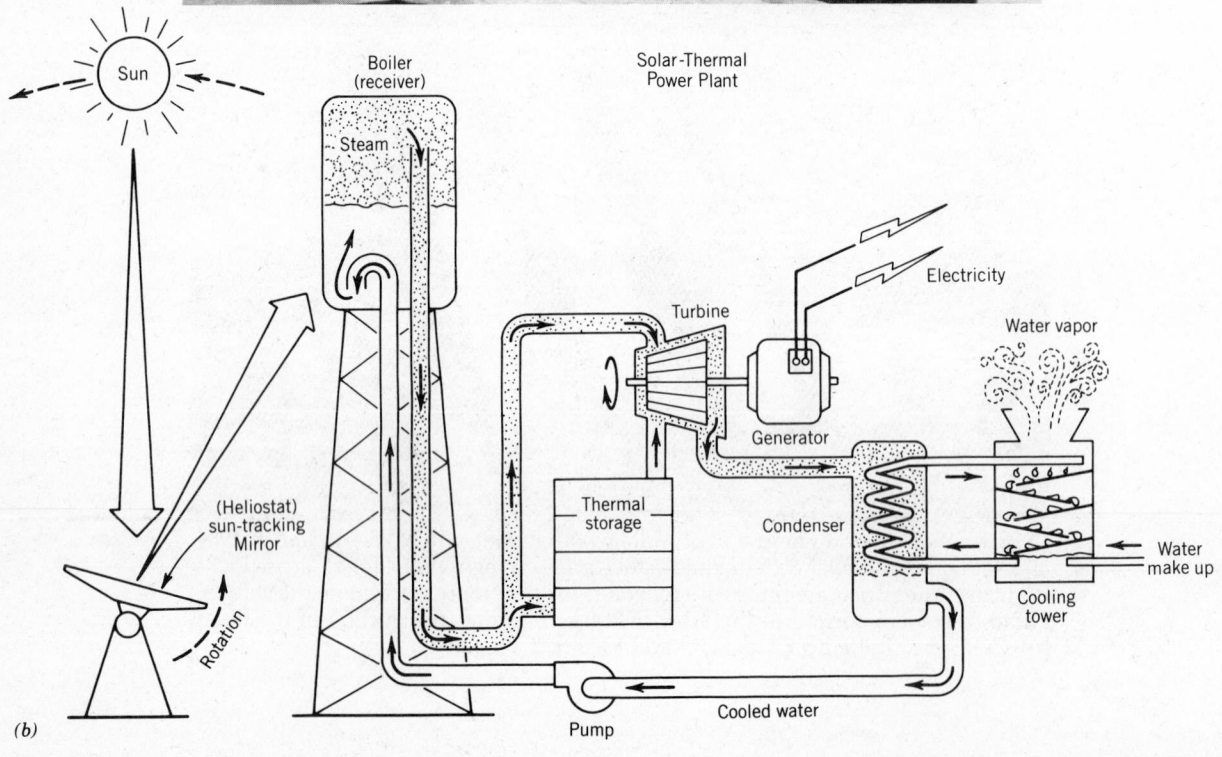

Sun

Boiler
(receiver)

Solar-Thermal
Power Plant

Steam

Electricity

Turbine

Water vapor

(Heliostat)
sun-tracking
Mirror

Generator

Rotation

Thermal
storage

Condenser

Water
make up

Cooling
tower

(b)

Pump

Cooled water

SOLAR ENERGIZED ELECTRIC POWER

Figure 18-30 *Solar One*—the world's largest solar-thermal electric generating plant. (*a*) The central tower with the receiver–boiler on top. Notice that it is so hot it glows even in the bright light of a desert day. Also shown is a small section of the heliostat array field. Note the closeup of the tracking mechanism in the center foreground. (*b*) A flow diagram illustrating the major processes in the operation of the *Solar One* power plant. (*c*) The central operations office, where *Solar One* systems are constantly controlled and monitored by computer. (Southern California Edison Company.)

The 34- by 36-ft arrays are mounted on two-axis, computer-controlled sun trackers in order to intercept maximum solar radiation at all hours of the day and all seasons of the year.

18-19   Solar-Thermal Electric Power

Another approach to solar-generated electric power is demonstrated by a 10-MW pilot plant operated by the Southern California Edison Company near Barstow, California, also on the high desert. A field of 1818 heliostats (see Fig. 1-6) reflects sunlight and focuses solar energy on a receiver at the top of a 300-ft tower (Fig. 18-30*a*). The intense heat in the receiver–boiler generates high-pressure superheated steam at 960°F, which flows to a conventional turbine generator at ground level. This plant, named *Solar One*, began operation in

April 1982, and it has been so successful that preliminary planning for a 100 MW plant of similar design has already begun.

Thermal storage for night and cloudy-day operation is provided by 6800 tons of crushed granite and sand contained in an underground steel tank 45 ft deep and 60 ft in diameter. The heat-exchange fluid (240,000 gal) is a special heat-stable oil. When fully charged, as at the end of an all-day sunny day, the thermal storage can provide enough heat on discharge for reduced-output operation throughout the night, or a maximum output of 7 MW for a four-hour period. Figure 18-30b is a flow diagram illustrating the operating principles of the plant.

All 1818 heliostats of the array field must track the sun all day long every day of the year and simultaneously reflect solar radiation to focus on the receiver–boiler at the top of the tower. This operation is extremely complex, of course, as is also the operation of the receiver–boiler where the daytime temperature is in the range of 1000°F to 1150°F with pressures exceeding 1500 psi. *Solar One* overall operation is computer controlled (Fig. 18-30c) and, since it is a pilot plant, it is monitored as a research project by Edison company personnel and by representatives of the U.S. Department of Energy and of Livermore Laboratories.

PROBLEMS

1. A remote water pumping station must lift 5000 gal of water per day from a well 125 ft deep to a surface cistern. If the pump and motor combined efficiency is 72 percent, and if there is an additional 15 percent loss in the batteries and BOS components of the PV power system, what must be the average daily energy output of the photovoltaic array in kilowatt-hours per day?

2. An electric energy load survey of a home is made with the following results, on an average per-day basis:

    750 W incandescent lamps operating for three hours
    1 color TV (ac) operating for five hours
    1 16-ft manual-defrost refrigerator
    1 electric range
    1 toaster for 0.5 hour
    1 washing machine (two loads)
    1 dishwasher (one load)
    1 gas dryer (two loads)
    1 vacuum cleaner for one hour
    1 electric iron for two hours
    1 furnace blower for three hours

Use the energy consumption values suggested in Table 2 of Fig. 18-27, and determine the average daily load in kilowatt-hours.

3. A National Park Service ranger station in northern Arizona is switching from a diesel generator to photovoltaic power. If the average energy budget is known to be 6200 W-hr/day, how many M-61 photovoltaic modules should be recommended? (See Fig. 18-27.)

4. A resort near Petoskey, Michigan, has an average electric energy budget of 68 kW-hr/day. If 40 percent of the load is to be met (on the average) by a stand-alone ac photovoltaic system, how many M-51 modules should be recommended? (Fig. 18-27.)

5. A remote ranch near Sheridan, Wyoming is planning an ac stand-alone PV system to provide an average daily energy budget of 12 kW-hr. How many M-53 modules should be installed? (Fig. 18-27.)

6. A summer home in the north woods is to have a stand-alone ac PV system. The location, at 48° N Lat, is in a region with 68 percent possible sunshine for August and a clearness factor of 0.92. There is an average daily electric energy

load for August of 12.5 kW-hr/day. The tilt angle of the PV array will be optimized for summer. The array will be made up of single-crystal silicon solar cells, with an average working efficiency for the array of 8.4 percent. A 16-percent allowance is made for losses in batteries, inverter, and other BOS equipment. (a) Calculate the required total area of PV array square feet to match the August load. (b) How many 48-by-12-in. modules will be required?

7. An office building in Hilo, Hawaii has an average daily electric energy load for December of 140 kW-hr/day. A line-tie ac system to meet 50 percent of this load is to be installed. The location has 65 percent possible sunshine and a clearness factor of 0.92 for December. The array will be tilted at an angle to optimize the winter sun. The PV array will have an average working efficiency of 8.4 percent, and a 14 percent loss is estimated for batteries, inverter, and other BOS equipment. What is the total area of the PV array that should be installed to meet one half of the average daily load for December?

Appendix

Selected References

The literature of solar energy would fill a good-sized library, ranging as it does from popular, how-to-do-it books and pamphlets to advanced solar engineering handbooks and scientific research reports published by engineering societies and university groups. From this vast literature, only a few selected titles have been chosen for inclusion here. The authors believe that a short list of references directly related to the content and level of the text will be of greater value than a long list of unrelated (and possibly unconsulted) works.

Each of the following titles should be of value to students for further reading and as a source of data and ideas for solar energy practice.

1. The following publications of the American Society of Heating, Refrigerating, and Air Conditioning Engineers (ASHRAE) are indispensable to students and practitioners in the solar energy field:

ASHRAE Handbook 1977 Fundamentals.

ASHRAE Handbook 1982 Applications.

ASHRAE Handbook 1979 Equipment.

Applications of Solar Energy for Heating and Cooling of Buildings (ASHRAE GRP 170). Richard C. Jordan and Benjamin Y. H. Liu, eds., 1977.

Cooling and Heating Load Calculation Manual (ASHRAE GRP 158).

William Rudoy and Joseph F. Cuba, eds., 1979.

All ASHRAE publications are available from the Society at 1791 Tullie Circle, N.E., Atlanta, Georgia 30329.

2. The handbooks published by the Solar Energy Group at the Los Alamos National Laboratory provide a wealth of research data and current ideas on passive solar design. These volumes include:

Pacific Regional Solar Heating Handbook. 1976
Energy Research and Development Administra-
tion (ERDA), San Francisco Operations Office,
1976.
Passive Solar Design Handbook, vol. I. Total Environment
Action, Inc. U.S. Department of Energy, March 1980.
Passive Solar Design Handbook, vol. II. J. D. Balcomb, ed.
U.S. Department of Energy, January 1980.
Passive Solar Design Handbook, vol. III. R. W. Jones, ed.
U.S. Department of Energy, July 1982.

All of these publications are available from the U.S. Government Printing Office, Washington D.C.

3. Anderson, Bruce, and Michael Riordan. *The Solar Home Book*. Harrisville, N.H.: Cheshire Books, 1976.

4. Beckman, W. A., S. A. Klein, and J. A. Duffie. *Solar Heating Design by the f-Chart Method*. New York: Wiley, 1977.

5. Butti, Ken, and John Perlin. *A Golden Thread—2500 Years of Solar Architecture and Technology*. New York: Van Nostrand Reinhold, 1980.

6. Di Certo, Joseph J. *The Electric Wishing Well—The Solution to the Energy Crisis*. New York: Macmillan, 1976.

7. Duffie, J. A., and W. A. Beckman, *Solar Energy Thermal Processes*. New York: Wiley, 1974.

8. Harris, Norman C. *Modern Air Conditioning Practice*, 3rd ed. New York: McGraw-Hill, 1983.

9. Howell, Yvonne, and Justin A. Bereny. *Engineer's Guide to Solar Energy*. San Mateo, Calif.; Solar Energy Information Services (SEIS), 1979.

10. Komp, Richard J. *Practical Photovoltaics—Electricity from Solar Cells*. Ann Arbor, Mich.: Aatec Publications, 1981.

11. Kreider, Jan F., and Frank Kreith. *Solar Energy Handbook*. New York: McGraw-Hill, 1981.

12. Maycock, Paul D., and Edward N. Stirewalt. *Photovoltaics—Sunlight to Electricity in One Step*. Andover, Mass.: Brick House Publishing, 1981.

13. Mazria, Edward. *The Passive Solar Energy Book*. Emmaus, Pa.: Rodale Press, 1979.

14. Stoeker, W. F. *Using SI Units in Heating, Air Conditioning, and Refrigeration*. Troy, Mich.: Business News, 1975.

15. *Climatic Atlas of the United States*. U.S. Department of Commerce. Asheville, N.C.: National Climatic Center, 1977.

16. *Energy Products Specifications Guide*, 1984. Solar Vision, Inc., Harrisville, N.H.

17. *Solar Energy Handbook*. Power Systems Group/Ametek, Inc. Radnor, Pa.: Chilton Publishing, 1979.

18. *Sunset Homeowner's Guide to Solar Heating*. Menlo Park, Calif.: Lane Publishing, 1978.

Journals and Magazines

1. *Air Conditioning and Refrigeration News*. Business News Publishing Company, Troy, Mich. A weekly trade paper with up-to-date news of the industry. Covers solar as well as conventional air conditioning and refrigeration.

2. *ASHRAE Journal*. The monthly journal of the American Society of Heating, Refrigerating, and Air Conditioning Engineers. Published by ASHRAE, Atlanta, Georgia.

3. *Passive Solar Journal*. American Solar Energy Society (ASES), the U.S. section of the International Solar Energy Society, 110 W. 34th St., New York, N.Y.

4. *Solar Age*. Solar Vision, Inc., Church Hill, Harrisville, N.H. A monthly journal with current articles on both passive and active solar energy system design. Includes product reviews, news of conventions, and so on.

5. *Solar Energy*. Pergamon Press, Fairview Park, Elmsford, N.Y. An international journal for scientists and engineers. Main emphasis is on research and scientific development. Papers presented at conferences are published.

6. *Solar Engineering and Contracting*. Business News Publishing Company, Troy, Mich. A bimonthly trade magazine.

Solar Economics

The design of solar systems should always be considered from an energy performance standpoint—that is, the performance of the solar energy system should be evaluated in terms of the energy that will be saved by the system. For passive systems, system performance is measured in terms of the Solar Savings (SS) and the Auxiliary Heat (AH) (see Chapters 6 and 9). The Solar Savings is the energy provided by the passive solar system toward the total space-heating load, and the Auxiliary Heat is the conventional energy required to supplement the passive solar heating. For active systems, the performance of a solar energy system is measured by the Solar Heating Fraction (SHF) (see section 13-7), which is the fraction of the conventional energy that is saved by the use of the active solar system.

Most clients or customers will ask two basic questions about solar energy systems. One question will relate to the energy savings the solar system will provide, and it can be answered in terms of either the Solar Savings (SS) or the Solar Heating Fraction (SHF). The second question is usually one that relates to the economics of the system. This question will often be phrased somewhat as follows: "How long will it be before the savings in conventional energy will pay for the solar energy installation?" Or, "What will be the return on my investment in the solar energy system?" To answer these questions, two sets of calculations are necessary—one set to determine the energy savings provided by the proposed solar system, and a second set to determine the economic performance (return on investment, or payback period) of the proposed system.

Several different mathematical procedures are used to make economic evaluations. The two that will be discussed here will be the *payback method* and the *life-cycle costing method*.

II-1   The Payback Method

This method determines the time required to amortize the initial system cost with the savings that result from the solar system. The payback period can be determined using the following simple equation, if the effects of inflation and the earning power of money are not considered:

$$PBP = \frac{I}{S_E} \qquad \text{(II-1)}$$

where

PBP = the payback period, years
I = the initial investment in the solar system in present-value dollars
S_E = the yearly energy savings in present-value dollars that result from the use of the solar system (assuming that the cost of conventional energy remains unchanged)

The yearly energy savings in dollars can be calculated as:

$$S_E = \frac{P \times Q_{SA}}{E} \qquad \text{(II-2)}$$

where

S_E = the yearly energy savings in present-value dollars
P = the present price of conventional energy, dollars/10^6 Btu (assumed constant)
Q_{SA} = the net quantity of heat energy supplied by the solar system per year, 10^6 Btu/year.
E = the overall efficiency of the conventional energy system, percent

Illustrative Problem II-1

A domestic solar water heater will provide a solar heating fraction of 0.70 for an annual domestic hot-water load of 14.4×10^6 Btu/year. The cost of electrical energy is $17.57/$10^6$ Btu ($0.06 kW-hr) and will be used with an assumed efficiency of 100 percent. The initial installed cost of the solar water heater system is $2600. Determine the payback period for this system, making no allowance for inflation or the earning power of money.

Solution

The annual energy supplied by the solar system can be calculated using Eq. (13-2), modified as follows:

$$Q_{SA} = SHF \times Q_{WA}$$

where

Q_{SA} = the annual energy supplied by the solar system, 10^6 Btu/year
SHF = the solar heat fraction
Q_{WA} = the annual domestic hot-water load, 10^6 Btu/year

Substituting,

$$Q_{SA} = 0.70 \times 14.4 \times 10^6 \text{ Btu/year}$$
$$= 10 \times 10^6 \text{ Btu/year}$$

The present-dollar value of energy savings is now calculated using Eq. (II-2):

$$S_E = \frac{P \times Q_{SA}}{E}$$
$$= \frac{(\$17.57/10^6 \text{ Btu}) \times 10 \times 10^6 \text{ Btu/year}}{1.0}$$
$$= \$175.70/\text{year}$$

The payback period (assuming constant dollars and making no allowance for the earning power of money) is then calculated using Eq. (II-1):

$$PBP = \frac{I}{S_E}$$

Substituting,

$$PBP = \frac{\$2600.00}{\$175.70}$$
$$= 14.8 \text{ years} \qquad \qquad Ans.$$

The simple payback period method as illustrated above does not consider several factors that will actually affect the overall economics of the solar system investment.

On the one hand, the cost of conventional energy will no doubt increase with time because of inflation; and on the other hand, there will be additional costs associated with the operation of the solar energy system, such as maintenance, insurance, and property tax. It is also necessary to relate system costs and savings to a common time and a common dollar measure. When an investment in a solar energy system is made, most of the expenditures occur in the present time, while savings occur in future time, and a common reference is necessary to make an accurate comparison. The value of a dollar will be different at some future time as compared to its present value. The time-dependent value of money is, of course, related to inflation, which reduces the purchasing power of a currency, but it is also dependent on the earning potential of money even if there were no inflation. The method used for adjusting the time value of money is termed *discounting*. Discounting is a procedure by which interest rate formulas are used to convert cash flows expected to occur at different future times to an equivalent value of dollars at the present time or at some stipulated future time.

Present Value of Future Savings. The *discount rate* represents the expected earning power of money. It is generally taken as the average return that the particular investor expects or could receive from typical investments. It should be pointed out that there is not a single discount rate that is appropriate for all calculations and that the selection of the discount rate is to some degree subjective. To discount a future payment or savings, an interest formula or discount formula must be used. The calculations are usually based on an annual period; that is, if a calculation is to cover a period of five years, there would be five annual calculation periods. The formula used to discount future values to a present value is the *single present value* formula:

$$SPV = \frac{1}{(1 + i)^N} \qquad \qquad \text{(II-3)}$$

where

i = the discount rate selected for use, %/year
N = the period for the calculation, years
SPV = the single present value factor for the period

Illustrative Problem II-2

If the savings during the fifth year of operation of a solar energy system are estimated to be $250 what is the present value of this savings if a discount rate of 0.08 is used?

Solution

Using Eq. (II-3) for the single present value factor, the present value can be calculated as:

$$PV = SPV \times \$250.00$$
$$= \frac{1}{(1 + 0.08)^5} \times \$250$$
$$= \frac{1}{(1.08)^5} \times \$250$$
$$= \frac{1}{1.4693} \times \$250$$
$$= \$170.15 \qquad \qquad Ans.$$

Another way to consider this is that if $170.15 in savings is invested now with an 8 percent rate of return on the investment, in five years the value of the investment with interest will be $250. Yet another way to consider it is that $170.15 in hand now is worth the same as a guaranteed payment of $250 five years from now.

Future Fuel Costs. To calculate the increase in fuel costs from inflation, the *single compound amount formula* can be used to determine the future value of a present amount:

$$SCA = (1 + i)^N \qquad \qquad (II-4)$$

where

i = the fuel cost inflation rate, %/year
N = the period for the calculation, years
SCA = the single compound amount factor for the period

Illustrative Problem II-3

If the present cost for fuel is $450 per year and fuel costs are increasing at a rate of 10 percent per year, determine what the future yearly fuel cost will be in 10 years.

Solution

Using Eq. (II-4) for the single compound amount factor, the future cost can be calculated as:

$$FC = SCA \times \$450$$
$$= (1 + i)^N \times \$450$$
$$= (1 + 0.10)^{10} \times \$450$$
$$= (1.10)^{10} \times \$450$$
$$= 2.5937 \times \$450$$
$$= \$1167.18 \text{ for the tenth year} \qquad Ans.$$

Thus, in 10 years, the fuel cost will more than double.

Tables of precalculated values for the present-value and compound-amount formulas are available in most standard references on the mathematics of finance and in accounting reference books.

Illustrative Problem II-4

For the domestic solar hot-water heating system described in Illustrative Problem II-1, determine the payback period for a fuel inflation rate of 15 percent and a discount rate of 8 percent.

Solution

From Illustrative Problem II-1, the present investment is $2600, the present fuel cost is $17.57/10^6$ Btu, and the annual solar heating fraction (SHF) is 0.70, which gave an annual energy savings of $Q_{SA} = 10 \times 10^6$ Btu/yr. For this calculation, it is helpful to set up a table like Table II-1 to aid in the calculation. First, the future fuel cost (single compound amount) factors are either calculated or obtained from tables of mathematical finance for a fuel inflation rate of 15 percent. The annual inflating fuel costs for future years are then calculated by multiplying the single compound amount factor (SCA) times the present fuel cost of $17.57/10^6$ Btu. For example, in the fifth year, the SCA from Eq. (II-4) is 2.0113, which is multiplied by the present fuel cost of $17.57/10^6$ Btu to obtain the fifth year inflated fuel cost of $35.34/10^6$ Btu. The annual energy cost savings in dollars is then calculated by multiplying the annual energy savings (in 10^6 Btu) by the annual inflating fuel cost in dollars/10^6 Btu.

There may be other savings resulting from the solar system also, such as incentive tax savings or energy credits. These dollar savings would then be added to the annual energy savings. But there may also be other *costs*, such as maintenance, repair, or operating costs, which are associated with the solar system, and these additional yearly costs would be subtracted from the dollar savings.

The net annual dollar savings are then converted to present value by multiplying the annual values by the single present value factor (SPV), found from Eq. (II-3). The payback period according to Table II-1 (column at far right) is 10+ years; that is, it is in the tenth year that the present value of accumulated savings becomes equal to the original investment of $2600.00. *Ans.*

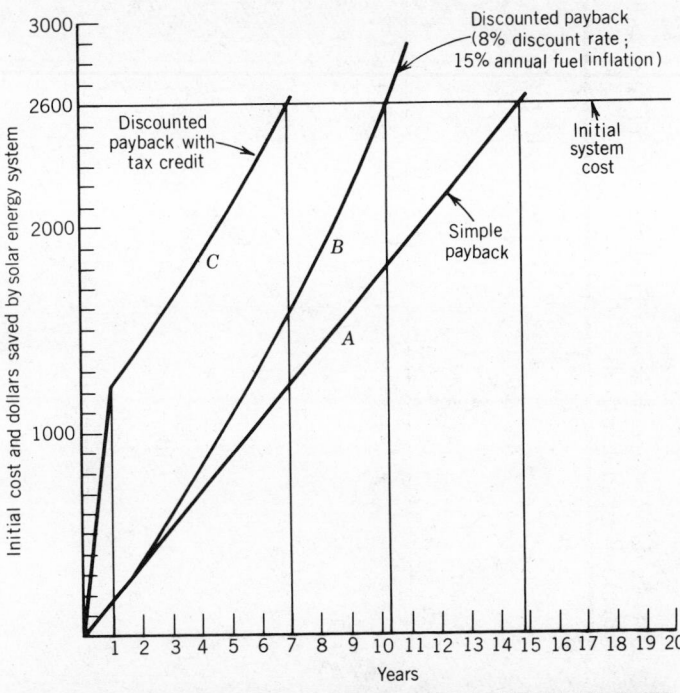

Figure II-1. Plot of results of simple payback method (curve A) of Ill. Prob. II-1, and the discounted payback method (curve B) of Ill. Prob. II-4 for comparison.

Table II-1   Calculation of Present Value of Accumulated Savings for a Discounted Payback Calculation (Illustrative Problem II-4)

Year of Calculation	Annual Energy Savings Q_{SA} ($\times 10^6$ Btu)	Single Compound Amount (SCA) Factor (Eq. II-4)	Annual Inflating Fuel Cost (dollars/10^6 Btu)	Annual Energy Savings (dollars)	Single Present Value Factor (SPV) (Eq. II-3)	Present Value of Annual Savings (dollars)	Present Value of Accumulated Savings (dollars)
1	10	1.1500	20.21	202.10	.92593	187.13	187.13
2	10	1.3225	23.24	232.40	.85734	199.25	386.38
3	10	1.5208	26.72	267.20	.79383	212.11	594.49
4	10	1.7490	30.73	307.30	.73503	225.87	824.36
5	10	2.0113	35.34	353.40	.68059	240.52	1064.88
6	10	2.3130	40.64	406.40	.63017	256.10	1320.98
7	10	2.6600	46.74	467.40	.58349	272.72	1593.71
8	10	3.0590	53.75	537.50	.54027	290.40	1884.10
9	10	3.5178	61.81	618.10	.50025	309.20	2193.31
10	10	4.0455	71.08	710.80	.46320	329.25	2522.55
11	10	4.6523	81.74	817.40	.42889	350.57	2873.12

The results of the simple payback method of Illustrative Problem II-1 and of the discounted payback method of Illustrative Problem II-4 are plotted in Fig. II-1, showing the increasing savings and the years when the initial cost will have been amortized, for both methods.

II-2 Incentives for Using Solar Energy

Government incentives for the use of solar energy have a very positive effect in reducing the payback period. There are several forms in which government incentives are offered:

1. Property tax incentives.
2. Income tax incentives.
3. Excise tax incentives.
4. Loan interest incentives.

The property tax reduction or exemption for solar energy equipment is an incentive that has been provided by many states. During the 1970s, 30 of the 50 states enacted some type of property-related tax incentive for solar systems. Many of these incentives are still applicable, and some have termination dates or renewal periods after 1984. The solar equipment exempted from property taxation varies widely from state to state. Some states limit their exemptions to active solar heating and/or cooling systems and domestic solar hot-water systems. Other states include passive solar and other renewable energy sources in their exemptions.

The income tax incentive for solar equipment provides a positive reward for an investment in solar equipment or in a total solar energy system. The income tax incentive currently exists at both state and federal levels of government. Today (1984), 29 states have some form of an income tax incentive, which may be either an income tax deduction or an income tax credit. The several states have different formulas for determining the deduction or credit, and the type of solar energy system that is eligible for an income tax incentive also varies from state to state.

The National Energy Act (Energy Tax Act of 1978) provides for a residential energy credit on the individual (federal) income tax. The income tax credit is allowed for any renewable energy source, including solar, wind, and geothermal energy equipment used in the principal residence of the taxpayer. These credits are in effect through 1985 and may be extended after 1985 by Congress.

Excise taxes include such taxes as sales, use, franchise, and transaction taxes. Eight states have provided some form of excise tax incentives.

Loan incentives in the form of subsidized low-interest loans or guaranteed loans can greatly reduce the first cost and life-cycle costs of solar energy systems. Presently, eight states have some type of loan incentive or grant program. These range from special loans for disaster victims, for low- or moderate-income persons, or for veterans, to programs that are open to any resident of the state.

The purpose of this brief discussion of government incentives is merely to point out that they exist and that they should be taken into account when costing out solar energy systems. Full coverage of the subject of incentives for solar energy would require a book-length treatment, and such a publication would have to be revised annually to keep up with changing legislation.

An excellent treatment of solar legislation at state levels, as of 1981, is contained in Chapter 11 of *Solar Heating and Cooling*, 2nd ed., by Jan F. Kreider and Frank Kreith, Washington, D.C.: Hemisphere Publishing Corporation, 1982.

Illustrative Problem II-5

For the $2600 domestic solar hot-water heating system cited in Illustrative Problem II-4, determine the payback period if a 40 percent income tax credit is received in the first year.

Table II-2 Calculation of Present Value of Accumulated
Savings with a Tax Credit (Illustrative Problem II-5)

Year of Calculation	Present Value of Annual Savings (dollars)	Tax Credit Received (dollars)	Present Value of Accumulated Savings (dollars)
1	187.13	1040.00	1227.13
2	199.25	0.00	1426.38
3	212.11	0.00	1638.49
4	225.87	0.00	1864.36
5	240.52	0.00	2104.88
6	256.10	0.00	2360.98
7	272.72	0.00	2633.70

Solution

First the income tax credit is calculated as:

$$\text{Tax credit} = 0.40 \times \$2600$$
$$= \$1040$$

The present value of accumulated savings for the first year will be the sum of the present value of the annual savings as calculated in Table II-1 for Illustrative Problem II-4 (which is $187.13) and the tax credit just calculated above, or

$$\$187.13 + \$1040.00 = \$1227.13$$

This value is entered in Table II-2 for the first year. The calculations for the present value of annual savings for the remaining years are the same as the calculations for Illustrative Problem II-4. The present value of accumulated savings for additional years is shown in Table II-2 through the seventh year, at which time the initial capital investment will have been amortized.   *Ans.*

The results of this payback calculation are also plotted on the chart of Fig. II-1, curve C.

One of the disadvantages of the payback method is that it does not provide information on an investment past the payback period, and therefore the total saving that a solar system could provide over its useful lifetime is not obtained. From payback analysis, it is not possible to compare several different systems and select the one that is the most cost-effective over equivalent "lifetime" periods.

The next section describes a method of cost analysis that allows for comparisons over system lifetime periods.

II-3   Life-Cycle Cost Analysis

Life-cycle cost analysis is a procedure by which the total cost of the system is calculated for the expected useful life of the system. This method takes into account all future expenses for both the solar energy system and an equivalent conventional energy system. These expenses are calculated on a yearly basis and then discounted to a present value. The total present value of the energy system is then the sum of the discounted yearly values. The yearly expenses for the system before taxes will consist of maintenance costs, repair or replacement costs, insurance costs, and fuel costs. The following discussion will not consider taxes or incentives, in order to keep the analysis simple. The discussion here is suited to residential systems only and not to the complexities of commercial systems.

The total present-value life-cycle cost for a system is defined as:

Total present-value life-cycle cost = Initial cost + Sum of the maintenance and insurance costs + Sum of repair and replacement costs + Sum of the fuel costs − Salvage value at the end of the calculation period.

Initial cost is the cost to install the system, including the design, equipment purchases, installation costs, value of floor space taken up by the equipment, and any building modification costs.

Maintenance and insurance costs are annual costs associated with the energy system. These costs will not remain constant and may increase or decrease from year to year. Ordinarily they increase at the general inflation rate.

Repair and replacement costs are associated with a major expense such as replacing a pump, collector, or other part of the system. These expenses are not annual expenses and will occur only when a part of the system must be replaced or repaired.

Fuel costs are an annual expense, and these costs have been increasing with an inflation rate greater than the average overall inflation rate.

Salvage value is the remaining value of the equipment at the end of the calculation period.

Illustrative Problem II-6

An active solar space-heating system costs $17,000 (present value), including all of the auxiliary heating system costs, as compared to $5000 for a conventional space-heating system for the same building. The annual heating load is 120×10^6 Btu/yr. The solar system has a solar heating fraction of 0.75, and the conventional system has an efficiency of 0.78. The conventional fuel cost is $6/$10^6$ Btu, and the fuel cost is estimated to inflate at 12 percent per year.

Maintenance and insurance costs for the solar system are estimated at $110 per year (present value) with an inflation rate of 5 percent per year. The conventional system will have a maintenance and insurance cost of $50 per year (present value), also at an inflation rate of 5 percent per year. The discount rate of money will be estimated at 8 percent per year. The active solar system is expected to have two $1000 repair jobs—one in the eighth year and one in the fifteenth year. The conventional system is expected to have one $1000 repair job in the fifteenth year. The estimated system life will be 20 years for both systems, and the salvage value will be assumed to be zero for both systems at the end of 20 years. Calculate the life-cycle costs for these two systems, and determine the net savings, if any, of the solar-heating system.

Solution

The annual fuel used by the conventional system is:

$$\text{Fuel} = \frac{120 \times 10^6 \text{ Btu/yr}}{0.78}$$
$$= 153.9 \times 10^6 \text{ Btu/year}$$

The first-year cost of fuel for the conventional system is, then:

$$\text{Fuel cost} = 153.9 \times 10^6 \text{ Btu/yr} \times \$6/10^6 \text{ Btu}$$
$$\text{(Conventional)} = \$923.40/\text{yr}$$

The annual Auxiliary Heat for the solar system is:

$$\text{AH} = 120 \times 10^6 \text{ Btu/yr} \times (1 - 0.75)$$
$$= 30 \times 10^6 \text{ Btu/yr}$$

The first-year fuel cost for the solar system is, then:

$$\text{Fuel cost} = \frac{30 \times 10^6 \text{ Btu/yr} \times \$6/10^6 \text{ Btu}}{0.78}$$
$$\text{(Solar)} = \$230.77/\text{yr}$$

The future costs for maintenance and insurance are found by multiplying the estimated maintenance and insurance costs by the future cost factor for inflation. The future cost for fuel is found by multiplying the first-year energy costs by the future fuel cost factor. The total future costs are then the sum of fuel, maintenance, and insurance costs. The discounted costs are found by multiplying the total future costs by the discount factor. The accumulated discounted costs are the sums of the annual discounted costs plus any repairs that occur.

Table II-3 shows these annual costs for the conventional system, with the future cost factor for inflation, the future fuel cost factors, and the discount factor applied to the costs, with the resulting accumulated discounted costs.

Table II-4 shows the annual costs for the active solar space-heating system, with the future cost factor for inflation, future fuel cost factors, and the discount factor applied to the costs, with the resulting accumulated discounted costs.

The total present-value life-cycle cost for the conventional system is, then:

$$
\begin{aligned}
\text{Total present-value} \\
\text{life-cycle cost,} \\
\text{(Conventional)} &= \$5000.00 \; + \; 29{,}409.07 \\
&= \$34{,}409.07 \qquad \text{Ans.}
\end{aligned}
$$

The total present-value life-cycle cost for the solar system is:

$$
\begin{aligned}
\text{Total present-value} \\
\text{life-cycle cost} \\
\text{(Solar heating system)} &= \$17{,}000 \; + \; 10{,}596.71 \\
&= \$27{,}596.71 \qquad \text{Ans.}
\end{aligned}
$$

In this case, the solar heating system will provide a savings of $6812.36 over a 20-year period.   Ans.

The major disadvantage of life-cycle cost analysis is that future costs must be predicted by using estimated inflation rates and discount rates. A change of even 1 to 2 percent in either the inflation rate or the discount rate will affect calculated life-cycle costs markedly.

A solar design can be optimized economically by calculating the life-cycle cost of several solar options based on their required solar collection areas. The option with the lowest life-cycle cost will represent the best economic option for the assumptions used in the analysis. Often, one calculation is based on a set of optimistic assumptions, and a second calculation is made based on a set of pessimistic assumptions. These calculations would then provide the client with a best-case and a worst-case result, from which major decisions could be made about the type and size of solar energy system to be designed.

CONCLUSION

Since this is an introductory text, the presentation of solar economics has been purposely held at an elementary level. Only the simplest analyses have been illustrated, and the treatment has been primarily from the viewpoint of residential systems, since commercial and industrial systems involve extremely complex accounting procedures. The concepts and methods of cost–benefit analysis of engineering, industrial, and mechanical systems is a mathematical science called *engineering economics*. Readers desiring further and more advanced discussions of solar economics may want to consult such sources as:

Ruegg, R. T. *Solar Heating and Cooling in Buildings: Methods of Economic Evaluation* (COM-75-11070). Springfield, Va.: National Technical Information Service, 1975.

Riggs, J. L. *Engineering Economics*. New York: McGraw-Hill, 1977.

Table II-3 Annual Life-Cycle Cost Calculations for a Conventional Space-Heating System (Illustrative Problem II-5)

Year of Calculation	Future Cost Factor for Inflation (5%)	Future Cost for Maintenance and Insurance (dollars)	Future Fuel Cost Factor (12%)	Future Fuel Cost (dollars)	Total Future Costs (dollars)	Discount Factor (8%)	Discounted Costs (dollars)	Repair Costs (dollars)	Accumulated Discounted Costs (dollars)
1	1.0500	52.50	1.1200	1034.21	1086.71	.9259	1006.18	0.0	1006.18
2	1.1025	55.13	1.2544	1158.31	1213.44	.8573	1040.28	0.0	2046.47
3	1.1576	57.88	1.4049	1297.28	1355.16	.7938	1075.73	0.0	3122.20
4	1.2155	60.78	1.5735	1452.97	1513.75	.7350	1112.61	0.0	4234.80
5	1.2762	63.81	1.7623	1627.31	1691.12	.6806	1150.97	0.0	5385.78
6	1.3401	67.01	1.9738	1822.61	1889.62	.6302	1190.84	0.0	6576.61
7	1.4071	70.36	2.2106	2041.27	2111.62	.5835	1232.13	0.0	7808.75
8	1.4775	73.88	2.4759	2286.25	2360.13	.5403	1275.18	0.0	9083.93
9	1.5513	77.57	2.7730	2560.59	2638.16	.5003	1319.87	0.0	10403.80
10	1.6289	81.45	3.1058	2867.90	2949.35	.4632	1366.14	0.0	11769.93
11	1.7103	85.52	3.4785	3212.05	3297.57	.4289	1414.33	0.0	13184.26
12	1.7959	89.80	3.8959	3597.47	3687.27	.3971	1464.22	0.0	14648.48
13	1.8856	94.28	4.3634	4029.16	4123.44	.3677	1516.19	0.0	16164.67
14	1.9799	99.00	4.8870	4512.66	4611.66	.3405	1570.27	0.0	17734.94
15	2.0789	103.95	5.4735	5064.23	5158.18	.3152	1625.86	1000.00	20360.79
16	2.1828	109.14	6.1303	5660.72	5769.86	.2919	1684.22	0.0	22045.01
17	2.2920	114.60	6.8659	6339.97	6454.57	.2703	1744.67	0.0	23789.68
18	2.4066	120.33	7.6899	7100.85	7221.18	.2503	1807.46	0.0	25597.14
19	2.5269	126.35	8.6126	7952.87	8079.22	.2317	1871.96	0.0	27469.10
20	2.6532	132.66	9.6462	8907.30	9039.96	.2146	1939.98	0.0	29409.07

Table II-4 Annual Life-Cycle Cost Calculations for an Active Solar Space-Heating System (Illustrative Problem II-5)

Year of Calculation	Future Cost Factor for Inflation (5%)	Future Cost for Maintenance and Insurance (dollars)	Future Fuel Cost Factor (12%)	Future Fuel Cost (dollars)	Total Future Costs (dollars)	Discount Factor (8%)	Discounted Costs (dollars)	Repair Costs (dollars)	Accumulated Discounted Costs (dollars)
1	1.0500	115.50	1.1200	258.46	373.96	.9259	346.25	0.0	346.25
2	1.1025	121.28	1.2544	289.48	410.75	.8573	352.14	0.0	698.39
3	1.1576	127.34	1.4049	324.21	451.54	.7938	358.44	0.0	1056.83
4	1.2155	133.71	1.5735	363.12	496.82	.7350	365.16	0.0	1421.99
5	1.2762	140.38	1.7623	406.69	547.07	.6806	372.33	0.0	1794.32
6	1.3401	147.41	1.9738	455.49	602.90	.6302	379.95	0.0	2174.28
7	1.4071	154.78	2.2106	510.14	664.92	.5835	387.98	0.0	2562.26
8	1.4775	162.53	2.4759	571.36	733.89	.5403	396.52	1000.0	3958.78
9	1.5513	170.64	2.7730	639.93	810.57	.5003	405.53	0.0	4364.31
10	1.6289	179.18	3.1058	716.73	895.90	.4632	414.98	0.0	4779.29
11	1.7103	188.13	3.4785	802.73	990.87	.4289	424.98	0.0	5204.27
12	1.7959	197.55	3.8959	899.06	1096.61	.3971	435.46	0.0	5639.73
13	1.8856	207.42	4.3634	1006.94	1214.36	.3677	446.52	0.0	6086.25
14	1.9799	217.57	4.8870	1127.77	1345.34	.3405	458.09	0.0	6544.34
15	2.0789	228.68	5.4735	1263.12	1491.80	.3152	470.21	1000.00	8014.56
16	2.1828	240.11	6.1303	1414.69	1654.80	.2919	483.04	0.0	8497.59
17	2.2920	252.12	6.8659	1584.44	1836.59	.2703	496.42	0.0	8994.01
18	2.4066	264.73	7.6899	1774.60	2039.32	.2503	510.44	0.0	9504.46
19	2.5269	277.96	8.6126	1987.53	2265.49	.2317	524.91	0.0	10029.37
20	2.6532	291.85	9.6462	2226.05	2517.91	.2146	540.34	0.0	10569.71

Glossary

Absorber plate   the element of a solar collector that absorbs solar radiation and converts it into heat. It is ordinarily black, and may be of metal, plastic, or rubberized materials.

Absorptance   the ratio of the solar radiation actually absorbed by a surface to the radiation incident on the surface. Its value is most often given for normal incidence and for all of the wavelengths of the solar spectrum. Absorptance is one of the key factors in solar collector efficiency.

Absorption refrigeration   vapor-cycle refrigeration without the use of a compressor. The refrigerant is first absorbed in a strong solution, then desorbed with an increase in pressure by an outside source of heat (solar-heated hot water, for example), and then recondensed at a high temperature so that the total heat added in the cycle can be rejected to outside air (usually by means of an atmospheric cooling tower).

Active solar system   a solar heating or cooling system that makes use of solar collectors, a solar thermal storage system, and mechanical equipment for distributing or moving heat.

Air conditioning   the science and practice of producing, within interior spaces, conditions of air that are conducive to human comfort. Temperature, humidity, air motion, and air purity are all controlled in a complete air-conditioning system.

Air-heating solar system   a solar heating system in which air is used as the heat transfer fluid. Air is heated in solar collectors, and the heated air, either by natural or forced convection, transfers heat to the structure or to thermal storage.

Air-to-air heat pump   a heat pump that extracts heat from outdoor air and transfers it to indoor air, raising the temperature of the air in the process.

Ambient air   the surrounding air in the atmosphere, or outdoor air.

Ambient temperature   the temperature of the air outside a building, as measured by an ordinary (dry-bulb) thermometer.

Angle of incidence   the angle at which the sun's rays strike a surface, measured from the perpendicular to the surface to the line of the rays.

Angstrom (Å)   a unit of length for measuring the wavelength of light. One angstrom equals 10^{-10}m.

Annual auxiliary heat   the total annual heating requirement from the backup or auxiliary heating system (conventional fuels) to balance the heating load of a solar building.

Annual solar savings   the total annual reduction in heating and/or cooling requirements of a building (or pool or spa) resulting from the use of solar energy.

Aperture   the surface, usually vertical, glazed, and south-facing, through which solar radiation enters a building. Also called the *solar collection area*.

Backflow valve   a valve in a liquid piping circuit that prevents an unintentional reversal of flow in the circuit. Backflow valves, or dampers, are also used in air distribution circuits.

Balance of system components (BOS)   the other elements, aside from photovoltaic cells (PVCs), that make up a residential or small commercial solar-electric power system.

Berm   a mound of earth positioned against one or more walls of a building, for the purpose of reducing the temperature variations in the structure.

Black body   a body or surface that is "perfectly" black; that is, it absorbs all radiation that strikes it and reflects none. It is also a perfect emitter of radiation at any temperature. Since such a body is theoretical only, the term *ideal black body* is often used.

Boron   the chemical element whose atomic number is 5. It is semimetallic, a borrower of electrons. When used as a dopant in silicon, it borrows electrons from silicon atoms, leaving positive charges as vacancies or holes.

British thermal unit   the amount of heat required to raise the temperature of 1 lb of water 1 F°. The British thermal unit (Btu) is the basic unit of heat in the English (engineering) system of measurement.

Building envelope   the walls, ceiling, floors, windows, and doors of a building taken together; all of the elements that enclose the living space. Sometimes called the *building skin*.

Building load coefficient (BLC)   the building heat loss caused by heat transmission through the building envelope and infiltration of outside air into the building. The units of BLC are Btu/F°-day or W-hr/C°-day.

Calorie   the amount of heat required to raise the temperature of 1 g of water 1 C°. A unit of heat in the metric system of measurement.

Clerestory window   a window high up on a wall that allows sunlight to enter rooms or parts of rooms that would not receive direct rays from windows of conventional height.

Closed-loop system   a flow circuit or piping circuit that is sealed as a unit, without losing or gaining additional fluid.

Coefficient of performance a ratio that indicates the performance of a heating or cooling device. It is the ratio of the heating (or cooling) output of the unit to the heat equivalent of mechanical energy input to the unit.

Collector (or solar collector) any of several types of devices that receive solar radiation and convert it into heat, usually in the form of hot water or hot air.

Collector efficiency the ratio of the collector heat output for a measured time interval to the solar radiation energy incident on the collector for the same period of time. Usually expressed as a percent.

Concentrating collector a solar collector that uses some means of focusing or concentrating the sun's rays onto a relatively small area of absorber surface in order to attain higher temperatures. Lenses or curved reflectors may be used.

Convection the transfer of heat in a fluid (liquid or air) by actual motion of the fluid. Natural convection occurs as a result of the differing densities of hot and cool fluids; forced convection is accomplished with pumps or fans.

Cooling season that portion of the year when the temperature–humidity condition of outdoor air is such that indoor cooling is necessary or desirable for human comfort. Varies from May to October in some U.S. regions to June to September in others. Some regions require no cooling.

Cost–benefit analysis as applied to solar energy systems, the determination of whether the long-term savings from the free solar energy justify the additional capital and operating costs of the solar system, when compared with a conventional system of the same capabilities.

Czochralski process a method of growing large-size silicon crystals by slowly drawing a seed crystal out of molten silicon in an electric furnace by computer-controlled methods.

Declination the angular position of the sun at any time of the year, measured north ($+$) or south ($-$) of the equator. Maximum declination is $\pm 23.5°$.

Deep discharge a situation in a photovoltaic power system where the storage batteries are operated continuously on heavy load to the point where they can no longer deliver current at their rated voltage.

Degree-Day a unit that is a measure of heating and cooling requirements. One degree-day represents a difference of 1° between the mean temperature for that day (the mean temperature is the average of the maximum and the minimum temperatures for the 24-hr period) and a fixed standard temperature (usually 65°F or 18°C).

    Cooling Degree-Days if the minimum and maximum temperatures for a given day are 67°F and 95°F, the number of cooling degree-days for that day is $81 - 65 = 16$.

    Heating Degree-Days if the minimum and maximum temperatures for a given day are 20°F and 40°F, the number of heating degree-days for that day is $65 - 30 = 35$.

Design cooling load the calculated maximum space-cooling load for the location, based on a stipulated indoor condition (usually 78°F, 50 percent relative humidity) and the design (i.e., worst case) outdoor condition for that locality in summer.

Design heating load the calculated maximum space-heating load, based on a stipulated indoor condition (usually 70°F to 72°F) and the design (i.e., worst case) outdoor condition for the locality in winter.

Design temperature summer and winter extremes of temperature listed in tables for hundreds of localities, giving the dry-bulb and wet bulb (summer) temperatures that are near the 20-year extremes for the stations named.

Design temperature difference the difference, in degrees, between the outdoor design temperature and the indoor design temperature.

Differential thermostat a thermostat that responds to temperature differences as indicated by sensors—for example, the difference in temperature of water as it leaves solar collectors from the temperature of water in the thermal storage tank.

Diffuse radiation   solar radiation that has been scattered as it passes through the atmosphere. It appears to come from the entire hemisphere of the sky, and it cannot be focused by lenses or reflectors.

Direct-beam radiation   radiation that comes in a direct line from the sun, without appreciable scattering. It casts sharp shadows and can be focused by lenses and reflectors.

Direct-gain solar energy system   a passive solar-heating system in which direct-beam solar radiation enters the living spaces and warms the interior directly as it is absorbed by interior surfaces.

Dopant   a substance added in trace amounts to a crystal to change its electrical properties. A dopant that has electrons to lend provides negative charges and is an N-dopant. One that borrows electrons creates electron vacancies or "holes" and is a P-dopant.

Draindown system   an open-loop solar water-heating system in which water is drained out of the system when freezing weather is impending.

Efficiency   in any system or machine, the ratio of useful energy output to the total energy input, usually expressed as a percent.

Emissivity   the property of materials that manifests itself in their emitting thermal radiation as their temperature increases.

Emittance   the ratio of the amount of radiant heat emitted from a surface to that which would be emitted from an ideal black body at the same absolute temperature.

Energy conversion   processes by which one form of energy is changed into another form, such as radiant energy to heat energy, heat energy to mechanical energy, or mechanical energy to electric energy.

Energy efficiency ratio (EER)   in mechanical cooling, the ratio of the useful cooling capacity of the unit to the required power input to the cooling equipment, in Btu/hr-W.

Eutectic mixtures   combinations of substances, usually metallic salts, that melt at relatively low temperatures (80°F to 100°F), storing large amounts of latent heat as they change phase, and then releasing this stored heat as they cool and revert to the solid phase.

Evaporative cooling   lowering the dry-bulb temperature of a stream of air by blowing it over wetted pads or through water sprays and then circulating this air through living spaces. Effective only in dry climates. Also called *adiabatic cooling*.

Fenestration   an architectural term that has to do with the design and placement of windows in a building.

Flat-black paint   a nonglossy black paint with a very high absorptance. Used to coat the surfaces of collector absorber plates.

Flat-plate collector   a flat absorber plate installed in a box or frame and usually covered with glazing. It collects solar radiation and converts it into heat. There is no concentration of the sun's rays by lenses or curved reflectors.

Flow rate   quantity (gallons or pounds or cubic feet) of fluid (liquid or air) that passes through a collector system or piping system or duct system in unit time, such as gal/min, gal/hr, lb/hr, or ft³/min (CFM).

Forced circulation   flow of a fluid in pipes or ducts energized by mechanical means, such as a pump or blower.

Fresnel lens   an optical element whose purpose is to bring incoming light to a focus, either at a point or, in some solar energy systems, on a photovoltaic cell or on a fluid-carrying pipe.

Gallium   a metallic element used in the manufacture of some types of photovoltaic cells.

Gigawatt   one billion watts (10^9W).

Greenhouse   a structure, either attached to a main building or free-standing, with large areas of glass in roof and walls, that collects large amounts of solar radiation. The glass, though transparent to short-wavelength solar radiation, is more or less opaque to the low-temperature, long-wavelength reradiation given off by surfaces inside. Heat is trapped, and temperature builds up in the greenhouse.

Grid   the total network of a public utility for the distribution of electric energy.

Heat exchanger   a device used for transferring heat contained in one fluid to a different fluid, without the two fluids coming into contact with each other.

Heat gain   the heat gained by a building from all possible sources—solar radiation, conduction through the building envelope, infiltration of warm, humid air, and heat generated by internal sources.

Heating season   that portion of the year when the temperature of the outdoor air makes indoor heating necessary or desirable for human comfort. Varies from October to May in some climates in the United States to no heating season at all in a few extreme southern regions.

Heat loss   the heat lost by a building from all possible processes—conduction outward through the building envelope, infiltration of cold outside air, convection, and radiation.

Heat pump   a refrigerating unit that can be reversed as desired to supply heat in the form of warm air to a building or space in winter, and extract heat from (i.e., cool) the space in summer. As described, this is an air-to-air heat pump. Water-to-air and water-to-water heat pumps are also used for a variety of purposes.

Heat source   a fluid or substance that contains large amounts of heat and serves as a source of supply of heat energy for a heat pump.

Heat storage   a body or substance that absorbs collected solar radiation and retains the heat energy for release later into a space or building.

Heat transfer   the process of moving heat from where it is collected or produced to some other location. This is often accomplished by using water or air as the heat-transfer medium.

Heliostat   a mirror or array of mirrors equipped with a sun-tracking mechanism for reflecting solar radiation to a specific point or surface at all hours of the day for every day in the year.

Hole   a location in a metallic crystal where there is an electron vacancy in the lattice. The hole acts as a positive charge.

Hour angle   the angular distance of the sun from its high point of the day at solar noon.

Humidity   a general term indicating moisture (water vapor) in the ambient air. *Absolute humidity* is the actual amount of water vapor in the air at a given time, in pounds of water vapor per pound of air. *Relative humidity* is the ratio of the absolute humidity at the given temperature to the humidity if the air were saturated with moisture at the same temperature. Relative humidity (RH) is a measure of the percentage of saturation.

HVAC system   a heating, ventilating, and air-conditioning system for a building.

Hybrid system   a solar-heating system that incorporates both passive solar and active solar elements.

Incident angle   the angle between a direct ray of sunlight and the perpendicular to the surface on which the ray is incident.

Indirect-gain system   a passive solar energy system wherein solar radiation is first absorbed by thermal storage and heat is later released into the living space. Direct-beam solar radiation does not enter the living space in this type of system.

Infiltration   air leakage from outdoors into a living space through cracks, through and around doors, or as doors or windows are opened. Wind pressure, especially in winter, increases infiltration.

Infrared radiation   electromagnetic radiations in the region just beyond the red end of the visible spectrum, beginning at a wavelength of about 7000 Å or 0.7 μ. So-called "heat waves" are infrared radiation.

Insolation   the total spectrum of radiation energy received from the sun, including infrared, visible, and ultraviolet. In solar energy practice, it is the rate at which solar energy is received (solar power) on a surface, measured in Btu/hr-ft^2 or in kW/m^2. Sometimes called intensity of solar radiation.

Insulator   any one of a number of materials that resist the flow of heat by conduction. Poor conductors of heat are good insulators.

Internal heat gain   heat generated in a living space by occupants, machinery, appliances, cooking, use of hot water—any source of heat within the space except the heating system itself.

Inverter   a device that converts dc to ac. An essential element of a photovoltaic power system that is to provide ac.

Kelvin (K)   a unit of temperature change on the Kelvin (Celsius absolute) scale of temperature. Temperatures are expressed as *kelvins* [no degree (°) sign].

Kilowatt-hour   a unit of electric energy expended or supplied when 1 kW of power is steadily applied for 1 hr. Abbreviated kW-hr.

Langley   a solar radiation intensity of 1 cal/cm^2. As a unit of insolation (solar power), 1 Ly/min = 221 Btu/hr-ft^2.

Latent heat   heat energy stored and/or released as a result of a change in phase. In air-conditioning practice, the latent heat of water vapor in the air constitutes a cooling load. Conversely, adding moisture to the air for winter air conditioning constitutes a heating load.

Life-cycle cost analysis   one method of economic analysis of a solar energy system. All capital, operating, and maintenance costs are estimated for the predicted lifetime of the system; and then this cost is compared to that for a conventional heating (and/or cooling) system.

Line-tie system   a photovoltaic power system that is tied in with utility power in such a way that not only can power be drawn from the utility when solar energy will not meet the system demand, but excess power can be supplied to the utility when demand is less than the PV system output. The excess power is metered and credited to the customer's account.

Liquid-heating system   an active solar energy system with collectors that heat a liquid, which is then used as the heat transfer medium to supply heat to the living space, pool, spa, or domestic hot water.

Load   in heating or cooling systems, the total amount of heat that is needed and is to be supplied by solar or conventional energy or a mix of both. In a photovoltaic system, the current that must be supplied at a given voltage to meet the electrical demand.

Load Collector Ratio (LCR)   a parameter used in predicting the performance of passive solar energy systems, obtained by dividing the Net Load Coefficient by the projected vertical-plane area A_p of the total solar collection area.

Local Solar Time (LSoT)   the actual time at a given meridian of longitude, based on solar noon being defined as the instant when the sun crosses that meridian.

Mean (average) daily temperature   the average of the minimum and the maximum temperatures during a 24-hr day.

Megawatt   one million watts (10^6 W or 10^3 kW).

Microclimate   the conditions of temperature, humidity, sunlight, wind, fog, rain, and so on, in the immediate surround or over a small area.

Micron   one millionth of a meter, 10^{-6} m (abbreviated μ). Another term is micrometer (abbreviated μm).

Movable insulation   insulating materials, often of the rigid-board type, that can be moved into place when heat transfer is not desired and moved aside when heat flow is desired or when natural lighting or a view out of a window is desired.

N-silicon   silicon containing a dopant such as phosphorus, resulting in the crystalline structure containing excess electrons in the crystal lattice.

Net Load Coefficient (NLC)   the daily heat loss from a solar building, calculated by excluding any gains or losses through the solar elements of the building. The solar elements are assumed to be replaced by a thermally neutral wall. Units of NLC are Btu/F°-day.

Normal   the direction that is perpendicular to a given surface.

Open-loop system   a flow circuit or piping circuit that is open to a continuous resupply of the fluid being heated or cooled.

Outgassing   the production of vapors from the chemical decomposition of materials exposed to high temperatures, such as insulating materials in solar collectors.

Outside air   air from outdoors that enters a building being heated or cooled, either by infiltration or by intentional introduction. Outside air usually contributes to the heating load in winter and to the cooling load in summer.

P-silicon   silicon containing a dopant, such as boron, which is a borrower of electrons. The crystal structure is then deficient in electrons.

Parabolic-trough collector   a solar collector that focuses solar radiation along the linear axis of a trough. The trough is formed by a reflector with a parabolic cross section.

Passive solar energy system   a solar heating and/or cooling system in which the energy is supplied with little or no mechanical equipment. The building itself acts as the heat energy collector and the distribution system.

Payback period   the number of years required for a solar heating or cooling system to pay for itself from savings realized by not having to buy fuel or electric energy.

Peak power   the power output of a solar system (in heat or electric energy) when there is peak solar radiation at perpendicular incidence on the collectors or photovoltaic cells. Peak solar radiation is also called 1 sun and is standardized at 1 kW/m² or 317 Btu/hr-ft².

Peak watt   essentially the same as peak power. More often used with photovoltaic systems to indicate the electric power that a solar cell will produce at noon with the cell directly facing the sun when the insolation is 1 kW/m² or 317 Btu/hr-ft².

pH   a measure of acidity or of alkalinity, important for corrosion control in piping circuits. A pH of 7 is neutral; pH less than 7 is acid; pH greater than 7 is alkaline.

Phase-change materials   substances that store heat as latent heat of fusion or latent heat of vaporization. Also called eutectic materials, they store heat as they melt (fusion), and when the phase change reverses, they release the stored heat on solidifying.

Phosphorus   a chemical element whose atoms have loosely bound electrons. Phosphorus is used as a dopant to create N-silicon.

Photon   according to the quantum theory, a particle of light energy. The photon is hypothesized as moving with the speed of light and as having the capability of reacting with atoms that it strikes to eject electrons from those atoms—the *photoelectric effect*.

Photovoltaic cell   a semiconductor device that converts light into a flow of electrons. A solar cell is a photovoltaic cell designed to produce electric energy from direct-beam solar radiation. Photovoltaic cells are called PVCs in industrial practice.

Photovoltaic module   a number of photovoltaic cells (quite commonly 36 to 45) mounted in a sealed panel and electrically connected to provide the desired output current characteristics. A photovoltaic *array* is an interconnected arrangement of modules designed to provide electric power for a specific application.

P–N junction   the interface region between *P*-type material and *N*-type material in semiconductors.

Power   the rate at which work is done or the rate at which energy is supplied or used. Power units are the ft-lb/sec, the horsepower (1 hp = 550 ft-lb/sec), the watt, and the kilowatt (1 kW = 1000 W).

Present value   the value in today's dollars of money or assets to be received at some future time.

Pressure   force on a unit of area. Units are lb/in.^{2}, lb/ft^2, or, in the metric system, pascals or kilopascals.

PV   the industry abbreviation for photovoltaic(s).

Pyranometer   an instrument for measuring total sky (direct and diffuse) solar radiation intensity (insolation).

Pyrheliometer   an instrument that intercepts and measures only direct-beam solar radiation.

Quad   one quadrillion (10^{15}) Btu. This is a vast amount of energy. Total annual energy use in the United States (mid-1980s) is in the range of 80 to 85 Quads.

Reflectance   the ratio of the radiation energy reflected from a surface to the radiation energy incident on the surface.

Refrigerant   any one of a number of low-boiling-point substances that accomplish cooling by means of a cyclic process involving two changes of state. Ammonia, methyl chloride, and the Freon family of refrigerants are examples.

Resistance to heat flow (R value)   A numerical parameter that expresses the ability of a material to impede the flow of heat by conduction. In the building construction industry, insulating materials are rated for their *R* values.

Resistance heating   the production of heat from electricity by forcing current through a resistance coil or rod. The conversion is accomplished at 100 percent efficiency, yielding 3410 Btu/kW-hr.

Retrofit   the application of solar heating elements or systems to a building not originally designed as a solar building, one that has originally had conventional heating and/or cooling equipment.

Rock bed   a bed or bin of stones used to store thermal energy while the energy is being collected, for release to living spaces at some later time. Rock beds are used with passive systems, active systems, and hybrid systems.

Selective coating   a coating material applied to solar collector absorber plates. These coatings have a high absorptance for solar (high-frequency, short-wavelength) radiation and a low thermal emittance for low-frequency, long-wavelength radiation. Such coatings increase collector efficiency if other factors are held constant.

Semiconductor   any one of a number of materials whose electrical conductivity lies between the class known as insulators (glass, mica, rubber) and the class known as good conductors (copper, aluminum, iron). Silicon, germanium, and gallium are semiconductors.

Sensible heat   heat energy that, when added to a substance, causes that substance's temperature to rise. This kind of heat is apparent to the human senses.

Sensor   any one of many devices that detect changes in temperature, pressure, humidity, and so on, and then relay this information to a gauge, thermostat, motorized control, or computer.

Shading angle   that elevation angle of the sun that begins to cause a shadow on a solar aperture, window, or thermal storage wall.

Shading coefficient (SC) a rating for solar control devices such as screens, blinds, or drapes. The coefficient for a window with solar controls is expressed as a fraction, compared to a baseline window with single glazing and no shading controls.

Silicon a chemical element and a metallic semiconductor. It is present in bountiful supply in the earth's crust in sand and quartz (SiO_2). The current workhorse semiconductor in both the electronics and the photovoltaics industries.

Solar altitude the angular distance of the sun above the horizon from a particular location, at a specified time of day and day of the year. Also called the *solar elevation angle*, and (sometimes) the *vertical angle*.

Solar azimuth the angular distance from due south to the sun's projection on the horizon.

Solar building a structure that relies on solar energy for a good portion of its heating and/or cooling requirements, and one that was designed with solar energy systems in mind.

Solar cell a photovoltaic cell designed for the specific purpose of converting solar radiation energy into electric energy.

Solar collection area the surface area on a building through which solar radiation energy flows. The area is most often glazed and is usually but not always, vertical and south-facing (northern hemisphere). Also called solar aperture, the area is measured in square feet or square meters.

Solar constant the intensity of solar radiation in near-earth space. The solar power incident normally on unit surface area just outside the earth's atmosphere. Solar constant = 429 Btu/hr-ft^2, or 1.35 kW/m^2.

Solar cooling the summer cooling of buildings by passive or active solar methods. Passive methods include night ventilation, earth-contact cooling, and night-sky radiation. Active solar cooling uses solar heat to energize conventional refrigerated cooling equipment.

Solar energy energy from the sun that arrives on earth as electromagnetic radiations in the spectrum range from about 0.25 to about 2.50μ.

Solar Load Ratio (SLR) a coefficient used in predicting the performance of a passive solar building. It is the monthly net solar energy absorbed by the building divided by the monthly net building heating load.

Solar noon the instant at which the sun, in its daily traverse across the sky, crosses the meridian of the observer.

Solar radiation energy from the sun. According to the wave theory, this energy is in the form of electromagnetic radiations (waves). The particle theory attributes solar energy to high-speed particles called photons that travel through space with the speed of light.

Solar spectrum the wavelength (or frequency) distribution of the electromagnetic radiations emanating from the sun.

Solar time the time of day as determined from the apparent motion of the sun across the sky. Distinguished from standard time or clock time, which is based on artificially determined time zones.

Space heating winter heating of living spaces. If solar energy is the heat source, either passive solar methods or active solar methods may be used.

Specific heat capacity the ratio of the thermal capacity of any substance to the thermal capacity of water.

Stagnation temperature the highest temperature reached inside a solar collector on a sunny day if there is no heat-transfer fluid circulating through the collector.

Stand-alone system a photovoltaic power system with no connection to a utility grid. Such systems usually have battery storage.

Supplementary (auxiliary) heat heat provided by a conventional source (furnace, boiler, or electric unit) when solar heat is insufficient to meet the load demand.

Temperature   an indication of heat intensity or degree of heat. As a hypothesis, temperature is a manifestation of the average kinetic energy of the molecules of a substance. Temperature determines the direction of heat flow, always from hot bodies to cooler bodies along a descending temperature gradient.

   Dry-bulb temperature   the temperature as read from an ordinary thermometer held in air.

   Effective temperature   a temperature scale that is a measure of human comfort, taking into account the effects of dry-bulb temperature, relative humidity, and air movement.

   Wet-bulb temperature   the air temperature indicated by a thermometer whose bulb is surrounded by a wick saturated with water. The thermometer must be either whirled or placed in a moving stream of air to maximize evaporation of the water, resulting in a cooling effect on the thermometer bulb. The wet-bulb temperature is an indication of the total heat content of the air (sensible and latent).

Thermal capacity   the amount of heat that must be added to or removed from unit mass (weight) of a substance to cause its temperature to change by 1°. Units are Btu/lb-F° or kcal/kg-C°. Thermal capacity is a measure of the heat storage capability of materials.

Thermal conductivity   the amount of heat transferred through unit cross-section area of a substance per unit time per unit temperature difference between two parallel surfaces unit distance apart. Units are Btu/hr-ft^2-F°/in., or W/m^2-C°/m.

Thermal efficiency   the ratio of the energy output of a device or system to the thermal energy input to the device or system. For example, the mechanical power output of a steam turbine divided by the heat power input to the steam (from solar or conventional sources) gives the thermal efficiency.

Thermal energy   energy of substances or materials resulting from the intensity of motion of their molecules and the internal energy of molecules and atoms.

Thermal storage   materials of considerable mass (weight) and with appreciable thermal capacity, used in solar buildings or in active solar energy systems to store thermal energy for later use. Water, stones, brick, adobe, concrete, and ceramic tile are all used for thermal storage.

Thermodynamics   the science of energy transformations involving mechanical work and heat. The First Law of Thermodynamics states that in any thermodynamic process, there is no gain or loss of energy—total energy remains constant. The Second Law of Thermodynamics concerns the availability of heat energy. Heat will not flow up a temperature hill unless energy from another source is supplied to force it to do so. Or—the corollary or inverse—mechanical energy cannot result from heat energy unless the source of heat energy (the "working substance") undergoes a drop in temperature.

Thermosiphon   circulation of a fluid by convection as warm, less dense fluid rises and cool, more dense fluid falls, under gravity forces.

Thermostat   a device for sensing the temperature of fluids (including air) and actuating controls so that the system will maintain the desired temperature within narrow limits.

Thin film   in photovoltaic cell manufacture, any one of several semiconductor materials (silicon, gallium arsenide, etc.) deposited in very thin coatings on a substrate material.

Tilt angle   the angle between a horizontal plane and the surface of a collector array or a photovoltaic array. Tilt angle is an important consideration in the year-round performance of the solar array.

Tracking array   a collector array or photovoltaic array equipped with a mechanism that automatically tracks the sun. Dual-axis tracking keeps the array pointed at the sun every minute of the day for all days of the year. Single-axis tracking follows the sun during each day but makes no adjustment for changes in the sun's declination during the year.

Transmittance the ratio of solar radiation energy transmitted through glazing to the incident radiation energy on its surface.

Trombe wall a thermal storage wall made of stone, brick, concrete, or adobe (for example) that has thermocirculation vents so that air from the living space will enter the space between the wall and the glazing (through bottom vents) and be heated, returning to the living space through top vents.

Ultraviolet radiation the spectrum band of radiations with wavelengths shorter than visible light but longer than x radiation.

Volumetric thermal capacity the thermal capacity per unit volume, as distinguished from thermal capacity per unit mass (weight). Volumetric thermal capacity is a more meaningful concept when designing thermal storage walls, since dimensions are more readily dealt with than masses.

Wafer a thin slice of single-crystal silicon (or other semiconductor) obtained by sawing from a cylindrical ingot.

Watt the metric unit of power, or work done or energy expended per unit time. It is the commonly used unit of electric power in both metric and English systems of measurement. One ampere (1 amp) of current at a potential of one volt (1 V) provides a power of one watt (1 W). A more practical unit of power is the kilowatt (1 kW = 1000 W).

Watt-hour the unit of electric energy. One watt of power provided for one hour results in 1 W-hr of electric energy. A more practical unit is the kilowatt-hour (kW-hr).

Index

A

Absolute pressure, 44
Absolute temperature, 78–79
    Kelvin scale, 78
    Rankine scale, 79
Absorbed solar radiation, 389
Absorber plate, 387ff
    surface coatings for, 421–425
Absorptance, of surface, 83
    of collector absorber plate, 422
Absorption cooling, solar energized,
        658–664
Absorption of radiant energy, 82–85
Absorption of solar radiation, by
        thermal storage, 256ff
Absorption refrigerating cycle,
        658–659
Active solar energy systems, defined,
        189, 385
    basic elements of, 385
    cooling configurations, 651–655
    heating configurations, 469–576
Air-based solar energy systems,
        586–590
    operating modes for, 588–590
Air conditioning, defined, 629–630
    solar-energized modes of,
        642–665

Air-heating flat-plate collectors,
        394–396, 418–421
Air mass attenuation, 96
Air spaces in walls, effect on
        conduction heat flow, 155–156
Alternative energy sources, 15–19
Angstrom unit, defined 77
Aperture (solar collection area),
        187–188
Aperture control strategies, 332–333
Apparent solar time (sun's hour
        angle), 105
Atmospheric pressure, 39, 42–43
Auxiliary Heat (AH), defined, 214
    domestic hot water systems,
        484–486
    pool and spa systems, 548–550
    space-heating systems, 584–585,
        586–590
Average monthly heating load, 172
    domestic hot water, 491–494
    pools and spas, 540–541
    space heating, 590–591

B

Back-draft damper, 290–291
Balance of system (BOS) equipment,
        699, 717–722

Baseboard radiator, hydronic
    system, 582
Base temperature, for degree-days
    (DD) determination, 591
Batch-type solar water heater,
    469–471
Bioclimatic Chart, 198–201
Biomass energy conversion, 14
Bourdon gauge, 44–45
Breadbox solar water heaters,
    471–472
British thermal unit (Btu), defined,
    54
Building envelope (skin), 154–168
Building heat gain, 231–238, 637–639
Building heating load, defined,
    149–151
    analysis and calculation of,
        150–178
    for space heating, 590–591
Building Load Coefficient (BLC),
    defined, 175–214
    for active solar space heating, 572

C
Calorie, metric unit of heat
    quantity, 54
Cavity radiation, 80
Change of state, basic principles,
    60–63
    of eutectic materials, 255–256
Check test and start-up procedures,
    526
Chromosphere, 90–91
Clear-day insolation, 116
    tables of values, 137–147
Clearness factor, 116–117
Clerestory windows, 243–246
Closed-loop system configurations:
    for domestic hot water heating,
        481–485
    for pool and spa systems, 548–550
    for process heat systems, 680–684
    for solar-energized cooling,
        658–663
    for space heating, 575–577
Coal, as energy source, 14–18
Coefficient of performance (COP),
    of refrigerating machines and
    heat pumps, 654–655,
    686–688
Collector absorber plates, 387–390,
    414–420
    for air-heating collectors, 418–421
    for liquid-heating collectors,
        413–418
Collector area requirements:
    for active solar cooling, 664–665
    for domestic hot water systems,
        495–497

for industrial process heating,
    680–683
for pool and spa heating, 539–546
for space heating, 572, 590–591
Collector assembly and details of
    construction, 438–443
Collector azimuth angle, 399
Collector connections and
    mounting, 439–443
Collector efficiency, defined,
    445–446
    curves of, 451, 456, 458, 460–461,
        462
    glazed collector, 449–454
    unglazed collector, 447–448
Collector energy production rate,
    452
Collector glazings, 432–434
Collector insulation, 435–437
Collector mounting procedures,
    515–523
Collector orientation angles, 399
Collector plumbing layouts, 502–511
Collector tilt angle, 399
Combination solar energy systems:
    domestic hot water and hot tub,
        565–566
    pool and domestic hot water, 567
    pool and hot tub (spa), 566
    space heating and pool, 620–623
Comfort Chart, 196
Comfort Zones, winter and
    summer, defined, 196–197
Commercial uses of solar energy,
    670ff
Components of flat-plate collectors,
    412–421
Compression refrigerating cycle,
    651–655
Concentrating (focusing) collectors,
    390–393
    for industrial process heat,
        674–680
    for solar cooling applications, 655,
        659
Concentrator solar cells, 710
Conductance (C) of materials, 70 ff
Conduction, basic principles of,
    66–70
    coefficient (K), 67–70
Conservation of energy, 57, 571–572,
    642, 669
Construction material, thermal
    properties, 155–158
Control systems, for domestic hot:
    for industrial process heating,
        685–686
    for pools and spas, 553–556
    for space heating, 618–619
    water heating, 512–514

Convection, basic principles, 72–76
Convection (air film) coefficient,
    73–74, 76, 155
  for losses from absorber plate, 388
  for losses from pool and spa
      surface, 542–544
Convective-loop systems, defined,
    192
Convector, fan-forced, for room
    heating, 582
Corona, 90–91
Cost-benefit analysis, solar energy
    systems, 743–754
  home photovoltaic (PV) system,
    728–729
Czochralski process, 704–706

D
Daylighting in solar buildings,
    358–364
  controlling artificial lighting, 364
  controlling glare, 363
Degree-days (DD), defined, 172, 214
  table of, 173
Density, defined, 41
Design building heating load, 171
Design cooling load, 636–642
Design temperature difference, 152,
    636
Dew-point temperature, 631
Differential linear expansion, 65–66
  as used in thermostats, 65
Differential thermostat, 396–397,
    512–513, 619
Diffuse radiation, 98–99
Dimensional analysis, in problem
    solving, 49–50
Direct-beam radiation, 98–99
Direct-gain passive systems, defined,
    189
  advantages and disadvantages,
    231
  essential features, 229–230
  steps in design, chart, 274
Discounting, in solar economics,
    745
"Doping" of semiconductors,
    703–705
Drainback system, for active solar:
  for domestic water heating,
    509–511
  space heating, 577–581
Draindown freeze protection,
    479–481
Dry-bulb temperature, defined, 151,
    630

E
Earth-contact cooling, 649–650
Earth's annual motion, 102–105

Economics of solar energy systems,
    743–754
Effective temperature, defined,
    196–197
  use of:
    in summer cooling analysis,
    635
    in winter heating analysis,
    196–202
Efficiency, defined, 59
  of solar cells, 707 ff
  of solar collectors, 445–460
    air-heating, 460–461
    liquid-heating, 445–458
Electrical load for PV systems,
    analysis of, 723–724
Electromagnetic spectrum, 76–77
  chart of, 77
Electrons and "holes" in crystal
    lattice, 702ff
Electron-volt, defined, 702ff
Emission of radiant energy, 82–85
Emissivity, of surface, 82–83
Emittance, of collector absorber
    plate, 422–423
Energy, defined, 47
Energy conservation, 150–151,
    571–572
Energy efficiency ratio (EER), 654
Energy flows, for glazed collectors,
    450
Energy Gap, 15–21
  charts illustrating, 17–18
Engineering economics, 754
Equation of time, 107
Equinoxes, spring (vernal), and fall
    (autumnal), 102–104
Eutectic salts, 62, 255
Evacuated-tube solar collectors,
    674–676
Evaporative cooling, 645–648
Extraterrestrial solar radiation, 93–95

F
Film (convection) coefficient, 73–75
Flat-black paint, 422–424
  use of on collector absorber
    plates, 422–425
Flat-plate solar collectors, 387–390,
    411–414
  air-heating type, 419–421
  efficiency of, 445–460
  heat absorbed by, 387–388
  heat losses from, 388–390
  liquid-heating type, 414–418
  for solar cooling, 659
Fluid-flow calculations, 505–509
Fluid pressure, 42–47
  calculations for static pressure,
    44–45

Fluid pressure (*continued*)
  measurement of, by gauges, 42–45
Force, defined, 38
  and mass relationships, 38–39
Forced convection, 72
Fossil fuels, 14–15
  unique values of, 15
Freeze protection, for open-loop
      liquid-heating systems,
      478–481
    for draindown systems, 479–481
    for outdoor pools and spas,
      555–556
Frequency, of electromagnetic
      waves, 77
Fundamental heat equation, 54

G
g, acceleration of gravity, 39
Gauge pressure, 44
Gauges, pressure:
  Bourdon type, 43
  U-tube type, 42–43
Geothermal energy, 16
Glazings:
  for flat-plate collectors, 390,
      425–435
    absorptance of, 425–426
    installation of, 432–435
    reflectance of, 425–426
    table of properties of, 427
    transmittance of, 425–426
  for thermal storage walls, 284–286
Glazing support systems, 283–284
Greenhouse effect, 82
Greenhouses as sunspaces and solar
      collection elements, 316–321
Grid-connected photovoltaic (PV)
      systems, 721

H
Heat, as form of energy, 32
  and change of state, 60–63
  and temperature, 33
Heat conductivity (K), 66–70
  table of, for selected materials, 69
Heat exchangers, for pools and spas,
      549–550
    for process heating, 681–685
    for solar hot water systems, 475,
      481–484
    for space heating, 575–577
Heat flow, as result of temperature
      difference, 53
Heat gains:
  from conduction, 640–641
  from infiltration, 641
  from radiation, 640
  from ventilation, 641

Heating load calculations for solar
      design, 171
Heat loss through building envelope,
      154–168
    from conduction, 154, 156–167
    from convection, 154
    through walls, ceilings, and floors,
      156–168
Heat measurement, 53–56
Heat pumps in solar energy systems,
      665ff
    coefficient of performance (COP),
      654, 687
    for auxiliary heat, 586–590
    high-temperature application,
      686–689
    solar-assisted configuration, 587
    solar-augmented configuration,
      588
Heat storage:
  latent heat, 255–256
  sensible heat, 254
Heat transfer, 66–85
  by conduction, 66–70
  by convection, 72–76
  by radiation, 76–85
Heat transmission overall coefficient
      (U), 71–72
Heat transmission losses, minimizing
      by insulation, 150
Heliostats, 732–734
Hertz, unit of frequency, 77
"Holes," in semiconductor crystal
      lattice, 703ff
Home electric (PV) systems, 719–727
Horsepower, English-system unit of
      power, 47
Hot tub solar-heating systems,
      560–565
    design criteria for, 560–562
Hot-water load, how to determine,
      491–494
    tables for average usage, U.S., 492
Hot water systems for domestic use,
      469–532
Human comfort, conditions for,
      195–200, 635
    in passive solar buildings, 195ff
Humidity, 197, 630–635
    relative, 196–198, 630, 642–651
    specific, 630
Hybrid photovoltaic systems,
      714–722
Hybrid solar energy systems,
      defined, 192–194
    other hybrid designs, 353
    using rock beds for storage,
      341–350
Hydroelectric energy, 15
Hydronic heating systems, 582–583

Hydrostatics, 45–47
  in collector piping systems,
      502–508

I
Ideal blackbody, 79–80
Incentives, for installing solar, 749
Incidence, angle of, 105, 421
Incident angle modifier, 452
Incident solar radiation, 389
Indirect-gain passive solar systems,
      defined, 189–190
  configurations for, 283–328
  using roof ponds, 310–314
  using sunspaces, 315–327
  using Trombe walls, 284–295
Indoor design conditions, 152, 635
Industrial uses of solar energy,
      669–687, 729–734
  for electric power production,
      729ff
  for generating steam, 680
  for process heat, 680–685
Infiltration air:
  component of heating load, 168
  component of total
      cooling load, 637
  estimating the volume of, 168,
      640–641
Infrared region of solar spectrum,
      76–77
Insolation, defined, 98–102
  component elements of, 98–99
  magnitude of, 99
  measurement of, 100–101
  total cooling season, 119–120
  total daily, 119
  total heating season, 119–120
  total monthly, 119
  units of, 98–100
Insolation available, 116–117
  calculation of, 102
  factors affecting, 116
  on photovoltaic arrays, 724
  on pools and spas, 542–546
  on solar collectors, 398
Insolation tables, 137–148
  use of, 117–120
Installation details:
  for pool and spa systems, 556–565
  for solar hot-water systems,
      515–525
  for space-heating systems, 609–626
Insulation:
  in buildings, 150, 159ff
  in solar collectors, 435–437
Integral collector storage (ICS)
      water heaters, 470–472
Intensity of solar radiation
      (insolation), 98

Internal energy of molecules, 33
Inverters, photovoltaic systems,
      720–722
Isolated-gain solar energy systems,
      defined, 191–192
  configurations and operating
      modes, 310–329

J
Joule, SI-metric unit of work,
      energy, and heat, defined, 47,
      59
Joule's constant, 58–60

K
Kelvin temperature scale, 78–79
Kilogram, SI-metric unit of mass, 35
Kilowatt, SI-metric unit of electrical,
      mechanical, and thermal
      power, 48
Kinetic-molecular hypothesis, 33

L
Latent heat:
  of fusion, 61
  of vaporization, 62
Latent heat gains, in air
      conditioning, 637
Latent heat thermal storage, 293–295
Life-cycle cost analysis, 750–754
Line-tie photovoltaic energy
      systems, 721–722
Liquid-heating solar collectors,
      387–394, 411–440, 445–465
Load:
  design cooling, 635–642
  design heating, 171ff
  monthly structure heating load,
      172
  total cooling season, 120, 642
  total heating season, 120, 171–175
Load Collector Ratio (LCR),
      defined, 218
  correlated with solar savings
      fraction (SSF), 219, 376
Load Collector Ratio (LCR) method
      of performance analysis,
      222–224, 336, 368–380
Local solar time (LSoT), defined,
      106
  calculation of, 107–109
  use of, in obtaining sun angles,
      105–109

M
Manometer gauges, 42–43
Masonry, concrete, and adobe as
      thermal storage walls, 286

Mazria's guidelines:
  for sizing solar collection area, 234
  for sizing thermal storage walls,
    297
Mean radiant temperature (MRT),
  198–201
Measurement, systems of, 34–45
  English and SI-metric units of,
    35–38
    tables of equivalents, 36–37
Mechanical equivalent of heat,
  57–60
Micron, defined, 77
Monthly and annual Auxiliary Heat,
  calculation of by SLR
  method, 375–380
Monthly and annual Solar Savings,
  calculation of by SLR
  method, 375–380
Monthly net reference thermal load
  (MNRTL), 215
Monthly net solar energy absorbed,
  215
Monthly Solar Load Ratio Method,
  advanced considerations,
  368–380
Monthly structure heating load,
  172–174
Multiple glazings, effect of on
  collector performance,
  428–432

N
Natural convection, 72
Natural gas as energy resource, 14,
  17–18
Near-earth space, 96
Net load coefficient (NLC), 175, 214
Net reference thermal load (NRTL),
  175, 214
Newton (N), SI-metric unit of force,
  38
Nuclear energy, 16–17

O
Ocean thermal energy conversion
  (OTEC), 13
Oil, as energy source, 15–17
Open-loop active solar systems,
  defined, 394–395
  for domestic hot water, 473–481
  for swimming pools, 548–549
Outdoor design conditions, 152–153,
  636
Outdoor design temperature, 151
  table of values for selected
  locations, 152–153
Overall coefficient of heat transmission (U), 154–168
  table of values, 162–164

use of in heating and cooling load
  calculations, 165–167

P
Parabolic-trough solar collectors, 387
  for commercial and industrial
  service, 678–680
  for solar-energized cooling
  systems, 655, 659
Pascal, SI-metric unit of pressure,
  40
Passive solar cooling, 642–651
  and human comfort, 643
  by earth contact, 649–651
  by evaporative methods, 645
  by radiation to night sky, 648
  by ventilation, 643
Passive solar elements, combinations
  of, 336–339
  effects on heating and cooling
  loads, 331–336
Passive solar energy, history of use,
  185–186
Passive solar energy systems,
  defined, 187–189
  illustrated and described, 190ff
Passive solar system design, basic
  steps in, 206–213
  construction documents phase,
  207
  design development phase, 206
  flow chart of, 208
  schematic design phase, 206
Passive solar systems, classified, 189
  advantages and disadvantages of,
  194–195
Payback method of economic
  analysis, 743–750
Percent of insolation, in solar access
  studies, 400–401
Percent of occlusion, in solar access
  studies, 401–402
Percent possible sunshine, 116
  table of values for selected
  locations, 136
Performance analysis of solar energy
  systems:
  the LCR method, 222–227,
  308–310, 324–328, 364–380
  the SLR method, 213–222,
  364–380
Petroleum, as energy source, 14–15,
  17–18
Phase-change, in water, 61–62
  in metallic salt hydrates
  (eutectics), 62
Phase-change materials, as thermal
  storage, 255–256
Phase-change solar water heaters,
  489–491

Photoelectric effect, 700–703
Photons, particles of light energy, 695
Photosphere, 90–91
Photovoltaic cells (PVCs), 693ff
Photovoltaic energy available from insolation, 724–728
Photovoltaics industries, 695ff
  cost trends in, 697
  development of, 696
  present status and future potential of, 697–698
Photovoltaic modules and arrays, sizing and installing, 712–714, 725–728
  short-cut methods for sizing, 727–728
Photovoltaic power, 711–722
  for electric utilities, 729–732
  system components (BOS equipment), 711–714
Photovoltaic systems, cost-benefit analysis, 728–729
Pipe and pump sizing for solar installations:
  domestic hot water systems, 502–509
  space-heating systems, 575–581
  swimming pools and spas, 550–552
Pool and spa controls, 553–555
Pool and spa covers, 546–548, 560–561
Pool heat gains, 542
  solar radiation absorbed, 544
Pool heating load, calculation of, 542–546
Pool heat losses, classified, 535–537
Pool solar energy system installation, 556–560
Power, defined, 47
Present value of future savings, 745
Pressure, defined, 39
  measurement of, 42–47
Pressure losses in fluid flow,
  in ducts, 347
  in pipes, 506–509
  through valves and fittings, 507
Prime Solar Fraction (PSF), 400
Process heat in industry, 680–685
Profile (shadow-line) angle, 114, 248
Psychrometric Chart, 630–632
  use of in determining indoor conditions, 633–635
Pyranometer, 100–101
Pyrheliometer, 100–101

Q
Quad, defined, 17
Quantum, unit increment of energy, 78

Quantum theory of radiant energy, 78

R
Radiant energy, 77–84
  from sun, 89–120
Radiation, 76–85
  and human body heat gain or loss, 84–85
  blackbody radiation, 79–81
  Stefan-Boltzmann law of, 80
  Wien's displacement law, 81–82
Radiation heat gains:
  by solar collectors, 387–392, 445–449
  by thermal storage walls, 284–288
  through glass, 231–238
  to earth, 92–100
Radiation heat losses:
  from outdoor pools, 542ff
  from roof ponds, 310–313, 648
  from solar collectors, 389–392
Radiative cooling, 648
Rankine-cycle solar cooling, 655–658
Rankine temperature scale, 79
Rating standards, solar hot water heaters, 528–530
Recirculation freeze protection, 478–479
Reflectance, of surface, 83–84
  of collector absorber plate, 421–423
  of collector glazing, 425–429
Relative humidity, 196–198, 630, 642–651
Renewable energy sources, 15–19
Retrofitting for passive solar, 353–358
  design procedures, 354–357
  south-oriented porches, 355–356
  sunspaces, 356
  thermal storage walls, 354–355
  windows and skylights, 354
Retrofit solar installations, defined, 354
Rock beds, as thermal storage, 339–353
  construction details, 352
  fan-forced type, 339–341
  for hybrid systems, 341–343
  heat transfer rate in, 346
  in space-heating systems, 625
  pressure drop in, 344–346
Roof overhang, as solar control device, 247–251
Roof ponds, 310–314
  construction details, 311–313
  sizing of, 313–314
Rules of thumb:
  for building orientation, 212

Rules of thumb (*continued*)
  for passive solar design, 208–213
  for sizing solar collection area,
    209–211
  for thermal storage, 211–212
R-values, indicators of thermal
    resistance, 70–72, 154–160

S
Saturated air, 151, 630
Seasonal heating and cooling loads,
    179
Seasons of year, 102–103
Secondary-irradiated thermal
    storage, defined, 264–265
Selective coatings for absorber
    plates, 422–425
  table of properties, 424
Semiconductors, 702–708
Sensible heat, defined, 61
  as part of cooling load, 637–641
Sensible heat storage, 211, 254
Shading and insulating devices,
    251–254
Shading coefficients (SC), 236–237
  table of, 236
Silicon, as semiconductor, 702–709
  N-type, 703
  P-type, 703
  wafers for solar cells, 705–706
Single-crystal silicon, 704–708
Siphon drainback system, 580
Site-built collectors, 441–443
Skylights, 243–244
Sling psychrometer, 632
Solar access, defined, 202
  for active solar energy systems,
    398–409
  devices for determining, 202–206
Solar access surveys, 400–409
Solar altitude (elevation angle), 248
Solar angle of incidence, 113
  tables of, 123–132
Solar angles, determination of,
    105–113
  from charts, 109–113
  from equations, 106–109
Solar azimuth angle, 105
Solar cells, 693, 698, 705–710
  concentrating (focusing) type, 710
  electrical response of, 707
  operating characteristics of, 699,
    705–708
Solar collection area (aperture),
    187–188
  design of, 242–247
  Mazria's rule for sizing, 234
  rule of thumb for design, 232
Solar collectors, 386–392, 411–465
  daily performance, 461–465

efficiency of, 445–452
  tests and test reports on, 454–465
Solar Constant, 96–97
  values of, table, 97
Solar declination, 103, 105–106
Solar economics, 743–754
Solar elevation (altitude) angle, 105,
    248
Solar-energized absorption cooling,
    658–664
Solar energy industry, 21–25
  careers and employment in, 23–25
  present status of and projected
    growth to year 2000, 21–22
Solar energy system performance,
    estimation of, 213–224
  the Load Collector Ratio (LCR),
    method, 222–224
  Solar Load Ratio (SLR) method,
    213–222
Solar energy systems, classified, 8–14
Solar flares, 90–91
Solar heat gain, control of, 247–254
  estimation of, 235–238
  providing for, 231–235
Solar heat gain factors (SHGF),
    235–238
Solar heating fraction (SHF), 602
Solar hour angle, 104–105
Solar-irradiated storage, 263
Solariums, 316–320
Solar Load Ratio (SLR), defined, 214
Solar Load Ratio (SLR) method of
    performance evaluation,
    213–222, 364–380
Solar position angles, table of for
    northern hemisphere, 133–135
Solar PV modules, 699–701, 712–714
Solar radiation energy, distribution
    among wavelengths of the
    spectrum, 95
Solar Savings (SS), defined and
    explained, 210–214
Solar Savings Fraction (SSF),
    defined, 216
Solar site audit worksheet, 403–404
Solar spectrum, 92–95
  chart of visible region, 94
Solar-thermal generation for
    network power, 732–734
Solar water heaters, 469–530
  classification of, 469
  forced-circulation type, 473 ff
  integral collector storage (ICS)
    type, 470–473
  thermosiphon type, 488
Solar water heating systems, pipe
    and pump sizing, 502–509
  sizing of collector bank, 497
  sizing the storage tank, 499

Solstices, winter and summer 102–103
Space-heating solar energy systems, 571–626
  air-based systems, 593
  collector area requirements, 573
  configurations for, 575–581
  controls for, 618–619
  design considerations, 590–594
  examples of installed systems, 620–626
  liquid-based systems, 592
  sizing guidelines, 572–573
  thermal storage requirements, 573–575
Space-heating system performance, 594–608
  solar collector considerations, 594
  Solar Load Ratio (SLR), method, 601–608
  thermal storage considerations, 599
Space-temperature swing 638
Spa heating systems, design of, 560–565
  losses from, 561–563
  sizing collectors for, 539
Specific heat capacity, 55
  of thermal storage materials, 254
  table of, for selected substances, 56, 254
Specific (absolute) humidity, 630
Spectral radiation intensity, 95
Spectroscope, 93–94
Stagnation temperature, defined, 387
  of solar collector, 448
Stand-alone photovoltaic systems, 714–720
Stefan-Boltzmann law of radiation, 80–81
Storage batteries for PV systems, 718–719
Summer cooling, 629–665
  active-solar methods, 651–665
  load analysis and determination, 631–642
  passive methods, 642–651
Summer vents, in Trombe wall, 290–292
Sun, as giant furnace, 90
  apparent motion of, 103–105
  layered structure of, 90
Sun charts, 109–112, 202–203
  use of in determining solar access, 203
Sunlight cycles in rooms, 239–241
  use of to plan thermal storage, 271
Sun's energy:
  origin of, 92

total magnitude of, 92
portion striking earth, 92
Sunspaces, 247, 315–329
  construction details, 320–323
  guidelines for sizing, 317–320
  interaction with main building, 315–317
  thermal performance of, 324–329
  thermal storage for, 323–324
Sun-tracking for solar collectors and PV arrays, 390–392, 711–712, 731–732
Surface conductance (film coefficient), 73–75
Surface-solar azimuth, 113
Swimming pool and spa solar energy systems, 535–566
  heat losses from, 535–537
  installation procedures, 556–560
  solar collectors for, 537–539
  system design criteria, 539–546
System efficiency of solar systems, defined, 446
System performance evaluation by advanced methods, 364–380
  Los Alamos reference designs, 365–367
  SLR-SS correlations, 367
  SSF-LCR-S/DD$_{month}$ correlations, 376–377

T
Temperature, 27–33
  and heat, 33–34
  Celsius scale of, 30
  Fahrenheit scale of, 28
  Kelvin scale of, 78
  Rankine scale of, 79
Temperature control in passive solar buildings, 201
Temperature difference and heat flow, 33, 53
Thermal capacity, 54–57
  of thermal storage mass, 254–269
Thermal expansion, 63–66
  linear expansion, 63–65
  coefficient of, 63
Thermal flywheel effect, 201
Thermal resistance (R), 70
  values for construction materials, 154–160
Thermal storage, defined, 188
  amount and distribution of, 262–269
  for active solar energy systems, 392–394
  for passive solar buildings, 254–274, 284–307
  for space heating, 599–600
  heat-flow time lag in, 259–262

Thermal storage (*continued*)
  latent heat materials for, 255–256
  placement of for best efficiency, 263–266
  sensible heat materials for, 254ff
  use of ordinary construction materials, 259–262
  use of water, 261–262
Thermal storage capacity, defined, 267
  rule of thumb for minimum capacity, 262–263
Thermal storage roof, 191
Thermal storage roof ponds, 310–314
Thermal storage walls, 284–304
  capacity requirements, 299–300
  construction details, 300–307
  operation of, on daily basis, 284–295
  performance of, 307–310
  sizing and construction, 295–307
    by Los Alamos rule of thumb, 296
    by Mazria's guidelines, 297–298
Thermocirculation vents, 289–292
Thermodynamics, 33
  First law of, 59
Thermometers, 28–30
Thermonuclear fusion, 16
Thermosiphon solar water-heating systems, 488
Thin-film deposition for solar cells, 709
Tidal power, 19
Tilt angle, 113
Time lag of heat flow through walls, 286–288
Total heat gain rate through glass, 237–238
Total outside air, estimation of, 170, 641
Total season cooling load, 642
Transmittance, of collector glazing, 450
Transmittance-absorptance product, 450
Trombe walls, 289–292. *See also* Thermal storage walls
Troubleshooting, solar hot-water systems, 526–528

U
Ultraviolet region of spectrum, 76–77

Units, of energy, work, and power, 47–48
  of Solar Constant, 97
  of solar radiation intensity (insolation), 99–100
Units and dimensions, 34–50
  as an aid to problem analysis and problem solving, 49–50
U-tube gauges, 42–43
U-valves, for heat transmission through structural elements, 71–72, 154–160
  table of 162–164
  use of in calculating building heat losses and gains, 165–168, 639–642

V
Ventilation air, 169
  for passive cooling, 643–645
Visible spectrum, 76–77
Volume, as measure of thermal storage capacity:
  masonry and water storage walls, 254–256
  rock beds, 344–352
Volumetric thermal capacity, 254ff

W
Water:
  as heat-transfer medium, 394–395
  as refrigerant, 660–661
  as thermal storage, 292ff
Water cooling tower, 661
Watt, SI-metric unit of power, defined, 47
Wavelength, of radiant energy, 77
Wet-bulb depression, 151
Wet-bulb temperature, defined, 151
  effect on cooling load, 630
Wien's displacement law, 81–82
Wind energy conversion, 11
Work, defined, 47
Work function, of metals, 702

Z
Zone air-handling unit, 581
Zones:
for hydronic heating, 582–583
  for radiant heating, 622
  for sunspace and solarium hybrid systems, 315–320

Conversion Factors
and Approximate Equivalents

A. Solar Energy Units

The Solar Constant $\qquad$ I_0 = 429 Btu/hr-ft^2
  (near-earth space) $\qquad$ = 1.353 kW/m^2

1 sun (clear-day insolation at the
earth's surface, normal incidence,
solar noon—average for mid-lati-
tudes) $\qquad$ = 317 Btu/hr-ft^2 = 1 kW/m^2
1 Btu/hr-ft^2 $\qquad$ = 3.155 W/m^2
1 W/m^2 $\qquad$ = 0.317 Btu/hr-ft^2
1 kW/m^2 $\qquad$ = 317 Btu/hr-ft^2
1 Langley (Ly) $\qquad$ = 1 calorie/cm^2
$\qquad$ = 11.63 W-hr/m^2
$\qquad$ = 3.69 Btu/ft^2
1 Ly/min $\qquad$ = 698 W/m^2
$\qquad$ = 221.2 Btu/hr-ft^2

B. Units of Heat, Energy, and Power

1 ft-lb $\qquad$ = 3.77×10^{-7} kW-hr
1 watt (W) $\qquad$ = 1 joule/sec
$\qquad$ = 3.41 Btu/hr
1 kilowatt (kW) $\qquad$ = 1000 W
$\qquad$ = 3410 Btu/hr
$\qquad$ = 1.34 horsepower (hp)
1 horsepower (hp) $\qquad$ = 0.746 kW
$\qquad$ = 746 W
$\qquad$ = 2545 Btu/hr
1 Btu $\qquad$ = 2.93×10^{-4} kW-hr = 778 ft-lb
1 kW-hr $\qquad$ = 3410 Btu
$\qquad$ = 2.655×10^6 ft-lb
1 ton (refrigeration) $\qquad$ = 12,000 Btu/hr
$\qquad$ = 3.52 kW
1 therm $\qquad$ = 10^5 Btu
$\qquad$ = 29.3 kW-hr
1 Quad $\qquad$ = 10^{15} (one quadrillion) Btu

C. Approximate Energy Yields of Fossil Fuels

1 barrel (bbl) of oil $\qquad$ = 42 U.S. gallons
1 Quad $\qquad$ $\cong$ 180 million bbls oil
$\qquad$ $\cong$ 40 million tons coal
$\qquad$ $\cong$ 1 trillion ft^3 natural gas
1 million bbl oil/day
(MB/DOE) for 1 year $\qquad$ = 2.02 Quads/year